Chemical Structure and Reactivity:
an integrated approach

Chemical Structure and Reactivity:

an integrated approach

James Keeler

Department of Chemistry and Selwyn College, University of Cambridge

Peter Wothers

Department of Chemistry and St Catharine's College, University of Cambridge

OXFORD

UNIVERSITY PRESS

OXFORD

UNIVERSITY PRESS

Great Clarendon Street, Oxford OX2 6DP

Oxford University Press is a department of the University of Oxford.
It furthers the University's objective of excellence in research, scholarship,
and education by publishing worldwide in

Oxford   New York

Auckland   Cape Town   Dar es Salaam   Hong Kong   Karachi
Kuala Lumpur   Madrid   Melbourne   Mexico City   Nairobi
New Delhi   Shanghai   Taipei   Toronto

With offices in

Argentina   Austria   Brazil   Chile   Czech Republic   France   Greece
Guatemala   Hungary   Italy   Japan   Poland   Portugal   Singapore
South Korea   Switzerland   Thailand   Turkey   Ukraine   Vietnam

Oxford is a registered trade mark of Oxford University Press
in the UK and in certain other countries

Published in the United States by
Oxford University Press Inc., New York

Typeset by the authors in LaTeX
Printed in Italy
on acid-free paper by
L.E.G.O S.p.A – Lavis TN

ISBN 978–0–19–928930–1

1 3 5 7 9 10 8 6 4 2

Preface

Our intention in writing this book was to produce a *single* text to accompany the first year, and somewhat beyond, of typical U.K. degree courses in chemistry. We also had it very much in mind that we wanted to treat the subject *as a whole* rather than dividing it up in the traditional way. This enables us to emphasise the important connections between different topics, to develop a unified view of the whole subject, and to present the material in rather a different way than is conventional.

Our presentation is very much based on the idea that both structure and reactivity can be understood, or at least rationalized, by thinking about the orbitals (atomic and molecular) involved, their energies and the way they interact. This is the unifying theme of the whole text, taking us right from the description of the simplest molecules, through to an understanding of reaction mechanisms. This central importance of an orbital description is reflected in the fact that three chapters right at the start of the book are devoted to establishing these ideas.

The quantitative and more theoretical aspects of chemistry are not ignored, but we have been at pains to present these in a way which emphasises their wider relevance. There are chapters covering the traditional physical chemistry topics of thermodynamics, kinetics and quantum mechanics, but these are interspersed throughout the book and are strongly connected to the rest of the discussion.

Content

The content of this book has been tailored to fit in with chemistry courses at U.K. universities. However, a sufficiently wide range of topics are covered that the text should also be useful in other countries.

Although this book contains a blend of inorganic, organic and physical chemistry it is emphatically not a 'general chemistry' text of the type produced for North American universities. Our approach, the content, and the starting point are all significantly different from general chemistry texts. In particular, we have included notably more organic chemistry than one typically finds in such books.

Given that we certainly did not want to produce a book of overwhelming size, we have had to be careful in the selection of the topics which are included and the level to which they are discussed. Our feeling is very much that at this level the most important thing is to understand the key ideas and concepts, and know how to use them. We have therefore devoted a lot of space to setting out these ideas carefully and in detail, so as to provide a sound basis for further study.

Space has not allowed us to include lengthy discussions of the chemistry of each group, nor to explore in detail the subtleties of synthetic organic chemistry, nor to delve deeply into the intricacies of quantum mechanics. Nevertheless we believe that the material which is presented here forms both a coherent story and a suitable basis on which to begin to get to grips with more advanced chemistry.

How the text is organized

The book is divided into two parts. The ten chapters in Part I are closely integrated with one another and are designed to be studied in the order they appear. Together they set out the fundamental ideas needed for the study of chemistry at this level. Although thermodynamics and kinetics appear in Chapters 6 and 10, they are treated with the absolute minimum of mathematics, and in particular without calculus. We have chosen this approach so that those students who are still developing their mathematical skills can nevertheless obtain a good grasp of these essential topics.

The chapters in Part II either introduce further topics, or take the discussion started in Part I on to a higher level. For example, with the aid of calculus Chapter 17 discusses thermodynamics in a somewhat more formal way than was the case in Chapter 6. Similarly, the quantum mechanical ideas which were used qualitatively in Part I are explored in more detail in Chapter 16. Generally speaking, the chapters in Part II can be studied in any order. However, Chapters 14 and 15 are so closely connected that they should be studied in this order; the same is true of Chapters 17 and 19.

Chapter 20 is a brief but self-contained exposition of the key mathematical ideas used in the rest of the book. While this chapter is certainly no substitute for an appropriate course in mathematics, we hope that it will be a convenient reference and handy refresher on the key ideas.

Each chapter concludes with a set of questions which are designed to test both a basic understanding of the material presented in the chapter, as well as the ability to apply the concepts in more unfamiliar situations. We have also listed in *Further Reading* some other texts which can be consulted for an alternative view or a more advanced or detailed discussion.

Acknowledgement

We are very grateful to many of our colleagues in the Department of Chemistry for finding the time and patience to answer our questions, direct us to relevant literature, or provide data. In particular we would like to thank both Ruth Lynden–Bell and Anthony Stone for greatly helping our understanding of the underlying theory. Duncan Howe and Paul Skelton from the physical methods section ran many NMR and mass spectra for us with great skill, for which we are most grateful. Alfa Aesar (Heysham, Lancs.) contributed all the samples used to run the NMR spectra, and the GCMS trace shown in Fig. 11.11 was provided by Phil Teale and Simon Hudson (HFL Ltd, Newmarket): we thank them for their contributions.

Our Cambridge colleagues John Kirkpatrick, Finian Leeper, Rob Paton, Mike Rogers and Steven Smith have read and commented on drafts of parts of the book (often at rather short notice). In addition, Michael Clugston (Tonbridge School), Ian Cooper (Newcastle University), Bridgette Duncombe (University of

Edinburgh), Jason Eames (University of Hull), David McGarvey (Keele University), Ruud Scheek (University of Groningen), Edward Smith (Imperial College, London), Patrick Steel (Durham University), David Worrall (Loughborough University), Rossana Wright (University of Nottingham) and Timothy Wright (University of Nottingham) have read parts of the first draft on behalf of the publishers. We are much indebted to all of these people for their careful reading and helpful comments which have undoubtedly improved the text.

Steven Smith has also prepared, with great skill and care, a significant part of the solutions manual. Stephen Elliott contributed a great deal to the development of the online resource centre, and we are very grateful to him for producing some really outstanding resources.

This book has been typeset by the authors using the LaTeX text processing system, in the implementation distributed by MiKTeX (http://miktex.org/). We acknowledge the exceptional and continuing effort of the many people throughout the world who have contributed to the LaTeX system and made it freely available.

We would like to record our especial thanks to Henneli Greyling for keeping the teaching office, and much else besides, running smoothly and efficiently while our attention was focused on completing the book. Jonathan Crowe, our editor at OUP, both commissioned the book and maintained his enthusiasm for the project throughout all of the stages of production. We thank him, and the rest of his team at OUP, for their unwavering support and professionalism.

Cambridge, April 2008

Online resource centre

The online resource centre provides both students and teachers with additional materials to complement and extend the text. It also provides a mechanism for all users of the book to feed back their comments and observations to the authors.

www.oxfordtextbooks.co.uk/orc/keeler

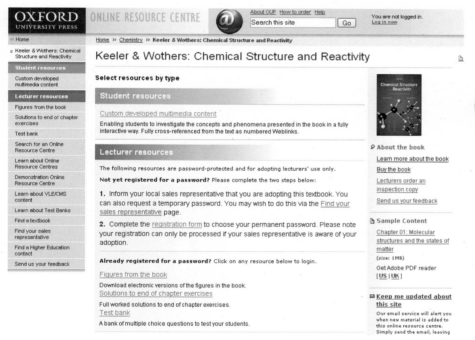

For students

Throughout the text, relevant online resources are indicated by *Weblink* boxes in the margin. On the website the links are arranged according to chapter, so you should easily be able to navigate to the particular link you are looking for.

The material on the web is not simply a repeat of a diagram in the text. Rather, it is an interactive resource in which you can alter the view of some three-dimensional object, or see how a graph changes in real time when you alter the parameters. There are four main kinds web resources

> *Weblink 1.1*
>
> Online resources are indicated by boxes like this one, located in the margin.

- Three-dimensional representations of molecules and lattices which you can rotate, zoom and view in different ways.

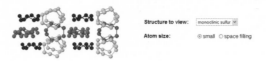

- Three-dimensional representations of orbitals which you can also rotate and zoom so that you can get a real feel for their shape.

- 'Movies' showing how orbitals change during reactions.

- Graphs or plots in which you can change the parameters (usually with real-time sliders), so that you can get a real feel for how the graph responds.

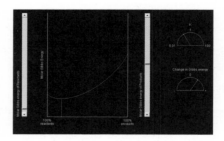

For teachers

Registered adopters of this book also have access to (password protected)

- All the figures, in a high-resolution digital format.

- Detailed worked solutions to all of the questions.

- A testbank of multiple-choice questions keyed to the book.

The *Weblink* resources can also be projected during lectures, and provide both excellent illustrations of the topics as well as helping to vary the presentation during the lecture.

Brief contents

Full contents

Part II Going further

Part I
The fundamentals

Contents

Molecular structures and the states of matter

Key points

- The basic types of bonding are covalent, ionic and metallic.
- There are weaker, non-covalent, interactions between molecules.
- Molecular structures are determined using X-ray diffraction.
- In drawing molecular structures, lines between atoms do not always represent bonds.
- The shapes of simple molecules can often be predicted using the valence shell electron pair repulsion model.
- For an ideal gas, the pressure, volume, temperature and amount in moles are related by the ideal gas equation.
- The Boltzmann distribution predicts that only those molecular energy levels with energies of the order of, or less than, $k_B T$ are occupied.

This book is concerned with trying to understand why molecules form, what determines their shape, and what it is about a particular molecule which makes it react in a certain way. This is undoubtedly an ambitious task which has already occupied chemists for hundreds of years, and will no doubt keep them employed for many years to come. However, a great deal of progress has been made and, through studying this book, you will see that with the aid of a relatively small number of fundamental principles it is possible to understand much about the chemical world of structures and reactivity.

The idea of a molecular structure – how we describe it, how we represent it on paper, and the different types of structures that occur – is absolutely central to the whole of this book. So this first chapter starts out by considering molecules, and assemblies of molecules, in a fairly general way before we start a detailed discussion of chemical bonding. We will also look at how molecular structures are drawn, and some details about the behaviour of gases which will be very useful to us in the remainder of the book.

1.1   Types of bonding

In this section we are going to take a broad over-view of the kinds of bonding seen in chemical compounds and the key characteristics of each type of bonding. At this stage we are not going to be concerned with why particular compounds exhibit certain types of bonding, or give a detailed description of the each type of bonding. These are topics which will be explored extensively throughout the rest of this book.

1.1.1   Covalent bonding

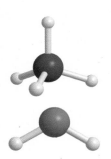

Fig. 1.1 Methane (CH_4, shown at the top) and water (H_2O, shown at the bottom), are examples of molecules held together by covalent bonds. Each such bond can be thought of as arising from the sharing of a pair of electrons, and is indicated by the grey cylinder joining the atoms. Carbon, oxygen and hydrogen atoms are coloured black, red and grey, respectively.

The key feature of covalent bonding is that it arises from the sharing of electrons between atoms, resulting in electron density between the atoms. The simplest examples of this kind of bonding are in molecules such as H_2O and CH_4, illustrated in Fig. 1.1, in which we equate each bond with a pair of electrons being shared between two atoms. Electrons can be shared between more than two atoms, giving rise to what is called delocalized bonding, as exemplified by the π system of benzene. It is also possible to form multiple covalent bonds in which two or three pairs of electrons are shared between atoms. We may think of a double bond, such as the C=C bond in ethene, as resulting from the sharing of two pairs of electrons, and a triple bond, such as the N≡N bond in N_2, as resulting from the sharing of three pairs of electrons.

Covalent bonds tend to be directional in the sense that the optimum interaction occurs when the bonds to a given atom have a particular spatial arrangement. For example, when a carbon is attached to four other atoms, the resulting bonds are usually arranged so as to create an approximately tetrahedral geometry around the carbon.

The number of atoms which are covalently bonded together to form a molecule can become very large, as in the case of biomolecules such as proteins, which contain thousands of atoms. It is also possible for the network of covalently bonded atoms to extend throughout a macroscopic *solid* sample: this gives rise to what is called a giant covalent structure. Examples of such systems are diamond (pure carbon) and quartz (SiO_2), shown in Fig. 1.2 on the next page.

Weblink 1.1

It is easier to appreciate the details of the structure of SiO_2 shown in Fig. 1.2 if you can rotate it in real time, and view it from different directions. Following the weblink will enable you to do just this.

The strength of covalent bonds

We can define a bond energy by imagining a process in which a molecule is broken apart into its constituent atoms, and then dividing up the energy needed to achieve this process between the bonds in the molecule. So, for example, the energy needed for the process $CH_4(g) \longrightarrow C(g) + 4H(g)$ can be taken to be four times the C–H bond energy (often called the bond strength).

The strengths of typical covalent single bonds vary between 150 kJ mol^{-1} and 350 kJ mol^{-1}. Double and triple bonds, where they are formed, are stronger. For example, a C=C double bond energy is around 600 kJ mol^{-1}, and the N≡N triple bond is around 950 kJ mol^{-1}.

1.1.2   Ionic bonding

Ionic bonding is essentially confined to solid materials which consist of a regular array of ions extending throughout the macroscopic sample. The most familiar example is solid sodium chloride, NaCl, which consists of a regular array of Na^+ and Cl^- ions, as shown in Fig. 1.3 on the facing page.

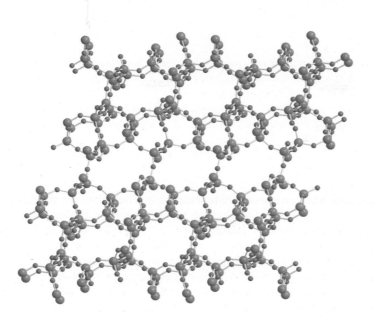

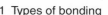

Fig. 1.2 A perspective view of part of the giant covalent structure of SiO_2. The network of covalent bonds between silicon (shown in blue) and oxygen (shown in red) extends throughout the entire solid.

In such solids there are strong electrostatic interactions between the ions, which are arranged in such a way that the attractive interactions between oppositely charged ions outweigh the repulsive interactions between ions with the same charge. In contrast to covalent bonds, the electrostatic interaction between ions is not directional i.e. all that matters is the distance between two ions, not the orientation.

When ions are present, the electrostatic interaction between them is so strong that it tends to be dominant. However, it is possible for there also to be significant covalent interactions between ions, so pure ionic bonding is best regarded as an idealization.

Sometimes you will see reference being made to an 'ionic bond' between two ions: such terminology is best avoided. The reason for this is that in an ionic solid each ion interacts with many other ions, both of the same and opposite charge to itself. The energy which binds the structure does not come from single interactions between ions (the so-called ionic bonds), but as a result of the interactions between all the ions in the sample.

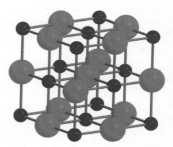

Fig. 1.3 Solid sodium chloride consists of a regular array of Na^+ and Cl^- ions (shown in blue and green, respectively) which are held together by electrostatic interactions. The lines joining the ions do not represent bonds, but are there to guide the eye.

Weblink 1.2

View and rotate in real time the section of the NaCl structure shown in Fig. 1.3.

The strength of ionic bonding

The strength of ionic bonding can be assessed by working out the energy required to take the ions in the solid and then separate them to infinity (in the gas phase), where they will no longer interact: this is often called the *lattice energy*. For NaCl this lattice energy is around 790 kJ mol^{-1}: clearly, the energy of interaction between the ions in an ionic structure is at least as great as that between atoms in a covalently bound molecule.

As well as varying inversely with their separation, the energy of interaction between two ions is also proportional to the product of their charges. This means that two doubly charged ions interact four times more strongly than two singly charged ions at the same separation. As a result, the lattice energy depends strongly on the charges of the ions. For example, MgF_2, in which the metal ion is doubly charged, has a lattice energy of 2905 kJ mol^{-1}, which is much greater than that for NaCl. Where both the cation and the anion are doubly charged, the lattice energy is higher again; for example, MgO has a lattice energy of 3850 kJ mol^{-1}. In Chapter 5 we will look in more detail at how lattice energies can be estimated.

1.1.3   Metallic bonding

As its name implies, metallic bonding is the kind of bonding seen in solid (and liquid) metals. In Chapter 5 we will have a great deal more to say about this kind of bonding, but the basic idea is that the structure is held together by electrons which can move freely throughout the macroscopic sample.

The strength of metallic bonding

We can obtain an estimate of the strength of metallic bonding by looking at the energy required to vaporize a metal i.e. turn it from a solid into gaseous atoms. These energies vary very widely, from as little as 100 kJ mol^{-1} for Na, to 330 kJ mol^{-1} for Al and 849 kJ mol^{-1} for tungsten. The values are comparable with those for covalent interactions, but not as strong as those for ionic interactions.

1.2   Weaker non-bonded interactions

It is found that between molecules there are a number of interactions which are considerably weaker than typical covalent bonds and so are usually described as *non-bonded interactions*. These interactions are also described as being *intermolecular* as they occur *between* molecules. Despite being weaker than covalent bonds, these intermolecular interactions play an important part in determining the physical and chemical properties of many molecules.

The two most important non-bonded interactions are the *dispersion interaction* (also called the *London interaction*) and *hydrogen bonding*. The latter only occurs in molecules in which a hydrogen atom is attached to an small highly electronegative element, such as nitrogen, oxygen or fluorine. In contrast, the dispersion interaction is a feature of all molecules, so we will describe it first. Collectively, intermolecular interactions (other than hydrogen bonds) are often referred to as *van der Waals* forces.

1.2.1   Dispersion interaction

It is easiest to describe this interaction if we think about the approach of two molecules which do not have dipoles, such as an N$_2$ molecule, or a cyclohexane molecule. Generally speaking, the reason why a molecule lacks a dipole is that it has certain kinds of symmetry (for example, a *centre of inversion*); however for the present purposes we can just as well imagine that there is no dipole as a result of the electron density being uniform across the molecule. This is represented in Fig. 1.4 (a) by an ellipse with a uniform colour.

It turns out that the electron density fluctuates over time. If at any instant the density becomes unsymmetrical a dipole will be generated, as shown in (b). Here, the electron density on the right has increased, giving a partial negative charge ($\delta-$, yellow), whereas the electron density on the left has decreased, giving a partial positive charge ($\delta+$, red).

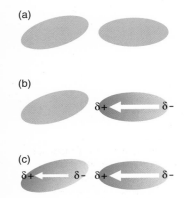

Fig. 1.4 Visualization of the dispersion interaction between two molecules, represented by ellipsoids in (a); the molecules have a symmetrical electron density, and so do not possess dipoles. However, the electron distributions will fluctuate, and this can result in the generation of a dipole, as shown in (b): yellow indicates an increase in electron density, and red a reduction. This dipole will distort the electron distribution of the left-hand molecule in such a way as to induce a dipole in it, as shown in (c). The direction of the induced dipole is always such that there is a favourable interaction between the two dipoles.

Fig. 1.5 Illustration of the arrangement of the dipole and the induced dipole for various orientations of the two molecules. In each pair, the induced dipole is represented by the smaller of the two arrows. Note that the closest contacts are always between red regions ($\delta+$) and yellow regions ($\delta-$), which means that the interaction between the two dipoles is always favourable.

The $\delta+$ end of the dipole is closest to the left-hand molecule, so the electron density of this molecule is pulled to the right, generating an excess of electron density on the right, and a reduction in electron density on the left. The result is the generation of an *induced dipole* on the left-hand molecule, as shown in (c).

As it was the dipole of the right-hand molecule which led to the induction of the dipole of the left-hand molecule, the two dipoles are necessarily arranged so that they have an energetically favourable interaction with one another i.e. the closest approach is between the $\delta+$ and $\delta-$ ends. It is this favourable interaction which is the origin of the dispersion interaction between molecules.

The key thing to realize is that the induced dipole will be aligned in such a way that it *always* has a favourable interaction with the first dipole. This point is illustrated in Fig. 1.5, in which the orientation of the two dipoles is shown for various different arrangements of the two molecules; in all cases the interaction is favourable.

The fluctuations which lead to the generation of a dipole, and the subsequent induction of a dipole in a nearby molecule, are very rapid processes compared to the rates at which the molecules are moving due to thermal motion. Thus, as the molecules move around, this favourable interaction is always maintained, regardless of the orientation of the two molecules.

If the molecules have *permanent* dipoles, these will also interact but the problem is that random thermal motion means that the dipoles will *not* always be aligned in such a way as to have a favourable interaction. As a result, when averaged across the sample, the interaction between permanent dipoles is usually much less that the dispersion interaction. Indeed, apart from hydrogen bonding, the dispersion interaction is usually the most significant interaction between molecules.

One of the factors which determines the strength of the dispersion interaction is the *size* of the induced dipole: the larger this dipole, the stronger the interaction. The size of the induced dipole depends on how much a given electric field will distort the electron distribution. The more easily the distribution is distorted, the more *polarizable* the molecule is said to be, and hence the greater the induced dipole.

If the electrons in molecules (and atoms) are held tightly, then the polarizability tends to be low. The most extreme example of this is the flguorine atom in which the high ionization energy reflects the fact that the electrons are tightly held. Thus, compounds containing many fluorine atoms tend to have low polarizabilities, and so the dispersion interaction is weak. This accounts for the fact that many fluoro compounds, such as SF_6, PF_5 and IF_7, occur as gases at room temperature. Even uranium hexafluoride UF_6, despite its high relative molecular mass of 352, sublimes at only 57 °C. This property is also exploited in non-stick plastics such as PTFE (polytetrafluoroethylene).

At first glance, it is somewhat counter-intuitive that the interaction between permanent dipoles is weaker than that between induced dipoles. The key point is that induced dipoles are always oriented so that their interaction is favourable; this is not true for permanent dipoles.

Fig. 1.6 Part of the crystal structure of solid ethanol showing, by dashed lines, the hydrogen bonds between different molecules; the molecules at the edge will form further hydrogen bonds with other molecules which are not shown. Note how each of the four molecules in the centre of the picture makes two hydrogen bonds, one to its oxygen and one to the hydrogen attached to the oxygen.

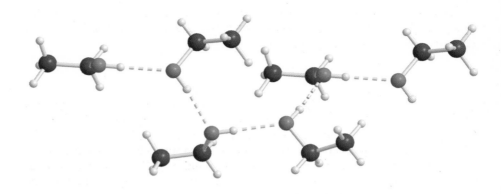

We can obtain an estimate of the strength of the dispersion interaction by looking at the energy needed to vaporize a liquid hydrocarbon, such as ethane. The dispersion interaction is the dominant one which causes such a molecule to liquefy at low temperatures, so the 15 kJ mol^{-1} needed to vaporize the liquid can be taken as an estimate of the size of the interaction. Clearly, this is much weaker than a covalent bond, but by no means insignificant.

1.2.2 Hydrogen bonding

The *hydrogen bond* is a very important non-bonded interaction that occurs between a hydrogen that is attached to a small highly electronegative atom, and a second electronegative atom. Figure 1.6 shows, as dashed lines, the hydrogen bonds which occur in the structure of solid ethanol. In this case, these bonds occur between the hydrogen atom which is attached to oxygen, and an oxygen atom on another molecule. These particular contacts are identified as hydrogen bonds as the O and H are only about 180 pm apart, which is much closer than is usual for atoms in separate molecules, but not as close as a full O–H bond which has a length of around 80 to 85 pm. In the liquid similar bonds are expected to form, although they will of course constantly be made and unmade as the molecules move around due to thermal motion.

There is likely to be a significant dipole in the bond between hydrogen and an electronegative atom, simply because the latter is drawing electron density to itself. The hydrogen therefore has a partial positive charge, which can interact favourably with the partial negative charge appearing on a electronegative atom of a second molecule; this illustrated in Fig. 1.7.

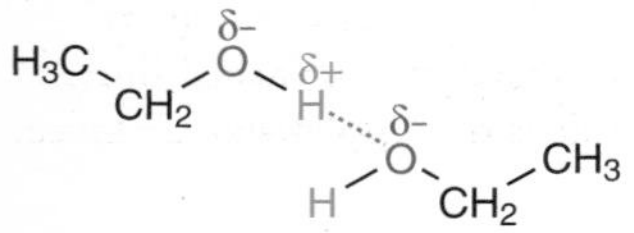

It is thought that this electrostatic interaction between dipoles is a significant contributor to the energy of a hydrogen bond. However, it is observed that such bonds are directional in nature, as is well illustrated in Fig. 1.6. It cannot be, therefore, that the interaction is purely electrostatic. Later on, we will return to a more detailed description of the hydrogen bond which will account for its directional properties.

We can obtain an estimate of the strength of the hydrogen bond by looking at the energy needed to vaporize ethanol, which turns out to be 44 kJ mol^{-1}. Of this, about 15 kJ mol^{-1} can be attributed to van der Waals interactions (mainly dispersion), leaving the remaining 29 kJ mol^{-1} attributable to the hydrogen bonding. Again, the interaction is not as large as for a covalent bond, but

🌐 *Weblink 1.3*

View and rotate the structure of solid ethanol shown in Fig. 1.6.

$$H_3C\diagdown_{CH_2}\diagup\overset{\delta-}{O}\diagdown_{\overset{\delta+}{H}}\overset{\delta-}{\cdots}\underset{CH_2}{\overset{O}{\diagup}}\diagdown CH_3$$

Fig. 1.7 Illustration of the formation of a hydrogen bond, shown in green, between two ethanol molecules. There is a partial positive charge on the hydrogen that is attached to the oxygen, on account of the high electronegativity of the latter. This partial positive charge can interact favourably with the partial negative charge on the oxygen of another molecule.

is certainly significant. As you may know, the three-dimensional structures of biological molecules such as proteins and DNA are maintained largely by an extensive network of hydrogen bonds.

1.2.3   Repulsive interactions

So far we have been talking about the favourable, attractive, interactions between atoms and molecules. Generally, the closer the molecules approach, the more favourable the interaction. This means that the energy due to the attractive interaction decreases as the molecules approach one another, as is illustrated by the blue curve in Fig. 1.8. Thinking of it the other way round, this favourable interaction means that we have to put energy in to break the molecules apart from one another.

It is found that when atoms and molecules approach one another closely there is in addition always a *repulsive* interaction which causes the energy to rise, as shown by the red curve. As the distance between the molecules decreases, the energy due to the repulsive interaction increases more quickly than the decrease due to the attractive interaction. Thus, at some distance, the total energy (shown by the black curve), which is the sum of the attractive and repulsive part, has a minimum value.

This distance at which the energy is a minimum is called the *equilibrium separation*, and is the point at which the attractive and repulsive forces are in balance. On average, we therefore expect to find that the two molecules are separated by this equilibrium separation. We say 'on average' as, due to thermal motion, the molecules will actually be moving around in the vicinity of this equilibrium distance.

These repulsive interactions come about when the electron density from one atom overlaps significantly with that of another. We will be in a position to understand why this leads to a repulsion when we have looked in more detail at the electronic structures of atoms and molecules, which is the topic of the next three chapters.

In a molecular solid (section 1.3), the way the molecules are arranged (e.g. the distances between them) is determined by this balance between the attractive and repulsive forces. Similarly, the spacing of the ions in an ionic lattice is determined by the balance between the attractive electrostatic interactions between oppositely charge ions, the repulsive electrostatic interactions between ions of the same charge, and the repulsive interactions between all the ions. Note that even oppositely charged ions, if they approach closely enough, will eventually repel one another.

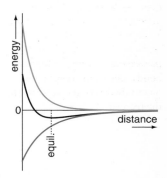

Fig. 1.8 When two molecules approach one another there is a favourable interaction which leads to a reduction in the energy, as shown by the blue curve. If we take the energy zero to be when the molecules are well separated, the energy due to this favourable interaction becomes more negative the closer the molecules approach. There is also a repulsive interaction which eventually leads to an increase in energy as the distance decreases, as shown by the red curve. The repulsive energy increases more quickly than the attractive energy decreases, so that at some point the total energy, shown by the black curve, is a minimum (marked by the dashed line); this is the *equilibrium distance*.

1.3   Solids

Solid materials can show the whole range of bonding interactions we have been describing above, and not surprisingly the properties of a particular solid reflect the kind of bonding involved.

Molecular solids are composed of discrete molecules which retain their individual identity, but are held together by non-bonded interactions. Since these interactions are relatively weak, molecular solids tend to have quite low melting points. Many organic compounds form molecular solids, as do elements such as iodine (I_2) and phosphorus. For example, as shown in Fig. 1.9 on the following page, the allotrope white phosphorus contains discrete P_4 molecules,

Fig. 1.9 The allotrope of phosphorus known as white phosphorus is a molecular solid, consisting of discrete P_4 molecules arranged in chains and layers. The view shown in (a) is taken looking down the chain of P_4 molecules. View (b) shows how the chains are arranged relative to one another within a layer.

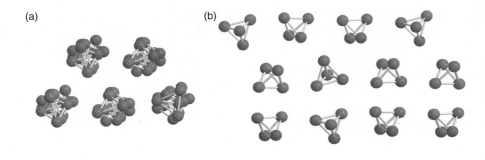

(a)   (b)

🌐 *Weblink 1.4*

View and rotate the structure of white phosphorus shown in Fig. 1.9.

arranged in chains and layers. Numerous inorganic compounds, such a P_4O_{10}, also form molecular solids.

Some solids consist not of discrete molecules, but are essentially formed as one giant covalently linked network which extends throughout of the sample. We mentioned this kind of bonding in section 1.1.1 on page 4, and illustrated it with the structure of SiO_2 shown in Fig. 1.2 on page 5. The strength of the covalent bonding tends to give such solids rather high melting points. Similarly, for *ionic solids*, which were described in section 1.1.2 on page 4 and illustrated in Fig. 1.3 on page 5, the strong interactions also lead to high melting points.

The final type of solid we need to distinguish is a *metallic solid*, where the bonding is of the type described in section 1.1.3 on page 6. Such solids are distinguished from all those mentioned so far by being good conductors of electricity on account of the mobile electrons they contain.

1.4 Structure determination by X-ray diffraction

The structure of a molecule is of immense importance to chemists, as so much depends on the way in which the atoms are joined up and the shape the molecule adopts. An enormous amount of effort has therefore been put into developing and refining methods for determining molecular structures. Of these methods, spectroscopy is the one in most common day-to-day use by chemists, and this topic is covered in detail in Chapter 11.

However, spectroscopy is rather a qualitative tool when it comes to structure determination: it can tell us the sorts of groups that are present and possibly how they are connected to one another, but it does not give precise structural information such as bond lengths and bond angles. When it comes to obtaining really detailed and definitive structural information one technique – that of *X-ray diffraction* – reigns supreme.

For an X-ray diffraction experiment we need a small crystal of the compound of interest. In this crystal the ions or molecules are arranged in an orderly way which is the same throughout the solid. We can think of the solid as being composed of the same structural motif (e.g. an array of ions or a molecule) repeated endlessly throughout the solid. This repeating motif is called the *unit cell*.

As a result of the orderly structure, a beam of X-rays which strikes the crystal is reflected in certain specific directions giving rise to what is known as a *diffraction pattern*. Figure 1.10 on the next page shows the key parts of

Fig. 1.10 A typical X-ray diffractometer. The crystal is mounted on the end of the fine glass rod 1, and is cooled by a stream of cold nitrogen gas from 2. This cooling helps to prevent the crystal from being damaged by the heat generated when the X-rays are absorbed, and the atmosphere of nitrogen gas also prevents chemical degradation of the crystal. The X-rays are generated in 3 by bombarding a target (here molybdenum) with high-energy electrons. The X-rays are focused along the tip 4 towards the crystal. The detector, 5, is placed behind the crystal, and a small TV camera, 6, is used to aid in the alignment process. The crystal can be rotated, under computer control, into different orientations by the goniometer, 7.

the typical X-ray diffractometer used to measure the diffraction pattern, and an example of such a pattern is shown in Fig. 1.11 on the following page. In fact, the X-rays are diffracted by the electrons, with the strongest diffraction coming from regions of high electron density. In molecules, most of the electron density is clustered around the atomic nuclei (i.e. from the filled shells), and so a diffraction experiment can be used to infer the positions of the atoms.

If the crystal is a molecular solid we will be able to determine from a diffraction experiment not only the precise spatial arrangement of the atoms which form the molecule, but also the way in which the molecules are arranged relative to one another. Similarly, for an ionic solid, we will be able to determine the positions of the ions.

It is not entirely a straightforward matter to go from the measured diffraction pattern to a three-dimensional map of the electron density in the unit cell. However, provided there are not too many atoms in the unit cell (say less than a thousand) it is possible to determine the electron density in a reasonable time using a computational approach known as the *direct method*.

Once we have determined a map of the electron density we can use it to identify where the atoms are. At the simplest level, we can expect there to be atoms at positions which have high electron density, but we have to decide which atom to place at each position. Hydrogen atoms can be difficult to locate as they have much lower electron density than most other atoms.

The usual process is to develop of model of the molecule and then calculate from it the expected electron density – something which is relatively easy to do. In building the model we take account of any other information we have, such as the molecular formula or the expected structure. For example, if the molecule has a benzene ring, we have a good idea where the six carbon atoms are relative

Fig. 1.11 A example of an X-ray diffraction pattern from a small crystal. The crystal is mounted at the centre of the picture and the X-ray beam is coming towards us. To stop the direct beam overloading the detector, a metal 'beam stop' is placed between the crystal and the detector; this beam stop, and the wire holding it, appear as the white shadow running from the middle to the top of the picture. X-rays are diffracted from the crystal at certain angles, and each diffracted beam gives rise to one of the black dots. In order to determine a molecular structure, many such diffraction patterns are taken in which the X-rays impinge on the crystal in different directions.

to one another, even though we do not know how to place these six atoms in the unit cell.

The electron density computed from the model is then compared to the experimental electron density; if the two differ at the position of one of the atoms, then that atom needs to be moved. The whole process is repeated over and over again until the best agreement between the experimental and calculated electron densities is obtained. This process is known as *refinement*.

With modern instruments to measure the diffraction pattern in an automatic way, and powerful computers to analyse it, it is possible to determine the structures of small- to medium-sized molecules in a more-or-less routine way. A very large number of structures have been determined, and these are deposited in various databases, as described in Box 1.1 on the next page.

1.4.1 Joining up the atoms in an X-ray structure

The final result of analysing the data from an X-ray diffraction experiment is the location of the atoms, but the experiment does not tell us how these atoms are joined up i.e. where the 'bonds' are. Typically the connections are made by assuming that atoms which are closer than some specified distance are bonded to one another. These distances are based on the bond lengths found in a large number of compounds whose structures have already been determined. In other words, we simply use the precedents set by existing, known structures.

The way these precedents are usually expressed is to assign a *covalent radius* to each type of atom i.e. a value for carbon, a different value for nitrogen and so on. If the distance between two atoms is comparable, within certain specified limits, to the sum of the covalent radii, then a bond is considered to exist between the two atoms.

For example, the covalent radius for carbon is usually taken as 68 pm, and that for chlorine as 99 pm. Typically, a tolerance of up to ±40 pm is permitted on any bond length computed from these values. So, a carbon and a chlorine atom approaching between 68 + 99 − 40 = 129 pm and 68 + 99 + 40 = 207 pm would be considered to be bonded, which is rather a wide range.

In joining up the atoms we also take account of the normal rules of valency, for example by making sure that each carbon has a valency of four, and each oxygen a valency of two. This approach works well for organic molecules, where the rules are rarely if ever broken, but becomes harder to apply once we move away from such well behaved molecules.

1.4.2 Caution with X-ray structures

Just about all the molecular structures which are described in this book have been determined by X-ray diffraction. The only exceptions are very small molecules in the gas phase, such as CO_2 and N_2O, for which there are spectroscopic methods capable of giving very detailed structural information. These methods are touched on in Chapter 16.

The information provided by X-ray diffraction is very precise and generally unambiguous. However, we need to be careful about a few things when using these data.

Box 1.1 Structural databases

A very large number of molecular structures have been determined using X-ray diffraction, and these have all been collected together into electronic databases, which are freely available to research scientists. The *Cambridge Structural Database* contains structural information on what are essentially organic molecules i.e. those with a carbon skeleton. At present there are over 360,000 entries in the database. The *Inorganic Crystal Structure Database* contains structural information on inorganic materials, be they molecular, giant covalent or ionic in form; there are presently over 90,000 entries. The structures of biological molecules, such a proteins and nucleic acids, are also collated in separate databases; over 35,000 structures of proteins have been deposited. These databases represent an enormous resource for those interested in patterns and trends in molecular structures.

- A diffraction experiment gives us the structure of the molecule *in the crystal*: when the compound is dissolved in a solvent, or passes into the gas phase, it may not have the same structure. For example, phosphorus pentachloride exists in the solid as an array of PCl_4^+ and PCl_6^- ions, whereas in the gas phase it exists as discrete PCl_5 molecules. More subtle changes of molecular geometry between the solid and solution phases are also commonly found.

- We need to remember that molecules do not necessarily have a single, fixed structure. All molecules are constantly flexing and bending due to vibrations, there may be rotation about (single) bonds, and some molecules readily interconvert between structural isomers. A crystal structure will not necessarily tell us about all of these possibilities, as the crystallization process tends to favour one particular form.

- X-ray diffraction tells us where the *atoms* are located in relation to one another. It does not tell us which atoms are *bonded* to one another; this is something we have to infer based on our chemical intuition and perhaps other experimental data.

We now turn to the important matter of how we represent molecular structures and paper, and what these representations imply.

1.5 Where are the bonds?

As soon as we learn about molecules we start drawing 'structures' in an attempt to explain and describe the way in which the atoms are joined up. This whole process becomes so second nature to a chemist that it is all too easy to forget that what we draw is just some representation of the bonding, which may or may not be adequate. In this section we will take a critical view of how we represent molecular structures.

Let us take the water molecule as an example. We know that water has the formula H_2O, and using either spectroscopy or diffraction experiments we can find out the locations of the three atoms relative to one another. Given the arrangement shown in Fig. 1.12 (a), our instinct is to join up the O with the two

(a)

(b)

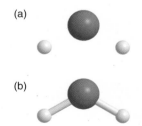

Fig. 1.12 Experiments tell us that the three atoms in water are arranged as in (a). Instinctively we would join these atoms up with two O–H bonds as shown in (b). In the case of water, it is pretty clear that this is the right approach, but for more unusual molecules it is not always clear which atoms are bonded to which.

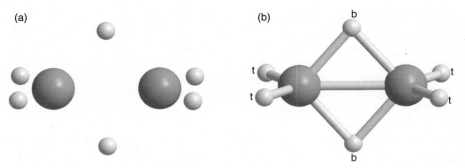

Fig. 1.13 Shown in (a) is the arrangement of the atoms in diborane, B_2H_6, as determined by X-ray diffraction. Boron atoms are shown in yellow, and hydrogen atoms in grey. Those atoms which are close enough to be considered bonded are joined together to give the structure shown in (b). There are two types of hydrogen atoms: terminal (marked 't') and bridging (marked 'b').

hydrogens, indicating the presence of two O–H bonds, as shown in (b). We do not, however, join up the two H atoms, as we do not expect a bond between them.

In drawing the structure in this way we have used our chemical knowledge that the usual valency of oxygen is two and that of hydrogen is one. Once we have made a covalent bond between the oxygen and each hydrogen, all of the valencies are satisfied so there is no 'need' for a bond between the two hydrogens.

In water, everything is clear cut, but as we look at more complex, and more exotic, molecules we will see that it becomes more and more difficult to say where the bonds are. A very good example of these problems is provided by the case of diborane, B_2H_6.

The arrangement of atoms in this molecule has been determined by X-ray diffraction, and is shown in the Fig. 1.13 (a). The task is now to join up the atoms in the same way we did for water. The two hydrogens on the left, and the two on the right, are 117 pm from the nearest boron atom, a distance which is certainly short enough to be considered as a bond, so we connect the nearby atoms to represent this. The hydrogen at the top is 132 pm from both boron atoms, which is again a short enough distance for it to be considered a bonded interaction. We therefore attach this hydrogen to both borons, and do the same for the hydrogen at the bottom.

Finally, the two boron atoms are 175 pm apart, which is just about close enough for a bonded interaction between two first-row elements, so we connect these two atoms by a bond, giving the complete structure shown in (b). There are two types of hydrogens in this structure: *terminal*, denoted 't' and *bridging*, denoted 'b'.

We can immediately see that there is a problem with this structure as it has nine bonds, each by implication comprising a pair of electrons giving a total electron count of 18. However, in B_2H_6 there are only 12 valence electrons. Put another way, in the structure the boron atoms have a valency of five, and the bridging hydrogens have a valency of two: both are unexpected.

It it clear from this discussion that the way the atoms are joined up in Fig. 1.13 (b) is inconsistent with both the electron count and the normal rules of valency. There are simply too many bonds.

🌐 *Weblink 1.5*

View and rotate the structure of diborane shown in Fig. 1.13.

The current understanding of the bonding in diborane is that the bonds between the boron atoms and the terminal hydrogens are of a conventional type in which a pair of electrons is shared between the two atoms. The bonding between the borons and the two bridging hydrogens is delocalized, in the sense that the electrons are shared over all four atoms, rather than between pairs of atoms. So, the lines in the figure are correct in that they show bonded interactions, but not all the lines represent simple bonds from the sharing of two electrons.

A second example is the lithium cluster compound, with formula $Li_4(CH_3CH_2)_4$, shown in Fig. 1.14. The core of the structure is four lithium atoms which are placed approximately at the corners of a tetrahedron. Each face of the tetrahedron is bridged by the CH_2 carbon of an ethyl group, such that there are close contacts between the carbon and three lithium atoms. As in the case of diborane, the atoms which are close enough to be considered bonded have been connected by lines. For the ethyl groups we can be sure that the line joining the two carbons does indeed represent a conventional C–C bond. As to the rest of the molecule, it is very much less clear where the bonds are. It is probably best simply to view the lines as a guide for the eye, indicating which atoms are close to one another, and in what spatial arrangement.

Our final example is the structure of solid NaCl, shown in Fig. 1.3 on page 5; here we see the familiar regular arrangement of sodium and chloride ions. Lines have been drawn which connect adjacent ions, but these lines certainly do not represent chemical bonds, as this structure is held together by non-directional electrostatic forces between ions. As in the lithium cluster, the lines are there simply to guide the eye and to help us appreciate the spatial arrangement of the ions.

In summary, we have seen in this section that we need to be careful about how we interpret the representations of molecular structures. For simple compounds, particularly organic-type molecules, there is probably a close correspondence between the lines connecting atoms and the presence of bonds. Once we move away from such structures, although we still connect nearby atoms by lines it is much less clear whether or not these lines are an accurate representation of the bonding in the molecule.

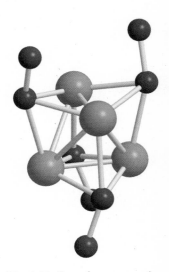

Fig. 1.14 Crystal structure of the compound $Li_4(CH_3CH_2)_4$, which contains a cluster of four lithium atoms (shown in orange); the faces of this cluster are bridged by carbon atoms (shown in dark grey). To avoid confusion, the hydrogen atoms are not shown.

1.6 How to draw molecules

Being able to draw representations of molecules on paper in a straightforward, clear and informative way is going to be very important for us in our study of chemistry. In this section we will look at the particular way in which organic molecules are drawn by practising chemists. It is important to get used to this method of drawing molecules as we will use it extensively throughout this book, and you will find that it is universally employed in more advanced texts.

Even a molecule as simple as methane already represents something of a challenge, as it is three-dimensional; various representations of CH_4 are shown in Fig. 1.15 on the next page. The so-called 'displayed' formula shown in (a) certainly represents the bonding correctly, but gives no sense of the three-dimensional shape which is shown in (b).

Although (b) is realistic, it is not convenient to draw by hand, so it is usually represented by the conventional representations shown in (c) and (d). In (c) the C–H bonds which are in the plane of the paper are drawn as single lines,

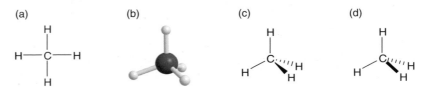

Fig. 1.15 Different representations of CH_4. The displayed formula shown in (a) does not represent the three-dimensional shape, shown in (b). Representations (c) and (d) are used to convey the three-dimensional shape. The C–H bonds in the plane of the paper are drawn as single lines, the bond coming out of the plane is represented by a solid wedge or thick line, and the bond going into the paper is represented by a dashed wedge or line.

Fig. 1.16 Shown in (a) is the three-dimensional shape of butane, C_4H_{10}; the conventional representation is given in (b). However, this is rather cluttered so the representation shown in (c) is generally preferred.

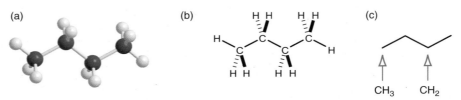

whereas the bond coming towards us out of the plane of the paper is shown by a solid wedge. The bond going into the plane of the paper is shown by a dashed wedge; note that the dashing runs across the wedge. Structure (d) is often used as an alternative to (c). Here the bond coming out of the paper is represented by a bold line, and the one going into the paper is represented by a dashed line.

Figure 1.16 (a) shows the three-dimensional structure of butane (C_4H_{10}). Note that the carbon chain lies in a plane and forms a zig-zag. In fact, there is plenty of experimental evidence that hydrocarbon chains often adopt this conformation, so the conventional representation shown in (b) is quite realistic. However, the problem with representation (b) is that it is very cluttered, and would take quite a long while to draw out.

To get round this problem, chemists use an abbreviated form of this structure, as shown in (c). In this representation all of the hydrogens attached to the carbons are left out, as are the letters 'C' representing the carbons themselves. Where two lines join, there is assumed to be a carbon, and in addition there is a carbon at the free end of a line. For a given carbon, it is simply assumed that there are sufficient hydrogens attached such that the valency of four is satisfied. So, the carbons on the far right and left are methyl groups, CH_3, and the two carbons on the middle are CH_2 groups.

This *framework* representation of molecules can be extended to other molecules by using the following rules:

1. Draw chains of atoms as zig-zags with approximately 120° angles.

2. Do not indicate carbon atoms by a 'C'.

3. Do not include any hydrogens, or bonds to hydrogens, which are *attached to carbon*.

4. Draw in all other atoms together with all of their bonds, and all the atoms to which they are attached

5. If a carbon atom is drawn in, include all the other atoms attached to it.

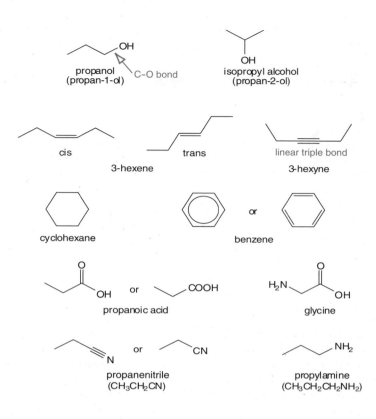

Fig. 1.17 Examples of organic molecules drawn as framework structures. There is often more than one way of drawing a structure, and which you choose depends on personal taste or on what part of the molecule you want to emphasise. For example, in the case of propanoic acid, if we are going to discuss the reactions of the carbonyl group, then the left-hand of the two representations is preferred.

Several examples of the applications of these rules are shown in Fig. 1.17. It is important not to be too rigid, though, about applying these rules, and to realise that often there are alternative ways of drawing the same structure. The key thing is to make sure that the structure is clear and unambiguous.

1.7 Common names and abbreviations

When drawing and talking about chemical structures there are quite a few abbreviations which are in common use for particular groups. The most important of these are given in Table 1.1 on the next page. For the more complex groups, framework structures are also given. In these, the wavy line indicates the point of attachment of the group.

As you will have learnt, there is a systematic way of naming chemical compounds. On the whole, most practising chemists will understand these names but they are also likely to use older, historical, names for some common compounds. A selection of these 'trivial names' are listed in Table 1.2 on page 19. You will simply have to get used to these as part of the everyday language of chemistry.

Table 1.1 Common abbreviations

abbreviation	name	formula	framework structure
Me	methyl	$-CH_3$	
Et	ethyl	$-CH_2CH_3$	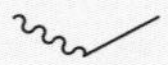
n-Pr	normal propyl	$-CH_2CH_2CH_3$	
i-Pr	iso propyl	$-CH(CH_3)_2$	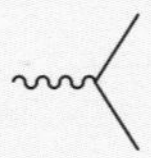
n-Bu	normal butyl	$-CH_2CH_2CH_2CH_3$	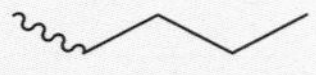
t-Bu	tertiary butyl	$-C(CH_3)_3$	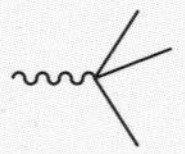
Ac	acetyl	$-COCH_3$	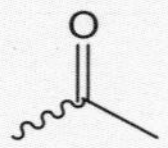
Ph	phenyl	$-C_6H_5$	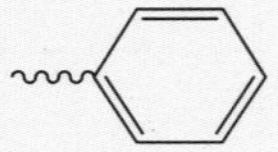
Ar	aryl	any aryl group (a substituted benzene ring)	
R	alkyl	any alkyl group	

1.8   Predicting the shapes of molecules using VSEPR

With modern theories and the aid of powerful computers we can predict, with quite good accuracy, the shapes of small- to medium-sized molecules. Often, however, we do not need to know all of the details of the molecular geometry (i.e. every bond length and bond angle), but simply require a general idea of the shape. The *valence shell electron pair repulsion* (VSEPR) model is a useful, non-computational, way of predicting the general shape of small molecules. We will look briefly at how this theory can be applied, its successes and its limitations.

The basic idea behind the VSEPR model is rather simple, and is best illustrated by considering the series of hydrides AH_n, where A is an element from the second period. All we do is count the number of *electron pairs* in the valence shell of A and then assume that these electron pairs will arrange themselves in such a way as to be as far apart from one another as possible. The process is illustrated in Fig. 1.18 on the next page

For example, in CH_4 the carbon contributes four electrons and each hydrogen contributes one, making eight in all. There are thus four pairs of electrons, one associated with each C–H bond. The idea is that these electron pairs, and

Table 1.2 Trivial names of some common compounds

trivial name	systematic name	framework structure
acetone	propanone	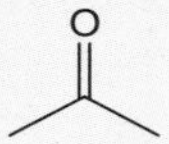
acetic acid	ethanoic acid	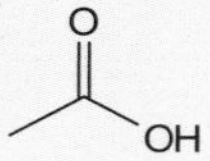
acetic anhydride	ethanoic anhydride	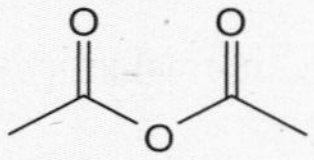
ether	ethoxyethane	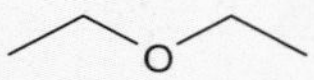
toluene	methyl benzene	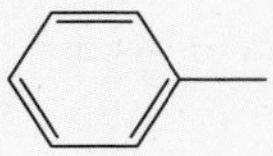
chloroform	trichloromethane	$CHCl_3$

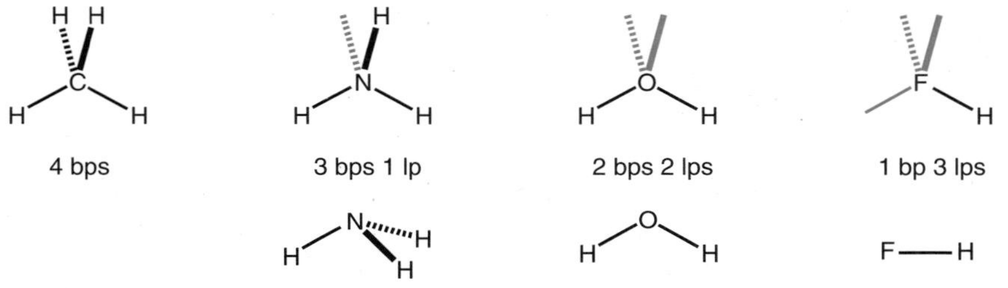

Fig. 1.18 Illustration of how the geometries of molecules with four electron pairs can be rationalized. CH_4 has four pairs of electrons in its valence shell, and if these arrange themselves as far apart as possible, we have the familiar tetrahedral arrangement shown on the left (a thick bond is coming out of the page, and a dashed one is going into the page). NH_3 also has a total of four electron pairs: three are bonded pairs (bps) and one is a lone pair (lp). As shown, these four pairs point towards the corners of a tetrahedron, with the lp occupying one position, indicated by the green line. The result is the familiar trigonal pyramidal shape drawn in a more conventional way beneath. OH_2 has two bps and two lps (indicated by green lines), and when these arrange themselves at the corners of a tetrahedron we obtain the familiar bent geometry. HF completes the picture with three lps and one bp.

of course the associated bonds, arrange themselves so as to be as far apart as possible, so as to minimize the repulsion between the electron pairs. In the case of four electron pairs, the repulsions are minimized by making a tetrahedral arrangement, which is of course in accord with the known tetrahedral geometry of CH_4.

For ammonia, NH_3, there are also four electron pairs, three involved in bonding to hydrogen and one *lone pair*. As in CH_4 these four pairs arrange themselves so as to point towards the corners of a tetrahedron, resulting in the familiar trigonal pyramidal geometry of ammonia.

Fig. 1.19 Illustration of the geometrical arrangements of two to six electrons pairs which minimize the repulsion between the pairs. The arrangement for three pairs is trigonal planar, four is tetrahedral, five is trigonal bipyramidal and six is an octahedral. For these last three arrangements, a picture of the solid from which the arrangement of ligands is derived is also shown. In a given shape, the positions occupied by the ligands are all equivalent, with the exception of the trigonal bipyramid; in this, the two *axial* positions at the top and bottom are different from the other three *equatorial* positions.

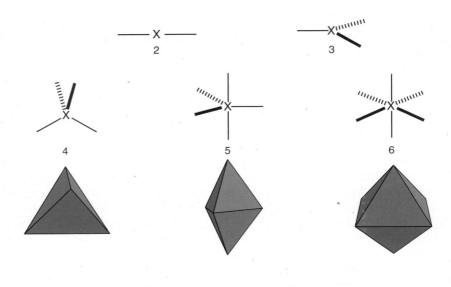

In fact experiment shows that the H–N–H bond angle in ammonia is 107°, somewhat less than the tetrahedral angle of 109.5°. We can refine the model to accommodate this by proposing that the repulsion between a bonded pair and a lone pair is greater than the repulsion between two bonded pairs. This difference in the repulsion would mean that N–H bonded pairs could get closer together, resulting in the required reduction in the H–N–H bond angle.

A similar argument applies to OH_2 where there are again four electron pairs, this time two bonded pairs and two lone pairs, resulting in a tetrahedral arrangement and hence a 'bent' geometry for water. The H–O–H bond angle in water is 104.5°, even smaller than the H–N–H bond angle in ammonia; this observation can be accommodated by supposing that the lone pairs repel one another even more than they repel bonded pairs.

A similar argument can be used for other numbers of electron pairs, and the optimum arrangements for between two and six lone pairs is shown in Fig. 1.19. For example, BeH_2 has just two pairs of electrons in its valence shell, and according to the figure these should be arranged at 180° to one another to give a linear geometry, which is exactly what is found experimentally.

BF_3 and the ion CH_3^+ (which is known in the gas phase) both have three electron pairs, which are arranged at 120° to one another, giving a trigonal planar geometry, again in agreement with experiment. The fluorides PF_5 and SF_6 have five and six bonded pairs, respectively, and their geometries are found to conform to the predictions of Fig. 1.19.

Multiply bonded species

The VSEPR model can also deal with multiply bonded species, such as CO_2 in which there are double bonds between each oxygen and the carbon. The valence shell of the carbon has eight electrons, four contributed by the carbon and, on account of the double bonds, *two* contributed by each oxygen. There are thus four electron pairs in the carbon valence shell, grouped into two 'pairs of pairs' of electrons associated with each C–O bond. By arranging the two bonds at 180° to one another, the repulsion between the two pairs of pairs of electrons is minimized, thus accounting for the linear geometry of CO_2.

A second example is sulfur trioxide, SO_3, which has six valence electrons contributed from the sulfur, and six in total from the three oxygens. These form three S–O double bonds, each comprising two electron pairs. Minimizing the repulsion between these 'pairs of pairs' gives a trigonal planar geometry, which is what is observed for SO_3.

A more complex example is the sulfate ion, SO_4^{2-}. Recognizing that the oxidation state of the sulfur is VI, we would normally draw the structure as having a double bond to two of the oxygens, and a single bond to the other two, as shown in Fig. 1.20 (a). The problem with this structure is that it is inconsistent with the known equivalence of the four oxygen atoms. The structure shown in (b), although rather less familiar, makes the four oxygens equivalent and, by giving the sulfur a 2+ charge, has the overall correct charge.

In structure (b) the electron count is six from the sulfur, less two for the 2+ charge, and a total of four from the oxygens. The total number of electrons is thus eight, giving four pairs which arrange themselves tetrahedrally. This is in accord with the known structure of the sulfate ion.

In more complex molecules, we can use the VSEPR approach to predict the spatial arrangement of bonds around a particular atom. For example, in organic molecules the four single bonds around a carbon are expected to be arranged tetrahedrally, and the two single bonds to an oxygen are expected to have an angle somewhat less than the tetrahedral angle between them. Similarly, the three single bonds to a nitrogen are expected to be arranged in a trigonal pyramid. This kind of approach is very useful for giving a rough idea of the arrangement of bonds around a particular atom.

1.8.1 Limitations of the VSEPR approach

The VSEPR approach is reasonably successful when applied to organic molecules and to other compounds consisting mainly of non-metallic elements. However, it is too simplistic a model for us to expect it to apply universally, and there are many cases where the predictions of the VSEPR theory are simply wrong.

For example, as shown in Fig. 1.21, in *amines* the three bonds around the nitrogen are found to be arranged in a trigonal pyramid, in accordance with the predictions of VSEPR. However, in *amides* the three bonds around the nitrogen are found to be arranged in a planar geometry, which is not what is expected from VSEPR. As we will see in later on, this planar geometry is associated with the bond between the nitrogen and the carbonyl carbon having partial double-bond character.

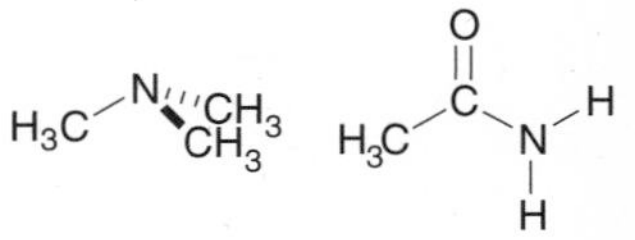

A further example is provided by the (gaseous) molecular fluorides of the Group 2 metals. As we have seen, the prediction is that these should be linear, but although this is so for BeF_2 and MgF_2, the bond angles in CaF_2, SrF_2 and BaF_2 are 145°, 120° and 108°, respectively. VSEPR can offer no explanation for this trend.

A final example of the failure of VSEPR is given by transition metal complexes. There are a large number of these of the type ML_6, all of which have an octahedral arrangement of ligands L, despite varying numbers of electrons in the valence shell of the metal.

It is important to realise that molecules do not necessarily adopt the shapes they do *because* of the repulsion between valence electrons, as implied by the VSEPR model. The real reasons why a shape is preferred are surely more

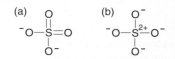

Fig. 1.20 Two different representations of the bonding in the sulfate anion. Representation (a) is more familiar, but is not consistent with the equivalence of all four oxygen atoms. In (b) these atoms are clearly equivalent, but note the 2+ charge on the sulfur.

Fig. 1.21 In an amine, shown on the left, the bonds around the nitrogen are found to be arranged in a trigonal pyramidal fashion, as predicted by VSEPR. However, in an amide, shown on the right, the three bonds to nitrogen lie in a plane, contrary to the predictions of VSEPR.

Fig. 1.22 Illustration of the pressure–volume relationship for a fixed quantity of an ideal gas at various temperatures. In (a) volume is plotted against pressure for various temperatures. The inverse relationship between these two quantities is best seen by plotting, for example, volume against the reciprocal of pressure, as shown in (b). The resulting straight lines are in accord with Boyle's Law.

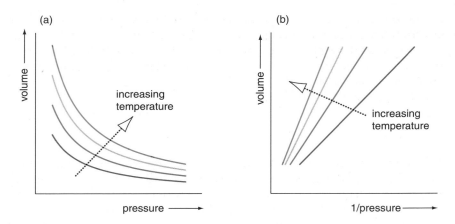

complex than the repulsion between electron pairs. The fact that the model sometimes predicts the correct shape does not mean that the model is correct – it might just be fortuitously giving the right answer. All in all, we need to approach the VSEPR model with a fair dose of scepticism. It does provide us with a useful guide, but we should not be too surprised if its predictions turn out to be incorrect.

1.9 The ideal gas

In the subsequent chapters we are quite often going to make use of the properties of gases, so this is a convenient moment to review these and introduce how they can be described in a quantitative way. Experimental work has shown that the pressure, volume and temperature of a fixed amount of a gaseous substance are related. For example, at a fixed temperature it is found that the volume is inversely proportional to the pressure applied i.e. the higher the pressure, the smaller the volume. This relationship is often known as *Boyle's Law*.

Similarly, for a fixed volume of gas, it is found that the pressure is proportional to the temperature i.e. the higher the temperature, the greater the pressure (know as *Charles' Law*). In fact these relationships are not followed precisely, with especially large deviations being seen for gases under high pressures or with higher densities. However, at sufficiently low pressures and densities, all gases follow Boyle's and Charles' Laws.

These observations led to the idea of an *ideal gas* (also called a *perfect gas*), which is one which obeys the *ideal gas equation*:

$$pV = nRT. \tag{1.1}$$

It is important to realize that the units of n are moles. n is often described as the 'number of moles'; although this usage is common, it is somewhat imprecise as n is not a number, but an amount.

In this equation, p is the pressure, V is the volume, T is the (absolute) temperature, n is the amount of gas in moles, and R is a universal constant known as the *gas constant*. In SI, the pressure is in N m^{-2}, the volume is in m^3 and R has the value 8.3145 J K^{-1} mol^{-1}.

If the amount of gas (i.e. n) and temperature are fixed, then the terms on the right-hand side of Eq. 1.1 are constant and so

$$pV = \text{const.} \qquad \text{or} \qquad V \propto \frac{1}{p}.$$

Thus the volume is inversely proportional to the pressure, which is Boyle's Law. In fact from Eq. 1.1 on the preceding page we can see that the exact relationship is $V = nRT/p$, so if we plot V against $1/p$ we will obtain a straight line whose slope is proportional to the temperature. This relationship is illustrated in Fig. 1.22 on the facing page.

Similarly, if we rearrange Eq. 1.1 to give

$$p = \frac{nRT}{V},$$

and then imagine keeping the amount of gas and the volume fixed, we have $p \propto T$, as illustrated in Fig. 1.23; this is Charles' Law. The last equation also tells us that, at fixed temperature and volume, the pressure is directly proportional to the amount of gas (i.e. n), which is also illustrated in the figure.

Equation 1.1 is called an *equation of state*, since it connects the variables p, V, n and T which describe the physical state of the system. To a good approximation, real gases obey the ideal gas equation provided that the pressure and density are not too high. For example, the behaviour of gases such as helium, nitrogen and methane at normal pressures and temperatures are well-approximated by the ideal gas equation. However, a dense gas, such as the vapour above a volatile liquid, or a gas such as ammonia at several atmospheres pressure, show significant deviations from ideal behaviour. Such gases are said to be *non-ideal*.

We will often need a relationship between the pressure, volume and temperature of a gas, such as that provided by the ideal gas law. For simplicity, we will simply *assume* that the gases we are dealing with obey the ideal gas law to a reasonable approximation. There are more complex laws which describe the behaviour of real gases to a better approximation (such as the van der Waals equation of state), but using such laws will make our calculations much more difficult, and so we will not go down this route.

At a molecular level, it can be shown that a gas in which there are negligible interactions between the molecules, and in which the molecules occupy a negligible part of the total volume, will obey the ideal gas law. It thus makes sense that ideal gas behaviour is seen at low pressures and densities, as this means that the molecules are far apart relative to their size. Similarly such an interpretation makes sense of the observation that helium, in which there are very weak interatomic interactions, behaves as an ideal gas over a much wider range of pressures than does ammonia, in which there are stronger intermolecular interactions.

1.9.1 Using the ideal gas equation

Equation 1.1 can be rearranged in the following way

$$pV = nRT$$
$$\frac{p}{RT} = \frac{n}{V}.$$

The quantity n/V is the moles per unit volume, in other words the *concentration*, which we will give the symbol c: $c = p/(RT)$. This shows that, at fixed temperature, the concentration of molecules in a gas is proportional to the pressure. So, in effect, the pressure of a gas is a measure of the concentration.

If you are unfamiliar with the use of SI units, refer to section 20.2 on page 882.

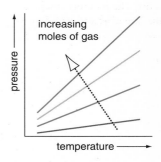

Fig. 1.23 Illustration of the pressure–temperature relationship for a fixed volume of an ideal gas; lines are plotted for amounts of gas in moles. These plots are in accord with Charles' Law.

Example 1.1 Using the ideal gas equation

Use the ideal gas equation to calculate the volume occupied by one mole of gas at atmospheric pressure (1.013×10^5 N m^{-2}) and 0 °C.

All we need to do is rearrange Eq. 1.1 on page 22 and then substitute in the values for n, p and T. We use SI units for all of the quantities (see section 20.2 on page 882), and are careful to use the absolute temperature, 273.15 K.

$$
\begin{aligned}
V &= \frac{nRT}{p} \\
&= \frac{1 \times 8.3145 \times 273.15}{1.013 \times 10^5} \\
&= 0.0224 \text{ m}^3.
\end{aligned}
$$

As we have used SI throughout, the answer comes out in m^3. This value may be familiar to you, as it is the molar volume at STP (standard temperature and pressure), often quoted as 22.4 dm^3.

The second example is to calculate the concentration, in molecules per cubic metre, of an ideal gas at STP. All we need to do is use Eq. 1.2 which gives the required concentration, N/V:

$$
\begin{aligned}
\frac{N}{V} &= \frac{pN_A}{RT} \\
&= \frac{1.013 \times 10^5 \times 6.022 \times 10^{23}}{8.3145 \times 273.15} \\
&= 2.69 \times 10^{25} \text{ molecules m}^{-3}.
\end{aligned}
$$

The amount in moles, n, is given by N/N_A, where N is the number of molecules (or atoms) and N_A is Avogadro's constant, which is the number of molecules (or atoms) per mole. Substituting this expression for n into the ideal gas equation gives

$$
pV = \frac{N}{N_A} RT.
$$

A simple rearrangement gives

$$
\frac{N}{V} = \frac{pN_A}{RT}. \tag{1.2}
$$

In this expression N/V is also the concentration, but this time in molecules (or atoms), rather than moles, per unit volume. As before, this measure of concentration is related to the pressure; we will find good use for both of these expressions for concentration later on.

Example 1.1 illustrates two typical calculations using the ideal gas equation.

1.9.2   Mixtures of gases: partial pressures

So far we have been talking about a single substance in the gas phase. However, it is fairly straightforward to extend the discussion to a mixture of gases by introducing the concept of the *partial pressure* of a gas. We will find this very useful later on when we discuss chemical equilibrium.

The concept of the partial pressure of a gas can be understood in the following way. Imagine that we have a mixture of gases, in a container of a particular volume and at a given temperature; the gas mixture exerts a pressure p_{tot}. Now imagine a thought experiment in which we remove all but one of the gases in the mixture. The pressure exerted by this remaining gas on its own is called its *partial pressure*.

Put another way, the partial pressure of a gas in a mixture is the pressure which that gas would exert *if* it occupied the whole volume on its own. This idea is illustrated in Fig. 1.24.

The total pressure exerted by the mixture is the sum of the partial pressures of the components of the mixture

$$p_{tot} = p_1 + p_2 + p_3 + \ldots,$$

where p_1 is the partial pressure of substance one in the mixture, p_2 that of substance two and so on. This relationship is sometimes referred to as *Dalton's Law*.

For a mixture of ideal gases, the partial pressure of component one is given by

$$p_1 = x_1 p_{tot},$$

where x_1 is the *mole fraction* of that substance. This mole fraction is defined as

$$x_1 = \frac{n_1}{n_{tot}},$$

where n_1 is the amount in moles of substance 1, and n_{tot} is the total amount (in moles) of all components in the mixture: $n_{tot} = n_1 + n_2 + n_3 + \ldots$.

Due to the way it is defined, the mole fraction of each component in a mixture is less than one, and the sum of the mole fractions of all the components is = 1

$$
\begin{aligned}
x_1 + x_2 + x_3 + \ldots &= \frac{n_1}{n_{tot}} + \frac{n_2}{n_{tot}} + \frac{n_3}{n_{tot}} + \ldots \\
&= \frac{n_1 + n_2 + n_3 + \ldots}{n_{tot}} \\
&= \frac{n_{tot}}{n_{tot}} = 1.
\end{aligned}
$$

If each gas in the mixture is ideal, we can work out the partial pressures by applying the ideal gas equation to each component. For example, for component one:

$$p_1 V = n_1 RT \qquad \text{hence} \qquad p_1 = \frac{n_1 RT}{V}. \tag{1.3}$$

Since $x_1 = n_1/n_{tot}$, it follows that $n_1 = x_1 n_{tot}$, and so equation above for p_1 can be written

$$p_1 = \frac{x_1 n_{tot} RT}{V}.$$

Example 1.2 demonstrates the practical application of these relationships.

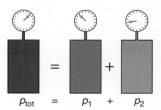

Fig. 1.24 Illustration of the concept of the partial pressure of a gas. The container on the left contains a mixture of two gases which together exert a total pressure p_{tot}. If the first gas occupied the total volume on its own, the pressure it would exert is called its partial pressure, p_1. Similarly, if the second gas occupied the total volume on its own, the pressure would be its partial pressure, p_2. The sum of the partial pressures is the total pressure.

Example 1.2 Partial pressures

A container of volume 100 cm^3 is filled with $N_2(g)$ to a pressure of 10^5 N m^{-2} at a temperature of 298 K; a second container of volume 200 cm^3 is filled with $O_2(g)$ to a pressure of 2×10^5 N m^{-2}, also at 298 K. The two containers are then connected so that the gases can mix. Calculate the partial pressure of the two gases in the mixture, assuming that the temperature is held constant at 298 K and that both gases are ideal.

All we need to know is the amount in moles of the two gases, then we can simply use Eq. 1.3 on the previous page to find the partial pressures. As we know the volume, temperature and pressure of each of the separate gases, the amount in moles of each is found be rearranging the ideal gas equation, $pV = nRT$, to give $n = pV/RT$

$$\begin{aligned} n(N_2) &= \frac{10^5 \times 100 \times 10^{-6}}{8.3145 \times 298} \\ &= 4.036 \times 10^{-3} \text{ mol.} \end{aligned}$$

Note that we had to covert the volume from cm^3 to m^3: 1 cm^3 = 10^{-6} m^3. Using this value in Eq. 1.3 on the preceding page we have

$$\begin{aligned} p(N_2) &= \frac{n(N_2)RT}{V} = \frac{4.036 \times 10^{-3} \times 8.3145 \times 298}{300 \times 10^{-6}} \\ &= 3.33 \times 10^4 \text{ N m}^{-2}. \end{aligned}$$

Note that this time, for the mixed gases, we had to use the volume of the mixture, which is 300 cm^3.

A similar calculation gives $n(O_2) = 1.614 \times 10^{-2}$ mol, and hence $p(O_2) = 1.33 \times 10^5$ N m^{-2}. In fact, the partial pressure of O_2 is *four* times that of N_2 as there are four times as many moles of O_2 as there are of N_2 (twice the initial volume, twice the initial pressure).

1.10   Molecular energy levels

One of the key ideas which underlies our understanding of chemistry at the molecular level is that atoms and molecules have available to them a set of *energy levels*. We will have a lot more to say about these energy levels throughout this book, and indeed in the next three chapters there is a lot of discussion of the energy levels available to the electrons in atoms and molecules. However, at this point it is useful just to have a quick overview of this very important topic.

When we say that molecules have energy levels available to them, what we mean is that the molecule cannot have any energy, but can only have an energy corresponding to one of these levels: we say that the molecule *occupies* a particular energy level. This rather strange result arises from *quantum mechanics*, which is the theory needed to describe the behaviour of microscopic objects, such as electrons, atoms and molecules. The following chapter, and Chapter 16, go into more detail about this theory.

It turns out that each kind of motion that an atom or molecule might undergo has associated with it a set of energy levels. These kinds of energy levels are:

translational energy levels: these are associated with the movement ('translation') of atoms and molecules through space. The separation between these energy levels is extremely small.

rotational energy levels: these are associated with the overall rotation of molecules (atoms do not rotate). The separation of these energy levels is very much greater than those for translation.

vibrational energy levels: these are associated with the vibration of the bonds in a molecule. The separation of these levels is typically two to three orders of magnitude greater than those for rotation.

electronic energy levels: these are associated with the electrons in atoms and molecules. The separation of these levels is usually much greater than for vibrational levels.

1.10.1 The Boltzmann distribution

Any one molecule, at any point in time, will occupy just one of the translational levels, one of the rotational levels, one of the vibrational levels and one of the electronic levels. As the molecules bump into one another, they exchange energy and so move between the different energy levels. Since collisions are very frequent, there is an incessant and relentless rearrangement of the molecules amongst the energy levels.

In any sample of material that we might handle in the laboratory (a *macroscopic* sample) there are an enormous number of molecules, so there is no way in which we can know which energy level each molecule is occupying. However, as a result of there being very many molecules, it turns out that we can specify the *average* number of molecules in any particular energy level: this average is called the *population* of the energy level.

At equilibrium, the populations of the levels are given by the *Boltzmann distribution*. This states that the population n_i of the energy level with energy ε_i is given by

$$n_i = n_0 \exp\left(\frac{-\varepsilon_i}{k_B T}\right), \tag{1.4}$$

where k_B is the Boltzmann constant (1.381×10^{-23} J K^{-1}), T is the absolute temperature, and n_0 is the population of the lowest level (which has energy $\varepsilon_0 = 0$).

As a result of the properties of the exponential function, the Boltzmann distribution says that the population of a level decreases as the energy of that level increases. Furthermore, the population of the higher energy levels increases as the temperature increases. This last point comes about because as T increases, $1/T$ decreases, and so $-\varepsilon/k_B T$ becomes less negative.

The predictions of the Boltzmann distribution are all summed up in the graph shown in Fig. 1.25. The vertical axis shows n_i/n_0, which is the population of the ith level, n_i, expressed as a fraction of the population of the lowest energy level, n_0; n_i/n_0 is called the fractional population. The maximum value that the fractional population can take is therefore one, and this occurs when all of the molecules are in the lowest level.

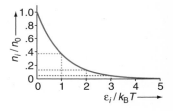

Fig. 1.25 Graph showing how the fractional population of the ith level, n_i/n_0, varies as a function of the energy of that level, expressed as a fraction of $k_B T$. The red, green and purple dashed lines show the fractional populations for $\varepsilon/k_B T = 1, 2$ and 3, respectively. The population is only significant if ε_i is of the order of, or less than, $k_B T$.

⇨ The properties of the exponential function are reviewed in section 20.4 on page 895.

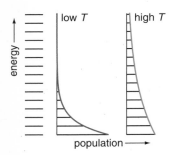

Fig. 1.26 Illustration of how the populations of a set of energy levels changes with temperature. The energy levels are shown on the left. In the two graphs on the right the population of each level is indicated by the *length* of the line, with the smooth curve showing the overall trend. At low temperatures, only the first few levels are occupied, but at high temperatures many more of the higher levels are occupied. Note that at the higher temperature the population of the lower levels decreases relative to what they were at the lower temperature.

🌐 *Weblink 1.6*

This link takes you to real-time version of Fig. 1.26 using which you can explore the effect of changing the energy level spacing and the temperature on the populations of the energy levels predicted by the Boltzmann distribution.

The horizontal axis is the energy of the level, expressed as a ratio of $k_B T$. Since the units of k_B are J K^{-1} and those of T are K, the product $k_B T$ has units of joules i.e. it is an energy. The ratio $\varepsilon/k_B T$ is therefore an energy divided by an energy, so it has no units and is said to be *dimensionless*. The graph illustrates what has already been explained: as the energy ε_i increases, the population decreases, and as the temperature increases the population of the levels above the ground state increases.

If the energy of the level is equal to $k_B T$, then $\varepsilon_i/k_B T = 1$, and from the graph the fractional population of this level is 0.37. Doubling the energy so that $\varepsilon_i/k_B T = 2$ reduces the fractional population to 0.13, and making $\varepsilon_i/k_B T = 3$ reduces the population still further to 0.05.

What this illustrates is that, as a result of the exponential function, the only levels which have significant populations are those whose energies are of the *order of* or *less than* $k_B T$. This is an exceptionally important result, which we will use frequently.

Figure 1.26 illustrates in a slightly different way how the populations change with temperature. This diagram shows in the *actual* values of the populations, rather than the fractional populations n_i/n_0 shown in Fig. 1.25 on the previous page. As the temperature increases, the higher energy levels become more populated, but since the number of molecules is fixed, this means that the population of the lower levels has to decrease. The overall effect of increasing the temperature is that the molecules are spread out more evenly, rather than being concentrated in the lower levels.

The thermal energy

The quantity $k_B T$ is often thought of as a measure of the thermal energy, as it is the size of this quantity which determines which energy levels will be populated at thermal equilibrium. At 298 K, $k_B T = 4.1 \times 10^{-21}$ J, which is rather a small number to comprehend. It can be made more tangible by multiplying by Avogadro's constant, so that we have an energy per mole; doing this gives the value 2.5 kJ mol^{-1}. We see that the thermal energy is a lot less than the energies associated with chemical reactions.

The spacing of translational energy levels is of the order of 10^{-41} J, which is incredibly much smaller than $k_B T$ at room temperature. There are therefore an extremely large number of translational energy levels which have significant populations.

Rotational energy levels are spaced by around 10^{-23} J, which is significantly smaller than $k_B T$, so typically there are hundreds of such energy levels which have significant populations. In contrast, vibrational energy levels are spaced by around 10^{-21} J, so only the first one or two levels above the ground state are populated to a significant extent. Finally, electronic energy levels are even more widely spaced, so it is unusual for any other than the ground state to be occupied.

Relation between the Boltzmann constant and the gas constant

In section 1.9 on page 22 the ideal gas equation, which relates pressure, volume and temperature, was introduced: $pV = nRT$. In this equation, the gas constant, R, is in fact related to the Boltzmann constant in the following way:

$$R = k_B N_A,$$

where N_A is Avogadro's constant i.e. the number of molecules (or atoms) per mole. It is because of this relationship between R and k_B that the gas constant appears in many expressions which, on the face of it, appear to have little to do with gases.

1.11 Moving on

We now need to move on to develop our understanding of bonding so that we can answer the question as to why it is that one compound consists of molecules held together by covalent bonds, whereas another exhibits ionic bonding. We will start by describing covalent bonding, which, as you know, is all about the 'sharing' of electrons.

The behaviour of small particles such as electrons is described using a theory known as quantum mechanics. We will introduce this theory by first using it to describe the behaviour of electrons in atoms, and then develop the theory further to describe the behaviour of electrons in molecules, and hence in chemical bonds.

You are already familiar to some extent with the quantum mechanical description of atoms, as the familiar atomic orbitals ($1s$, $2s$, $2p$ etc.) are a consequence of this theory. In the following chapter we will say much more about where these orbitals come from, and what they actually represent.

QUESTIONS

1.1 The energy released by the complete combustion of gaseous methane to CO_2(g) and H_2O(g) is 803 kJ mol^{-1}. Given that the O–O bond energy in O_2 is 498 kJ mol^{-1}, the C–O bond energy in CO_2 is 805 kJ mol^{-1} and the O–H bond energy in H_2O is 497 kJ mol^{-1}, estimate the C–H bond energy in methane. (Hint: write a balanced chemical equation for the combustion of methane and then think about the number and type of bonds broken or made as the reaction proceeds).

1.2 What is the distinction between a *molecular* solid and an *ionic* solid? Account for the following observations:

(a) Solid $PbBr_2$ does not conduct electricity, but when molten the salt is a good conductor.

(b) Neither solid naphthalene nor molten naphthalene conduct electricity.

(c) Metallic gold, both when solid and molten, conducts electricity.

naphthalene

1.3 As we go down Group 18, the noble gases, the atoms become more polarizable. Explain what you understand by this statement. Also explain how this trend in polarizability can be used to explain the observation that the boiling points of the liquefied noble gases increase as you go down the group.

1.4 Explain the following trends in the boiling points of the following two sets of hydrides:

set (a)	boiling point / °C	set (b)	boiling point / °C
H_2O	100.0	CH_4	−161.5
H_2S	−59.6	NH_3	−33.3
H_2Se	−41.3	H_2O	100.0
H_2Te	−2		

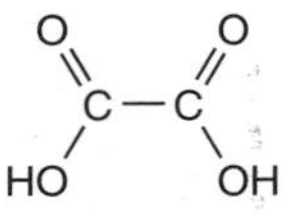

1.5 What types of intermolecular forces are present in the following molecules: (a) butane C_4H_{10}; (b) CH_3F; (c) CH_3OH; (d) CF_4?

1.6 In the gas phase, ethanoic acid is thought to exist as a dimer, held together by *two* hydrogen bonds. Suggest a structure for the dimer.

In the solid, oxalic acid, $(COOH)_2$, forms extended chains, also held together by hydrogen bonds. Sketch a likely structure for such a chain.

oxalic acid

1.7 The following framework structures are poorly drawn or simply implausible. Point out the errors in each, and re-draw them correctly.

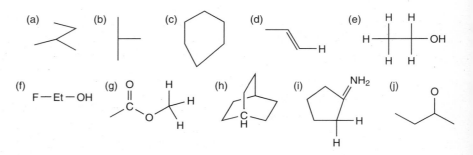

1.8 Find the molecular formula (i.e. $C_aH_b...$) of each of the following framework structures:

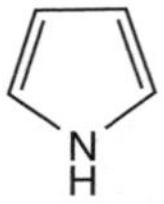

pyrrole

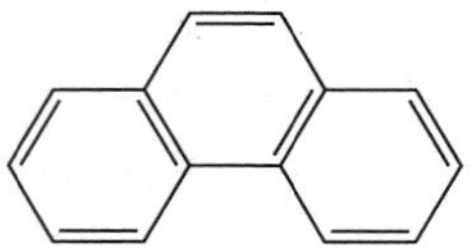

phenanthrene

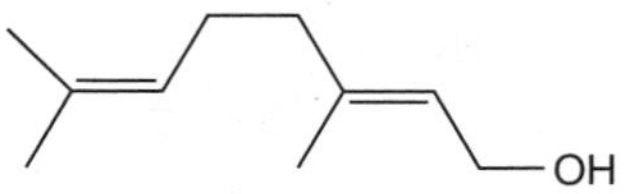

geraniol

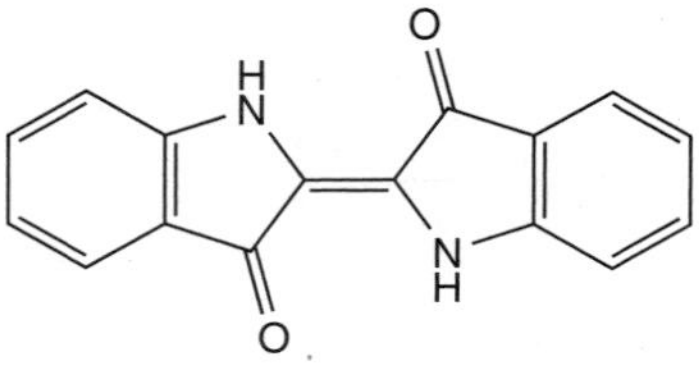

indigo

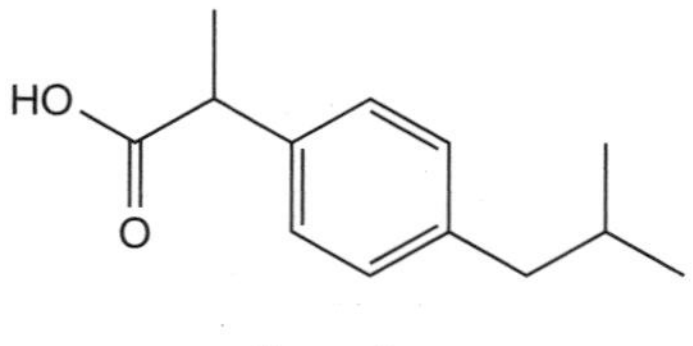

ibuprofen

1.9   Draw framework structures of the following molecules:

t-BuOH   i-PrOEt   EtOAc   EtAc

n-BuCl   AcOH   PhMe   PhH

Ac₂O   EtCN   Et₂O   n-PrNHAc

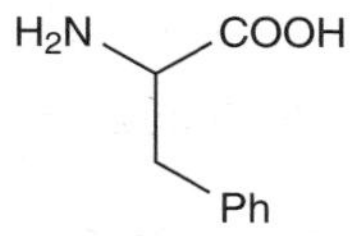

(Phenylalanine) 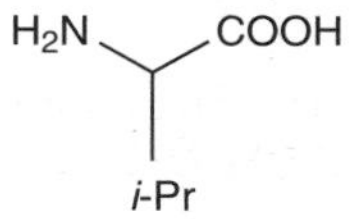   (Valine) 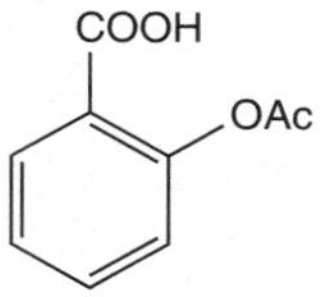   (Aspirin)

1.10   Draw framework structures of the following molecules:

(a)   CH₃CH₂CH₂NHCH₂CH₃

(b)

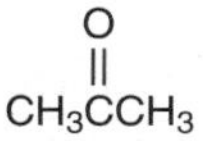

(c)

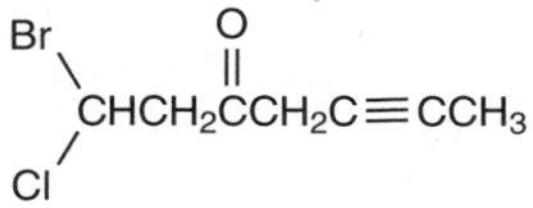

(d)   HOCH₂CH₂CO₂Me

(e)   (CH₃)₃CNH₂

(f)

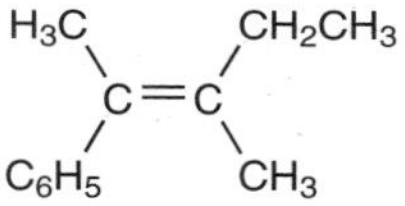

(g)

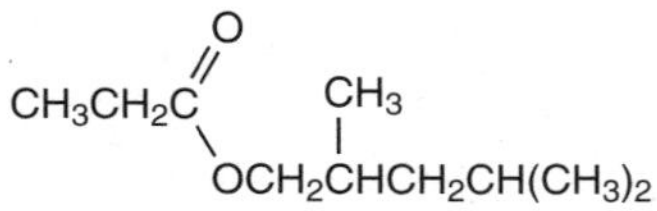

(h)
A triacyl glycerol 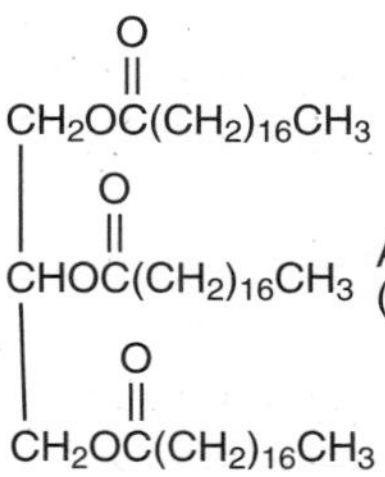
(a component of saturated fat)

(i)

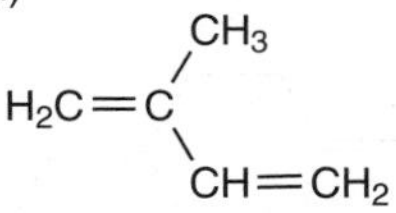

Isoprene
(a component of natural rubber)

(j)   C(CH₂ONO₂)₄

PETN
(an extremely powerful explosive)

Hint: the nitro groups can be written simply as
-NO₂, or drawn out in full as:

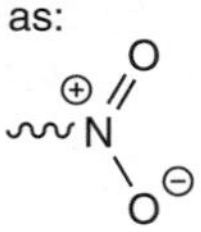

(k)

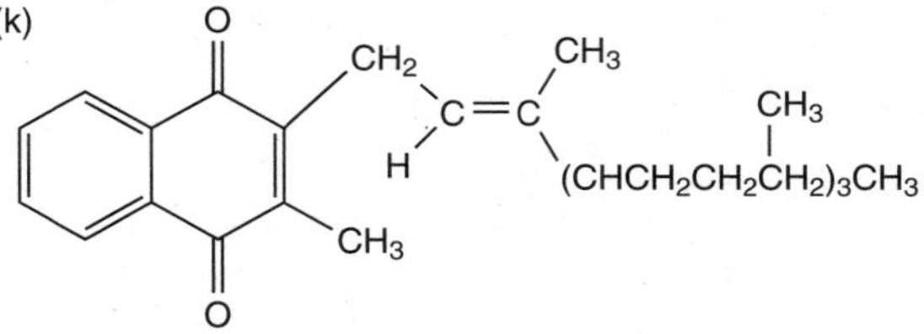

Vitamin K₁

1.11 The boron hydride with formula B_4H_{10} is often drawn as

(a) If each line represents a conventional bond in which two electrons are shared between two atoms, how many electrons are indicated by this structure?

(b) Assuming that each boron contributes three electrons, and each hydrogen contributes one, how many valence electrons are there in B_4H_{10}?

(c) How can you reconcile your answers to (a) and (b)?

1.12 Use the VSEPR model to predict approximate structures for the following species: (i) BH_3, (ii) BH_4^-, (iii) H_3O^+, (iv) CH_5^-, (v) PCl_5, (vi) PCl_4^+, (vii) PCl_6^-, (viii) NO_3^- (see opposite). (Hint: for the charged species, first work out the number of electrons in the valence shell ignoring the charge, and then reduce this total by one for a positive overall charge, or increase it by one for an overall negative charge).

1.13 ClF_3 is a highly reactive but nevertheless well-characterized volatile liquid used (among other things) to produce UF_6 in the processing of nuclear fuels. It has the following T-shaped structure

(a) Use the VSEPR theory to show that the structure of ClF_3 can be expected to be based on a trigonal bipyramid.

(b) The T-shaped structure can be considered to be a distorted trigonal bipyramid in which two 'equatorial' positions are occupied by lone pairs. Draw a diagram to illustrate this, and suggest why the bond angle in ClF_3 is not 90° as it would be in a regular trigonal bipyramid.

1.14 Explain why, at normal pressures and temperatures, $MgCl_2$ is a solid, SiO_2 is a solid, CO_2 is a gas and Ar is a gas.

1.15 Calculate the concentration, in moles m^{-3} and molecules m^{-3}, of nitrogen gas at a pressure of 0.1 atmospheres and a temperature of 298 K. You may assume that the gas behaves ideally. (1 atmosphere is 1.013×10^5 N m^{-2})

1.16 What pressure will one mole of an ideal gas exert at 298 K if it is confined to a volume of (i) 1 m^3, (ii) 1 dm^3, and (iii) 1 cm^3?

1.17 A container of volume 100 cm^3 contains 1.0×10^{-4} moles of H_2 and 2.0×10^{-4} moles of N_2, such that the total pressure is 0.1 atmospheres. Calculate the mole fraction and partial pressure of each species. Also, calculate the temperature of the mixture.

1.18   Two containers of equal volume, and held at the same temperature, each contain the same amount in moles of an ideal gas. Explain why the pressure in each container is the same.

Suppose that one of these containers is filled with gas A and one with an equal amount in moles of gas B. Now, we connect the two containers such that the gases A and B mix. On mixing, what happens to the total pressure? How are the partial pressures of A and B related to the total pressure?

1.19   Suppose we take a container of fixed volume and maintain it at a constant temperature. In the container there is a mixture of three gases, which can be considered to be ideal. Suppose that the amount in moles of one of the gases is now increased. What happens to: (i) the total pressure and (ii) the partial pressures of each gas?

If the amount in moles of each gas is doubled, what happens to (i) the total pressure, (ii) the mole fraction of each gas?

1.20   Calculate the fractional population (i.e. n_i/n_0) of the following energy levels of carbon monoxide, at 298 K and at 2000 K:

(a)   A rotational level at energy 8.0×10^{-23} J above the ground level.

(b)   A vibrational level at energy 4.3×10^{-20} J above the ground level.

(c)   An electronic level 1.3×10^{-18} J above the ground level.

Comment on your answers.

Electrons in atoms

Key points

- Quantum mechanics provides a complete description of the behaviour of electrons in atoms and molecules.

- The energy of an electron is quantized – that is, it can only have certain values.

- Each electron is described by a wavefunction; the square of the wavefunction gives the probability distribution of the electron.

- In the hydrogen atom, the familiar orbitals ($1s$, $2s$, $2p$ ...) are the electron wavefunctions.

- The energy and shape of each orbital is specified by three quantum numbers.

- In multi-electron atoms, the electrons can be assigned to hydrogen-like orbitals.

- The energies and sizes of the orbitals in multi-electron atoms are affected by electron–electron repulsion in a way which can be understood using the concepts of screening and penetration.

- Trends in ionization potentials can be rationalized using orbital energies.

As chemists, our ambitious aim is to understand the structures of molecules, and the reactions which they undergo. To a large extent, this all boils down to understanding 'what the electrons are doing' in a molecule, for it is the electrons which are forming the bonds, the electrons which are rearranged when bonds are made and broken, and the transfer of electrons which leads to ions.

The next three chapters are devoted to developing an understanding of the behaviour of electrons first in atoms, and then in molecules. In the remainder of the book, we will go on to show how chemical structures and reactivity can be understood in terms of the electronic structure of the molecules involved. You can therefore see why the ideas which will be developed in these early chapters are so important, and why so much space is devoted to them, as they are the bedrock on which our understanding will be built.

This chapter starts the discussion by thinking about the behaviour of electrons in atoms – a topic which you are already quite familiar with. If asked to describe 'where the electrons are' in lithium, the chances are that as part of your answer you would explain that two of the electrons are in the $1s$ orbital,

---

Box 2.1 Classical mechanics

Classical mechanics is the theory, based on Newton's Laws, used to describe the behaviour of *macroscopic* objects. By 'macroscopic' we mean an everyday object which has a significant mass and size, such that we can observe its behaviour in a straightforward way. So a lump of metal of mass 1 g and the earth are macroscopic objects, but individual electrons and molecules are not. The behaviour of macroscopic objects is predicted to essentially arbitrary precision by classical mechanics. So, for example, we can guide a space craft across our solar system and engineer a soft landing on a planet simply using Newton's Laws to describe the motion. However, these laws do not apply to microscopic objects such as electrons.

---

and the third is in the $2s$ orbital – in other words, the electronic configuration of lithium is $1s^2 2s^1$. In this chapter we will look in more detail at what exactly these orbitals are, how they arise, their energies and what they tell us about the electronic properties of atoms. In the next chapter, we will extend this concept of an orbital to molecules, and see how *molecular orbitals* can be used to describe covalent bonding.

Atomic orbitals arise from *quantum mechanics*, which is the theory needed to describe the behaviour of individual atoms, molecules and the electrons they contain. This theory is very powerful and provides a complete description of the behaviour of atoms and molecules. However, quantum mechanics is quite subtle and involved, and the theory has to be framed in mathematical language.

At this stage we do not want to get bogged down in all the details of quantum mechanics, but want to get on and use the results it provides to help us understand the behaviour of electrons in atoms and molecules. Therefore what we are going to do is to introduce the key ideas of quantum mechanics in an essentially non-mathematical way, and then go on to look at what the theory predicts about the behaviour of electrons in atoms. If you are interested in more of the mathematics of quantum mechanics, then you can study Chapter 16 in Part II.

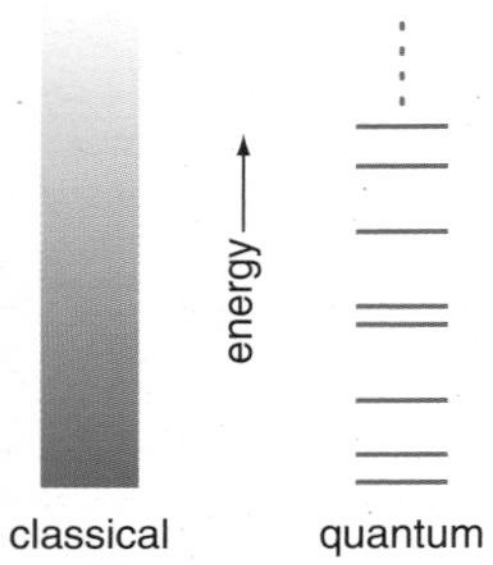

2.1 Introducing quantum mechanics

One of the most surprising and important results arising from quantum mechanics is that the energy of a particle, such as an electron, is *quantized*. This means that, rather than the energy being able to take any value, it is restricted to a certain set of values, usually called *energy levels*, as illustrated in Fig. 2.1.

In many ways this is a very surprising result, as our experience of the everyday world is that the energy can vary smoothly from one value to another i.e. it can vary *continuously*. For example, when riding a bicycle we can increase our speed – and hence our kinetic energy – smoothly from one value to another. It is not the case that our speed (energy) can only have certain values which we jump between. However, this is exactly what quantum mechanics predicts for the energy of an electron.

Quantum mechanics only applies to very small objects, such as electrons: it does not apply to the macroscopic objects we experience in the everyday world.

Fig. 2.1 In classical mechanics – which is what we experience directly in our day-to-day lives – energy can vary smoothly from one value to another. In quantum mechanics, which applies to very small particles, the energy can only have certain values, called energy levels.

For such objects, Newtonian or classical mechanics applies (Box 2.1 on the facing page), and in this description the energy is not quantized. This is why the prediction that the energy is quantized is initially an unfamiliar idea.

The second key idea which arises in quantum mechanics is that we cannot say precisely where a particle is located – all we can do is give the *probability* that a particle will be at a particular position. Different positions (coordinates) have different probabilities associated with them, so as a result it appears that the particle is 'spread out' over a region of space, as illustrated in Fig. 2.2. Like quantization, this idea simply does not accord with our experience of the everyday world, in which we expect to be able to specify exactly where an object is located.

Both the quantization of energy, and the idea that we can only talk about the probability of a particle being at a particular location, are closely related to the *wavefunction*, which is the next topic.

2.1.1 The wavefunction

In quantum mechanics, a particle has associated with it a *wavefunction* which, within the theory, specifies everything there is to know about the particle. For example, if we know the wavefunction we can work out the energy of the particle and determine the probability of finding it at a particular position. In this section we are going to discuss what the wavefunction is and how it is interpreted.

In mathematics a function is something which, for given values of the appropriate variables, can be evaluated to give a number. An example of a function is $x^2 - x + 2$, which can be evaluated for any value of the variable x to give a number e.g. if $x = 1$, $x^2 - x + 2 = 1^2 - 1 + 2 = 2$, so the function evaluates to 2. By evaluating the function over a range of values of the variable x we can make a plot, as shown in Fig. 2.3.

Such a function of x is usually written $f(x)$, where f is a label for the function and the '(x)' indicates that the function depends on the variable x. So, the notation $f(x)$ tells us that 'f is a function of x'. A function may depend on more than one variable, and if this is the case we simply add these to the bracket. So, a function of x, y and z would be written $f(x, y, z)$ e.g. $f(x, y, z) = ax + by + cz^2$, where a, b and c are constants.

In quantum mechanics, the wavefunction is simply a mathematical function of the relevant variables. For example, for the single electron in the hydrogen atom, the wavefunction is a function of the coordinates x, y and z which specify the position of the electron i.e. x, y and z are the variables. If there are two electrons, as in a helium atom, then the wavefunction would depend on the coordinates of each electron i.e. there would be six variables, x, y and z for each electron. The wavefunction is usually given the Greek letter 'psi', ψ, so for the electron in hydrogen it would be written as $\psi(x, y, z)$ to indicate that it is a function of x, y and z.

2.1.2 Probability interpretation of the wavefunction

One of the really important properties of the wavefunction is that it gives us a way of working out the *probability* of finding the particle (electron) in a particular region of space.

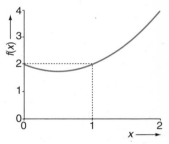

classical   quantum

Fig. 2.2 In classical mechanics the position of an object can be specified exactly. In quantum mechanics, we can only talk about the probability of a particle being at a particular location: some regions have higher probability, and some lower.

Fig. 2.3 A plot of the function $f(x) = x^2 - x + 2$ against the variable x. For any value of x, the function can be evaluated to give a number, as illustrated here for $x = 1$.

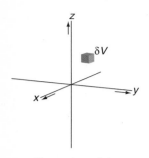

Fig. 2.4 Illustration of the probability interpretation of the wavefunction. The probability of finding a particle in a small volume δV, here indicated by a cube, centred at coordinates (x, y, z), is proportional to the *square* of the wavefunction at that point.

If the wavefunction at position (x, y, z) is $\psi(x, y, z)$, then the probability of finding the particle in a small volume δV located at this position is given by $[\psi(x, y, z)]^2 \, \delta V$. This is illustrated in Fig. 2.4. Note that this probability depends on the *square* of the wavefunction.

We can describe $[\psi(x, y, z)]^2$ as the probability per unit volume, more commonly known as the *probability density*. The reason for this name is that when we multiply the probability density, $[\psi(x, y, z)]^2$, by the volume, δV, we obtain the probability of being in that volume:

$$\text{prob. of being in volume } \delta V \text{ at position } (x, y, z) = \underbrace{[\psi(x, y, z)]^2}_{\text{prob. density}} \; \underbrace{\delta V}_{\text{volume}} .$$

This is analogous to multiplying the density (in kg m^{-3}) by the volume (in m^3) to obtain the mass (in kg).

If the wavefunction is that for an electron, the square of the wavefunction tells us how the probability density of the electron varies from place to place, in other words it tells us the spatial variation of the *electron density*. As we shall see, knowing this electron density is very important when it comes to understanding the behaviour of electrons in atoms and molecules.

You will often hear it said that 'the probability is proportional to the square of the wavefunction': this is almost, but not quite, true. The correct statement is that it is the *probability density* which is given by the square of the wavefunction. For most purposes, the subtle distinction between these two statements is not that important, but it is as well to be aware that only the latter is correct.

If you go on to study quantum mechanics in more detail you will find that the wavefunction can be complex, that is it can have a real and an imaginary part. If this is the case, then the probability density is given by $\psi^{\star}(x, y, z)\,\psi(x, y, z)$, where the $\star$ indicates the complex conjugate of the wavefunction. All of the wavefunctions we are going to be dealing with are strictly real, in which case the complex conjugate is the same as the original function, so the statement that the probability density is given by $[\psi(x, y, z)]^2$ is correct.

2.1.3 Energy

For much of the time, we will be using quantum mechanics to determine and describe the energy levels available to atoms and molecules, but before we start down this road it is a good idea to remind ourselves about the different types of energy, and the concept of the conservation of energy.

Energy comes in several different forms, but the two we are most concerned with here are *kinetic* and *potential*. Kinetic energy is associated with motion: a particle of mass m moving with speed (velocity) v, has kinetic energy $\frac{1}{2}mv^2$ i.e. the faster it is moving, the greater the kinetic energy.

Potential energy can be described as energy which has been 'stored up'. For example, if a mass is raised up it gains gravitational potential energy as its height above the earth increases. Similarly, when a spring is stretched, potential energy is stored up in the spring. Another example is the energy of interaction between charged particles, which is often called the electrostatic potential energy.

Finding the wavefunction: the Schrödinger equation

It is all very well to talk about energy levels and wavefunctions, but how do we find out what these are? Quantum mechanics tells us that they are found by solving the *Schrödinger equation*. We are not going to go into the details of how this is done, but will simply look at the solutions to this equation which have already been worked out. If you are interested in precisely what the Schrödinger equation is, and how it can be solved in some simple cases, you should refer to Chapter 16 in Part II.

The form of the Schrödinger equation depends on the particular system whose wavefunctions we are trying to find. For example, the Schrödinger equation for hydrogen will be different to that for helium, as there is an extra electron involved in the latter atom. It turns out that the key thing we need to know in order to construct the Schrödinger equation is how the potential energy experienced by the particles (e.g. the electrons) varies with position. As we are about to see, in the hydrogen atom this variation takes a simple well-known form.

Having set up the Schrödinger equation, we then need to solve it. This is where we come across a substantial difficulty with quantum mechanics, which is that an *exact* solution to the Schrödinger equation can only be obtained in a few very simple cases. Luckily, one of the cases is the hydrogen atom.

2.2   Introducing orbitals

In this section, and the one which follows, we are going to look at the simplest atom, which is one with a single electron – the hydrogen atom. For this system it is possible to solve the Schrödinger equation exactly, and so obtain the wavefunctions and the associated energy levels. In fact, you already know a lot about these wavefunctions as they are the familiar orbitals such as $1s$, $2s$, $2p$ and so on. We will look in more detail at the shapes and energies of these orbitals, the way they can be represented on paper, and on their mathematical forms.

We are interested in the behaviour of the electron, not the overall motion of the whole hydrogen atom. So we will imagine the nucleus to be stationary and positioned at the origin, and then let the electron move about this nucleus. In fact, because the electron is so much lighter than the nucleus, it turns out to be an excellent approximation to think of the motion of the electron in this way.

The electron is held to the nucleus as a result of the favourable electrostatic interaction between the positively charged nucleus and the negatively charged electron. Due to this interaction, the potential energy of the electron varies as $-1/r$, where r is the distance between the electron and the nucleus; the form of this *Coulomb potential* is shown in Fig. 2.5. When the electron is very far away from the nucleus the interaction between the two charges, and hence the potential energy, goes to zero. As the electron moves in closer, there is a favourable interaction resulting in a decrease in the potential energy i.e. the energy becomes negative.

The Coulomb potential only depends on the distance r and not on the direction – a situation which is described as *spherically symmetric*. It turns out that the wavefunctions themselves are not necessarily spherically symmetric, and can depend on both the distance from the nucleus and the direction.

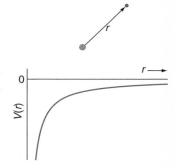

Fig. 2.5 In the hydrogen atom, there is a favourable interaction between the electron and the nucleus, on account of their opposite charges. This interaction results in the potential energy of the electron, $V(r)$, varying as $-1/r$, where r is the distance from the nucleus. The potential is zero at large values of r, where there is no interaction between the electron and the nucleus, and at shorter distances becomes negative as a result of the interaction being favourable.

The wavefunctions which are solutions to the Schrödinger equation for the hydrogen atom are so important that they have their own name – they are called *orbitals* or more specifically *atomic orbitals*. So we can interchangeably use the terms 1s orbital, 1s atomic orbital and 1s wavefunction.

The 1s orbital

The lowest energy wavefunction, which turns out to be the 1s orbital, depends only on the distance r between the electron and the nucleus. The function is very simple, taking the form of an exponential:

$$\psi_{1s}(r) = N_{1s} \exp(-r/a_0). \tag{2.1}$$

The constant a_0 is a length and is called the *Bohr radius*; it has the value 52.9 pm. N_{1s} is a constant called the *normalization constant*; for our discussion, its value is of no particular importance, so we will not worry about it further. Figure 2.6 shows a plot of this wavefunction as a function of the distance r. The wavefunction has its maximum value when $r = 0$ i.e. at the nucleus, and then falls off as the distance increases. Due to the properties of the exponential function, the wavefunction falls to half its initial value at about $r = a_0$, and to half again at about $r = 2a_0$.

Although this wavefunction only depends on the distance r, it is nevertheless three-dimensional, as the distance r can be measured in any direction from the nucleus. In other words, the 1s wavefunction is spherically symmetric. Imagining such a three-dimensional wavefunction in our minds, and representing it on paper can be quite difficult, so we will spend some time introducing different ways of representing such functions.

2.2.1 Representing orbitals on paper

The important point to grasp about a wavefunction is that at any point in space it has a value; what we need to do is develop a way of representing these values so that we can see how they vary with position.

Figure 2.7 on the next page is one such representation of the 1s wavefunction. What we have done here is to take a box, centred on the nucleus, and divide it up into small cubes. Each cube is then shaded according to the value of the wavefunction in the centre of that cube, with darker shading representing a larger value of the wavefunction.

So that we can see what is going on inside the box, three views are shown in which a different number of layers of cubes have been lifted up. On the surface that is exposed, each cube is labelled with a number which is proportional to the value of the wavefunction in that cube. For convenience, the maximum has been set to 100; you can see that the larger numbers go with darker shading.

View (a) exposes a layer of cubes which are well away from the nucleus, and so the wavefunction has a rather low value at these points. In view (b) the exposed face is somewhat closer to the nucleus, so the value of the wavefunction is larger; note that the maximum value is in the middle of the layer, as this is the point closest to the nucleus. Finally, in view (c) the nucleus is in the centre of the exposed face, so the central cube shows the maximum value (100) of the wavefunction. Due to the spherical symmetry of the wavefunction, we would obtain a similar picture if we lifted off layers from any of the faces of the box.

Another way of representing the three-dimensional shape of this orbital is to take a cross section through the wavefunction and then make a *contour plot*

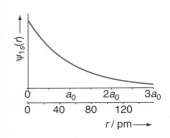

Fig. 2.6 Plot of the wavefunction of the 1s orbital as a function of the distance r from the nucleus. Two scales are given: one in terms of the Bohr radius, a_0, and one in pm.

⇨ The properties of the exponential function are reviewed in section 20.4 on page 895.

🌐 *Weblink 2.1*

Follow the weblink to see an animated version of Fig. 2.7.

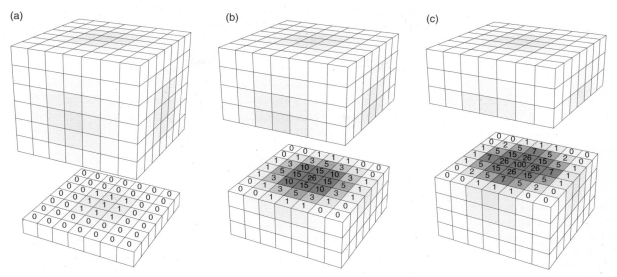

Fig. 2.7 Three-dimensional representation of the wavefunction of the $1s$ orbital plotted in Fig. 2.6 on the facing page. The space around the nucleus is divided up into small cubes, each of which is shaded according to the value of the wavefunction within that cube: the greater the value of the wavefunction, the darker the colour. The stack of cubes is split apart by lifting off several upper layers. On the exposed face, the numbers represent the value of the wavefunction in each cube. Arbitrarily, the maximum has been set to 100; the whole box is of side approximately eight Bohr radii.

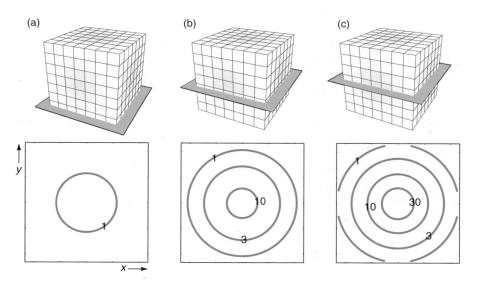

Fig. 2.8 Contour plots of three different cross sections taken through the $1s$ wavefunction at the positions indicated by the green planes; these cross sections correspond to the exposed faces shown in Fig. 2.7. The contour lines join points which have the same value of the wavefunction; the value for each line is shown. As the wavefunction is spherically symmetric, all of the contours are circular.

of the resulting values of the wavefunction. In a contour plot, lines are drawn between points which have the same value, just in the same way that contours are drawn on a map to represent the height of the land. What we see in the contour plot depends crucially on how many contours we draw and the values at which they are drawn.

Figure 2.8 on the preceding page shows contour plots of the cross sections corresponding to the exposed faces in Fig. 2.7. Arbitrarily, we have drawn contours at values of the wavefunction of 1, 3, 10 and 30 (recall that the maximum is set to 100). Section (a) is well away from the nucleus, so the wavefunction has decayed away and we only see one contour at a value of 1. In section (b) the wavefunction has higher values on the exposed face, so contours are seen at values of 1, 3 and 10. Finally, section (c) is taken right through the nucleus so the wavefunction has its maximum value and all four contours are seen. Since the wavefunction is spherically symmetric, the contours are all circular.

Contour plots are very useful ways of representing three-dimensional wavefunctions, but we do have to remember that what we see in such a plot depends crucially on where the cross section is taken, and the values of the contours which are drawn.

An alternative to drawing contours is to represent the value of the wavefunction using shading, with a deeper shade indicating a higher value of the wavefunction. Figure 2.9 shows such a representation of the 1s orbital for the same cross section as in Fig. 2.8 (c). Again, the spherical form of the wavefunction is clear, as is the fact that it is a maximum at the nucleus.

The final representation of the wavefunction we will use is a three-dimensional version of a contour plot. Imagine throwing a net over the function and then pulling the net tight until at all points on the net the wavefunction has the same value. The surface which this net creates is called an *iso-surface*, which is the three-dimensional analogue of a contour.

As with contours, just what the iso-surface looks like depends crucially on the value at which we choose to draw it. This point is illustrated in Fig. 2.10 where we have drawn three iso-surfaces for the 1s wavefunction, taken at the values 30, 10 and 3, where the maximum value (at the nucleus) has been set to 100. The computer program which was used to draw these iso-surfaces attempts to make them look solid by adding highlights and shading. The effect is reasonably convincing, and we certainly gain the impression that the wavefunction has spherical symmetry. However, the apparent 'size' of the orbital depends entirely on the value of the wavefunction at which we choose to draw the iso-surface.

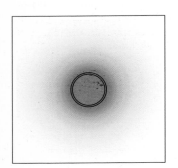

Fig. 2.9 A representation of a cross section taken through the 1s orbital in which the value of the wavefunction is indicated by the depth of the shading. The cross section is the same as in Fig. 2.8 (c). The ring indicates the radius at which the electron density in a thin spherical shell is a maximum.

2.2.2 The radial distribution function

In section 2.1.2 on page 37 we introduced the idea that the square of the wavefunction is the probability density, so that $\psi(x, y, z)^2 \, \delta V$ is the probability

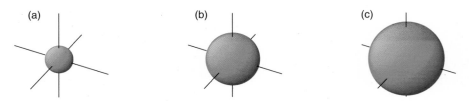

Fig. 2.10 Iso-surface representations of the wavefunction of the 1s orbital. Each plot is for the same range of the x, y and z coordinates as used in Fig. 2.7 on the preceding page, but in (a), (b) and (c) the iso-surface is drawn at values of the wavefunction of 30, 10 and 3. The maximum value, at the nucleus, is 100.

of finding the electron in a small volume δV at position (x, y, z). When we are thinking about atomic orbitals it is often useful to think not about the probability of the finding the electron in a small volume δV, but the probability of finding the electron in a thin shell of radius r and thickness δr, as illustrated in Fig. 2.11.

The reason that this approach is useful is that it adds up the probability in all directions, and so gives us a measure of the probability of finding the electron at a particular distance from the nucleus, regardless of the direction. Remember that if we are talking about an electron, this probability density is the electron density.

The *radial distribution function*, $P_{1s}(r)$, for the $1s$ orbital is defined as

$$P_{1s}(r) = 4\pi r^2 \, [\psi_{1s}(r)]^2.$$

With this definition, $P_{1s}(r) \, \delta r$ is the probability of finding the electron in a shell of radius r and thickness δr.

We can rationalize the form of the radial distribution function by noting that a sphere has surface area $4\pi r^2$, so that the volume of a thin shell of thickness δr will be $4\pi r^2 \times \delta r$. Thus $4\pi r^2 \, \delta r$ is a small volume δV, so $4\pi r^2 \, [\psi_{1s}(r)]^2 \, \delta r$ is the probability of the electron being in this volume.

Figure 2.12 shows the steps in constructing the radial distribution function (RDF) for the $1s$ orbital. Plot (a) is of the wavefunction, which we have seen before. Plot (b) shows $[\psi_{1s}(r)]^2$ and the rising function r^2; note that at larger values of r, $[\psi_{1s}(r)]^2$ is decreasing faster than r^2 is increasing. The product of these two functions, which is the RDF, is shown in plot (c).

Since r^2 is zero at $r = 0$, the RDF also goes to zero at this point. To start with, the RDF rises on account of the increase in the function r^2, but eventually the decreasing function $[\psi_{1s}(r)]^2$ wins out, and the RDF starts to fall. As a result, the RDF shows a maximum, which turns out to be at a distance of one Bohr radius. The position of this shell of maximum density is shown by the ring drawn in Fig. 2.9 on the facing page.

The presence of this maximum in the RDF can be rationalized by thinking about the volume of the thin shell, which is $4\pi r^2 \delta r$. At $r = 0$, the shell has zero radius and so has no volume. Thus, although the wavefunction is a maximum at this point, the probability of finding the electron in a shell of zero volume is nevertheless zero. This is why $P_{1s}(r)$ is zero at $r = 0$.

As r increases, the volume of the shell increases, but at the same time the wavefunction is decreasing. The two effects therefore work against one another.

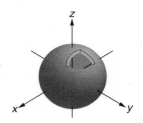

Fig. 2.11 Visualization of a thin shell surrounding the nucleus; the small section cut away allows us to see into the interior. If the radius of the shell is r, and its thickness is δr, the volume of the shell is $4\pi r^2 \delta r$.

(a)

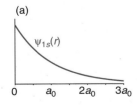

(b)

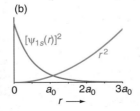

(c)

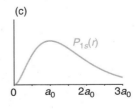

Fig. 2.12 Illustration of how the radial distribution function (RDF) for the $1s$ orbital is constructed. Plot (a) shows the wavefunction. Plot (b) shows the square of the wavefunction (in green) and the function r^2 (in red). Plot (c) shows the product $4\pi r^2 \times [\psi_{1s}(r)]^2$, which is the RDF, $P_{1s}(r)$. The vertical scales for each function are different. The maximum in the RDF results from the competition between the rising function r^2 and the falling function $[\psi(r)]^2$

To start with, the volume increases faster than the wavefunction decreases, so the RDF increases. Eventually the decrease in the wavefunction dominates over the increase in the volume, and the RDF starts to decrease. The result is that at some distance the RDF reaches a maximum.

We will often use RDFs to compare the distribution of electrons between different orbitals.

2.3 Hydrogen atomic orbitals

Having looked at the $1s$ orbital in some detail, and introduced the different ways of representing the orbital, we now need to turn to the other atomic orbitals, and look at their properties.

2.3.1 Quantum numbers, orbitals and energies

When we solve the Schrödinger equation we often find that the wavefunctions and energy levels are characterized by a particular set of numbers, called *quantum numbers*. By 'characterised' it is meant that if we know the quantum numbers we can fairly easily work out the mathematical form of the wavefunction and the associated energy. Quantum numbers are usually restricted to be integers $(0, 1, 2 \ldots)$, or half integers $\frac{1}{2}, \frac{3}{2} \ldots$.

For the hydrogen atom, it turns out that there are *three* quantum numbers which characterize the orbitals:

(a) The *principal* quantum number, n, which takes values $1, 2, 3 \ldots$

(b) The *orbital angular momentum* quantum number, l, which takes values from $(n - 1)$ down to 0, in integer steps.

(c) The *magnetic* quantum number, m_l, which takes values from $+l$ to $-l$ in integer steps; there are thus $(2l + 1)$ different values of m_l (i.e. l positive values, l negative values, plus a value of zero).

Each separate orbital has a unique set of values of the three quantum numbers n, l and m_l. A knowledge of these quantum numbers enables us to specify the energy and three-dimensional shape of the orbital, and it is this relationship which we are going to look at in detail here.

It turns out that the energy of an orbital depends *only* on the value of the principal quantum number n, so it is common to group orbitals with the same value of n together into *shells*. Traditionally, orbitals with $n = 1$ are called the K shell, those with $n = 2$ the L shell, and those with $n = 3$ the M shell. We will start by identifying the number of orbitals in each of these shells, and the associated quantum numbers.

The K shell ($n = 1$)

Recall that the orbital angular momentum quantum number l takes values from $(n - 1)$ down to 0 in integer steps. Thus, if $n = 1$ the only value for l is zero. The magnetic quantum number m_l takes values between $+l$ and $-l$ in integer steps, so as $l = 0$ there is again only one value, which is $m_l = 0$. So, in the K shell there is just one orbital with $n = 1$, $l = 0$ and $m_l = 0$.

Traditionally this orbital is denoted $1s$. The number gives the value of n and the letter gives the value of l according to Table 2.1. This orbital is spherically symmetric, as has already been discussed in detail in section 2.2 on page 39.

The L shell ($n = 2$)

This shell has $n = 2$, so the maximum value of l is $(n − 1)$ which is $(2 − 1) = 1$. Remember that l can take integer values from this maximum down to zero, so that $l = 0$ is also possible. As we had before, for $l = 0$ there is only one value of m_l, which is zero. This orbital is denoted $2s$, the '2' being the value of n, and the s indicating that $l = 0$.

For $l = 1$, there are more values of m_l. Recall that the rule is that m_l can take values between $+l$ and $−l$ in integer steps, so there are three m_l values: $+1$, 0 and $−1$. These three orbitals are all denoted $2p$: the '2' giving the value for n, and the 'p' being the letter for $l = 1$, according to Table 2.1. We will see shortly that the three $2p$ orbitals have different shapes, and they are often denoted $2p_x$, $2p_y$ and $2p_z$.

The M shell ($n = 3$)

For this shell $n = 3$, so l can take the values 2, 1 and 0. When $l = 2$, m_l can take the values $+2$, $+1$, 0, $−1$ and $−2$: these correspond to the five $3d$ orbitals, d being the letter used to represent $l = 2$.

For $l = 1$ we have as before three orbitals, denoted $3p$, and in addition for $l = 0$ there is one orbital denoted $3s$. In total there are nine orbitals in the M shell.

Orbital energies

The energy of an atomic orbital (AO) in hydrogen depends *only* on the principal quantum number n

$$E_n = -\frac{Z^2 R_H}{n^2}. \tag{2.2}$$

In this expression E_n is the energy of the orbital with principal quantum number n, Z is the nuclear charge (here = 1), and R_H is the *Rydberg constant*, which takes the value 2.180×10^{-18} J, or 1312 kJ mol^{-1} in molar units. The Rydberg constant can be expressed in terms of other fundamental constants in a way which arises from the solution of the Schrödinger equation.

We will find it most convenient to express the orbital energies in electron-volts (eV), in which case R_H takes the value 13.61 eV; the reason for choosing to use these units is that the numbers have a convenient size. If you are unfamiliar with eV, Box 2.2 on the following page gives some background information.

These energies are measured on a scale whose zero point corresponds to the electron being infinitely removed from the nucleus. At such large distances the interaction with the nucleus is zero, so it makes sense to call this point zero energy. As the electron moves in towards the nucleus there is a favourable interaction, so the energy falls and becomes negative – this is why the energies in Eq. 2.2 are negative. The energy levels are illustrated in Fig. 2.13.

Orbitals which have $n = 1$ thus have the lowest energy (of $−R_H$), those with $n = 2$ have the next lowest (of $−\frac{1}{4}R_H$) and so on. As n approaches infinity, the energy goes to zero, which corresponds to the electron being very far away from the nucleus, in other words, when the electron is ionized.

Table 2.1 Letters used to represent l values

l	letter
0	s
1	p
2	d
3	f

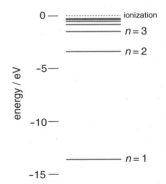

Fig. 2.13 Illustration of the energy levels of the hydrogen atom. The energy is characterized by the principal quantum number n, and is measured downwards from a zero point which corresponds to ionization.

Box 2.2 Energy units: J, kJ mol^{-1} and eV

The SI unit for energy is the joule (symbol J), but it is not always convenient to use this unit, especially when we are dealing with atoms and molecules. The energy of the electron in the 1s orbital in hydrogen is, from Eq. 2.2 on the previous page, -2.180×10^{-18} J: an inconveniently small and rather unmemorable number.

If we consider a mole of hydrogen atoms, then the energy is obtained simply by multiplying the energy of one atom by Avogadro's constant, N_A. So the energy of the 1s electron is $N_A \times -2.180 \times 10^{-18}$, which is -1312 kJ mol^{-1}. This is a more memorable number, and one we can compare with the energies of chemical reactions, which are also quoted in kJ mol^{-1}.

It is very common to use another unit, called the *electron-volt* (symbol eV) to express the energies of atoms and molecules. One eV is the energy which a single electron would acquire by being accelerated through a potential difference of one volt. In practice, we convert from joules to eV by dividing by the charge on a single electron, which is 1.602×10^{-19} coulombs.

So, in eV the Rydberg constant is $2.180 \times 10^{-18}/1.602 \times 10^{-19}$, which is 13.61. Therefore, the energy of the electron in the 1s orbital is -13.61 eV, a nicely sized number that we can grasp.

You can read more about non-SI units in section 20.2.3 on page 886.

Degeneracy

A rather special feature of these orbitals for hydrogen is that the energy depends *only* on the value of the principal quantum number. This means that the 2s and 2p orbitals have the same energy, as do the 3s, 3p and 3d orbitals. In quantum mechanics, wavefunctions which are different but have the *same* energy are described as being *degenerate*. So the 3s, 3p and 3d are described as degenerate orbitals or wavefunctions.

In quantum mechanics, degeneracy is always associated with symmetry, and the spherical symmetry of the hydrogen atom gives it unusually high degeneracy. For example, in the M shell there are nine orbitals (one 3s, three 3p and five 3d) all of which are degenerate. We will see later on, though, that in multi-electron atoms this degeneracy is lost.

2.3.2 The shapes of the 2s and 2p orbitals

In this section we are going to describe the shapes of the 2s and 2p orbitals using different representations of the orbitals on paper in order to illustrate their key features. Once we have got to grips with the shapes of these orbitals, we will look at the mathematical form of the wavefunctions and see how the shapes arise from these.

2s

The 2s wavefunction, like that for the 1s, is spherically symmetric, but unlike the 1s orbital, the 2s has both a positive and a negative part. Figures 2.14, 2.15 and 2.16 show different representations of the 2s orbital just like those used in section 2.2 on page 39 for the 1s orbital.

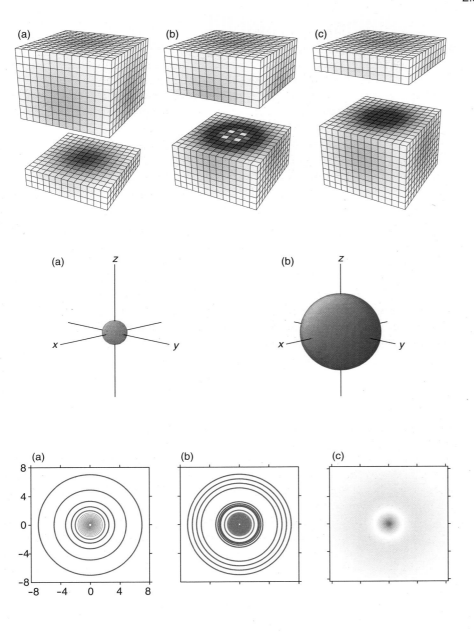

Fig. 2.14 Representation of the wavefunction of the 2*s* AO using the same method as in Fig. 2.7 on page 41. The depth of shading of each small cube is proportional to the value of the wavefunction in that cube; red indicates positive values, and blue negative. The nucleus is in the centre of the box, and each side is of length sixteen Bohr radii.

Fig. 2.15 Iso-surface representations of the 2*s* AO. The surface shown in red in (a) is drawn at a positive value of the wavefunction, whereas that shown in blue in (b) is drawn at the same value, but negative. The spherical symmetry of the wavefunction is immediately evident. The region of space plotted here is the same as in Fig. 2.14.

Fig. 2.16 contour plots (a) and (b) are of a cross section taken through the 2*s* AO at the nucleus; red and blue contours are for positive and negative values. The green contour indicates where the wavefunction is zero. The contours are spaced more closely in (b) than in (a). Plot (c) represents the same cross section using red shading for positive values, and blue for negative. The scale is in Bohr radii.

Figure 2.14 uses shading to convey the value of the wavefunction in each small cube. Positive and negative values are indicated by red and blue shading, respectively, and the nucleus is located in the centre of the box. As can be seen from cut-away view (b), close to the nucleus the wavefunction is positive (red shading). As we we move away from the nucleus the value of the wavefunction falls quickly, crossing zero and becoming negative (blue shading). This negative part of the wavefunction is much more extensive than the positive part. Eventually, at large enough distances, the wavefunction tends to zero (white).

🌐 *Weblink 2.1*
Follow the weblink to see an animated version of Fig. 2.14.

The sides of the box in Fig. 2.14 on the previous page are twice the size of those used to illustrate the 1s orbital in Fig. 2.7 on page 41. The 2s orbital thus extends over a significantly greater range than the 1s.

Figure 2.15 on the previous page shows iso-surface representations of the 2s wavefunction, similar to those shown in Fig. 2.10 on page 42 for the 1s. The spherical symmetry of the orbital is immediately apparent. Plot (a) shows, in red, an iso-surface taken at a positive value of the wavefunction and, as expected from Fig. 2.14, this positive region is close in to the nucleus. Plot (b) shows, in blue, an iso-surface taken at the same value as that used in (a), but negative. The negative part of the wavefunction extends much further from the nucleus than does the positive part, again as we expect from Fig. 2.14.

Figure 2.16 (a) and (b) are contour plots (like those shown in Fig. 2.8 on page 41) of a cross section taken through the 2s orbital at the position of the nucleus. Positive values are indicated by red contours, negative blue and the zero contour is green. Due to the spherical symmetry of the orbital, the contours are all circular.

The values of the wavefunction at which the contours are drawn are evenly spaced, so you can see from the way that the red contours are closely packed together that the positive part of the wavefunction goes to higher values, and is much more compact, than the negative part. Note the position of the zero contour (green) which separates the positive and negative parts of the wavefunction. As the orbital is spherically symmetric, this green contour corresponds to a distance from the nucleus (a certain radius) at which the wavefunction is zero in all directions. This is termed a *radial node*.

Plot (c) represents the the same cross section but this time using shading, with positive values in red and negative in blue. The compact central positive part of the wavefunction is clear, as is the white ring near to the radius where the wavefunction goes to zero (the radial node).

2p

Weblink 2.2

This link takes you to a web page on which you can display, and rotate in real time, iso-surface representations of all of the hydrogen AOs. It is well worth spending some time looking at the shapes of the various AOs so that you can really get to grips with their key features. One of the options you will have there is to display the nodal planes: try this for the 2p AOs.

There are three 2p orbitals, usually denoted $2p_x$, $2p_y$ and $2p_z$. The iso-surface plots, shown in Fig. 2.17 on the facing page, give the most immediate impression of the shapes of these orbitals. Each has a positive part (red) and a symmetrically placed negative part (blue), arranged along one of the axes. For example, the $2p_z$ wave function is positive at positive values of z, and negative for negative values. The two 'lobes' are related by reflection in the xy-plane, and the whole wavefunction has rotational symmetry about the z-axis, meaning that it is unaltered by rotation through any angle about this axis. The $2p_x$ and $2p_y$ orbitals are just versions of $2p_z$ which have been rotated onto the x- and y-axes, respectively. We often say that '$2p_z$ points along z', meaning that the lobes are aligned along this direction.

Figure 2.18 on the next page shows a representation of the $2p_z$ orbital using the shaded cubes. In the figure the z-axis is vertical and, as before, the nucleus is in the centre of the box. Cut-away view (a) shows the negative values of the wavefunction for negative z values, whereas view (c) shows the positive values of the wavefunction for positive z values i.e. the negative and positive lobes. On the exposed face we see that the values of the wavefunction are distributed symmetrically about the centre, reflecting the rotational symmetry that the wavefunction has about the z-axis.

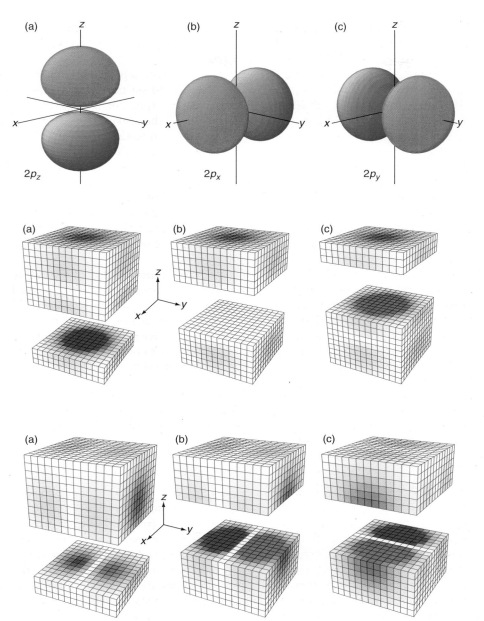

Fig. 2.17 Iso-surface plots of the three $2p$ orbitals: positive values of the wavefunction are indicated by red, and negative by blue. The three orbitals shown in (a), (b) and (c) are $2p_z$, $2p_x$ and $2p_y$, respectively. Each orbital has the same shape, all that is different about them is the axis along which they are aligned.

Fig. 2.18 Representations of the $2p_z$ orbital using shaded cubes; as before the nucleus is in the centre of the box, red indicates positive values, and blue negative. The wavefunction takes negative values for $z < 0$, as shown in cut-away view (a), and positive values for $z > 0$, as shown in (c). At $z = 0$, the xy-plane, the wavefunction is zero, as shown by the white face exposed in cut-away view (b).

Fig. 2.19 Views (a) and (b) are of the $2p_y$ orbital. This orbital is similar to $2p_z$ shown in Fig. 2.18, except that it points along the y-axis, and the nodal plane, where the wavefunction is zero, is the xz-plane. The $2p_x$ orbital, which points along the x-axis, is shown in (c); this time the nodal plane is the yz-plane.

Cut-away view (b) is taken through the nucleus, and there the exposed face (the xy-plane) is white, indicating that the wavefunction is zero. Such a plane where the wavefunction is zero is called a *nodal plane*. Referring to Fig. 2.17 (a) you can see that as the wavefunction is positive in the 'northern hemisphere' and negative in the 'southern hemisphere', it must go through zero as we cross the equator. Thus, the xy-plane is a nodal plane.

Figure 2.19 (a) and (b) show two cut-away views of the $2p_y$ orbital. This time, the positive values of the wavefunction are for positive y values, and the negative values for negative y values. The vertical layer of white cubes indicates

Fig. 2.20 Contour plot (a), and shaded plot (b), taken through the $2p_z$ orbital at the position of the xz-plane. The zero contour (shown in green) corresponds to the xy-plane which is a nodal plane. The axes are labelled in units of the Bohr radius, a_0.

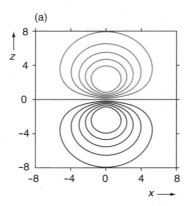

(a)

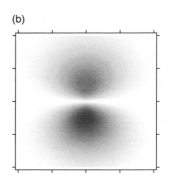

(b)

the nodal plane, which in this case is the xz-plane. Cut-away view (c) is of the $2p_x$ orbital, and this time the layer of white cubes shows us that the nodal plane is the yz-plane. These nodal planes can also be envisaged on the corresponding iso-surface plots of Fig. 2.17 (b) and (c).

Finally Fig. 2.20 shows a contour plot and a shaded plot of a cross-section taken through the $2p_z$ orbital in the xz-plane. The contour plot, (a), shows clearly the positive and negative lobes, pointing along $+z$ and $-z$. The zero contour (in green) is at the position of the nodal plane (the xy-plane), and separates the two lobes. The shaded plot, (b), gives the same impression, with the white line indicating the nodal plane. On account of the cylindrical symmetry of the orbital about the z-axis, any cross section which contains the z-axis and passes through the nucleus will look exactly the same as that shown in Fig. 2.20.

Cross sections taken through $2p_x$ and $2p_y$ orbitals and containing the x- and y-axes, respectively, will look just the same as those shown in Fig. 2.20.

Figure 2.21 shows a very commonly used representation of a $2p$ orbital. The shading correctly indicates the positive and negative lobes, but the shape is just wrong. The problem is that this representation makes it look as if the $2p$ orbital is like an hour glass or a figure of eight – in particular, it shows the wavefunction coming to a point at the nucleus. If you look at the various pictures we have given of the $2p$ orbital you will see that it is certainly *not* hour-glass shaped, but it better described as two slightly flattened spheres, which are close but not touching.

The problem is that, wrong as the representation shown in Fig. 2.21 is, it is used very widely, from text books right though to research-level publications. So pervasive is this picture that some chemists even refuse to believe that it is *not* correct! We will avoid this representation, but you should be aware that it is something you are likely to comes across frequently.

Fig. 2.21 A common, but entirely erroneous, representation of a $2p$ orbital. Compare this representation with the plots shown in Fig. 2.17 on the preceding page and Fig. 2.20.

Summary

It is useful to summarize what we have discovered so far about the orbitals for the K and L shells:

- The $1s$ orbital is spherically symmetric and has no radial node.

- The $2s$ orbital is spherically symmetric, but has a radial node.

• There are three 2p orbitals which point along the x-, y- and z-axes. Each has a nodal plane, but there is no radial node.

We can see a pattern developing here which will become more evident as we consider the orbitals with $n = 3$. However, before we do this we will look at the mathematical form of the 1s, 2s and 2p orbitals. Appreciating these will help us to understand more clearly the way the various orbitals relate to one another.

2.3.3 Mathematical form of the 1s, 2s and 2p orbitals

We expect the hydrogen atom wavefunctions to be functions of x, y and z, but because of the spherical symmetry of the atom it turns out to be far more convenient to express the wavefunctions in *spherical polar coordinates* rather than in terms of x, y and z coordinates (cartesian coordinates).

In spherical polar coordinates the position of a point in space is specified by three coordinates: r, θ and ϕ, as illustrated in Fig. 2.22. The coordinate r is simply the distance from the origin to the point: r is always positive. The angle θ is measured downwards from the z-axis, and can be likened to a latitude. The angle ϕ is measured round from the x-axis, and is like a longitude.

All points on a sphere can be reached by allowing ϕ to range between 0° and 360°, and θ to range between 0° and 180°. Note that it is not necessary for both angles to cover a complete revolution of 360°.

If we express the hydrogen wavefunctions in this coordinate system we discover a great simplification, which is that each wavefunction can be expressed as a product of two functions, the first of which only depends on r, and the second of which only depends on the two angles

$$\psi_{n,l,m_l}(r, \theta, \phi) = \underbrace{R_{n,l}(r)}_{\text{radial part}} \times \underbrace{Y_{l,m_l}(\theta, \phi)}_{\text{angular part}} .$$

The function which only depends on r is called the *radial function* or *radial part*. It turns out that this radial function depends on the two quantum numbers n and l, so it is written $R_{n,l}(r)$. The function which depends only on the two angles is called the *angular part* or the *angular function*; this depends on the two quantum numbers l and m_l, and so is written $Y_{l,m_l}(\theta, \phi)$. In mathematics the $Y_{l,m_l}(\theta, \phi)$ are called spherical harmonics.

The reason why this separation offers a great simplification is that we can think about the radial and angular parts entirely separately. This simplifies the discussion of the orbitals as, for example, all three of the 2p orbitals have the same radial dependence, as the radial part does not depend on m_l. Furthermore, as the angular part does not depend on n, all orbitals with the same l and m_l quantum numbers have the same angular dependence, regardless of the shell in which they occur.

We will start out by discussing the radial parts of the wavefunction and showing how these are related to the radial distribution functions.

The radial parts of the wavefunction

The 1s orbital is spherically symmetric, meaning that it does not depend on the angles θ and ϕ, so the wavefunction already given in Eq. 2.1 on page 40 is the radial part. Using the notation we have introduced above, the radial function is

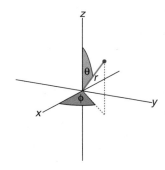

Fig. 2.22 Illustration of the spherical polar coordinate system. The position of a point is specified by its distance r from the origin, and two angles, θ and ϕ. θ is measured downwards from the z-axis, and ϕ is measured round from the x-axis. θ ranges from 0 to π radians, and ϕ from 0 to 2π radians.

Fig. 2.23 Plots of: (a) the radial parts of the 2s and 2p AOs, (b) the corresponding RDFs. For comparison the 1s radial part and RDF is also shown, on the same vertical scale (the upper range of the 1s functions have had to be clipped off). The horizontal scale is given in units of Bohr radii.

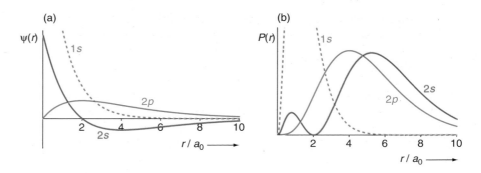

written $R_{1,0}(r)$, with the subscripts indicating that $n = 1$ and $l = 0$

$$1s: \quad R_{1,0}(r) \;=\; N_{1,0} \exp\left(-\frac{r}{a_0}\right).$$

In this expression a_0 is the Bohr radius and $N_{1,0}$ is a normalization constant. The value of this constant depends on n and l, so we have added these as subscripts: $N_{n,l}$.

The radial parts of the 2s ($n = 2$, $l = 0$) and 2p ($n = 2$, $l = 1$) orbitals are

$$2s: \quad R_{2,0}(r) \;=\; N_{2,0}\left(2 - \frac{r}{a_0}\right)\exp\left(-\frac{r}{2a_0}\right)$$

$$2p: \quad R_{2,1}(r) \;=\; N_{2,1}\left(\frac{r}{a_0}\right)\exp\left(-\frac{r}{2a_0}\right).$$

Figure 2.23 (a) shows plots of these radial parts, with that of the 1s orbital included for comparison. In the case of the 2s orbital there is no angular dependence, so it is the radial function which determines the shape of the orbital. We can therefore compare our various representations of the 2s orbital give in Figs. 2.14, 2.15 and 2.16 on page 47 directly with the plot of the radial part.

The first thing we note from the plot of the radial part of the 2s AO ($R_{2,0}(r)$) is that the function starts positive, falls quickly to zero at $r = 2a_0$, and then goes negative. The function then returns to zero quite slowly compared to its initial decay from $r = 0$. This behaviour is exactly that which we identified from the plots of page 47.

The value of r at which the wavefunction crosses the axis, and so has the value zero, gives the position of the radial node. From the mathematical form of $R_{2,0}(r)$ you can see this node occurs when $r = 2a_0$; when this value is substituted for r the radial function goes to zero.

Figure 2.24 shows how the radial function is related to the 2s contour plot (as shown in Fig. 2.16 on page 47). Above the contour plot is shown a cross section taken through the nucleus and at the position indicated by the dashed line. Since the orbital is spherically symmetric, if we move away from the nucleus in any direction the wavefunction varies in just the same way as the radial function. So, the cross section is simply the radial function extending in both directions from the nucleus.

The zero crossing at $r = 2a_0$ gives rise to a radial node which appears as the green zero contour. The minimum value of $R_{2,0}(r)$ occurs at $r = 4a_0$, and from

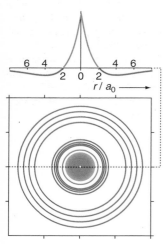

Fig. 2.24 Shown above the contour plot of the 2s orbital is a cross section taken at the position indicated by the dashed line; on each side of zero this cross section is the radial function $R_{2,0}(r)$. The node at $r = 2a_0$ gives rise to the green zero contour. The minimum at $r = 4a_0$ corresponds to the area where the contours are sparse.

Fig. 2.24 on the preceding page we can see that at this radius the contours are widely spaced on account of the function changing slowly around its minimum.

Returning to Fig. 2.23 (a), we see that the behaviour of the radial part of the $2p$ orbital, $R_{2,1}(r)$, is somewhat different to that for $2s$. Firstly, as the function contains a factor of r, $R_{2,1}(r)$ is zero at $r = 0$. However, it is not usual to count $r = 0$ as a radial node. As the radius increases the factor of r increases, but the exponential term $\exp(-r/2a_0)$ decreases. So, the function rises to start with, but is then brought back to zero by the exponential term.

The three-dimensional shapes of the $2p$ orbitals illustrated on page 49 depend on both the radial part and the angular part of the wavefunction. However, if we move away from the nucleus in a particular direction (i.e. at fixed angles), then the way in which the wavefunction varies is determined solely by the radial part. The idea is illustrated in Fig. 2.25 which shows a contour plot of the $2p_y$ orbital (as in Fig. 2.20 on page 50) and a cross section taken along the y-axis. For positive values of y, the variation in the wavefunction is just that of the radial function $R_{2,1}(r)$: the function rises quickly to a maximum and then falls away more slowly. This behaviour is reflected in the contours which are more closely spaced in the region close to the nucleus.

For negative values of y the wavefunction has become negative, but apart from this sign change the variation along a particular direction is just the same as that for $R_{2,1}(r)$. We shall see presently that this sign change is due to the angular part of the wavefunction.

For both the $2s$ and the $2p$ it is the exponential term, $\exp(-r/2a_0)$, which eventually drives the function to zero at large values of r. For the $1s$ the exponential term is $\exp(-r/a_0)$, which decays more quickly. Thus the $1s$ orbital spreads into a smaller region of space than does the $2s$ or $2p$, as is clear from Fig. 2.23 (a). In other words, the $1s$ orbital is 'smaller' than the $2s$ or $2p$.

Radial distribution functions

On page 42 we introduced the concept of the radial distribution function (RDF) which gives the total electron density in a thin shell i.e. summed over all angles. It turns out that the RDF depends *only* on the radial part of the wavefunction, which should not be a surprise as computing the RDF involves summing over the angular part. For an orbital with principal quantum number n and orbital angular momentum quantum number l, the RDF, $P_{n,l}(r)$, is given by

$$P_{n,l}(r) = r^2 \times [R_{n,l}(r)]^2.$$

Figure 2.23 (b) on page 52 shows plots of the RDFs for the $2s$ and $2p$ orbitals, with that of the $1s$ orbital shown for comparison. As we commented on before, the RDF always goes to zero at $r = 0$ and, due to the exponential term, the RDF will decay towards zero at large values of r. For the $2s$ orbital the RDF shows a minimum at the position of the node ($r = 2a_0$), since at this point the radial function is zero, so the RDF must also be zero. Since there is no node in the radial function for the $2p$, there is no minimum in the RDF.

It is interesting to compare the RDFs of the $2s$ and $2p$ orbitals with that of the $1s$ orbital. The thing we see immediately is how much closer in to the nucleus, and more tightly bunched, is the electron density for the $1s$ orbital when compared to the other two orbitals. This is consistent with the fact that the $1s$ orbital is lower in energy than $2s$ or $2p$ orbitals, so we might reasonably expect the electron density to be closer to the nucleus, where it experiences a greater (favourable) interaction.

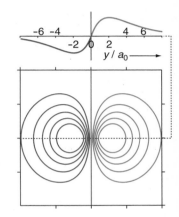

Weblink 2.3

This link takes you to a page where you can examine and compare graphs of the radial parts of the wavefunctions for the hydrogen AOs.

Fig. 2.25 Shown above the contour plot of the $2p_y$ orbital is a cross section taken along the y-axis at the position indicated by the dashed line. The variation in the wavefunction as we move away from the nucleus in a particular direction is determined solely by the radial function $R_{2,1}(r)$.

🌐 *Weblink 2.3*

This link takes you to a page where you can examine and compare graphs of the RDFs of the hydrogen AOs.

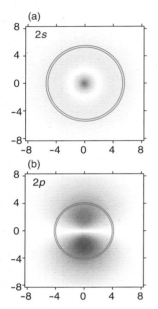

Fig. 2.26 Shaded plots of cross sections through: (a) the 2s and (b) one of the 2p AOs; the scales are given in units of Bohr radii. The green ring indicates the radius at which the principal maximum in the RDF for each orbital occurs.

If we compare the RDFs for the 2s and 2p orbitals we see that the principal maximum for the 2s orbital occurs at 5.2 a_0, whereas for 2p orbital the maximum occurs at 4.0 a_0. In Fig. 2.26 the positions of these maxima are indicated on shaded plots of the wavefunctions. It is perhaps surprising that the maximum in the RDF for the 2s orbital is so far from the nucleus, given that the wavefunction itself is a maximum at the nucleus. However, it is important to recall that the RDF gives the probability of being in a *thin shell*, and that the volume of this shell increases with the radius. Thus, although the 2s wavefunction is much smaller at $r = 5.2\,a_0$ than it is at $r = 0$, this is compensated for by the increase in the volume of the shell.

It is important to remember that despite the principal maxima in the RDFs for the 2s and 2p orbitals being at different radii, the two orbitals nevertheless have the *same* energy as they have the *same* principal quantum number. It is tempting, but wrong, to look at the plots of the RDFs and say that 'the 2p orbital must be lower in energy as the maximum in its RDF is closer to the nucleus than for 2s orbital'.

Angular parts of the wavefunction

Having looked at the radial parts of these wavefunctions, we can now turn to the angular parts. Remember that these only depend on the quantum numbers l and m_l, so they are the same functions for *all* s orbitals, and the same set of functions for all p orbitals.

For an s orbital, which has $l = 0$ and $m_l = 0$, the situation is very simple as there is no angular dependence, as we have already seen for the 1s and 2s orbitals. The same is true for all s orbitals, regardless of the value of the principal quantum number, n.

There are three 2p orbitals which all have $l = 1$ but are distinguished by having different m_l values of 0, +1 and −1. The two functions $Y_{1,\pm 1}(\theta,\phi)$ turn out to be *complex*, which means that they have a 'real' and an 'imaginary' part. It is inconvenient to use these complex functions as we would have to plot the real and imaginary parts separately.

Luckily, quantum mechanics gives us a way round this problem. It turns out that as the three 2p orbitals are degenerate (i.e. have the same energy), any linear combination of the wavefunctions is also a valid wavefunction. By careful choice of the combination we can generate a set of real wavefunctions, which are much more convenient to use. In the case of the p orbitals, the combination of the angular parts gives the following three real functions, which we will use from now on:

$$Y_z(\theta,\phi) = \cos(\theta) \qquad Y_x(\theta,\phi) = \sin(\theta)\cos(\phi) \qquad Y_y(\theta,\phi) = \sin(\theta)\sin(\phi).$$

Let us look first at the angular function $Y_z(\theta,\phi) = \cos(\theta)$, which is illustrated in various ways in Fig. 2.27 (a). It is helpful also to refer to Fig. 2.22 on page 51 to remind ourselves that the angle θ is measured downwards from the z-axis, and that it takes the range from 0 to π (180°). The angle ϕ is measured round from the x-axis, and takes the range 0 to 2π (360°).

At the top of Fig. 2.27 (a) we see a plot of $\cos(\theta)$ in the range 0 to π. The function has the value +1 when $\theta = 0$, and then decreases steadily towards zero as the angle reaches $\pi/2$ (90°). Beyond this, $\cos(\theta)$ becomes increasingly negative, heading towards −1 when the angle is π (180°).

In Fig. 2.27 (a) the behaviour of this function is also represented in a slightly different way. Shown below the graph of $\cos\theta$ is a set of axes in which we have

⇨ The key properties of trigonometric functions are reviewed in section 20.3 on page 890.

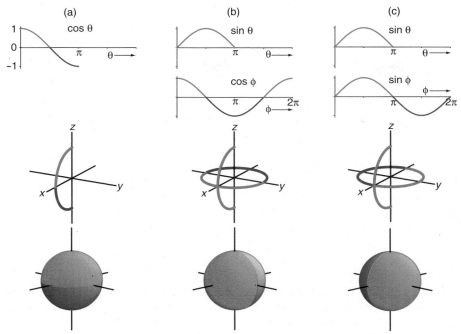

Fig. 2.27 Illustration of the three angular functions associated with the p orbitals. The plots represent: (a) $Y_z(\theta, \phi) = \cos\theta$; (b) $Y_x(\theta, \phi) = \sin\theta\cos\phi$; and (c) $Y_y(\theta, \phi) = \sin\theta\sin\phi$. At the top are shown separate plots of the sine and cosine functions; recall that in spherical polar coordinates the angle θ only goes from 0 to π. Below are drawn colour-coded arcs representing the behaviour of these functions (red for positive, blue for negative): the arc going from $+z$ to $-z$ is for θ, and that in the xy-plane is for ϕ. At the bottom are shown three-dimensional representations of the overall function, obtained by rotating the $+z$ to $-z$ arc through 360° about z, taking into account the sign of the function indicated by the arc in the xy-plane. In each case, there is a positive and a negative hemisphere.

drawn an arc from $+z$ to $-z$ which represents the path taken as θ goes from 0 to π. In addition, the arc has been colour-coded red for positive values, and blue for negative.

We need to remember that this function $Y_z(\theta, \phi) = \cos\theta$ is three dimensional, so although it does not depend on the angle ϕ, we nevertheless have to plot the function over the full range of ϕ. Essentially what this means is rotating the arc which represents $\cos(\theta)$ by 2π (360°) about the z-axis to give the sphere shown at the bottom of Fig. 2.27 (a).

From this sphere we see that the angular function $Y_z(\theta, \phi) = \cos(\theta)$ is positive in the northern hemisphere (positive z values), and negative in the southern hemisphere (negative z values). For $\theta = \pi/2$ (90°) the function goes to zero, and this results in the xy-plane being a *nodal plane*. In fact, a better description is to call this an *angular node* as it occurs at a particular angle, here $\theta = \pi/2$.

At this point we should refer back to the various representations of the $2p_z$ orbital given in Fig. 2.17 on page 49, Fig. 2.18 on page 49, and Fig. 2.20 on page 50. The positive and negative lobes we saw in these diagrams, and the nodal plane, are all due to the angular part of the wavefunction, $Y_z(\theta, \phi) = \cos(\theta)$. It is the radial part of the wavefunction which causes it to decay away to zero at large distances from the nucleus.

Fig. 2.28 Shown in (a) are the radial parts of the 3s, 3p and 3d AOs, and in (b) the corresponding RDFs. For comparison the 2s RDF is also shown, on the same vertical scale (the upper part of the RDF has been clipped off). Note how the 2s RDF is much closer in to the nucleus than are the RDFs for the orbitals with $n = 3$.

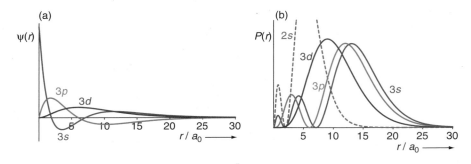

Now let us turn to the function $Y_x(\theta, \phi) = \sin\theta\cos\phi$, which is represented in Fig. 2.27 (b). In the range $\theta = 0$ to π (180°), $\sin\theta$ is always positive, as is illustrated on the graph and by the red arc going from $+z$ to $-z$. On the other hand, $\cos\phi$ is positive for $\phi = 0$ to $\pi/2$ (90°), then negative until $\phi = 3\pi/2$ (270°), and finally positive up to $\phi = 2\pi$ (360°). This is illustrated in the graph at the top of (b) and also by the colour-coded arc drawn in the xy-plane of the axes beneath.

We now have to construct the three-dimensional form of $Y_x(\theta, \phi)$, and as before this can be done by imagining that the arc which goes from $+z$ to $-z$ is rotated about the z-axis through 360°. However, we have to take into account the sign change which results from the $\cos(\phi)$ part of the function, represented by the coloured arc in the xy-plane. The result of this is to make the hemisphere which has positive x values positive, and the other hemisphere negative, as shown at the bottom of the plot. There is an angular node at $\phi = \pi/2$ (90°) or $\phi = 3\pi/2$ (270°) for all values of θ i.e. the yz-plane. Clearly, this is the angular part of a p_x orbital.

The third function $Y_y(\theta, \phi) = \sin(\theta)\sin(\phi)$, which is represented in Fig. 2.27 (c), can be treated in a similar way. The difference here is that the function $\sin(\phi)$ is positive for $\phi = 0$ to π (180°), and is then negative for the rest of the range. As a result, it is the hemisphere for positive y values which is positive, and the opposite hemisphere is negative. The angular node is at $\phi = 0$ or $\phi = \pi$ (180°) for all values of θ i.e. the xz-plane. This is the angular part of a p_y orbital.

Clearly it is the the angular parts of the wavefunction which are responsible for both the lobed structure of the p orbitals, along with the associated nodal planes or angular nodes.

2.3.4 The 3s, 3p and 3d orbitals

The M shell, which has $n = 3$, contains one 3s, three 3p and five 3d orbitals. Before we look at the full three-dimensional shapes of these functions, it is useful to consider the radial parts and the corresponding radial distribution functions.

Radial parts and RDFs

Figure 2.28 shows the radial parts of the 3s, 3p and 3d orbitals, which are the functions $R_{3,0}(r)$, $R_{3,1}(r)$ and $R_{3,2}(r)$, respectively. The mathematical form of these are

$$3s \ (n = 3, l = 0): \quad R_{3,0}(r) \ = \ N_{3,0}\left(27 - 18\left(\frac{r}{a_0}\right) + 2\left(\frac{r}{a_0}\right)^2\right)\exp\left(-\frac{r}{3a_0}\right)$$

$$3p \ (n = 3, l = 1): \quad R_{3,1}(r) \ = \ N_{3,1}\left(6\left(\frac{r}{a_0}\right) - \left(\frac{r}{a_0}\right)^2\right)\exp\left(-\frac{r}{3a_0}\right)$$

$$3d \ (n = 3, l = 2): \quad R_{3,2}(r) \ = \ N_{3,2}\left(\frac{r}{a_0}\right)^2\exp\left(-\frac{r}{3a_0}\right).$$

The radial part of the $3s$ is non-zero at the nucleus ($r = 0$). This is a feature of all s orbitals, regardless of the shell to which they belong. The function $R_{3,0}(r)$ is zero when the term in the large bracket is zero, and it is easy to find the values of r at which this occurs by solving the quadratic equation $27 - 18x + 2x^2 = 0$, where $x = r/a_0$. These nodes occur at $r = 1.9\,a_0$ and $r = 7.1\,a_0$. Note that this time there are two radial nodes, whereas for $2s$ there was only one, and for $1s$ there were none.

The radial part of the $3p$, $R_{3,1}(r)$, is zero when $r = 0$, as is the case for the $2p$ orbital; as before, we do not count $r = 0$ as a node. The function also goes to zero when $r = 6\,a_0$ i.e. there is a node at this radius, which is in contrast to the $2p$ orbitals which have no radial nodes. Finally, the radial part for the $3d$, $R_{3,2}(r)$, starts from zero at $r = 0$, but has no radial nodes.

All three radial functions decay away to zero at large values on r on account of the exponential term $\exp(-r/3a_0)$. This decay is slower than for the orbitals with $n = 2$, so the $n = 3$ orbitals are generally larger.

Figure 2.28 on the preceding page shows the corresponding RDFs. As we found before, all the RDFs are zero at the nucleus, and a node in the radial part of the wavefunction gives rise to a minimum (of zero) in the RDF. The greater size of the orbitals with $n = 3$ when compared to $2s$ is also evident from the plot, and we can make sense of this by noting that the $n = 3$ orbitals are higher in energy than those with $n = 2$, and so on average further away from the nucleus.

Three-dimensional shapes: 3s and 3p

Recall that the angular parts of the wavefunctions depend only on l and m_l, and *not* on the value of the principal quantum number n. So, we expect the $3s$ to have the same angular properties as $2s$, and the $3p$ to have the same angular properties as the $2p$.

An s orbital has no angular dependence, so the $3s$ wavefunction has spherical symmetry. Figure 2.29 (a) and (b) show contour plots of a cross section taken through the wavefunction and passing through the nucleus. The two radial nodes, which we identified in the $3s$ radial function plotted in Fig. 2.28 (a), give rise to the two green zero contours. As with the other s orbitals, the central part, close in to the nucleus, has the largest value and is rather compact. A plot of the wavefunction as we move away from the nucleus i.e. as a function of r, is also shown in (a). From this we can identify the two radial nodes, the minimum and the maximum, and see how these relate to the contour plot.

Figure 2.30 on the following page shows iso-surface plots of the three $3p$ orbitals, and Fig. 2.31 shows plots of a cross section taken through one of the orbitals. The angular parts of these orbitals are just the same as for $2p$. For example, for $3p_z$ the angular part is positive for positive values of z, and negative for negative values; the two hemispheres are separated by a nodal plane, the

⊕ *Weblink 2.2*

This link takes you to a web page on which you can display, and rotate in real time, iso-surface representations of all of the hydrogen AOs. It is very instructive to use the option for displaying the angular nodes (nodal planes) on the $3p$ and $3d$ AOs.

Fig. 2.29 Plots of a cross section through the 3s orbital taken at the position of the nucleus. In (b) the contours are more closely spaced than they are in (a). The positions of the two radial nodes are indicated by the green zero contours visible in (a) and (b). Also shown above (a) is a plot of the wavefunction taken along the dotted line. As can be seen by comparison with Fig. 2.28 (a), this is the radial part of the wavefunction. The shaded plot (c) illustrates clearly the positive, compact nature of the part of the wavefunction close to the nucleus.

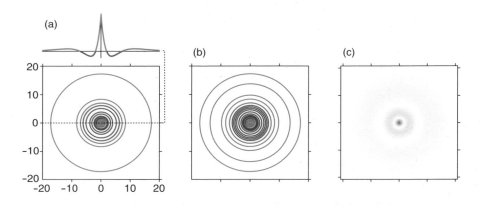

Fig. 2.30 Iso-surface plots of the three 3p orbitals. As for the 2p, one 3p orbital is aligned along each axis, and each orbital has a nodal plane (or angular node). For example, $3p_z$ points along the z-axis and the nodal plane is the xy-plane. The radial node cuts through each lobe, dividing them into two.

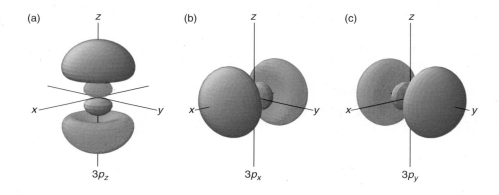

Fig. 2.31 Plots of a cross section containing the x-axis, and passing through the nucleus, taken through the $3p_x$ orbital. In the contour plot, (a), the green zero contour identifies the radial node at $r = 6a_0$, and the nodal plane (the yz-plane). The two lobes of opposite sign are aligned along the x-axis, but each lobe is cut into two by the radial node. In the shaded plot, (b), the radial and angular nodes are clearly identified by the white regions which divide the wavefunction. Shown above (a) is a plot of the wavefunction taken along the dotted line: this is the radial part.

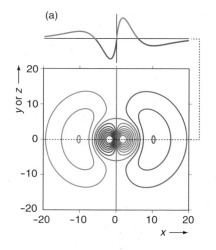

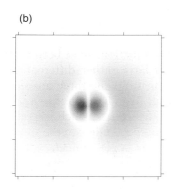

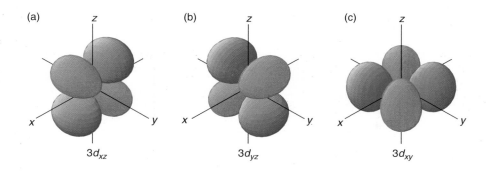

Fig. 2.32 Iso-surface plots of three of the 3d orbitals. Each of these orbitals has two nodal planes (or angular nodes) giving rise to four lobes which point *between* the axes.

xy-plane. However, in the 3p there is also a radial node at $r = 6\,a_0$ which cuts each lobe into two parts, as can clearly be seen from the contour plot, Fig. 2.31 (a).

Above Fig. 2.31 (a) we see a plot of the wavefunction taken along the dotted line. As explained before, since this is taken in a fixed direction, the way the wavefunction varies is determined only by the radial part ((a) in Fig. 2.28 on page 56). The radial node at $r = 6\,a_0$ gives rise to the green zero contour, and the sign change as we cross this node is also evident. The sign change as we go from positive to negative x-coordinates results from the angular part of the wavefunction.

Three-dimensional shapes: 3d

For $n = 3$ we have a set of 3d orbitals all with $l = 2$ and with five possible values of m_l: -2, -1, 0, $+1$ and $+2$. As we saw in Fig. 2.28 on page 56, these orbitals have no radial nodes, but do go to zero at the nucleus. As for the p orbitals, the corresponding angular functions $Y_{2,m_l}(\theta, \phi)$ are complex, so it is usual not to consider these but a set of combinations of these five functions which are real. These functions are

$$Y_{xz}(\theta, \phi) = \cos(\theta) \sin(\theta) \cos(\phi) \qquad Y_{z^2}(\theta, \phi) = 3 \cos^2(\theta) - 1$$

$$Y_{yz}(\theta, \phi) = \cos(\theta) \sin(\theta) \sin(\phi) \qquad Y_{x^2-y^2}(\theta, \phi) = \sin^2(\theta) \cos(2\phi)$$

$$Y_{xy}(\theta, \phi) = \sin^2(\theta) \sin(2\phi)$$

The functions have been arranged deliberately into two columns, as the two groups have rather different properties. Iso-surface plots of the 3d orbitals deriving from the three angular functions in the left-hand column are shown in Fig. 2.32. These three orbitals all have a common structure: in each there are *two* nodal planes which split the orbital into four lobes; these four lobes lie in a plane perpendicular to the two nodal planes, and point between the axes.

For example, for $3d_{xz}$ the xy- and yz-planes are the nodal planes, and the four lobes lie in the xz-plane. Similarly, $3d_{yz}$ lies in the yz-plane, with the xy- and xz-planes as nodal planes, whereas the $3d_{xy}$ lies in the xy-plane.

As we did before, we can locate the nodal planes by looking at the form of the angular functions. For example, $Y_{xz}(\theta, \phi) = \cos(\theta) \sin(\theta) \cos(\phi)$ is zero for $\theta = \pi/2$: this is the xy-plane. The function is also zero for $\phi = \pi/2$ and $3\pi/2$, which corresponds to the yz-plane.

The angular function also explains the signs of the four lobes. Looking at the form of $Y_{xz}(\theta, \phi)$, we see that $\sin(\theta)$ is positive in the $+z$ hemisphere, and

🌐 *Weblink 2.2*

This link takes you to a web page on which you can display, and rotate in real time, iso-surface representations of all of the hydrogen AOs. Use the option for displaying the angular nodes to investigate their location for the 3d AOs.

Fig. 2.33 Plots of a cross section taken at the position of the xz-plane through the $3d_{xz}$ orbital. The contour plot, (a), shows the four lobes which alternate in sign and point between the axes. The green zero contours correspond to the xy- and yz-planes, which are nodal planes. The shaded plot, (b), shows the same features. Also shown above (a) is a plot of the wavefunction taken along the diagonal dotted line; this is the radial part of the wavefunction, as can be seen by comparison with Fig. 2.28 on page 56.

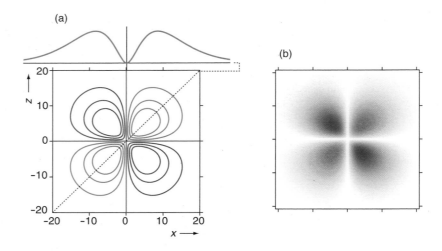

negative in the $-z$ hemisphere; further, $\cos(\phi)$ is positive in the $+x$ hemisphere, and negative in the $-x$. Putting these two together means that the function is divided into four quadrants, of alternating sign, as we see in Fig. 2.32 (a).

Figure 2.33 shows contour and shaded plots of a cross section taken through the $3d_{xz}$ orbital at the position of the xz-plane. The four lobes, pointing between the axes, are clearly visible, as are the green zero contours arising from the two nodal planes. Also shown above Fig. 2.33 (a) is a plot of the wavefunction along the dotted line: this corresponds to the radial part of the function, as shown in Fig. 2.28 on page 56.

Figure 2.34 on the facing page shows iso-surface plots of the two remaining $3d$ orbitals, with angular functions $Y_{x^2-y^2}(\theta, \phi)$ and $Y_{z^2}(\theta, \phi)$, and contour plots of cross sections through these two orbitals are shown in Fig. 2.35. These orbitals look rather different to the d orbitals shown in Fig. 2.32.

The orbital $3d_{x^2-y^2}$, shown in Fig. 2.34 (a) and Fig. 2.35 (a) and (b), has four lobes, just like the other $3d$ orbitals, but this time they are pointing *along* the x- and y-axes. There are two nodal planes lying at $45°$ to the xz- and yz-planes, something which is particularly clear from the position of the green zero contours on Fig. 2.35 (a).

We can see the origin of these nodal planes by looking at the corresponding angular function $Y_{x^2-y^2}(\theta, \phi) = \sin^2(\theta)\cos(2\phi)$. This is zero for $\phi = \pi/4$, $3\pi/4$, $5\pi/4$ and $7\pi/4$, which gives rise to the two vertical planes angled round by $\pi/4$ ($45°$) and $3\pi/4$ ($135°$) from the x-axis.

The orbital $3d_{z^2}$, shown in Fig. 2.34 (b) and Fig. 2.35 (c) and (d), has quite a different structure. It has cylindrical symmetry about the z-axis but, unlike the other orbitals, is not divided into four lobes. The best way to appreciate what is going on here is to start with the corresponding angular function $Y_{z^2}(\theta, \phi) = 3\cos^2(\theta) - 1$. We can identify the angular nodes by working out at what angles

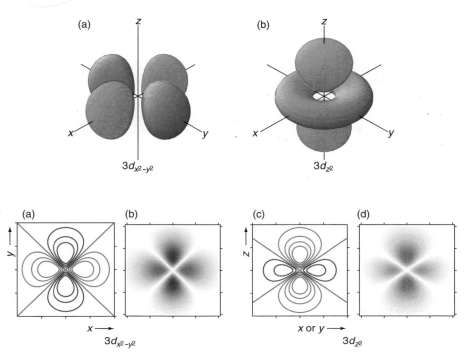

Fig. 2.34 Iso-surface plots of the remaining two 3*d* orbitals. In contrast to those shown in Fig. 2.32 on page 59, these orbitals point *along* the axes.

Fig. 2.35 Plots (a) and (b) are of a cross section taken through the $3d_{x^2-y^2}$ orbital at the position of the *xy*-plane. In contrast to the orbitals shown in Fig. 2.33 on the preceding page, the four lobes point *along* the axes. The green zero contours correspond to the two nodal planes which lie *between* the axes. Plots (c) and (d) are of a cross section taken through the $3d_{z^2}$ orbital; the cross section contains the *z*-axis and the nucleus. For this orbital, the angular nodes form two cones.

θ this function will be zero:

$$
\begin{aligned}
3\cos^2(\theta) - 1 &= 0 \\
\cos^2(\theta) &= \frac{1}{3} \\
\theta &= \cos^{-1}\left(\frac{\pm 1}{\sqrt{3}}\right).
\end{aligned}
$$

A little work with a calculator shows that θ is 54.7° or 125.3° (recall that θ only varies between 0 and 180°, and that $\cos\theta$ is negative for θ greater than 90°).

The wavefunction therefore has angular nodes for $\theta = 54.7°$ and for $\theta = 125.3°$, for *all* values of the angle ϕ. Therefore, these two angular nodes take the form of two cones, one inclined at 54.7°, and one at 125.3° to the *z*-axis, as shown in Fig. 2.36.

When we look at a contour plot of a cross section taken through this orbital, as shown in Fig. 2.35 (c), it is these two angular nodes (in the form of cones) which give rise to the green zero contours inclined at the corresponding angles.

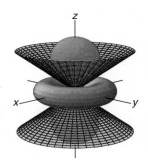

Fig. 2.36 Iso-surface of the $3d_{z^2}$ orbital showing, in green, the two angular nodes. These take the form of cones at 54.7° and 125.3° to the *z*-axis.

2.3.5 The general picture

Now that we have looked in detail at the orbitals in the K, L and M shells, some trends begin to become clear. For example, we have seen that *s* orbitals have no nodal planes or angular nodes, but the number of radial nodes increases

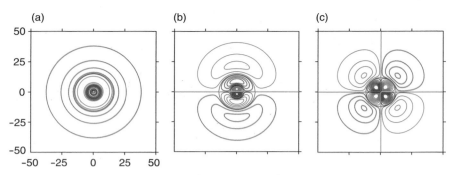

Fig. 2.37 Contour plots of three typical orbitals from the shell with $n = 4$. Plot (a) is of the $4s$: the orbital has spherical symmetry and has three radial nodes, indicated by the green zero contour, the innermost of which is just discernible. Plot (b) is of one of the $4p$ orbitals: we see that two radial nodes, indicated by the green circles, cut the two lobes which are separated by a nodal plane. Plot (c) is on one of the $4d$ orbitals: here, one radial node cuts the four lobes which are separated by two nodal planes.

as the principal quantum number increases. Similarly, as we go from s to p to d orbitals, the number of nodal planes (angular nodes) increases steadily. The number of these radial and angular nodes is important as their presence determines the overall shape of an orbital. We have also seen that, for hydrogen, the orbital energy goes as $-1/n^2$ and the size of the orbital increases with n.

In fact, the total number of nodes, and the numbers of the radial and angular nodes, obey the following simple set of rules. For an orbital with quantum numbers n and l:

- The total number of angular and radial nodes is $(n - 1)$.

- The number of angular nodes (nodal planes) is l.

- The number of radial nodes is $(n - 1 - l)$.

For example, for a $3p$ orbital, there are a total of $(3 - 1) = 2$ nodes. As $l = 1$ there is one angular node, and $(3 - 1 - 1) = 1$ radial node.

The other interesting point is the relationship between the number of nodes and the energy. In hydrogen, orbitals with the same n have the same energy and the same number of nodes. As n increases the energy increases, and so does the number of nodes. There is thus a strong connection between the number of nodes and the energy: more nodes means higher energy, a connection which will see repeated when we look at orbitals in molecules.

Using these rules we can have a good guess at the form of orbitals from higher shells. For example, for the shell with $n = 4$ possible values of l are 3, 2, 1 and 0. Applying the rules we can deduce the following about the AOs in this shell.

(a) The $4s$ orbital will have no angular nodes (i.e. it will be spherically symmetric) and three radial nodes.

(b) The $4p$ will have one angular node (nodal plane), and so have a shape similar to the $2p$ and $3p$ consisting of two lobes. However, these lobes will be cut by two radial nodes, just like the $3p$ was cut by one radial node.

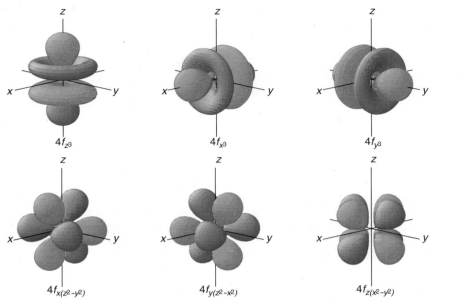

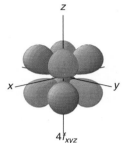

Fig. 2.38 Iso-surface plots of the seven $4f$ orbitals, each of which has three angular nodes. For example, for the $4f_{z^3}$ orbital one of the angular nodes is the xy-plane, and the other two are in the form of cones about the z-axis. For the $4f_{xyz}$ orbital the angular nodes are formed by the xy-, xz- and yz-planes.

(c) The five $4d$ orbitals will, like the $3d$, have two angular nodes (nodal planes) but the resulting lobes will also be cut be a radial node.

(d) The seven $4f$ orbitals (with $l = 3$) will have no radial nodes, but *three* angular nodes, or nodal planes.

For the $4s$, $4p$ and $4d$ orbitals, these features can be seen in the selection of contour plots shown in Fig. 2.37 on the facing page. The seven $4f$ AOs are shown as iso-surface plots in Fig. 2.38. It is clear that these have no radial nodes, and in each we can identify three angular nodes, which may take the form of cones (as in the $3d_{z^2}$ AO) or planes.

⊕ *Weblink 2.2*

This link takes you to a web page on which you can display, and rotate in real time, iso-surface representations of all of the hydrogen AOs. You can look at the shapes of all of the orbitals with $n = 4$.

2.4 Spin

So far, we have just been concentrating on the wavefunction which describes the spatial distribution of the electron i.e. the electron density. However, electrons have a further property – that of *spin* – which we need to add to our description. Spin is a kind of *angular momentum* possessed by the electron, so it is useful to start the discussion by describing what angular momentum is.

Angular momentum

In classical mechanics, the *momentum* of a particle of mass m is given by the product $m \times v$, where v is the velocity. Velocity is a *vector quantity*, meaning that it has both a magnitude (the speed) and a direction. Therefore momentum is also a vector quantity, having a magnitude and a direction.

If a mass is moving around a circular path (an orbit) it has a special kind of momentum, called *angular momentum*, which is associated with this circular motion. Angular momentum, like the linear momentum already described, is a vector quantity.

Quantum mechanics predicts that an electron in an atom can have angular momentum, and that the size of this angular momentum is related to the angular momentum quantum number l. This is, of course, why l is given this name. An electron in an s orbital with $l = 0$ has no orbital angular momentum, and higher values of l $(1, 2, \ldots)$ correspond to increasing amounts of angular momentum.

It is tempting, but entirely wrong, to think of this orbital angular momentum as arising from the electron orbiting around the nucleus in the same way that the Earth orbits the Sun. The whole point of quantum mechanics is that the electron is *not* a localized particle, but is described using a wavefunction and a probability density. It just so happens that one of the properties of the electron, as described by a wavefunction, is that it possesses angular momentum.

Spin angular momentum

Experimental work in the first decades of the 20th century led to the idea that, in addition to its orbital angular momentum, an electron also possesses another kind of angular momentum, which was dubbed *spin angular momentum*. Subsequent theoretical work showed that electron spin arises when relativistic ideas are incorporated into quantum mechanics.

Calling this kind of angular momentum 'spin angular momentum' is rather problematic as it tempts us to think of the electron as a little sphere which is spinning about its own axis; this is simply not a correct way of describing spin. Rather, we simply have to recognize that an electron possesses an intrinsic source of angular momentum which, for historical reasons, has become known as spin angular momentum.

Each electron turns out to have one half of a unit of spin angular momentum. This is specified by the *spin angular momentum quantum number*, s, which always takes the value $\frac{1}{2}$. We saw before that associated with the orbital angular momentum quantum number l there is another quantum number m_l which takes values from $+l$ to $-l$ in integer steps. In a similar way, associated with s there is another quantum number m_s which takes values from $+s$ to $-s$ in integer steps. In fact, as $s = \frac{1}{2}$, the only two values of m_s are $+\frac{1}{2}$ and $-\frac{1}{2}$.

These two values of m_s are usually referred to as *spin states*, and are depicted and described in a variety of different ways. $m_s = +\frac{1}{2}$ is often called 'spin up' or 'the α spin state'; it is often depicted by an up-arrow: ↑. $m_s = -\frac{1}{2}$ is called 'spin down' or 'the β spin state', and is denoted ↓. In the absence of magnetic or electric fields, these two spin states have the same energy i.e. they are degenerate.

To give a complete description of an electron in a hydrogen atom, we need to specify four quantum numbers. Three of these refer to the spatial part of the wavefunction: the principal (n), orbital angular momentum (l) and magnetic (m_l) quantum numbers. The fourth, m_s, refers to the spin. Note that it is not necessary to specify the value of s as it is always $\frac{1}{2}$.

The inclusion of spin has profound consequences when it comes to multi-electron atoms, which we will come to shortly.

2.5   Hydrogen-like atoms

The hydrogen atom has a very special place in our discussions as it has just one electron, but if we include ions as well as atoms then there are other one-electron systems such as He^+ and Li^{2+}. The only difference between these ions and hydrogen is that the nuclear charge is greater. For example, whereas in hydrogen the nuclear charge is $+1$, in He^+ it is $+2$, and in Li^{2+} it is $+3$.

This increase in nuclear charge means that the electron experiences a greater attraction to the nucleus. We might therefore expect that the orbitals are lowered in energy, and pulled in towards the nucleus (i.e. they are smaller).

In fact, the energies and the wavefunctions change in rather a simple way as the charge on the nucleus, Z, increases. As we saw in Eq. 2.2 on page 45, the energies go as Z^2, where Z is the nuclear charge:

$$E_n = -\frac{Z^2 R_H}{n^2}.$$

Thus, as the nuclear charge increases, the energies of the orbitals all decrease (become more negative). For example, the $1s$ orbital of He^+ has an energy of $-4R_H$ compared to $-R_H$ for hydrogen.

The radial functions $R_{n,l}(r)$ all scale with Z in a simple way. For example:

$$1s: \quad R_{1,0}(r) \quad = \quad N_{1,0} \exp\left(-\frac{Zr}{a_0}\right)$$

$$2s: \quad R_{2,0}(r) \quad = \quad N_{2,0}\left(2 - \frac{Zr}{a_0}\right)\exp\left(-\frac{Zr}{2a_0}\right).$$

The factor of Z in the exponential term means that the exponential function decays faster as Z increases. In other words, increasing the nuclear charge decreases the size of the orbitals. This point is well made in Fig. 2.39 where we see the RDFs for $1s$ orbitals with nuclear charges of 1, 2 and 3. As Z increases, the electron density is pulled into the nucleus and becomes more compact.

Later on, we will have much more to say about these hydrogen-like atoms, as they are much used as useful approximations when it comes to multi-electron atoms.

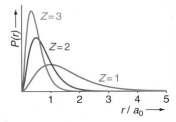

Fig. 2.39 Plots of the RDFs for a $1s$ electron with increasing nuclear charge, Z. As Z increases the orbital is pulled in towards the nucleus and so becomes more compact.

🌐 *Weblink 2.3*

This link takes you to a page where you can see how the RDFs of the hydrogen AOs are affected by increasing the nuclear charge.

2.6   Multi-electron atoms

We have spent a long time looking at the simplest possible atom which has just one electron for the good reason that the ideas which arise when we look at hydrogen are immediately transferable to more complex atoms. So, you will find in this section that we are already well along the way to understanding all atoms, not just hydrogen.

However, we will see that, in contrast to our detailed and exact description of hydrogen, we can only develop a more approximate picture for multi-electron atoms. The reasons for this is that in such atoms we need to take account of the repulsion between electrons, something which does not, of course, occur in hydrogen. We will see that it is this electron–electron repulsion that is the source of all the difficulties when it comes to describing multi-electron atoms.

Let us start by thinking about the helium atom, which has a nuclear charge of $+2$ and two electrons; the interactions between the charged particles are shown

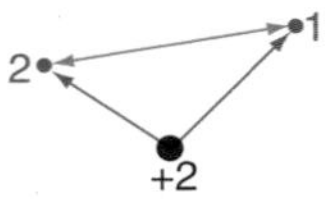

Fig. 2.40 Illustration of the interactions present in the helium atom. The two electrons (blue) both have a favourable (attractive) interaction with the positive nucleus (black). However, there is an unfavourable (repulsive) interaction between the two electrons.

in Fig. 2.40. Just as in hydrogen, each electron is attracted to the nucleus on account of the opposite charges. However, the new feature is that the two electrons repel one another, on account of them both being negatively charged: this is termed *electron–electron repulsion*.

The attraction between each electron and the nucleus lowers the potential energy of the electron, whereas the electron–electron repulsion increases the potential energy. The contribution due to the repulsion between electrons is significant when compared to the contribution due to attraction to the nucleus.

Although we can easily write down the Schrödinger equation for such a two-electron atom, it is quite impossible to solve the equation *analytically* so as to find the wavefunctions and associated energy levels. By 'solve analytically' we mean find mathematical functions, such as those quoted in section 2.3.3 on page 51, which are solutions to the Schrödinger equation. However, it is possible to solve the equation *numerically*. A numerical solution to the problem gives us a wavefunction, but not in the form of a simple mathematical function. Rather, we obtain a set of numbers which define the wavefunction at each point in space. Computer programs are very good at determining such numerical solutions to the Schrödinger equation, but the results are not always so easy to understand or describe in the way we can for exact solutions.

Mathematically, it is the terms in the potential which arise from the electron–electron repulsion which are causing the problem. If these terms were absent, an analytical solution to the Schrödinger equation could easily be found. What we need to find is a way of 'dealing with' these awkward terms. The approach which is most convenient for chemists is to use the *orbital approximation*, which is described in the next section. In fact, as you will shortly see, you have been using this approximation for a while now, without knowing it.

2.6.1 The orbital approximation and electronic configurations

If there were no electron–electron repulsion, each electron in an atom would simply experience the nuclear charge Z. Solving the Schrödinger equation for any of the electrons would give a set of wavefunctions and energy levels which are just like those for hydrogen, adjusted for a nuclear charge Z in the way described in section 2.5 on the preceding page. Thus, each electron has available to it the usual set of hydrogen-like orbitals: $1s$, $2s$ etc.

The question then arises as to which orbital wavefunction to assign to each electron. As you know, this process is governed by the *Pauli exclusion principle* which states that 'no two electrons can have all four quantum numbers the same'. In practice, this means that each orbital, specified by given n, l and m_l can be 'occupied by' two electrons whose spins must be opposed i.e. arranged as ↑↓.

So, the lowest energy arrangement for helium is to assign both electrons to $1s$ orbitals, with spins opposed, giving the *electronic configuration* $1s^2$. As a result of the Pauli principle, no more electrons can be assigned to the $1s$ orbital, so when it comes to lithium, which has three electrons, the lowest energy orbitals available for the third electron are those with $n = 2$. It turns out, for reasons that will be described in section 2.6.2 on page 69, that the $2s$ orbital is lower in energy than the $2p$, so the electronic configuration of lithium is $1s^2 2s^1$.

There are four electrons in beryllium, so the configuration is $1s^2 2s^2$; the $2s$ orbital is now fully occupied. As we move across the remainder of the second period from boron to neon, the additional electrons are added to the $2p$ orbitals.

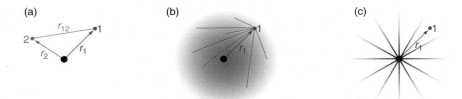

(a)   (b)   (c)

Fig. 2.41 Illustration of the central field approximation. In (a) we see depicted the interactions between each electron and the nucleus (shown in black), and between the two electrons. In (b) electron 2 is assigned to an orbital, and the average electron–electron repulsion between electron 1 and the electron density due to electron 2 (shown in green) is computed. Averaging the repulsion in this way results in electron 1 experiencing a potential which depends *only* on the distance r_1 from the nucleus, as is illustrated in (c) by the shaded lines radiating from the nucleus.

As there are three of these, they can contain a maximum of six electrons, so by the time we reach neon they are full and the configuration is $1s^2 2s^2 2p^6$.

This simple way of describing the structure of atoms in terms of electronic configurations is all very well, but remember that we started out by supposing that the electrons do not repel one another, which is simply not true. However, we can retain much of the simplicity of this approach by using the *orbital approximation*. In this, we assume that each electron experiences the nuclear charge and an *average* repulsion from all of the other electrons in the atom. This approximation results in a set of energy levels for each electron which are closely related to those for hydrogen-like atoms. Thus, all of the knowledge we have built up about the hydrogen-atom wavefunctions can be carried over to multi-electron atoms.

To understand how the orbital approximation works we first need to look at how the average electron–electron repulsion is calculated, and then how the wavefunctions for each electron are found. The first step involves the *central field approximation* and the second uses what is called a *self consistent field*.

The central field approximation

Figure 2.41 illustrates the way in which we go about calculating the average electron–electron repulsion. In (a) we see a picture of the interactions involved in the helium atom: each electron is attracted to the nucleus, and the two electrons repel one another. It is the complexity of this three-way interaction which makes solution of the Schrödinger equation so difficult.

The situation is simplified by *assuming* that electron 2 occupies a $1s$ orbital, so that the electron density (i.e. the square of the wavefunction) due to this electron is known. We then compute the electron–electron repulsion energy between electron 1 and the electron density due to electron 2, as illustrated in (b). This repulsion is averaged over the density of electron 2, and so does not depend on the position of that electron but only on its wavefunction.

In this example, as the electron density for electron 2 is spherically symmetric, the average repulsion experienced by electron 1 only depends on its distance r_1 from the nucleus, and not on the direction. As a result, the combined effect of the attraction to the nucleus and the averaged electron–electron repulsion is that the potential energy of electron 1 depends *only* on its distance r_1 from the nucleus, as illustrated in (c).

In more complex cases, where the electron density due to the other electrons is not spherically symmetric, the average repulsion will depend on the distance and direction from the nucleus. This makes solving the Schrödinger equation rather difficult, so what is normally done is to take the average of the repulsion over all angles (i.e. a spherical average) so that the potential energy simply depends on the distance. Such an approach is called the *central field approximation*.

The outcome of all of this is that electron 1 experiences a potential which depends only on the distance r_1. This is the same situation that we have in a one-electron atom, but the difference is that whereas in such an atom the potential varies simply as $-1/r_1$, in a multi-electron atom the form of the potential is more complex. However, we can calculate what this potential is numerically, and using this we can solve (numerically) the Schrödinger equation for electron 1. As a result, we obtain a wavefunction and an associated energy for electron 1. The important thing about this wavefunction is that it takes into account the (average) repulsion to electron 2, and so will be different from a simple $1s$ orbital.

Self consistent field orbitals

The method we have just described for dealing with the effects of electron–electron repulsion in helium is related to a more general procedure which can be used to find the orbitals for multi-electron atoms.

This procedure starts by assigning the electrons into approximate orbitals (for example the orbitals for hydrogen), taking into account the Pauli principle. The average repulsion between one of the electrons and all of the others is then computed, and a spherical average taken. This gives a potential which depends only on the distance from the nucleus. The somewhat surprising thing about this potential is that it is the *same* for all of the electrons; we are not in a position here to discuss why this is the case, so we will simply have to accept this point.

The Schrödinger equation is then solved numerically for this potential, giving a new set of orbitals analogous to the $1s$, $2s$ etc. These are *not* the same as the ones we started with as the potential is different. We then assign the electrons to these new orbitals, calculate the average repulsion between an electron and all the others, take the spherical average, and hence work out a new potential. We then solve the Schrödinger equation for this new potential, thus generating another set of orbitals, to which the electrons are assigned. The process is carried on until the orbitals do not change significantly: when this point is reached we have what are called the *self consistent field orbitals*. The name arises from the fact that placing the electrons in these orbitals give rise to a potential which, via the Schrödinger equation, leads to the same set of orbitals.

As a result of the central field approximation, which forces the potential to depend only on r, the self consistent field orbitals (the solutions to the Schrödinger equation) turn out to be factorizable into a radial part and an angular part, just as they were for hydrogen. As before, these wavefunctions are characterized by three quantum numbers

$$\psi_{n,l,m_l}(r,\theta,\phi) = \underbrace{C_{n,l}(r)}_{\text{radial part}} \times \underbrace{Y_{l,m_l}(\theta,\phi)}_{\text{angular part}}.$$

The angular parts, $Y_{l,m_l}(\theta,\phi)$, are *exactly* the same as for hydrogen, but the radial parts, $C_{n,l}(r)$, are different. However, these differences are more in matters of

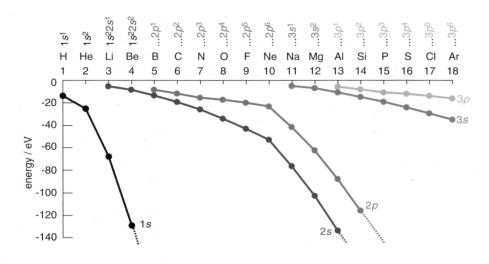

Fig. 2.42 Energies, in eV, of the occupied AOs for the electronic ground states of the elements H to Ar. Above each element is also shown its electronic configuration, with … indicating that the lower energy orbitals are full. Data for orbitals with energies lower than −140 eV are not shown.

detail than in the overall form. For example, these radial parts have the same number of radial nodes as the corresponding functions in hydrogen, and they have associated RDFs which are quite similar in form. What is substantially different is the overall size of the orbitals: as we will see in section 2.6.3 on page 74, the orbitals in multi-electron atoms are generally more contracted than those in hydrogen.

Summary

At the beginning of this section it was stated that in the orbital approximation each electron experiences an average repulsion due to the other electrons. We have now seen is that this approximation means that each electron can be assigned to a wavefunction which is closely related to the hydrogen-like atomic orbitals. The electronic structure of a multi-electron atom can therefore be described in terms of the occupation of these orbitals, which is precisely what you have been doing for some time now.

It is important to realize that, when we describe the electron configuration of lithium as $1s^2 2s^1$, the orbitals to which the electrons assigned are *not* the same as those for hydrogen. So, the $2s$ in lithium is *not* the same function as the $2s$ in hydrogen. However, the functions are sufficiently related that it is useful to use the same labels.

A second important point is that although an orbital in a multi-electron atom is a wavefunction for one electron, its energy is influenced by all of the other electrons as a result of the electron–electron repulsion.

Our next task is to understand the way in which the energies of these orbitals in multi-electron atoms are different from those in hydrogen. This is a very important point as the orbital energies determine the way in which they are occupied, and hence the form of the Periodic Table. Later on, we will also see that orbital energies are very important in determining bonding.

2.6.2   Orbital energies

You have been taught that the order in which the atomic orbitals are filled is first $1s$, then $2s$, followed in sequence by $2p$, $3s$ and then $3p$. What we are going

Fig. 2.43 Energies, in eV, of the occupied AOs for the electronic ground states of the elements H to Ne. These are the same data as in Fig. 2.42 on the previous page, but plotted on an expanded scale for the first two periods.

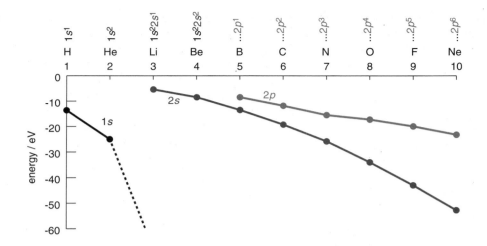

to do in this section is to try to rationalize the order in which these orbitals are filled. In doing so, we will introduce the important concepts of electron *screening* and *penetration*.

Figure 2.42 on the preceding page shows the energies of the occupied orbitals for the ground states of the atoms in the first three periods; Fig. 2.43 shows the same data from just the first two periods and on an expanded scale. The ground state is the electronic configuration which gives the lowest overall energy; this configuration is also shown on the plot. The orbital energies have been computed using the self-consistent method described in section 2.6.1 on page 66.

The data in the plot are consistent with the order in which you have been told to fill the orbitals. So, for example, we see that the 2s is filled before the 2p, and the n = 2 shell is filled before any electrons are placed in n = 3. The plot also shows how, once they start to be filled, the energy of a set of orbitals with the same n and l (e.g. 3s or 2p) falls steadily as the nuclear charge increases. In addition, once a shell is full its energy falls rather rapidly. So, for example, the energy of the 1s drops steeply from He to Li, and similarly the 2s and 2p drop more steeply as we move beyond Ne.

The orbital energy of 1s in He is −25.0 eV. If the two electrons in He did not repel one another, each would experience just the nuclear charge of Z = 2, and so we could use Eq. 2.2 on page 45 to compute the orbital energy:

$$E_n = -\frac{Z^2 R_H}{n^2}. \tag{2.3}$$

Putting Z = 2 and n = 1 we find that the orbital energy is −54.4 eV, which is substantially different from the actual value of −25.0 eV. We can attribute the difference to the electron–electron repulsion, which is clearly a very significant contribution to the orbital energy.

Moving on to Li, the highest energy electron is now in a 2s orbital. If we look at the radial distribution functions for the 2s and 1s orbitals in hydrogen, plot (b) in Fig. 2.23 on page 52, we see that the majority of the electron density of the 2s is outside the area occupied by the 1s electrons. This leads to the idea that the 1s electrons might form a *screen* between the 2s electron and the nuclear charge, thus reducing the attraction felt by the 2s electron.

Screening is simply a consequence of electron–electron repulsion. The distant $2s$ electron experiences an attraction due to the nuclear charge of $+3$, and a repulsion from each $1s$ electron due to their charge of -1. If the latter are clustered close enough to the nucleus, the repulsion due to the two electrons cancels out completely two units of the positive charge on the nucleus, leaving an *effective nuclear charge*, Z_{eff}, of $+1$. The idea is illustrated in Fig. 2.44.

If the screening by the two $1s$ electrons is perfect, then the $2s$ electron in Li is essentially in a one-electron atom with a nuclear charge $+1$; we can therefore use Eq. 2.3 on the facing page to compute the orbital energy. Remembering that $n = 2$, the energy is $E_2 = -(1^2 \times R_{\text{H}})/(2^2) = -3.40$ eV. The actual energy of the $2s$ in Li is -5.34 eV, which is significantly lower than the energy we have just computed. We interpret this result by saying that the screening by the two $1s$ electrons is not perfect, so the $2s$ electron experiences an effective nuclear charge which is greater than $+1$.

If we refer again to the RDFs for the $1s$ and $2s$ shown in (b) from Fig. 2.23 on page 52, we see that a small part of the electron density for the $2s$ does indeed fall inside the region occupied by the $1s$. In other words, there is some probability that the $2s$ electron 'gets inside' the screen formed by the $1s^2$, thus experiencing more of the nuclear charge. This is usually described by saying that the $2s$ electron *penetrates* to the nucleus. One useful approach to quantifying the extent of shielding and penetration is to rewrite Eq. 2.3 on the facing page using not the actual nuclear charge but an effective nuclear charge, Z_{eff}

$$E_n = -\frac{Z_{\text{eff}}^2 R_{\text{H}}}{n^2}.$$

We then rearrange this so that given the orbital energy, E_n, we can compute Z_{eff}

$$Z_{\text{eff}} = \sqrt{\frac{-n^2 E_n}{R_{\text{H}}}}. \tag{2.4}$$

Using Eq. 2.4 we find Z_{eff} for $2s$ in Li to be 1.25. That this value is greater than 1 indicates that the screening is not perfect i.e. the $2s$ has penetrated the $1s^2$ shell somewhat. However, Z_{eff} is much less than the actual nuclear charge of $+3$, indicated that the $1s$ electrons are pretty effective at screening the nucleus.

Penetration

We now turn to the question as to why the ground state configuration of Li is $1s^2 2s^1$ and not $1s^2 2p^1$. Recall that in a one-electron atom the $2s$ and $2p$ orbitals are degenerate i.e. have the same energy, but this is no longer the case in Li, where the configuration of the ground state shows that occupation of the $2s$ gives a lower energy than occupation of $2p$.

The form of the RDFs for the $2s$ and $2p$ provide a way of rationalizing their relative energies in the Li atom. These RDFs, along with that for $1s$, are shown in Fig. 2.45; the RDFs are approximate as they have been computed using the hydrogen-like one-electron wavefunctions. For the $2s$ and $2p$ we have assumed that the nuclear charge is $+1$ i.e. the $1s$ electrons form a perfect screen, but it has been assumed that the $1s$ experiences the full nuclear charge of $+3$. As a result the $1s$ RDF is significantly contracted in towards the nucleus in the way described in section 2.5 on page 65.

Looking at the plots of these RDFs we can see that, as expected, most of the electron density of the $2s$ and $2p$ orbitals is outside the region occupied

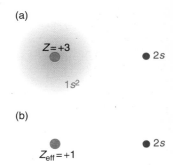

Fig. 2.44 Illustration of the idea of electron screening. In Li, the two $1s$ electrons cluster closely in to the nucleus so that from the point of view of the more distant $2s$ electron the effective nuclear charge is reduced. If the screening effect of the two $1s$ electrons is perfect, the effective nuclear charge seen by the $2s$ is just $+1$, as shown in (b).

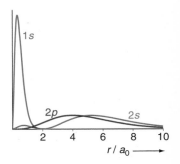

Fig. 2.45 Approximate RDFs for the occupied orbitals in Li; it has been assumed that the wavefunctions are those for a one-electron hydrogen-like atom. The $1s$ RDF has been computed for a nuclear charge of $+3$, whereas the RDFs for the $2s$ and $2p$ have been computed for a nuclear charge of $+1$. Note how the extent of penetration of the $2s$ into the region occupied by the $1s$ is greater than for $2p$.

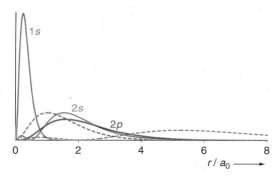

Fig. 2.46 RDFs for the occupied AOs in boron. The solid lines show RDFs computed using a numerical solution to the Schrödinger equation. For comparison, the dashed lines show the RDF for the $1s$ (green) and $2s$ (red) AO in hydrogen; note how much the orbitals in boron have contracted compared to those in hydrogen. The computed RDFs are similar in form to those for hydrogen e.g. the $2s$ shows a small subsidiary maximum, while the $2p$ has a single maximum.

● *Weblink 2.3*

This link takes you to a page where you can make plots of the RDFs of the various AOs and alter the effective nuclear charges experienced by each; you can then see how screening and penetration come about.

by the $1s$. However, there is a small amount of electron density from $2s$ and $2p$ which appears inside the $1s$ shell. The subsidiary maximum of the $2s$ (at about $r = 0.7\,a_0$) is mostly inside the region occupied by the $1s$, whereas for the $2p$ we just have the tail of the RDF. As a result, the $2s$ electron has greater probability of being inside the $1s$ screen than does than does the $2p$ electron, and so experiences, on average, a greater effective nuclear charge. Thus, the overall energy of Li is lower if the $2s$, rather than the $2p$, orbital is occupied.

The language which is often used to describe this different behaviour of the $2s$ and $2p$ is to say that the $2s$ electron *penetrates* to the nucleus, or penetrates the $1s$ shell, more than does the $2p$. The $2s$ electron is described as being 'more penetrating' than the $2p$. It is important to keep in mind, though, that what these various phrases are trying to describe is the result of the electron–electron repulsion.

Figure 2.45 on the preceding page also illustrates why the $2s$ electron has rather little effect on the nuclear charge experienced by the $1s$ electron. The electron density due to the $2s$ is mostly well outside the region occupied by the $1s$, and so has little influence on the interaction between the $1s$ and the nucleus. This is why the energy of the $1s$ falls rather steeply as we go from He to Li. The nuclear charge increases by one, as does the number of electrons, but the extra electron goes into the $2s$ and so does not shield the core $1s$ electrons from the increase in nuclear charge.

Moving across the second period

Increasing the nuclear charge to 4 brings us to Be, which has the ground state configuration $1s^2\,2s^2$. Referring to Fig. 2.43 on page 70, we see that the orbital energy of the $2s$ is -8.41 eV, significantly lower than the energy of -5.34 eV for Li; computing Z_{eff} using Eq. 2.4 on the preceding page, we find a value of 1.57 for Be. Note that although in going from Li to Be the nuclear charge has increased by one, the Z_{eff} experienced by the $2s$ electrons has only increased from 1.25 to 1.57. We interpret this by saying that the $2s$ electrons are shielding one another to some extent, so that they do not experience the full increase in the nuclear charge. Put another way, the addition of an extra electron in Be increases the amount of electron–electron repulsion, which offsets the increased attraction to the nucleus.

The $2s$ electrons do not shield one another as well as the $1s$ electrons shield the $2s$. This is because the $2s$ electrons occupy the same region of space as one another, whereas the $1s$ electrons generally lie between the $2s$ and the nucleus. Generally speaking, electrons in the same shell offer only a limited amount of screening of one another.

Moving on to boron, the extra electron has to go into the $2p$ because the $2s$ is full. As expected, in boron the $2s$ lies somewhat lower than the $2p$ (-13.5 eV as compared to -8.43 eV) which we can rationalize as before by noting that the $2s$ is more penetrating than the $2p$. The $2s$ in boron is lower in energy than the equivalent orbital in Be on account of the increase in nuclear charge not being entirely offset by the increase in electron–electron repulsion.

Figure 2.46 on the preceding page shows the RDFs for the occupied AOs in boron. These RDFs are computed from the wavefunctions found by a numerical solution to the Schrödinger equation, as described in section 2.6.1 on page 66. For comparison the RDFs for the $1s$ and $2s$ AOs in hydrogen are shown. These plots clearly show the greater penetration of the $1s$ by the $2s$ when compared to the $2p$.

The first thing we note about the RDFs in boron is that they are very much contracted compared to those in hydrogen; this is on account of the much larger nuclear charge in boron. However, apart from being contracted, the form of the boron RDFs is quite similar to those for hydrogen. For the $2s$, both show two maxima, with one being much greater than the other; for the $2p$, both show a single maximum. This gives us confidence that the hydrogen AOs, suitably scaled, are good guides to the actual orbitals in multi-electron atoms.

As we move across the remainder of the period, the $2p$ sub-shell is steadily filled, becoming complete at Ne with the configuration $1s^2 2s^2 2p^6$. From Fig. 2.43 on page 70 we can see that there is a steady downward trend in the energies of the $2s$ and $2p$ AOs, with the $2s$ falling considerably more steeply.

These trends can be rationalized by noting that for each successive element the nuclear charge increases by one, and one extra electron is added. The increase in nuclear charge will result in a decrease in the energy of the orbitals, but the extra electron results in an increase in electron–electron repulsion which increases the orbital energies. As we have already noted, electrons in the same shell do not shield one another particularly well, so the effect of the increase in nuclear charge dominates, accounting for the general fall in the orbital energies.

The $2s$ electron is more penetrating that the $2p$ and so feels the influence of the nuclear charge more than does the $2p$. As a result, the increase in nuclear charge has a greater effect on the $2s$ than the $2p$, and this explains why the $2s$ falls more steeply as the nuclear charge increases.

Moving across the third period

On moving from Ne to Na the additional electron has to go into the $n = 3$ shell, and it is found that the lowest energy arrangement is when the electron goes into the $3s$. The argument about why it is the $3s$ rather than the $3p$ which is occupied is precisely the same as for why $2s$ is filled before $2p$. The $3s$ is more penetrating than the $3p$, as can be seen from the RDFs shown in Fig. 2.28 on page 56.

The energy of the $3s$ in Na is -5.00 eV, somewhat higher than for the $2s$ electron in Li (-5.34 eV). If we use Eq. 2.4 on page 71 with $n = 3$ we can compute that Z_{eff} for the $3s$ electron is 1.8, which can be contrasted with the

Fig. 2.47 Computed average distances r_{av} of the electron from the nucleus for occupied orbitals of the elements of the first three periods; the distance is given in Bohr radii and pm. Note that, once occupied, an orbital contracts steadily as the nuclear charge increases. Particularly pronounced is the large increase in r_{av} of the outer orbital when a new shell is occupied e.g. Li compared to He, and Na compared to Ne. The orbitals are all smaller than the corresponding orbitals in hydrogen, an effect which is especially marked for the $3s$ and $3p$ orbitals.

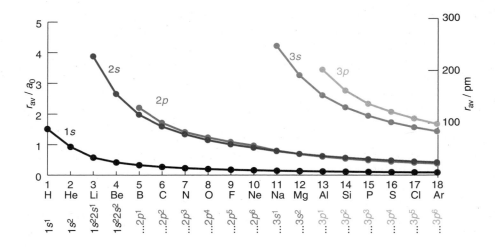

value of 1.25 for $2s$ in Li. The higher value for Na $3s$ is a result of the much higher nuclear charge and the fact that the $3s$ electron penetrates well to the nucleus. However, despite the higher Z_{eff}, the fact that the electron is in the $n = 3$ shell results in an increase in its energy relative to the $2s$ in Li.

As we go across the third period, we see similar trends to those seen for the second period, which can be rationalized in the same way. The energies of the $3s$ and $3p$ orbitals fall steadily, with the $3s$ falling faster than $3p$. For atoms in the same group in the periodic table, the energies of outer electrons are always higher if these are in the $n = 3$ shell than if they are in the $n = 2$ shell. For example, the energy of the $3s$ orbital in Si is higher than the $2s$ in C, and similarly the $3p$ is higher than the $2p$. This indicates that it is the change in principal quantum number n which dominates the difference in energy between $2p$ in C and $3p$ in Si; any change in the effective nuclear charge is of secondary importance.

2.6.3 Trends in the size of the orbitals

The radial distribution function tells us how the electron probability, or density, varies with distance from the nucleus, and from this we can work out the average distance of the electron from the nucleus, r_{av}. This average is rather meaningless as the whole point about quantum mechanics is that is tells us that the electron is not at a particular place, but can only be described by a probability distribution. However, r_{av} can be used as a rough indication of the 'size' of an orbital.

The maximum in the RDF of the hydrogen $1s$ orbital occurs at $r = a_0$; this is the radius at which we are most likely to find the electron. However, the *average* value of r is *not* the same as the value with greatest probability.

In hydrogen, this average distance for a $1s$ electron is $1.5\,a_0$, for $2s$ it is $6\,a_0$, and for $2p$ it is $5\,a_0$. You can make sense of these values by looking at the plots of the RDFs in Fig. 2.23 on page 52. When we move to the shell with $n = 3$, the hydrogen orbitals become much larger, with r_{av} being $13.5\,a_0$ and $12.5\,a_0$ for $3s$ and $3p$, respectively (see Fig. 2.28 on page 56).

Figure 2.47 shows the values of r_{av} for the occupied orbitals of the elements from the first two periods. These data are computed using the wavefunctions found by a numerical solution of the Schrödinger equation, as described in section 2.6.1 on page 66.

There are several points to note from these data. Firstly, there is a large increase in r_{av} the first time the outer electron goes into the next shell e.g. on going from He to Li, and on going from Ne to Na. We know from what we have seen for hydrogen that as the principal quantum number n increases the orbitals become much larger, and this is the explanation of the effect seen in the plot. However, the $2s$ electron in Li is significantly contracted when compared to a $2s$ electron in hydrogen ($r_{av} = 3.9\,a_0$ as compared to $6\,a_0$). This observation can be rationalized by noting that the $2s$ electron in Li penetrates somewhat to the nucleus, and so experiences an effective nuclear charge of more than +1. As has been commented on before in section 2.5 on page 65, this leads to a contraction of the orbital. This contraction of the $2s$ is also illustrated, for boron, in Fig. 2.46 on page 72.

The contraction is even more pronounced for the $3s$. In hydrogen this orbital has an r_{av} of $13.5\,a_0$, whereas in Na the value is $4.2\,a_0$. Clearly, the penetration by the $3s$ results in a significant contraction of the orbital. If we compare elements in the same group e.g. Li and Na, Ne and Ar, we find that if the outer electron occupies a $3s$ or $3p$ r_{av} is greater than for the case where the outer electron is in $2s$ or $2p$. However, the increase in size on going from $n = 2$ to $n = 3$ orbitals is nothing like so large as one would expect from the hydrogen wavefunctions.

The second thing we see in Fig. 2.47 on the facing page is that once an orbital is filled, it contracts steadily as the nuclear charge increases. This exactly mirrors the trend in orbital energies, and has the same explanation: the extra attraction due to increasing the nuclear charge by one is not compensated for by the increased electron–electron repulsion, and so the orbital contracts. This effect is particularly pronounced for the $1s$ orbital which quickly contracts to a very small radius. It is for this reason, as we shall see, that this orbital takes no part in the bonding for atoms beyond He.

The final point to note is that whereas r_{av} is much the same for $2s$ and $2p$, it is more different between $3s$ and $3p$, with $3p$ being larger. There appears to be no simple rationalization of this behaviour.

The take-home message from these data is that the occupied orbitals which are available for covalent bonding have sizes of between 50 and 250 pm for the first period, and between 100 and 250 pm for the second. These are useful numbers to keep in mind when we come to think about bond lengths.

2.6.4 The effective nuclear charge

The concept of the effective nuclear charge was introduced on page 71. To recap, the idea is that if we know the orbital energy (from a numerical calculation), we can work backwards to find what nuclear charge would give rise to a hydrogen (one-electron) orbital with that energy. The expression for Z_{eff} was given in Eq. 2.4 on page 71

$$Z_{eff} = \sqrt{\frac{-n^2 E_n}{R_H}},$$

where n is the principal quantum number and E_n is the orbital energy.

What the value of Z_{eff} tells us is the extent to which the orbital energy deviates from that of a corresponding orbital in hydrogen. If there is no penetration (or, put another way, the screening is perfect) we expect $Z_{eff} = 1$, as

Fig. 2.48 Effective nuclear charges, Z_{eff}, computed from the orbital energies used in Fig. 2.42 on page 69, for the occupied orbitals of the elements of the first three periods; values of Z_{eff} greater than 5 are not plotted. Once an orbital is occupied, its Z_{eff} increases steadily as the nuclear charge increases.

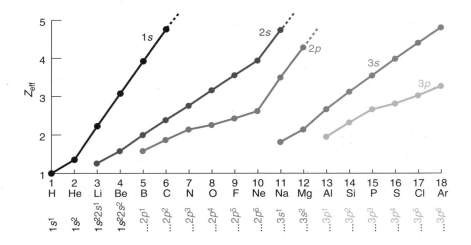

in hydrogen. The greater the degree of penetration (the poorer the screening), the larger the effective nuclear charge will become.

Figure 2.48 is a plot of Z_{eff} for the occupied orbitals of the elements of the first three periods. The orbital energies used to compute the nuclear charges are those used to generate Fig. 2.42 on page 69.

As expected Z_{eff} for $1s$ in H is 1, and for He it is somewhat greater than 1. After that, the effective nuclear charge for the $1s$ orbital rises steeply on account of it being ineffectively screened by the electrons in the $n = 2$ and $n = 3$ shells. For the $2s$ and $2p$ the effective nuclear charges increase across the second period, and then make a sharp upturn as we move to the next period in which the $n = 3$ shell starts to be filled. Z_{eff} for $2s$ is always greater than for $2p$, and the separation between the two increases as we go across the second period.

The reasons for this behaviour of the effective nuclear charge are essentially the same as those given for the behaviour of the orbital energies. The $2s$ is more penetrating than the $2p$, and so experiences a greater effective nuclear charge; in addition, the greater penetration by the $2s$ means that it feels the increase in the nuclear charge more than does the $2p$, so Z_{eff} of the $2s$ increases more rapidly. The general rise in Z_{eff} for both $2s$ and $2p$ is as a result of the increase in the nuclear charge not being offset by the increase in electron–electron repulsion. It is interesting to note that by the end of the period (Ne), Z_{eff} has risen to 2.6 for $2p$, indicating that the shielding of $2p$ electrons by one another is far from perfect.

When we move to Na, and the $3s$ is occupied, the effective nuclear charge drops back down to 1.8, indicating that the $n = 2$ shell is quite an effective screen for $3s$. Then, as we move across the period Z_{eff} values follow a very similar trend to those already seen for the $n = 2$ orbitals. The explanation for this behaviour is identical.

It is interesting to compare Z_{eff} values between the outer electrons for elements in the same group e.g. Li with Na, N with P etc. What we see is that Z_{eff} is always greater for the $n = 3$ orbital than the corresponding $n = 2$ orbital. We can probably attribute this to the fact that the actual nuclear charge is higher for those atoms whose outer shell contains $3s$ electrons when compared to those whose outer shell contains $2s$ electrons. When compared to hydrogen, the greater contraction of the $n = 3$ orbitals than the $n = 2$ orbitals, noted in section 2.6.3 on page 74, can be attributed to the larger effective nuclear charge felt by the $n = 3$ orbitals.

Slater's rules

The power of modern computers means that it is possible to solve Schrödinger's equation to high precision for atomic systems. However, in the past such calculations were by no means so quick and easy, so some approximate methods were developed for estimating orbital energies. Although these methods are no longer needed, they can provide a useful way of rationalizing various trends, and so are worthwhile considering.

Here we will look at *Slater's rules*, which are an approximate method for estimating the effective nuclear charge. Slater proposed that Z_{eff} could be written as

$$Z_{\text{eff}} = Z - S,$$

where Z is the actual nuclear charge and S is a *shielding constant*, which is computed in the following way.

First we divide up the orbitals into groups (indicated by the brackets) and order them as follows

$$(1s)\,(2s, 2p)\,(3s, 3p)\,(3d)\,(4s, 4p)\,(4d)\,(4f)\,\ldots$$

In this grouping, s and p orbitals with the *same* principal quantum number are grouped together, but d and f orbitals are separated. The value of S of a particular electron (in a particular group) is computed using the following rules:

1. There is no contribution to S from electrons in groups to the right of the one being considered.

2. A contribution of 0.35 is added to S for each electron in the same group as the one being considered, except in the $(1s)$ group where the contribution is 0.30.

3. If the electron being considered is in an ns or np orbital, then electrons in the next lowest shell (i.e. that with $(n-1)$) each contribute 0.85 to S. Those electrons in lower shells (i.e. $(n-2)$ and lower) contribute 1.00 to S.

4. If the electron being considered is in an nd or nf orbital, all electrons below it in energy contribute 1.00 to S.

These rules encapsulate the ideas we have been exploring in this section. Rule 1 says that electrons in higher orbitals do not contribute to the screening e.g. $2s$ electrons do not screen $1s$ electrons. Rule 2 says that electrons in the same shell shield one another to some extent. Rule 3 says that electrons in the next lower shell are quite good at screening the nuclear charge, and those in the shell below that (if there is one) form a perfect screen e.g. for a $3s$ electron the $2s$ form a good screen, and the $1s$ form a perfect screen. We shall return to Rule 4 when we consider transition elements in more detail.

Let us use carbon to illustrate how these rules are applied. The electronic configuration, grouped in the way suggested, is

$$(1s^2)\,(2s^2, 2p^2).$$

Fig. 2.49 Effective nuclear charges, Z_{eff}, estimated using Slater's rules for the occupied orbitals of the elements of the first three periods. Note that these rules give the same Z_{eff} values for s and p orbitals with the same n. These data should be compared with those presented in Fig. 2.48 on page 76. The overall trend is the same, but the numerical values are not in good agreement.

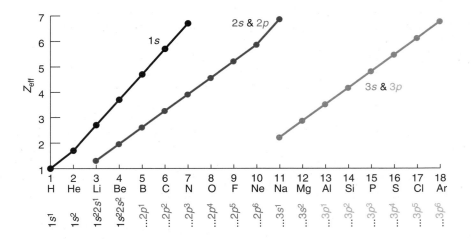

For a $1s$ electron, the $2s$ and $2p$ make no contribution to S (Rule 1), but according to Rule 2 we have a contribution of 0.30 to account for the effect of the other electron. So, remembering that the nuclear charge is 6, $Z_{\text{eff}} = 6 - 0.30 = 5.70$; clearly, the $1s$ electrons feel most of the nuclear charge.

For a $2s$ electron, Rule 1 does not apply and Rule 2 says that we have a contribution of 0.35 for the other three electrons in the same group. Rule 3 says that each $1s$ electron, as it is in the next lowest shell, contributes 0.85. So $Z_{\text{eff}} = 6 - 3 \times 0.35 - 2 \times 0.85 = 3.25$. As the $2s$ and the $2p$ are in the same group, they have the same Z_{eff}.

The values for Z_{eff} obtained from modern calculations are 4.8 for $1s$, 2.4 for $2s$ and 1.9 for $2p$. Salter's rules give values which are in the right ball park, but the numerical agreement is not good. In particular, these rules predict Z_{eff} to be the same for $2s$ and $2p$, which is not the case. However, we should not be too critical of Slater's efforts: the interactions involved in multi-electron atoms are subtle and complex, so we cannot reasonably expect an accurate prediction of Z_{eff} based on a simple set of rules.

For Si ($Z = 14$), the grouped electronic configuration is

$$(1s^2)\,(2s^2, 2p^6)\,(3s^2, 3p^2).$$

For a $1s$ electron $Z_{\text{eff}} = 14 - 0.30 = 13.70$; again, this electron feels most of the nuclear charge. For a $2s$ or $2p$ electron, the electrons in the $n = 3$ shell make no contribution, the seven electrons in the same shell contribute 0.35, and the $1s$ electrons contribute 0.85 each.

$$Z_{\text{eff}} = 14 - \underbrace{7 \times 0.35}_{2s\ \text{and}\ 2p} - \underbrace{2 \times 0.85}_{1s} = 9.85$$

The $2s$ and $2p$ feel a significant part of the nuclear charge.

For the $3s$ and $3p$ the calculation is

$$Z_{\text{eff}} = 14 - \underbrace{3 \times 0.35}_{3s\ \text{and}\ 3p} - \underbrace{8 \times 0.85}_{2s\ \text{and}\ 2p} - \underbrace{2 \times 1.00}_{1s} = 4.15$$

Z_{eff} is higher for $3s$ and $3p$ in Si than for $2s$ and $2p$ in C, which is what we found above.

Figure 2.49 on the preceding page shows the effective nuclear charges, estimated using Slater's rules, of the elements of the first three periods. This plot shows the same general trends as the data shown in Fig. 2.48 on page 76, but the numerical agreement between the two sets is poor. Note that it is a feature of Slater's rules that ns and np orbitals have the same Z_{eff} values.

2.6.5   Orbital energies of excited states and of empty orbitals

In a hydrogen-like atom, which contains just one electron, we have seen that there is a set of orbitals whose energies depend only on the value of the principal quantum number n. In the ground state the $1s$ orbital is occupied, as this has the lowest energy. However, it is also possible for the electron to occupy a higher energy orbital, such as $2s$ or $3p$, leading to what are called *excited states* of the atom.

The mental picture we have of a hydrogen-like atom is that the electron has 'available' to it a set of energy levels (orbitals) whose energies are characterized by the principal quantum number n. At any particular time, the electron occupies one of these orbitals. All of these other orbitals are 'empty', but if the electron moved to one of these, then the energy of the electron would simply be determined by the value of n of the occupied orbital. What we have therefore is a fixed set of energy levels which the electron can hop between.

For a multi-electron atom the situation is quite different, as in such atoms the energy of an electron in a particular orbital depends on the repulsion between this electron and *all* the other electrons on the atom. To add further complication, the amount of repulsion depends on exactly which other orbitals are occupied.

For example, in the case of the ground state of carbon, $1s^2\,2s^2\,2p^2$, the energy of the electron in the $2p$ orbital is determined by its repulsion with the other electron in the $2p$, the two electrons in the $2s$ and the two electrons in the $1s$. If we go to an excited configuration $1s^2\,2s^1\,2p^3$, the energy of the $2p$ electron is now determined by the repulsion with *two* other $2p$ electrons, *one* $2s$ electron and two $1s$ electrons. As a result, the energy of $2p$ in this excited state is *different* to the energy of $2p$ in the ground state.

In fact, the same is true of *all* of the orbitals in the atom, since rearranging the electrons will alter the details of the electron–electron repulsion for each electron. This point is illustrated in Fig. 2.50 which shows the approximate orbital energies for different configurations of the carbon atom. The key thing to note is that the orbital energies change as the occupation of the orbitals change – that is, the orbital energies depend on the configuration.

Empty orbitals

Strictly speaking we can only talk about the energy of an *occupied* orbital. An empty orbital cannot have an energy, since if there is no electron in the orbital there is nothing there to have any energy; there would be no energy of interaction with the nucleus. However, in practice we often talk about the energy of an unoccupied orbital. What we really mean by such an energy is the energy that an electron *would* have *if* it were to occupy that orbital.

In a hydrogen-like atom, the energies of these unoccupied orbitals are simply determined by the value of the principal quantum number, n. However, in multi-electron atoms the situation is altogether more complicated, and a full discussion of this is well beyond the level of this text. For now we will have

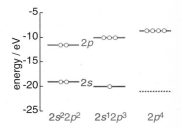

Fig. 2.50 Illustration of the approximate orbital energies of the $2s$ and $2p$ orbitals in carbon for three different configurations; the arrangement of the electrons is shown by the red dots. In the configuration $2p^4$, the $2s$ is unoccupied (a virtual orbital); its energy is therefore shown by a dashed line. Note that as there are three $2p$ orbitals, up to six electrons can be accommodated in them.

to content ourselves with noting that the method described in section 2.6.1 on page 66, as well as giving us the wavefunctions and energies of the occupied orbitals, also generates a set of *virtual* orbitals which are not occupied. Like the occupied orbitals, these virtual orbitals can be recognized as being related to the familiar hydrogen orbitals. For example, such a calculation on the lithium atom will give occupied 1*s* and 2*s* orbitals, as well as virtual orbitals which are recognizable as 2*p*, 3*s*, 3*p* and so on.

We can think of these virtual orbitals as being the unoccupied orbitals to which an electron could move. However, it is important to recognize that moving an electron to one of these orbitals will result in changes to the energy of this and all the other orbitals on account of the changes in electron–electron repulsion.

This point is illustrated in Fig. 2.50 on the previous page, which shows on the far right the orbital energies for an excited state of carbon in which all four valence electrons are in the 2*p*. This leaves the 2*s* empty, so 2*s* is one of the virtual orbitals. However, if this orbital is then occupied, to give the configuration $2s^1 2p^3$, its energy changes, as shown in the figure.

2.7 Ionization energies

In this section we are going to explore the relationship between ionization energies (ionization potentials) and the orbital energies we have been discussing so far. It is very useful to understand the trends in ionization energies as such values are often used to rationalize the occurrence of particular ions or oxidation states.

The ionization energy of a particular electron in an atom A is defined as the energy change for the following process

$$A(g) \longrightarrow A^+(g) + e^-.$$

We have already defined the energy of the electron when it is infinitely removed from the atom as zero, so the ionization energy, IE, is simply

$$IE = \text{energy of } A^+ - \text{energy of } A. \tag{2.5}$$

If we assume that A and A^+ are at rest, then the energy of each is just due to the electrons. So, to find the ionization energy we need to know the electronic energies of A and A^+.

2.7.1 One-electron atoms

In the case of a one-electron atom with nuclear charge Z, the calculation of the ionization energy is very simple. If the electron is in an orbital with principal quantum number *n*, the energy of the electron in the atom is $-R_H Z^2/n^2$. There is no electron in the ion, so its energy is zero, so the calculation of the IE is very simple

$$
\begin{aligned}
IE_n &= \text{energy of } A^+ - \text{energy of } A \\
&= 0 - (-\frac{R_H Z^2}{n^2}) \\
&= \frac{R_H Z^2}{n^2}.
\end{aligned}
$$

The IE is given the subscript n to indicate that the electron being ionized has this value of the principal quantum number. There is thus a very simple relationship between the IE and the orbital energy:

$$\text{one-electron atom}: \quad \text{IE} = -\text{ orbital energy.}$$

2.7.2 Multi-electron atoms

Once the atomic orbitals have been determined using the self-consistent method described in section 2.6.1 on page 66, it is a simple matter for the computer program to calculate the total electronic energy of an atom or ion. The difference of these energies gives the ionization energy, according to Eq. 2.5 on the preceding page.

It is important to understand that the total electronic energy of an atom is *not* the sum of the energies of the individual electrons (occupied AOs). The reason for this is that in the self-consistent approach, the energy of each electron depends on the average repulsion to all the other electrons. Therefore, if we add up the energies of all the electrons we will overcount the amount of electron–electron repulsion.

This is best understood by thinking about a simple example, such as helium. The energy of electron one includes a contribution from the repulsion by electron two; similarly, the energy of electron two includes a contribution from the repulsion by electron one. So, if we add up the energy of the two electrons, the contribution from the repulsion comes in twice, whereas in reality the total electronic energy has a single contribution from the electron–electron repulsion.

To put some specific numbers on this, the $1s$ orbital energy in helium is -25.0 eV; there are two $1s$ electrons, so the sum of their orbital energies is -50.0 eV. This value is much greater that the computed total electronic energy of helium which is -77.9 eV; counting in the electron–electron repulsion twice clearly has a very large effect.

Our discussion of multi-electron atoms has focused very much on orbital energies, and not on total electronic energies. It would therefore be very useful if we could use this understanding of orbital energies to rationalize the trends in ionization energies. The simplest approach is to use the following approximation, which mirrors the exact relationship found for one-electron atoms

$$\text{multi-electron atom}: \quad \text{IE} \approx -\text{ orbital energy.}$$

For example, the orbital energy of the outer electron in lithium, the $2s$, is -5.34 eV, so we approximate the ionization energy of this electron as 5.34 eV. The experimental value for the ionization energy is 5.39 eV, so the agreement is quite good.

This approach to computing the IE, sometimes known as *Koopmans' theorem*, is valid if the orbitals of the ion are the *same* as the orbitals of the neutral atom. A moment's thought will show that this condition is never true: removing an electron to make the ion will change the electron–electron repulsion experienced by all the remaining electrons, so their orbital energies *must* change.

However, it is found that the ionization energies determined in this way are not too different from the experimental values, as is clearly demonstrated by Fig. 2.51 on the following page. This graph compares the experimental first ionization energy with minus the energy of the highest energy AO. With a few

Fig. 2.51 A comparison of the experimental (first) ionization energy of the elements in the first three periods with minus the orbital energy of the ionized electron. Generally speaking the two values agree quite well, although there are particularly pronounced differences for O, F and Ne, and to a lesser extent for S, Cl and Ar. Along the bottom is shown the ground state configuration of the atom; it is assumed that the electron which is ionized is the last mentioned in these configurations.

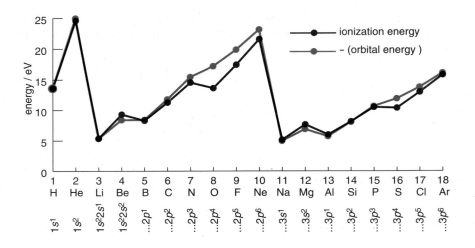

exceptions, the trend in the ionization energies is reproduced faithfully, as are the numerical values. What this is telling us is that the AOs of the ion are not that much different from those of the neutral atom.

There are three elements which stand out as having significantly different ionizations energies to those estimated from the orbital energies: O, F and Ne. These atoms have the electronic configurations $\ldots 2p^4$, $\ldots 2p^5$ and $\ldots 2p^6$. To a lesser extent, the elements S, Cl and Ar (with configurations $\ldots 3p^4$, $\ldots 3p^5$ and $\ldots 3p^6$) also show significant deviations. In all these cases, the experimental ionization energy is *lower* than the prediction from the orbital energy.

It turns out that these differences are associated with the way in which the electrons fill the p orbitals, leading to changes in the quantum-mechanical *exchange energy*. The details of what is going on are described in the next section.

2.7.3 Exchange energy

Recall that each orbital can contain at most two electrons, and that if there are two electrons in the same orbital, they must have their spins opposed i.e. one spin up ↑, and the other down ↓. So for Be there is no choice but for the two electrons in the $2s$ to have their spins opposed.

As we carry on along the third period and start to fill the $2p$ orbitals, there are many more possibilities as to how the electrons can fill the orbitals. This is because there are *three* degenerate $2p$ orbitals, each of which can be occupied by up to two electrons.

Figure 2.52 shows the *experimentally* determined arrangement of the electrons in the $2p$ orbitals for the ground states of B to Ne. Our task is to rationalize why the electrons occupy the orbitals in this way.

Looking at this picture we can see that in all cases the electrons are arranged in such a way as to *maximize the number of parallel spins*. So, for example, for the configuration $2p^2$ we place the electrons in separate $2p$ orbitals so that the electron spins can be parallel; the spins cannot be parallel if the electrons are placed in the same orbital. Similarly, for the configuration $2p^3$, the third electron is placed in a different $2p$ orbital to the other two, so that all the spins

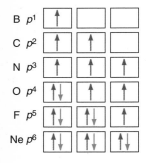

Fig. 2.52 Illustration of the ground state electronic configurations of the elements B to Ne. Each box represents a separate $2p$ orbital, and the arrows show the alignment of the electron spins.

can be parallel. For the configurations $2p^4$ to $2p^6$ we progressively have to pair up more and more electrons, but note that as we do this the number of (parallel) spin-down electrons increases.

The reason for this behaviour is associated with a contribution to the energy of the electrons called the *exchange interaction*. This interaction arises from the details of the quantum mechanical treatment of the electrons, and has no analogy in classical mechanics or electrostatics. Most importantly, the exchange interaction has nothing to do with the fact that the electrons are charged.

Although we cannot easily explain the origin of the exchange term, we can understand its effect in a simple way. Quantum mechanics indicates that the result of the exchange interaction is that *each pair* of electrons with parallel spins (both up, or both down) leads to a *lowering* of the electronic energy of the atom. Using this idea, we can rationalise the configurations shown in Fig. 2.52 on the facing page.

Let us first consider the configuration $2p^3$, for which Fig. 2.53 shows several possible arrangements of the electrons. All of the arrangements satisfy the Pauli principle but, as we shall see, they have different energies.

If we number the electrons 1, 2 and 3, we can see that arrangement (a) has *three pairs* of electrons with parallel spins: these pairs are 1–2, 1–3 and 2–3. Arrangement (b) has just *one pair* of electrons with parallel spins, and the same is true of arrangement (c), but in this case it is the spin-down electrons which are parallel. Arrangement (d), even though two electrons are paired up in one orbital, still has *one pair* of electrons with parallel spins i.e. the two with spin up.

Clearly, arrangement (a) has the greatest number of pairs of electrons with parallel spins, and so its energy is lower than the other arrangements as a result of it having the greatest contribution from the exchange interaction. A similar line of argument for the other configurations in Fig. 2.52 on the facing page will show that each of these has the greatest exchange interaction possible for that number of $2p$ electrons.

You will often come across a somewhat different explanation as to why the configurations shown in Fig. 2.52 are the lowest in energy. The explanation generally comes in two parts. Firstly, it is stated that as electrons are negatively charged, the repulsion between them is minimized by placing them in different orbitals. The argument is that such orbitals occupy different regions of space, and so 'keep the electrons as far apart as possible', thus minimising the repulsion. Secondly, it is stated that 'electrons with parallel spins tend to avoid one another', thus having the spins parallel further minimizes the repulsion.

This explanation sounds quite convincing, but it is simply not in accord with the picture which arises from a detailed quantum mechanical analysis of electrons in atoms. In contrast, the explanation we have given in terms of the exchange interaction is firmly grounded in the theory.

Effect on the ionization energies

We now turn to what effect this exchange energy has on the ionization energies. We will discover that the contribution made by the exchange interaction offers a good explanation for the deviations between the ionization energies and minus the orbital energies found for O, F and Ne. The key point to realize here is that this exchange contribution to the energy will be *different* for the atom and the ion, on account of the fact that the number of electrons is reduced by one in the ion.

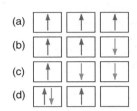

Fig. 2.53 Illustration of different arrangements of three electrons in the three $2p$ orbitals. The exchange interaction results in (a) having the lowest energy as it has the greatest number of pairs of parallel electron spins.

Fig. 2.54 Graph (a) shows the exchange contribution to the energy of the atom (in blue) and corresponding ion (in red) for the series $2p^1$–$2p^6$. Graph (b) shows the difference in exchange energy of the ion and the atom: this is the exchange energy contribution to the ionization energy. Plot (c) shows the experimental ionization energies for these atoms. The orange dotted line connects the ionization energies of B and O, which have no exchange contributions: this line is this an indicator of the trend in the absence of exchange effects. The deviations of the ionization energies from the orange line follow the exchange contributions shown in (b).

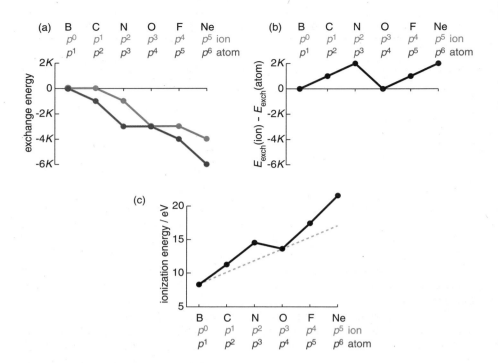

The first thing we want to do is to quantify the exchange contribution to the energy of the configurations $2p^1$–$2p^6$. A simple and reasonably accurate way of doing this is to say that each *pair* of electrons with parallel spins contributes $-K$ to the electronic energy of the atom, where K is a constant. The contribution from each pair is written as *minus K* since the interaction is favourable and so *lowers* the energy.

Referring to Fig. 2.52 on page 82 the contribution for the configuration $2p^1$ is clearly zero, whereas $2p^2$ has one pair of parallel spins and so the contribution is $-K$. As we have already explained, there are three pairs of parallel spins for $2p^3$, so the contribution is $-3K$.

For $2p^4$ there are still three pairs of spin-up electrons, so the contribution is $-3K$. For $2p^5$ there are three pairs of spin up electrons and one pair of spin down electrons, so the contribution is $-4K$. Finally, for $2p^6$ there are three pairs of spin-up and three pairs of spin-down electrons, making a contribution of $-6K$.

To compute the ionization energy, we need to know the electronic energy of both the atom and the ion (Eq. 2.5 on page 80)

$$\text{IE} = \text{energy of A}^+ - \text{energy of A}.$$

What we are going to work out is the exchange energy contribution to the ionization energy; the details of the calculation are set out in Table 2.2 on the next page. Here, we list the configuration of the atom and the exchange contribution to its energy $E_{\text{exch}}(\text{atom})$; also given is the configuration of the corresponding ion and the exchange contribution to its energy $E_{\text{exch}}(\text{ion})$.

The right-hand column gives $E_{\text{exch}}(\text{ion}) - E_{\text{exch}}(\text{atom})$, which will be the exchange contribution to the ionization energy. The same data are also presented graphically in Fig. 2.54.

Table 2.2 Exchange contribution to the ionization energy

element	atom		ion		
	configuration	$E_{\text{exch}}(\text{atom})$	configuration	$E_{\text{exch}}(\text{ion})$	$E_{\text{exch}}(\text{ion}) - E_{\text{exch}}(\text{atom})$
B	$2p^1$	0	$2p^0$	0	0
C	$2p^2$	$-K$	$2p^1$	0	K
N	$2p^3$	$-3K$	$2p^2$	$-K$	$2K$
O	$2p^4$	$-3K$	$2p^3$	$-3K$	0
F	$2p^5$	$-4K$	$2p^4$	$-3K$	K
Ne	$2p^6$	$-6K$	$2p^5$	$-4K$	$2K$

For boron, with atomic configuration $2p^1$, there is no exchange contribution to either the atom or the ion, and exchange contribution to the ionization energy is zero. As we move to the atomic configurations $2p^2$ and $2p^3$, the exchange contribution of the atom increases and is always greater than that of the ion. The difference $E_{\text{exch}}(\text{ion}) - E_{\text{exch}}(\text{atom})$ increases, and so the exchange contribution results in an increase in the ionization energy, as can be seen in Fig. 2.54 (b).

When we reach oxygen, with atomic configuration $2p^4$, the exchange energy is the *same* for the atom and ion, so there is no exchange contribution to the ionization energy. This is in contrast to the previous atom, nitrogen, which has the maximum contribution. As a result, the ionization energy of oxygen is less than that of nitrogen, despite oxygen having the larger nuclear charge. The next two atoms with the configurations $2p^5$ and $2p^6$ again show an increase in the exchange energy contribution to the ionization energy, as shown in Fig. 2.54 (b).

This exchange contribution therefore provides us with a good explanation of the slightly erratic behaviour of the ionization energies of B–Ne, as shown in Fig. 2.54 (c). Boron and oxygen have no exchange contribution to their ionization energies. The orange dotted line, which connects the ionization energies of B and O, can be regarded as an indication of the trend in the absence of exchange contributions. The ionization energies of C, N, F and Ne all lie above this line on account of the exchange contribution; in addition, N and Ne are further above the line than C and F, which mirrors the size of the exchange contribution shown in Fig. 2.54 (b).

You may have heard before of this idea that a 'half-filled shell is unusually stable'. The evidence often cited for this is that the ionization energy of N (configuration $2p^3$) is larger than that for O ($2p^4$), even though O has a greater nuclear charge than N. As we have just seen, the real explanation for this effect is considerably more complicated than just attribution of some special stability to $2p^3$. Rather, it rests on an appreciation that the ionization energies reflect *differences* between the exchange contribution to the atom and the ion.

2.8 Moving on

The ideas introduced in this chapter are very important for the rest of the book. We will constantly be referring to the energies, shape and size of atomic orbitals when we are discussing chemical bonding and reactivity, so it is very important to get a good grasp of these essential ideas: they are the bedrock of our understanding of chemistry.

In the next chapter we move on to extend the idea of orbitals from atoms to molecules. This will enable us to develop a detailed understanding of what a covalent bond is, the factors that influence its strength, and how the shapes of covalent molecules are determined by their electronic structure.

FURTHER READING

Elements of Physical Chemistry, fourth edition, Peter Atkins and Julio de Paula, Oxford University Press (2005).

QUESTIONS

2.1 The probability of finding an electron in a small volume δV located at a particular point is given by $\psi(x, y, z)^2\, \delta V$, where $\psi(x, y, z)$ is the wavefunction at that point.

For an electron in a $1s$ orbital, describe how this probability varies as the distance from the nucleus is increased. At what point is the probability (electron density) a maximum?

2.2 Explain what is meant by the *radial distribution function*. For an electron in a $1s$ orbital, how does the RDF vary with distance from the nucleus?

Explain why it is that although the $1s$ wavefunction is a maximum at the nucleus, the corresponding RDF goes to zero at the nucleus. Also, explain why the RDF shows a maximum, and why the RDF goes to zero for large values of the distance r.

2.3 **This question requires a knowledge of calculus.** The RDF for a $1s$ orbital is $4\pi r^2 [\psi_{1s}(r)]^2$. Given that the $1s$ wavefunction is $\psi_{1s}(r) = N_{1s} \exp\left(-r/a_0\right)$, show that the RDF is given by

$$P_{1s}(r) = 4N_{1s}^2 \pi r^2 \exp\left(-2r/a_0\right).$$

We can find the maximum in this RDF by differentiating it with respect to r, and then setting the derivative to zero. Show that the required derivative is

$$\frac{dP_{1s}(r)}{dr} = 8N_{1s}^2 \pi r \exp\left(-2r/a_0\right) - 8N_{1s}^2 \pi \frac{r^2}{a_0} \exp\left(-2r/a_0\right).$$

Further show that this differential goes to zero at $r = a_0$, and use a graphical argument to explain why this must correspond to a maximum.

For a hydrogen-like atom with nuclear charge Z, the $1s$ wavefunction is $\psi_{1s}(r) = N_{1s} \exp\left(-Zr/a_0\right)$. Show that the corresponding RDF has a maximum at $r = a_0/Z$.

2.4 The N shell has the principal quantum number n equal to four. Determine the quantum numbers (l and m_l) of all the possible AOs in this shell. What new feature arises in this shell that is not present in the M shell? How do the energies of these orbitals in the N shell compare with one another?

How many electrons could be accommodated in the N shell?

Orbitals for which the orbital angular momentum quantum number, l, takes the value 4 are given the letter g. In which shell would you expect g orbitals first to appear?

2.5 For a hydrogen atom (with $Z = 1$), calculate the energies, in kJ mol^{-1}, of the four lowest energy levels. Do the same for the He$^+$ ion, which has $Z = 2$. Plot the levels of He$^+$ to scale in a similar way to Fig. 2.13 on page 45, labelling each with the value of the principal quantum number n. Discuss the choice of energy zero you have used.

2.6 What do you understand by the terms *radial node* and *nodal plane*, as applied to AO wavefunctions? Illustrate your answer using the $2s$ and $2p$ AOs.

Explain why radial nodes arise from the radial part of the wavefunction, whereas nodal planes arise from the angular part of the wavefunction.

2.7 The radial part of the $3s$ AO wavefunction is

$$R_{3,0}(r) = N_{3,0} \left[27 - 18\left(\frac{r}{a_0}\right) + 2\left(\frac{r}{a_0}\right)^2 \right] \exp\left(-\frac{r}{3a_0}\right).$$

This function will go to zero, i.e. have a radial node, when the term in the large square bracket goes to zero

$$\left[27 - 18\left(\frac{r}{a_0}\right) + 2\left(\frac{r}{a_0}\right)^2 \right] = 0.$$

Finding out the values of r at which this is the case is made easier if we substitute $x = r/a_0$. Show that the above equation then becomes

$$27 - 18x + 2x^2 = 0.$$

This is a simple quadratic whose roots can be found using the standard formula. Show that these roots are $x = 1.9$ and $x = 7.1$. Hence show that there are radial nodes at $r = 1.9\,a_0$ and $7.1\,a_0$. Compare these answers with the plot shown in Fig. 2.28 on page 56.

Using a similar approach, determine the position of the radial node in the $3p$ orbital; the radial part of the wavefunction is

$$R_{3,1}(r) = N_{3,1} \left[6\left(\frac{r}{a_0}\right) - \left(\frac{r}{a_0}\right)^2 \right] \exp\left(-\frac{r}{3a_0}\right).$$

2.8 Draw up a table showing the number of radial nodes, the number of angular nodes (nodal planes), and the total number of radial and angular nodes for $1s$, $2s$, $2p$, $3s$, $3p$ and $3d$ orbitals. Confirm that the numbers in your table are in accord with the general rules given on page 62.

Describe the nodal structures of $5s$, $5p$, $5d$, $5f$ and $5g$ orbitals ($5g$ has $l = 4$).

2.9 The contour plot shown below is of one of the $4p$ orbitals: positive intensity is indicated by red contours, negative by blue, and the zero contour is indicated in green.

Sketch how the wavefunction will vary along the dotted line a, and along the two circular paths b and c (for the latter two, this means making a plot of the wavefunction as a function of an *angle* which specifies how far we have moved around the circle).

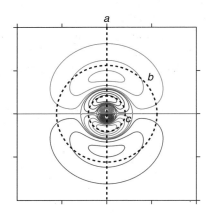

2.10 Using the concepts of electron *screening* and *penetration*, and with reference to the relevant RDFs, explain the following observations concerning the orbital energies of $3s$ and $3p$ shown in Fig. 2.42 on page 69:

(a) The ground state configuration of Na is $\ldots 3s^1$ and not $\ldots 3p^1$.

(b) In Al, the $3s$ is lower in energy than the $3p$.

(c) As we go along the series Na $\ldots$ Ar, the energy of both the $3s$ and the $3p$ orbitals falls steadily; however, the energy of the $3s$ falls more rapidly than does that of the $3p$.

What would you expect to happen to the energies of the $3s$ and $3p$ orbitals for elements with $Z > 18$ i.e. along the fourth period?

2.11 Calculate the effective nuclear charges of the following orbitals for Al and Ar. Comment on the differences of Z_{eff} values seen between these two elements, and between the two different types of orbital.

element	Z	energy of $3s$ (eV)	energy of $3p$ (eV)
Al	13	−10.7	−5.71
Ar	18	−34.8	−16.1

Write down the full electronic configurations of Al and Ar. User Slater's rules to compute Z_{eff} for a $3p$ electron and a $2p$ electron in each atom. Comment on the values you obtain.

2.12 The electron in hydrogen can be promoted from one orbital to another by the absorption of a photon whose energy matches the energy difference between the two orbitals. This process is described as a transition.

If the electron starts in an orbital with principal quantum number n_1 and moves to an orbital with principal quantum number n_2 ($n_2 > n_1$), show that the energy change $\Delta E(n_1 \rightarrow n_2)$ is given by

$$\Delta E(n_1 \rightarrow n_2) = E_{n_2} - E_{n_1}$$
$$= R_{\text{H}} \left(\frac{1}{n_1^2} - \frac{1}{n_2^2} \right).$$

The energy of a photon with frequency v is hv, where h is Planck's constant (6.626×10^{-34} J s). Show that it follows that the frequency of a photon which can cause a transition from n_1 to n_2, $v_{n_1 \rightarrow n_2}$, is

$$v_{n_1 \rightarrow n_2} = \frac{R_{\text{H}}}{h} \left(\frac{1}{n_1^2} - \frac{1}{n_2^2} \right).$$

Using $R_H = 2.180 \times 10^{-18}$ J, work out the frequency (which will be in Hz), of the transitions $2 \to 3$, $2 \to 4$ and $2 \to 5$.

Convert these frequencies to wavelengths using $c = \nu\lambda$, where c is the speed of light $(2.998 \times 10^8$ m s$^{-1})$ and λ is the wavelength in m. Give your answers in m and nm.

What region of the electromagnetic spectrum do these transitions appear in?

2.13 Using a similar approach to that in the previous question, show that for a hydrogen-like (one-electron) atom with nuclear charge Z, the frequency of the photon needed to cause a transition from an orbital with principal quantum number n_1 to one with quantum number n_2 is

$$\nu_{n_1 \to n_2} = \frac{Z^2 R_H}{h} \left(\frac{1}{n_1^2} - \frac{1}{n_2^2} \right).$$

Super-nova remnant E0102-72 is located some 200,000 light years from the Earth, and is 20 light years across. The matter contained in the remnant is at extremely high temperatures, up to millions of degrees. At such extremes, very highly ionized atoms are found, such as O^{7+} (which is a hydrogen-like atom).

In the light from this remnant, astronomers have observed transitions attributed to this ion. The transitions are thought to be $1 \to 2$, $1 \to 3$ and $1 \to 4$. Work out the frequency (in Hz) and wavelength (in nm) of the light which would cause these transitions. What part of the electromagnetic spectrum does this appear in?

2.14 Explain why it is that in a hydrogen-like atom the ionization energy is exactly equal to minus the orbital energy, whereas this is only approximately true for multi-electron atoms.

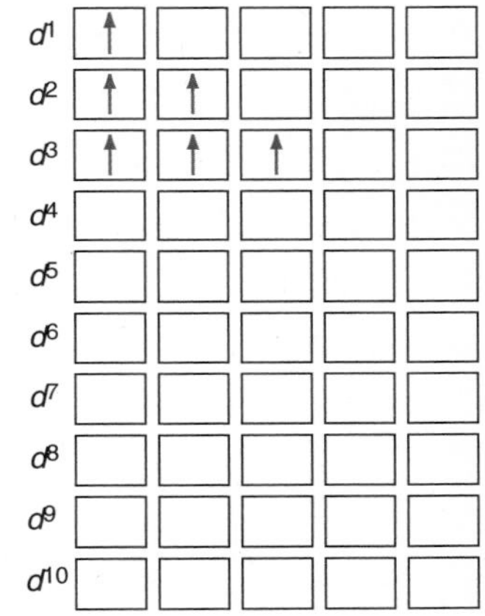

2.15 There are five $3d$ orbitals, which can accommodate a maximum of ten electrons. The lowest energy arrangement of the electrons in these five orbitals is shown opposite for the configurations d^1, d^2 and d^3. Explain why these are the lowest energy arrangements, and go on to complete the figure showing the lowest energy arrangements for d^4 to d^{10}.

For each configuration, work out the number of pairs of electrons with parallel spins, and hence draw up a table similar to Table 2.2 on page 85 showing the exchange contribution to the energy of the atom, the corresponding ion, and the difference between the two.

Finally, plot the difference of these exchange energies in a similar way to Fig. 2.54 (b) on page 84. What do these data imply about the variation in ionization energies of atoms with the configurations $d^1 \ldots d^{10}$?

Electrons in molecules: diatomics

Key points

- In molecules, electrons can be assigned to molecular orbitals which are analogous to the atomic orbitals introduced in the previous chapter.
- Molecular orbitals are constructed from a linear combination of atomic orbitals (AOs).
- The combination of two atomic orbitals gives two molecular orbitals: (1) a bonding molecular orbital which places electron density between the atoms and contributes to the formation of a bond; (2) an antibonding molecular orbital which pushes electron density away from the internuclear region and does not favour bond formation.
- There are simple rules governing the way in which MOs are formed from AOs: of these, the most important are that AOs have to have the correct symmetry and similar energies in order to interact to form MOs.
- The relationship between the MOs and atomic orbitals from which they are formed is conveniently expressed using an MO diagram.
- Electronegativity is a consequence of orbital energies.
- Photoelectron spectroscopy provides an experimental way of probing the energies and other characteristics of MOs.

In the previous chapter we saw how the atomic orbitals predicted by quantum mechanics give a good description of the energies and spatial distributions of electrons in atoms. Our task now is to apply quantum mechanics to molecules so that we can develop an orbital description of covalent bonding. We will discover that electrons in molecules can be assigned to *molecular orbitals* which are analogous to the atomic orbitals we have encountered so far. The shapes and energies of these molecular orbitals are crucial in determining the strength and directional properties of bonds.

The simple picture of a covalent bond is that it comes about due to the sharing of electrons between atoms. The molecular orbital approach confirms this key idea, but enables us to be more precise about why sharing leads to a lowering of energy, how the energy of a particular bond is related to the nature of the sharing, and why the sharing might not be equal between the two atoms.

Using molecular orbitals we will be able to give straightforward explanations of a number of observations, such as:

- H_2 exists as a stable molecule, He_2 is unknown, but He_2^+ has been detected.

- The bond dissociation energy of Be_2 is less than one tenth of that of either Li_2 or B_2.

- N_2 has a stronger bond than does O_2.

- The bond in N_2^+ is weaker than that in N_2, but the bond in O_2^+ is stronger than that in O_2.

- In Li–H, the hydrogen has a partial negative charge, whereas in F–H, the hydrogen has a partial positive charge.

- O_2 is *paramagnetic*, meaning that it is drawn into a magnetic field, but N_2 is not.

No doubt you can think of explanations for some of these observations but, as we shall see, molecular orbital theory gives a single framework within which all of these observations can be understood in a satisfying way.

In this chapter we are going to restrict ourselves to describing the bonding in diatomics. To start with we will look at *homonuclear* diatomics, which are those in which both atoms are the same, such as H_2 and N_2. Then we will go on to look at *heteronuclear* diatomics, in which the two atoms are different, such as HF and CO.

However, the whole discussion starts with the simplest possible molecule, H_2^+. Like the hydrogen atom, this molecule contains only one electron, so there is no electron–electron repulsion, which simplifies things a great deal. We will see that the ideas we develop for this one-electron system can quickly be transferred to multi-electron molecules, just as we did for atoms.

3.1 Introducing molecular orbitals

H_2^+ is perhaps not such a familiar molecule, but it exists in the gas phase where its properties have been studied by spectroscopic methods. Dissociating this molecule to give a hydrogen atom and a proton (H^+) requires 256 kJ mol^{-1}, so H_2^+ is lower in energy than a hydrogen atom and a proton; our initial task is to understand why this is so.

Let us start by thinking about the different contributions to the energy of the separated atoms and then of the molecule. For the H atom we have the potential energy due to the favourable interaction between the electron and the nucleus. As was described before, this energy is negative as it is measured from a zero point corresponding to the electron being far removed from the nucleus. In addition, there is the kinetic energy of the electron, a contribution we can associate with its orbital motion. As there is no electron in H^+, there are no further contributions to the energy from this atom.

Fig. 3.1 Diagrammatic representations of the electrostatic interactions in H_2^+. The two positive nuclei (shown in red) repel one another leading to an increase in the potential energy. In contrast, the negatively charged electron (shown in blue) is attracted to each of the positively charged nuclei, leading to a decrease in the potential energy.

In H_2^+ the situation is a little more complicated. There are two different contributions to the potential energy, as illustrated in Fig. 3.1 on the preceding page. The first contribution is from the favourable interaction between the electron and the nuclei; just as in an atom, this potential energy is negative. It is found that this contribution decreases (i.e. becomes more negative) as the internuclear separation R decreases, as is shown by the blue line (marked e–n) in Fig. 3.2. We can make sense of this by noting that when R is large the electron can only be close to, and so interact with, one nucleus at a time, whereas when R is small the electron can interact with both nuclei.

The second contribution to the potential energy comes from the unfavourable interaction between the two positively charge nuclei i.e. a repulsion. This contribution varies as $1/R$, so it is zero when R is large and becomes increasingly positive as R decreases, as is shown by the red line (marked n–n) on Fig. 3.2.

The final contribution to the energy is the kinetic energy of the electron associated with its motion about both nuclei. It is found that as R decreases the kinetic energy at first decreases by a small amount from the value it has for the separated atoms, and then starts to increase more rapidly. This is shown by the green line (marked ke) on Fig. 3.2. The explanation for this behaviour is rather complex and is beyond the scope of this chapter.

The total energy of the molecule is a sum of these three contributions, and is shown by the black line (marked tot) on Fig. 3.2. Although it is not very marked, there is a minimum in this total energy at a particular value of R, known as the equilibrium separation R_e. The prediction is therefore that H_2^+ will form with this bond length.

In this plot, the total energy does not go to zero when R becomes large. This is because what we have at large values of R is a well-separated hydrogen atom and a proton (H^+), so the energy is just the energy of the electron in H. Similarly, the kinetic energy and the electron–nuclear energy are not zero at large values of R, but rather tend to the values of the H atom.

The value of R_e results from a subtle balance between all three different contributions to the energy. However, it is the case that it is the electron–nucleus interaction which is in large part responsible for overcoming the nucleus–nucleus repulsion.

The next part of our task is to see how quantum mechanics can be used to understand the form of Fig. 3.2, and hence explain why H_2^+ forms.

3.1.1 Linear combination of atomic orbitals

As there is only one electron in H_2^+, it proves to be possible to solve the Schrödinger equation for this molecule in a fairly straightforward way so as to obtain the wavefunction of the electron. In analogy to atomic orbitals, a wavefunction which describes an electron in a molecule is called a *molecular orbital* (MO). For a given molecule there will be a series of MOs of ever increasing energy, just like the AOs in an atom. The square of the MO wavefunction gives the probability density of the electron, again just as for an AO.

One way of constructing the molecular orbitals is to add together atomic orbital wavefunctions which are located on the various atoms in the molecule. This is called the *linear combination of atomic orbitals* (LCAO) approach. We can put forward a number of arguments as to why this is a sensible way to proceed.

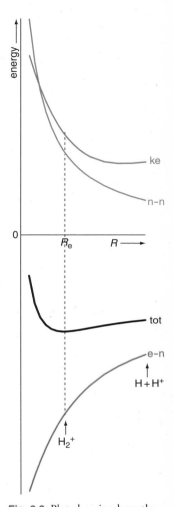

Fig. 3.2 Plot showing how the different contributions to the energy of H_2^+ vary with the internuclear distance R. The green line (marked ke) gives the kinetic energy of the electron, the red line (marked n–n) gives the nuclear repulsion energy, and the blue line (marked e–n) gives the electron–nuclear energy. The total energy is shown by the black line marked tot. This shows a minimum at the equilibrium separation, R_e.

Firstly, if we are really close to a particular nucleus in a molecule, the dominant interaction will be with that nucleus, so we can expect the electron wavefunction to be very similar to that of an AO centred on that nucleus. Secondly, if we imagine moving the nuclei in a molecule further and further apart, we will eventually end up with separated atoms, for which the wavefunctions must be AOs. Since the wavefunctions have to change steadily as the nuclei are pulled apart, the MOs will steadily transform into AOs: it therefore makes sense to construct the MOs from AOs in the first place. Finally, we know a lot about AOs, so constructing MOs from them will be convenient way to proceed.

Quantum mechanics gives us a way of working out precisely which combination of AOs will be the best approximation to each MO. As before, we are not going to go into all the mathematical details of how this best combination is found. Rather, we will look at the results of such calculations and see how we can rationalize and understand them. We will also find that there are quite simple qualitative rules for how MOs are constructed from AOs, so without making any calculations we can have a good guess at the form of the MOs. In many cases, such a guess will be sufficient.

In H_2^+ it makes sense to construct the MOs from a $1s$ AO on one hydrogen atom (which we will label A), and a $1s$ AO on the other hydrogen atom (which we will label B). As the two atoms are identical we cannot really distinguish between them, but the A and B labels are nevertheless useful for keeping track of what is going on. The MO wavefunction is constructed in the following way

$$\text{MO} = c_A \times (1s \text{ AO on atom A}) + c_B \times (1s \text{ AO on atom B}).$$

c_A and c_B are the *orbital coefficients*; they are just numbers, whose values can be determined by quantum mechanics. As you can see, the values of these coefficients determine how much of each atomic orbital is present in the MO.

For H_2^+ the detailed quantum mechanical calculation reveals a very simple result arising from the combination of these two AOs. We find that there are *two* MOs: the first (ψ_+) has $c_A = 1$ $c_B = 1$, and the second (ψ_-) has $c_A = 1$ $c_B = -1$. These MOs can be written

$$\psi_+ = N_+ [(1s \text{ AO on atom A}) + (1s \text{ AO on atom B})]$$
$$\psi_- = N_- [(1s \text{ AO on atom A}) - (1s \text{ AO on atom B})],$$

where N_+ and N_- are two normalizing factors whose values need not concern us.

The energies of these two MOs can also be found using quantum mechanics, and these energies are plotted in Fig. 3.3 as a function of the internuclear distance R. Here E_+ is the energy of the MO ψ_+, and E_- is the energy of the MO ψ_-; note that these energies include the contribution from the repulsion of the two nuclei. Arbitrarily we have defined the energy of the well-separated atoms (i.e. at large R) as zero.

As the atoms approach one another the energy of ψ_+ falls at first, reaching a minimum at $R = 2.5\,a_0$ (145 pm) before rising steeply. If the electron is placed in this orbital, then it is favourable for the two atoms to come together until $R = 2.5\,a_0$, at which point the energy is a minimum. In other words, it is favourable for a bond to form, and so ψ_+ is called a *bonding* MO.

In contrast, the energy of ψ_- simply rises as the atoms come closer together. If the electron were placed in this orbital it would be disadvantageous, in energy

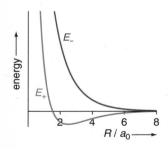

Fig. 3.3 Plots of the energies, as a function of the internuclear distance R (in Bohr radii), of the two MOs arising from the combination of two $1s$ AOs in H_2^+. E_+ and E_- are the energies of the MOs ψ_+ and ψ_-, respectively. At large values of R the energies tend to the same value (the energy of a hydrogen atom and a proton) which has arbitrarily been taken as zero.

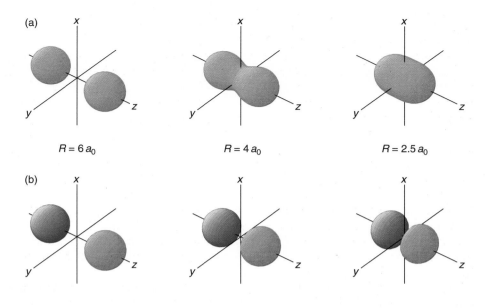

(a)

$R = 6\,a_0$ $R = 4\,a_0$ $R = 2.5\,a_0$

(b)

Fig. 3.4 Iso-surface representations of (a) the bonding and (b) the antibonding MOs in H_2^+, formed from the overlap of two $1s$ AOs. The orbitals are shown for three different values of the internuclear distance, R; red indicates a positive, and blue a negative, value of the wavefunction. The two nuclei are located along the z-axis, symmetrically placed about $z = 0$. In (a) we see the in-phase (constructive) overlap which leads to the bonding MO. In contrast the antibonding MO shown in (b) has out-of-phase (destructive) overlap between the two $1s$ AOs. In the antibonding MO there is a nodal plane (here xy) between the two nuclei.

terms, for the the atoms to come together to form a molecule. For this reason, ψ_- is called an *antibonding* MO.

Figure 3.4 shows iso-surface representations of the bonding and antibonding MOs for various values of the internuclear separation. For the bonding MO, shown in (a), the orbital coefficients c_A and c_B have the same sign, and so the two $1s$ AOs add together and reinforce one another. This is called *constructive* or *in-phase* overlap. For the largest internuclear separation there is little overlap of the two AOs, and so the diagram appears to show two separate $1s$ AOs. However, as the internuclear separation decreases, the two AOs start to overlap more, eventually giving an MO which encompasses both nuclei.

In contrast, for the antibonding MO shown in (b), the orbital coefficients have *opposite* signs, so the two AOs cancel one another out in the region where they overlap. This is called *destructive* or *out-of-phase* overlap. The antibonding MO always has a nodal plane between the two atoms; in this case the xy-plane is the nodal plane.

The contour plots shown in Fig. 3.5 on the next page perhaps give a clearer impression of what is happening to the MO wavefunctions as the internuclear distance decreases. Looking first at the bonding MO, shown in (a), at the largest internuclear distance ($R = 6.0\,a_0$), most of the wavefunction is located close to the nuclei. However, as the nuclei approach one another, the wavefunction begins to spread away from the nuclei, moving into the region between the two nuclei (the internuclear region).

In contrast, for the antibonding MO shown in (b), the destructive overlap ensures that there is a nodal plane between the two nuclei. As the internuclear separation decreases, the wavefunction is pushed away from the internuclear region.

Remember that the square of the wavefunction gives the probability density of the electron. So, what we see for the bonding MO is a build up of electron density in the internuclear region, whereas for the antibonding MO the electron density is pushed away from this region. This behaviour gives us an explanation

🌐 *Weblink 3.1*

View, and rotate in real time, iso-surface representations of the bonding and antibonding MOs of H_2^+.

🌐 *Weblink 3.2*

Follow this link to go to a real-time version of Fig. 3.5 in which you can vary the distance between the nuclei and see how the bonding and antibonding MO wavefunctions respond.

Fig. 3.5 Contour plots of cross-sections taken through (a) the bonding MO, and (b) the antibonding MO of H_2^+; plots are shown for three different values of the internuclear separation R. Positive values of the wavefunction are shown by red contours, and negative by blue, with the zero contour in green. The position of the two nuclei is indicated by the maroon dots joined by a line. Above each contour plot is shown a graph of how the MO wavefunction varies along the line of the nuclei. Recalling that the electron density depends on the square of the wavefunction, these plots show the way in which the electron density due to the bonding MO accumulates in the internuclear region as the separation of the two nuclei decreases. In contrast, for the antibonding MO there is a nodal plane between the two nuclei, and the electron density is pushed away from the internuclear region.

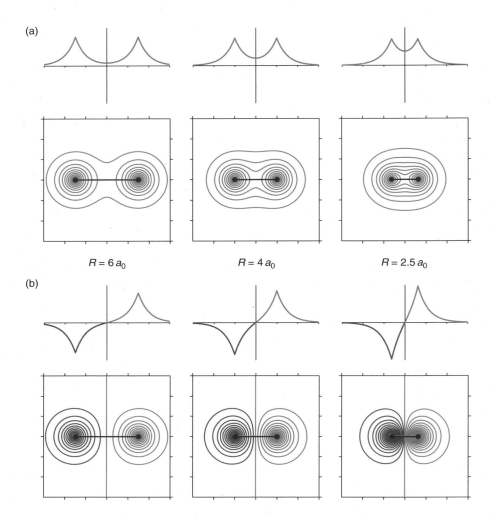

Weblink 3.2

Follow this link to go to a real-time version of Fig. 3.6 in which you can vary the distance between the nuclei and see how the difference electron density of the MOs change.

of why the energies of the MOs behave so differently as the internuclear separation changes, as shown in Fig. 3.3 on page 94.

The way in which the distribution of the electron density changes between the bonding and antibonding MOs is well illustrated in Fig. 3.6 on the next page. These plots show the *difference* between the electron density due to the MO and that due to two *non-interacting* 1s orbitals placed at the same internuclear separation. The blue shading indicates an increase in electron density compared to the non-interacting AOs, whereas the orange shading indicates a reduction in the electron density.

At large separations, where the interaction is small, we expect this difference in electron density to be small. However, as the internuclear separation decreases, the interaction between the AOs will increase and so there will be changes in the electron density which will show up in the difference plot.

Looking first at the bonding MO, shown in (a), we see that as the internuclear separation decreases there is a gradual build up of electron density in the internuclear region, whereas the electron density to the left and right of the molecule is decreased. In contrast, for the antibonding MO, shown in (b),

(a)

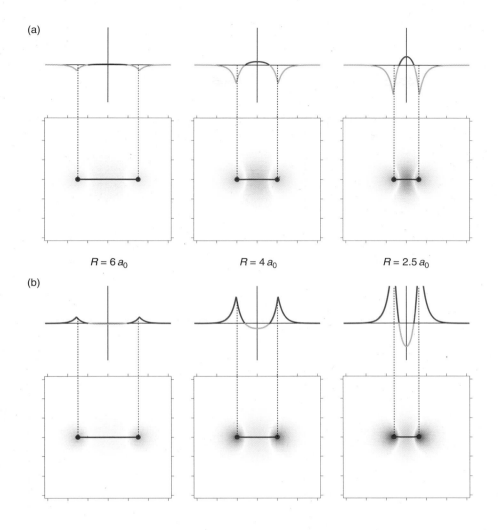

$R = 6\,a_0$ $R = 4\,a_0$ $R = 2.5\,a_0$

(b)

Fig. 3.6 Shaded plots showing the *difference* in electron density between that from an MO and that from two non-interacting $1s$ AOs located on the two atoms. These plots therefore show how the electron density is redistributed as a result of the interaction between the AOs. Blue shading indicates an *increase* in electron density, whereas orange shading indicates a *decrease*. The graphs along the top of each plot show the variation in this difference density along the internuclear axis; the vertical dashed lines are at the positions of the nuclei. The bonding MO, shown in (a), shows a build-up of electron density in the internuclear region as R decreases. In contrast, for the antibonding MO shown in (b), there is a reduction of electron density in this region. In this case, the electron density is pushed away from the nuclei, to the left and right of the molecule. This redistribution of electron density is a large part of the explanation for why the MO energies vary in the way shown in Fig. 3.3 on page 94.

electron density is built up to the left and right of the molecule, and is decreased in the internuclear region.

We can argue that the initial fall in energy of the bonding MO in part arises from this accumulation of electron density in the internuclear region. When the electron is located there it has a favourable interaction with *both* nuclei, whereas away from this region the electron can only have a significant interaction with one of the nuclei. Thus, the increased electron density in this internuclear region lowers the energy of the MO.

For the antibonding MO the electron density is reduced in the internuclear region, indeed careful inspection of the plots in Fig. 3.6 (b) for $R = 2.5\,a_0$ will show that the electron density is also pushed away from the nuclei themselves.

These changes in the distribution of the electron density primarily affect the electron–nuclear part of the potential energy which was discussed in section 3.1 on page 92. We have to remember that the *total* energy will also be affected by the electron kinetic energy and the nuclear repulsion energy. It is the balance between these three contributions which result in a minimum in the energy of the bonding MO. In contrast, for the antibonding MO, there is no reduction in the electron–nuclear energy to compensate for the nuclear repulsion, so no minimum is seen in the energy.

3.1.2 What makes an orbital bonding or antibonding?

The existence of, and differences between, bonding and antibonding orbitals is so crucial in our discussion of bonding that it is well worth while pausing to review the key differences between these orbitals, and how these arise.

Figure 3.3 on page 94 shows how the energies of the MOs varies with the internuclear distance. The energy of the bonding MO (E_+) initially falls as the internuclear separation is decreased, reaches a minimum, and then increases. As a result, if the electron occupies this MO there is an energetic advantage for the atoms to come together to form a molecule, with the lowest energy being achieved at the distance corresponding to the minimum.

On the other hand, for the antibonding MO (E_-), the energy simply rises as the internuclear separation decreases. If the electron occupies this orbital, there is no energy advantage for the atoms to come together – on the contrary, the lowest energy arrangement will be when the two atoms are separate.

The reason why the energies of the orbitals behave in this way can be rationalized by looking at the distribution of electron density which each gives rise to, as shown in Fig. 3.6 on the previous page. In the bonding MO, there is a build-up of electron density in the internuclear region. This results in a lowering of the energy of the electron, since in this region it has a favourable interaction with both nuclei. Provided that the internuclear separation is not too small, this lowering of the energy due to the arrangement of the electrons exceeds the increase in the energy due to the repulsion between the nuclei.

In contrast, for the antibonding MO there is a build-up of electron density outside the internuclear region; this electron density on the periphery of the molecule can be thought of as 'pulling apart' the two nuclei. This, combined with the nuclear repulsion, means that the energy of the antibonding MO simply increases as the internuclear separation decreases.

When we were discussing atomic orbitals we noted that the more nodes there were in an orbital, the higher its energy. The MOs behave in a similar way, with the higher energy antibonding MO having a node. This node arises from the destructive overlap of the two AOs, which is also the underlying reason why the MO does not favour bonding. There is thus a strong connection between the presence of nodes and antibonding behaviour, and we will see many examples of this as we move on to look at more complex MOs.

> For the bonding MO, there is also a small decrease in the electron kinetic energy as we approach the equilibrium separation.

3.1.3 Symmetry labels

As we saw in the previous chapter, each atomic orbital has a label, such as $2s$ or $3p$, which tells us the values of the n and l quantum numbers. Once we know the values of these quantum numbers, we can say something about the shape of the orbital, such as the number and type of nodes which are present.

In a similar way, MOs are also given labels to identify them, but in contrast to AOs, the labels used for MOs tell us about the symmetry properties of the wavefunction. Symmetry turns out to be a very powerful tool for dealing with all sorts of problems in molecular structure and spectroscopy, but in order to appreciate how symmetry can help us we would need to spend some time getting to grips with *group theory*. There is not the time to do this here, so we will for the present limit our discussion to just a few key ideas needed to understand the symmetry labels of the MOs in a diatomic.

If we take a diatomic molecule and rotate it through any angle about the internuclear axis, the molecule ends up looking indistinguishable to its starting position. The rotation though any angle is therefore described as a *symmetry operation*. Although this rotation has no net effect on the molecule, the MOs can be affected, and we use the way in which they respond to a particular rotation in order to classify them.

Referring to the iso-surface plots of the bonding and antibonding MOs shown in Fig. 3.4 on page 95, it is clear that these wavefunctions are unaffected by a rotation through *any angle* about the internuclear axis. MOs with this property are given the symmetry label σ, the lower case Greek letter 'sigma'.

If the MO wavefunction changes sign on rotation by 180° about the internuclear axis, the symmetry label given is π, and if there is a sign change on rotation by 90° the symmetry label is δ ('delta'). Later on, we will come across examples of MOs which have these symmetry labels.

Figure 3.7 illustrates another way of assigning these σ, π and δ labels. We imagine traversing a circular path in any plane perpendicular to the internuclear axis; such a path is shown by the red circle. If in traversing this path we do not encounter any nodal planes in the wavefunction, the label σ is given. If we encounter one nodal plane, the label is π, and if there are two nodal planes the label is δ.

The second symmetry label used for these MOs is connected with the *centre of inversion*, which is a symmetry element possessed by a homonuclear diatomic. Figure 3.8 illustrates the meaning of this symmetry element. The letter 'Z', shown in (a), possesses a centre of inversion at the point indicated by the bull's eye (red dot in a black circle). If we start at any position and move in a straight line directly towards the centre of inversion, and then carry on in the same direction for the same distance, we end up at an equivalent position.

Starting at the green point at the top left, and moving through the centre of inversion to the green point at the bottom right brings us to an equivalent position. The same is true of the two blue, and two orange, points, and indeed of any point.

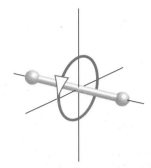

Fig. 3.7 Illustration of how the symmetry labels σ, π and δ can be assigned. The red circle shows a path which lies in a plane perpendicular to the internuclear axis. In traversing this path, if no nodal planes are encountered the symmetry label is σ. If one or two nodal planes are encountered, the label is π or δ, respectively.

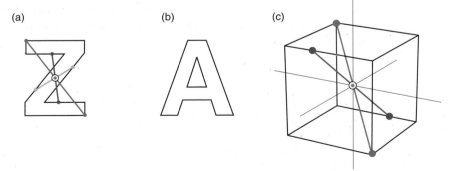

(a)   (b)   (c)

Fig. 3.8 Illustration of the meaning of a the *centre of inversion* symmetry element. The letter 'Z' possesses such an element, shown by the bull's eye. If we start at any point (e.g. the top left-hand green dot), move towards the centre of inversion and then carry on in the same direction for the same distance (as shown by the green line), we arrive at an equivalent point (e.g. the bottom right-hand green dot). The letter 'A' does not possess a centre of inversion, but the cube shown in (c) does.

(a) (b)

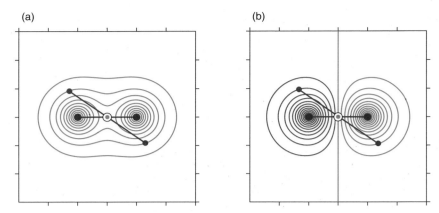

Fig. 3.9 Illustration of the behaviour of the bonding and antibonding MOs under the inversion operation; the centre of inversion is shown by the bull's eye. For the bonding MO shown in (a), if we start at some arbitrary point shown by the purple dot and then pass through the centre of inversion to an equidistant point we end up at a position where the wavefunction has the same value. The MO is thus given the label g. In contrast, the same process carried out on the antibonding orbital shown in (b) results in a change in sign of the wavefunction. The MO is thus given the label u.

The letter 'A', shown in (b), does not have a centre of inversion (although it does have other kinds of symmetry). It is impossible to find a point in this letter which has the same properties as the point identified for the letter 'Z'.

Three-dimensional objects can also possess a centre of inversion, for example the cube shown in (c). Here, the centre of inversion is at the very centre of the cube. A homonuclear diatomic also possesses a centre of inversion, located at the exact mid-point between the two atoms.

The MO wavefunctions of a homonuclear diatomic can be classified according to whether they remain the same or change sign under this inversion operation. As shown in Fig. 3.9 (a) for the bonding MO, the inversion operation leaves the wavefunction unaltered. The wavefunction is therefore described as being *symmetric* with respect to the inversion operation, and is given the symbol g. The g is for 'gerade', the German word for 'even'.

In contrast, under the inversion operation the antibonding MO has the same value but different sign, as illustrated in (b). The MO is therefore described as being *antisymmetric*, and is given the symbol u, standing for 'ungerade', German for 'odd'.

Expressed in mathematical terms, the inversion operation simply reverses the sign of the x-, y- and z-coordinates which the wavefunction (orbital) depends on. A wavefunction with g symmetry has the property that $\psi(-x, -y, -z) = +\psi(x, y, z)$, whereas one with u symmetry has the property $\psi(-x, -y, -z) = -\psi(x, y, z)$.

The bonding MO derived from the two $1s$ orbitals is therefore labelled σ_g (pronounced 'sigma gee'), and the antibonding MO is labelled σ_u (pronounced 'sigma you'); note that the g/u label is given as a subscript. Sometimes, a superscript $\star$ is added to the label for the antibonding MO, giving $\sigma_u^\star$ (pronounced 'sigma you star').

In more complex molecules it is possible that there will be more than one σ_g orbital, so to avoid confusion they are just numbered $1\sigma_g$, $2\sigma_g$ as they increase

The *g/u* symmetry labels only apply to *homonuclear* diatomics. Heteronuclear diatomics do not have a centre of inversion, so these labels *cannot* be applied.

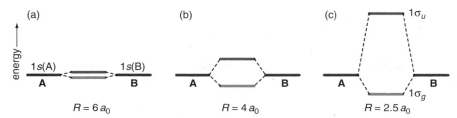

Fig. 3.10 MO diagrams, drawn to scale, for H_2^+ as a function of the internuclear separation R; the vertical scale is energy. The energy of the AOs are indicated by the maroon lines, those of the bonding MOs by red lines, and blue lines are used for the antibonding MOs. The dashed lines show which AOs contribute to which MOs. Note that, as shown in Fig. 3.3 on page 94, the energy of the bonding MO falls as R is reduced towards $2.5\,a_0$. The energy of the antibonding MO is raised by more than the lowering in energy of the bonding MO.

in energy. It is also quite common to give the AOs from which the MO is formed, so the σ_g MO might be written as $1s\sigma_g$. This notation can get rather cumbersome and is sometimes ambiguous, so we will not use it.

3.1.4 Molecular orbital diagrams

A convenient way of illustrating the relationship between the energy of the MOs and the AOs from which they are formed is to construct a *molecular orbital diagram*. A series of such diagrams for H_2^+ for different internuclear separations is shown in Fig. 3.10.

The vertical scale is energy, and the coloured horizontal lines indicate the energies of the orbitals. The maroon lines indicate the energies of the AOs, and the labels **A** and **B** distinguish the two atoms. Of course, in the case of H_2^+ the atoms and AOs are the same, but we retain the labels for convenience later on. The red line indicates the energy of the $1\sigma_g$ bonding MO, and the blue line indicates the energy of the $1\sigma_u$ antibonding MO. Finally, the dashed lines indicate which AOs contribute to which MOs – in this case both AOs contribute to both MOs.

The three MO diagrams (a), (b) and (c) in Fig. 3.10 are drawn to scale for the indicated values of the internuclear separation, R. In (a), R is so large that there is rather little interaction between the two AOs, so the MOs are close in energy to the AOs. As R decreases the interaction increases, and the MOs move further away in energy from the AOs. Note that, as shown in Fig. 3.3 on page 94, the antibonding MO goes up in energy more than the bonding MO goes down. The MO diagram shown in (c) is for the case where the bonding MO has the minimum energy. Here, the antibonding MO is raised by much more than the bonding MO is lowered. It is usual to indicate on these MO diagrams the symmetry labels of the MOs, and the usual labels for the AOs.

3.1.5 Significance of the sign of the wavefunction

Since the *square* of the wavefunction gives the probability density of the electron, when it comes to determining the distribution of the electrons the sign of the wavefunction is of no consequence. It turns out that all physical properties similarly depend on the square of the wavefunction.

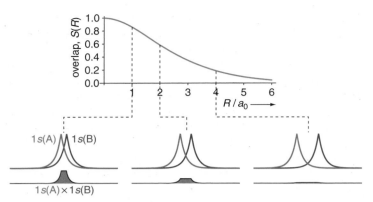

Fig. 3.11 Plot of the overlap integral $S(R)$ for two $1s$ orbitals as a function of the internuclear separation R. Beneath the plot is shown illustrations of how the integral is computed. $S(R)$ is the integral of the product of the two wavefunctions, which in this case is the *area* under the graph of this product. The two AO wavefunctions are shown in green and purple for the two atoms, and beneath these is plotted the product of these two functions. The overlap integral is the area shown in red.

However, when we start combining AOs to form MOs, the sign of the AO wavefunction starts to be of significance. It is because the $1s$ AO is positive that adding two such AOs will lead to constructive interference, and hence the σ_g bonding MO. Similarly, in this case subtracting the AOs leads to destructive interference. Later on we will see other examples where the constructive interference comes about when two AOs are subtracted.

A question which often arises when MOs are encountered for the first time is just how the destructive interference which leads to the antibonding MO comes about. After all, the $1s$ AO wavefunction on each atom is positive, so why should the sign of one of the functions apparently 'flip' in order to give destructive interference?

The solution to this possibly perplexing problem is to realize that the form of the bonding and antibonding MOs arise from a quantum mechanical analysis of the problem. When we look at these MOs, we can see that they arise from constructive and destructive interference between the AOs. However, this is just our interpretation of the result – it does not mean that in physical reality one of the AOs had to flip sign in order to form the antibonding MO. Rather, the whole idea of MOs arising from the constructive and destructive interference between AOs is simply a convenient way for us to rationalize the appearance of the MOs.

3.1.6 Overlap and the overlap integral

As we have seen, as the internuclear separation decreases, the MOs shift in energy further and further away from the energy of the AOs. The calculation of the energy shift at a particular internuclear separation is not particularly straightforward, even for the simplest case of two interacting $1s$ orbitals. However, a useful guide to the strength of the interaction between the AOs, and hence the energy shift of the MOs, can be obtained by looking at the *overlap integral*.

The overlap integral between two AOs is found by multiplying together the two AO wavefunctions, and then taking the *integral* of this product. If you are familiar with elementary calculus you will recall that the integral of a function is the *area* under a graph of that function. So the overlap integral is the area under a plot of the product of the two AOs.

Figure 3.11 on the preceding page illustrates how the overlap integral $S(R)$ varies with internuclear distance R for two $1s$ orbitals. We see that $S(R)$ has a value of 1 when $R = 0$, and then falls steadily as R increases, dropping to a small value by the time $R \approx 6\,a_0$. Due to the way AO wavefunctions are defined, $S(R)$ will always have the value 1 for the case of two identical AOs at a separation of zero i.e. on top of one another.

The diagrams underneath the graph of $S(R)$ offer an explanation for the behaviour of this overlap integral. For three different values of R (1, 2 and 4 a_0), we see plots of the $1s$ wavefunctions on atom A (shown in green), and on atom B (shown in purple). Beneath these are shown plots of the *product* of the two wavefunctions. The area under these curves, shaded in red, is the value of the overlap integral.

As R increases the two $1s$ wavefunctions move apart, so the range over which the two functions overlap when they both have significant amplitude decreases. As a result the *product* of the two functions is smaller, and therefore the area under the product, and hence the overlap integral, decreases.

From this graph we can see that when $R = 6\,a_0$ the overlap integral is small, so we would not expect the MOs to differ very much in energy from the AOs. If R is decreased to $4\,a_0$, the overlap integral is significant ($S(R) = 0.19$), so the MOs will shift in energy from the AOs. Finally, at $R = 2.5\,a_0$, the point at which the bonding MO has its minimum energy, $S(R) = 0.46$, and we expect an even greater shift of the MOs. This is precisely the behaviour seen in Fig. 3.10 on page 101.

As the internuclear separation decreases below $2.5\,a_0$, the overlap integral goes on increasing, but as we have seen the energy of the bonding MO begins to increase on account of the repulsion between the nuclei. The overlap integral does not therefore give the whole story, since it does not take into account the nuclear repulsion.

The size of the overlap integral at a particular value of R is clearly going to depend on the size of the AOs. The smaller the AOs, the smaller will be the value of $S(R)$ at a particular value of R on the grounds that the wavefunctions overlap to a smaller extent. This is illustrated in Fig. 3.12, which shows in blue the overlap integral for two $1s$ AOs appropriate for a nuclear charge of +1, and in red the same but for a nuclear charge of +2. As described is section 2.5 on page 65, increasing the nuclear charge to 2 shrinks the $1s$ AOs, and so they have to get closer together to have the same overlap as for the larger AOs with a nuclear charge of +1.

For example, as shown by the dashed lines in Fig. 3.11, to achieve an overlap integral of 0.5 the separation of the two $1s$ orbitals with $Z = 1$ needs to be about $2.3\,a_0$, but for the smaller orbitals with $Z = 2$, the separation has to be reduced to about $1.2\,a_0$.

Later on, when we come to look at the overlap of other kinds of orbitals, we will again use the overlap integral as a guide to how well the AOs are interacting. Before leaving the topic of the overlap integral, it is just worth noting that as the orbitals are three dimensional, the integral is computed over all three dimensions. For simplicity, the plots shown in Fig. 3.11 just show the wavefunction along the internuclear axis.

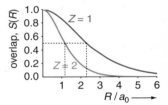

Fig. 3.12 Illustration of how the overlap integral depends on the size of the AOs. Shown in blue is $S(R)$ for two $1s$ AOs appropriate to a nuclear charge $Z = 1$. The red line is $S(R)$ for $Z = 2$. Increasing the nuclear charge shrinks the orbitals, so to have the same amount of overlap they have to get closer together. For example, as shown by the dashed line, to have $S = 0.5$ the orbitals with $Z = 1$ have to be separated by about $2.3\,a_0$, whereas when Z is increased to 2, the separation has to be $1.2\,a_0$.

3.1.7 Summary

We have covered quite a few important new ideas in this section, so it is a good idea to summarize these before we move on to consider diatomics more complex than H_2^+.

- Molecular orbitals are formed from the linear combination of atomic orbitals on different atoms.

- In the case of H_2^+ the combination of two $1s$ AOs on the two atoms gives two MOs: one bonding and the other antibonding.

- The energies of these MOs depend on the internuclear separation: the bonding MO shows a minimum in its energy at a certain separation, whereas the energy of the antibonding MO simply increases as the internuclear separation decreases.

- The bonding MO arises from an in-phase or constructive combination of the AOs; this leads to a concentration of electron density in the internuclear region.

- The antibonding MO arises from an out-of-phase or destructive combination of the AOs; this leads to electron density being excluded from the internuclear region and being concentrated on the periphery.

- The interaction between the AOs can be described in terms of the overlap integral.

3.2 H_2, He_2 and their ions

We have seen in the case of H_2^+ that if the single electron goes into the $1\sigma_g$ bonding MO, the energy reaches a minimum at $R = 2.5\,a_0$, and at this point the energy is lower than for the separated atoms. This implies that H_2^+ is stable with respect to dissociation into an atom and an ion, which is in accord with experiments which have detected and characterized this molecular ion. The electronic arrangement in H_2^+ is illustrated in Fig. 3.13.

What then of H_2? This has two electrons, so we immediately realize that things are going to be much more complicated on account of the electron–electron repulsion, just as we found for atoms. However, we can follow the same route that we used for multi-electron atoms and simply assign electrons to the MOs, just as we assigned them to AOs. In doing so, we need to recognize that we are using the orbital approximation as described in section 2.6.1 on page 66 – that is, we are assuming that the electron experiences an *average* repulsion from all of the other electrons.

Using this approach, we can assign both electrons in H_2 to the $1\sigma_g$ MO, making sure that the spins are opposed, just as we would do for two electrons in the same AO. The resulting configuration is written $1\sigma_g^2$, and is illustrated in the MO diagram shown in the middle of Fig. 3.13. Since both electrons are in a bonding orbital, we expect H_2 to be stable with respect to dissociation into two hydrogen atoms, which indeed it is.

The experimental dissociation energy of H_2 is 432 kJ mol^{-1}, whereas that for H_2^+ is 256 kJ mol^{-1}. Our MO picture can make sense of this difference by noting

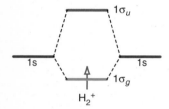

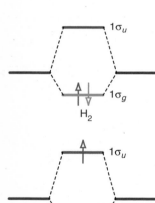

Fig. 3.13 Illustrative MO diagrams for the species H_2^+, H_2, H_2^- and He_2^+, showing which orbitals are occupied. It is important to realize that, on account of the electron–electron repulsion, the MOs will change when their occupancy changes. In addition, for He_2^+ the nuclear charges are increased to 2, which will have a further effect on the MOs. The diagrams are therefore only indicative.

Table 3.1 Properties of H_2, He_2 and their ions

molecule	configuration	dissociation energy / kJ mol^{-1}	bond length / pm
H_2^+	$1\sigma_g^1$	256	106
H_2	$1\sigma_g^2$	432	74
He_2^+	$1\sigma_g^2 1\sigma_u^1$	241	108
He_2	$1\sigma_g^2 1\sigma_u^2$	not observed	

that whereas H_2^+ has just *one* electron in the bonding MO, H_2 has *two*. So, we might reasonably expect that H_2 will have a stronger bond than H_2^+. Doubling the number of electrons in the bonding MO does not simply double the bond dissociation energy, though, since in H_2 there is electron–electron repulsion to take account of.

H_2^- has three electrons, so the third would have to go into the antibonding $1\sigma_u$ orbital, to give the configuration $1\sigma_g^2 1\sigma_u^1$, illustrated at the bottom of Fig. 3.13 on the facing page. We might reasonably expect the effect of the two bonding electrons to outweigh the effect of the one antibonding electron, thus making H_2^- stable with respect to dissociation into atoms. In fact, there is little experimental data for this ion as it decays very quickly by ejecting an electron.

He_2 has four electrons, leading to the configuration $1\sigma_g^2 1\sigma_u^2$. We know that the antibonding MO is raised more in energy that the bonding MO is lowered, so we can expect that the two antibonding electrons will outweigh the two bonding ones. Our prediction is that He_2 is not lower in energy than two helium atoms, so we would not expect He_2 to form. This is indeed in line with experiment, which has produced no evidence for this molecule.

However, He_2^+ has three electrons, and so will have the same configuration as H_2^-, i.e. $1\sigma_g^2 1\sigma_u^1$. We might therefore expect He_2^+ to be stable with respect to dissociation, and indeed this is the case. Experimental work indicates that the bond dissociation energy is 241 kJ mol^{-1}. Table 3.1 summarizes the properties of these molecules and their ions, giving in addition to the bond dissociation energies the experimentally determined bond lengths.

The bond length for H_2^+ given in the table is different from the value of 2.5 a_0 (132 pm) for the position of the energy minimum of the bonding MO we quoted above. The reason for this difference is that it is only the first approximation to form the MOs from just two $1s$ AOs. To obtain better agreement between theory and experiment we need to include more AOs in the calculation, and if we do this we can reproduce the experimental value of the bond length.

We have reached a very satisfying position. With one simple MO diagram we have been able to rationalize the trends in the bond strengths of H_2, He_2 and their ions, and provide a good explanation for failure to observe He_2 experimentally. We now turn to the homonuclear diatomics of the second period ($Li_2 \ldots Ne_2$) where we will need a more complex MO diagram to explain what is going on.

3.3 Homonuclear diatomics of the second period

As we move along the second period, the $2s$ and $2p$, as well as the $1s$, orbitals are occupied. It is not unreasonable to expect to have to use all of these AOs to form MOs for the diatomic species we are interested in. The way in which these AOs combine to form MOs is governed by five 'rules' which come from a full quantum mechanical analysis of the problem. We will first look at these rules and their consequences, and then go on to use them to understand the MO diagrams of the second period homonuclear diatomics.

3.3.1 Rules for forming MOs

The five 'rules' can be summarized as:

(a) The combination of a certain number of AOs produces the same number of MOs e.g. combining four AOs gives four MOs.

(b) Only AOs of the correct symmetry will interact to give MOs.

(c) The closer in energy the AOs, the larger the interaction when MOs are formed.

(d) In general, each MO is formed from a different combination of AOs; AOs which are close in energy to the MO contribute more than those which are further away in energy.

(e) The size of the AOs must be compatible for there to be a strong interaction when MOs are formed.

We will look at each of these rules in turn.

There is not much to say about the first rule other than it sounds reasonable. If we imagine taking a molecule and then moving the nuclei further and further apart we will end up with atoms, and in the process the MOs will have transformed into AOs. Since the MOs must evolve steadily into AOs, the number of orbitals cannot change, which is what the first rule says.

Rule 2: symmetry

The easiest way to illustrate what is meant by AOs having to have the correct symmetry to interact is to give an example of two AOs which do not have the correct symmetry: for example a $1s$ and a $2p_x$ orbital. It is usual to take the internuclear axis as the z direction, so a $2p_x$ AO points perpendicular to the axis.

What we are going to do is to think about the overlap integral between these two orbitals. Recall from section 3.1.6 on page 102 that the overlap between two AOs is found by multiplying the two AOs together and then taking the area under this product. As can be seen in Fig. 3.14, the $1s$ wavefunction and the upper lobe of the $2p_x$ wavefunction are both positive (red), so in this region their product is positive. We call this 'positive overlap', as it makes a positive contribution to the overlap integral.

In contrast, the lower lobe of the $2p_x$ wavefunction is negative (blue) and of opposite sign to the $1s$ wavefunction. The product of the two wavefunctions is therefore negative, and so the area under the product is negative. We call this 'negative overlap' as it makes a negative contribution to the overlap integral. You can see from the diagram that the positive and negative overlaps are equal and opposite, and so the net overlap is zero. As a result, these two AOs *do not* interact to form MOs.

We can also use a symmetry argument to show that the overlap integral is zero. The $1s$ orbital is *symmetric* with respect to the yz-plane, meaning that

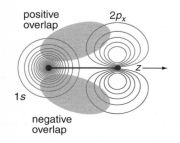

Fig. 3.14 Illustration of why the $1s$ and $2p_x$ orbitals do not overlap to give MOs. The positive overlap between the upper lobe of the $2p_x$ orbital and the $1s$ orbital exactly cancels the negative overlap between the lower lobe and the $1s$. As a result, the overlap integral is zero, and the AOs do not interact to form MOs.

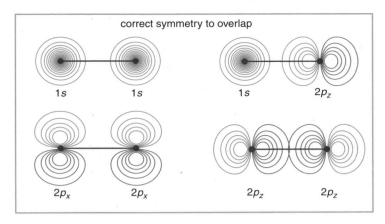

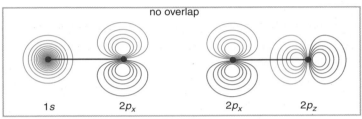

Fig. 3.15 Illustration of different possible combinations of $1s$ and $2p$ AOs. In the first four combinations the AOs have the correct symmetry to overlap, and so can lead to the formation of MOs. In the last two combinations, there is positive and negative overlap, leading to an overlap integral of zero; the AOs do not therefore have the correct symmetry to form MOs. Replacing the $1s$ AOs by $2s$ will gave the same result. Similarly, the $2p_x$ can be replaced by $2p_y$ throughout. The signs of the AOs have been adjusted so that there is net positive overlap in the first group.

when the wavefunction is reflected in this plane there is no change of sign. Reflection of the $2p_x$ orbital in this plane does, however, result in a sign change, so the $2p_x$ orbital is said to be *antisymmetric*. As the two orbitals have different symmetry with respect to this plane, their overlap integral is necessarily zero.

Various possible combinations of the $1s$ and $2p$ AOs are shown in Fig. 3.15. By applying the same ideas as we used in Fig. 3.14 on the facing page, we can see that whereas the first four pairs have the correct symmetry to overlap, the two pairs at the bottom do not. This diagram has been drawn using $1s$ AOs, but the same conclusions would arise for $2s$ AOs and these have the same spherical symmetry. In addition, the $2p_x$ orbital can be replaced by $2p_y$ (which also points perpendicular to the internuclear axis) without altering the conclusions.

The orbitals in Fig. 3.15 have been drawn with the signs which lead to the formation of bonding MOs as shown i.e. in-phase overlap. Do not forget though that by reversing the sign of *one* of the AOs we will obtain the antibonding MO in each case.

Rules 3 and 4: energy match and the contributions from different AOs

Figure 3.16 on the following page illustrates what happens to the energies of the MOs as the energy gap between the two AOs is increased. To start with, when the AOs are closely matched in energy, the antibonding MO (blue) and the bonding MO (red) lie significantly above and below the energies of the AOs from which they are formed.

However, as the energy gap between the two AOs is increased, the bonding MO ends up closer and closer to the lower energy AO (here **B**), whereas the antibonding MO ends up close in energy to the higher energy AO (here **A**). When the energy separation becomes large, the two MOs have essentially the same energy as the two AOs.

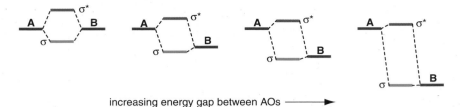

increasing energy gap between AOs ⟶

Fig. 3.16 Series of MO diagrams showing the effect of increasing the energy separation between the two AOs (**A** and **B**) which are interacting to form the MOs (blue: antibonding, red: bonding). Across the series, the energy of **A** is held constant while the energy of **B** falls. When the AO energies are different, the bonding MO lies below the lower energy of the AOs, while the antibonding MO lies above the higher energy of the two AOs. In the limit that the energy separation of the AOs is large, the bonding MO is at the same energy as the lower of the two AOs, and the antibonding MO is at the same energy as the higher energy AO.

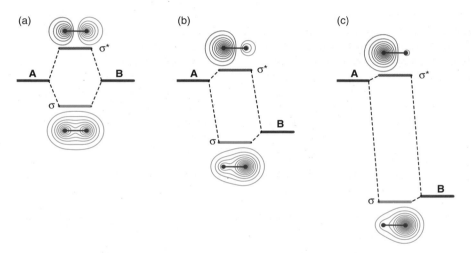

Fig. 3.17 Illustration of how the form of the bonding and antibonding MOs are affected by increasing the energy separation of the two AOs from which they are formed. In (a) the AOs **A** and **B** have the same energy, and so the MOs, whose form is shown by contour plots, have equal contributions from the the two AOs. As the energy of the AO from **B** decreases, the contribution from AO **B** to the bonding MO increases, while the contribution from **A** decreases. In contrast, for the antibonding MO, it is the contribution from **A** which increases, while that from **B** decreases.

Figure 3.17 shows the effect on the MO wavefunctions of increasing the separation of the AO energies. In (a) we see the situation in which the two AOs (**A** and **B**) have the same energy. As we have described above, for this arrangement the bonding MO has equal contributions from the AOs on **A** and **B**, and likewise the antibonding MO has equal and opposite contributions from the two AOs.

If the energy of the AO on **B** decreases, as shown in (b), the two AOs no longer contribute equally to the MOs. From the contour plots of the MOs we can see that the bonding MO has a greater contribution from the AO on atom **B** than it does from the AO on **A**. In contrast, the antibonding MO has a greater contribution from **A** than from **B**.

● *Weblink 3.3*

Follow this link to go to real-time versions of Fig. 3.16 and Fig. 3.17 in which you can vary the energy separation between the two AOs and see how the energy and form of the two MOs are affected.

(a) 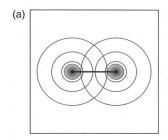 (b)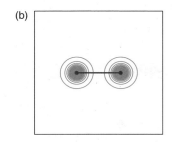

Fig. 3.18 Contour plots of (a) two $2s$ AOs and (b) two $1s$ AOs separated by a typical internuclear distance for a diatomic of the second period; the same contour levels have been used for both. The $1s$ AO is contracted as a result of the increased effective nuclear charge for such an orbital in the second period. There is significant overlap between the two $2s$ AOs, but at this internuclear separation there is no overlap between the compact $1s$ AOs.

In (c) we see the situation where the energy of the AO on **B** has fallen even more. Compared to (b), the bonding MO is even more dominated by the AO from **B**, and the antibonding MO is even more dominated by the contribution from **A**. When the separation between the AO energies becomes very large, the bonding MO is essentially the AO from **B**, and the antibonding MO is the AO from **A**. In this limit we say that the AOs are so far apart in energy that there is no interaction between them, and so the MOs are identical to the AOs.

We can summarize these effects in the following way.

As the energy separation between the AOs increases:

- the bonding MO lies closer and closer in energy to that of the lower energy AO.

- the antibonding MO lies closer and closer in energy to that of the higher energy AO.

- the contribution to the bonding MO from the lower energy AO increases, while that from the higher energy AO decreases.

- the contribution to the antibonding MO from the higher energy AO increases, while that from the lower energy AO decreases.

The last two points can also be expressed by saying that the major contribution to an MO is from the AO which is *closest to it in energy*.

Rule 5: size

In order to interact to form MOs, the AOs must overlap to a significant extent, and as we have seen in section 3.1.6 on page 102 a convenient way of quantifying this overlap is to look at the overlap integral, $S(R)$. Only when the internuclear separation is small enough that the overlap integral starts to rise do the AOs interact significantly to form MOs.

This point is illustrated in Fig. 3.18. Contour plot (a) is of two $2s$ AOs separated by a typical bond length for a diatomic of the second period. The outer contours of the orbitals clearly overlap quite a lot, so we can expect that the overlap integral will be significant. The resulting bonding and antibonding MOs will therefore be significantly shifted away from the energy of the AOs.

Contour plot (b) shows two $1s$ AOs, separated by the same distance as in (a). By the time we reach the second period, these $1s$ AOs will be rather contracted on account of the increased effective nuclear charge. Thus, although the two orbitals have the correct symmetry to overlap, the overlap integral will be essentially zero as the orbitals do not intersect. In this situation, bonding and antibonding MOs will not be formed.

In summary, whether or not there will be a significant interaction between two AOs depends on their size relative to the separation between the orbitals. Core orbitals, which are those in shells below the outer shell, are likely to be so contracted that at typical bond lengths there will be no interaction between them.

3.3.2 Types of MOs from $2s$ and $2p$ AOs

For the elements of the second period, it is the MOs formed from the overlap of the $2s$ and $2p$ AOs which are going to be important in determining the bonding. The reason for this is that in the atom these AOs are occupied, so in turn the MOs formed from them will also be occupied. We can ignore the $1s$ AOs since, as we have just seen, these are too contracted to interact significantly at typical bond lengths.

As we saw in section 2.3.1 on page 45, the $2s$ orbital is lower in energy than the $2p$, and the energy separation between these two orbitals increases as we go across the period. To start with, therefore, we will simplify the discussion by assuming that the energy separation between $2s$ and $2p$ is sufficient that we do not need to consider the overlap between these orbitals (Rule 3). Later on, we will relax this restriction.

MOs from $2s$–$2s$ overlap

Two $2s$ orbitals can overlap just in the same way as two $1s$ orbitals in order to form a bonding σ_g MO and an antibonding σ_u MO. Both the $1s$ and $2s$ AOs have spherical symmetry, but the difference is that the $2s$ has a radial node, close in to the nucleus (see Fig. 2.16 on page 47). In hydrogen, the radial node in the $2s$ AO occurs at $2\,a_0$, which is 106 pm. However, in a second period atom the $2s$ will be significantly contracted on account of the increased effective nuclear charge felt by the electron. For example, in boron the radial node comes at around $0.5\,a_0$ (26 pm). Since typical bond lengths are in the range 100–300 pm, the region of the $2s$ orbital inside the radial node is simply not going to be involved in overlap with other orbitals. Thus, for the purposes of forming MOs, the presence of the radial node in the $2s$ orbital is of no significance.

Figure 3.19 on the facing page shows contour plots of the bonding σ_g and antibonding σ_u MOs formed from the overlap of two $2s$ AOs. These $2s$ orbitals have been contracted to account for the effective nuclear charge typical of the second period.

As we saw in Fig. 2.16 on page 47, the $2s$ AO wavefunction has a positive part close in to the nucleus and then as we move away from the nucleus there is a radial node followed by a more extended negative part of the wavefunction. Thus, when two such AOs interact to form an MO by constructive, or in-phase, overlap, the outer part of the MO wavefunction is negative, as shown in Fig. 3.19 (a). The contour plot also shows the positive part of the wavefunction which is close in to the nucleus.

Weblink 3.4

View, and rotate in real time, iso-surface representations of all of the MOs described in this section.

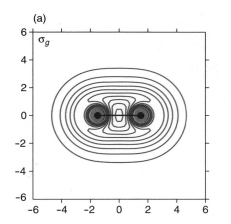
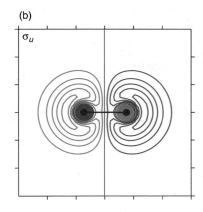

Fig. 3.19 Contour plots of the bonding σ_g and antibonding σ_u MOs formed by the overlap of two $2s$ AOs. The plots are taken in a plane which passes through the two atoms and the bond, which are indicated by the purple dumb-bell. The $2s$ orbitals used to form these MOs have been contracted to account for the effect of the effective nuclear charge felt by a typical second period element; similarly, the internuclear separation of $3\,a_0$ (170 pm) is typical for homonuclear diatomics of this period. The scale is given in Bohr radii.

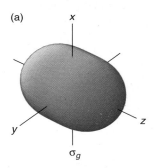
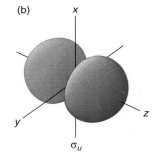

Fig. 3.20 Iso-surface plots of the bonding σ_g and antibonding σ_u MOs formed by the overlap of two $2s$ AOs, as shown in Fig. 3.19. Both wavefunctions are cylindrically symmetric about the internuclear axis (the z-axis), and so have the symmetry label σ.

The σ_u antibonding MO shown in Fig. 3.19 (b) arises from destructive, or out-of-phase, overlap. As we saw before for the $1s$–$1s$ overlap, this antibonding MO has a nodal plane between the two nuclei, and the electron density is pushed away from the internuclear region. In contrast, in the bonding MO electron density is pushed into this internuclear region, which in part accounts for the lowering in energy of this MO.

Figure 3.20 shows iso-surface plots of these two MOs. It is evident from these pictures that the wavefunctions do not change sign as we traverse a circular path perpendicular to the internuclear axis (they have cylindrical symmetry) so according to the discussion in section 3.1.6 on page 102, the appropriate symmetry label is σ. The assignment of the g or u label, which gives the symmetry under the inversion operation, is clear from the contour plots of Fig. 3.19.

It is often useful to have a simple 'cartoon-like' representation for how these MOs are constructed from AOs, both to help us remember the form of the MOs and also as a sketch which we could reproduce quickly on paper. Figure 3.21 on the next page shows such a sketch for the construction of MOs from $1s$ or $2s$ AOs. The sign of the AO wavefunction is indicated by shading. At the top

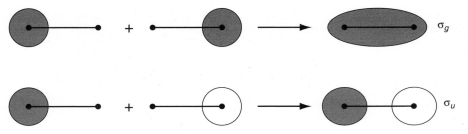

Fig. 3.21 Cartoons showing the formation of the bonding σ_g and antibonding σ_u orbitals from the in-phase and out-of-phase overlap of two $1s$ or $2s$ AOs. On the left are shown the AOs on the individual atoms, with the location of the nuclei being shown by the dumb-bell. On the right are shown the MOs. Positive and negative signs for the wavefunctions are indicated by shaded and open areas.

Fig. 3.22 Iso-surface plots of the bonding σ_g and antibonding σ_u orbitals formed from the 'head on' overlap of two $2p_z$ orbitals. Note that the bonding MO has a concentration of electron density in the internuclear region, which is not the case for the antibonding MO.

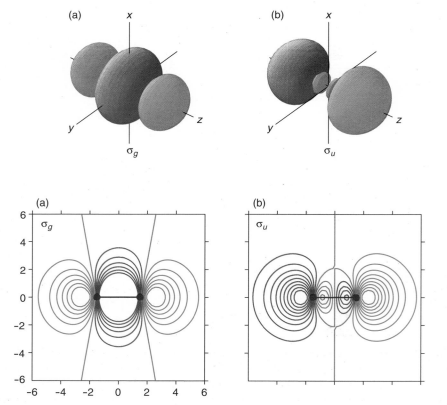

Fig. 3.23 Contour plots of the same two MOs as shown in Fig. 3.22. These plots are taken in the plane containing the two atoms and the bond, which are indicted by the purple dumb-bell. The scale is given in units of Bohr radii.

is shown how in-phase overlap leads to the σ_g bonding MO, whereas at the bottom we have out-of-phase overlap leading to the σ_u antibonding MO.

It is important to realize that in constructing the MOs the *absolute* sign of the AO wavefunctions, and hence the sign of the MO wavefunction, is not important. All that matters is that there will be in-phase overlap to give a bonding MO, and out-of-phase overlap to give an antibonding MO. Whether two positive AO wavefunctions overlap to give an overall positive MO wavefunction, or two negative wavefunctions overlap to give a negative MO

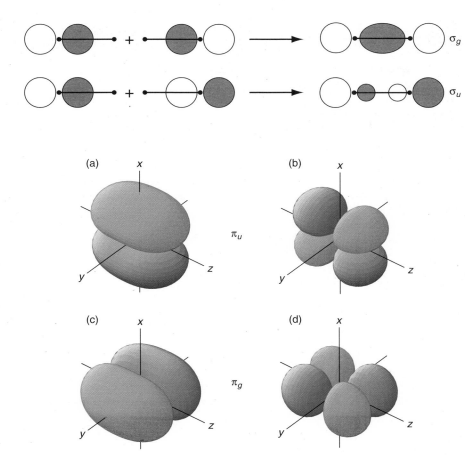

Fig. 3.24 Cartoons, like those in Fig. 3.21 on the preceding page, showing the formation of a σ bonding and antibonding MO from the head-on overlap of two $2p_z$ AOs. The $2p$ orbitals are represented by two spheres, barely touching.

Fig. 3.25 Iso-surface plots of the MOs formed from the overlap of $2p$ orbitals which point perpendicular to the internuclear axis. Overlap of two $2p_x$ AOs gives the π_u bonding MO shown in (a) and the corresponding antibonding π_g MO shown in (b). Note that the yz-plane forms a nodal plane for both MOs, hence their symmetry label is π. The two $2p_y$ MOs form a similar pair of MOs, shown in (c) and (d).

wavefunction is not important. What is important is for the AO wavefunctions to have the *same* sign, so that the overlap is constructive, leading to a bonding MO.

MOs from 2p–2p σ overlap

Two $2p_z$ orbitals, where z is the internuclear axis, can overlap 'head on' to give σ-type bonding and antibonding MOs. These two MOs are shown as iso-surface plots in Fig. 3.22, and as contour plots in Fig. 3.23 on the facing page. Both MOs have cylindrical symmetry about the internuclear axis and so have the symmetry label σ. From the contour plots it is clear that the bonding MO is symmetric under the inversion operation and so has the label g, whereas the antibonding MO is antisymmetric, and has the label u.

Figure 3.24 shows our cartoon representations of how these σ MOs are formed from $2p_z$ AOs. As with all the bonding MOs we have seen so far, there is a concentration of electron density in the internuclear region, whereas in the antibonding MO, there is a nodal plane between the two nuclei.

Fig. 3.26 Contour plots of the π_u and π_g MOs formed by overlap of two $2p_x$ AOs. The plots are taken in the xz-plane and the scale is given in units of Bohr radii. The horizontal green zero contour indicates the nodal yz-plane, and in addition the antibonding π_g MO has a nodal xy-plane. For the bonding MO, electron density is concentrated in the region between the two nuclei, but not actually along the internuclear axis.

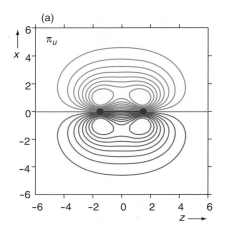

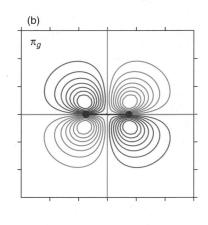

MOs from 2p–2p π overlap

The $2p_x$ and $2p_y$ AOs point in perpendicular directions to the internuclear axis and so cannot overlap in the same way as do the $2p_z$ AOs. However, two $2p_x$ orbitals can overlap 'sideways on' to give the bonding and antibonding MOs shown in Fig. 3.25 (a) and (b). As before, the bonding MO comes about from in-phase overlap of the two AOs, and the antibonding MO comes about from the out-of-phase overlap.

For these two MOs, the yz-plane is a nodal plane. Therefore, as we go round a circular path perpendicular to the internuclear axis we cross a nodal plane. From the discussion in section 3.1.3 on page 98, it follows that the orbitals are given the symmetry label π.

Figure 3.26 shows contour plots of these two MOs. These plots show clearly how both have the yz-plane as a nodal plane, and that the antibonding MO also has the xy-plane as a nodal plane. The bonding MO, shown in (a), changes sign on inversion through the centre so has the label u; in contrast, the antibonding MO, (b), does not change sign and so has the label g.

The two $2p_y$ MOs can also overlap sideways on to give a π_u bonding MO and a π_g antibonding MO shown in Fig. 3.25 (c) and (d). Apart from being rotated about the z-axis by 90°, these orbitals are identical to the π MOs formed from the $2p_x$ AOs; in particular they have the same energy. What this means is that there are *two* degenerate π_u MOs, and *two* degenerate π_g MOs.

The π_u bonding MO concentrates electron density in the region between the two nuclei but, in contrast to a σ bonding MO, the concentration is above and below the internuclear axis, rather than along it. For this reason, a π bonding MO is generally less strongly bonding than is a σ bonding MO formed from the same kind of orbital. Figure 3.27 on the facing page shows a cartoon of the formation of these two π MOs.

3.3.3 Idealized MO diagram

If we assume that there is no overlap between the $2s$ and $2p_z$ AOs, then for the homonuclear diatomics we can draw up the idealized MO diagram as shown in Fig. 3.28 on page 116. On the left and the right are shown the energies of the AOs, and the MO energies are shown in the middle. Which AOs contribute to which MOs are indicated by dashed lines.

⊕ *Weblink 3.4*

View, and rotate in real time, iso-surface representations of all of the MOs described in this section.

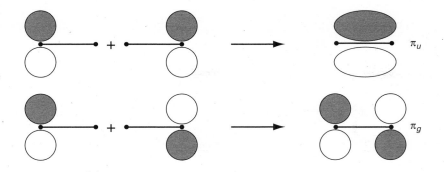

Fig. 3.27 Cartoons, like those of Fig. 3.21 on page 112, showing the formation of a π bonding and antibonding MO from the sideways overlap of two $2p$ AOs.

The $2s$ is shown considerably lower than the $2p$, since this is the condition under which we can ignore the overlap between the $2s$ and the $2p$. For example, in oxygen the energy of the $2s$ AO is -1.2 eV, whereas that of the $2p$ AOs is -0.63 eV. Also, we have shown three energy levels for the $2p$, as there are three such orbitals which are degenerate (i.e. have the same energy).

Not shown on this diagram are the $1s$ orbitals, which lie at much lower energy than the $2s$ AOs; for example, in oxygen the energy of the $1s$ AO is -21 eV. If placed on the same scale as the $2s$ and $2p$ AOs in Fig. 3.28, the $1s$ AOs would be 2.8 m off the bottom of the page! We have already noted that these AOs are too contracted to overlap significantly, but in principle they give rise to a σ_u MO and a σ_u MO. These are given the labels $1\sigma_g$ and $1\sigma_u$ as they are the first MOs with this symmetry. Occupation of these MOs has no effect on the bonding as the MOs have the same energy as the AOs.

The two $2s$ orbitals overlap to give a bonding σ_g and an antibonding σ_u MO, which are labelled $2\sigma_g$ and $2\sigma_u$ as these are the second MOs with these symmetries. The antibonding MO is raised in energy by more than the bonding MO is lowered, as has been noted before.

The two $2p_z$ AOs overlap to give the third set of σ MOs, labelled $3\sigma_g$ and $3\sigma_u$. The two $2p_x$ AOs overlap to give a π bonding and a π antibonding MO, and the same is true for the $2p_y$ AOs. The result is two degenerate π_u (bonding) MOs, and two degenerate π_g (antibonding) MOs. Generally speaking, π overlap is less effective than σ overlap between p orbitals, which is why the $3\sigma_g$ is shown below the $1\pi_u$.

Now that we have prepared this MO diagram we can feed in the electrons, filling the MOs in the same way that we are used to doing for atoms. Having done this, we can see what consequences the resulting electronic configuration has for the strength of bonding.

It turns out that we can only ignore the overlap between $2s$ and $2p$ for the elements O, F and Ne which are on the far right of the second period. Recall from Fig. 2.42 on page 69 that the energy separation between the $2s$ and $2p$ AOs increases as we go across the period. The extent to which such AOs can interact to form MOs thus decreases as we go across the period, and it turns out that by the time we reach oxygen, it is reasonable to ignore the $2s$–$2p$ overlap. So, using the idealized MO diagram we can discuss the bonding in O_2, F_2 and Ne_2, but not in the other diatomics of the second period.

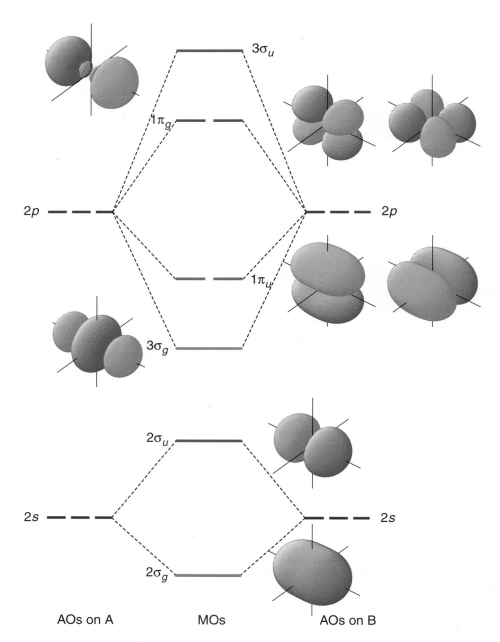

Fig. 3.28 Idealized MO diagram for the homonuclear diatomics of the second period; overlap between $2s$ and $2p_z$ has been ignored. Not shown on the diagram are the $1\sigma_g$ and $1\sigma_u$ MOs which arise from the overlap of the $1s$ AOs. These AOs lie at much lower energies than the $2s$ and $2p$ AOs. Also, there is little interaction between the $1s$ AOs on account of them being much contracted by the effect of the nuclear charge. Note that there are three degenerate $2p$ AOs on each atom, and that the $1\pi_u$ and $1\pi_g$ MOs are both doubly degenerate. An iso-surface representation of each MO is also shown.

Bonding in O_2, F_2 and Ne_2

An oxygen atom has eight electrons, so in O_2 there are a total of sixteen electrons. If we fill the MOs with these in the usual way we come up with the configuration

$$1\sigma_g^2\, 1\sigma_u^2\, 2\sigma_g^2\, 2\sigma_u^2\, 3\sigma_g^2\, 1\pi_u^4\, 1\pi_g^2.$$

Note that the first four electrons go into the 1σ orbitals (formed from the $1s$ AOs) not shown in Fig. 3.28 on the preceding page. Also, the two degenerate orbitals which form the $1\pi_u$ are filled by four electrons – two, spin paired, in each. The arrangement of electrons in MOs is illustrated in Fig. 3.29.

In the configuration, the electrons in bonding MOs are indicated in red, while those in antibonding MOs are indicated in blue. The 1σ MOs are not shifted significantly from the energies of the $1s$ AOs, and so are neither bonding nor antibonding; they are shown in black. There are a total of eight electrons in bonding MOs, and four in antibonding MOs.

Recall from our discussion of H_2^+ that occupation of a bonding MO lowers the energy as a bond is formed, whereas occupation of an antibonding MO has the opposite effect. As the number of bonding electrons in O_2 exceeds the number of antibonding electrons, we expect the molecule to be stable with respect to dissociation into atoms.

From this configuration we can work out a quantity called the *bond order*, BO:

$$\text{BO} = \tfrac{1}{2}\,(\text{no. of bonding electrons} - \text{no. of antibonding electrons}) \qquad (3.1)$$

The factor of one half arises because we normally consider a single bond as arising from a shared pair of electrons, and a double bond from the sharing of four electrons. For O_2 the bond order is $\tfrac{1}{2}(8 - 4) = 2$; such a value implies the presence of a double bond.

For F_2 we have a total of eighteen electrons, giving the configuration

$$1\sigma_g^2\, 1\sigma_u^2\, 2\sigma_g^2\, 2\sigma_u^2\, 3\sigma_g^2\, 1\pi_u^4\, 1\pi_g^4.$$

There are now eight bonding and six antibonding electrons, so we would again predict that this molecule is stable with respect to dissociation into atoms.

However, as there are more antibonding electrons in F_2 than in O_2, the expectation is that the bond dissociation energy of F_2 will be less than that of O_2. Experiment bears this out: the dissociation energies are 494 kJ mol^{-1} for O_2 and 154 kJ mol^{-1} for F_2. For F_2 the BO is $\tfrac{1}{2}(8 - 6) = 1$, which we can take as indicating a 'single bond'. Compared to O_2, the reduced BO of F_2 is reflected in its reduced bond dissociation energy.

Finally, for Ne_2 the configuration is

$$1\sigma_g^2\, 1\sigma_u^2\, 2\sigma_g^2\, 2\sigma_u^2\, 3\sigma_g^2\, 1\pi_u^4\, 1\pi_g^4\, 3\sigma_u^2.$$

There are now eight bonding and eight antibonding electrons, giving a bond order of zero. The prediction is that there is no energetic advantage for two Ne atoms to form a molecule. Experimentally, Ne_2 has been detected, but the bond dissociation energy is certainly less than 1 kJ mol^{-1}, so it can hardly be called a 'stable' molecule.

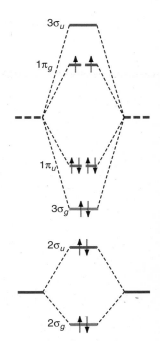

Fig. 3.29 MO diagram for O_2 showing how the electrons are arranged. Note that the last two electrons go into separate $1\pi_g$ MOs, and have their spins parallel.

Fig. 3.30 On the left is shown a vacuum flask containing liquid oxygen (which is pale blue in colour). If a strong magnet is brought up to the edge of the flask, as shown on the right, the liquid is attracted towards the magnet. This illustrates that liquid oxygen is paramagnetic.

Paramagnetism

Some substances show a physical property called *paramagnetism*, which means that they are drawn *into* a magnetic field. At a molecular level, paramagnetism is associated with the presence of *unpaired* electron spins.

Molecular oxygen is known to be paramagnetic, something which can be demonstrated clearly by the way in which liquid oxygen is attracted to a magnet, as shown in Fig. 3.30. Our MO diagram provides a simple explanation for this paramagnetism. As is shown in Fig. 3.29 on the preceding page, the final two electrons go into the doubly degenerate $1\pi_g$ MOs and, as was discussed for AOs in section 2.7.3 on page 82, we expect to obtain the lowest energy by putting these electrons in *separate* $1\pi_g$ orbitals with their spins *parallel*. We therefore predict that O_2 has unpaired electron spins, and so will be paramagnetic.

In contrast, in F_2 the $1\pi_g$ MOs contain four electrons, and so all the spins are paired up. We do not therefore expect F_2 to be paramagnetic, which is indeed the case.

Ions of O_2

Table 3.2 gives experimental data on various ions of O_2, all of which can be understood using the MO diagram. Removing an electron from O_2 to give O_2^+, which has the configuration $\ldots 1\pi_g^1$, reduces the number of antibonding electrons: the bond is therefore stronger and the dissociation energy greater than for O_2. It is interesting to note that, as usual, the strengthening of the bond is associated with a decrease in its length.

Table 3.2 Properties of O_2 and its ions

species	bond length / pm	dissociation energy / kJ mol^{-1}	bond order	paramagnetic?
O_2^+	112	643	2.5	yes
O_2	121	494	2	yes
O_2^-	135	395	1.5	yes
O_2^{2-}	149	204	1	no

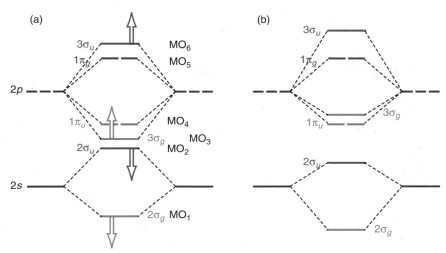

Fig. 3.31 Illustration of the effect of $2s$–$2p_z$ mixing on the MO diagram of a homonuclear diatomic. Shown in (a) is the MO diagram for the case where there is no interaction (mixing) allowed between $2s$ and $2p_z$. So, MO_1 is composed of just $2s$, whilst MO_3 is composed of just $2p_z$ AOs. However, these two MOs have the same symmetry (σ_g), and so can combine further to form new MOs. As a result, MO_1 falls in energy and MO_3 rises, as indicated by the red arrows. A similar interaction occurs between MO_2 and MO_6, both of which have symmetry σ_u, resulting in the new arrangement of MOs shown in (b). Note that $3\sigma_g$ now lies *above* $1\pi_u$.

In contrast, adding an electron to give O_2^- (configuration $\dots 1\pi_g^3$) increases the number of antibonding electrons, and so the bond is weaker and longer than in O_2. The doubly charged anion has one more antibonding electron giving the configuration $\dots 1\pi_g^4$, so its bond bond length is increased further.

Note that O_2^+, O_2 and O_2^- all have unpaired electrons, and so are paramagnetic. In contrast O_2^{2-}, which has the electronic configuration $\dots 1\pi_g^4$, has no unpaired electrons, and so is not paramagnetic.

This simple MO diagram therefore provides a straightforward explanation of all these properties of O_2 and its ions, as well as helping us to understand the relative bond strengths of O_2, F_2 and Ne_2. We now move on to modifying our MO diagram so that it can be used to describe the other homonuclear diatomics of the second period.

The data in Table 3.2 all refer to gas-phase species, but the ions are more commonly found in ionic solids. O_2^- occurs in potassium superoxide, KO_2; O_2^{2-} occurs in sodium peroxide, Na_2O_2; O_2^+ occurs in the compound $O_2[AsF_6]$.

3.3.4 Allowing for *s*–*p* mixing

We now turn to how our simple MO diagram can be modified to allow for interaction (often called *mixing*) between $2s$ and $2p_z$ AOs. In principle, we have a total of four AOs which can overlap to produced σ-type MOs: two $2s$ AOs, and two $2p_z$ AOs. We therefore expect the overlap of these four AOs to produce four MOs (Rule 1), and that each MO will be a linear combination of all four AOs. The contribution of each AO to a given MO will depend on details such as the relative energies and sizes of the AOs. This is not something we can easily guess at, so will have to use numerical (computer) calculations to determine the precise outcome.

However, a useful way of understanding the effect of this *s*–*p* mixing is illustrated in Fig. 3.31. In (a) we see an MO diagram drawn up assuming there is no overlap between the $2s$ and $2p_z$ (just like Fig. 3.28 on page 116). Thus,

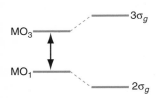

Fig. 3.32 Illustration how two MOs with the same symmetry, here σ_g, can interact and mix in such a way as to 'push them apart' in energy.

in this diagram, the $2\sigma_g$ (MO$_1$) is composed entirely of $2s$ AOs, while the $3\sigma_g$ (MO$_3$) is composed of entirely $2p_z$ AOs.

As MO$_1$ and MO$_3$ have the *same* symmetry (σ_g), they can combine with one another in just the same way that two AOs of appropriate symmetry combine. Combining two AOs gives two MOs, one lower in energy then the lower energy AO, and one higher in energy than the higher energy AO. The same thing happens when we combine two MOs: we end up with a new MO which is lower in energy than the lower of the two original MOs, and one which is higher in energy than the higher of the two original MOs, as is illustrated in Fig. 3.32.

As a result of this interaction, MO$_1$ moves down in energy, and MO$_3$ moves up, as indicated by the red arrows on Fig. 3.31 (a). The final position of the new MOs are shown in (b). In a similar way there is an interaction between the σ_u MOs (MO$_2$ and MO$_6$) which results in MO$_2$ moving down in energy, and MO$_6$ moving up, as indicated by the blue arrows.

Figure 3.33 shows how this mixing affects the form of the MOs. Just as when we combine AOs to give MOs, the mixing process gives two new MOs, one in which the original MOs are combined in phase, and one in which they are combined out of phase. The in-phase mixing of the two σ_g MOs (MO$_1$ and MO$_3$) to give the $2\sigma_g$ MO is shown in (a). Since this MO is closest in energy to MO$_1$, the $2\sigma_g$ MO is formed from mainly MO$_1$ plus a small contribution from MO$_3$. The contours plots show how adding in some MO$_3$ increases the electron density between the two nuclei, and also reduces the electron density on the periphery of the molecule. It makes sense, therefore, that the $2\sigma_g$ which results from this mixing is more bonding than MO$_1$.

The out-of-phase mixing of the σ_g MOs is shown in Fig. 3.33 (b). The new $3\sigma_g$ is closest in energy to MO$_3$, so this is the major contributor to the new MO. This time we *subtract* a small amount of MO$_1$ to give the new MO, and we can see that this results in an MO in which the electron density on the periphery of the molecule has been increased. It is thus less bonding than MO$_3$.

Fig. 3.33 Illustration of the effect of mixing between the σ_g MOs. In (a) we see the in-phase mixing of MO$_1$ (which is from the overlap of $2s$ AOs) and MO$_3$ (which is from the overlap of $2p_z$ AOs). A small amount of MO$_3$ is added to MO$_1$, so as to increase the bonding character of the resulting $2\sigma_g$ MO; note how the electron density in the internuclear region has increased. The out-of-phase mixing is shown in (b). This time, a small amount of MO$_1$ is subtracted from MO$_3$, resulting in an decrease in the bonding character of the $3\sigma_g$ MO.

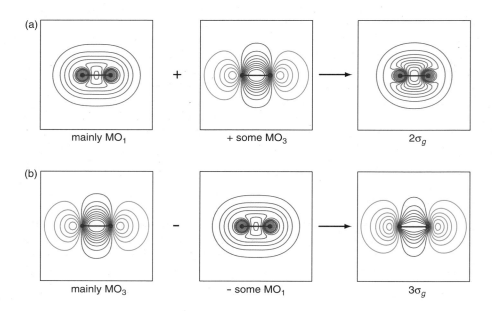

Table 3.3 Properties of the homonuclear diatomics of the second period

species	configuration	bond length / pm	dissociation energy / kJ mol^{-1}	paramagnetic?
Li_2	$2\sigma_g^2$	267	105	no
Be_2	$2\sigma_g^2 2\sigma_u^2$	245	approx. 9	no
B_2	$2\sigma_g^2 2\sigma_u^2 1\pi_u^2$	159	289	yes
C_2	$2\sigma_g^2 2\sigma_u^2 1\pi_u^4$	124	599	no
N_2	$2\sigma_g^2 2\sigma_u^2 1\pi_u^4 3\sigma_g^2$	110	942	no
O_2	$2\sigma_g^2 2\sigma_u^2 3\sigma_g^2 1\pi_u^4 1\pi_g^2$	121	494	yes
F_2	$2\sigma_g^2 2\sigma_u^2 3\sigma_g^2 1\pi_u^4 1\pi_g^4$	141	154	no
Ne_2	$2\sigma_g^2 2\sigma_u^2 3\sigma_g^2 1\pi_u^4 1\pi_g^4 3\sigma_u^2$	310	< 1	

A consequence of the s–p mixing is that $2\sigma_u$ becomes less antibonding, and $3\sigma_g$ becomes less bonding than is the case without such mixing. Depending on the degree of mixing, it is possible for the $3\sigma_g$ MO to be pushed above the $1\pi_u$. Calculations indicate that for the diatomics $Li_2 \ldots N_2$ the order of the MOs is as shown in Fig. 3.31 (b), whereas for O_2 and F_2 the order is as in (a). As was commented on before, it is not possible to say what the precise ordering of the orbital energies is without resorting to detailed calculations.

3.3.5 Properties of the homonuclear diatomics

We are now in a position to use our MO diagrams to interpret the properties of the homonuclear diatomics listed in Table 3.3. Generally speaking we see from this table that the greater the dissociation energy, the shorter the bond i.e. 'stronger bonds are shorter bonds'. Recall that for Li_2 to N_2 the order of the MOs is as shown in Fig. 3.31 (b) i.e. with the $3\sigma_g$ lying above the $1\pi_u$, whereas for O_2 and F_2, the diagram is as shown in (a) i.e. with $3\sigma_g$ lying below $1\pi_u$.

The electronic configurations given in Table 3.3 do not include the two electrons in the $1\sigma_g$ and the two in the $1\sigma_u$ MOs formed from the $1s$ AOs which do not contribute to the bonding. The table is concerned with what are called the *valence electrons*, which are the ones responsible for bonding. The $1s$ electrons would be classified as *core electrons*, and are not involved in bonding.

Li_2 has two valence electrons which occupy the bonding $2\sigma_g$ MO, creating a 'single' bond between the two atoms which is analogous to the single bond in H_2. Be_2 has four valence electrons which occupy the bonding $2\sigma_g$, and the antibonding $2\sigma_u$. At first sight, this seems analogous to the case of He_2 in which both the $1\sigma_g$ and $1\sigma_u$ MOs are occupied. However, the difference here is that on account of the s–p mixing described in the previous section, the $2\sigma_u$ MO is lowered in energy, as shown in Fig. 3.31 (b). The antibonding effect of the $2\sigma_u$ therefore does not outweigh the bonding effect of the $2\sigma_g$, so there is net bonding in Be_2. However, as we can see from the table, the bond dissociation energy is only 9 kJ mol^{-1}, indicating an exceptionally weak bond.

For B_2 there are six valence electrons, and the last two of these go into the bonding $1\pi_u$ MO, which lies below the $3\sigma_g$. As there are two degenerate $1\pi_u$ MOs, the lowest energy arrangement is to put one electron in each and to have the spins parallel. So, just like O_2, B_2 has unpaired spins and is therefore paramagnetic.

Fig. 3.34 Representation of the orbital energies of the *occupied* MOs for the molecules $Li_2 \ldots F_2$. Unless otherwise indicated, each orbital shown is filled by two electrons in the case of σ MOs, or by four electrons in the case of degenerate π MOs. The exceptions are in B_2 and O_2 where the highest energy pair of degenerate π MOs are occupied by just two electrons. Each of these electrons occupies a different orbital and their spins are parallel, giving rise to paramagnetism. Note how the energy ordering of the $1\pi_u$ and $3\sigma_g$ MOs change over as we go from N_2 to O_2.

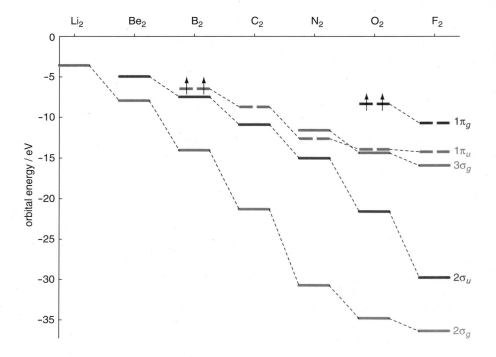

Moving on to C_2 gives two further electrons which fill the $1\pi_u$ MOs. If we assume that bonding effect of the $2\sigma_g$ and the antibonding effect of the $2\sigma_u$ MOs more or less cancel one another out, then B_2 has two bonding electrons, leading to a bond order of one i.e. a single bond between the two atoms. In contrast, C_2 has four bonding electrons and hence a bond order of two (a double bond). This increase in bond order accounts for the strengthening and shortening of the bond as we go from B_2 to C_2.

In N_2 the final two electrons go into the bonding $3\sigma_g$ MO, giving a bond order of three – in other words a triple bond. This molecule has the largest bond order, which correlates with the shortest and strongest bond. We have already discussed the bonding in O_2, F_2 and Ne_2 on page 117.

Overall, our MO diagrams provide a convenient and concise explanation of the pattern of bond strengths and bond lengths, and also the occurrence of paramagnetism. However, it has to be remembered that in order to explain the data we have had to invoke a reordering of the $3\sigma_g$ and $1\pi_u$ MOs as we went from N_2 to O_2. There is no way of predicting, other than by a full calculation, that this reordering occurs at this point.

Figure 3.34 shows the energies of occupied orbitals for the diatomics $Li_2 \ldots F_2$. These energies have been computed using a numerical approach similar to that used to find the AOs in multi-electron atoms. As we go across the period, we see that the orbital energies fall steadily. This is simply a result of the AO energies falling as the nuclear charge experienced by the electrons increases. The other feature which is clear is the swap over in energy of the $1\pi_u$ and $3\sigma_g$ MOs when we go from N_2 to O_2.

3.3.6 Limitations of qualitative MO diagrams

In this section we have seen how powerful the MO approach is, and how it provides a convenient way of rationalizing trends in the bonding of homonuclear diatomics. However, we have already begun to see that even for these simple molecules it is not always possible to 'guess' the detailed form of the MO diagram. For example, the ordering of the $3\sigma_g$ and $1\pi_u$ MOs is not something that we can predict without resorting to detailed calculations.

Even for the simplest molecule H_2^+ the situation is not quite as simple as we have implied. If the MOs are constructed from only $1s$ AOs, the predicted bond length is $2.5\,a_0$ (132 pm) and the predicted bond dissociation energy is 170 kJ mol^{-1}. These values are considerably different from the experimental values of 106 pm and 268 kJ mol^{-1}. In order for the calculation to reproduce these experimental values, many more orbitals than just the $1s$ need to be included. The contributions from these other orbitals are small, so the MOs which arise from these more sophisticated calculations look much the same as from our simple picture e.g. the $1\sigma_g$ bonding MO looks pretty much like that shown in Fig. 3.4 on page 95. Nevertheless, these small contributions are vital if quantitative agreement between theory and experiment is required.

The same considerations apply to other homonuclear diatomics. The MO diagram and representations of the orbitals shown in Fig. 3.28 on page 116 were described as 'idealized' because not only was the σ overlap between $2s$ and $2p$ ignored, but also the MOs were only constructed from a limited set of AOs (just the $2s$ and the $2p$). A more realistic calculation needs to include contributions from many more AOs. As with H_2^+, the qualitative result is much the same, as can be seen by comparing the computed MOs for O_2 shown in Fig. 3.35 on the next page with those in Fig. 3.28 on page 116, and the contour plots of Fig. 3.19 on page 111, Fig. 3.23 on page 112 and Fig. 3.26 on page 114.

When we first drew MO diagrams, like those of Fig. 3.10 on page 101, we emphasized the relationship between the energies of the AOs and MOs. In the simple case of H_2^+, the bonding MO lies below the energy of the AOs from which it is composed, and the antibonding MO lies above the energy of the AOs. This nice simple arrangement can also be seen in the idealized MO diagram for the homonuclear diatomics shown in Fig. 3.28 on page 116.

However, in reality, the relationship between the AO and MO energies is often not so simple as these idealized pictures imply. The reasons for this lie in the complicated interactions which can occur in a molecule containing many electrons and nuclei, and the way in which these interactions change when a molecule is formed from atoms. As a result, we cannot expect our simple MO diagrams to be anything more than a good guess at the true situation. This is not to imply that such diagrams are not useful – on the contrary, we will use them to great effect throughout the rest of this book. However, it is important to understand their limitations.

If we need to know the detailed form of the MOs in a particular molecule, we can always compute them using one of the computer programs which have been developed for this purpose. Such is the power of modern computers, and the sophistication of the programs, that MO calculations on small to medium-sized molecules can readily be performed. Throughout the book, we will make use of the results of such calculations. However, even when we have calculated the MOs, we will often interpret their shapes and energies using simple qualitative arguments like those introduced in this chapter.

Weblink 3.5

View, and rotate in real time, iso-surface representations of all of the filled MOs of O_2, illustrated in Fig. 3.35.

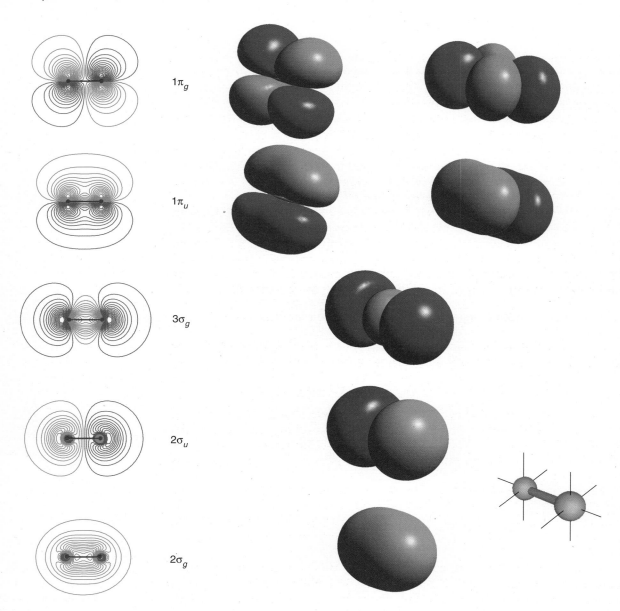

Fig. 3.35 Contour plots and iso-surface plots of the occupied MOs for O_2 obtained from a computer calculation which constructs the MOs from many more AOs than the $2s$ and $2p$ used to produce the idealized MO diagram shown in Fig. 3.28 on page 116. Comparison of the form of the MOs shown here with those in the idealized diagram (and the contours plots shown in Fig. 3.19 on page 111, Fig. 3.23 on page 112 and Fig. 3.26 on page 114) shows that although there are some differences of detail between the calculated and idealized MOs, the basic shapes of the MOs are much the same. You will notice that the overall signs of these MOs are not always the same as in the previous pictures: this is of no significance, as changing the overall sign of the wavefunction makes no difference to the electron density.

3.4   Photoelectron spectra

Photoelectron spectroscopy gives us an excellent experimental way of probing the energies and other characteristics of molecular orbitals, thus complementing the theoretical calculations which we have been discussing so far. Such spectra can be recorded in a straightforward way.

The sample is irradiated with an intense source of ultraviolet (UV) light. These short-wavelength photons are sufficiently energetic to cause ionization of the higher energy electrons in molecules – a process analogous to the *photoelectric effect* in which electrons are ejected from a metal when sufficiently energetic photons strike the surface. The energy and number of the ejected electrons is then measured, and presented in the form of a spectrum.

The UV light is often generated by passing an electric discharge through low-pressure helium gas. Under the right conditions, such a lamp generates UV light at a single wavelength of 58.4 pm, which corresponds to a photon energy of 21.2 eV. As we know the energy of the photon, and have also measured the energy of the ejected electron, we can work out the ionization energy (or ionization potential) of that electron simply from the conservation of energy:

$$\text{energy of photon} = \text{energy of ejected electron} + \text{ionization energy}.$$

As we saw in section 2.7.2 on page 81, to a good approximation the ionization energy of an electron is simply minus its orbital energy

$$\text{ionization energy} \approx -\text{ orbital energy}.$$

Using this, we can interpret each peak in the photoelectron spectrum as being due to the ionization of an electron from a particular orbital. In addition, the energy of the electron can be used to compute the ionization energy of the orbital. Thus, such spectra give a rather direct picture of the energies of the molecular orbitals.

Figure 3.36 shows the photoelectron spectrum of H_2, along with the corresponding MO diagram; the horizontal axis gives the ionization energy, computed in the way described above. What we see is a series of lines, called a *band*, *all* of which arise from the ionization of an electron from the bonding $1\sigma_g$ MO.

The reason that there are several lines, even though the electron is being ionized from a single orbital, is that removing this *bonding* electron causes a significant change in the strength of bonding in the molecule. In particular, with only one bonding electron remaining, we expect the molecular ion to have a greater equilibrium bond length than does the neutral molecule, which has two bonding electrons. This change in the bond length leads to the excitation of vibrations in the molecular ion H_2^+, and each line in the band corresponds to the molecular ion being excited to a different vibrational energy level.

Figure 3.37 on the following page shows the photoelectron spectrum of N_2, along with the corresponding MO diagram. There are three bands which show varying amounts of vibrational structure, and which have been colour coded so that it is easy to see which line goes with which band.

The band with the lowest ionization energy (about 15.5 eV) shows little vibrational structure, in contrast the band seen in the spectrum of H_2. A lack of extensive vibrational structure is associated with the ionization of an electron which does not have a strong influence on the extent of bonding. As a result,

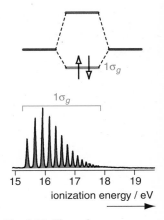

Fig. 3.36 Photoelectron spectrum of H_2 recorded using 21.2 eV UV radiation; this spectrum can be interpreted using the MO diagram shown above. In the spectrum there is a single band which corresponds to the ionization of the $1\sigma_g$ electron. The fine structure in the band is due to different vibrational energy levels of the molecular ion being excited by the ionization process.

Fig. 3.37 Photoelectron spectrum of N_2 recorded using 21.2 eV UV radiation, along with the MO diagram used in its interpretation. The spectrum shows three bands, coloured red, green and blue, which can be attributed to ionization from the $3\sigma_g$, $1\pi_u$ and $2\sigma_u$ MOs, respectively. The lack of vibrational structure for the red and blue bands indicates that the corresponding MOs are neither strongly bonding nor strongly antibonding.

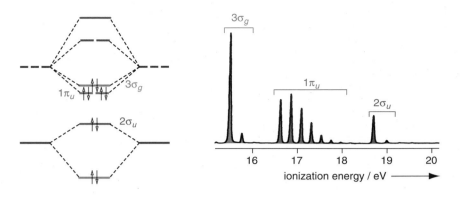

there is little vibrational excitation of the molecular ion, and so rather than seeing a band with several lines, we see little vibrational structure. The form of the 15.5 eV band is thus consistent with this being due to the ionization of an electron from the $3\sigma_g$ MO, since we have already noted that this orbital is only weakly bonding.

In contrast, the next highest energy band (coloured green) shows extensive vibrational structure, which implies that removal of the electron causes a significant change in the extent of bonding. This is entirely consistent with the green band being due to the ionization of an electron from the bonding $1\pi_u$ MO.

Finally, the highest energy band (coloured blue) shows some vibrational structure, but not as much as for the green band. This again points to the ionization of an electron from an MO which is not strongly influencing the extent of bonding. As we discussed above, the $2\sigma_u$, on account of the s–p mixing, is only weakly antibonding, which is consistent with the appearance of the blue band. The UV photons are insufficiently energetic to ionize electrons from any of the lower lying MOs.

The photoelectron spectrum (PES) thus gives us a very direct way of probing both the energies of the MOs, and from the vibrational structure we can determine whether or not an MO has a strong influence on the extent of bonding. We will look at the PES of other molecules later on in this chapter.

3.5 Heteronuclear diatomics

Having been successful in rationalizing the properties of homonuclear diatomics, the next step is to move on to heteronuclear diatomics such as HF and CO. The new feature which will arise with such molecules is that they can have dipole moments as a result of the electrons not being evenly distributed between the two atoms. We will start our discussion by considering LiH and HF which illustrate how this polarization of the molecule can come about.

Box 3.1 Dipole moments

The dipole moment of a molecule can be measured using either spectroscopic techniques or by measuring bulk properties, such as the dielectric constant. For small molecules in the gas phase, spectroscopic measurements are preferred.

The value of the dipole moment is usually given in *debye* (symbol D), a unit named in honour of the Dutch chemical physicist Peter Debye. 1 debye is equivalent to 3.336×10^{-30} C m in SI.

Homonuclear diatomics do not, of course, have dipole moments on account of their symmetry. A dipole moment of 2 D is considered quite large, and values over 5 D are rarely encountered.

3.5.1   LiH and HF: two typical hydrides

LiH is most commonly encountered as a crystalline solid, but the molecular species Li–H also exists in the gas phase where it has been studied experimentally. The dissociation energy of LiH is 234 kJ mol^{-1}, its bond length is 160 pm, and it has a large *dipole moment* of 5.9 D (see Box 3.1 for details on dipole moments), the hydrogen atom being $\delta-$ and the lithium $\delta+$.

HF also exists in the gas phase, with a somewhat shorter bond length of 91.7 pm. Like LiH, the molecule has a dipole moment (1.9 D), but this time it is the fluorine atom which is $\delta-$ and the hydrogen which is $\delta+$.

You can probably already think of an explanation, using the concept of *electronegativity*, as to why the hydrogen is $\delta-$ in LiH but $\delta+$ in HF. We would argue that in a bond the electron density is drawn towards the more electronegative atom. Therefore, since H is more electronegative than Li, the H atom is $\delta-$ in LiH; however, F is more electronegative that H, so the H atom is $\delta+$ in HF.

Using an MO description of the bonding in these two hydrides we will be able to explain why they are polarized in the way they are, and along the way it will become clearer how this very important concept of electronegativity is related to orbital energies.

LiH

The first thing we do when drawing up an MO diagram is to think about the energies of the AOs which we might be involved. Referring to Fig. 2.42 on page 69, we see that for H we have the $1s$ AO with an energy of -14 eV, and for Li we have the $2s$ with an energy of -5 eV. We can safely ignore the Li $1s$ as being too low in energy and too contracted to have significant interaction with any other orbital.

There are thus just two AOs which will interact, the H $1s$ and the Li $2s$, and these will give rise to a bonding and an antibonding MO. The bonding MO will lie below the lower energy of the AOs, which is the H $1s$, and the antibonding MO will lie above the higher energy of the AOs, which is the Li $2s$; this is depicted in the MO diagram of Fig. 3.38 on the next page.

The overlap of two s orbitals gives, as before, MOs with cylindrical symmetry, so they have the label σ. However, unlike a homonuclear diatomic, a heteronuclear diatomic does not possess a centre of inversion, so we cannot

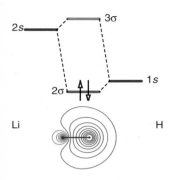

Fig. 3.38 Approximate MO diagram for LiH. The Li $2s$ overlaps with the H $1s$ to give two σ MOs, labelled 2σ and 3σ; a contour plot of the occupied 2σ MO, found by a computer calculation, is also given. The 2σ MO is closest in energy to the lower energy of the AOs, the H $1s$, so this AO is therefore the major contributor to the MO, as can clearly be seen from the contour plot. Not shown on the diagram is the Li $1s$ which is too low in energy to have any significant interaction with the H $1s$; this Li AO also has σ symmetry, and is given the label 1σ.

assign a g or u symmetry label. The two MOs are therefore labelled 2σ and 3σ; we could also denote the 3σ as $3\sigma^\star$, to indicate that it is antibonding.

The reason that the bonding orbital is labelled 2σ is that, strictly speaking, it is the *second* lowest energy σ-type orbital. The lowest such orbital is the Li $1s$ which, although not involved in the bonding, can nevertheless be classified as having σ symmetry; it is therefore given the label 1σ.

According to rule 4 from section 3.1 on page 92, the 2σ MO has an unequal mixture of the two AOs, with the major contribution coming from the H $1s$ as this is closest in energy to the MO. In contrast, it is the Li $2s$ which is the major contributor to the 3σ MO.

The two Li $1s$ electrons are classified as core electrons and so are not involved in the bonding. This leaves LiH with just two valence electrons which are assigned to the bonding MO, to give the configuration $2\sigma^2$. With both electrons in the bonding MO, we predict that the molecule is stable with respect to dissociation into atoms.

The new feature we have here is that the electron density due to these two bonding electrons is not equally distributed between the Li and H. Recall that the square of the wavefunction gives the probability density, so simply by looking at the contour plot of the 2σ MO in Fig. 3.38 we can see that the majority of the electron density in this orbital will be on the hydrogen. It is therefore not surprising that the bond has a dipole moment, with the H being $\delta-$.

Previously, we argued that we expected that the hydrogen would be $\delta-$ on the grounds that it is more electronegative than lithium. Our MO picture tells us that the reason the hydrogen is $\delta-$ is that the hydrogen AO has lower energy than the Li AO, resulting in the bonding MO being polarized towards the hydrogen.

It is the fact that the hydrogen $1s$ AO has lower energy than the $2s$ AO in the lithium which results in the bond being polarized towards the hydrogen. In other words, hydrogen is more electronegative than lithium *because* the hydrogen AO is lower in energy than the AO on lithium.

We will see more examples of this as this section proceeds, but the general idea which comes out of this discussion is that the lower the energy of the AOs of an atom, the more electronegative that atom becomes. This statement needs slight qualification by adding the proviso that we are talking about the AOs which are involved in bonding i.e. the valence electrons. Electronegativity is thus a consequence of orbital energies.

It is tempting to say that the orbital energies are lower in more electronegative atoms, but this is the wrong way round to make the connection. Rather, we should say that atoms with high electronegativity are those with low energy orbitals.

HF

In HF the AOs available for bonding are the H $1s$ at -14 eV, the F $2s$ at -43 eV, and the three F $2p$ at -20 eV. As before, we can safely ignore the F $1s$, and we can also ignore any interaction with the F $2s$ as this orbital is very much lower in energy than the H $1s$. These two F orbitals have σ symmetry, and so are given the labels 1σ and 2σ.

If we take the H–F bond to be along the z-axis then, as shown in Fig. 3.15 on page 107, the H $1s$ and the F $2p_z$ have the correct symmetry to overlap to

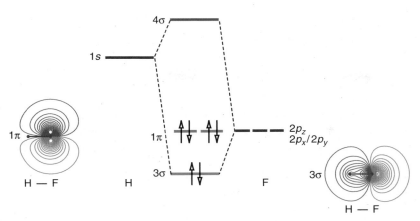

Fig. 3.39 Approximate MO diagram for HF, along with contour plots of the 3σ and one of the degenerate 1π MOs, found by a computer calculation. The F $2p_x$ and $2p_y$ AOs do not have the correct symmetry to overlap with the H $1s$ and so are not involved in the formation of MOs, and so are called *nonbonding* orbitals. These orbitals have π symmetry. The F $1s$ and $2s$ AOs are too low in energy to interact with the H $1s$; these two fluorine AOs have σ symmetry and so form the 1σ and 2σ orbitals in the molecule.

give σ type MOs. These two AOs give the 3σ bonding and 4σ antibonding MO. The F $2p_x$ and $2p_y$ do not have the correct symmetry to overlap with the H $1s$, again as shown in Fig. 3.15 on page 107, and so are not involved in the bonding; their energies are thus unchanged.

The resulting MO diagram, along with contour plots of the occupied MOs, is shown in Fig. 3.39. There are eight valence electrons in HF, two are located in the F $2s$ (not shown), two in the 3σ and one pair in each of the F $2p_x$ and $2p_y$ orbitals, which in the molecule have the symmetry label 1π as they have a nodal plane containing the internuclear axis. Overall, we therefore expect HF to be stable with respect to dissociation into atoms.

The F $2p_x$ and $2p_y$ are often described as *nonbonding* orbitals as they are neither bonding nor antibonding. The electron pairs in these orbitals are described as nonbonding electrons or, more commonly, *lone pairs*. The F $2s$ (2σ) is also occupied by a pair of electrons which could be described as a third lone pair.

As can be seen from the contour plots, the major contributor to the bonding 3σ MO is the fluorine $2p_z$ as this AO is closest in energy to that of the MO. This asymmetry of the 3σ MO, combined with the electron density due the lone pairs, results in the F being $\delta-$. Following the previous argument, we see that the fluorine is more electronegative than the hydrogen as it has lower energy valence AOs.

PES of HCl

An experimental PES of HF is not available, presumably due to the difficulties of handling such a highly corrosive gas. However, a PES spectrum of HCl is shown in Fig. 3.40 on the next page. The MO diagram for this molecule is essentially the same as for HF with the exception that it is the $3p$ orbitals which are involved in the bonding.

The spectrum shows two bands. The lowest energy band (highlighted in green) shows little vibrational structure, and can be associated with ionization

Fig. 3.40 Photoelectron spectrum of HCl recorded using UV radiation with a photon energy of 21.2 eV; the corresponding MO diagram is shown on the left. The band highlighted in green is from the nonbonding π orbitals i.e. the $3p_x$ and $3p_y$ on the chlorine. The band highlighted in red is from the 3σ MO; the presence of significant vibrational structure is indicative of the bonding nature of this MO.

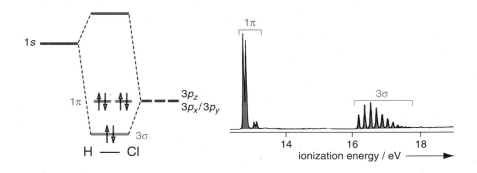

of the nonbonding electrons from the 1π orbitals (the $3p_x$ and $3p_y$). In contrast, the band highlighted in red shows significant vibrational structure and can be assigned to ionization of the bonding 3σ MO. As with N_2, the PES provides experimental evidence to corroborate our MO diagram.

3.5.2 NO, CO and LiF

Finding the MOs for diatomics between two second-period atoms is rather more complex than for hydrides as many more orbitals are involved. Generally, we will need to consider the $2s$ and the $2p$ AOs on both atoms. The interaction of the two $2s$ and the two $2p_z$ AOs will lead to four σ type MOs (Rule 1), each of which will in principle be a mixture of all four AOs. It is only by doing a computer calculation that we will be able to identify the exact form of these four MOs.

In addition, the two $2p_x$ AOs and the two $2p_y$ AOs will form a degenerate pair of π bonding MOs, and a degenerate pair of π antibonding MOs, just as we found in the homonuclear case. The difference is that the π MOs will not have equal contributions from the AOs on each atom, as these AOs do not have the same energy.

As before, the $1s$ AOs are too contracted to interact significantly. These AOs both have σ symmetry and, in the molecule, are labelled 1σ and 2σ: they will not be shown on our MO diagrams.

NO

The valence orbitals on nitrogen are the $2s$ at -26 eV and the $2p$ at -15 eV; for oxygen the orbitals are $2s$ at -34 eV and $2p$ at -17 eV. There is no way to guess the detailed form of the MOs, so we will have to resort to a computer-based calculation and see if we can make sense of the results. Figure 3.41 (a) shows the energies of the MOs, plotted to scale, and contour plots of the occupied MOs are shown in Fig. 3.42 on page 132.

There are eleven valence electrons in NO, so the configuration is $3\sigma^2 \, 4\sigma^2 \, 1\pi^4 \, 5\sigma^2 \, 2\pi^1$. As expected for a molecule with an odd number of electrons, there is one unpaired electron (in the 2π).

It is interesting to compare the MOs for NO with those of N_2, and to facilitate this the MO diagram for N_2 is shown in Fig. 3.41 on the next page, while contour plots of the MOs are shown in Fig. 3.42 on page 132. Overall, the MO diagram, and the MOs themselves, for NO and N_2 are remarkably

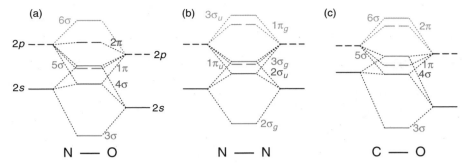

Fig. 3.41 MO diagrams, based on computer calculations, for NO, N_2 and CO. The occupied MOs are shown in full colours, whereas the unoccupied ones (6σ in NO, $1\pi_g$ and $3\sigma_u$ in N_2, and 2π and 6σ in CO) are shown in pale colours. The energies of the AOs and the occupied MOs are shown to scale, but the energies of the unoccupied MOs are not well defined. The dashed lines show which AOs contribute to which MOs. For these molecules, the $2s$ and the $2p_z$ contribute to all of the σ-type MOs.

similar. If we were looking to construct a qualitative diagram for NO, then simply reusing the existing one for N_2 would be a pretty good start. All we need to account for is the fact that the lower energy of the oxygen AOs will mean that they make a larger contribution to the bonding MOs than do the nitrogen AOs.

The MOs in NO are not strongly polarized towards either atom, so we should not be surprised to find that the dipole moment in NO is only 0.15 D. If we ionize NO to give NO^+ we might reasonably expect the strength of the bonding to increase on the grounds that an antibonding electron has been removed. Experiment bears this out: for NO the dissociation energy is 627 kJ mol^{-1} and the bond length is 115 pm, whereas for NO^+ the corresponding values are 1048 kJ mol^{-1} and 106 pm.

CO

The valence orbitals on carbon are the $2s$ at -19 eV and the $2p$ at -12 eV; for oxygen the orbitals are $2s$ at -34 eV and $2p$ at -17 eV. Compared to NO, there is a greater difference in the energies of the AOs, so we can expect the MO diagram for CO to depart somewhat more from that for N_2. This is borne out by the MO diagram shown in Fig. 3.41 (c) on page 131. Nevertheless, we see that the N_2 diagram is still a pretty good starting point for drawing up a qualitative description of CO.

There are ten valence electrons in CO, giving the electronic configuration $3\sigma^2 \, 4\sigma^2 \, 1\pi^4 \, 5\sigma^2$; in contrast to NO, all of the electrons are paired up. Contour plots of the occupied MOs are shown in Fig. 3.42 on the next page.

The degenerate pair of 1π MOs are clearly bonding and have a larger contribution from the oxygen, which we can rationalize as being the result of the orbitals on this atom being lower in energy. The highest energy occupied orbital is the 5σ; this MO is more unsymmetrical than the corresponding orbital in NO, with the carbon being the major contributor. Again, comparing the 5σ MO to the $3\sigma_g$ MO in N_2 seems to indicate that the 5σ is weakly bonding, perhaps less so that in N_2.

Figure 3.43 on page 133 shows the PES of CO, alongside the MO diagram. The lowest energy band (highlighted in red) is from the 5σ, and the lack of

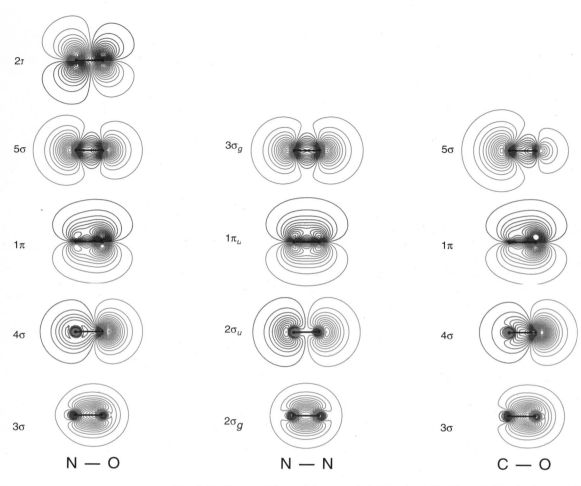

Fig. 3.42 Contour plots of the occupied MOs for NO, N_2 and CO; the form of these MOs have been found using a computer calculation. For each molecule, the energies of the MOs increase as you go up the page, but a more detailed picture of their relative energies can be seen in Fig. 3.41 on the preceding page.

vibrational structure indicates that this MO is essentially nonbonding. The band highlighted in green can be assigned to ionization of the 1π, and the vibrational structure is consistent with the bonding nature of this MO. The band highlighted in blue has some vibrational structure, and can be assigned to the 4σ.

The molecules CO and N_2 are said to be *isoelectronic*, meaning that they both have the same number of valence electrons (ten, in this case); NO^+ is also isoelectronic with N_2. It is interesting to note the similarity between the photoelectron spectra of CO and N_2 (Fig. 3.37 on page 126). Given that these two molecules are isoelectronic, the similarity of the spectra is perhaps not so surprising.

LiF

Our last example is rather an extreme one, as it involves two atoms at opposite ends of the second period. Of course, LiF is well known as a crystalline ionic

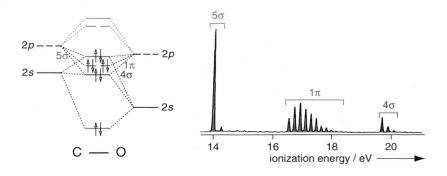

Fig. 3.43 Photoelectron spectrum and MO diagram of CO. The red band is from the 5σ MO; the lack of vibrational structure indicates that this MO is essentially nonbonding. The green band shows significant vibrational structure, and is from the bonding 1π MOs. The blue band is from the 4σ; it shows some vibrational structure, consistent with the MO being weakly antibonding.

solid, in which there is a regular array of Li$^+$ and F$^-$ ions, but what we are talking about here is the discrete diatomic molecule Li–F which exists in the gas phase. Its bond length is 160 pm, the dissociation energy is 574 kJ mol^{-1}, and it has a large dipole moment of 6.3 D.

The occupied AOs of Li and F are very mismatched in energy. For the Li we have the 2s with an energy of −5 eV, and for the F the 2s and 2p are at −43 eV and −20 eV, respectively. As it is so low in energy compared to the other AOs, we can safely assume that the F 2s is not involved in bonding, and so becomes the nonbonding 3σ MO in the molecule (recall that the even lower energy F 1s and Li 1s will become the 1σ and 2σ). There are no AOs on the Li of the correct symmetry to overlap with the fluorine $2p_x$ and $2p_y$ (the internuclear axis is along z), so these AOs become the nonbonding 1π MOs.

In principle, the Li 2s has the correct symmetry to overlap with the F $2p_z$, and this interaction gives rise to the bonding 4σ and antibonding 5σ MOs. However, since the energy separation between the AOs is so large, the 4σ MO is made up of the F $2p_z$ with only a very small contribution from the Li 2s. The resulting MO diagram is shown in Fig. 3.44; note how closely the 4σ MO resembles a $2p_z$ AO, indicating that there is very little interaction with the Li 2s. In LiF there are eight valence electrons which occupy the orbitals as shown.

It is useful to think about this molecule in terms of how the electrons which were in the Li and F AOs relate to the electrons in the MOs. The F atom

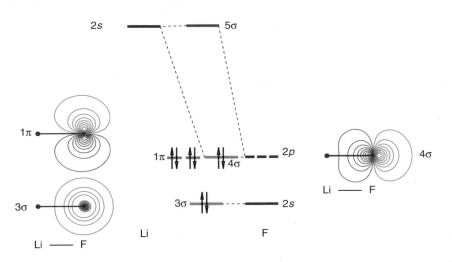

Fig. 3.44 Approximate MO diagram for LiF, and contour plots on the occupied MOs obtained from a computer calculation; only one of the degenerate 1π MOs is shown. There is such a large mismatch between the energy of the Li and F AOs that there is hardly any mixing between them. Thus, the 3σ and 4σ are essentially the fluorine 2s and $2p_z$ AOs, whereas the 5σ is the Li 2s (the internuclear axis is along z). The 1π are the F $2p_x$ and $2p_y$ AOs, and are nonbonding.

originally had seven electrons, two in the $2s$ and five in the $2p$. When LiF forms the $2s$ becomes the 3σ, and the two of the $2p$ AOs become the 1π, without change. The $2p_z$ becomes the 4σ, but there is so little mixing with the Li $2s$ AO that this MO is dominated by the contribution from the fluorine. Therefore, to a good approximation the electrons which started out on the fluorine atom remain localized on that atom when the molecule forms.

The story for the Li is quite different. Originally the single valence electron was in the $2s$, but on forming LiF this electron ends up in the much lower energy 4σ MO, to which the F $2p_z$ is the major contributor. On moving from the Li $2s$ to the 4σ there is clearly a significant reduction in energy, and it is to this that we can attribute the bonding in LiF. Essentially what has happened is that the $2s$ electron which was associated with the Li has been transferred to a fluorine orbital. As a result, the molecule might reasonable be represented as Li^+F^-.

Such a situation is often described by saying that there is an 'ionic contribution' to the bonding in LiF. In contrast to the molecules we have been discussing so far, in LiF the bonding arises not from a sharing of electron density *between* the atoms, but from a more complete transfer of an electron from a high energy orbital on one atom to a lower energy orbital on the other: there is little electron density shared between the atoms. It is important to realize that this transfer of electron density from one atom to the other arises quite naturally in the MO picture, and would be included in any numerical calculations. It is not some additional term we have to add in to our calculation.

We will return to this question of the extent to which bonding arises from the sharing of electrons ('the covalent contribution') or the more complete transfer of electrons ('the ionic contribution') in section 7.9.2 on page 298 where we discuss trends in bonding across the periodic table.

3.5.3 The HOMO and the LUMO

As we go on to make more use of MOs to discuss chemical reactions we will discover that there are two particular MOs which tend to be particularly important in determining the way in which a molecule reacts. These are the *highest occupied molecular orbital* (HOMO) and the *lowest unoccupied molecular orbital* (LUMO): the names are self explanatory.

So, for example, in CO the HOMO is the 5σ and the LUMO is the 2π. In N_2 the HOMO is the $3\sigma_g$ and the LUMO is the $1\pi_g$. Not surprisingly, for many molecules the LUMO is often an antibonding MO.

3.6 Moving on

We have seen in this chapter that the MO approach provides a good explanation for, and rationalization of, the properties of simple diatomic molecules. The way in which MOs are formed are governed by some simple rules which we can use to construct qualitative MO diagrams. Alternatively, we can use these rules to help us interpret the results of computer calculations.

The MO picture encompasses the full range of molecules all the way from a homonuclear diatomic, where there is an equal sharing of the electrons between the two atoms, through a molecule like NO where the sharing is unsymmetrical, to an extreme case like LiF where there is more or less complete transfer from one atom to another.

Our task is now to extend the MO approach to larger molecules. What we will find is that the same general rules apply, although the increase in the number of atoms and orbitals involved means that it gets more and more difficult to guess at the detailed form of an MO diagram. Of course, we can always use computer calculations to find the MOs, but this is not always going to be convenient.

For many molecules a more localized description of the bonding is often satisfactory, and we will see how such an approach can be combined with the MO approach by introducing the concept of hybrid atomic orbitals. However, parts of the bonding of some molecules, such as the π bonds in benzene, can really only be described using the delocalized MO approach.

FURTHER READING

Elements of Physical Chemistry, fourth edition, Peter Atkins and Julio de Paula, Oxford University Press (2005).

QUESTIONS

3.1 Explain how the bonding and antibonding MOs are formed in H_2^+, and rationalize why it is that the energy of the bonding MO shows a minimum when plotted as a function of the internuclear separation R, whereas the energy of the antibonding MO simply increases as R decreases.

Construct a MO diagram for H_2^+, and make rough sketches of the form of the MOs. Explain how symmetry labels are assigned to the two MOs.

3.2 Explain what is meant by the overlap integral between two AOs, and also explain why this quantity goes to zero for large distances between the two AOs. How would you expect the plot of $S(R)$ to compare between two $1s$ AOs and two $2s$ AOs?

Shown below is a plot of the overlap integral between two $2p$ orbitals. The blue line is for the sideways-on overlap to give π MOs, whereas the red line is for the head-on overlap to give σ MOs. Explain the form of these two curves.

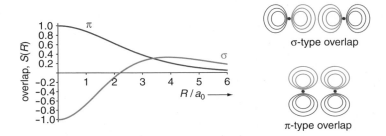

3.3 Use MO diagrams to rationalize why He_2 is an unknown species, but the ion He_2^+ has been observed. Make what predictions you can about the stability of the molecules He_2^{2+} and H_2^{2-} with respect to dissociation.

3.4   Suppose that there is overlap between the $2p_z$ AOs from two atoms A and B, where z is the internuclear axis. Sketch the form of the MO diagram, and the form of the bonding and antibonding MOs, for the following cases: (a) A and B of very similar electronegativity; (b) A somewhat more electronegative than B; (c) A much more electronegative than B.

   If the resulting bonding MO is occupied by a pair of electrons, what consequences would this have for the polarity of the A–B bond in these three cases?

3.5   Consider the overlap between a $3d_{xy}$ AO on one atom and a $2p$ AO on a second atom as these two approach one another along the x-axis. Explain, using sketches, whether or not there is overlap between the $3d_{xy}$ and each of the three $2p$ orbitals in turn. If there is overlap, classify the resulting MO as having σ or π symmetry.

   Do the same for a $3d_{x^2-y^2}$ AO interacting with each of the $2p$ AOs.

3.6   Explain why the *overall* sign of the MO wavefunction is not important, whereas the relative signs of two AOs which are forming an MO is of significance.

3.7   In O_2 the lowest energy MO is given the label $1\sigma_g$ and the next lowest is $1\sigma_u$. Explain how these MOs are formed (i.e. from which AOs); sketch the form of the MOs.

   Explain why it is that although both of these MOs are occupied, they make little contribution to the bonding in O_2.

3.8   The dissociation energy of N_2 is 942 kJ mol^{-1}, whereas that for N_2^+ is 842 kJ mol^{-1}; the dissociation energy of O_2 is 494 kJ mol^{-1}, whereas that for O_2^+ is 642 kJ mol^{-1}. rationalize these data.

3.9   Sketch an MO diagram for BeH. On the basis of your diagram, would you expect this molecule to be stable with respect to dissociation into atoms? Use your MO diagram to predict any other properties you can.

3.10   Construct an MO diagram for the diatomic molecule CH, label the MOs, indicate which orbitals are occupied, and sketch the form of the occupied orbitals. The relevant orbital energies are: H $1s$ -14 eV; C $2s$ -19 eV; C $2p$ -12 eV. To a first approximation, you can ignore the interaction between the H $1s$ and the C $2s$ (why?).

   What does your MO diagram predict about CH and its ion CH$^+$?

3.11   Explain what you understand by the statement that CN$^-$ is *isoelectronic* with CO.

   Construct an approximate MO diagram for CN$^-$, indicate which MOs are occupied and explain what the diagram predicts about this molecule.

3.12   We argued that in H_2 the atoms are held together as a result of the build-up of electron density along the internuclear axis. However, in LiF there is practically no such build up. What, then, holds the atoms in LiF together, and in what way is this different to H_2?

Electrons in molecules: polyatomics

Key points

- The use of symmetry helps us in constructing qualitative MO diagrams for larger molecules.
- The detailed form and energies of MOs can be found using computer calculations.
- Hybrid atomic orbitals provide a convenient localized description of bonding in larger molecules.
- The description of parts of the bonding in some molecules requires a delocalized MO approach.
- The delocalized molecular orbitals of simple π systems can be found using a geometric construction.
- Resonance structures are a useful way of describing some aspects of delocalized bonding.

Our task is now to extend the molecular orbital approach from diatomics to the much larger molecules we are going to encounter in our study of chemistry. We have already seen that even for molecules as simple as diatomics there is no way we can construct a detailed MO diagram with pen and paper: rather, we need to resort to computer calculations in order to find out the exact shape and energy of the MOs. However, we will often find that a simple qualitative MO diagram, of the sort we can construct without resort to detailed calculations, is sufficient. Even in cases where we have calculated the MOs, we still have to interpret the result of the calculation, which we usually do by referring back to qualitative MO diagrams. So, although it has its limitations, there is definitely a role for the qualitative MO approach we will develop in this chapter.

We will start by showing how symmetry considerations make it possible to construct MO diagrams for simple triatomics, such as H_3^+ and H_2O. What we will find is that in these molecules the MOs extend over all the atoms – that is, the bonding is delocalized. In fact, it is a general feature of the MO approach that AOs on many atoms contribute to each MO. Unfortunately, this means that as the number of atoms (and hence orbitals) increases, it rapidly becomes almost impossible to draw up useful qualitative MOs diagrams: another approach is needed.

The one we will explore is a description of bonding in which the overlap of *hybrid atomic orbitals*, rather than atomic orbitals, leads to the formation of bonds. Due to the directional nature of the hybrid atomic orbitals, their overlap gives bonds which are localized between two atoms. This greatly simplifies the description of the bonding in larger molecules, as it can be broken down into a set of interactions between just two atoms at a time. This approach is very successful in describing the bonding in organic molecules, and in fact you have, most probably without knowing it, already been using it.

In some cases, the localized description using hybrids is not adequate to describe all of the bonding in a molecule. For parts of a structure, the delocalized MO approach is more appropriate, for example in describing the π electrons in benzene, and in the carboxylate anion. Typically, what we do is use the localized approach as far as we can (as it is simple), and reserve the more complex MO approach for the delocalized part of the molecule.

By the end of this chapter, you will be able to draw up a good description of the bonding in a wide range of molecules, and will be able to use this to explain some of the physical and chemical properties of the molecule. In later chapters, you will see that this understanding of the orbitals involved in the bonding will be crucial to understanding the way in which molecules react.

4.1 The simplest triatomic: H_3^+

H_3^+ is not a familiar molecule to the earth-bound chemist, but in galactic terms it is ubiquitous. For example, it is found in the vast 'molecular clouds' which exist in interstellar space. In these clouds it is thought that H_3^+ plays an important part in the series of reactions which lead to the formation of many quite complex molecules. H_3^+ has also been observed somewhat nearer to home in the atmosphere of gas giant planets, such as Jupiter. The light given off by these remote objects is analysed spectroscopically, and the resulting data are compared with the spectra of known molecules which have been recorded in the laboratory. In this way, it is possible to verify the identity of the molecules, even though they are at vast distances from the Earth.

If we think about constructing the MO diagram for this molecule, the natural choice is to create the MOs by a linear combination of three $1s$ orbitals, one on each atom. This is exactly analogous to the way we approached H_2^+. As we are combining *three* AOs, we expect to form *three* MOs, but so far we have no way of working out the form of these MOs, nor any way of finding their relative energies. In this section, we will see that by exploiting the *symmetry* of the molecule, we can extend the ideas used in diatomics to enable us to form the MOs of more complex molecules.

4.1.1 Constructing the MOs using symmetry

The first thing we need to do is decide what the geometry of H_3^+ is. A reasonable guess is that it is linear, with the central H atom equidistant from the two end H atoms i.e. $(H–H–H)^+$. In fact, it turns out that this is not the correct structure, but it is nevertheless instructive to look at the form of its MOs. Later on we will look at the MOs of the correct structure.

In this linear structure, there are two different kinds of hydrogen atoms: (1) the central atom; (2) the atoms at the ends. The two end atoms are clearly the same as one another, and if we look at the molecule we realize that this is because of the particular symmetry which the molecule possesses. As is shown in Fig. 4.1, the molecule has a mirror plane, perpendicular to the internuclear axis and passing through the central atom. Reflection in this plane swaps over the two end nuclei (1 and 3), so by symmetry they are equivalent. The central atom, 2, is unaffected by reflection in the plane. The molecule also possesses many other elements of symmetry, but this mirror plane is the crucial one for the present discussion as its action differentiates the end atoms from the central atom.

When we were thinking about the way in which $2p$ and $2s$ orbital could overlap to form MOs, we came across the idea that in order for two AOs to overlap, they must have the same symmetry. For example, as shown in Fig. 3.15 on page 107, in a diatomic (with the internuclear axis along z), a $2p_x$ AO cannot overlap with a $2s$, whereas a $2p_z$ can. This is because the $2s$ and the $2p_z$ are unaffected by rotation about the z-axis, whereas the $2p_x$ changes sign when rotated by 180°.

What we are going to do now is to classify the three $1s$ AOs in H$_3^+$ according to the mirror plane which has just been identified as a key element of symmetry. Having done this, we will invoke the rule that only orbitals with the same symmetry can overlap to form MOs. The simplification which results from this application of symmetry will make it possible to construct both the MOs and the energy level diagram.

Figure 4.2 shows the effect of the mirror plane on the three AOs. The AO on atom 2 is simply reflected onto itself and is therefore unaffected: the orbital is said to be *symmetric* with respect to reflection in the mirror plane. On the other hand, the AOs on the two atoms (1 and 3) are interchanged by the reflection.

For an orbital to be classified under a symmetry operation, there can only be two possible outcomes when the operation is applied to the orbital. Either the orbital remains the same, in which case it is classified as *symmetric*, or the orbital changes sign, in which case it is classified as *antisymmetric*. So, the AO on atom 2 is clearly symmetric, but we cannot classify the AO on atom 1 as it neither remains the same nor changes sign.

However, suitable *combinations* of the AOs on the two end atoms can be classified as symmetric or antisymmetric, as is illustrated in Fig. 4.3 on the next page. If we simply add the two $1s$ AOs on atoms 1 and 3, the resulting combination is simply reflected onto itself by the mirror plane: it can therefore be classified as symmetric. On the other hand, if we take the AO on atom 1 and *subtract* that on atom 3, the resulting combination is antisymmetric, since reflection in the plane results in a sign change.

Since symmetry is going to be very important in this section, the orbitals are going to be colour-coded according to symmetry. Symmetric orbitals, or combinations of orbitals, will be orange, whereas antisymmetric orbitals will be green; orbitals yet to be classified will be grey. We also need to be able to indicate the sign of the orbital. A positive lobe of an orbital will be filled in with the appropriate colour, whereas a negative lobe will be left white and outlined with the appropriate colour.

An orbital, or combinations of orbitals, which has been classified according to symmetry is called a *symmetry orbital* (SO) or sometimes a *symmetry adapted orbital*. So, the two combinations shown in Fig. 4.3 are symmetry orbitals, one

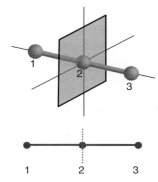

Fig. 4.1 A linear symmetrical molecule possess a mirror plane which swaps over the two end atoms, 1 and 3, but leaves the central atom, 2, in place. In the side view, shown at the bottom, the mirror plane is indicated by the green dashed line.

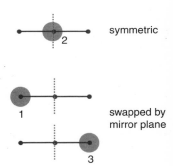

Fig. 4.2 Illustration of the effect of the mirror plane (green dashed line) on the three $1s$ AOs. The AO on atom 2 is unaffected by reflection in the mirror plane: it is said to be symmetric. The AOs on the end atoms are swapped over by the effect of reflection in the mirror plane.

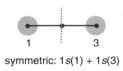

symmetric: $1s(1) + 1s(3)$

antisymmetric: $1s(1) - 1s(3)$

Fig. 4.3 Illustration of the symmetric and antisymmetric combinations of the $1s$ AOs on the end atoms. In the symmetric combination the two AOs have the same sign; reflection in the mirror plane results in the same orbital. In the antisymmetric combination, reflection gives an overall sign change. Coloured-in circles indicate a positive wavefunction whereas open circles indicate negative parts. Symmetric combinations are coloured orange, and antisymmetric ones are coloured green.

symmetric and one antisymmetric. In addition, the $1s$ AO on atom 2 is also a symmetric SO.

We are now ready to construct the MO diagram, which is shown in Fig. 4.4. On the left we have the single $1s$ orbital from the central atom, labelled SO1. As has been explained, this orbital is symmetric, and so is coloured orange. On the right we have the two SOs formed from the AOs on the end atoms. SO2 is the symmetric SO, and is therefore coloured orange, whereas SO3 is the antisymmetric SO, coloured green.

SO2 is an in-phase combination of AOs, whereas SO3 is an out-of-phase combination; therefore, SO2 is expected to be lower in energy than a $1s$ AO, whereas SO3 is expected to lie higher in energy. However, the two interacting AOs are not on adjacent atoms, but are separated by two bonds. There will therefore only be a weak interaction between the AOs, so the energy separation of SO2 and SO3 is not large.

The two symmetric SOs, SO1 and SO2, interact to give a bonding MO (MO1), which is lowered in energy, and an antibonding MO (MO3), which is raised in energy. Just as we saw before, the bonding MO arises from the in-phase (constructive) overlap of the two SOs, whereas the antibonding MO arises from the out-of-phase (destructive) overlap of the SOs. The sketches of the MOs shown in Fig. 4.4 really only show the contributions (and in particular the *sign* of the contributions) from each AO; they are not supposed to be accurate representations of the form of the MOs. Note that MO1 is bonding and has no nodes, whereas MO3 is antibonding and has two nodes. These nodes take the form of planes between atoms 1 and 2, and between atoms 2 and 3.

The antisymmetric SO3 has no other orbitals to overlap with, and is therefore unchanged as it forms MO2. This MO is weakly antibonding, on account of the destructive overlap of the two AOs. Contour plots of the computed MOs for this linear geometry are shown in Fig. 4.5 on the next page. These plots are in good agreement with the the cartoons of the MOs shown in Fig. 4.4, especially in regard to the relative signs of the contributions from the various AOs. For MO1 and MO3 the detailed calculation shows that the AO from atom 2 makes a larger contribution than the other two. This is not something we can predict simply from symmetry considerations.

Fig. 4.4 Illustration of the use of symmetry to construct the MOs of a linear H_3^+. On the right are the two SOs, one symmetric (SO2) and one antisymmetric (SO3) formed from the AOs on the end atoms. On the left we have the AO on the central atom, which is also a symmetric SO, SO1. The two symmetric SOs overlap to give a bonding MO (MO1) and an antibonding MO (MO3), shown in cartoon form. The antisymmetric SO3 has no other orbitals to overlap with, and so is unaltered, becoming MO2.

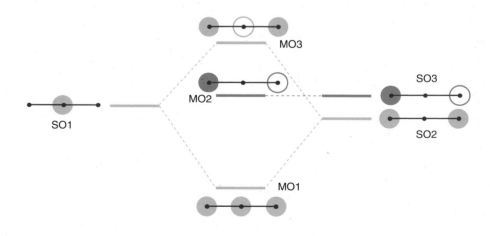

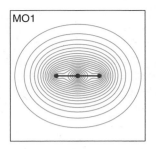
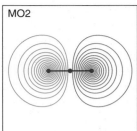
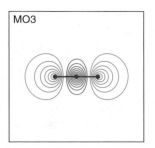

Fig. 4.5 Contour plots of the MOs computed for linear H_3^+. These MOs match closely those shown in Fig. 4.4 which were predicted using symmetry. MO1 is bonding, but MO3 is antibonding; note the nodes between atoms 1 and 2, and between atoms 2 and 3, which are a feature of antibonding MOs.

The two electrons in H_3^+ are allocated to the bonding MO, MO1, so we predict that the molecule is stable with respect to dissociation into atoms. What we have here is an example of delocalized bonding: there are two bonding electrons in an MO which is delocalized over three atoms. With two electrons spread over two bonds, this arrangement could be described as there being 'half a bond' between each adjacent hydrogen atom.

The energy ordering of the MOs reflects the number of nodes. MO1 has no nodes and is the lowest in energy, MO2 has one node (coincident with atom 2) and is next highest in energy, and finally MO3 has two nodes and is highest in energy. As this chapter proceeds, we will see this pattern over and over again: more nodes means higher energy, and more antibonding character.

4.1.2 Constructing the MOs for triangular H_3^+

We started out making the reasonable guess that H_3^+ is linear. However, spectroscopic studies have determined unambiguously that in H_3^+ the three atoms lie at the corners of an equilateral triangle of side 87 pm. Our MO diagram for the linear geometry predicts that the molecule is stable with respect to dissociation into atoms, but this does not mean that the linear geometry has the lowest overall energy.

What we are now going to do is to use the symmetry approach to construct an MO diagram for triangular H_3^+, and see if this gives us a clue as to why this is the preferred geometry.

In an equilateral triangle, all of the vertices are equivalent, so all three atoms in triangular H_3^+ are the same. This is in contrast to the linear case where it was clear that the two end atoms were in some way different from the central atom. The equilateral triangle also has many symmetry elements, so we are faced with the dilemma of which to choose for the classification of our orbitals. We will choose a mirror plane which passes through the upper vertex and the mid point of the bottom edge.

Figure 4.6 shows that under this symmetry operation, the AO on atom 1 is unaffected, whereas the AOs on atoms 2 and 3 are swapped over. Just as we did in the linear case, we can form combinations of the AOs on atoms 2 and 3 which are either symmetric or antisymmetric: these are shown in Fig. 4.7 on the next page.

SO2 is the symmetric combination, and SO3 is the antisymmetric combination. In addition, the AO on atom 1 is also a symmetric SO, denoted SO1. In SO2 there is clearly constructive overlap between adjacent atoms, so we expect

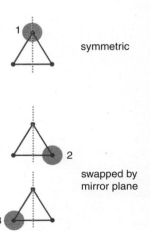

Fig. 4.6 Illustration of the effect of a mirror plane (indicated by the green dashed line) on the three $1s$ AOs at the corners of an equilateral triangle. The AO on atom 1 is reflected onto itself, and so is symmetric. However, the AOs on atoms 2 and 3 and swapped over, and so are neither symmetric for antisymmetric.

the energy of this orbital to be significantly below the energy of a $1s$ AO. In contrast, SO3 has destructive overlap between the AOs, so it will be higher in energy.

Having found the symmetry orbitals, we can now use them to construct the MO diagram; the process is shown in Fig. 4.8. On the right we have the two SOs constructed using the AOs on atoms 2 and 3, whereas on the left we have the AO on atom 1, which is also an SO. Note that SO2 is shown lower in energy, and SO3 higher in energy, than the $1s$ AO.

The two symmetric SOs, SO1 and SO2, overlap constructively to give the bonding MO1, and destructively to give the antibonding MO3. The antisymmetric SO3 has no orbitals to overlap with and so is unchanged as it becomes the antibonding MO2. Figure 4.9 on the next page shows the computed form of the MOs for this triangular geometry. Our qualitative sketches in Fig. 4.8 are a good match with the computed form of the orbitals. However, as before, our simple symmetry argument does not reveal details such as the larger contribution of the AO on atom 1 to MO3, and the fact that MO2 and MO3 turn out to have the same energy (i.e. they are degenerate).

The two electrons on H_3^+ are allocated to the bonding MO1, and so we predict that the molecule is stable with respect to dissociation into atoms. As with the linear geometry, the MO is delocalized over all three atoms. However, with two electrons and three bonding interactions (along the three sides of the triangle), we have 'one third' of a bond between each adjacent atom.

4.1.3 The optimum geometry of H_3^+

Comparing the form of the occupied bonding MOs (MO1 in both Fig. 4.9 and Fig. 4.5), we can see that in the triangular case there is bonding (electron density)

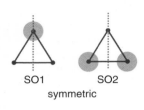

Fig. 4.7 Illustration of the symmetry orbitals of triangular H_3^+. SO1 is just the AO on atom 1, and is clearly symmetric. SO2 and SO3 are the symmetric and antisymmetric combinations of the AOs on atoms 2 and 3.

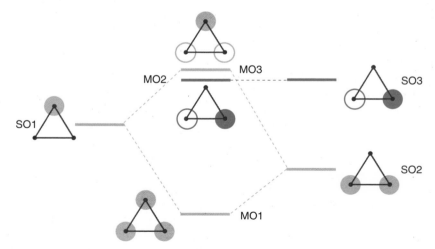

Fig. 4.8 Qualitative MO diagram for triangular H_3^+ constructed using symmetry. On the right are the two SOs, one symmetric (SO2) and one antisymmetric (SO3), formed from the AOs on atoms 2 and 3 (see Fig. 4.7). On the left we have the AO from atom 1, which is also a symmetric SO, SO1. The two symmetric SOs overlap to give a bonding MO (MO1) and an antibonding MO (MO3), shown in cartoon form. The antisymmetric SO3 has no other orbitals to overlap with, and so is unaltered as it becomes MO2. It is not possible from these simple considerations to determine the relative energies of MO2 and MO3.

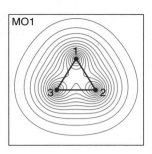 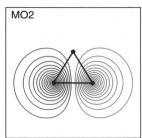 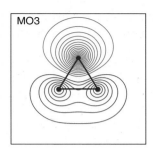

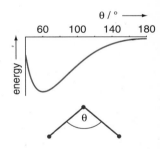

Fig. 4.9 Contour plots of the MOs computed for triangular H_3^+. These MOs match closely the predictions of our symmetry-based analysis, shown in Fig. 4.8 on the preceding page. MO1 is clearly bonding; it turns out from the calculation that MO2 and MO3 are both antibonding and also degenerate.

along all three sides of the triangle. In contrast, for the linear geometry there are bonding interactions between atoms 1 and 2, and between 2 and 3. However, on account of the large separation of atoms 1 and 3, there is little bonding interaction between these atoms. So, we might conclude that MO1 is 'more bonding' for the triangular geometry than for the linear case, and hence this is why the triangular geometry is preferred.

Of course, there is no way of telling without resorting to a detailed calculation whether or not a 'bent' geometry intermediate between the linear case and the equilateral triangle might in fact have the lowest energy. Furthermore, for any given geometry the bond lengths also needs to be adjusted in order to minimize the energy.

The programs used to compute the MOs and their energies usually also have an option to optimise the geometry of a molecule. What this means is that the program computes the energies of the occupied MOs, and hence the total electronic energy, for a series of different geometries in which the arrangement of the atoms are changed. The program then tries to 'home in' on the arrangement with the lowest total energy. For a few atoms, finding the optimum geometry is not too laborious, but the task rapidly becomes more difficult as the number of atoms, and hence the number of possible geometries, increases. Things can be helped along by starting with a reasonable guess as to the geometry, based on experiments or the precedents from related molecules.

Figure 4.10 shows a plot of the computed total electronic energy of H_3^+ as a function of the angle θ, as shown. It is clear that $\theta = 60°$ gives the lowest energy, and so is indeed the preferred geometry.

Geometry of H_3

Neutral H_3 has one more electron than H_3^+, and this electron has to be placed in MO2, as MO1 is already full. For the linear geometry, MO2 is only weakly antibonding, whereas for the triangular geometry, this MO is antibonding. As a result, we might expect that for H_3 the linear geometry is preferred, and indeed this is borne out by computer calculations.

Furthermore, the H–H bond length in linear H_3 is calculated to be 93 pm, somewhat longer than the 87 pm found for H_3^+. This is what would be expected for a system with an additional electron which is not contributing to the bonding. These speculations have to remain theoretical, though, as there is no experimental data available for neutral H_3.

Fig. 4.10 Plot of the computed total electronic energy of H_3^+ as a function of the bond angle θ, as shown; the H–H bond length on the short sides of the triangle is held fixed. The minimum energy geometry is when $\theta = 60°$ i.e. an equilateral triangle.

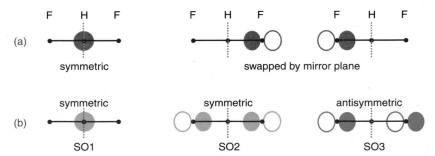

Fig. 4.11 Construction of the SOs for linear FHF⁻. The top row (a) shows the AOs: a $1s$ on the H, and a $2p_z$ on each F. The $1s$ on the central atom is already symmetric with respect to the mirror plane (shown by the dashed green line); however, the two $2p_z$ AOs are interchanged by the mirror plane. Shown in (b) are the symmetry orbitals. SO2 and SO3 are the symmetric and antisymmetric combinations of the F $2p_z$ AOs; SO1 is just the H $1s$, which is also symmetric.

4.2 More complex linear triatomics

The approach we have illustrated for H_3^+ of using symmetry to identify which combinations of AOs will overlap can be extended to more complex molecules which involve rather more orbitals. We will look at how this approach can be used to describe the bonding in FHF⁻.

The anion FHF⁻ consists of the linear arrangement F–H–F, with the hydrogen atom placed centrally between the two fluorines. The H–F bond length is 113 pm. This anion has not actually been observed in the gas phase, but it is known to exist as a discrete molecular entity in solutions of hydrofluoric acid, and in the ionic solid $K^+(HF_2)^-$.

In constructing the MO diagram for this molecule, we need to take into account the relative energies of the AOs. The F $2s$ are at −43 eV, whereas the $2p$ are at −20 eV. Given that the H $1s$ is at −14 eV, we can safely ignore any interaction between the F $2s$ and the H $1s$ on the grounds that they are too far apart in energy. The fluorine $2s$ orbitals will therefore remain unaffected by the formation of the molecule, and will become nonbonding σ-type MOs.

If we take the F–H–F internuclear axis to be the z-axis, then the F $2p_x$ and $2p_y$ will not have any interaction with the H $1s$ as they have the wrong symmetry, just as we noted for diatomic molecules in Fig. 3.14 on page 106. Therefore, the $2p_x$ and $2p_y$ on both F atoms are also nonbonding, and have π symmetry. There are thus a total of six nonbonding orbitals, three on each fluorine.

Just as we did for linear H_3^+, the remaining orbitals – the H $1s$ and the two F $2p_z$ – need to be classified according to how they behave under the mirror plane which passes through the central atom. Figure 4.11 shows the by now familiar procedure. On the top row, (a), we see the three AOs, and in (b) we see the symmetry orbitals. The two F $2p_z$ AOs combine to give SO2 and SO3, which are symmetric and antisymmetric, respectively. As in H_3^+, the overlap in SO2 is constructive, but the two atoms are quite distant from one another. So, SO2 will be a little lower in energy than a F $2p_z$ AO. Similarly, on account of the destructive overlap in SO3, it will be higher in energy.

Figure 4.12 on the next page shows the qualitative MO diagram constructed using these SOs. As before, on the left we have the orbitals of the central H atom, and on the right we have the SOs formed from the F $2p_z$ AOs. In addition, there are the six nonbonding orbitals located on the fluorines. The

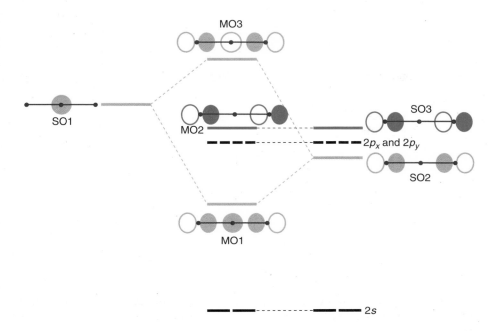

Fig. 4.12 Qualitative MO diagram for linear FHF⁻ constructed using symmetry. On the right are the two SOs, one symmetric (SO2) and one antisymmetric (SO3), formed from the $2p_z$ AOs on the fluorine atoms (see Fig. 4.11 on the preceding page), along with the nonbonding $2s$, $2p_x$ and $2p_y$ AOs. On the left we have the $1s$ AO from the central hydrogen, which is the symmetric SO1. The two symmetric SOs overlap to give a bonding MO (MO1) and an antibonding MO (MO3), shown in cartoon form. The antisymmetric SO3 has no other orbitals to overlap with, and so is unaltered as it becomes MO2. The six nonbonding orbitals are also unaffected.

two symmetric SOs, SO1 and SO2, overlap constructively and destructively to form the bonding MO, MO1, and the antibonding MO, MO3. SO3 has nothing to overlap with, and so is unaltered as it forms MO2.

There are a total of sixteen valence electrons in FHF⁻: seven from each F, one from the H, and one for the negative charge. Two of these are allocated to the bonding MO, MO1, and a further twelve are allocated to the six nonbonding orbitals on the fluorines which were described above. The final pair of electrons are allocated to the weakly antibonding MO2. We therefore predict that the molecule is stable with respect to dissociation into atoms.

Figure 4.13 shows the computed form of the bonding MO1, and the HOMO, MO2. The overall form of these orbitals corresponds closely to our qualitative picture of the MOs given in Fig. 4.12.

In this molecule, there is just *one* pair of electrons (the pair in MO1), shared across all three atoms, which is responsible for the bonding. As in H_3^+ the bonding is delocalized, and there is 'half a bond' between the H and each of the F atoms.

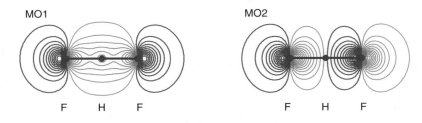

Fig. 4.13 Computed MOs for linear FHF⁻. On the left is the σ bonding MO, which corresponds to MO1. On the right is the HOMO, which corresponds to MO2 from our qualitative MO picture of Fig. 4.12.

Fig. 4.14 Construction of the symmetry orbitals for water; the molecule lies in the yz-plane, with the z-axis bisecting the H–O–H bond angle. The top row shows the valence orbitals, and beneath them are the SOs, classified according to the mirror plane shown by the green dashed line. Each oxygen AO can be classified on its own, but as before we have to form a symmetric and an antisymmetric SO from the hydrogen $1s$ AOs.

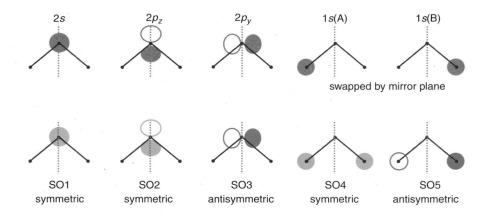

This example begins to show us the strength of the MO approach. Using the simple rules of valency it is hard to 'understand' the bonding in F–H–F⁻ as the hydrogen appears to have a valency of 2, which is not what we normally expect. However, the MO approach has no difficulties with describing this molecule and tells us that the bonding is due to an orbital which is spread across all three atoms. In other words, the bonding is delocalized.

4.3 MOs of water and methane

Our last two examples are more complex for different reasons. Water is more difficult to deal with due to its geometry, and methane is complicated on account of the larger number of atoms.

4.3.1 Using symmetry to construct an MO diagram of water

Water is 'bent', with an H–O–H bond angle of 104.5°. As we have been doing with the other triatomics, we can use a symmetry argument to help us to construct the MO diagram. In water, the two hydrogens are equivalent, and the element of symmetry that swaps them is a mirror plane that passes though the oxygen atom, and lies perpendicular to the plane of the molecule. We will therefore classify the AOs according to this mirror plane.

The valence orbitals we need to consider are the $2s$ and $2p$ on the oxygen, and a $1s$ on each of the hydrogens. It is usual to take the z-axis as passing through the oxygen atom and bisecting the H–O–H bond angle. The other two axes can be placed as we wish, and we will choose to place the molecule in the yz-plane, so that the x-axis is coming out of the plane of the molecule.

The oxygen $2p_x$ points out of the plane of the molecule, and for this orbital the yz-plane is a nodal plane. The two hydrogen $1s$ AOs lie in this nodal plane and so do not have the correct symmetry to overlap with the $2p_x$; this orbital is therefore nonbonding.

Figure 4.14 shows the classification of the remaining orbitals according to the mirror plane. The oxygen $2s$ and $2p_z$ are symmetric, whereas the $2p_y$ is antisymmetric. As before, the two hydrogen $1s$ AOs are swapped by the mirror plane, so we have to form a symmetric and an antisymmetric SO. The symmetric combination, SO4, will lie lower in energy than the antisymmetric SO5 on account of the in-phase overlap in the former.

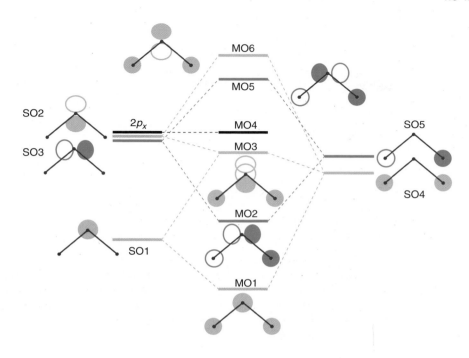

Fig. 4.15 Approximate MO diagram for water. On the left are shown the SOs for the oxygen, and on the right are the SOs for the two hydrogens. The oxygen $2p_x$ points out of the plane of the molecule, and so cannot overlap with any of the SOs: it is therefore nonbonding. There are three symmetric SOs (SO1, SO2 and SO4) which overlap to give three MOs (MO1, MO3 and MO6). MO1 is formed mainly from SO1 and SO4, and MO6 is formed mainly from SO2 and SO4; however, MO3 has significant contributions from all of the symmetric (orange) SOs. The two antisymmetric (green) SOs overlap to form the bonding MO2 and the antibonding MO5.

Using the SOs, we can draw up the approximate MO diagram shown in Fig. 4.15. The oxygen $2p_x$ cannot overlap with any of the SOs, and so is a nonbonding orbital. The two antisymmetric SOs, SO3 and SO5, overlap to give the bonding MO2 and the antibonding MO5.

There are three symmetric SOs (SO1, SO2 and SO4), which will overlap to give three symmetric MOs. This situation is a little more complex than the cases we have encountered so far in which only two SOs were overlapping. It is not possible to determine the exact form of the three MOs, or their relative energies, without resorting to a full computer calculation. However, we can make some general statements about the form of these MOs.

Firstly, there will be a bonding MO (MO1) which will lie *below* the lowest energy SO. In this case, the lowest energy symmetry orbital is SO1, so we have placed MO1 below it. This MO will have unequal contributions from all three SOs, but the major contributions will be from those which are closest to it in energy. So, MO1 will mostly be composed of an in-phase combination of SO1 and SO4.

Secondly, there will be an antibonding MO (MO6) which lies *higher* in energy than the highest energy SO. For water, SO2 is the highest energy symmetric SO, so MO6 is placed above it. Again, this MO is composed of all three SOs, but the major contributors are SO2 and SO4, as these are closest in energy to MO6. The MO is antibonding, so SO4 and SO2 are combined out-of-phase.

Finally, the third MO (MO3) lies 'somewhere in the middle', and is likely to contain significant contributions from all three SOs. The details of this MO really can only be found from a computer calculation. For H_2O, it turns out that MO3 is weakly bonding.

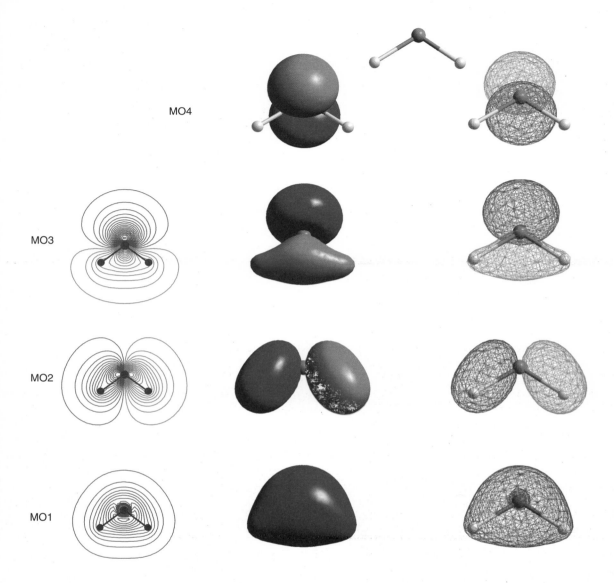

Fig. 4.16 Different representations of the occupied MOs computed for H_2O. On the left are contour plots of those MOs which lie in the plane of the molecule. In the middle are shown iso-surface representations in which the surface forms a solid object. On the right is a slightly different representation in which the iso-surface is shown as a net. For these iso-surface representations, the molecule is shown using a 'ball and stick' representation, as shown at the top.

There are eight valence electrons in H_2O: six from the oxygen and one from each of the hydrogens. All the MOs up to MO4 are thus occupied by pairs of electrons. There are thus two pairs of electrons, in MO1 and MO2, which are responsible for the bonding between the two hydrogens and the oxygen. There is one nonbonding pair in the $2p_x$ and further pair of electrons in MO3, which is weakly bonding.

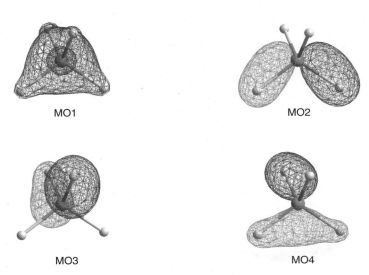

MO1

MO2

MO3

MO4

Fig. 4.17 Iso-surface representations of the occupied MOs for methane, obtained from a computer calculation. MO1 is the lowest energy of these MOs; the other three are all degenerate.

One of the interesting features of this MO picture of H_2O is that, contrary to the simple expectations, there is only one (strictly) nonbonding pair of electrons. It is true that there is another pair of electrons in MO3 which is only weakly bonding, but this pair is in quite a different orbital from the nonbonding pair in MO4. As with all of the other molecules we have been looking at, the MOs which are responsible for bonding (MO1 and MO2) are spread over all three atoms i.e. they are delocalized.

Figure 4.16 on the facing page shows the form of the occupied MOs in water which arise from a computer calculation. The MOs are represented both as iso-surfaces and, for those orbitals which lie in the plane of the molecule, as contour plots. Two different ways of representing the iso-surface are used: the first is to use the surface to create a solid object, and the second is to make the surface a net. The advantage of this latter representation is that we can see through the net to the representation of the molecule.

Our simple predictions about the MOs, shown in Fig. 4.15 on page 147, match the computed MOs quite well. MO1 is a totally in-phase orbital which encompasses all three atoms, whereas MO2 has a nodal plane down the centre. MO3 appears to be an in-phase combination of the oxygen $2p_z$ and the symmetric SO4. Finally, MO4 is the oxygen $2p_x$ AO, which lies out of the plane of the molecule.

4.3.2 MOs of methane

We are more or less at the point where increasing the complexity of the molecule much beyond something like H_2O will make our qualitative approach grind to a halt. For example, to construct an MO diagram for OF_2 we will have to deal with four valence AOs on each atom which will result in twelve MOs. Using symmetry we will be able to simplify things somewhat, but we will then be left with the problem of trying to make an intelligent guess as to the energy ordering of the MOs.

Weblink 4.1

View, and rotate in real time, iso-surface representations of the MOs of H_2O illustrated in Fig. 4.16.

🌐 *Weblink 4.2*

View, and rotate in real time, iso-surface representations of the MOs of CH_4 illustrated in Fig. 4.17.

Methane, CH_4, although it has more atoms than OF_2, has fewer orbitals. What is more, it has much higher symmetry than H_2O, and we can exploit this to help us construct the MO diagram. However, to do this we need the apparatus of *Group Theory* which you may well learn about later on in your study of chemistry, but is beyond the scope of this text. So, as the title of this section says, we have come to the end of the line when it comes to constructing MO diagrams.

Of course, we can always use a computer program to find the MOs, and this is what we have done for CH_4 to obtain the MOs shown in Fig. 4.17 on the preceding page. MO1 is the lowest in energy, which makes sense is it has no nodal planes. The calculation reveals that the other three MOs are degenerate, a fact which is not entirely obvious from looking at the pictures. Notice, too, that all the MOs are spread over several atoms i.e. the bonding is delocalized, just as we have found in every molecule we have looked at so far.

The 'problem' with these MOs for methane is that they imply a picture for the bonding in this molecule which is at odds with the simple view which you have been taught, and which just about every chemist would use on a day-to-day basis. The straightforward view of the bonding in CH_4 is that each hydrogen is bonded to the carbon via a single bond in which a pair of electrons is shared. The MO picture, with its delocalized orbitals, does not seem to fit in with this simple description.

A further difficulty is that if we have to resort to computer calculations to find the MOs, and hence a description of the bonding, in something as simple as methane, what are we going to do if we want to describe the bonding in molecules with 20, 50 or a 100 atoms? Clearly, we need a way of describing the bonding which is much simpler than full-blown MO theory, and something which fits in with the tried and tested electron-sharing view of bonding.

In the next section, we will see how the introduction of *hybrid atomic orbitals* allows us to create a picture of the bonding which is straightforward to use, and which also sits well with the simple picture in which there are distinct bonds between atoms. However, it is important to realize that along with the simplification which hybrids offer comes a considerable loss of sophistication when compared with the full MO approach. At best, a description of the bonding made using hybrids will be a rough, but nevertheless useful, guide as to what is going on.

All the detailed work we have done with MOs is not wasted though: firstly, the hybrid approach grows out of the MO picture, and secondly we will see that a complete description of some kinds of molecules is best achieved by a combination of the hybrid approach with a delocalized MO approach.

4.4 Hybrid atomic orbitals

Part of the 'problem' with the MO approach to CH_4 is that the AOs on carbon do not point towards the four hydrogen atoms. The $2s$ is spherical, so it points equally in all directions, whereas the three $2p$ AOs point at right-angles to one another.

The idea we are now going to pursue is that, since we know that CH_4 is tetrahedral, we should construct some new orbitals on the carbon that point towards the corners of a tetrahedron. Each one of these new orbitals could then overlap with just *one* of the hydrogen AOs to give a bonding and an antibonding

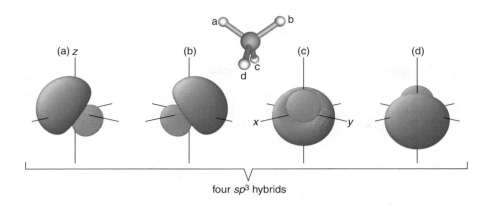

Fig. 4.18 Iso-surface plots of the four sp^3 hybrids formed from the $2s$ and the three $2p$ AOs. The four HAOs point towards the corners of a tetrahedron, as indicated by the schematic molecule at the top.

MO. Placing a pair of electrons in the bonding MO would then constitute a single bond between the C and that particular H, just as in our very simple picture.

These new orbitals on the carbon are constructed by combining the existing atomic orbitals, much in the same way that MOs are constructed. The important distinction is that we are combining AOs from the *same* atom, in contrast to when MOs are formed where we combine AOs from *different* atoms.

An orbital constructed from a linear combination of AOs on a particular atom is called a *hybrid atomic orbital*, HAO. As with MOs, combining a certain number of AOs always gives rise to the same number of HAOs. The energies and shapes (i.e. directional properties) of these HAOs depend on the precise combination of AOs used to form the HAO.

An important property of the HAOs constructed on a particular atom is that they are chosen to be *orthogonal* to one another. Orthogonality refers to a mathematical property of the HAO wavefunctions, the practical consequence of which is that different HAOs have no net overlap with one another, and so tend to point in different directions.

We will not go into the mathematical details of how HAOs are constructed, but will look at some particular sets of HAOs which are useful for describing the bonding in commonly encountered molecules, particularly organic molecules.

4.4.1 sp^3 hybrids

The $2s$ and the three $2p$ AOs can be combined to form four *equivalent* HAOs, known as sp^3 hybrids. These hybrids are equivalent in the sense that they all have the same energy and the same overall shape: all that is different between them are the directions in which they point. Iso-surface plots of these four hybrids are shown in Fig. 4.18, and a contour plot taken in a plane through one of these hybrids is shown in Fig. 4.19.

Each HAO has a major lobe and a minor lobe, aligned along a particular direction. To have the best overlap with this HAO, an orbital must approach the major lobe head on, so the HAO is 'more directional' than the AOs from which it is formed. The four hybrids point towards the corners of a tetrahedron, with an angle of 109.5° between the any two hybrids. The energy of these HAOs lies between that of the $2s$ and $2p$ AOs from which they are formed, and since the proportion of $2p$ in the hybrid is greater than that of $2s$, the HAO lies closer in energy to the $2p$.

Weblink 4.3

View, and rotate in real time, iso-surface representations of the four sp^3 HAOs illustrated in Fig. 4.18.

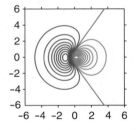

Fig. 4.19 Contour plot of a cross section taken through one of the sp^3 hybrids shown in Fig. 4.18; the scale is in units of Bohr radii. The HAO is directional in the sense that its major lobe is pointing in a particular direction (here to the left). We therefore expect the strongest bonding to be with appropriate orbitals on an atom located on the left.

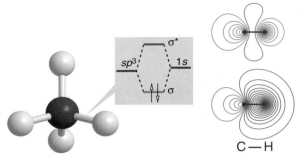

Fig. 4.20 Description of the bonding in CH_4 using sp^3 HAOs. Each sp^3 hybrid overlaps with the H $1s$ AO towards which it points, thus giving rise to σ bonding and antibonding MOs, as shown in the tinted box. Contour plots of the two MOs are also shown. Two electrons are placed in the bonding MO, giving what is called a two-centre two-electron bond between the carbon and one of the hydrogens. An identical description is used for each of the C–H bonds.

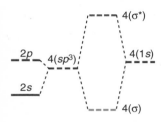

Fig. 4.21 MO diagram for CH_4. On the left we see the carbon AOs combining to give four sp^3 hybrids. These overlap, one on one, with the hydrogen $1s$ AOs to give four σ and four $\sigma^\star$ MOs.

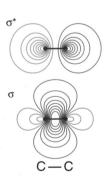

Fig. 4.22 Contour plots of the σ and $\sigma^\star$ MOs formed from the overlap of two sp^3 HAOs. Occupation of the σ MO by a pair of electrons results in a C–C single bond.

We can immediately put these HAOs to good use to describe the bonding in methane, CH_4. Each HAO overlaps with the $1s$ AO of the hydrogen to which it points, giving rise to a bonding and an antibonding MO. The key thing about these bonding MOs is that they are localized between the carbon and just one of the hydrogens. Placing two electrons in each of these bonding MOs thus creates four localized bonds, as is illustrated schematically in Fig. 4.20. All eight valence electrons in CH_4 are located in bonding MOs, which accords with the known stability of this structure.

Figure 4.21 shows an alternative way of depicting the relationship between the AOs, the HAOs and the MOs. On the left we have the $2s$ and $2p$ AOs of carbon; these combine together to give the four sp^3 hybrids. The energy of these hybrids falls between the energy of the AOs from which they are formed, but lies closer to the energy of the $2p$ AOs as these make a greater contribution to the hybrids. Each hybrid overlaps with a hydrogen $1s$ AO to give a σ and a $\sigma^\star$ MO. There are therefore four equivalent bonding, and four equivalent antibonding MOs.

In contrast to the MO descriptions of polyatomics we have seen so far, these bonding MOs are only spread across *two* atoms, rather than across all the atoms in the molecule. The MOs are thus localized, and we can associate each bonding MO, and the pair of electrons it contains, with a single bond between the carbon and one of the hydrogens. Such a bond is called a *two-centre two-electron* (2c–2e) bond.

These sp^3 hybrids can be used to describe the bonding in larger hydrocarbons. For example, in ethane (CH_3–CH_3), three of the hybrids on each carbon overlap with H $1s$ AOs to give six 2c–2e bonds. The remaining hybrid on each carbon overlap with one another to form a bonding and an antibonding MO, contour plots of which are shown in Fig. 4.22. By placing two electrons in the bonding MO, we obtain another 2c–2e bond between the two carbons. You can see how this idea can be extended to describe any hydrocarbon network.

The MOs formed from the overlap of sp^3 hybrids have cylindrical symmetry about the bond axis, and so are termed σ. The bonds which result from filling these MOs are therefore called σ bonds.

4.4.2 Doubly-bonded carbon: sp^2 hybrids

Carbon atoms to which there is a double bond generally adopt a planar geometry in which the angles between adjacent bonds around the carbon are 120°. The simplest example of this is ethene C_2H_4, shown as a three-dimensional view and also drawn in the conventional way in Fig. 4.23.

To describe the bonding in this molecule we need a set of hybrids that point at 120° to one another, and lie in a plane. We can see immediately that these hybrids will be formed from the $2s$ and two of the $2p$ AOs which lie in the plane. The third $2p$ AO points out of the plane, and has the wrong symmetry to overlap with the hydrogen $1s$ AOs.

The required set of HAOs formed from the $2s$ and *two* of the $2p$ AOs are called sp^2 hybrids. Figures 4.24 and 4.25 show pictures of these sp^2 hybrids formed from the $2s$, $2p_x$ and $2p_y$ AOs. The three HAOs lie in the xy-plane and are directed to the corners of an equilateral triangle, as required. The $2p_z$ AO is not involved in the formation of the HAOs, and lies perpendicular to the plane of the hybrids.

We can use these HAOs to describe the bonding in ethene in the following way, as illustrated in Fig. 4.26 on the next page. Each C–H bond is formed from the overlap of one sp^2 HAO with the hydrogen $1s$ towards which it points: this is just the same as in CH_4 except that a different hybrid is involved. The remaining HAOs on each carbon overlap with one another to form a bonding and an antibonding σ MO; putting a pair of electrons in this MO results in a C–C σ bond. Occupation of the C–C σ bonding MO, and the four C–H σ bonding MOs accounts for ten out of the total of twelve valence electrons.

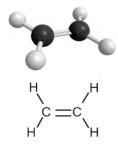

Fig. 4.23 Ethene, C_2H_4 is a planar molecule, with H–C–H and H–C–C bond angles of 120°. At the top is shown a three-dimensional representation, and below is shown the usual way in which this molecule is drawn.

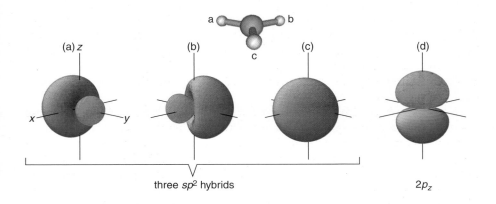

three sp^2 hybrids $2p_z$

Fig. 4.24 Iso-surface representations of the three sp^2 HAOs formed from the $2s$, the $2p_x$ and $2p_y$ AOs; the $2p_z$ AO which is not involved in the hybridization, is also shown on the far right. The three HAOs all lie in the xy plane and point at 120° to one another, as shown by the schematic molecule at the top.

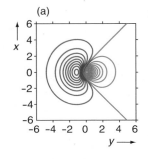
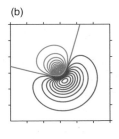
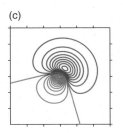

Fig. 4.25 Contour plots, taken in the xy-plane, of the three sp^2 HAOs shown in Fig. 4.24. The major lobe of each HAO points in a particular direction, with an angle of 120° between these directions.

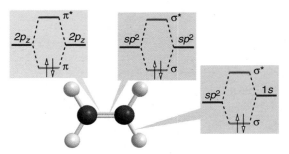

Fig. 4.26 Schematic description of the bonding in ethene using HAOs. Each C–H bond arises from the occupation of the σ bonding MO formed from an sp^2 hybrid and a hydrogen $1s$. Another σ bond is formed from the overlap of an sp^2 HAO on each carbon. Finally, a π bond arises from the occupation of the π bonding MO formed from the overlap of the $2p_z$ AOs on each carbon.

🌐 *Weblink 4.4*

View, and rotate in real time, iso-surface representations of the three sp^2 HAOs and one unhybridized $2p$ AO illustrated in Fig. 4.24.

The two out-of-plane $2p_z$ AOs overlap to give a π bonding and a π anti-bonding MO, just like the π MOs we saw in diatomics. Placing two electrons in the bonding MO results in a π bond between the two carbon atoms, and accounts for all twelve valence electrons.

As we generally expect σ bonding orbitals to lie lower in energy that π bonding orbitals (between the same atoms), the HOMO is likely to be the C–C π bonding MO. Similarly, as $\sigma^\star$ (antibonding) orbitals are generally higher in energy than $\pi^\star$ orbitals, the LUMO is likely to be the C–C π antibonding MO. As we shall see later on, the reactivity of this molecule is thus very much focused on the π system, as this is where both the LUMO and HOMO are located.

One of the important differences between the carbon-carbon bonds in ethane and in ethene is that it is easy to rotate about the C–C bond in ethane, whereas in ethene the two CH_2 groups are held in the same plane. We can now see why this is. In ethane, the bond between the carbons is the result of the occupation of the σ-type MO, which has cylindrical symmetry about the bond axis. Rotation about this bond does not, to a first approximation, alter the degree of overlap between the two sp^3 hybrids.

In contrast, in ethene one of the C–C bonds arises from the occupation of a π MO, which is formed from the overlap of two out-of-plane $2p_z$ AOs. As is shown in Fig. 4.27, if we attempt to twist one of the CH_2 groups out of the

Fig. 4.27 Illustration of the effect of rotation about the C–C bond of ethene on the π MO formed from the overlap of the $2p_z$ AOs. In the planar form, (a), there is good overlap between the $2p$ AOs, but if one of the CH_2 groups is twisted by 45°, as shown in (b), there is clearly poorer overlap. Twisting the group by 90°, shown in (c), results in essentially no overlap between the AOs: the π bond is therefore broken.

(a)   (b)   (c)

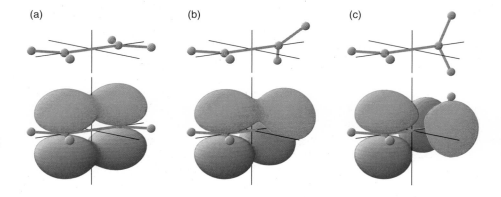

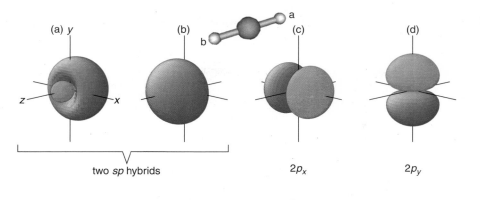

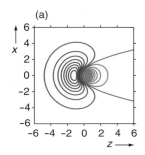

Fig. 4.28 Iso-surface representations of the two *sp* HAOs formed from the 2*s* and the 2*p_z* AOs; the other two 2*p* AOs which are not involved in the hybridization are also shown. The two HAOs point at 180° to one another; the unhybridized 2*p* orbitals point at right angles to the hybrids.

Fig. 4.29 Contour plots, taken in the *xz*-plane, of the two *sp* HAOs shown in Fig. 4.28, along with one of the unhybridized 2*p* AOs. The major lobe of the HAOs point in opposite directions.

plane, then the lobes of the two $2p_z$ AOs will no longer be parallel, and so they will not overlap so well. In the limit that the CH_2 group is rotated by 90°, there will be no overlap between the 2*p* AOs, and so no π bond will be present.

Thus, rotating about the C–C double bond in ethene is very difficult as it leads to the progressive breaking of the π bond between the two carbons. Experimental work indicates that the barrier to rotation about the C–C double bond in ethene is around 270 kJ mol^{-1}. Given that so much energy is needed, to all intents and purposes we can assume that there is no rotation about a C–C double bond.

4.4.3 Triply-bonded carbon: *sp* hybrids

In a carbon with a triple bond, the remaining single bond is found to point in the opposite direction to the triple bond i.e. the bond angle is 180°. The simplest example of this is ethyne (acetylene), HCCH, which is a linear molecule containing a triple bond between the two carbons.

To describe the bonding in this molecule we need hybrids on the carbon which point at 180° to one another, and these are made by hybridizing the 2*s* with just the $2p_z$ (we are assuming that the internuclear axis is along *z*). The resulting two hybrids are called *sp* hybrids, and are shown as iso-surface plots in Fig. 4.28 and as contour plots in Fig. 4.29. The major lobes of the two *sp* hybrids point at 180° to one another. The $2p_x$ and $2p_y$ AOs are not involved in the hybridization, and point perpendicular to the two hybrids.

Using these hybrids the bonding in ethyne can be described in the following way (illustrated in Fig. 4.30 on the next page). The C–H σ bonds arise from the overlap of an *sp* HAO with a hydrogen 1*s*. Between the two carbons there is a

Weblink 4.5

View, and rotate in real time, iso-surface representations of the π MOs illustrated in Fig. 4.27.

Weblink 4.6

View, and rotate in real time, iso-surface representations of the two *sp* HAOs and two unhybridized 2*p* AOs illustrated in Fig. 4.28.

Fig. 4.30 Schematic description of the bonding in ethyne using HAOs. The two C–H bonds arise from the occupation of the σ bonding MO formed from an sp hybrid and a hydrogen $1s$. Another σ bond is formed from the overlap of an sp HAO on each carbon. There are two π bonds which arise from the occupation of the π bonding MOs formed from the overlap of the $2p_x$ AOs, and the $2p_y$ AOs.

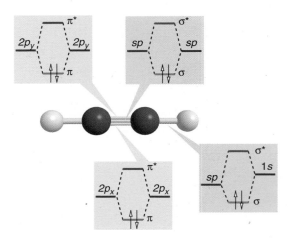

triple bond: a σ bond arising from the overlap of the two remaining sp hybrids, and two π bonds arising from the overlap of the $2p_x$ and $2p_y$ AOs. Ethyne has ten valence electrons, all of which can therefore be accommodated in bonding MOs.

As with ethene, the HOMO is the C–C π bonding MO, and the LUMO is the $\pi^\star$ MO. So, the reactivity of this molecule is centred on the π system once more.

4.4.4 Summary

We have seen in this section that by a suitable choice of HAOs we can describe the bonding in simple hydrocarbons in terms of 2c–2e bonds. The choice of HAO is determined by the bond angles we require around a particular carbon:

- For bond angles of 109.5°, a tetrahedral arrangement, choose sp^3 hybrids.

- For bond angles of 120°, a trigonal planar arrangement, choose sp^2 hybrids.

- For bond angles of 180°, a linear arrangement, choose sp hybrids.

4.5 Comparing the hybrid and full MO approaches

Before we carry on with using the HAO approach, it is useful to compare the picture that it generates of the bonding with that provided by a full MO treatment. We will make this comparison for N_2, whose MOs have already been described in detail: the energy level diagram is given in Fig. 3.31 (b) on page 119, and the MOs themselves are depicted in Fig. 3.42 on page 132.

If the nitrogen is sp hybridized, then we realise that on each of the nitrogen atoms one of the hybrids points towards the other nitrogen, whereas one points in the opposite direction. The two hybrids which point towards one another will overlap to give a σ MO and a $\sigma^\star$ MO. The two hybrids which point away from one another will not overlap significantly, and so are nonbonding orbitals. The $2p$ orbitals which were not involved in forming the hybrids will overlap to give a pair of π MOs and a pair of $\pi^\star$ MOs, just as in ethyne.

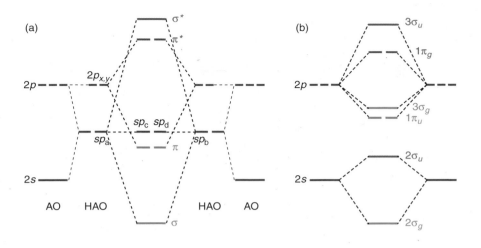

Fig. 4.31 Two different MO diagrams for N_2. MO diagram (a) is constructed by first forming sp HAOs on the nitrogen atoms, and then letting these overlap. The two HAOs which point towards one another (sp_a and sp_b) overlap to give a σ and a $\sigma^\star$ MO. The other two HAOs (sp_c and sp_d) remain nonbonding. The unhybridized $2p$ AOs overlap to give a degenerate pairs of π and $\pi^\star$ MOs. Diagram (b) is the same as that from Fig. 3.31 (b) on page 119. The relative energies of the MOs in the two diagrams is not significant.

Figure 4.31 (a) shows the resulting MO diagram. On the extreme right and left, we have the nitrogen AOs. Next, moving toward the centre we have the sp hybrids; note that, not surprisingly, the energy of each hybrid lies midway between that of the $2s$ and $2p$ AOs from which it is formed. The two hybrids which point towards one another are denoted sp_a and sp_b, whereas those which point away from one another are denoted sp_c and sp_d. In the middle we have the MOs, which have already been described. The relative energies of the MOs are just guesses, and no great significance should be attached to them.

There are ten valence electrons in N_2, and these occupy the σ MO, the two degenerate π MOs, and the two nonbonding HAOs sp_c and sp_d. The picture we obtain of the bonding is therefore that there is a σ bond and two π bonds between the two atoms, together with a nonbonding lone pair on each nitrogen. The HOMO is one of these nitrogen lone pairs, and the LUMO is the $\pi^\star$ MO.

Figure 4.31 (b) is the MO diagram of N_2 which we have described before. The occupied orbitals are the $2\sigma_g$ (bonding), the $2\sigma_u$ (weakly antibonding), the $1\pi_u$ (bonding) and the $3\sigma_g$ (weakly bonding – the HOMO).

There are some similarities between the two descriptions of the bonding in N_2. Both predict the presence of two π bonds, and that the LUMO is a π^* orbital. However, there are more significant differences when it comes to the σ bonding, and in the prediction by the hybrid approach that there are two lone pairs. You will recall that the photoelectron spectrum (Fig. 3.37 on page 126) provides good experimental evidence in support of the MO picture, so we are forced to the conclusion that the hybrid picture of the bonding is deficient, certainly in matters of detail. This simple example serves to highlight the important point that the simplicity of the hybrid approach comes at the price of a loss of sophistication when compared to the full MO treatment.

You may be wondering why we chose sp hybrids on the nitrogen atoms, rather than sp^2 or sp^3 hybrids. Had we chosen sp^2 hybrids we would have ended up with a σ bond, a single π bond, and then six electrons to distribute amongst four nonbonding sp^2 hybrids. Presumably these would be arranged as two pairs in two of the HAOs, and then one electron in each of the remaining HAOs. In other words, we would predict there to be unpaired electrons in N_2. This is clearly not correct, and probably therefore a good reason not to choose sp^2 hybrids, but had we not know that N_2 had no unpaired electrons, we would not have been able to chose between sp and sp^2 hybrids.

Throughout the rest of this chapter, and indeed the rest of the book, we will frequently use HAOs to describe the bonding in molecules. However, we have to be aware that the resulting descriptions will only be approximate, and may not be adequate for some purposes. If we need something more precise, we can always make a full MO calculation, but this will require access to a computer, rather than the pencil and paper we need for the hybrid approach.

4.5.1 More about lone pairs

This comparison between the HAO and MO pictures of the bonding in N_2 prompts us to think a little more about how the familiar concept of a lone pair can be reconciled with the MO picture. In this picture some orbitals are strictly nonbonding, usually as a result of not having the correct symmetry to overlap with other orbitals e.g. the out-of-plane $2p_x$ AO in H_2O (Fig. 4.15 on page 147). A pair of electrons occupying such an orbital is certainly a lone pair in the sense that it localized on an atom and is not involved in bonding.

However, there is another way in which something akin to a lone pair can arise. Suppose that *both* a bonding MO and its antibonding partner are occupied by pairs of electrons. As far as the bonding between the atoms goes, these two pairs of electrons essentially cancel one another out, so in effect both pairs of electrons are nonbonding. We could therefore describe the occupation of the bonding MO and its antibonding partner as giving rise to two lone pairs.

In a diatomic, each MO is spread across both atoms, and the contributions from the AOs on each atom may well not be the same. However, if atom A is the major contributor to the bonding MO, then atom B will be the major contributor to the antibonding MO. As a result, if both MOs are occupied there is an even electron distribution between the two atoms. It is therefore reasonable to describe this situation as resulting in two lone pairs.

Applying this idea to the MO diagram of N_2, we would argue that the occupied bonding $2\sigma_g$ and antibonding $2\sigma_u$ MOs (which originate from the same AOs) give rise to two lone pairs. The pair of electrons occupying the $3\sigma_g$ are bonding, as are the two pairs occupying the $1\pi_u$. We thus have the familiar picture for N_2 a triple bond (from the three bonding pairs) and one lone pair on each nitrogen.

In O_2 there is an additional pair of electrons in the $1\pi_g$ antibonding MOs, and we can say that these cancel out the effect of one of the pairs of electrons in the $1\pi_u$ bonding MOs, giving a lone pair on each oxygen. Recall that there is already a lone pair on each atom arising from the occupation of the $2\sigma_g$ and $2\sigma_u$ MOs, so in total there are two lone pairs on each oxygen. There is one bonding pair in the $3\sigma_g$ and one pair in the $1\pi_u$ which has not been 'cancelled' by a pair in the $1\pi_g$, giving a σ bond and a π bond. This is in line with the simple picture we have of the bonding in O_2.

Finally, for F_2 there is a further pair of electrons in the $1\pi_g$ antibonding MOs. The four electrons in the $1\pi_g$ cancel out the bonding effect of the four electrons in the $1\pi_u$, giving rise to two lone pairs on each fluorine. There is already one lone pair from the occupation of the $2\sigma_g$ and $2\sigma_u$, so the overall picture is of three lone pairs on each fluorine, and just a single σ bond between the atoms.

We will come back to this idea of lone pairs arising from pairs of occupied bonding and antibonding MOs when we consider the bonding in the elements in section 7.6.3 on page 284.

4.6   Extending the hybrid concept

The hybrids we have described so far are ideal for describing the bonding in hydrocarbons as the angles between the hybrids match the usual bond angles found at carbon. What we now want to do is to extend the hybrid approach so that we can describe the bonding around atoms which adopt bond angles other than 109.5°, 120° or 180°. We will see that by choosing the correct ratio of $2s$ to $2p$ we can arrange for the angle between two hybrids to have any value between 90° and 180°.

The hybrids that have been described above are all examples of *equivalent hybrids*. What this means is that, apart from the fact that they point in different directions, the four sp^3 HAOs all have exactly the same shape. This comes about because the relative proportions of $2s$ and $2p$ orbitals is the same in each hybrid. The same is true for the three sp^2, and the two sp hybrids.

As we have seen, the angle between the hybrids is 109.5° for sp^3, 120° for sp^2 and 180° for sp. There is a clear trend here, which we can describe by introducing the concept of the 's character' of a hybrid.

Recall that sp^3 hybrids are, as their name implies, formed from the $2s$ and the three $2p$ AOs; therefore the $2s$ and $2p$ contribute in the ratio 1:3. This means that 25% of a hybrid comes from the $2s$, which we describe by saying that the hybrid has 25% s character. In a similar way we can describe an sp^2 hybrid as having 33% s character, as these hybrids are formed from one $2s$ and two $2p$ AOs. An sp hybrid has 50% s character.

As we move along the series of hybrids from sp^3 to sp^2 to sp both the angle between the hybrids and the s character increase. This relationship can be made quantitative, as is illustrated in Fig. 4.32. This graph also indicates that simply by selecting the correct s character we can achieve any angle we like between the hybrids, a point which we are about to exploit.

When we form four HAOs from the $2s$ and three $2p$ AOs, we do not have to make all four of the resulting HAOs equivalent. We can, if we wish, have a different proportion of $2s$ in each hybrid, and therefore have different angles between different hybrids. The only restriction on the HAOs is that, when taken together, the overall proportion of $2s$ must be 25%, and the hybrids must be orthogonal to one another. We will look at how this approach can be used to describe the bonding in two simple cases: water and ammonia.

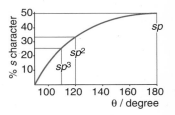

Fig. 4.32 Plot showing the percentage s character needed in a pair of hybrids as a function of the angle θ between them. As the angle increases, the required s character increases. The values for the equivalent sp^3, sp^2 and sp hybrids are shown.

4.6.1   Water

The H–O–H bond angle in water is 104.5°. To make a localized description of the bonding we will need two HAOs on the oxygen with this angle between them so that they point towards the hydrogens. The angle between equivalent sp^3 hybrids is 109.5°, which is a little too large, but we can reduce this simply by decreasing the s character of *two* of the hybrids. In fact, according to Fig. 4.32, we need to reduce the s character to 21% to give the desired angle between the hybrids.

Figure 4.33 on the following page shows how the two resulting hybrids, HAO1 and HAO2, point directly at the hydrogen atoms. Each of these hybrids can then overlap with the hydrogen $1s$ to which they point, giving a bonding and an antibonding σ MO. Two electrons are placed in each of the bonding MOs, thus forming the two 2c–2e O–H bonds.

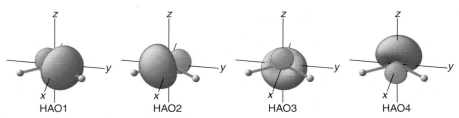

Fig. 4.33 Iso-surface plots of the four oxygen HAOs needed for a localized description of the bonding in water; the water molecule is shown schematically, and lies in the xy plane. These hybrids are very similar to the (equivalent) sp^3 hybrids shown in Fig. 4.18 on page 151, except that the s character of HAO1 and HAO2 has been decreased so that the angle between them is $104.5°$; they therefore point directly at the hydrogens. The s character of the other two HAOs has been increased, so that the angle between them is greater than the tetrahedral angle. Note that these latter two HAOs point away from the hydrogens.

Weblink 4.7

View, and rotate in real time, iso-surface representations of the four approximate sp^3 HAOs, illustrated in Fig. 4.33, which have been adjusted to match the bond angle in H_2O.

As we have decreased the s character in two of the hybrids, the s character of the other two hybrids must be increased in such a way that the total proportion of $2s$ in the HAOs is 25%. According to Fig. 4.32 on the previous page, this means that the angle between the other two hybrids will increase. As can be seen from Fig. 4.33, these hybrids point away from the hydrogens, and so are not involved in bonding. Each is occupied by a pair of electrons, which can be classified as lone pairs.

In summary, our localized description of the bonding in water is that the oxygen is *approximately* sp^3 hybridized: two of the hybrids overlap with the hydrogen $1s$ to form O–H bonds, and the other two hybrids are occupied as lone pairs. This accounts for all eight valence electrons. The hybridization is described as being 'approximately' sp^3 on account of the small adjustment needed to the hybrids in order to make them point in the right directions.

The O–H σ bonding MO will lie below the energy of the HAOs on the oxygen, so the HOMO will be one of the HAOs which has remained nonbonding i.e. one containing a lone pair. The LUMO will be the O–H $\sigma^\star$ antibonding MO. As we expect the oxygen orbitals to be lower in energy than the hydrogen, the major contributor to the σ bonding MO will be from the oxygen, whereas the major contributor to the $\sigma^\star$ MO (the LUMO) will be the hydrogen.

An alternative view of water

This localized description of the bonding in water starting from sp^3 hybrids is not the only possibility. Another approach is to start with the three sp^2 hybrids which lie in the plane of the molecule, and then increase the s character of two of these hybrids until the angle between them in the required $104.5°$. These two hybrids can then overlap with the hydrogen $1s$ to form the two O–H bonds.

The remaining hybrid has increased s character, and points away from the hydrogens. This HAO is occupied by two electrons to form a lone pair. Finally, a pair of electrons is placed in the unhybridized $2p$ orbital which lies out of the plane of the molecule: this forms the second lone pair.

What differs between these two descriptions of water is the nature of the orbitals in which the lone pairs are located. In the first description, based on sp^3 hybrids, the lone pairs are in two identical HAOs. In the second description, based on sp^2 hybrids, one lone pair is in an out-of-plane $2p$ orbital, and one is in an HAO which lies in the plane.

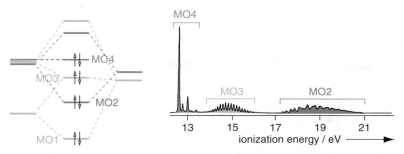

Fig. 4.34 Photoelectron spectrum of water, recorded using UV photons with an energy of 21.2 eV. The MO diagram from Fig. 4.15 on page 147 is also shown. The lowest energy band is assigned to MO4, and the absence of vibrational structure indicates that this MO is nonbonding. The orange band is assigned to MO3, and the presence of vibrational structure indicates that this MO is bonding. The green band, with its extensive vibrational structure, is assigned to the bonding MO2. The 21.2 eV photons are insufficiently energetic to ionize the electrons from MO1.

Which scheme is correct?

We now have three different descriptions of the bonding in water: the MO approach, given in section 4.3.1 on page 146, and two localized approaches based on different hybridization schemes. The question is, which of these descriptions is 'correct', or at least which is 'the best'?

One way of answering this question is to look at some experimental evidence in the form of the PES, which is shown in Fig. 4.34, along with the MO diagram from Fig. 4.15 on page 147. The lowest energy band can be assigned to MO4, and the lack of vibrational structure is consistent with MO4 being nonbonding, just as we expected.

The next band (coloured orange) shows significant vibrational structure, confirming that MO3 has some bonding character. Finally, the green band, which shows extensive vibrational structure, is assigned to the bonding MO2. There is no band corresponding to the ionization of the electrons in MO1 because the binding energy of these electrons is greater than the energy of the photons used to record the photo electron spectrum.

The PES is therefore in accord with the MO diagram we drew up earlier. However, the spectrum is *not* in accord with the description of the bonding using either sp^3 or sp^2 hybrids, as these schemes both predict that there will be two lone pairs (equivalent in the case of sp^3 hybrids).

As we saw before for the case of N_2, the picture of the bonding produced by the hybrid approach does not agree in detail with the more sophisticated MO treatment. Nevertheless, the hybrid approach does capture the key points about the bonding in water, which we can appreciate by looking at the pictures of the computed MOs shown in Fig. 4.16 on page 148.

MO1 and MO2 are responsible for the most of the bonding between the oxygen and the hydrogen. If we add MO1 and MO2, we end up with an orbital which is mainly located between the oxygen and the left-hand hydrogen: this is equivalent to the bonding MO formed when the oxygen HAO which points to the left overlaps with the $1s$ orbital on the left-hand hydrogen. Similarly, if we subtract MO1 and MO2 we end up with an orbital between the oxygen and the right-hand hydrogen. This orbital is analogous to the bonding MO formed between the other oxygen HAO and the right-hand hydrogen AO.

Taken together MO1 and MO2 can be regarded as being equivalent to the two localized bonding MOs.

In the MO approach, MO4 is nonbonding and MO3 is only weakly bonding. So although we do not have the two lone pairs predicted by the hybrid approach, we do have two pairs of the electrons which are not making a large contribution to the bonding.

The question arises as to which hybridization scheme we should choose to describe the bonding in water: sp^3 or sp^2? In a sense it does not matter too much as both will only give an approximate picture when compared to the full MO treatment. Using sp^3 hybrids is attractive as it predicts that there are two equivalent lone pairs, which fits in with the simple picture of the bonding in water which most chemists first encounter. However, we must not let this familiarity blind us to the fact that both the full MO treatment, and the photoelectron spectrum, clearly indicate that there are *not* two equivalent lone pairs in water.

4.6.2 Ammonia

Ammonia, NH_3, has a trigonal pyramidal structure, with an H–N–H bond angle of 107.8°; there are a total of eight valence electrons. Just as we did with water, we can make a localized description of the bonding by starting with sp^3 hybrids on the nitrogen. The angle between these hybrids is too large, so we need to increase the s character in three of them until the angle is the required 107.8°.

These three HAOs overlap with the hydrogen $1s$, and two electrons are placed in the resulting σ bonding MOs to give the three N–H bonds. The fourth hybrid points away from the other three and does not interact with the hydrogens. It is occupied by two electrons to give a lone pair.

As with water, the HOMO is the lone pair on the nitrogen, and the LUMO is one of the the N–H $\sigma^\star$ MOs, to which the hydrogen is the major contributor.

4.7 Bonding in organic molecules

In this section we will illustrate how the hybrid approach can be used to describe the bonding in organic molecules. As we have seen for simpler molecules, we have to realize that the resulting picture of the bonding will only be approximate. Nevertheless, it will be a useful start.

When we move on to looking at the reactions of molecules with one another we will find that the form of the HOMO and LUMO are quite important, so in each case we will try identify these important orbitals.

4.7.1 Halides, alcohols and amines

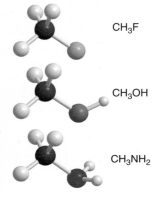

CH₃F

CH₃OH

CH₃NH₂

Three simple examples of these compounds which have only single bonds are shown opposite: all have a methyl group, CH_3, attached to a functional group which contains a heteroatom (i.e. an atom other than carbon or hydrogen). In methanol the C–O–H bond angle is 108.5°, and the coordination around the nitrogen in methylamine is trigonal pyramidal.

In all of these compounds, the arrangement of atoms around the carbon is close to tetrahedral, so we will assume that the carbon is sp^3 hybridized. Three of these hybrids overlap with the three hydrogen $1s$ AOs to give three localized

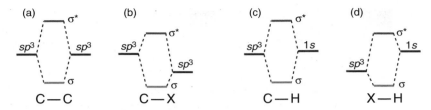

Fig. 4.35 MO diagrams showing the comparison between the MOs formed between different combinations of carbon, an electronegative heteroatom (X), and hydrogen. On account of the lower energy of the hybrid orbitals on X, the C–X $\sigma^\star$ MO lies below the corresponding MO from C–C. Similarly, the X–H $\sigma^\star$ is lower in energy than the corresponding C–H orbital. These diagrams indicate that it likely that the LUMO will be one of the $\sigma^\star$ orbitals involving a bond to X.

C–H bonds. It is probably most convenient to imagine that the heteroatom X (which may be F, O or N) is approximately sp^3 hybridized. One of these hybrids overlaps with the remaining sp^3 hybrid on the carbon, to form the C–X localized bond.

In CH_3F the remaining three sp^3 hybrids on the fluorine are occupied by pairs of electrons to give the expected three lone pairs. In CH_3OH, one of the sp^3 hybrids on the oxygen overlaps with the hydrogen $1s$ to give an O–H localized bond; the remaining two hybrids contain the lone pairs. Finally, in CH_3NH_2, two of the hybrids on nitrogen are used to form localized N–H bonds, and the third is occupied to give a lone pair.

We could just as well have assumed that the oxygen or the fluorine was sp^2 hybridized. However, the trigonal pyramidal geometry around the nitrogen indicates that sp^3 hybridization is most appropriate for this atom.

In all of these molecules, the HOMO is clearly one of the heteroatom HAOs which is not involved in bonding i.e. one which contains a lone pair. As to the LUMO, there are three $\sigma^\star$ antibonding MOs associated with the C–H, X–H and C–X bonds which are all possible candidates.

Figure 4.35 shows schematic MO diagrams for localized C–C, C–X, C–H and X–H bonds, where X is an electronegative atom. The HAO on the X atom lies lower in energy than either the HAO on the carbon, or the hydrogen $1s$. As a consequence, the MOs for X–H and C–X are not shifted so far from the HAO energies as they are for the case of C–H and C–C. It therefore follows, as is shown in the diagram, that the C–X $\sigma^\star$ lies at lower energy than the corresponding MO for C–C; similarly, the X–H $\sigma^\star$ is lower in energy than the corresponding orbital in C–H.

In CH_3F the LUMO must therefore be the C–F $\sigma^\star$. Since this MO is closest in energy to the carbon sp^3 HAO, the major contributor to the MO is the carbon HAO. In CH_3OH and CH_3NH_2 there are two possible candidates for the role as LUMO: either the C–X or X–H $\sigma^\star$ (where X is O or N). It is not possible to say, without a detailed calculation, which of these MOs is the LUMO in a particular case. The major contributor to the X–H $\sigma^\star$ is the hydrogen $1s$, and to the C–X $\sigma^\star$ is the carbon HAO.

4.7.2 Aldehydes, ketones and imines

Next we turn our attention to organic compounds in which there is a double bond to a heteroatom. This group includes ketones and aldehydes which both

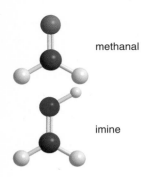

methanal

imine

contain the carbonyl group, C=O, and imines which contain a doubly-bonded C=N linkage. Two simple examples are given in the margin: methanal (CH_2O) and the corresponding imine (CH_2NH).

Methanal is planar with bond angles of approximately 120° around the carbon: it is therefore appropriate to use sp^2 hybrids on the carbon. Two of these HAOs overlap with the hydrogen $1s$ AOs to give the two C–H σ bonds.

In order to satisfy the valence on carbon we need to form a double bond to the oxygen and, as in ethene, this bond between the C and O comprises a σ and π bond. We therefore need to leave at least one of the $2p$ AOs on oxygen unhybridized so that we can form π MOs with this orbital; we will choose sp^2 hybrids on the oxygen, although sp HAOs would be equally suitable.

The oxygen HAO which points towards the carbon forms a σ bond, and the two remaining oxygen HAOs are occupied by pairs of electrons to give lone pairs. Finally, the out-of-plane $2p$ AOs on the carbon and oxygen overlap to give bonding and antibonding π MOs. Two electrons are placed in the bonding MO to create the second bond between the C and O.

The HOMO is either of the hybrids on the oxygen which are occupied by lone pairs. The LUMO will be the $\pi^\star$ MO since, as is usually the case, $\pi^\star$ MOs lie below $\sigma^\star$ MOs. The major contributor to this $\pi^\star$ MO is the $2p$ on the carbon, as this AO is higher in energy than the oxygen $2p$.

The imine is also planar, and so can be described in a very similar way, using sp^2 hybrids on both the carbon and the nitrogen. The only difference to the carbonyl compound is that one of the sp^2 hybrids on the nitrogen is used to form a σ bond to a hydrogen, and one becomes a lone pair. The HOMO is this lone pair, and the LUMO is the $\pi^\star$ MO.

4.7.3 Cyanides and other triply-bonded species

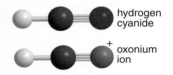

hydrogen cyanide

oxonium ion

The only commonly encountered organic group with a triple bond to a hetero-atom is the cyano group, C≡N, as illustrated opposite for the simplest case of HCN. The bond angles around the carbon are 180° i.e. it is linear, so sp hybridization is appropriate for the carbon, and in order to form the triple bond to nitrogen, this atom also needs to be sp hybridized.

One sp hybrid from the carbon overlaps with the hydrogen $1s$ to give the C–H σ bond, and the other hybrid overlaps with an sp hybrid on nitrogen to give a C–N σ bond. The second nitrogen sp hybrid points away from the carbon and is filled to give a lone pair. The two pairs of $2p$ orbitals from the C and the N, which lie perpendicular to the internuclear axis, overlap to give two π and two $\pi^\star$ MOs. The bonding MOs are occupied to create two π bonds. The HOMO is the lone pair on the nitrogen, and the LUMO is the $\pi^\star$ MO, to which the carbon will be the major contributor.

Rather less familiar as a triply bonded species is the oxonium ion shown opposite. This is in fact isoelectronic with HCN, as substituting N by O increases the number of electrons by one, but then the total is reduced by one due to the positive charge. Overall the two molecules have the same number of electrons, so the bonding can be described in exactly the same way.

4.7.4 The energy ordering of orbitals

A computer-based calculation will tell us the energy ordering of the MOs, and hence make it possible to identify the HOMO and the LUMO. However, in the absence of such a calculation, we will often need to make an 'educated guess' as to the ordering of the orbitals, and to do this the following rough guide is useful:

highest energy	σ^* antibonding
	π^* antibonding
	nonbonding orbitals (including lone pairs)
	π bonding
lowest energy	σ bonding.

This ordering is not inviolable: it is simply a guide. As we have already seen in the case of molecules such as CH_3X (X is a heteroatom), we may have to use other criteria to choose between different σ orbitals. Another useful point to remember is that generally speaking the more nodes an MO has, the higher its energy. This rule of thumb is useful for working out the energy ordering of a set of MOs formed from the same AOs.

4.8 Delocalized bonding

For very many molecules, the localized description of bonding using HAOs proves to give a useful picture of the electronic structure. However, there are some cases – and they are by no means uncommon – in which the localized picture fails to account for some key properties. The most well known of these is benzene, C_6H_6, a conventional representation of which is shown opposite.

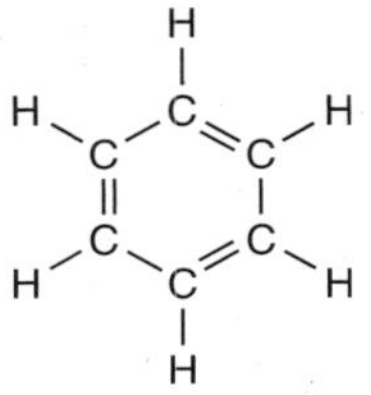

The experimental evidence points clearly to the fact that benzene has a planar hexagonal structure in which all of the C–C bonds have the *same* length. This is at odds with the picture we have drawn opposite which shows alternating single and double bonds: double bonds are usually significantly shorter than single bonds.

Another familiar example is the caboxylate ion, shown in Fig. 4.36 (b), which is formed by the de-protonation of a carboxylic acid, shown in (a). In ethanoic acid, the C–O bond length to the carbonyl oxygen is 122 pm, whereas that to the hydroxy oxygen is 132 pm; as expected, the C–O double bond in shorter than the single bond. In the ethanoate ion the two C–O bond lengths are equal at 125 pm, which is a value intermediate between that for C–O double and single bonds.

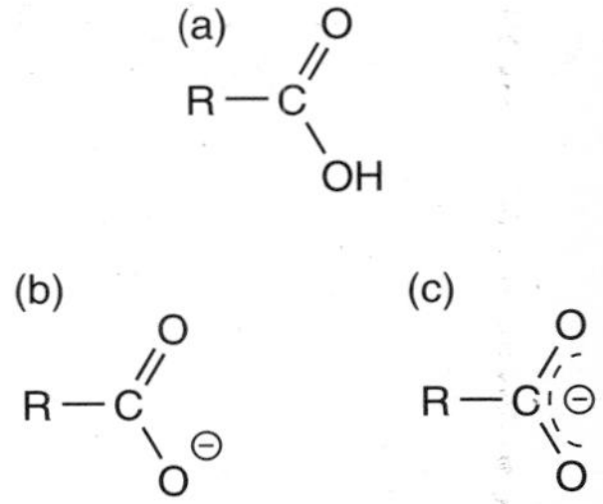

These data indicate that in a carboxylate the two oxygens are equivalent, and that the bonding between then and the carbon is intermediate between and single and a double bond. The structure shown in (b) is simply not consistent with these observations, so it is common to draw the anion with an extended π bond running across both oxygens and the carbon, as shown in (c). This structure emphasizes the equivalence of the two oxygens and that the negative charge is not localized on just one of these atoms.

Rather more subtle is the case of butadiene (shown below), in which there is experimental evidence for a significant barrier to rotation, of about 30 kJ mol^{-1}, about the central C–C bond. This is significantly greater than the barrier for rotation about a typical C–C single bond (around 15 kJ mol^{-1}), but much less than the barrier to rotation about a C=C double bond. This representation of butadiene, in which there is a single bond between the central two carbon atoms, does not account for this increased barrier to rotation.

benzene

Fig. 4.36 Deprotonation of the carboxylic acid (a) gives the anion (b). The C–O bond lengths in (b) are the same, which is inconsistent with structure (b). A better representation is (c) in which there is a delocalized π bond running across all three atoms.

These molecules, and many like them, show what is usually called *delocalized* π bonding. This is bonding involving out-of-plane $2p$ orbitals (hence the π) from *more than two atoms*.

For example, in the case of benzene each carbon can be thought of as being sp^2 hybridized, and so each has one $2p$ orbital pointing perpendicular to the plane of the ring. The π molecular orbitals are formed from *all six* of these $2p$ AOs, rather than just from a pair of adjacent AOs, as is implied by the simple representation of benzene with alternating single and double bonds. As we shall see, it is this delocalized π bonding which makes all of the C–C bonds equivalent.

Similar considerations apply to butadiene. Each carbon is sp^2 hybridized, leaving a single $2p$ AO from each carbon, and the π MOs are formed from all four of these AOs. We will show that the resulting π MOs extend over the whole molecule in such a way that the central C–C bond has partial π character.

In both benzene and butadiene there is a distinct separation between the σ bonded framework and the orbitals which are involved in the delocalized π system. The σ framework is formed by the overlap of the HAOs on carbon with one another and with the hydrogen $1s$ AOs; these interactions can be described pretty well using a localized approach. It is only the π system for which the localized picture is not sufficient.

Of course, if we made a full computer-based MO calculation, all of the bonding would be delocalized, as such calculations do not force the interactions to be between only two orbitals. However, in the results of such a calculation there is still a separation between the σ and π MOs, simply because the AOs which form the π MOs have the wrong symmetry to overlap with any of the AOs which form the σ MOs. So, what we can usually do is use a simple localized description for the σ bonded framework of the molecule, and then reserve the full power of the MO approach for just the π interactions. This is the approach we will illustrate in this section.

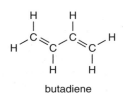

butadiene

4.8.1 Orbitals in a row

We have seen earlier in this chapter that when more than two or three orbitals are involved, finding the form of the MOs using just pen and paper is not so straightforward. However, there are some special situations in which the form of the MOs can be found in quite complex systems using a graphical argument. One of these situations is when we have a *row* of (more or less) identical orbitals. We will see that for such an arrangement, it is easy to find the form of the MOs without resorting to detailed calculations.

There are not many molecules which literally have orbitals arranged in a row, but there are many molecules where the arrangement of the orbitals can usefully be approximated in this way. One example is butadiene, in which the four out-of-plane $2p$ orbitals can be thought of as being in a row – we will start with this system, explaining first how the MOs can be constructed, and then how they explain the restricted rotation about the central C–C bond.

The overlap of four $2p$ orbitals will generate four MOs, and in general each AO will make a contribution to each MO. Our task is to work out the contribution of each AO, and this can be done using the following graphical procedure, illustrated in Fig. 4.37 on the facing page.

First, we draw out evenly spaced dots on a line to represent the positions of the atoms. In this case there will be four such dots, numbered 1 to 4, as shown in (a). We then add a dot (with the same spacing) to the left of number 1, and a dot to the right of number 4. These two extra dots are numbered 0 and 5, respectively, as shown in (b). Of course, there are no atoms at these positions, but we need these for the graphical construction.

Now we inscribe the first half of a sine wave starting at position 0 and finishing at position 5, as is shown in (c). Recall that the first half of a sine wave starts at zero (position 0), goes up to +1 and then falls back to zero (position 5). It turns out that the contribution of the AO from atom 1 to the MO is proportional to the *height* of this sine wave at the position of atom 1. Similarly, the contribution of the AO from atom 2 is proportional to the height of the sine wave at the position of atom 2 and so on for atoms 3 and 4. These heights, and hence the contributions of the AOs, are shown by the arrows in (d).

From the diagram we can see that the AOs at positions 2 and 3 contribute equally, as do those at positions 1 and 4, but that the contributions from atoms 2 and 3 are greater than those from atoms 1 and 4. The construction shown in Fig. 4.37 is for the lowest energy MO. The next highest energy MO is constructed in a similar way, except this time we inscribe a complete sine wave (i.e. *two* half sine waves) between positions 0 and 5, and use this to determine the contributions from each AO. The next MO is found by inscribing *three* half sine waves, and the fourth and final MO is found by inscribing *four* half sine waves. Figure 4.38 on the following page shows the contributions which each AO makes to the four MOs. In this figure, the size of each $2p$ AO has been made proportional to the contribution which that AO makes to the MO.

In the 1π MO, each AO contributes in the same sense, so there is constructive overlap (bonding) between each adjacent AO; it is not therefore surprising that this is the lowest energy of the MOs.

The sine wave which describes the form of the 2π MO is positive at positions 1 and 2, but negative at positions 3 and 4. The AOs from atoms 1 and 2 thus contribute in the positive sense, whereas those from atoms 3 and 4 contribute in the negative sense, as can be seen from the reversal of the sign of the AOs. There is a bonding interaction between 1 and 2, and between 3 and 4, but an antibonding interaction between 2 and 3 i.e. there is a node between these two atoms. Overall, the orbital has two bonding and one antibonding interaction, so the net result is one bonding interaction. Not surprisingly, therefore, the 2π MO is higher energy the 1π MO, and overall less strongly bonding.

The form of the 3π MO is determined by three half sine waves, and as a result the AOs at positions 1 and 4 contribute in the positive sense, whereas those at positions 2 and 3 contribute in the negative sense. There are therefore two nodes between atoms 1 and 2, and between atoms 3 and 4. This time, there are two antibonding interactions (1–2 and 3–4), and one bonding interaction (2–3). As a result the MO is net antibonding.

Finally, in the 4π MO are three nodes, one between each adjacent pair of atoms. All of the interactions are antibonding, so this MO is the highest in energy and the most antibonding of the four. The pattern of the nodes here is just as we have come to expect. As we go along the series $1\pi \ldots 4\pi$ the number of nodes increases as the energy of the MOs increases.

The MOs for any number of p orbitals in a row can be found using this method. It is important to remember that we must add the fictitious atoms to the left and right of the chain, and the the sine waves are inscribed *between*

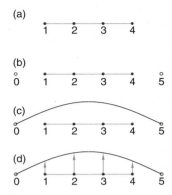

Fig. 4.37 Illustration of a graphical method for determining the form of the MO from four orbitals in a row, (a). Two extra atomic positions, 0 and 5, are added at the same spacing as the real atoms, as shown in (b). A half sine wave is drawn between positions 0 and 5, as shown in (c). The contributions of each AO to the MO is given by the height of the sine wave at the position of that atom, as illustrated in (d). The MO shown here is the one with the lowest energy.

This pattern in which the number of nodes in the wavefunction increases as the energy increases is a recurring these in quantum mechanics. We will look in more detail at the origin of this in Chapter 16.

Fig. 4.38 Schematic representations of the contribution
of each AO to the four MOs
arising from the overlap of
four *p* orbitals in a row. 1π is
the lowest in energy, followed
by 2π and so on. The relative
contribution of each AO to
a particular MO is found
by inscribing an appropriate
sine wave over the atoms, as
illustrated in Fig. 4.37 on the
preceding page for the lowest
energy MO, 1π. The higher
energy MOs are found by
inscribing two, three and four
half sine waves. The size of the
p orbitals is made proportional
to their contribution to the MO.
Nodal planes are indicated by
the green dashed lines; note
how the number of nodes
increases as the energy of the
MO increases.

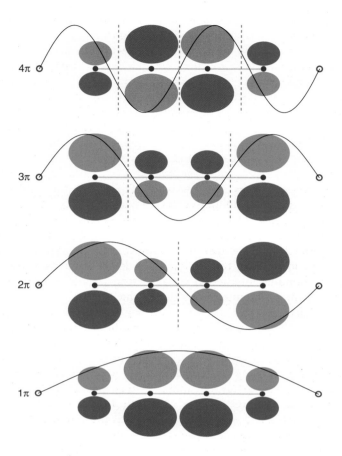

these extra atoms. To construct the various sine waves it is useful to remember
that the number of nodes will increase by one for each successive MO, and that
the spacing between these nodes is given by

$$\frac{N+1}{n+1} \times R,$$

where N is the total number of atoms, n is the number of nodes and R is the
spacing between the atoms.

For example, with $N = 4$ the MO with one node (the 1π) has a spacing of
$5/2 \times R$ i.e. the node is located at a distance $2.5\,R$ from the fictitious atom at
position zero. The next MO has $n = 2$ so the spacing is $5/3 \times R$ i.e. the nodes
are $1.667 \times R$ and $3.333 \times R$ from the position of atom zero.

Butadiene

We can put the orbitals we have just devised to immediate use in describing
the π system in butadiene, the localized structure of which is shown opposite.
There are a total of twenty-two valence electrons in this molecule (four from
each carbon, and a total of six from the hydrogens). If we assume that the
carbon atoms are sp^2 hybridized, and then use these hybrids to form the six
C–H bonds, and the three C–C bonds, this accounts for eighteen electrons; this

butadiene

'σ bonded' framework, as it is usually known, is illustrated in Fig. 4.39. There are thus four electrons to be accommodated in the π MOs.

If we assume that we can approximate the interaction between the four out-of-plane 2p orbitals using the 'four orbitals in a row' picture, two electrons are assigned to the 1π MO and two to the 2π MO shown in Fig. 4.38 on the facing page. The key thing about these MOs is that they extend over all four carbons, and in particular the 1π MO has a bonding interaction between all four atoms. In other words, there is some π bonding between the *central* two atoms, 2 and 3. This is completely at variance with the simple localized picture, which has a single σ bond between these two atoms.

The 2π MO is also occupied, and contributes to bonding between atoms 1 and 2, and between atoms 3 and 4, which fits in with the localized picture. However, there is an antibonding interaction between atoms 2 and 3. This detracts from the bonding interaction between these atoms due to the occupation of the 1π MO. However, as can be seen from Fig. 4.38, in the 1π MO the two orbitals on atoms 2 and 3 which are responsible for the bonding interaction make a greater contribution to the MO than do the same two AOs in the 2π MO. Thus, the bonding interaction between atoms 2 and 3 due to the 1π MO outweighs the antibonding interaction due to the 2π MO.

The overall picture is that there is π bonding between 1 and 2, and between 3 and 4, due to both the 1π and 2π MOs. In addition there is some π bonding between atoms 2 and 3, but this is considerably less than that between the end atoms. We can say that there is a *partial* π bond between the middle two atoms, and this in turn accounts for the restricted rotation about this bond.

Figure 4.40 shows iso-surface plots of the four π MOs computed for butadiene; of these the HOMO and the next lowest MO (HOMO−1) are occupied. These orbitals compare very well with the predictions of Fig. 4.38. The simple graphical argument is thus a good guide to the actual form of the MOs determined by a much more sophisticated calculation.

Fig. 4.39 Schematic illustration of the σ bonded framework in butadiene; the carbon atoms are assumed to be sp^2 hybridized. Eighteen electrons occupy the σ bonding MOs in this framework, leaving four electrons for the out-of-plane π system.

⟲ *Weblink 4.8*

View, and rotate in real time, iso-surface representations of the four π MOs of butadiene illustrated in Fig. 4.40.

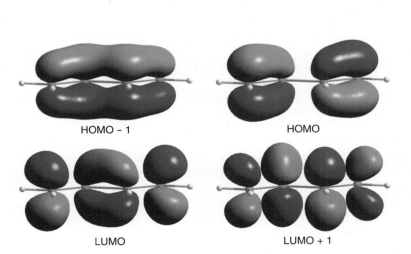

HOMO − 1

HOMO

LUMO

LUMO + 1

Fig. 4.40 Iso-surface plots of the computed π MOs for butadiene. Comparison of these with Fig. 4.38 shows that the simple graphical argument is a good guide to the overall form of the MOs. The HOMO is the 2π MO, and the next lowest MO (denoted HOMO−1) is the 1π MO. The LUMO is the 3π MO, and the highest energy orbital (denoted LUMO+1) is the 4π MO.

Fig. 4.41 On the left are shown the contributions which each AO makes to the MOs for three p orbitals in a line, constructed using the same graphical method as was used for four orbitals. The 1π MO is bonding across all the atoms, but the 2π MO is nonbonding since the middle AO makes no contribution (the sine wave is zero at this point). The 3π MO is antibonding. On the right is shown the occupation of the MOs for the allyl anion and cation.

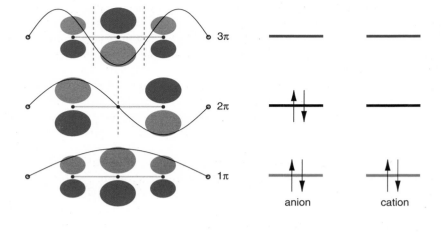

allyl cation

allyl anion

The allyl cation and anion

The allyl cation and anion, illustrated opposite, are somewhat exotic looking species but which have nevertheless been studied both in solution and in the solid state. Both species can be thought of as deriving from propene: removing H^+ from this molecule gives the allyl anion, and removing H^- gives the allyl cation, although a more realistic reaction would be by the loss of Br^- from allyl bromide ($CH_2{=}CHCH_2Br$).

The experimental evidence is that both of these molecules are planar and symmetrical, in the sense that the two end carbons are the same. The localized structures shown opposite make it seem as if the end carbons are different, as one is doubly bonded, and the other is only singly bonded. We will show in this section that these ions have a delocalized π system which accounts for the symmetry, and also explains several other properties of the molecules.

The allyl anion has eighteen valence electrons: four from each of the three carbons, five from the hydrogens and one for the negative charge. If we imagine that the carbons are sp^2 hybridized, and then use these hybrids to form the two C–C and five C–H σ bonds, this accounts for fourteen of the electrons, leaving four to go into the π MOs. For the allyl cation there are two fewer electrons, leaving just two to go into the π system.

Each carbon has an out-of-plane $2p$ AO, and these interact to form the π MOs. As before, we can model these using three orbitals in a row, and can construct the form of the MOs using the same graphical argument as we used before.

The contributions of the AOs to the MOs is shown on the left of Fig. 4.41. As before, we have added an imaginary atom to the left and to the right of the three atoms, and then inscribed one, two and three half sine waves across these five positions. The 1π MO is bonding across all atoms, but the AO from the central atom makes a larger contribution to the MO than do the AOs on the end atoms.

The 2π MO has one node, which is spaced by $(N+1)/(n+1)\times R = (4/2)\times R = 2R$ from the fictitious atom on the left. This places the node on the central atom, so the AO from this atom makes no contribution. The AOs from the end atoms make equal, and opposite, contributions. This MO is nonbonding as the two

AOs on the end atoms are too far apart for there to be a significant interaction between them.

The 3π MO has two nodes which are spaced by $(N+1)/(n+1)\times R = (4/3)\times R = 1.333\,R$. So there are nodes at $1.333\,R$ and $2.667\,R$ from the fictitious atom on the left. The contribution from each AO is in the opposite sense to its neighbours, so all the interactions are antibonding. As before, the energy of the MOs increase as the number of nodes increases.

In the allyl anion, there are four electrons in the π system so these occupy the 1π and 2π MOs. The occupation of the 2π MO makes no contribution to the bonding, so π bonding is entirely due to the 1π MO. Since the two electrons in this MO are spread across all three atoms, so we have a partial π bond between adjacent atoms. We note that the 1π MO is symmetric, so the bonding likewise has to be symmetric, in contrast to the impression given by the localized picture.

Although the two electrons in 2π MO do not contribute to the bonding, they do increase the electron density on the end atoms. A full MO calculation shows that there is a partial (negative) charge of -0.62 on the end atoms, with slight positive charges on all the other atoms. This can be compared to ethene in which there is a charge of -0.35 on each carbon, and $+0.17$ in each hydrogen. In the allyl anion, there is clearly a concentration of negative charge on the end carbon atoms.

In the cation, only the 1π MO is occupied, so again there is a partial π bond across all three atoms. The form of this MO tells us that the electron density will be greatest on the central atom, in contrast to the anion where the density is greatest on the end atoms.

4.8.2 Resonance structures

As we have seen, the delocalized picture of the bonding in butadiene and in the allyl cation and anion gives a picture which is in accord with experimental evidence about these molecules, whereas the localized picture does not. However, we can improve on the localized picture somewhat by introducing the concept of *resonance*. How this applies to the molecules we have been discussing so far is illustrated in Fig. 4.42 on the next page.

Allyl cation and anion

In Fig. 4.42 (a) two localized bonding structures are given for the allyl cation, one with the positive charge on the right-most carbon, and one with the charge on the left-most carbon. These two structures are entirely equivalent, and are both equally acceptable. However, as we have seen, neither of these structures on their own is adequate to explain the observed properties of the cation, because they do not show that the two end carbons are the same.

The idea of resonance is that if we take these two structures *together*, they are a better representation of what is going on than either on its own. It is important not to get the wrong idea here. We are *not* saying that the molecule is flicking back and forth between these two structures. What we are saying is that the true electronic structure is best represented by a mixture of these two. So, the π bond is neither located on the left nor on the right, but across all three atoms. Similarly, the positive charge is not localized, but is spread out.

These different localized representations of the bonding are called *resonance structures*, and traditionally they are connected by a double-headed arrow. This symbol is chosen so as not to be confused with a reaction arrow ($\longrightarrow$) or the

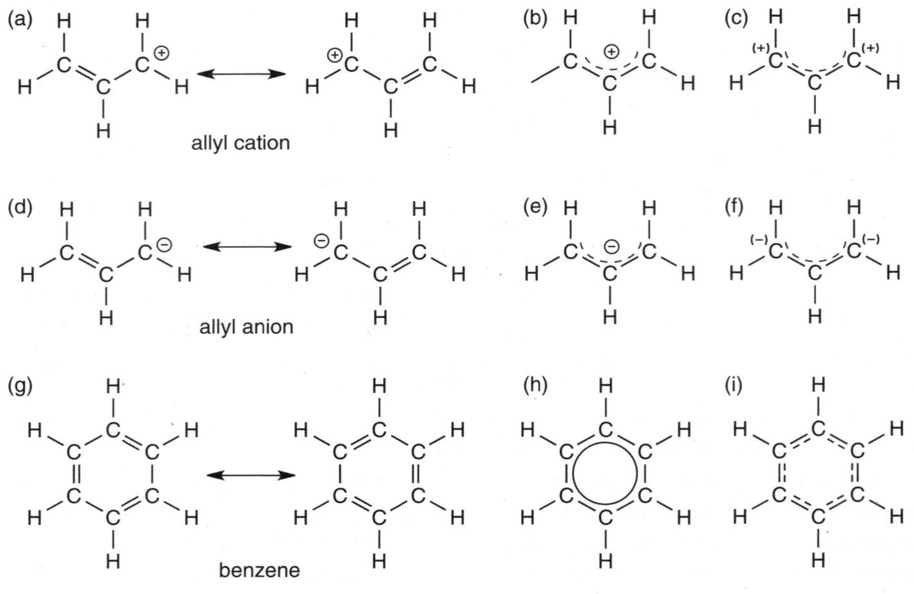

Fig. 4.42 Illustration of different ways of representing delocalized π bonding. In (a) we see two *resonance structures* of the allyl cation, connected by a double headed arrow. These are two equivalent localized representations of the bonding, but neither is adequate on its own. Shown in (b) and (c) are alternative ways of representing the bonding: a dashed line indicates a delocalized π bond, and the '(+)' represents a possible location of the positive charge in one of the resonance structures. The equivalent representations of the allyl anion are shown in (d)–(f). Benzene can be represented as resonance structures, shown in (g), or in the ways shown in (h) and (i).

symbol for equilibrium ($\rightleftharpoons$). In different resonance structures the atoms are all in the same positions, but the electrons (i.e. the bonds) are arranged differently. So these resonance structures are different representations of the *same* molecule.

Figure 4.42 (b) and (c) show two commonly used alternative ways of representing the delocalized bonding in the allyl cation. In (b) a dashed line is used to represent the idea that the π bond is spread over all three atoms, and the positive charge is shown as being associated with the whole molecule. Representation (c) also uses a dashed line, but the positions at which the positive charge may be localized, as shown by the resonance structures in (a), are indicated as '(+)'. The purpose of the bracket around the plus sign is to remind us that a full positive charge is not present, but that this is a position at which a positive charge is present in one resonance structure. Structure (c) is more useful than (b) as the former shows the possible locations of the positive charge.

Figure 4.42 (d) shows the resonance structures of the allyl anion, and further alternative representations are shown in (e) and (f). These are closely analogous to those used for the allyl cation. Finally, (g) gives the resonance structures for benzene, which differ in the placement of the double bonds. This delocalized bonding is often represented in the ways shown in (h) and (i).

Butadiene

Butadiene is rather a different case to benzene and the allyl ions. The delocalized picture tells us that there is a partial π bond between the two central carbons, so if we are to use the resonance concept to explain this then we need to include some resonance structures in which there is a double bond between the middle two carbons. Figure 4.43 on the facing page shows two such resonance structures, A and C, along with the usual form, B. What is significantly different about structures A and C is that in order to place the double bond in the middle we have had to create a positive charge on one of the end carbons, and a negative charge on the other. Structures A and C only differ in the placement of these two charges.

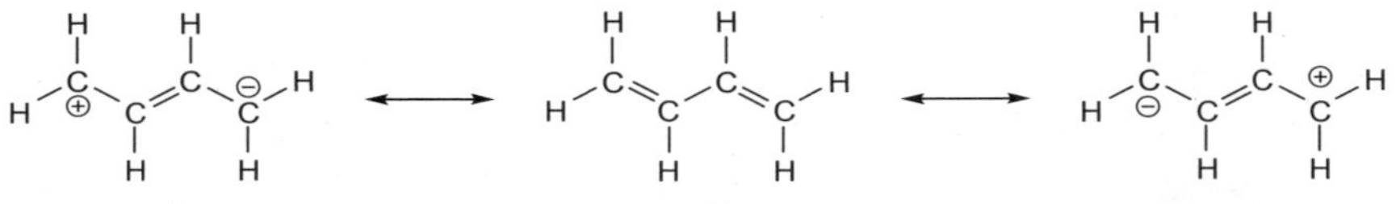

Fig. 4.43 Illustration of resonance for butadiene. The usual representation, B, fails to account for the partial π bond between the central carbons, but this can be remedied by including small contributions from forms A and C.

For benzene and the allyl ions, the two resonance structures contribute equally i.e. the electronic arrangement is a 50:50 mixture of the two. However, for butadiene the central double bound is only partial, so A and C are only *minor contributors* to the structure, whereas B is the major contributor. There must be equal contributions from A and C as otherwise we predict the molecule to have a net dipole moment, which it certainly does not.

The concept of resonance has to be used with great care. It is especially important to understand that the resonance structures are just different *localized* representations of the bonding, and that the true electronic structure is represented by some combination of these resonance structures: most emphatically *the molecule does not 'flicker' between these resonance structures*. We have a tendency to want to represent the bonding in molecules using localized two-centre two-electron bonds, simply as this is both familiar and convenient. However, we have to accept that such representations are, for many molecules, neither accurate nor adequate.

The MO picture avoids all of these difficulties as it naturally produces a delocalized picture of bonding. The concept of resonance is a way of trying to introduce delocalized bonding using localized structures.

4.9 Delocalized structures including heteroatoms

So far we have only considered delocalized structures in which all of the atoms are the same (carbon). However, much the same considerations apply when there is a mixture of carbon and heteroatoms (nitrogen, oxygen). The MOs we have developed for orbitals in a line will not be quite right as the constituent AOs will not all have the same energy, but as we shall see these MOs are still a useful guide.

4.9.1 Carboxylate anion

At the start of this section we used the carboxylate ion, whose structure is shown in Fig. 4.36 on page 165, as an example of a delocalized bonding. You will recall that both C–O bond lengths in such an ion are the same, and intermediate between those for double and single bonds. Using the concept of resonance, we can rationalise this observation by proposing that there are two equally contributing resonance structures, as shown opposite.

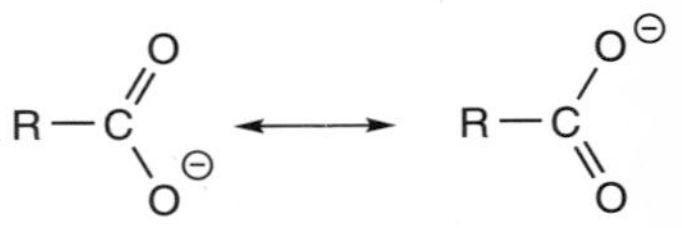

carboxylate anion

An alternative explanation is to use an MO description of the π system. If we allow the carbon and both of the oxygens to be sp^2 hybridized, then each has a $2p$ orbital pointing out of the plane. These three AOs will combine to form three π MOs, in which we have to accommodate four electrons – one from each atom, plus one for the negative charge.

HOMO − 3

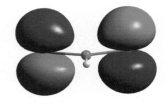

HOMO

LUMO

Fig. 4.44 Iso-surface plots of the computed form of the π MOs in the methanoate anion (HCO_2^-); the hydrogen atom is coming towards us. The HOMO and LUMO compare well with the form of the 2π and 3π MOs predicted in Fig. 4.41 on page 170; the lowest energy π MO, labelled HOMO−3, compares well with the 1π MO.

To a rough approximation we can model this π system as three $2p$ orbitals in a line, and so can use the MOs illustrated in Fig. 4.41 on page 170. Two of the electrons occupy the 1π MO, which is π bonding across all atoms, and two occupy the nonbonding 2π MO, which localizes electron density on the end atoms. This is of course just the same description that we used for the allyl anion.

Figure 4.44 shows the computed form of the occupied π MO in the methanoate anion (HCO_2^-). These agree well with the simple MOs computed for three p orbitals in a line. Given that the pair of electrons in MO1 is responsible for the bonding across all three atoms, what we have is a partial π bond between the C and each O. Often this is represented by a dashed line, in the way shown opposite.

4.9.2 Enolates

If we take a ketone or an aldehyde and treat it with a strong base, it is possible to ionize one of the hydrogens on the carbon *adjacent* to the carbonyl carbon to give what is known as an *enolate*. The process is illustrated in Fig. 4.45 for ethanal.

The two carbons and the oxygen lie in a plane, so just as in the allyl anion and the carboxylate anion there is the possibility of forming a delocalized π system involving these three atoms. We can model this using the simple MOs from Fig. 4.41 on page 170, placing two electrons in the 1π MO and two in the 2π MO. There is thus partial π bonding across all three atoms. However, this enolate lacks the symmetry of the carboxylate anion, so we must expect the actual π MOs to look somewhat different from those shown in Fig. 4.41.

Figure 4.46 on the facing page shows the computed form of the occupied π MOs for the enolate anion from ethanal. The lowest energy π MO, labelled HOMO−2, compares well with the 1π MO in Fig. 4.41. However, rather than the orbitals from all three atoms contributing equally, as they do in the 1π MO, the orbitals from the oxygen atom and the carbonyl carbon are the major contributors. The next highest energy π MO is also the HOMO; broadly speaking this is analogous to the 2π MO from Fig. 4.41. However, in contrast to the 2π MO, there is a contribution from the carbonyl carbon, and the contributions from the other carbon and the oxygen are not the same. When we come to look at the reactions of this enolate, we will see that the form of the HOMO is important in determining how this species reacts.

Weblink 4.9

View, and rotate in real time, iso-surface representations of the three π MOs of the methanoate ion illustrated in Fig. 4.44.

Fig. 4.45 A base may remove H^+ from the carbon adjacent to a carbonyl group to give an enolate anion. Two resonance forms of the enolate are shown: in A, the negative charge is on the carbon, and in B it is on the oxygen.

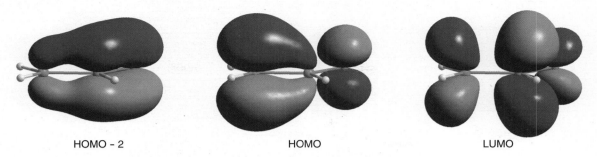

HOMO − 2   HOMO   LUMO

Fig. 4.46 Iso-surface plots of the computed form of the π MOs in the enolate formed from ethanal (structure shown in Fig. 4.45 on the preceding page). The CH_2 group is to the left and the carbonyl oxygen is to the right.

The enolate can also be described using the two resonance structures A and B shown in Fig. 4.45 on the preceding page. Both of these forms are significant contributors, as is indicated by a detailed MO calculation which shows that the partial charges are −0.81 on the oxygen, +0.34 on the carbonyl carbon, and −0.62 on the other carbon.

4.9.3 Amides

Amides, such as methanamide (shown opposite), have a rather unexpected structural feature which is illustrated in Fig. 4.47. This shows the three-dimensional structure of the amide obtained by X-ray diffraction on a crystal. The special thing to note is that the oxygen, the two carbon atoms, the nitrogen *and* the two hydrogens attached to it all lie in a plane.

At first sight this is unexpected. The nitrogen has three single bonds and a lone pair, so we would expect the bonds to be arranged in a trigonal pyramid, just as they are in NH_3, or in an amine RNH_2.

What is going on here becomes clear once we look at the form of the π MOs of methanamide ($HCONH_2$) shown in Fig. 4.48 on the next page. These clearly show that there is an interaction between $2p$ AOs on the nitrogen, carbon and oxygen. In particular, the lowest energy π MO (HOMO−2) shows that there is a significant bonding interaction across all three of these atoms.

We could describe this situation by saying that the nitrogen is approximately sp^2 hybridized, which makes available an out-of-plane $2p$ AO which can then be involved in π interactions with $2p$ AOs on the carbon and the oxygen. The overlap of these $2p$ AOs is optimized when they all point in the same direction, and this accounts for the planarity of the molecule.

These orbitals indicate that there is a partial π bond between the nitrogen and the carbon. In terms of resonance structures, this can be accommodated by saying that there is some contribution from the resonance structure B shown opposite. Further evidence that there is a partial π bond between the N and the C comes from spectroscopic measurements which indicate that the energy barrier for rotation about the C–N bond is around 70 kJ mol^{-1}, significantly higher than that expected for a single bond.

In energetic terms, we can rationalize this planar geometry of the amide by noting that by adopting this geometry the π system can be extended over three atoms. The most bonding π MO spread over the three atoms is lower in energy than the C–O π bonding MO, thus by forming the localized system the energy of the molecule as a whole is lowered.

Weblink 4.10
View, and rotate in real time, iso-surface representations of the three π MOs of the enolate ion (from ethanal) illustrated in Fig. 4.46.

methanamide

Fig. 4.47 Three-dimensional structure of methanamide obtained by X-ray diffraction. Note that all of the atoms lie in a plane.

A   B

HOMO − 2

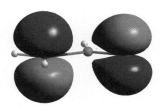
HOMO

LUMO

Fig. 4.48 Iso-surface plots of the computed form of the occupied π MOs in methanamide. The NH_2 group is to the left and the carbonyl oxygen is to the right.

4.10 Moving on

Weblink 4.11

View, and rotate in real time, iso-surface representations of the three π MOs of methanamide illustrated in Fig. 4.48.

Between this chapter and the previous one we have shown how molecular orbitals can be used to describe the bonding in molecules. Although the detailed form and energies of the MOs can only be found from a computer calculation, we can nevertheless construct useful qualitative MO diagrams without resort to such detailed calculations. As we shall see when we start to look at chemical reactions, the form of the HOMO and the LUMO are particularly important, and these can often be identified by the kind of qualitative arguments outlined in this chapter. We have thus developed a powerful tool which we will use extensively in the rest of the book.

Our focus has so far been on small molecules which are held together with covalent bonds. In the next chapter we consider the extended structures which often occur in solid materials. There are two extremes of bonding in such materials: at one end we have giant covalent structures in which there is a continuous network of covalent bonds throughout the sample. At the other extreme are ionic structures, which are held together not by covalent interactions but by the forces between charges.

FURTHER READING

Introduction to Molecular Symmetry, J. S. Ogden, Oxford University Press (2001).

QUESTIONS

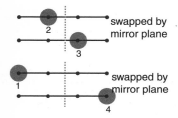
swapped by mirror plane

swapped by mirror plane

4.1 Consider a hypothetical molecule in which four hydrogen atoms lie in a row. Like H_3^+, this molecule has a mirror plane down the middle. As shown opposite, atoms 2 and 3, and atoms 1 and 4 are swapped by the mirror plane. We therefore need to consider the AOs on atoms 2 and 3 together, and the AOs on atoms 1 and 4 together.

Draw sketches of the symmetric and antisymmetric combinations of AOs on atoms 2 and 3 (i.e. the symmetry orbitals), and then do the same for the AOs on atoms 1 and 4. Using the rule that only AOs of the same symmetry overlap, sketch the form of the four MOs, two symmetric and two antisymmetric, which are formed from the overlap of the SOs.

Arrange these MOs in order of increasing energy by looking at the number of nodes in the MOs. What does your MO diagram predict about the hypothetical molecule H_4?

4.2   Construct an MO diagram for a hypothetical linear water molecule, H–O–H, using a similar approach to that used for FHF^-; take the z-axis to be along the axis of the molecule. Sketch the form of the MOs, and indicate which will be occupied.

[Hints: The orbitals can be classified according to a mirror plane, as we did in FHF^-. The two hydrogen $1s$ form a symmetric and an antisymmetric SO, and the oxygen $2s$ and $2p_z$ can be classified individually. The oxygen $2p_x$ and $2p_y$ have the wrong symmetry to overlap with the hydrogen $1s$ (why?), and so remain nonbonding.]

Compare your MO diagram with that for 'bent' water. Using these two diagrams, rationalize why it is that water adopts a bent, rather than a linear, geometry.

4.3   Using appropriate HAOs, described the bonding in each of the following molecules, being sure to specify the form of each occupied MO, and also making sure that you have accounted for all the electrons. Identify the HOMO and the LUMO for the first three molecules.

4.4   Using a similar approach to that used in section 4.5 on page 156, construct an MO diagram for F_2 assuming that (i) the two fluorine atoms are sp hybridized and (ii) the two fluorine atoms are sp^3 hybridized. State which orbitals are occupied and compare the resulting descriptions of the bonding with that from the full MO picture (Fig. 3.28 on page 116).

Use a similar approach to construct an MO diagram for CO using sp hybrids on both atoms; remember to place the oxygen AOs and HAOs somewhat lower in energy than the corresponding orbitals on carbon. What does your diagram predict about the form of the HOMO and the LUMO? Compare your predictions about the MOs with the computed orbitals shown in Fig. 3.42 on page 132.

4.5   The MO diagram for the linear form of H_3^+ can be constructed by realizing that it is just three $1s$ orbitals in a row. The resulting MOs have the same form as the π MOs in the allyl cation, except that they are composed of $1s$ AOs. Compare the MOs deduced using this approach to those found in section 4.1.1 on page 138 using symmetry arguments.

4.6   Using the graphical method described in section 4.8.1 on page 166, deduce the form of the π MOs of hexatriene ($CH_2=CH-CH=CH-CH=CH_2$). Which MOs are occupied?

Identify the contribution (in terms of bonding or antibonding) that each occupied MO makes to the π bonding between the end two carbons (numbered 1 and 2). Do the same for the bonding between carbons 2 and 3, and between 3 and 4. Based on your results, which of these bonds has the greatest net bonding interaction, and which has the least?

4.7   Using the graphical method described in section 4.8.1 on page 166, deduce the form of the π MOs of the cation ($CH_2=CH-CH=CH-CH_2^+$). Which MOs are occupied?

Adding one electron to this species gives a radical, and adding a further electron gives an anion. Explain what effect you would expect adding one or two extra electrons to have on the strength of the π bonding.

4.8 The azide ion, N_3^-, has a symmetrical linear structure. Describe the bonding in this ion by using sp hybrids on each atom, and by forming delocalized π MOs from the $2p$ orbitals which point perpendicular to the long axis (the z-axis) of the molecule. Note that there will be two separate π systems: one formed form the $2p_x$, and one from the $2p_y$ AOs. Identify the HOMO and the LUMO.

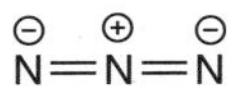

Compare your MO picture with the bonding shown in the localized structure of azide, shown opposite.

The molecules CO_2 and NO_2^+ are both symmetrical and linear, and can be described as being *isoelectronic* with N_3^-. What do you understand by this term? How does your MO description of the bonding in the azide ion need to be modified to describe CO_2 and NO_2^+?

4.9 Using appropriate HAOs, describe and compare the bonding in the following three molecules.

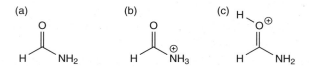

4.10 In the cyclopropenyl cation (shown below) a π system is formed from the overlap of an out-of-plane $2p_z$ AO from each carbon. The MOs formed from these AOs have exactly the same form as the MOs for triangular H_3^+ found in section 4.1.2 on page 141, except that they are formed from $2p_z$ AOs, rather than $1s$ AOs.

Carefully count up how many electrons there are in this molecule, and hence determine which of the π MOs are occupied. Compare the form of the occupied π MOs with those for the allyl cation, commenting on any relevant points of difference.

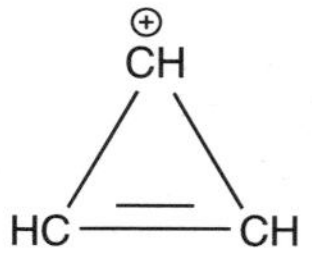

5

Bonding in solids

Key points

- In solids, the overlap between orbitals on different atoms can give rise to crystal orbitals which extend throughout the material; these orbitals are analogous to delocalized MOs.
- The crystal orbitals which arise from a particular set of atomic orbitals form a band, which can hold a certain number of electrons.
- The electrical conductivity of metals is the result of partially filled bands; insulators have full bands, but in semiconductors there is a small gap between a filled and an empty band.
- The lattice enthalpy of an ionic solid can be estimated using a simple electrostatic model.
- The lattice enthalpy depends on the size (radii) of the ions and their three-dimensional arrangement in the crystal.

In section 1.3 on page 9 we discussed briefly how different types of solids could be distinguished on the basis of the bonding they contain. These types are:

- molecular solids, which contain discrete molecules, held together by weak interactions, such as hydrogen bonds;

- giant covalent solids, in which there is a network of covalent bonds extending throughout the entire structure;

- metallic solids, in which there is extensive delocalization of the electrons;

- ionic solids, in which it is the interactions between discrete ions which hold the structure together.

Now that we have developed the molecular orbital approach, we are in a position to be more precise about the nature of the bonding in metals and in giant covalent structures. We will see that such structures can be described using the concept of *bands* which are rather like giant molecular orbitals which extend throughout the material.

Having done this, we will turn our attention to ionic solids, and show how a simple electrostatic model can account for energetics of such structures.

An important question which arises in these discussions is why a given chemical species adopts a particular form. For example, why is it that under normal conditions we find lithium as the solid metal, but nitrogen as N_2 gas? It is, as we have seen, perfectly possible for lithium to form the molecule Li_2, but it must be that the metallic form is preferred as it is 'more stable'. Similarly, although we could conceive of a metallic form of nitrogen, the molecular form is 'more stable'.

Precisely what we mean by 'more stable' will have to wait until the next chapter when we look at the topic of *chemical thermodynamics*. For the moment we will not enquire as to *why* a particular form is preferred, but simply content ourselves with describing the bonding in a given structure.

5.1   Metallic bonding: introducing bands

A defining feature of metals is that they are excellent conductors of electricity, which we take to imply that they contain electrons which are free to move throughout the solid. So, rather than electrons being closely associated with a particular nucleus, they are free to move from atom to atom.

This description of the behaviour of electrons in a metal is reminiscent of our discussion of delocalized molecular orbitals, particularly those from extended π systems described in section 4.8 on page 165. We saw there that an electron occupying an MO in such a system is not localized on one or two atoms, but is spread out over all of the atoms whose AOs contribute to the MO.

The picture we will develop of a metal is that some of the electrons occupy orbitals which are delocalized *throughout the entire structure*. These orbitals are formed in just the same way as MOs, but instead of being derived from a few AOs, they are formed from AOs on every atom in the whole solid. These orbitals, which extend throughout the entire structure, lead to what are known as *bands*, whose properties we can use to describe metallic bonding.

We will start by considering a one-dimensional band, which is the easiest to visualize, and then generalize this to more dimensions.

5.1.1   Bands in one dimension

Let us start by imagining a chain of lithium atoms, and the molecular orbitals they might give rise to. As we did when thinking about discrete molecules, we can ignore the two $1s$ orbitals as these are so contracted that they are unlikely to overlap with other orbitals. The $1s^2$ are the core electrons. What we are left with is a single (valence) electron in the $2s$ AO on each atom.

The MOs formed from a chain of $2s$ AOs can be constructed in exactly the same way as was described in section 4.8.1 on page 166 for a chain of $2p$ orbitals. Figure 5.1 on the facing page shows the resulting MOs for chains of between two and eleven atoms. The MOs are placed vertically according to their energies.

In each case, the lowest energy MO is one in which there is constructive interference between all the constituent AOs i.e. the AOs all have the same sign. The highest energy MO is when there is destructive interference between adjacent AOs i.e. the signs alternate. From the diagram it can be seen that as

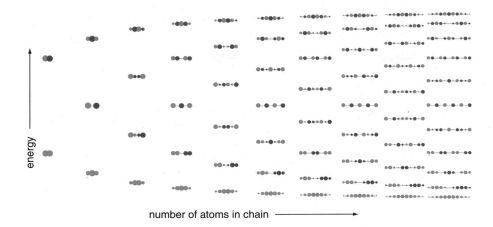

number of atoms in chain ⟶

Fig. 5.1 Representation of the form of the MOs, and their energies, for 2–11 s orbitals in a row. The lowest energy MO has constructive overlap between all adjacent AOs, whereas in the highest energy MO there is destructive overlap between adjacent AOs. As the number of orbitals increases the energy range between the lowest and highest energy MO tends towards a constant value, and the MOs cluster towards the ends of this range.

the number of atoms increases, the energy range which the MOs span increases at first but then starts to level out, so that the energy of the most bonding MO becomes independent of the number of atoms in the chain. The same is true of the most antibonding MO.

The reason for this is that, for a chain of N atoms, in the most bonding MO there are $(N-1)$ bonding interactions between AOs on adjacent atoms. The total number of bonding interactions per atom in the chain is therefore $(N-1)/N$, and in the limit that the number of atoms is large, this ratio becomes one. As a result, for chains with many atoms, the number of bonding interactions per atom is independent of the number of atoms, and therefore the energy of the most bonding MO is independent of the number of atoms. A similar argument applies to the most antibonding MO, in which the number of antibonding interactions per atom tends to one for large N.

We also note that for four or more atoms, the MOs are not evenly spread across the energy range, but tend to cluster at the top and bottom of the range. This point is made even more clearly in Fig. 5.2 which shows the energy levels for a chain of 200 atoms. As we approach the lowest and highest energies, the levels become more and more densely packed, to the point where the separation between levels is no longer clear from the diagram. A convenient way of describing the way in which the levels are spread out is to define a quantity known as the *density of states*. This is the *number* of energy levels in a small range of energies centred around a particular energy. As is shown in Fig. 5.2, for this chain the density of states is quite low in the middle of the energy range, but increases sharply at the highest and lowest energies.

In a macroscopic sample of a metal, there will not be twenty or 200 atoms in the chain, but more like 10^{20} atoms. Nevertheless, the pattern of the energies of the resulting MOs will be of the same form as we have described: they will cover a certain energy range, which is independent of the number of atoms, and there will be a clustering of the MOs at the extremities of this energy range. As these MOs encompass the whole sample, they are usually called *crystal orbitals* (COs).

The COs which result from the overlap of a particular AO form what is called a *band*. The band has a *width* which is the energy separation between the lowest energy CO and the highest energy CO.

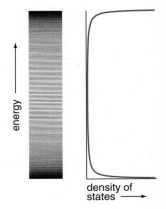

density of states ⟶

Fig. 5.2 On the left is shown the energies of the MOs for 200 atoms in a chain. On the right is shown the density of states, which is the number of energy levels in a small range of energies. Note how the density of states increases dramatically as we approach the lowest or highest energies.

A useful way of thinking about the formation of a band is to imagine a thought experiment in which we start out with a large separation between the atoms in our chain. At such large separations there is no interaction between the AOs, and so each of the N AOs is unaffected by being in the chain. There are thus N energy levels, but they are all the same.

As the atoms are moved closer together, the orbitals start to interact, giving rise to COs with a range of energies, from the most bonding to the most antibonding. In other words, a band is formed. The closer the atoms become, the stronger the interaction between the AOs, and so the most bonding CO drops further in energy, and the most antibonding CO goes up further in energy i.e. the width of the band increases, as is illustrated in Fig. 5.3. The width of the band therefore reflects the strength of the interaction between the AOs. Of course, if the atoms get too close together repulsive interactions will start to come into play, and the width of the band will stop increasing or may even decrease. Just as with a molecule, there will be a separation at which the optimum interaction occurs.

In the case of a chain of N lithium atoms, each atom contributes one electron to the band: there are thus N electrons to be accommodated in the band. The band itself contains N COs, as the number of COs must be the same as the number of AOs. As with MOs, two electrons can occupy each CO, so half the COs are occupied, starting from the lowest energy and working up to the non-bonding level in the middle of the band.

We therefore predict that the energy is lowered when the chain of atoms comes together, as this results in a lowering in the energy of the electrons as they occupy bonding COs. It should be remembered that the COs cluster at the lower energy end of the band, so the majority of the electrons are in COs with a significant bonding character.

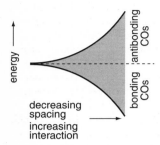

Fig. 5.3 Illustration of how the width of a band initially increases as the separation between the AOs decreases i.e. as the interaction between the AOs increases. The COs in the bottom half of the band are bonding, while those in the top half are antibonding. If the separation becomes too small, repulsive terms will start to be important and the band will cease to increase in width.

5.1.2 Conduction of electricity

One of the key features of a metal is its ability to conduct electricity, and the concept of a band gives us a ready explanation for this phenomenon. The key idea is illustrated in Fig. 5.4 on the facing page for the two cases of a partly filled band, and a filled band.

Let us start by considering the case of a partly filled band, shown in the upper part of the diagram. In (a) we see, in schematic form, the levels which comprise the band in three different parts of the sample; the electrons are shown as green dots. Of course in practice there would be vastly more COs in the band, and many more electrons, but these few levels shown here will suffice.

When an electric field is applied, as shown in (b), the electrons close to the negative end are raised in energy, and those close to the positive end are lowered in energy. We can think of this as leading to a change in the energies of the COs, as shown in (b). The result of applying the field is that the energies of the COs within the band vary across the sample.

With the field applied, there are now *empty* COs on the right of the sample which are *lower* in energy than *filled* COs on the left. Some electrons can therefore drop down into lower energy COs, as shown by the arrows in (c), thus effectively moving from left to right across the sample. This flow of electrons across the sample results in an electric current, so the sample is therefore a conductor of electricity.

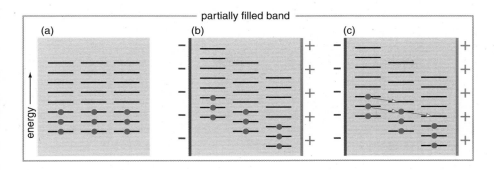

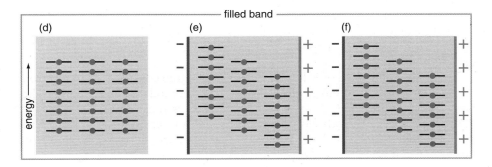

Fig. 5.4 Illustration of how a partially filled band leads to conduction. In (a) we see 'snap shots' of the COs from a band at three different locations in the sample; electrons are shown in green. Applying an electric field leads to a shift in the energies of the COs, as shown in (b). Electrons can then flow from left to right by dropping down into lower energy unoccupied COs, as shown by the arrows in (c). The material is therefore a conductor. If the bands are full, as shown in (d)–(f), no such flow is possible as there are no empty COs on the right for the electrons to move into.

Now consider the case where the band is full, shown in the lower part of the diagram. Just as before, applying an electric field shifts the energies of the COs to those shown in (e). However, as there are no empty COs on the right-hand side, there are no spaces for the electrons to flow into. Thus there is no current, and the sample does not therefore conduct electricity. We come to the very important conclusion that to conduct electricity a sample must have a *partially filled band*.

The picture shown in Fig. 5.4 is slightly deceiving as the COs extend throughout the sample rather than being localised as shown. In practice the effect of the electric field is to cause the energy of the COs to vary continuously across the sample. So, what we have in this figure is 'snap shots' of the COs at three different parts of the sample.

5.1.3   Bands in three dimensions

Although we have introduced the concept of a band by thinking about a one-dimensional chain of atoms, you can see in principle how the same idea can be extended to three dimensions. The AOs on each atom interact with those on all of the neighbouring atoms, giving rise to COs which extend in all directions through the solid. Just as in the one-dimensional case, the interaction of the AOs leads to the formation of bands which contain COs ranging from strongly bonding to strongly antibonding.

The detailed form of the bands depends on the exact three-dimensional arrangement of the atoms in the solid. As a result, the width of the band can be different in different directions in the solid. Overall, the behaviour of the bands in three dimensions is rather a complex matter, which is well beyond the scope of this text.

Overlapping bands

In the previous section, we considered the case of a chain of lithium atoms, and showed that the $2s$ AOs give rise to a band which is half full. This predicts that there is net bonding, and that there will be conduction along the chain.

Moving to three dimensions does not really alter this picture in a significant way. The overlap of N AOs gives rise to a band containing N COs, which vary between strongly bonding and strongly antibonding. Each lithium atom contributes one electron, so the band is half full, which explains why the metallic solid is lower in energy than the gaseous atoms, and also why the material is a conductor.

Moving to beryllium, which has the configuration $2s^2$, the same kind of band arises but this time there is a total of $2N$ electrons from N atoms, so the band is completely filled. This means that for every bonding CO that is occupied, an equivalent antibonding CO is also occupied. As a result we predict that there is no reduction in energy when the gaseous atoms form a solid. Furthermore, as the band is full, we do not expect the sample to conduct electricity.

These predictions are, of course, completely wrong. The stable form of beryllium under normal conditions is a metallic solid, which in fact has a *higher* enthalpy of vaporization than does lithium (324 kJ mol^{-1} as compared to 159 kJ mol^{-1}); in addition, beryllium is an excellent conductor of electricity.

In fact, what is going on here is that in addition to the band formed from the $2s$ AOs, there is another band formed from the overlap of the $2p$ AOs. This should come as no surprise, since we have seen in simple molecules that MOs are formed from the overlap of all the available AOs.

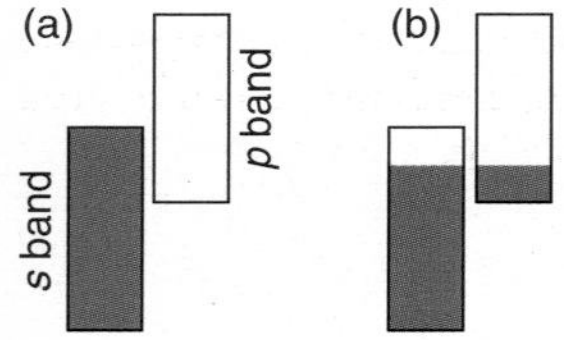

If this p band overlaps the s band, as shown in Fig. 5.5 (a), then the strength of bonding can be increased if some of the electrons at the *top* of the s band are transferred into the *bottom* of the p band, as shown in (b). The reason why this increases the strength of bonding is that the electrons from the top of the s band will be coming from antibonding COs and going into bonding COs which are at the bottom of the p band.

Therefore, we can rationalize the observation that beryllium is a metallic solid by supposing that there is a p band which overlaps in this way, thus allowing for there to be more bonding electrons than antibonding ones. This also results in partially filled bands which account for the high conductivity.

Just as with MOs, the details of the bands, such as their energies and widths, and whether or not they will overlap, are not something that we can predict from simple considerations. Therefore, we will be using bands to rationalize observations rather than make predictions. For example, in the case of beryllium we conclude that the p band must overlap the s band, but there is no simple way of predicting that this will be so, or the extent of the overlap.

5.1.4 Bands formed from MOs

Bands can be formed from the overlap of other kinds of orbitals than AOs, and in many cases this is a more natural way to approach the problem. In a metal, the atom is the fundamental object which repeats throughout the whole sample, but in other solids the repeating unit may consist of more than one atom.

A simple example of this is the case of solid hydrogen, which consists of discrete H_2 molecules. The repeating motif is therefore the hydrogen molecule, not a hydrogen atom. Given this, it is natural to form bands from the MOs of H_2, rather than the AOs on H.

Fig. 5.5 Illustration of the consequences of two bands overlapping. In (a) we see a full s band (shaded blue) overlapping with an empty p band. The energy of the electrons is decreased if some move from the top of the s band to the bottom of the p band, as shown in (b). Since the electrons have moved from antibonding COs to bonding COs, the strength of the bonding is increased.

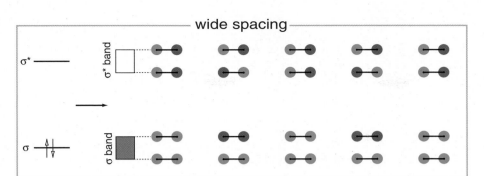

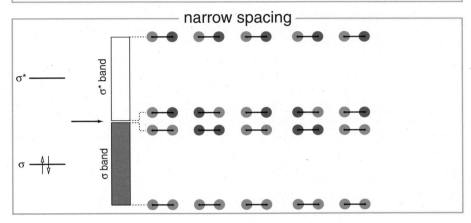

Fig. 5.6 Illustration of the bands formed from the σ and $\sigma^\star$ MOs of a linear chain of H_2 molecules (indicated by the black lines). The upper part shows the bands formed when the separation of the H_2 molecules is relatively large, giving a weak interaction between the MOs. For each band, the form of the CO with the lowest and highest energy is shown schematically. The lower part shows the case where the molecules are much closer together, so that the interactions between the MOs are larger; note the greater width of the bands. The highest energy CO of the σ band, and the lowest energy CO of the σ^* band have, on average, the same degree of bonding and antibonding, and so lie close in energy. As the separation of the molecules is decreased the two bands may touch or even overlap.

Just as with lithium, it is easiest to start out thinking about a linear chain of hydrogen molecules, and the one-dimensional bands that the orbitals from these molecules form. The process is visualized in Fig. 5.6.

First, look at the upper part of the diagram, which is for the case that the spacing between the H_2 molecules is large compared to the H–H bond length. A band is formed from the H_2 σ bonding MOs, and a separate band is formed from the $\sigma^\star$ antibonding MOs. For each of these bands, the COs with the lowest and highest energy are shown schematically.

For the σ band, the lowest energy CO is formed by having all of the σ MOs with the same sign; there is thus constructive overlap between the AOs from the two bonded hydrogen atoms and also, to a lesser extent, between MOs on adjacent H_2 molecules. The highest energy CO in this band is still formed from σ MOs, but this time the sign alternates between adjacent MOs. There is still bonding between the AOs from the two bonded hydrogen atoms, but there is antibonding between adjacent H_2 molecules.

In the σ^* band, the COs are formed from σ^* MOs. For the lowest energy CO in this band, the MOs alternate in sign so that there is a bonding interaction between adjacent MOs. For the highest energy CO, there is no such alternation, so there is an antibonding interaction between adjacent MOs. The amount of antibonding thus increases from the bottom to the top of the whole diagram.

When the spacing between the H_2 molecules is large compared to the H–H bond length, the interaction between MOs on adjacent atoms is small. Whether these interactions are bonding or antibonding thus only makes a small difference, and so the bands are narrow. The σ band is full, and does not overlap the empty $\sigma^\star$ band: the material is therefore an insulator.

Now turn to the bottom part of Fig. 5.6 on the preceding page, which is appropriate for the case that the separation between the H_2 molecules is much smaller. In fact, this part of the diagram has been drawn for the case where the hydrogen atoms are all evenly spaced. The form of the COs at the top and bottom of each band are still the same, but as the H_2 molecules are now so much closer, the bonding or antibonding interactions *between* the MOs is now much greater. As a result, the bands become wider.

A key point is that the highest energy CO from the σ band is now very similar to the lowest energy CO from the $\sigma^\star$ band. Both have interactions which go bonding–antibonding–bonding–antibonding as you go along the chain. If the spacing between the hydrogen molecules becomes small enough, the σ and $\sigma^\star$ bands will touch, and they then may even overlap. At this point the solid will become a conductor.

Although our picture is one-dimensional, it serves as a useful model for discussing the behaviour of solid H_2, which can be formed at sufficiently low temperatures (below 14 K). It is found that the solid is an insulator.

This fits in well with the picture given in the upper part of Fig. 5.6, where the σ band is filled, and the empty band is not overlapping. Also, since the band is filled, bonding and antibonding COs are occupied equally, so there is no bonding advantage to be gained by the individual molecules forming the solid. The fact that solid H_2 only forms at very low temperatures is indicative of very weak interactions between the molecules.

Interestingly, if solid hydrogen is subject to very high pressures (around 1.4×10^6 times atmospheric pressure) there is a phase change and the solid becomes highly conductive. We can interpret this by saying that the pressure forces the molecules together so that the interaction between them increases, as shown in the lower part of Fig. 5.6. Presumably at these high pressures the σ and $\sigma^\star$ bands must overlap, so electrons can move into the $\sigma^\star$ leading to partially filled bands, and hence electrical conductivity.

5.1.5   Band gaps and semiconductors

The separation between a full band and an empty band with which it does not overlap is called the *band gap*, as is illustrated in Fig. 5.7. We have already seen an example of such a band gap in the case of the chain of H_2 molecules, where the gap occurs between the σ and $\sigma^\star$ bands.

Often the filled band is called the *valence band*, as this is the band which has been filled by what are nominally the valence electrons. The empty band is called the *conduction band*, since if electrons were to be promoted to this band, the result would be electrical conduction, due to the presence of partially filled bands.

If the band gap is sufficiently small, electrons can be promoted from the valence band to the conduction band as a result of the thermal energy which the electrons acquire. This thermal energy is of the order of 2.5 kJ mol^{-1} at room temperature, or about 0.03 eV. For there to be significant numbers of electrons promoted to the conduction band, the band gap must be less than or comparable to this thermal energy. It is also possible for the absorption of sufficiently energetic photons of light to cause electrons to be promoted to the conduction band.

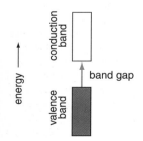

Fig. 5.7 Illustration of the concept of a band gap. This is the energy separation between the filled valence band and the empty conduction band.

Materials that conduct as a result of electrons being thermally promoted to the conduction band are called *semiconductors*. Typically, such materials are much poorer conductors of electricity than are metals, but they do nevertheless conduct to a significant extent. As the temperature is raised, the thermal energy is increased and so more electrons are promoted to the conduction band, resulting in an increase in conductivity. This behaviour is in contrast to metals which generally show a reduction in conductivity as the temperature rises.

Table 5.1 gives data on the band gaps for the elemental solids from Group 14. Of these elements, lead and the β form of tin have no measurable band gaps and are metallic conductors. The band gap in diamond is so much larger than the thermal energy that no electrons can be promoted to the conduction band, making diamond an insulator. The other elements have somewhat smaller band gaps, and so it is possible for some thermally excited electrons to enter the conduction band: these solids are therefore semiconductors.

The band gap clearly decreases as we go down the group. We can rationalize this trend by thinking about the the orbitals which form the band, and the strength of the interaction between these orbitals.

In diamond, we can imagine that adjacent carbon atoms are bonded together as a result of the occupation of the σ bonding MO formed when two sp^3 hybrids interact. At the same time, a σ^* MO is formed, but this is unoccupied. The sp^3 derived σ MOs can overlap to form the valence band, which will be filled. The σ^* MOs overlap to form the empty conduction band. A similar description is appropriate for the solid Si, Ge and α Sn, each of which has the same structure as diamond. In each case the sp^3 hybrids are formed from the orbitals in the valence shell (i.e. $n = 3$ for Si, $n = 4$ for Ge, and so on).

The band gap depends on two factors: firstly, the width of the valence and conduction bands; secondly, the energy separation of the σ and σ^* MOs from which these bands are formed. The stronger the interaction between the sp^3 HAOs, the larger the separation between the resulting σ and σ^* MOs. Generally speaking, the strength of the interaction between HAOs decreases as we go down a group. This is because the orbitals become larger and more diffuse, and so overlap less effectively. Thus, the separation between the σ and σ^* MOs decreases as we go down the group. This is illustrated in Fig. 5.8, in which the energy of the MOs is shown by the dashed lines.

When it comes to the formation of the bands from these MOs, similar considerations apply. We expect the strength of interaction to decrease as we go down the group, and so the width of the bands will decrease as we go down the group; again, this is illustrated in Fig. 5.8. The two effects therefore work in opposite directions to one another, so we cannot say which will win out. The fact that the band gap is found to decrease indicates, though, that the dominant factor is the separation of the σ and σ^* MOs.

5.1.6 Graphite

Graphite, which is an allotrope of carbon, represents an interesting contrast to the allotrope diamond. Whereas diamond is an insulator, graphite is quite a good conductor – not quite to the extent of a metal, but significantly greater than a typical semiconductor. We can find a rationalization of this behaviour using bands.

Graphite has a layered structure, and in each layer the carbon atoms form a hexagonal net. We can describe the bonding within a layer by assuming that

Table 5.1 Band gaps for the elements of Group 14

element	band gap	
	/ eV	/ kJ mol^{-1}
C (diamond)	6.0	580
Si	1.1	107
Ge	0.67	64.2
α Sn	0.08	7.7
β Sn	0	0
Pb	0	0

α Sn and β Sn are two *allotropes* tin, in the same way that diamond and graphite are allotropes of carbon.

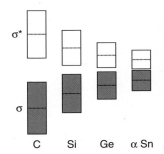

Fig. 5.8 Schematic diagram showing the band structure of the Group 14 elemental solids. In each there is a filled band arising from the overlap of σ type MOs, and an empty band arising from the overlap of $\sigma*$ MOs. The dashed lines show the energies of the MOs formed by the interaction of HAOs on adjacent atoms i.e. the approximate centre of the band. As we go down the group, the separation between the energies of these MOs decreases, but at the same time the width of the bands decreases. The overall result is that the band gap decreases.

Fig. 5.9 In graphite there are layers of carbon atoms which form a hexagonal net, as shown in (a). This pattern can be replicated by tiling a plane with the unit cell shown in (b). This cell contains just two atoms; for convenience the edges are shown in blue. Tiling a plane with the unit cell gives the arrangement shown in (c). Connecting adjacent carbon atoms gives (d), in which the hexagonal net is revealed.

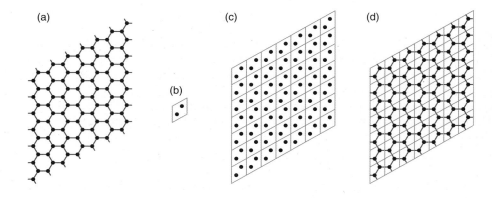

the carbons are sp^2 hybridized, so that each carbon forms three σ bonds with its neighbours. The out-of-plane $2p$ AOs form an extended π system which we can describe using band theory.

The hexagonal net of carbon atoms which forms each layer is shown in Fig. 5.9 (a). This pattern can be generated by 'tiling' a plane with a *unit cell*, shown in (b), which contains just *two* carbon atoms. By tiling we mean placing each unit cell next to another simply by moving it along or up (or both), but without rotating the cell. Due to their shape (a rhombus) the unit cells will entirely cover the plane giving the arrangement shown in (c). If we then join up adjacent carbon atoms the hexagonal net becomes clear, as shown in (d).

As there are just two atoms in the unit cell, it makes sense to construct the bands from the π and $\pi^\star$ MOs which are formed by the interaction of the out-of-plane $2p$ AOs on the two carbons. The overall picture is therefore rather similar to the situation for our linear array of H_2 molecules, shown in Fig. 5.6 on page 185, except that instead of the overlap between $1s$ AOs, with have the side-on overlap between $2p$ AOs.

The π MOs interact to form a band, as do the $\pi^\star$ MOs. If the spacing between adjacent carbons is the same – which it is in the graphite layer – the π and $\pi^\star$ bands just touch, as shown in the lower part of Fig. 5.6 on page 185. Each carbon atom contributes one electron via its $2p$ AO. Therefore, each unit cell contributes two electrons, so the π band is just filled. What we have is therefore a filled band in contact with an empty band: it is therefore easy for electrons to be promoted to the $\pi^\star$ band and so graphite is a conductor. It is found that the conductivity of graphite is much larger than a typical semiconductor, but somewhat less than a metal; for this reason, graphite is often called a semimetal. This fits in with the picture we have developed of the π and $\pi^\star$ bands just touching.

As we have described them, the bands are confined to the layers of carbon atoms. So, although we expect electrons to be able to move along the layers, we do not expect there to be a flow of electrons from one layer to another. The conductivity is thus *anisotropic*, meaning that it is not the same in all directions. This anisotropy has been verified experimentally for graphite.

The difference in electrical properties between diamond and graphite is a result of the different band structures, which in turn arises from the different three-dimensional structures the two allotropes adopt.

5.2   Ionic solids

An ionic solid is held together by the electrostatic forces between the charged ions. Of course, there may also be covalent interactions between such ions, but what we are going to be concerned with here are solids in which the electrostatic interactions are dominant.

For an ionic solid we can define a useful quantity called the *lattice enthalpy*, which is the enthalpy change when the gaseous ions are brought together, from infinite separation, to form the lattice e.g.

$$M^+(g) + X^-(g) \longrightarrow MX(s)$$

At infinite separation, the energy of interaction is zero. If a lattice forms, then there must be a reduction in energy when it is formed, so the lattice enthalpy must be negative. Typical values are -787 kJ mol^{-1} for NaCl, and -3850 kJ mol^{-1} for MgO. Some books define the lattice enthalpy as the enthalpy change when the ions in the lattice are separated to infinity. Using this definition, rather than the one above, simply results in a sign change in the quoted values.

As you probably know, lattice enthalpies can be determined using a *Born–Haber* cycle in which experimentally determined enthalpy changes for other processes are manipulated in a Hess' Law cycle. What we will see in this section is that it is possible to develop a simple model for ionic lattices which we can use to make a good estimate of the lattice energy. This is somewhat in contrast to covalently bonded molecules and solids, where no such 'back of the envelope' calculations are possible.

5.2.1   The ionic model for lattice enthalpies

There is an attractive force between two opposite charges which varies as $1/r^2$, where r is the distance between the charges i.e. the force increases as the distance between the ions decreases. If this were the only force acting, then two oppositely charged ions would simply accelerate into one another.

In an ionic solid the ions occupy fixed positions so we conclude that there must be a repulsive force acting which, at a particular separation of the ions, exactly balances the attractive force due to their charges. Our task is to work out the separation of the ions at which this balance lies, and what the energy of this arrangement is.

Since ultimately we want to work out the energy of a lattice, it is convenient to work in terms of the *energy* of interaction of the ions, rather than the force between them. The (potential) energy of interaction of two oppositely charged ions goes as $-1/r$. As r decreases, the potential energy becomes more negative i.e. it is favourable to bring the two ions closer and closer together. The blue curve plotted in Fig. 5.10 shows the form of this favourable interaction.

The repulsive force between the ions must result from a contribution to the potential energy which *increases* as the distance between the ions decreases i.e. making it unfavourable to bring the ions together. It turns out that there is such a contribution to the energy of interaction of all atoms, ions and molecules which results from the overlap of filled orbitals. When a filled orbital on one species interacts with a filled orbital on another the result is a filled bonding and a filled antibonding MO. As we have seen, the energy of the antibonding MO goes up by more than the energy of the bonding MO is lowered, so this interaction leads to an overall increase in energy. The closer we push the species

The lattice enthalpy is the energy released when the lattice is formed under constant pressure conditions. You will also see reference to the lattice *energy*, which is not quite the same thing as the lattice enthalpy, but the difference is so small as to be irrelevant for our purposes.

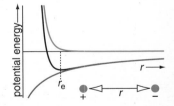

Fig. 5.10 There is a favourable interaction between two oppositely charged ions which varies as $-1/r$ (the blue curve). In addition, there is a unfavourable (repulsive) interaction which increases steeply as r decreases (the red curve). The total energy of interaction, shown by the black curve, thus shows a minimum, the position of which is indicated by the dashed line.

together the more the orbitals will overlap, and at very short distances even core orbitals are involved. Thus, the energy of interaction goes on rising.

Typically, it is found that the potential energy due to this interaction between filled orbitals varies as $1/r^n$, where the exponent n varies between 9 and 12. This contribution to the potential energy therefore rises very steeply as the atoms, molecules or ions come together, as shown by the red curve in Fig. 5.10 on the previous page. A value for n can be determined from experimental measurements on the compressibility of ionic solids. However, we will see below that for the present purposes the exact value of n does not matter too much.

The total potential energy, shown by the black line in Fig. 5.10 on the preceding page, is the sum of the favourable attractive and unfavourable repulsive contributions. At some separation of the two ions this total energy will be a minimum, a point which corresponds to the equilibrium separation, r_e. At this separation there is no net force on the ions.

Our aim is to calculate the energy of interaction of a regular three-dimensional array of cations and anions at their equilibrium separations, such as is formed in many ionic compounds. The energy of such an array will depend on the charges of the ions, the spacings between them, and the details of their spatial arrangement. It is easiest to see how such a calculation can be made by starting out with a simple one-dimensional arrangement of ions.

Energy of a chain of ions

Imagine an infinite chain of ions in which cations with charge $+z$ alternate with anions with charge $-z$, and the separation between adjacent ions is r; the arrangement is shown in Fig. 5.11.

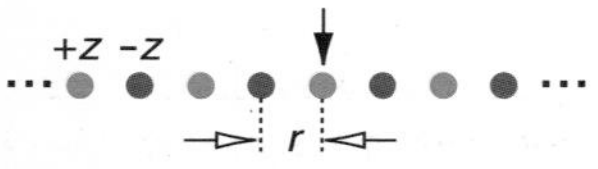

The potential energy (in joules) of two ions with charges z_1 and z_2 separated by a distance r is given by

$$\frac{z_1 z_2 e^2}{4\pi\varepsilon_0 r}, \tag{5.1}$$

where e is the elementary charge (the charge on the electron, 1.602×10^{-19} C), and ε_0 is a physical constant called the permittivity of vacuum (8.854×10^{-12} F m^{-1}). This energy of interaction between two charges is called the *electrostatic* or *Coulomb* energy.

We are going to first concentrate on the cation marked by the arrow in Fig. 5.11, and work out its energy of interaction with all the other ions in the chain. Although there are in principle a very large number of other ions, because they get further and further away the interaction gets weaker and weaker. It is therefore possible to work out the total interaction in a relatively straightforward way.

The ion to the right of the arrow is an anion and is separated from the cation by a distance r, so the energy of interaction is

$$\frac{-z^2 e^2}{4\pi\varepsilon_0 r}.$$

This expression has been obtained by putting $z_1 = +z$ and $z_2 = -z$ into Eq. 5.1.

The next ion to the right is a cation, and it is separated by $2r$ from the arrowed position, so the energy of interaction is

$$\frac{+z^2 e^2}{4\pi\varepsilon_0 (2r)}.$$

Do not confuse the exponent n with the principal quantum number.

Fig. 5.11 A linear chain of ions consisting of cations with charge $+z$ alternating with anions with charge $-z$. The spacing between adjacent ions is r.

This was obtained from Eq. 5.1 on the facing page with $z_1 = +z$ and $z_2 = +z$; note that the energy is positive as two cations are interacting. The next ion to the right is an anion separated from the arrowed cation by $3r$, so the contribution to the energy is negative and is $(-z^2 e^2)/(4\pi\varepsilon_0(3r))$. We can carry on in this way, adding a term for each ion, taking into account its separation and its change.

We must not forget that to the left of the arrowed cation there are also an equivalent set of ions to those on the right, so the total energy of interaction with the arrowed cation must take these into account. By symmetry, the interaction with the ions to the left is exactly the same as with those to the right.

The total energy is therefore the sum

$$2\left[-\frac{z^2 e^2}{4\pi\varepsilon_0 r} + \frac{z^2 e^2}{4\pi\varepsilon_0(2r)} - \frac{z^2 e^2}{4\pi\varepsilon_0(3r)} + \frac{z^2 e^2}{4\pi\varepsilon_0(4r)} \cdots\right],$$

where the factor of 2 is to account for the ions on the left and on the right of the arrowed cation.

Tidying this up somewhat, the total Coulomb or electrostatic energy is

$$\varepsilon_{\text{Coulomb}} = -2\frac{z^2 e^2}{4\pi\varepsilon_0 r}\left[1 - \frac{1}{2} + \frac{1}{3} - \frac{1}{4} \cdots\right]$$

After a large number of terms, the quantity in the square bracket converges to a value of $\ln 2 = 0.693$. It is usual to combine this term with the factor of 2 into a single constant $\mathcal{A}$, called the *Madelung constant*. In this case $\mathcal{A}$ takes the value 1.386.

So far we have just computed the energy of one cation in the chain. To find the energy per mole of the hypothetical ionic compound $M^{+z}X^{-z}$ with a linear chain of ions all we need to do is to multiply $\varepsilon_{\text{Coulomb}}$ by Avogadro's constant N_A to give

$$E_{\text{Coulomb}} = -N_A\mathcal{A}\frac{z^2 e^2}{4\pi\varepsilon_0 r}.$$

You might think that we should compute the electrostatic energy of the anions as well, and add this to the energy of the cations. However, to do this would count each ion–ion interaction twice over, and so would be incorrect.

An important point about this expression is that because the ions are regularly spaced at a distance r, the energy depends only on this distance and the Madelung constant, which is itself simply a function of the geometrical arrangement of the ions. Note that the charge on the ions is *not* included in the value of $\mathcal{A}$, but appears as a separate factor.

We now need to take into account the energy due to the repulsion between the ions. As we have seen, this is very short range, going as $1/r^9$ or an even higher power. Thus, it is sufficient to just consider the repulsion between neighbouring ions. Following the same line of argument as we used above, the energy, per formula unit, due to these interactions can be written as $N_A C/r^n$, where C is some constant.

The total energy is thus

$$E = -N_A\mathcal{A}\frac{z^2 e^2}{4\pi\varepsilon_0 r} + \frac{N_A C}{r^n}.$$

As was explained above, this energy will be a minimum at some particular value of r, the equilibrium separation r_e, and this is the separation we would expect to find in practice.

Example 5.1 Energy of a chain of ions

By evaluating Eq. 5.2 for some typical parameters we can see what kind of energies are involved in the formation of a one-dimensional array of ions. In solid LiF (which is of course three-dimensional), the distance between the ions is 201 pm: we will take this as the separation of the ions in the chain. Taking n to have a typical value of 9, we can evaluate Eq. 5.2 as follows:

$$
\begin{aligned}
E_{\text{equil.}} &= -N_A \mathcal{A} \frac{z^2 e^2}{4\pi\varepsilon_0 r_e}\left(1 - \frac{1}{n}\right) \\
&= -6.022 \times 10^{23} \times 1.386 \frac{1^2 \times (1.602 \times 10^{-19})^2}{4\pi \times 8.854 \times 10^{-12} \times 201 \times 10^{-12}}\left(1 - \frac{1}{9}\right) \\
&= -8.52 \times 10^5 \text{ J mol}^{-1} \\
&= -852 \text{ kJ mol}^{-1}.
\end{aligned}
$$

The bond energy of a discrete LiF molecule (in the gas phase) is found to be 574 kJ mol^{-1}, so you can see that the energy of the array of ions is easily comparable with the bond energy.

Altering the value of n from 9 to 12 increases the energy to 878 kJ mol^{-1}, a change by only 3%; thus, as was noted above, the value of n is not critical in the calculation.

Table 5.2 Madelung constants for various crystal structures

structure	$n_C : n_A$	$\mathcal{A}$
linear	2:2	1.386
square net	4:4	1.613
wurtzite, ZnS	4:4	1.641
NaCl	6:6	1.748
CsCl	8:8	1.763
rutile, TiO_2	8:4	2.408
fluorite, CaF_2	8:4	2.519
cuprite, Cu_2O	2:4	2.221

If you are familiar with elementary calculus you can easily work out the value of r at which the minimum in the energy occurs: this is explored in one of the exercises. The value of r will depend on the various constants, including C and n, both of which are unknown. The usual procedure is to assume that we know a value for r_e, for example from experimental data as will be explained below, and that n is also known. We can then use the energy minimization step to find the value of C, thus eliminating it from the expression for the total energy. The result of these manipulations is the following expression for the lattice energy at the equilibrium separation

$$
E_{\text{equil.}} = -N_A \mathcal{A} \frac{z^2 e^2}{4\pi\varepsilon_0 r_e}\left(1 - \frac{1}{n}\right). \tag{5.2}
$$

This is a rather straightforward expression which allows us to estimate the lattice energy, provided we have information about the typical values of r_e and n. Such a calculation is given in Example 5.1.

Generalization to three dimensions

If we move from a line of ions to a three-dimensional lattice, the only thing that changes in the calculation is the value of the Madelung constant, as this is the part that depends on the geometrical arrangement of the ions. Working out the value of $\mathcal{A}$ is not entirely straightforward as the series involved tend to converge only rather slowly. However, values of the Madelung constant have been computed for the common crystal structures, and these values are listed in Table 5.2. Several of the commonly adopted crystal structures are shown in Fig. 5.12 on the next page.

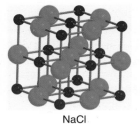
NaCl

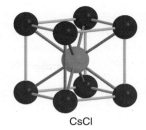
CsCl

ZnS

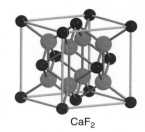
CaF₂

Fig. 5.12 Four commonly found structures of ionic crystals. In NaCl (Cl⁻ in green, Na⁺ in blue) the closest neighbours to each Na⁺ are six Cl⁻, and similarly there are six Na⁺ around each Cl⁻: the coordination number of both ions is six. In contrast, in the CsCl structure (Cl⁻ in green, Cs⁺ in blue) the Cl⁻ is eight coordinate, and the same is true for the Cs⁺, although this is not evident from the fragment shown here. In the wurtzite structure of ZnS (Zn²⁺ in blue, S²⁻ in yellow) the coordination number of each ion is four. In the fluorite structure, exemplified by CaF₂, each Ca²⁺ (blue) is surrounded by eight F⁻ (green), whereas each F⁻ is surrounded by four Ca²⁺. Nearest neighbour contacts are indicated by the grey cylinders joining the ions; the pale blue cylinders are simply there as a guide to help in visualizing the arrangement of the ions in a cube. It is important to realize that what are shown here are simply illustrative parts of the entire structure.

The table also gives the coordination numbers for the cation, n_C, and anion, n_A. The coordination number of the cation is the number of nearest-neighbour anions, and similarly the coordination number of the anion is the number of nearest-neighbour cations. For the linear chain, n_C and n_A are both two, and for the square net these values increase to four. In the NaCl structure the coordination numbers of both the cations and anions is six, whereas for the CsCl structure the coordination numbers are both eight. In the wurtzite structure, adopted by ZnS, the coordination number is four. For salts where the ratio of anions to cations is not one to one, the coordination numbers of the two ions are different. For example, in CaF₂ (which adopts the fluorite structure) the coordination number of the Ca²⁺ is eight, whereas that of the F⁻ is four.

From the tabulated values we can see that the value of $\mathcal{A}$ increases on going from one to two then to three dimensions. Furthermore, for ionic salts with the formula MX, increasing the coordination number also results in a modest increase in the Madelung constant. We can rationalize these trends by noting that as we increase the number of dimensions, or the coordination number, there are more ions with which a given ion can interact.

Table 5.2 also lists the values of the Madelung constant for the fluorite and rutile structures, which can be adopted by salts with the formula MX₂, and for the cuprite structure which can be adopted by salts with the formula M₂X. We note that these values are considerably larger than those for the the salts MX. Recall that the value of $\mathcal{A}$ does not take into account the charges on the ions, so this increase in $\mathcal{A}$ has to be due to the geometrical arrangement of the ions. In a salt MX₂ there are twice as many anions as cations, and so the number of favourable interactions is increased when compared to a salt MX. As a result, the Madelung constant is increased.

By using the appropriate value of the Madelung constant, the lattice energy can be calculated from

$$E_{\text{lattice}} = -N_A \mathcal{A} \frac{z_+ z_- e^2}{4\pi\varepsilon_0 r_e}\left(1 - \frac{1}{n}\right). \tag{5.3}$$

◉ *Weblink 5.1*
View and rotate in real time the structures shown in Fig. 5.12.

This is a minor modification of Eq. 5.2 on page 192 in which instead the charges on the ions being z, the charges are generalized to z_+ for the cation and z_- for the anion. Note that z_+ and z_- are *not* signed quantities e.g. for Mg^{2+} $z_+ = 2$, and for O^{2-} $z_- = 2$.

Equation 5.3 gives the energy of interaction of the ions when they are formed into the lattice. We can assume that when the ions are infinitely separated the energy of interaction is zero, so that the energy change of the reaction

$$a M^{z+}(g) + b X^{z-}(g) \longrightarrow M_a X_b(s)$$

is $E_{lattice}$.

The enthalpy change for this reaction is the *lattice enthalpy*, $\Delta H^\circ_{lattice}$. At 0 K $\Delta H^\circ_{lattice} = E_{lattice}$, but at finite temperatures there is a small term to add to $E_{lattice}$. However, as this correction is much less than 1% of the value of $E_{lattice}$, we can safely ignore it and simply write

$$\Delta H^\circ_{lattice} = -N_A \mathcal{A} \frac{z_+ z_- e^2}{4\pi\varepsilon_0 r_e}\left(1 - \frac{1}{n}\right). \tag{5.4}$$

5.2.2   Ionic radii

In an ionic solid, the closest approach is between two ions of opposite sign i.e. an anion and a cation, and this distance can be determined from X-ray diffraction experiments. A large number of such distances have been measured, and a study of these data reveals the following particularly interesting facts.

The difference between anion–cation spacings in MCl and MF, i.e. $r_{MCl} - r_{MF}$, turns out to be more or less independent of the nature of M. For example, as we go along the series M = Na, K, Rb, Cs, the difference varies between just 46 pm and 50 pm. Similarly, the difference $r_{KX} - r_{NaX}$, where X = F, Cl, Br, I only varies between 30 pm and 36 pm.

These observations suggest that the interionic spacing can be thought of as having a contribution from the cation, r_+, and a contribution from the anion, r_-, such that $r = r_+ + r_-$. Furthermore, the value of r_+ for a particular cation is more or less independent of the anion which it is paired within, and likewise for the anion. The distances r_+ and r_- are called the *ionic radii* of the cation and anion, respectively.

It is possible to draw up a table of these radii such that, to some reasonable approximation, the interionic distance for any ionic compound can be predicted simply by adding together the appropriate values for the ions present. We can then use this value for r_e in Eq. 5.3 on the previous page in order to compute the lattice energy, provided that the structure, and hence the Madelung constant, is known.

The optimum values of the ionic radii are found to depend on the coordination number. Most commonly, the tabulated values of these radii are for sixfold coordination, as is the case for the data given in Table 5.3 on the facing page. From this table we can see a number of trends.

- 2+ ions have smaller ionic radii than 1+ ions in the same period.

- The ionic radii of anions are generally much larger than those of cations in the same period.

- Within the same period, 2− anions have similar radii to 1− anions.

- The ionic radius increases in size as we go down a group.

Table 5.3 Ionic radii for sixfold coordination

ion	r_+ / pm	ion	r_+ / pm	ion	r_- / pm	ion	r_- / pm
Li$^+$	68			O^{2-}	142	F$^-$	133
Na$^+$	100	Mg^{2+}	68	S^{2-}	184	Cl$^-$	182
K$^+$	133	Ca^{2+}	99	Se^{2-}	197	Br$^-$	198
Rb$^+$	147	Sr^{2+}	116	Te^{2-}	217	I$^-$	220
Cs$^+$	168	Ba^{2+}	134				

These trends are relatively easy to rationalize by thinking about the behaviour of the electrons in an atom as electrons are added or removed. Removing an electron will result in a reduction in electron–electron repulsion, and so the effective nuclear charge will increase, pulling the remaining electrons in towards the nucleus. Removing two electrons simply causes a larger effect, and in addition, Mg has a higher nuclear charge than Na, further increasing the effective nuclear charge in Mg^{2+}, and hence contracting the ion further.

Conversely, adding an electron increases the electron–electron repulsion, and so decreases the effective nuclear charge, causing the the electrons to move out from the nucleus. On going from S^{2-} to Cl$^-$ the nuclear charge in increased by one, which causes a contraction, but one additional electron is added which causes an expansion. The two effects seem to more or less cancel one another out, resulting in the two anions having rather similar sizes. Finally, we generally expect the size of an atom or ion to increase as we go down a group simply because the outer electrons are being placed in higher and higher shells.

The values of the ionic radii are not unique, but depend on the details of the criteria used to find a set of values which are a 'best fit' to the experimental data. You will therefore find different values quoted in different places (a quick trawl of the web gave values for the ionic radius of Na$^+$ varying from 95 pm to 116 pm, although values close to 100 pm are the commonest).

5.2.3 The Kapustinskii equation

If we just require a fairly rough and ready estimate of the lattice energy, then the approach suggested by the Russian chemist Anatolii Fedorovich Kapustinskii can be very useful. Kapustinskii noticed that the fraction

$$\frac{\text{Madelung constant, } \mathcal{A}}{\text{number of ions in one molecule, } n_{\text{ions}}}$$

was roughly constant at about 0.87 for a wide range of compounds. So, in Eq. 5.3 on page 193, we can replace $\mathcal{A}$ with $0.87\,n_{\text{ions}}$. In addition, setting $n = 9$ is a reasonable compromise since, as we have already seen, variations in n do not affect the lattice energy strongly. The lattice energy can therefore be approximated in the following way

$$
\begin{aligned}
E_{\text{lattice}} &= -N_A \mathcal{A} \frac{z_+ z_- e^2}{4\pi\varepsilon_0 r_e}\left(1 - \frac{1}{n}\right) \\
&\approx -N_A\, 0.87\, n_{\text{ions}} \frac{z_+ z_- e^2}{4\pi\varepsilon_0(r_+ + r_-)}\left(1 - \frac{1}{9}\right).
\end{aligned}
$$

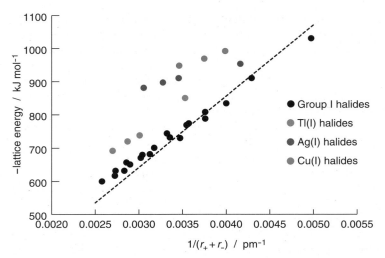

Fig. 5.13 Plot showing the correlation between the experimentally determined lattice energy and $1/(r_+ + r_-)$, where r_+ and r_- are the ionic radii of the cation and anion, for a series of salts MX. The dashed line is the prediction of the Kapustinskii equation. The lattice energies of the Group I halides show a good correlation with $1/(r_+ + r_-)$, whereas for the halides of Tl(I), Cu(I) and Ag(I) the experimental values are significantly greater in size than the predictions of the ionic model. This is attributed to covalent contributions to the lattice energy for these salts.

Note that we have replaced r_e with the sum of the ionic radii. Substituting in for all the constants we find

$$E_{\text{lattice}} \; / \; \text{kJ mol}^{-1} \approx \frac{-1.07 \times 10^5 \, n_{\text{ions}} z_+ z_-}{(r_+ + r_-)/\text{pm}}. \tag{5.5}$$

Note that in this expression the ionic radii are in pm and the lattice energy is in kJ mol^{-1}. This is the Kapustinskii equation for estimating lattice energies, and as noted above $\Delta H^\circ_{\text{lattice}} \approx E_{\text{lattice}}$.

For example, using the radii given in Table 5.3 on the preceding page the Kapustinskii equation predicts a lattice energy for NaCl of -762 kJ mol^{-1}; the experimentally derived value is -787 kJ mol^{-1}. There is a deviation of around 3% between these two values, which is hardly significant.

Sometimes, the values of the ionic radii are adjusted so that the Kapustinskii equation gives values as close as possible to the experimental lattice energies. These radii, known as *thermochemical radii*, can differ significantly from those obtained from crystal structures. This approach is generally used for complex anions, such as SO_4^{2-}, for which it is hard to assign a 'radius' based on a crystal structure.

5.2.4   Validity of the ionic model

The ionic model assumes that the lattice energy is simply a consequence of an electrostatic interaction between the ions, together with a short-range repulsive term. It is perhaps remarkable that such a simple model is capable of reproducing lattice energies which are so close to the experimental values, and indeed one might be forgiven for being somewhat suspicious of the level of agreement. Remember that if a theory predicts values which are in good agreement with experiment it does not *necessarily* mean that the theory is correct.

The truth of the matter is that it is likely that in many 'ionic' solids there are significant covalent interactions between the atoms. Just as was the case for simple molecules, the strength of these covalent interactions depends on the energy gap between, and overlap of, the AOs involved.

The greater the covalent contribution becomes, the less charge separation there is between M and X, and so the smaller the electrostatic contribution. So, to some extent, these two contributions compensate for one another, which is probably one of the reasons why the simple electrostatic model is so successful.

Figure 5.13 on the preceding page illustrates what happens when the co-valent contribution starts to become more significant. Here we see plotted, as black dots, the lattice energies of the Group I halides (LiF ... LiI through to CsF ... CsI) against $1/r$, where r is the sum of the radii of the anion and cation. If we ignore variations in the Madelung constant and the repulsion parameter n, then we expect the lattice energy for these salts to vary as $1/(6r_+ + r_-)$.

The graph shows that there is indeed a good correlation with $1/(r_+ + r_-)$. Furthermore, the prediction of the Kapustinskii equation, shown by the dashed line, is in remarkably good agreement with the experimental data, given the assumptions involved.

However, if we look at Tl(I), Ag(I) and Cu(I) halides (shown by red, blue and green dots, respectively), there is much poorer agreement between experiment and theory. In all cases, these halides have lattice energies which are *greater* in magnitude than the values predicted by the ionic model, a difference which we can attribute to the presence of significant covalent contributions to the lattice energy.

Presumably, compared to the Group I metals, the orbital energies of Cu, Ag and Tl are significantly lower and so are closer to the energies of the halide AOs. As a result there is a greater interaction, and hence a greater covalent contribution to the lattice energy.

5.3 Moving on

It seems as if we have spent a lot of time and effort describing bonding, but this effort is both necessary and well spent as the ideas we have developed over these initial chapters form the bed rock of our understanding of structure and reactivity.

Our next step is to look at precisely what it is that determines whether one species is more 'stable' than another: this is the domain of thermodynamics.

FURTHER READING

Inorganic Chemistry, fourth edition, Peter Atkins, Tina Overton, Jonathan Rourke, Mark Weller and Fraser Armstrong, Oxford University Press (2006).

QUESTIONS

5.1 In Fig. 5.1 on page 181 are shown the MOs (or COs) formed from a chain of s orbitals. We can just as well form such COs from chains of $2p$ AOs. Assuming that the chain is aligned along the z-axis, sketch the form of the *lowest* and *highest* energy COs for a chain composed of: (a) $2p_x$ AOs, and (b) $2p_z$ AOs. Comment on any differences between the COs you have drawn.

5.2 Use the idea of overlapping bands to rationalize why both lithium and beryllium are metallic conductors, and why the enthalpy of vaporization of beryllium (found to be 324 kJ mol^{-1}) is significantly greater than that of lithium (found to be 159 kJ mol^{-1}).

5.3 The bond length on the discrete (gaseous) Li$_2$ molecule is 267 pm, and the bond dissociation energy is 105 kJ mol^{-1} (i.e. 52.5 kJ is required to create one mole of Li atoms). For metallic lithium, the Li–Li spacing is 304 pm, and the enthalpy of vaporization is 159 kJ mol^{-1} (i.e. 159 kJ is required to create one mole of Li atoms). Discuss these data in the light of the different type of bonding in Li$_2$(g) and Li(m).

5.4 What is the difference between a metallic conductor and a semiconductor, and how can this difference be explained using band theory? Explain why diamond is an insulator, whereas silicon is a semiconductor.

If a small fraction of the Si atoms in solid silicon are replaced by phosphorus atoms, it is observed that the conductivity of the material increases significantly. Rationalize this observation (hint: how many electrons does a P atom have compared to a Si atom?).

5.5 What do you understand by the statement 'BN is isoelectronic with C$_2$'?

Solid boron nitride (BN) can exist in two forms. The α-form consists of hexagonal layers, is a soft material, and is a electrical insulator with a band gap of around 5.2 eV. The β-form is a very hard, abrasive material, which is also an insulator. By comparison with the allotropes of carbon, propose a structure for the β-form of BN, and explain why the α-form is an insulator.

5.6 Graphite can be 'doped' by exposing the material to potassium vapour. In the resulting materials the potassium atoms are found to fit *between* the layers of carbon atoms in the original structure, and it is also found that doping increases the electrical conductivity greatly. Use band theory to explain why doping with potassium increases the conductivity (hint: the outer electron from a K atom is easily removed: where might it be transferred to?).

Similarly, doping with bromine vapour leads to the incorporation of Br atoms between the layers, again accompanied by an increase in the conductivity. Explain why this is so.

5.7 Use the standard enthalpy changes given for the following processes to compute the lattice enthalpy of $MgBr_2(s)$:

$$Mg(s) \longrightarrow Mg(g) \quad 147 \text{ kJ mol}^{-1}$$
$$Br_2(l) \longrightarrow 2\,Br(g) \quad 224 \text{ kJ mol}^{-1}$$
$$Mg(g) \longrightarrow Mg^+(g) \quad 738 \text{ kJ mol}^{-1}$$
$$Mg^+(g) \longrightarrow Mg^{2+}(g) \quad 1451 \text{ kJ mol}^{-1}$$
$$Br(g) \longrightarrow Br^-(g) \quad -325 \text{ kJ mol}^{-1}$$
$$Mg(s) + Br_2(l) \longrightarrow MgBr_2(s) \quad -524 \text{ kJ mol}^{-1}$$

5.8 (Requires a knowledge of elementary calculus) In section 5.2.1 on page 189 it was shown that the energy of a chain of ions with alternating charges is given by

$$E = -N_A \mathcal{A} \frac{z^2 e^2}{4\pi\varepsilon_0 r} + \frac{N_A C}{r^n}.$$

The value of r at which this energy is a minimum, the equilibrium separation r_e, is found by computing the derivative with respect to r, and then setting this derivative equal to zero:

$$\frac{dE}{dr} = 0$$

Assuming that all of the other parameters do not depend on r (i.e. are constants), find the derivative and, by setting it equal to zero, show that the constant C is given by:

$$C = \frac{\mathcal{A} z^2 e^2 r^{n-1}}{4\pi\varepsilon_0 n}.$$

Substitute this expression for C into the original expression for the energy E and hence, after some tidying up, obtain Eq. 5.2 on page 192.

5.9 Use the Kapustinskii equation, Eq. 5.5 on page 196, along with the radii listed in Table 5.3 on page 195 to estimate the lattice enthalpy of $MgBr_2$. Compare your answer to the value obtained in question 5.7.

Calculate the lattice enthalpy using the full expression given in Eq. 5.4 on page 194 using the following values for the parameters: $r_e = 270.7$ pm, $\mathcal{A} = 2.355$ and $n = 9$ (recall that $e = 1.602 \times 10^{-19}$ C and $\varepsilon_0 = 8.854 \times 10^{-12}$ F m^{-1}). You need to be careful with units: r_e must be in m, and the energy will come out in J mol^{-1}.

5.10 (a) The discrete diatomic molecule MgO is known in the gas phase and has a bond length of 175 pm. If we *assume* that the this molecule consists of Mg^{2+} and O^{2-} ions held together by an electrostatic interaction then we can work out the energy change for

$$Mg^{2+}(g) + O^{2-}(g) \longrightarrow Mg^{2+}O^{2-}(g)$$

by using Eq. 5.2 on page 192 with $\mathcal{A} = 1$:

$$E_{\text{dimer}} = -N_A \frac{z^2 e^2}{4\pi\varepsilon_0 r_e}\left(1 - \frac{1}{n}\right).$$

Taking $n = 7$ and $z = 2$, and being careful with the units, compute E_{dimer} in kJ mol^{-1}.

Hence, given the following standard enthalpy changes

$$Mg(g) \longrightarrow Mg^{2+}(g) \quad 2188 \text{ kJ mol}^{-1}$$
$$O(g) \longrightarrow O^{2-}(g) \quad 703 \text{ kJ mol}^{-1},$$

determine the enthalpy change for

$$Mg(g) + O(g) \longrightarrow Mg^{2+}O^{2-}(g).$$

(b) Solid MgO is a crystalline solid with an Mg–O spacing of 210.2 pm. Compute its lattice enthalpy using Eq. 5.4 on page 194 taking $\mathcal{A} = 1.7476$ and $n = 7$. Hence determine the enthalpy change for

$$Mg(g) + O(g) \longrightarrow MgO(s).$$

(c) Compare and comment on your answers to (a) and (b).

Thermodynamics and the Second Law

Key points

- All spontaneous processes are accompanied by an increase in the entropy of the Universe; this is the Second Law of Thermodynamics.
- Entropy is a property of matter which can be associated with the way in which molecules are distributed amongst energy levels.
- Entropy can also be defined in terms of the heat involved in a process.
- The Gibbs energy decreases in a spontaneous process.
- The equilibrium constant is related to the standard Gibbs energy change for a reaction, $\Delta_r G^\circ$: $\Delta_r G^\circ = -RT \ln K$.
- $\Delta_r G^\circ$ can be computed from tabulated enthalpies of formation and absolute entropies.
- Thermodynamics can be used to predict the position of equilibrium, and how this responds to external changes e.g. pressure, temperature and concentration.

We now come to a very important topic in our study of chemistry, which addresses one of the key questions which chemists seek to answer. This is, what determines whether a reaction will 'go' or not, or put more subtly, when a reaction comes to equilibrium, what determines whether the equilibrium will favour the products or the reactants, and to what extent. Understanding the answer to this question is not only of fundamental scientific importance, but is also of great practical significance as it helps us to determine and understand the factors which will lead to the greatest yield in a chemical reaction.

We will see that the approach to chemical equilibrium is governed by the celebrated *Second Law of Thermodynamics*, which involves the physical property known as *entropy*. These ideas arise within the physical theory generally known as *thermodynamics*. The name is something of a misnomer, as the theory is concerned with much more than heat and temperature.

Thermodynamics is a theory which describes the behaviour of *bulk* matter, in contrast to quantum mechanics which focuses in on the behaviour of *individual* atoms and molecules. Of course, it is certainly the case that the behaviour of a mole of hydrogen molecules must be intimately connected to the properties of a single hydrogen molecule, and this connection can be made in a precise way using *statistical thermodynamics*. However, the principles of thermodynamics are not dependent on an understanding of the microscopic behaviour of atoms and molecules.

In this chapter we will tread a path which lies somewhere between the molecular and thermodynamic views of chemical equilibria. We will develop the key ideas using thermodynamics, but we will also appeal frequently to our understanding of the behaviour of atoms and molecules in order to interpret the results we derive. Nowhere is this dual approach more evident than when it comes to getting to grips with the somewhat mysterious concept of entropy.

Thermodynamics is, by scientific standards, a rather old theory, having its roots in the early nineteenth century, and being more or less fully developed by the beginning of the twentieth. It has, however, stood the test of time very well, so we can have confidence in its predictions. Einstein is quoted as having said

> [Classical thermodynamics] is the only physical theory of universal content, which I am convinced, that within the framework of applicability of its basic concepts will never be overthrown.

Praise indeed, from the man who overthrew classical physics!

For chemists, the most important result in thermodynamics is the relationship between the equilibrium constant for a reaction, K, and a quantity known as the *standard Gibbs energy change*, $\Delta_r G^\circ$

$$\Delta_r G^\circ = -RT \ln K.$$

In turn $\Delta_r G^\circ$ is related to the standard enthalpy change of the reaction, $\Delta_r H^\circ$, and the standard entropy change, $\Delta_r S^\circ$

$$\Delta_r G^\circ = \Delta_r H^\circ - T\Delta_r S^\circ.$$

These relationships allow us to understand the factors which determine the size of the equilibrium constant i.e. whether a reaction will favour products or reactants. It will take us quite a while to develop these key relationships from more fundamental ideas. However, the effort is well worthwhile as they are a key part of our understanding of chemical processes.

6.1 Spontaneous processes

If we drop a lump of sodium metal into water, there is a violent reaction leading to the formation of NaOH and hydrogen gas. If we ignite a mixture of hydrogen and oxygen, they combine very rapidly to form water. If we mix the gases NH_3 and HCl, clouds of NH_4Cl form at once. All of these chemical reactions can be described as *spontaneous*, meaning that once initiated they 'go' on their own, without further intervention from us.

There are also physical processes which can be described as spontaneous. For example, solid NaCl dissolving in water, the mixing of ethanol and water, and the freezing of liquids at low temperatures.

The reverse of these spontaneous processes do not take place on their own. Water does not break apart into hydrogen and oxygen, NH_4Cl does not dissociate back to NH_3 and HCl, and a mixture of water and ethanol does not separate out. Of course we can force these changes to take place, for example water can be converted back to hydrogen and oxygen by electrolysis, and heating to high temperatures will dissociate NH_4Cl. However, these changes are fundamentally different from the spontaneous processes which happen 'on their own'.

The spontaneous reactions mentioned above are rather extreme examples in which the reaction goes almost exclusively to products. In general, chemical reactions go to an equilibrium position in which there are appreciable amounts of both reactants and products. The remarkable thing is that at equilibrium there is a particular relationship, given by the equilibrium constant, between the concentrations of the reactants and products. For example, dissolving ethanoic acid in water sets up the equilibrium

$$CH_3COOH + H_2O \rightleftharpoons CH_3COO^- + H_3O^+.$$

The equilibrium constant, K, for this reaction is defined as

$$K = \frac{[CH_3COO^-][H_3O^+]}{[CH_3COOH][H_2O]},$$

where the square brackets indicate concentration. The key point is that, at a given temperature, the value of K has a particular fixed value.

When we add ethanoic acid to water, there is a spontaneous reaction in which some of the acid reacts with the water to give H_3O^+ and ethanoate ions. However, this reaction does not continue until all of the acid has dissociated, but only until the point where the concentrations are in the ratio set by the value of K. Once this point is reached, no further changes in the concentrations are observed.

Put in more general terms, there is a spontaneous process which leads to the establishment of equilibrium. Once this point is reached there is no further change.

Is energy minimization the criterion for a spontaneous process?

When we think of spontaneous processes, most of the examples which come readily to mind – such as combustion – are reactions in which there is a significant release of energy in the form of heat. Such processes are described as being *exothermic*. In an exothermic process the products are lower in energy than the reactants, so as the reaction proceeds the reduction in the energy of the chemical species appears as heat, as is illustrated in Fig. 6.1.

It is thus tempting to say that these reactions are spontaneous *because* the products are lower in energy than the reactants. However, a moment's thought will reveal that this interpretation is *entirely wrong*.

If the products have to be lower in energy than the reactants, then all spontaneous reactions would have to be exothermic. However, there are examples of spontaneous *endothermic* processes e.g. when solid NH_4NO_3 dissolves in water or when ice melts. There are also processes which are spontaneous even though they are not accompanied by any energy change e.g. the mixing of gases.

The establishment of chemical equilibrium, such as that between NO_2 and N_2O_4, also provides an example of how both exothermic and endothermic processes can be spontaneous. At low temperatures nitrogen dioxide, NO_2,

Later on in this chapter we will see that this is not quite the right way to write the equilibrium constant for this reaction.

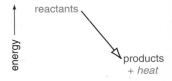

Fig. 6.1 If the products are lower in energy than the reactants, then in going from reactants to products there will be a reduction in energy. This energy will appear as heat, making the reaction *exothermic*.

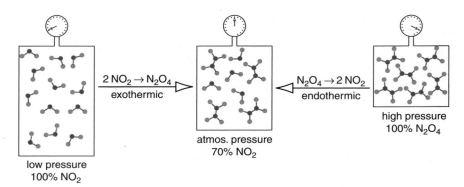

Fig. 6.2 At low pressure, NO_2 exists mostly as the monomer, as shown on the left. Increasing the pressure to atmospheric pressure causes some of the NO_2 to dimerize to give N_2O_4: this is a spontaneous exothermic process. At high pressures, dimerization is almost complete and the sample is almost entirely N_2O_4, as shown on the right. Decreasing the pressure results in some of the dimers dissociating to NO_2: this is a spontaneous endothermic process. The approach to equilibrium can therefore involve both exothermic and endothermic processes.

dimerizes to form dinitrogen tetroxide, N_2O_4. As a bond is formed in this reaction, the process is exothermic

$$\text{exothermic:} \quad 2\,NO_2(g) \longrightarrow N_2O_4(g)$$

The reverse reaction, in which a bond is broken, is therefore endothermic

$$\text{endothermic:} \quad N_2O_4(g) \longrightarrow 2\,NO_2(g).$$

In general the monomer and dimer are in equilibrium:

$$2\,NO_2(g) \rightleftharpoons N_2O_4(g).$$

As is illustrated in Fig. 6.2, at low pressures the monomer NO_2 is favoured, but if we then compress the gas a new equilibrium is established in which some of the dimer is formed: this is an exothermic process. On the other hand, at high pressures the dimer N_2O_4 is favoured, but if we then release the pressure, equilibrium is established by some of the dimer dissociating to give NO_2: this is an endothermic process. So, depending on where we start from, equilibrium can be established by either endothermic or exothermic processes.

It is thus clear that energy changes are not a predictor of whether or not a process will be spontaneous. We will see in the next section that it is in fact changes in a quantity called entropy which we need to consider in order to decide whether or not a process will be spontaneous. Later on, we will see that energy changes *are* important, but only to the extent that they influence entropy changes.

6.2 Properties of matter: state functions

Before we start on our discussion of entropy and the Second Law, it is useful to pause for a moment and introduce the idea that there are certain quantities which can be described as the 'properties of matter'.

A familiar example of such a property is *density*, the mass per unit volume. Each pure substance, at a given temperature and pressure, has a particular density which is entirely independent of how that substance has been prepared or previously treated. The density is thus some fundamental property which a substance has: it is therefore a 'property of matter'.

In thermodynamics, such properties of matter are often described as *state functions* because the value that they take only depends on the state of the substance, and not on how it was brought to that state. The 'state' is specified by variables such as temperature and pressure.

It is very useful to identify these properties of matter, or state functions, as their values can be measured and tabulated for future use. As we will see, we can use such tabulated data to compute useful quantities, such as equilibrium constants.

Another state function which you have probably already come across is the *enthalpy*, probably in the form of enthalpies of reaction, $\Delta_r H°$. These $\Delta_r H°$ values are actually the change in the enthalpy when going from reactants to products. For example, for the reaction

$$H_2(g) + \tfrac{1}{2}O_2(g) \longrightarrow H_2O(g)$$

$\Delta_r H°$ is -242 kJ mol^{-1}. This means that the *change* in enthalpy on going from one mole of H_2 and half a mole of O_2 to one mole of H_2O is -242 kJ. Such a change has a definite value since the enthalpy, being a state function, has definite values for the reactants and the products.

It is important to understand that when the value of $\Delta_r H°$ is quoted as -242 kJ mol^{-1}, the specification 'per mole' does not mean that the enthalpy change is for one mole of product or one mole of reactant. Rather, 'per mole' means that the quoted enthalpy change is for when the reaction proceeds according to a specified balanced chemical equation in which the numbers of moles of reactants and products are given by the stoichiometric coefficients. If we wrote the above reaction as

$$2H_2(g) + O_2(g) \longrightarrow 2H_2O(g),$$

then $\Delta_r H°$ would be *twice* the value quoted above i.e. $2\times(-242) = -484$ kJ mol^{-1}. For the reaction

$$CH_4(g) + 2H_2O(g) \longrightarrow CO_2(g) + 4H_2(g)$$

$\Delta_r H°$ is 165 kJ mol^{-1}, which means that when one mole of CH_4 reacts with two moles of H_2O to give one mole of CO_2 and four moles of H_2, the change in enthalpy is 165 kJ.

There are quite a lot more of these properties of matter or state functions in thermodynamics, but the ones we are going to be most interested in are enthalpy, entropy and Gibbs energy. They will all be introduced in this chapter.

This reaction is used commercially for the production of hydrogen from natural gas (methane).

6.3 Entropy and the Second Law

Whether or not a process will be spontaneous, i.e. whether or not it will happen, is governed by the Second Law of Thermodynamics

Second Law: In a spontaneous process, the entropy of the Universe increases.

This is a rather simple statement, but it will take us quite a while to work out firstly what it means, and then how to apply it in a practical way. Our first task will be to discuss what entropy is, and how it can be determined. Having done this, we will then go on to look at how the entropy change of something as large as the Universe can be found, and thus how we can apply the Second Law.

Before we start with our discussion of entropy, it is worthwhile noting that, like all physical laws, the Second Law cannot be 'proved'. However, its validity is established by comparing its predictions with experimental observations. It has been found that this law leads to predictions which are backed up by experiment, and so we take this as evidence that the law is valid. Do not forget, though, that all physical laws have their limitations. Thermodynamics applies only to bulk matter, not to individual molecules, just in the same way that Newton's Laws apply to macroscopic objects, and not to individual atoms and molecules.

6.3.1 A microscopic view of entropy

If you have ever heard about entropy, then the one idea you are likely to have encountered is that entropy is something to do with 'randomness': apparently, the more 'random' a substance is, the higher its entropy. So, gases have higher entropy than solids, as the chaotic movement of the molecules in a gas is more random than the orderly arrangement one finds in a solid.

This idea of entropy being associated with randomness is at first attractive, but some further thought reveals significant problems. For example: what is 'randomness'? How can it be determined in a quantitative way? Is one gas more random than another? Clearly, to answer any of these questions we are going to have to be much more precise than simply saying that entropy is 'randomness'.

Entropy is a property of matter, or a state function. There are a number of ways of defining entropy, but to start with by far the easiest approach is to take a microscopic view and think about how the behaviour of the individual molecules and atoms in a sample affects its entropy.

In a macroscopic sample there are an enormous number of molecules, and in turn these molecules have available to them a very large number of *energy levels*. We came across the idea of energy levels when discussing the behaviour of electrons in atoms and molecules. You will recall that quantum mechanics restricts the energy of the electron to certain discrete values: the energy is not allowed to vary continuously as it does in the classical world. As was discussed in section 1.10 on page 26, in addition to electronic energy levels, molecules also have available to them many more energy levels associated with translation, rotation and vibration.

When we were looking at the way in which electrons occupy energy levels (orbitals), we always placed the electrons in the lowest energy levels available, subject to the rule that only two electrons can occupy any one orbital. Although it is possible to move the electrons to higher energy orbitals, this usually involves a large increase in energy as the orbitals are quite widely spaced in energy, and as a result it is not likely to take place. However, compared to the electronic energy levels, the energy levels due to vibration, rotation and (especially) translation are much more closely spaced. As a result, many more of these energy levels than just the lowest ones are occupied.

The question is, which of the many energy levels available to it will a particular molecule occupy? Clearly, with so many molecules and energy levels involved, it is quite impossible for us to answer this question – and in any case, as the molecules bump into one another they are constantly changing which energy level they are in. However, as the numbers of molecules and levels involved is very large, it turns out that we can work out *on average* the number of molecules which are occupying each energy level. We will see that from this knowledge we can then determine the entropy.

A simple example

It it helpful at this point to focus on a system containing rather a small number of molecules and energy levels, so that we can actually look at the way in which the molecules are arranged in the energy levels. Although the system we are going to look at is very small, it will nevertheless show the key features which also apply to the enormously larger number of molecules in a macroscopic system.

In our simple system we will imagine that we have just 14 molecules, and that each molecule has available to it a set of energy levels with energies 0, 1, 2, 3 ... in arbitrary units. It will make things simpler if we assume that the molecules are distinguishable from one another, for example by having labels A, B, C .... Of course, in practice this is not going to be the case, but the final conclusions of this section are not really altered by this somewhat unrealistic assumption.

Let us suppose that the *total* energy possessed by these 14 molecules is 10 units. Clearly there are a large number of different ways of arranging the molecules amongst the energy levels such that their total energy is 10. One such arrangement, usually called a *distribution*, is

$$\left|\begin{array}{l|cccccc} \text{energy, } \varepsilon_i & 0 & 1 & 2 & 3 & 4 & 5 \\ \text{population, } n_i & 8 & 3 & 2 & 1 & 0 & 0 \end{array}\right|$$

The upper row of the table shows the energy of each level, ε_i, and the lower row shows the number n_i in each level, usually called the *population* of the level. This distribution is also illustrated in Fig. 6.3 (a).

The total energy is found by multiplying the population of each level by its energy, and then summing over all the occupied levels

$$\begin{aligned} E &= n_0\varepsilon_0 + n_1\varepsilon_1 + n_2\varepsilon_2 + n_3\varepsilon_3 \ldots \\ &= 8 \times 0 + 3 \times 1 + 2 \times 2 + 1 \times 3 \\ &= 10. \end{aligned}$$

As we required, the total energy is 10 units.

Since we decided that the molecules are distinguishable, there are many different ways in which this distribution of molecules amongst energy levels can be achieved. For example, the molecule in level 3 could be any one of the 14 molecules. Likewise, the two molecules in level 2 could be any two of the remaining 13 molecules, and so on. Working out the number of ways in which distinguishable objects can be arranged in a particular distribution is quite a simple problem in statistics which you may already have come across. If you have not, then you might like to refer to Box 6.1 on the next page.

If the population of the *i*th level is n_i, and the total number of molecules is N, then the number of ways, W, in which the distribution can be achieved is

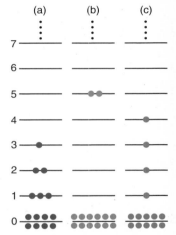

Fig. 6.3 Three possible distributions of 14 molecules amongst a set of evenly spaced energy levels; only the first few energy levels are shown. Each distribution has the same total energy of 10 units.

Box 6.1 Ways of arranging balls in boxes

Suppose we have three boxes and 7 differently coloured balls, and want to arrange the balls in the boxes in the following way:

first box	second box	third box
4	2	1

There are 7 balls to choose from when we select the first ball to go into the first box, then 6 as we select the second ball, 5 as we select the third, and 4 as we select the fourth. The total number of ways of choosing the four balls for the first box is therefore $7 \times 6 \times 5 \times 4$. Now we fill the second box. For the first ball we have 3 left to choose from, and for the second we have 2 to choose from. The total number of ways of filling the second box is thus 3×2. Finally, there is only one ball left to choose for the third box. The total number of ways of filling the boxes is thus $(7 \times 6 \times 5 \times 4) \times (3 \times 2) \times (1)$, which is written 7! (spoken 'seven factorial').

However, we have counted too many ways, as it does not matter in which *order* we choose the 4 balls to go into the first box. To compensate for this we need to divide the number of ways computed so far by the number of ways of arranging the 4 balls in the first box. As we can choose any of the 4 balls first, then any of the remaining 3 balls, and so on, the number of ways of arranging the balls in the box is $4 \times 3 \times 2 \times 1 = 4!$. Similarly, the number of ways of arranging the balls in the second box is 2!, and in the third box it is 1!. So, the total number of ways of arranging the balls in the boxes is

$$\frac{7!}{4! \times 2! \times 1!} = 105.$$

In general, the number of ways, W, of arranging N distinguishable objects in a number of boxes such that there are n_1 in the first, n_2 in the second and so on is given by

$$W = \frac{N!}{n_1! \times n_2! \times n_3! \dots}.$$

We take $0! = 1$, so that empty boxes have no effect on the calculation.

$$W = \frac{N!}{n_0! \times n_1! \times n_2! \dots}.$$

In this case $N = 14$, $n_0 = 8$, $n_1 = 3$, $n_2 = 2$ and $n_3 = 1$, so W is 1.8×10^5.

There are many other ways of arranging these 14 molecules such that the total energy is 10 units. Another possible distribution (shown in Fig. 6.3 (b)) is

energy, ε_i	0	1	2	3	4	5
population, n_i	12	0	0	0	0	2

The number of ways this distribution can be achieved is (recall that $0! = 1$):

$$W = \frac{14!}{12!\,2!}.$$

This is 91, considerably less than for the first distribution.

Another possibility (shown in Fig. 6.3 (c)) is

$$\left|\begin{array}{l} \text{energy, } \varepsilon_i \\ \text{population, } n_i \end{array}\right.\left|\begin{array}{cccccc} 0 & 1 & 2 & 3 & 4 & 5 \\ 10 & 1 & 1 & 1 & 1 & 0 \end{array}\right|$$

for which W is 2.4×10^4.

It can be shown mathematically that the distribution with the largest value of W, called the *most probable distribution*, is the one in which the populations obey the Boltzmann distribution (introduced in section 1.10.1 on page 27)

$$n_i = n_0 \exp\left(\frac{-\varepsilon_i}{k_B T}\right). \tag{6.1}$$

In this expression T is the absolute temperature, and k_B is a fundamental constant called Boltzmann's constant, which takes the value 1.381×10^{-23} J K^{-1}. The Boltzmann distribution predicts that as the energy of a level increases, its population decreases in a particular way which depends on the temperature. In addition, the population of a level with energy ε_i will only be significant if $\varepsilon_i \leq k_B T$. If $\varepsilon_i > k_B T$ the exponential term $\exp(-\varepsilon_i/k_B T)$ will be very small, thus making the population small.

The factor $k_B T$ has units of energy (J), and can be regarded as a measure of the average thermal energy available to a molecule. Thus, only those levels whose energies are less than or comparable with $k_B T$ will be populated significantly.

If the number of molecules is very large, such as would be the case in a macroscopic sample, it is found that the most probable distribution has an overwhelmingly larger value of W than any other distribution. So, to all practical intents and purposes, we can assume that the molecules are *always* distributed in this way.

The argument behind this assumption is that the molecules are constantly rearranging themselves amongst the energy levels, thereby creating different arrangements. However, as the most probable distribution can be achieved in an overwhelming larger number of ways than any other distribution, virtually all of the arrangements correspond to the most probable distribution.

Now comes the really crucial part. Boltzmann hypothesized the following connection between the entropy, S, and the number of ways, W_{max}, in which the *most probable* arrangement can be achieved

$$S = k_B \ln W_{max}, \tag{6.2}$$

where as before k_B is the Boltzmann constant. This equation is arguably one of the most profound in the whole of physical science as it relates the entropy, which is a bulk property, to W which is determined by the energy levels available to the molecules. We can use this relationship to understand the way in which entropy responds to heating, changes in physical state, and pressure.

The units of entropy

From Eq. 6.2 we can see that, since W_{max} is just a number, the units of the entropy are the same as the units of the Boltzmann constant i.e. J K^{-1}. As with other thermodynamic quantities such as enthalpy, it is common to quote the entropy as a molar quantity, meaning the entropy of one mole of the substance. In this case, the units of S are J K^{-1} mol^{-1}.

The first distribution we looked at of our 14 molecules with 10 units of energy is the one predicted by the Boltzmann distribution.

🌐 *Weblink 6.1*

This link takes you to a page where you can explore the relationship between the populations, the spacing of the energy levels and the value of $k_B T$, as predicted by the Boltzmann distribution.

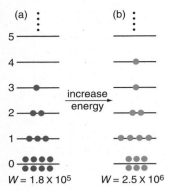

Fig. 6.4 In both (a) and (b), the molecules are distributed according to the Boltzmann distribution, but (a) has a total of 10 units of energy whereas (b) has 15 units. As a result, the molecules are distributed over more energy levels in (b), and so the number of ways, W, is increased. As a consequence, the entropy increases.

In section 6.5.1 on page 215 we will look at how actual values of the entropies of substances are determined. A typical value is that for helium gas which has an entropy of 126 J K^{-1} mol^{-1} at atmospheric pressure and 298 K.

6.3.2 Using Boltzmann's hypothesis

Heating the sample

If we heat the sample so as to raise its temperature, its energy must increase and so the molecules have to move up to higher energy levels. As before, the populations are given by the Boltzmann distribution, Eq. 6.1 on the previous page, using the appropriate temperature. For example, for the simple system we considered above, increasing the temperature so that the energy is increased by 50% to 15 units gives the following distribution (illustrated in Fig. 6.4):

energy, ε_i	0	1	2	3	4	5
population, n_i	6	4	2	1	1	0

W for this distribution is 2.5×10^6, significantly greater than the value of 1.8×10^5 for the original distribution. The reason why W has increased is that the molecules are spread out more amongst the energy levels. Since $S = k_B \ln W_{max}$, the entropy has therefore increased. We conclude therefore that heating a sample (thus raising its temperature) causes the molecules to move to higher energy levels, and thereby increases the entropy.

For an ideal gas, such as helium, doubling the temperature (but keeping the pressure constant) leads to an increase in the entropy of 14 J K^{-1} mol^{-1}, which is about a 10% change on the value at 298 K and 1 atmosphere.

Effect of temperature

We have already shown that the entropy of a sample is increased when its energy is increased by heating. What we will now show is that for a fixed amount of energy, the increase in entropy depends on the *initial* temperature of the sample.

Let us consider three distributions, each of which conforms to the Boltzmann distribution, and has 14 particles. The first distribution has 5 units of energy, the second has 10, and the third has 15. As a result, the temperature of the third is greater than the second, which in turn is greater than the first. These distributions are illustrated in Fig. 6.5, and are tabulated below.

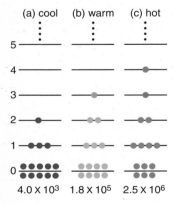

Fig. 6.5 All three distributions have 14 molecules, but the energy increases from 5 to 10 to 15 units as we go from left to right. As a result, the value of W (given beneath each distribution) increases. Note, however, that W increases much more in going from (a) to (b), than it does on going from (b) to (c).

	ε_i	0	1	2	3	4	5	W	$\ln W$
lowest temp. ($E = 5$)		10	3	1	0	0	0	4.0×10^3	8.29
intermediate temp. ($E = 10$)		8	3	2	1	0	0	1.8×10^5	12.1
highest temp. ($E = 15$)		6	4	2	1	1	0	2.5×10^6	14.7

As expected, W (and hence $\ln W$) increases when the energy increases, that is when the temperature is increased. However, adding 5 units of energy to the lowest temperature system causes a much greater increase in W than does adding the same amount of energy to the intermediate temperature system. As a result, the increase in entropy is greater in the first case.

This turns out to be a general effect. The *lower* the temperature of the sample, the *greater* the increase in entropy when a set amount of energy is supplied to it.

Expanding a gaseous sample

It was described in section 1.10 on page 26 that for gases the vast majority of the accessible levels are those associated with translation. These energy levels can be modelled in a simple way using elementary quantum mechanics, and what we find is that the spacing of these levels varies *inversely* with the volume occupied by the gas. Therefore, expanding a gas results in the energy levels moving closer together.

We can see what effect this has on the calculation of W by returning to our example with 14 molecules and 10 units of energy. If we now make the spacing of the energy levels $\frac{1}{2}$ of one unit of energy, the most probable distribution becomes

energy, ε_i	0	0.5	1.0	1.5	2.0	2.5
population, n_i	5	4	2	1	1	1

which is illustrated in Fig. 6.6. The number of ways of distributing the particles in the more closely spaced levels is 1.5×10^7, which is much larger than the value of 1.8×10^5 found when the level spacing is one unit. Bringing the energy levels closer together has resulted in the molecules being spread out more, and thus W has increased.

We conclude therefore that expanding a gaseous sample (i.e. increasing its volume) *increases* its entropy. Since pressure is inversely proportional to volume, it follows that increasing the pressure of a gas *decreases* its entropy.

For example, doubling the volume (but keeping the temperature constant) of an ideal gas such as helium increases the entropy by 5.8 J K^{-1} mol^{-1}, which is about 5% of the value at 298 K and 1 atmosphere pressure.

Increasing the molecular mass

The same piece of elementary quantum mechanics which tells us how the spacing of the energy levels changes with volume also tells us that the spacing varies *inversely* with the *mass* of the molecules. Thus increasing the mass, like increasing the volume, decreases the spacing of the energy levels and so increases the entropy.

For example, the entropies of He, Ne, Ar, Kr, and Xe (all at 298 K and 1 bar pressure) are 126, 146, 155, 164 and 170 J K^{-1} mol^{-1}; there is a steady increase as the mass of the atoms increase.

Changes of state

Gases are significantly different from solids and liquids in that they have available to them a very large number of translational energy levels. We should therefore expect that W for a gas, and hence the entropy, will be much greater than that for a solid or liquid.

Although the molecules in a liquid are much more constrained than those in a gas, they are certainly freer to move than they are in a solid. Modelling the energy levels available in a liquid is not at all easy, due to the strong interactions between the molecules. Nevertheless, we can reasonably assume that on account of the freer motion in a liquid there will be more energy levels available than there are in a solid. We therefore expect that the entropy of a substance in the liquid state will be higher than in the solid state.

For example, the entropies at 298 K of solid water, liquid water and gaseous water are 38, 70 and 189 J K^{-1} mol^{-1} respectively. As expected, the entropy of the gas is much greater than that of either the solid or the liquid.

▷ In Chapter 16 we will see how simple quantum mechanics can be used to model these translational energy levels of a gas. What we will find is that the spacing of the levels decreases both as the volume increases and as the mass of the molecules increases.

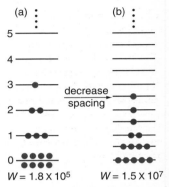

$W = 1.8 \times 10^5$ $W = 1.5 \times 10^7$

Fig. 6.6 Both distributions have 14 molecules and 10 units of energy, but in (b) the spacing of the energy levels is halved compared to (a). As the molecules are spread over more levels, the value of W is increased.

🌐 *Weblink 6.1*

This link takes you to a page where you can explore how the entropy is affected by the spacing of the energy levels and the value of $k_B T$. The applet illustrates very clearly how decreasing the energy level spacing, or increasing the temperature, increases the entropy.

Under normal conditions solid water (ice) does not exist at 298 K; the value of the molar entropy quoted here at 298 K has been extrapolated from the value at a lower temperature.

6.3.3 Summary

Boltzmann's hypothesis shows us that the entropy of a substance is related to the way in which the molecules are distributed amongst the energy levels. Using this, we have been able to predict that the entropy of a substance will behave in the following way:

- The entropy increases with temperature i.e. as energy is supplied to the system, its entropy will increase.

- Absorption of a given amount of energy gives rise to a larger increase in entropy the lower the initial temperature.

- The entropy increases as a gas is expanded, and decreases as a gas is compressed.

- The entropy of a gas increases as the mass of the atoms/molecules increases.

- The entropy of the gaseous state of a substance is greater than that of the liquid state, which in turn is greater that the entropy of the solid state.

There is no mention of 'randomness' in this list, and the idea that there is a connection between entropy and randomness is really best abandoned at this point. The connection which is useful to us is that between entropy and the distribution of molecules amongst energy levels.

It is possible to develop this view of entropy as deriving from the distribution of molecules amongst energy levels in such a way that we can actually calculate the value of the entropy: this is the theory called *statistical thermodynamics*. However, we are not going to follow this topic any further here, partly because the mathematics involved is rather more complicated than we wish to use at the moment, and partly because we want to introduce an alternative view of entropy which does not rest on an understanding of molecular energy levels. This latter view will turn out to be most useful for the discussion of chemical equilibrium.

You must not think that this description of the molecular basis of entropy has been a pointless diversion. On the contrary, we will constantly refer back to this microscopic interpretation of entropy in order to interpret the results we are about to derive from a more 'classical' point of view.

6.4   Heat, internal energy and enthalpy

In the next section we are going to define entropy in terms of the heat involved in a process, so before doing that it is as well to refresh our understanding of what heat is. Having done this, we will introduce an important concept in thermodynamics which is that of a *reversible* process. We need to understand what such processes are in order to understand the definition of entropy in terms of the heat.

Finally, we will look in a little more detail at precisely how the familiar enthalpy change is related to the heat, and introduce another thermodynamic variable called the internal energy.

6.4.1 What is heat?

Heat is a familiar term from everyday life: we are used to talking about 'heating things up', 'letting the heat out', 'turning on the heater' and so on. However, its familiarity can be something of a problem when it comes to pinning down a more exact definition, which is what we need in thermodynamics.

Heat is a form of energy, in particular that form of energy which is involved in bringing a hot object placed in contact with a cooler one to a common temperature, as is illustrated in Fig. 6.7. Although we talk about heat 'flowing' from the hot object to the cooler one, it is important to realize that heat is not some kind of mysterious fluid or other substance. Rather, heat is just our word to describe a certain kind of energy transfer which takes place between objects. The word 'flow' should not really be used when when it comes to talking about heat, but if we abandon this common usage the phrasing becomes so awkward and tortuous as to be quite unusable. So we will continue to talk about heat flow, but keep in the back of our minds that this is rather inaccurate terminology.

Heat is not a property of matter, that is it is not one of the state functions described in section 6.2 on page 204. We *cannot* talk about objects having a certain amount of heat in the same way that we can talk about them having a certain amount of entropy. Rather, heat is something that 'happens' during a particular process.

As a result of heat not being a state function, it follows that the amount of heat involved in going from one state to another depends on precisely how that change is brought about. There is one way of bringing about the change which is of particular significance in thermodynamics: this is to bring about the change *reversibly*. In the next section, we explore what this means.

Reversible heat

In thermodynamics, a *reversible* process is one whose direction can be changed by a very small (*infinitesimal*) change in some variable. For example, imagine that we have two objects A and B which are at the same temperature and in contact with one another. If the temperature of A is raised by a very small amount, then there will be a flow of heat from A to B in order to equalize the temperatures. On the other hand, if the temperature of A is lowered by the same small amount, then there will be a flow of heat from B to A, as shown in Fig. 6.8. The transfer of heat is reversible since an *infinitesimal* change in the temperature caused a change in the direction of the flow.

As there is only an infinitesimal temperature difference between the two blocks, the amount of heat which flows between them will also be infinitesimal. If we made the temperature difference much larger – say ten degrees – then a finite amount of heat would flow. However, the flow would *not* be reversible since a infinitesimal change in the temperature will not cause the direction of flow to reverse. To change its direction we would need to alter the temperature by a much larger, finite, amount (at least ten degrees). Such a heat transfer is said to be *irreversible*.

If we want to transfer a finite amount of heat from A to B in a reversible way, then we need to keep adjusting the temperature of A so that it is always just infinitesimally above that of B. At any point the direction can be changed by an infinitesimal lowering of the temperature of A, so the process is reversible. However, as A is kept infinitesimally hotter than B, there is a steady flow of heat from A to B.

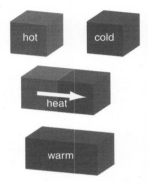

Fig. 6.7 When a hot object is placed in contact with a cold object, energy is transferred from the hot to the cold object in order to equalize their temperatures. This transfer of energy is the process we call heat.

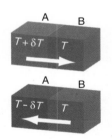

Fig. 6.8 Illustration of the concept of a reversible process. If object A is infinitesimally hotter than object B, a small amount of heat will flow from left to right. However, the direction of flow can be reversed by making object A infinitesimally cooler than B. A process whose direction can be altered by an infinitesimal change in a variable (here the temperature) is said to be *reversible*.

As you can see from this description, a reversible process is not something that can be achieved in practice – rather it is an idealization or limit of a real process. Any real process which is actually observed to take place cannot, more of less by definition, be reversible.

6.4.2 Internal energy and enthalpy

In thermodynamics it is possible to define two quantities called the *internal energy* (given the symbol U), and the *enthalpy* (given the symbol H). Like entropy, the internal energy and the enthalpy are properties of matter (state functions, as described in section 6.2 on page 204), so we can talk about a substance having a certain amount of internal energy or enthalpy.

The really useful feature of these two new quantities is that they are related to the heat under certain special circumstances. If an amount of heat is transferred to the system under conditions in which the *volume* remains constant, it can be shown that the heat is equal to the change in the internal energy, ΔU. Likewise, if the heat is transferred to the system under conditions of constant *pressure*, the amount of heat is equal to the change in the enthalpy, ΔH.

The internal energy and the enthalpy are state functions, i.e. they are properties of matter. ΔU and ΔH are therefore independent of the way in which a change is achieved, and, in particular, their values are the *same* for both reversible and irreversible processes.

> ⇨ We will prove these assertions in Chapter 17.

6.5 Entropy in terms of heat

Entropy can also be defined in a way which makes no reference to energy levels, but nevertheless leads to the same properties as we have already described. Such a definition could be described as being 'classical' in the sense that is does not rely on results from quantum mechanics.

The definition is as follows: if a small amount of heat δq_{rev} is supplied *under reversible conditions* to a system at temperature T_{sys}, the change in entropy of the system is given by

$$\text{change in entropy} = \frac{\delta q_{rev}}{T_{sys}}. \tag{6.3}$$

A lower case Greek 'delta', δ, is used to indicate a small amount or a small change. So δq indicates a small amount of heat. A larger, finite, change is indicated by a capital delta, Δ.

The qualification 'reversible conditions' is important since, as we have just seen, the amount of heat depends on the way in which the change is brought about.

We can see immediately that this definition leads to two conclusions about the entropy change. Firstly, it predicts that if heat is supplied to the system (an endothermic process, δq is positive), the entropy of the system increases. Secondly, it predicts that the increase in entropy for a fixed amount of heat is greater the lower the temperature. It is encouraging that both of these predictions are in accord with what we have already deduced about entropy in section 6.3.2 on page 210 by taking a molecular point of view.

We will now go on to explore how this definition of entropy can be used to determine the actual value of the entropy, and then how it can be used in conjunction with the Second Law to determine whether or not a process will be spontaneous.

Box 6.2 Heat capacities

When a certain amount of heat δq is supplied to a substance, its temperature rises in direct proportion to the amount of heat supplied. The relationship between the heat supplied and the temperature rise δT is

$$\delta q = C \times \delta T.$$

The constant of proportion between the heat and the temperature rise is called the *heat capacity*, C.

As described in section 6.4.2 on the preceding page, under constant pressure conditions, the heat is equal to the enthalpy change i.e. $\delta q = \delta H$. So, the above relationship can be written

$$\delta H = C_p \times \delta T,$$

where the subscript p has been added to the heat capacity to remind us that this is the value appropriate for constant pressure conditions.

The value of the heat capacity is relatively simple to measure. All we have to do is to measure the temperature rise of the sample as a known amount of energy is supplied, usually by electrical heating. We then plot a graph of the heat input against temperature, and the slope at any temperature T gives the heat capacity directly. It is found that the heat capacity of substances varies with temperature, so we really should write it as $C_p(T)$ to remind ourselves of this fact.

The amount of heat needed to increase the temperature of substance by a given amount increases as the amount of that substance increases. It is therefore usual to quote the heat capacity as a *molar* quantity, so that if we have n moles the enthalpy change is $\delta H = nC_{p,m}\,\delta T$. From now on, we will assume that all C_p values are molar.

Heat capacity data for a great many substances have been measured and is available in tabulated form.

6.5.1 Measuring entropy

By starting from Eq. 6.3 on the facing page we can develop a practical method by which the entropy of a substance can be determined. A key part of this method is the use of heat capacities, which are discussed in Box 6.2.

If a small amount of heat δq is supplied to a substance with heat capacity $C(T)$, then the resulting temperature rise δT is related to δq and $C(T)$ via

$$\delta q = C(T)\,\delta T. \tag{6.4}$$

Generally, the heat capacity varies with temperature, so we have written the heat capacity as $C(T)$, which is the mathematical way of indicating that 'C is a function of T' i.e. it varies with temperature.

If we make our measurements under constant pressure conditions, then as explained in section 6.4.2 on the facing page, the heat is equal to the enthalpy change, δH. We can therefore rewrite Eq. 6.4 as

$$\delta H = C_p(T)\,\delta T,$$

where $C_p(T)$ is the constant pressure heat capacity. Furthermore, as enthalpy is a state function, the value of δH is the same for both reversible and irreversible processes, so we can write

$$\text{constant pressure} \qquad \delta q_{\text{rev}} = C_p(T)\,\delta T.$$

This expression for δq_{rev} can be used to evaluate the entropy change by substituting it in the definition of δS, Eq. 6.3 on page 214:

$$\begin{aligned}
\delta S &= \frac{\delta q_{\text{rev}}}{T} \\
&= \frac{C_p(T)}{T} \times \delta T.
\end{aligned} \qquad (6.5)$$

This expression leads to a graphical method of finding the entropy.

If we plot a graph of $C_p(T)/T$ against T, then $C_p(T)/T \times \delta T$ can be interpreted as the *area* of a rectangle of height $C_p(T)/T$ and width δT, as is shown Fig. 6.9 (a). Equation 6.5 tells us that this area is the small change in the entropy, δS, when the temperature is increased by δT at temperature T.

Now we imagine adding up very many of these small rectangles placed side by side to cover the range from T_1 to T_2, as shown in Fig. 6.9 (b). The sum of the areas of these little rectangles is the *change* in entropy on going from T_1 to T_2. As the rectangles get narrower and narrower, we see that the sum of their areas is the same thing as the *area under the curve* between T_1 and T_2.

Therefore, by taking the area under a plot of $C_p(T)/T$ against T we can work out the entropy change between any two temperatures. Since we can measure the heat capacity quite easily, this gives a practical method for measuring the entropy change between two temperatures.

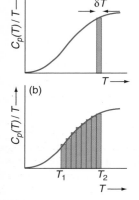

Fig. 6.9 If we make a plot of $C_p(T)/T$ against T, Eq. 6.5 implies that the area of a small rectangle of height $C_p(T)/T$ and width δT will be equal to the small entropy change when the temperature is increased by δT. One such typical rectangle is shown in (a). By joining together many such rectangles between T_1 and T_2 we can see that the sum of their areas, which is in effect the area under the curve, gives the entropy change between T_1 and T_2, as shown in (b).

Absolute entropies

If we return to our molecular interpretation of entropy, we can deduce that at absolute zero (0 K) all perfect crystals have zero entropy. The argument is that at this temperature the thermal energy has gone to zero, so all of the molecules must be in the lowest possible energy state, and arranged in a perfectly ordered way in the solid. There is only one way of achieving this arrangement, so $W = 1$ and hence the entropy is zero since $S = k_B \ln 1$, which is zero.

Therefore, if we use our graphical method to measure the increase in entropy between 0 K and some finite temperature T, this increase in entropy will be the entropy at temperature T, since at 0 K the entropy is zero. Entropies determined in this way are called *absolute entropies*, or sometimes *Third Law entropies*.

There are experimental difficulties with making heat capacity measurements very close to absolute zero, so in practice we make measurements as close to absolute zero as we can and then extrapolate to 0 K. The resulting errors in the absolute entropies are not large.

From the definition of entropy, Eq. 6.3 on page 214, it follows that its units are $J\,K^{-1}$ or, for the case of a molar quantity, $J\,K^{-1}\,mol^{-1}$; this fits in with what we found before by working from the Boltzmann definition of entropy, Eq. 6.2 on page 209. Extensive tabulations of the absolute entropies of elements and compounds have been compiled. Generally, these data are tabulated at 298 K. Entropies do vary with temperature somewhat, but for temperatures not too far from 298 K it is often acceptable simply to use the tabulated value and ignore the temperature variation.

6.6 Calculating the entropy change of the Universe

At last we are in a position to use the Second Law in a practical way. Recall that this Law says that in a spontaneous process the entropy of the Universe increases. It sounds like it is going to be a tall order to compute the entropy change of the Universe, it being so large, but in fact it is not quite as difficult as it seems.

The first thing we do is to divide the Universe into two parts, as illustrated in Fig. 6.10. The first part is the *system*, which is the thing we are interested in e.g. our reacting species in a beaker on the bench. The second part is the *surroundings*, which is *everything else* in the Universe which is not part of the system. The system and the surroundings comprise everything in the Universe. We only allow heat to be transferred between the system and the surroundings.

With this separation, we can compute the entropy change of the Universe, ΔS_{univ}, as the sum of the entropy changes of the surroundings, ΔS_{surr}, and of the system, ΔS_{sys}

$$\Delta S_{univ} = \Delta S_{surr} + \Delta S_{sys}. \tag{6.6}$$

We can determine the entropy change of the system by using tabulated values of absolute entropies.

The entropy change of the surroundings is computed in rather a different way. Since we have already decided that the only thing which is transferred between the system and the surroundings is heat, the entropy change of the surroundings will simply result from this heat flow. As is illustrated in Fig. 6.11, if heat flows into the system, the process is endothermic from the point of view of the system, so δq_{sys} is positive. However, this heat must be flowing out of the surroundings, so from the point of view of the surroundings the process is exothermic, and so δq_{surr} is negative. In other words, the direction of heat flow is opposite from the point of view of the system and surroundings i.e.

$$\delta q_{surr} = -\delta q_{sys}.$$

The surroundings comprise the rest of the Universe, and so are very large indeed. We can therefore safely assume that their temperature, T_{surr}, will not be altered by absorbing or giving out the amounts of heat typical for chemical processes. Therefore, rather than dealing with the entropy change resulting from a small amount of heat being transferred, as involved in the definition of Eq. 6.3 on page 214, we can go straight to the entropy change resulting from the transfer of all of the heat, $q_{rev,surr}$

$$\Delta S_{surr} = \frac{q_{rev,surr}}{T_{surr}}.$$

Again, since the surroundings are very large, we can also assume that they remain at a constant pressure when heat is transferred to or from the system. As we saw in section 6.4.2 on page 214, under constant pressure conditions the heat is equal to the enthalpy change, and since enthalpy is a state function its value is the same regardless of whether the heat is transferred reversibly or irreversibly. We can therefore drop the 'rev' subscript and write the entropy change of the surroundings as

$$\Delta S_{surr} = \frac{q_{surr}}{T_{surr}}.$$

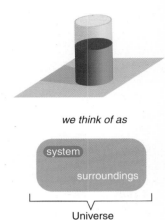

we think of as

Fig. 6.10 Illustration of the division of the Universe into two parts: the object of interest (e.g. the reacting species contained in a beaker), called the *system*, and the rest of the Universe, called the *surroundings*. Only heat is allowed to move between the system and the surroundings.

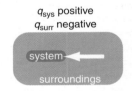

q_{sys} positive
q_{surr} negative

Fig. 6.11 In an endothermic process, heat flows into the system so q_{sys} is positive. This heat can only come from the surroundings, so from the point of view of the surroundings the process is exothermic i.e. q_{surr} is negative. It follows that $q_{surr} = -q_{sys}$.

If the system is also at constant pressure, then $q_{sys} = \Delta H_{sys}$. We have already seen that $q_{surr} = -q_{sys}$, so it follows that $q_{surr} = -\Delta H_{sys}$. We can therefore write the entropy change of the surroundings as

$$\begin{aligned} \Delta S_{surr} &= \frac{q_{surr}}{T_{surr}} \\ &= \frac{-\Delta H_{sys}}{T_{surr}}. \end{aligned}$$

Now that we have an expression ΔS_{surr} we can now go back to Eq. 6.6 on the preceding page and compute the entropy change of the Universe as

$$\Delta S_{univ} = \underbrace{\frac{-\Delta H_{sys}}{T_{surr}}}_{\text{surroundings}} + \underbrace{\Delta S_{sys}}_{\text{system}}. \tag{6.7}$$

This is an exceptionally important result as it enables us to compute the entropy change of the Universe simply by knowing the entropy change and enthalpy change of the *system*. The Second Law tells us that the entropy of the Universe must increase in a spontaneous process, which means that ΔS_{univ} must be positive. Now that we have a way of computing this entropy change, we can put the Second Law to work and find out whether or not a process will be spontaneous.

6.6.1 Putting the Second Law to work: how to make ice

We all know that you make ice by putting water into a refrigerator. If the temperature of the refrigerator is set below 0 °C, then ice will form, but if the temperature is any higher we will not get any ice. The Second Law gives a very neat explanation of why this is so.

As was explained on page 211, when a liquid freezes to form a solid there is a reduction in entropy. This means that for liquid water going to ice ΔS_{sys} is *negative*. Going from ice to liquid water is endothermic since we have to put in energy to break up the hydrogen bonds in the structure. Therefore, the reverse process in which liquid water goes to ice must be *exothermic* since these bonds are being formed. The entropy change of the surroundings, given by $\Delta S_{surr} = -\Delta H_{sys}/T_{surr}$, is therefore *positive* when ice is formed from liquid water. This makes sense as heat is transferred to the surrounding as the ice is formed. ΔS_{univ} can thus be calculated using

$$\Delta S_{univ} = \underbrace{\frac{-\Delta H_{sys}}{T_{surr}}}_{\text{positive for water} \rightarrow \text{ice}} + \underbrace{\Delta S_{sys}}_{\text{negative for water} \rightarrow \text{ice}}.$$

In order for the formation of ice to be spontaneous, ΔS_{univ} must be positive, so ΔS_{surr} has to be more positive than ΔS_{sys} is negative. As the temperature is lowered, ΔS_{surr} becomes more and more positive until eventually it outweighs the negative ΔS_{sys}. At this point ΔS_{univ} becomes positive, and the freezing of liquid water to ice becomes spontaneous according to the Second Law. The overall process is illustrated in Fig. 6.12 on the next page.

We can put some numbers to all of this. The enthalpy of freezing of liquid water is -6.01 kJ mol^{-1} (-6.01×10^3 J mol^{-1}), and the corresponding entropy

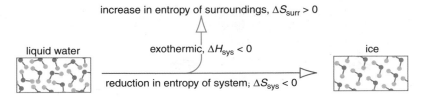

Fig. 6.12 When liquid water freezes to ice there is a reduction in the entropy of the system, but as the process is exothermic, the entropy of the surroundings increases. If the temperature is low enough, the entropy increase of the surroundings will be sufficient to overcome the entropy decrease of the system. As a result, ΔS_{univ} will be positive and the process will be spontaneous.

change is -22.0 J K^{-1} mol^{-1}. Let us first compute ΔS_{univ} at $-10\ ^\circ$C, which is 263 K, noting that as ΔS_{sys} is in J K^{-1} mol^{-1}, we must use ΔH_{sys} in J mol^{-1}

We are ignoring any variation in the values of ΔS and ΔH with temperature.

$$\Delta S_{univ} = \frac{-\Delta H_{sys}}{T_{surr}} + \Delta S_{sys}$$
$$= -\frac{-6.01 \times 10^3}{263} - 22.0$$
$$= 22.9 - 22.0$$
$$= +0.9 \text{ J K}^{-1}\text{ mol}^{-1}.$$

Since ΔS_{univ} is positive, the process is spontaneous according to the Second Law, and this is indeed in line with our expectation that liquid water will freeze to ice at $-10\ ^\circ$C. At this low temperature the increase in entropy of the surroundings, resulting from the exothermic process, is large enough to compensate for the decrease in entropy of the system.

Let us repeat the calculation at $+10\ ^\circ$C (283 K)

$$\Delta S_{univ} = \frac{-\Delta H_{sys}}{T_{surr}} + \Delta S_{sys}$$
$$= -\frac{-6.01 \times 10^3}{283} - 22.0$$
$$= 21.2 - 22.0$$
$$= -0.8 \text{ J K}^{-1}\text{ mol}^{-1}.$$

At this temperature ΔS_{univ} is negative, and so the process is *not allowed* by the Second Law: as we expected, ice will not form at $+10\ ^\circ$C. At this higher temperature the increase in entropy of the surroundings is not large enough to compensate for the decrease in entropy of the system.

Repeating the calculation once more at a temperature of 273.15 K we find $\Delta S_{univ} = 0.0$; this is of course the temperature at which liquid water and ice are in equilibrium. At this temperature there is no change in the entropy of the Universe for conversion of liquid water to ice, or *vice versa*, so both can exist together. This is what we mean by equilibrium.

Spontaneous endothermic processes

If we have a process which is endothermic e.g. ice melting to water, then ΔH_{sys} is positive, so the entropy change of the surroundings ($-\Delta H_{sys}/T$) will be negative. Such a process can therefore only be spontaneous if the entropy of the system increases sufficiently to overcome the entropy reduction of the surroundings. In this case increasing the temperature makes the entropy change of the surroundings less negative. This means that, even if ΔS_{univ} is not positive at a certain temperature, then it may become so if the temperature is raised.

In summary, an endothermic process can be spontaneous provided that it is accompanied by an increase in the entropy of the system, and that the temperature is high enough.

6.6.2   Summary

The really key idea in this section is that we have to think not only about the entropy of the system, but also about the entropy of the surroundings, in order to compute the entropy change of the Universe, which is what the Second Law refers to.

To summarize this very important section:

- $\Delta S_{\text{univ}} = \dfrac{-\Delta H_{\text{sys}}}{T_{\text{surr}}} + \Delta S_{\text{sys}}$.

- $\Delta S_{\text{univ}} > 0$ for a spontaneous process.

- $\Delta S_{\text{univ}} = 0$ at equilibrium.

- A process in which the entropy of the system decreases can be spontaneous provided that ΔH_{sys} is sufficiently negative and T_{surr} is low enough.

- An endothermic process can be spontaneous provided that ΔS_{sys} is sufficiently positive and T_{surr} is high enough.

6.7   Gibbs energy

As we have seen, to work out whether or not a process is spontaneous, all we need to do is inspect the sign of ΔS_{univ}. We have also seen that we can compute this entropy change of the Universe from a knowledge of the entropy and enthalpy changes of the system (Eq. 6.7 on page 218). Rather than computing ΔS_{univ} it turns out to be convenient to have some property of the system whose value will tell us whether or not a process is spontaneous. The *Gibbs energy*, G (also known as the free energy or the Gibbs free energy), turns out to be the property we require.

Formally, G is defined in terms of the enthalpy, entropy and temperature as follows

$$\text{definition of } G: \quad G = H - TS.$$

What is going to concern us most is the change in G, ΔG. If this takes place at constant temperature, then from the above definition it follows that

$$\Delta G_{\text{sys}} = \Delta H_{\text{sys}} - T_{\text{sys}} \Delta S_{\text{sys}}. \tag{6.8}$$

The subscript 'sys' is added to remind us that the quantities refer to the system; there is no 'ΔT' term as we have imposed the condition of constant temperature.

We can see why the Gibbs energy change is a useful quantity by dividing both sides of Eq. 6.8 by $-T_{\text{sys}}$, to give

$$\frac{-\Delta G_{\text{sys}}}{T_{\text{sys}}} = \frac{-\Delta H_{\text{sys}}}{T_{\text{sys}}} + \Delta S_{\text{sys}}.$$

We now compare this with the expression given in Eq. 6.7 on page 218 for ΔS_{univ}

$$\Delta S_{univ} = \frac{-\Delta H_{sys}}{T_{surr}} + \Delta S_{sys}.$$

Recall that Eq. 6.7 applies under constant pressure conditions.

If we assume that the system and the surroundings are at the same temperature T_{sys} (i.e. they are in thermal equilibrium), then the right-hand sides of the two previous equations are the same. It therefore follows that the left-hand sides are the same, so that

$$\frac{-\Delta G_{sys}}{T_{sys}} = \Delta S_{univ}.$$

What we have shown is that, under conditions of constant temperature and pressure, $-\Delta G_{sys}/T_{sys}$ is the *same* as ΔS_{univ}. It follows that as ΔS_{univ} is positive in a spontaneous process, ΔG_{sys} must be negative. Similarly, at equilibrium both ΔG_{sys} and ΔS_{univ} are zero.

What we have done is to re-express the Second Law in terms of the Gibbs energy change of the system, rather than the entropy change of the Universe. By looking at the sign of ΔG_{sys}, which is just a property of the system, we can therefore determine whether or not a process will be spontaneous.

From now on we will frame our discussion in terms of the Gibbs energy change, rather than the entropy change of the Universe. We will also drop the 'sys' subscript, and simply remember that all of the quantities refer to the system. Thus, ΔG for a process at constant temperature is given by

$$\Delta G = \Delta H - T\,\Delta S.$$

The units of ΔG are the same as for ΔH i.e. $J\ mol^{-1}$, but given the typical values of ΔG they are more usually quoted in $kJ\ mol^{-1}$.

6.7.1 Making ice – again

We can illustrate the use of the Gibbs energy by revisiting the example of liquid water freezing to ice, discussed previously in section 6.6.1 on page 218. Recall that for the process liquid water → ice, ΔH is $-6.01\ kJ\ mol^{-1}$, and ΔS is $-22.0\ J\ K^{-1}\ mol^{-1}$.

Ignoring the slight variation of ΔH and ΔS with temperature, ΔG at 263 K can be computed as

$$\begin{aligned}
\Delta G &= \Delta H - T\,\Delta S \\
&= -6.01 \times 10^3 - 263 \times (-22.0) \\
&= -224\ J\ mol^{-1}.
\end{aligned}$$

At this temperature, ΔG is negative, so the process is spontaneous – just as we found before.

At the higher temperature of 283 K, the calculation of ΔG gives

$$\begin{aligned}
\Delta G &= -6.01 \times 10^3 - 283 \times (-22.0) \\
&= +216\ J\ mol^{-1}.
\end{aligned}$$

Now we see that ΔG is positive, so the process is not spontaneous. Finally, at 273.15 K, ΔG is zero, indicating that the solid and liquid are in equilibrium.

This example shows that it is rather more straightforward to discuss whether or not a process will be spontaneous in terms of the Gibbs energy rather than in terms of the entropy change of the Universe.

6.7.2 Summary

As we have seen, ΔG is computed using

$$\Delta G = \Delta H - T\,\Delta S.$$

A spontaneous process requires ΔG to be negative, which can come from having either a *negative* ΔH, or a *positive* ΔS (or both).

It is important to realize that both the ΔH and ΔS terms in the expression for ΔG are about *entropy changes*: $-\Delta H/T$ is the entropy change of the surroundings, and ΔS is the entropy change of the system. A negative ΔH may result in a process being spontaneous *not* because the energy is lowered but because it leads to an increase in the entropy of the surroundings.

We are going to use the Gibbs energy a lot, so let us summarize its important properties:

- For a process taking place at constant temperature $\Delta G = \Delta H - T\,\Delta S$, where all of the quantities refer to the system.

- $-\Delta G/T$ is the entropy change of the Universe.

- For a process to be spontaneous, the Gibbs energy must decrease i.e. ΔG must be negative.

- At equilibrium, there is no further change in Gibbs energy i.e. $\Delta G = 0$.

The last three points all refer to processes which are at constant pressure and temperature.

6.8 Chemical equilibrium

We now come to what is the most important part of this chapter, which is how the ideas which we have developed can be used to describe how reactions come to equilibrium. The key result in this discussion is the following relationship

$$\Delta_r G^\circ = -RT \ln K, \qquad (6.9)$$

in which R is the gas constant (8.314 J K^{-1} mol^{-1}) and T is the temperature. This expression relates the equilibrium constant K to a quantity known as the standard Gibbs energy change for the reaction, $\Delta_r G^\circ$. $\Delta_r G^\circ$ can be found using

$$\Delta_r G^\circ = \Delta_r H^\circ - T\Delta_r S^\circ,$$

where $\Delta_r H^\circ$ is the standard enthalpy change for the reaction, and $\Delta_r S^\circ$ is the standard entropy change. We will define all of the these quantities later on in this section, but the key point to appreciate here is that the values of $\Delta_r H^\circ$ and $\Delta_r S^\circ$ can be determined using tabulated data. Thus, having found $\Delta_r G^\circ$ using these data, we can find the equilibrium constant using Eq. 6.9. What this means is that from these tabulated data we can predict the equilibrium constant of more or less any reaction without any more effort than flicking open a book.

The derivation of Eq. 6.9 is rather involved and also needs some ideas we have not yet introduced, so we will delay this until Chapter 17. Our discussion here is going to be focused on understanding in outline where this relationship comes from, and then how it can be used in a practical way.

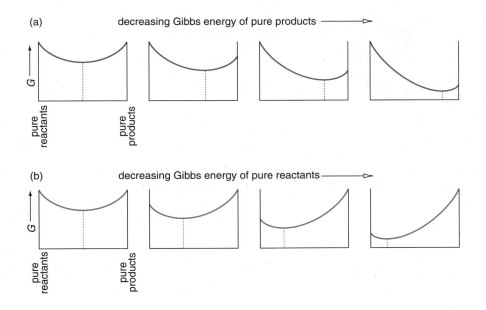

Fig. 6.13 Plots showing how the Gibbs energy of a reacting mixture varies as we go from pure reactants (on the left of each plot) to pure products (on the right). The plots shown in (a) are for steadily decreasing Gibbs energies of the pure products, whereas for those shown in (b) it is the Gibbs energy of the pure reactants which is decreasing. All the plots have the same general form, with a single minimum at the position indicated by the dashed line: this is the equilibrium point. The equilibrium composition is closer to whichever of the reactants or products has the lower Gibbs energy.

6.8.1 The approach to equilibrium

In the previous section we saw that it was convenient to use the Gibbs energy to discuss whether or not a physical process, like the freezing of liquid water to ice, is spontaneous. A process will be spontaneous if it is accompanied by a decrease in the Gibbs energy, and when the Gibbs energy reaches a minimum, no further changes are allowed. The equilibrium position is defined by this minimum in the Gibbs energy.

Exactly the same principles apply to chemical reactions. If we start with the reactants, there will be some conversion to products *provided* such a process is accompanied by a reduction in the Gibbs energy. The process will carry on until the Gibbs energy of the mixture reaches a minimum, at which point the reaction has come to equilibrium.

As you know, depending on the reaction, the position of equilibrium may favour the products, or the reactants, or indeed at equilibrium there may be significant amounts of both reactants and products present. The *composition* of the equilibrium mixture, by which we mean the ratio of products to reactants, is specified by the familiar equilibrium constant, K.

To work out the equilibrium composition we need to know how the Gibbs energy of the mixture of reactants and products varies as the ratio between them changes i.e. as the composition changes. In Chapter 17 we will see in detail how this can be done, but at this stage we will simply look at some graphs of Gibbs energy against composition and discuss the consequences for chemical equilibrium.

Figure 6.13 shows a series of plots of Gibbs energy against composition. On each graph the horizontal axis runs from pure reactants on the left, to pure products on the right; as we go from left to right the fraction of products increases steadily. The vertical axis is the Gibbs energy of the mixture.

Along the series of plots shown in (a) the Gibbs energy of the pure products is decreasing, whereas for those shown in (b) it is the Gibbs energy of the pure

Weblink 6.2

This link takes you to a version of Fig. 6.13 in which you can alter the Gibbs energy of the reactants and products and see how this affects the position of equilibrium.

reactants which is decreasing. Although each plot is different, all the curves have the same general shape, and in particular they all have a single minimum, at the composition indicated by the dashed line.

These plots all tell the same story. If we start at any composition other than that indicated by the dashed line, moving towards the composition indicated by the dashed line will result in a decrease in the Gibbs energy. The process is therefore spontaneous.

When the composition reaches the value indicated by the dashed line, no further change is possible. This is because, starting from this minimum, increasing the amount of products (moving to the right) or increasing the amount of reactants (moving to the left) results in an increase in the Gibbs energy, which is not allowed. The dashed line thus indicates the equilibrium composition. The shape of the curve therefore 'funnels' the composition of the reaction mixture towards the equilibrium position.

If the Gibbs energies of the pure reactants and pure products are the same, the equilibrium composition lies midway between pure reactants and pure products. As the Gibbs energy of the pure products decreases (the series of plots shown in (a)), the equilibrium composition moves closer and closer to pure products i.e. the fraction of the products at equilibrium increases. On the other hand, as the Gibbs energy of the pure reactants decreases (plots (b)), the position of equilibrium moves closer and closer to the reactants i.e. the fraction of reactants at equilibrium increases.

In fact, the position of equilibrium, and hence the value of the equilibrium constant K, can be shown to be *only* a function of the difference in Gibbs energy between pure products and pure reactants. Before we explore the exact relationship between these two quantities, we need to look at how we are going to write the equilibrium constant.

6.8.2 Writing the equilibrium constant

In a general balanced chemical equation, such as

$$a\text{A} + b\text{B} \rightleftharpoons c\text{C} + d\text{D},$$

A and B are the reactants, and C and D are the products. a is the *stoichiometric coefficient* for reactant A, b is that for B and so on. These stoichiometric coefficients are the numbers needed to balance the chemical equation. For example, in the reaction between methane and water to give carbon dioxide and hydrogen

$$\text{CH}_4(\text{g}) + 2\text{H}_2\text{O}(\text{g}) \rightleftharpoons \text{CO}_2(\text{g}) + 4\text{H}_2(\text{g})$$

The stoichiometric coefficient of CH_4 is 1, that of H_2O is 2, that of CO_2 is 1 and that of H_2 is 4.

The equilibrium constant for this general reaction is written as 'products over reactants' with the concentrations of each being raised to the power of the corresponding stoichiometric coefficient

$$K = \frac{[\text{C}]^c[\text{D}]^d}{[\text{A}]^a[\text{B}]^b}, \tag{6.10}$$

where [A] represents the *equilibrium* concentration of species A, and so on.

If written in this way, the units of the equilibrium constant depend on the units used to express the concentration and on the values of the stoichiometric coefficients. For example, for the reaction

$$I^-(aq) + I_2(aq) \rightleftharpoons I_3^-(aq),$$

the equilibrium constant is written

$$K = \frac{[I_3^-]}{[I^-][I_2]}.$$

If the concentrations are expressed in mol dm^{-3}, then this equilibrium constant has units (mol dm^{-3})$^{-1}$.

To relate the value of the equilibrium constant to thermodynamic parameters, such as the Gibbs energy, we need to write it in a way such that it is *dimensionless*. This is achieved by dividing each of the concentration terms by the *standard concentration*, c°. Using this approach, the equilibrium constant for the general reaction is written

$$K = \frac{([C]/c^\circ)^c([D]/c^\circ)^d}{([A]/c^\circ)^a([B]/c^\circ)^b}. \tag{6.11}$$

Provided we choose the same units for [A] and c°, the ratio ([A]/c°) is dimensionless, and so, therefore, is K.

There is a well established convention that for species in solution the concentrations are expressed in mol dm^{-3}, and that c° is taken as 1 mol dm^{-3}. Therefore, in Eq. 6.11 each concentration, expressed in mol dm^{-3}, is divided by the number 1. You would be forgiven for thinking that this is a bit perverse since it makes absolutely no difference to the value of K. Nevertheless it is important for what follows that we ensure that the equilibrium constant is written in a way which makes it dimensionless.

You will often come across the expression for the equilibrium constant written without the c° terms i.e. as Eq. 6.10 on the preceding page, and indeed you will also find values of K which are quoted with units. Strictly speaking, to do so is incorrect. However, it is acceptable to write the equilibrium constant in the form of Eq. 6.10 *provided* that we remember to write the concentrations in mol dm^{-3}, and that '[A]' is really a shorthand for '[A]/c°'.

If gases are involved in the equilibrium, it is usual to express their concentrations not in mol dm^{-3} but as *partial pressures* – a concept which was introduced in section 1.9.2 on page 24. The partial pressure of a gas in a mixture is simply the pressure that the gas would exert if it occupied the whole volume on its own. As we saw, this pressure is proportional to the concentration.

Just as with concentrations, when writing the equilibrium constant in terms of the partial pressure we have to divide each pressure by the standard pressure p° in order to make the equilibrium constant dimensionless. So, for the general equilibrium the expression for K is

$$K = \frac{(p_C/p^\circ)^c(p_D/p^\circ)^d}{(p_A/p^\circ)^a(p_B/p^\circ)^b}.$$

The convention is that the standard pressure is 1 bar, which is exactly 10^5 N m^{-2}; this is close to, but not the same pressure as, 1 standard atmosphere. When writing the equilibrium constant in terms of partial pressures, we have to be sure to use the same units for the partial pressures and for p°.

The equilibrium constant expressed in terms of partial pressures is sometimes denoted K_p, whereas that expressed in terms of concentrations is sometimes denoted K_c.

6.8.3 The standard Gibbs energy change

Provided that the equilibrium constant is written in the way described above, its value is related to the standard Gibbs energy change $\Delta_r G^\circ$ for the reaction via

$$\Delta_r G^\circ = -RT \ln K.$$

For the general equilibrium

$$a\text{A} + b\text{B} \rightleftharpoons c\text{C} + d\text{D},$$

the standard Gibbs energy change is defined as the change in Gibbs energy when a moles of A react with b moles of B to give c moles of C and d moles of D, with all of the substances being in their *standard states* and at the stated temperature.

The standard states are different for gases, liquids, solids and solutions. They are defined as follows:

phase	definition of standard state
gas	the *pure* gas at a pressure of 1 bar and at the stated temperature
liquid	the *pure* liquid at the stated temperature
solid	the *pure* solid at the stated temperature
solution	an (ideal) solution with the solute at the standard concentration of 1 mol dm^{-3}, and at the stated temperature

It is important to note that for gases, liquids and solids the standard state refers to the *pure* substance. This means that when the reactants are in their standard states they are *not mixed*, but separate; the same applies to the products. So when we talk about the standard Gibbs energy change for

$$2\text{H}_2(g) + \text{O}_2(g) \rightleftharpoons 2\text{H}_2\text{O}(g)$$

it is for two moles of pure H_2 and one mole of pure O_2, unmixed, going to two moles of pure H_2O, all species at 1 bar pressure and at the stated temperature. $\Delta_r G^\circ$ is really a hypothetical quantity, rather than one referring to a practically realisable reaction.

For solutions we have specified that the standard state is 1 mol dm^{-3} with the solution behaving ideally. This means that there must be no interactions between the solute molecules, something which is not likely to be true for a real solution at such a concentration. Dealing with non-ideal solutions is well beyond the scope of this text, so for the present we will simply have to assume that the solutions we are dealing with are ideal.

The value of $\Delta_r G^\circ$ always refers to a particular balanced chemical equation, which must be specified: this is what the subscript 'r' is there to remind us. In addition, the value of $\Delta_r G^\circ$ is strongly temperature dependent, so the temperature must be stated. It is a common misconception that 'standard' implies a temperature of 298 K: this is *not* the case.

6.8.4 Equilibria involving solids and liquids

If an equilibrium involves a solid or pure liquid, then no 'concentration' term for such species is included in the expression for the equilibrium constant. For example, for the reaction

$$CaCO_3(s) \rightleftharpoons CaO(s) + CO_2(g),$$

the equilibrium constant is written as

$$K = p(CO_2)/p^\circ,$$

where $p(CO_2)$ is the equilibrium partial pressure of CO_2.

Similarly for the acid dissociation equilibrium in water

$$HA(aq) + H_2O(l) \rightleftharpoons A^-(aq) + H_3O^+(aq).$$

the equilibrium constant is written as

$$K = \frac{([A^-]/c^\circ)([H_3O^+]/c^\circ)}{[HA]/c^\circ}.$$

No term is included for the water since, at modest acid concentrations, it is essentially in its pure form, that is in its standard state. Note that the fact there is no term for the concentration of the water *does not* mean that pure water has a concentration of 1 mol dm^{-3}.

It is important to realize that there are two different standard states involved here. For the solute the standard state is a solution with concentration 1 mol dm^{-3}; for the solvent the standard state is the pure solvent. More details of why equilibrium constants are written in this way, and the choice of standard states, are given in Chapter 17 in Part II.

6.9 Finding the standard Gibbs energy change

Now that we have seen how to write the equilibrium constant appropriately, and how $\Delta_r G^\circ$ is defined, our next task is to see how we can use tabulated data to determine the value of $\Delta_r G^\circ$.

For the general equilibrium

$$aA + bB \rightleftharpoons cC + dD,$$

$\Delta_r G^\circ$ is given by

$$\Delta_r G^\circ = \underbrace{c \times G^\circ_{m,C} + d \times G^\circ_{m,D}}_{\text{products}} - [\underbrace{a \times G^\circ_{m,A} + b \times G^\circ_{m,B}}_{\text{reactants}}], \qquad (6.12)$$

where $G^\circ_{m,A}$ is the molar standard Gibbs energy of A, and so on. This expression for $\Delta_r G^\circ$ is in line with the definition given above i.e. the Gibbs energy change when reactants go to products according to the stated chemical equation, with all species being in their standard states. Note that the stoichiometric coefficients come into the expression for $\Delta_r G^\circ$ as it is written in terms of molar quantities e.g. $G^\circ_{m,A}$.

The Gibbs energy itself is defined as $G = H - TS$, so $G^\circ_{m,A}$ can be written

$$G^\circ_{m,A} = H^\circ_{m,A} - TS^\circ_{m,A},$$

where $H_{m,A}^\circ$ is the standard molar enthalpy of A, and $S_{m,A}^\circ$ is the standard molar entropy of A; similar expressions can be written for the other species. Using these expressions for $G_{m,A}^\circ$ etc. in Eq. 6.12 on the previous page we have

$$
\begin{aligned}
\Delta_r G^\circ &= cG_{m,C}^\circ + dG_{m,D}^\circ - aG_{m,A}^\circ - bG_{m,B}^\circ \\
&= c\left(H_{m,C}^\circ - TS_{m,C}^\circ\right) + d\left(H_{m,D}^\circ - TS_{m,D}^\circ\right) \\
&\quad -a\left(H_{m,A}^\circ - TS_{m,A}^\circ\right) - b\left(H_{m,B}^\circ - TS_{m,B}^\circ\right) \\
&= \left[cH_{m,C}^\circ + dH_{m,D}^\circ - aH_{m,A}^\circ - bH_{m,B}^\circ\right] \\
&\quad -T\left[cS_{m,C}^\circ + dS_{m,D}^\circ - aS_{m,A}^\circ - bS_{m,B}^\circ\right]
\end{aligned}
\tag{6.13}
$$

The quantity in the first square bracket is defined as the standard enthalpy change for the reaction, $\Delta_r H^\circ$:

$$
\Delta_r H^\circ = \underbrace{\left(cH_{m,C}^\circ + dH_{m,D}^\circ\right)}_{\text{products}} - \underbrace{\left(aH_{m,A}^\circ + bH_{m,B}^\circ\right)}_{\text{reactants}},
\tag{6.14}
$$

and the quantity in the second square bracket is defined as the standard entropy change for the reaction, $\Delta_r S^\circ$:

$$
\Delta_r S^\circ = \underbrace{\left(cS_{m,C}^\circ + dS_{m,D}^\circ\right)}_{\text{products}} - \underbrace{\left(aS_{m,A}^\circ + bS_{m,B}^\circ\right)}_{\text{reactants}}.
\tag{6.15}
$$

These quantities are defined in exactly the same way as $\Delta_r G^\circ$ i.e. for reactants in their standard states going to products in their standard states, according to the balanced chemical equation.

Using these definitions of $\Delta_r S^\circ$ and $\Delta_r H^\circ$ in Eq. 6.13 we can see that

$$
\Delta_r G^\circ = \Delta_r H^\circ - T\Delta_r S^\circ.
\tag{6.16}
$$

This is an exceptionally important relationship, which we will use frequently.

The values of $\Delta_r H^\circ$, $\Delta_r S^\circ$ and $\Delta_r G^\circ$ are all quoted 'per mole' because they are defined in terms of molar quantities (e.g. S_m°). As discussed in section 6.2 on page 204, the 'per mole' *does not* refer to a mole of reactants or products.

6.9.1 Finding the standard enthalpy change, $\Delta_r H^\circ$

Although values of standard molar entropies can be found by the method already described, it is not possible to measure the standard enthalpies of compounds (H_m°), but only the changes in such enthalpies. As a result, we cannot use Eq. 6.14 directly. However, we can find $\Delta_r H^\circ$ using *standard enthalpies of formation*, $\Delta_f H^\circ$, extensive tabulations of which are available.

The standard enthalpy of formation of a compound is defined as the enthalpy change when *one mole* of that compound is formed from its constituent elements in their *reference states*, all species being present in their standard states.

The *reference state* of an element is its most stable form at the stated temperature and at the standard pressure, 1 bar. For example, at 298 K the reference state of hydrogen is $H_2(g)$, that of carbon is solid graphite, and that of mercury is the liquid. The definition of the standard enthalpy of formation means that $\Delta_f H^\circ$ of the elements *in their reference states* is necessarily zero.

$\Delta_r H°$ for any reaction can be computed from the standard enthalpies of formation using

$\Delta_r H°$ = (sum of $\Delta_f H°$ of products) − (sum of $\Delta_f H°$ of reactants)

For example, for the reaction

$$CH_4(g) + 2H_2O(g) \longrightarrow CO_2(g) + 4H_2(g)$$

$\Delta_r H°$ is computed as

$$\Delta_r H° = [\Delta_f H°(CO_2(g)) + 4\Delta_f H°(H_2(g))] - [\Delta_f H°(CH_4(g)) + 2\Delta_f H°(H_2O(g))].$$

Note that each $\Delta_f H°$ value has to be multiplied by the corresponding stoichiometric coefficient in the balanced chemical equation.

From data tables we can find the required values of $\Delta_f H°$ at 298 K, and so can compute the required $\Delta_r H°$ as

$$
\begin{aligned}
\Delta_r H° &= [-393.51 + 4 \times 0] - [-74.81 + 2 \times (-241.8)] \\
&= 164.9 \text{ kJ mol}^{-1}.
\end{aligned}
$$

Note that by definition the standard enthalpy of formation of $H_2(g)$ is zero.

Put slightly more formally, for the general reaction

$$aA(g) + bB(g) \longrightarrow cC(g) + dD(g),$$

$\Delta_r H°$ is given by

$$\Delta_r H° = [c \times \Delta_f H°(C) + d \times \Delta_f H°(D)] - [a \times \Delta_f H°(A) + b \times \Delta_f H°(B)]. \quad (6.17)$$

The reason why this expression works is demonstrated in Fig. 6.14. The enthalpy change for the reaction can be determined using a Hess's Law cycle in which we first break apart the reactants into their elements (in their reference states), and then reassemble the products from the elements. The first step is the reverse of the formation process, so the enthalpy changes are minus the enthalpies of formation, and for the second step the enthalpy changes are just those of formation.

All of the quantities on the right-hand side of Eq. 6.17 are molar (i.e. have units of kJ mol^{-1}), so $\Delta_r H°$ is also in molar units (usually kJ mol^{-1}).

6.9.2 Finding the standard entropy change $\Delta_r S°$

We have already seen in section 6.5.1 on page 215 that absolute entropies can be determined from heat capacity data. Extensive tabulations of such $S°_m$ values are available, and these can be used directly in Eq. 6.15 on the facing page.

For example, for the reaction

$$CH_4(g) + 2H_2O(g) \longrightarrow CO_2(g) + 4H_2(g)$$

$\Delta_r S°$ is computed as

$$
\begin{aligned}
\Delta_r S° &= [S°_m(CO_2(g)) + 4 \times S°_m(H_2(g))] - [S°_m(CH_4(g)) + 2 \times S°_m(H_2O(g))] \\
&= [213.74 + 4 \times 130.68] - [186.3 + 2 \times 188.8] \\
&= 172.56 \text{ J K}^{-1} \text{ mol}^{-1}.
\end{aligned}
$$

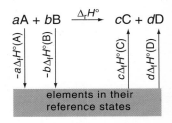

Fig. 6.14 The enthalpy change for a reaction can be computed using a Hess's Law cycle in which the reactants are first broken apart into their constituent elements in their reference states (the downward pointing arrows), and then these elements are reassembled into the products (the upward pointing arrows). The enthalpy changes for the downward pointing arrows are minus the enthalpies of formation of the reactants, and for the upward pointing arrows are the enthalpies of formation of the products.

6.9.3 Finding $\Delta_r G°$ and hence K

Now that we have $\Delta_r H°$ and $\Delta_r S°$ we can find $\Delta_r G°$ using $\Delta_r G° = \Delta_r H° - T\Delta_r S°$. For example, for the reaction we have been considering

$$
\begin{aligned}
\Delta_r G° &= \Delta_r H° - T\Delta_r S° \\
&= 164.9 \times 10^3 - 298 \times 172.56 \\
&= 1.135 \times 10^5 \text{ J mol}^{-1} \\
&= 113.5 \text{ kJ mol}^{-1}.
\end{aligned}
$$

Note that we had to be careful in this calculation to account for the fact that the value of $\Delta_r H°$ is given in kJ mol^{-1}, but $\Delta_r S°$ is in J K^{-1} mol^{-1}. It is therefore necessary to express $\Delta_r H°$ in J mol^{-1} before making the calculation. However, on the last line the value of $\Delta_r G°$ has been converted back to kJ mol^{-1}, since this is the usual unit for Gibbs energy changes.

Now that we know $\Delta_r G°$ we can compute K using $\Delta_r G° = -RT \ln K$; once more, we have to be careful to express the value of $\Delta_r G°$ in J mol^{-1} before making the calculation

$$
\begin{aligned}
\ln K &= \frac{-\Delta_r G°}{RT} \\
&= \frac{-113.5 \times 10^3}{8.314 \times 298} \\
&= -45.8 \\
\text{therefore } K &= 1.27 \times 10^{-20}.
\end{aligned}
$$

The value of K is extremely small which means that at 298 K the equilibrium mixture consists of essentially pure reactants; in other words this reaction simply does not go to products at all. We were able to work this out just with the aid of tabulated data. No experiments on our part were required.

6.9.4 Which reaction?

The values of $\Delta_r H°$ and $\Delta_r S°$ (and hence $\Delta_r G°$) relate to a particular balanced chemical equation, which must either be specified or implied by a well-known convention. For example, if we specify that the reaction is

$$
\tfrac{1}{2}N_2(g) + \tfrac{3}{2}H_2(g) \rightleftharpoons NH_3(g),
$$

then from tables we can find that (at 298 K) $\Delta_r H°$ is -46.1 kJ mol^{-1} and $\Delta_r S°$ is -99.1 J K^{-1} mol^{-1}, giving the value of $\Delta_r G°$ as -16.6 kJ mol^{-1}; from this value of $\Delta_r G°$ we can compute that the equilibrium constant is 812. For this reaction the equilibrium constant is given by

$$
K = \frac{p(NH_3)/p°}{(p(N_2)/p°)^{\frac{1}{2}} (p(H_2)/p°)^{\frac{3}{2}}}.
$$

If, however, we double the number of moles on each side of the equation to give

$$
N_2(g) + 3H_2(g) \rightleftharpoons 2NH_3(g),
$$

$\Delta_r H°$, $\Delta_r S°$ and $\Delta_r G°$ are all doubled in value compared to what they were before. This is because these values now refer to a reaction which has twice as many moles on the left and right as it did before.

The value of $\Delta_r G°$ is therefore -33.2 kJ mol^{-1}, which corresponds to $K' = 6.6 \times 10^5$. This equilibrium constant refers to the doubled reaction, and is given by

$$K' = \frac{(p(NH_3)/p°)^2}{(p(N_2)/p°)\,(p(H_2)/p°)^3}.$$

It is not surprising, therefore, that the equilibrium constants K and K' have different values.

On account of the logarithmic relationship between $\Delta_r G°$ and K, *doubling* the value of $\Delta_r G°$ results in the equilibrium constant being *squared*, which is precisely what we have here: $K' = K^2$. You can see, therefore, that it is very important to be clear about which chemical equilibrium your value of $\Delta_r G°$ refers to. Indeed the subscript 'r' is there to remind us that the value refers to a particular reaction.

6.9.5 Temperature dependence

The values of $\Delta_r H°$ and $\Delta_r S°$ are temperature dependent, so if we want to work out $\Delta_r G°$ (and hence K) at a particular temperature, then we need to be sure to use the values of $\Delta_r H°$ and $\Delta_r S°$ appropriate to this temperature. Most commonly, these values are tabulated at 298 K, which presents something of a problem if we want to find $\Delta_r G°$ at a temperature other than this.

In section 17.8.1 on page 785 and in section 17.8.2 on page 787 we will show how the values of $\Delta_r H°$ and $\Delta_r S°$ can be converted from one temperature to another using heat capacity data. What we will find when we do this is that $\Delta_r H°$ and $\Delta_r S°$ are not strongly temperature dependent. Provided that we are prepared to sacrifice some accuracy in our calculations, it is often acceptable to use the tabulated values at temperatures other than 298 K.

Note that because the temperature appears explicitly in the calculation of $\Delta_r G°$

$$\Delta_r G° = \Delta_r H° - T\Delta_r S°,$$

$\Delta_r G°$ has a strong temperature dependence, except in the special case where $\Delta_r S°$ is small. Often the temperature dependence of $\Delta_r G°$ due to the factor of T appearing in its definition far outweighs the temperature variation of $\Delta_r H°$ and $\Delta_r S°$.

As an example, let us look again at the reaction we were discussing above

$$CH_4(g) + 2H_2O(g) \longrightarrow CO_2(g) + 4H_2(g).$$

At 298 K we found that $\Delta_r H°$ is 164.9 kJ mol^{-1} and $\Delta_r S°$ is 172.56 J K^{-1} mol^{-1}, which gives the value of $\Delta_r G°$ as 113.5 kJ mol^{-1}. This large positive value indicates that, at equilibrium, essentially no products are present i.e. the reaction does not 'go'.

To work out $\Delta_r G°$ at 1000 K we will assume that $\Delta_r H°$ and $\Delta_r S°$ are the same as at 298 K, so

$$
\begin{aligned}
\Delta_r G° &= \Delta_r H° - T\Delta_r S° \\
&= 164.9 \times 10^3 - 1000 \times 172.56 \\
&= -7.66 \times 10^3 \text{ J mol}^{-1} \\
&= -7.66 \text{ kJ mol}^{-1}.
\end{aligned}
$$

Note that, as before, when computing $\Delta_r G°$ the value of $\Delta_r H°$ is converted to J mol^{-1}.

The value of $\Delta_r G°$ is substantially different to that at 298 K, simply because of the $-T\Delta_r S°$ term in the calculation of $\Delta_r G°$.

The corresponding value of K is found as before

$$
\begin{aligned}
\ln K &= \frac{-\Delta_r G°}{RT} \\
&= \frac{-(-7.66 \times 10^3)}{8.314 \times 1000} \\
&= 0.92
\end{aligned}
$$

therefore $K = 2.5$.

At this higher temperature the equilibrium now favours the products, although not to a great extent. In fact, the yield of H_2 at this high temperature is sufficient for this reaction to be used in the commercial production of hydrogen from methane.

6.9.6   The standard Gibbs energy and entropy of formation

The standard enthalpy of formation is defined as the enthalpy change when one mole of a substance is formed from its elements in their reference states, with all species in their standard states. In a similar way we can define the *standard Gibbs energy of formation* of a substance, $\Delta_f G°$, as the Gibbs energy change for the same process, and the *standard entropy change of formation*, $\Delta_f S°$, as the entropy change for the same process.

As a result of these definitions, it follows that both the standard Gibbs energy of formation, and the standard entropy change of formation, of an element in its reference state are zero:

for elements in their reference states: $\Delta_f G° = 0$ and $\Delta_f S° = 0$.

6.9.7   Summary

We have covered a lot of ground in our discussion of chemical equilibrium, so this is a good time to summarize the key points:

- The equilibrium constant must be written in a way which makes it dimensionless.

- In writing the equilibrium constant as 'products over reactants', no terms are included for solids or liquids.

- The equilibrium constant K is related to the standard molar Gibbs energy change via $\Delta_r G° = -RT \ln K$.

- $\Delta_r G°$ can be computed using $\Delta_r G° = \Delta_r H° - T\Delta_r S°$.

- $\Delta_r H°$ can be found using the tabulated values of standard enthalpies of formation of compounds.

- $\Delta_r S°$ can be found from tabulated values of absolute standard molar entropies.

- $\Delta_r H°$ and $\Delta_r S°$ have a weak temperature dependence.

6.10   Interpreting the value of $\Delta_r G^\circ$

In this section we are going to look at how to interpret the value of $\Delta_r G^\circ$ in the light of its relationship with the equilibrium constant

$$\Delta_r G^\circ = -RT \ln K.$$

If we divide both sides by $-RT$ we obtain

$$\frac{-\Delta_r G^\circ}{RT} = \ln K,$$

and then taking the exponential of both sides gives

$$K = \exp\left(\frac{-\Delta_r G^\circ}{RT}\right). \qquad (6.18)$$

This equation says that if $\Delta_r G^\circ$ is positive, on the right we will have the exponential of a negative number, which is less than one. Therefore, the equilibrium constant will be less than one, meaning that at equilibrium the reactants are favoured.

On the other hand, if $\Delta_r G^\circ$ is negative, on the right-hand side of Eq. 6.18 we will have the exponential of a positive number, which is greater than one. Therefore, the equilibrium constant will be greater than one, meaning that the products are favoured. The way in which the equilibrium constant varies with $\Delta_r G^\circ$ is illustrated in Fig. 6.15.

Another way of looking at how the position of equilibrium varies with $\Delta_r G^\circ$ is to work out what fraction of the equilibrium mixture is products. For the simple $A \rightleftharpoons B$ equilibrium it is easy to show that the fraction of product B at equilibrium is $K/(1 + K)$, and Fig. 6.16 shows a plot of this, expressed as a percentage, as a function of $\Delta_r G^\circ/RT$, graph (a), and as a function of $\Delta_r G^\circ$ at 298 K, graph (b).

The graph shows that the fraction of product B changes rather rapidly as $\Delta_r G^\circ/RT$ goes through zero. However, by the time $\Delta_r G^\circ/RT$ is significantly negative, the equilibrium mixture is essentially 100% B, and similarly a significantly positive value of $\Delta_r G^\circ/RT$ means that there is very little product formed. There is only a small range of values of $\Delta_r G^\circ/RT$ which results in significant amounts of both reactants and products at equilibrium.

It is important to realize that a positive $\Delta_r G^\circ$ does *not* mean that the reaction will 'not go' – all that it means is that at equilibrium the reactants are favoured

▷ The properties of the exponential function and natural logarithms are reviewed in section 20.4 on page 895.

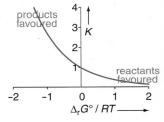

Fig. 6.15 Graph of K as a function of the dimensionless parameter $\Delta_r G^\circ/RT$. If $\Delta_r G^\circ/RT = 0$ the equilibrium constant is 1; if $\Delta_r G^\circ/RT > 0$ the equilibrium constant is less than 1, meaning that the reactants are favoured. If $\Delta_r G^\circ/RT < 0$, $K > 1$ indicating that the products are favoured.

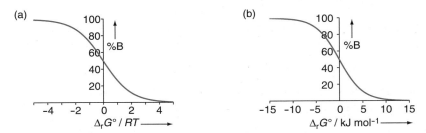

Fig. 6.16 Graphs showing how the percentage of product B for the $A \rightleftharpoons B$ varies (a) as a function of the dimensionless parameter $\Delta_r G^\circ/RT$, and (b) as a function of $\Delta_r G^\circ$ at 298 K. There is only a small range of values of $\Delta_r G^\circ$ which result in a significant amount of both products and reactants at equilibrium.

Fig. 6.17 Graph showing how the equilibrium constant varies with $\Delta_r G°$ at three different temperatures; note that the equilibrium constant is plotted on a logarithmic scale. The lower the temperature, the more rapidly K varies with $\Delta_r G°$. At 298 K, values of $\Delta_r G°$ greater than 30 kJ mol^{-1} give such small equilibrium constants that in effect no reaction takes place. In contrast, values more negative than -30 kJ mol^{-1} give such large equilibrium constants that the reaction is essentially complete.

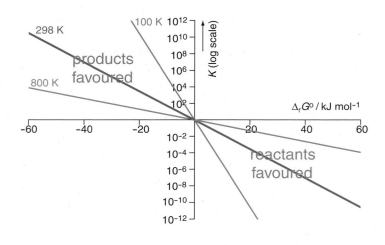

over the products. This point was illustrated in the series of plots (b) shown in Fig. 6.13 on page 223. For these the Gibbs energy of the pure products is greater than that of the pure reactants, meaning that $\Delta_r G°$ is positive. Overall in going from pure reactants to pure products there is an increase in Gibbs energy, so this process is not allowed. However, as we move away from pure reactants the Gibbs energy initially decreases, so that the formation of *some* amount of products is spontaneous. The only consequence of the positive $\Delta_r G°$ is that the equilibrium position favours the reactants i.e. the equilibrium constant is less than 1.

Since the value of K varies exponentially with $\Delta_r G°$, rather small changes in $\Delta_r G°$ result in large changes in the equilibrium constant. For example, at 298 K a $\Delta_r G°$ of -10 kJ mol^{-1} corresponds to $K = 57$; doubling $\Delta_r G°$ to -20 kJ mol^{-1} increases K to 3205, which is rather a large change.

Figure 6.17 shows how the equilibrium constant varies as a function of $\Delta_r G°$. This time K has been plotted on a logarithmic scale, so the graph is a straight line ($\ln K$ is proportional to $-\Delta_r G°$). As expected, for positive $\Delta_r G°$ we find that the equilibrium constant is less than 1, and for negative $\Delta_r G°$ we see that the constant is greater than 1. Lines are shown on the graph for three different temperatures (100 K, 298 K and 800 K); these show that the strongest variation of K with $\Delta_r G°$ occurs at the lowest temperature.

At a temperature of 298 K, by the time $\Delta_r G°$ reaches 30 kJ mol^{-1}, the equilibrium constant has become so small that, to all intents and purposes, there are essentially no products present at equilibrium. Therefore values of $\Delta_r G°$ greater than about 30 kJ mol^{-1} can be taken to imply that the reaction simply does not take place to a significant extent.

On the other hand, by the time $\Delta_r G°$ reaches -30 kJ mol^{-1}, the equilibrium constant has become so large that we can assume that only products are present at equilibrium i.e. the reaction goes to completion.

A reaction which has a negative value of $\Delta_r G°$ is said to be *exergonic*; if $\Delta_r G°$ is positive the reaction is said to be *endergonic*. These terms are not widely used in chemistry, but are quite common in discussions of biological reactions and processes.

6.11 $\Delta_r H°$ and $\Delta_r S°$ for reactions not involving ions

It is very helpful in our study of chemistry to have some understanding of the factors which influence the values of $\Delta_r H°$ and $\Delta_r S°$, as well as some idea of typical values for these quantities. We are going to look at this in two parts. Firstly, we will look at reactions that do not involve ions in solution, and then we will look separately at ions in solution. The reason for making this separation is that there are some rather different issues involved for ions in solution.

6.11.1   Enthalpy changes

In a typical chemical reaction as we go from reactants to products some bonds are broken and others are formed. The value of $\Delta_r H°$ is therefore very much influenced by the balance between the energy needed to break the bonds and the energy released as a result of forming new bonds.

In a simple dissociation reaction such as

$$N_2O_4(g) \longrightarrow 2NO_2(g) \qquad \Delta_r H° = +57.2 \text{ kJ mol}^{-1}$$

$\Delta_r H°$ is positive since a bond is broken and none are made. On the other hand, the following reaction has a negative $\Delta_r H°$ as a bond is made, but none are broken

$$NO_2(g) + NO_3(g) \longrightarrow N_2O_5(g) \qquad \Delta_r H° = -93.1 \text{ kJ mol}^{-1}$$

In general, where bonds are both made and broken $\Delta_r H°$ can have either sign. For example, in the combustion of methane

$$CH_4(g) + 2O_2(g) \longrightarrow 2H_2O(g) + CO_2(g) \qquad \Delta_r H° = -803 \text{ kJ mol}^{-1},$$

four C–H and two O=O bonds are broken, but four O–H and two C=O bonds are made. The fact that the reaction is exothermic implies that taken together the bonds which are made are stronger than those which are broken.

The physical state of the reactants and products also has an influence. For example, the $\Delta_r H°$ values for forming water as a gas and as a liquid are significantly different

$$H_2(g) + \tfrac{1}{2}O_2(g) \longrightarrow H_2O(g) \qquad \Delta_r H° = -242 \text{ kJ mol}^{-1}$$
$$H_2(g) + \tfrac{1}{2}O_2(g) \longrightarrow H_2O(l) \qquad \Delta_r H° = -286 \text{ kJ mol}^{-1}.$$

The enthalpy change is more negative when liquid water is formed as a result of the heat released when gaseous water condenses

$$H_2O(g) \longrightarrow H_2O(l) \qquad \Delta_r H° = -44 \text{ kJ mol}^{-1}.$$

Bond enthalpies

In trying to understand the values of $\Delta_r H°$ it is useful to introduce the concept of a *bond enthalpy* (often called bond energy). The bond enthalpy for a diatomic A–B is defined as the enthalpy change in going from A–B to well separated atoms A and B. For CH_4, we define the C–H bond enthalpy as one quarter of the enthalpy change of the reaction

$$CH_4(g) \longrightarrow C(g) + 4H(g).$$

Table 6.1 Average bond enthalpies at 298 K

bond	enthalpy / kJ mol^{-1}	bond	enthalpy / kJ mol^{-1}	bond	enthalpy / kJ mol^{-1}
C–H	414	C–C	346	C=C (double)	614
N–H	391	C–N	286	C=N (double)	615
O–H	463	C–O	358	C=O (double)	804
P–H	322	C–F	485		
S–H	364	C–Cl	327	O=O (double)	498
Si–F	597	C–Br	285		
P–F	490	C–I	178	C≡C (triple)	839
P–Cl	322			C≡N (triple)	890

More complex molecules have more than one type of bond, so the enthalpy change on going to separate atoms has to be interpreted accordingly. For example, for ethane the enthalpy change for

$$C_2H_6(g) \longrightarrow 2C(g) + 6H(g)$$

can be thought of as the bond enthalpy of a C–C single bond and six C–H bonds. For ethene, the corresponding process

$$C_2H_4(g) \longrightarrow 2C(g) + 4H(g)$$

is for one C=C double bond and four C–H bonds.

By looking at the enthalpy changes of such processes for a range of different molecules it is possible to come up with a set of average bond enthalpies for different types of bonds. These have to be average values as it is found that there is some variation in the bond enthalpies for the same type of bond in different compounds.

For example, the C–H bond enthalpy in CH_4 is 439 kJ mol^{-1}, but in C_2H_6 it is 423 kJ mol^{-1}, and in CH_3OH it is 402 kJ mol^{-1}. The value usually quoted for the average bond enthalpy of C–H is between 412 and 414 kJ mol^{-1}.

Table 6.1 gives a selection of average bond enthalpies for some commonly encountered bonds. These can be used to estimate $\Delta_r H°$ for reactions, for example the addition of Cl_2 to ethene:

$$CH_2CH_2(g) + Cl_2(g) \longrightarrow CH_2ClCH_2Cl(g).$$

We have broken a Cl–Cl bond, which takes 242 kJ mol^{-1} (the bond enthalpy of Cl_2), and one of the bonds between the two carbons, but have made two C–Cl bonds. From the table the C=C double bond enthalpy is 614 kJ mol^{-1}, and that for the C–C single bond is 346 kJ mol^{-1}, so breaking one bond from the C=C double bond takes (614 − 346) = 268 kJ mol^{-1}. The calculation of $\Delta_r H°$ is therefore

$$\Delta_r H° = \underbrace{(242 + 268)}_{\text{bonds broken}} - \underbrace{(2 \times 327)}_{\text{bonds made}} = -144 \text{ kJ mol}^{-1}.$$

The experimental value is found to be −179 kJ mol^{-1}. You might think that this is not very good agreement, but as the table only gives average values of bond strengths, we really cannot expect a very accurate result. At best, what we have here is a useful guide.

6.11.2   Entropy changes

It was explained in section 6.3.2 on page 210 that the entropy of a gas is greater than that of a liquid, which in turn is greater than that of a solid. In fact, a gas has so many more energy levels available to it than either a solid or a liquid that the entropy of a gas is *much* greater than it is for the other phases. For example, for water the standard molar entropy (at 298 K) of the gas is 189 J K^{-1} mol^{-1}, for the liquid it is 70 J K^{-1} mol^{-1}, and for the solid it is 38 J K^{-1} mol^{-1}.

It therefore follows that in a chemical reaction the value of $\Delta_r S^\circ$ is dominated by the *change* in the number of moles of gaseous species in going from reactants to products. For example, in the decomposition

$$CaCO_3(s) \longrightarrow CaO(s) + CO_2(g)$$

we can expect $\Delta_r S^\circ$ to be significantly positive as one of the products, but none of the reactants, is gaseous. $\Delta_r S^\circ$ turns out to be 159 J K^{-1} mol^{-1}.

The following reaction also has a positive $\Delta_r S^\circ$ as one mole of gas becomes two

$$N_2O_4(g) \longrightarrow 2NO_2(g) \qquad \Delta_r S^\circ = +176 \text{ J K}^{-1} \text{ mol}^{-1}.$$

On the other hand, a reaction in which there is a reduction in the number of moles of gaseous species has a negative $\Delta_r S^\circ$, for example

$$NO_2(g) + NO_3(g) \longrightarrow N_2O_5(g) \qquad \Delta_r S^\circ = -137 \text{ J K}^{-1} \text{ mol}^{-1}.$$

The following reaction has the same number of moles of gas on either side, so $\Delta_r S^\circ$ is expected to be small, which indeed it is

$$CH_4(g) + 2O_2(g) \longrightarrow 2H_2O(g) + CO_2(g) \qquad \Delta_r S^\circ = -4 \text{ J K}^{-1} \text{ mol}^{-1}.$$

However, if the water in this reaction is formed in the liquid rather than gaseous state, the entropy change becomes significantly negative as going from gaseous water to liquid water involves a reduction in the entropy

$$CH_4(g) + 2O_2(g) \longrightarrow 2H_2O(l) + CO_2(g) \qquad \Delta_r S^\circ = -48 \text{ J K}^{-1} \text{ mol}^{-1}.$$

Ionic solids

As we have seen in section 5.2.1 on page 189, the enthalpy change accompanying the formation of an ionic solid from gaseous ions is called the lattice enthalpy. This is the enthalpy change for the reaction

$$M^+(g) + X^-(g) \longrightarrow MX(s).$$

Since two moles of gas are going to one mole of solid, we can expect there to be a significant reduction in the entropy. For example, for NaCl(s) the entropy change on forming the lattice is -229 J K^{-1} mol^{-1}.

6.12 $\Delta_r H°$ and $\Delta_r S°$ for reactions involving ions in solution

The thermodynamics of a reaction involving ions in solution, such as when an ionic solid dissolves (a process called *dissolution*)

$$NaCl(s) \longrightarrow Na^+(aq) + Cl^-(aq),$$

is somewhat more complicated than the reactions we have looked at so far as there are strong interactions between the ions and the solvent water. These interactions affect both the enthalpy and entropy changes of reactions involving ions.

When thinking about the thermodynamics of ions in solution it is useful to think about the standard enthalpy and standard entropy of *hydration* of an ion. These are the $\Delta_r H°$ and $\Delta_r S°$ values for a reaction in which one mole of the ion (at standard pressure) is taken from the gas phase into solution, to give a final concentration of 1 mol dm^{-3}:

$$M^{z+ \text{ or } z-}(g) \longrightarrow M^{z+ \text{ or } z-}(aq, \ 1 \text{ mol dm}^{-3}).   \Delta_{hyd}H° \text{ and } \Delta_{hyd}S°$$

6.12.1   Enthalpies of hydration

Given that water is a very polar molecule, we can expect that there will be strong favourable interactions between the ion and solvent water. Studies indicate that for most ions between four and eight water molecules are quite tightly held to the ion, forming what is known as the *primary hydration sphere*. At somewhat greater distances, a larger number of water molecules form the *secondary hydration sphere*, as shown in Fig. 6.18. These molecules are not as tightly held as those in the primary sphere, but they are significantly affected by the presence of the ion.

For cations, the negative end of the dipole of the water molecule is oriented towards the ion, as this orientation gives the favourable interaction. For anions, the water molecules are oriented the other way round.

The primary form of interaction between the water molecules and the ions is the electrostatic interaction between a charge and a dipole. However, for anions there is also the possibility of hydrogen bonding to the water molecules. In addition, water molecules in the secondary hydration sphere are certainly hydrogen bonded to those in the primary sphere.

Table 6.2 gives the values $\Delta_{hyd}H°$ for a selection of cations and anions. The enthalpy of hydration is large and negative, in line with the expected strong interactions between the ions and the solvent water. What is also clear from the table is that $\Delta_{hyd}H°$ is much more negative for 2+ ions than for 1+ ions. Again, this makes sense as we would expect a doubly charged ion to have a stronger interaction with the solvent than a singly charged ion. We also note that singly charged anions and cations have comparable values of $\Delta_{hyd}H°$.

As we go down a group (e.g. Li$^+$ → Cs$^+$), for both cations and anions the $\Delta_{hyd}H°$ values become less negative. This seems to correlate with the increase in the ionic radius, values of which are also shown in the table. This trend can be explained using a simple electrostatic model for hydration which is described in the next section.

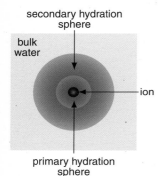

secondary hydration sphere

bulk water

ion

primary hydration sphere

Fig. 6.18 Schematic of a hydrated ion. Close to the ion there is a layer of four to eight water molecules which are quite tightly held: these form the primary hydration sphere. Further out there is a less well-defined layer of water molecules which are not so tightly held: this is the secondary hydration sphere. At large distances from the ion, we simply have unperturbed bulk water.

Ionic radii were introduced in section 5.2.2 on page 194.

Table 6.2 Standard enthalpies and entropies of hydration of ions at 298 K

ion	$\Delta_{hyd}H^\circ$/ kJ mol^{-1}	$\Delta_{hyd}S^\circ$/ J K^{-1} mol^{-1}	ionic radius / pm
Li$^+$	-538	-141	76
Na$^+$	-424	-110	102
K$^+$	-340	-72.5	138
Rb$^+$	-315	-63.8	152
Cs$^+$	-291	-57.9	167
Be^{2+}	-2524	-308	45
Mg^{2+}	-1963	-329	72
Ca^{2+}	-1616	-250	100
Sr^{2+}	-1483	-203	118
F$^-$	-504	-139	133
Cl$^-$	-359	-76.3	181
Br$^-$	-328	-60.0	196
I$^-$	-287	-41.9	220

6.12.2 A simple model for enthalpies of hydration

The enthalpy of hydration of an ion can be estimated using a simple electrostatic model similar to that used for estimating lattice energies. The details of the calculation are beyond the scope of this book, but the result is quite straightforward to use and interpret.

The model assumes that the ion is a sphere of radius r, having a charge z. The energy change when the ion goes from a vacuum into water turns out simply to depend on the bulk physical properties of the water, the radius of the ion and its charge. We end up with the following simple expression for $\Delta_{hyd}H^\circ$ at 298 K

$$\Delta_{hyd}H^\circ(298 \text{ K}) = -\frac{z^2}{r\,(\text{pm})} \times (6.98 \times 10^4) \quad \text{kJ mol}^{-1}. \qquad (6.19)$$

Note that r must be in pm; the value of the constant reflects the bulk physical properties of the solvent water.

This equation predicts that $\Delta_{hyd}H^\circ$ is negative, and that increasing the charge z, or decreasing the radius r, results in the enthalpy change becoming more negative. We can rationalize this behaviour by noting that ions with higher charge, or smaller radius, will have a higher charge density and hence a stronger interaction with the solvent.

Equation 6.19 predicts that a plot of $\Delta_{hyd}H^\circ$ against z^2/r should be a straight line. This relationship is tested in Fig. 6.19 on the following page, where experimental values of $\Delta_{hyd}H^\circ$ for some monovalent cations and anions, as well as for some divalent cations, are plotted in this way. In each plot, the dashed line is the prediction of Eq. 6.19.

Looking firstly at the monovalent ions, we see that the data follow the predicted trend quite well in that as z^2/r increases, so does the size of $\Delta_{hyd}H^\circ$. The experimental values for the anion hydration enthalpies fall quite close to the dashed line. However, for the cations the experimental values of $\Delta_{hyd}H^\circ$ lie significantly below the prediction of Eq. 6.19, shown by the dashed line. For the divalent cations once again the trend is correct, but the experimental values are similarly well below the predicted values.

We cannot expect too much of this very simple electrostatic model as it takes no account of the details of the interaction between the ion and the solvent at

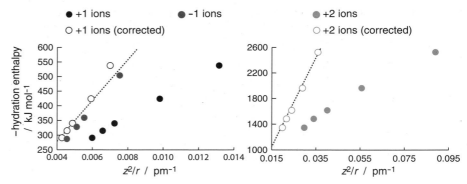

Fig. 6.19 Plots of experimental values of enthalpies of hydration against z^2/r for monovalent ions (left) and divalent cations (right). There is a reasonable linear trend between these two quantities. The prediction of the simple electrostatic model is shown by the dashed line. Whereas the anions fall quite close to this, the experimental values for the cations are smaller in size than the values predicted by the model. If 68 pm is added to the radii of the cations, the correlation with the model is much better, as shown by the open circles.

a molecular level. Probably the biggest difficulty is over the choice of the ionic radius. To draw Fig. 6.19 we simply used the values obtained from crystal structures, and these may indeed not be appropriate for ions in solution.

In fact, for the cations if we simply add 68 pm to the radius obtained from crystal structures, the values of $\Delta_{hyd}H°$ predicted by Eq. 6.19 are in quite close agreement with the experimental values, as shown by the open circles in Fig. 6.19. However, this correction is simply empirical i.e. without a theoretical basis.

6.12.3   Entropies of hydration of ions

Some typical values of the entropy of hydration of ions are given in Table 6.2 on the previous page; the values are all negative i.e. there is a reduction in entropy on going from the gas to the solution. Two other trends are clear: firstly, 2+ ions have more negative values of $\Delta_{hyd}S°$ than do 1+ or 1− ions; secondly, as the (crystal) ionic radius decreases, the value of $\Delta_{hyd}S°$ becomes more negative.

We can use a molecular argument to explain these data. When the ion moves into solution it is certainly more constrained than when it was in the gas phase, simply due to strong interactions with the polar water molecules; this accounts for the generally negative values of $\Delta_{hyd}S°$. The greater the charge on the ion becomes, or the more 'concentrated' that charge becomes by being localized on a smaller ion, the more tightly held are the water molecules in the hydration spheres. Constraining the water molecules in this way results in a reduction in their entropy, and so further reduces the entropy of the solution. These negative values of $\Delta_{hyd}S°$ are thus in part due to the reduction in the entropy of the solvent due to the presence of the ion.

6.12.4   Summary

Before we look at some applications, this is a good point to summarize what we have found out about typical $\Delta_r H°$ and $\Delta_r S°$ values

- For reactions not involving ions, the value of $\Delta_r H°$ is mainly determined by the type and number of bonds being made or broken. The physical state (e.g. gas, liquid or solid) of the species involved also affects the value of $\Delta_r H°$.

- The value of $\Delta_r S°$ is dominated by the change in the number of moles of gaseous species on going from reactants to products.

- In solution, the solvation of ions is important. The enthalpy of solvation of an ion is determined primarily by the size and charge of the ions. The smaller or more highly charged an ion, the more negative the enthalpy of solvation.

- The entropies of solvation of ions in solution are unfavourable (negative), and become more so as the charge on the ion increases or its size decreases.

6.13 Applications

We will look at two examples in which the ideas and data presented in the previous section can be used to rationalize differences in the behaviour of related compounds.

6.13.1 Solubility of ionic solids

When an ionic solid MX dissolves in water, the following equilibrium is set up

$$MX(s) \rightleftharpoons M^+(aq) + X^-(aq),$$

where we have taken the example of monovalent ions. If $\Delta_r G°$ for this reaction is large and negative, then K will be much greater than one, and the equilibrium will be very much on the side of the products (the dissolved ions). In such a situation, we would describe MX as being very soluble. On the other hand, if $\Delta_r G°$ for this reaction is large and positive, K will be much less than 1, the reactants (the solid) will be favoured, and so MX will be very insoluble.

We can think about the values of $\Delta_r H°$, $\Delta_r S°$ and $\Delta_r G°$ for this process using the simple Hess' cycle shown in Fig. 6.20 in which the dissolution of the solid, reaction 4, is broken down into three steps. The value of $\Delta_r H°$ for reaction 4 is found by adding together the $\Delta_r H°$ values for steps 1, 2 and 3; similarly, the value of $\Delta_r S°$ for step 4 is found by adding together the $\Delta_r S°$ values of steps 1–3.

For step 1, $\Delta_r H°$ is minus the lattice enthalpy of MX(s). For steps 2 and 3 the $\Delta_r H°$ and $\Delta_r S°$ values are the enthalpies and entropies of hydration of the appropriate ions.

Values of $\Delta_r H°$, $\Delta_r S°$ and $\Delta_r G°$ for all four steps for NaCl and LiF are presented in graphical form in Fig. 6.21 on the next page. Since $\Delta_r G° = \Delta_r H° - T\Delta_r S°$, it makes more sense to indicate not the value of $\Delta_r S°$, but the product $-T\Delta_r S°$, as it is this value which can be added directly to $\Delta_r H°$.

Bar charts (a) and (b) show the data for NaCl; chart (b) is a twelve-fold vertical expansion of (a). Firstly we see that $\Delta_r H°$ for step 1 (shown in black), the breaking apart of the lattice, is, as expected, large and positive. The enthalpies of solvation of the two ions are however quite large and negative (steps 2 and

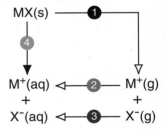

Fig. 6.20 Hess' Law cycle for analysing the dissolution of an ionic solid MX in water.

Fig. 6.21 Bar charts indicating the sizes of $\Delta_r H^\circ$, $-T\Delta_r S^\circ$ and $\Delta_r G^\circ$ for the four steps shown in Fig. 6.20. The bars are coloured according to the which step they represent, and the vertical scale is in kJ mol^{-1}. Charts (a) and (b) show the data for NaCl; the vertical scale for (b) is expanded twelve-fold so that the small data values are more easily seen. Similarly, graphs (c) and (d) are for LiF, with (d) being a twelve-fold scale expansion of (c).

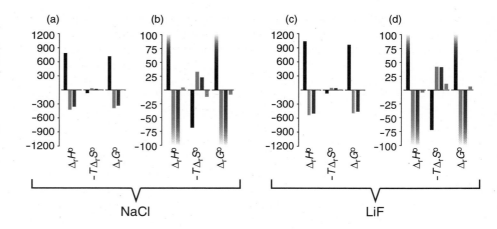

3, shown in green and turquoise), and pretty much compensate for the large positive value of step 1. Overall, the value of $\Delta_r H^\circ$ for reaction 4, shown in red, turns out to be just +4 kJ mol^{-1}, a value which is only just visible on the twelve-fold vertical scale expansion shown in (b).

Since step 1 involves going from solid to gas, $\Delta_r S^\circ$ is positive, so $-T\Delta_r S^\circ$ is negative. On the other hand, for steps 2 an 3, $-T\Delta_r S^\circ$ is positive, as the entropy decreases when an ion is solvated. What is clear from charts (a) and (b) is that the values of $-T\Delta_r S^\circ$ for steps 1, 2 and 3 are much smaller than those for $\Delta_r H^\circ$. The value of $-T\Delta_r S^\circ$ for reaction 4 (shown in red) turns out to be -12.9 kJ mol^{-1}. In other words, the entropic contribution to $\Delta_r G^\circ$ is favourable.

This favourable entropic term is sufficiently negative to overcome the unfavourable (positive) enthalpy term, resulting in $\Delta_r G^\circ$ for NaCl dissolving (reaction 4) being negative at -8.9 kJ mol^{-1}. This means that the process of solid NaCl dissolving in water is a favourable one.

Charts (c) and (d) in Fig. 6.21 present the data for LiF. On account of the much smaller size of both ions, the lattice energy (step 1) is significantly greater than for NaCl. However, for the same reason the enthalpies of solvation of Li$^+$ and F$^-$ (steps 2 and 3) are more negative than for Na$^+$ and Cl$^-$. It turns out that the solvation terms win out, so $\Delta_r H^\circ$ for the dissolution process (step 4) is just negative, in contrast to the case of NaCl, where $\Delta_r H^\circ$ is just positive.

As before, the $-T\Delta_r S^\circ$ terms are much smaller than the enthalpy terms. On account of the smaller size of the ions in LiF, the ions are more strongly solvated and so the entropy of solvation terms are more negative, resulting in an overall positive value of $-T\Delta_r S^\circ$ for step 4 (the dissolution process): this is in contrast to the case of NaCl, for which this term is negative. The data for the two salts can be contrasted by comparing charts (b) and (d).

Overall, $\Delta_r G^\circ$ for the dissolution of LiF is positive, indicating that the salt is not soluble to a significant extent. From the charts we can see that this result is a consequence of the unfavourable $-T\Delta_r S^\circ$ term outweighing the favourable $\Delta_r H^\circ$ term. The solubility of LiF (at 298 K) is found to be just 1.3 g dm^{-3} (equivalent to ≈ 0.005 mol dm^{-3}), considerably less than for NaCl which has a solubility of 360 g dm^{-3} (≈ 6 mol dm^{-3}).

These two examples reveal that whether or not a salt will be soluble is the result of a rather subtle balance between conflicting terms. Small ions give high lattice energies, and are strongly solvated. This means that the enthalpies of hydration are large and negative, which may compensate for the large lattice enthalpies. However, the strong solvation also results in more negative entropies of solvation. Which one of these terms will win is hard to predict in advance.

Figure 6.22 shows similar data for $CaCO_3$ which is rather insoluble. The double positive charges on the ions result in a much higher lattice energy than for NaCl or LiF, but in compensation the enthalpies of solvation are much more negative. Overall, $\Delta_r H°$ for the dissolution process turns out to be just negative, as it was in LiF.

The big difference for $CaCO_3$ is that the $-T\Delta_r S°$ terms are much more significant – to the extent that they can be seen on the chart without using a scale expansion. Doubling the charges on both the cation and the anion makes the entropies of solvation much more negative than for singly charged ions, and as a result $-T\Delta_r S°$ for the dissolution process is significant and positive. In fact it is sufficiently positive to overcome the negative value of $\Delta_r H°$, thus making $\Delta_r G°$ positive. As a result, the dissolution process is not favoured and the salt is insoluble. Experimentally, the solubility of $CaCO_3$ is found to be around 10^{-4} g dm^{-3} at 298 K ($\approx 10^{-6}$ mol dm^{-3}); the salt is very insoluble.

6.13.2   Acid strengths of HF and HCl

When an acidic species HX is present in water, the following equilibrium is set up

$$HX(aq) \rightleftharpoons H^+(aq) + X^-(aq).$$

If $\Delta_r G°$ for this reaction is significantly negative, then the equilibrium lies in favour of the products (i.e. the dissociation of the acid is more or less complete), and we have what is known as a *strong acid*. In contrast, if $\Delta_r G°$ is significantly positive, the equilibrium lies well the left, and there is little dissociation, giving what is known as a *weak acid*.

All of the hydrogen halides (HX) turn out to be very strong acids, with the exception of HF which turns out to be rather weak. For HCl, $\Delta_r G°$ for the above reaction is -43 kJ mol^{-1}, but for HF the value is dramatically different at $+17$ kJ mol^{-1}. In this section we will use our understanding of the thermodynamic properties of ions to rationalize this difference.

The Hess' Law cycle shown at the top of Fig. 6.23 on the following page is a useful way of discussing the overall process, which is step 5. In step 1 *undissociated* HX is taken from the aqueous phase to the gas phase. In step 2 the gaseous HX molecule is dissociated into atoms, and in step 3 the H atom loses an electron, while the X atom gains one. Finally, the ions go from the gas phase to the aqueous phase in step 4.

Charts (a) and (b) present for comparison the values of $\Delta_r H°$ and $-T\Delta_r S°$ for each step; data for HF are shown in green, while those for HCl are shown in blue. As we have seen before, the values of $\Delta_r H°$ are considerably larger than those of $-T\Delta_r S°$.

Looking first at the $\Delta_r H°$ values, chart (a), we see that for step 1 HF has a somewhat larger value than does HCl. This can probably be attributed to the strong hydrogen bonding between HF and solvent water. Step 2, the dissociation of the H–X bond, is also larger for HF, which we can rationalize as resulting from the better overlap between the AOs in HF. For step 3, there is little difference between the two halogens, but for step 4 the value is much more negative for HF than for HCl. This can be attributed to the greater enthalpy of solvation of the smaller F^- anion.

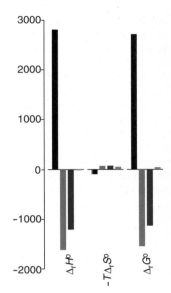

Fig. 6.22 Data, presented in the same way as in Fig. 6.21, for the dissolution of $CaCO_3$. $\Delta_r H°$ for step 4 (shown in red) is just negative, but $-T\Delta_r S°$ for this step is sufficiently positive that $\Delta_r G°$ becomes positive. As a result of the unfavourable entropy terms, the salt is insoluble. The vertical scale is in kJ mol^{-1}.

In water H^+ exists as solvated H_3O^+ ions, but for simplicity here we will just write the ion as H^+(aq).

Fig. 6.23 Shown at the top is a Hess' Law cycle for discussing the acid dissociation of HX (step 5). The values of $\Delta_r H^\circ$ for each step are shown in chart (a), with green bars for HF and blue for HCl. Similarly, the values of $-T\Delta_r S^\circ$ are shown in chart (b): note the different scale. Chart (c) compares the values of $\Delta_r H^\circ$ and $-T\Delta_r S^\circ$ for the overall process, step 5. For HF, the positive $-T\Delta_r S^\circ$ term dominates, resulting in a positive value for $\Delta_r G^\circ$ and hence little dissociation. In contrast, for HCl the negative $\Delta_r H^\circ$ dominates, resulting in a negative $\Delta_r G^\circ$, and hence a strong acid.

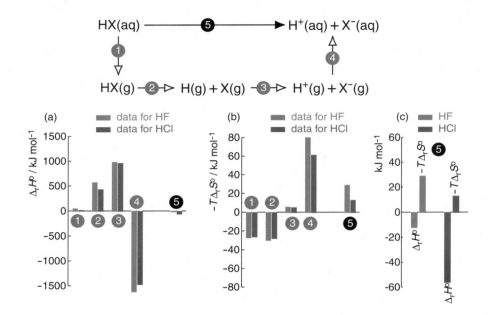

Overall, $\Delta_r H^\circ$ for step 5 is just negative for HF (-12 kJ mol^{-1}), but significantly negative (-56 J K^{-1} mol^{-1}) for HCl. We can attribute this difference to fact that steps 1 and 2 are more endothermic for HF, and that this is not compensated for by the more exothermic step 4.

The $-T\Delta_r S^\circ$ values shown in chart (b) do not reveal any great differences between HF and HX except for step 4. Solvation is more unfavourable for HF on account of the large reduction in entropy when the small F$^-$ ion is solvated. Overall, the $-T\Delta_r S^\circ$ values are both positive (unfavourable), but the value for HF is significantly more positive than that for HCl.

Chart (c) compares the values of $\Delta_r H^\circ$ and $-T\Delta_r S^\circ$ for the overall process, step 5. For HF, $\Delta_r H^\circ$ is negative, but this is exceeded in size by the positive value of $-T\Delta_r S^\circ$. As a result, $\Delta_r G^\circ$ is positive, and the equilibrium is to the left, meaning that the majority of HF molecules in the solution are undissociated. In contrast, for HCl $\Delta_r H^\circ$ is so much more negative that it is greater in size than the positive $-T\Delta_r S^\circ$ value, which in any case is rather small. As a result, $\Delta_r G^\circ$ is significantly negative, and the equilibrium is far to the right, meaning that virtually all of the HCl molecules are dissociated to give H$_3$O$^+$ and Cl$^-$.

We see that there several factors which influence the acid strength of HX, and that these are finely balanced. Overall, the low acidity of HF can be attributed to: the strong hydrogen bonding of undissociated HF in solution, the strong H–F bond, and the unfavourable entropy of solvation of F$^-$. Note that both enthalpy and entropy effects are important in determining the acid strength. For HCl, the hydrogen bonding is weaker, the H–Cl bond is weaker, and the solvation of Cl$^-$ has a less unfavourable entropy change: taken together these result in dissociation being much more favourable.

6.14 Acidity, basicity and pK_a

Equilibria involving acids and bases play an important part in many chemical processes and, as we will see later on, such equilibria also have an important influence on reactivity. In this section we will introduce the key thermodynamic ideas used to describe equilibria involving acids and bases.

6.14.1 Acids and bases

The simplest definition of an *acid* is that it is a substance which can give up a proton (H^+) to another species. For example, if the acid is AH the equilibrium under discussion is

$$\underbrace{AH}_{acid} + \underbrace{B}_{base} \rightleftharpoons \underbrace{A^-}_{conjugate\ base} + \underbrace{BH^+}_{conjugate\ acid}.$$

The species which accepts the proton, here B, is described as a *base*.

Having given up its proton, AH becomes the species A^- which is called the *conjugate base* of the acid AH. The reason for this name is that A^- can accept a proton to give AH, which means that A^- is itself a base. It is described as the conjugate base of AH since it is the base which derives from the dissociation of AH. In the same way, having accepted the proton the base B becomes the species BH^+, which is described as the *conjugate acid* of the base B.

If we dissolve an acid in water, then it is the solvent H_2O itself which acts as the base in the following equilibrium:

$$\underbrace{AH}_{acid} + \underbrace{H_2O}_{base} \rightleftharpoons \underbrace{A^-}_{conjugate\ base} + H_3O^+.$$

The species which is formed when water accepts a proton, H_3O^+, is called the *hydronium ion.*

As equilibria involving acids and bases in water are so important, we will restrict our discussion to these from now on. Writing this equilibrium out slightly more formally, so that the state of all of the species is indicated, gives

$$AH(aq) + H_2O(l) \rightleftharpoons A^-(aq) + H_3O^+(aq).$$

The equilibrium constant associated with this is called the *acid dissociation constant*, and is given the symbol K_a

$$K_a = \frac{[H_3O^+]/c^\circ\ [A^-]/c^\circ}{[AH]/c^\circ}.$$

Note that, in line with the discussion in section 6.8.2 on page 224, each concentration term has been divided by the standard concentration c°, and no term appears for the solvent water since (for dilute solutions) it is essentially in its pure form i.e. in its standard state.

The standard concentration is 1 mol dm^{-3}, so if we remember to write the concentrations in mol dm^{-3}, then it is permissible to write K_a as

$$K_a = \frac{[H_3O^+]\ [A^-]}{[AH]}.$$

It is important to remember that, despite appearances to the contrary, this equilibrium constant is dimensionless.

When a base B is dissolved in water it can remove a proton from H_2O to give the hydroxide ion OH^-

$$\underbrace{B(aq)}_{\text{base}} + \underbrace{H_2O(l)}_{\text{acid}} \rightleftharpoons \underbrace{BH^+(aq)}_{\text{conjugate acid}} + OH^-(aq).$$

Note that this time the solvent water is acting as an acid, and that the resulting species BH^+ is the conjugate acid of the base B. Using a similar approach to that for K_a, the equilibrium constant for this process, K_b, can be written

$$K_b(B) = \frac{[BH^+]\,[OH^-]}{[B]}. \tag{6.20}$$

We have written the equilibrium constant as $K_b(B)$ to remind ourselves that it is for the base B.

Of course, the conjugate acid BH^+ could also transfer a proton to water, setting up the equilibrium

$$BH^+(aq) + H_2O(l) \rightleftharpoons B(aq) + H_3O^+(aq),$$

for which the acid dissociation constant is

$$K_a(BH^+) = \frac{[B]\,[H_3O^+]}{[BH^+]}. \tag{6.21}$$

Again we have written this as $K_a(BH^+)$ to remind ourselves that this is the acid dissociation constant for BH^+.

6.14.2 Self ionization of water

We have already seen that water can act as an acid and as a base, and in fact it can do this 'with itself' to give the following equilibrium

$$\underbrace{H_2O(l)}_{\text{base}} + \underbrace{H_2O(l)}_{\text{acid}} \rightleftharpoons H_3O^+(aq) + OH^-(aq).$$

As with K_a and K_b, the concentration terms in K_w are in fact divided by $c°$ so that the equilibrium constant is dimensionless.

The equilibrium constant for this is denoted K_w, and is called the *autoprotolysis constant* for water

$$K_w = [H_3O^+]\,[OH^-].$$

At 298 K it takes the value 1.008×10^{-14}.

It is easy to show that the values of $K_b(B)$, $K_a(BH^+)$ and K_w are related. From Eq. 6.20 and Eq. 6.21 we can compute the product $K_b(B)K_a(BH^+)$ as

$$\begin{aligned}
K_b(B)\,K_a(BH^+) &= \frac{[BH^+]\,[OH^-]}{[B]}\,\frac{[B]\,[H_3O^+]}{[BH^+]} \\
&= [OH^-][H_3O^+] \\
&= K_w.
\end{aligned}$$

Since at a given temperature K_w is fixed, the last line tells us that the values of $K_b(B)$ and $K_a(BH^+)$ are not independent of one another. As we shall shortly see, this means that it is possible to discuss the basicity of B in terms of the acidity of the conjugate acid BH^+.

6.14.3   The acidity scale: pK_a

The values of the acid dissociation constant cover many orders of magnitude, so it is common to quote not the constant K_a but a quantity called the pK_a, defined as

$$pK_a = -\log_{10} K_a.$$

Note that the logarithm is to the base 10. For example, the value of K_a for ethanoic acid in water at 298 K is 1.78×10^{-5}, so

$$
\begin{aligned}
pK_a &= -\log_{10}(1.78 \times 10^{-5}) \\
&= 4.75
\end{aligned}
$$

An acid which only dissociates to a small extent will have a value of K_a which is much less than one and hence, on account of the minus sign in the definition of pK_a, a positive value of the pK_a. Such acids are described as *weak* e.g. ethanoic acid, with a pK_a of 4.75 . An acid which to all intents and purposes is completely dissociated in solution is described as a *strong acid*. Since such acids have $K_a \gg 1$, they have negative values of the pK_a. However, as we shall see in section 6.14.6 on page 250, for such acids the actual value of the pK_a has little effect of the acidity of the solution.

If we have the value of pK_a and want to work back to the value of K_a we simply use

$$K_a = 10^{-pK_a}.$$

Interpretation of pK_a values

The pK_a value refers to the the equilibrium

$$AH(aq) + H_2O(l) \rightleftharpoons A^-(aq) + H_3O^+(aq).$$

If the pK_a is negative, K_a is greater than 1, so the above equilibrium is over to the right i.e. the products are favoured, such that the concentrations of A^- and H_3O^+ will be greater than the concentration of undissociated acid AH. The more negative the pK_a becomes, the further to the right the equilibrium lies, meaning that more and more of the AH is dissociated. For strong acids, such as HBr, the equilibrium is so far to the right that dissociation is essentially complete.

On the other hand, positive pK_a values mean that K_a is less than 1, so that the equilibrium lies to the left. The concentration of the undissociated AH will be greater than that of A^- or H_3O^+, and as the pK_a becomes more positive, less and less of the AH will be dissociated. By the time the pK_a exceeds 10, to all intents and purposes none of the AH will have dissociated.

6.14.4   Basicity in terms of pK_a

It was shown above that the value of $K_b(B)$ of a base B, and the acid dissociation constant, $K_a(BH^+)$, of its conjugate acid BH^+ were related in the following way

$$K_b(B)\, K_a(BH^+) = K_w. \tag{6.22}$$

We will now show how this relationship makes it possible to discuss the strength of bases in terms of the acidity of the conjugate acids.

▷ The properties of logarithms and powers are reviewed in section 20.4 on page 895.

Fig. 6.24 Graphical illustration of how the pK_a value of an acid AH can be interpreted. Strong acids have large negative values of pK_a. If A^- is a strong base, the pK_a of its conjugate acid AH will be large. The conjugate acids of very strong bases will have pK_a values in excess of 15.

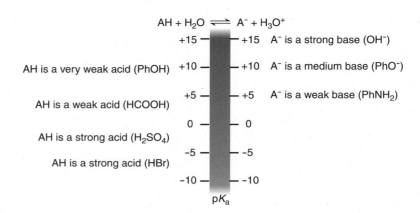

For example, to discuss the base NH_2^- we need to consider the pK_a of its conjugate acid, NH_3, which has a value of around 33. This pK_a refers to the following equilibrium

$$NH_3(aq) + H_2O(l) \rightleftharpoons NH_2^-(aq) + H_3O^+(aq),$$

The very large positive value for the pK_a indicates that the equilibrium is essentially completely to the left. This tells us that NH_2^- is a very strong base, as the reaction in which it removes the proton from H_3O^+ (right to left in the above) is essentially complete.

To put this discussion in a more quantitative form, we start with Eq. 6.22 on the preceding page, take $\log_{10}$ of both sides, and multiply by -1 to give

$$pK_b(B) + pK_a(BH^+) = pK_w,$$

where

$$pK_b = -\log_{10} K_b \quad \text{and} \quad pK_w = -\log_{10} K_w.$$

At 298 K $K_w = 1.008 \times 10^{-14}$, so pK_w is 14.0. So, in the case of NH_3

$$
\begin{aligned}
pK_b(NH_2^-) &= pK_w - pK_a(NH_3) \\
&= 14.0 - 33 \\
&= -19
\end{aligned}
$$

The equilibrium constant K_b refers to the equilibrium

$$NH_2^-(aq) + H_2O(l) \rightleftharpoons NH_3(aq) + OH^-(aq).$$

The very large negative value for pK_b tells us that the equilibrium is completely to the right i.e. NH_2^- is a very strong base. This example illustrates the general principle that the conjugate acids of strong bases have large positive pK_a values.

Generally we will not use pK_b values, but will characterize both acids and bases using pK_a values. To summarize the conclusions of this section:

- A strong acid has a large negative pK_a.

- The conjugate acid of a strong base has a large positive pK_a.

These conclusions are illustrated graphically in Fig. 6.24.

6.14.5 Competition between two acids

Suppose we make an aqueous solution containing a mixture of two acids, AH and XH. Both will dissociate to protonate the water, but the extent to which each dissociates will not necessarily be the same. Intuitively, we would expect that it will be the strongest acid which is more dissociated: we are now in a position to prove that this is so using pK_a values.

Individually, the two acids dissociate according to the equilibria

$$AH(aq) + H_2O(l) \rightleftharpoons A^-(aq) + H_3O^+(aq) \qquad (6.23)$$
$$XH(aq) + H_2O(l) \rightleftharpoons X^-(aq) + H_3O^+(aq) \qquad (6.24)$$

K_a for these two equilibria are given by

$$K_a(AH) = \frac{[H_3O^+]\,[A^-]}{[AH]} \quad \text{and} \quad K_a(XH) = \frac{[H_3O^+]\,[X^-]}{[XH]}. \qquad (6.25)$$

If we subtract Eq. 6.24 from Eq. 6.23, the terms in H_2O and H_3O^+ cancel to give

$$AH(aq) + X^-(aq) \rightleftharpoons A^-(aq) + XH(aq). \qquad (6.26)$$

The equilibrium constant for this reaction is given by

$$K = \frac{[A^-]\,[XH]}{[AH]\,[X^-]}. \qquad (6.27)$$

By comparing Eq. 6.27 with Eq. 6.25, we can see that

$$K = \frac{K_a(AH)}{K_a(XH)}. \qquad (6.28)$$

Taking $\log_{10}$ of each side and multiplying by -1 gives

$$pK = pK_a(AH) - pK_a(XH), \qquad (6.29)$$

where $pK = -\log_{10} K$. As before, a negative value of pK implies that the equilibrium is to the right, whereas a positive value implies that the equilibrium is to the left.

We can now use Eq. 6.29 to interpret the position of equilibrium in the competition reaction (Eq. 6.26)

$$AH(aq) + X^-(aq) \rightleftharpoons A^-(aq) + XH(aq).$$

If AH is a stronger acid than XH, then pK_a(AH) will be less than pK_a(XH), and so from Eq. 6.29 pK will be negative. This tells us that the equilibrium is over to the right i.e. AH is more dissociated than XH

$$AH(aq) + X^-(aq) \rightleftharpoons \underbrace{A^-(aq) + XH(aq).}_{\text{favoured if } pK_a(AH) < pK_a(XH)}$$

On the other hand, if XH is a stronger acid than AH, then pK_a(XH) will be less than pK_a(AH), and so pK will be positive. This means that the equilibrium is to the left i.e. XH is more dissociated than AH

$$\underbrace{AH(aq) + X^-(aq)}_{\text{favoured if } pK_a(XH) < pK_a(AH)} \rightleftharpoons A^-(aq) + XH(aq).$$

As an example, consider making up a solution containing methanoic acid (AH, $pK_a = 3.75$) and ethanoic acid (XH, $pK_a = 4.76$). The pK for

$$HCOOH(aq) + CH_3COO^-(aq) \rightleftharpoons HCOO^-(aq) + CH_3COOH(aq)$$

is $3.75 - 4.76 = -1.01$, and so the equilibrium constant is 10.2 i.e. the right-hand side is favoured, meaning that the methanoic acid is more dissociated than the ethanoic acid.

On the other hand, if XH is a strong acid such as HCl ($pK_a = -7$), then $pK = 3.75 - (-7) = 10.75$, meaning that the equilibrium

$$HCOOH(aq) + Cl^-(aq) \rightleftharpoons HCOO^-(aq) + HCl(aq)$$

is essentially completely over to the left. As we would expect, the much stronger acid HCl is the one which dissociates.

6.14.6   Levelling by the solvent

If we dissolve a strong acid in water (one with a large negative pK_a), then the equilibrium

$$AH(aq) + H_2O(l) \rightleftharpoons A^-(aq) + H_3O^+(aq)$$

will essentially lie all the way to the right i.e. all of the acid has dissociated, and the concentration of H_3O^+ is equal to the initial concentration of the acid AH. Making the pK_a more negative has no effect on the concentration of H_3O^+, since all strong acids essentially dissociate completely.

The result is that dissolving any strong acid in water, regardless of how negative its pK_a is, simply gives a solution in which the concentration of H_3O^+ is equal to the initial concentration of the acid. H_3O^+ is itself an acid, with a pK_a of -1.74 at 298 K, so we describe this situation by saying that in water all strong acids are *levelled* to the acidity of H_3O^+ in the sense that this species is the most acidic that can exist in water. So, although the pK_a of HCl is -7 and that of HBr is -9, dissolving either acid in water gives the same concentration of the acidic species H_3O^+.

A similar situation exists for strong bases, which deprotonate water (which has a pK_a of 15.74) to give OH^-. In other words, in water all strong bases are levelled to the basicity of OH^-.

The solvent water therefore imposes a limit on the acidity and basicity of the species it can support. A similar effect exists for other solvents, but the limits will be different. For example, much stronger bases can exist in hydrocarbon solvents than in water because hydrocarbons have very much lower acidity than water. Dissolving the very strong base butyllithium in water simply gives OH^-, but the butyllithium can exist in a solvent such as cyclohexane.

6.14.7   pH

Like the pK_a, the concentration of H_3O^+ is also specified on a log scale by defining a quantity known as the pH

$$pH = -\log_{10}[H_3O^+].$$

In pure water, H_3O^+ is produced by the autoprotolysis equilibrium

$$H_2O(l) + H_2O(l) \rightleftharpoons H_3O^+(aq) + OH^-(aq),$$

for which

$$K_w = [H_3O^+][OH^-].$$

If this is the only equilibrium present, then since the formation of each H_3O^+ also results in the formation of an OH^-, it follows that the concentrations of these two species are equal $[H_3O^+] = [OH^-]$. Thus, K_w can be written

$$K_w = [H_3O^+]^2.$$

At 298 K K_w is 1.008×10^{-14}, so from the above $[H_3O^+] = 1.004 \times 10^{-7}$ mol dm^{-3}, and hence the pH is 7.00.

This value of the pH for pure water defines a neutral solution. Higher concentrations of H_3O^+ give acidic solutions with values of the pH below 7; lower concentrations of H_3O^+ give basic solutions with pH values greater than 7.

If we dissolve a strong acid in water, the acid dissociates completely resulting in a concentration of H_3O^+ equal to the initial concentration of the acid AH. So the pH is simply $-\log_{10}[AH]_{initial}$.

6.14.8 The pH of solutions of weak acids

indexpH!solutions of weak acids If we dissolve a weak acid in water we expect that the resulting solution will be acidic due to the partial ionization of the acid. We are now in a position to work out exactly what the pH of the resulting solution is.

As usual we start with the equilibrium

$$AH(aq) + H_2O(l) \rightleftharpoons A^-(aq) + H_3O^+(aq).$$

for which K_a is defined as

$$K_a = \frac{[H_3O^+][A^-]}{[AH]}.$$

When we make up our solution of the acid we add a known number of moles of AH to a known volume of water to give an *initial concentration* of AH, $[AH]_{init}$. This is called the initial concentration as it is the concentration of AH before any of it has dissociated.

If the acid is weak, rather little of the AH will actually dissociate, so we can safely assume that when we have come to equilibrium the actual concentration of AH will be very similar to the initial concentration of the acid i.e. $[AH] \approx [AH]_{init}$. Furthermore, since each AH molecule which does dissociate gives an A^- ion and an H_3O^+ ion, the concentration of these two latter species is the same: $[A^-] = [H_3O^+]$. Under these conditions, the expression for the equilibrium constant can be written

$$K_a \approx \frac{[H_3O^+]^2}{[AH]_{init}}.$$

It therefore follows that

$$[H_3O^+] = \sqrt{K_a[AH]_{init}}.$$

Taking the logarithm to the base 10 of each side, and multiplying by -1 gives

$$-\log_{10}[H_3O^+] = \tfrac{1}{2}\left(-\log_{10}K_a - \log_{10}[AH]_{init}\right).$$

The term on the left is the pH, and the first term in the bracket on the right is the pK_a so

$$pH = \tfrac{1}{2}\left(pK_a - \log_{10}[\text{AH}]_{\text{init}}\right).$$

With this expression we can work out the pH of a solution of a weak acid.

For example, ethanoic acid has a pK_a of 4.76, so a 0.1 mol dm^{-3} solution of the acid has pH

$$pH = \tfrac{1}{2} \times 4.76 - \tfrac{1}{2}\log_{10}(0.1),$$

which evaluates to 2.88 – a moderately acidic solution.

Note that, since it is the case that $[\text{A}^-] = [\text{H}_3\text{O}^+]$, the concentration of A^- is easily computed from the pH. In this case the pH of 2.88 corresponds to an H_3O^+ concentration of $10^{-2.88} = 1.32 \times 10^{-3}$ mol dm^{-3}. The A^- concentration is therefore the same: only just over 1% of the acid has dissociated.

6.15 How much product is there at equilibrium?

In this chapter we have so far learnt how to compute the equilibrium constant using tabulated data, but we have not really addressed the question of how, given the value of the equilibrium constant, we can work out exactly how much of each of the reactants and products is present at equilibrium. This is the topic of this section. We will look at two examples of how such calculations are made. The first is a simple dissociation equilibrium which we will analyse using the concept of the *degree of dissociation*. The second is a more complex equilibrium, but which can nevertheless be analysed in a similar way.

6.15.1 Dissociation equilibria

Let us imagine a simple equilibrium between a dimer A_2 and the two monomers from which it is composed

$$A_2(g) \rightleftharpoons 2\,A(g).$$

Examples of this might be the dissociation of a diatomic into atoms (e.g. $I_2 \rightarrow 2I$), the dissociation of a molecule into radicals (e.g. $C_2H_6 \rightarrow 2CH_3$), or the dissociation of a dimer such as Al_2Cl_6 into two $AlCl_3$ molecules.

Let us suppose that we start out with n_0 moles of pure A_2 and introduce this into a vessel with volume V which is held at temperature T in a thermostat. What then happens is that some fraction α of the A_2 molecules dissociate until the equilibrium position is reached, at which point the partial pressures of A and A_2, p_A and p_{A_2}, are related to the value of the equilibrium constant

$$K = \frac{(p_A/p^\circ)^2}{p_{A_2}/p^\circ} = \frac{p_A^2}{p_{A_2}p^\circ}, \tag{6.30}$$

where p° is the standard pressure (1 bar, 10^5 Pa).

We can draw up a table showing the amount in moles, mole fractions and partial pressures of A_2 and A before and after the equilibrium has been established:

line	quantity	$A_2(g)$	$\rightleftharpoons$	$2A(g)$
1	initial moles	n_0		0
2	moles at equil.	$n_0(1 - \alpha)$		$2 \times n_0\alpha$
3	mole fractions	$(1 - \alpha)/(1 + \alpha)$		$(2\alpha)/(1 + \alpha)$
4	partial pressures	$p \times (1 - \alpha)/(1 + \alpha)$		$p \times 2\alpha/(1 + \alpha)$

Let us go through the lines one by one. Line 1 gives the initial situation as described above in which there are n_0 moles of A_2.

We then imagine that, in coming to equilibrium, a *fraction* α of the A_2 molecules have dissociated; it is important to realize the α is a fraction, not a number of moles. The amount in moles of A_2 which has dissociated is $n_0\alpha$, leaving behind $n_0 - n_0\alpha = n_0(1 - \alpha)$ moles of undissociated A_2. Since the dissociation of each mole of A_2 gives *two* moles of A, the total amount in moles of A is twice $n_0\alpha$. These values of the amounts in moles at equilibrium are shown on line 2.

At equilibrium, the total amount in moles present is $n_0(1 - \alpha) + 2n_0\alpha$, which is $n_0(1 + \alpha)$. Using this we can compute the mole faction of A_2 at equilibrium as

$$\frac{n_0(1 - \alpha)}{n_0(1 + \alpha)} = \frac{(1 - \alpha)}{(1 + \alpha)}.$$

The mole fraction of A can be computed in a similar way in order to give the value shown on line 3. Finally, we compute the partial pressures using the relationship $p_i = x_i p$, where p is the total pressure and x_i is the mole fraction: these values are shown on line 4.

The values of the partial pressure can then be substituted into the expression for K given in Eq. 6.30 on the facing page:

$$
\begin{aligned}
K &= \frac{p_A^2}{p_{A_2} p^\circ} \\
&= \left(\frac{2\alpha p}{1 + \alpha}\right)^2 \left(\frac{1 + \alpha}{(1 - \alpha)p}\right) \frac{1}{p^\circ} \\
&= \frac{4\alpha^2}{(1 + \alpha)(1 - \alpha)} \frac{p}{p^\circ}.
\end{aligned}
\tag{6.31}
$$

We can simplify this somewhat if we assume that the gases are ideal so that at equilibrium the total pressure p is related to the volume and the *total* amount in moles n_{tot} according to the ideal gas equation

$$pV = n_{tot}RT.$$

We have already worked out that $n_{tot} = n_0(1 + \alpha)$ so

$$p = \frac{n_0(1 + \alpha)RT}{V}.\tag{6.32}$$

Using this value for p in Eq. 6.31 leads to the cancellation of $(1 + \alpha)$ between top and bottom of the fraction to give the final expression

$$K = \frac{4\alpha^2}{1 - \alpha} \frac{n_0 RT}{p^\circ V}.\tag{6.33}$$

Using this relationship we can work out α, the fraction of A_2 which has dissociated provided that we know K, the initial amount in moles, the temperature and the volume. The parameter α is often known as the *degree of dissociation*.

Example 6.1 Calculation of the degree of dissociation

In the gas phase, bicyclopentadiene dissociates to cyclopentadiene (this is the reverse reaction of a Diels–Alder reaction)

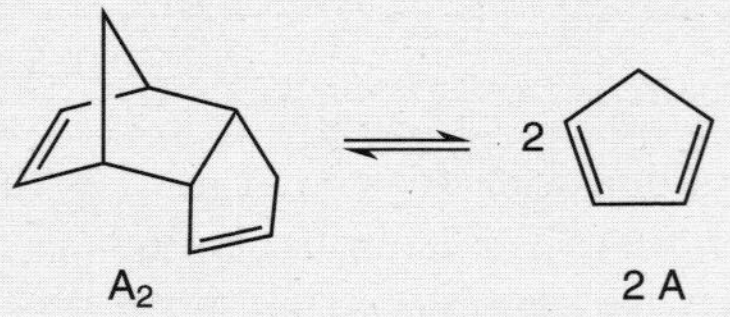

K for the above equilibrium is 0.035 at 373 K (100 °C). Suppose that 0.01 moles (about 1.3 g) of the dimer are allowed to come to equilibrium in a volume of 1 dm^3. What is the degree of dissociation, α?

All we need to do is to use Eq. 6.34, but we have to be very careful with the units. The best approach is to use the appropriate SI unit for all the quantities (SI units are described in section 20.2 on page 882). The SI unit of pressure is N m^{-2} (or Pa), so the standard pressure $p°$ of 1 bar has to be put as 1×10^5 N m^{-2}. The SI unit of volume is m^3, so the volume V of 1 dm^3 has to be put at 1×10^{-3} m^3. The calculation of α is thus

$$\left(\frac{Kp°V}{4n_0RT} \right)^{\frac{1}{2}} = \left(\frac{0.035 \times 1 \times 10^5 \times 1 \times 10^{-3}}{4 \times 0.01 \times 8.3145 \times 373} \right)^{\frac{1}{2}}$$

which evaluates to 0.168.

The degree of dissociation of the dimer is quite small. From the value of α we can work out that the amount in moles of A (cyclopentadiene) at equilibrium is $2n_0\alpha = 3.4 \times 10^{-3}$, which is just 0.022 g.

Simplification for small degrees of dissociation

If the equilibrium constant is much less than 1, then we can expect that α will also be much less than 1, in which case we can simplify Eq. 6.33 on the preceding page somewhat. In this limit, $(1 - \alpha)$ can be approximated to 1, so that

$$K = \frac{4\alpha^2 \times n_0RT}{p°V}.$$

Rearranging this gives a nice simple expression for α

$$\alpha = \left(\frac{Kp°V}{4n_0RT} \right)^{\frac{1}{2}}. \tag{6.34}$$

Example 6.1 illustrates the use of this expression to determine α.

Effect of pressure

If α is small, then $(1 + \alpha) \approx 1$, and the equilibrium pressure in the reaction vessel, given by Eq. 6.32 on the previous page, can be approximated as

$$p = \frac{n_0RT}{V}.$$

This pressure can be altered by changing the volume or the amount in moles (changing the temperature will change the equilibrium constant).

By inverting both sides of this expression for the equilibrium pressure we obtain $1/p = V/(n_0 RT)$. This can then be used to rewrite Eq. 6.34 on the facing page as

$$\alpha = \left(\frac{Kp^\circ}{4p}\right)^{\frac{1}{2}}.$$

The equation tells us that *increasing* the equilibrium pressure *decreases* the degree of dissociation α.

This result can be interpreted using Le Chatelier's principle. If the pressure is increased, the system tries to respond by shifting the equilibrium in such a way as to decrease the pressure. This is done by reducing the amount in moles present, i.e. some of the A molecules recombining to give A_2, which corresponds to a decrease in α, as predicted.

The general case

If we cannot assume that α is small, then we have to solve Eq. 6.33 as a quadratic equation in α. Rearranging somewhat gives us

$$\frac{\alpha^2}{1-\alpha} = \underbrace{\frac{p^\circ V K}{4 n_0 RT}}_{F}.$$

If we define the quantity on the right to be F, then multiplying out the above and bringing all the terms to the left gives

$$\alpha^2 + F\alpha - F = 0,$$

which can be solved in the usual way (i.e. using the quadratic formula) to give a tidy expression for α

$$\alpha = \tfrac{1}{2}\left(-F \pm \sqrt{F^2 + 4F}\right). \tag{6.35}$$

It turns out that the plus sign gives values of α which are between 0 and 1; the minus sign gives physically nonsensical values. Example 6.2 illustrates the use of this expression, and Fig. 6.25 shows how α varies with F.

6.15.2 More complex equilibria

More complex equilibria can be analysed in a similar way to that used for dissociation. Let us take as an example the production of hydrogen from methane, which we analysed above

$$CH_4(g) + 2H_2O(g) \longrightarrow CO_2(g) + 4H_2(g).$$

Imagine that we start by mixing n_0 moles of CH_4 with $2n_0$ moles of H_2O, and then allow the reactants to come to equilibrium. We can draw up a table just as we did before:

line	$CH_4(g)$	$2H_2O(g)$	$\rightleftharpoons$	$CO_2(g)$	$4H_2(g)$
1	n_0	$2n_0$			
2	$n_0(1-\alpha)$	$2n_0(1-\alpha)$		$n_0\alpha$	$4n_0\alpha$
3	$(1-\alpha)/(3+2\alpha)$	$2(1-\alpha)/(3+2\alpha)$		$\alpha/(3+2\alpha)$	$4\alpha/(3+2\alpha)$

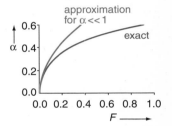

Fig. 6.25 Graph showing how the degree of dissociation α varies as a function of the parameter F, defined as $(p^\circ VK)/(4n_0 RT)$. The blue line gives the exact result predicted by Eq. 6.35, whereas the green line gives the value of α predicted by Eq. 6.34 on the facing page, which is only correct for $\alpha \ll 1$.

Example 6.2 Calculation of the degree of dissociation in the general case

The equilibrium

$$N_2O_4(g) \rightleftharpoons 2\,NO_2(g),$$

has $K = 3.97$ at 348 K (75 °C). Calculate the degree of dissociation when 0.01 moles (about 0.9 g) of N_2O_4 come to equilibrium in a volume of 1 dm^3.

First, we have to find F, and as in Example 6.1 on page 254 we need to be sure to put all of the quantities in SI units. Thus the standard pressure is 1×10^5 N m^{-2} and the volume is 1×10^{-3} m^3. Using these values F is computed as

$$\frac{p^\circ VK}{4n_0 RT} = \frac{1 \times 10^5 \times 1 \times 10^{-3} \times 3.97}{4 \times 0.01 \times 8.3145 \times 348} = 3.43$$

This value of F of 3.43 can be substituted into Eq. 6.35 on the previous page so as to find α:

$$\tfrac{1}{2}\left(-F + \sqrt{F^2 + 4F}\right) = \tfrac{1}{2}\left(-3.43 + \sqrt{3.43^2 + 4 \times 3.43}\right) = 0.81\,.$$

The resulting value of $\alpha = 0.81$ implies that, under these conditions, there is a significant amount of dissociation. There are $2n_0\alpha = 0.016$ moles of NO_2 present at equilibrium.

Line 1 shows the initial amounts in moles. We then imagine that at equilibrium a fraction α of the initial amount in moles of CH_4 has reacted to give products i.e. αn_0 moles of the methane react. This leaves behind $n_0 - \alpha n_0 = n_0(1-\alpha)$ moles of CH_4, as shown on line 2. Since two moles of H_2O react with one mole of CH_4, the amount in moles of H_2O is reduced by $2n_0\alpha$, leaving $2n_0(1-\alpha)$ moles behind. The amount in moles of CO_2 formed equals the amount in moles of CH_4 lost ($n_0\alpha$), but four times as much H_2 is generated ($4n_0\alpha$).

By adding up all the entries on line 2 we find that the total amount in moles at equilibrium is $n_0(3 + 2\alpha)$; using this, we can work out the mole fractions on line 3. In turn, by multiplying these mole fractions by the equilibrium pressure p_{eq} we can find the partial pressures of all of the species.

Using these, the equilibrium constant can be written in the following way

$$
\begin{aligned}
K &= \frac{p_{CO_2} p_{H_2}^4}{p_{CH_4} p_{H_2O}^2} \frac{1}{(p^\circ)^2} \\[2mm]
&= \frac{\alpha\,(4\alpha)^4}{(1-\alpha)\,[2(1-\alpha)]^2} \frac{1}{(3+2\alpha)^2} \frac{p_{eq}^2}{(p^\circ)^2}.
\end{aligned}
\tag{6.36}
$$

This looks, and is, pretty complicated. It is certainly not going to be possible to find α in a simple way as this expression includes powers up to α^5. However, we can make some progress if we assume that the equilibrium constant is small so that $\alpha \ll 1$. Under these circumstances we can approximate $1 - \alpha \approx 1$ and $3 + 2\alpha \approx 3$ to give the much simpler form

$$K = \frac{64\alpha^5}{9} \frac{p_{eq}^2}{(p^\circ)^2}.$$

This can be rearranged to give the following expression for α

$$\alpha = \sqrt[5]{\frac{9K(p^\circ)^2}{64p_{eq}^2}}. \tag{6.37}$$

When used commercially, this reaction is run at around 800 K, at which temperature the equilibrium constant is 0.018 . If the reaction is run so that the equilibrium pressure is 10 bar, then we find that $\alpha = 0.12$, which is a modest conversion to products. Equation 6.37 tells us that *decreasing* the pressure will increase α, a result which is consistent with the application of Le Chatelier's principle to this reaction in which there is an increase in the number of moles of gas in going from left to right. However, in a commercial process quite a high pressure is needed in order for the reaction to proceed at a reasonable rate (increasing the pressure increases the rate of collisions between molecules), so it may not be practicable to reduce the pressure too much.

As we did before, we can use the ideal gas law to compute the equilibrium pressure in terms of the total amount in moles

$$p_{eq} = \frac{n_0(3 + 2\alpha)RT}{V}.$$

Using this, Eq. 6.36 on the preceding page becomes

$$K = \frac{\alpha\,(4\alpha)^4}{(1 - \alpha)\,[2(1 - \alpha)]^2}\left(\frac{n_0RT}{p^\circ V}\right)^2.$$

6.16 Moving on

Our study of thermodynamics provides us with two very important ideas. The first is the very fundamental point that reactions are driven by an increase in the entropy of the Universe or, equivalently, a decrease in the Gibbs energy. The second is that thermodynamics gives us a practical way for predicting equilibrium constants from tabulated data.

We have also seen how we can understand how the position of equilibrium is affected by changes in chemical structure by thinking about the effect this has on the enthalpy and entropy changes.

In these first six chapters we have developed all the key ideas which we will now go on to use to understand and make sense of chemical structures and reactivity. We will start this exploration by taking an overview of the Periodic Table, how the properties of the elements vary across the periods and down the groups, and how this plays out in their chemistry.

FURTHER READING

Why Chemical Reactions Happen, James Keeler and Peter Wothers, Oxford University Press (2004).
Thermodynamics of Chemical Processes, Gareth Price, Oxford University Press (1998).
The Second Law, P. W. Atkins, Scientific American Books (1994).

QUESTIONS

The gas constant, R, has the value 8.3145 J K^{-1} mol^{-1}. 1 bar is 10^5 N m^{-2}, and 1 atmosphere is 1.01325×10^5 N m^{-2}.

6.1 Following the discussion on page 207, devise some more arrangements of the 14 molecules amongst the given energy levels such that the total energy is 10 units. For each distribution, compute the value of W. Were you able to find a distribution with a value of W greater than distribution (a) from Fig. 6.3 on page 207?

6.2 Imagine that we have two large copper blocks, A and B, which are isolated from the surroundings. Block A is at a temperature of 250 K, and block B is at a temperature of 300 K. As the blocks are large, we can assume that small amounts of heat flowing into or out of them is reversible, and that such a process will not change the temperature of the block. It therefore follows that for an amount of heat q the entropy change of the block is q/T, where T is the temperature of the block.

Compute the entropy change of each block when 10 J of heat flows: (a) from block A to block B; (b) from block B to block A. Using these results, compute the entropy change of the Universe in each case, and hence determine which direction of heat flow is spontaneous.

6.3 Solid elemental tin can exist as two allotropes, called β-tin and α-tin (sometimes these are called white and grey tin, respectively). For the interconversion process

$$\text{Sn}(\beta) \longrightarrow \text{Sn}(\alpha)$$

the enthalpy change is -2.1 kJ mol^{-1}, and the entropy change is -7.1 J K^{-1} mol^{-1}, both at 298 K. For the purposes of this question you may assume that these values are independent of temperature.

By computing the entropy change of the Universe for the above process, determine which allotrope is the stable form at: (a) -10 °C; (b) $+40$ °C. At what temperature will the two allotropes be in equilibrium with one another?

Calculate ΔG for the above process at the two temperatures, and comment on what the resulting values tell you about which allotrope is the stable form.

6.4 Solid elemental sulfur exists in two allotropes, α-sulfur (rhombic) and β-sulfur (monoclinic). The molar entropies of the two allotropes are 31.8 J K^{-1} mol^{-1} for α-sulfur, and 32.6 J K^{-1} mol^{-1} for β-sulfur (at 298 K). The enthalpy change for the conversion of α-sulfur to β-sulfur is 330 J mol^{-1}.

Determine the temperature at which the two allotropes are in equilibrium, and the allotrope which is favoured at temperatures higher than this.

6.5 Explain why it is that an endothermic process can be spontaneous provided (a) it is accompanied by an increase in the entropy of the system, and (b) the temperature exceeds a particular value.

6.6 This question and the one which follows concern the equilibrium

$$\text{CO(g)} + 2\,\text{H}_2\text{(g)} \rightleftharpoons \text{CH}_3\text{OH(g)},$$

which we are going to investigate as a viable commercial method for the production of methanol. The following data are provided (all at 298 K)

	CO(g)	H$_2$(g)	CH$_3$OH(g)
$\Delta_f H°$ / kJ mol^{-1}	-110.53		-200.66
S_m° / J K^{-1} mol^{-1}	197.67	130.68	239.81

Using these data, determine $\Delta_r H°$, $\Delta_r S°$, $\Delta_r G°$ and hence K, all at 298 K. On the basis of your answer, comment on the viability of the reaction as a method for the production of methanol.

6.7 In practice, it is found that the reaction in the previous question only proceeds at a viable rate at 600 K. Assuming that the values of $\Delta_r H°$ and $\Delta_r S°$ are the same at 600 K as they are at 298 K, find the value of the equilibrium constant at 600 K. Qualitatively, is your answer in accord with Le Chatelier's principle?

6.8 Consider the equilibrium in which solid calcium carbonate decomposes to the oxide plus carbon dioxide

$$CaCO_3(s) \rightleftharpoons CaO(s) + CO_2(g).$$

Write down an expression for the equilibrium constant of this reaction in terms of the partial pressure of CO_2.

The standard enthalpies of formation of $CaCO_3(s)$, $CO_2(g)$ and $CaO(s)$ are -1207.6 kJ mol^{-1}, -393.5 kJ mol^{-1} and -634.9 kJ mol^{-1} respectively, and the standard entropies are 91.7 J K^{-1} mol^{-1}, 213.8 J K^{-1} mol^{-1} and 38.1 J K^{-1} mol^{-1} (all at 298 K). Assuming that these values are independent of temperature, compute $\Delta_r H°$, $\Delta_r S°$ and $\Delta_r G°$ at 800 K; hence find the equilibrium pressure of carbon dioxide at this temperature.

6.9 The following data are all at 298 K

reaction	$\Delta_r H°/$ kJ mol^{-1}	$\Delta_r S°/$ J K^{-1} mol^{-1}
$C(s) + \frac{1}{2}O_2(g) \longrightarrow CO(g)$	-110.53	89.36
$Pb(s) + \frac{1}{2}O_2(g) \longrightarrow PbO(s)$	-218.99	-100.9

Assuming that these values are independent of temperature, compute $\Delta_r G°$ for the reduction of PbO by carbon

$$PbO(s) + C(s) \rightleftharpoons CO(g) + Pb(s)$$

at 298 K and at 700 K. Also, compute the temperature at which $\Delta_r G°$ is zero, and comment on the significance of this value.

6.10 Using the approach described in section 6.15.2 on page 255, complete the following table for the equilibrium $CO(g) + 2 H_2(g) \rightleftharpoons CH_3OH(g)$

line	CO(g)	$2H_2(g)$	$\rightleftharpoons$	CH$_3$OH(g)
1	n_0	$2n_0$		0
2	$n_0(1-\alpha)$			
3				

Line 1 is the initial amount in moles, line 2 is the amount in moles after a fraction α of the CO has reacted, and line 3 gives the mole fractions.

Hence show that the equilibrium constant is given by

$$K = \frac{\alpha(3-2\alpha)^2}{4(1-\alpha)^3} \frac{(p°)^2}{p_{eq}^2},$$

where p_{eq} is the equilibrium pressure.

If it can be assumed that the equilibrium constant is such that $\alpha \ll 1$, show that the equilibrium constant can be approximated as

$$K = \frac{9\,\alpha(p°)^2}{4p_{eq}^2},$$

hence obtain an expression for α in terms of K and p_{eq}.

Using your value of K from exercise 6.7, find α at 600 K for the case where the total pressure is (a) 1 bar and (b) 50 bar. Comment on your answers in the light of Le Chatelier's principle. Would it be advantageous to run the reaction at high pressure?

6.11 At high temperatures molecular iodine dissociates to iodine atoms

$$I_2(g) \rightleftharpoons 2\,I(g).$$

At 298 K, the standard enthalpies of formation of $I_2(g)$ and $I(g)$ are 62.44 kJ mol^{-1} and 106.84 kJ mol^{-1}, respectively, and the standard molar entropies are 260.69 J K^{-1} mol^{-1} and 180.79 J K^{-1} mol^{-1}, respectively. Assuming that these values are independent of temperature, compute $\Delta_r H^\circ$, $\Delta_r S^\circ$ and $\Delta_r G^\circ$ for the dissociation reaction at 600 K, and hence find the value of the equilibrium constant. 0.008 moles of I_2 are allowed to come to equilibrium in a vessel of volume 1 dm^3 at a temperature of 600 K. Using Eq. 6.34 on page 254 determine the degree of dissociation α at this temperature, and hence the amount in moles of iodine atoms present.

6.12 Hydrogen cyanide, HCN, has a pK_a of 9.21 at 298 K. Compute the pH of a 0.1 mol dm^{-3} solution of HCN. Also, find the ratio of the concentrations of CN$^-$(aq) to HCN(aq) in such a solution.

Methanoic acid has a pK_a of 3.75 at 298 K. If we prepared an equimolar mixture of hydrogen cyanide and methanoic acid in water, which out of the two conjugate bases CN$^-$ and HCO$_2^-$ would you expect to be present at the higher concentration? Give reasons for your answer.

6.13 Imagine preparing a solution of a weak acid AH and its sodium salt NaA. The acid dissociates according to the usual equilibrium

$$AH(aq) + H_2O(l) \rightleftharpoons A^-(aq) + H_3O^+(aq),$$

and we can assume that the sodium salt dissociates completely in water. Under these conditions, explain why it is possible to write the acid dissociation constant as

$$K_a = \frac{[H_3O^+]\,[NaA]_{init}}{[AH]_{init}},$$

where $[NaA]_{init}$ is the initial concentration of NaA in the solution, and $[AH]_{init}$ is the initial concentration of AH used to make up the solution.

Hence show that the pH of the solution is given by

$$pH = pK_a + \log \frac{[NaA]_{init}}{[AH]_{init}}.$$

Propanoic acid has a pK_a of 4.87. Compute the pH of a solution with initial concentrations of 0.1 mol dm^{-3} propanoic acid and 0.1 mol dm^{-3} sodium propionate. Such a solution is called a *buffer* as it has the property that its pH is largely unaltered by the addition of small amounts of H$_3$O$^+$ or OH$^-$ ions. Explain, in qualitative terms, why the presence of a reservoir of A$^-$ ions and undissociated AH leads to this buffering action.

6.14 Use the following data (all at 298 K) to calculate $\Delta_r H^\circ$, $\Delta_r S^\circ$ and $\Delta_r G^\circ$ for the dissolution of BaSO$_4$(s) at 298 K:

$$BaSO_4(s) \rightleftharpoons Ba^{2+}(aq) + SO_4^{2-}(aq)$$

reaction	$\Delta_r H^\circ$ / kJ mol^{-1}	$\Delta_r S^\circ$ / J K^{-1} mol^{-1}
BaSO$_4$(s) $\rightarrow$ Ba^{2+}(g) + SO$_4^{2-}$(g)	2469	297
Ba^{2+}(g) $\rightarrow$ Ba^{2+}(aq)	−1346	−202.5
SO$_4^{2-}$(g) $\rightarrow$ SO$_4^{2-}$(aq)	−1099	−183

To what do you attribute the insolubility of BaSO$_4$?

Trends in bonding

Key points

- To understand the trends in the bonding between the elements we need to look at how the energies and sizes of atomic orbitals vary across the periodic table.

- The variation in orbital energies across the periodic table can be understood in terms of the increase in nuclear charge, the extent to which different electrons screen one another from the nucleus, and the shell to which a particular orbital belongs.

- The concept of the effective nuclear charge experienced by an electron is a useful way of rationalizing trends in the sizes of orbitals, ionization energies and electron affinities.

- Anomalies are observed in the properties of the elements from Period 2 due to their small sizes, and in the very heaviest elements from Period 6 due to relativistic effects.

- The structures adopted by the elements can be rationalized in terms of the number of valence electrons available and the size of the orbitals occupied by these electrons.

- The structures of metals can be understood by considering how hard spheres stack together.

- To understand the bonding in compounds, we also need to consider the relative energies of the AOs which combine to give MOs.

In this chapter we will see how an understanding of the way in which orbital energies and sizes vary across the periodic table enables us to rationalize the physical properties of the elements, and the nature of the bonding between the elements. Our discussion starts by considering the electronic configuration of the elements, and the energies of the occupied orbitals. Using the concepts introduced in Chapter 3, we will be able to rationalize the trends in orbital energies and sizes, as well as in ionization energies and electron affinities, as we move across the periodic table.

Having completed this ground work, our attention turns to the bonding between atoms of the same element, and how the nature of this bonding affects the structure and physical properties of the element. We will see that whether an element exists as discrete small molecules, or as a giant molecular structure, or as a metal, depends on the energies and sizes of the valence orbitals.

The chapter closes with a discussion of the bonding between atoms of different elements. In this case, the energy match between the valence orbitals needs to be considered. We will see that, depending on the energy separation of these orbitals, the bonding can vary on a spectrum from purely covalent to ionic.

The concepts and ideas introduced in this chapter will form the framework for our understanding of the structures and properties of the chemical elements and their compounds.

7.1   Electronic configuration and the periodic table

The chemical elements were first arranged in a system resembling our modern periodic table by Dmitri Mendeleev in 1869. His arrangement was based on atomic masses, and on the chemical and physical properties of the elements. It is now realized that the elements should be arranged according to their atomic number, that is the number of protons in the nucleus (which is, of course, equal to the number of electrons in the neutral atom), and according to the arrangement of the electrons in various shells. It is the arrangement of the electrons, specifically the highest energy valence electrons, that determines the chemical and physical properties of the elements.

Figure 7.1 on the next page shows a periodic table of the first 102 elements, with the atomic number of each element shown beneath the symbol, and the electronic configuration of the valence electrons shown in blue or green. The configuration is only given for the highest energy (valence) electrons; core electrons are not included. The full electronic configuration for potassium, for instance, is that of argon $[1s^2 2s^2 2p^6 3s^2 3p^6]$ followed by $4s^1$. The configuration for an element is given in green where the configuration differs from the general trend in that period or group. Thus the electronic configurations of chromium and copper are highlighted in green since these are the only two transition metals in Period 4 where the $4s$ orbital contains just one electron rather than two.

Divisions – blocks, periods and groups

There are a number of important features of this periodic table we need to appreciate.

Blocks Four blocks are distinguished in the periodic table: the s-, p-, d-, and f-blocks. These names reflect the orbitals used to accommodate additional electrons as we move across a given block, although there are a few exceptions where electrons are added into different shells.

Periods The periods are numbered according to the principal quantum number of the s orbital most recently filled. For example, in the period which starts with potassium (K) and ends with krypton (Kr) the $4s$ orbital is the most recently filled, so these elements form Period 4. The lanthanide and actinide elements are in Periods 6 and 7, respectively.

Groups The recommended group numbers are shown in red at the top of each group. For Groups 1–12, the group number gives the total number of valence electrons for any atom in that group. For example ruthenium (Ru) is in Group 8 and has a total of 8 valence electrons: one in the $5s$ shell, and seven in the $4d$ shell. For the elements of Groups 13–18, the number

Palladium (Pd) is the unique exception to this rule.

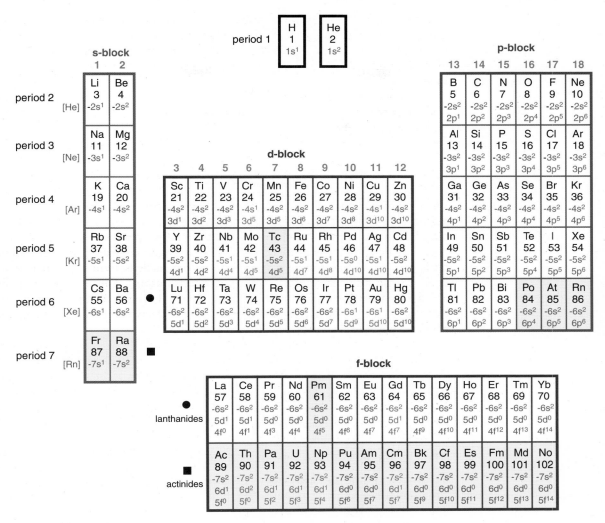

Fig. 7.1 A periodic table of the first 102 elements showing the configuration of the valence electrons for each atom. The atomic number is given in black beneath the symbol of each atom. The electronic configurations are shown in blue or green, with green being used to highlight where, on moving across a given period, a regular trend is broken. Elements which only occur as unstable radioactive isotopes are shown with a yellow background. The recommended group numbers are shown in red. A number of elements with atomic number greater than 102 have been made artificially, but in each case only a few atoms have been synthesized, and even then the mean lifetime of these is just a fraction of a second.

of valence electrons is given by the group number minus ten. Later, in section 7.10 on page 303, we will see how these numbers relate to the possible oxidation states of the elements.

A further classification which is often made is to identify elements in the *s*- and *p*-blocks, together with those in Group 12 (which are not considered to be transition metals), as the *main-group* elements.

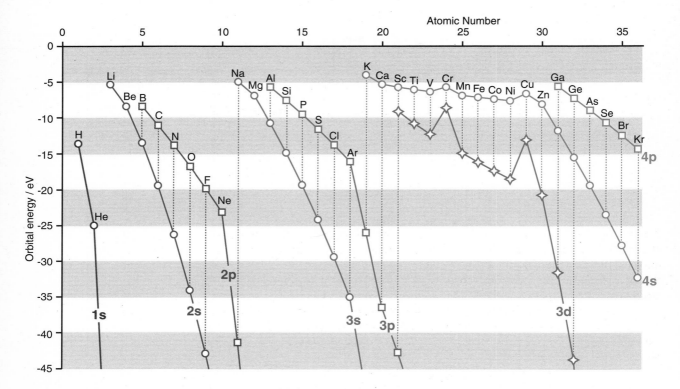

Fig. 7.2 A graph of the atomic orbital energies for the elements hydrogen to krypton. Only the energies of the highest energy occupied orbitals are shown. The graph is coloured according to the principal quantum number of the orbital.

The placing of hydrogen, helium, and the lanthanides and actinides

Hydrogen and helium are somewhat difficult to place in the periodic table and are sometimes shown, as here, 'floating' in a less than satisfactory fashion. On the grounds of electron configuration, with one and two electrons in their *s* orbital, hydrogen and helium ought to be placed at the top of Groups 1 and 2. However, with a filled outermost shell, helium could also be placed above the noble gases which it most resembles in properties, i.e. at the top of group 18. However, this would then suggest that helium is part of the *p*-block, which does not make sense as it has no electrons in *p*-orbitals.

The exact placing of the lanthanides and actinides also varies in different versions of the periodic table. In many versions the element lanthanum (La) is placed in Group 3 which then means that elements 58 (cerium, Ce) to 71 (lutetium, Lu) must be squeezed in between Groups 3 and 4. There are persuasive arguments for putting lutetium in Group 3, as we have done in Fig. 7.1, which means the lanthanides and actinides fit in between the *s*-block and *d*-block.

The electron configurations of the elements play a crucial role in their chemical properties: for instance, they may be used to predict whether an atom forms singly or doubly charged ions. We shall explore how the observed oxidation states of atoms relate to the electronic configurations, but before we do so, we must revisit another topic of equal importance to electronic configuration. This is the energies of the occupied orbitals.

7.2 Orbital energies and effective nuclear charges

In Chapters 3 and 4 we have already seen the importance of the energies of atomic orbitals when considering how well orbitals from different atoms combine to form molecular orbitals. Consequently, we need to have a feel for how the energies of the AOs vary across the periodic table if we want to understand how the elements bond to one another. A graph showing the orbital energies of all the elements up to radon (Rn) is shown on page 925; the pattern of these orbital energies is so important that you will have cause to refer to this graph rather often.

In order to explore the AO energies in more detail Fig. 7.2 on the preceding page shows just the data for the first 36 elements. In section 2.6.2 on page 69 we looked at these data for the first three periods and introduced the concept of *screening* (also called shielding) as an explanation for the observed trends. The key idea is that as the nuclear charge, and consequently the number of electrons, increases by one, the electrons experience a greater attraction to the nucleus, but also there is an increase in electron–electron repulsion. We describe this by saying that the electrons screen one another from the nucleus so that the electrons do not experience the full increase in the nuclear charge.

For example, for the elements potassium (K) to zinc (Zn) the energy of the 4s AO only decreases by a total of about four eV over the twelve elements. An explanation for this is that between scandium (Sc) and copper (Cu) electrons are being added to 3d orbitals, and such electrons form an effective screen for the 4s electrons. To put it another way, as far as the 4s electrons are concerned, on moving from one element to the next between scandium and copper, the effect of the extra proton added is largely cancelled out by the addition of the electron to the lower 3d orbital. Consequently the energy of the 4s falls rather slowly.

From gallium (Ga) onwards, extra electrons are being added to the 4p orbitals. These do not screen the electrons in the 4s orbitals as efficiently as do electrons in the 3d. Hence, from gallium to krypton (Kr), the energy of the 4s orbital drops far more rapidly as the 4s electrons experience more of the effect of the additional proton. We see from the graph that the energy of the 4s orbital drops by more than 20 eV over the series of just six elements.

The energies of the 2s and 2p orbitals drop quite rapidly as these orbitals start to become occupied (i.e. between Li and Ne); the same is true of the 3s and 3p AOs between Na and Ar. Across these two series, each added electron is going into the same shell and so not providing an efficient screen for the other electrons in that shell. Note that the energy of the s orbital drops slightly faster than that of the p orbitals. This is due to the fact that electrons in the s orbital penetrate to the nucleus to a greater extent than electrons in the p orbitals, as was discussed in section 2.6.2 on page 69.

7.2.1 The effective nuclear charge

A useful way of thinking about the trends in these orbital energies is to calculate the effective nuclear charge, Z_{eff}, experienced by an electron in a particular orbital. We have already seen in section 2.6.4 on page 75 how Z_{eff} can be calculated from the orbital energy E_n by rearranging the expression for E_n

$$E_n = \frac{-Z_{eff}^2 R_H}{n^2}, \qquad (7.1)$$

In Eq. 7.1 R_H is the Rydberg constant and n is the principal quantum number of the orbital. Recall from section 2.3.1 on page 45 that orbital energies are *negative*.

Fig. 7.3 A periodic table showing the element symbol, atomic number, and two sets of effective nuclear charges for the highest occupied orbital. Those determined using Slater's rules are shown in blue and the more reliable values of Clementi are shown in red. Note how on descending a group, Slater's Z_{eff} initially increase, but then remain constant. In contrast, Clementi's values continue to increase down the group.

Key to each cell:
Symbol / Atomic Number / Z_{eff} (Slater) / Z_{eff} (Clementi)

Element	Z	Z_{eff} (Slater)	Z_{eff} (Clementi)
H	1	1.0	1.0
He	2	1.7	1.69
Li	3	1.30	1.28
Be	4	1.95	1.91
B	5	2.60	2.42
C	6	3.25	3.14
N	7	3.90	3.83
O	8	4.55	4.45
F	9	5.20	5.10
Ne	10	5.85	5.76
Na	11	2.20	2.51
Mg	12	2.85	3.31
Al	13	3.50	4.07
Si	14	4.15	4.29
P	15	4.80	4.89
S	16	5.45	5.48
Cl	17	6.10	6.12
Ar	18	6.75	6.76
K	19	2.20	3.50
Ca	20	2.85	4.40
Sc	21	3.00	4.63
Ti	22	3.15	4.82
V	23	3.30	4.98
Cr	24	2.95	5.13
Mn	25	3.60	5.23
Fe	26	3.75	5.43
Co	27	3.90	5.58
Ni	28	4.05	5.71
Cu	29	3.70	5.84
Zn	30	4.35	5.97
Ga	31	5.00	6.22
Ge	32	5.65	6.78
As	33	6.30	7.45
Se	34	6.95	8.29
Br	35	7.60	9.03
Kr	36	8.25	9.77
Rb	37	2.20	4.98
Sr	38	2.85	6.07
Y	39	3.00	6.26
Zr	40	3.15	6.45
Nb	41	2.80	6.70
Mo	42	2.95	6.98
Tc	43	3.60	7.23
Ru	44	3.25	7.45
Rh	45	3.40	7.64
Pd	46	6.85	7.84
Ag	47	3.70	8.03
Cd	48	4.35	8.19
In	49	5.00	8.47
Sn	50	5.65	9.10
Sb	51	6.30	9.99
Te	52	6.95	10.81
I	53	7.60	11.61
Xe	54	8.25	12.42
Cs	55	2.20	6.36
Ba	56	2.85	7.58
Lu	71	3.00	8.80
Hf	72	3.15	9.16
Ta	73	3.30	9.53
W	74	3.45	9.85
Re	75	3.60	10.12
Os	76	3.75	10.32
Ir	77	3.90	10.57
Pt	78	3.55	10.75
Au	79	3.70	10.94
Hg	80	4.35	11.15
Tl	81	5.00	12.25
Pb	82	5.65	12.39
Bi	83	6.30	13.34
Po	84	6.95	14.22
At	85	7.60	15.16
Rn	86	8.25	16.08

to give

$$Z_{eff} = \sqrt{\frac{-n^2 E_n}{R_H}}. \tag{7.2}$$

If an electron is perfectly screened from the nucleus $Z_{eff} = 1$, but as the screening becomes less than perfect Z_{eff} increases. The value of the effective nuclear charge is a measure of the extent to which an electron in a given orbital is more tightly held than an equivalent electron in hydrogen.

On page 77 we saw how approximate values for Z_{eff} can also be obtained using a set of simple rules devised by Slater; these values are not based directly on orbital energies. Values of Z_{eff}, such as the set calculated by Clementi, which are based on orbital energies in the way we have described are more reliable.

Trends in effective nuclear charges

Figure 7.3 shows a periodic table annotated with effective nuclear charges for the highest energy electrons. Two values are given: those calculated using Slater's rules are shown in blue, and the values determined by Clementi are shown in red. Clementi's values are also shown in a more graphical form in Figure 7.4 on the next page.

The Clementi values for Z_{eff} increase *across* a period and *down* a group. Broadly speaking the Slater values show the same trends, although there are some significant differences. For example on descending any particular group, Slater's effective nuclear charges increase up to Period 4, and then remain constant for the remainder of the group (Groups 1 and 2 are an exception as for these Z_{eff} remains constant from Period 3 onwards). This is in contrast to the Clementi values which show a steady increase down all the groups.

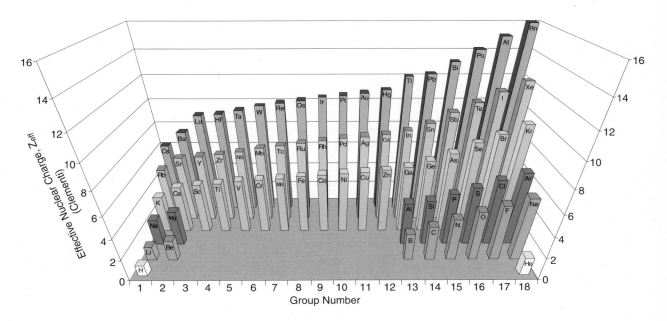

Fig. 7.4 A periodic table showing the effective nuclear charge (Clementi values) of the highest occupied orbital of each element. Each period of the table is given a different colour.

7.2.2 Variation in orbital energies down a group

Figure 7.5 on the following page shows the orbital energies of the elements arranged according to their group number, from which it can clearly be seen that the orbital energies *increase* down a group. At first this can seem a bit puzzling, since we have just noted that the effective nuclear charge *increases* down the group, and we might therefore (erroneously) have expected the orbital energies to become more negative. The error in our thinking arises because we have forgotten that, according to Eq. 7.1 on page 265, the orbital energy also depends on $-1/n^2$. As we descend the group, n increases and so $-1/n^2$ becomes less negative; this term therefore leads to an increase in the orbital energy going down the group. Looking at the data in Fig. 7.5, it is clear that this term dominates over the effect of the increase in Z_{eff}.

It is important to remember that the value of Z_{eff} is determined from the orbital energy using Eq. 7.2. The two quantities are not independent of one another, therefore.

The second period anomaly

An important feature to notice from Fig. 7.5 is that the energies of the *s* and *p* orbitals for the second period elements seem to be lower than expected when compared to the energies of *s* and *p* orbitals for the later members of the groups. For example, whilst there is a gradual increase in the *s* and *p* orbital energies for the series S–Se–Te–Po, there is a much sharper increase between oxygen and sulfur.

The reason why the *s* and *p* orbitals for the elements in the second period are so low in energy is usually attributed to the particularly small value of the principal quantum number ($n = 2$). Since the orbital energy is proportional to $-1/n^2$, a small value of n (either one or two) means that the energy ends up rather large and negative.

This effect is illustrated in Fig. 7.6 on the following page which shows a plot of $-1/n^2$ against n. The biggest jump is from when $n = 2$ to $n = 3$, with a more

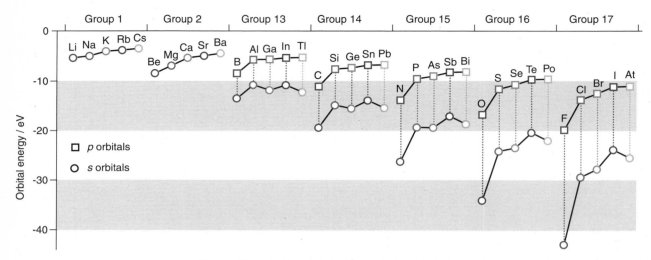

Fig. 7.5 The orbital energies of the *s*- and *p*- block elements arranged in groups. The energies of the *s* orbitals are indicated with circles; *p* orbitals with squares. On the whole, on moving down a group, the orbital energies tend to increase. There are a couple of irregularities in the energies of the *s* orbitals which will be discussed later.

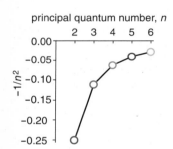

Fig. 7.6 A graph to show how $-1/n^2$ varies with *n*. The biggest jump is between $n = 2$ and $n = 3$. The trend mimics that seen in the graph of the orbital energies.

gradual rise from then on. Even though we have not included any effective nuclear charges, we see that the graph mimics the increases in energy seen in Fig. 7.5.

We shall see later that there are a number of other anomalous properties associated with the second period, all of which ultimately have the same origin.

7.2.3 The effects of the *d*-block

The fact that the graph shown in Fig. 7.6 does not exactly mirror the trend in orbital energies shown in Fig. 7.5 should not be too surprising since Fig. 7.6 only looks at the effects of *n*. However, there is a significant deviation in the orbital energies that we need to explain, and that is why there is a downwards kink in the orbital energy on moving from the 3*s* (shown with the green circles) to the 4*s* (shown with the red circles). This is especially noticeable in Groups 13 and 14. The origin of this downwards kink is quite simple. For any two elements from the same group, the element in Period 3 always has eight more protons than the element in Period 2. For instance, aluminium has eight protons more than boron; silicon has eight more than carbon. However, for pairs of elements from Periods 3 and 4, this is only true for the Group 1 elements sodium and potassium, and the Group 2 elements magnesium and calcium. Comparing any other two elements from the same Group, the element from Period 4 now has 18 more protons than the element from Period 3. For instance, gallium has 18 protons more than aluminium.

This difference is due to the filling of the 3*d* orbitals between calcium and gallium. The effects of all the extra protons in the nucleus is to cause an additional lowering in the energy of the valence orbitals for the *p*-block elements in Period 4. However, the effect is not really drastic: the 4*s* and 4*p* orbitals in gallium do not seem to have experienced the full effect of the ten extra protons. We interpret this by saying that the effect of the extra protons on the orbital energies of the 4*s* and 4*p* has, to a large extent, been cancelled out by the electrons that have been added in the 3*d* orbitals.

We could rephrase all of this by looking at the problem from the perspective of effective nuclear charges rather than orbital energies. The increase in Z_{eff} for gallium is slightly greater than might have been expected by comparison with boron and aluminium because of the filling of the *d*-block. This is illustrated in Fig. 7.7 which shows Z_{eff} for Ga, As, and Br all being slightly greater than the previous trend might have predicted.

We shall see shortly how this same jump in Z_{eff} is the cause of a number of other irregularities in the trends down groups, such as in sizes and ionization energies. Before discussing this, there is a further effect on the orbital energies of the heavier elements which we must consider.

7.2.4   Relativistic effects in the heavy elements

In order to understand fully the complete orbital energy graph on page 925, we need to understand how the energies of electrons can be affected by relativistic effects. It is really only the heavy elements for which these effects are really significant but, as we shall see, relativistic effects provide an explanation for some of the anomalous properties of these heavy elements.

Notice in the graph of orbital energies that the energy of the 6*s* orbital for the later transition metals such as osmium (Os), iridium (Ir), platinum (Pt), and gold (Au) is somewhat *lower* than the energies of the 5*s* orbitals from the previous row of transition metals which ends at silver (Ag). The data are also presented in Fig. 7.8 on the following page in a slightly different way. This plot shows the energies of the 5*s* and 5*p* orbitals (as purple circles and squares respectively) of the elements from Period 5, together with the energies of the 6*s* and 6*p* orbitals (as orange circles and squares) for the elements in Period 6; the data are plotted against the group number. With the exception of caesium and barium, we see that the energy of the 6*s* orbital for the elements in the Period 6 is actually *lower* than the energy of the 5*s* electrons for the elements from the Period 5. This is in contrast to what we have previously seen, where the energy of the highest occupied orbital generally *increases* down a group.

A consequence of the 6*s* orbital being lower in energy than expected is that the effective nuclear charge for electrons in the 6*s* shell is rather large. This is seen to be the case for Clementi's Z_{eff} in Fig. 7.3 on page 266 and Fig. 7.4 on page 267.

Part of the reason for this unexpected increase is due to the filling of the first row of the *f*-block. So far as the 6*s* electrons are concerned, as the 4*f* orbitals are filled, the addition of the extra electron into the 4*f* shell almost completely cancels out the effects of the additional proton. However there is a very slight increase in the effective nuclear charge experienced by the 6*s* electrons. This in part explains the larger Z_{eff} experienced by the 6*s* electrons in the third row of the transition metals. A further contribution to the increased Z_{eff} felt by the 6*s* electrons (or, equivalently, to the lowering of their energy) is due to *relativistic effects*.

In order to understand this, we need to look back at the equation for the energy, E_n, of an electron in a one-electron system in an orbital with principal quantum number n

$$E_n = -R_H \times \frac{Z^2}{n^2}$$

Whilst this relationship is most conveniently written using the Rydberg constant, R_H, this constant is just a collection of the other fundamental constants.

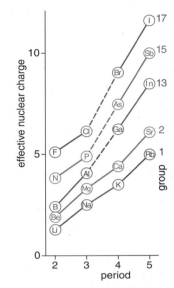

Fig. 7.7 A graph showing how Z_{eff} changes on descending a group. The effective nuclear charge for gallium (Ga) and the following *p*-block elements in Period 4 is greater than expected based on the previous trend. This is due to the filling of the *d*-block in between calcium and gallium.

Fig. 7.8 A graph comparing the energies of the valence s and p orbitals of the elements from Periods 5 and 6. The lanthanides are not included. From Group 3 onwards, rather than being higher in energy, the energy of the $6s$ electrons for the elements from Period 6 (shown by orange circles) turn out to be *lower* than the energy of the $5s$ electrons for the elements from Period 5 (shown by the purple circles). The energies of the corresponding p electrons (shown by the squares) are approximately the same in the two periods.

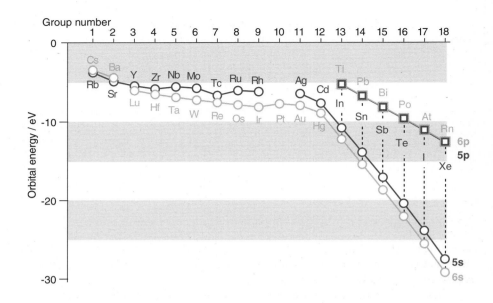

The equation may be rewritten in terms of these fundamental constants as

$$E_n = -\frac{m_e e^4}{8\varepsilon_0^2 h^2} \times \frac{Z^2}{n^2}$$

where m_e is the mass of the electron, e is the charge on the electron, ε_0 is the vacuum permittivity, and h is Planck's constant.

The important point is that the orbital energy depends on the *mass* of the electron, m_e. According to the theory of relativity, this mass is not actually constant but varies with the speed of the electron.

In hydrogen, the speed at which the electron moves in the $1s$ orbital is about $1/137$ times the speed of light. Increasing the nuclear charge increases the speed at which the electron moves, and it can be shown that for a one-electron atom with charge Z the electron moves at $Z/137$ times the speed of light.

The relativistic mass of the electron

Einstein's theory of relativity states that the mass m_{rel} of an object moving at velocity v is related to its rest mass m_{rest} by

$$m_{rel} = \frac{m_{rest}}{\sqrt{1 - (v/c)^2}},$$

where c is the speed of light.

Using this equation, we calculate that the electron in the $1s$ orbital of a hydrogen atom has a relativistic mass of only 1.00003 times the rest mass. However, for a heavier element such as mercury ($Z = 80$) the velocity of the $1s$ electron is approximately $80/137$ or 58% of the speed of light, and so in this case the mass of the electron is 1.23 times the rest mass. Taking into account this increase in the mass, the relativistic energy of the $1s$ electron in mercury is lower than would otherwise be predicted.

It is not just the orbital energy which is affected by the mass of the electron. The expression for the Bohr radius also includes the mass of the electron

$$a_0 = \frac{\varepsilon_0 h^2}{m_e \pi e^2 Z}.$$

Recall that in section 2.2.2 on page 42 we saw that the maximum in the radial distribution function for a $1s$ orbital is at $r = a_0$. From the above expression it follows that the relativistic increase in the mass causes a_0 to decrease and hence the orbital to contract. The same contraction occurs in other orbitals for which relativistic effects are significant.

As the wavefunction for an s orbital is non-zero at the nucleus, electrons in s orbitals experience the effects of the nuclear charge more than electrons in other orbitals. Consequently, relativistic effects are most significant for s electrons in the heavier atoms. Furthermore, since the s electrons are most effective in shielding the other electrons in the atom, their being closer to the nucleus shields the other electrons more than would otherwise be expected leading to a notable *expansion* of the valence d and f orbitals.

These relativistic effects are only significant for the heaviest elements, but nonetheless, as we shall see, their effects can be important. Figure 7.9 shows how the separation between the valence shell s and p orbital energies in Group 14 vary when relativistic corrections are taken into account, and when they are ignored. There is essentially no difference for the lightest members of the group, carbon and silicon, but for the heaviest member, lead, the calculated separation between the s and p orbitals is greater than for carbon once relativistic effects are taken into account.

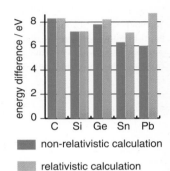

non-relativistic calculation

relativistic calculation

Fig. 7.9 A comparison of the calculated energy difference between the valence s and p orbitals of the elements of Group 14 using relativistic and non-relativistic methods.

7.2.5 Electronegativity and orbital energies

Electronegativity is a rather nebulous concept which is widely used, but difficult to quantify. Pauling defined it as 'the power of an atom when in a molecule to attract electrons to itself'. The 'when in a molecule' phrase is necessary to distinguish the concept from the more precisely defined electron affinity (also called the electron attachment energy) which is a measure of the attraction of an isolated atom towards an electron.

Pauling's scale

The problem with electronegativity is how to turn its description in words into a numerical scale. Pauling chose to define electronegativity in terms of bond strengths. He noticed that the strength of a heteronuclear bond A–B is always greater than the average of the homonuclear bonds A–A and B–B. For example, the H–Cl bond strength is 432 kJ mol^{-1}; this is greater than the average of the strengths of the H–H bond (436 kJ mol^{-1}) and the Cl–Cl bond (243 kJ mol^{-1}), which is 340 kJ mol^{-1}.

Pauling argued that the heteronuclear bond was stronger than the average of the two homonuclear bonds due to some ionic contribution which can be quantified by a difference in electronegativities. The exact relationship he settled on was

$$E_{A-B} - \sqrt{E_{A-A} \times E_{B-B}} = C \times [\chi_A - \chi_B]^2$$

where E_{A-A} is the bond strength of the diatomic A–A, E_{A-B} is the bond strength of the diatomic A–B, C is a constant of proportionality, and χ_A and χ_B are

Fig. 7.10 A graph showing the correlation between the Pauling electronegativity of main group elements (i.e. excluding the transition metals, lanthanides, and actinides) with orbital energy. In this plot we have taken the weighted average of the orbital energy of the valence *s* and *p* orbitals. Some of the more common elements are marked on the graph. The red line is the line of best fit; there is clearly a strong correlation between the quantities plotted.

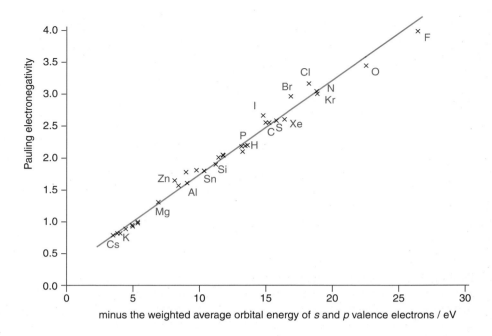

the electronegativities of A and B. Note also that in this equation Pauling used the geometric mean (the square root of the product) of the heteronuclear bond strengths rather than the arithmetic mean (half the sum). Pauling arbitrarily chose the constant C so that the most electronegative element fluorine had an electronegativity value of 4, or 3.98 in his later work.

Other electronegativity scales

Since Pauling, many others have developed different electronegativity scales. These include: the Allred–Rochow scale, based on the force exerted by an atom on its *s* and *p* electrons; the Mulliken absolute electronegativity scale, based on the arithmetic mean of the first ionization energy and the electron affinity of an atom; and the Allen configuration energy or spectroscopic electronegativity based on the weighted average of the ionization energies of the *s* and *p* valence electrons of an atom.

Electronegativity *vs.* orbital energy

Electronegativities are often used to predict the polarity of a bond, with the more electronegative element having the partial negative charge. However, we have seen in section 3.5 on page 126 that a proper explanation of bond polarity comes from considering the energies of the AOs which combine to form MOs. The atom which contributes the lowest energy AO to the MO ends up with the greater electron density. We will much prefer to talk about bond polarity and related effects in terms of orbital energies which can now be computed reliably, rather than the nebulous concept of electronegativity which is so hard to define in a quantitative way.

A reliable set of orbital energies were not available to Pauling when he devised the concept of electronegativity, but it turns out that there is a good correlation between the two. This is clearly shown in Fig. 7.10 which plots the

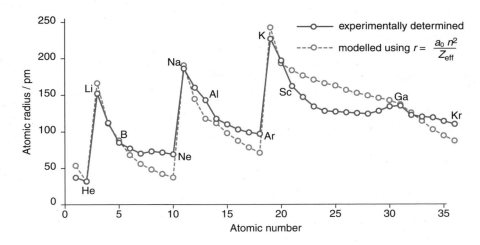

Fig. 7.11 The trends in the radii of atoms on moving across the periodic table. The experimentally determined radii of the elements are shown in blue; predicted radii in red.

Pauling electronegativity against the weighted average orbital energy of the s and p valence electrons of the main group elements; the main group is usually defined as the elements in the s and p-blocks together with Group 12 which are not considered to be transition metals. The correlation between the two quantities is impressive considering the different data on which the two plotted quantities are based.

Having explored how the orbital energies and effective nuclear charges of the elements vary across the periodic table, we are now in a position to understand the trends of many other properties, such as atomic sizes, ionization energy and electron affinities. Understanding these will be useful when we come to look at the bonding and chemical reactions of the elements.

7.3 Atomic sizes across the periodic table

As we have seen in Chapter 2, an electron is not confined to be at a particular distance from the nucleus but is spread out in a way described by the radial distribution function. Talking about the 'size' of a particular orbital therefore presents some difficulties. Equally, the size of an *atom* is also a difficult concept to pin down experimentally. We could take the distance between neighbouring atoms in the solid element as being twice the radius, but there is then the difficulty of which allotrope to use for this measure, as well as what to do for elements which do not have readily accessible solid form. As a result, the quoted values are subject to considerable variation depending on the choices made by the compilers of the data.

A simple, and surprisingly successful, model for these atomic radii proposes that they are approximated by

$$r = \frac{a_0 n^2}{Z_{\text{eff}}} \tag{7.3}$$

where a_0 is the Bohr radius, equal to 53 pm, n is the principal quantum number of the highest occupied orbital in the atom, and Z_{eff} is the effective nuclear charge for an electron in that orbital.

Figure 7.11 plots the experimentally determined atomic radius for each element (shown in blue), together with the value predicted using this simple

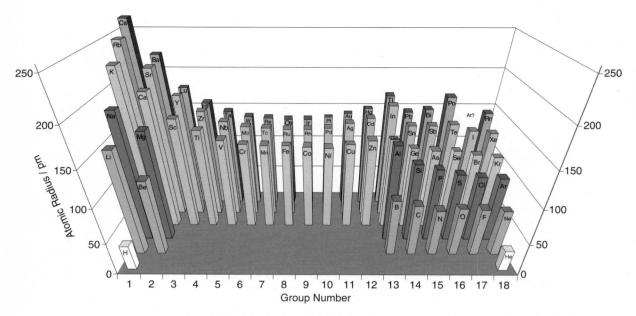

Fig. 7.12 A periodic table showing the experimental atomic radius of each element. Each period of the table is given a different colour.

model (in red). We can see that while there are many disagreements between the predicted and the observed radii, the model does predict the periodic trends with a fair degree of accuracy. Considering how difficult it is to decide on a meaningful radius for an atom in the first place, together with the simplicity of the model used, the correlation is surprisingly good.

Trends across the periodic table

Figure 7.12 shows the observed radii of the elements in the form of a three-dimensional periodic table. From this plots we see that, for a given period, the atomic radii are at a maximum for the Group 1 metals, and then decrease steadily across the period, reaching a minimum at Group 18 (we shall look at the trends within the transition metals later in the chapter). This steady decrease we can attribute to the increase in Z_{eff} across a period, which we noted in section 7.2.1 on page 265.

On moving down a group, n increases and this, according to Eq. 7.3 on the previous page, should lead to an increase in the size. However, recall from section 7.2.1 that Z_{eff} also increases down a group and according to Eq. 7.3 this will lead to a decrease in the size. The two effects thus work in opposite directions. From Fig. 7.12 is is clear that the size increases down a period, so we conclude that the effect of increasing n is dominant over the effect of the increase in Z_{eff}.

On moving down a group there is a particularly significant increase in size on going from Period 2 to 3. This is evident from Fig. 7.12 but is also shown for the members of Groups 13, 14 and 18 in Fig. 7.13. For each of these groups, the largest increase is between the element from Period 2 and that from 3. Another way to put this is that the elements from Period 2 are especially small. We will see a number of important consequences of this later in the chapter.

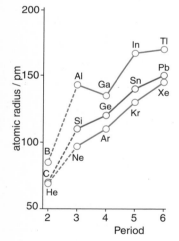

Fig. 7.13 Atomic radii of the elements from Groups 13, 14 and 18 plotted against the period number. Note the large increase in the radius on going from the Period 2 to the corresponding Period 3 element.

Notice also from Fig. 7.13 on the preceding page that, on descending Group 13, gallium in Period 4 is *smaller* than aluminium in the Period 3. This is yet another consequence of the Z_{eff} for gallium being greater than that of aluminium, as has been previously discussed in section 7.2.3 on page 268.

Decreasing radii of the heavier elements

With the effects of n^2 and Z_{eff} opposing one another, it would be difficult to predict exactly how the size might change on descending a group. For instance, it turns out that the third row transition metals are approximately the same size as the second row. For these elements, the increase in n^2 must be more or less balanced by the increase in Z_{eff} together with a relativistic contraction of the outermost $6s$ orbital (see section 7.2.4 on page 269).

Relativistic effects become more and more important as we descend a group, and these effects result in the very heaviest members of some groups actually being smaller than the previous members. Consider the data for Group 2, as an example

element	Be	Mg	Ca	Sr	Ba	Ra
radius / pm	112	160	197	215	222	215

We can explain the fact that radium is actually smaller than barium as being due to the greater value of Z_{eff} for radium, together with a contribution from relativistic effects as mentioned above.

7.3.1 Size, overlap and bond strengths

It is important to understand how the size of orbitals vary across the periodic table since the size plays an important role in determining bond strengths. The crucial point is that overlap of smaller valence orbitals tends to give stronger bonds than the overlap of larger orbitals. This is best illustrated with an example.

Unlike hydrogen, the Group 1 elements all exist as metallic solids at room temperature and pressure. Nonetheless, like hydrogen they can form diatomic molecules in the gas phase. The bonding in each of these arises from the overlap of the valence s orbitals to form a σ bonding molecular orbital and $\sigma^\star$ antibonding MO. For each diatomic, the two valence electrons occupy the bonding MO.

The observed bond strengths for these molecules in the gas phase are

molecule	H_2	Li_2	Na_2	K_2	Rb_2	Cs_2
bond strength / kJ mol^{-1}	436	100	71	50	47	43

There is a clear trend here with the overlap of the two $1s$ orbitals of hydrogen giving a much stronger bond than the overlap of the two $2s$ orbitals of lithium, and so on down the group. As was described in section 3.1.6 on page 102, it is possible to calculate the degree of overlap (the overlap integral) between the two valence s orbitals as the separation between the atoms is varied. The results for H_2, Li_2, and K_2 are shown in Fig. 7.14 on the next page. At their equilibrium bond lengths, it is seen that the overlap between the two $1s$ orbitals in H_2, indicated by the black arrow, is considerably greater than that between the two $2s$ orbitals in Li_2 (the blue arrow), which in turn is greater than that between the two $4s$ orbitals in K_2 (the red arrow).

Fig. 7.14 Plot of the overlap integral between the two valence s orbitals in the dimers H_2, Li_2, and K_2 as a function of separation between the atoms. At the experimentally observed bond lengths for these dimers, indicated by the dashed lines, the overlap integral in H_2 is greater than that in Li_2, which in turn is greater than that in K_2.

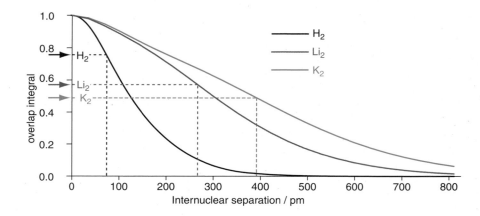

We could say that the poorer overlap of the larger valence orbitals at the equilibrium separations results in the bond strengths of these homonuclear diatomics decreasing down the group. Another way to put this, is that as we descend the group, the increased repulsion due to the 'core' electrons (i.e. the electrons other than the valence electrons) means the atoms cannot approach each other sufficiently well to achieve the same degree of overlap as is possible for the lighter atoms. We shall return to trends in bonding across the periodic table later in the chapter.

7.4 Ionization energies and electron affinities

As we have seen in section 2.7.2 on page 81, the energy required to ionize an atom is reasonably well approximated by minus the energy of the orbital from which the electron is being removed. The trends we have seen in the effective nuclear charges may also therefore be used to understand the trends in ionization energies (IE).

Trends in ionization energies across the periodic table

Figure 7.15 on the next page shows the trends in the first ionization energies across the periodic table. The IE generally increase on moving across a period due to the increasing Z_{eff}. The decrease in the IE on going from nitrogen to oxygen has been discussed in section 2.7.3 on page 82 and is due to the effects of the exchange interaction.

On descending a group, even though Z_{eff} continues to increase, the ionization energies *decrease* due to the fact that electrons are being removed from shells with higher principal quantum numbers. However, this decrease in IE begins to level off, and for some groups the IE may even begin to increase again. For instance, the IE of the third row transition metals, in which the $6s$ electron is removed, are all greater than those of the second row due to the increase in Z_{eff} after filling the f-block, and also due to relativistic effects.

Trend reversal for the heaviest elements

The ionization energies for Group 1 are shown in Table 7.1 on the next page; note that the ionization energy of francium (Fr) is actually *greater* than that

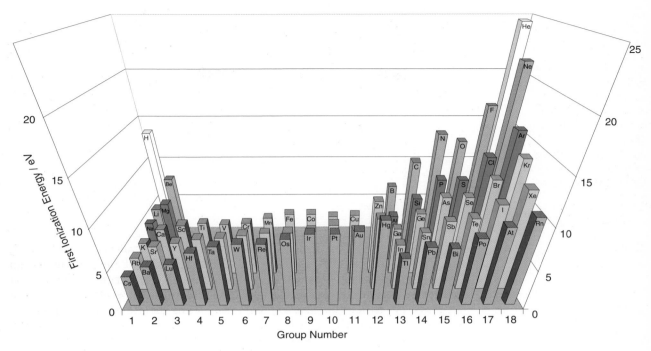

Fig. 7.15 A periodic table showing the first ionization energies of the elements. The IE generally increases across a period due to the increase in Z_{eff}, and generally decreases down a group as electrons are removed from shells with higher principal quantum numbers.

of caesium (Cs). The increase for francium is essentially due to relativistic effects (see section 7.2.4 on page 269), which lowers the energy of the $7s$ orbital. This may be seen by comparing the calculated ionization energies which take relativistic corrections into account with those which ignore them. When relativistic effects are taken into account, the calculated value for francium is in good agreement with experiment. However, when these effects are not considered, the predicted IE of Fr is *lower* than that of caesium. Interestingly, the calculated value for element 119, which has yet to be synthesized, predicts this reversal in the trend continues with an ionization energy more like that of sodium.

Table 7.1 First ionization energies of Group 1 elements (eV).

element	experimental IE	calculated IE (relativistic)	calculated IE (non-relativistic)
Li	5.39	–	–
Na	5.14	–	–
K	4.34	–	–
Rb	4.18	4.18	4.11
Cs	3.89	3.90	3.75
Fr	4.07	4.08	3.61
element 119	?	4.80	3.41

Fig. 7.16 A graph to show the average ionization energies for the removal of the valence electrons for the members of Groups 13, 14, and 15. For Group 13, this is the average of the first three IEs; for Group 14, the average of the first four IEs; for Group 15, the average of the first five IEs. In each group the change on moving from Period 2 to 3 is significantly greater than any subsequent change.

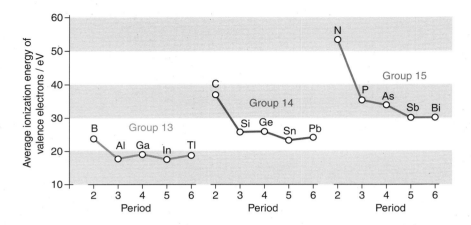

7.4.1 The second period anomaly

Figure 7.16 shows how the average ionization energy for successively removing the valence electrons varies on moving down Groups 13, 14, and 15. For the members of Group 13 (shown in red), this is the average IE for successively removing the first three electrons; for Group 14 (shown in blue), the average is for removing the first four electrons; for Group 15 (shown in green), the average is for removing the first five electrons.

For each group, the average IE of the first member (i.e. boron, carbon, and nitrogen) is considerably greater than for the remaining group members. For Group 13, there is a sharp drop from Period 2 to 3 (boron to aluminium), but then a rise from Period 3 to 4 (aluminium to gallium). Similarly, whilst there is also a sharp drop in Groups 14 and 15 between the elements from Period 2 and Period 3, the difference on moving from Period 3 to 4 is much less than expected. This behaviour on going from Period 3 to Period 4 is attributed to the effects of filling the *d*-block as outlined in section 7.2.3 on page 268.

The large (average) ionization energies for the first members of each group mean that these elements are unlikely to form positive ions and hence compounds with considerable ionic character. However, we could have come to this conclusion with just the aid of the orbital energy chart. Indeed, the group trends in the ionization energies mirror the orbital energies shown in Fig. 7.5 on page 268. The ionization energies of boron, carbon, and nitrogen are large since their orbital energies are relatively low. This means their orbital energies will be better matched with those of electronegative elements such as oxygen or fluorine than the other members of groups 13–15. The closer the match in orbital energy, the more covalent the bonding, as we have seen. We shall return to issues of bonding later in the chapter.

7.4.2 Electron affinities

Definition

The *electron attachment enthalpy* is defined as the enthalpy change for adding an electron to an atom

$$X(g) + e^- \longrightarrow X^-(g).$$

period 1

H	He
73	≤0

s block

	1	2
period 2	Li 60	Be ≤0
period 3	Na 53	Mg ≤0
period 4	K 48	Ca 2
period 5	Rb 47	Sr 5
period 6	Cs 46	Ba 14

d block

	3	4	5	6	7	8	9	10	11	12
period 4	Sc 18	Ti 8	V 51	Cr 64	Mn 0	Fe 16	Co 64	Ni 112	Cu 118	Zn 0
period 5	Y 30	Zr 41	Nb 86	Mo 72	Tc 53	Ru 101	Rh 110	Pd 54	Ag 126	Cd 0
period 6	Lu ??	Hf 0	Ta 31	W 79	Re 14	Os 106	Ir 151	Pt 205	Au 223	Hg 0

p block

	13	14	15	16	17	18
period 2	B 27	C 122	N ≤0	O 141	F 328	Ne ≤0
period 3	Al 43	Si 134	P 72	S 200	Cl 349	Ar ≤0
period 4	Ga 29	Ge 130	As 78	Se 195	Br 325	Kr ≤0
period 5	In 29	Sn 116	Sb 103	Te 190	I 295	Xe ≤0
period 6	Tl 19	Pb 35	Bi 91	Po 183	At 270	Rn ≤0

Fig. 7.17 A periodic table showing electron affinities in kJ mol^{-1}. The more positive the value, the more favourable it is for an electron to be added to the atom.

For most elements, this process is exothermic. It is also common to talk about the *electron affinity*, which is simply *minus* the electron attachment enthalpy

$$\text{electron affinity} = -\text{ electron attachment enthalpy}$$

The *more positive* the electron affinity, therefore, the *more favourable* energetically it is for the electron to be attached to the atom. There is, unfortunately, a considerable degree of confusion of the sign of these quantities.

Electron affinities (EA) are notoriously difficult to measure experimentally and as a result are often determined by indirect methods. There is certainly a large degree of error in most values quoted. Despite these shortcomings, the idea of the affinity an atom has for adding an extra electron is important and Fig. 7.17 shows a set of values and how they vary across the periodic table. The general trends may be understood by considering how the relevant orbital energies vary, and by considering the electron configurations of the elements.

Filled and half-filled shells

Trying to add an extra electron to an atom with a filled shell or filled subshell is not favourable. In the case of the elements from Group 18, the extra electron would have to be added to a higher energy orbital starting a new shell. This is particularly unfavourable, and so the EA for all the noble gases are negative. The elements of Group 2 all have a filled *s* orbital and so the extra electron would have to be accommodated in a *p* orbital, which as we have seen, is slightly higher in energy. This results in the EA for these elements being small or negative. Similar behaviour is seen for zinc (Zn), cadmium (Cd), and mercury (Hg) in Group 12 which all have filled *ns* and $(n-1)d$ subshells. The additional electron would have to go into the slightly higher energy *np* subshell.

Groups 7 and 15 have a half-filled subshell. In the case of manganese (Mn), technetium (Tc), and rhenium (Re) from Group 7, the *d* subshell is half filled;

for Group 15, it is the *p* subshell that is half filled. As we have seen from the discussion of ionization energies in section 2.7.3 on page 82, adding an electron to a partially filled subshell usually causes an increase in the favourable exchange energy. However, there is an exception to this for half filled subshells in which case there is no increase in exchange energy. This explains why elements in Group 7 and 15 have lower electron affinities than those in groups on either side.

The influence of the effective nuclear charge

Ignoring the special cases of elements with filled and half-filled subshells, there is a general increase in EA across a period. This is due to the lowering in energy of the orbital into which the added electron is accommodated. Another way to put this is to say that the electron affinities generally increase across a period as the effective nuclear charge increases. It is not surprising, therefore, that the halogens, Group 17, have the largest electron affinities of all the elements.

What is surprising is that, after the halogens, the element with the next largest EA is gold. This can be understood by recalling that the electronic configuration of gold ([Xe] $6s^1 5d^{10}$) has a filled *d* subshell but a vacancy in the $6s$ shell. We have already seen how the energy of the $6s$ orbital for the third row transition metals is lowered due to the large Z_{eff} and relativistic effects (see section 7.2.4 on page 269). Gold is the element that has the $6s$ orbital lowest in energy and still able to accommodate an extra electron. The *auride* ion, Au^-, has been detected in a number of salt-like compounds.

The second period anomaly

One further feature apparent from Fig. 7.17 on the previous page is that the electron affinities for the *p*-block elements in Period 2 are all less than the values for the corresponding elements in the same group from Period 3. Since the valence orbital energies of oxygen and fluorine are lower than those of any elements other than the noble gases, we might have expected these to have particularly large electron affinities. It comes as a surprise, therefore, to find that the EA of chlorine is greater than that of fluorine, and even more of a surprise to find that oxygen has the smallest EA of all of the Group 16 elements.

It seems reasonable to expect that there will be a correlation between the electron affinity and the energy of the orbital into which the extra electron is added. Such a plot for the elements of Groups 16 and 17 is shown in Fig. 7.18 on the facing page. With the exception of the first element in each group, the correlation is very good. However, the EA for both oxygen and fluorine are much less than predicted by the trend. Indeed, we can use the graph to estimate much lower than expected the electron affinities are: approximately 132 kJ mol^{-1} for fluorine and 85 kJ mol^{-1} for oxygen.

The reason for the low electron affinities for the Period 2 elements has been put down to their small sizes. As we have already seen, these elements are particularly small, fluorine especially so. Adding an extra electron results in more electron–electron repulsion than is the case for the larger atoms, and it is this that results in their small electron affinities.

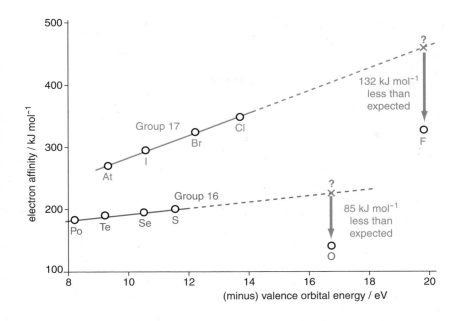

Fig. 7.18 A graph showing a correlation between electron affinity and the valence orbital energy for Group 16 and Group 17. The graph shows a good correlation between the two energies with the notable exception of the first members of each group. Thus the electron affinities of fluorine and oxygen are considerably less than might have been predicted. This is thought to be due to the extra electron repulsion that would arise on adding a further electron to these particularly small atoms.

7.5   Summary of the trends in orbital energies and sizes

Before we look at the bonding in the elements, it is worthwhile summarizing the important points so far.

- The energies of orbitals follow a fairly regular pattern across the periodic table, decreasing across a period, and increasing down a group.

- Effective nuclear charges, Z_{eff}, are determined from orbital energies and these can give us an alternative perspective to help rationalize trends. Generally Z_{eff} increases across a period and down a group.

- Understanding why the orbital energies (or effective nuclear charges) vary in the way they do gives a way of understanding other trends in the periodic table, such as the sizes of orbitals (which may be related to the sizes of atoms), ionization energies, and electron affinities.

- Atomic sizes generally decrease across a period and increase down a group.

- The Period 2 elements often have anomalous properties when compared to their other group members. Their orbital energies are significantly lower and their sizes are smaller than would be expected on the basis of the trend in the other periods.

- Relativistic effects need to be taken into account when trying to rationalize the properties of the elements from Period 6.

7.6   Bonding in the elements – non-metals

We are now in a position to look at the bonding in the elements and see how this varies across the periodic table. We shall start with the bonding in the main

Fig. 7.19 A graph showing how the bond strengths of the second row homonuclear diatomics vary across the periodic table. The bond strength reaches a maximum at nitrogen which has all of the bonding MOs filled but only one antibonding MO occupied.

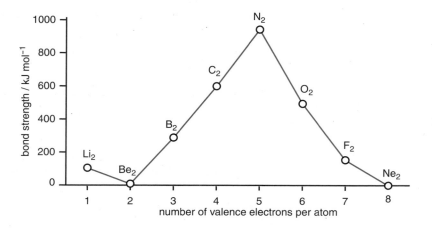

group elements, particularly the non-metals from the *p*-block, and then move on to the remainder of the periodic table.

7.6.1 Orbital sizes and numbers of electrons

When considering the bonding of an element with itself, we do not need to worry about the relative energies of the valence orbitals since these will be the same and thus are perfectly matched. However, as we saw in section 7.3.1 on page 275, the sizes of the orbitals are important: generally, the overlap of smaller orbitals gives rise to stronger bonding.

To illustrate how important the *number* of valence electrons is in bonding, we shall remind ourselves of the simplest case, that of the bonding in homonuclear diatomics. This was discussed in section 3.3 on page 105 where we saw how to construct molecular orbitals suitable for all the homonuclear diatomics from Period 2 using the $2s$ and $2p$ orbitals. We also saw how the strength of the bond depends on the number of electrons in bonding and antibonding orbitals. Generally speaking, the greater the excess of electrons in bonding MOs over electrons in antibonding MOs, the stronger the bonding.

Figure 7.19 shows how the bond strengths for the homonuclear diatomics from Period 2 vary across the periodic table. The nitrogen molecule has the strongest bond since it has enough electrons to fill all the bonding MOs while only one antibonding MO is filled (see section 3.3.5 on page 121 and Table 3.3 on page 121). The molecules B_2 and C_2 do not have enough electrons to fill all the bonding MOs; O_2 and F_2 have all the bonding MOs filled, but also have more electrons in antibonding MOs.

Of course, carbon does not exist as C_2 molecules under normal conditions, but is a solid. The solid forms of carbon, such as diamond and graphite, with their giant covalent structures, are considerably more stable than the diatomic molecule would be. Similarly, lithium, beryllium, and boron all exist as solids. It is possible to quantify the bonding in these different structures by looking at the amount of energy needed to produce a mole of *atoms* from the solid element. This quantity is the standard enthalpy of atomization, ΔH°_{atom}, and is plotted for the elements from the *s*- and *p*- blocks in Fig. 7.20 on the facing page.

The first point that stands out from Fig. 7.20 is that it is *carbon* that has the greatest standard enthalpy of atomization and not nitrogen. Indeed, in moving

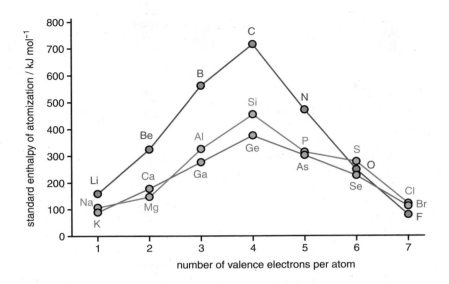

Fig. 7.20 A graph showing the enthalpies of atomization of the elements from the *s*- and *p*-blocks for Periods 2 (shown in blue), 3 (in red) and 4 (in green). In each case the strongest bonding occurs with the elements from Group 14. In this group each atom contributes four electrons to the bonding MOs.

across the main groups of each period, $\Delta H^{\circ}_{\text{atom}}$ rises steadily from Group 1, reaches a maximum at Group 14, and then decreases again.

The second point is that, on the whole, the enthalpies of atomization for the elements of Period 2 are greater than those of Period 3, which are in turn greater than those of Period 4. This is a reflection of the trend, noted in section 7.3.1 on page 275, that stronger bonds arise from the overlap of smaller orbitals. However, Fig. 7.20 shows that there are a few exceptions to this general trend.

The final point to note from Fig. 7.20 is that the graph is roughly symmetrical about the elements of Group 14. In other words, in each period $\Delta H^{\circ}_{\text{atom}}$ for Group 13 is approximately the same as for Group 15, Group 12 is approximately the same as for Group 16, and Group 11 as for Group 17.

To understand why the bonding reaches a maximum for the Group 14 elements, we must return again to the MO diagram for the diatomics shown in Fig. 3.28 on page 116. There we see that whilst it is true that nitrogen is the first diatomic with all of its bonding orbitals filled, there is still one antibonding orbital filled, the $2\sigma_u$. For *diatomic molecules* it is always the case that there is one relatively low energy antibonding MO which must be filled before all the bonding MOs can be filled. However, this is *not* the case for a three-dimensional structure.

If we ignore the *d* orbitals for the time being, each atom has four valence orbitals (one *s* and three *p* orbitals) which can combine to form molecular orbitals. For a structure consisting of N atoms, there are a total of $4N$ AOs which interact to form $4N$ MOs. Half of these MOs are bonding, and half are antibonding. Each Group 14 element has four valence electrons which are just enough to fill all the bonding MOs and none of the antibonding MOs. The elements prior to Group 14 do not have enough electrons to fill all the bonding MOs and so the bonding has not been maximized. The elements after Group 14 have more than four valence electrons per atom which means that some antibonding MOs are filled which weakens the bonding. This explains why the elements of Group 14 have the greatest enthalpies of atomization.

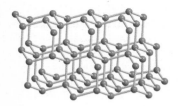

Fig. 7.21 The structure of diamond. Each carbon atom is bonded to four other atoms which are arranged in a three-dimensional network of tetrahedra.

7.6.2 The structure of the elements of Group 14

The structure of the diamond form of carbon is shown in Fig. 7.21. Each carbon atom in diamond is surrounded by four other carbon atoms arranged at the corners of a regular tetrahedron.

Each carbon atom in the solid bonds using its four valence orbitals: the $2s$ and the three $2p$. Since the carbon atoms are arranged tetrahedrally, the simplest way to view the bonding is by imagining that each carbon is sp^3 hybridized and uses these HAOs to form MOs. For a solid made up of N atoms, there are $4N$ HAOs in the solid which interact to form $4N$ MOs: half of these MOs are bonding and half are antibonding. Since each carbon atom contributes four valence electrons, there are just enough electrons to fill all the bonding MOs and none of the antibonding MOs.

All the elements from Group 14 with the exception of lead have an allotropic form analogous to the diamond structure. Lead has a metallic structure, as does one form of tin, 'white tin' or β-tin. The other common form of tin, 'grey tin' or α-tin, has this diamond structure in which all of the atoms arranged are tetrahedrally.

7.6.3 The structure of the elements of Group 15

Each element from Group 15 has the same number of orbitals as those from Group 14 but now contributes *five* valence electrons rather than four. In the solid with N atoms there are therefore $5N$ valence electrons. Since only $4N$ electrons are needed to fill completely the bonding MOs, the remaining N electrons must occupy antibonding MOs. Having electrons in antibonding MOs decreases the bond order from a value of four per atom to three.

As we discussed in section 4.5.1 on page 158, a convenient way to think of this is that, for each atom, the extra electron in the antibonding MO 'cancels out' the effect of one electron in a bonding MO to leave just three electrons per atom in bonding MOs, together with two electrons which make no net contribution to the bonding, i.e. a lone pair of electrons. In the earlier section we used this approach to explain how the MO picture of N_2, which has four bonding and one antibonding MOs filled, can be thought of as giving rise to three N–N bonds and one lone pair per atom.

Fig. 7.22 The layer structure of hexagonal-rhombohedral α-As. Each atom has three bonds.

Fig. 7.23 The structure of α-As, α-Sb, and α-Bi The layers shown in Fig. 7.22 stack with a repeating pattern every three layers: ABCABC etc. View (a) shows five layers viewed from the side. View (b) shows the same layers viewed from above. Atoms which lie directly over one another are coloured the same. The six-membered rings are clearly visible.

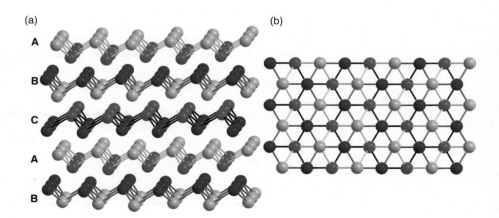

With the exception of nitrogen, there are a number of allotropes for each member of Group 15. In all of these structures we can regard each atom as having three bonds and one lone pair. One of the forms, the hexagonal-rhombohedral structure, consists of layers of puckered six-membered rings as shown in Fig. 7.22 on the preceding page. The most common forms of arsenic, antimony, and bismuth all adopt this structure. Phosphorus can adopt this structure but is more usually found in either the red or white phosphorus structure which are both based on P_4 units; the structure of white phosphorus is shown in Fig. 1.9 on page 10. Black phosphorus also consists of layers of puckered six-membered rings but arranged in a slightly different manner to those shown in Fig. 7.22.

Note how the structure shown in Fig. 7.22 strongly resembles part of the diamond structure but with each atom only having three bonds rather than four. In this layer structure we could imagine a lone pair taking the place of the fourth bond present in the diamond structure.

Figure 7.23 on the preceding page shows how these layers stack together; atoms which lie directly over one another are coloured the same. View (a) is from the side and view (b) shows the same arrangement viewed directly from above. The stacking pattern repeats every three layers.

7.6.4 The structure of the elements of Group 16

Moving on to Group 16, each atom contributes *six* valence electrons. With $4N$ electrons needed to fill completely the bonding MOs, the remaining $2N$ electrons must occupy antibonding MOs.

As before, we can imagine that each antibonding electron cancels out the effect of a bonding electron. The result is that $2N$ electrons are bonding, giving two bonds per atom, and $4N$ electrons make no net contribution to the bonding, i.e. there are two lone pairs per atom.

At room temperature and pressure, the stable form of sulfur is rhombic or α-sulfur. Above about 96 °C the stable form is monoclinic or β-sulfur. Both forms consist of puckered eight-membered rings, but the packing patterns differ between the two structures. A single ring is shown in Fig. 7.24 and Fig. 7.25 shows three views illustrating how the rings stack together in α-sulfur. Other

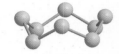

Weblink 7.1

View and rotate in real time the three-dimensional structures of solid α-As, α-Sb, and α-Bi.

Fig. 7.24 A puckered eight-membered ring found in the solid structure of sulfur. Each sulfur atom has two bonds.

Weblink 7.2

View and rotate in real time the three-dimensional structures of solid rhombic or α-sulfur.

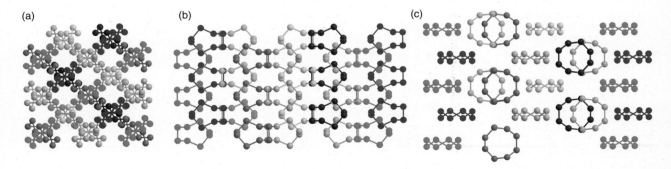

(a)   (b)   (c)

Fig. 7.25 Three views showing how the eight-membered rings stack together in rhombic or α-sulfur. S_8 rings in different environments are shown in different colours so the repeating pattern may be seen more easily. Each view contains the same number of S_8 rings but some lie directly over one another.

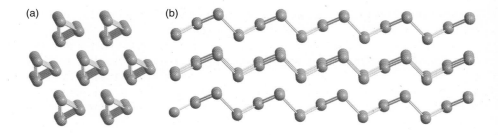

Fig. 7.26 Two views showing how the spirals in tellurium stack together. View (a) is end-on whereas view (b) has been rotated through 90 °.

Fig. 7.27 The spiral structure of solid tellurium. The spiral repeats every fourth atom.

> *Weblink 7.3*
>
> View and rotate in real time the three-dimensional structure of solid tellurium.

> *Weblink 7.4*
>
> View and rotate in real time the three-dimensional structure of solid iodine.

Fig. 7.28 Two views of the structure of iodine. View (a) is from the side and shows the layer structure. View (b) is from above and shows how alternate layers are staggered. The I–I bond length is 271.3 pm; the next closest atoms are 353.6 pm away and are shown using the hashed bonds.

allotropes of sulfur exist with rings containing from six to thirteen or more sulfur atoms.

Selenium also exists in various forms, including rings. Tellurium exists in a single crystalline form consisting of a network of spiral chains. A single spiral is shown in Fig. 7.27 where it can be seen that the spiral repeats every fourth atom; Fig. 7.26 shows how the spirals stack together. There are also spiral forms of sulfur and selenium with similar structures. Once again, each atom in the structure has just two bonds.

7.6.5 The structure of the elements of Group 17

For Group 17, each atom contributes seven electrons, filling the $4N$ bonding MOs and $3N$ of the antibonding MOs. There is thus only one bond per atom, and three lone pairs.

All the members of Group 17 exist as diatomic molecules. At room temperature and pressure, fluorine and chlorine are gaseous, bromine is a liquid, and iodine is a solid. The structure of solid iodine is shown in Fig. 7.28. The structure consists of layers which can most easily be seen in view (a); the layers are coloured differently to show how adjacent layers do not lie directly on top of one another, but are staggered. This is clear when the structure is viewed from above as in view (b).

7.6.6 The second period anomaly

One of the most noticeable anomalies for the Period 2 elements is the fact that oxygen and nitrogen exist as diatomic gases whereas the other members of the group exist as solids. On first inspection, it might seem perfectly reasonable for oxygen to have structures similar to sulfur with rings and chains; every oxygen would still have the necessary two bonds. Similarly, we might have thought that

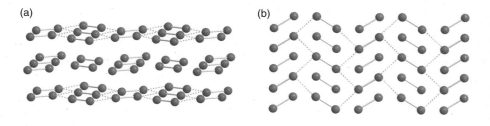

nitrogen could exist in sheets like phosphorus and the other Group 15 elements; each nitrogen would still have the three bonds it needs.

In order to understand why these structures are not observed for nitrogen and oxygen, we first need to look at another anomaly in Period 2 which is why the bond strength in F_2 (at 159 kJ mol^{-1}) is weaker than that in Cl_2 (243 kJ mol^{-1}) and Br_2 (193 kJ mol^{-1}) whereas generally bond strengths decrease on descending a group. We need to look again at the MO diagram for F_2, shown in Fig. 7.29. Whilst the simplest view of the F_2 molecule is that it has a single bond and three lone pairs on each of the fluorine atoms, the more accurate view from MO theory tells us that there are four occupied bonding MOs ($2\sigma_g$, $3\sigma_g$, and two $1\pi_u$ orbitals) and three occupied antibonding MOs ($2\sigma_u$, and two $1\pi_g$ orbitals).

As we saw in section 3.1.4 on page 101, when two AOs combine to form two MOs, the antibonding MO is raised more in energy than the bonding orbital is lowered. Unfortunately, things are further complicated by the interactions of more orbitals as we saw when we looked at the effects of s–p mixing in section 3.3.4 on page 119. This has the effect of lowering the energies of the $2\sigma_g$ and the $2\sigma_u$ orbitals; it may not then be the case that the antibonding MO is raised more in energy than the bonding is lowered. However, it remains true that the antibonding $1\pi_g$ orbital is raised more in energy than the bonding $1\pi_u$ orbital is lowered since these orbitals are not affected by s–p mixing.

The antibonding effect

In F_2, both of the $\pi^\star$ antibonding MOs are occupied. This *more than compensates* for the fact that the π bonding MOs are also occupied. Not only do the electrons in the $\pi^\star$ MOs cancel out the bonding due to the electrons in the π bonding MOs, they also partially cancel out some of the σ bonding in F_2. This effect is sometimes called the *antibonding effect* and it is this that makes the F–F bond so weak. The key idea is illustrated in Fig. 7.30 (a) where we see two $2p$ AOs interacting to form a π bonding and antibonding MO. The dashed purple line indicates where the energy of the π_g MO would be if the orbital was raised by the same amount that the bonding π_u was lowered. In fact it is raised more that this, with the difference in energy being labelled ΔE. It is this difference that weakens the bonding in the F_2 molecule.

The MO diagram for Cl_2 is rather similar to that for F_2 so we might have expected the same argument to apply for Cl_2. However, this is where the second

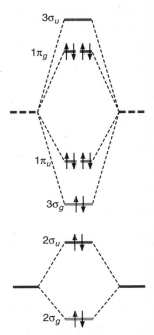

Fig. 7.29 A schematic MO diagram for F_2.

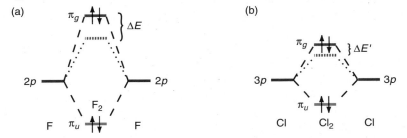

Fig. 7.30 In (a) two $2p$ orbitals interact to form a bonding π_u MO and an antibonding π_g MO. The dashed purple line shows the energy the antibonding orbital would have if it were raised by the same amount as the bonding MO was lowered. In fact it is higher in energy than this by an amount ΔE. The interaction between two $3p$ orbitals in (b) is weaker and so $\Delta E'$ is smaller.

Table 7.2 Bond strengths in the second and third period diatomics.

second period diatomic	B_2	C_2	N_2	O_2	F_2
bond strength / kJ mol^{-1}	297	607	945	498	159
third period diatomic	Al_2	Si_2	P_2	S_2	Cl_2
bond strength / kJ mol^{-1}	133	327	490	425	243

row anomaly really comes in. The key point is that there is a much stronger interaction between the small $2p$ orbitals in fluorine than there is between the larger $3p$ orbitals in Cl_2. In Cl_2 *all* the orbitals interact less strongly than in F_2, but in both molecules, the σ bonding interactions are stronger than the π (see section 3.3.2 on page 110). The key point is that the destabilizing interaction due to the filled π_g orbitals is less significant in Cl_2 than in F_2, and so less of the σ bonding is cancelled out. This is illustrated in Fig. 7.30 (b) where the destabilizing contribution, $\Delta E'$, arising from the overlap of two $3p$ orbitals is significantly less than that from two $2p$ orbitals, ΔE in (a).

The 'lone pair / lone pair' repulsion model

An alternative explanation sometimes given for F_2 having a weaker bond than Cl_2 is to attribute this to the more significant lone pair repulsion in the smaller F_2 molecule as compared to the Cl_2 molecule. The MO picture shows that there are no lone pairs in the F_2 molecule but, as was discussed in section 4.5.1 on page 158, we can think of lone pairs as arising from electrons in antibonding MOs 'cancelling out' the effects of electrons in bonding MOs. This leads to a picture of the bonding in F_2 as consisting of one bonded pair in the $3\sigma_g$ MO and a total of three lone pairs on each F atom arising from the cancellation of the $2\sigma_g$ with the $2\sigma_u$ MOs, and the $1\pi_u$ with the $1\pi_g$ MOs.

As has been discussed, in F_2 the antibonding π MOs more than cancel out the bonding π MOs, and it is this which is the origin of the 'strong lone pair repulsion' in this molecule.

p-block diatomics revisited

The antibonding effect also helps to explain the otherwise rather odd trend seen when the bond strengths of the p-block diatomics from Periods 2 and 3 are compared. These are shown in Table 7.2. Comparing B_2 with Al_2, C_2 with Si_2, and N_2 with P_2, we see that in each case the bond strength of the third period diatomic is approximately *half* that of the second period diatomic. Each of these diatomics has electrons in one or more of the π_u bonding MOs, but none has any electrons in the π_g antibonding MOs. The bonds of the second period diatomics are considerably stronger than those of the third period due to the better overlap of the smaller orbitals.

However, once the π_g antibonding MOs start to be occupied in O_2 and F_2 the situation changes around. As we have just seen, due to the effect of occupying the π_g antibonding MOs, the bond strength in F_2 turns out to be less that that in Cl_2. The effect for O_2 is not as great as for F_2, but nonetheless the bond in O_2 is nowhere near twice as strong as the bond in S_2. Once again, this is because of the two electrons in the π_g orbitals which are particularly destabilizing in the case of O_2, but less so in S_2.

Fig. 7.31 A linear chain of oxygen atoms. Each atom has two σ bonds and two filled p orbitals (shown in red and blue). The p orbitals interact to form bonding and antibonding bands, both of which are filled completely.

The structures of nitrogen and oxygen

We are now in a position to understand why oxygen does not form structures with rings and chains like sulfur, and why nitrogen does not form solids with layer structures like the other members in its group. Let us consider the case of a linear chain of oxygen atoms as shown in Fig. 7.31 on the preceding page. We can consider each oxygen as being *sp* hybridized and as having two σ bonds formed from the overlap of the *sp* HAOs. This leaves two 'lone pairs' on each oxygen atom; in this linear structure, each lone pair is in a *p* orbital.

The simplest answer to explain why this structure is unstable is that the lone pair/lone pair repulsion between all the adjacent oxygen atoms is just too severe. In fact, these two sets of *p* orbitals on each oxygen overlap to form bands in the way described in section 5.1.1 on page 180. Each set of *p* orbitals gives a band in which half of the crystal orbitals (COs) are bonding, and half are antibonding; as with MOs, the antibonding COs are more antibonding than the bonding COs are bonding. In this chain of oxygen atoms these bands will be full, giving a net antibonding interaction which would destabilize the structure. This effect is less severe in the case of sulfur, although in reality the structure is not linear but bent which helps reduce the destabilizing interaction still further.

In the case of nitrogen, we might imagine a planar sheet of nitrogen atoms, with each atom having three bonds. Every nitrogen atom would also have a lone pair in a *p* orbital as shown in Fig. 7.32. Once again, we could simply say that the lone pair/lone pair repulsion between the neighbouring atoms would be too great and this destabilizes the structure. A slightly more sophisticated argument would be that the *p* orbitals form a band but, as in the case of the chain of oxygen atoms, this band will be full and so destabilizing to the structure. With the larger phosphorus atoms, the π interactions are weaker and not sufficient to break the molecule apart (although the structure does pucker to reduce the destabilization).

Finally, note that for carbon with one fewer electron per atom, such a planar system shown in Fig. 7.32 only has enough electrons to half fill the band formed from the *p* orbitals, resulting in a net bonding interaction. In contrast to the cases of oxygen and nitrogen, the delocalized bonding is *favourable* to the structure, and indeed this structure is that adopted by graphite, the bonding in which was discussed in section 5.1.6 on page 187.

7.7 Metallic structures

Most of the elements exist as metals. Unlike non-metals, where a great variety of different structures are shown, most metals adopt a structure based around one of just three patterns: hexagonal close-packed, cubic close-packed and body centred cubic. We shall look in detail at these structures shortly, but the key feature of every metallic structure is the number of 'bonds' associated with each atom. We need to be slightly careful with the word 'bond' here. It would be hard (if not impossible) to rationalize these metallic structures in terms of two-centre two-electron bonds that we are more used to dealing with. This is because in most cases there simply are not enough valence electrons for this to be the possible. Sometimes the term 'metallic bond' is used to distinguish the interactions between the atoms in a metal from the more classical bonds that are present in most of the non-metallic structures we have looked at so far. What

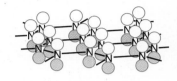

Fig. 7.32 A planar sheet of nitrogen atoms. Each atom has the necessary three bonds and each atom has a lone pair in *p* orbital.

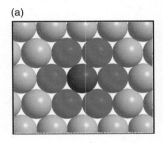

(a)

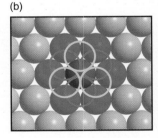

(b)

Fig. 7.33 The most efficient packing of a single layer of spheres has each sphere surrounded by six others. In view (a) the sphere coloured red is in contact with six spheres picked out in dark green. Three spheres from the next layer could be placed in contact with the red sphere. These could be placed in either of two ways as indicated by the yellow and blue circles in (b).

Fig. 7.34 The two commonly observed close-packed structures. Both structures consist of layers in which every atom touches six others. In the hexagonal close-packed structure shown in (a) and (b), every third layer is placed directly over the first. In the cubic close-packed structure shown in (c) and (d), the structure repeats every fourth layer. In both structures each atom in in contact with twelve others, as indicated by the red bonds made to the red atom.

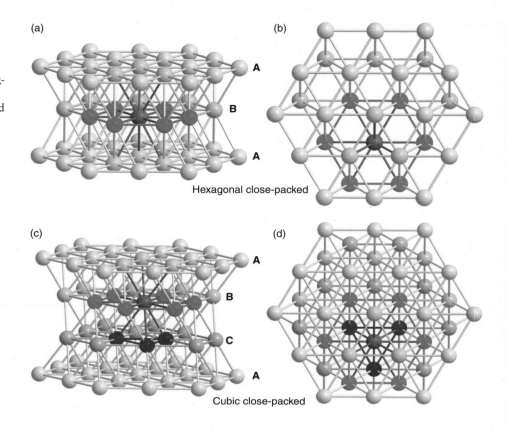

(a)

(b)

Hexagonal close-packed

(c)

(d)

Cubic close-packed

we really mean when we say the 'number of bonds associated with each atom' is the number of *nearest neighbours* each atom has in the structure. In most metallic structures every atom has either eight or twelve nearest neighbours.

7.7.1 Close-packed structures

We can model the structures of metals by considering the different ways it is possible to pack together hard spheres, such as ball bearings. Let us first consider just one layer of spheres, as shown in (a) in Fig. 7.33 on the preceding page. This layer has the closest possible packing of hard spheres, and each sphere is in contact with six others. For example, the sphere picked out in red is in contact with the six spheres picked out in a darker shade of green.

When it comes to adding the next layer, the closest packing is achieved by placing it such that the spheres in the second layer fit over the holes (or interstices) between the spheres in the first. If we consider the red sphere, then just three spheres in the next layer can make contact with this red sphere. However, there are two possible ways of arranging these, as shown in Fig. 7.33 (b). The new spheres could either be placed in the positions indicated by the yellow circles, or in the positions shown by the blue circles. For just two layers it makes no difference which we choose since the two arrangements are related by a 180° rotation around the red atom. When structures are built up with three or more layers, the arrangements do become important. In theory many different arrangements might be possible for a multi-layered structure; however in practice, just two structures are commonly observed.

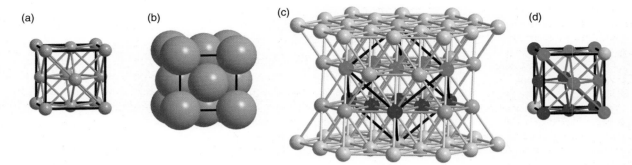

Fig. 7.35 The face-centred cube motif present in the cubic close-packed structure. A single cube is shown in (a); (b) shows the same arrangement but illustrating how the spheres touch one another in the structure. Structure (c) shows how the cube may be found within the cubic close-packed structure from Fig. 7.34 (c). A single cube is shown in (d) with the spheres coloured to indicate the close-packed layers.

Hexagonal close-packing

In the simplest packing arrangement, the structure repeats every third layer, as is shown in (a) and (b) from Fig. 7.34 on the preceding page. View (a) is from the side, and shows all three layers. Twelve red bonds show the contacts between the red sphere and its nearest neighbours; six within the same layer coloured green, three above and three below coloured yellow. Since the third layer is equivalent to the first the packing may be labelled ABAB to signify the alternating pattern of layers. View (b) shows the same structure viewed from above. The two yellow layers are equivalent and so lie on top of one another when viewed from above. This structure is called the *hexagonal close-packed* (*hcp*) structure. Magnesium, titanium and cobalt are examples of metals which adopt this structure.

Cubic close-packing

The alternative packing that is commonly observed has a structure which repeats every *fourth* layer, i.e. ABCABC. This is shown in Fig. 7.34 (c) and (d). View (c) is from the side showing the four layers. Once again, twelve red bonds show the contacts between the red sphere and its nearest neighbours: six within the same layer coloured green, three in the layer above coloured yellow, and three in the layer below coloured blue.

View (d) is taken from above, and all twelve of the nearest neighbours to the red sphere may still be seen. This structure is known as *cubic close-packed* (*ccp*) or *face-centred cubic* (*fcc*) for reasons we shall see shortly. Aluminium, copper, silver, and gold are examples of metals which adopt the *ccp* structure.

The *ccp* structure may be generated in a number of different ways; one is by combining a number of repeating units with a so-called *face-centred cubic* structure. A single cube is shown in Fig. 7.35 (a). The cube is made up from eight spheres placed at the corners of a cube and a further six placed in the centres of the sides. View (b) shows the same structure but illustrates which spheres touch one another. Structure (c) shows the layer structure from Fig. 7.34 (c) with the face-centred cube picked out; this cube is shown isolated in (d).

🌐 *Weblink 7.5*
View and rotate in real time a section of a hexagonal close-packed structure.

🌐 *Weblink 7.6*
View and rotate in real time a section of a cubic close-packed structure.

7.7.2 Non close-packed structures

Primitive cubic

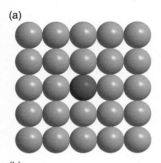

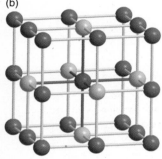

Fig. 7.36 View (a) shows a single cubic-packed layer of spheres. Each sphere is in contact with just four others within the layer. If many such layers are stacked on top of one another, the primitive cubic structure is generated. This is shown in view (b) where it is seen that each sphere is contact with six others.

If instead of packing a layer of spheres so that each row lies staggered from the previous row, each row is arranged exactly as the previous, a single layer ends up with every sphere touching just *four* others. This is shown in Fig. 7.36 (a) where the red sphere is seen to be in contact with just four others. If each successive layer is placed directly over the first, a very simple structure known as a primitive cubic structure is generated. The repeating unit in this structure consists of a cube with an atom at each corner. Eight such cubes are shown in Fig. 7.36 (b) where it is seen that each atom is in direct contact with only six others.

With only six nearest neighbours rather than the twelve found in the close-packed structures, it is not surprising that the packing is less efficient in the primitive cubic structure. It may be shown the spheres occupy only 52% of the space within this structure. This is considerably less than the 74% occupied within the close-packed structures. Despite its simplicity, this structure is extremely rare; the only element found to adopt this structure under normal conditions is polonium in the allotrope α-Po.

Body-centred cubic

The final important structure observed in metals is known as *body-centred cubic*, *bcc*. This structure is made up of repeating units in which each sphere is in contact with eight others placed at the corners of a cube, as shown in Fig. 7.37. Note in this structure the grey spheres are *not* in contact with each other but only with the red sphere.

Figure 7.38 on the facing page shows how these units stack together. We can picture the structure as having layers of spheres arranged on a square grid but not touching one another. Each successive layer is placed over the spaces of the layer below. The red sphere in (a) is touching eight neighbouring spheres, coloured grey. There are six next-nearest neighbours coloured bright yellow. The structure is perhaps best viewed showing the spheres at their full sizes as in (b). The spheres occupy 68% of the structure; once again this is less than the maximum 74% achieved in the close packed structures.

Fig. 7.37 The packing arrangement in body-centred cubic. The red sphere is in contact with the eight grey spheres which are arranged on the corners of a cube. Note the grey spheres are *not* in contact with each other but only with the red sphere.

7.7.3 Bonding in the metals

Now we have seen the structures adopted by typical metals, we should look at how the atoms actually bond together in these structures. In section 7.6.1 on page 282 we looked at the trends in the enthalpies of atomization on moving across the *p*-block. We found that we could understand the trend by considering how the valence orbitals overlap to give MOs, together with the number of electrons available to put into these orbitals. We can consider metallic bonding in just the same way. Figure 7.39 on the facing page shows how the enthalpies of atomization vary across the periodic table for Groups 1–12.

Let us first focus on the trend for the elements from Period 6, shown in orange in Fig. 7.39. The trend here is a gradual rise to a maximum at Group 6, followed by a drop to Group 12. This parallels the trend in Fig. 7.20 on page 283 where we saw the highest enthalpies of atomization were for the Group 14 elements which have just enough valence electrons to fill completely the bonding MOs formed from the overlap of the valence orbitals.

(a) (b)

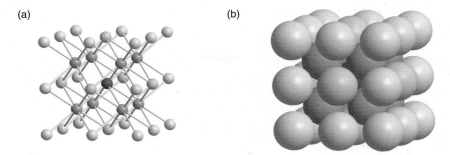

Fig. 7.38 The body-centred cubic structure. The structure is best thought of as having spheres arranged on a square grid within layers. The layers are here coloured yellow and grey. Each layer fits over the spaces from the layer below. The sphere coloured red is in direct contact with eight spheres coloured grey. The next nearest neighbours are the six spheres coloured bright yellow. View (b) shows how the spheres within each layer do not touch each other.

For the metallic elements shown in Fig. 7.39, we now need to consider how the *d* orbitals are involved in bonding. Unlike in section 7.6.1 on page 282, it is no longer appropriate to consider the atoms as being simply sp^3 hybridized. We have seen in section 5.1 on page 180 how the orbitals overlap to form bands in metals. Let us imagine that for these elements from Period 6, the 5*d* orbitals (which are lowest in energy) overlap to form a band, as do the 6*s* orbitals, and the 6*p* orbitals. However, for the moment, let us assume that while the bands from the 5*d* and 6*s* overlap, the band formed from the higher-energy 6*p* orbitals does not overlap with the other two bands. This will give an arrangement as shown in Fig. 7.40 on the next page.

For a solid made up from *N* atoms, *N* 6*s* AOs overlap to form a band of *N* crystal orbitals (COs). As usual, each orbital can accommodate two electrons so the band requires 2*N* electrons in order to fill it completely. Since each atom has five *d* AOs, the 5*d* band is formed from 5*N* AOs and requires 10*N* electrons to

Weblink 7.7
View and rotate in real time a section of a primitive cubic structure.

Weblink 7.8
View and rotate in real time a section of a body-centred cubic structure.

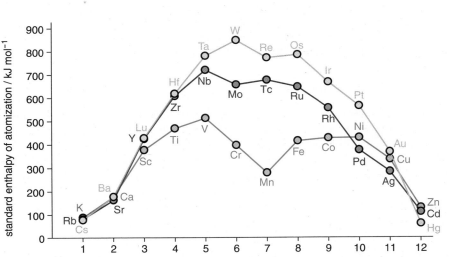

Fig. 7.39 A graph showing how the enthalpy of atomization of metals varies on moving across the periodic table from Group 1 to 12. Elements from the Period 4 are shown in red, those from Period 5 in purple, and those from the Period 6 in orange.

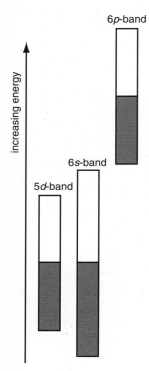

Fig. 7.40 A hypothetical arrangement of bands for Period 6. The shaded halves represent the crystal orbitals which are net bonding; the unshaded halves those which are net antibonding.

fill it. If we assume for a moment that the bands remain as indicated in Fig. 7.40, we see that on moving across Period 6, initially the electrons are going into the most bonding parts of the $5d$ and $6s$ bands. In order to fill completely the bonding COs of these two bands, each atom needs to contribute six electrons. This occurs for the element in Group 6, tungsten (W). On continuing across the period, the additional electrons are now filling the antibonding COs of the $6s$ and $5d$ bands. To fill both these bands requires a total of twelve electrons to be contributed by each atom. This occurs with mercury from Group 12. We can therefore see how this simple arrangement of bands can explain the trend observed in Fig. 7.39 on the previous page.

The trend in the atomic sizes of the d-block elements, as determined from the structures of the metals, mimics the pattern in the bond strengths. The strongest bonds from the elements in the centre of the d-block give rise to shorter bonds. On crossing the d-block, the bond lengths therefore decrease slightly until the middle, then increase again. This trend is seen in the sizes of the atoms in Fig. 7.12 on page 274 where the sizes of the metal atoms are taken as being half the bond length in the metal structure.

The arrangement of bands shown in Fig. 7.40 is rather over-simplified. As we have seen in section 7.2 on page 265, the energy of the d orbitals drop sharply as a period is traversed. The widths of the bands will vary depending on the degree of overlap, which in turn will vary as Z_{eff} changes. The band from the p orbitals will overlap, to some degree, with the s and d bands. It may also be the case that some antibonding parts of certain bands overlap with bonding parts of others. We should therefore not be too surprised by the erratic behaviour in the trend shown in Fig. 7.39 on the previous page on moving across Period 4. Nonetheless, the simple picture of valence electrons first contributing to bonding COs, and then antibonding COs is a useful one and helps us to explain why the strongest bonds are seen towards the middle of the d-block. By the time the p-block is reached, the d orbitals have become so low in energy that the only valence orbitals are the s and p orbitals which can be used for bonding as discussed in section 7.6.1 on page 282.

7.8 The transition from metals to non-metals

One of the interesting features of the periodic table is the division of the elements between metals and non-metals. This is shown in Fig. 7.41 on the next page where we see that the non-metals are found towards the top right of the periodic table, the metals towards the left, and that there is a rather ill-defined transition between the two involving semiconductors and semimetals. Before looking at *why* the change from metals to non-metals takes place, we should be clear about the different classifications used in Fig. 7.41.

Metals have delocalized bonding with each atom typically having between eight to twelve nearest neighbours. The presence of partially filled bands or overlapping bands allow metals to conduct electricity (see section 5.1 on page 180).

Semimetals typically have structures with fewer nearest neighbours than metals and they are usually poorer conductors of electricity than true metals. This is either due to the presence of a very small band gap (typically less

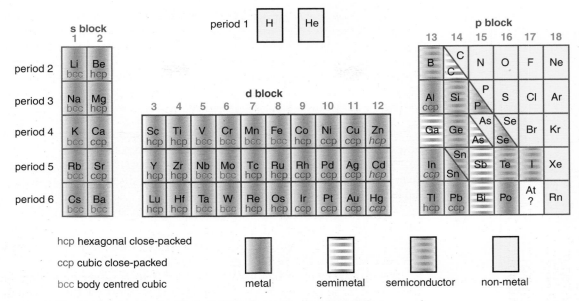

Fig. 7.41 A periodic table showing the distribution of metals, semimetals, semiconductors, and non-metals. The structure adopted by the metal under normal conditions is shown beneath the symbol of the metal. Where this is given in italics, the structure is slightly distorted from the ideal.

than 0.1 eV), or bands which just touch or overlap to a very small degree. The electrical conductivity of both metals and semimetals *decreases* as the temperature is increased.

Semiconductors conduct electricity poorly but the conductivity increases with temperature and when irradiated with light of a suitable wavelength. A band gap of between about 0.1 and 4 eV separates the filled valence band and the vacant conduction band.

Non-metals are electrical insulators and hence do not conduct electricity. The filled valence band and vacant conduction band are typically separated by more than 4 eV. Their structures consist of more localized bonding with four or fewer bonds per atom.

Orbital sizes and overlap

The transition from metallic structures to non-metallic follows the trend in the sizes of the valence orbitals. On crossing a period, Z_{eff} increases and hence the sizes of the orbitals decrease; consequently the elements tend towards non-metallic structures with more localized bonding and fewer nearest neighbours (typically four or fewer). Conversely, on descending a group, the orbital sizes increase due to the increase in the principal quantum number of the valence shell, and at the same time the trend towards metallic structures is seen with more delocalized bonding and many more nearest neighbours (typically eight or twelve).

Let us consider lithium which in the solid form adopts a body centred cubic structure where each lithium atom has eight nearest neighbours 304 pm away, and a further six at 351 pm (see Fig. 7.38 on page 293). In the gas phase, discrete Li_2 molecules can be detected in which the Li–Li bond length of 267 pm

Table 7.3 Ratios of different bond lengths in the structures of α-As, Sb, and Bi

Group 15 element	As	Sb	Bi
distance to nearest three atoms within layer (r_1) / pm	252	288	306
distance to nearest three atoms from adjacent layer (r_2) / pm	312	335	351
ratio r_2/r_1	1.24	1.16	1.15

is considerably less than the Li–Li spacing in the metal. However, the enthalpy of atomization, $\Delta H^\circ_{\text{atom}}$ of Li_2 is 51 kJ mol^{-1} whereas for metallic Li, $\Delta H^\circ_{\text{atom}}$ is 159 kJ mol^{-1}. In other words, having many slightly longer bonds in the solid gives stronger interactions per atom than having a single closer partner in the dimer.

We may wonder then why the same is not true for hydrogen for which H_2 molecules have a lower energy than does the solid. We have already seen in section 7.3.1 on page 275 how the bond strength in H_2 is considerably greater than that in Li_2, and this we attributed as being due to the better overlap of the $1s$ AOs in H_2 as compared with the $2s$ AOs in Li_2. Figure 7.14 on page 276 shows the degree of overlap as a function of internuclear separation for hydrogen, lithium, and potassium. The key point is that the overlap drops far more quickly with the smaller orbitals in hydrogen than with the larger orbitals in lithium and potassium. For lithium and potassium, even at the separations found in the metallic structures, there is still considerable overlap present. In other words, the larger orbitals are able to overlap with many neighbours over longer distances which gives rise to the metallic structures and delocalized bonding. Smaller orbitals overlap with their immediate neighbours only to form more localized bonding.

Intermediate structures: semimetals

The transition from metallic to non-metallic structures is not abrupt; as Fig. 7.41 on the preceding page shows, the border between the two includes semiconductors and semimetals with properties in between those of typical metals and non-metals. Like metals, semimetals conduct electricity, but their structures usually show more localized bonding like the non-metals. However, these structures gradually become more metal-like as the group is descended. For example, α-arsenic, antimony, and bismuth all have similar structures as shown in Fig. 7.23 on page 284, but as the group is descended, the differences in separation between any atom and its three nearest neighbours *within* each layer, and the next three nearest neighbours from *adjacent* layers becomes less and less.

As Table 7.3 shows, the ratio of these distances decreases as the group is descended. At the bottom of the group it is almost the case that bismuth has six equally spaced nearest neighbours rather than just three. The higher coordination number of six rather than just three is what we might expect of something with more metallic character. Under higher pressures, the distances to the nearest six neighbours become the same and these elements adopt the α-polonium structure shown in Fig. 7.36 on page 292.

Similar trends are observed down any group in the p-block. For instance, in the crystal structures of the solid halogens, on descending the group the X–X bond length (r_1) begins to approach the closest distance between one X_2 molecule and the next (r_2). The ratio of r_2/r_1 for solid chlorine, bromine, and

Table 7.4 Bond strengths (kJ mol^{-1}) of various diatomics involving Group 1 elements.

	H	Li	Na	K	Rb	Cs
H	436	238	186	175	167	175
Li		100	83	72	–	–
Na			71	62	62	59
K				50	–	–
Rb					47	46
Cs						43

iodine are 1.68, 1.46, and 1.29 respectively, again indicating that each iodine atom very nearly has many equally spaced neighbours rather than just the I_2 unit. It should not be too surprising that iodine has the most metallic character of all of the halogens in their solid forms. What is more, at pressures greater than 28 GPa, iodine adopts a distorted cubic close-packed structure and is a metal.

7.9 Bonding between the elements

7.9.1 General trends

We have already seen in section 7.3.1 on page 275 that the strength of the bond between two atoms of the *same* element is strongly influenced by the size of the interacting orbitals. When two *different* atoms combine, we have to consider not only the size of the orbitals but also the energy match between them.

The importance of size is illustrated in Table 7.4 where the bond strengths of hydrides and mixed diatomics from Group 1 are compared. With the exception of hydrogen, the orbital energies of the Group 1 metals are comparable, and for each of the diatomics given, it seems that the smaller one of the combining orbitals is, the stronger the bond formed. This turns out to be a general result: if two valence orbitals of approximately the same energy combine, the smaller the orbitals, the stronger the bond.

Of course, the reason for the stronger bonds might not necessarily be due to the fact that the orbitals are smaller *per se*, but another (correlated) reason such as the fact that the the repulsion between core electrons is smaller, at a given bond length, for the smaller atom. Nonetheless, the fact remains that bonds resulting from two orbitals from Period 2 combining tend to be stronger than bonds from two orbitals from Period 3. Similarly, a bond involving a $2s$ orbital to another s orbital is stronger than a bond involving a $3s$ orbital, and so on.

π bonding

The overlap of two p orbitals to form a π bonding MO is even more susceptible to the sizes of the orbitals, as illustrated in Table 7.5 on the following page. This shows the energy needed to break various π bonds in order to rotate the two halves of the molecules (see section 4.4.2 on page 153). The data show that whilst considerable energy is needed to break the π bonds resulting from the interaction of two $2p$ orbitals, much less is required to break a bond formed from one $2p$ and one $3p$ AO, and still less to break the π bond formed from two $3p$ AOs.

Table 7.5 Calculated barriers to rotation (kJ mol^{-1}) about various π bonds

2p–2p overlap	barrier	2p–3p overlap	barrier	3p–3p overlap	barrier
$H_2C=CH_2$	274	$H_2C=SiH_2$	149	$H_2Si=SiH_2$	105
$H_2C=NH$	265	$H_2C=PH$	180	$H_2Si=PH$	120
$HN=NH$	251	$HN=PH$	185	$HP=PH$	142

This rapid fall-off in the strength of π bonding accounts for the very clear trend that multiple bonding is common, and indeed often preferred, between Period 2 elements, but is much less common when elements from later periods are involved. For such elements there is a clear preference for forming single bonds over multiple bonds.

Orbital energies

Although size is important, when it comes to bonding between two different elements, the effect of the different energies of the two interacting orbitals is perhaps the dominant effect on the strength of the bond. Generally speaking, when the orbital energies are closely matched, we expect to find a strong bond in which there is more or less equal sharing of electron density between the two atoms. This is what we describe as a 'covalent bond'. However, strong bonds can still arise as a result of the transfer of an electron from one atom to another giving what is sometimes called an 'ionic bond'. It is important that we understand these two different contributions to bonding, which is best approached by considering the simple case of the diatomics of Period 2.

7.9.2 Covalent *vs* ionic bonding

In section 3.3.5 on page 121 and in section 7.6.1 on page 282 we looked at how the numbers of electrons in bonding and antibonding MOs affected the strengths of bonds in the diatomic molecules. The bonding reached a maximum in N$_2$ since this had ten valence electrons – enough to fill all the bonding MOs

Fig. 7.42 Bond strengths, measured in the gas phase, of the Period 2 homonuclear diatomics, fluorides, and oxides as a function of the number of valence electrons in the diatomic. Whilst the bond strengths for the fluorides and oxides broadly follow the same trend as for the homonuclear diatomics, this hides the fact that the nature of the bonding changes substantially across the series.

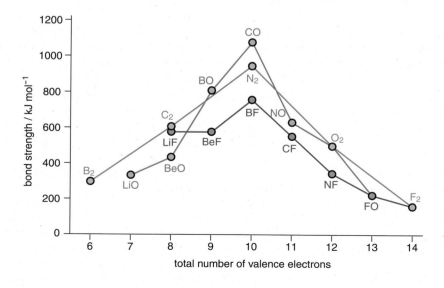

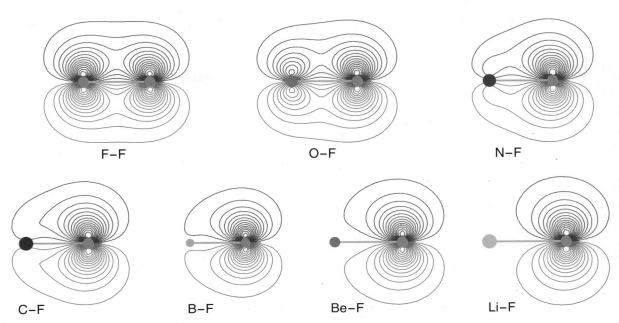

Fig. 7.43 Contour plots of one of the π bonding MOs of the fluorides from Period 2. In F_2 the contours are symmetrical over both atoms, showing the electrons to be equally distributed in this orbital. At the other end of the series, LiF, there are no contours over the lithium indicating no π bonding between the two elements. For the fluorides in between, the contribution that the π MO makes to the bonding steadily decreases.

but only one antibonding MO. Figure 7.42 on the facing page shows a similar plot comparing the strengths of homonuclear diatomics with the oxides and fluorides from Period 2. The trend seems to suggest that once again the number of electrons is important, with each series reaching a maximum when there are ten valence electrons. However, what the graph does not reveal is how the nature of the bonding changes across each series.

The MOs for carbon monoxide, CO, were compared with those of N_2 in Fig. 3.41 on page 131. This showed that although the AOs from carbon and oxygen no longer made equal contributions to the MOs, the overall form of the MOs is very similar. In particular, the π MOs of CO make a significant contribution to the bonding, so that we can talk of there being a triple bond between the two atoms. Like N_2, and CO, BF also has ten electrons occupying an equivalent set of orbitals, but does this mean that we should expect BF to have a triple bond?

The π bonding MOs for the series of diatomic fluorides are shown in Fig. 7.43. In F_2 the orbital is symmetrical with each fluorine atom making an equal contribution to the MO. In this molecule the electrons are distributed equally between the two atoms. At the other end of the series, the π MO in LiF is noticeably different. As we have already seen on page 132, this orbital has virtually no contribution from the lithium and is, in effect, simply a fluorine $2p$ AO. The transition from F_2 at one extreme to LiF at the other is gradual. This means that for the diatomics in between these extremes, such as CF and BF, there is a degree of π bonding in the molecule but this is clearly much less than in N_2 or CO. Quantifying the contribution that these π MOs make to the bonding is difficult, but it is clear that although there may be some contribution in BF, by the time we reach LiF, the π bonding is negligible.

Fig. 7.44 MO diagrams for LiF and N_2. The electrons present in the atoms have also been included and are shown as red arrows. (a) In N_2, some electrons are lowered in energy relative to those in the AOs, others are raised. However, there is a net lowering in energy and it is this that gives rise to the bonding in N_2. (b) In LiF the electron that was initially in the $2s$ AO of lithium ends up much lower in energy in the 4σ MO. It is this lowering of energy of the electron that gives rise to the bond between the Li and F.

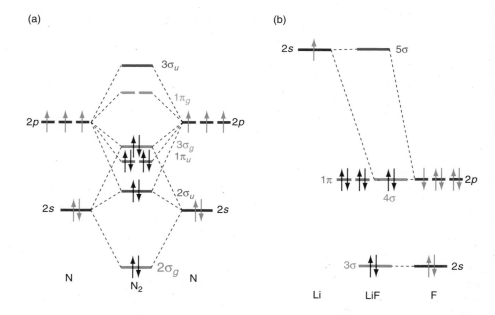

The question which then arises is if BF does not have a full triple bond like the isoelectronic N_2, and LiF does not have a double bond like C_2, then just why are these bonds quite as strong as they are? The answer to this is that the nature of the bonding between the atoms is changing. As the amount of π bonding decreases across the series, the separation in charge between the two atoms increases. As we go across the series F_2 to LiF, the mismatch in the interacting AOs results in an increasing imbalance of electron density between the two atoms, i.e. the bond becomes increasingly polar. Calculations indicate that in LiF the fractional charge on the two atoms is around +0.7 and −0.7, indicating a substantial transfer from lithium to fluorine.

The MO perspective

It is often said that this charge separation leads to an 'electrostatic or ionic contribution' to the bonding in a molecule like LiF. However, this is not some new effect which is not included in the MO picture. Figure 7.44 illustrates how the energies of the MOs in N_2 and LiF relate to the energies of the combining AOs. In N_2, shown in (a), when the molecule forms some electrons are lowered in energy and others are raised relative to their energies when in the atom. However, there is a net lowering of energy of the electrons in the molecule as compared to the atoms, and it is this that gives rise to the bond.

For LiF, shown in (b), only one electron is significantly lower in energy in the molecule than in the uncombined atoms. Lithium has one high energy electron in the $2s$ orbital and in the molecule this electron ends up in the 4σ orbital which is largely made up of a fluorine $2p$ orbital. This significant lowering in energy is responsible for the bond between the lithium and the fluorine.

The MO diagram for LiF shows that the $2s$ electron from lithium ends up in the 4σ MO, to which the F $2p$ AO is the major contributor. If it were the case that the 4σ MO had no contribution from the lithium $2s$ then the transfer of the electron to the fluorine is complete, and the molecule could reasonably be represented as Li$^+$F$^-$. However, as we have seen above, this is probably not the case and there is some participation from the Li 2s AO in the σ bonding MO, and hence some degree of covalency to the bond.

This discussion shows that the MO picture can be used to explain why the bonding in LiF is strong. It is not necessary to invoke some extra 'ionic contribution' to the bonding; rather the lowering in the energy which arises from the near complete transfer of the electron from lithium to fluorine has a natural explanation within the MO picture.

7.9.3 The classification of compounds as ionic or covalent

Trying to classify a compound as 'ionic' or 'covalent' is in many ways meaning-less. As we have seen above, in the gas phase even lithium fluoride does not exist as discrete lithium cations and fluoride anions and hence is not entirely ionic. We like to think that solid LiF *is* composed of a lattice of discrete ions but even in this case there is evidence that a small amount of electron density is shared between the ions. In solution, ions do not float around free, but instead have strong interactions with the solvent (section 6.12.1 on page 238). However, whilst we must remember that there really is no such thing as a pure electrostatic bond, there is nonetheless a meaningful distinction between compounds such as lithium fluoride which we would label as being 'ionic' and compounds such as N_2 or F_2 which we would label as being covalent.

The properties associated with *ionic* compounds include:

- high melting points and boiling points;

- the ability to conduct electricity when molten, but not when solid;

- generally high solubility in polar solvents such as water.

In contrast, the properties associated with *covalent* compounds include:

- often low melting points and boiling points;

- are unable to conduct electricity when solid or molten;

- generally more soluble in non-polar solvents such as toluene or hexane;

- often have structures with discrete molecular units.

The correlation between orbital energy and bond type

Perhaps the best prediction of whether the bonding in a compound is going to be more 'ionic' or 'covalent' is how closely matched in energy its valence orbitals are. A large difference results in more ionic compounds with the properties mentioned above, a small difference results in more covalent bonding. However, just as there is a gradual transition in the relative contributions of the AOs to the π MOs as shown in Fig. 7.43 on page 299, the transition in the nature of the bonding, such that it exists at all, is also a continuum.

Figure 7.45 shows the boiling points for the halides of sodium, aluminium, and silicon. The number in brackets under each formula is the difference in eV between the average orbital energies of the two atoms. We think of all the sodium halides as being ionic and this is supported by their high boiling points. In addition, their crystal structures reveal lattices of ions rather than discrete

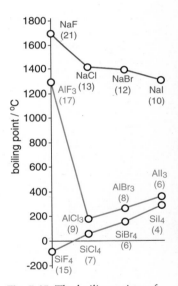

Fig. 7.45 The boiling points of the halides of sodium, aluminium, and silicon. The numbers beneath the formulae are the differences between the orbital energies of the combining elements, in eV. Generally speaking, the compounds with high boiling points (the more ionic compounds) also have large differences in their orbital energies, whereas those with lower boiling points (the more covalent molecules) tend to have smaller differences.

Fig. 7.46 A Van Arkel diagram. For a species composed of just two elements X and Y, the difference between their valence orbital energies is plotted against the average of their valence orbital energies. The baseline therefore lists the elements in order of increasing orbital energy. Metallic compounds tend to be found in the lower left hand corner of the triangle (coloured blue), more ionic compounds in the upper corner (coloured red), and more covalently bonded compounds in the lower right hand corner (coloured yellow). Semiconducting materials are generally found in the area in between the metals and non-metals (coloured green). There are no sharp divisions between the different regions – the transitions are seamless.

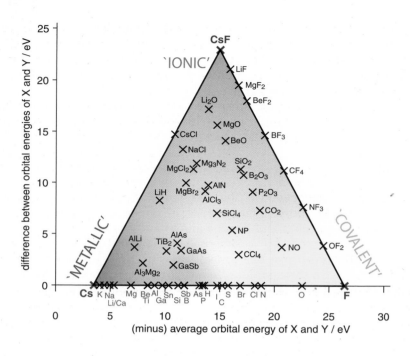

molecular units. The downwards trend in their boiling points is probably an indication of the decreasing lattice energies due to the increasing size of the halide ion (section 5.2.2 on page 194). For these compounds, the difference in orbital energies between the sodium and the halide is large, at least ten eV.

The low boiling points of the silicon halides suggests that the bonding in these molecules is predominantly covalent, and the upward trend in these boiling points as we go from SiF_4 to SiI_4 can be attributed to the increase in the dispersion interaction as the size of the halide increases (see section 1.2.1 on page 6). In the solid, individual tetrahedral SiX_4 units are found. With the exception of silicon fluoride, the difference in orbital energies for these compounds is rather small.

The aluminium halides show a different trend. The very high boiling point of AlF_3 suggests the bonding in this species to be ionic, whereas the low boiling points and their increasing trend suggests that the bonding in $AlCl_3$, $AlBr_3$, and AlI_3 to be more covalent. These ideas are supported by the crystal structrues of the halides; in AlF_3 each aluminium is surrounded by six fluoride ions, whereas $AlBr_3$ and AlI_3 exist as Al_2X_6 dimers packed together in the solid.

The Van Arkel diagram

A useful aid correlating difference in orbital energies (sometimes under the guise of electronegativities) with bond type is a diagram first proposed by Van Arkel and Ketelaar in the 1940s, and later modified by Sproul in the 1990s. For species composed of just two elements, the difference between their orbital energies is plotted against the average of these energies, as shown in Fig. 7.46. The baseline indicates the average valence orbital energies of the elements, and these are shown for a selection of elements in the figure. These are arranged in order of decreasing orbital energies, caesium is found at the extreme left, and fluorine at the extreme right. The compound with the greatest difference in the orbital energies of its constituent atoms is caesium fluoride, and this forms the uppermost point of the triangle.

As we have seen in this chapter, there is a correlation between orbital energies, Z_{eff} and the sizes of orbitals. It should therefore come as no surprise that on moving across the baseline, the elements show a gradual transition from metallic to covalent bonding. Metals in general tend to be found in the lower left-hand corner of the triangle, and compounds with more covalent bonding tend to be found in the lower right-hand corner. The top of the triangle contains the compounds with the greatest differences in orbital energies – the ionic compounds. The transitions between these different regions is seamless.

7.10 Oxidation states

Even though there is no such thing as a pure ionic compound, it is often useful to think of substances such as the alkali metal halides as containing discrete ions. In fact, since there are no elements with occupied orbitals higher in energy than the alkali metals, whenever these elements are bonded with another element, they will always be polarized with a degree of positive charge. In almost all cases, (alloys excepted) it might be helpful to think of any compound which contains an alkali metal, M, as containing an M^+ ion.

Similarly, since fluorine is the most electronegative atom, in compounds it will always bear some partial negative charge; very often it might be convenient to think of the compound as containing the F^- ion. This line of thought has lead to the idea of *oxidation states* for elements in compounds. This is a purely artificial construction, where we imagine a given compound as being made up entirely of ions. Whilst this is clearly never the case, the idea is still extremely useful and forms an essential part of the nomenclature used for very many compounds.

7.10.1 Definition of oxidation state

The oxidation state of an element in a molecule or ion may be defined as the charge that the element would have *if* the molecule were composed entirely of ions, with the ions in their 'closed shell electron configurations'.

It follows from this definition that if we take each element in the species of interest, multiply its oxidation state by the number of atoms of that element present, and then sum these terms over all the elements present, the result must equal the overall charge on the species (i.e. zero for a neutral species)

$$\sum_{\text{all elements } i \text{ present}} (\text{oxidation state of element } i) \times (\text{number of atoms of element } i)$$

$$= \text{overall charge on the species} \tag{7.4}$$

In assigning oxidation states we usually make some assumptions about the oxidation states of the least and most electronegative elements in the species. These assumptions are set out here in three steps:

(a) Identify the *least* electronegative element in the species. If it is in Group 1 or Group 2, assign it an oxidation number of +1 or +2 respectively.

(b) *If* an element was assigned an oxidation state in step 1, and *if* there is only one other element present in the species, its oxidation state can be determined using Eq. 7.4 on the preceding page; this finishes the whole procedure.

(c) Identify the *most* electronegative element in the species. If this element is in Group 17 assign it an oxidation state of −1, if it is in Group 16 assign it an oxidation state of −2, and if it is nitrogen assign a value of −3.

(d) Determined the oxidation state of the remaining element using Eq. 7.4 on the previous page.

Two examples

Consider first the anion SO_4^{2-}. In step 1 we identify the least electronegative species as sulfur, but as this is neither in Group 1 nor in Group 2, we cannot assign it an oxidation state as yet. Step 2 therefore does not apply. In Step 3 we identify the most electronegative element as oxygen, which is in Group 16, and so it is assigned an oxidation state of −2. In Step 4 we use Eq. 7.4, noting that the overall charge is −2 and letting the oxidation state of sulfur be x

$$\underbrace{1 \times x}_{\text{sulfur}} + \underbrace{3 \times (-2)}_{\text{oxygen}} = -2$$

It therefore follows that the oxidation state of sulfur in SO_4^{2-} is +4.

As our second example we will work out the oxidation state of iodine in potassium ortho-periodate, K_5IO_6. In Step 1 we identify potassium from Group 1 as the least electronegative element, and so assign it an oxidation state of +1. Step 2 does not apply as we have more than one other element present. In Step 3 we identify oxygen from Group 16 as the most electronegative element and therefore assign it an oxidation state of −2. In Step 4 we use Eq. 7.4, noting that the species is not charged and letting the oxidation state of iodine be x

$$\underbrace{5 \times (+1)}_{\text{potassium}} + \underbrace{6 \times (-2)}_{\text{oxygen}} + \underbrace{x}_{\text{iodine}} = 0$$

The oxidation state of the iodine is therefore +7.

When used in the name of a species, by convention the oxidation state is given in roman numerals in brackets after the name of element, or the species containing the element. For example, iron(III) chloride, $FeCl_3$ contains iron in the +3 oxidation state; potassium manganate(VII), $KMnO_4$ contains manganese in the +7 oxidation state. Note that no space is included between the name and the oxidation state. It must be remembered that these compounds *do not* contain ions with such high charges; the oxidation state is just a formal counting device.

7.10.2 Trends in oxidation states across the periodic table

Figure 7.47 on the facing page shows the oxidation states exhibited by the elements and their trends on moving across the periodic table. The trends are

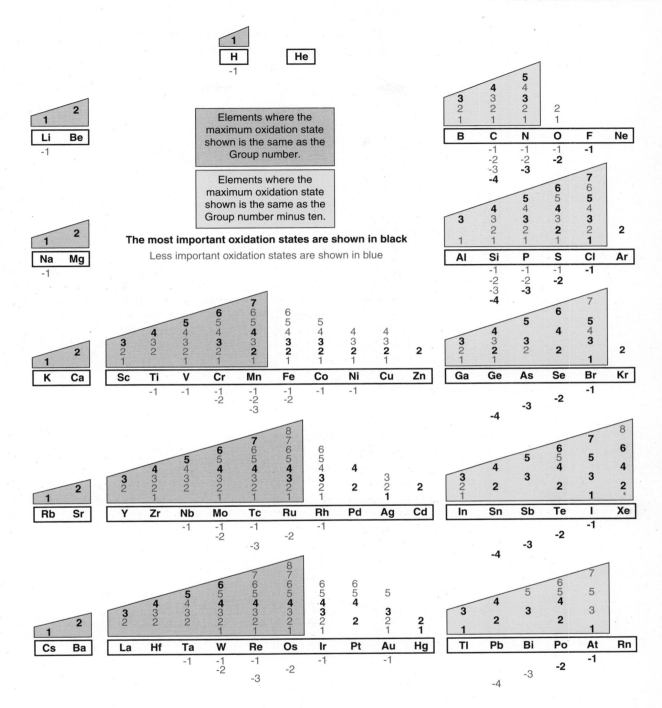

Fig. 7.47 The oxidation states exhibited by the elements, arranged according to the periodic table. The legend gives details of the colour coding used.

relatively easily understood in light of the earlier sections of this chapter. By definition, the oxidation states of Groups 1 and 2 are I and II respectively. Up to manganese, the first row of the transition metals can form oxides or oxyanions where the oxidation state is the same as the group number. Thus we see, scandium(III) in Sc_2O_3, titanium(IV) in TiO_2, vanadium(V) in V_2O_5, chromium(VI) in CrO_3, and manganese(VII) in the MnO_4^- ion.

Beyond manganese, as the energies of the $3d$ orbitals drop (see Fig. 7.2 on page 264), it becomes increasingly difficult to use these electrons in bonding. As a result, the maximum oxidation state shown gradually decreases again until for Group 12, the maximum shown is +2. With the valence s and p electrons all available for bonding in the p-block elements, the highest oxidation states here (except for oxygen and fluorine) are the group number minus 10. Thus we see oxidation states varying from +3 in aluminium(III) oxide, Al_2O_3, up to +8 in xenon tetroxide, or xenon(VIII) oxide, XeO_4.

7.11   Moving on

In this chapter we have shown how the broad trends in bonding can be rationalized using the fundamental concept of orbital energy, and the related concepts of effective nuclear charge and orbital size. Chemical reactions are often driven by the changes in energy that accompany the formation and destruction of bonds. In the next chapter we will look at how an understanding of the forms and energies of the molecular orbitals of a given species enable us to rationalize how that species reacts with another.

FURTHER READING

Inorganic Chemistry, fourth edition, Peter Atkins, Tina Overton, Jonathan Rourke, Mark Weller and Fraser Armstrong, Oxford University Press (2006).

QUESTIONS

7.1   Explain the following features and trends of the orbital energies shown in Fig. 7.2 on page 264.

(a)   In going across the second period from Li to Ne the energies of both the $2s$ and $2p$ AOs fall, but the $2s$ AO falls more steeply than the $2p$.

(b)   On going from Ne to Na there is a large decrease in the energy of the $2p$ AO.

(c)   The energies of the $2s$ AO in Li, the $3s$ AO in Na and the $4s$ AO in K show a gentle upward trend but in going from Li to Na, and from Na to K the nuclear charge increases by eight.

(d)   On going from Sc to Zn the energy of the $4s$ AO decreases rather little, but the energy of the $3d$ AO shows a greater fall in energy.

7.2   The atomic radii of the lanthanide elements (Ce–Lu) are remarkably similar, varying only by a few pm across all fourteen elements. Use Slater's rules (page 77) to determine the effective nuclear charge experienced by the outer electrons (the $6s$) of these elements, and hence explain why the atomic radii vary so little.

7.3 Explain why there is a general increase in atomic radii on moving from the first row transition metals (those in period 4) to the second row (those in period 5), but there is very little change on moving to the third row (those in period 6).

7.4 Rationalize the trends in the atomic radii of the Group 2 elements given in the table (the radius given for element 120 is based on a calculation, rather than experiment).

element	Be	Mg	Ca	Sr	Ba	Ra	element 120
radius / pm	112	160	197	215	222	215	200

7.5 When metallic gold is exposed a caesium vapour a solid compound with formula CsAu is formed; the solid is a poor conductor of electricity, but the conductivity rises sharply when the solid is molten. Discuss the likely nature of the bonding in this compound. Would you expect to be able to form a similar compound from the reaction of silver with caesium?

7.6 Explain the following trends in the bond strengths (in $kJ\ mol^{-1}$) of the Group 1 and Group 11 diatomics.

molecule	bond strength	molecule	bond strength
K_2	50	Cu_2	177
Rb_2	47	Ag_2	160
Cs_2	43	Au_2	225

7.7 (a) Construct an MO diagram for Be_2 and use it to explain why, although the bond order is zero, s–p mixing results in the the molecule having a weak bond.

(b) Explain the following observations: (i) the bond strength of Li_2 is 100 $kJ\ mol^{-1}$, a value much greater than that for Be_2; (ii) the enthalpy of vaporization of solid beryllium is greater than that for solid Li.

(c) Rationalize the following bond strengths of the Group 12 homonuclear diatomics:

molecule	bond strength / $kJ\ mol^{-1}$
Zn_2	29
Cd_2	7
Hg_2	≈ 0

(d) Suggest why metallic mercury has the lowest enthalpy of vaporization of any metal, and hence why it is a liquid at room temperature.

7.8 The common oxidation states of mercury are +1 and +2. While Hg^{2+} ions are found in solution, mercury(I) exists in solution as Hg_2^{2+}. The Hg–Hg bond length in mercury(I) salts is considerably less than the Hg–Hg distance in Hg(l).

(a) By considering the $6s$ orbitals only, draw a simple MO diagram to explain the bonding in Hg_2^{2+}.

(b) Would you expect mercury(I) salts to be paramagnetic?

(c) Identify the species present in the ionic compound $Hg_3(AlCl_4)_2$, and describe the bonding in each. [Hint: consider the ions present in the compound.]

7.9 The following table gives the bond strengths (kJ mol^{-1}) for the homonuclear diatomics of the p-block.

group	13	14	15	16	17
	B_2	C_2	N_2	O_2	F_2
	297	607	945	498	159
	Al_2	Si_2	P_2	S_2	Cl_2
	133	327	490	425	243
	Ga_2	Ge_2	As_2	Se_2	Br_2
	112	264	382	333	193
	In_2	Sn_2	Sb_2	Te_2	I_2
	100	187	299	260	151
	Tl_2	Pb_2	Bi_2	Po_2	At_2
	64	87	200	187	80

Rationalize the trends in the bond strengths for both crossing the periods and descending the groups. Your answer should include a consideration of the bond orders present in each group.

Why are the bond strengths in Groups 16 and 17 generally greater than those in Groups 14 and 13 respectively? Explain any exceptions to this trend.

7.10 (a) Discuss the factors which lead to the following trend in bond energies

diatomic	F_2	Cl_2	Br_2	ClF
bond energy / kJ mol^{-1}	158	242	193	297

(b) Discuss the factors which lead to the following trend in bond lengths

bond	C–Cl	Si–Cl	Ge–Cl	Sn–Cl
bond length / Å	1.76	2.08	2.13	2.20

7.11 Discuss the following observations.

Solid carbon dioxide sublimes at temperatures above -78 °C to give a gas containing discrete CO_2 molecules.

Solid silicon dioxide has a giant covalent structure. It melts at around 1500 °C, and finally boils at temperatures over 2800 °C; discrete SiO_2 molecules can be detected in the gas phase.

If CO_2 is subject to extremely high pressures a solid material is produced which appears to have a giant covalent structure in which there are C–O single bonds.

7.12 (a) Using the data on orbital energies given below, place the following compounds on the van Arkel diagram: $CaCl_2$, TiB_2, S_4N_4 and AlN.

element	B	C	N	Al	S	Cl	Ca	Ti
minus valence orbital energy / eV	11.8	15.2	18.8	9.05	15.8	18.2	5.34	8.44

(b) The following descriptions refer to the properties of the compounds listed above, but not necessarily in the same order.

W brown solid, melting point 2980 °C, good conductor

X orange crystals, mp 170 °C, soluble in organic solvents

Y   white solid, mp 1418 °C, insulator, moderately soluble in water

Z   white solid, mp > 2400 °C, band gap 6 eV, insulator.

With the aid of the van Arkel diagram, and from your general knowledge, deduce which compound corresponds to which letter, giving brief reasons for your answers.

7.13   Pure calcium carbide, CaC_2, is a colourless solid (melting point 2300 °C) which does not conduct electricity in the solid state. The structure of CaC_2 is essentially an NaCl-type lattice with alternating Ca^{2+} ions and C_2^{2-} ions. The C–C bond length in these ions is 119.1 pm.

(a)   Use the van Arkel diagram to predict the sort of bonding you might expect to find in calcium carbide (the necessary orbital energies are given in the previous question).   Are the properties given above consistent with the position of CaC_2 in the van Arkel diagram?

(b)   Construct an MO diagram for C_2 and use it to explain why the C–C bond length in CaC_2 is considerably shorter than that in gaseous C_2 (131.2 pm).

(c)   Lanthanum carbide, LaC_2, is thought to contain the La^{3+} ion and the C_2^{3-} ion; the C–C bond length is this ion is 130.3 pm. Use your MO diagram to explain why the C–C bond length in LaC_2 is greater than that in CaC_2.

7.14   At room temperature *white phosphorus* is an insulating solid. It boils at 280 °C to give a vapour which is a mixture of P_4 and P_2 molecules. If the vapour is condensed, white phosphorus reforms. The P–P bond length in P_2 is 1.90 Å, and in P_4 it is 2.21 Å.

If white phosphorus is held at elevated temperatures for an extended period, various other solid allotropes are formed, none of which boil under 600 °C. One of these allotropes, *black phosphorus*, is a semiconductor.

The α allotropes of arsenic, antimony and bismuth all have similar structures to that of black phosphorus.

(a)   Sketch the structures of P_4 and P_2, and explain their relative bond lengths.

(b)   How would you expect the equilibrium between P_4 and P_2 to vary with temperature?

(c)   Suggest the common structure adopted by black phosphorus and the α allotropes of the other Group 15 elements.

(d)   Explain the differences in the physical properties of the black and white allotropes of phosphorus.

(e)   Rationalize the trends in the bond lengths and bond angles observed in the solid form of the Group 15 elements

allotrope	black P	α-As	α-Sb	α-Bi
bond length / Å	2.23	2.52	2.91	3.07
bond angle / degree	98.2	96.7	96.6	95.5

7.15   (a)   The Si–Si bond lengths in $(R_2Si)_2$ and the *cyclic* compound $(R_2Si)_3$ are 2.14 Å and 2.40 Å, respectively. Draw the structures of these two compounds and comment on the Si–Si bond orders in each.

R =

(b)   Suggest why naturally occurring carbon may be found in either of two allotropes, whereas silicon exists in only one allotropic form in which the Si–Si bond length is 2.35 Å.

(c) The standard enthalpies of sublimation of graphite, diamond and silicon are 715 kJ mol^{-1}, 710 kJ mol^{-1} and 456 kJ mol^{-1}, respectively. Comment on these values.

(d) Explain why silicon is a semiconductor whereas diamond is an electrical insulator.

(e) Explain why heating silicon with small quantities of arsenic leads to a new material with higher electrical conductivity then pure silicon.

7.16 At high temperatures, calcium metal reacts with silicon to give a salt with empirical formula $CaSi_2$, and with carbon to give a salt with empirical formula CaC_2. X-ray crystallography reveals that the structures of these two salts are very different: CaC_2 contains discrete C_2^{2-} ions, whereas in $CaSi_2$ the silicon atoms form sheets of $(Si^-)_n$ with each silicon having three Si–Si single bonds.

(a) Which element occurs in a form which is isoelectronic and isostructural with the C_2^{2-} anion, and which element occurs in a form which is isoelectronic and isostructural with the $(Si^-)_n$ sheets?

(b) Why is there such a difference between the structures of the carbon and silicon anions?

(c) Calcium and silicon can also react to give a compound with empirical formula CaSi, whose structure is found to contain helical chains of Si atoms. How may the structure of these chains be rationalized?

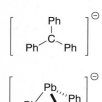

7.17 (a) In the anion Ph_3C^- the Ph–C bonds have a trigonal planar arrangement about the carbon, with the Ph–C–Ph bond angle being 120°. In contrast, in the anion Ph_3Pb^- the Pb–Pb bonds have a trigonal pyramidal arrangement about the Pb, with a Ph–C–Ph bond angle of 91°. Comment on the reasons why these two ions have such different structures.

(b) From your answer to (a), explain why Ph_3C^- is a far stronger base than Ph_3Pb^-.

7.18 Determine the oxidation state of oxygen in the following compounds: (a) Li_2O, (b) Na_2O_2, (c) KO_2, (d) MgO, (e) $Ba(O_3)_2$.

7.19 Determine the oxidation state of phosphorus in the following compounds or ions: (a) PCl_3, (b) PF_5, (c) P_2O_5, (d) PO_4^{3-}.

7.20 Determine the oxidation state of chromium in the following compounds or ions: (a) CrF_6, (b) $CrCl_4^-$, (c) CrO_4^{3-}, (d) $Cr_2O_7^{2-}$.

7.21 Discuss the likely reasons for the observation that in going from Sc to Mn the maximum oxidation state shown by successive elements increases steadily, but as we carry on from Fe to Zn there is a steady decrease in the maximum oxidation state for successive elements.

Describing reactions using orbitals

Key points

- During a reaction, as bonds are formed and broken, the electrons are redistributed.
- The redistribution of electrons may be followed by calculating how the molecular orbitals change as the reaction proceeds.
- 'Curly arrow mechanisms' summarize the way in which electrons are redistributed during a reaction.
- The way in which two reactants may react can often be understood by considering the interaction between the highest energy occupied MO (the HOMOO) and the lowest energy unoccupied MO (the LUMO).

So far we have seen how molecular structures can be understood using an orbital description of the bonding: in this chapter we are going to extend these ideas to cover chemical reactions. We will discover that by looking at the orbitals involved it is possible to understand why molecules react in a particular way. The key ideas which grow out of this orbital description provide a framework for rationalizing a great deal about chemical reactivity, and they will be applied in many of the following chapters, particularly those concerned with organic chemistry.

During a chemical reaction bonds are broken and new bonds are made as the reactants are transformed into the products. As a result the electrons are redistributed during the reaction, meaning that the shapes and energies of the orbitals involved must change. Our focus in this chapter is therefore on understanding how and why the orbitals change during a chemical process.

8.1 The redistribution of electrons in a reaction

The first reaction we are going to describe is just about the simplest one imaginable: the reaction between a proton H^+ and a hydride ion H^- to give H_2.

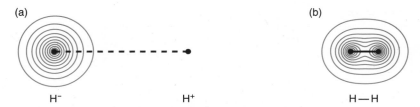

Fig. 8.1 Contour plots of the occupied MO for the H^- + H^+ system. In (a) the ions are separated by a large distance and the contours indicate that the electrons are only associated with the hydride. In (b) the nuclei are at their equilibrium separation and the contours indicate that the electrons are shared equally by both nuclei.

We will think about the reaction taking place in the gas phase so that we do not have to worry about the role of the solvent.

Initially, as the H^+ and H^- approach one another there will be an electrostatic interaction between the ions, due to their opposite charges. The product of the reaction is H_2, which is of course not H^+H^- but a covalently bound molecule in which electrons are *shared* between the two atoms. It therefore follows that in this reaction there has been a redistribution of the electrons between the two hydrogen atoms.

This redistribution of electrons is immediately apparent from the contour plots of the orbitals shown in Fig. 8.1. In (a) the two ions are separated by a considerable distance and the contours reveal that the electron density is associated with only one nucleus, that of the hydride. At the equilibrium separation shown in (b), the electron density is shared equally over the two nuclei. There is thus a redistribution of electron density as the reactants go to products.

8.1.1   Curly arrows

We can summarize the redistribution of electrons that occurs during this simple reaction using a *curly arrow*. This is a convention, used extensively in organic chemistry, in which an arrow represents the movement of a *pair* of electrons. The arrow starts from where the electrons originate, and the head of the arrow indicates where the electrons end up.

In the trivial example of the reaction between a hydride ion and a proton, the arrow starts at the hydride and ends up in the region in between the two nuclei where we draw the bond in the final structure, as shown in Fig. 8.2. Positioning the head of the arrow in this way can be rather ambiguous in more complex reactions, so to avoid this problem it is common to draw the head of the arrow pointing *at* the atom to which the new bond is formed.

Since the arrow indicates the motion of an electron pair, it would be most correct to indicate the pair from which the arrow starts. Often this will be a lone pair, but it could also be a bonding pair. However where the lone pair is on an anion it is common to show the arrow starting from the minus charge. Although this is strictly not correct, it is nevertheless the normal practice.

We should not be misled into thinking that the curly arrows indicate *how* a reaction takes place. Rather, all the arrows do is summarize the net redistribution of the electrons which results from the reaction. In the H^- + H^+ reaction the arrow shows that the two electrons which were associated with the hydride end up being shared between the two nuclei, and thus forming the bond.

Fig. 8.2 The redistribution of electrons in the H^- + H^+ reaction can be described using a *curly arrow*. This is a convention in which an arrow (here shown in blue) indicates the movement of a *pair* of electrons. The arrow starts where the electrons originate from and finishes where the electrons end up.

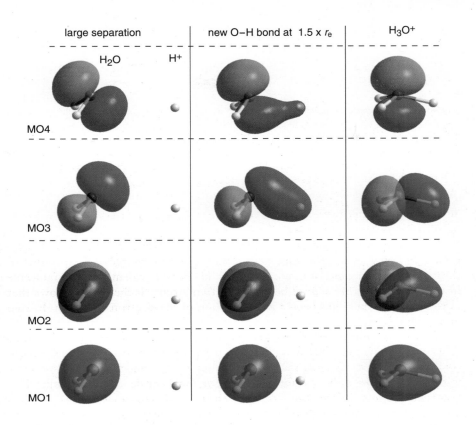

Fig. 8.3 Iso-surface plots of the occupied MOs of water showing how they change during protonation. The left-hand column shows the orbitals in water, the right-hand column the corresponding orbitals in H_3O^+. The middle column show the orbitals at an intermediate stage when the developing O–H bond is still 50% longer than it is in H_3O^+. During the protonation, all the orbitals of the water change to some degree. MO1 and MO2 are initially responsible for bonding the two hydrogens to the oxygen in H_2O, but in H_3O^+ they are involved in bonding all three hydrogens to the water. The most significant change takes place in MO3 which is weakly bonding in water, but strongly bonding in H_3O^+. r_e is the equilibrium O–H bond length in the product.

8.1.2   The protonation of water

A more interesting reaction than $H^- + H^+$ is the protonation of water to form the hydronium ion, H_3O^+. The MOs of water were discussed in section 4.3.1 on page 146 and are illustrated in Fig. 4.15 on page 147. We saw that there are two strongly bonding MOs (MO1 and MO2), one weakly bonding MO (MO3) and one nonbonding MO (MO4). Figure 8.3 shows what happens to these orbitals as a proton is brought up to the water molecule. On the left we have the situation in which the H^+ is so far away from the water that there is no interaction, and on the right we have the MOs of the product hydronium ion. In the middle are shown the MOs for the case where the incoming ion is $1.5 \times r_e$ from the oxygen, where r_e is the equilibrium distance.

In H_2O the bonding MO1 arises from the overlap of the oxygen $2s$ with the two hydrogen $1s$ AOs. In H_3O^+ the bonding is simply extended to cover all three hydrogens. MO2 arises from the overlap of one of the $2p$ AOs on oxygen with the two hydrogen $1s$ AOs and, just as for MO1, in H_3O^+ the overlap is extended to include all of the hydrogen AOs.

MO3 undergoes a more significant change; in water this orbital is only weakly bonding, whereas in H_3O^+ it is strongly bonding. Finally, MO4 is non-bonding in water, but becomes weakly bonding in H_3O^+. The electrons in this orbital are those which, in a simple description of H_3O^+, would be described as a lone pair.

The important point is that *all* of the orbitals change to some extent during this reaction, but some change more than others. The most significant change is in MO3 which changes from weakly bonding in water, to strongly bonding in H_3O^+.

Weblink 8.1

Follow this link to see a 'movie' version of Fig. 8.3 in which you can see how the MOs change as the reaction proceeds.

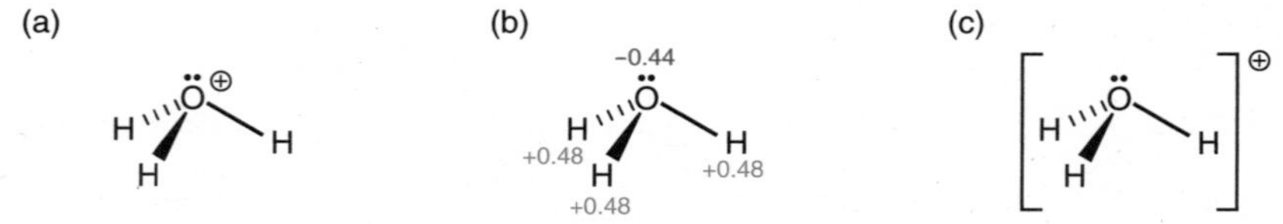

(a)   (b)   (c)

Fig. 8.4 The hydronium ion is often represented as in (a) with a positive charge on the oxygen. However, the calculated charge distribution, shown in (b), clearly shows that the oxygen has a negative charge. Representation (c) is less misleading but more cumbersome to use.

Curly arrow mechanism

As before, the net change that takes place in the reaction may be summarized using curly arrows

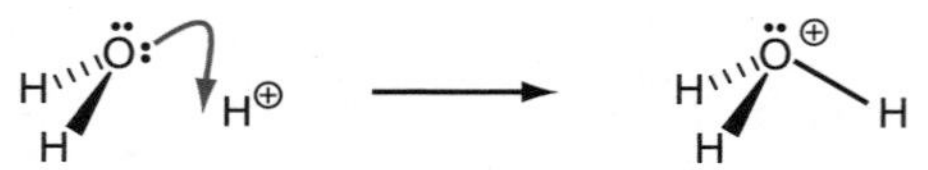

The arrow starts on one of the lone pairs in water, and ends where the new bond is formed. We have drawn the arrow head closer to the proton to emphasize that the new bond is forming to this proton.

Charge distribution in H_2O and H_3O^+

Notice that in the above mechanism we have represented the hydronium ion as having a positive charge on the oxygen. However, calculations on this ion show that, far from being positive, the oxygen bears a *negative* charge of -0.44, as shown in Fig. 8.4 (b). In water the charge on the oxygen is -0.67, so it is true that protonation has reduced the charge of the oxygen, but it has certainly not acquired a positive charge. Structure (c) is a more realistic representation of the hydronium ion in that it indicates that the *whole* ion bears a positive charge; however this representation is somewhat cumbersome to use.

8.1.3 Formal charges

The positive charge shown on the oxygen in Fig. 8.4 (a) is a *formal charge* which arises from a particular way of assigning the electrons to the atoms. These formal charges arise from a kind of 'electron accountancy' which is done using the rules set out in Box 8.1 on the next page. A key feature of these rules is the assumption that the two electrons in a bond are shared *equally* between the two atoms.

To compute the formal charge on the oxygen in the hydronium ion we note that the oxygen has three bonds and so is counted as having a half share of three pairs of electrons i.e. three electrons, one from each bond. In addition, the oxygen has a lone pair, making a total of *five* electrons. An isolated oxygen atom needs *six* valence electrons in order to be neutral, but in the hydronium ion the oxygen only has five, so it follows that the formal charge on the oxygen is +1.

The problem with this kind of electron accountancy is the assumption that each bond represents a pair of electrons shared *equally* between two atoms. We have seen many times how the MO picture tells us that, whilst it is convenient to think of bonds in this way, it is not realistic. However, if we insist of thinking

---

Box 8.1 Calculating formal charges

The formal charge on any atom may be calculated in the following manner:

(a) Count the number of bonds to the atom. Double bonds count as two, triple bonds as three.

(b) Add to the number of bonds the number of electrons in lone pairs, and any single electrons if the species is a radical. This gives the total number of electrons associated with the atom, N_e.

(c) Calculate the number of valence electrons, N_v, associated with the neutral atom. If the atom is from Groups 1–12 in the periodic table, this number is the group number. If the atom is from Groups 13–18, this number is the group number minus ten.

(d) The formal charge is given by $N_v - N_e$.

The formal charge may be positive, negative, or zero. If the charge on an atom is already known, the method can also be used in reverse to calculate the number of lone pairs (or free electrons) on that atom.

Example – carbon monoxide

Carbon monoxide may be represented in one of two ways

(a) $:\overset{\ominus}{C}\!\equiv\!\overset{\oplus}{O}:$ (b) $:C\!=\!\overset{\cdot\cdot}{\underset{\cdot}{O}}:$

In structure (a), carbon has three bonds and one lone pair, so $N_e(\text{carbon}) = 5$. Carbon is in Group 14 which means a neutral carbon atom has $(14 - 10) = 4$ valence electrons i.e. $N_v(\text{carbon}) = 4$. Therefore the formal charge on carbon is $(4 - 5) = -1$ as shown in the structure.

The oxygen also has three bonds and one lone pair, so $N_e(\text{oxygen}) = 5$. Oxygen is in Group 16 and so the neutral atom has $(16 - 10) = 6$ valence electrons i.e. $N_v(\text{oxygen}) = 6$. Therefore the formal charge on oxygen is $(6 - 5) = +1$, as shown in structure (a).

In structure (b), each atom has the required number of electrons to be neutral (four for carbon, six for oxygen) and hence no formal charges are needed. Whilst neither representation is ideal, (a) is preferable since we would normally think of CO as consisting of a triple bond rather than a double bond (recall that CO is isoelectronic with N_2).

---

of bonds in this way, and draw curly arrow mechanisms which represent the redistribution of pairs of electrons, we will end up with these formal charges. We need to keep in the back of our minds the limitations of this approach.

8.2 HOMO–LUMO interactions

As reactants go to products all of the orbitals change to some extent, but it is often the case that the most significant changes, particularly in the early stages of a reaction, can be thought of as arising from an interaction between the highest occupied MO (the HOMO) of one reactant and the lowest unoccupied

Fig. 8.5 Illustration of different outcomes from the interaction of two orbitals. In (a) a filled orbital interacts with an empty one resulting in a decrease in the energy of the electrons. In (b) two filled orbitals interact: one pair of electrons goes down in energy and one up, resulting in no net lowering in energy. In (c) neither orbital is occupied, so no electrons change energy.

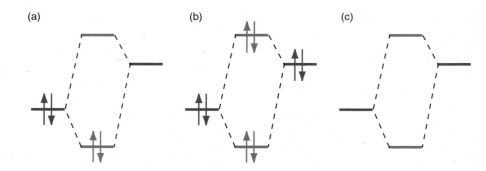

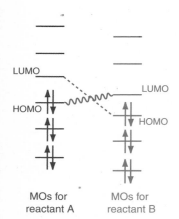

Fig. 8.6 The HOMO–LUMO interaction with the smallest energy gap is between the HOMO from A with the LUMO from B, as shown by the wavy green line. The other possibility, the interaction between the HOMO from B with the LUMO from A, shown by the dashed green line, is not so strong as the MOs are further apart in energy.

MO (the LUMO) of the other. As we will explain, this HOMO–LUMO interaction results in a lowering in the energy of some electrons, and this can be identified as the driving force for the reaction. In addition, it is useful to classify reactions according to the type of HOMO and LUMO involved e.g. nonbonding, antibonding etc.

Figure 8.5 illustrates the outcome of different types of orbital interactions. The blue lines represent typical orbitals of the well-separated reactants, and the green lines show how the energies of these orbitals change as a result of the interaction between them. According to the by now familiar pattern, the mixing between these orbitals results in two new orbitals, one lower in energy than either of the original orbitals, and one higher in energy.

In (a) the interaction between a filled orbital and an empty orbital is depicted. As a result of this interaction the two electrons move to a lower energy orbital, thus lowering the overall energy. In contrast, in (b) the two interacting orbitals are both full: one pair of electrons moves to lower energy and one to higher, resulting in no net lowering in the overall energy. Finally, in (c) the interaction is between two empty orbitals: since these orbitals are unoccupied, their interaction has no effect on the electron energy.

The HOMO–LUMO interaction is between a filled and an empty orbital, and so is like (a) and leads to a lowering of the energy. Of course there are other possible interactions between filled and empty MOs, but the HOMO–LUMO interaction is likely to be the most significant as these two orbitals will be most closely matched in energy.

For a given pair of reactants A and B there are always two possible choices of HOMO–LUMO interaction, as illustrated in Fig. 8.6. Either the HOMO of A can interact with the LUMO of B, as shown by the wavy line, or the HOMO of B can interact with the LUMO of A, as shown by the dashed line. The strongest interaction will be the one in which the HOMO and LUMO are closest in energy, which in this case is the one indicated by the wavy line.

8.2.1 Identifying the HOMO and LUMO

When trying to work out how two species might react, a useful approach is first to identify the HOMO and LUMO of each of the species since it is likely that a HOMO–LUMO interaction will be particularly important in determining the nature of the reaction. The rough energy ordering of MOs, given in section 4.7.4 on page 164 and repeated here for convenience, can be a useful guide for identifying the HOMO and LUMO.

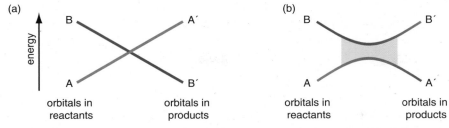

Fig. 8.7 Illustration of the effects of mixing between MOs. Orbitals A and B in the reactants become A′ and B′ in the products, and it is assumed that at some intermediate geometry the energies of the two MOs become equal. If the MOs do not have the same symmetry, this *crossing* has no consequences, as shown in (a). If, however, the MOs have the *same* symmetry, there is extensive mixing between them and a crossing is not permitted, as shown in (b). As a result of the mixing, orbital A takes on some of the character of B, and *vice versa*.

> *highest energy* σ^* antibonding
> π^* antibonding
> nonbonding orbitals (including lone pairs)
> π bonding
> *lowest energy* σ bonding.

We will often find that the highest energy electrons are lone pairs on heteroatoms such as oxygen or nitrogen. The LUMO is typically a nonbonding orbital (e.g. a vacant p orbital), a π^*, or a σ^* orbital. In choosing between different antibonding MOs it is useful to recall the discussion in section 4.7.1 on page 162 where we saw that the antibonding orbitals between carbon and a heteroatom (e.g. chlorine or oxygen) are usually *lower* in energy than those between two carbon atoms, or between a carbon and a hydrogen atom.

In a given molecule, there may be several candidates for the HOMO, which in turn may mean that there are a number of ways in which the molecule could react under certain circumstances. The same is true for the LUMO. Identifying these possible HOMOs and LUMOs is thus the first step in understanding reactions.

8.2.2   How orbitals change during a reaction

Using computer calculations, we can follow the way in which a particular MO in a reactant molecule changes during the reaction until it finally becomes an MO in a product molecule (for example, as shown in Fig. 8.3 on page 313). When we do this, we often see a steady and understandable change in the form of the orbital, but sometimes we see more 'unexpected' changes which can be hard to understand. This behaviour can be put down to *mixing* between the MOs which is occurring at a geometry intermediate between reactants and products.

We have already come across such mixing in section 3.3.4 on page 119 when we were looking at the MOs of homonuclear diatomics. What we saw there was when two orbitals of the *same symmetry* came close in energy there was an interaction between them which resulted in the energy of the lower energy orbital becoming even lower, and that of the higher energy orbital becoming even higher. In addition, mixing causes one orbital to take on some of the character (i.e. shape) of the other orbital.

Figure 8.7 on the preceding page illustrates how this mixing affects the MOs as they change during the course of a reaction. In (a) MO A in the reactant becomes MO A′ in the product, and similarly B becomes B′. The energies are such that at some intermediate geometry between reactants and products the MOs cross (i.e. have the same energy). If the two MOs have different symmetry, the crossing has no consequences, which is what is shown here.

On the other hand, if the two MOs have the same symmetry then it turns out that this crossing is 'forbidden', and what we see is the behaviour shown in Fig. 8.7 (b). In the zone shaded green there is extensive mixing between the two MOs, pushing A to lower energy and B to higher. In addition, MO A takes on some of the character of MO B, and *vice versa*. As a result, when we look at how MO A changes over the course of the reaction we will see changes due to the mixing in of B. Until we recognize that this is what is going on, this behaviour can be somewhat perplexing. If we do see 'unexpected' changes, then mixing is likely to be the culprit.

8.3   Interactions involving nonbonding LUMOs

In this section we are going to look at reactions in which two species come together to form a single compound, or adduct, and in which the LUMO is a nonbonding orbital. The species providing the LUMO in such a situation is typically a *Lewis acid*.

Lewis acids

A Lewis acid is a substance that can form a bond by accepting a pair of electrons. A simple example is H^+ which, as we have seen, can bond to water by the $1s$ AO (the LUMO) interacting with a lone pair on the oxygen. More complex species such as $AlCl_3$, BH_3, and BF_3 can also be classified as Lewis acids.

The LUMO of BH_3, illustrated in Fig. 8.8 (a), is a vacant p orbital on the boron. For molecules such as $AlCl_3$ and BF_3, the p orbital from the central atom is the major contributor to the LUMO, but there are some smaller contributions from the halogen p orbitals, as is illustrated by the LUMO of BF_3 shown in Fig. 8.8 (b). However, for the purposes of simplicity, we shall just treat the LUMO of $AlCl_3$ and BF_3 as being a vacant p orbital.

If a Lewis acid such as $AlCl_3$ or BF_3 combines with a species containing a high-energy pair of electrons (such as a lone pair), then a new bond may be formed. Initially, the high-energy electrons are localized on one atom, called the donor atom. However, in the adduct, the electrons help contribute towards the bonding between the donor atom and the acceptor, the Lewis acid.

8.3.1   The reaction of BH_3 with H^-

A simple example of the reaction between a donor and a Lewis acid is that between a hydride ion, H^-, and borane, BH_3. The highest energy electrons of either the hydride or the borane are the electrons in the $1s$ AO of the hydride. The negative charge on this ion means these electrons are higher in energy than the $1s$ orbital of a neutral hydrogen atom. The lowest energy unoccupied MO of either species is the vacant p AO on the boron. As the hydride approaches the borane, the electrons that were initially localized on the hydride become shared between the hydrogen and the boron.

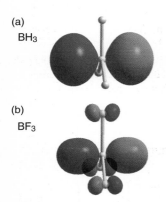

(a)
BH_3

(b)
BF_3

Fig. 8.8 Surface plots of (a) the LUMO of BH_3 and (b) the LUMO of BF_3. In BH_3 the LUMO is a p orbital on the boron, but in BF_3 there are some small contributions from the halogen p orbitals.

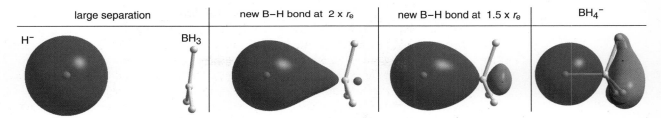

Fig. 8.9 Illustration of how the highest occupied orbital in the BH_3 / H^- system gradually changes as the hydride approaches the borane. At a large separation, the HOMO is the $1s$ AO of the hydride. In BH_4^- the HOMO is a bonding orbital in which a p orbital of the boron interacts with the four hydrogens. There is a gradual transition between the two orbitals as the reaction proceeds from reactants to products, as shown by the two intermediate geometries.

Figure 8.9 shows how the highest occupied orbital changes form during this interaction. When the hydride ion and BH_3 are separated by a large distance, the orbital is essentially a $1s$ AO on the hydrogen. As the two species come together, electron density becomes shared between the hydrogen and the boron. In the final product, BH_4^-, the electrons are now in an orbital which bonds the boron atom to all four of the hydrogens.

The early stages of this interaction can be thought of in terms of the interaction between the HOMO, the $1s$ on the hydride, and the LUMO, the $2p$ on the borane; these orbitals are shown in cartoon form in Fig. 8.10. The interaction between these two orbitals gives a bonding MO in which a small amount of the $2p$ AO has overlapped in-phase with the $1s$ AO. This is precisely what we see in the second and third frames of Fig. 8.9.

The LUMO of the borane is just a p orbital, which in BH_3 makes no contribution to the B–H bonding. However, in the product, BH_4^-, this p orbital does contribute to the overall bonding between the boron and hydrogens. During the course of the reaction, we can say that the hybridization of the boron changes from sp^2 in BH_3 to sp^3 in BH_4^-.

Curly arrow mechanism

The curly arrow mechanism for the reaction between H^- and BH_3 is shown below.

This mechanism suggests that the electrons initially associated with the hydride form the new B–H bond in the product. Whilst this is not entirely consistent with our delocalized view of the bonding, it is certainly a useful way to think of the reaction, and is consistent with the overall outcome.

Note that in the product the negative charge is shown associated with the boron. This is merely a formal charge – with four bonds, the boron is counted as having four electrons, one more than it needs to be neutral. However, calculations suggest the bulk of the charge is associated with the four hydrogens rather than with the boron. A more realistic, but cumbersome representation is shown in Fig. 8.11 (b).

Weblink 8.2
Follow this link to see a 'movie' version of Fig. 8.9.

hydride
HOMO

borane
LUMO

Fig. 8.10 A cartoon showing the HOMO–LUMO interaction between hydride and borane.

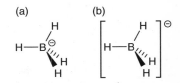

Fig. 8.11 Two different representations of BH_4^-. In (a) the negative charge is shown associated with the boron. However, this is just a formal charge and is misleading in that calculations show the bulk of the charge is spread over the hydrogens. A more accurate representation is shown in (b) which emphasizes that the charge is spread over the whole ion.

(a)

(b)

Fig. 8.12 Illustration of the HOMO–LUMO interactions for two different angles of attack by H^- on BH_3. In (a) the hydride approaches the face of the borane, resulting in positive overlap between the orbitals. In (b) the approach is in the plane of the B–H bonds; this results in no net interaction as the positive and negative overlap cancel one another out.

Angle of attack

One last point that comes out of a consideration of the orbitals involved during the reaction between H^- and BH_3 concerns the direction from which the hydride approaches the borane. This is usually described as the direction or angle 'of attack'.

Since the early stages of the reaction can be thought of in terms of the HOMO–LUMO interaction, the preferred angle of attack is generally the one which gives the best overlap between the HOMO and LUMO. Figure 8.12 illustrates the orbital overlap for two different angles of attack. In (a) the H^- approaches one face of the BH_3 resulting in positive overlap of the two orbitals. In (b) the H^- approaches sideways on i.e. in the plane of the existing bonds in the BH_3 molecule. Now there is positive overlap with one lobe of the p orbital, and negative overlap with the other, resulting in no net overlap, and hence no net interaction between the two orbitals. It is clear, therefore, that (a) is the preferred direction of attack.

We shall explore the importance of orbital interactions in determining the optimum angle of approach between two reactants in more complicated reactions later in the chapter.

8.3.2 Nucleophiles, bases, acids and electrophiles

In the reaction between hydride and borane, we described the borane as a Lewis acid – that is a species capable of accepting a pair of electrons. The hydride, the donor of the pair of electrons, could likewise be described as a *Lewis base*. However, it would be more usual to describe the hydride as a *nucleophile*. A nucleophile (which literally means 'nucleus loving') is a species which takes part in reactions by donating a pair of electrons. Similarly, the borane could be described as an *electrophile* ('electron loving').

Although there is little distinction between the terms electrophile and Lewis acid, the latter tends to be used when the species in question forms an adduct. Lewis acids are often used in organic chemistry to enhance the reactivity of a given species in the same way that a proton, H^+, may be used.

A proton is, not surprisingly, just a special example of a Lewis acid. An electron donor which reacts specifically with a proton is usually called a base, and the term nucleophile is reserved for when the electron donor reacts with other atoms, especially carbon.

In inorganic chemistry, the term *ligand* is used for electron donors which bond to metal ions.

8.3.3 The protonation of ethene

In the reaction between hydride ion and borane, both the HOMO of the hydride and the LUMO of the borane were nonbonding orbitals. The interaction between the two gave rise to a new bonding orbital. In the next example, the protonation of ethene, the HOMO is a bonding orbital. When this interacts with the LUMO we shall see that as the new bond forms, an existing bond breaks.

In strong aqueous acid, alkenes may be hydrated to give alcohols. The first step in this reaction is the protonation of the alkene as shown below. The reaction forms a new bond between carbon and hydrogen, but breaks the C=C π bond.

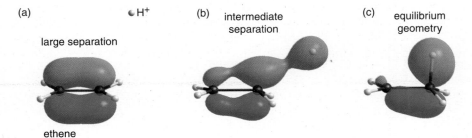

Since the proton has no electrons, the HOMO can only be one of the ethene MOs. In section 4.4.2 on page 153 we saw how we can describe the bonding in ethene using hybrid atomic orbitals. In this description, the HOMO is the π bonding MO. The LUMO of the proton is the vacant $1s$ orbital.

Figure 8.13 shows the how the π bonding MO of ethene changes as a proton approaches. In (a), the proton is sufficiently far away from the ethene not to affect it. The orbital shown is just the π MO of ethene; note that this orbital does not contribute to the C–H bonding in ethene. As the proton approaches the alkene, electron density from the π MO begins to envelope the proton, as shown in (b). The electron density between the two carbons decreases, but that between the right-hand carbon and the incoming proton increases. Finally, in the resulting ion shown in (c) there is little electron density between the two carbon atoms. The MO can be viewed as arising from the overlap of a carbon p orbital with the $1s$ AOs on the three attached hydrogens; this MO contributes to the bonding between these atoms.

> ● *Weblink 8.3*
>
> Follow this link to see a 'movie' version of Fig. 8.13.

(a)   ● H⁺   (b)   intermediate separation   (c)   equilibrium geometry

large separation

ethene

Fig. 8.13 Iso-surface plots of the HOMO during the reaction between H⁺ with ethene. In (a) the proton is far away from the ethene, so the HOMO is simply the π MO of ethene. At the intermediate separation, (b), the proton draws electron density away from the region between the two carbons. In the resulting ion (c), the HOMO is involved in bonding between the right-hand carbon and the three hydrogens attached to it.

Fig. 8.14 A curly arrow mechanism for the protonation of ethene. The arrow in (a) shows the electrons from the π bond moving to form a new bond to the proton. The alternative representation in (b) emphasizes that the electrons form a new bond between the right-hand carbon and the proton.

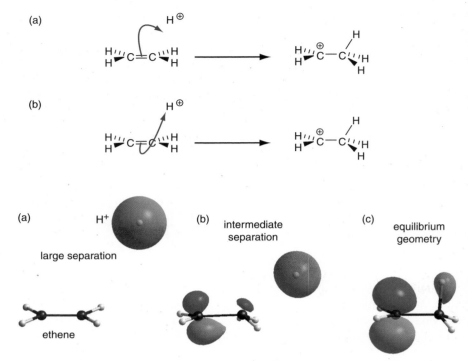

Fig. 8.15 Iso-surface representations of the LUMO in the H$^+$ + ethene reaction. Initially, as shown in (a), the LUMO of either reactant is the vacant $1s$ AO of the incoming proton. In the product, shown in (c), the LUMO is essentially just a vacant p orbital on the left-hand carbon. In the intermediate stages, such as that shown in (b), the LUMO has contributions from both the carbon and the hydrogen.

Curly arrow mechanism

Once again, the redistribution of electron density may be summarized using curly arrows, as shown in Fig. 8.14. The arrow in (a) shows a pair of electrons from the π bond moving to form the new bond to the incoming proton. The alternative curly arrow shown in (b) is drawn through the right-hand carbon to emphasize that it is this atom to which the proton becomes attached. Either representation is acceptable.

Note that in Fig. 8.14 the product has a formal positive charge on the left-hand carbon atom. This arises as the carbon has three bonds and no lone pairs, resulting in an electron count of three – one less than is needed to make the atom neutral. We could view the incoming proton as taking electron density away from this carbon, using the electrons from the π bond to form the new C–H σ bond.

As it only has three bonds, we can think of the left-hand carbon as being sp^2 hybridized with a vacant p orbital. In fact the LUMO of the ion is essentially the vacant p orbital centred on this carbon. Figure 8.15 shows how the LUMO gradually changes during the protonation of ethene. Initially, the LUMO is the vacant $1s$ AO of the incoming proton as shown in (a), and in the product it is essentially just a vacant $2p$ orbital on carbon as shown in (c). In the intermediate stages of the reaction, exemplified by (b), the orbital has contributions from both the carbon and the incoming hydrogen. The LUMO of the product is shown in cartoon form in Fig. 8.16.

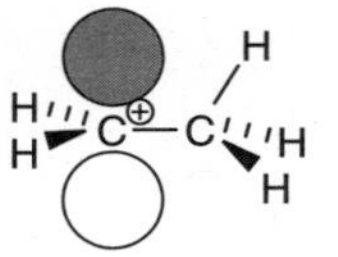

Fig. 8.16 A cartoon representation of the LUMO in protonated ethene. It is essentially a vacant p orbital.

8.4   Interactions involving π antibonding LUMOs

In the previous section, the reactions involved the donation of electrons into nonbonding vacant orbitals of electrophiles, resulting in the formation of a new bond to the electrophile. If the LUMO is *antibonding* rather than nonbonding, the interaction will result in a bond in the electrophile being broken. We will illustrate this first by showing how the interaction with a π antibonding MO results in the breaking of the π bond.

8.4.1   Nucleophilic attack on a carbonyl

A very important reaction in organic chemistry involves the attack of a nucleophile on a carbonyl group. The nucleophile might be an anion such as cyanide, or hydroxide, or it could be a neutral species such as water or an amine. The general reaction can be summarized as

The key points of this reaction are as follows.

- A new bond is formed between the nucleophile and the carbon of the carbonyl.

- The C=O double bond becomes a single bond.

- The geometry about the central carbon changes from trigonal planar to tetrahedral.

- The nucleophile loses a negative charge (either going from negative to neutral, or from neutral to positive).

- The oxygen becomes negatively charged.

Methanal and H$^-$

The simplest possible example of the reaction between a nucleophile and a carbonyl is the reaction between a hydride ion and methanal (formaldehyde)

The carbonyl bond is polarized with a partial negative charge on the oxygen and a partial positive charge on the carbon. As the hydride ion approaches the methanal, there is an electrostatic attraction between the negatively charged hydride and the carbon of the methanal. However, as the two species get closer, orbitals start to interact. The highest energy orbital which is occupied in either the methanal or the hydride is the $1s$ orbital of the hydride. The unoccupied orbital with the lowest energy is the $\pi^\star$ of the methanal, pictured in Fig. 8.17.

(a)

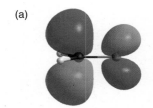

Fig. 8.17 Iso-surface representations of the vacant $\pi^\star$ orbital of methanal; this is the LUMO.

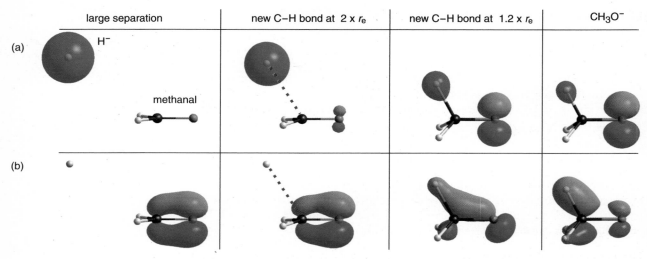

| large separation | new C–H bond at 2 x r_e | new C–H bond at 1.2 x r_e | CH_3O^- |

(a) H⁻

methanal

(b)

Fig. 8.18 The evolution of two key orbitals during the attack of hydride ion on methanal. Shown in (a) is the HOMO at various stages of the reaction. At a large separation the orbital is just the hydride $1s$ AO, whereas in the product it is mainly just a p orbital on the oxygen. At intermediate stages of the reaction it changes between these two extremes. Shown in (b) is the evolution of the C=O π bonding MO of methanal into what is essentially a C–H σ bonding MO in the product.

🌐 *Weblink 8.4*

Follow this link to see a 'movie' version of Fig. 8.18.

Figure 8.18 (a) shows how the HOMO of the whole system changes during the course of the reaction. Initially, with a large separation between the two species, this orbital is just the $1s$ orbital of the hydride. When the forming C–H bond is at twice the distance that it ends up at in the product, the orbital is still mainly the hydrogen $1s$, but there is also a small contribution from the $\pi^\star$ MO of methanal. This contribution is most visible on the oxygen atom. As the hydride approaches closer still, the contribution on the oxygen increases. In the product, this orbital has most electron density around the oxygen atom, and so we would describe it as essentially being a lone pair on oxygen.

Figure 8.18 (b) shows the evolution of the orbital which starts off as the π MO of methanal. We see that during the reaction electron density shifts from this orbital to help form the new C–H bond. It might seem rather strange that the π MO, which initially has a greater contribution from the oxygen p orbital, ends up essentially giving rise to the C–H bond, whereas in the HOMO the electrons that were initially associated with the hydride end up being mainly associated with the oxygen as a lone pair. The reason for this is that in addition to the HOMO–LUMO interaction there is also strong mixing between different MOs, as described in section 8.2.2 on page 317. The extent of this mixing is only revealed by a full calculation of the type used to produce Fig. 8.18 – it is not easy to predict using qualitative arguments.

Note that the initial interaction between the hydride and methanal can be seen as resulting in electrons being placed into the $\pi^\star$ MO of the carbonyl. We should expect that putting electrons in the antibonding orbital will 'cancel out' the effects of the occupied bonding orbital, thus breaking the π bond. This is found to be the case, although the form of the orbitals changes so much during the reaction it is no longer sensible to describe the product in terms of the original π and $\pi^\star$ orbitals.

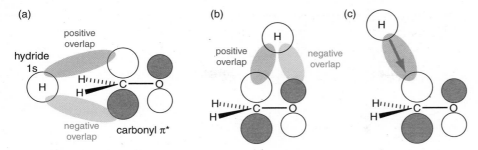

Fig. 8.19 If the hydride approaches the methanal in the plane of the molecule, as in (a), there is no net overlap between the hydride $1s$ AO and the $\pi^{\star}$ MO of the methanal. Similarly, an approach as shown in (b) leads to positive and negative overlap. The most favourable line of approach is as shown in (c) which maximizes the positive overlap between the two MOs.

Curly arrow mechanism

The net changes that take place during the reaction are conveniently summed up using curly arrows

The arrow starting on the negative charge of the hydride and finishing on the carbonyl carbon suggests that the electrons initially associated with the hydride go towards the formation of the new C–H bond. The arrow starting in the middle of the C=O bond and ending up on the oxygen suggests that the electrons in the π bond end up associated on the oxygen as a lone pair. This is not consistent with how the MOs change during the course of the reaction since these indicate that the electrons that were initially associated with the hydride end up on the oxygen, whereas the electrons that were in the π bonding MO end up contributing to the new bond to the hydrogen. However, the net result is the same and ultimately, since the electrons are indistinguishable, it is not meaningful to say that one pair of electrons ends up in any particular place; the electrons are delocalized over the entire molecule.

Angle of attack

An appreciation of the orbital interactions involved also helps us to understand the best angle of attack for the hydride to take as it approaches the carbonyl. Three different possibilities are illustrated in Fig. 8.19. In (a) the hydride approaches in the plane of the molecule and towards the carbonyl carbon. This approach is not favoured since the positive and negative overlap of the HOMO and LUMO cancel one another out leading to no net interaction. Similarly, if the hydride approaches onto the face of the molecule, and between the carbon and oxygen, as shown in (b), there is again little net interaction due to the cancellation between positive and negative overlap. It is clear that positive overlap will be maximized by approaching at an angle, such as that shown in (c).

Detailed calculations and experimental evidence suggest the most favourable angle of attack is along a line inclined at approximately 109° from the carbonyl, as shown in Fig. 8.20. This maximizes the degree of positive overlap, and

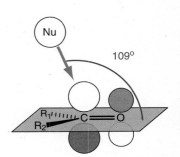

Fig. 8.20 The most favourable angle of attack for an incoming nucleophile reacting with a carbonyl is at approximately 109° as shown. This gives the most favourable overlap between the $\pi^{\star}$ and the HOMO of the nucleophile, and also minimizes the electronic repulsion between the π bond and the nucleophile.

also minimizes the repulsion between the filled π orbital and the incoming nucleophile.

8.5 Interactions involving σ antibonding LUMOs

In the previous section we saw how the donation of electrons into a $\pi^\star$ MO results in the C–O π bond being broken, but the σ bond remaining intact. In this section we shall see how the donation of electrons into a $\sigma^\star$ MO leads to bond breaking such that atoms or groups separate from each other. The type of reaction we shall look at is called *nucleophilic substitution* and is outlined below.

The following are the key points of this reaction.

- A new bond is formed between the nucleophile and the central carbon.

- As the C–X σ bond is broken, the leaving group X⁻ is released.

- The geometry around the carbon undergoes an inversion; in some sense it is 'turned inside-out' (like an umbrella).

To keep things as simple as possible, we will once again use H⁻ as the nucleophile, and investigate its reaction with CH_3Cl to give $CH_4 + Cl^-$. The overall HOMO of the two reactants is the $1s$ AO of the hydride and, as we saw in section 4.7.1 on page 162, the LUMO of an alkyl halide RX is the C–X $\sigma^\star$ orbital. Three different representations of this MO in chloromethane (methyl chloride) are shown in Fig. 8.21.

Figure 8.22 on the next page shows how two of the key MOs change during the course of the reaction. The behaviour of the HOMO, which starts out as the $1s$ AO on the hydride, is shown in (a). As the hydride approaches the methyl chloride the HOMO starts to include a small contribution from the LUMO of the methyl chloride, whilst the contribution from the $1s$ decreases. As the hydride approaches closer still we note that the HOMO is more and more dominated by the contribution from the chlorine until eventually, when we reach the product, the HOMO is just a $3p$ AO on the chloride ion.

Curly arrow mechanism

In the reaction, a new σ bond forms between the hydrogen and the carbon, and at the same time the C–Cl σ bond breaks. This is illustrated by the evolution of the MO shown in Fig. 8.22 (b). In the reactants, the MO is responsible for the C–Cl bond, but as the hydride approaches we see the build-up of electron density between the carbon and hydrogen, and the reduction of electron density between the carbon and chlorine. In other words, the making of the C–H bond and the breaking of the C–Cl bond.

(a)

(b)

(c)

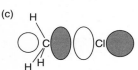

Fig. 8.21 Three different representations of the LUMO of chloromethane, which is the C–Cl $\sigma^\star$ MO: (a) a contour plot, (b) an iso-surface plot, and (c) a simple cartoon sketch.

🌐 *Weblink 8.5*

Follow this link to see a 'movie' version of Fig. 8.22.

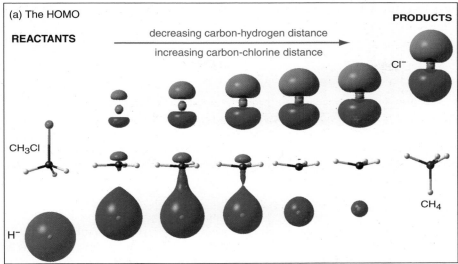

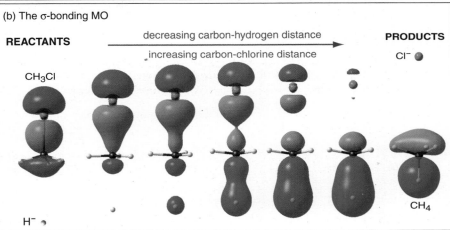

Fig. 8.22 Iso-surface plots of two key occupied MOs involved in the H^- + CH_3Cl reaction. The HOMO during the reaction is shown in (a). At the start of the reaction, this is the $1s$ AO of the hydride ion and at the end it is a $3p$ orbital on the chloride ion; there is a gradual transition between these two extremes. In the early stages of the reaction we can recognize that a small contribution from the LUMO, the C–Cl $\sigma^\star$ MO depicted in Fig. 8.21, is mixed into the HOMO. The evolution of the MO which starts out as the C–Cl σ bonding MO is shown in (b). Again we can see a gradual change of this MO from being a C–Cl bonding MO to being a C–H bonding MO. As a result, the C–Cl bond is broken, and a C–H bond is formed.

As usual, we can sum up the redistribution of electrons using curly arrows

The first curly arrow starts at the electrons of the hydride ion and ends up near the carbon. This arrow represents the electrons that are contributing to the new C–H bond. The second curly arrow starts in the middle of the C–Cl σ bond and ends up on the chlorine. This represents both the electrons that were in the C–Cl bond moving off with the chlorine to form the chloride ion.

As we have seen before, the redistribution of the electrons implied by the curly arrows is not the same as that predicted by the behaviour of the orbitals. However, the curly arrows do give an accurate representation of the outcome of the reaction, which is the formation of a new C–H bond and the breaking of a C–Cl bond.

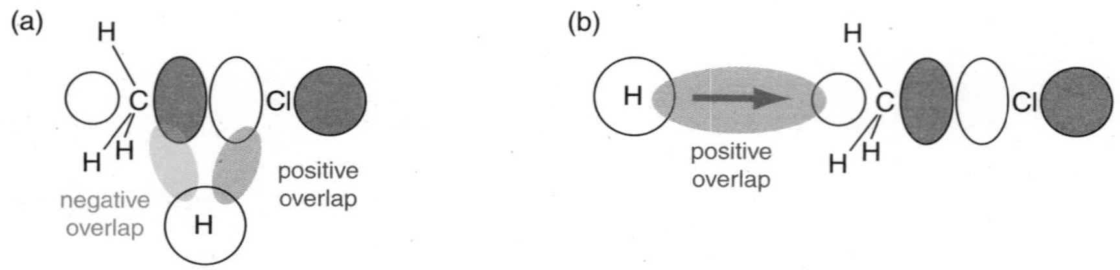

Fig. 8.23 Illustration of the HOMO–LUMO interactions for two possible angles of attack of H⁻ on CH₃Cl. In (a) the hydride attacks from the side, and as a result there is positive and negative overlap between the $1s$ AO and the C–Cl $\sigma^\star$ LUMO; the overall interaction is therefore weak. In contrast, approach from the rear, as shown in (b), leads to positive overlap and hence a favourable interaction.

Angle of attack

As with previous examples, a consideration of the orbitals involved helps us to understand part of the reason why the reaction proceeds with the geometry it does. In the initial stages of the reaction, the negatively charged hydride is attracted to the carbon which bears a partial positive charge due to the effects of the electron-withdrawing chlorine. Figure 8.23 (a) shows that if the hydride were to approach side-on to the C–Cl bond there would be both positive and negative overlap between the HOMO and LUMO, leading to an overall weak interaction. If, on the other hand, the hydride approaches the carbon from the opposite side to the chlorine, as shown in (b), there is net positive overlap. This geometry is therefore preferred.

8.6 Summary of the effects of different HOMO–LUMO interactions

Before moving on, it is useful to summarize the important points from the discussion so far.

- During the course of a reaction, the electrons which bond the atoms together become rearranged. This rearrangement may be visualized by following how the molecular orbitals change.

- Only the highest energy valence electrons play a significant part in a reaction; core electrons are not involved. In many cases, a useful starting point is to identify the electrons which have the highest overall energy in either reactant. Lowering the energy of these can provide a strong driving force for the reaction to occur.

- The interaction of a filled orbital with a vacant orbital lowers the energy of the electrons in the filled orbital. A particularly important interaction is between the HOMO of one reactant with the LUMO of the other.

- New bonds may be formed when electrons are donated from nonbonding or bonding molecular orbitals into vacant nonbonding or antibonding orbitals. Donation from a bonding orbital breaks an existing bond as the new bond forms. Donation from a nonbonding MO can form a new bond without breaking any existing ones. Putting electrons into an antibonding orbital breaks an existing bond, whereas putting electrons into a nonbonding orbital does not.

- A curly arrow mechanism provides a convenient way of summarizing the net redistribution of electrons between the reactants and products. However, it is important to realize that each arrow does not represent the behaviour of the electrons in a particular MO.

8.7 The role of protonation in reactions

Many reactions only proceed at reasonable rates after the starting material has been protonated or combined with a Lewis acid. For example the hydrolysis of ethyl ethanoate to ethanol and ethanoic acid occurs several thousand times faster as the acidity of the solution is increased from pH 7 to pH 1. The initial step in this reaction is the attack of the oxygen lone pair of a water molecule into the C=O $\pi^\star$ of the ester

This step is considerably faster if the carbonyl oxygen has first been protonated

The acid speeds up the rate of the reaction by increasing the reactivity of the starting material, the ester. In this section we shall briefly explore why this should be so.

It is easier for a water molecule to attack a protonated carbonyl than a neutral carbonyl for two reasons: firstly, the electrostatic interaction which brings the electron-rich oxygen of the water and the carbonyl carbon together is greater if the carbonyl is protonated; secondly, the LUMO into which the water is donating electron density, the carbonyl $\pi^\star$, is lower in energy if the carbonyl has been protonated, thus leading to a stronger interaction.

Adding a proton to the oxygen takes some electron density away from the oxygen, resulting in an increase in its effective nuclear charge. Despite the fact that we draw a protonated carbonyl with a positive charge on the oxygen, this is just a formal charge. The actual charge on the oxygen is still negative, but less so than for the neutral carbonyl.

Lowering the energy of the C–O $\pi^\star$ MO

We can understand why protonating the oxygen gives rise to a lower energy $\pi^\star$ MO by considering the MO diagram in Fig. 8.24. Increasing the Z_{eff} for oxygen lowers the energy of its atomic orbitals (shown in green) relative to the neutral oxygen (shown in purple). This means that there is a poorer energy match between the carbon and oxygen AOs, which in turn means that the energy of

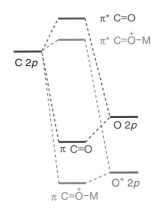

Fig. 8.24 Adding either a proton or metal cation (denoted by M) to the oxygen atom increases the effective nuclear charge of the oxygen, which lowers the energy of the oxygen AOs (shown in green) relative to what they were originally (shown in purple). This means there is an even greater difference between the energies of the carbon and oxygen orbitals, which results in the $\pi^\star$ orbital (red) being lower in energy than in the neutral carbonyl (blue).

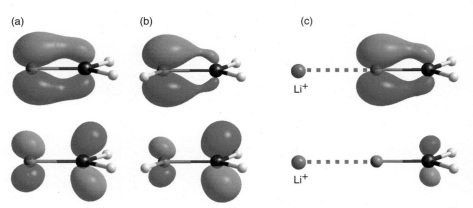

Fig. 8.25 Illustration of how the π and $\pi^\star$ MOs of methanal are affected by protonation or coordination to a cation. The upper diagram in (a) shows the π MO for methanal, and the lower diagram the $\pi^\star$ orbital. In (b) the oxygen has been protonated. The bonding MO now shows an increased contribution from the oxygen, whilst the contribution from the carbon has decreased. In the $\pi^\star$ orbital beneath, the contribution from the carbon has increased and that from the oxygen has decreased. A similar trend is seen in (c) when the oxygen is coordinated to a Li$^+$ ion.

the $\pi^\star$ MO is not raised as much above that of the carbon AO as in the case of the neutral carbonyl.

In a carbonyl group we expect that the π MO will have a greater contribution from the oxygen than from the carbon, and that the opposite will be the case for the $\pi^\star$ MO. This comes about because the oxygen AO is lower in energy than carbon AO. Protonating the oxygen further lowers the energy of its AO, and so we expect that the π MO will have an even greater contribution from the oxygen, and the $\pi^\star$ MO an even greater contribution from the carbon. The plots of the π MOs shown in Fig. 8.25 illustrate this point by comparing the π MOs of methanal, shown in (a), with those of protonated methanal, shown in (b). Coordination of the oxygen by a Li$^+$ ion has a similar effect to protonation, as shown in (c).

The important point is that when trying to understand or predict how two reagents react together, protonation or binding to a Lewis acid may increase the reactivity of a carbonyl quite significantly. It does this by increasing the charge separation, and also by lowering the energy of vacant antibonding orbitals.

Representations of the protonated carbonyl

We can also think about the effect of protonating the carbonyl oxygen using resonance structures, as is illustrated in Fig. 8.26 on the facing page. For methanal the two important resonance structures are shown in (a) and (b). Structure (a) has a π bond in which the electrons are shared *equally* between the carbon and oxygen. In structure (b) the two electrons which were in the π bond have now been localized on the oxygen, giving it a negative charge and leaving a positive charge on the carbon. It is important to remember that neither of these structures is an adequate description: the 'real' structure lies somewhere between the two.

The MO picture of the bonding in methanal (Fig. 8.25) tells us that the π bonding MO has a greater contribution from the oxygen, and that there is therefore greater electron density on the oxygen. In terms of the resonance

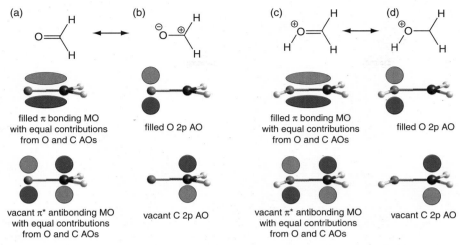

Fig. 8.26 Resonance structures, and the corresponding orbitals, for methanal and protonated methanal. For methanal, structure (a) implies an equal sharing of the electrons in the π bond, whereas structure (b) is the opposite extreme in which there is no π bond and the electrons are localized on the oxygen. The real structure lies between these two extremes. The fact that we know that the π MO in methanal is polarized towards the oxygen implies that there must be some contribution from (b). The two resonance structures for protonated methanal are shown in (c) and (d). The greater polarization of the π MO in this ion implies that (d) makes a larger contribution for protonated methanal than (b) does for methanal i.e. the structure of protonated methanal lies more towards (d) than the structure of methanal lies towards (b).

structures, this implies that there is some contribution from structure (b). Following on this line of thought, we can imagine that the π MO arises from a combination of the symmetrical π MO shown beneath structure (a) with a small contribution from the oxygen $2p$ AO shown beneath (b). Likewise, we can think of the $\pi^\star$ MO as arising from a combination of the symmetrical $\pi^\star$ MO shown beneath (a) with a small contribution from the carbon $2p$ AO shown beneath (b).

The corresponding two resonance structures for protonated methanal are shown in Fig. 8.26 (c) and (d). The MO picture shows that the π bond is more polarized towards the oxygen than in methanal itself, and this implies that structure (d) makes a greater contribution in protonated methanal than does (b) in methanal. Spectroscopic evidence confirms these predictions, since it is found that on protonation the C–O bond becomes weaker and the carbon becomes even more positively charged.

8.8 Intramolecular orbital interactions

In the previous reactions we have seen how the HOMO from one species can interact with the LUMO from another. For example, to understand the reaction between methylamine, CH_3NH_2, and chloroethane, CH_3CH_2Cl, we identify the HOMO to be the nitrogen lone pair in the methylamine, as shown in Fig. 8.27 (a), and the LUMO to be the C–Cl $\sigma^\star$ in chloroethane, as shown in (b). The reaction between the two involves the nitrogen lone pair attacking into C–Cl $\sigma^\star$. This is a nucleophilic substitution reaction, of the same type as the H^- + CH_3Cl

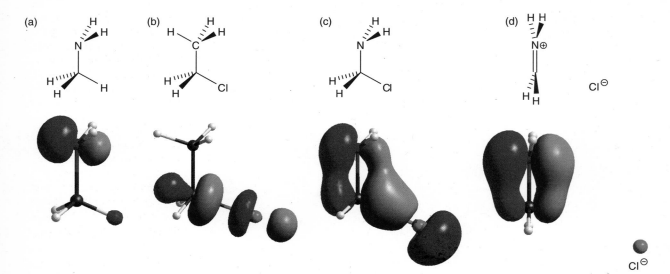

Fig. 8.27 Iso-surface plots of (a) the HOMO of methylamine, which is a lone pair on the nitrogen, and (b) the LUMO of chloroethane, which is the C–Cl $\sigma^\star$ MO. When an NH_2 group and chlorine atom are attached to the *same* carbon atom the HOMO, shown in (c), can be thought of as arising from a combination of the two orbitals shown in (a) and (b). The interaction between these two MOs results the C–N bond being strengthened and the C–Cl bond being weakened. Such an interaction may eventually lead to the formation of a C=N double bond and the elimination of a chloride ion; (d) shows the HOMO of the product.

reaction. We will now look at what happens if *both* these components are in the *same* molecule.

With both the NH_2 group and Cl attached to the same carbon, the nitrogen lone pair and the C–Cl $\sigma^\star$ can interact with each other, as shown in Fig. 8.27 (c). The result of this interaction is to strengthen the C–N bond, and at the same time weaken the C–Cl bond. This orbital shows considerable π character and in fact the molecule is well on its way to forming a full double bond between the carbon and nitrogen, whilst breaking the C–Cl bond to give the products shown in (d). We can show this interaction with a curly arrow mechanism

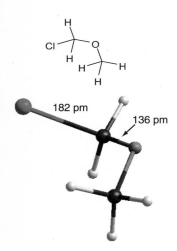

Fig. 8.28 The crystal structure of CH_3–O–CH_2Cl shows that the C–Cl bond is somewhat longer than typical values, whereas the indicated C–O bond is somewhat shorter than expected. Both effects can be attributed to an interaction between the oxygen lone pair and the C–Cl $\sigma^\star$ MO.

In fact, it turns out that no compounds can be isolated with both an NH_2 group and a chlorine attached to the same carbon. The interaction shown above results in any such compounds falling apart and forming a double bond between the carbon and nitrogen, and eliminating a chloride ion.

It is certainly not always the case that this kind of interaction between MOs within a molecule will lead to it falling apart. However, a consideration of which orbitals can interact within a molecule can often help us understand the bonding in that molecule. For example, in chloromethyl methyl ether (CH_3–O–CH_2Cl),

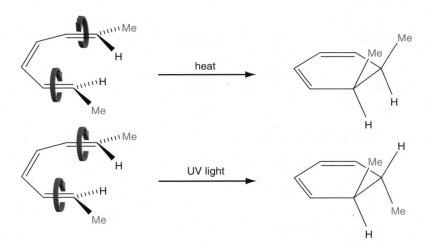

Fig. 8.29 When heated, *trans-cis-trans*-octatriene gives the substituted cyclohexadiene with the methyl substituents on the same face of the ring, whereas when irradiated with UV light, it forms the compound with the two methyl groups on opposite faces of the ring. In the first case, the product is formed by rotating the two bonds indicated in opposite directions (one clockwise, one anticlockwise); in the second case it is formed by rotating them in the same direction.

shown in Fig. 8.28 on the facing page, the length of the indicated C–O bond is 136 pm, significantly shorter than the typical value of 143 pm for such bonds, whereas the C–Cl bond length of 182 pm is longer than the typical value of 177 pm. These changes can be attributed to an interaction between the oxygen lone pair and the C–Cl $\sigma^\star$ MO leading to a shortening and strengthening of the C–O bond, and a lengthening and weakening of the C–Cl bond.

8.9   Rearrangement reactions

Sometimes it is not really appropriate to think of a reaction in terms of HOMO–LUMO interactions between two species. This is particularly true when molecules undergo rearrangements. However, the reaction can still be followed, and indeed is best understood, by considering how the molecular orbitals change during the course of the reaction.

One class of reaction where the reactions are best understood by following the changes in the relevant MOs are *pericyclic* reactions. These are reactions in which all the bonds being formed or broken lie in a ring. The reaction may involve one species or two, and need not necessarily involve a cyclic compound so long as all the forming and breaking bonds make up a ring. As an example, we shall look at the cyclization of *cis*-1,3,5-hexatriene to form 1,3-cyclohexadiene

heat or UV light

cis-1,3,5-hexatriene   1,3-cyclohexadiene

During this reaction, a new σ bond is formed between the end carbons and the π bonds become rearranged. The particularly interesting thing about the reaction is that when there are different substituents on the terminal carbons which allow us to follow what is happening more closely, we see that the reaction proceeds in a subtly different manner depending on whether it is promoted by heat or UV light. Figure 8.29 shows this different behaviour for the case where there are methyl groups on the terminal carbons. We see that when the starting material is heated, as in (a), the product is formed by rotating the two bonds shown in *opposite* directions so the two methyl groups end up on the *same* face of the six-membered ring. In contrast, when the starting material is irradiated with UV light, as in (b), the product is formed by rotating the two bonds in the *same* direction so the methyl groups end up on *different* faces in the product.

Before molecular orbital theory had been fully developed, such different outcomes were impossible to explain. However, by considering how the orbitals change during the reaction it is possible to understand what is going on here.

The π MOs of 1,3,5-hexatriene

In section 4.8 on page 165 you saw how to predict the form of the π MOs of conjugated hydrocarbons using a simple rule; using this we can easily find the π MOs of the six *p* orbitals in a row for 1,3,5-hexatriene; these are shown in Fig. 8.30. Figure 8.31 shows how these π MOs of hexatriene change during the heat-induced cyclization reaction. The MOs of the starting material *cis*-1,3,5-hexatriene are shown on the left and the corresponding MOs of the product, 1,3-cyclohexadiene, on the right. The orbitals for the intermediate stages of the reaction are shown in between. For the starting material, the three MOs are arranged in order of increasing energy, whereas for the product, the MO B′ is now the HOMO, MO C′ is the second highest occupied MO and MO A′ the third highest.

Twisting the end two carbons towards each other brings the end *p* orbitals towards one another. In MO A and MO C, the contributions from the *p* AOs on the end carbons interact in-phase, whereas for MO B, they interact out-of-phase. There is a strong mixing interaction between MO A and MO C during the

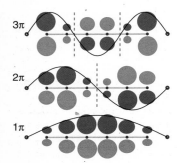

Fig. 8.30 A schematic representation of the occupied π MOs of 1,3,5-hexatriene as predicted using the sine-wave rule. The computer calculated orbitals are shown on the left of Fig. 8.31. Note that the contributions from the central two carbons in MO2 are so small, they no longer show up.

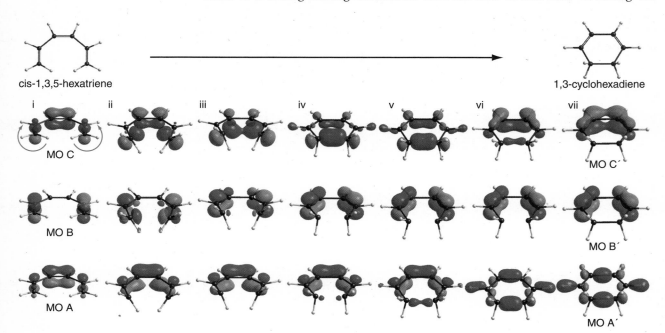

Fig. 8.31 The evolution of the occupied π orbitals during the thermal cyclization of hexatriene. The π MOs of the starting material are shown on the left and the corresponding orbitals in the product are shown on the right. The orbitals at several key stages during the reaction are shown in between.

transformation: whereas it initially appears that MO C will give rise to the new σ bond between the two terminal carbons, the mixing during stages (iv), (v), and (vi) shows that it is MO A that finally gives rise to the new σ bond in the product. Figure 8.32 shows how the energies of the MOs change during the reaction and also illustrates the mixing that takes place between MOs A and C.

Curly arrow mechanism

Whilst we can draw a curly arrow mechanism which summarizes the redistribution of electrons during the course of the reaction, the MOs show how inappropriate they are for this reaction, even more so than in the previous reactions we have looked at.

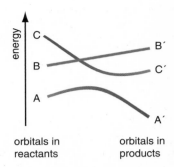

Fig. 8.32 A schematic graph to show how the energies of the MOs shown in Fig. 8.31 change during the reaction. MOs A and C have the same symmetry and so rather than crossing, interact with each other with MO A taking the form of MO C and *vice versa*. MO B does not have the same symmetry and crosses over the energy of MO C to become the HOMO in the product.

Since the reaction is symmetrical, we could equally well draw the curly arrows in a clockwise or anti-clockwise direction. It is not the case that one particular π bond forms the σ bond – the electrons in the whole molecule are rearranged simultaneously. While somewhat misleading in detail, the curly arrows are still useful in summarizing the overall changes that take place.

8.10 Moving on

In this chapter we have seen how orbitals change during the bond making and breaking which takes place in some simple reactions. We have also seen how identifying key orbitals can help in understanding how two molecules may react. In the next chapter we shall use these ideas to help rationalize some key reactions in organic chemistry.

QUESTIONS

8.1 By following the implications of the curly arrows, determine the products of the following reactions

(a)

$Cl^{\ominus}$ $H^{\oplus}$ ⟶

(b)

$\overset{\cdot\cdot}{N}H_3$ $H^{\oplus}$ ⟶

(c)

$F^{\ominus}$ BF_3 ⟶

(d)

$H-C\equiv C-H$ $H^{\oplus}$ ⟶

(e)

$N\equiv C$ ⟶

(f)

$Cl^{\ominus}$ H_3C-Br ⟶

(g)

$:OH$ ⟶

(h)

⟶

(i)

⟶

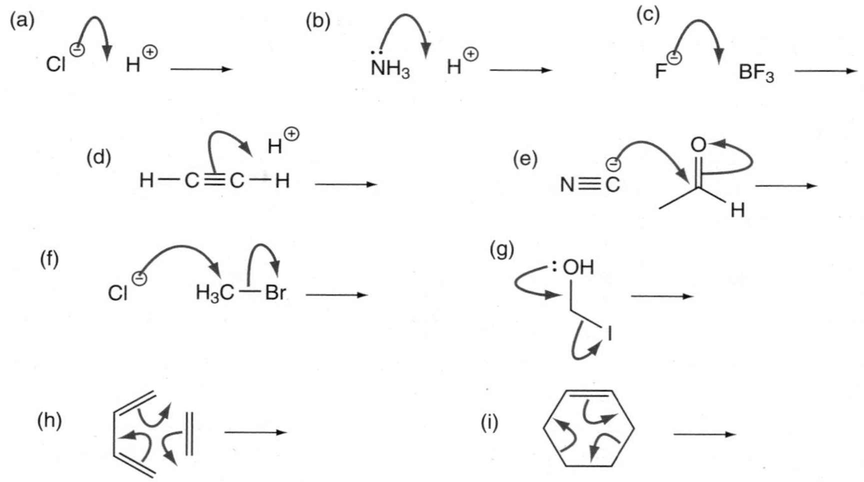

8.2 Draw the appropriate curly arrows for the following reactions (if your arrow originates from a lone pair, indicate this)

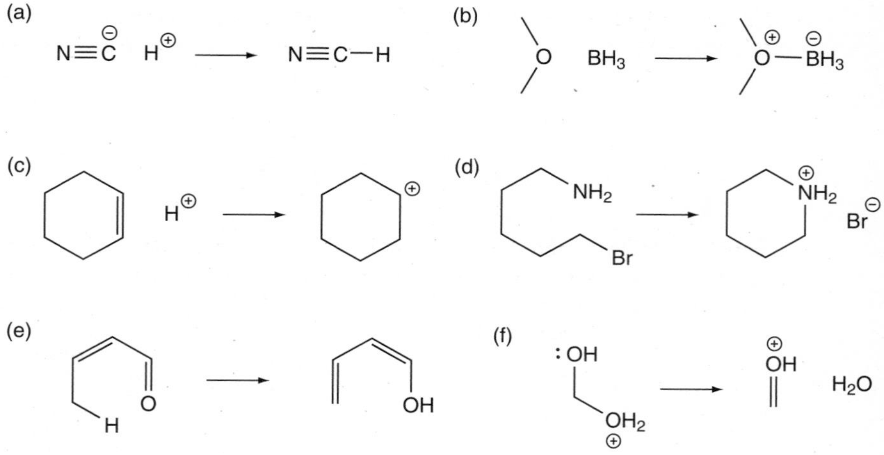

(a)

$N\equiv \overset{\ominus}{C}$ $H^{\oplus}$ ⟶ $N\equiv C-H$

(b)

O BH_3 ⟶ $\overset{\oplus}{O}-\overset{\ominus}{B}H_3$

(c)

$H^{\oplus}$ ⟶ $\oplus$

(d)

NH_2 Br ⟶ $\overset{\oplus}{N}H_2$ $Br^{\ominus}$

(e)

O H ⟶ OH

(f)

$:OH$ $\overset{\oplus}{O}H_2$ ⟶ $\overset{\oplus}{O}H$ H_2O

8.3 Explain how the formal charges on the oxygen and boron arise in the product of reaction (b) in the previous question, and how the formal charges on the oxygen atoms arise in both the reactant and product of reaction (f).

8.4 Assuming that there is no charge on the sulfur atom, use the approach described in Box 8.1 on page 315 to determine the number of lone pairs on the sulfur in the following molecules: (a) SF_2, (b) SF_4, (c) SF_6, (d) SO_3.

8.5   Dimethyl sulfoxide $(CH_3)_2SO$ is often depicted as

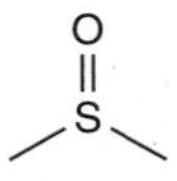

Use the fact that there is no formal charge on the sulfur to determine the number of lone pairs on that atom. Hence predict the shape of the molecule.

Draw an alternative structure in which there is a single bond between the sulfur and the oxygen, and determine any formal charges present.

8.6   Explain why it is likely that the most favourable interaction between two molecules will be between the HOMO of one and the LUMO of the other.

Identify the HOMO in each of these molecules and hence suggest the position at which each would be most easily protonated. Draw a curly arrow mechanism for the protonation in each case.

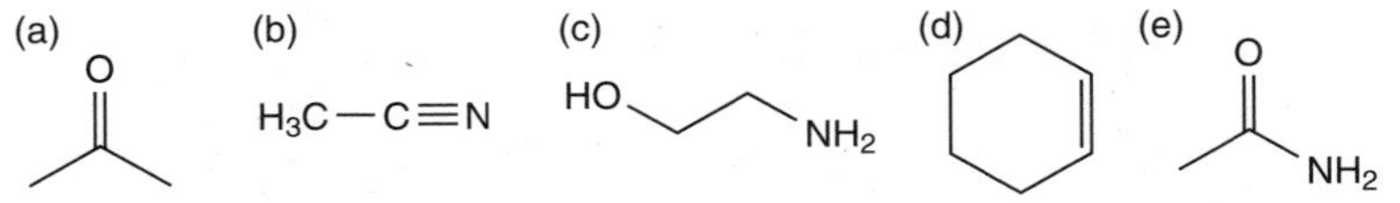

8.7   Under normal conditions pure BF_3 is a gas, but it can be purchased from chemical suppliers as a solution in ethoxyethane (diethyl ether, Et_2O). Describe the interaction between BF_3 and the solvent, and draw a curly arrow mechanism for the interaction you suggest. Do you think that BH_3 might also be transported in this way?

8.8   At high temperatures in the gas phase aluminium trichloride exists as discrete $AlCl_3$ molecules which have a trigonal planar structure. At lower temperatures, a dimer Al_2Cl_6 is formed, and this species is also found in molten aluminium trichloride. The structure of the dimer is illustrated below (any formal charges are not shown)

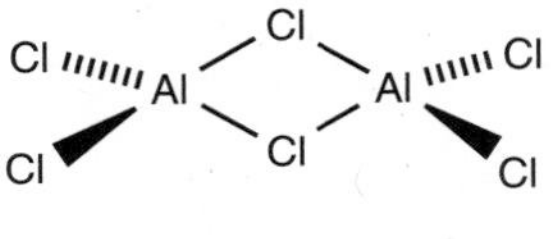

Identify the HOMO and the LUMO in an $AlCl_3$ molecule and hence explain how a reaction between two such molecules can give rise to a dimer with the structure shown. Assign any formal charges required, and explain why the dimerization must be an exothermic process.

8.9   Protonation of cyclohexene gives and ion **A** which reacts with water to give a species **B**

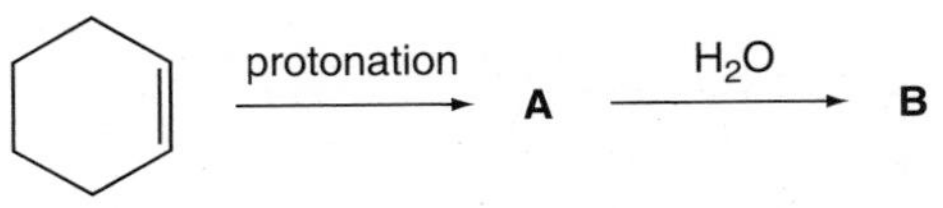

Identify the HOMO in cyclohexene and hence determine the structure of **A**; draw a curly arrow mechanism for its formation. Consider the possible HOMO/LUMO interactions between **A** and H_2O, and hence predict the structure of **B**, drawing a curly arrow mechanism for its formation.

8.10 (a) Assuming the C and N to be *sp* hybridized, draw up a description of the bonding in CN⁻ and hence identify the HOMO.

(b) By considering the likely orbital interactions, predict the initial product of the reaction between CN⁻ and methanal. Draw a curly arrow mechanism for your proposed reaction.

8.11 Cyanides (nitriles) R–C≡N react hardly at all with H_2O, but under acid conditions their reactivity is greatly enhanced. By considering the orbitals involved, explain why protonation of a nitrile enhances its reactivity towards nucleophiles. In your answer be sure to specify the site of protonation, and give a curly arrow mechanism for the initial reaction of the protonated nitrile with H_2O.

8.12 Compound (a) is readily available and is stable under normal conditions, whereas compound (b) is not listed by any chemical supplier. Why is this?

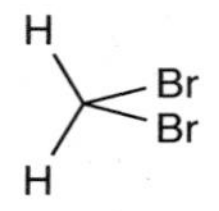

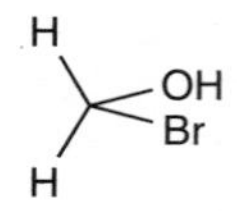

Organic chemistry 1: functional groups

Key points

- Functional groups can be classified according to the number of bonds a given carbon has to elements of greater electronegativity; the functional group level is the number of these bonds.
- Reactions can be classified according to the change in functional group level.
- The outcome of a reaction can be rationalized by considering where the highest energy electrons are to be found, and what low energy vacant MOs are available.
- Within a particular functional group level, there is often a general order of reactivity for different functional groups.
- Transformations from one functional group to another may take place via different mechanisms depending on the particular reagents and reaction conditions.

Organic chemistry is concerned with molecules which are based on a framework of carbon atoms, typically in combination with hydrogen, oxygen and nitrogen. Despite this rather limited set of elements, there are more organic compounds known than there are compounds not containing carbon. This vast number of organic compounds is a tribute to the extraordinary range of practical uses to which such substances are put, and the ingenuity of chemists in producing them.

Very many of the the chemical compounds which make our modern lives possible, such as pharmaceuticals (drugs), pesticides, herbicides, polymers (e.g. plastics), pigments and dyes can be classed as organic. Great skill goes into designing and synthesizing organic molecules which have the required properties, and it is this desire to make particular compounds which drives the field of organic chemistry.

The name 'organic chemistry' reminds us that historically these carbon-containing compounds were identified as being the chemical components of living systems. The kinds of reactions which take place in biological systems are essentially the same as those which an organic chemist will do in the laboratory, but in biology such reactions are facilitated by enzymes (as catalysts) and are carried out under rather different conditions.

Box 9.1 Common level-one functional groups

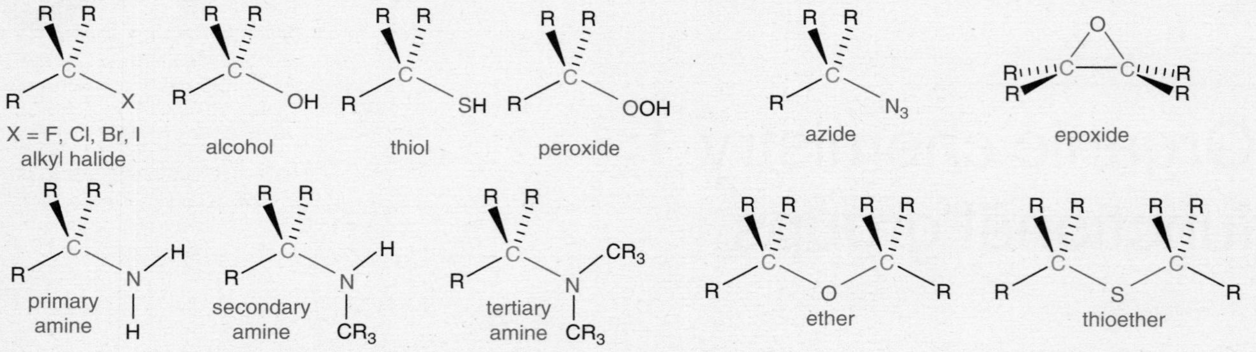

In each functional group, the carbon atom(s) shown in red is attached to *one* electronegative element, shown in blue.

R is either an alkyl group, an aryl group, or a hydrogen. When there is more than one *R* group in a given molecule, they need not be the same.

In organic chemistry a great deal of effort has been put into devising ways of extending and elaborating simple molecules to generate more complex ones. This is the area of *organic synthesis*. A vast repertoire of reactions, reagents and strategies have been developed, and so one of the difficulties we face is finding a way of rationalizing and classifying the reactions so as to be able to make sense of them. We will start on this process in this chapter.

The approach we will adopt is very much based on understanding orbitals and orbital interactions, using the ideas which has been introduced in the previous chapter. In this way we will find that for most reagents and conditions we can predict pretty well what will happen, understand why particular products are formed, and rationalize trends in reactivity. This approach brings some order to the seemingly endless number of organic reagents and reactions.

Before we start to look at specific reactions we will look at a way of classifying functional groups which we will find very useful in understanding and classifying many organic reactions.

9.1 Functional groups

One of the ways that we can begin to rationalize so vast a subject is by recognizing certain units of atoms that always behave in similar manner no matter what molecule they are in. These units are known as *functional groups*. The convention used for drawing organic molecules helps to emphasize the functional groups by allowing the carbon–hydrogen backbone to fade into the background. Hydrocarbons with no double or single bonds are not terribly reactive, indeed, the old name for this class of compound was *paraffins* which was derived from the observation that they had little affinity for other compounds.

Apart from double and triple bonds, functional groups involve the heteroatoms in a molecule – that is to say, atoms other than carbon and hydrogen. As we have already seen, heteroatoms are often involved in reactions since they

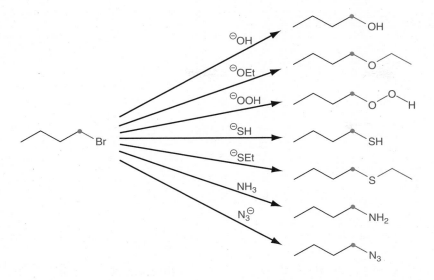

Fig. 9.1 The functional group in bromobutane can easily be converted to the different functional groups shown by reaction with the appropriate nucleophile. In each reaction the carbon atom highlighted with a red dot remains at the same functional group level since it has one attached heteroatom.

often have high energy lone pairs, and because the energy and size mismatch of their orbitals with carbon means they give rise to low energy vacant antibonding orbitals. For the moment we shall ignore groups with C–C π bonding (either isolated double bonds, triple bonds or aromatic systems) and just focus on individual carbon atoms with varying numbers of heteroatoms attached.

9.1.1 Functional group level

A useful way to divide up the many different functional groups is by considering how many bonds a particular carbon atom has to elements of greater electronegativity. The heteroatoms we will usually be concerned with include oxygen, nitrogen, the halogens, and sulfur. Functional groups with just one heteroatom attached to a carbon we can classify as level-one functional groups; with two bonds to heteroatoms (either as a double bond or two single bonds) a group is classified as a level-two functional group, and so on. Some examples of functional groups from level one are shown in Box 9.1 on the preceding page.

The point about classifying functional groups in this way, is that it is often easy to replace one carbon–heteroatom bond with another and thereby convert one functional group to another within the same level. For example, the functional group in bromobutane may be converted to different functional groups by reaction with the appropriate nucleophile, as shown in Fig. 9.1.

Each of the reactions shown in Fig. 9.1 proceeds via the high-energy electrons of the nucleophile attacking into the LUMO of the bromobutane, the C–Br $\sigma^\star$ MO, as outlined in section 8.5 on page 326. We will look at such reactions in more detail later in this chapter where we shall see that alternative mechanisms are possible for conversions between different functional groups within level one.

9.1.2 Level-two functional groups

Box 9.2 on the next page shows some common level-two functional groups. Each functional group has a carbon atom with *two* bonds to an electronegative element. The bonds can either be a double bond to the electronegative element,

Box 9.2 Common level-two functional groups

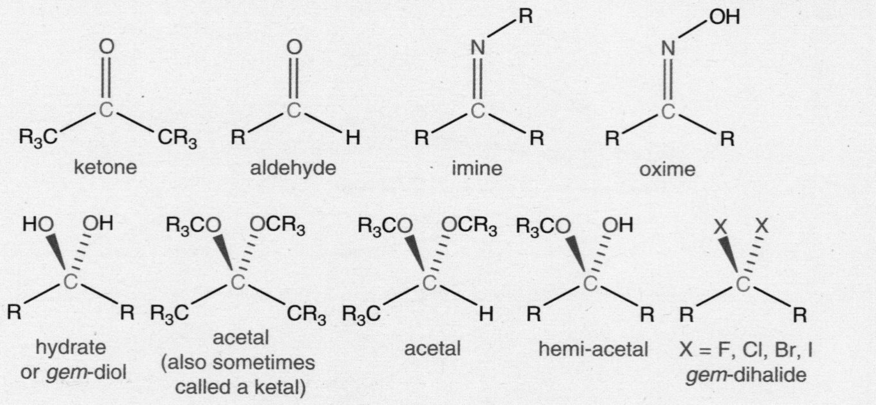

In each functional group, the carbon atom shown in red has *two* bonds to an electronegative element, shown in blue.

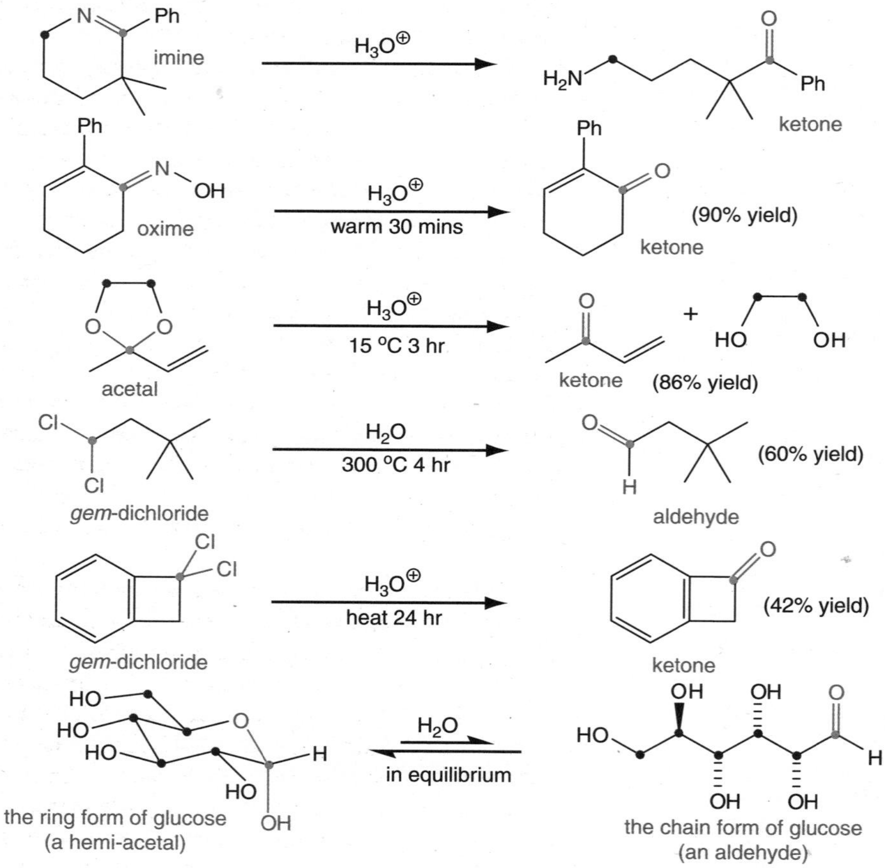

Fig. 9.2 Examples of the formation of aldehydes or ketones by the hydrolysis of other level-two functional groups. In each case, the carbon atom in the functional group with two bonds to an electronegative atom or atoms is marked with a red dot. It is this carbon that ends up as the aldehyde or ketone in each reaction, thus remaining at level two. Several of the compounds also include level-one functional groups; such carbons are indicated with a purple dot. These carbons remain at level one during the reaction.

Box 9.3 Common level-three functional groups

carboxylic acid   ester   amide   acyl chloride
                                      or acid chloride

orthoester   trihalide   cyanide   acid anhydride
                           or nitrile

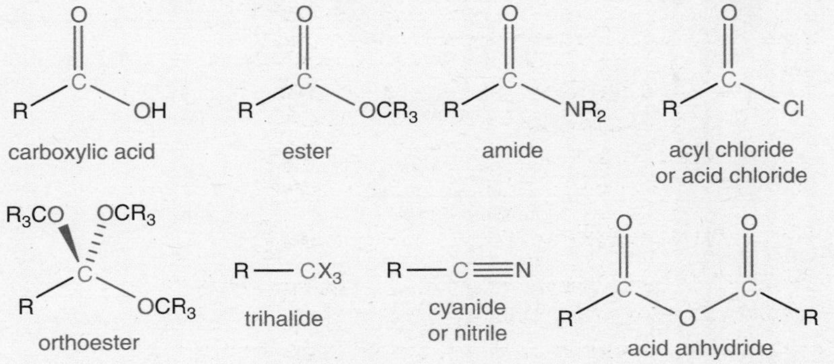

In each functional group, the carbon atom shown in red has *three* bonds to an electronegative element, shown in blue.

or two σ bonds. There are many possible choices for the two electronegative elements which are attached to the same carbon, but not all of these are stable. As we saw in section 8.8 on page 331, there are no stable compounds with both an $-NH_2$ group and a chlorine on the same carbon. Such compounds would rapidly fall apart to form an imine containing a carbon–nitrogen double bond. However, both of these functional groups are at level two. All that has happened is that one carbon–heteratom bond has been replaced by another.

As we shall see shortly, two of the functional groups shown in Box 9.2 – the hydrate and the hemi-acetal – are not usually stable. These groups rapidly form the ketone (or aldehyde) by eliminating water. All of the other functional groups shown in Box 9.2, whilst usually stable, can be converted to the corresponding ketone (or aldehyde) by treating with aqueous acid. Some examples of such conversions are shown in Fig. 9.2 on the preceding page.

9.1.3   Level-three functional groups

Some common level-three functional groups are shown in Box 9.3. Most of these have a C=O double bond and one other electronegative element attached to the same carbon, but it is also possible to have three single bonds (as in the trihalide and orthoester) or a triple bond to nitrogen (as in the nitrile). Each of these functional groups may be hydrolysed to give the corresponding carboxylic acid; some examples are given in Fig. 9.3.

The conditions needed to hydrolyse the different function groups vary considerably. acyl chlorides hydrolyse extremely rapidly in water whereas trichlorides are only hydrolysed with difficulty. The nitrile may be partially hydrolysed to yield an amide, or more fully to give the carboxylic acid. We will explore the mechanisms for these reactions, and the order of reactivity of the different groups, later in this chapter.

In addition to those shown in Box 9.3 there are many other ways of arranging for a carbon to have three bonds to an electronegative element. However, not all of these will be stable compounds. For example, no stable compounds have been isolated with three $-OH$ groups on one carbon atom. Such groups rapidly eliminate water and form the carboxylic acid as shown in Fig. 9.4 on the next page.

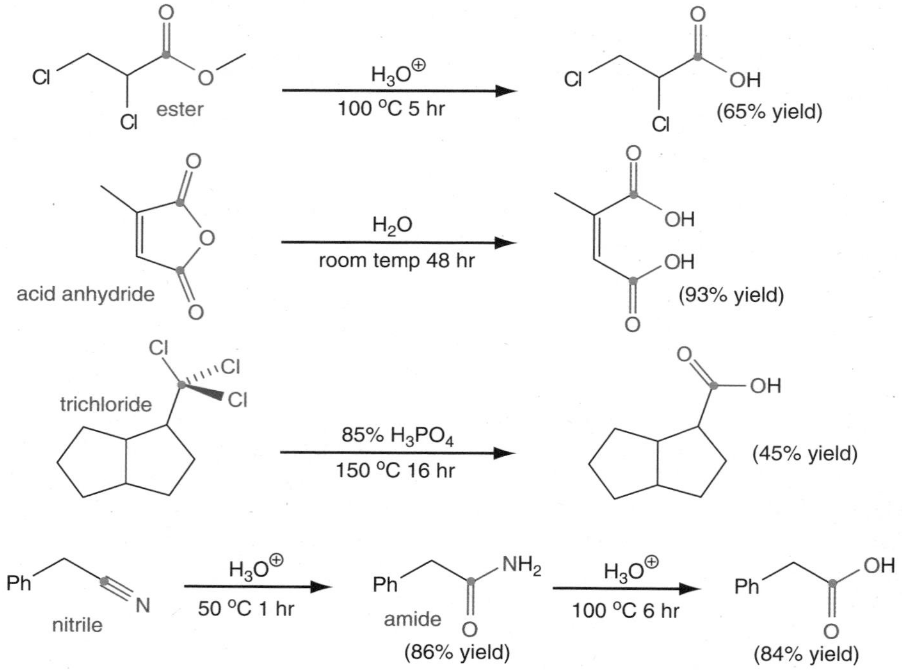

Fig. 9.3 Examples of the hydrolysis of different level-three functional groups to form the corresponding carboxylic acid. In each example, the level-three carbon of the group is marked with a red dot. Note that some groups are more easily hydrolysed than others, as indicated by the conditions required for the hydrolysis. The nitrile may be partially hydrolysed to form an amide, which on further hydrolysis yields the carboxylic acid.

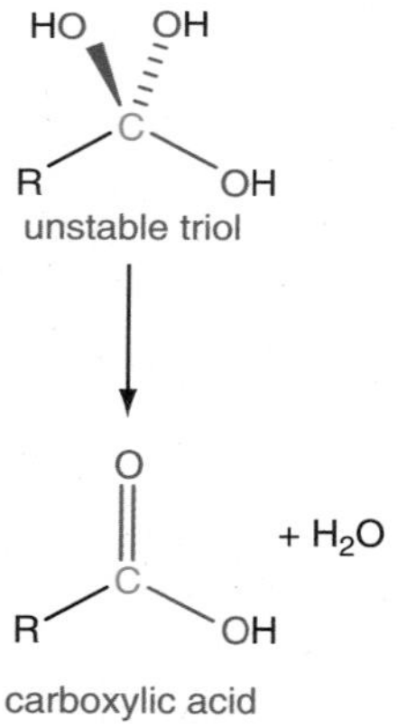

9.1.4 Level-four functional groups

Functional groups which contain carbons with four bonds to heteroatoms are rather less common than the other groups we have seen. However, some examples are shown in Box 9.4 to illustrate the principle. There are many small molecules in which carbon has four bonds to electronegative elements, and some examples of such molecules are shown in Fig. 9.5 on the next page. Perhaps the most common molecule is the extremely stable carbon dioxide molecule; its hydrated forms, orthocarbonic acid and carbonic acid, are not stable and readily lose water.

Fig. 9.4 Having three hydroxyl groups on the same carbon gives an unstable group which rapidly loses water to form a carboxylic acid.

Box 9.4 Level-four functional groups

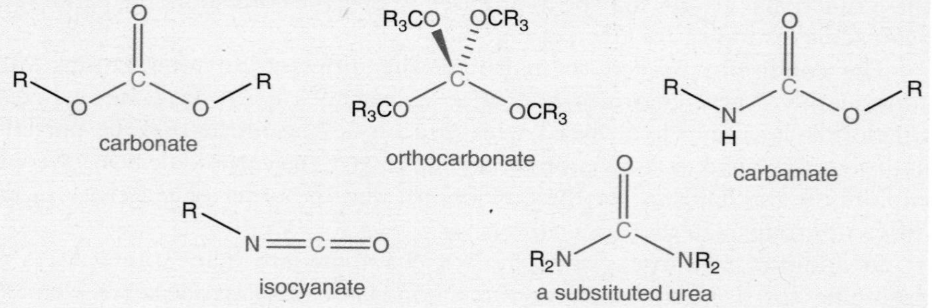

In each functional group, the carbon atom shown in red has *four* bonds to an electronegative element, shown in blue.

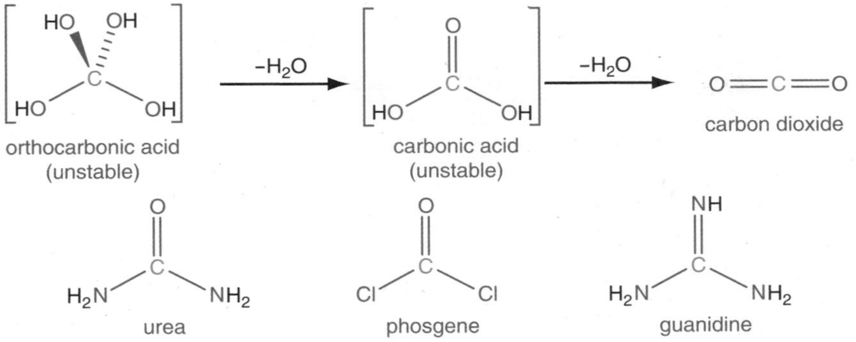

Fig. 9.5 Examples of small molecules which contain a level four carbon atom. Orthocarbonic acid and carbonic acid are unstable and rapidly lose water to form carbon dioxide. The other molecules are stable, but react with water to form carbon dioxide.

All of the functional groups shown in Box 9.4 and the molecules in Fig. 9.5 can be hydrolysed. In each case, the carbon–heteroatom bonds of the level-four carbon are swapped for C–O bonds and ultimately carbon dioxide is formed. Figure 9.6 shows an example where a carbamate functional group is hydrolysed. Every carbon atom in the molecule remains at the same functional group level during the reaction.

9.2 Changing functional group level

Whilst it is relatively easy to change from one functional group to another in the same level simply by swapping one carbon–heteroatom (C–X) bond for another, it is also possible to move up or down a level.

To move *down* a level involves the formation of a new C–C bond or a new C–H bond at the expense of a C–X bond. Forming a new C–H bond requires the use of a *reducing agent*, whereas forming a new C–C bond requires the use of reactant with a *nucleophilic carbon*. To move *up* a level involves the loss of a C–C bond or C–H bond in order to form a new C–X bond. This requires an *oxidizing agent*.

When carrying out such reactions, more than one step may often be necessary to obtain the desired product. Box 9.5 on the next page illustrates the conventional way in which this is indicated in a reaction scheme.

9.2.1 Moving down a functional group level using a reducing agent

Many different reducing agents are used to move down a functional group level, and some examples of such reactions are shown in Fig. 9.7 on page 347. In

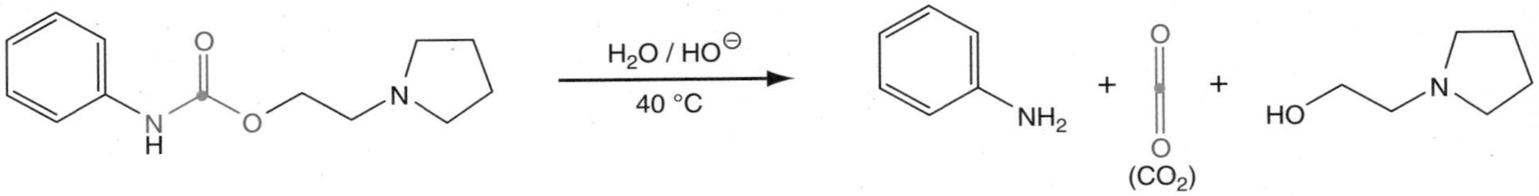

Fig. 9.6 An example of the hydrolysis of a carbamate group. During this reaction, every carbon maintains the same functional group level. The carbamate carbon (marked with a red dot) ends up as carbon dioxide.

Box 9.5 Reaction steps

When a reaction is written down it is common to number the reagents indicated above or below the reaction arrow. This notation means that the reaction must be carried out in a series of steps, each following the other. As an example, consider the reaction below.

This notation means that the ketone is first reacted with the reducing agent lithium aluminium hydride (LiAlH$_4$) in the solvent diethyl ether, and then *after* the reduction, aqueous acid is added. The reaction is carried out in this order because the first step must be absolutely free of any acidic protons since these would destroy the reducing agent. After the reduction step has been carried out, the oxygen will be bound with the aluminium and this needs to be removed in order to obtain the desired alcohol. This is the role of the aqueous acid in the second step.

This second step of freeing up the product and purifying it is an essential part of any organic reaction. It is called a 'work-up' and since it is essential to most reactions, it is often omitted from the reaction scheme altogether.

---

each example, the reducing agent supplying the hydrogen is shown in green and the functional group level of the carbon which undergoes the reduction is indicated in red. At this stage do not worry too much about the exact details of the reagents and conditions used for each reaction. The important thing is to recognize which carbon has changed functional group level and which hydrogens have been supplied by the reducing agent.

In example (a) from Fig. 9.7 an imine is reduced from level two to one using the reducing agent sodium borohydride, NaBH$_4$. Lithium aluminium hydride, LiAlH$_4$, is used in an analogous reaction reducing a ketone to an alcohol in (b). The same reducing agent is used in (c) to reduce an ester from level three to one.

It is possible, though harder, to partially reduce a functional group: in (d) an ester is reduced from level three to two using the reducing agent DIBAL (also called DIBAL-H, *DiIsoButylALuminium Hydride*). Similarly, in (e) a nitrile (cyanide) is reduced using the same reducing agent from level three to two, initially yielding an imine anion which is then hydrolysed (keeping at the same functional group level as usual) to the aldehyde.

Reaction (f) shows the reduction of an amide using the reducing agent sodium aluminium hydride. The major product is the amine which results from the amide being reduced from level three down to level one. A minor product is the alcohol – also reduced down two levels. A small amount of the aldehyde is formed from a partial reduction.

The systematic names for NaBH$_4$ and LiAlH$_4$ are sodium tetrahydridoborate and lithium tetrahydridoaluminate. However, in the chemical literature these reagents are universally known as sodium borohydride and lithium aluminium hydride; we will therefore continue to use these widely recognized names.

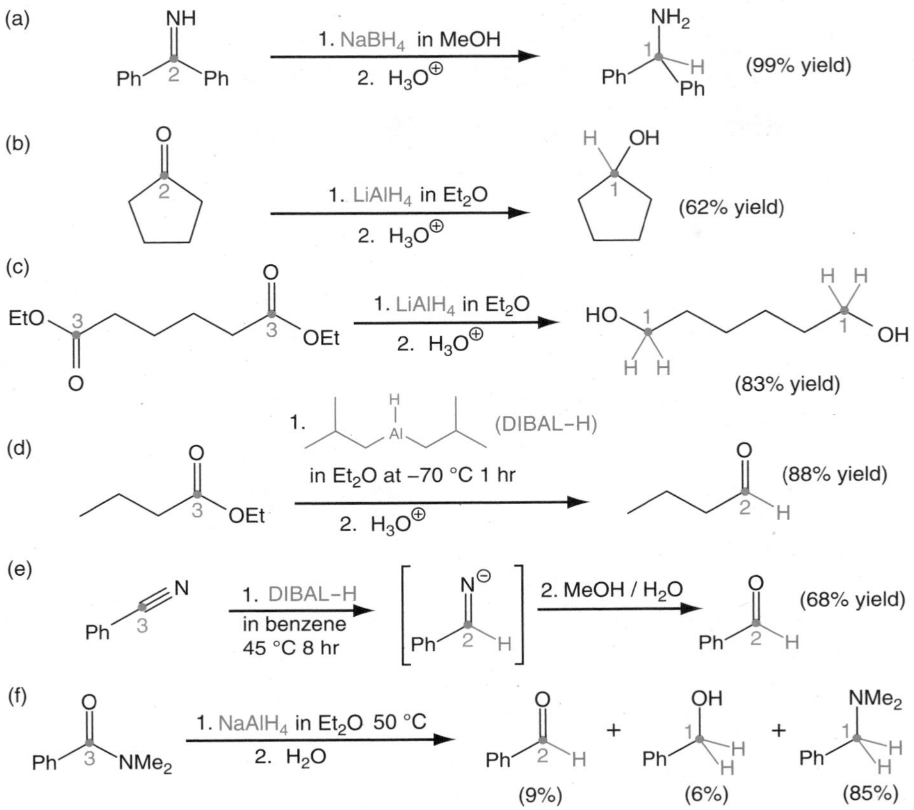

Fig. 9.7 Some examples of reactions in which reduction is taking place. In each example, the carbon atom with the red dot moves down one or two functional group levels as indicated by the red numbers. A variety of reducing agents are used and these are highlighted in green. In the product, the hydrogens which have been added by the reducing agent are shown in green.

9.2.2   Moving down a functional group level using a carbon nucleophile

In order to replace a bond between a carbon atom and an electronegative element with a new C–C bond, a nucleophilic carbon is used to replace the heteroatom; this results in moving down a functional group level. A number of carbon nucleophiles are available including cyanide (CN⁻), organolithium reagents (R–Li), and Grignard reagents (R–MgBr). Some examples of such reactions are shown in Fig. 9.8 on the next page. Once again, at this stage do not worry too much about the details of the reagents and conditions, but focus on which carbon has changed functional group level and which are the newly formed C–C bonds.

In example (a) from Fig. 9.8, the carbon nucleophile is the cyanide anion, CN⁻. The carbonyl carbon of the ketone at functional group level two gives rise to an alcohol at level one.

In (b) the carbon nucleophile is allyllithium. The C–Li bond is strongly polarized with a partial positive charge on the lithium and a partial negative charge on the carbon. This takes the ketone carbon from level two to level one in the resulting alcohol. A similar reaction takes place in example (c) but the carbon nucleophile is butylmagnesium bromide, an example of a Grignard reagent. In this the C–Mg bond is strongly polarised with a partial positive charge on the magnesium and a partial negative charge on the carbon.

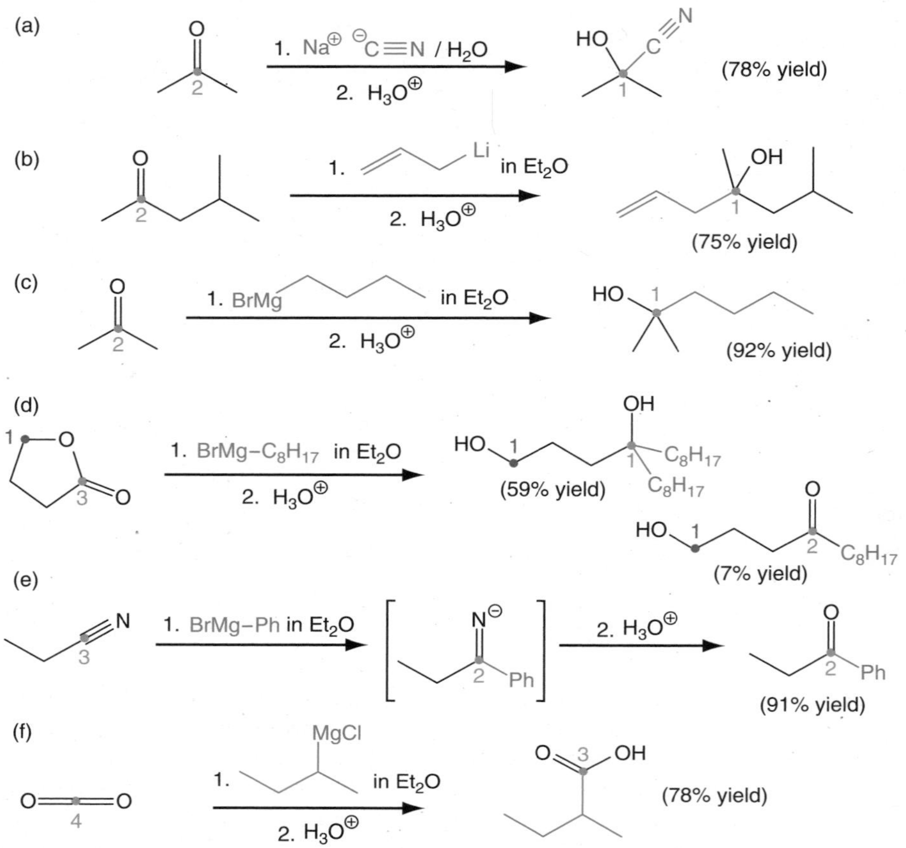

Fig. 9.8 Some examples of moving down one or two functional group levels by forming a new C–C bond at the expense of a C–X bond. In each example, the functional group level of the carbon concerned is shown in red. The carbon nucleophile reagent is shown in green.

Grignard reagents are also used in examples (d) and (e). In (d) two molecules of the Grignard reagent react with one of the ester to take the carbonyl carbon from level three to level one in the alcohol. Note that the carbon highlighted with a purple dot is at level one in the starting material, and remains at level one during the reaction.

In (e), a nitrile is taken from level three to level two. Initially the imine anion is formed as an intermediate, but this is then hydrolysed in aqueous acid to the ketone. As usual, the functional group level stays the same, at level two, during the hydrolysis.

In (f), the Grignard reagent reacts with carbon dioxide to form a carboxylic acid. The carbon atom in CO_2 starts out at level four and ends up as level three in the carboxylic acid after the formation of the new C–C bond.

9.2.3 Moving up a functional group level

To move up a functional group level involves losing a C–C or C–H bond and replacing it with a C–X bond. This is usually accomplished with an oxidizing agent. Numerous different oxidizing agents are routinely used, often involving transition metals in high oxidation states. Picking the right oxidizing agent for the specific task means finding the one which will raise the functional group level of the group in question by the desired amount whilst leaving any other functional groups unaffected: Fig. 9.9 on the facing page gives some examples.

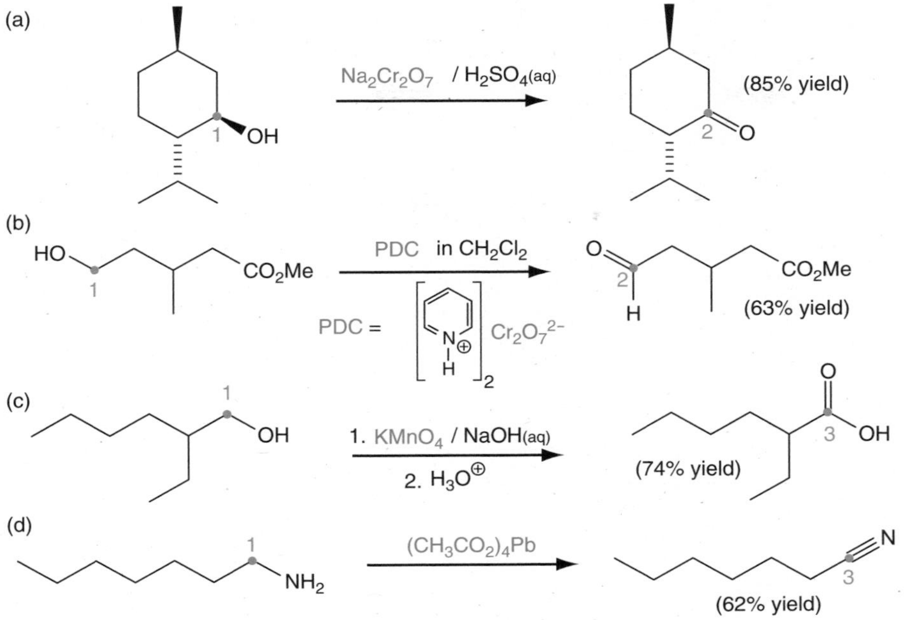

Fig. 9.9 Some examples of moving up one or two functional group levels using an oxidizing agent (shown in green). In each example, the carbon atom which is oxidized is marked with a red dot and its functional group level is also shown in red.

In example (a), an alcohol is oxidized using sodium dichromate to the corresponding ketone. In the oxidizing agent, the chromium is Cr(VI), i.e. it is in its highest oxidation state (see section 7.10 on page 303). A slightly different oxidizing agent, pyridinium dichromate (PDC), is used in (b). In this reagent, the sodium cation is replaced by the pyridinium ion. The chromium is still in its highest oxidation state, but the reagent is more gentle and allows the partial oxidation of the primary alcohol from level one to level two in the aldehyde.

In contrast, example (c) uses potassium manganate(VII) to oxidize the primary alcohol at level one to the carboxylic acid at level three. As in the previous examples, the metal in the oxidizing agent is in its highest oxidation state.

Example (d) shows a primary amine at level one being oxidized to a nitrile at level three. Once again, the oxidizing agent, this time lead(IV) ethanoate, has the metal in its highest oxidation state.

9.2.4 Summary of functional group transformations

Before investigating exactly how one functional group may be changed into another, it is useful to summarize the important points so far.

- Reactions are usually centred around the heteroatoms in molecules since these have high energy electrons and give rise to low energy vacant orbitals. Different arrangements of heteroatoms on a given carbon atom are called functional groups.

- Apart from when there is π bonding between carbon atoms, the carbon framework of a molecule is usually less reactive and often remains unchanged during a reaction. Carbon–carbon π bonds, in the form of isolated double or triple bonds, conjugated systems, or aromatic systems, are often relatively reactive and provide means of introducing more functional groups. This will be discussed in Chapter 13.

- A useful classification of a functional group is according to the number of bonds that a given carbon atom has to elements of greater electronegativity (denoted X). This number is the functional group level.

- It is relatively easy to move within a functional group level by swapping one C–X bond for another. In the presence of excess water (and possibly acid or base), most level-one functional groups will be hydrolysed to form alcohols, most level-two groups will form aldehydes or ketones, most level-three groups will form carboxylic acids, and most level-four functional groups will form carbon dioxide.

- To move *down* a functional group level requires a reducing agent to form a new C–H bond at the expense of a C–X bond, or a nucleophilic carbon reagent to form a new C–C bond at the expense of the C–X bond.

- To move *up* a functional group level requires the use of an oxidizing agent to form a new C–X bond, at the expense of a C–H or C–C bond.

9.3 Level two to level one – carbonyl addition reactions

Perhaps the simplest reaction of a carbonyl group is when a nucleophile with high-energy electrons attacks into the LUMO of the carbonyl, the CO $\pi^\star$, breaking the π bond to give a tetrahedral product as shown in Fig. 9.10. The orbital interactions in this reaction were discussed in section 8.4.1 on page 323. If the nucleophile is an electronegative atom such as nitrogen or oxygen, then the functional group remains at level two, and it may be possible for a π bond to reform. If the nucleophile is carbon or hydrogen, then the functional group level drops from two to one, and it is this class of reaction we shall look at now.

9.3.1 Addition of cyanide: cyanohydrins

One of the simplest addition reactions to a carbonyl is the addition of cyanide to form a group called a *cyanohydrin* which has an –OH group and a –CN group attached to the same carbon. An important cyanohydrin is acetone cyanohydin (also called 2-hydroxy-2-methyl-propanenitrile) which is used as a precursor in the production of acrylic plastic. Acetone cyanohydrin is prepared industrially by the reaction between acetone (propanone) and hydrocyanic acid, HCN, with a catalytic amount of potassium cyanide. A mechanism for this reaction is shown in Fig. 9.11.

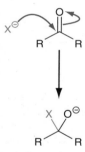

Fig. 9.10 The addition of nucleophile X⁻ to an aldehyde or ketone to give a tetrahedral product.

Fig. 9.11 A mechanism for the formation of acetone cyanohydrin.

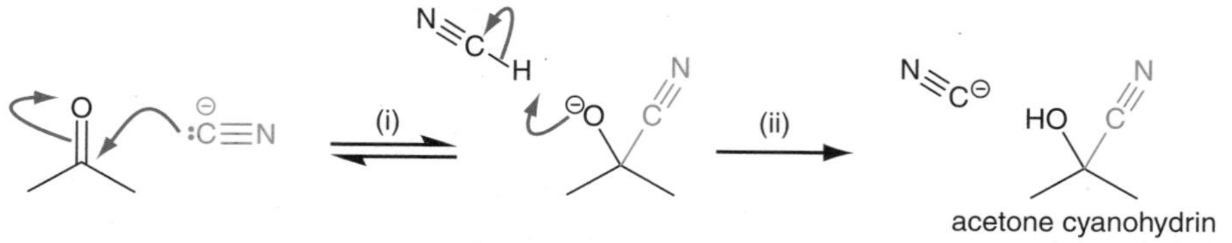

acetone cyanohydrin

The highest energy electrons of the reactants are the HOMO of the cyanide anion. Cyanide is isoelectronic with carbon monoxide, and its HOMO is a σ MO which can be treated as a carbon lone pair (see section 3.5.2 on page 130 and section 4.5 on page 156). It is these high-energy electrons that attack into the CO $\pi^\star$ MO to give the tetrahedral anion, as shown in step (i) in Fig. 9.11.

The reverse of step (i) is also possible; the high-energy electrons on the negatively-charged oxygen can re-form the carbonyl and push out cyanide ion, as shown in Fig. 9.12. However, the oxygen lone pair can also be protonated by HCN, which is a weak acid with a pK_a of 9.2. This proton transfer lies well over in favour of the neutral acetone cyanohydrin and cyanide, as shown in step (ii). The cyanide ion formed in the second step can go on to attack another molecule of acetone.

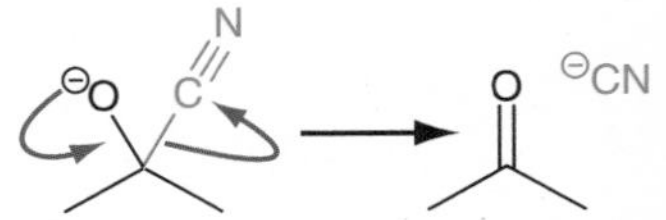

Fig. 9.12 The tetrahedral anion can return to the carbonyl by pushing out the cyanide ion.

Acetone cyanohydrin as a source of cyanide

Acetone cyanohydrin is also used in reactions as a convenient source of cyanide ions. The neutral compound is stable under normal conditions since the oxygen lone pair on the –OH group is not sufficiently high in energy to push out cyanide. However, if a base is added, the –OH group is deprotonated and the negatively charged oxygen can now push out cyanide as it re-forms the C=O π bond, as shown in Fig. 9.12.

9.3.2   Reaction with organometallic reagents

Another extremely useful reaction is the reaction between a carbonyl such as an aldehyde or ketone, with an *organometallic* reagent. These are reagents that have a carbon–metal bond. Many metals are used in organic chemistry, but in this section, we shall look at organomagnesium reagents (usually called Grignard reagents after the French chemist who discovered them), and organolithium reagents. These are prepared by reacting magnesium or lithium metal with an organic halide, R–X, in a solvent such as diethyl ether (often simply called 'ether') or tetrahydrofuran. The structures of these solvents are shown in Fig. 9.13. The reaction proceeds by a rather complex mechanism, much of which takes place on the surface of the metal and involves the transfer of single electrons from the metal. The overall reactions may be summarized

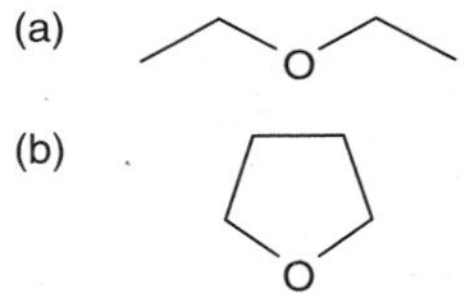

Fig. 9.13 The structures of (a) diethyl ether, also called ethoxyethane, or just simply ether, and (b), tetrahydrofuran, often abbreviated to THF.

$$2Li(s) + R–X \longrightarrow R–Li + LiX$$

$$Mg(s) + R–X \longrightarrow R–MgX$$

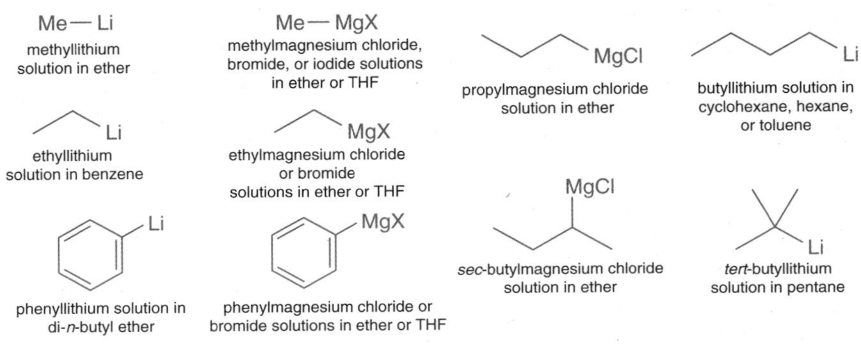

Me — Li
methyllithium
solution in ether

Me — MgX
methylmagnesium chloride,
bromide, or iodide solutions
in ether or THF

propylmagnesium chloride
solution in ether

butyllithium solution in
cyclohexane, hexane,
or toluene

ethyllithium
solution in benzene

ethylmagnesium chloride
or bromide
solutions in ether or THF

sec-butylmagnesium chloride
solution in ether

tert-butyllithium
solution in pentane

phenyllithium solution in
di-*n*-butyl ether

phenylmagnesium chloride or
bromide solutions in ether or THF

Fig. 9.14 A selection of some commercially available organolithium and Grignard reagents. Organolithium reagents tend to be more reactive than Grignard reagents. Sometimes they react with ethers and hence are supplied as solutions in hydrocarbons instead.

Organolithium and organomagnesium reagents are often made as required and used in solution without isolating them, but many are also commercially available. A few examples are shown in Fig. 9.14 on the preceding page.

Bonding in organolithium and Grignard reagents

The simple structures shown in Fig. 9.14 on the previous page are slightly misleading since it has been shown that these compounds tend to form aggregates such as dimers, tetramers, and hexamers both in solution and in the solid phase. Ethyllithium, for example, exists as a tetramer, $Li_4(CH_3CH_2)_4$, and its structure is shown in Fig. 1.14 on page 15. For the sake of simplicity, however, we shall use the simple monomeric structures in our mechanisms.

The important point about both organolithium and Grignard reagents is that the carbon–metal bond is strongly polarized with increased electron density on the carbon. This is due to the fact that the AOs of carbon are lower in energy than the orbitals of the metals, as shown in Fig. 9.15. The strong interaction between two carbon AOs is shown in (a) which leads to a rather low-energy σ bonding MO. The interaction between the AOs of carbon and lithium is weaker due to the poor energy match between the AOs, as shown in (b). The C–Li σ bonding MO is higher in energy than the C–C MO, and it has a greater contribution from the carbon than from the lithium. We could also say that the C–Li bond has considerable ionic character.

It is the high energy electrons in the C–Li and C–Mg σ bonds that are utilized in their reaction with a carbonyl. A general mechanism is shown in Fig. 9.16. In the first step the high-energy electrons from the carbon–metal bond attack into the $\pi^\star$ MO of the carbonyl, forming a new C–C bond and breaking the C=O double bond, to give the usual tetrahedral anion. It may be that the metal also coordinates to the oxygen of the carbonyl in this first step, making it even more favourable (see page 329). It has also been suggested that more than one molecule of the organometallic reagent is needed in this step. Be this as it may, the essence of the reaction is as shown in Fig. 9.16.

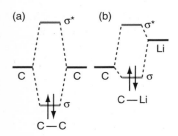

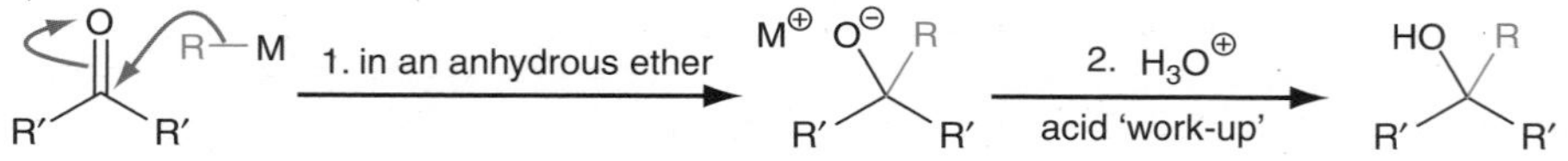

Fig. 9.15 The good energy match between the AOs of carbon gives rise to a low energy bonding MO and high energy antibonding MO. The interaction between carbon and lithium is weaker due to the difference in energy of their AOs, and this results in a high-energy C–Li σ bonding MO.

Fig. 9.16 A simple mechanism for the reaction between an organometallic reagent and an aldehyde or ketone.

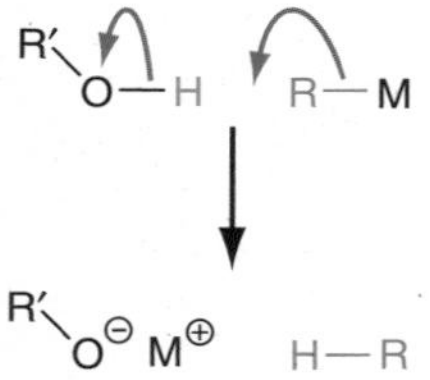

Fig. 9.17 Organometallic reagents can also act as strong bases and easily remove any acidic protons, such as those in an alcohol or water.

What is extremely important when using organometallic reagents is that water, or indeed anything with acidic hydrogens, is totally absent from the reaction mixture. Organometallics are destroyed by water since the reagent can also act as a base and pick up a proton as shown in Fig. 9.17. This is why in Fig. 9.16, there are two separate stages: the first being the reaction between the carbonyl and the Grignard reagent or organolithium reagent, the second stage being the subsequent protonation of the tetrahedral anion to give the alcohol. This step is known as the 'work-up'.

Some examples of the addition of organolithium and Grignard reagents to aldehydes and ketones are given in Fig. 9.18 on the facing page.

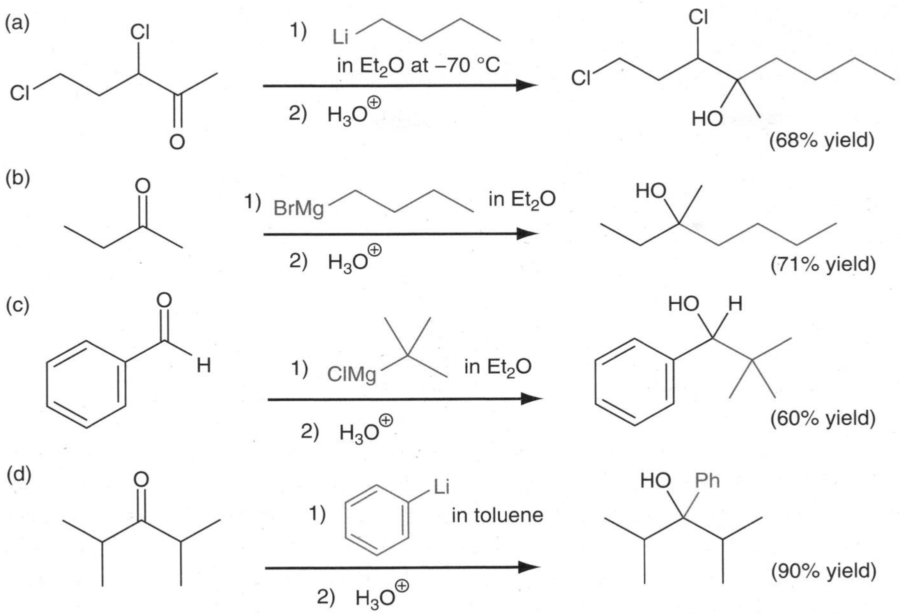

(a)
Cl
Cl
O
1) Li
in Et2O at −70 °C
2) H3O⊕
Cl
Cl
HO
(68% yield)
(b)
O
1) BrMg
in Et2O
2) H3O⊕
HO
(71% yield)
(c)
O
H
1) ClMg
in Et2O
2) H3O⊕
HO H
(60% yield)
(d)
O
Li
1)
in toluene
2) H3O⊕
HO Ph
(90% yield)

Fig. 9.18 Some examples of the addition of organolithium and Grignard reagents to aldehydes and ketones. Each reaction takes place in two separate stages: first the reaction with the organometallic reagent under strictly anhydrous conditions; second, an acid work-up. The added alkyl group is shown in blue.

9.3.3 Reaction with metallic hydride reducing agents

Aldehydes and ketones are also readily reduced using metal hydrides – two of the most common being sodium borohydride, $NaBH_4$, and lithium aluminium hydride, $LiAlH_4$. Each of these reagents may be treated as consisting of an alkali metal cation, Na^+ or Li^+, and a tetrahedral anion, BH_4^- or AlH_4^-, in which the central boron or aluminium is surrounded by four hydrogens.

Since the hydrogens are arranged at the corners of a tetrahedron around either the boron or the aluminium, we can describe the central atom as being sp^3 hybridized. The BH_4^- anion is isoelectronic with methane, CH_4. However, in CH_4 the sp^3 HAO of carbon and the $1s$ AO of hydrogen are well matched in energy (see section 4.4.1 on page 151) and so the C–H σ bonding MO is relatively low in energy, as shown in Fig. 9.19 (a).

In contrast, an sp^3 AO of boron is considerably higher in energy than the $1s$ AO of hydrogen, and hence the resulting B–H σ bonding MO is also high in energy as shown in Fig. 9.19 (b). What is more, whereas the electrons in the C–H bonds in methane are shared more or less equally between the carbon and hydrogen, in BH_4^- the electrons are more associated with the hydrogens.

(a)
σ*
C sp3
HAO
H 1s
AO
σ
C—H
(b)
σ*
B sp3
HAO
H 1s
AO
σ
B—H

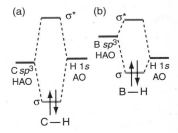

Fig. 9.19 MO diagram for (a) methane, and (b) BH_4^-. Whereas an sp^3 HAO of carbon and a hydrogen $1s$ AO have approximately the same energy, an sp^3 HAO of boron is higher in energy than a hydrogen $1s$ AO and so the resulting σ bonding MO is higher in energy.

The reactivities of $LiBH_4$ and $NaBH_4$ compared

The mismatch in orbital energy is even greater between *aluminium* and hydrogen than for *boron* and hydrogen. What is more, the larger $3s$ and $3p$ AOs of aluminium do not interact so strongly with the $1s$ AO of hydrogen as do the $2s$ and $2p$ AOs of boron. Both of these factors mean the Al–H σ bonding MO is even higher in energy than the corresponding B–H MO. Since it is the electrons in the high-energy B–H bonds and Al–H bonds that participate in reactions, it should come as no surprise to learn that the AlH_4^- ion is more reactive than the BH_4^- ion. For example, whereas $NaBH_4$ reacts only slowly with water, $LiAlH_4$ reacts vigorously to form hydrogen gas in a reaction analogous to that between water and an organometallic reagent.

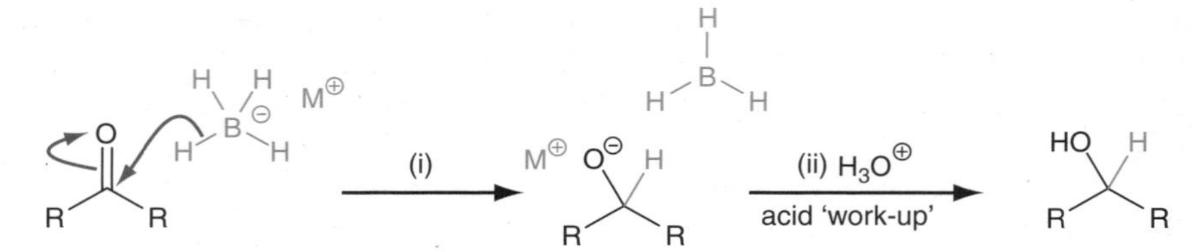

Fig. 9.20 A mechanism for the reduction of an aldehyde or ketone with borohydride.

Mechanism for the reduction of an aldehyde or ketone

A mechanism for the reduction of an aldehyde or ketone with $NaBH_4$ is shown in Fig. 9.20. In the first step, the LUMO of the aldehyde or ketone is attacked by the high-energy electrons in one of the B–H bonds to form the alkoxide **A**. Note that the curly arrow comes from the B–H bond and not from the negative charge. There is no lone pair on any atom in BH_4^- – all the electrons are in bonding MOs.

The smaller lithium ion interacts more strongly with the oxygen than does a sodium ion, which helps make the lithium reagents more reactive than the sodium ones.

This first step has been simplified by neglecting the role of the alkali metal cation which will interact with the oxygen of the carbonyl. This has the effect of both lowering the energy of the $\pi^\star$ MO (see section 8.7 on page 329), and also of helping to bring the BH_4^- ion and the carbonyl together; if the cation is absent, the reduction does not take place.

A further simplification in the mechanism shown in Fig. 9.20 is that only one of the hydrogens from the BH_4^- ion has been used, whereas it is possible for all four to be used to reduce further molecules of the aldehyde or ketone. It is worthwhile looking at how this occurs. The initial products after step (i) of the mechanism shown in Fig. 9.20 can react further: the overall HOMO is now one of the lone pairs of the oxygen on alkoxide **A**, and the overall LUMO is the vacant boron $2p$ AO of the BH_3. There is a simple interaction between these two orbitals to form the adduct **B** as shown in Fig. 9.21.

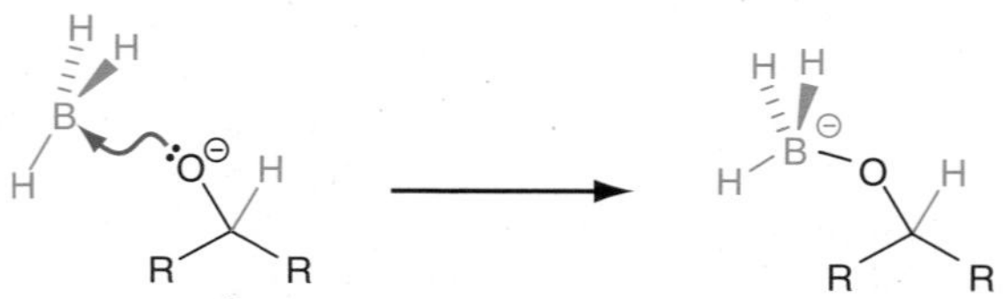

Fig. 9.21 The interaction between an alkoxide ion **A** and BH_3.

The adduct **B** once again has a tetrahedral boron and the high-energy electrons from one of the remaining B–H σ bonds can reduce a further molecule of the carbonyl as shown in Fig. 9.22.

Fig. 9.22 The tetrahedral boron anion acts as a reducing agent a second time.

This reaction produces more of the alkoxide **A**, and a new species **C** which has a trigonal boron with a vacant $2p$ AO. This LUMO can once again be attacked by a further oxygen lone pair from the alkoxide allowing the process to be repeated until, theoretically, all four of the original hydrogens in the BH_4^- ion have been lost and replaced by four of the alkoxide ions to give the species $B(OR')_4^-$ shown in Fig. 9.23.

The $B(OR')_4^-$ ion is hydrolysed in acid to give the alcohol and boric acid, $B(OH)_3$, according to the equation

$$B(OR')_4^\ominus + 2H_2O + H_3O^\oplus \longrightarrow B(OH)_3 + 4R'OH$$

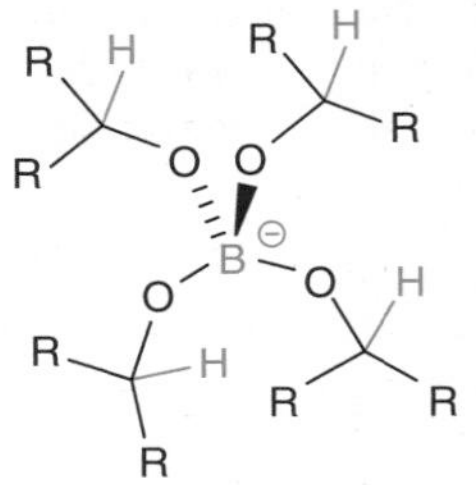

Fig. 9.23 The anion formed if all four hydrogens from BH_4^- have reacted with a ketone R_2CO and so been replaced by four alkoxide ions. We will denote this as $B(OR')_4^-$ where $R' = R_2CH$.

Other reductions

It is also possible to reduce the CN π bond of an imine using either $NaBH_4$ or $LiAlH_4$ in an analogous mechanism to that for the reduction of a CO π bond. Examples of the reduction of an imine, and of a ketone, are shown in (a) and (b) from Fig. 9.7 on page 347.

Note that in these examples the less reactive $NaBH_4$ reducing agent is used in a solution of methanol. It can even be used in neutral or alkaline solitions in water but not in acid, when it would rapidly be protonated. In contrast, the more reactive $LiAlH_4$ must be used with no acidic protons or water present.

We will come back to these reducing agents when we see their reactions with level-three functional groups later in the chapter (section 9.6.2 on page 388), but first we will look at some reactions which convert one level-two functional group to another.

9.4 Transformations within functional group level two

In this section we will look at reactions which the functional group stays within level two. First we shall look at reactions in which the double bond of the carbonyl is replaced, then at reactions in which a new double bond is made between carbon and an electronegative element.

9.4.1 Hydrates and hemiacetals

When an aldehyde is dissolved an alcohol, a *hemiacetal* is formed. This group has an –OH and and –OR group on the same carbon atom.

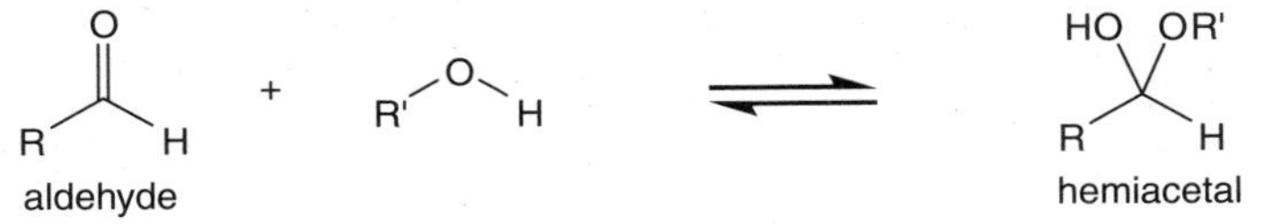

Similarly, when an aldehyde is dissolved in *water*, hydrates are formed. These are also sometimes known as *gem*-diols which means a diol with both alcohol groups on the same carbon.

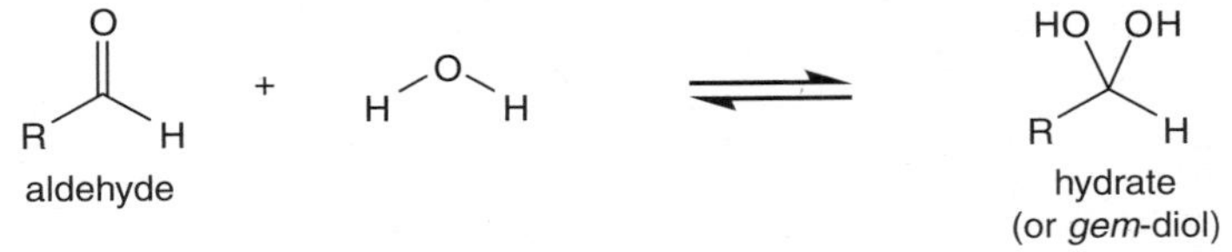

Mechanism for the formation of hydrates and hemiacetals

In these reactions, the high energy electrons of the lone pair on either the alcohol or the water attack into the LUMO of the aldehyde, the C=O $\pi^\star$ MO (see section 8.4 on page 323). A curly arrow mechanism for the formation of the hydrate from ethanal (acetaldehyde) is shown below

The above mechanism is slightly misleading since it seems to suggest that after the initial attack of the water on the aldehyde, a negative charge is formed on the carbonyl oxygen and a positive charge is formed on the oxygen from the attacking water. In order to form the hydrate it is therefore necessary to transfer a proton from the water oxygen to the carbonyl oxygen.

What the mechanism fails to take into account is that the incoming water will be hydrogen-bonded to many other water molecules in the solvent, as will the oxygen of the carbonyl. Whilst it *looks* as if a proton needs to be removed from one oxygen and be added onto the other, this need not be the case.

Figure 9.24 (a) shows the initial product after the addition of water to ethanal. In (b) a number of water molecules have been included to show the sort of hydrogen bonding that might exist in aqueous solution. The hydrogen bonds are shown as the red dashed lines between oxygen and hydrogen.

In (c), no atom has moved its position to any significant degree, but all the bonds that were hydrogen bonds in (b) have become full bonds, and some of the full O–H bonds in the water molecules in (b) have now become hydrogen bonds. The result is the desired hydrate – still hydrogen bonded with the solvent. The hydrate is shown without the solvent in (d). The process which we initially described as proton transfer can therefore be achieved with only a rearrangement of the bonding – no atoms have to move.

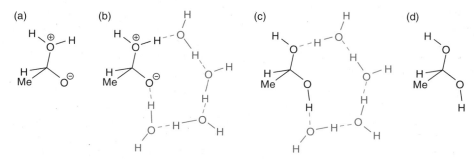

Fig. 9.24 The initial charged species formed after the addition of water to ethanal is shown in (a). This is shown hydrogen bonded to a number of water molecules in (b), with the hydrogen bonds in red. With a slight redistribution of the O–H bonds and the hydrogen bonds, the hydrogen-bonded hydrate shown in (c) is formed. This is shown without the solvent in (d).

Table 9.1 The percentage of hydrate present in aqueous solutions of aldehydes

aldehyde	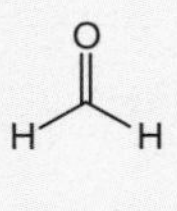	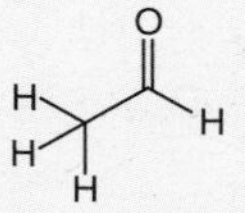	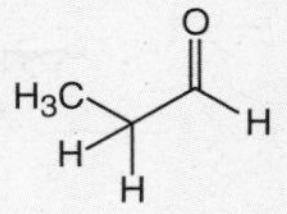	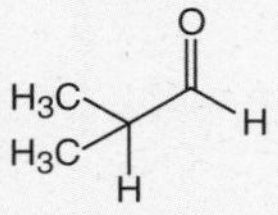	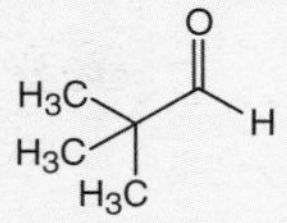
% hydrate	99.95	59	42	30	20

Table 9.2 The percentage of hemiacetal present in solutions of aldehydes in alcohol

aldehyde	alcohol	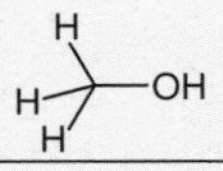	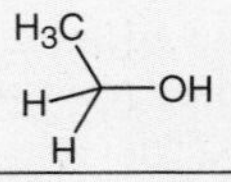	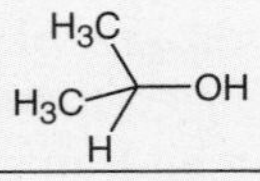	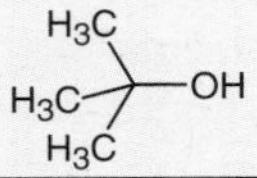
		97	91	71	12
		95	87	58	–
		90	81	42	–

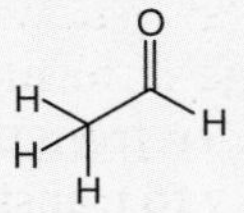

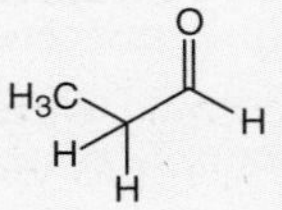

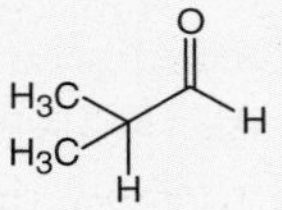

The carbonyl to hydrate / hemiacetal equilibrium

During the formation of either a hydrate or a hemiacetal from an aldehyde, the crowding around the level-two carbon atom increases. In the aldehyde the three groups are at approximately 120° to one another, whereas in the hydrate or hemiacetal the four groups are closer together at approximately 109°. It should not be too surprising to learn that the sizes of any side-groups on the reactants is one factor which affects the position of the equilibrium between the aldehyde and the hydrate or hemiacetal.

The amount of hydrate present when an aldehyde is dissolved in water also depends on the structure of the aldehyde. There are two factors which affect the equilibrium strongly: the size of the group R attached to the carbonyl carbon, and whether this group is electron withdrawing or not.

In general terms, the larger the group on the aldehyde, the smaller the amount of hydrate. Table 9.1 shows the percentage of the hydrate form present at equilibrium when aldehydes with increasingly large R groups are dissolved in water. In these examples, the aldehyde solution is made up to a concentration of between 2 and 3 mol dm^{-3}. The clear trend is that as the size of R increases, the fraction of the hydrate form decreases.

Similarly, Table 9.2 shows the percentage of the hemiacetal form present at equilibrium when different aldehydes are dissolved in alcohols R′OH in which R′ is increasing in size. In these examples, the aldehyde solution is made up to a concentration of 0.1 mol dm^{-3}. Once again, the trend is that as the size of the R group in either the aldehyde or the alcohol increases, the fraction of the hemiacetal present at equilibrium decreases.

Table 9.3 The percentage of hydrate present in aqueous solutions of aldehydes and ketones.

aldehyde or ketone	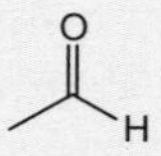	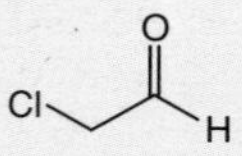	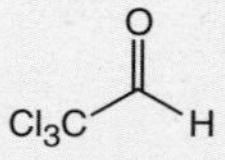	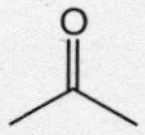	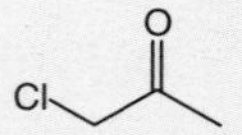	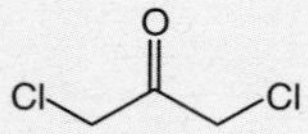
% hydrate	59	97.4	99.996	0.2	10	91

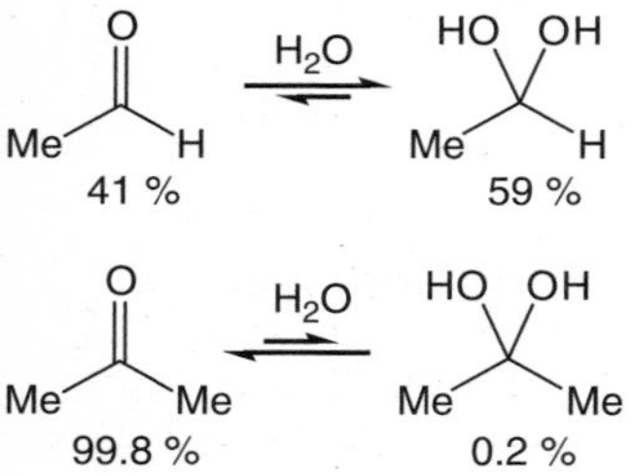

Fig. 9.25 Whilst ethanal in water exists largely as the hydrate form, acetone (propanone) exists largely as the ketone form.

Given that the sizes of the groups attached to the carbonyl carbon influence the fraction of the hydrate or hemiacetal so strongly, it is not surprising that replacing the hydrogen of the aldehyde with an alkyl group to form a ketone should result in even smaller equilibrium proportions of the hydrates or hemiacetals. Thus whilst an aqueous solution of ethanal exists mainly in the hydrate form, a solution of acetone (propanone) exists almost exclusively as the ketone as shown in Fig. 9.25.

Electronic factors

Size is not the only factor which determines the amount of hydrate or hemiacetal present at equilibrium. If the aldehyde has an electron-withdrawing group attached to the carbonyl carbon, the amount of hydrate or hemiacetal increases significantly.

Table 9.3 shows how the amount of hydrate present at equilibrium drastically increases as successive hydrogen atoms are replaced by electron-withdrawing chlorine atoms for both the aldehyde ethanal, and the ketone acetone (propanone). We shall see in section 9.5.3 on page 370 that there is a favourable interaction between the C=O π system and any C–H or C–C σ bonds from the carbon adjacent to the carbonyl carbon. These interactions are not possible when the C–H or C–C is replaced by a C–Cl bond. The chlorine atoms therefore destabilize the carbonyl form which results in the lower energy hydrate being favoured at equilibrium.

9.4.2 Acetals

Simply dissolving an aldehyde or ketone in an alcohol gives rise to a proportion of the hemiacetal which has one –OH group and one –OR group on the level-two carbon atom. However, the majority of hemiacetal compounds are not stable and they cannot be isolated pure. Any such attempts usually result in the elimination of water and the reforming of the carbonyl compound as shown in Fig. 9.26 (a).

In contrast to hemiacetals, compounds which have *two* –OR groups on the same carbon, shown in (b), are more stable and are usually easy to isolate in a pure state. These are called *acetals* when prepared from aldehydes (the

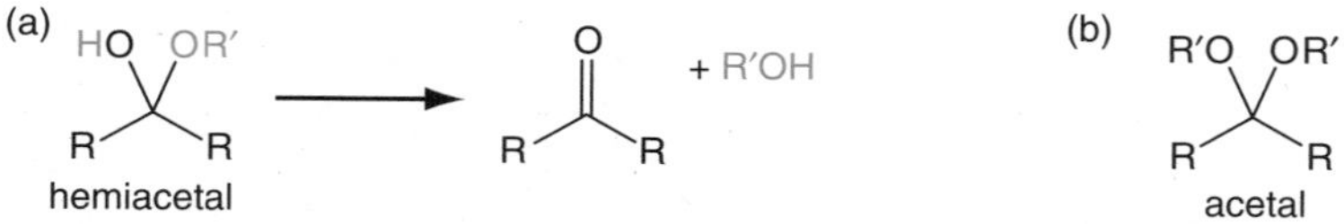

Fig. 9.26 Most hemiacetals are only present in equilibrium in alcohol. Any attempt to isolate them results in them eliminating the alcohol and reforming the carbonyl as shown in (a). In contrast, with two –OR groups, actetals, shown in (b), are stable and can readily be purified.

level-two carbon has a hydrogen atom attached) or *ketals* when prepared from ketones (in which the level-two carbon has no hydrogen attached). In everyday use it is becoming increasingly common to call both functional groups acetals irrespective of whether the level-two carbon has a hydrogen attached to it or not. We shall follow this convention from now on.

The formation of acetals

Acetals are formed when aldehydes or ketones react with alcohols in the presence of acid. The reaction is an equilibrium and may be forced to reactants or products simply by altering whether water or alcohol is present in excess, as shown in Fig. 9.27.

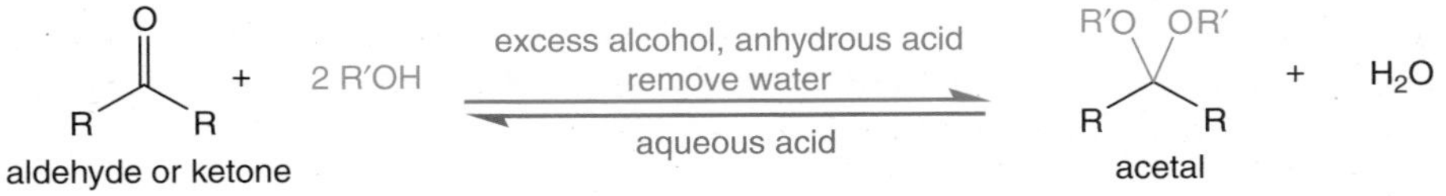

Fig. 9.27 The equilibrium between a carbonyl and an acetal may be driven in either direction simply by having an excess of either the alcohol or water.

In order to prepare the acetal, an anhydrous acid is used together with a large excess of the alcohol. Very often, the water is also removed as it is formed, for example by distilling it out or adding a drying agent, in order to drive the reaction to the right. To hydrolyse the acetal, an excess of aqueous acid is used.

We shall first of all explore the role of the acid in the formation of an acetal. In fact acids play a similar role in many reactions in organic chemistry, so it is well worth spending some time on this now to ensure we understand fully the concepts involved.

The role of the acid in acetal formation

In order to prepare acetals in good yields via the reaction between alcohols and aldehydes or ketones, an acid catalyst must be present. The acid must also be *anhydrous* in order to avoid the reverse reaction – the hydrolysis of the acetal to the carbonyl and alcohol. This reverse reaction is also acid catalysed but requires the presence of water. Typical anhydrous acids used in the preparation of acetals might include HCl or toluenesulfonic acid, which is often just called *tosic acid* and given the symbol TsOH. The structure of tosic acid is shown in Fig. 9.28.

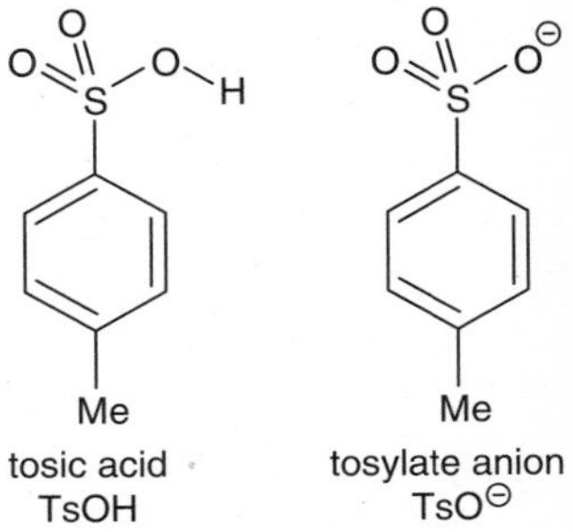

In a mixture of a ketone (or aldehyde), an alcohol and acid, the acid will protonate some of the alcohol, as shown in Fig. 9.29 (a), and also some of the ketone as shown in (b). The proton transfers to and from oxygen all occur very rapidly and so the mixture will contain both the neutral alcohol and ketone, together with the protonated species.

In the reaction between the alcohol and the ketone, the lone pair of the alcohol oxygen attacks into the vacant CO $\pi^\star$ of the ketone. Whilst the

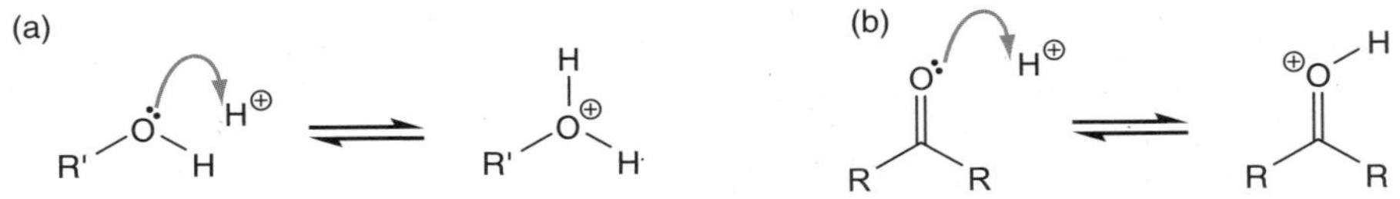

Me
tosic acid
TsOH

Me
tosylate anion
TsO$^{\ominus}$

Fig. 9.28 The structure of toluenesulfonic acid, often called tosic acid, and its conjugate base, the tosylate anion.

Fig. 9.29 In a mixture of acid, an alcohol and a ketone, the acid will protonate some of the alcohol, as shown in (a), and some of the ketone, as in (b).

protonated alcohol still has a lone pair on the oxygen, it is *much* lower in energy than that of the unprotonated alcohol. This is because protonating the oxygen increases its effective nuclear charge and hence lowers the energy of its orbitals. As a result, the interaction between the lone pair of the protonated oxygen and the carbonyl $\pi^\star$ MO is unfavourable and so does not occur.

The mechanism of acetal formation

Protonating the lone pair of the ketone oxygen also increases the effective nuclear charge of the oxygen, and this has the effect of lowering the energy of the C=O $\pi^\star$ as outlined in section 8.7 on page 329. Lowering the energy of the $\pi^\star$ MO makes the interaction with the lone pair of the alcohol oxygen even more favourable than the interaction with the unprotonated ketone. The attack of the lone pair of the alcohol into the C=O $\pi^\star$ of the protonated ketone gives a protonated hemiacetal, **A**, and is shown in Fig. 9.30. Note that the oxygen atoms from the alcohol and ketone have been labelled O_a and O_k respectively.

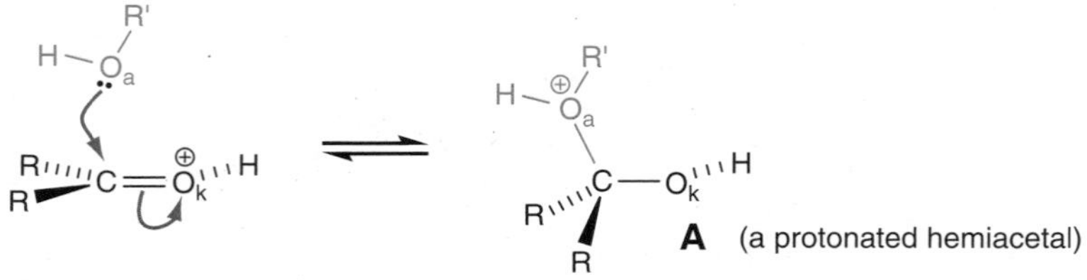

Fig. 9.30 Attack of the lone pair of the alcohol into the C=O $\pi^\star$ of the protonated ketone to give a protonated hemiacetal, **A**.

As we have already seen, hemiacetals tend to lose water to reform the carbonyl, and indeed this is precisely what can happen to **A**. The HOMO of **A** is the lone pair on the oxygen without the positive charge, O_k. There is good orbital overlap between this lone pair and the C–O_a $\sigma^\star$ MO, and this MO is particularly low in energy due to the increased effective nuclear charge of the protonated O_a. The mechanism for the reformation of the ketone is shown in Fig. 9.31. Note that this reaction is exactly the reverse of the reaction shown in Fig. 9.30.

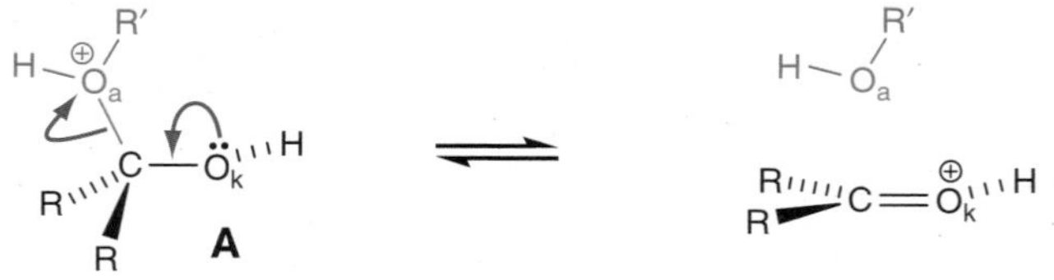

Fig. 9.31 The reformation of the ketone from the protonated hemiacetal **A**.

However **A** can readily undergo another reaction in which it loses a proton from oxygen O_a. It is also possible for O_k to pick up a proton (recall that such proton transfers to and from oxygen are extremely rapid). Rather than showing both of these steps separately, we shall just write '$\pm H^+$' although we should keep it in mind that the protons are being added to and removed from all oxygen atoms very rapidly in a dynamic equilibrium. Figure 9.32 on the facing page shows how proton transfers give species **B**, which is still a protonated acetal, but one protonated on O_k rather than O_a.

This proton transfer is an extremely important step in the formation of the desired acetal because whereas the HOMO of **A** is the oxygen lone pair on O_k,

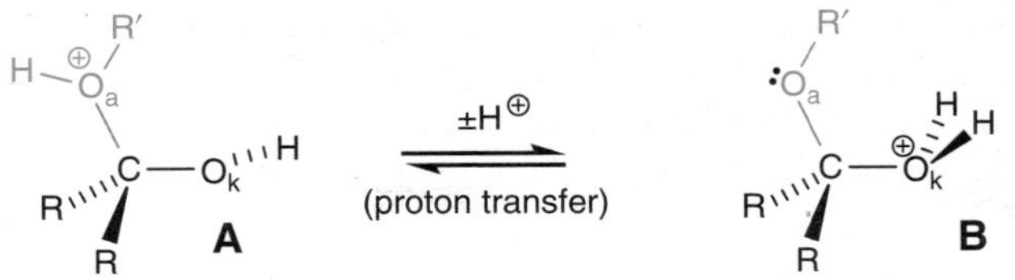

Fig. 9.32 Protons are rapidly coming on and off different oxygen atoms giving an equilibrium between the protonated hemiacetals, **A** and **B**.

the HOMO of **B** is the oxygen lone pair on O_a. This lone pair has good overlap with the C–O_k $\sigma^\star$ MO. Once again, this MO is particularly low in energy since O_k has a formal positive charge. The result of this orbital interaction is the *water* is now eliminated from the protonated hemiacetal rather than alcohol. We say that the protonation of oxygen O_k makes it a better *leaving group* than oxygen O_a. The mechanism for the loss of O_k as water is shown in Fig. 9.33.

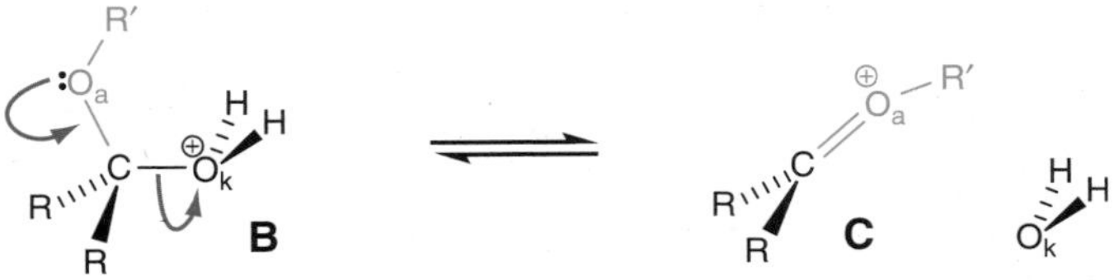

Fig. 9.33 The interaction between the lone pair of O_a and the C–O_k $\sigma^\star$ MO results in the elimination of water and the formation of **C**.

The species formed in this reaction, **C**, is rather like a protonated ketone except that instead of having a *proton* on the oxygen atom, there is an *alkyl group* instead. Whereas the hydrogen bonding in the solution means that a proton would be removed rapidly from the oxygen of the ketone, this alkyl group remains firmly attached.

Like a protonated ketone, **C** also has a very low energy C=O $\pi^\star$ MO which is readily attacked by nucleophiles. The lone pair on the oxygen atom in *water* could attack into the C=O $\pi^\star$ which would be the reserve reaction to that shown in Fig. 9.33. This may well happen, but due to the greater concentration of alcohol rather than water that is present, it is far more likely that the lone pair from an oxygen atom in an *alcohol* molecule will attack instead. This interaction forms a protonated acetal molecule, **D**, as shown in Fig. 9.34, which rapidly loses a proton to give the desired neutral acetal.

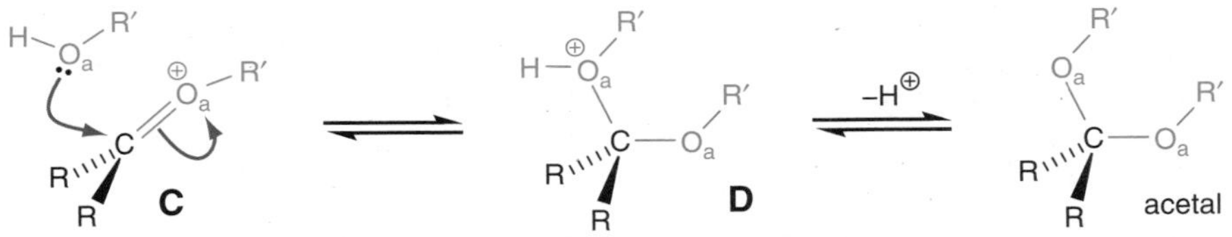

Fig. 9.34 The attack of the oxygen lone pair from an alcohol into the C=O $\pi^\star$ MO of **C** gives rise to the protonated acetal **D**. This rapidly loses a proton to give the neutral acetal.

Note that, in the reaction, the oxygen atom O_k that started out as the ketone oxygen is ultimately lost as water. Both the oxygen atoms in the acetal come from the alcohol.

Hydrolysis of acetals

The acetal formation reaction is easily reversible, and acetals are hydrolysed back to the ketone (or aldehyde) by reaction with aqueous acid. The initial step is the protonation of the acetal to give species **D**, followed by the removal of water to give **C**. In aqueous solution, **C** is now far more likely to be attacked

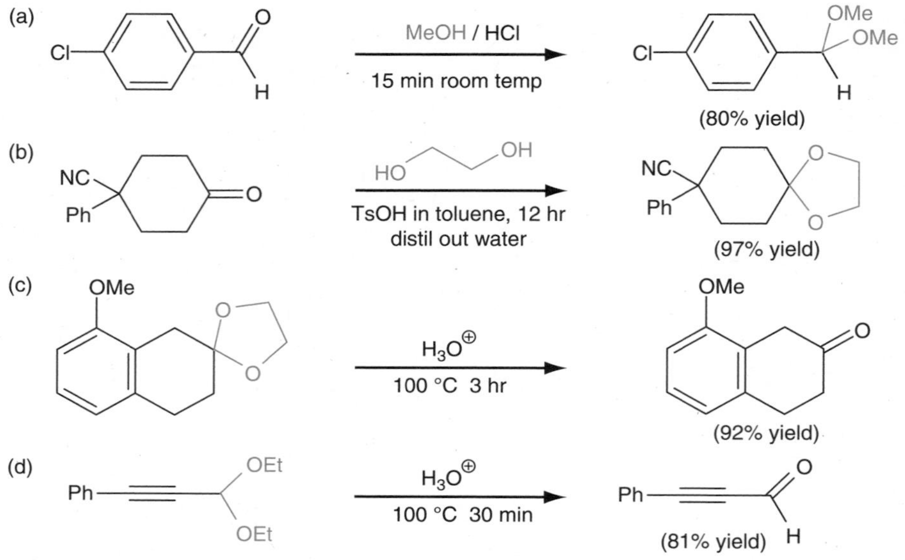

Fig. 9.35 Reactions (a) and (b) show examples of the formation of acetals from an aldehyde or a ketone using alcohol and anhydrous acid. Reactions (c) and (d) show the hydrolysis of acetals in aqueous acid.

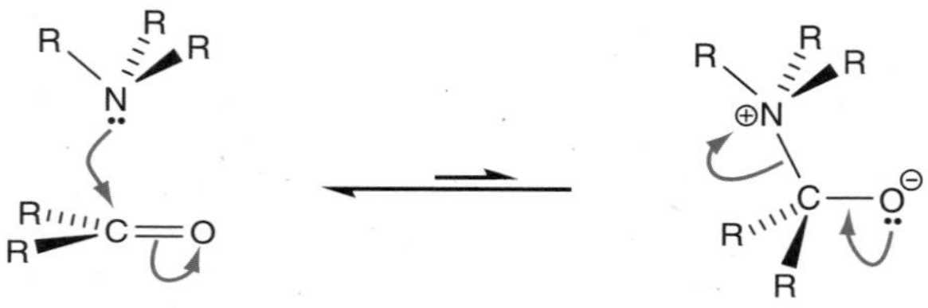

Fig. 9.36 If a tertiary amine attacks a carbonyl to form a tetrahedral intermediate, the tetrahedral intermediate will readily decay by the reverse reaction in which the amine is 'pushed out' by the electrons from the oxygen.

by a *water* molecule rather than an *alcohol* and so the reverse of the reaction shown in Fig. 9.33 on the preceding page takes place to give **B**. After a proton transfer **A** is formed, which then loses the second alcohol as shown in Fig. 9.31 on page 360 to give the protonated ketone.

Some examples of the formation and hydrolysis of acetals are shown in Fig. 9.35.

9.4.3 The reaction of amines with aldehydes and ketones

A primary amine has two hydrogens attached to the nitrogen, RNH_2; a secondary amine has one attached hydrogen, R_2NH; a tertiary amine has no attached hydrogens, R_3N. The R groups need not be the same.

The way in which amines react with aldehydes and ketones depends on the particular carbonyl and on whether the amine is primary, secondary, or tertiary i.e. the number of hydrogens attached to the nitrogen of the amine.

Tertiary amines R_3N are the easiest to deal with: there is essentially no net reaction. If a tertiary amine did attack a carbonyl, this would form a tetrahedral intermediate. This intermediate would have high energy oxygen lone pairs and, due to the positive formal charge on the nitrogen, a particularly low energy C–N σ^* MO. The intermediate would quickly fall apart by pushing out the amine and reforming the carbonyl as shown in Fig. 9.36.

Just because it is not possible to isolate the tetrahedral intermediate, it does not mean that it does form. Later in the chapter (section 9.5.6 on page 381) we shall see how this initial reaction helps tertiary amines to catalyse substitution reactions at level-three carbonyls.

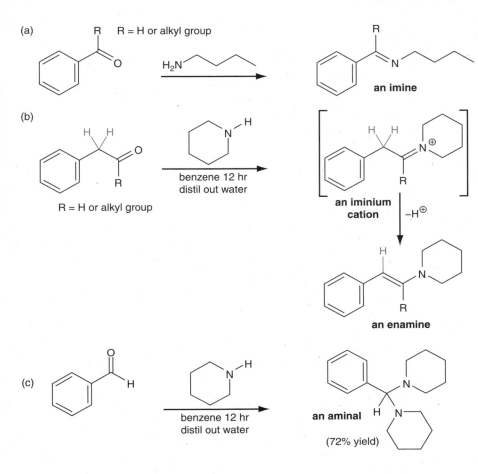

Fig. 9.37 The reaction between a primary amine and an aldehyde or ketone gives an imine as shown in (a). Secondary amines react with most ketones to give enamines, as shown in (b). However, these can only form if the carbonyl has at least one hydrogen (shown in green) on the adjacent carbon. When no such hydrogens are present, aldehydes can react with secondary amines to give aminals, as shown in (c). The increase in crowding as the carbonyl carbon goes from trigonal planar to tetrahedral generally means that aminals do not form from ketones.

In contrast to tertiary amines, primary and secondary amines form very different looking products as shown in Fig. 9.37. *Primary amines* react with aldehydes and ketones to form *imines* as shown in (a). In the formation of an imine, the oxygen from the carbonyl is lost together with the two protons from the amine nitrogen.

When the amine has just one proton on the nitrogen (i.e. a *secondary amine*), it is not possible to form a neutral imine. Instead a positively charged *iminium* ion is initially formed, as shown in (b). However, it is not usually possible to isolate this reactive intermediate. The protonated C=N increases the acidity of any hydrogens attached to the adjacent carbon and one of these may be lost to form a new group called an *enamine*. These hydrogens are shown in green in Fig. 9.37 (b). The name of the enamine is derived from the fact that this functional group is made up of an alk*ene* and an *amine*. We shall cover these groups in more detail in Chapter 13 where we look at the formation and reaction of π systems.

Only in special cases, such as the reaction between a secondary amine and an aldehyde with no hydrogens on the carbon next to the carbonyl, is it possible to make the amine equivalent of an acetal, an *aminal*, as shown in Fig. 9.37 (c).

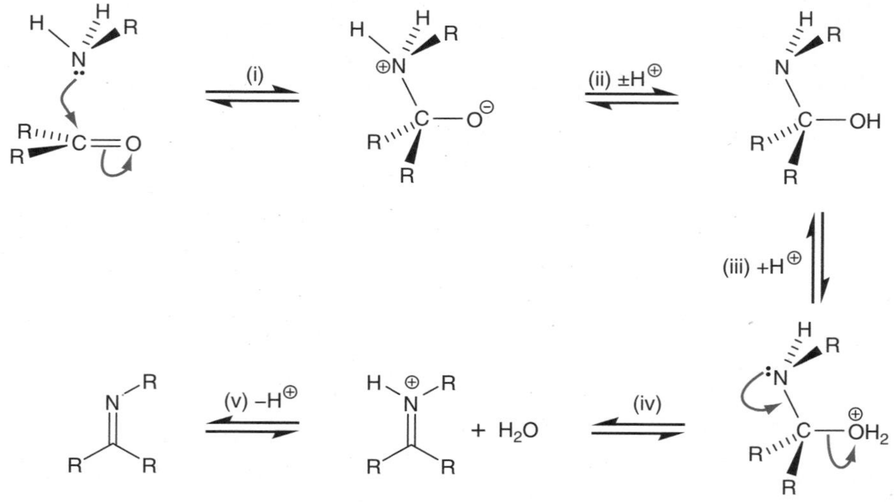

Fig. 9.38 A mechanism for the formation of an imine from a primary amine and an aldehyde or ketone.

The mechanism for the formation of an imine

A mechanism for the formation of an imine from a carbonyl and amine is shown in Fig. 9.38. Step (i) is the attack by the high-energy lone pair of the nitrogen of the amine into the LUMO of the carbonyl, the C=O $\pi^\star$. Remember that even before this attack, any hydrogen bonding to the oxygen of the carbonyl will increase the Z_{eff} of the oxygen and hence lower the energy of the C=O $\pi^\star$ MO still further; it may be the case that no full negative charge ever appears on the oxygen. The formal protonation of the oxygen and the deprotonation of the nitrogen in step (ii) are both extremely rapid and may already have occurred to some extent during the first step.

Step (iii) shows a further protonation of the oxygen. This protonation lowers the energy of the C–O $\sigma^\star$ MO and makes the oxygen a much better leaving group. The oxygen is lost as water in step (iv) and a CN π bond is formed.

Whether or not the oxygen is fully protonated depends on the pH of the solution. However, the effects of the pH on the rate of the reaction are not straightforward since some steps are easier in acid (such as step (iii)), whereas others are made harder (such as step (i), which cannot proceed at all if the nitrogen is fully protonated and hence has no lone pair to attack with).

The loss of water in step (iv) leaves a protonated imine which is rapidly deprotonated, as in step (v), to give the neutral imine.

Enamines and aminals

With secondary amines, the mechanism is the same up to the formation of the iminium cation, but then the nitrogen has no hydrogen to lose in order to become neutral. Instead the ion can lose a proton from a neighbouring carbon to form the enamine as shown in Fig. 9.39. We shall look at this reaction more fully in Chapter 13.

In the formation of an aminal, the iminium ion is attacked by further amine. As we noted in the analogous formation of acetals, moving from the trigonal carbon to tetrahedral carbon increases the steric interactions around the carbon. This is even worse in the case of aminals since an $-NR_2$ group is larger than an $-OR$ group. Consequently, aminals can only usually be made from aldehydes, and even then only when the formation of an enamine is not possible.

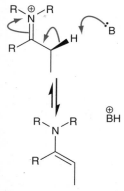

Fig. 9.39 The mechanism for the loss of a proton from an iminium ion to form an enamine.

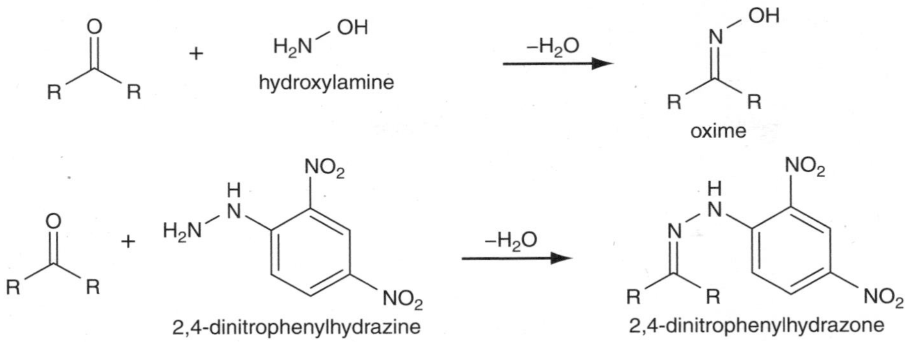

Fig. 9.40 The $-NH_2$ groups from a number of other compounds can form CN π bonds in an analogous manner to the formation of an imine.

9.4.4 Oximes, hydrazones and related compounds

Aldehydes and ketones readily react with a number of other compounds which contain $-NH_2$ groups to form groups with a C=N in a reaction analogous to the formation of an imine. Common reactions include the formation of oximes (from hydroxylamine) and hydrazones (from hydrazines) as shown in Fig. 9.40.

9.4.5 Other transformations

In this section we have looked at transforming one level-two functional group to another. Other transformations are possible, such as the hydrolysis of the *gem*-dichlorides illustrated in Fig. 9.2 on page 342 to give a carbonyl. However, whilst possible, not all transformations are common, or indeed useful. For instance, it is harder to synthesize *gem*-dichlorides in the first place than it would be to make an aldehyde or a ketone so this reaction is not a sensible synthetic method.

9.5 Transformations within functional group level three

In Fig. 9.3 on page 344 we saw examples of different level-three functional groups being hydrolysed to carboxylic acids. The conditions needed to hydrolyse the different functional groups vary enormously and, as we shall see, this reflects the reactivity of the functional groups. Sometimes acid or base is needed to bring about the effective hydrolysis of the group, and we will see why this is so by looking at the mechanism for the hydrolysis of the ester group.

9.5.1 The hydrolysis of esters

How fast an ester is hydrolysed to the alcohol and carboxylic acid depends very strongly on the pH of the solution. The rate of hydrolysis for an ester such as ethyl ethanoate at constant pH may be written

$$\text{rate of hydrolysis of ester} = k_{obs} \times [\text{ester}]$$

where [ester] is the concentration of the ester in solution, and k_{obs} is the *rate constant* for the reaction. The value of k_{obs} depends on the concentration of H_3O^+ and OH^- ions and hence the pH of the solution, as shown in Fig. 9.41 on the next page. For a given concentration of ester, the reaction is fastest *either*

Fig. 9.41 A graph showing how k_{obs} for the hydrolysis of ethyl ethanoate in aqueous solution varies with the pH of the solution at 25 °C. For a given concentration of ester, the reaction is fastest where there is a high concentration of H_3O^+ ions (low pH) *or* where there is a high concentration of OH^- ions (high pH).

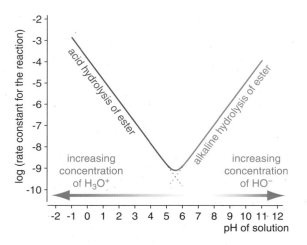

where there is a high concentration of H_3O^+ ions (low pH) *or* where there is a high concentration of OH^- ions (high pH). The mechanisms for the hydrolysis of ethyl ethanoate under acidic conditions and under alkaline conditions are different.

The alkaline hydrolysis of an ester

A mechanism for the alkaline hydrolysis of ethyl ethanoate is shown in Fig. 9.42.

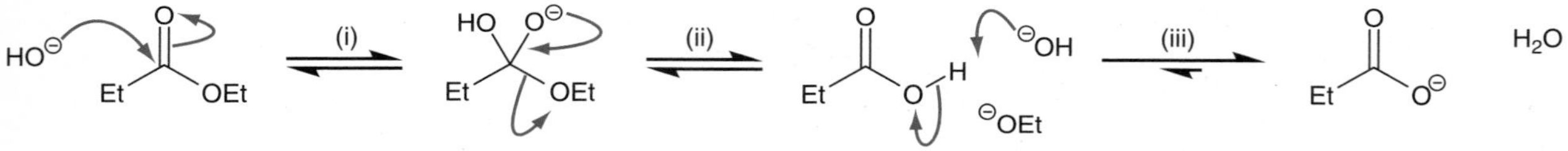

Fig. 9.42 A mechanism for the alkaline hydrolysis of ethyl ethanoate.

The highest energy electrons of the reactants are those of the oxygen lone pair in the hydroxide ion. The LUMO of the reactants is the CO $\pi^\star$ of the ester. The initial interaction is the attack of the oxygen lone pair of hydroxide into the CO $\pi^\star$ MO of the ester to give the tetrahedral intermediate as shown in step (i). Note it is not possible to protonate the oxygen of the carbonyl since the concentration of acid present under alkaline conditions is tiny. Furthermore, if the carbonyl oxygen was protonated, the first thing that would happen in base is that hydroxide would deprotonate it.

Two things can now happen to the tetrahedral intermediate: it can either lose hydroxide ion in the reverse of step (i) to re-form the ester, or it can lose an ethoxide ion as shown in step (ii). There is very little to chose between these two alternatives. In each case a C–O bond is being broken and an anion (HO^- or EtO^-) is produced in which the negative charge is localized on a single oxygen atom. The ethoxide ion can rapidly pick up a proton from water to form ethanol and more hydroxide.

The relative energies of the species at each stage of the reaction are shown in Fig. 9.43 on the next page. The tetrahedral intermediate is highest in energy since there is more crowding round the central carbon atom, but there is very little difference in energy between the hydroxide and ester on the left of the tetrahedral intermediate, and the carboxylic acid and ethoxide ion on the right.

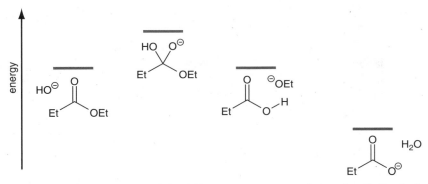

Fig. 9.43 The relative energies of the different species present during the alkaline hydrolysis of ethyl ethanoate. The tetrahedral intermediate is highest in energy since it has a negative charge localized on one oxygen, and it is also crowded with four groups attached to the central carbon. There is little to choose between the ester and hydroxide ion on one hand, and the carboxylic acid and ethoxide ion on the other; in both cases there is a negative charge localized on one oxygen atom. The solution of ethanoate ion is lowest in energy since this has the negative charge delocalized over two oxygen atoms.

However, the reaction does not end here after step (ii). The carboxylic acid is then deprotonated in these alkaline conditions as shown in step (iii) in Fig. 9.42 on the facing page. It might be the ethoxide ion that has just left the tetrahedral intermediate in step (ii) that does the deprotonation, but it is more likely to be some of the hydroxide that is present in larger concentrations. Either way, deprotonating the carboxylic acid leaves the carboxylate ion which is considerably lower in energy due to the charge now being delocalized over two oxygens rather than concentrated on just one.

Overall, the reaction is driven to the lowest energy species, the carboxylate ion and the alcohol. Note that during the reaction, hydroxide ion is consumed. This means that, strictly speaking, the reaction cannot be said to be catalysed by hydroxide as is sometimes stated.

The acid hydrolysis of an ester

A mechanism for the hydrolysis of ethyl ethanoate under *acidic* conditions is shown in Fig. 9.44 on the next page. The highest energy electrons in the reactants are the lone pair in water. These are not as high in energy as the lone pair in hydroxide, and the reaction where the oxygen lone pair from neutral water attacks the $\pi^\star$ MO of the ester only occurs very slowly. However, the carbonyl $\pi^\star$ is lowered in energy upon protonation of the carbonyl oxygen (see section 8.7 on page 329), which is a rapid process. Either of the oxygen atoms can be protonated, but only when the carbonyl oxygen is protonated, as in step (i), can further reaction take place.

With the carbonyl oxygen protonated, the CO $\pi^\star$ MO is sufficiently low in energy for the lone pair from water to attack as shown in step (ii) and form the tetrahedral intermediate labelled **A** in Fig. 9.44.

The LUMO of **A** is the $\sigma^\star$ MO between the carbon atom and the oxygen atom with the formal positive charge. The highest energy electrons are those on one of the neutral oxygen atoms, either the –OH group or the –OEt group. The interaction between these two orbitals re-forms the CO π bond, pushing out water as a leaving group in the reverse of step (ii).

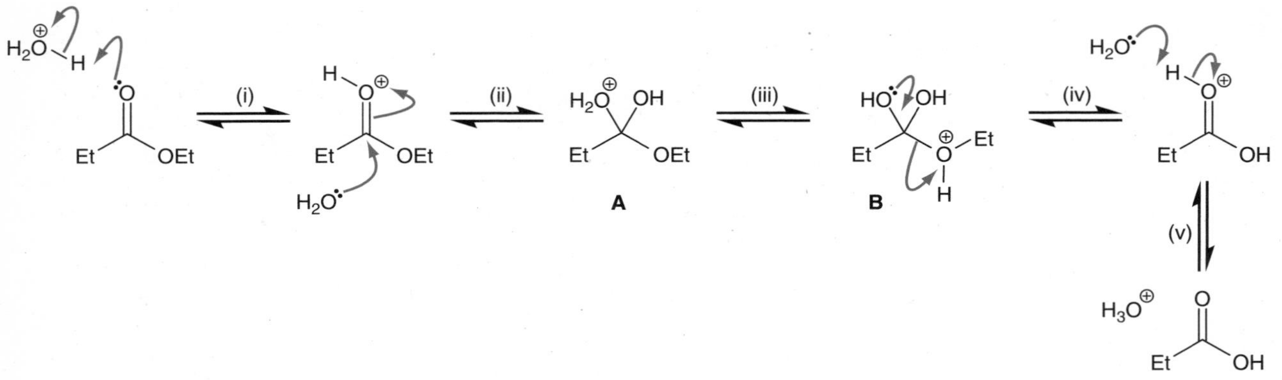

Fig. 9.44 A mechanism for the acid catalysed hydrolysis of ethyl ethanoate.

However, since proton transfer to oxygen is extremely rapid in aqueous solution, protons can be added to and removed from any of the oxygens in the tetrahedral intermediate. When an –OH group is protonated, the reverse of step (ii) takes place, but when the –OEt group is protonated to give **B**, step (iv) can take place. In this neutral ethanol is expelled as the carbonyl π bond is re-formed giving the protonated carboxylic acid.

The protonated carboxylic acid and water are in equilibrium with the neutral carboxylic acid and H_3O^+ as shown in step (v). This equilibrium regenerates the H_3O^+ used in step (i) and so this acid hydrolysis is catalysed by acid, unlike the alkaline hydrolysis in which hydroxide ions are consumed.

Driving the equilibrium

Notice that unlike in the alkaline hydrolysis, every step of the acid hydrolysis mechanism shown in Fig. 9.44 is reversible. With excess water and an acid catalyst, esters are hydrolysed to the carboxylic acid and the alcohol; an example is shown in (a) in Fig. 9.45 on the facing page. Esters can be prepared from the carboxylic acid and alcohol by using an excess of alcohol and driving out the water as it is formed in the reaction, or by adding a drying agent to remove the water; examples are shown in (b) and (c) in Fig. 9.45. The formation of the ester is also catalysed by acid and often concentrated sulfuric acid is used since this is also effective in helping to remove the water by strongly solvating with it.

9.5.2 Preparation of carboxylic acids by the hydrolysis of other level-three functional groups

Whilst all the level-three functional groups can be hydrolysed to give carboxylic acids as shown in Fig. 9.3 on page 344, it is not common to make carboxylic acids in this way since the carboxylic acids themselves are often the starting materials used to make these functional groups. Ester hydrolysis, both under alkaline and acidic conditions, is routinely used to make carboxylic acids, and another useful route is the hydrolysis of nitriles. This latter route is useful because the nitrile group may easily be introduced into a molecule by means of the substitution of, for example, bromide by cyanide. Figure 9.46 on the facing page shows such an example, used in one of the early steps in a synthesis of flunitrazepam, the active component in the drug Rohypnol.

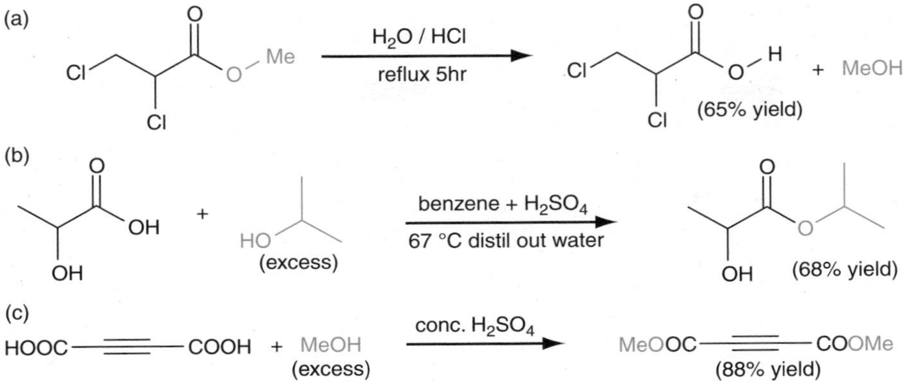

(a)

H₂O / HCl
reflux 5hr

(65% yield)   + MeOH

(b)

benzene + H₂SO₄
67 °C distil out water

(68% yield)

(excess)

(c)

HOOC———————COOH + MeOH   conc. H₂SO₄   MeOOC———————COOMe
(excess)

(88% yield)

Fig. 9.45 Reaction (a) is an example of the acid-catalysed hydrolysis of an ester using aqueous acid. Reactions (b) and (c) are examples of the formation of esters using an excess of alcohol and removing the water either by distilling it out, or solvating it with concentrated sulfuric acid.

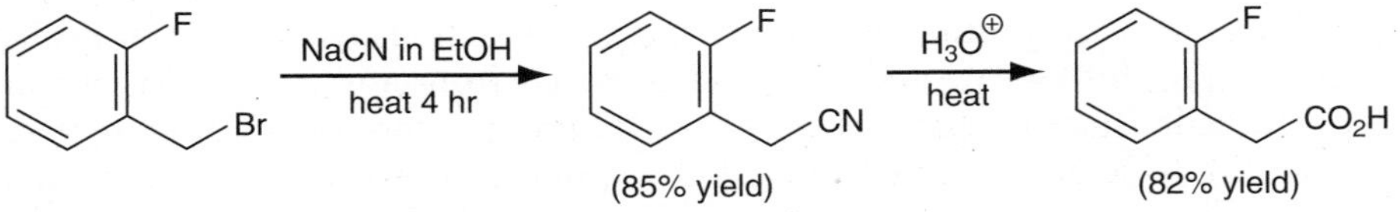

NaCN in EtOH
heat 4 hr

(85% yield)

H₃O⊕
heat

(82% yield)

Fig. 9.46 In the first step, a nucleophilic substitution reaction replaces the bromine atom with cyanide to form a nitrile. This reaction extends the number of carbon atoms in the molecule by one. In the second step, the nitrile is hydrolysed to give the carboxylic acid.

Nitrile hydrolysis mechanism

A mechanism for the hydrolysis of a nitrile in aqueous acid to give a carboxylic acid is shown in Fig. 9.47. The first step is the protonation of the nitrile by the HOMO (the nitrogen lone pair) interacting with H^+. It is relatively difficult to protonate the nitrile, but this lowers the energy of the CN $\pi^\star$ MO in an analogous manner to the protonation of the oxygen of a carbonyl as outlined in section 8.7 on page 329. Once protonated, the LUMO of the nitrile, the CN $\pi^\star$ is attacked by the high energy lone pair of the oxygen in water to give intermediate **B**, as shown in step (ii).

The HOMO of **B** is the nitrogen lone pair. This has good overlap with the low energy C–O $\sigma^\star$ MO and this interaction results in water being lost from

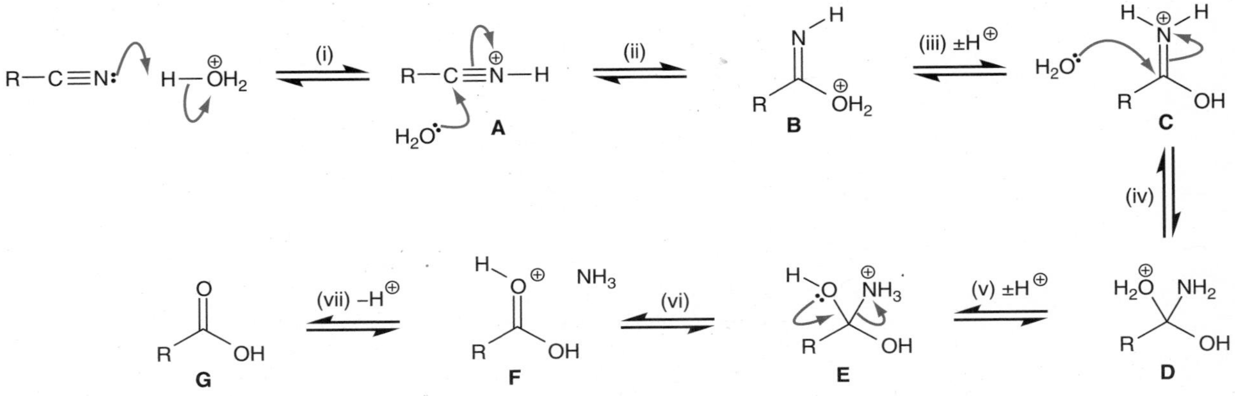

Fig. 9.47 A mechanism for the acid-catalysed hydrolysis of a nitrile.

the intermediate, re-forming **A**. However, removing a proton from the oxygen and adding one to the nitrogen in a rapid proton transfer, gives **C**. The LUMO for this intermediate is the $\pi^\star$ MO and this can readily be attacked by the high energy electrons from an oxygen lone pair, as shown in step (iv), to give the tetrahedral intermediate **D**.

In **D**, the high energy nitrogen lone pair can expel the protonated oxygen (as H_2O) to re-form **C**. However, after the rapid swapping of protons in step (v), one of the oxygen lone pairs can expel the protonated nitrogen (as NH_3) to give **F**. Species **F** is simply a protonated carboxylic acid; this can readily lose a proton to give the neutral carboxylic acid **G**.

Partial hydrolysis of a nitrile to an amide

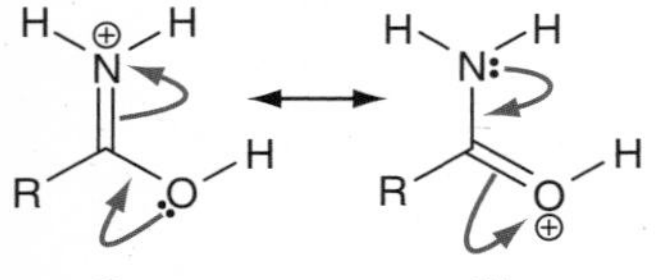

In the hydrolysis of a nitrile shown in Fig. 9.3 on page 344, the amide was formed first before being further hydrolysed to the carboxylic acid. The amide is also present in the mechanism shown in Fig. 9.47, but in a disguised form. Intermediate **C** is actually a protonated amide, which is shown in its more usual form **C′** in Fig. 9.48. The curly arrows show how these two resonance structures are related but both forms are valid representations of a protonated amide.

We saw in section 4.9.3 on page 175 that the π system in an amide extends over the oxygen, carbon, and nitrogen, and this is what the resonance structures also show. It makes no difference which resonance structure is used when writing the mechanism of the hydrolysis; in Fig. 9.47 on the preceding page structure **C** was used, but we could equally well have drawn a mechanism using structure **C′**. Both mechanisms involve the attack on the level-three carbon atom and, as Fig. 4.48 on page 176 shows, this is the atom in the LUMO of an amide that has the largest coefficient.

Fig. 9.48 Different representations of a protonated amide.

9.5.3   Carbonyl reactivity series

The partial hydrolysis of a nitrile to give an amide requires less extreme conditions than the full hydrolysis to the carboxylic acid, as may be seen in Fig. 9.3 on page 344. This is an important point: different functional groups hydrolyse at very different rates, reflecting the reactivity of the functional group. This is illustrated in Fig. 9.49 for various carboxylic acid derivatives.

In the hydrolysis of each of the acid derivatives shown in Fig. 9.3, the first step (ignoring any possible protonation of the carbonyl oxygen), is the attack of the oxygen lone pair into LUMO, the $\pi^\star$ of the carbonyl. The relative energies of the $\pi^\star$ MOs for these common acid derivatives are shown in Fig. 9.50 on the facing page, together with the approximate energy of the $\pi^\star$ MO of methanal

Fig. 9.49 Acyl chlorides are the most reactive derivatives of carboxylic acids, and are rapidly hydrolysed in water. The next most reactive are anhydrides. Amides are the the least reactive acid derivative and are only hydrolysed by prolonged treatment with strong hot acid or alkali.

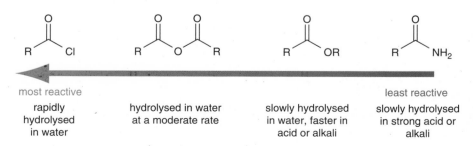

for comparison. The most reactive acyl chloride has the lowest energy LUMO, the least reactive amide has the highest energy LUMO.

We can understand this order by considering how we can combine an additional p AO with the π and $\pi^\star$ MOs of an isolated C=O double bond such as in methanal. The approximate form and energy of the π and $\pi^\star$ MOs of methanal are shown on the left hand side of Fig. 9.51, and on the right hand side is shown the additional p AO from the atom which will take the place of one of the hydrogens in methanal. The interaction of these orbitals generates three π MOs which are shown in the centre of the figure. We have already seen how to predict the *form* of these MOs in section 4.8.1 on page 166 but here we want to look at the relative *energies* of these orbitals.

The most important points are that the lowest energy orbital before the interaction, the CO π MO, is lowered further on forming the new 1π MO, whereas the highest energy orbital, the CO $\pi^\star$ MO, is raised in energy on forming the new 3π MO. It is not easy without carrying out detailed calculations to predict whether the 2π MO is higher or lower in energy than the isolated p orbital.

Conjugating an extra p orbital with an existing π bonding MO and anti-bonding MO results in a net lowering in energy of electrons since the 1π MO is lower in energy than the original CO π MO. However, the vacant antibonding 3π MO is raised in energy because of the interaction. This makes no difference to the *stability* of the molecule, but it does make it less susceptible to attack by a nucleophile.

Esters *vs* aldehydes and ketones

Figure 9.51 thus predicts that a functional group such as an ester, which has the p orbital of a second oxygen atom conjugated with the carbonyl, will have a higher energy LUMO than a functional group which has just the C=O carbonyl such as an aldehyde or ketone. Having a higher energy LUMO, the $\pi^\star$ MO, means the ester will be less susceptible to nucleophilic attack than the aldehyde or ketone.

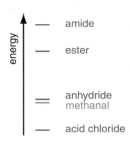

Fig. 9.50 The approximate relative energies of the $\pi^\star$ MOs of some carboxylic acid derivatives.

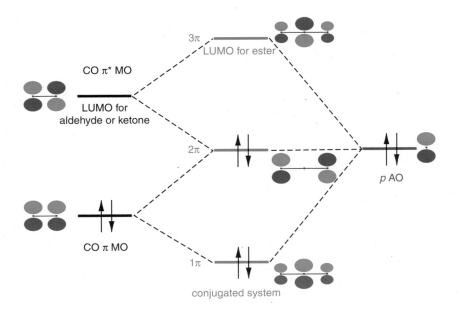

Fig. 9.51 A schematic MO diagram to show how conjugating an extra p orbital with an existing π and $\pi^\star$ MO alters the relative energies of the orbitals. The energies and forms of the CO π and $\pi^\star$ MOs are shown on the left, whilst on the right is the extra p orbital which is to interact with the existing π MOs. The three resulting MOs from this interaction are shown in the centre.

Fig. 9.52 A schematic MO diagram to compare the relative energies of the π MOs of an ester and an amide. Since a nitrogen p AO is higher in energy than an oxygen p AO, there is a stronger interaction between the nitrogen AO and the CO π* MO. This results in the LUMO for an amide being higher in energy than the LUMO for an ester.

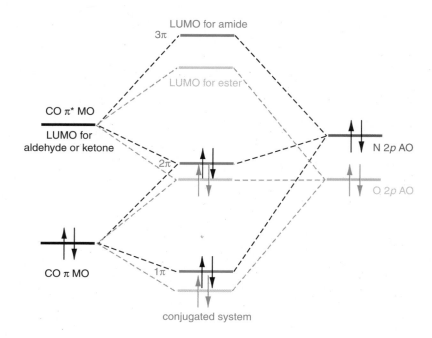

Esters *vs* amides

In constructing the MOs for a conjugated π system from a carbonyl and an extra p AO, the energy match between the p orbital and the π MOs is important. Figure 9.52 shows what happens when the p orbital is changed from an oxygen p AO to a nitrogen p AO. The nitrogen p orbital is higher in energy than the oxygen p orbital. This means there is a worse match in energy between the nitrogen p orbital and the CO π MO, but a better match between the nitrogen p orbital and the π* MO. The net result is that the LUMO for the amide, MO 3π, is higher in energy for an amide than for an ester.

The MO diagrams shown in Fig. 9.51 and Fig. 9.52 are purely schematic and show only the effects of π conjugation. Other factors are important too, such as the size of the extra substituent attached to the carbonyl: the larger the group, the harder it would be for a nucleophile to attack the carbonyl. The substituent may also attract electron density towards its nucleus, altering the effective nuclear charge of the carbonyl carbon.

Acyl chlorides *vs* amides

We have now rationalized why amides are less reactive than esters, but why are acyl chlorides *more* reactive? There is very little difference in *energy* between the valence p AOs of nitrogen and chlorine. However, the valence orbital of chlorine is a 3p AO whereas the valence orbital of nitrogen is a 2p AO. We saw in section 7.9.1 on page 297 that there is a stronger interaction between two 2p AOs than between a 2p AO and a 3p AO. This means whilst there is a strong interaction between the nitrogen 2p AO with the carbon and oxygen 2p AOs in an amide, there is a weaker interaction with the chlorine 3p AO with the carbon and oxygen 2p AOs in an acyl chloride. This is shown schematically in Fig. 9.53 on the facing page.

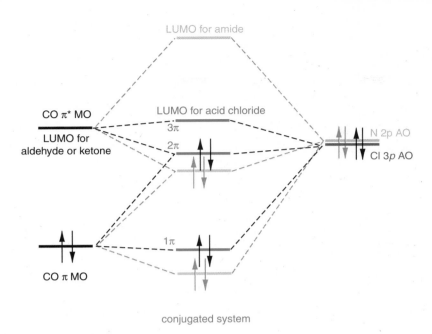

conjugated system

Fig. 9.53 A schematic MO diagram to compare the relative energies of the π MOs of an amide and an acyl chloride. Even though the p AOs of chlorine and nitrogen are of similar energy, there is a much weaker interaction between the chlorine $3p$ AO with the $2p$ AOs of carbon and oxygen in the acyl chloride than there is between the $2p$ AO of nitrogen with the $2p$ AOs of carbon and oxygen in the amide. This results in the LUMO for an acyl chloride being much lower in energy than the LUMO for an amide.

Since there is not a strong interaction between the chlorine AO and the carbonyl MOs, the LUMO for the acyl chloride (the 3π MO) is largely the CO $\pi^\star$ MO, the 1π MO is largely the CO π MO, and the 2π MO is largely an unchanged $3p$ AO of chlorine. This can be seen in Fig. 9.54 on the next page which compares the isosurface plots of the π MOs of an amide and an acyl chloride. Whereas the nitrogen makes a significant contribution to 1π and 3π MOs in the amide, there is little contribution from the chlorine to these orbitals and hence they appear as just a normal CO π MO and $\pi^\star$ MO respectively. In contrast, whereas the non-bonding 2π MO in the amide has significant contributions from both the nitrogen and the oxygen (with a node on the carbon), for the acyl chloride this MO has the greatest contribution from the chlorine $3p$ AO.

Acyl chlorides *vs* aldehydes and ketones

Figure 9.53 suggests that the since there is still some interaction between the chlorine $3p$ AO and the carbonyl π system, the LUMO for the acyl chloride should be *slightly* higher in energy than that of the aldehyde or ketone which has no such interaction. However, as Fig. 9.50 on page 371 shows, the acyl chloride LUMO is *lower* in energy than that of an aldehyde. This is because whilst the chlorine atom does not efficiently conjugate with the π system, it does still attract electron density towards itself since it has a larger Z_{eff} than carbon. Hence the carbonyl carbon in an acyl chloride has a greater partial positive charge than in aldehyde or ketone, and this in turn helps to lower the energy of the $\pi^\star$ MO further.

Aldehydes and ketones *vs* methanal

It is worthwhile finishing this section by comparing the energies of the π MOs for aldehydes, ketones and methanal even though these are level-two functional

(a) amide MOs

(b) acid chloride MOs

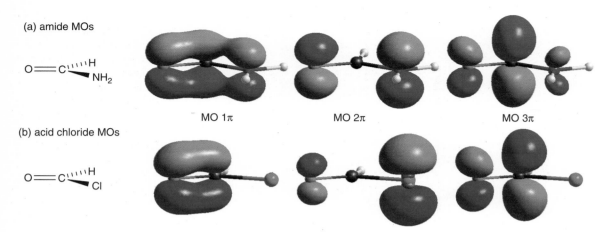

Fig. 9.54 The π MOs for (a) an amide, and (b) an acyl chloride. For the amide, the $2p$ AO of the nitrogen makes a significant contribution to all of the orbitals. For the acyl chloride, the larger size of the chlorine $3p$ AO means it does not conjugate to the same extent with the carbon and oxygen $2p$ AOs. The 1π and 3π MOs have no significant contribution from the chlorine and appear as normal CO π and $\pi^\star$ MOs. The nonbonding 2π MO is largely a chlorine $3p$ AO.

groups. We rationalized the energies of amides and esters by combining an extra p orbital with the π system of a carbonyl. The two hydrogens attached to the carbonyl of methanal cannot interact with the π system since they lie in the nodal plane of the π system and hence there are equal amounts of positive and negative overlap between the orbitals.

However, any group other than a hydrogen will have a p orbital which can overlap to some degree with the π system of the carbonyl. In the above examples, we looked at the interaction of a high-energy lone pair with the carbonyl, but even σ MOs can interact in an analogous way, although the interaction is not so strong.

Figure 9.55 on the facing page shows a schematic MO diagram illustrating the interaction between a C–H σ bonding MO with the π and $\pi^\star$ MOs of a carbonyl. The first point to note, in contrast to the previous examples, is that the C–H σ bonding MO is *lower* in energy than both the CO π and $\pi^\star$ MOs. Nonetheless, there is still a weak interaction between it and the CO π MOs. MO 1 is still largely a C–H bonding MO, but it is now a little lower in energy and has a small contribution from the CO π bonding MO in it. Similarly, MO 2 is still largely the CO π bonding MO, but now with a small contribution from the σ MO. Finally, MO 3 is still mainly the CO $\pi^\star$ MO, but now it also has a small contribution from the σ MO.

The form of these MOs are shown as isosurface plots in Fig. 9.56 on the next page. The plots in (a) are made with a higher value of the wavefunction than those in (b). Thus the plots in (a) show where the electron density is most concentrated, whereas those in (b) also show the atoms on which smaller amounts of electron density are to be found.

MO 1 made up of the $2p$ AO of the methyl carbon overlapping with two of the hydrogens attached to that carbon. The third hydrogen of the methyl group lies in the nodal plane of the p orbital and so does not interact with the p orbital. View (a) shows that MO 1 contributes to the C–H bonding in the methyl group. However, view (b) shows that there is some interaction of this orbital with the CO π system.

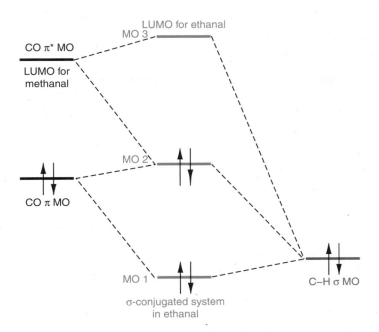

Fig. 9.55 An adjacent C–H σ MO can interact with the CO π and $\pi^\star$ MOs to some degree. This weak interaction lowers the energy of the σ MO slightly, but raises the energy of the $\pi^\star$ MO. Such interactions are not possible in methanal since the hydrogens are in the nodal plane of the π system.

View (a) of MO 2 shows that this orbital is essentially just the CO π bonding MO, but again view (b) shows that there is also some interaction with the methyl group. Even MO 3, which is essentially just the CO $\pi^\star$ MO, also has a small contribution from the hydrogens of the methyl group.

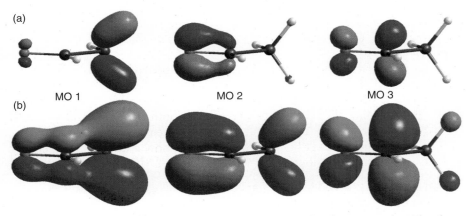

Fig. 9.56 Isosurface plots of three MOs of ethanal. In (a) the plots are made with a large value of the wavefunction and thus show where the electron density is most concentrated. In (b) a smaller value of the wavefunction has been chosen and hence these plots show the atoms which also have smaller amounts of electron density. MO 1 essentially just contributes to the bonding between the methyl carbon and its attached hydrogens, MO 2 is essentially the CO π bonding MO, and MO 3 is essentially the CO $\pi^\star$ MO. However, the plots shown in (b) illustrate that each orbital is delocalized to some extent over the whole molecule.

Fig. 9.57 Acyl chlorides readily react with carboxylate ions, water, alcohols, and amines to form acid anhydrides, carboxylic acids, esters, and amides respectively.

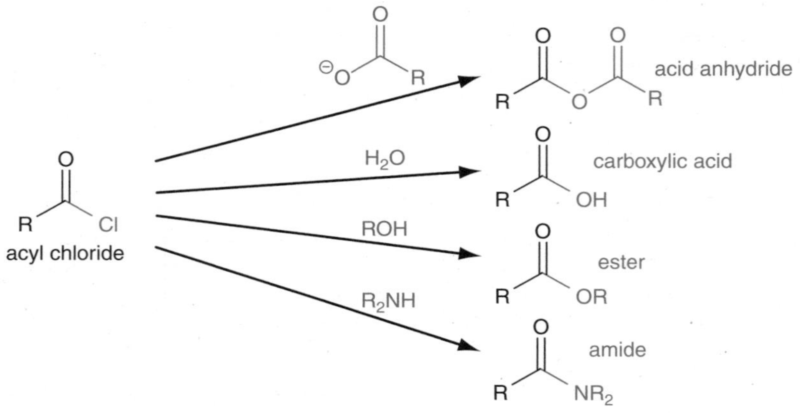

σ conjugation

This conjugation of the C–H σ bonding MO with the π system is sometimes called σ conjugation. This has a number of important consequences for the ketone group as we shall see later, but for the moment the key point is how this interaction affects the energies of the LUMOs in aldehydes and ketones. Ketones such as acetone (propanone) have C–H σ bonds on both sides of the carbonyl that can interact in this fashion, and this raises the energy of the CO $\pi^\star$ MO (the LUMO) relative to the energy of the $\pi^\star$ in methanal which has no such interaction. Aldehydes have σ conjugation from one side only and the LUMO of an aldehyde is in between that of a ketone and methanal.

Of course, steric factors also play an important role in the reactivity of aldehydes and ketones: it is much easier for a nucleophile to approach an aldehyde with its small hydrogen attached to the carbonyl than it is for the nucleophile to attack a ketone which has two larger groups attached to the carbonyl.

Summary

We have now rationalized the ordering of the $\pi^\star$ LUMOs of the common carbonyl groups as shown in Fig. 9.50 on page 371. This in part explains the order of reactivity of these groups as shown in Fig. 9.49 on page 370: acyl chlorides are the most reactive and amides are the least. The other important feature of the acyl chloride which makes it so reactive is the weak C–Cl bond which means the group readily loses chloride ion in its reactions.

We can also include aldehydes and ketones in the reactivity series. Generally, these are more reactive than esters, but less reactive than acyl chlorides. In the remainder of this chapter we shall see how it is possible to exploit this order of reactivity.

9.5.4 Acyl chlorides

Acyl chlorides are particularly useful level-three functional groups since, being so reactive, they can be used to make all the other acid derivatives, as shown in Fig. 9.57. The mechanism for each of the reactions is the same and is shown in Fig. 9.58 on the next page.

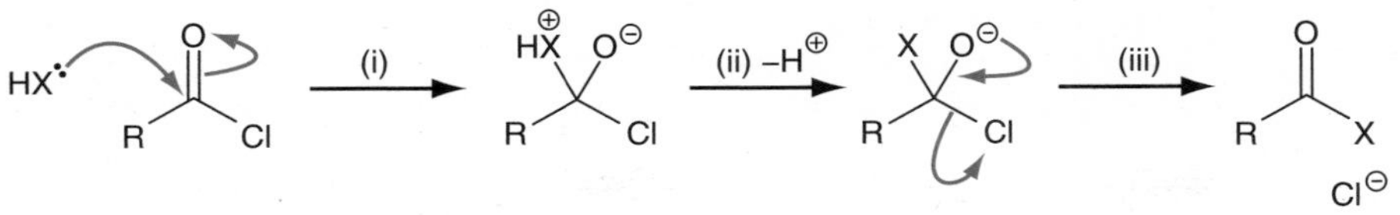

Fig. 9.58 The mechanism for the nucleophilic attack on an acyl chloride.

In step (i) the high-energy lone pair of the nucleophile attacks into the CO $\pi^\star$ MO, the LUMO of the acyl chloride. This gives rise to a tetrahedral intermediate. In step (ii) this loses a proton to become neutral, and then in step (iii) the high energy oxygen lone pair in the tetrahedral intermediate pushes out chloride ion to re-form the carbonyl.

Note that whilst it might be tempting to draw a mechanism in which the nucleophile replaces the chloride directly, as shown in Fig. 9.59, this is incorrect. For this reaction to occur, the nucleophile would need to attack into the C–Cl $\sigma^\star$ MO, thus breaking the C–Cl σ bond. However, the LUMO of the acyl chloride is not the C–Cl $\sigma^\star$ MO but the CO $\pi^\star$ MO. Attacking into this orbital breaks the C=O π bond and forms the tetrahedral intermediate as shown in Fig. 9.58.

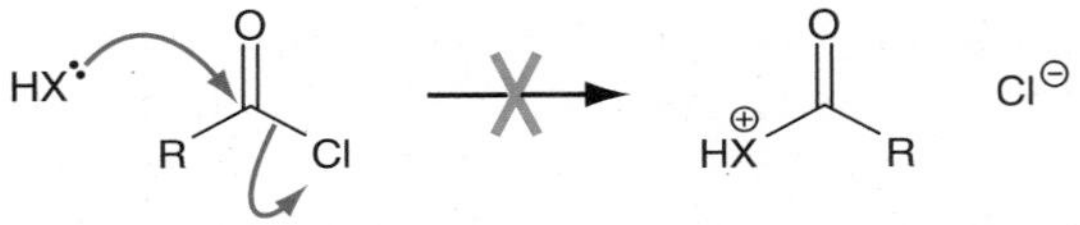

Fig. 9.59 Direct replacement of chloride ion by the nucleophile would mean attacking into the C–Cl $\sigma^\star$ MO. This does not occur since this MO is not the LUMO of the acyl chloride.

Making an acyl chloride

It is not possible to make an acyl chloride by the reaction of a chloride ion with one of the other acid derivatives. For example, a chloride ion cannot displace hydroxide ion according to the plausible-looking mechanism shown in Fig. 9.60.

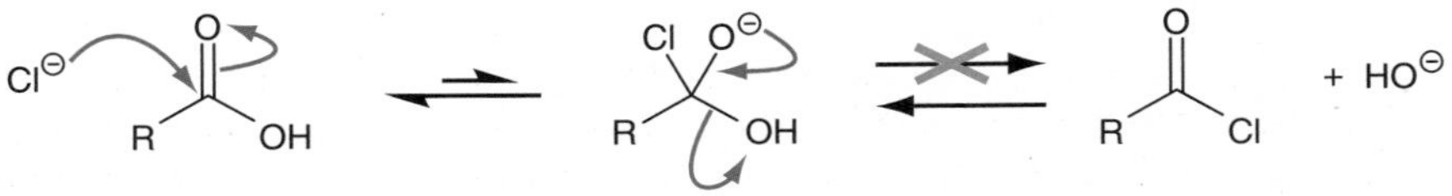

Fig. 9.60 If a chloride ion attacks a carboxylic acid to form a tetrahedral ion, this will not then eliminate hydroxide ion to form the acyl chloride, but will instead eliminate the chloride ion to re-form the carboxylic acid.

If a chloride ion did attack into the CO $\pi^\star$ MO of a carboxylic acid (which is not impossible), the tetrahedral intermediate so formed would not push out the hydroxide ion to give the acyl chloride. Instead it would push out the chloride ion to re-form the carboxylic acid. In fact, the reverse of the reaction – the reaction of an acyl chloride with an hydroxide ion – is extremely fast and the chloride ion is readily substituted by the hydroxide ion. We say that the chloride ion is a much better leaving group from the tetrahedral intermediate than is the hydroxide ion.

Calculating $\Delta_r G^\circ$ and K for the reaction

It is worthwhile – and good revision – to calculate $\Delta_r G^\circ$ and K as outlined in section 6.9 on page 227 for the hypothetical reaction between HCl and a carboxylic acid to give the acyl chloride. The reaction is

$$CH_3COOH(l) + HCl(g) \longrightarrow CH_3COCl(l) + H_2O(l)$$

Δ_rH° can be calculated from the standard enthalpies of formation

$$
\begin{aligned}
\Delta_rH^\circ &= [\Delta_fH^\circ(CH_3COCl(l)) + \Delta_fH^\circ(H_2O(l))] - [\Delta_fH^\circ(CH_3COOH(l)) + \Delta_fH^\circ(HCl(g))] \\
&= [-273.8 - 285.8] - [-484.5 - 92.3] = +17.2 \text{ kJ mol}^{-1}.
\end{aligned}
$$

Similarly Δ_rS° is computed as

$$
\begin{aligned}
\Delta_rS^\circ &= [S_m^\circ(CH_3COCl(l)) + S_m^\circ(H_2O(l))] - [S_m^\circ(CH_3COOH(l)) + S_m^\circ(HCl(g))] \\
&= [200.8 + 70.0] - [159.8 + 186.9] \\
&= -75.9 \text{ J K}^{-1} \text{ mol}^{-1}.
\end{aligned}
$$

We can use these values to find Δ_rG° and hence K for the reaction

$$\Delta_rG^\circ = \Delta_rH^\circ - T\Delta_rS^\circ = 17.2 \times 10^3 - 298 \times (-75.9) = 3.98 \times 10^4 \text{ J mol}^{-1}.$$

$$\ln K = \frac{-\Delta_rG^\circ}{RT} = \frac{-3.98 \times 10^4}{8.314 \times 298} = -16.1$$

$$\text{therefore } K = 1.05 \times 10^{-7}.$$

Such a small value for the equilibrium constant essentially means this reaction does not go at all. The reason the reaction is unfavourable is partly due to negative entropy change (which is mainly due to the loss of the HCl gas), and partly due to the enthalpy change of the reaction: the reaction is endothermic. This is in turn largely due to the fact that, in order to form the acyl chloride, a strong C–O bond needs to be broken but only a weak C–Cl bond is formed in its place. These bond strengths are shown in Fig. 9.61. The C–O bond is particularly strong in the acid due to the conjugation of the oxygen lone pair with the CO π system.

Given that an acyl chloride cannot be formed by the direct substitution with chloride ion on an acid, the question now arises of how can the acyl chloride be made? The answer is to make use of entropy.

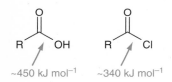

~450 kJ mol⁻¹   ~340 kJ mol⁻¹

Fig. 9.61 A comparison of the bond strengths in a carboxylic acid and an acyl chloride.

A more efficient method of making acyl chlorides

A more effective way to prepare an acyl chloride from a carboxylic acid is to react the acid with thionyl chloride, $SOCl_2$, the structure of which is shown in Fig. 9.62. Let us consider the thermodynamics of this reaction in the gas phase for which the overall equation for the reaction is

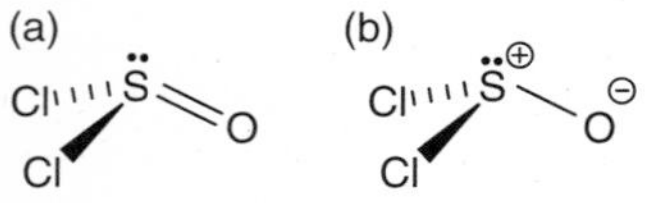

$$CH_3COOH(l) + SOCl_2(l) \longrightarrow CH_3COCl(l) + SO_2(g) + HCl(g)$$

Δ_rH° and Δ_rS° can be calculated for this reaction in the same way as before. The results are:

$$\Delta_rH^\circ = +15.5 \text{ kJ mol}^{-1} \qquad \Delta_rS^\circ = +256 \text{ J K}^{-1} \text{ mol}^{-1}$$

These give a value for Δ_rG° of -61 kJ mol^{-1} and hence a value for K of 4.5×10^{10}. The equilibrium now lies strongly towards favouring the products. However, the reaction is still endothermic; Δ_rG° is negative because Δ_rS° is so large and positive.

(a)   (b)

Fig. 9.62 The structure of thionyl chloride may either be drawn as in (a) which suggests a double bond between the sulfur and the oxygen, or as in (b) with formal charges on these atoms. The formal charges turn out to be fairly close to the actual charges on these atoms.

Ethanoic acid, thionyl chloride and ethanoyl chloride are all liquids at room temperature. Going from two liquids to one liquid and two gases gives a massive increase in the entropy of the system and it is this that drives the reaction.

The reaction is carried out in the laboratory by essentially using the thionyl chloride as a solvent. The carboxylic acid is dissolved in neat thionyl chloride, stirred whilst allowing the gases to escape, and then the excess thionyl chloride is removed by distillation to leave the acyl chloride.

Mechanism for the formation of an acyl chloride

It is possible to draw the mechanism for the formation of an acyl chloride from a carboxylic acid and thionyl chloride in a number of ways; one way which summarizes the key features of the reaction is shown in Fig. 9.63.

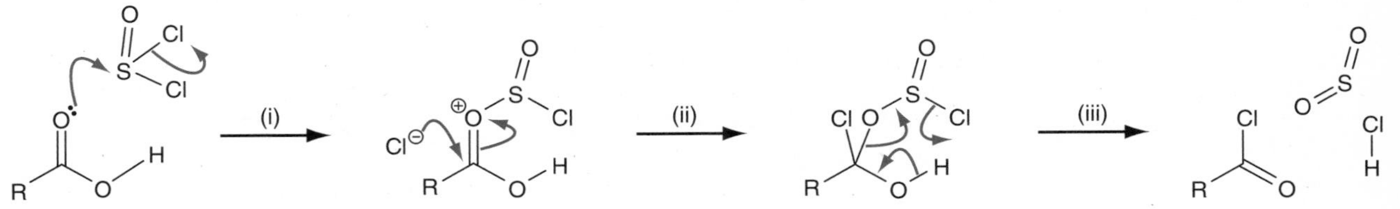

Fig. 9.63 A mechanism for the formation of an acyl chloride from a carboxylic acid and thionyl chloride.

The first step in the reaction is the attack by the HOMO of the carboxylic acid (the oxygen lone pair) into the LUMO of the thionyl chloride (the S–Cl $\sigma^\star$). This interaction results in a new strong O–S bond at the expense of the weak S–Cl bond. Note the formal positive charge on the oxygen which helps to lower the energy of the C=O $\pi^\star$ MO. Step (ii) is the attack of the high-energy lone pair of the chloride ion into the LUMO of the carbonyl species, the C=O $\pi^\star$, to give the usual tetrahedral species.

The curly arrows shown in step (iii) are probably not at all realistic in terms of describing *how* the next step takes place but nonetheless summarize the key changes. It is quite possible that a further chloride ion is eliminated as shown in Fig. 9.64. This again results in a formal positive charge on the oxygen which makes the C–O $\sigma^\star$ MO lower in energy. The lone pair from the other oxygen atom interacts with this C–O $\sigma^\star$ MO, pushing out SO₂.

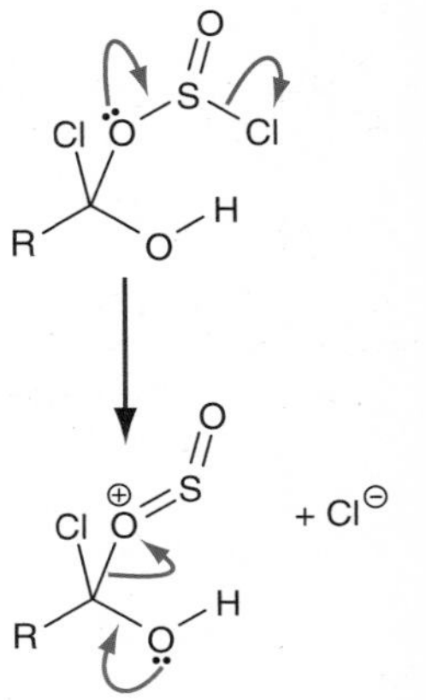

Other methods for making acyl chlorides

We can summarize the role of the thionyl chloride in the preparation of acyl chlorides as essentially making the oxygen into a good leaving group; whereas OH⁻ is not a feasible leaving group, neutral SO₂ is an excellent one. The thionyl chloride allows the formation of a new, strong S–O bond at the expense of a weak S–Cl bond.

Other reagents can perform a similar role to the thionyl chloride; one of the most commonly used is phosphorus pentachloride, PCl₅. The reaction is analogous to the reaction with thionyl chloride but now a new strong P–O bond is formed at the expense of a weak P–Cl bond, and the oxygen from the carboxylic acid leaves not as OH⁻ but as the neutral molecule phosphorus oxychloride, POCl₃. POCl₃ is a liquid under normal conditions (boiling point 106 °C) and may be distilled off at the end of the reaction.

Fig. 9.64 It is possible that a double bond is formed between the oxygen and sulfur, whilst at the same time breaking the S–Cl bond. This forms an even better leaving group, neutral SO₂.

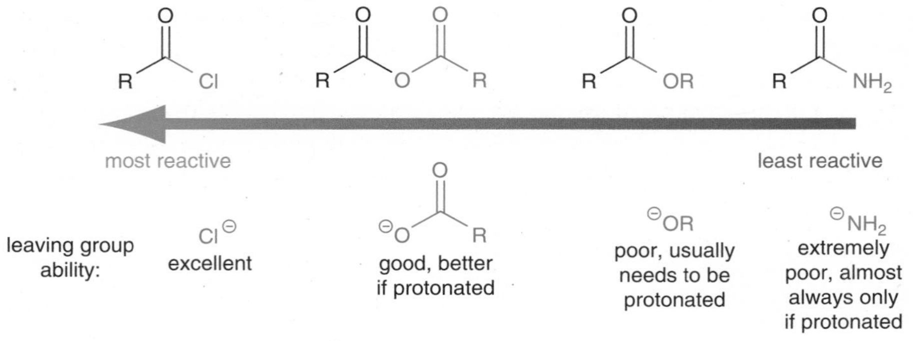

Fig. 9.65 The order of the reactivity of different acid derivatives can also be rationalized by considering the different leaving groups from each were they to be attacked by a nucleophile. The chloride ion is an excellent leaving group which helps to make acyl chlorides so reactive. In contrast, the NH_2^- of an amide is an extremely poor leaving group.

9.5.5 The leaving groups in acid derivatives

In the formation of the acyl chloride above, we saw that it was not possible to react a carboxylic acid with a chloride ion to form a tetrahedral intermediate and then to lose an hydroxide ion (see Fig. 9.60 on page 377). We said that the chloride ion is a much better leaving group than the hydroxide ion.

There are a number of different ways of looking at what makes a good leaving group. When considering the possible reactions of the tetrahedral intermediate in Fig. 9.60 on page 377, we could simply note that the C–Cl $\sigma^\star$ MO is lower in energy than the $\sigma^\star$ MO for the C–OH bond and hence it is not surprising that the chloride is therefore expelled in preference to the hydroxide ion.

The low-energy C–Cl $\sigma^\star$ MO is a result of the poor orbital overlap between the carbon and chlorine. Such poor overlap also results in a weak C–Cl bond as compared with the C–OH bond, and this could be another way of looking at why the chloride ion leaves in preference to the hydroxide ion.

A third way that you may see in other text books is to say that the chloride ion is 'more stable' in some way than the hydroxide ion. This is an argument we need to be rather careful with and one that we shall return to later in section 9.7.2 on page 393. However, it is certainly the case that some anions are higher in energy than others. For example, a high energy anion, such as CH_3^-, will never act as a leaving group.

When we considered the order of reactivity of common acid derivatives (see Fig. 9.49 on page 370), we looked at how easy it was for a nucleophile to attack into the $\pi^\star$ MO of the carbonyl to form the tetrahedral intermediate. If this intermediate is to then re-form a carbonyl, one of the groups attached to the central carbon must leave. This can only happen if there is a good leaving group present.

When attacking an acyl chloride there is always a good leaving group present – the chloride ion – and this leads to the many nucleophilic substitution reactions illustrated in Fig. 9.57 on page 376. For other acid derivatives there is not always a good leaving group present. Figure 9.65 indicates the relative ease with which the different substituents can act as leaving groups from the common acid derivatives; it can be seen that the leaving group ability corresponds to the general reactivity of the whole functional group. Of course, in these reactions the leaving group only leaves from the tetrahedral intermediate formed after the acid derivative is attacked by a nuecleophile, not from the acid derivative itself.

Part of the reason that acyl chlorides are so reactive is because Cl^- is such a good leaving group. In contrast, NH_2^- will virtually never act as a leaving group from an amide which is the least reactive of the acid derivatives. In order for the nitrogen from an amide to leave, it will usually have to be protonated first and leave as a neutral amine (as in the hydrolysis of the amide). However, whether or not this is possible depends on whether the reaction is carried out under acidic or basic conditions.

9.5.6 Improving the leaving group: nucleophilic catalysis

It is also possible to speed up a reaction by temporarily replacing one leaving group with another to make a more reactive functional group. As an example, the reaction between acid anhydrides and alcohols to form esters is rather slow but proceeds easily in the presence of pyridine as shown in Fig. 9.66.

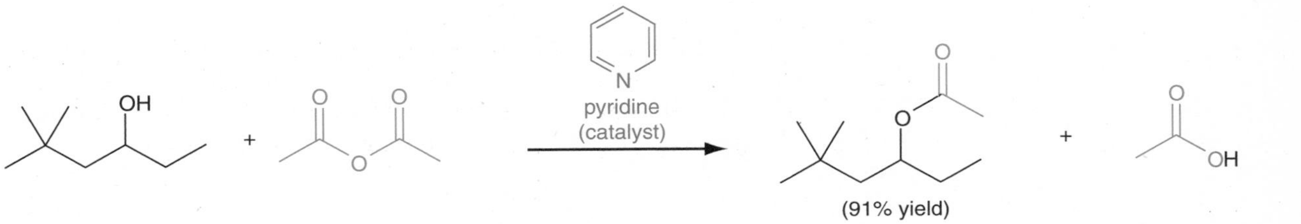

Fig. 9.66 The preparation of an ester from an alcohol and anhydride.

As usual, the LUMO of the reactants, is the CO $\pi^\star$ MO of the acid anhydride. In the absence of the pyridine, the HOMO is the lone pair of the alcohol oxygen. These two orbitals can interact, but the reaction is slow. When pyridine is included, there is a source of higher-energy electrons on the nitrogen in the pyridine. We can consider the nitrogen in pyridine to be sp^2 hybridized with a filled sp^2 orbital forming the lone pair. This orbital is shown in Fig. 9.67. This can interact with the CO $\pi^\star$ MO of the anhydride to form the tetrahedral intermediate **A** as shown in step (i) of Fig. 9.68 on the following page.

Most of the time the tetrahedral intermediate **A** will simply re-form the acid anhydride by pushing out neutral pyridine, which is a good leaving group, in the reverse of step (i). However, occasionally ethanoate ion can be lost as the leaving group, as shown in step (ii), leaving the reactive species **B**.

The positive charge on the nitrogen adjacent to the already polarized carbonyl carbon makes **B** considerably more reactive than the acid anhydride. The lone pair of the alcohol quickly attacks into the $\pi^\star$ of **B** to form a new intermediate **C** as shown in step (iii) in Fig. 9.69 on the next page.

Tetrahedral intermediate **C** can re-form the carbonyl either by loss of the neutral alcohol in the reverse of step (iii), or by the loss of pyridine. However, the proton can easily be removed from the alcohol, as in step (iv) to give **D**. In this step, some more of the pyridine could act as the base, as could the ethanoate ion formed in step (ii). There is only one good leaving group on **D** and that is the pyridine. This is lost, thereby regenerating the catalyst, and the desired ester is formed as shown in step (v).

The pyridine speeds us this reaction by being a better nucleophile than the alcohol, and forming reactive species **C**, but then also being an excellent leaving group.

Fig. 9.67 Iso-surface representation of the nitrogen 'lone pair' MO of pyridine.

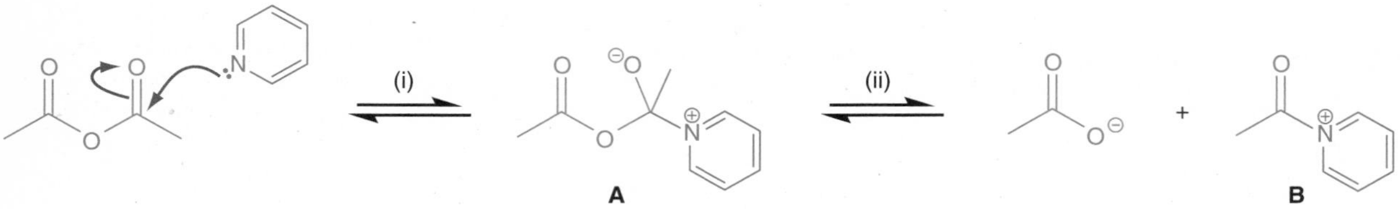

Fig. 9.68 The acid anhydride is first attacked by the high-energy nitrogen lone pair of pyridine. The tetrahedral intermediate **A** that results can sometimes push out ethanoate ion as a leaving group to leave the reactive species **B**.

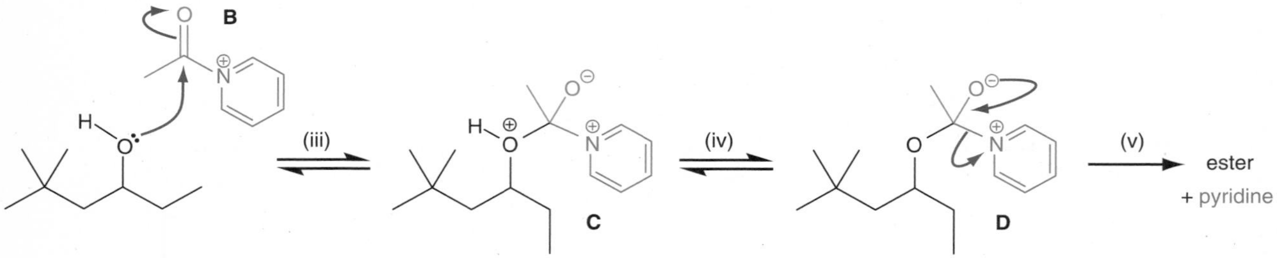

Fig. 9.69 The formation of the ester from the reactive intermediate **B**.

Summary

There are many other possible transformations possible moving within functional group level three, and some of these are explored in the *Questions* at the end of the chapter. Whilst it is not possible to cover every possible permutation in detail, you should now possess the skills to appreciate how and why a given reaction takes place. In each case, the principles are the same: identify the high-energy electrons in the nucleophile, attack into the CO $\pi^\star$ MO to form a tetrahedral intermediate, possibly re-form the CO π bond if there is a realistic leaving group present. Attention must be paid to whether the reaction is carried out in acid or in base, and how this may affect the reaction.

9.6 Moving down from functional group level three

The order of reactivity of the common acid derivatives is extremely important when it comes to understanding how they react with reducing agents and organometallic reagents. As we shall see, sometimes the reducing agent or organometallic can take the functional group from level three to level two, other times from level three to level one. We shall first consider the reactions with organometallic reagents.

9.6.1 Reaction with organometallic reagents

There are a number of different outcomes from the reactions between organometallic reagents and the different acid derivatives depending on the particular derivative, the relative amount of the organometallic added, and the reaction conditions. Figure 9.70 on the facing page outlines the most likely outcomes when either *one equivalent* or *an excess* of organometallic reagent are added to the different acid derivatives and the reaction mix is worked up in

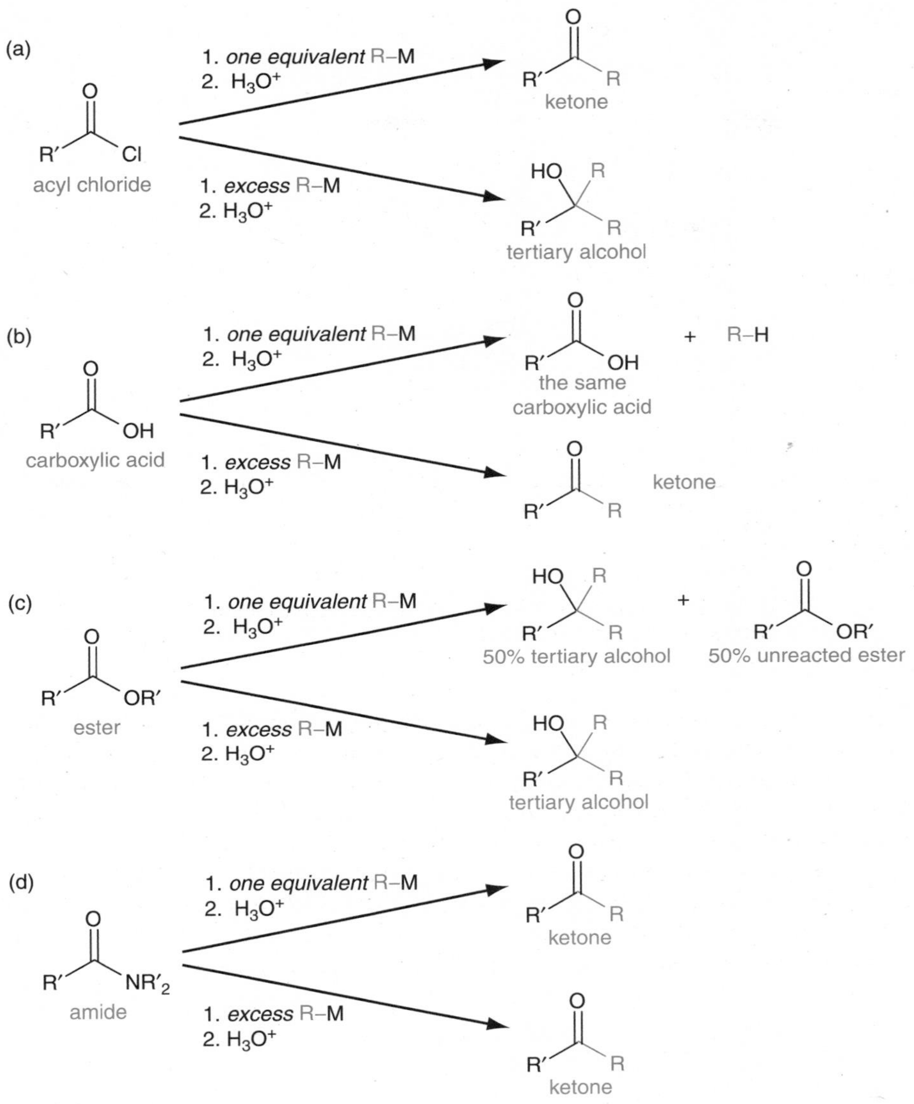

Fig. 9.70 Illustration of the most likely outcomes when adding either one equivalent or an excess of an organolithium or Grignard reagent to the different carboxylic acid derivatives.

the usual way with aqueous acid. Adding one equivalent of the organometallic means adding the same number of moles as the carbonyl compound.

Whilst the outcome for the different cases may seem rather confusing at first glance, we can understand what happens by considering the relative reactivities of the different carbonyl groups. We will consider each reaction in turn.

Acyl chlorides and organometallics

When an organometallic reagent, such as a Grignard reagent, is gradually added to an acyl chloride, the initial interaction is between the high-energy electrons in the metal–carbon bond from the organometallic, with the LUMO of the acyl chloride, the CO $\pi^\star$, as shown in step (i) in Fig. 9.71 on the next page.

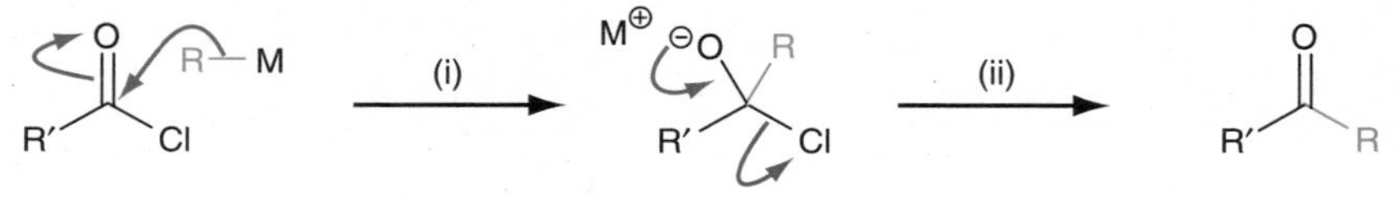

Fig. 9.71 The initial reaction between an organometallic and an acyl chloride.

The resulting tetrahedral intermediate rapidly re-forms the carbonyl by elimination of the excellent leaving group Cl⁻, as shown in step (ii). The question is, what happens next when more of the organometallic solution is added? There are now *two* carbonyl species present which the organometallic could react with: more of the acyl chloride, and the ketone that has just formed from the initial reaction.

Since the acyl chloride is more reactive that the ketone, it is this that reacts with the organometallic in preference to the ketone. Hence if just one equivalent of an organolithium or Grignard reagent is added to an acyl chloride, the major product is the ketone, as shown in Fig. 9.70 (a). When excess of the organometallic is added, the ketone goes on to react further as we have already seen in section 9.2.2 on page 347, ultimately yielding the tertiary alcohol.

In practice, this is not usually a clean way to prepare ketones since some further addition to the ketone will inevitably occur. As we shall see, there are also better methods for preparing the alcohol, so this method is also seldom used in practice.

Other organometallic reagents, notably organocadmium reagents (CdR$_2$) and organocopper reagents are somewhat less reactive than their lithium and magnesium counterparts and these are more often used in the preparation of ketones from acyl chlorides as they do not add to the product ketone.

Some examples of the preparation of ketones from acyl chlorides using different organometallic reagents are given in Fig. 9.72. In example (a) the Grignard reagent is added to a cold solution of the acyl chloride in order to minimize further reaction with the ketone. Example (b) illustrates the use of an organocadmium reagent, prepared by adding one equivalent of cadmium chloride to two equivalents of the corresponding Grignard reagent. This organocadmium reagent is less reactive than the organomagnesium reagent and will only react with the most reactive carbonyls. Notice in this reaction that the less reactive ester group is not attacked.

The orbital energies of cadmium are lower in energy than either lithium or magnesium which means the carbon–cadmium bond is more covalent. This is reflected in the properties of these reagents: dimethylcadmium, for example, exists as discrete, linear molecules and melts at −4.5 °C and boils at 105.5 °C. In contrast, methyllithium is a crystalline solid which decomposes above 200 °C.

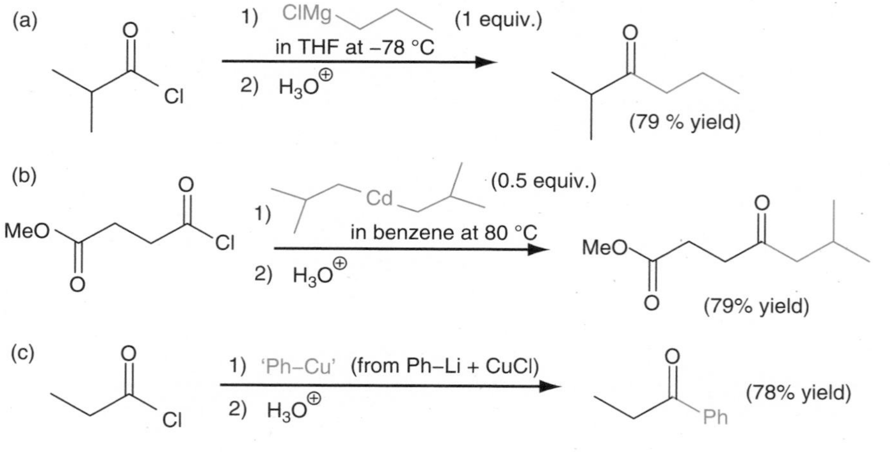

Fig. 9.72 The preparation of ketones from acyl chlorides using organometallic reagents. When using organomagnesium reagents the reaction must be carried out at low temperatures in order to obtain good yields, as shown in (a). More reliable results are usually obtained using either organocadmium or organocopper reagents as shown in (b) and (c).

Example (c) illustrates the use of an organocopper reagent, prepared by adding copper(I) chloride to the phenyllithium reagent. Once again, the organocopper reagent is less reactive and produces a good yield of the ketone.

Carboxylic acids and organometallics

Returning to Fig. 9.70 on page 383, we come to the reaction of carboxylic acids and organometallics. When one equivalent of a organometallic reagent such as a Grignard or organolithium reagent, all that happens is that the organometallic reagent reacts with the acidic proton of the carboxylic acid as shown in Fig. 9.73. The carboxylate anion formed is then re-protonated in the work up of the reaction as shown in example (b) in Fig. 9.70.

However, when two equivalents of the organometallic reagent are used (organolithium reagents being the most commonly used in practice), ketones can be formed in good yields. The first equivalent reacts with the acid as before, the second equivalent adds to form the ketone.

Rather than wasting the first equivalent to make the lithium salt of the carboxylic acid, it is also possible (often giving better yields), to start from the lithium salt of the carboxylic acid. Figure 9.74 shows some examples of ketones prepared from carboxylic acids or their lithium salts.

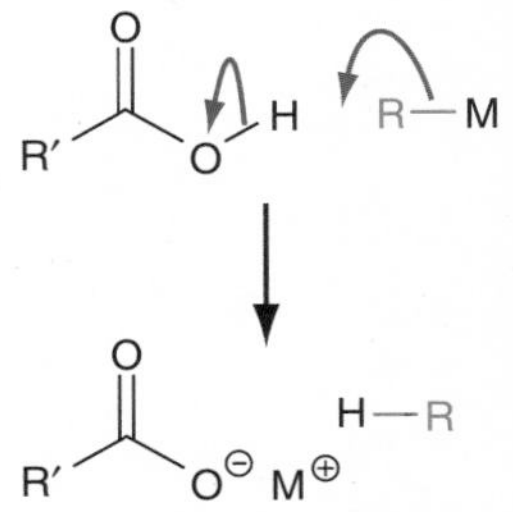

Mechanism of reaction

By now you should be able to predict the mechanism for the formation of a ketone from a carboxylic acid and organolithium reagent. The first step is the formation of the lithium salt of the acid, as shown in Fig. 9.73. This salt is then attacked by a second equivalent of the organolithium to give the usual tetrahedral intermediate as shown in step (i) of Fig. 9.75 on the following page.

There is no leaving group attached to the central carbon of the tetrahedral intermediate and so it remains as this dilithium salt until acid is added. Protonation of the oxygens gives a hydrate, which, as we have seen in section 9.4.1 on page 355, readily forms the ketone with which it is in equilibrium.

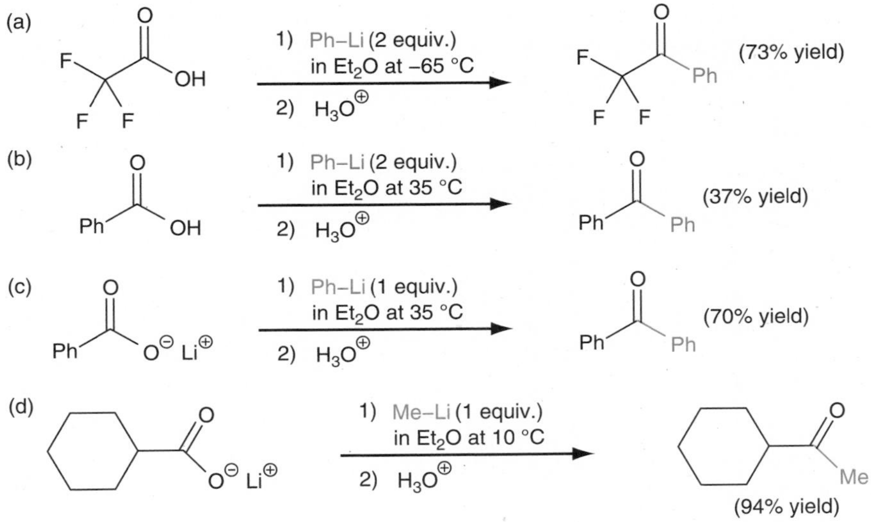

Fig. 9.73 The initial reaction between a carboxylic acid and an organometallic reagent.

Fig. 9.74 The formation of ketones from carboxylic acids using organolithium reagents. In examples (a) and (b), the ketones are prepared directly from the carboxylic acids with two equivalents of the organolithium reagent. In examples (c) and (d), the ketone is prepared from the lithium salt of the acid with just one equivalent of organolithium reagent. Comparison of reactions (b) with (c) shows the latter route often gives better yields.

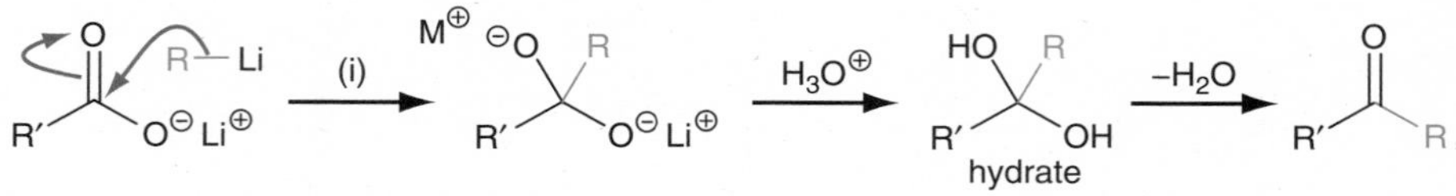

Fig. 9.75 The reaction between carboxylate anion and an organolithium reagent gives the usual tetrahedral intermediate which, in this case, has no leaving group. Addition of acid gives the hydrate, and subsequently the ketone is formed by dehydration.

Notice that this reaction takes the carbon of the carboxylic acid from level three to level two. It is not possible for further organolithium to react with the tetrahedral intermediate since it has no low energy vacant orbital – there is no longer a CO $\pi^\star$. The level-two ketone is only formed in the second stage of the reaction, when acid is added. The acid would destroy any remaining organolithium reagent present in the mix before it protonates the intermediate to form the hydrate; the resulting ketone is never in contact with the organolithium reagent.

Esters and organometallics

In Fig. 9.70 on page 383 we saw that, when two equivalents of the organometallic are used, esters react to give tertiary alcohols. The mechanism for this reaction is analogous to the reaction with acyl chlorides. The organometallic attacks in the CO $\pi^\star$ MO as usual to give the tetrahedral intermediate as shown in step (i) in Fig. 9.76.

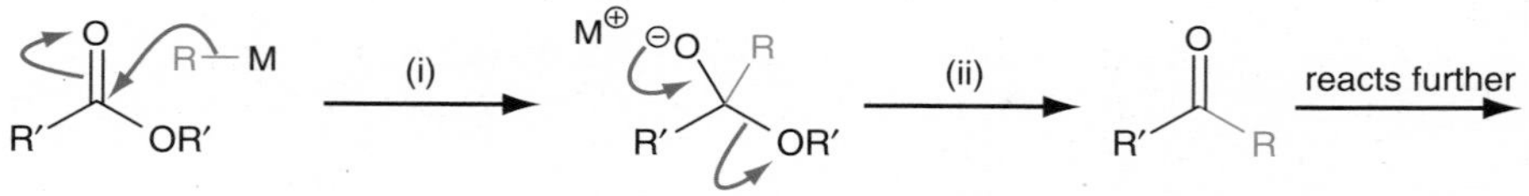

Fig. 9.76 The initial reaction between an ester and an organometallic reagent. The tetrahedral intermediate can expel an alkoxide ion to form the ketone, but this quickly reacts with more of the organometallic reagent.

Whilst an alkoxide ion, RO^-, is not a good leaving group, it is being expelled by the other negatively charged oxygen in the tetrahedral intermediate. This is similar to the alkaline hydrolysis of an ester in which one $-O^-$ pushes out another. In such a process there is a reduction in crowding as the central carbon goes from tetrahedral to trigonal, but otherwise there is little difference in energy. The key point in this reaction is that the ketone which is formed is *more* reactive towards nucleophilic attack than the ester from which it was made. Consequently, even if there is only a limited amount of the organometallic reagent present, it reacts with whatever ketone has already been made in preference to any remaining ester.

Some ketones have been prepared from esters – generally in hindered systems or where the alkoxide leaving group is particularly bad or held in place by a ring structure. However, by far the most important reaction is the preparation of alcohols by the reaction of two equivalents of the organometallic with one of the ester. Some examples are shown in Fig. 9.77 on the facing page.

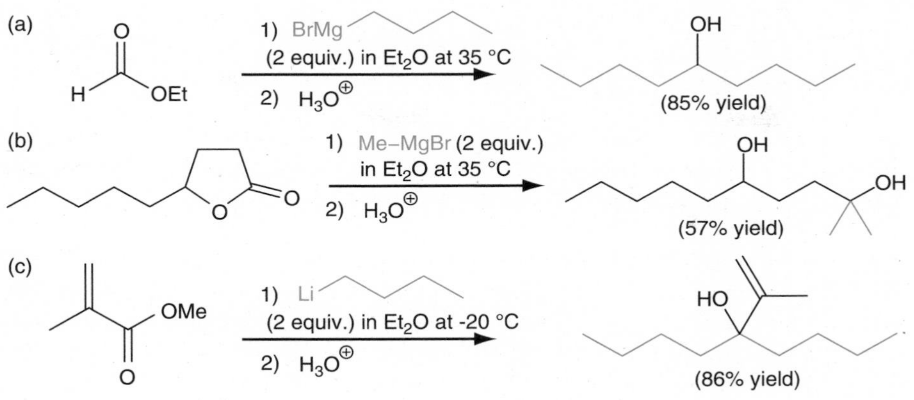

Fig. 9.77 Examples of the formation of alcohols by the reaction of two equivalents of either Grignard reagent or organolithium reagent with esters.

Amides and organometallics

The final acid derivative to consider from Fig. 9.70 on page 383 is the amide and its reaction with organometallic reagents. If there are any protons attached to the nitrogen, the organometallic reagent will simply remove these and be destroyed. The pK_a of an amide, at around 17, is about the same as an alcohol, and so a proton is easily removed by an alkyllithium or Grignard reagent. Provided there are no protons on the amide, the reaction with a Grignard or organolithium reagent yields an aldehyde or ketone. Some examples are given in Fig. 9.78.

The first step in the mechanism, is as usual, the attack by the organometallic reagent into the CO $\pi^\star$ MO to form the tetrahedral intermediate, as shown in step (i) in Fig. 9.79 on the next page.

The tetrahedral intermediate has no feasible leaving group; it is not possible for the $-O^-$ to push out $-NR'_2$. Hence the intermediate stays like this until acid is added. In acid, the $-O^-$ is rapidly protonated to give the unstable level-two species with both an alcohol and amine group on the same carbon atom. In acid, the amine group is readily lost and pushed out by the alcohol to give the ketone.

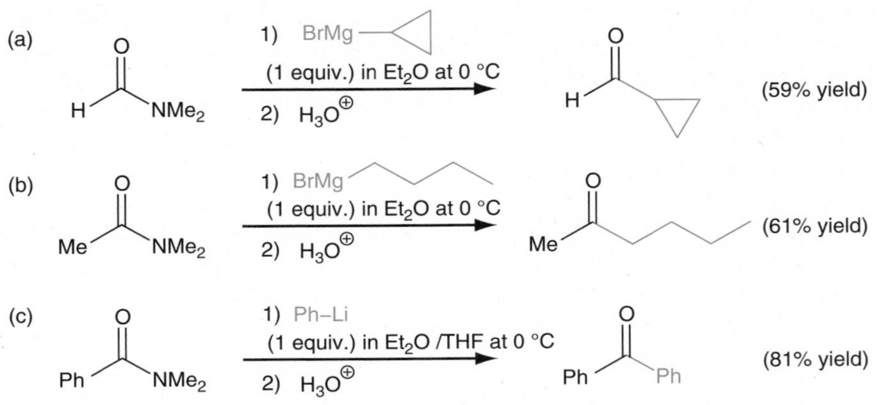

Fig. 9.78 Examples of the formation of aldehydes and ketones by the reaction of a Grignard reagent or organolithium reagent with amides.

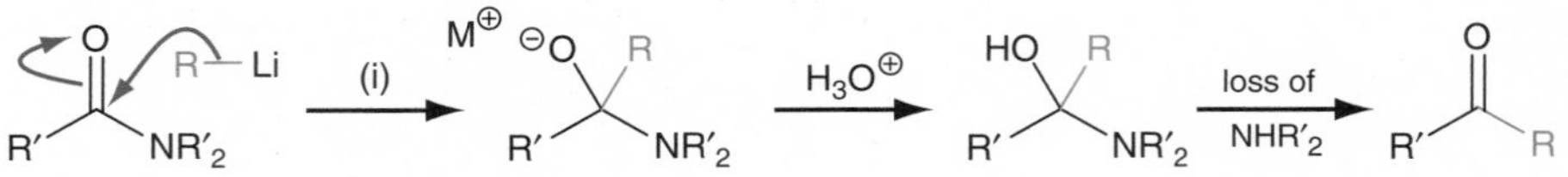

Fig. 9.79 The reaction of an organometallic reagent with an amide to form an aldehyde or ketone.

9.6.2 Reaction with metal hydride reducing agents

Two of the most commonly used reducing agents are $LiAlH_4$ and $NaBH_4$. As described in section 9.3.3 on page 353, $LiAlH_4$ is significantly more reactive than $NaBH_4$ and this is reflected in their reactions with level-three functional groups. $LiAlH_4$ will reduce *all* of the acid derivatives, but $NaBH_4$ will only reduce the most reactive acyl chloride.

These reactions take place in a similar manner to the reactions of the acid derivatives with organometallic reagents but with one major difference: in each case, when a group is reduced, it is reduced from level three to level one i.e. two hydrogens are added. Unlike with the organometallic reagents, it is usual to prepare aldehydes and ketones by the partial reduction of acid derivatives using $LiAlH_4$ or $NaBH_4$.

The reactions of $LiAlH_4$ and $NaBH_4$ with the common acid derivatives are summarized in Fig. 9.80. The mechanisms for the reactions with acyl chlorides and esters are straightforward and mirror the reaction of these groups with the organometallic reagents. The reactions which differ from their organometallic counterparts are the reactions of $LiAlH_4$ with either carboxylic acids or amides. With the organometallic reagents, both carboxylate anion and amides reacted once to give the tetrahedral intermediate but then there was no good leaving group which could enable the CO π bond to re-form. The different reaction in the case of $LiALH_4$ is due to the presence of the aluminium.

The orbital energies of aluminium are lower than those of either lithium or magnesium. Whereas we drew the initial tetrahedral intermediate formed from the organometallic in Fig. 9.79 as an *ionic* species with an $-O^-$ and either Li^+ or MgX^+, the interaction between oxygen and aluminium is far more covalent

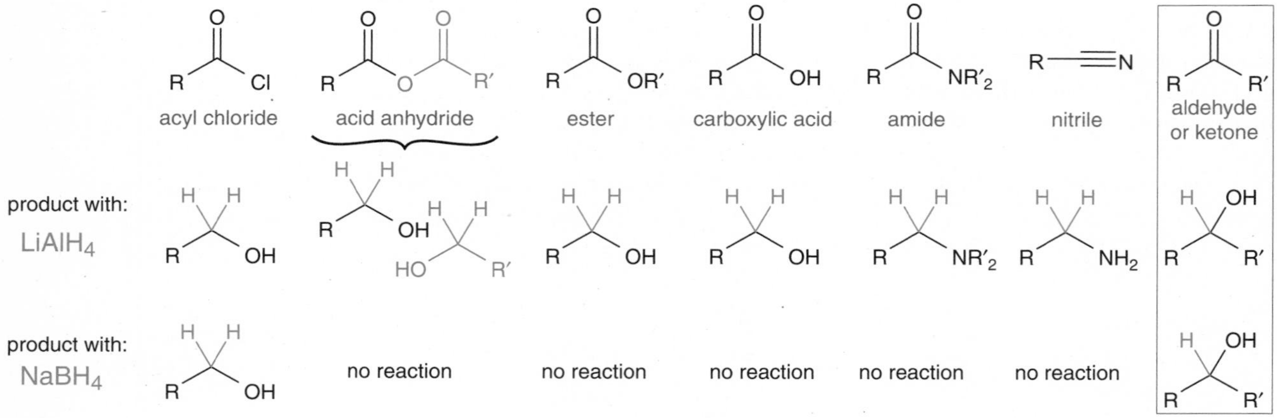

Fig. 9.80 The reactions of the common acid derivatives with $LiAlH_4$ and $NaBH_4$. $LiAlH_4$ reduces every group from level three to level one whereas $NaBH_4$ only reduces the most reactive acyl chloride. The reaction with aldehydes and ketones is included to complete the table.

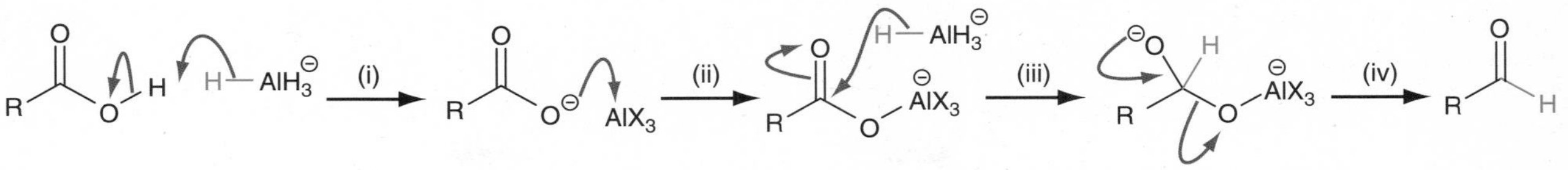

Fig. 9.81 The first part of a mechanism for the reduction of a carboxylic acid by LiAlH$_4$. An X is used where it could be either a hydrogen atom or an oxygen atom bonded to the aluminium.

in nature. An oxygen strongly bonded to aluminium can act as a leaving group and be pushed out by an –O$^-$; an oxygen bonded to lithium cannot.

The first part of a mechanism for the reduction of a carboxylic acid by LiAlH$_4$ is shown in Fig. 9.81. First of all, the acid proton reacts with the hydride to give hydrogen gas. In step (ii) the resulting carboxylate anion can form a bond with the AlH$_3$ that has just been formed. However, in later stages of the reaction, these hydrogens could have been replaced by other alkoxide ions, so we have written AlX$_3$ in step (ii) since it is not important exactly what the other groups on the aluminium actually are.

In step (iii), a second equivalent of hydride attacks, this time into the CO $\pi^\star$ MO to give a tetrahedral intermediate, as usual. The important point is now that the oxygen bonded to the aluminium is a viable leaving group which can be expelled to re-form the CO π bond, as shown in step (iv). The resulting aldehyde is rapidly reduced by a third equivalent of hydride as we have already seen in section 9.3.3 on page 353.

The reduction of an amide is similar, except that when the oxygen–aluminium species is expelled, an imine is formed instead of the aldehyde.

9.6.3   Reduction from level three to level two

In Fig. 9.7 on page 347 there were two examples of the partial reduction of level-three groups to level two. One of the most useful reagents for such transformations is DIBAL, diisobutylaluminium hydride, and it is worthwhile for us to consider briefly how this reagent works.

The way in which DIBAL differs from LiAlH$_4$ is that there are only three groups bonded to the aluminium in DIBAL. This means in a mixture of DIBAL and ester, the LUMO is the vacant $3p$ AO on the aluminium so that the initial interaction is between this vacant orbital and the high-energy electrons of the oxygen lone pair from the ester, as shown in step (i) in Fig. 9.82.

On coordination to the oxygen, the aluminium now has four groups bonded to it, together with a formal negative charge. The high-energy electrons in the Al–H σ bond now interact with the CO $\pi^\star$ MO of the carbonyl in the usual manner to give the tetrahedral intermediate as shown in step (ii). At low

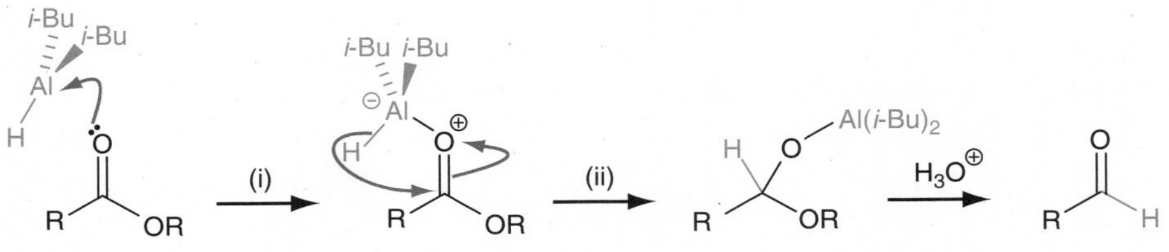

Fig. 9.82 A mechanism for the partial reduction of an ester to an aldehyde using DIBAL.

temperatures, this is where the reaction usually stops. On work up with aqueous acid, the level-two tetrahedral species is hydrolysed to the aldehyde in the usual manner.

9.6.4 Summary

There are many other reactions of level-three functional groups, but we hope that by now you have a good feel for how these might work. For the remainder of the chapter, we shall look at substitution reactions within level one.

9.7 Transformations within level one

The reactions we have looked at so far in this chapter have all involved an initial interaction between a source of high-energy electrons and a vacant $\pi^\star$ MO. This interaction resulted in a breaking of the C–O π bond and the formation of a tetrahedral intermediate. What we now want to do in order to transform one level-one functional group to another, is replace one C–X single bond with another. This is called *nucleophilic substitution* and may be summarized

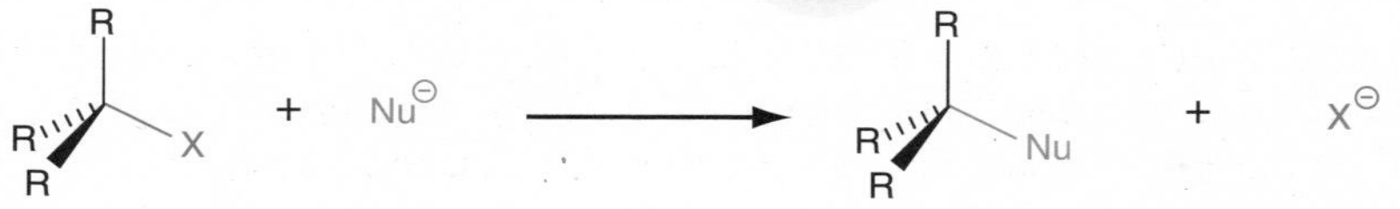

where the nucleophile Nu may be a charged or neutral species.

9.7.1 The S$_N$1 and S$_N$2 mechanisms

Perhaps the simplest way for this substitution to take place, is for the high energy electrons of the nucleophile to attack directly into the C–X $\sigma^\star$ MO as shown in Fig. 9.83. The orbital interactions involved in this reaction have been discussed in detail in section 8.5 on page 326.

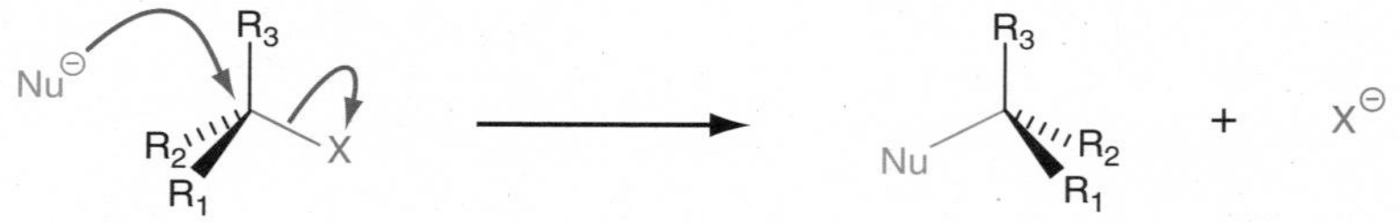

Fig. 9.83 The S$_N$2 mechanism.

This mechanism is known as the S$_N$2 mechanism; where the S in S$_N$2 stands for substitution, since one group is substituted for another, the subscript N for *nucleophilic* since it is a nucleophile which is involved in the attack, and the 'two' indicates that both species are involved in the *rate-limiting* step of the reaction, a concept which is discussed in detail in the following chapter (section 10.7 on page 422). All the reactions in Fig. 9.1 on page 341 proceed via this mechanism, and one of these examples is shown again in full in Fig. 9.84 on the next page.

We have also come across a slightly different mechanism for bringing about the substitution of one group for another. In section 9.4.2 on page 358 we looked at the equilibria between acetals and aldehydes or ketones. In the

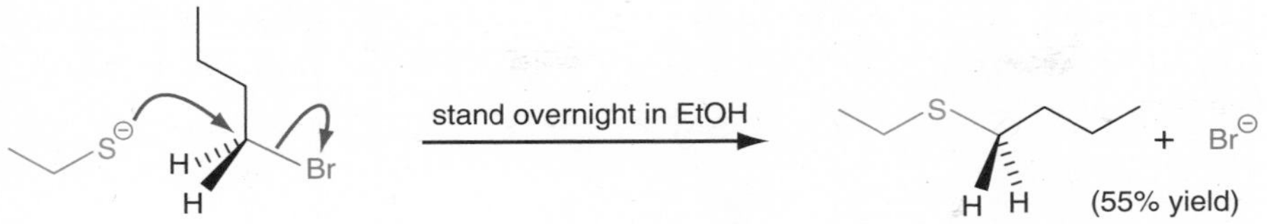

Fig. 9.84 An example of a reaction that proceeds via the S_N2 mechanism.

formation of the carbonyl compound from an acetal, one of the –OR groups needs to be replaced by an –OH group from water. The mechanism for this process is the reverse of that shown in Fig. 9.34 on page 361, and this is written out in the correct sequence in Fig. 9.85.

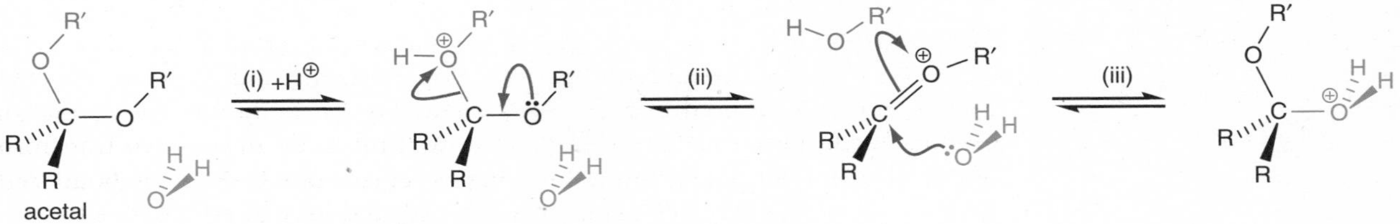

Fig. 9.85 The initial steps in the hydrolysis of an acetal.

In step (i) one of the –OR′ groups is protonated to make it into a better leaving group. In step (ii) the high-energy lone pair from the other –OR′ oxygen pushes out the protonated –OR′ group as the alcohol, forming the CO π bond in the process. Finally, in step (iii) the lone pair of the water oxygen attacks into the CO $\pi^\star$ MO. The net result is that the H_2O has substituted for one of the –OR′ groups. The important point is that this reaction does not take place by the water simply replacing the protonated –OR′ group directly as shown in Fig. 9.86.

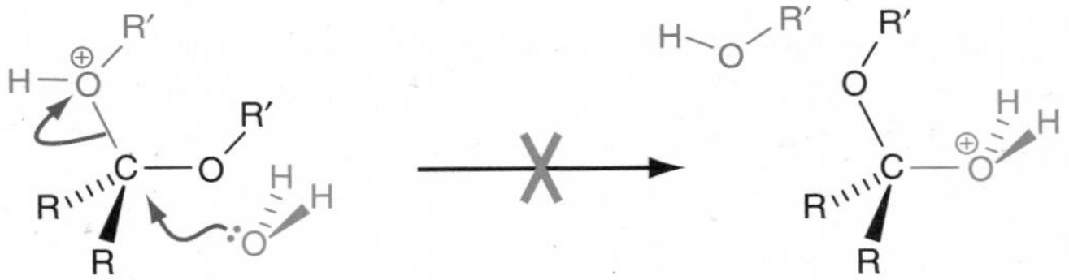

Fig. 9.86 This S_N2 mechanism is *not* that followed for the hydrolysis of an acetal.

The correct mechanism for the substitution step in the acetal hydrolysis is called the S_N1 mechanism and is shown in a more general way in Fig. 9.87 on the next page. This mechanism proceeds in two steps: the first is where the leaving group departs, the second is where the nucleophile bonds to the carbon. Making the bond in the second step is much faster than breaking the bond in the first, and the first step is the one that limits the speed of the reaction – i.e. it is the rate-limiting step. Only one species is involved in this step and this is what the '1' in S_N1 signifies.

The consequences that the S_N1 and S_N2 mechanisms have for the overall rate law of the reaction are discussed in detail in the following chapter, section 10.8.2 on page 430.

Intermediates and transition structures

In the S_N1 mechanism, the bond to the leaving group breaks, to give an intermediate cation – a carbenium ion – in which the central carbon now only has three groups attached. The most favourable geometry for such a species is

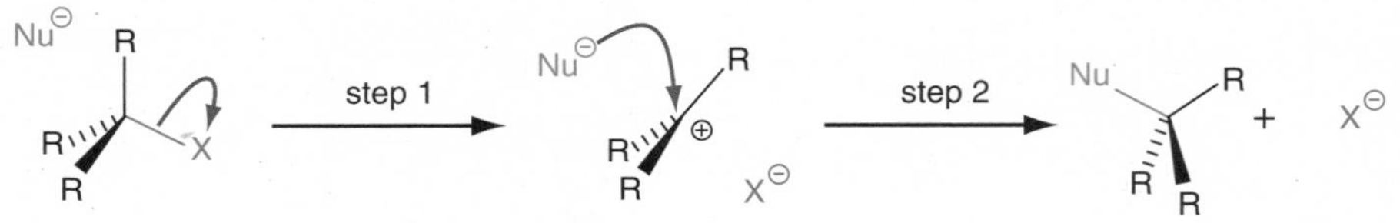

Fig. 9.87 The S_N1 mechanism.

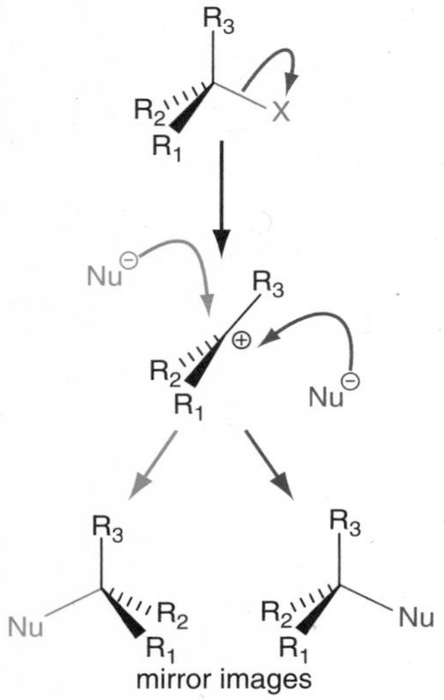

planar. When the nucleophile attacks in the second step, it could attack from either side, with both possibilities being equally likely.

If the central carbon has different groups attached, the incoming nucleophile attacking from one side or the other gives rise to products which are mirror images, as shown in Fig. 9.88. Such molecules which are not the same but mirror images of each other are called *optical isomers*, or *enantiomers*; these are considered in more detail in section 12.2 on page 523.

In contrast, in the S_N2 mechanism, no intermediate is formed and the reaction takes place by the new bond forming as the old bond breaks. As we have seen in section 8.5 on page 326, in this reaction the nucleophile must approach from directly behind the leaving group. This results in the central carbon essentially turning inside-out during the substitution as it proceeds through the *transition structure*, as shown in Fig. 9.89.

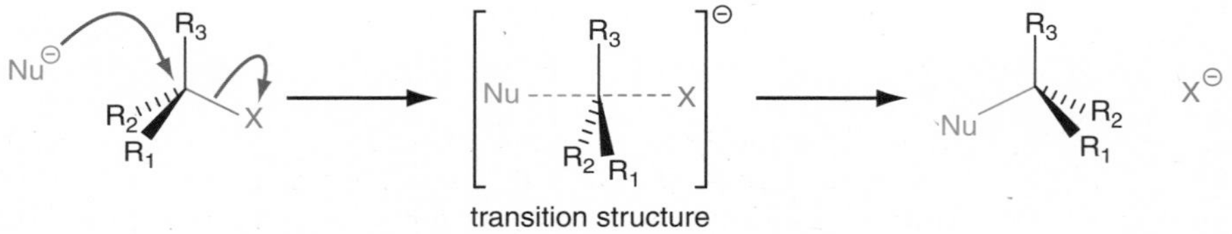

mirror images

Fig. 9.88 In an S_N1 mechanism, a trigonal planar intermediate is formed which can be attacked on either side by the incoming nucleophile as shown by the red and purple arrows. If there are different groups attached to the central carbon, the resulting products are not identical but are mirror images of each other.

transition structure

Fig. 9.89 The transition structure in the S_N2 mechanism. The R groups all lie in the same plane, at right angles to the plane of the paper. The bond to the nucleophile is half formed, and the bond to the leaving group is half broken.

In the transition structure, the bond between the central carbon and the incoming nucleophile is half formed, and the bond between the central carbon and the leaving group is half broken. All three of the R groups and the central carbon lie in the same plane, at right angles to the plane of the paper.

If the central carbon has different groups attached, only one product is produced as shown in Fig. 9.89.

S_N1 *vs* S_N2

By careful inspection of the products it may be possible in some cases to tell whether a reaction has taken place by an S_N1 or an S_N2 mechanism. However, *why* a reactant favours one mechanism over the other depends on a number of factors:

- the leaving group

- the nucleophile

- the other groups attached to the level-one carbon

- the reaction conditions, such as the solvent.

We will look at each of these factors in turn.

leaving group	compound	pK_a of conjugate acid (R = H)
N_2	$R\!-\!N_2^{\oplus}$	
$TsO^{\ominus}$	$R\!-\!OTs$	-6.5
$I^{\ominus}$	$R\!-\!I$	-10
$Br^{\ominus}$	$R\!-\!Br$	-9
H_2O	$R\!-\!\overset{\oplus}{O}H_2$	-1.7
$Cl^{\ominus}$	$R\!-\!Cl$	-7
NR_3	$R\!-\!\overset{\oplus}{N}R_3$	10
$F^{\ominus}$	$R\!-\!F$	3
$R'COO^{\ominus}$	$R'COOR$	$3\text{-}5$
NH_3	$R\!-\!\overset{\oplus}{N}H_3$	9
$OH^{\ominus}$	$R\!-\!OH$	15.7
$R'O^{\ominus}$	$R\!-\!OR$	$16\text{-}17$
$H^{\ominus}$	$R\!-\!H$	35
$NH_2^{\ominus}$	$R\!-\!NH_2$	38
$CH_3^{\ominus}$	$R\!-\!CH_3$	48

Increasingly better leaving groups ↑

very poor leaving groups

Fig. 9.90 The left-hand column lists a number of common groups arranged in order of their ability to act as leaving groups. The central column shows how each group occurs in a compound. N_2 is almost unstoppable as a leaving group, but to all intents H^-, NH_2^- and CH_3^- never act as leaving groups. The right-hand column gives the approximate pK_a value of the conjugate acid of the leaving group, i.e. the acid formed when R=H.

9.7.2   The effect of the leaving group

We have already seen the importance of the leaving group when looking at the reactivities of the acid derivatives in section 9.5.5 on page 380. For the nucleophilic substitution reactions the leaving group is even more important, and unless there is a good leaving group present, the reaction simply will not occur.

Generally speaking, if the C–X $\sigma^\star$ MO is low in energy, X^- will be a good leaving group. Such a situation will arise when the AOs of X are much lower in energy than those of carbon, and also when there is a poor match in size between the orbitals. The same factors also give rise to weak C–X bond strengths. What is more, if atom X has low-energy orbitals, the X^- anion should also be relatively stable.

Rather than simply considering the energy of the C–X $\sigma^\star$ MO, it is also possible to rationalize how good a leaving group something X^- is by considering the C–X bond strength, and the 'stability' of the X^- anion.

By studying a large number of reactions it is possible to draw up an approximate order of how good particular species are as leaving groups, and such an order is shown in Fig. 9.90.

The very best leaving group is neutral nitrogen, N_2. This is a very stable molecule, and once it has been eliminated there is very little chance that it will add back again. Further, when bonded in a molecule, the N_2 group bears a positive charge which draws electrons towards it. This results in a weakening of the C–N bond, making it easier to break, thereby further enhancing the leaving group ability.

The halide ions appear in the list in the order $I^- > Br^- > Cl^- > F^-$, meaning that iodide in the best leaving group and fluoride the worst. This order reflects the halogen–carbon bond strengths as exemplified by the series CH_3I, CH_3Br, CH_3Cl and CH_3F in which the bond strengths are 237, 293, 352 and 472 kJ mol^{-1} respectively.

The order $F^- > OH^- > NH_2^- > CH_3^-$ reflects the 'stability' of the anions themselves. The elements formally bearing the negative charge are all in the second period and so the effective nuclear charge experienced by the valence electrons in these atoms increases in the order $C < N < O < F$. Consequently the lowest energy orbitals are in F^- and the highest in CH_3^-, which means that F^- is lower in energy (i.e. 'more stable') than O^-, and so on for the others.

The order $TsO^- > RCOO^- > RO^-$ also reflects the stability of the anion which is related to the degree of delocalization that is possible in these ions. The negative charge in the tosylate ion can be delocalized over *three* oxygen atoms, which confers greater stability when compared to the carboxylate anion in which the charge can only be delocalized over *two* oxygen atoms. Both of these ions are lower in energy than an alkoxide, RO^- (e.g. $CH_3CH_2O^-$), in which the charge essentially remains on the single oxygen atom.

From the list we see that it is always harder for an anion to act as a leaving group than the protonated form of the same group. For example, OH^- is a much poorer leaving group than neutral H_2O, and NH_2^- is a much poorer leaving group than neutral NH_3. Thus one way to transform an –OH group into a much better leaving group is to protonate it to form $-OH_2^+$. Now, rather than the anion OH^- leaving, it is the neutral species H_2O which leaves.

Finally, it is worth noting that there is often a reasonable correlation between how good a leaving group something is, and the pK_a of its conjugate acid. This is explored in Box 9.6 on the next page.

Making a better leaving group

When it comes to actually trying to do a substitution, it may be necessary to convert a poor leaving group into a better one. We have already seen that one way to do this is to protonate a group. We have also seen in section 9.5.4 on page 376 that in the synthesis of acyl chlorides the –OH group can be made into a better leaving group with the use of thionyl chloride, $SOCl_2$.

Another common method used to make an –OH group into a good leaving group is to convert it to the tosylate ion. We have already come across this ion in Fig. 9.28 on page *359* where we used the conjugate acid, tosic acid, as an anhydrous acid. To convert an –OH group into tosylate, we react the alcohol with tosyl chloride, shown in Fig. 9.91.

When this reagent is mixed with an alcohol and a base such as pyridine, the high-energy oxygen electrons of the alcohol attack into the S–Cl $\sigma^\star$ MO and a new S–O bond is formed at the expense of a weak S–Cl bond (the same strategy that was used in the synthesis of acyl chlorides in section 9.5.4 on page 376). The oxygen from the alcohol is now bonded strongly to the sulfur and can leave with that group as the tosylate ion.

An example of this approach is shown in Fig. 9.92 on page 396. In this scheme the net result is the replacement of an –OH group by an –NH_2 group. In the first stage the tosylate is prepared by reacting the alcohol with tosyl chloride in pyridine. In the second stage, the tosylate is reacted with sodium azide, resulting in the azide nucleophile displacing the tosylate group. Finally, the azide is reduced using $LiAlH_4$ to give the amine.

9.7.3 The effect of the nucleophile

In the S_N1 mechanism, the rate-limiting step is the initial breaking of the bond to the leaving group to form a reactive intermediate (the carbenium ion). The

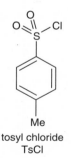

Fig. 9.91 Tosyl chloride, a reagent used to convert the poor leaving group OH^- into the excellent leaving group TsO^-.

tosyl chloride
TsCl

Box 9.6 The correlation between leaving group ability and pK_a.

How well an ion (X^-) or neutral species (Y) can act as a leaving group is to some extent reflected by the acidity of HX or HY^+.

The connection between acidity of the conjugate acid and leaving group ability comes about because similar factors affect both. Acidity is determined by the H–X bond strength and the stability of X^-. As we have seen, the same factors determine how good X^- is as a leaving group, although we should note that for a leaving group it is the C–X, rather than H–X, bond strength which is relevant.

As we saw in section 6.14 on page 245, we usually quantify the strength of an acid using pK_a values. A negative pK_a corresponds to an acid which readily dissociates in water, whereas a positive pK_a corresponds to a weak acid which only partly dissociates.

The pK_a values for the conjugate acids of the leaving groups are given in Fig. 9.90 on page 393. It can be seen that the conjugate acids of good leaving groups, such as TsO^-, I^- and H_2O, have negative pK_a values whereas the conjugate acids of groups which do not ordinarily act as leaving groups, such as H^-, NH_2^- and CH_3^-, have large positive pK_a values.

The same trends that we see for leaving group ability are found in the strengths of the conjugate acids. For example, the acidity of the hydrogen halides increases in the order HI > HBr > HCl > HF which mirrors exactly the trend in leaving group ability.

The acidity of hydrides increases as we move across the Periodic Table: CH_4 is not at all acidic and hence has an extremely high pK_a (around 48), NH_3 ($pK_a = 38$) can only be deprotonated by very strong bases, H_2O is a very weak acid ($pK_a = 15.7$) and HF is acidic, but still only weakly so ($pK_a = 3$). As we saw for leaving groups, this order reflects the 'stability' of the ions formed on dissociation of the acid, with F^- being the most stable due to the large effective nuclear charge on fluorine, and CH_3^- being the least stable as a result of the smaller effective nuclear charge of carbon.

Whilst we see the same general trends in both leaving group ability and acid strength, the order of leaving group ability from carbon does not match exactly the order of the acidity of the conjugate acids. For example, H_2O is a better leaving group than Cl^-, yet HCl is a stronger acid than H_3O^+. Such discrepancies are due to the fact that how good something is as a leaving group depends on its bond strength to *carbon*, whereas how strong its conjugate acid is depends on its bond strength to *hydrogen*; the two are not necessarily the same.

---

subsequent reaction between this intermediate and the nucleophile is very fast. Since the intermediate is so reactive, it does not really matter how good the nucleophile is for the second step – the intermediate will react with whatever nucleophile is nearby.

By contrast, in the S_N2 mechanism the bond to the nucleophile forms as the bond to the leaving group breaks. Whatever factors affect how easy it is to form the new bond will affect the rate of the reaction. Such factors include:

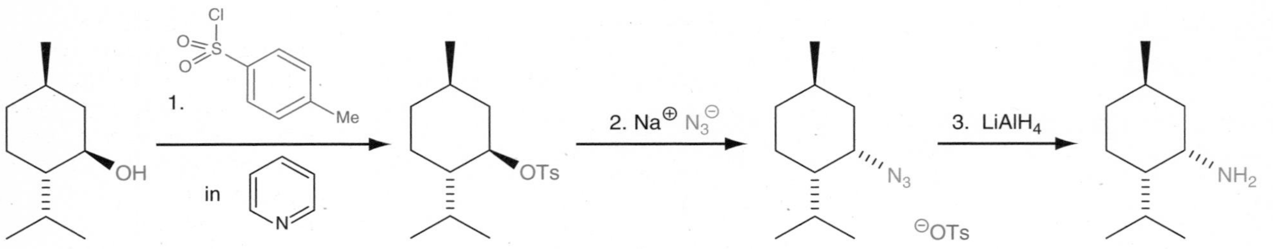

Fig. 9.92 The three-stage conversion of an alcohol into an amine. The poor leaving group –OH is first converted into the much better leaving group tosylate before being substituted by the azide ion. In the final stage of the reaction the azide is reduced to the amine using $LiAlH_4$.

- the relative energy of the electrons on the nucleophile;
- how bulky the nucleophile is;
- how solvated the nucleophile is.

It should not be too surprising that the higher the energy of the electrons of the nucleophile, the better it is as a nucleophile. We have already seen in the alkaline hydrolysis of esters that hydroxide is more effective as a nucleophile than neutral water. Generally, any anion is a better nucleophile than its neutral conjugate acid.

With caution, we can use pK_a as a guide to predicting how good a nucleophile a particular species is: any anion whose conjugate acid has a *high* pK_a will usually be a *good nucleophile* (provided it is not too bulky). However, it does *not* follow that any anion whose conjugate acid has a low pK_a will be a poor nucleophile. Iodide, for example, is an extremely good nucleophile, but its conjugate acid, HI, has one of the lowest pK_a values of around -10.

Figure 9.93 on the next page plots how effective different groups are as nucleophiles, their *nucleophilicity*, against the pK_a of the conjugate acid. The nucleophilicity of a particular species is measured by comparing the rate at which it displaces I^- from CH_3I with the rate at which methanol causes this displacement. For the series of nucleophiles which have an oxygen as the reactive centre (shown in red), there is a correlation between the pK_a and nucleophilicity, with the stronger bases acting as better nucleophiles. Note, for example, that the methoxide ion, MeO^- is around six orders of magnitude more effective as a nucleophile than neutral methanol.

However, there is certainly not a good overall correlation between basicity and nucleophilicity. This is because the two look at rather different things. The pK_a of the conjugate acid of the species X^- looks at how easy it is to break the X–H bond and protonate water; or, looking at the equilibrium in reverse, how easy it is for X^- to remove a proton from H_3O^+. The nucleophilicity looks at how easy it is for X^- to form a new bond to carbon. These are not the same things and so we should be cautious about trying to relate the two.

The solvent can also play a significant role in how good a nucleophile is, since polar solvents like water, which are capable of hydrogen-bonding to anions, will hinder the action of a nucleophile. Iodide is a much better nucleophile than the other halides (at least in such polar solvents) partly because it is less solvated, and partly because it has higher energy electrons.

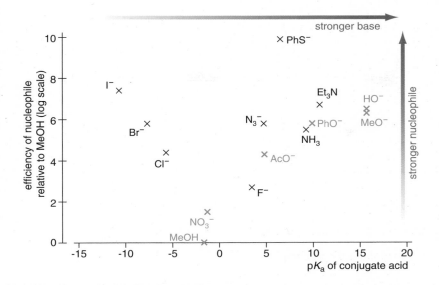

Fig. 9.93 When a group has the same element attacking either carbon (when it acts as a nucleophile), or hydrogen (when it acts as a base), there may be a reasonable correlation between how good a nucleophile the group is, and how good a base the group is. This is shown by the entries in red in the graph of the efficiency of the nucleophile against the pK_a of its conjugate acid. However, there is no general correlation between these two quantities with poor bases, such as iodide, also being very good nucleophiles.

9.7.4 The effects of the other groups attached to the reactive centre

Given that there is a good leaving group and reasonable nucleophile, the main factor which determines whether a subsutitution reaction will proceed via an S_N1 mechanism or an S_N2 mechanism, is what other groups are attached to the level-one carbon atom.

Steric factors in the S_N2 mechanism

In the S_N2 mechanism, the incoming nucleophile interacts with the C–X $\sigma^\star$ MO and so has to approach from directly behind the C–X bond. Any groups attached to the level-one carbon can partially block this approach and slow the reaction down. For the S_N2 mechanism to be efficient, two (or three) of the groups on the level-one carbon need to be hydrogens. This point is illustrated by Fig. 9.94 which shows the relative rates at which various chlorides are substituted by iodide via the S_N2 mechanism.

When the methyl group in **A** is replaced by a smaller hydrogen to give methyl chloride (chloromethane), the substitution proceeds almost a hundred times faster. When the methyl group is replaced by an ethyl group to give **C**, the rate is just three times slower. Replacing the methyl group by other alkyl groups also has rather little effect.

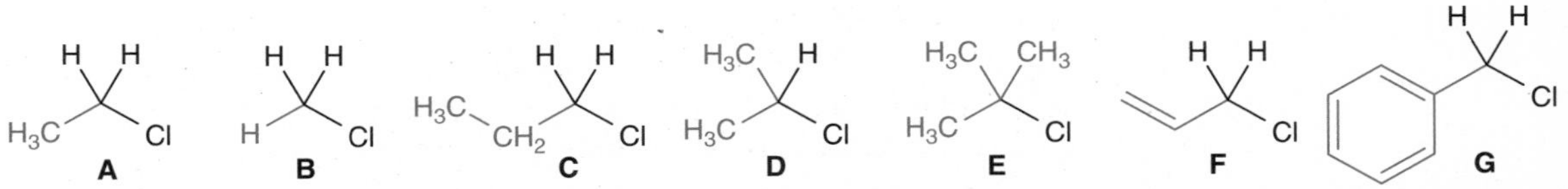

	A	**B**	**C**	**D**	**E**	**F**	**G**
	EtCl	MeCl	*n*-PrCl	*i*-PrCl	*t*-BuCl	allyl chloride	benzyl chloride
relative rate of substitution with iodide	1	93	0.37	0.0076	negligible	33	93

Fig. 9.94 Relative rates of nucleophilic substitution by iodide by the S_N2 mechanism at 60 °C in acetone (propanone) solution.

(a)

(b)

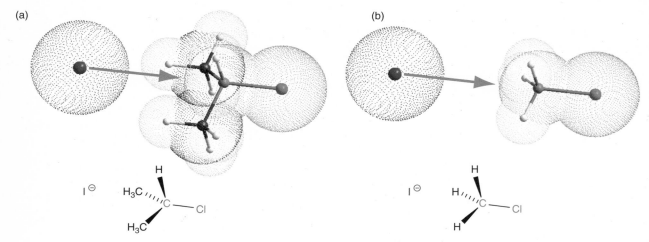

Fig. 9.95 Illustration of how in the S$_N$2 mechanism the groups attached to the level-one carbon hinder the approach of the nucleophile. The dotted surfaces indicate the approximate sizes of the atoms. The line of approach of the nucleophile must be from directly behind the bond to the leaving group in order for there to be an interaction with the C–X $\sigma^\star$ MO. With two methyl groups attached to the level-one carbon atom (shown in red), this line of approach is hindered, as shown in (a). When the level-one carbon has only hydrogens attached, i.e. when it is a methyl derivative, the approach is clear, as shown in (b).

However, the rate decreases by a factor of about a hundred if the level-one carbon has only *one* hydrogen attached, as in **D**. Figure 9.95 shows how the line of approach of the nucleophile is hindered when two of the hydrogens attached to the level-one carbon are replaced by methyl groups. With no hydrogens on its level-one carbon, **E** does not undergo any significant degree of substitution by the S$_N$2 mechanism.

Stabilization of the S$_N$2 transition structure by π systems

Given that larger groups attached to the level-one carbon generally slow down a substitution by an S$_N$2 mechanism, it is initially surprising to note in Fig. 9.94 that when the methyl group in **A** is replaced by a phenyl ring, the reaction proceeds about one hundred times *faster*. Similarly, compound **F** (allyl chloride) also reacts significantly faster than **A**.

The explanation for this observation is thought to be due to a stabilization of the transition structure (shown in Fig. 9.89 on page 392) through which the reaction passes during the substitution of the leaving group by the nucleophile. To describe the bonding in the transition structure, we note that there are still three full bonds to the R groups and that these bonds are all in the same plane, and at 120° to one another. We can consider the central carbon as being sp^2 hybridized, with one sp^2 HAO overlapping with an orbital from each R group to give rise to the three C–R bonds. This leaves a $2p$ AO on the central carbon which can overlap with an orbital from both the nucleophile and the leaving group to form a delocalized three-centre bond; this orbital interaction is shown in Fig. 9.96.

The key point is that if one of the R groups can also interact with the $2p$ AO of the central carbon, and so allow electron density to be further delocalized, then this will stabilize the structure. This can happen when one of the R groups

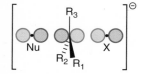

Fig. 9.96 The bonding in the transition structure of the S$_N$2 mechanism. A p AO from the central carbon interacts with a $2p$ orbital from the nucleophile and from the leaving group to form a three-centre bond.

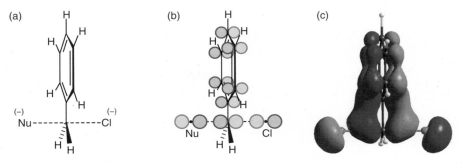

Fig. 9.97 The transition structure in the nucleophilic substitution of benzyl chloride. The benzene ring is aligned at right angles to the plane of the paper, as shown in (a). This alignment allows an overlap of the p AOs in the benzene ring with the $2p$ AO of the central carbon and the p orbitals of the nucleophile and leaving group. This is shown in cartoon form in (b) and as an isosurface plot in (c).

relative rate of substitution with iodide			
1	2 800	33 000	97 000

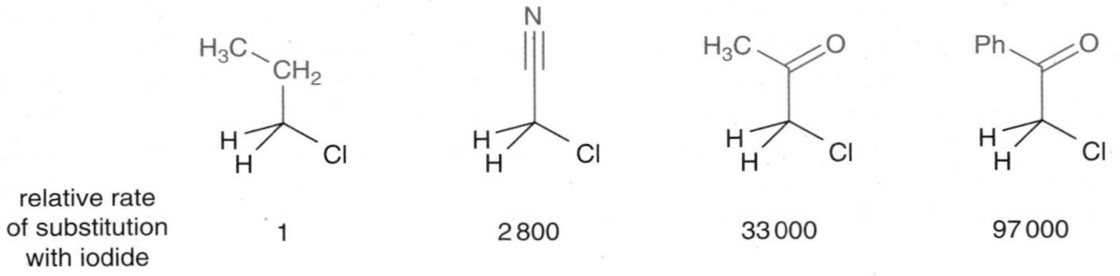

Fig. 9.98 The rate of substitution of chloride by iodide is greatly enhanced when there is a C=O or C=N π system attached to the level-one carbon. Each reaction is carried out in acetone (propanone) at 50 °C.

is a benzene ring, or a C=C double bond as in examples **F** and **G** from Fig. 9.94 on page 397. The lower in energy the transition structure, the faster the reaction occurs.

Figure 9.97 (a) shows the the transition structure in the nucleophilic substitution of benzyl chloride by a nucleophile. A delocalized MO formed from the interaction of the p orbitals on the central carbon, the carbons from the benzene ring, and the p AOs of the nucleophile and leaving group is shown in cartoon form in (b) and as an isosurface plot in (c).

The transition structure of an S_N2 mechanism can be stabilized even further when the C=C π system is replaced by a C=O or C=N π system. This is because the oxygen or nitrogen help to withdraw more electron density from the reaction centre in the transition structure. Some examples of the substitution of various chlorides by iodide are given in Fig. 9.98.

Stabilization of the S_N1 intermediate: (i) by a neighbouring lone pair

For the S_N1 mechanism to be efficient, the groups attached to the level-one carbon must be able to help stabilize the trigonal planar intermediate cation or, looking at it from a different perspective, the groups must be able to help this intermediate form in the first place. This was the role of the oxygen shown in black in the acetal hydrolysis mechanism in Fig. 9.85 on page 391 – the oxygen helped to stabilize the carbenium ion.

This interaction between the vacant $2p$ AO of the central carbon and the filled $2p$ AO of the oxygen is shown as energy level diagram in Fig. 9.99. It simply represents the formation of a π bond from an oxygen and a carbon $2p$ AO.

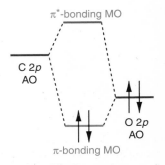

Fig. 9.99 The interaction between a filled oxygen $2p$ AO and a vacant carbon $2p$ to form a π bond.

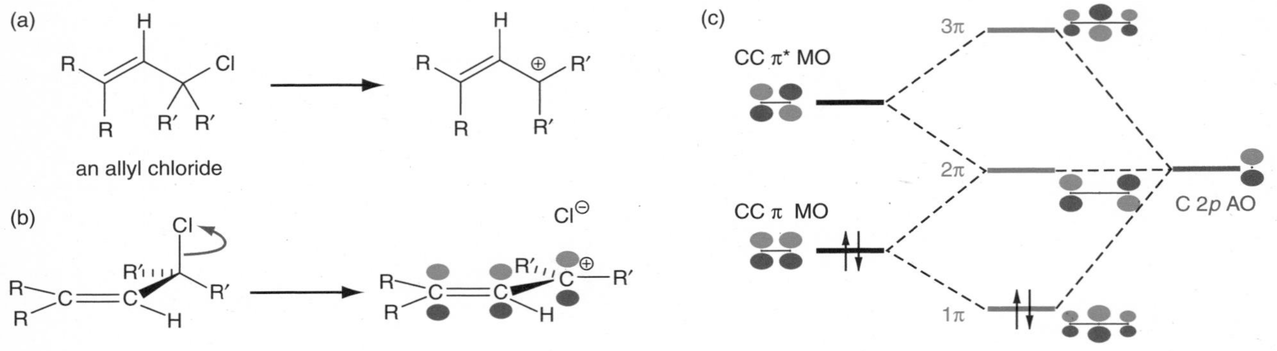

(a)

an allyl chloride

(b)

(c)

CC π* MO

CC π MO

3π

2π

C 2p AO

1π

Fig. 9.100 It is possible for an allyl chloride to undergo substitution by the S_N1 mechanism in which the first step is the formation of the intermediate as shown in (a). The neighbouring π system can interact with the vacant p orbital to form the delocalized π MOs of the allyl cation as shown in (b) and (c).

Stabilization of the S_N1 intermediate: (ii) by a neighbouring π system

Other groups can also interact with the vacant carbon p AO of the intermediate in the S_N1 mechanism to give rise to a lowering in energy of electrons, and hence some net stabilization. The allyl chloride shown in Fig. 9.100 (a) has a C=C double bond as one of the groups attached to the level-one carbon. If the chloride leaves, the neighbouring π bond can help to stabilize the cation formed. How the p orbitals overlap in the planar intermediate is more clearly seen in (b). This interaction is also shown as an energy level diagram in (c). The interaction between the CC π MOs and the vacant carbon $2p$ AO gives rise to the MOs for the allyl cation that we have seen before in section 4.8.1 on page 166.

It is the lowering in energy of the π electrons that stabilizes the structure. Whilst it might be tempting to say that the positive charge is delocalized, the positive charges remain at the nuclei of the atoms and it is the *electrons* that move.

It is interesting to consider why the chloride 'falls off' in the first place. In the hydrolysis of the acetal, we saw it was the interaction between the oxygen lone pair and the vacant C–O $\sigma^\star$ MO that helped the C–O bond break. In the reaction of the allyl chloride in Fig. 9.100, it is the interaction between the filled C=C π MO and the vacant C–Cl $\sigma^\star$ that helps break the C–Cl bond.

Whether an allyl chloride such as that shown in Fig. 9.100 (a) actually undergoes a nucleophilic substitution by the S_N1 mechanism as shown depends on the other substituents in the molecule. If the two groups R′ are both hydrogens, an S_N2 mechanism is also possible (remember the C=C π system also helps to stabilize the transition structure in the S_N2 mechanism).

Stabilization of the S_N1 intermediate: (iii) by neighbouring alkyl groups

Finally, and in many ways perhaps most important of all, the intermediate cation formed in an S_N1 mechanism can also be stabilized by neighbouring alkyl groups.

Figure 9.101 shows the methyl cation that would have to form if bromomethane (methyl bromide) underwent an S_N1 mechanism. The three C–H bonds lie in the nodal plane of the unoccupied $2p$ orbital of the carbon atom, and hence have the wrong symmetry for any interaction. Given there is no

Fig. 9.101 The methyl cation; CH_3^+. The three hydrogens are in the nodal plane of the unoccupied carbon $2p$ AO and so cannot interact with it.

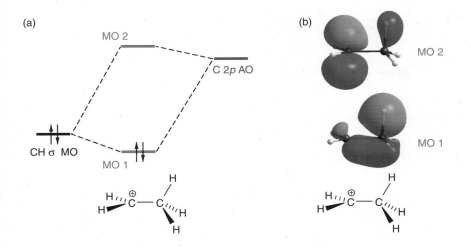

(a)

MO 2

C 2p AO

CH σ MO

MO 1

(b)

MO 2

MO 1

Fig. 9.102 In the ethyl cation, $CH_3CH_2^+$, the C–H σ bonding MOs of the methyl group can interact with the vacant $2p$ AO on the adjacent carbon. The schematic MO diagram, shown in (a), shows that this interaction lowers the energy of the electrons in the C–H σ bonds. Iso-surface plots of the resulting MOs are shown in (b).

stabilizing interaction present in the methyl cation, it does not readily form and methyl bromide will not undergo nucleophilic substitution by the S_N1 mechanism. It will, of course, readily undergo nucleophilic substitution by the S_N2 mechanism.

If one of the hydrogens is replaced by a methyl group, the C–H σ bonds from the methyl group can interact to some degree with the vacant carbon $2p$ AO, as shown in Fig. 9.102 (a). The resulting MOs from this interaction are shown in (b). MO 1 is still largely a C–H σ bonding MO, and the vacant MO 2 is largely a carbon $2p$ orbital. However, the important point is that as a result of the interaction, the C–H σ bonding MO is lowered in energy slightly. The interaction between a neighbouring σ bond and a vacant orbital is known as σ *conjugation*.

It turns out that the σ conjugation from one neighbouring alkyl group is generally not sufficient to stabilize the intermediate cation enough for the S_N1 mechanism to be fast enough to be seen. What is more, with just two hydrogens on the level-one carbon, the S_N2 mechanism is very fast and so nucleophilic substitution reactions proceed by this mechanism. Even with two neighbouring alkyl groups, there is not usually enough stabilization for the S_N1 mechanism to be competitive. However, with *three* groups, the intermediate cation that must be formed in the S_N1 mechanism is stabilized enough for this mechanism to occur readily. As an example, Fig. 9.103 shows the hydrolysis of *t*-butyl chloride to form *t*-butyl alcohol, a reaction which occurs readily in alcohol/water mixtures (the alcohol help the two otherwise immiscible liquids to mix).

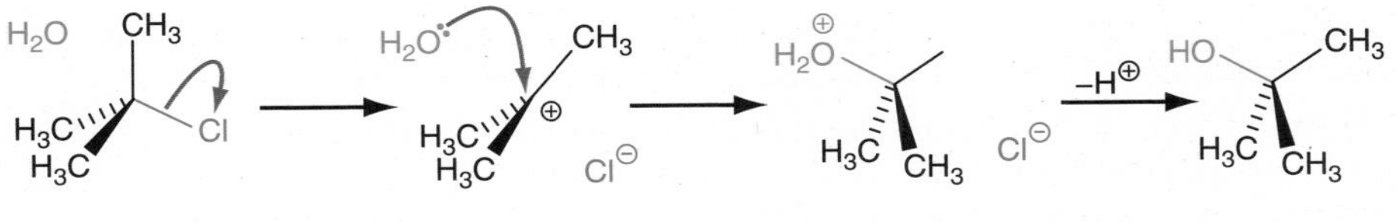

Fig. 9.103 The hydrolysis of *t*-butyl chloride to *t*-butyl alcohol. The reaction proceeds by an S_N1 mechanism with the intermediate cation being stabilized by the three methyl groups.

Table 9.4 Relative rates of reaction of different bromoalkanes with hydroxide

	MeBr	EtBr	*i*-PrBr	*t*-BuBr
number of hydrogens on level-one carbon	3	2	1	0
rate by S_N1 mechanism	–	–	0.83	3500
rate by S_N2 mechanism	74	5.8	0.17	–
overall rate of reaction	74	5.8	1	3500

The swap over between the S_N1 and S_N2 mechanisms

The two mechanisms by which nucleophilic substitution can occur are in some ways complimentary. With *two or three hydrogens* attached to the level-one carbon the S_N2 mechanism is favourable since the incoming nucleophile can easily approach from behind the leaving group, but the S_N1 mechanism is unfavourable since no stable cation can be formed. In contrast, with *three alkyl groups* and no hydrogens attached to the level-one carbon, the approach of the nucleophile is hindered and so the S_N2 mechanism is unfavourable, but now there is sufficient stabilization of the intermediate cation for the S_N1 mechanism to be favourable.

This swap over in the mechanism is shown in Table 9.4 which shows the relative rates of reaction of different bromoalkanes (alkyl bromides) with 0.01 M hydroxide in aqueous ethanol. Neither bromomethane nor bromoethane react at all by the S_N1 mechanism, but do react by the S_N2 mechanism. In contrast, *t*-butyl bromide does not react at all by the S_N2 mechanism, but exclusively by the S_N1 mechanism. The rate of reaction of *t*-BuBr is particularly large since the intermediate cation reacts with the solvent rather than hydroxide. The solvent is at a much higher concentration than the hydroxide, and so the reaction occurs extremely quickly.

With one hydrogen attached to the level-one carbon, the rate of substitution of *i*-PrBr is the slowest of the alkyl bromides shown in the table. Neither the S_N1 nor the S_N2 mechanism is particularly good, and both mechanisms contribute to some degree to the reaction.

Other substitution mechanisms

For some molecules, nucleophilic substitution is not possible by either the S_N1 or the S_N2 mechanism. An example is bromobenzene. It is impossible for a nucleophile to attack directly behind the C–Br σ bond, since this would mean an approach through the benzene ring itself, as shown in Fig. 9.104 (a). This means the S_N2 mechanism is not possible for bromobenzene.

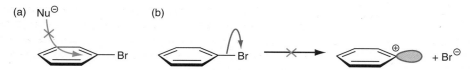

Fig. 9.104 Bromobenzene cannot undergo nucleophilic substitution by the S_N2 mechanism, since this necessitates a line approach from behind the C–Br bond as shown in (a). This is blocked by the benzene ring. It cannot undergo substitution by the S_N1 mechanism since the cation that would have to be formed has a vacant orbital at right angles to the π system of the benzene ring, as shown in (b), and so cannot be stabilized by the π system.

Even though at first glance it looks as if a stable cation might be formed on breaking the C–Br bond to give bromide ion, this is not the case. The C–Br σ bond is in the nodal plane of the benzene π system. Consequently, the vacant orbital that would result from the C–Br bond breaking is also in the nodal plane as shown in (b). This means no delocalization using the π electrons is possible, and hence with no stabilization of the cation. The S_N1 mechanism is also ruled out for bromobenzene. However, it is possible for aromatic systems like bromobenzene to undergo nucleophilic substitution but by a different mechanism involving the π system.

FURTHER READING

There is a wide selection of organic chemistry texts to choose from, but be aware that none of them uses quite the same approach we have adopted here. Rather, they tend to discuss the reactions of each functional group in turn.

March's Advanced Organic Chemistry, sixth edition, Michael B. Smith and Jerry March, Wiley (2007).
Organic Chemistry, Jonathan Clayden, Nick Greeves, Stuart Warren and Peter Wothers, Oxford University Press (2001).

QUESTIONS

9.1 Identify the functional groups present in the following molecules.

(a)

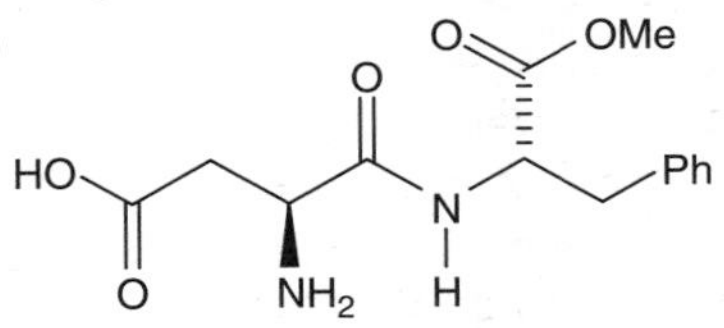

Aspartame – an artificial sweetener

(b)

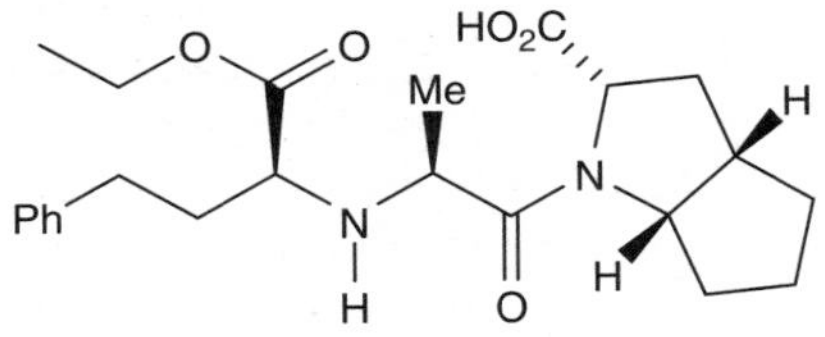

Ramipril – used to treat hypertension and congestive heart failure

(c)

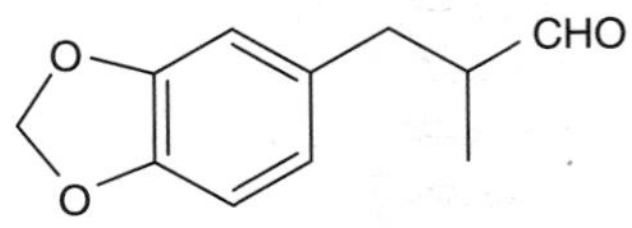

Tropional – a key ingredient in Chanel's perfume *Allure*

(d)

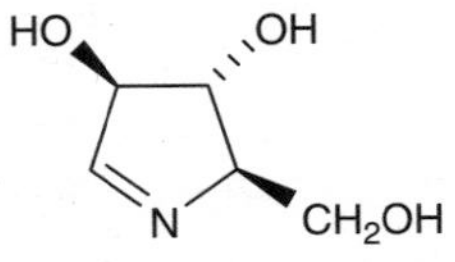

Nectrisine – an antibiotic

9.2   Copy the structures of each of the following molecules and identify the functional group level of each of the carbon atoms in them.

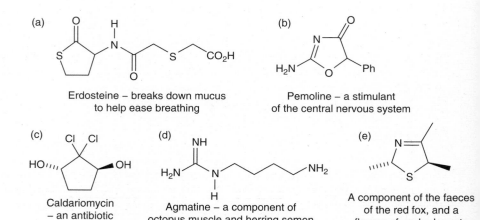

(a)

Erdosteine – breaks down mucus
to help ease breathing

(b)

Pemoline – a stimulant
of the central nervous system

(c)

Caldariomycin
– an antibiotic

(d)

Agmatine – a component of
octopus muscle and herring semen

(e)

A component of the faeces
of the red fox, and a
flavour of cooked meats

9.3   Draw the structures of the products formed in the following hydrolysis reactions.

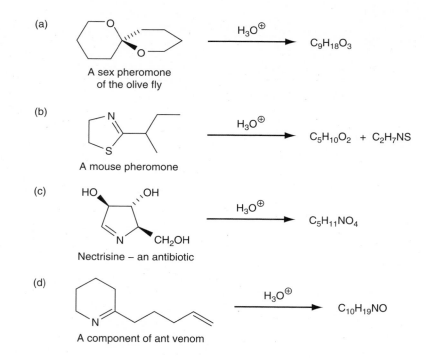

(a)

A sex pheromone
of the olive fly

$H_3O^{\oplus}$

$C_9H_{18}O_3$

(b)

A mouse pheromone

$H_3O^{\oplus}$

$C_5H_{10}O_2$ + C_2H_7NS

(c)

Nectrisine – an antibiotic

$H_3O^{\oplus}$

$C_5H_{11}NO_4$

(d)

A component of ant venom

$H_3O^{\oplus}$

$C_{10}H_{19}NO$

9.4 The drug *Fenipentol*, shown below, is synthesized by the reaction between butyl-magnesium bromide and an aldehyde. Draw the structure of the aldehyde and give the mechanism for the reaction.

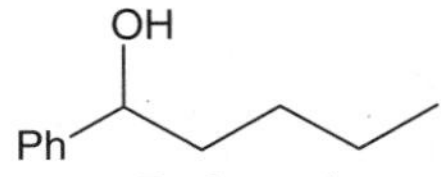

Fenipentol

9.5 A synthesis of the anti-inflammatory drug *Felbinac* is shown below. Suggest a structure for the intermediate **A** and give a mechanism for its formation. Give a mechanism for the hydrolysis of **A** to Felbinac.

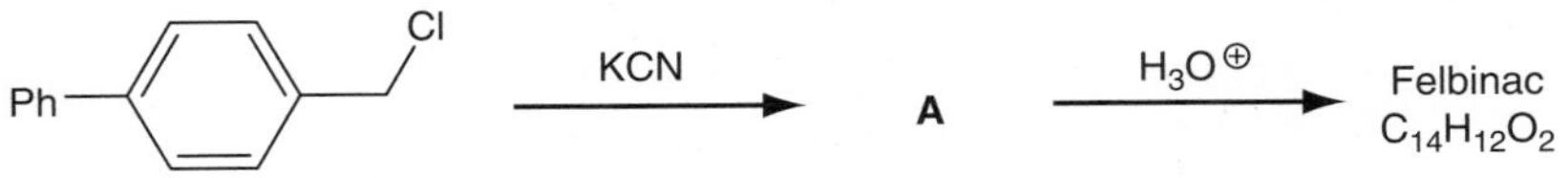

9.6 Consider the following scheme for the synthesis of the drug *Cicletanine*, which is used as a diuretic and antihypertensive.

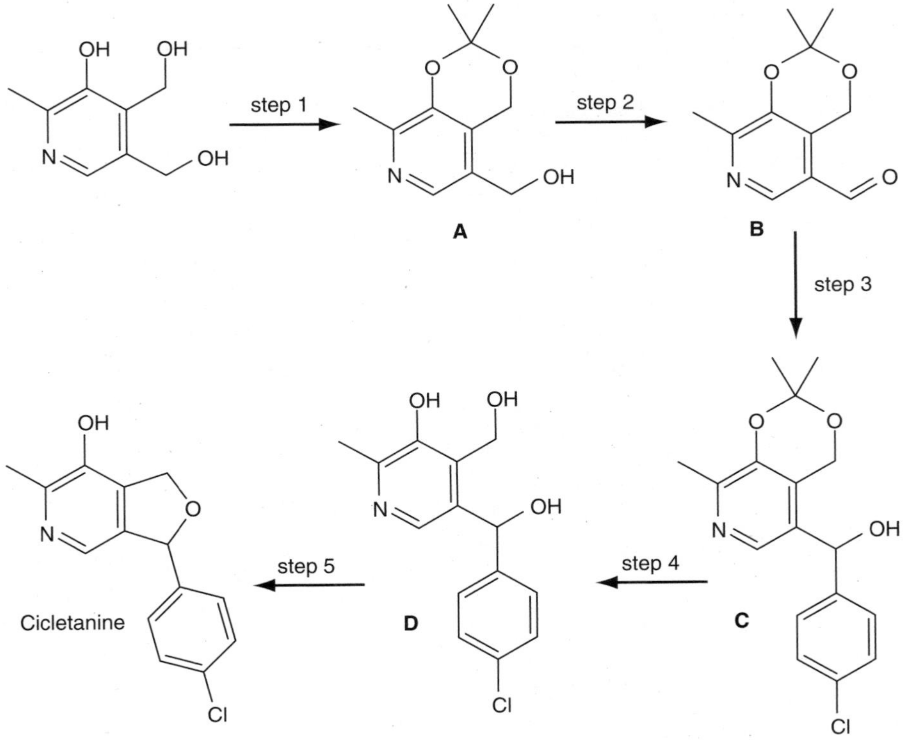

(a) Give the reagents needed to form acetal **A** in step 1 and draw a mechanism for this step.

(b) What sort of reaction is step 2: a reduction; an oxidation; or a nucleophilic substitution?

(c)   Give the reagent needed for step 3 and draw a mechanism for this step.

(d)   Steps 4 and 5 occur under the same conditions. What conditions are needed for the hydrolysis of acetal **D**? Give a mechanism for this step.

(e)   What sort of reaction is step 5: a reduction; an oxidation; or a nucleophilic substitution?

9.7   The synthesis of the antihistimine drug *Bamipine* is given below.

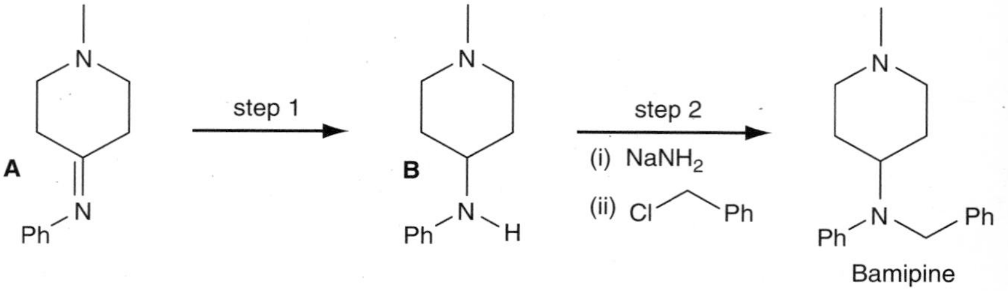

(a)   What sort of reaction is step 1: a reduction; an oxidation; or a nucleophilic substitution?

(b)   What is the role of the $NaNH_2$ in step 2? Draw mechanisms for the reactions in step 2.

(c)   The imine **A** is formed by the reaction of $PhNH_2$ with another reagent. Identify this reagent, and give a mechanism for its reaction with $PhNH_2$ to form **A**.

9.8   A synthesis of the antiparkinsonian drug *Pridinol* is outlined below.

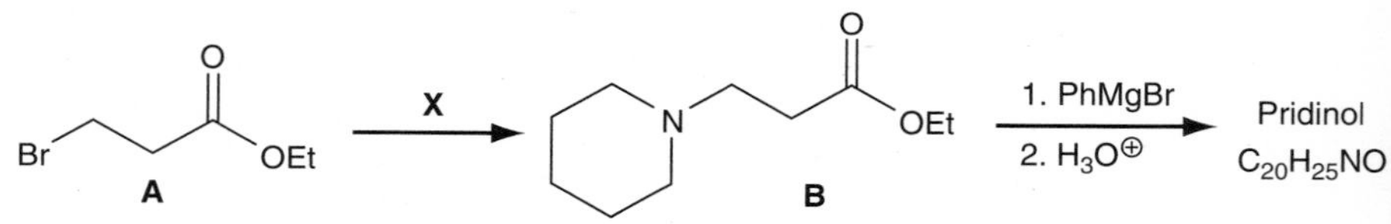

(a)   At which *two* sites could bromoester **A** react with a nucleophile? Name the key orbitals involved in each case.

(b)   Suggest a reagent **X** to form **B** from **A** and give a mechanism for the reaction. Explain why **X** reacts in the manner shown, rather than at the alternative site identified in (a).

(c)   Intermediate **B** reacts with excess phenylmagnesium bromide to give Pridinol. Suggest a structure for Pridinol and give a mechanism for its formation.

9.9   Part of the synthesis of the antiasthmatic drug *Fenspiride* is shown below. Identify the structure of the intermediate **B** and give a mechanism for its formation. Give a mechanism for the reduction of **B** to **C**.

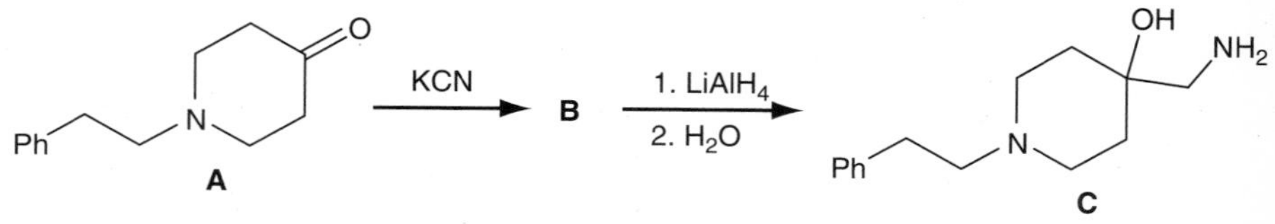

The drug Fenspiride is prepared from intermediate **C** by reacting it with the base NaOEt and compound **X**. **X** is one of the compounds shown in the box below. By considering the functional group levels, suggest which one of the compounds **D–H** is **X** and give a mechanism for the formation of Fenspiride.

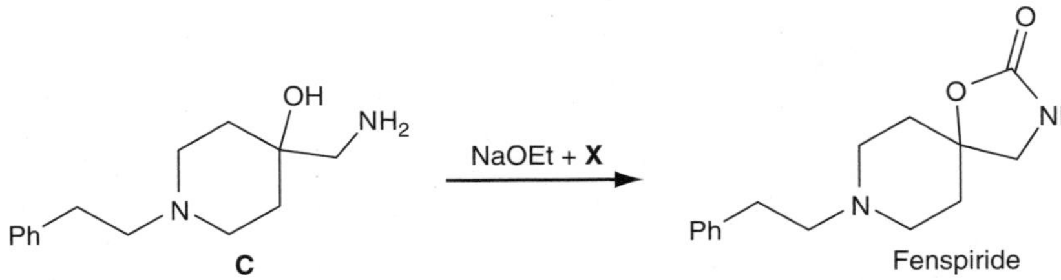

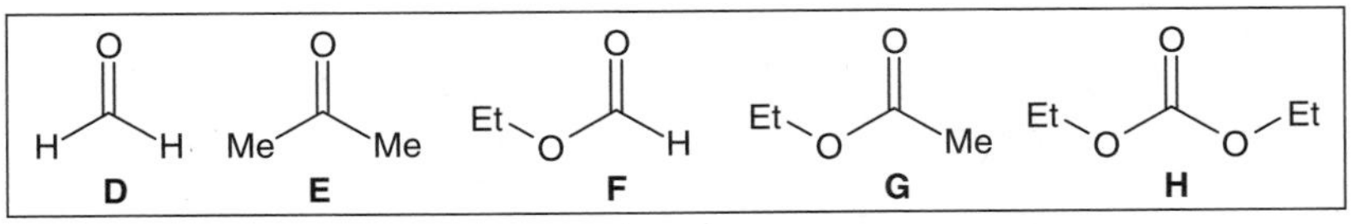

9.10 A synthesis of the sedative *Ethinamate* is shown below. Suggest a structure for the intermediate **A** and draw a mechansim for its formation. Give mechanisms for the formation of Ethinamate from **A**.

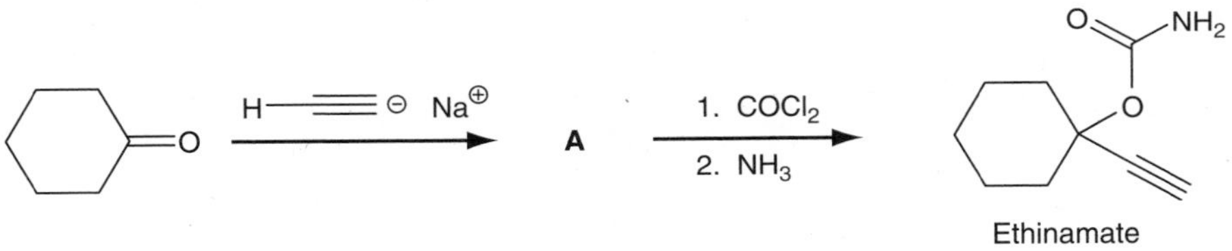

9.11 A synthesis of the mite and spider killer *Chlorobenzilate* is shown below. When the Grignard reagent **A** is added slowly to a solution of diethyl oxalate **B** at low temperatures, the intermediate **C** is first formed, but this reacts with further Grignard reagent to form, after work-up in aqueous acid, Chlorobenzilate. Suggest a structure for the intermediate **C** and give a mechanism for its formation. Give a mechanism for the reaction of the intermediate **C** with the Grignard reagent **A** and explain why the addition of the second Grignard reagent occurs at the position it does.

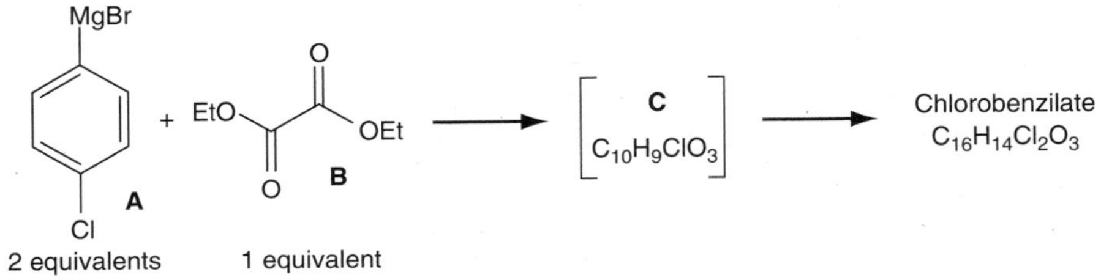

9.12 Explain the following. When sodium borohydride is added to a solution of an ester, as shown in (a), no reduction takes place. However, on addition of aluminium trichloride, as shown in (b), the reduction readily takes place.

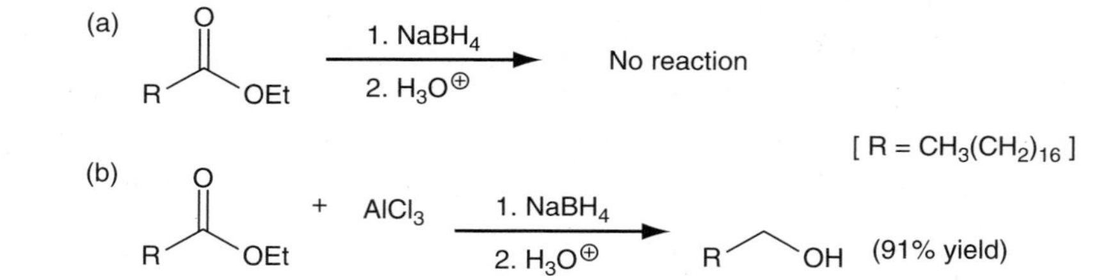

(a)

1. NaBH$_4$
2. H$_3$O$^\oplus$

No reaction

[R = CH$_3$(CH$_2$)$_{16}$]

(b)

+ AlCl$_3$

1. NaBH$_4$
2. H$_3$O$^\oplus$

R⌃OH   (91% yield)

9.13 A synthesis of the muscle relaxant *Nefopam* is shown below.

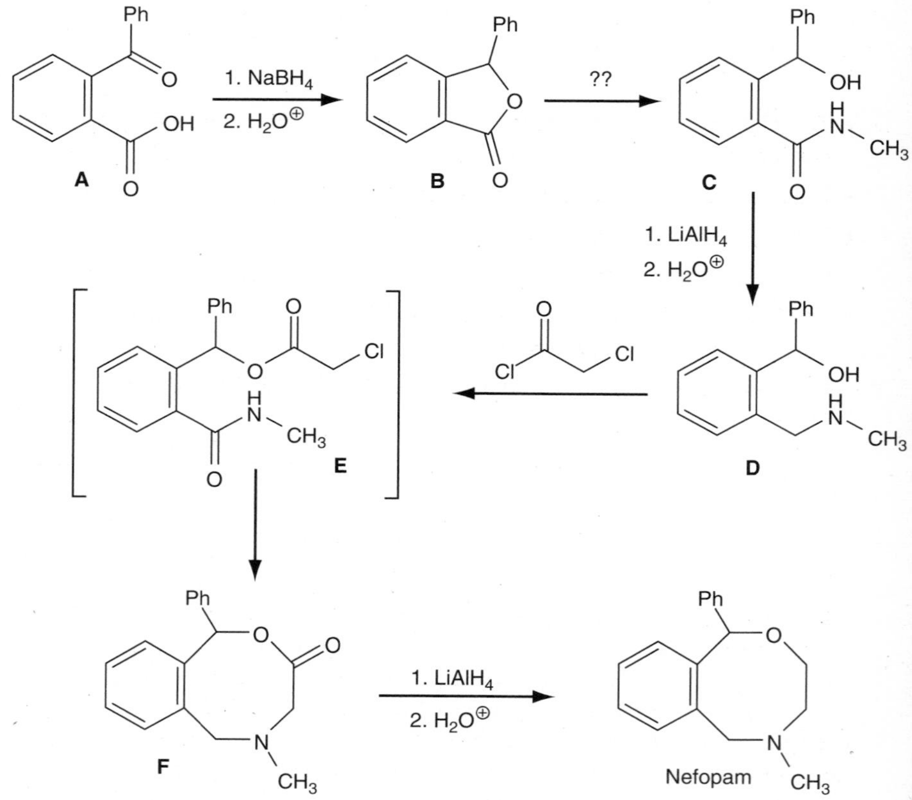

(a) Which of the carbonyls in compound **A** is reduced by the borohydride? Draw a mechanism for this reduction and for the subsequent formation of **B**.

(b) Suggest a suitable reagent for the conversion of **B** to **C** and draw a mechanism for this step.

(c) Give a mechanism for the reduction of **C** to **D**.

(d) The cyclyzation of intermediate **E** to give **F** is an example of an intramolecular nucleophilic substitution reaction. Will this proceed via an S$_N$1-like mechanism or an S$_N$2- like mechanism? Explain your answer.

(e) Draw a mechanism for the reduction of **F** and the subsequent formation of Nefopam which takes place in acid.

(f) Explain why NaBH$_4$ is the reducing agent of choice for the first step, but LiAlH$_4$ for the later two reductions.

The rates of reactions

Key points

- The experimentally determined rate law gives the dependence of the rate on the concentrations of the reactants.
- Rate constants have a strong temperature dependence given by the Arrhenius law.
- There is an energy barrier to most reactions.
- Complex reactions proceed by a series of elementary steps, called a mechanism.
- The kinetics of complex reactions can sometimes be simplified using the pre equilibrium or steady-state approximations.
- In a multi-step reaction, one step can often be identified as the rate-limiting (or rate-determining) step.

This chapter is concerned with how we describe and understand the rate at which a chemical reaction proceeds. Not only is this of fundamental importance to our understanding of the nature of chemical reactions, but it is also of great practical significance. Many industrial and commercial processes involve chemical reactions – for example the cracking of petroleum, the setting of cement, the formation of polymers and the drying of paint – so that understanding what determines the rates of the relevant reactions, and how they can be influenced, has very direct commercial consequences.

In the Earth's atmosphere there are many chemical species which are involved in an elaborate network of chemical reactions, driven by the energy from the sun. Attempting to model what will happen to the atmosphere over time, for example to understand the effect of certain kinds of pollution, requires a detailed knowledge of not only the reactions involved, but their rates and how they respond to changes in the conditions, such as temperature.

Life itself, reduced to its basics, involves a set of carefully controlled and interdependent chemical reactions. Living organisms need to be able to control the rates of different chemical processes to ensure that the required amounts of certain molecules are generated in the right place at the right time. Again, an understanding of how the rates of these reactions are regulated in living systems is crucial to understanding how a living organism works.

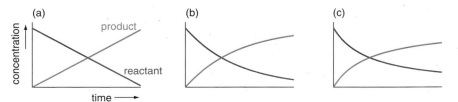

Fig. 10.1 Typical plots showing how the concentration of a reactant (the blue line) and a product (the red line) may change during a reaction. In (a) the reactant decays linearly, in (b) the decay is according to an exponential, and in (c) the decay curve has a different shape, in fact varying as $1/(t + C)$, where C is a constant. In each case, the rise of the concentration of the product mirrors the decay of the reactant.

In this chapter we are first going to look at how we describe the rates of reactions, and in particular how these rates can be expressed in terms of a *rate law*. We will then introduce the idea of the energy barrier to reaction and show how this leads to the characteristic temperature dependence of the rate.

The remainder of the chapter is devoted to looking at how we can understand the kinetics of complex reactions which involve several steps. To do this we will develop two important techniques: the *pre-equilibrium hypothesis* and the *steady-state approximation*.

In Chapter 18 in Part II we will return to chemical kinetics and revisit some of the topics introduced here, but taking a more mathematical approach. In that chapter we will also look at how reaction rates and rate laws are determined experimentally.

10.1 The rate of a reaction

Once a reaction has been initiated we expect that the concentration of the reactants will fall and that of the products will rise. Precisely how these concentrations vary with time depends on the details of the reaction we are studying, but Fig. 10.1 shows some typical ways in which the concentrations of products and reactants can change.

In (a) the concentration of the reactant falls linearly with time, and this fall is mirrored by a linear rise in the concentration of the product. Plot (b) shows a case where there is an exponential decay of the concentration of the reactant, which is mirrored by an exponential rise in the concentration of the product. Finally plot (c) also shows a decay curve, but the shape is subtly different to exponential shown in (b). In fact in (c) the concentration of the reactant is decaying as $1/(t + \text{const.})$, where t is the time.

No matter how the concentration is changing, at a particular time we can determine the *rate* of the reaction, which is defined as

$$\text{rate} = \frac{\text{change in concentration during the interval } \Delta t}{\Delta t}.$$

The rate therefore has units of (concentration time^{-1}). Figure 10.2 (a) shows how this measurement of the rate can be visualized for the case where the reactant A is decaying according to a curve.

What is immediately obvious from this plot is that the rate of reaction decreases in size as time goes on i.e. it is not constant. This being the case, it is

If you are familiar with calculus you will recognize this rate is the *derivative* of concentration with respect to time: rate = d conc./dt. In this chapter we will not be using the language of calculus, but later on in Chapter 18 we will return to the topic of rates of reactions and show how these can be described using calculus.

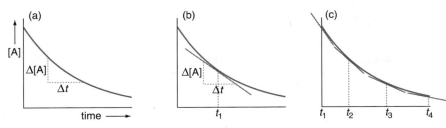

Fig. 10.2 The rate of a reaction is given by the change in concentration of A, Δ[A], during an interval of time Δt, divided by that time interval; this process is visualized in (a). For this curve, the rate is varying with time, so it is best to define the rate as the instantaneous slope of the curve, as illustrated in (b); the grey line is a tangent to the curve at time t_1. As shown by the tangents in (c) at times t_1, t_2 etc, the rate decreases in size as time progresses.

better to define the rate as the (instantaneous) *slope* of the curve at a particular time. As shown in Fig. 10.2 (b) this slope can be found by drawing a tangent (the grey line) to the curve, and then measuring the slope of the tangent. For different times, t_1, t_2 ..., as shown in (c), the tangents have different slopes, so the rate is changing.

Figure 10.3 shows that the slope, and hence the rate, is negative for a reactant, and positive for a product. This turns out to be rather inconvenient, so we will in fact define the rate is a slightly more subtle way, as described in the next section.

10.1.1 Defining the rate

Consider the reaction

$$2H_2(g) + O_2(g) \longrightarrow 2H_2O(g).$$

As we have already seen, if we define the rate as the slope of the graph of concentration against time, the rate for the two reactants H_2 and O_2 will be negative, and that for the product H_2O will be positive.

Furthermore, as two moles of H_2O are produced for each mole of O_2 consumed, the rate of *formation* of H_2O will be twice the rate of *loss* of O_2. Similarly, as two moles of H_2 react with one mole of O_2, the rate of *loss* of H_2 will be twice that of O_2. This is all rather messy.

These difficulties are neatly avoided by *defining* the rate for species A as

$$\text{rate of reaction} = \frac{1}{\nu_A} \times \text{slope of a plot of [A] against time,}$$

where: (a) ν_A is the stoichiometric coefficient of species A in the balanced chemical equation; and (b) the stoichiometric coefficients of reactants are taken as being *negative*, and those of products as being *positive*. The consequence of the first point is that the rate will be the *same* for all species involved in the reaction, and the second point ensures that this rate will *always* be positive.

For example, in the above reaction the stoichiometric coefficient for H_2 is −2, as it is a reactant, and likewise the coefficient for O_2 is −1. For the product H_2O the coefficient is +2. The rate of reaction can therefore be written in terms of any of the three species

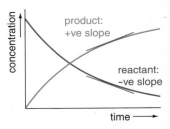

Fig. 10.3 Illustration of how the slope, and hence the rate of reaction, has opposite signs for a reactant and a product.

in terms of H_2:   rate $= \dfrac{1}{-2} \times$ slope of a plot of $[H_2]$ against time

in terms of O_2:   rate $= \dfrac{1}{-1} \times$ slope of a plot of $[O_2]$ against time

in terms of H_2O: rate $= \dfrac{1}{+2} \times$ slope of a plot of $[H_2O]$ against time.

As these slopes are in the ratio $-2{:}-1{:}+2$, all three rates given above have the same value. From now on, we will be careful to use this approach when we write the rate of a reaction.

10.2   Rate laws

Experimental work has revealed that the rates of reactions are strongly influenced by the concentration of the reactants. This in itself is hardly a surprise, but what is more interesting is the precise way in which the rate depends on the concentrations.

It is often found that the way in which the experimentally measured rate varies with concentration has a relatively simple mathematical form which can be expressed in the form of a *rate law*. A typical form for such a law is

$$\text{rate} = k\,[A]^a\,[B]^b \dots \tag{10.1}$$

In this expression $[A]$ represents the concentration of species A, and the power to which this concentration is raised, *a*, is called the *order* with respect to A. Similarly *b* is the order with respect to B; there may be more terms in the rate law, each with its own order. The overall order is the sum of the orders for each species: in this case the overall order is $(a + b + \dots)$.

The species A and B are usually reactants, but in some complex reactions the rate law may also depend on the concentration of products. Typically, the orders *a* and *b* are integer values, but again in more complex reactions they can be fractional (e.g. $\frac{1}{2}$) and can also be negative.

k is the *rate constant* (sometimes called rate coefficient) for the particular reaction we are considering; as its name implies, the value of *k* is independent of the concentrations of the species in the rate law. The units of the rate constant depend on the overall order of the reaction and, as we shall see later on, the rate constant is strongly temperature dependent.

It is very important to realise that the order is an experimentally determined quantity. Except for elementary reactions, discussed in section 10.5 on page 419, it is *not* the same as the stoichiometric coefficient in the balanced chemical equation which describes the overall reaction.

10.2.1   Some examples of rate laws

Rate laws can take many forms, and in this section we will look at a selection of reactions which illustrate the variety of rate laws that are found. The details as to how such rate laws are determined experimentally are discussed in Chapter 18.

The nucleophilic substitution reaction between the amine DABCO and benzyl bromide (in solution)

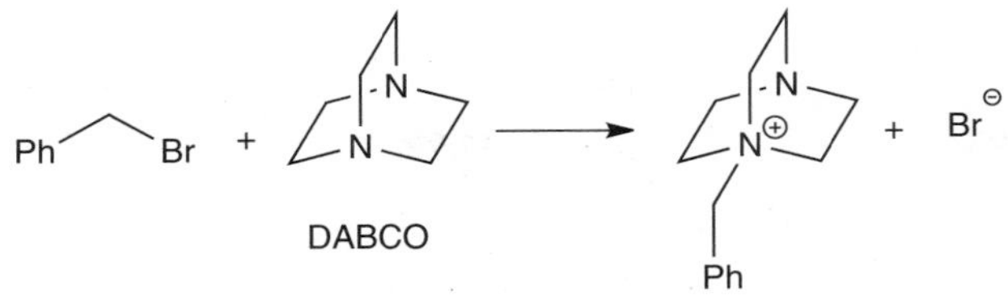

DABCO is effectively a tertiary amine, R_3N, in which the R groups close to form a ring.

has the rate law

$$\text{rate} = k\,[\text{PhCH}_2\text{Br}]\,[\text{DABCO}].$$

The reaction is said to be first order in PhCH_2Br and is also first order in DABCO; the overall order is therefore two.

The bromination of propanone (acetone) in aqueous solution under acid conditions has the overall reaction

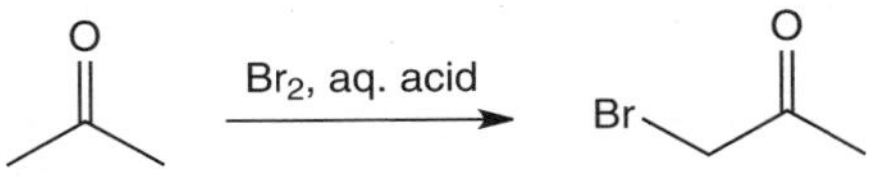

The rate law is found to be

$$\text{rate} = k\,[\text{CH}_3\text{COCH}_3]\,[\text{H}^+],$$

which is first order in both propanone and in acid. The somewhat unexpected feature here is that the reactant Br_2 does not appear in the rate law, whereas H^+ does, despite not obviously being a reactant. This is the first of several examples we will see where the relationship between the rate law and the balanced chemical equation is not straightforward.

The free-radical chlorination of an alkane RH in the gas phase

$$\text{RH} + \text{Cl}_2 \longrightarrow \text{RCl} + \text{HCl}$$

has the rate law

$$\text{rate} = k\,[\text{RH}]\,[\text{Cl}_2]^{\frac{1}{2}}.$$

This is the first example of a non-integer order, which are common for radical reactions. Another example is the gas phase decomposition of ethanal

$$\text{CH}_3\text{CHO} \longrightarrow \text{CH}_4 + \text{CO}$$

which has the rate law

$$\text{rate} = k\,[\text{CH}_3\text{CHO}]^{\frac{3}{2}}.$$

The decomposition of ozone to oxygen in the gas phase

$$2\text{O}_3 \longrightarrow 3\text{O}_2,$$

has the rate law

$$\text{rate} = k\,\frac{[\text{O}_3]^2}{[\text{O}_2]}.$$

The reaction is second order in O_3, but has an order of -1 with respect to oxygen. What this means is that the rate goes *down* as the concentration of oxygen goes up i.e. the reaction is *inhibited* by oxygen.

Our final example is the gas phase reaction between H_2 and Br_2 (to give HBr), which has been studied in great detail and has the rate law

$$\text{rate} = \frac{k_a[H_2][Br_2]^{\frac{3}{2}}}{[Br_2] + k_b[HBr]}.$$

This complex rate law does not conform to the pattern of Eq. 10.1 on page 412, so we cannot define an order with respect to each species.

These examples serve to illustrate that the rate law does not always have a simple relation to the overall reaction, and can be surprisingly complex.

10.2.2 Units of the rate constant

The rate law for a simple *first-order* reaction is

$$\text{rate} = k_{1st}[A],$$

which can be rearranged to

$$k_{1st} = \frac{\text{rate}}{[A]}.$$

The rate is defined as the change in concentration divided by time, so it has units of concentration/time (we are not being specific about what units either of these are measured in). Putting in the units of the term on the right of the last equation we have

$$\frac{\text{concentration/time}}{\text{concentration}} = \frac{1}{\text{time}}.$$

It follows that the units of the first-order rate constant, k_{1st} are simply time^{-1} e.g. s^{-1}.

For a simple *second-order* reaction the rate law is

$$\text{rate} = k_{2nd}[A]^2.$$

A similar argument to that used above shows that the units of the second-order rate constant k_{2nd} are $\text{concentration}^{-1}\,\text{time}^{-1}$ e.g. $\text{mol}^{-1}\,\text{dm}^3\,\text{s}^{-1}$.

10.2.3 Reactions involving the solvent

Sometimes the solvent in which the reaction is being carried out is involved chemically in the reaction i.e. the solvent molecules are themselves reactants. However, in such cases we usually do not find a term in the rate law for the concentration of the solvent. The explanation for this is that, compared to the other reactants, the concentration of the solvent is very high. Therefore, to a good approximation, this concentration does not vary during the reaction, so the effect of the solvent on the rate is constant throughout the reaction.

For example, the reaction in which I^- is displaced from an alkyl iodide by solvent water

$$RI + H_2O \longrightarrow ROH + H^+ + I^-,$$

is expected to be first order in RI and in the nucleophile H_2O, and hence second order overall

$$\text{rate} = k_{2nd}[RI][H_2O].$$

However, as H_2O is the solvent, its concentration is effectively constant and so the product $k_{2nd}[H_2O]$ is constant. The rate law then becomes first order overall

$$\text{rate} = k_{1st}[\text{RI}],$$

where $k_{1st} = k_{2nd}[H_2O]$. Sometimes such a reaction is described as begin *pseudo first order* overall, and k_{1st} is described as a *pseudo first-order rate constant*. The prefix 'pseudo' is added to remind us that the reaction and the rate constant are not really first order, but only apparently so because of the high concentration of one of the reactants (the solvent).

10.3 Temperature dependence

Experimentally, it is found that the rate constant often has a strong temperature dependence, increasing as the temperature increases. For very many reactions, the temperature variation of the rate constant k can be represented to a good approximation by the *Arrhenius equation*

$$k = A \exp\left(\frac{-E_a}{RT}\right). \tag{10.2}$$

A is called the *pre-exponential factor* or sometimes the *A factor*, E_a is the *activation energy* and R is the gas constant.

The product RT has the dimensions of energy per mole, as does E_a, making the fraction E_a/RT dimensionless. The activation energy therefore has units $J\,mol^{-1}$, and the pre-exponential factor has the same units as the rate constant, which, as we have seen, depend on the order of reaction.

If we take the natural logarithm of both sides of Eq. 10.2 we obtain

$$\underbrace{\ln k}_{y} = \underbrace{\frac{-E_a}{R}}_{m} \times \underbrace{\frac{1}{T}}_{x} + \underbrace{\ln A}_{c}.$$

This is of the form $y = mx + c$, so if we plot $\ln k$ against $1/T$ we will obtain a straight line of slope $-E_a/R$ and intercept (when $1/T = 0$) $\ln A$, as shown in Fig. 10.4. Such a graph is called an *Arrhenius plot*, and it is the method by which the values of the activation energy and pre-exponential factor (called the *Arrhenius parameters*) are determined.

Table 10.1 on the next page gives some values of the Arrhenius parameters for a range of different types of reactions; note that activation energies are usually quoted in $kJ\,mol^{-1}$. Activation energies can be up to $300\ kJ\,mol^{-1}$, but values around $100\ kJ\,mol^{-1}$ are more typical. Some reactions, particularly those involving radicals, can have much lower activation energies.

The exponential term in the Arrhenius expression means that the rate constants for reactions with significant activation energies will vary strongly with temperature. For example, using the data from the table for the reaction $EtBr + OH^- \rightarrow EtOH + Br^-$ in solvent ethanol, at 273 K the rate constant is computed as

$$4.3 \times 10^{11} \times \exp\left(\frac{-89.5 \times 10^3}{8.314 \times 273}\right) = 3.2 \times 10^{-6}\ dm^3\,mol^{-1}\,s^{-1},$$

> ⇨ Dimensional analysis is discussed in section 20.1 on page 876.

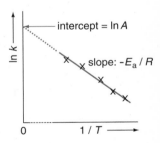

Fig. 10.4 Example of analysing the temperature dependence of rate-constant data using an Arrhenius plot. The activation energy E_a can be determined from the slope, and the pre-exponential factor A from the intercept.

Table 10.1 Some typical Arrhenius parameters

reaction	phase	pre-exponential factor	E_a / kJ mol^{-1}
cyclopropane $\rightarrow$ propene	gas	3×10^{15} s^{-1}	275
t-butyl bromide $\rightarrow$ isobutene + HBr	gas	1×10^{14} s^{-1}	177
$2N_2O_5(g) \rightarrow 4NO_2(g) + O_2(g)$	gas	4.9×10^{12} s^{-1}	113
$NO + O_3 \rightarrow NO_2 + O_2$	gas	7.9×10^8 dm^3 mol^{-1} s^{-1}	10.5
$NO + Cl_2 \rightarrow NOCl + Cl$	gas	4.0×10^9 dm^3 mol^{-1} s^{-1}	84.9
$2ClO \rightarrow Cl_2 + O_2$	gas	6.3×10^7 dm^3 mol^{-1} s^{-1}	0
EtONa + MeI $\rightarrow$ EtOMe + NaI	ethanol	2.4×10^{11} dm^3 mol^{-1} s^{-1}	81.6
EtBr + OH$^-$ $\rightarrow$ EtOH + Br$^-$	ethanol	4.3×10^{11} dm^3 mol^{-1} s^{-1}	89.5

where we have remembered to convert the given value of the activation energy into J mol^{-1}. Repeating the calculation at 298 K gives a rate constant of 8.8×10^{-5} dm^3 mol^{-1} s^{-1}, which is an increase by a factor of almost 30 for a rise in temperature of just 25 K. As a result of the $\exp(-E_a/RT)$ term in the Arrhenius expression, the higher the activation energy, the more quickly the rate constant changes with temperature.

10.4 The energy barrier to reaction

As we have seen in section 8.4.1 on page 323, for molecules to react they must encounter one another in the correct orientation.

The fact that the Arrhenius expression describes the temperature dependence of many rate constants is strong evidence that there is an *energy barrier* to reaction. The idea is that when two reactant molecules collide with the correct orientation for reaction, they will only react to give products if the energy of the collision exceeds a certain minimum. Collisions whose energies are less than this threshold do not lead to products – the reactant molecules simply separate, unchanged, after the collision.

For reactions taking place in the gas phase, we can use the simple gas kinetic theory to model the frequency and energy of collisions between reactant molecules. This theory indicates that the fraction of collisions with energy greater than or equal to E is given by

$$(\text{fraction of collisions with energy} \geq E) = \exp\left(\frac{-E}{RT}\right),$$

where T is the temperature and R is the gas constant.

Imagine that we have a simple reaction in which two reactants P and Q come together to give products, and that the rate law is simply

$$\text{rate} = k[\text{P}][\text{Q}].$$

If we use the Arrhenius expression, Eq. 10.2 on the previous page, for the rate constant, the rate law can be written

$$\text{rate} = A \exp\left(\frac{-E_a}{RT}\right)[\text{P}][\text{Q}]$$

$$= \underbrace{A\,[\text{P}][\text{Q}]}_{\text{rate if no barrier}} \times \underbrace{\exp\left(\frac{-E_a}{RT}\right)}_{\text{frac. with energy} \geq E_a}$$

As is indicated, the expression for the rate can be separated into two terms, each of which has a distinct physical meaning. A [P][Q] is rate that the reaction would have *if there were no barrier* i.e. if $E_a = 0$ so that $\exp(-E_a/RT) = 1$. Under these circumstances, each collision in which the molecules have the correct orientation leads to reaction, so A [P][Q] can be thought of as the rate of collisions with this correct orientation.

Not surprisingly, this rate is proportional to the concentrations of the two reactants since if there are more molecules present, the chances of them colliding must increase. The value of the pre-exponential factor A must depend on other factors which affect the rate of collision, such as how fast the molecules are moving and how large they are, and must also depend on the fraction of these collisions which have the correct orientation.

The second term is interpreted as the fraction of these collisions with energy greater than or equal to E_a. The activation energy term in the Arrhenius expression can therefore be identified as the energy barrier to the reaction.

Figure 10.5 shows a typical way in which this idea of an energy barrier to reaction is represented. The vertical scale is energy, and the green and red lines indicate the energies of the reactants and the products. The blue line shows the energy barrier which the reactants have to surmount in order to become products. The energy separation between the reactant and the top of the barrier is therefore the activation energy, E_a. The *energy profile* on the left is for an endothermic reaction, in which the products are higher in energy than the reactants, whereas the profile on the right is for an exothermic reaction, with the products being lower in energy than the reactants.

The horizontal axis in Fig. 10.5 is often referred to as the *reaction coordinate*, a somewhat loosely defined quantity which maps out the path between the reactants and the products. The shape of the blue curve is not significant: all that is important is the energy at which it starts and ends, and the energy at its peak.

For a typical activation energy of 50 kJ mol^{-1}, the fraction of collisions which have sufficient energy to surmount this barrier at 298 K is

$$\exp\left(\frac{-50 \times 10^3}{8.314 \times 298}\right) = 1.7 \times 10^{-9}.$$

Only a tiny fraction, of the order of one in 10^9, of the collisions are therefore sufficiently energetic to lead to reaction.

10.4.1 The transition state

As we move along the reaction coordinate the reactants are steadily transformed into products: bonds are beginning to be broken, new bonds are starting to be made, and no doubt parts of the molecules are being rearranged. The resulting arrangements of atoms are only transitory, and these arrangements certainly do not correspond to 'normal' stable molecules.

The arrangement which occurs at the very top of the energy barrier is of particular significance, and is called the *transition state*. It is the energy separation between this transition state and the reactants which determines the value of the activation energy, and hence the rate constant for this reaction. Chemists therefore spend a lot of time speculating about the nature of the transition state, and how its energy might be affected for example by changing

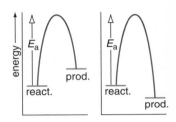

Fig. 10.5 Diagrammatic representation of the energy barrier between reactants and products. The energy of the reactants and products are represented by the green and red lines, respectively. The blue line represents the pathway which the reaction takes. The energy separation between the reactants and the highest energy point on the pathways is the activation energy. The energy profile on the left is for an endothermic reaction, and that on the right is for an exothermic reaction.

The arrangement which occurs at the top of the energy barrier is also referred to as the *transition structure*.

the structure of the reacting species, or the polarity of the solvent. In later chapters we will often use these types of arguments to understand the rates of reactions.

10.4.2 Reversible reactions and equilibrium

In principle, all reactions are reversible, meaning that the products can revert back to reactants. It is assumed that in going from products to reactants the molecules follow the same energy profile as when going from reactants to products, but in the opposite direction (this is rather grandly called the *principle of microscopic reversibility*).

This being the case, we can use the profiles shown in Fig. 10.5 to discuss both the forward and the reverse reactions. As is shown in Fig. 10.6, the activation energy of the reverse reaction, $E_{a,rev}$, is the energy separation between the top of the barrier and the products. In general, this activation energy will be different to that for the forward reaction, $E_{a,fwd}$. It is also clear from the diagram that the difference in these activation energies is the energy separation between the reactants and products, ΔE

$$\Delta E = E_{a,fwd} - E_{a,rev}. \tag{10.3}$$

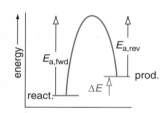

Fig. 10.6 Energy profile showing the activation energies of the forward and reverse reactions, $E_{a,fwd}$ and $E_{a,rev}$, and the relationship between these quantities and the energy difference ΔE between reactants and products.

When a reaction has come to equilibrium the concentration of the products and reactants are no longer changing. However, as you know, this does not mean that no reactions are taking place. On the contrary, what it means is that the rates of the forward and reverse reactions are *equal*. This situation is sometimes called *dynamic equilibrium*.

If we have an equilibrium between A and B as reactants, and P and Q as products

$$A + B \rightleftharpoons P + Q,$$

the equilibrium constant K_{eq} is written

$$K_{eq} = \frac{[P]_{eq}[Q]_{eq}}{[A]_{eq}[B]_{eq}}, \tag{10.4}$$

where $[A]_{eq}$ are the equilibrium concentrations.

If we assume that the rate law for the forward reaction is

$$\text{forward rate} = k_{fwd}[A][B],$$

and that the reverse reaction has the rate law

$$\text{reverse rate} = k_{rev}[P][Q],$$

then at equilibrium these two rates are equal

$$k_{fwd}[A]_{eq}[B]_{eq} = k_{rev}[P]_{eq}[Q]_{eq}.$$

Rearranging this gives

$$\frac{k_{fwd}}{k_{rev}} = \frac{[P]_{eq}[Q]_{eq}}{[A]_{eq}[B]_{eq}}.$$

Comparing this with Eq. 10.4 shows us that the equilibrium constant is equal to the ratio of the rate constants for the forward and reverse reactions

$$K_{eq} = \frac{k_{fwd}}{k_{rev}}.$$

We can push this relationship a bit further if we write each rate constant using the Arrhenius law, Eq. 10.2 on page 415:

$$K_{eq} = \frac{A_{fwd} \exp(-E_{a,fwd}/RT)}{A_{rev} \exp(-E_{a,rev}/RT)}$$

$$= \frac{A_{fwd}}{A_{rev}} \exp(-[E_{a,fwd} - E_{a,rev}]/RT)$$

$$= \frac{A_{fwd}}{A_{rev}} \exp(-\Delta E/RT),$$

where ΔE is defined in Eq. 10.3 on the preceding page and is the energy difference between the reactants and products.

This equation shows that the equilibrium constant is related to the *difference* in energy between the reactants and the products. If the products are lower in energy than the products i.e. ΔE is negative, then the exponential term is greater than one. The more negative ΔE becomes, the greater the equilibrium constant.

However, we have to be careful here as the equilibrium constant does not just depend on ΔE but also on the ratio of pre-exponential factors. We have no way of knowing, other than by experiment, whether this ratio is large or small.

10.5 Elementary reactions and reaction mechanisms

The basic view as to how a chemical reaction takes place at a molecular level is that first the two reactants have to collide with the correct orientation and then, if the collision is sufficiently energetic, products will be formed. For an encounter involving two reactants P and Q, the number of collisions will be proportional to the concentration of each, and so we expect that the rate law will be

$$rate = k[P][Q].$$

The reaction is first order in each reactant, and second order overall.

A reaction is described as being *elementary* if it takes place just as we have described, that is in a single reactive encounter between the molecules. We can simply write down the rate law for such a reaction as it will be first order in each reactant. For example, the gas-phase reaction

$$NO + O_3 \longrightarrow NO_2 + O_2,$$

is thought to be elementary i.e. it takes place when a molecule of NO collides with a molecule of O_3. The rate law for the reaction is therefore

$$rate = k[NO][O_3].$$

If a reaction is not elementary, it is *not* possible to write down the rate law just by inspecting the overall reaction. We saw several examples of this in section 10.2.1 on page 412 where the form of the rate law has little to do with the overall stoichiometry of the reaction.

A reaction which is not elementary proceeds via a series of interconnected elementary reactions which are collectively called the *mechanism* of the reaction. For example, the reaction (in aqueous solution) between Br_2 and dicyanomethane, $CH_2(CN)_2$, results in the substitution of H by Br according to the overall reaction

$$CH_2(CN)_2(aq) + Br_2(aq) \longrightarrow CHBr(CN)_2(aq) + HBr(aq).$$

The reaction is thought to take place via a mechanism involving the following three elementary reactions (all the species are aqueous)

$$CH_2(CN)_2 \longrightarrow CH(CN)_2^- + H^+$$
$$CH(CN)_2^- + H^+ \longrightarrow CH_2(CN)_2$$
$$CH(CN)_2^- + Br_2 \longrightarrow CH(CN)_2Br + Br^-.$$

The anion $CH(CN)_2^-$ is called an *intermediate*. It does not appear as a reactant or product in the overall reaction, but is involved in the reactions which ultimately take reactants through to products. Complex reactions usually involve the generation of such intermediates, and their detection is an important part of the experimental evidence for a proposed mechanism. Often these intermediates are very reactive chemically and so only have a fleeting existence while the reaction is taking place.

If the experimentally determined rate law is complex, or does not match the overall stoichiometry, then this is a strong indication that the reaction is not elementary. For example, as we saw in section 10.2.1 on page 412, the reaction between hydrogen and bromine has a rather complex rate law. Detailed experimental work has shown that the mechanism for this reaction involves no less than five significant elementary reactions in which H and Br atoms are intermediates

$$Br_2 \longrightarrow 2Br \qquad\qquad Br + Br \longrightarrow Br_2$$
$$Br + H_2 \longrightarrow HBr + H \qquad H + HBr \longrightarrow H_2 + Br$$
$$H + Br_2 \longrightarrow HBr + Br.$$

Even if the experimentally determined rate law is simple, it does not necessarily follow that the reaction proceeds in a simple way. For example, the gas-phase decomposition of N_2O_5

$$2N_2O_5 \longrightarrow 4NO_2 + O_2$$

has the simple rate law

$$\text{rate} = k[N_2O_5].$$

Nevertheless, this reaction is thought to involve several steps and a number of intermediates.

Notation for elementary reactions

The number of species which come together in an elementary step is called the *molecularity* of that step. For example, in the elementary reaction

$$Br + H_2 \longrightarrow HBr + H$$

two species come together so the molecularity is two. Such a reaction is described as being *bimolecular*. On the other hand the elementary reaction (in solution)

$$Me_3Cl \longrightarrow Me_3C^+ + I^-$$

involves just one species so the molecularity is one: the reaction is described as being *unimolecular*.

It is important to distinguish the molecularity from the order. The molecularity refers to the number of molecules coming together *in an elementary step*. The order is the power to which the concentration of a species is raised in the *experimentally* determined rate law. In the special case of an elementary reaction, the overall order will be equal to the molecularity.

10.6 Reactions in solution

In the gas phase molecules are rushing around at rather high speeds and as a result they frequently collide with one another. For example, at a pressure of 1 bar and at 298 K, a single molecule might experience between 10^9 and 10^{10} collisions per second, depending on its size.

In solution, the picture is rather different. The density is much higher than in a gas, so the molecules are packed together much more tightly – there is not a lot of 'space' between them. At modest concentrations, the bulk of the solution is solvent, so solute molecules are entirely surrounded by solvent molecules. As a result of thermal motion there are frequent collisions between a solute molecule and the solvent molecules surrounding it.

Experiments and computer simulations indicate that a solute molecule is likely to undergo many collisions with the solvent molecules that are in the immediate vicinity. These molecules are said to form a *solvent cage*. Eventually, as a result of thermal agitation, the solute 'breaks through' the surrounding solvent, but of course in doing so simply finds itself surrounded by another group of solvent molecules. The solute molecule then experiences many collisions with the solvent molecules which form the new cage. By jumping from cage to cage in this way, a solute molecule is able to move through the bulk solution, a process which is called *diffusion*.

If two solute molecules are going to react they first need to get close enough to interact, which means that they must encounter one another in the *same* solvent cage. Once they are in the same cage, they will experience many collisions with one another (and with the solvent molecules). If the collisions are sufficiently energetic, then the molecules may react, just as in the gas phase.

There are therefore two distinct stages to a reaction in solution. Firstly, by the process of diffusion, the molecules have to come together into the same solvent cage. Secondly, they have to react.

For most of the reactions we are going to be discussing the slow step is the actual reaction. However, there are some reactions which are so fast once the molecules are in the same solvent cage that their overall rate is limited by how fast the reacting molecules diffuse together. Such reactions are said to be *diffusion controlled*.

The most well known diffusion controlled reaction is that between H^+ and OH^- in aqueous solution

$$H^+(aq) + OH^-(aq) \longrightarrow H_2O(l).$$

This reaction has no energy barrier, which is why it is so fast.

Although the picture of what is happening in a solution phase reaction is rather different to what is happening in the gas phase, the resulting kinetics are essentially the same. For an elementary reaction involving two species P and Q, the chance of them encountering one another in solution is proportional to their concentrations, just as is the case in the gas phase. The expected rate law is therefore rate = k[P][Q], just as before.

10.7 Sequential reactions

In a reaction mechanism it is often the case that the product of one elementary reaction is a reactant for another elementary reaction. If this is the case, then the rate of the second reaction must be dependent on the rate of the first, as this determines the concentration of the reactant for the second reaction. In this section we are going to look at how these rates are connected and what the consequences are for the overall rate of reaction. This will lead us to introduce the important concept of the *rate-limiting step* of a complex reaction.

To make things as simple as possible we are going to consider two sequential first-order reactions. In the first reactant A becomes B, which then reacts to give C

$$A \xrightarrow{k_1} B \xrightarrow{k_2} C.$$

In this scheme, B can be described as an intermediate. The rate constants for the two steps are k_1 and k_2.

One example of this kind of arrangement is radioactive decay in which A is a radioactive nucleus which decays into another radioactive nucleus B, which then decays into C. A more chemical example is the acetylation of 1,4-dihydroxybenzene, using ethanoic anhydride (acetic anhydride) $(CH_3CO)_2O$. The OH groups are acetylated in two sequential steps

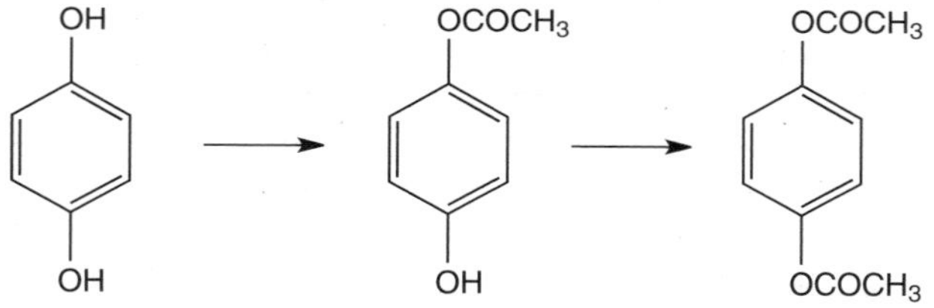

If the reaction is carried out using ethanoic anhydride as the solvent, so that it is in excess, then the two reactions are pseudo first order.

We will not go into the mathematical details here, but simply note that it is relatively straightforward to compute how the concentrations of A, B and C vary with time for given values of the rate constants k_1 and k_2. Figure 10.7 on the facing page shows some graphs of how these concentrations change for three different combinations of rate constants; in all cases it is assumed that only A is present at the start of the reaction.

Graph (a) is for the case where the rate constant for the first step, A → B, is ten times that for the second step, B → C. At the start of the reaction the concentration of A quickly drops to zero, a decay which is mirrored by an increase in the intermediate B. Then, over a longer time scale, the concentration of B falls back to zero as the concentration of the product C rises steadily.

Our interpretation of what is happening here is that, since step 1 has a much larger rate constant than step 2, A is more or less completely converted to B before there is time for much of B to be converted to C. Once this initial phase

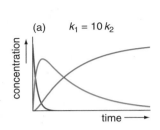
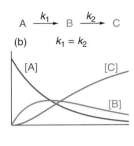
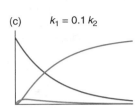

Fig. 10.7 Graphs showing how the concentrations of A, B and C vary with time for the case of sequential first-order reactions with rate constants k_1 and k_2. Graph (a) is for the case $k_1 = 10 k_2$, graph (b) is for the case where $k_1 = k_2$, and (c) is for the case $k_1 = 0.1 k_2$. In (a) the growth of the product C mirrors the decay of the intermediate B; step 2 (from B to C) is rate limiting. In (c) the growth of C mirrors the decay of A; step 1 (from A to B) is rate limiting.

is over, what we have is just the slower conversion of B to C. Note from the graph how at longer times the decay of B is mirrored by the rise in C. In this case, step 2 (B → C) is described as the *rate-limiting* or *rate-determining* step. It is described as this since it is the slowest step in the sequence and so sets a limit on the rate of formation of the product C.

In graph (c) we have the opposite situation in which the rate constant for step 2 is much larger than that for step 1. We see that rather little of the intermediate B is formed, and that the rise of the final product C mirrors the decay of the reactant A.

What is going on here is that the moment any of the intermediate B is formed it rapidly reacts to give the final product C. As a result, little B accumulates, and the fall of A is mirrored by the rise of C. In this case, it is the first reaction (A → B) which is the rate-limiting step.

Graph (b) is for the case where the two rate constants are the same. We can clearly see how the way in which the concentrations vary is intermediate between the behaviour in (a) and (c). Note that the rise of the product C does not simply mirror the fall in the concentration of either A or B.

Figure 10.8 shows two possible possible energy profiles for these sequential reactions. Profile (a) is for the case where $k_1 \gg k_2$, which means that the activation energy for the first step, $E_{a,1}$, must be less than that for the second step, $E_{a,2}$. Profile (b) is for the opposite case $k_2 \gg k_1$, which means $E_{a,1}$ must be greater than $E_{a,2}$, as shown.

Weblink 10.1

This link takes you to a 'real time' version of Fig. 10.7 in which you can alter the two rate constants and explore how this affects the way in which the concentrations of A, B and C vary with time.

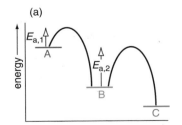
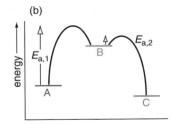

Fig. 10.8 Energy profiles for two sequential first-order reactions A → B → C. In (a) step 1 has a lower activation energy than step 2, and so the latter is the rate-limiting step; this corresponds to graph (a) in Fig. 10.7. In (b) step 2 has the lower activation energy, so step 1 is rate limiting; this corresponds to graph (c) in Fig. 10.7.

10.7.1 Pre equilibrium

Although it is useful to start out by thinking about the simple scheme in which there are two sequential reactions A → B → C, it has to be admitted that such a scheme is not very realistic. The key thing that is missing is the possibility of the reverse reactions i.e. B → A and C → B.

The profiles shown in Fig. 10.8 on the preceding page were carefully drawn so that the activation energy of the reverse reaction B → A is considerably greater than for the forward reaction B → C. As a result, the rate at which the intermediate B becomes C is much greater than the rate at which it reverts to A. Under these circumstances we can reasonably ignore the back reaction B → A.

Figure 10.9 shows a profile in which the activation energy $E_{a,-1}$ of the reverse reaction B → A is *less* than that for the forward reaction B → C, $E_{a,2}$. Under these circumstances, we cannot ignore the reverse reaction, so the scheme has to be modified to

$$A \underset{k_{-1}}{\overset{k_1}{\rightleftharpoons}} B \xrightarrow{k_2} C,$$

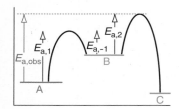

Fig. 10.9 Reaction profile in which the reverse reaction B → A has a lower activation energy ($E_{a,-1}$) than the reaction B → C ($E_{a,2}$). As a result B is more likely to return to A than go on to C, thus establishing an equilibrium between A and B. The activation energy $E_{a,obs}$ for the overall reaction is the energy separation between the reactant A and the highest energy point on the profile.

where the rate constant for the reverse reaction B → A is k_{-1}.

If the rate at which the intermediate B reverts to reactant A is faster than the rate at which it goes on to product C i.e. $k_{-1} \gg k_2$, we can argue that while the reaction is taking place A and B will come to equilibrium with one another. Recall from section 10.4.2 on page 418 that equilibrium is a dynamic state in which the forward and back reactions have the same rate. Since in this case the interconversion between A and B is so fast, even if some of the B goes on to product C, the equilibrium between A and B will quickly be re-established.

Assuming that this equilibrium has been established we can write

$$K_{eq} = \frac{[B]}{[A]},$$

where, as explained in section 10.4.2 on page 418, $K_{eq} = k_1/k_{-1}$. It follows that the concentration of B is given by

$$[B] = K_{eq}[A]. \tag{10.5}$$

The rate at which C is formed can therefore be written in terms of the concentration of A

$$
\begin{aligned}
\text{rate of formation of C} &= k_2[B] \\
&= k_2 K_{eq}[A] \\
&= \frac{k_1 k_2}{k_{-1}}[A].
\end{aligned}
$$

This is a nice simple result. We have shown that if the rate of the reverse reaction B → A is faster than the rate of the reaction in which B becomes products, then the overall rate law is simply first order in A with an observed rate constant $k_{obs} = k_1 k_2/k_{-1}$. We call this the observed rate constant as it is the value which would be determined experimentally.

If we write each rate constant using the Arrhenius expression then we can express k_{obs} in the following way

$$
\begin{aligned}
k_{obs} &= \frac{k_1 k_2}{k_{-1}} \\
&= \frac{A_1 \exp(-E_{a,1}/RT) \times A_2 \exp(-E_{a,2}/RT)}{A_{-1} \exp(-E_{a,-1}/RT)} \\
&= \frac{A_1 A_2}{A_{-1}} \exp\left(\frac{-(E_{a,1} + E_{a,2} - E_{a,-1})}{RT}\right).
\end{aligned}
$$

The final line shows that the activation energy, $E_{a,obs}$, associated with the observed rate constant is $(E_{a,1} + E_{a,2} - E_{a,-1})$. From Fig. 10.9 on the facing page it is clear that this activation energy is the energy separation between the reactant A and the top of the energy profile between B and C i.e. between A and the *highest energy point on the profile*.

It is interesting to compare the rate of step 2 (B → C), in which the product C is formed, to the rate of step 1 (A → B)

$$
\begin{aligned}
\frac{\text{rate of B} \rightarrow \text{C}}{\text{rate of A} \rightarrow \text{B}} &= \frac{k_2[B]}{k_1[A]} \\
&= \frac{(k_2 k_1/k_{-1})[A]}{k_1[A]} \\
&= \frac{k_2}{k_{-1}}.
\end{aligned}
$$

On the second line we have used Eq. 10.5 on the preceding page to substitute for the concentration of B. Referring to the profile in Fig. 10.9 on the facing page we see that the rate constant k_2 is less than k_{-1}, and so it follows that in this situation the rate for B → C is less than that for A → B. In other words, B → C is the rate-limiting step – a conclusion which is independent of the value of the rate constant k_1 for A → B.

This kinetic scheme is an example of a what is generally called *pre-equilibrium* kinetics in which an intermediate is assumed to be in equilibrium with the reactants. We will look at several more examples of this later on in this chapter. Generally, pre-equilibrium kinetics will apply if an intermediate reverts to the reactants from which it is formed faster than it goes on to products. Analysing the kinetics in such a situation is simplified as a result of us being able to relate the concentration of the intermediate to that of the reactants from which it is formed using an equilibrium expression. The equilibrium constant for this can be expressed as a ratio of rate constants.

10.7.2 The steady-state approximation

Let us return now to reaction profile (b) shown in Fig. 10.8 on page 423, for which a typical variation in the concentrations of A, B and C is shown in plot (c) from Fig. 10.7 on page 423. In this scheme, the rate constant for step 2 (B → C) is much greater than that for step 1 (A → B). As a result, the moment any B is formed by step 1, it more or less immediately reacts to give C. It follows that, to a good approximation,

rate of formation of B = rate of loss of B.

Using this, we will be able to work out the overall rate of reaction in quite a straightforward way.

This assumption about the behaviour of B is called the *steady-state approximation*; B is said to be *in the steady state*. It will apply when B is a reactive, high energy species whose formation has a high activation energy, but whose reactions generally have lower activation energies.

In our simple scheme

$$A \xrightarrow{k_1} B \xrightarrow{k_2} C,$$

the rate of formation of B is $k_1[A]$, and the rate of loss of B is $k_2[B]$, so applying the steady-state approximation gives

$$k_1[A] = k_2[B]_{SS},$$

where we have added the subscript 'SS' to the concentration of B to remind ourselves that this is the steady-state concentration. Rearranging this gives $[B]_{SS} = (k_1/k_2)[A]$, which we can use to work out the rate of formation of the product C as

$$
\begin{aligned}
\text{rate of formation of C} \ &= \ k_2[B]_{SS} \\
&= \ k_2 \frac{k_1}{k_2}[A] \\
&= \ k_1[A].
\end{aligned}
$$

In words, the rate of formation of the product C is just the rate of the first reaction A → B.

Our interpretation of what is going on here is the following. The rate constant for B → C (k_2) is very much larger than that for A → B (k_1). This means that *if* the concentrations of A and B were equal, then B → C would proceed at a much greater rate than A → B. However, as has been explained, the concentrations of A and B are *not* equal, since the moment any B is formed it reacts to give C, resulting in the concentration of B being very small. It is this low concentration of B which results in the rate of B → C being reduced so that it is equal to that of the A → B.

Under these circumstances, the rate at which C is formed is limited by the rate of the A → B step, which makes this the rate-limiting step. Increasing the rate constant of the B → C step will not increase the rate of formation of C, but increasing the rate constant for A → B will. This is what we mean by saying that the first step is rate limiting.

Since the rate of formation of the product is just $k_1[A]$, the rate constant for the overall reaction is simply k_1. It follows that the observed activation energy will just be that associated with k_1, i.e. $E_{a,1}$. Looking back at profile (b) in Fig. 10.8 on page 423 we see that the energy separation between the reactants and the highest point on the profile is $E_{a,1}$. So, just as we had in the pre-equilibrium case, the observed activation energy is determined by the highest energy point on the pathway.

Validity of the steady-state approximation

Figure 10.10 on the facing page compares the prediction of the steady-state approximation (dashed lines) with the exact behaviour (solid lines) of the concentrations of A, B and C. In (a), which is for the case $k_2 = 20 k_1$, the steady-state approximation gives excellent agreement. In (b), where $k_2 = 5 k_1$

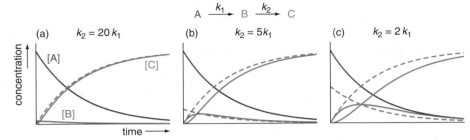

Fig. 10.10 Plots showing how the concentrations of A, B and C vary for different combinations of the rate constants k_1 and k_2. The solid lines give the exact behaviour, the dashed lines give the behaviour predicted using the steady-state approximation. Except at short times, the steady-state approximation is quite good in (a) and (b), where k_2 is significantly greater than k_1. However, in (c) k_2 is not sufficiently larger than k_1 for the steady-state approximation to apply, so that dashed lines are a poor approximation to the real behaviour of the [B] and [C]. The behaviour of [A] is identical in the exact and steady-state treatments.

the agreement is not as good, but it is nevertheless close enough to be useful. This plot also clearly shows that the steady-state prediction for [B] is completely wrong at short times, but that after a while the agreement is much better. The steady-state approximation is only expected to apply once the reaction has been running for a while so that the intermediate B has had time to reach its steady concentration. Finally, plot (c) is for the case $k_2 = 2k_1$, which is an insufficient difference in the rate constants for the steady-state approximation to apply.

Steady-state analysis of reversible reactions

We can also use the steady-state approximation to analyse the slightly more complex case where B is allowed to revert to A

$$A \underset{k_{-1}}{\overset{k_1}{\rightleftharpoons}} B \xrightarrow{k_2} C.$$

The steady-state approximation can still be used provided that the sum of the rate constants k_{-1} and k_2 are larger than k_1. This means that the moment any of the intermediate B is formed it reacts either to go back to reactant A or on to product C.

For this scheme the rate of loss of B is

$$\text{rate of loss of B} = k_{-1}[B] + k_2[B].$$

There are two terms since the reactions B → C and B → A both result in a loss of B. As before, the rate of formation of B is $k_1[A]$. Therefore, in the steady state

$$k_1[A] = k_{-1}[B]_{SS} + k_2[B]_{SS},$$

from which we find $[B]_{SS} = k_1[A]/(k_{-1}+k_2)$. The rate of formation of the product C is therefore

$$
\begin{aligned}
\text{rate of formation of C} &= k_2[B]_{SS} \\
&= \frac{k_1 k_2}{k_{-1}+k_2}[A].
\end{aligned}
$$

⊕ *Weblink 10.1*

This link takes you to a 'real time' version of Fig. 10.10 in which you can alter the two rate constants and compare the time variation of the concentrations with that predicted by the steady-state approximation.

The overall reaction is first order in A, but the observed rate constant now depends on all three rate constants, $k_{obs} = k_1 k_2 / (k_{-1} + k_2)$.

If $k_2 \gg k_{-1}$, then the bottom of the fraction can be approximated to k_2 and so the observed rate constant becomes k_1, which is what we found when we applied the steady-state approximation to the simple A $\rightarrow$ B $\rightarrow$ C scheme. This makes sense, as $k_2 \gg k_{-1}$ means that we can ignore the reverse reaction B $\rightarrow$ A.

If $k_{-1} \gg k_2$, then the bottom of the fraction can be approximated to k_{-1}, and so the observed rate constant becomes $k_1 k_2 / k_{-1}$, which is what we found when we used a pre-equilibrium analysis. The assumption $k_{-1} \gg k_2$ means that B is more likely to revert to A than to go on to C, which is precisely the situation in which the pre-equilibrium approach is valid. We can see from this that pre-equilibrium can be regarded as a special case of the steady-state approximation.

10.7.3 More complex cases

We have used the simple A $\rightarrow$ B $\rightarrow$ C scheme to introduce the ideas of pre-equilibrium and the steady-state approximation, but now we want to go on to apply these to some real reactions, which are inevitably more complicated. In particular, the reactions involved are not all going to be first order.

Why this leads to difficulties is best illustrated with an example. Suppose that we have a bimolecular reaction between A and B to give an intermediate Q, which can dissociate back to A and B. Q goes on to react with C to give the products. The overall scheme is therefore

$$
\begin{array}{lccc}
\text{step 1} & \text{A} + \text{B} & \xrightarrow{k_1} & \text{Q} \\
\text{step } -1 & \text{Q} & \xrightarrow{k_{-1}} & \text{A} + \text{B} \\
\text{step 2} & \text{Q} + \text{C} & \xrightarrow{k_2} & \text{products.}
\end{array}
$$

If we want to use the pre-equilibrium hypothesis, then this will require that the rate of step −1 is faster than that of step 2. However, the relative rate of these two steps will not only depend on the size of the rate constants k_{-1} and k_2, but also on the concentration of the reactant C.

If we wanted to apply the steady state approximation to the intermediate Q, then we would have to assume that the rate of its formation is equal to the rate of its loss

$$k_1[\text{A}][\text{B}] = k_{-1}[\text{Q}]_{\text{ss}} + k_2[\text{C}][\text{Q}]_{\text{ss}}.$$

Again, whether or not this is true will depend not only on the rate constants, but also on the concentrations of A, B and C. The essence of the problem is that we are having to compare the rates of first-order (step −1) and second-order (steps 1 and 2) reactions.

This example shows that for a complex reaction whether or not we can use the pre-equilibrium or steady-state approaches depends on *both* the rate constants *and* the concentrations of the species involved.

10.8 Analysing the kinetics of complex mechanisms

In this section we are going to look at how we can predict the overall rate law expected for a given mechanism. The reason we are interested in this is that, as the rate law can be determined experimentally, we can check whether or not a proposed mechanism is consistent with experiment. This is a useful check as to whether or not a proposed mechanism is acceptable.

It is important to realize that the fact that a proposed mechanism gives a rate law which is in agreement with experiment does not mean that the proposed mechanism is correct. All that it means is that the mechanism is *consistent* with the experimentally determined rate law. It is necessary to perform other experiments, such as the direct detection of intermediates, in order to build up evidence for a proposed mechanism.

If a mechanism has many elementary reactions, and several intermediates, it is by no means simple to work out the overall rate law. This is because the rate of each elementary reaction depends on the concentration of the species involved, which in turn depend on the rates of other elementary reactions. It may also be the case that a number of different mechanisms contribute in varying degrees to the overall reaction, further complicating the rate law.

However, if we can make some assumptions about the relative rates of different elementary reactions we may be able to use the pre-equilibrium approach or the steady-state approximation. Assuming that this is the case, we will be able to find the overall rate law without too much difficulty.

10.8.1 Oxidation of methanoic acid

Under acid conditions, and in aqueous solution, methanoic acid is oxidized by bromine to give CO_2 according to the following overall reaction (all species are aqueous)

$$HCOOH + Br_2 + 2H_2O \longrightarrow CO_2 + 2Br^- + 2H_3O^+.$$

The experimentally determined rate law is

$$\text{rate} = k_{obs} \frac{[Br_2][HCOOH]}{[H_3O^+]}, \tag{10.6}$$

where k_{obs} is the observed rate constant. The somewhat unusual feature is that the rate *decreases* as the concentration of H_3O^+ increases. We will use the pre-equilibrium approach to show how this rate law comes about from a simple reaction scheme.

The mechanism proposed for this reaction is

$$HCOOH + H_2O \underset{k_{-1}}{\overset{k_1}{\rightleftarrows}} HCOO^- + H_3O^+$$

$$HCOO^- + Br_2 + H_2O \overset{k_2}{\longrightarrow} CO_2 + H_3O^+ + 2Br^-.$$

The first reaction is the reversible dissociation of methanoic acid to give methanoate ($HCOO^-$) and H_3O^+. In the second reaction the methanoate is oxidized by the bromine to give CO_2 and the other products.

If we assume that the equilibrium between the methanoic acid and methanoate is largely undisturbed by the reaction of the latter with bromine, then we can write the usual equilibrium constant between these as

$$K_{eq} = \frac{[HCOO^-][H_3O^+]}{[HCOOH]},$$

where $K_{eq} = k_1/k_{-1}$; as usual, we do not include the solvent water in our expression for the equilibrium constant. This can be rearranged to give an expression for $[HCOO^-]$

$$[\text{HCOO}^-] = \frac{K_{eq}[\text{HCOOH}]}{[\text{H}_3\text{O}^+]}.$$

The rate of consumption of Br_2 is simply the rate of the second reaction, and if we assume that this is elementary it has the rate law

$$\text{rate of consumption of } Br_2 = k_2[\text{HCOO}^-][\text{Br}_2].$$

We have not included a term for the concentration of the solvent H_2O as this is so large that it is effectively constant and so is taken into the value of k_2. Substituting in the above expression for $[\text{HCOO}^-]$ gives

$$\text{rate of consumption of } Br_2 = k_2 K_{eq} \frac{[\text{Br}_2][\text{HCOOH}]}{[\text{H}_3\text{O}^+]}.$$

Our predicted rate law is of precisely the same form as the experimentally determined rate law, Eq. 10.6 on the previous page, with $k_{obs} = k_2 K_{eq}$.

All that we are really saying here is that it is the *methanoate* anion which is reacting with the bromine, and the concentration of methanoate is depressed by adding H_3O^+. Thus the rate goes down as the concentration of H_3O^+ is increased.

For the pre-equilibrium approach to be valid the rate at which the methanoate reacts with H_3O^+ must be faster than the rate of reaction with Br_2

$$k_{-1}[\text{HCOO}^-][\text{H}_3\text{O}^+] \gg k_2[\text{HCOO}^-][\text{Br}_2].$$

This is likely to be true since step -1 is just the abstraction of H^+ from H_3O^+ by a negatively charged base – a reaction which is likely to be fast.

10.8.2 Nucleophilic substitution

The general form of a nucleophilic substitution reaction at carbon is

$$\text{R--X} + \text{Nu}^- \longrightarrow \text{R--Nu} + \text{X}^-,$$

where R is an alkyl group, Nu^- is the nucleophile and X^- is the leaving group. Typical examples of such reactions (both in solution) are

$$(1): \quad CH_3Br + I^- \quad \longrightarrow \quad CH_3I + Br^-$$
$$(2): \quad (CH_3)_3CBr + SH^- \quad \longrightarrow \quad (CH_3)_3CSH + Br^-$$

The rate laws of such reactions have been studied in great detail, and it is found that, broadly speaking, the laws fall into two categories

⯈ The S_N1 and S_N2 mechanisms were first introduced in section 9.7.1 on page 390.

$$S_N1, \text{ first order overall:} \quad \text{rate} = k_{1st}[\text{RX}]$$
$$S_N2, \text{ second order overall:} \quad \text{rate} = k_{2nd}[\text{RX}][\text{Nu}^-].$$

Reaction (1) is found to have a second-order rate law, whereas reaction (2) is found to be first order.

The second-order rate law is thought to arise straightforwardly from a single-step mechanism in which the nucleophile attaches itself to the carbon at the same time as the leaving group departs. We are going to be concerned here with the first-order rate law (S_N1) which has the feature that the concentration of the nucleophile does not appear in the rate law.

It is generally thought that such a rate law arises because the reaction proceeds via a more complex mechanism involving a carbenium ion R^+ as an intermediate. The simplest mechanism which can explain the observed kinetics involves just two steps. In the first step, R^+ is formed, and in the second step this goes on to react with the nucleophile

$$R-X \xrightarrow{k_1} R^+ + X^- \qquad R^+ + Nu^- \xrightarrow{k_2} R-Nu.$$

Using a steady-state analysis we will show that this mechanism give overall first-order kinetics, provided that the first step is rate limiting.

Since R^+ is likely to be a reactive species, it seems reasonable to assume that the steady-state approximation will be applicable. Setting the rate of formation and of consumption of R^+ to be equal gives

$$k_1[RX] = k_2[R^+]_{SS}[Nu^-].$$

Rearranging this gives us an expression for the steady-state concentration of R^+

$$[R^+]_{SS} = \frac{k_1[RX]}{k_2[Nu^-]}.$$

The rate of formation of RNu is just the rate of step 2 i.e. $k_2[R^+]_{SS}[Nu^-]$; using the above expression for $[R^+]_{SS}$ in this gives

$$\begin{aligned} \text{rate of formation of RNu} \quad &= \quad \frac{k_2 k_1[RX][Nu^-]}{k_2[Nu^-]} \\ &= \quad k_1[RX]. \end{aligned}$$

This predicted rate law is first-order overall, as for an S_N1 reaction, so our proposed two-step mechanism is consistent with the observed rate law. The key assumption we made was that R^+ was in the steady state, which means that it must be consumed the moment it is formed, implying that step 1 is rate limiting.

This simple two-step mechanism accounts for the basic observation of an overall first-order rate law. However, more detailed work has shown that things are rarely this simple. A good example is the classic study published in 1940 by Bateman, Hughes and Ingold, who made careful measurements on the following reaction

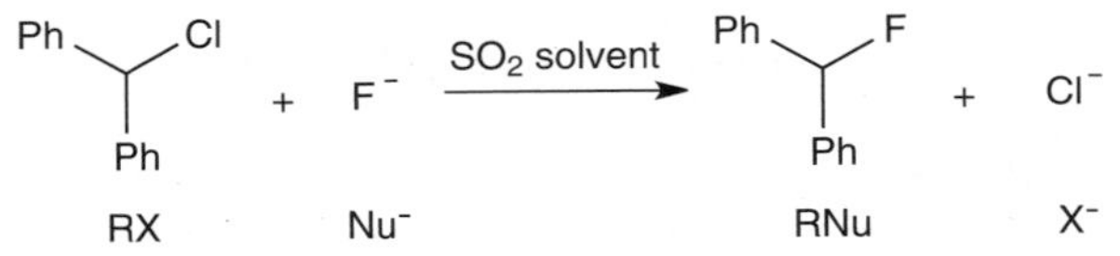

Liquid sulfur dioxide was chosen as the solvent since it is polar enough to dissolve the ions, and also because SO_2 is not a good nucleophile and so will not compete with F^-.

It was found that whereas in the early stages of the reaction the rate law was simply first order in RX, as the reaction proceeds the rate fell more quickly than expected for a simple first-order reaction. Furthermore, deliberately adding Cl^- (a product) was found to depress the rate further.

These observations can be accounted for by supposing that R^+ can not only react with the nucleophile but can also react with the leaving group to regenerate the starting material

$$R^+ + X^- \xrightarrow{k_{-1}} RX.$$

Adding this step, gives the following mechanism

$$RX \quad \overset{k_1}{\underset{k_{-1}}{\rightleftharpoons}} \quad R^+ + X^-$$

$$R^+ + Nu^- \quad \overset{k_2}{\longrightarrow} \quad RNu.$$

We can analyse this mechanism as before by putting R^+ into the steady state. Assuming that the rate of formation and consumption of R^+ are equal gives

$$k_1[RX] = k_{-1}[R^+]_{SS}[X^-] + k_2[R^+]_{SS}[Nu^-].$$

The steady-state concentration of R^+ is therefore given by

$$[R^+]_{SS} = \frac{k_1[RX]}{k_{-1}[X^-] + k_2[Nu^-]}.$$

Using this, we can find the rate of formation of RNu:

$$\begin{aligned} \text{rate of formation of RNu} \quad &= \quad k_2[R^+]_{SS}[Nu^-] \\ &= \quad \frac{k_1 k_2[RX][Nu^-]}{k_{-1}[X^-] + k_2[Nu^-]}. \end{aligned} \qquad (10.7)$$

This rate law is more complex than the one we found for the simple two-step mechanism. The key new feature that it predicts is that the rate will go *down* as the concentration of the leaving group X^- increases. Let us explore what this rate law predicts in two special cases.

R^+ reacts more rapidly with Nu^- than with X^-.

In the first special case, let us suppose that rate of step 2 is much greater than that of step -1, so we can write

$$k_2[R^+][Nu^-] \gg k_{-1}[R^+][X^-].$$

What this means is that the rate at which the carbenium ion R^+ reacts with the nucleophile Nu^- is much faster than the rate at which R^+ recombines with the leaving group X^-. The inequality can be simplified by cancelling $[R^+]$ from each side to give

$$k_2[Nu^-] \gg k_{-1}[X^-].$$

If this is the case, the term $(k_{-1}[X^-] + k_2[Nu^-])$ in the denominator of the predicted rate law, Eq. 10.7, can be approximated as $k_2[Nu^-]$, which allows us to simplify the rate law in the following way

$$\begin{aligned} \text{rate of formation of RNu} \quad &= \quad \frac{k_1 k_2[RX][Nu^-]}{k_{-1}[X^-] + k_2[Nu^-]} \\ &\approx \quad \frac{k_1 k_2[RX][Nu^-]}{k_2[Nu^-]} \\ &= \quad k_1[RX]. \end{aligned} \qquad (10.8)$$

In this special case, the rate law is first order overall, just as we found for the simple two-step mechanism. This is not a surprise, since we assumed that the rate of step 2 is much greater than that of step -1, which is tantamount to ignoring step -1 completely.

We expect that the rate law of Eq. 10.8 on the facing page will apply if we arrange things so that the concentration of the nucleophile is high and confine our attention to the early part of the reaction when the concentration of X^- is low. It will then be more likely that R^+ will react with Nu^- rather than X^-. The mechanism is then simply the formation of R^+ in the rate-limiting step, followed by the rapid reaction of the carbenium ion with Nu^-. If our aim is to produce a good yield of RNu, then these are the conditions under which we should run the reaction.

R^+ reacts more rapidly with X^- than with Nu^-.

The second special case is the other extreme in which we assume that the rate at which R^+ reacts with X^- is much faster than the rate at which R^+ reacts with Nu^-

$$k_{-1}[R^+][X^-] \gg k_2[R^+][Nu^-].$$

Cancelling $[R^+]$ as before gives

$$k_{-1}[X^-] \gg k_2[Nu^-].$$

This time, the term $(k_{-1}[X^-] + k_2[Nu^-])$ in the denominator of Eq. 10.7 can be approximated as $k_{-1}[X^-]$, which allows us to simplify the rate law in the following way

$$
\text{rate of formation of RNu} = \frac{k_1 k_2 [RX][Nu^-]}{k_{-1}[X^-] + k_2[Nu^-]}
$$
$$
\approx \frac{k_1 k_2 [RX][Nu^-]}{k_{-1}[X^-]}. \tag{10.9}
$$

This rate law is very different to the one we found previously. Firstly, it predicts that the rate depends on the concentration of the nucleophile, whereas previously there was no such dependence. Secondly, it predicts that the rate goes *down* as the concentration of the leaving group X^- increases. In other words, the reaction is *inhibited* by the product.

This behaviour arises because R^+ is more likely to react with the leaving group X^- than it is to react with the nucleophile Nu^-. If we keep the concentration of the nucleophile low, or deliberately add more of the leaving group X^-, then we are creating the conditions in which it is more likely that R^+ will react with X^- than Nu^-, which will lead to a rate law of the form of Eq. 10.9.

Looking at this rate law we recognize that k_1/k_{-1} is the equilibrium constant for

$$RX \rightleftharpoons R^+ + X^-.$$

In this case we can interpret the mechanism as a pre-equilibrium step in which R^+ is formed, followed by the slow rate-limiting step in which R^+ reacts with Nu^-.

The general case

Bateman, Hughes and Ingold found that their experimental data were a good fit to the full rate law of Eq. 10.8 on the facing page. They deliberately added more of the Cl^- leaving group so that they could explore its effect, and found it to be in accord with the predictions of this three-step mechanism. It is interesting to see that such a simple mechanism can give rise to relatively complex kinetics in which the form of the rate law, and the rate-limiting step, changes with the concentration of the species involved.

10.8.3 Overview

These few mechanisms for which we have looked at the kinetics in detail begin to give us a flavour of the relationship between the mechanism, the assumptions we make about it, and the overall rate law. One key idea which comes out of these case studies is that the species whose concentrations appear in the experimental rate law must be involved in the reaction *up to and including* the rate-limiting step. Species which are involved after the rate-limiting step do *not* appear in the rate law.

For example, the experimental rate law for an S_N1 reaction does not include a concentration term for the nucleophile, which tells us straight away that the nucleophile is involved *after* the rate-limiting step. If you look through the examples, you will see that a similar interpretation can be made in each case.

The second important point is that the rate-limiting step can change with concentration, as we saw in the case of the S_N1 reaction, and the ligand substitution reaction of transition metal complexes.

10.9 Chain reactions

Chain reactions are a particular class of complex reactions in which the initial generation of a small amount of an intermediate can result in the formation of much larger amounts of products. Examples of such reactions are polymerization (for example of ethene to give polyethene), explosions (for example, the reaction between oxygen and hydrogen), and the destruction of ozone by CFCs in the stratosphere.

We will take this latter process as an example to illustrate the key ideas. In the stratosphere, CFCs such as $CFCl_3$ can be broken down by UV light to give chlorine atoms and other products

$$CFCl_3 \xrightarrow{\text{UV light}} Cl + \text{other products.}$$

The chlorine atom can react with ozone to give an oxygen molecule and the ClO radical

$$\text{reaction 1} \qquad Cl + O_3 \longrightarrow ClO + O_2.$$

In the stratosphere there is a significant amount of oxygen atoms present, and these can react with the ClO in the following way

$$\text{reaction 2} \qquad ClO + O \longrightarrow Cl + O_2.$$

The key thing is that in reaction 2 a chlorine atom is generated. This can then react with another ozone molecule in reaction 1, and the resulting ClO can then react with O in reaction 2 to regenerate the Cl atom again.

Reaction 1 followed by reaction 2 results in the destruction of an ozone molecule, but the regeneration of the initial chlorine atom. These two reactions thus form a chain in which a single chlorine atom can destroy many ozone molecules.

Terminology for chain reactions

The intermediates which are responsible for carrying the chain forward, and hence for the consumption of reactants, are called *chain carriers*. In the above example, Cl and ClO are chain carriers. It is usually the case that chain carriers are reactive intermediates, such as free radicals.

The reaction in which the chain carriers are initially generated is called an *initiation* reaction. As we saw above, UV light might be responsible for such an initiation step, or it might be simple thermal dissociation.

Reactions in which the chain carriers lead to the formation of products, but are themselves ultimately regenerated, are called *propagation* reactions. Finally, reactions in which the chain carriers are consumed are called *termination* reactions. Typically these involve the recombination of radicals.

Each of the reactions we have mentioned in conjunction with the destruction of ozone can be classified as follows

$$\text{initiation} \quad CFCl_3 \xrightarrow{\text{UV light}} \mathbf{Cl} + \text{other products}$$
$$\text{propagation} \quad \mathbf{Cl} + O_3 \longrightarrow \mathbf{ClO} + O_2$$
$$\mathbf{ClO} + O \longrightarrow \mathbf{Cl} + O_2$$
$$\text{termination} \quad \mathbf{Cl} + \mathbf{Cl} \longrightarrow Cl_2.$$

The chain carriers are shown in bold.

> A free radical is a species with a single unpaired electron. Radicals are often created by breaking a bond e.g. breaking the C–C bond in ethane CH_3–CH_3 creates two methyl radicals, CH_3. Radicals tend to be rather reactive, and hence short lived, as they can readily combine with other radicals, or abstract atoms from stable molecules, creating a different radical e.g.
> $CH_3 + H_2 \rightarrow CH_4 + H$

> In the stratosphere there are several termination reactions for Cl involving the formation of HCl and nitrogen containing species.

10.9.1 The hydrogen + bromine reaction

One of the best understood chain reactions is that between H_2 and Br_2 to give HBr. As has already been noted, the experimentally determined rate law is quite complex

$$\text{rate} = \frac{k_a[H_2][Br_2]^{\frac{3}{2}}}{[Br_2] + k_b[HBr]}. \tag{10.10}$$

We will show in this section that this rate law can be explained by a chain reaction involving five elementary steps

type	reaction	step
initiation	$Br_2 \xrightarrow{k_1} 2\mathbf{Br}$	1
propagation	$\mathbf{Br} + H_2 \xrightarrow{k_2} HBr + \mathbf{H}$	2
	$\mathbf{H} + Br_2 \xrightarrow{k_3} HBr + \mathbf{Br}$	3
inhibition	$\mathbf{H} + HBr \xrightarrow{k_4} H_2 + \mathbf{Br}$	4
termination	$\mathbf{Br} + \mathbf{Br} \xrightarrow{k_5} Br_2$	5

As before, the chain carriers H and Br are indicated in bold.

The initiation step is just the thermal dissociation of Br_2 to give bromine atoms. A Br atom reacts with H_2 in step 2 to generate the product HBr and an H atom. This then reacts with Br_2 in step 3 to give another product molecule and regenerate the original Br atom. Together, steps 2 and 3 result in the production of HBr without the loss of any chain carriers.

Step 5 is a termination reaction as it results in the loss of the chain carrier Br. Step 4 is called an *inhibition* reaction as it leads to the destruction of product, but not to the overall reduction in the number of chain carriers.

You may be asking yourself why we are limiting ourselves to these five reactions, and excluding others which, on the face of it, look equally plausible, such as

$$\text{initiation} \quad H_2 \longrightarrow 2\mathbf{H}$$
$$\text{inhibition} \quad \mathbf{Br} + HBr \longrightarrow Br_2 + \mathbf{H}$$
$$\text{termination} \quad \mathbf{H} + \mathbf{H} \longrightarrow H_2.$$

The reason that these reactions are excluded is that it is not necessary to include them in order to explain the observed kinetics. What this implies is that although these reactions may be taking place, they are doing so at rates which are insignificant compared to the competing process which comprise the original list.

To simplify things we are going to think about the rate of the reaction in its early stages when the concentration of HBr is low. Looking at the rate law, Eq. 10.10 on the previous page, we see that when [HBr] is low the term $k_b[\text{HBr}]$ can be ignored, so the rate law simplifies to

$$\text{rate} = k_a[\text{H}_2][\text{Br}_2]^{\frac{1}{2}}. \tag{10.11}$$

When the concentration of HBr is low, we can ignore the inhibition reaction, step 4, as its rate is proportional to [HBr]; this simplifies the analysis considerably.

The approach we are going to adopt is to assume that the intermediate H and Br atoms are in the steady state. This leads to the following expression for Br

To save space and remove clutter we will not add the subscript 'SS' to [Br] and [H], but simply recall that these are the steady-state concentrations.

$$\text{rate of formation of Br} \;=\; \text{rate of loss of Br}$$
$$\underbrace{2k_1[\text{Br}_2]}_{\text{step 1}} + \underbrace{k_3[\text{H}][\text{Br}_2]}_{\text{step 3}} \;=\; \underbrace{k_2[\text{Br}][\text{H}_2]}_{\text{step 2}} + \underbrace{2k_5[\text{Br}]^2}_{\text{step 5}}. \tag{10.12}$$

On the left we have the rates of those processes which lead to the formation of Br. For step 1, the rate is $2k_1[\text{Br}_2]$ because *two* bromine atoms are produced in this reaction. We are assuming that each step is elementary so we can simply write down that that rate of step 3 is $k_3[\text{H}][\text{Br}_2]$. On the right, we have the rates of those reactions which lead to the consumption of Br; the factor of 2 in the rate of step 5 arises because two Br atoms are consumed in this reaction.

Doing the same for H (and still ignoring step 4) gives

$$\text{rate of formation of H} \;=\; \text{rate of loss of H}$$
$$\underbrace{k_2[\text{Br}][\text{H}_2]}_{\text{step 2}} \;=\; \underbrace{k_3[\text{H}][\text{Br}_2]}_{\text{step 3}}. \tag{10.13}$$

Things are beginning to look a little complicated, but a considerable simplification comes about if we add Eq. 10.12 to Eq. 10.13 since doing this results in the terms in k_2 and k_3 cancelling to give

$$2k_1[\text{Br}_2] = 2k_5[\text{Br}]^2.$$

This rearranges into a simple expression for [Br]

$$[\text{Br}] = \left(\frac{k_1}{k_5}\right)^{\frac{1}{2}} [\text{Br}_2]^{\frac{1}{2}}. \tag{10.14}$$

We can interpret this expression by noting that step 5 is just the reverse of step 1, so the ratio k_1/k_5 which appears in this expression is the equilibrium constant for

$$Br_2 \underset{k_5}{\overset{k_1}{\rightleftarrows}} Br + Br.$$

Assuming that these two reactions do set up an equilibrium, we can write the usual expression for K_{eq} $(= k_1/k_5)$

$$\frac{k_1}{k_5} = \frac{[Br]^2}{[Br_2]}.$$

This can be rearranged to give exactly the same expression for [Br] as in Eq. 10.14 on the facing page. The reason that it is acceptable to assume that steps 1 and 5 set up an equilibrium is that although steps 2 and 3 involve Br atoms, together they do not alter the concentration of Br.

The rate of formation of HBr is the sum of the rates of steps 2 and 3

$$\text{rate of formation of HBr} = k_2[Br][H_2] + k_3[H][Br_2],$$

but from Eq. 10.13 on the preceding page we see that the two terms on the right are equal, so we can rewrite the rate as

$$\text{rate of formation of HBr} = 2k_2[Br][H_2]. \tag{10.15}$$

It is now simply a matter of substituting in the expression for [Br] from Eq. 10.14 on the facing page into Eq. 10.15 to give

$$\text{rate of formation of HBr} = 2k_2 \underbrace{\left(\frac{k_1}{k_5}\right)^{\frac{1}{2}} [Br_2]^{\frac{1}{2}}}_{[Br]} [H_2].$$

The predicted rate law is precisely of the form of the experimentally observed rate equation, Eq. 10.11 on the facing page, for low HBr concentrations. In this law, the terms grouped by the brace are simply the concentration of Br, as determined by the equilibrium set up between steps 1 and 5, and expressed in Eq. 10.14 on the preceding page.

Interpretation of the rate law

The form of the overall rate law, and the reasons why only certain reactions are included in the reaction scheme, can be interpreted with the aid of the data given in Table 10.2 on the following page. Looking at these data for reactions 2, 3 and 4, we see that the two strongly exothermic reactions (3 and 4) have rather small activation energies, whereas reaction 2, which is endothermic, has a much larger activation energy. From Fig. 10.5 on page 417 it follows that for an *endothermic* reaction the activation energy has to be *at least as large* as the energy difference ΔE between reactants and products: this explains why for reaction 2 the activation energy is greater than 70 kJ mol^{-1}. For exothermic reactions, there is no necessary relationship between the activation energy and the energy change, but for these radical reactions it is usually the case that the activation energies are rather low. We can thus use the ΔE values as a guide to the activation energies.

Table 10.2 Parameters for some elementary reactions involved in the hydrogen/bromine reaction

type	step	reaction	E_a / kJ mol^{-1}	ΔE / kJ mol^{-1}
initiation	1	$Br_2 \longrightarrow 2Br$		193
initiation		$H_2 \longrightarrow 2H$		436
propagation	2	$Br + H_2 \longrightarrow HBr + H$	82	70
propagation	3	$H + Br_2 \longrightarrow HBr + Br$	4	−173
inhibition	4	$H + HBr \longrightarrow H_2 + Br$	12	−70
inhibition		$Br + HBr \longrightarrow Br_2 + H$		173

ΔE is the energy difference between reactants and products

The fact that k_2 appears in the rate law tells us that step 2, $Br + H_2 \longrightarrow HBr + H$, is the rate-limiting step. Referring to Table 10.2 we can see this makes sense as the activation energy for step 2 is much greater than for step 3 on account of the fact that in the former reaction it is an H–H bond which is being broken as opposed to the weaker Br–Br bond in the latter.

The factor of two in the rate law arises in the following way. Each bromine atom which is consumed in the rate-limiting step 2 leads to the formation of one HBr molecule and an H atom. The only possible fate of this H atom is to react in step 3 to give another HBr and regenerate the Br atom; since step 2 is rate limiting, the H atom is consumed in step 3 almost immediately that it is formed. Thus, each Br atom which is consumed by step 2 ultimately leads to the formation of *two* HBr molecules, hence the factor of two in the rate law.

We can now see why certain other elementary steps, although perfectly plausible in themselves, are not significant in the overall reaction. The initiation step $H_2 \longrightarrow 2H$ involves breaking the H–H bond which is much stronger than the Br–Br bond broken in step 1; the rate of step 1 will therefore be much greater, and so it is the dominant initiation process.

The termination step $H + H \longrightarrow H_2$ is not likely to be competitive with step 5 since the concentration of H atoms is much lower than that of Br atoms. This is because the H atoms are rapidly consumed in step 3.

Once a Br atom is formed in step 1 it can feed into step 2 thus producing HBr and an H atom. In step 3 the H atom produces more HBr and regenerates the Br atom. So, steps 2 and 3 together form a chain in which a single Br atom can lead to the production of many HBr molecules. The process is only stopped by step 5 in which the Br atoms are removed. Under typical conditions, each Br atom produced in step 1 goes on to generate around 10^{13} HBr molecules before it is terminated in step 5.

Inhibition

Introducing the inhibition reaction, step 4 $H + HBr \longrightarrow H_2 + Br$, adds some more complexity to the calculation; the details are explored in the Exercise 10.12 at the end of the chapter. The predicted form of the rate law becomes

$$\text{rate of formation of HBr} = \frac{2k_2 \left(\frac{k_1}{k_5}\right)^{\frac{1}{2}} [H_2][Br_2]^{\frac{3}{2}}}{[Br_2] + \left(\frac{k_4}{k_3}\right)[HBr]}.$$

The important new feature is that the rate is predicted to decrease as the concentration of the product HBr increases. This is because the rate of step 4 is proportional to the concentration of HBr, and also because this step actually consumes the product HBr.

We see from Table 10.2 that whereas step 4 is exothermic and has an activation energy of 4 kJ mol^{-1}, the alternative inhibition step $Br + HBr \longrightarrow Br_2 + H$ is strongly endothermic. We can rationalize this by noting that the alternative step involves making a Br–Br bond, whereas in step 4 the much stronger H–H bond is made. Given these data, it is not surprising that $Br + HBr \longrightarrow Br_2 + H$ is not a significant source of inhibition.

10.9.2 The hydrogen + oxygen reaction

The reaction between H_2 and O_2 is rather more complex than the H_2/Br_2 reaction, and so we will not go into all of the details here. The chain carriers are the radicals OH, H and O, and it is thought that the initiation reaction is

$$\text{initiation} \qquad H_2 + O_2 \longrightarrow 2\mathbf{OH}.$$

The main propagation step involves OH, and generates H

$$\text{propagation} \qquad H_2 + \mathbf{OH} \longrightarrow \mathbf{H} + H_2O.$$

The H atoms produced in this step are involved in what are called *branching* reactions in which one chain carrier produces two

$$\text{branching} \qquad \mathbf{H} + O_2 \longrightarrow \mathbf{OH} + \mathbf{O}$$
$$\mathbf{O} + H_2 \longrightarrow \mathbf{OH} + \mathbf{H}.$$

Together these two reactions keep the number of H atoms the same and generate two OH radicals. The number of chain carriers, and consequently the rate of reaction, therefore increases. If unchecked, there will be a runaway increase in the rate of reaction and hence an explosion.

If the termination reactions, which remove the chain carriers, are fast enough they can prevent the uncontrolled increase in the number of chain carriers, and hence prevent an explosion. Whether or not we see a steady reaction or an explosion depends on a delicate balance between the branching and termination reactions.

It turns out that at low pressures we see a steady reaction, at intermediate pressures there is an explosion, but at high pressures a steady reaction is seen once again. The reason for this behaviour is that the rates of the various termination and branching reactions respond differently to changes in the overall pressure.

FURTHER READING

Modern Liquid Phase Kinetics, B. G. Cox, Oxford University Press (1994).

QUESTIONS

10.1   The thermal decomposition of N_2O_5 in the gas phase has the overall reaction

$$2N_2O_5 \longrightarrow 4NO_2 + O_2.$$

Using the approach described in section 10.1.1 on page 411, write down the value of the stoichiometric coefficients of the reactant and each product. Then, write down the expression for the rate of reaction in terms of the slope of a plot of $[N_2O_5]$ against time, $[NO_2]$ against time, and $[O_2]$ against time.

10.2   By writing the units of concentration as 'conc' and the units of time as 'time', determine the units of the rate constants in the following rate laws

$$\text{rate} = k_1\,[N_2O_5] \qquad \text{rate} = k_2\,[PhCH_2Br]\,[DABCO]$$

$$\text{rate} = k_3\,[RH]\,[Cl_2]^{\frac{1}{2}} \qquad \text{rate} = k_4\,[CH_3CHO]^{\frac{3}{2}}$$

$$\text{rate} = k_5\,\frac{[O_3]^2}{[O_2]} \qquad \text{rate} = k_6.$$

10.3   The thermal decomposition of N_2O_5 in the gas phase is found to be first order in $[N_2O_5]$. The first order rate constant, k_{1st}, has been measured as a function of temperature as follows

$T\,/\,K$	349	368	378	396
$k_{1st}\,/\,s^{-1}$	0.0068	0.041	0.13	0.50

Make an Arrhenius plot of these data, and hence determine a value for the activation energy and the pre-exponential factor; state the units of each quantity.

10.4   Methanoic acid is oxidized in acid solution by Br_2 according to the following stoichiometric equation

$$HCOOH + Br_2 + 2H_2O \longrightarrow CO_2 + 2Br^- + 2H_3O^+.$$

It is found that the rate of reaction depends on the concentration of HCOOH and Br_2, and is also influenced by the concentration of added H_3O^+. If H_3O^+ and HCOOH are in excess, then it is found that the reaction is first order in $[Br_2]$. The following rate law was therefore proposed

$$\text{rate} = k_{obs}[Br_2][HCOOH]^a[H_3O^+]^b,$$

where a and b are the orders with respect to HCOOH and H_3O^+, respectively.

If H_3O^+ and HCOOH are in excess the rate law can be written

$$\text{rate} = k_{1st}[Br_2],$$

where k_{1st} is a pseudo first-order rate constant.

(a)   Write down an expression for k_{1st}.

(b)   The following values of k_{1st} were measured for different excess concentrations of HCOOH and H_3O^+:

[HCOOH] / mol dm^{-3}	0.10	0.10	0.12	0.22
[H$_3$O$^+$] / mol dm^{-3}	0.05	0.12	0.15	0.15
k_{1st} / s^{-1}	7.00×10^{-3}	2.92×10^{-3}	2.80×10^{-3}	5.13×10^{-3}

By comparing the data in the first two columns, determined the value of b. Similarly, determine the value of a by comparing the data in the third and fourth columns.

(c) Use all of the data in the table to determine an average value of k_{obs}; state the units of this rate constant.

10.5 Consider the following two energy profiles for the reaction scheme (all reactions are first order)

$$A \underset{k_{-1}}{\overset{k_1}{\rightleftharpoons}} B \overset{k_2}{\longrightarrow} C.$$

(a)

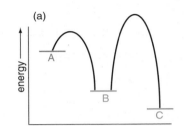

(b)

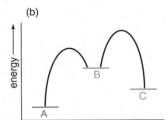

In each case, discuss whether or not the pre-equilibrium and/or the steady-state approaches can be used to analyse the overall kinetics. Using the appropriate approximation(s) in each case, obtain expressions for the rate of formation of C , identify the rate-limiting step, and mark the apparent activation energy on the energy profiles.

10.6 Consider the following reaction scheme, in which all of the reactions are first order and reversible.

$$A \underset{k_{-1}}{\overset{k_1}{\rightleftharpoons}} B \underset{k_{-2}}{\overset{k_2}{\rightleftharpoons}} C.$$

By equating the rates of formation and loss of B, show that the steady-state concentration of B is given by

$$[B]_{SS} = \frac{k_1[A] + k_{-2}[C]}{k_2 + k_{-1}}.$$

Hence show that the overall rate of change of the concentration of C is given by

$$\text{rate of change of } [C] = \frac{k_1 k_2[A] - k_{-1}k_{-2}[C]}{k_2 + k_{-1}}.$$

Under what conditions will this steady-state analysis be valid? Draw an energy profile for such a situation, marking on it all of the activation energies.

10.7 The reaction (in solution) between Br$_2$ and dicyanomethane, CH$_2$(CN)$_2$, has the overall stoichiometry.

$$CH_2(CN)_2 + Br_2 \longrightarrow CHBr(CN)_2 + Br^- + H^+.$$

The following mechanism, involving the intermediate CH(CN)$_2^-$, is proposed

$$CH_2(CN)_2 \overset{k_1}{\longrightarrow} CH(CN)_2^- + H^+$$

$$CH(CN)_2^- + H^+ \overset{k_{-1}}{\longrightarrow} CH_2(CN)_2$$

$$CH(CN)_2^- + Br_2 \overset{k_2}{\longrightarrow} CH(CN)_2Br + Br^-.$$

By putting the intermediate in the steady state, show that

$$\text{rate of formation of } CH(CN)_2Br = \frac{k_1 k_2 [CH_2(CN)_2][Br_2]}{k_{-1}[H^+] + k_2[Br_2]}.$$

Under what conditions will this analysis be valid?

What simplification of the expression for the rate law occurs if: (a) the rate of reaction 2 is much greater than that of reaction -1; (b) the rate of reaction -1 is much greater than that of reaction 2? In each case, identify the rate-limiting step and the apparent activation energy.

10.8 In aqueous solution the oxidation of Fe^{II} by Pb^{IV} is though to proceed via the intermediate species Pb^{III} according to the following mechanism

$$Fe^{II} + Pb^{IV} \xrightarrow{k_1} Fe^{III} + Pb^{III}$$

$$Fe^{III} + Pb^{III} \xrightarrow{k_{-1}} Fe^{II} + Pb^{IV}$$

$$Fe^{II} + Pb^{III} \xrightarrow{k_2} Fe^{III} + Pb^{II}.$$

Assuming that Pb^{III} can be placed in the steady state, determine the overall rate law for the rate of change of the concentration of Fe^{III}.

10.9 Under acid conditions and in aqueous solution, *t*-butanol undergoes a substitution reaction with iodide to give *t*-butyl iodide

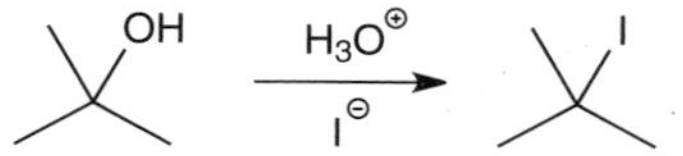

Experimentally the rate law for this reaction is found to be first order in the concentration of the *t*-butanol and also first order in H_3O^+, but not to have a dependence on the concentration of the iodide. The fact that the rate depends on the concentration of the acid is a clue that protonation is involved at some stage. A possible mechanism is

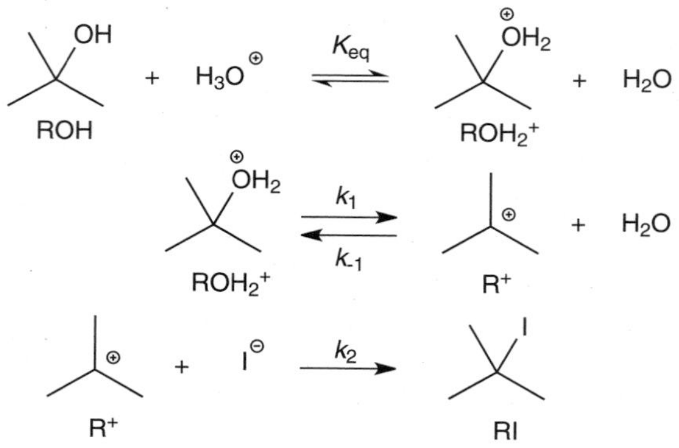

The reason for proposing the initial protonation as the first step is that the dissociation of ROH_2^+ to give R^+ and H_2O is thought to be considerably faster than the dissociation of ROH to give R^+ and OH^-.

(a) Assuming that ROH and ROH_2^+ are in equilibrium, show that

$$[ROH_2^+] = K_{eq}[ROH][H_3O^+]. \tag{10.16}$$

(b) Apply the steady-state approximation to R^+ and hence shown that

$$[R^+]_{ss} = \frac{k_1 K_{eq}[ROH][H_3O^+]}{k_{-1} + k_2[I^-]}.$$

You will need to use the expression for $[ROH_2^+]$ from Eq. 10.16 on the facing page.

(c) The rate of formation of the alkyl iodide is simply $k_2[R^+][I^-]$. Using you expression for $[R^+]$, show that

$$\text{rate of formation of RI} = \frac{k_1 k_2 K_{eq}[ROH][H_3O^+][I^-]}{k_{-1} + k_2[I^-]}.$$

If it can be assumed that the rate at which R^+ reacts with iodide is much faster than the rate at which the carbenium ion reacts with H_2O, show that the rate of formation of RI can be approximated by

$$\text{rate of formation of RI} = k_1 K_{eq}[ROH][H_3O^+].$$

(d) Give an interpretation of this rate law, identifying the rate-limiting step.

10.10 The destruction of ozone by Cl atoms in the stratosphere can be modelled using the following four reactions

$$\begin{aligned}
\text{initiation} \qquad & Cl_2 \xrightarrow{k_1} 2Cl \\
\text{propagation} \qquad & Cl + O_3 \xrightarrow{k_2} ClO + O_2 \\
& ClO + O_3 \xrightarrow{k_3} Cl + 2O_2 \\
\text{termination} \qquad & Cl + Cl \xrightarrow{k_4} Cl_2.
\end{aligned}$$

(a) Write down an expression equating the rate of formation and loss of Cl; do the same for ClO.

(b) Compare these two equations carefully, and hence deduce that, in the steady state,

$$[Cl] = \sqrt{\frac{k_1[Cl_2]}{k_4}}.$$

(c) Using your equation equating the rate of formation and loss of ClO, together with the above expression for [Cl], to show that

$$[ClO] = \frac{k_2}{k_3} \sqrt{\frac{k_1[Cl_2]}{k_4}}.$$

(d) Hence show that the rate of loss of ozone is given by

$$\text{rate of loss of ozone} = 2k_2[O_3] \sqrt{\frac{k_1[Cl_2]}{k_4}}.$$

(e) Explain why this final expression does not include the rate constant k_3. Also, discuss the origin of the term in the square root.

10.11 This question concerned the mechanism of the $H_2 + Br_2$ reaction discussed in section 10.9.1 on page 435. In that section it was assumed that the concentration of HBr was low, so that the inhibition step (step 4) could be omitted. This question explores the effect of including this step.

Taking into account *all five* steps in the mechanism, show that equating the rate of formation and loss of Br gives

$$2k_1[Br_2] + k_3[H][Br_2] + k_4[H][HBr] = k_2[Br][H_2] + 2k_5[Br]^2.$$

Further, by equating the rate of formation and loss of H show that

$$k_2[Br][H_2] = k_3[H][Br_2] + k_4[H][HBr]. \tag{10.17}$$

By adding together these two equations show that

$$[Br] = \left(\frac{k_1}{k_5}\right)^{\frac{1}{2}} [Br_2]^{\frac{1}{2}}.$$

Substitute this expression for [Br] into Eq. 10.17 and rearrange the result to obtain an expression for [H].

The rate of formation of HBr is

$$\text{rate of formation of HBr} = k_2[Br][H_2] + k_3[H][Br_2] - k_4[H][HBr].$$

Use Eq. 10.17 to show that this can be rewritten

$$\text{rate of formation of HBr} = 2k_3[H][Br_2],$$

and then substitute your expression for [H] into this to obtain, after some rearrangement

$$\text{rate of formation of HBr} = \frac{2k_2\left(\frac{k_1}{k_5}\right)^{\frac{1}{2}} [H_2][Br_2]^{\frac{3}{2}}}{[Br_2] + \left(\frac{k_4}{k_3}\right)[HBr]}.$$

Compare this rate law with the experimentally determined form.

10.12 It is possible that the gas phase reaction between NO and O_2 to give NO_2

$$2NO + O_2 \longrightarrow 2NO_2$$

proceeds via the following mechanism involving the intermediate N_2O_2

$$2NO \xrightarrow{k_1} N_2O_2$$
$$N_2O_2 \xrightarrow{k_{-1}} 2NO$$
$$N_2O_2 + O_2 \xrightarrow{k_2} 2NO_2.$$

Show how this mechanism can be consistent with the experimentally observed rate law

$$\text{rate of formation of NO}_2 = k_{obs}[O_2][NO]^2,$$

and give an expression for k_{obs}.

Part II
Going further

Contents

Spectroscopy

Key points

- Mass spectrometry essentially weighs molecules and fragments of molecules.
- Light from different parts of the electromagnetic spectrum causes different processes to occur in atoms and molecules.
- IR spectroscopy looks at molecular vibrations, and helps to identify functional groups.
- NMR looks at the environments of nuclei within molecules, and is able to detect which nuclei are close to one another on the bonding network.

Every time a chemist does a reaction, or isolates a compound, the first task is to identify the molecular structure of what they have made or isolated. Modern chemists, unlike their forebears from only a few decades ago, now have available to them a powerful array of instrumental techniques for determining molecular structures, so that for small- to medium-sized molecules it is possible to determine structures relatively easily and with confidence. Understanding molecular structure is not only about finding out what we have made, but also about determining the three-dimensional shape of the molecule. As we have seen, shape plays an important role in determining physical properties and reactivity.

The ultimate technique in structural determination is undoubtedly X-ray diffraction, which was discussed in Chapter 1. This technique reveals the precise positions of atoms, allowing us to determine the three-dimensional shape, along with bond angles and bond lengths. The drawback of this method is that it requires a single crystal of the pure compound: unfortunately not all compounds crystallize in the way required. Even if crystals are available, we might nevertheless be more interested in the structure in solution, for example in the case of biological molecules which are active in solution. Fortunately there are many other techniques available which can provide structural information from molecules in solution, although generally not with the precision available from diffraction experiments.

In this chapter we will look at just three techniques: mass spectrometry, infrared (IR) spectroscopy, and nuclear magnetic resonance (NMR) spectroscopy. Of these, NMR is by far the most important on a day-to-day basis for most chemists.

Fig. 11.1 A mass spectrum shows the relative numbers of ions of a given mass:charge ratio, m/z. In this spectrum of N, N'-dimethylethylenediamine the peak at 88.0995 units is the *molecular ion* formed by the loss of a single electron from the original molecule. The other labelled peaks are due to fragmentation of the molecular ion.

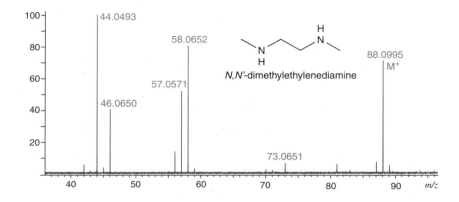

N, N'-dimethylethylenediamine

11.1 Mass spectrometry

In the late nineteenth and early twentieth century, it was discovered that the ions of atoms and molecules could be separated according to their mass and charge by the application of electric and magnetic fields. The early experiments looked at the ions of atoms and some very simple molecules containing two or three atoms. The technique has developed to such an extent that in modern instruments even proteins with masses of several thousand times that of a hydrogen atom are routinely studied.

The basic function of a mass spectrometer is to measure the mass of a molecule or, more correctly, an ion. A mass spectrometer can *only* separate and detect charged ions – neutral molecules cannot be detected. Both positive and negative ions can be detected, depending on the technique used.

Ions are separated according to their mass:charge ratio, m/z. This means a singly charged ion with a mass of 100 will be indistinguishable from a doubly charged ion with a mass of 200. A mass spectrum consists of a series of peaks corresponding to ions of differing mass:charge ratios. A typical mass spectrum is shown in Fig. 11.1. The horizontal axis shows the m/z ratio, and the vertical axis shows the relative numbers of ions of a given m/z ratio reaching the detector.

Sometimes it may be possible to observe a peak corresponding to the original molecule just missing an electron. This is known as the *molecular ion* and is denoted M^+. In other cases, the ion detected may correspond to the original molecule with an extra ion attached; such an ion is sometimes referred to as a *pseudo-molecular ion*. Ions such as H^+, Na^+, NH_4^+, MeO^-, or AcO^- are commonly attached to give pseudo-molecular ions. The pseudo-molecular ions are labelled to reflect the added ions e.g. $[M+Na]^+$ and $[M+MeO]^-$, indicating that the detected ion is the original molecule with an attached sodium cation or with an attached methoxide anion.

Rather than seeing just a single peak corresponding to the molecular ion, it is usually the case that many peaks are observed due to the molecule splitting up into various fragments within the instrument. The degree of fragmentation depends on how the spectrum was obtained. Very often, the peak in the spectrum with the greatest mass:charge ratio will be the molecular ion although this is not always the case. In Fig. 11.1 we see the mass spectrum of a diamine compound, showing the molecular ion at 88.0995 and various fragments.

Before we look at some spectra in more detail, we need to outline some of the different techniques used to record them, since this affects the nature of the spectra.

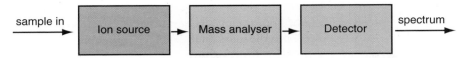

Fig. 11.2 The basic components of a mass spectrometer: a means of ionizing the sample (the ion source), a means of separating the ions according to their different *m/z* ratios (the mass analyser) and a means of detecting the ions.

11.1.1 Instrumentation

There are many different kinds of mass spectrometry in use in the modern laboratory. Whilst the details vary, each includes three main components, as illustrated in Fig. 11.2. There is an ionization source, a method for the separation of ions (a mass analyzer), and a method for detecting ions. The ions are created, analysed, and detected in a high vacuum so as to avoid collisions between the ions which would decrease the resolution of the instrument.

Ionization methods

The sample may be introduced into the spectrometer either as a solution or as a solid, and it is then vaporized in the vacuum within the instrument. Perhaps the simplest method used to ionize the sample is to fire high-energy electrons at it. In this so-called *electron ionization* (EI) technique, the high-energy electrons are able to knock out electrons from the gaseous sample, thereby forming positively charged ions of the original sample. The electrons removed tend to be those which are least tightly held, for example those from lone pairs or multiple bonds. These require less energy to remove compared to lower energy bonding electrons, or the lower-still 'core-like' electrons.

The positively charged ions formed are then repelled out into the mass analyzer by means of a positively charged plate. The main disadvantage of this ionization technique is that it is rather harsh, and can lead to the molecular ion fragmenting before it reaches the detector.

Sometimes a reagent gas such as methane is present in the ionization chamber before the sample is introduced. The reagent gas is at a pressure of about a thousand times that of the sample (but still much less than atmospheric pressure). Due to this difference in concentration, the electron beam is far more likely to ionize the reagent gas than the sample. However, collisions between the ionized reagent gas and the sample can lead to the production of ions, for example by the transferral of a proton. This technique, known as *chemical ionization* (CI), is a more gentle method for ionization, and so allows the parent molecular ion to be seen before fragmentation can occur. In contrast to the EI method, the parent ion produced using CI will be the neutral sample molecule with an extra ion (usually a proton) attached, i.e. $[M+H]^+$.

In *fast atom bombardment* ionization (FAB) the sample is mixed with a viscous liquid, such as glycerol, to form a 'matrix'. Ions are produced by bombarding the matrix with very high energy atoms or ions of, for example, krypton or xenon. The sample ions are then accelerated into the mass analyzer by an electric field. This ionization technique is quite gentle ('soft') and is particularly valuable for compounds which are hard to volatilize.

A further refinement of this method is the *matrix-assisted laser desorption ionization* (MALDI) technique. In this, the sample is co-crystallized with an organic compound such as benzoic acid to give a matrix. A UV laser, the light

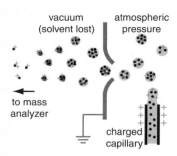

Fig. 11.3 Illustration of the electrospray ionization method. A solution of the sample is sprayed through a charged capillary to produce charged aerosol droplets. Once inside the insrument, the solvent evaporates, and the droplets eventually explode to leave molecules of the sample with ions stuck to them.

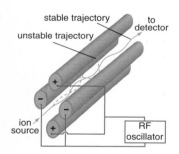

Fig. 11.4 A quadrupole analyzer consists of four rods, charged as shown. A radio-frequency signal applied to the rods causes ions of different mass-charge ratios to adopt either stable or unstable trajectories as they pass between the rods.

from which is strongly absorbed by the organic compound, is then focused onto this matrix. Ions consisting of both the matrix compound and the sample then break off and are accelerated into the mass analyser. This technique can be used to produce ions of large proteins with relative molecular masses of 100,000 or more.

The final ionization method we shall mention is *electrospray ionization* which is rapidly becoming the technique of choice in modern laboratories. The process is shown schematically in Fig. 11.3. A solution of the sample, sometimes with an added ionic compound such as a sodium alkoxide or carboxylate, is sprayed through a fine capillary held at a potential of several thousand volts. This produces an aerosol consisting of charged droplets of solution. The region where the aerosol is created is at atmospheric pressure but this chamber is separated from the high vacuum of the spectrometer by a small orifice. Due to the pressure difference the droplets are swept through the orifice into the vacuum. In the vacuum, the solvent evaporates, and the droplets become even smaller. Since each droplet may contain many like-charged ions, as the droplets become smaller they also become unstable and eventually explode, giving a molecule of the sample with a few attached ions. This is then accelerated in the mass analyzer.

The mass analyzer

Early mass spectrometers used a series of magnetic and electric fields to deflect the stream of ions in an arc. Just how far the ions deflect depends on their masses and charges. More modern instruments either make use of a *time of flight* (TOF) mass analyzer, or a *quadrupole* mass analyzer. The TOF instrument is perhaps the simplest to understand. By applying an electric field, a packet of ions from the ion source are first accelerated to the same kinetic energy. The ions are then allowed to drift through a field-free region, typically 30–100 cm long. Since the ions have the same kinetic energy, given by $\frac{1}{2}mv^2$, the lighter ions must have a greater velocity and hence reach the end of the field-free region before the heavier ions. Thus the different times at which the ions reach the detector give a measure of the mass:charge ratio of the ions.

The quadrupole mass analyzer (shown schematically in Fig. 11.4) consists of four parallel metal rods, usually between 5 and 10 cm long. Opposite pairs of rods are connected to the same polarity of a DC voltage source. A radio-frequency signal is also applied to the rods which has the effect of accelerating ions alternately towards them and away from them. By controlling the applied voltages and radio-frequencies, only ions of a desired mass:charge ratio can pass all the way through the analyzer – other ions have an unstable trajectory and collide with a rod and are neutralized. Simultaneously ramping the voltage and RF signal separates the ions according to the mass:charge ratios and allows a spectrum to be recorded.

Sometimes two mass analyzers are used in tandem. The first allows the selection of an ion of a given mass/charge ratio, the second analyzer allows the fragmentation of this one ion to be followed. This technique is called *tandem mass spectrometry* or MS–MS.

The detector

The earliest instruments used photographic plates to detect the ions. Most instruments now use either electron multipliers or photomultipliers to detect

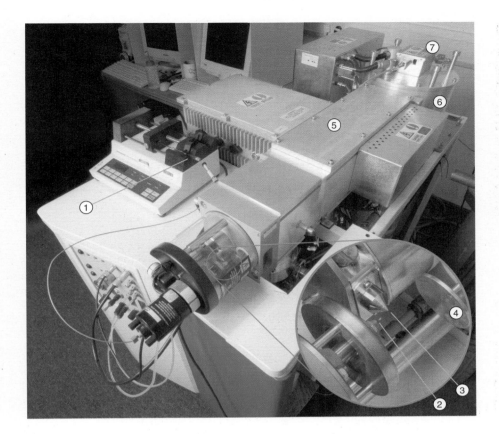

Fig. 11.5 A typical electrospray ionization mass spectrometer. (1) The microsyringe used to inject a solution of the sample into the instrument. (2) The charged capillary through which the solution containing the sample is sprayed. (3) The inlet orifice. (4) A sputtering plate to capture most of the sprayed sample solution. (5) A quadrupole mass analyzer. (6) A second mass analyzer (a time-of-flight mass analyzer). (7) The detector.

ions as these detectors give an electrical output which can most easily be recorded. It is also possible to use array detectors in which ions with a range of masses are detected simultaneously. The details of these detectors are not as important as the methods of producing the ions or separating them, and so we shall dwell on them no further. The various components of a typical mass spectrometer are illustrated in Fig. 11.5.

11.1.2 Isotope ratios

Figure 11.6 on the following page shows an EI mass spectrum of butyl bromide, $CH_3CH_2CH_2CH_2Br$. Using a table of relative atomic masses you can calculate that the relative molecular mass of C_4H_9Br is 137.02. However, in the spectrum instead of a peak at 137 we see a large peak at 136, and a second peak with almost equal intensity at 138. Closer inspection reveals that there is actually a very small peak at 137 together with a similar sized peak at 139.

The key to understanding this pattern of peaks is to realize that we need to consider the *specific isotopes* of the atoms involved. Whilst the relative atomic mass of bromine is 80, naturally occurring bromine contains no ^{80}Br atoms. Instead, naturally occurring bromine contains almost equal amounts of ^{79}Br and ^{81}Br. When we record the mass spectrum of butyl bromide, half of the M^+ ions detected contain a ^{79}Br atom, and so give rise to the peak at 136. The other half contain a ^{81}Br and give the peak at 138. The small peak at 137 is due to a

Fig. 11.6 An EI mass spectrum of butyl bromide. The presence of bromine is indicated by the 1:1 ratio of peaks separated by 2 mass units. These arise due to the 1:1 proportion of ^{79}Br and ^{81}Br isotopes in the sample.

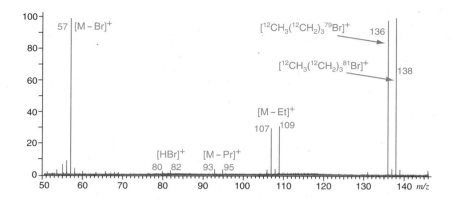

It is the slight variation in the exact abundances of various isotopes that leads to the relative atomic masses of different elements being quoted to different numbers of significant figures. For instance, the RAM for fluorine, with its single isotope ^{19}F, may be quoted accurately with up to seven decimal places. In contrast, since the proportions of the six naturally occurring isotopes of selenium vary slightly depending on the origin of the selenium containing compound, the RAM for selenium is usually quoted to only two decimal places.

molecular ion containing a ^{79}Br but also with one of the carbon atoms being a ^{13}C atom instead of the more usual ^{12}C. Similarly, the peak at 139 is due to an M$^+$ ion with a ^{81}Br atom and one ^{13}C atom. The mass spectrometer is easily able to distinguish between these M$^+$ ions containing different isotopes – indeed, it was the very first spectrographs recorded at the turn of the twentieth century which revealed the existence of isotopes.

Relative atomic masses of specific isotopes and of mixtures

The important point is that when working out the masses of the ions that we might expect to see in a spectrum, we cannot use the normal relative atomic masses that we use in most chemical calculations. The typical values we would find on a periodic table, for example, are actually *weighted averages* of the masses of all the naturally occurring isotopes of a given element. In order to understand the appearance of a mass spectrum, we require the relative atomic masses of *each specific isotope* and the proportions, or abundances, that the isotopes are found in a naturally occurring sample.

Some common isotopes and their masses are shown in Table 11.1 on the next page together with the relative atomic masses (RAM) used for the elements when we are refering to the mixtures of isotopes that occur in a natural sample.

Of the elements shown in the table, only fluorine and sodium have the same relative masses for the element and isotope. This is because these two elements only have one naturally occurring isotope. In contrast, the RAM of naturally occurring hydrogen, at 1.00794, is just a fraction greater than that for the ^{1}H isotope (at 1.007825032) due to the presence of a tiny percentage of the heavier deuterium isotope, ^{2}H (with a RAM of 2.0141011778). Similarly, the masses for elemental nitrogen and oxygen are slightly greater than that of their most abundant isotopes due to traces of heavier isotopes. Some elements have many naturally occurring isotopes – ruthenium, for example, has seven with proportions varying from 1.88% to 31.6%. Elemental tin holds the record with ten isotopes.

Isotopic signatures

Far from being a nuisance in mass spectrometry, the distributions of the different isotopes of an element, its 'isotopic signature', can be very useful. Looking back at the spectrum of butyl bromide in Fig. 11.6, we see that there are pairs of peaks at 107 and 109, at 93 and 95, and at 80 and 82; in each pair, the two

Table 11.1 Relative atomic masses for elements consisting of the mixtures of isotopes found in a naturally occurring sample, and the relative atomic masses of specific isotopes

element	RAM of element	isotope	RAM of isotope	% abundance
hydrogen	1.00794	^{1}H	1.007825032	99.985
carbon	12.011	^{12}C	12 (exactly)	98.89
		^{13}C	13.00335484	1.11
nitrogen	14.00674	^{14}N	14.00307401	99.634
oxygen	15.9994	^{16}O	15.99491462	99.762
fluorine	18.9984032	^{19}F	18.9984032	100
sodium	22.989770	^{23}Na	22.989770	100
sulfur	32.066	^{32}S	31.9720707	95.03
		^{34}S	33.9678669	4.21
chlorine	35.4527	^{35}Cl	34.96885271	75.77
		^{37}Cl	36.9659026	24.23
bromine	79.904	^{79}Br	78.918338	50.69
		^{81}Br	80.916291	49.31

peaks have the same height. Each of these pairs is due to an ion that contains a single bromine atom. For instance, the peaks at 107 and 109 are due to a positive ion corresponding to a molecule of butyl bromide that has lost an ethyl group. Such an ion is denoted $[M - CH_2CH_3]^+$. The peaks at 93 and 95 correspond to a molecule of butyl bromide that has lost a propyl group, i.e. $[M - CH_2CH_2CH_3]^+$. The peaks at 80 and 82 can only be due to a molecule of hydrogen bromide that has lost an electron, i.e. $[HBr]^+$.

Even the 1.1% of naturally occurring ^{13}C can be useful. If a compound contained just one carbon atom, the ratio of the peaks due to the M$^+$ ion (where the compound contains the more abundant ^{12}C isotope) to the M+1 peak (where the compound contains a ^{13}C atom instead) should be approximately 100 : 1.1. If a compound contains n carbon atoms, then the ratio will be approximately 100 : $(n \times 1.1)$. This means that for an organic compound, by comparing the heights of a peak M$^+$ to $[M+1]^+$ we can estimate the number of carbon atoms in the ion. As an example, returning to the spectrum of butyl bromide in Fig. 11.6 on the preceding page, we see that the molecular ion only contains around four carbon atoms since the ratio of the M$^+$ to the $[M+1]^+$ ions is approximately 100 : 4.4.

Note if the compound contains more than 90 carbon atoms, it is *more* likely that the molecule contains at least one ^{13}C atom than none at all.

More complex isotopic patterns

When the molecule contains more than one atom of an element with multiple isotopes, then distinctive peak intensity patterns arise. As an example, consider the EI mass spectrum of 1,3-dichlorobenzene shown in Fig. 11.7 on the following page. The molecular ions are the peaks between 146 and 150. As shown in Table 11.1, naturally occurring chlorine consists of ^{35}Cl and ^{35}Cl in a 76:24 ratio. The chance that *both* chlorine atoms in the dichlorobenzene are ^{35}Cl is therefore $0.76 \times 0.76 = 0.578$. The chance of having one of each isotope are $2 \times 0.76 \times 0.24 = 0.365$ (the factor of 2 arising since there are two ways of achieving this – either chlorine could be the ^{35}Cl with the other being the ^{37}Cl). Finally, the chance of both chlorine atoms being ^{37}Cl is $0.24 \times 0.24 = 0.058$. So the ratio of the three peaks should be 0.578 : 0.365 : 0.058 or 100 : 63 : 10

Fig. 11.7 The EI mass spectrum of 1,3-dichlorobenzene. The 3:1 ratio of $^{35}Cl{:}^{37}Cl$ gives rise to a characteristic pattern of intensities.

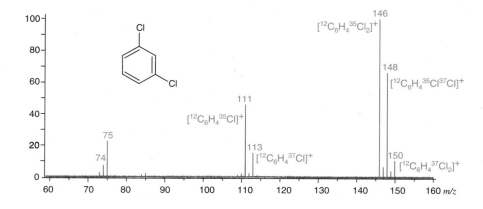

as seen in the spectrum. The peaks at 111 and 113 correspond to the molecular ion having lost one of its two chlorine atoms. Here the 3:1 ratio is clearly seen, as expected for an ion containing one chlorine atom.

The examples used so far have been relatively straightforward. When more isotopes are present, the patterns can rapidly become very complicated. For example, three atoms of ruthenium (which exists as a mixture of seven isotopes) give rise to the pattern of peaks shown in Fig. 11.8. Recognizing such isotope patterns plays an important part in the analysis of spectra – fortunately, computer programs are available to assist with this task.

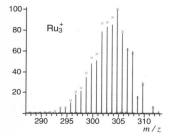

Fig. 11.8 The experimental isotope pattern for an Ru_3^+ ion. The red crosses represent the values predicted using simple statistics. These agree remarkably well with the experimental values.

11.1.3 'Accurate mass' spectrometry

Determination of molecular formulae

A well set-up mass spectrometer is easily capable of measuring the mass of an ion to within five parts per million or better. To put this into context, this would be like measuring the mass of an 80 kg person to within 400 mg. Given that as a person breaths in and out their mass will change by around 600 mg, this degree of accuracy is really rather impressive. Such accurate measurements are useful since it makes it possible to determine the *molecular formulae* of ions from their exact masses. As a simple example, consider the gases nitrogen, N_2, carbon monoxide, CO, and ethene, C_2H_4. Each gas has an approximate relative molecular mass of 28. However, using Table 11.1 on the previous page we see that the accurate mass of $^{14}N_2$ (the most common form of nitrogen) is 28.0062 (to within 5 ppm), for $^{12}C_2{}^1H_2$ it is 28.0313 and for $^{12}C^{16}O$ it is 27.9949. Thus if we measured the mass of the gas accurately to within 5 ppm, we could easily distinguish between these three possibilities.

As a more realistic example, let us suppose that in our high-resolution mass spectrum (HRMS), we observed a peak with a relative molecular mass of 88.1000. Computer programs are available which will produce best fit formulae for any desired mass – example programs may be found on the internet. Table 11.2 on the facing page shows the five closest match formulae for singly charged cations composed of carbon, hydrogen, nitrogen and oxygen only. The best match is the molecular ion formed by losing an electron from a molecule with formula $C_4H_{12}N_2$. This is not the *very* best match – that turns out to be an ion with the formula $[H_{15}Be_2MgP]^+$. However, this formula makes no chemical sense and may be discarded. The spectrum in Fig. 11.1 on page 448 is

Table 11.2 The five closest match CHNO only compounds of mass 88.1000

formula	RMM	error in ppm	example structure
$C_4H_{12}N_2^+$	88.0995	−5.7	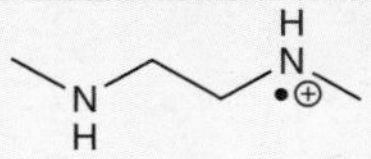
$C_5H_{12}O^+$	88.0883	−133	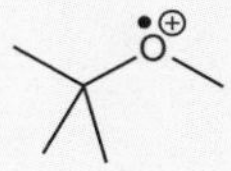
$C_3H_{10}N_3^+$	88.0869	−148	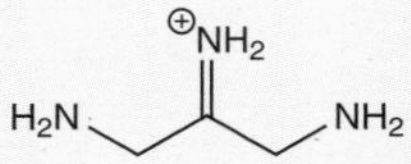
$C_5H_{14}N^+$	88.1121	+137	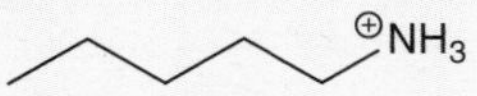
$C_4H_{10}NO^+$	88.0757	−276	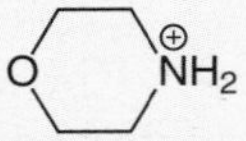

In each case, the masses used for C, H, N and O have been those of the most abundant isotopes.

of a compound with the formula $C_4H_{12}N_2$. The accurate mass of the molecular ion in the spectrum is given as 88.0995, and is in exact agreement with the calculated value.

Whilst high-resolution mass spectrometry can give a good indication as to the molecular formula for a compound, there may still be many structural isomers possible that fit that formula. However, it may be possible to learn something about the structure by careful study of the fragmentation pattern, which we consider next.

11.1.4 Fragmentation

We have already seen that ions may fragment in the mass spectrometer, and often this gives rise to a very complicated spectrum. By careful analysis of the spectrum it may be possible to learn something about the structure of the compound, although such information is usually more easily obtained from other spectroscopic techniques, such as NMR.

Figure 11.9 on the following page shows the EI mass spectrum for a ruthenium carbonyl compound, $Ru_3(CO)_{12}$. In this spectrum we can see peaks corresponding to the consecutive loss of each of the twelve carbon monoxide molecules. Note the repetition of the characteristic isotopic pattern for Ru_3 that we saw before in Fig. 11.8 on the preceding page.

Multiply charged ions

An interesting feature of Fig. 11.9 is the collection of peaks around 150, which are shown in more detail in Fig. 11.10 on the following page. We see the same isotopic pattern for Ru_3 as before (see Fig. 11.8), but now the separation between each of the peaks is no longer one unit, but one third of a unit. Since we cannot be losing a third of a nucleon at a time, this fractional spacing is indicative of multiply charge ions. In this case, each envelope of peaks is from a *triply* charged ion. The difference between one peak and the next is one mass unit, but the x-axis actually shows the mass:charge ratio, m/z, and not just mass. For a triply charged ion, differences of one mass unit correspond to differences

Fig. 11.9 The structure of $Ru_3(CO)_{12}$ has the three ruthenium atoms in a triangle with four carbon monoxide ligands attached to each metal atom. In the EI mass spectrum we see same isotope pattern for Ru_3 repeated as each successive CO molecule is lost.

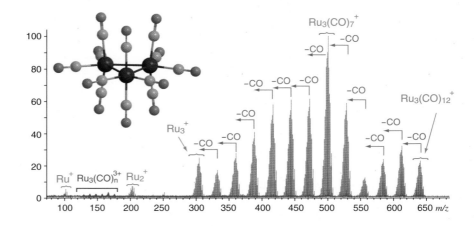

Fig. 11.10 An expansion of the EI mas spectrum of $Ru_3(CO)_{12}$ between 155 and 180 shows a set of triply charged ions. The spacing between each of the isotope peaks is now one third.

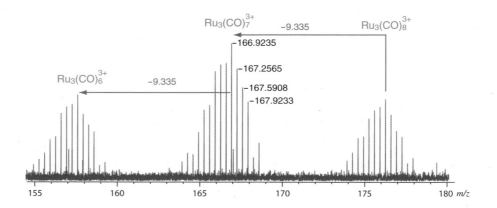

of one third on the m/z scale. As we move from one envelope of peaks to the next, a carbon monoxide ligand is again lost, but the change in the m/z value is now 9.335, i.e. the mass of CO divided by three.

Fragmentation patterns as 'molecular fingerprints'

The fragmentation of ions often results in rather complicated spectra. Whilst it is possible to work out what some fragments may correspond to, and to use this information to help piece together some structural information, on the whole, there are better techniques available for elucidating structures. In practice, the fragmentation pattern is often used as a kind of 'fingerprint' to help identify substances in analysis.

For example, mass spectrometry is routinely used in the testing of athletes for drugs. Typically, a urine sample is taken from the subject, and then passed through a gas or liquid chromatography column. On the chromatography column, the different substances present in the sample pass through the column at different speeds but the sample may contain many hundreds of compounds which will therefore only be partially separated. The output from the chromatography column is fed directly into a mass spectrometer and a spectrum is obtained every second or so.

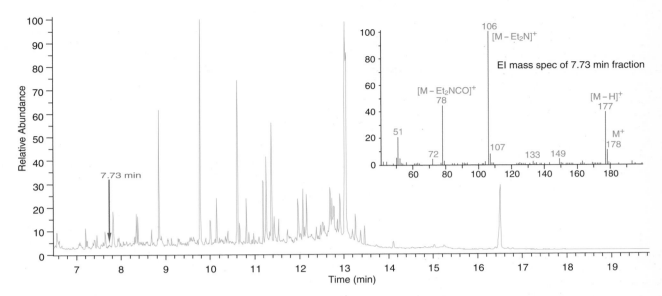

Fig. 11.11 The red trace is a gas chromatograph of a sample of horse urine containing a suspected illegal drug. The output from the chromatography column is fed directly into a mass spectrometer and automatically analysed every second or so. The inset is the EI-MS of the fraction passing through after 7.73 min and reveals the presence of the illegal stimulant *Nikethamide*.

The analyst knows how certain suspected drugs behave on the column and hence how long it takes for them to pass through. The mass spectrum of the appropriate fraction from the column is collected and then compared with the known spectrum of that drug. In this way its presence can be detected.

This combined technique using either (high-performance) liquid or gas chromatography with mass spectrometry is known as (HP)LC–MS or GC–MS. What makes the technique so valuable is the vast range of substances it is possible to detect and the extreme sensitivity – it is capable of detecting concentrations as low as nanograms per ml. Figure 11.11 shows a GC–MS trace from a sample of horse urine. The inset shows the EI mass spectrum of the fraction eluting at 7.73 min and reveals the presence of the illegal stimulant *Nikethamide*, whose structure is shown opposite.

11.1.5 Summary

- Mass spectrometry essentially weighs the ions of molecules and fragments of molecules.

- Molecular formulae can often be obtained from a knowledge of the precise mass of a molecular ion.

- Care must be taken to use the correct isotopic masses for the elements when interpreting mass spectra.

- Isotopic patterns can be a useful aid when interpreting spectra.

- Fragmentation in the mass spectrum may be of use in determining structure, but is often more useful by providing a 'molecular fingerprint' which may help to identify compounds by comparison with the spectra of authentic samples.

- Mass spectrometry is the most sensitive analytical technique, capable of detecting trace amounts of compounds.

11.2 Spectroscopy and energy levels

11.2.1 Energy levels in atoms and molecules

We have seen in the early chapters of this book that electrons in atoms and molecules may only possess certain energies, in other words that the energy is *quantized*. This means that to promote an electron from one energy level to another requires a fixed amount of energy. What is perhaps more surprising, is that other sorts of energy that a molecule may possess are also quantized. For example, there are only certain amounts of vibrational energy that a molecule can have – i.e. there exists a series of vibrational energy levels. Even rotational energy is quantized which means that molecules can only rotate at certain speeds.

It is often convenient to apportion the total energy that a molecule has into separate components – electronic, vibrational, rotational, and translational. Figure 11.12 on the facing page shows some of the different kinds of energy levels for carbon monoxide.

The electronic energy levels for carbon monoxide are shown in blue in Fig. 11.12. The lowest (the electronic *ground state*) is defined to be at zero energy. The minimum energy needed to promote an electron from the ground state to a higher energy level is shown in the diagram as ΔE_{elec}. This requires almost 50000 cm^{-1} (about 6 eV or 580 kJ mol^{-1}).

However, no matter what amount of electronic energy the molecule has, it is always vibrating and rotating. Some of the possible vibrational energies that the molecule may have when in the electronic ground state are shown in Fig. 11.12 on the next page by the red lines. These vibrational energy levels are shown expanded in (b). As can be seen in the figure, the vibrational energy levels are much closer together than the electronic energy levels. The minimum energy, ΔE_{vib}, now needed to promote the molecule from the lowest vibrational energy level (labelled $v = 0$) to the next highest vibrational energy level ($v = 1$) is around 2200 cm^{-1}.

Irrespective of which vibrational energy level is occupied, the carbon monoxide molecule may be rotating. The rotational energy levels for the lowest vibrational energy level are shown in green and are expanded in (c). The spacing between these rotational energy levels is typically just a few wavenumbers which is *much* less than the spacing between the vibrational energy levels.

11.2.2 Transitions between energy levels

It is possible for a molecule or atom to change the amount of energy it has by moving from one energy level to another. This might mean promoting an electron from one orbital to another, or it might mean promoting the molecule

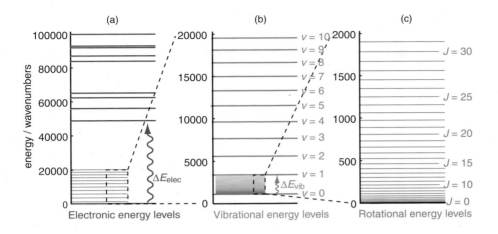

Fig. 11.12 The electronic, vibrational and rotational energy levels of carbon monoxide. The vibrational levels shown in red in (a) are expanded in (b). The rotational levels shown in green in (b) are shown expanded in (c).

from one vibrational energy level to another. Irrespective of the change, moving from one energy level to a higher energy level requires an input or absorption of energy, whereas moving from a higher energy level to a lower one releases or emits energy.

The energy absorbed or emitted can come from light, not necessarily just visible light, but other forms of electromagnetic radiation, such as microwave, infrared, ultraviolet, and X-ray. When light causes a transition between energy levels, a discrete, quantized amount of light energy, called a *photon*, is either absorbed or emitted. Figure 11.13 (a) shows the absorption of a photon (represented by the wavy red arrow) as a molecule is promoted from energy level E_1 to the higher energy E_2. Figure 11.13 (b) shows the emission of a photon as a molecule drops to the lower energy level E_1 from E_2.

A photon of light has a specific energy, and is related to a specific frequency, v, by the equation

$$E = hv$$

The constant of proportionality, h, is Planck's constant. If the energy, E, is in joules and v is in hertz, then h is 6.626×10^{-34} J s. We can therefore either talk about the energy or frequency of a photon.

The electromagnetic spectrum

The frequency (and hence energy) of electromagnetic radiation spans a huge range covering many orders of magnitude – arranging the electromagnetic radiation in order of frequencies gives us the *electromagnetic spectrum* shown in Fig. 11.14 on the following page. Whilst the spectrum is actually a seamless continuum, we arbitrarily split it up into different regions and give each region a distinct name. These divisions (radiowaves, microwaves etc) are useful since, as we shall see, the different regions of radiation tend to bring about different changes in an atom or molecule. However, we must remember that these divisions are purely artificial – the boundaries between one region and another are ill defined, and you will certainly encounter different boundaries for the divisions.

Note that a number of different units are used in Fig. 11.14. The relationship between these is given in Box 11.1 on the next page.

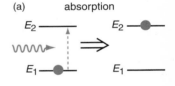

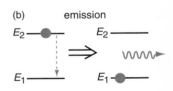

Fig. 11.13 In (a) the absorption of a photon of light (represented by the wavy red arrow) promotes the molecule from one energy level E_1 to a higher one, E_2. In (b) a photon is emitted as the molecule drops from energy level E_2 down to E_1.

Fig. 11.14 The approximate divisions of the electromagnetic spectrum arranged in order of increasing energy. The different regions of the spectrum are often referred to in different ways, for example by expressing the wavelength (λ), wavenumber ($\tilde{\nu}$), frequency (ν), or energy (E) associated with the radiation. The lower figure shows an expansion centred around the visible region of the spectrum.

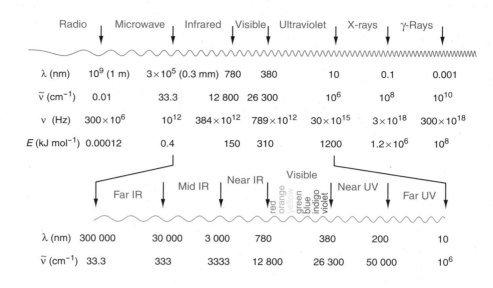

	Radio	Microwave	Infrared	Visible	Ultraviolet	X-rays	γ-Rays
λ (nm)	10^9 (1 m)	3×10^5 (0.3 mm)	780	380	10	0.1	0.001
$\tilde{\nu}$ (cm^{-1})	0.01	33.3	12 800	26 300	10^6	10^8	10^{10}
ν (Hz)	300×10^6	10^{12}	384×10^{12}	789×10^{12}	30×10^{15}	3×10^{18}	300×10^{18}
E (kJ mol^{-1})	0.00012	0.4	150	310	1200	1.2×10^6	10^8

	Far IR	Mid IR	Near IR	Visible red orange yellow green blue indigo violet	Near UV	Far UV
λ (nm)	300 000	30 000	3 000	780 · 380	200	10
$\tilde{\nu}$ (cm^{-1})	33.3	333	3333	12 800 · 26 300	50 000	10^6

Box 11.1 Units used in spectroscopy

The wavelength, λ, of a given type of electromagnetic radiation is the distance from the crest of one wave to the next, as shown opposite

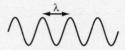

As a distance, the wavelength is measured in metres, or some submultiple of metres such as nanometres, nm. Wavelength (in nm) is commonly used when referring to different regions of the visible spectrum. Occasionally you may also come across micrometres, or microns, usually just given the symbol μ.

The *wavenumber*, $\tilde{\nu}$, is defined as the number of waves in one centimetre, and is the reciprocal of the wavelength in cm

$$\tilde{\nu} \text{ (in cm}^{-1}) = \frac{1}{\lambda} \quad \text{(where } \lambda \text{ is in cm.)}$$

The tilde ˜ is used to indicate that the quantity is a reciprocal wavelength. The wavelength (in m) and frequency (in Hz) of the light are related via

$$\nu = \frac{c}{\lambda}$$

where c is the speed of light in m s^{-1}. It follows that the wavenumber (in cm^{-1}) is also proportional to frequency (in Hz)

$$\tilde{\nu} = \frac{\nu}{c}$$

Note that in this expression, c must be in cm s^{-1} rather than the more usual m s^{-1}. Wavenumbers are commonly used in infrared spectroscopy. Hertz (as MHz) are commonly used when referring to regions of the radiowave region.

The transitions in different regions of the electromagnetic spectrum

The energy of the photon of light that is absorbed during a transition between two energy levels is equal to the difference between the two energy levels. However, as we saw in Fig. 11.12 on page 459, a typical separation between two electronic energy levels is much greater than the separation between two vibrational energy levels, which in turn is much greater than that between two rotational energy levels. This means that the energy needed to cause an electronic transition is much greater that that needed to cause a vibrational transition, which in turn is much greater than that needed for a rotational transition.

The energies are so different that they usually occur in quite separate parts of the electromagnetic spectrum. Transitions between electronic energy levels occur in the X-ray, ultraviolet or visible regions of the spectrum. Transitions between vibrational energy levels occur in the infrared region, whilst transitions between rotational energy levels involve microwave or radiowave radiation.

By analysing the exact frequencies of electromagnetic radiation absorbed, it is possible to find out about the different energy levels present in molecules, and so calculate parameters such as bond lengths and angles. In this chapter, however, we shall see how to interpret the spectra qualitatively to infer information about the structure of a compound. In the following section, we shall see how infrared (IR) spectroscopy may be used to identify the different functional groups present in a compound.

11.3 IR spectroscopy – introduction

In IR spectroscopy, IR radiation is shone through a sample and certain frequencies (energies) are absorbed as molecules move to higher vibrational energy levels. Historically, IR spectra are shown as transmission spectra: 100% transmission means all the light passes straight through, none is absorbed. If a certain frequency is absorbed, then a dip is shown in the trace. The resulting spectrum is shown schematically in Fig. 11.15 (a). Alternatively, spectra may be shown as absorption spectra in which case peaks are seen in the spectra pointing upwards as shown in Fig. 11.15 (b).

11.3.1   Sample preparation

The practicalities of recording an IR spectrum depend on the nature of the sample, especially whether it is a solid, liquid, or gas. Fig. 11.16 on the following page shows a modern IR spectrophotometer with a number of accessories.

For gaseous samples, a glass gas cell is filled and placed in the path of the IR beam. However, since glass absorbs IR light, a different material is used for the windows of the cell through which the IR passes. Usually, simple inorganic salts are chosen since they do not absorb in the usual region of interest. Sodium chloride is probably the cheapest and most commonly used material, but potassium bromide, calcium fluoride or even caesium iodide are sometimes used for particular purposes. Whatever the material, a single crystal of the salt is taken, cut into the appropriate shape for the ends of the gas cell, and then polished.

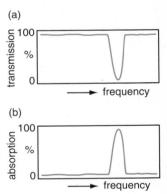

Fig. 11.15 In an IR spectrophotometer, IR light is shone through the sample. If plotted as a transmission spectrum, the absorption of light of a particular frequency is shown as a dip in the trace, as shown in (a). When plotted as an absorption spectrum, the absorption appears as a peak in the trace as shown in (b).

Fig. 11.16 A typical IR spectro-photometer and accessories. (1) A glass gas cell with NaCl windows. (2) A plate holder with NaCl plates. (3) An ATR diamond window and press.

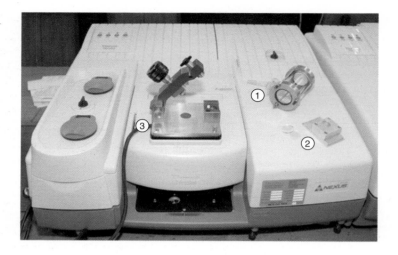

Liquid samples are often just applied as a thin film between two sodium chloride plates. Solid samples are sometimes prepared by grinding them up with a viscous hydrocarbon oil called *nujol* to make what is known as a *mull*. This toothpaste-like mull is then applied as a very thin layer sandwiched between two sodium chloride plates. Alternatively, the solid may be ground up with dry potassium bromide and then the mix put into a mould and compressed in a hydraulic press with a force equivalent to a ten tonne weight. Under such conditions the KBr exhibits *cold flow* and forms a transparent disc but with the sample incorporated. Whilst this method is slightly more laborious than the nujol mull method, it has the advantage of not introducing any extra substance that absorbs IR in the way that nujol does.

A more modern method used for liquids or solids is to record what is called an *attenuated total reflectance* (ATR) spectrum. In this technique, the sample is pressed against a diamond window, and rather than the IR light passing through the sample, it is reflected off the sample making use of total internal reflectance. ATR is a very quick, easy method to use: the sample is simply placed on top of the window and then the press is screwed down. The disadvantage is that the equipment is rather more expensive than the conventional apparatus.

11.3.2 The IR spectrum of a diatomic molecule

We have already seen the energy levels for carbon monoxide in Fig. 11.12 on page 459. IR radiation has the right energy to bring about transitions between vibrational energy levels. What this means is that when an IR photon is absorbed, the molecule gains vibrational energy i.e. it vibrates more vigorously.

Figure 11.17 on the facing page shows a high resolution IR spectrum of carbon monoxide gas. Closely spaced sets of absorptions, known as bands, are seen at just over 2000, 4000 and 6000 cm^{-1}. These correspond to transitions from the lowest vibrational energy level, (the vibrational ground state, labelled $v = 0$), to the first, ($v = 1$), second ($v = 2$), and third ($v = 3$) excited states. These transitions are called the *fundamental*, the *first overtone* and *second overtone* respectively; they are shown on the right of Fig. 11.17. Rather than a sharp peak, we see that each transition consists of an envelope of many peaks. The peaks within the envelope are all due to the same vibrational transition, but

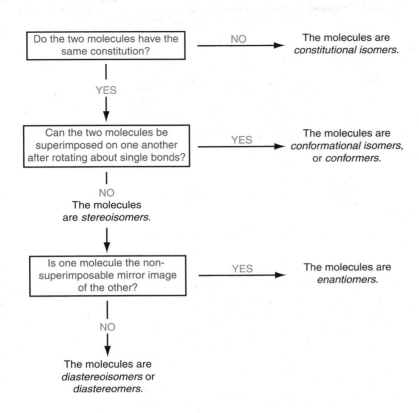

Fig. 12.6 A flow chart for determining the kind of isomerism shown by two molecules.

Fig. 12.5 (a) **A** and **B** are diastereoisomers because they are not mirror images of one another. The two enantiomers shown in (b) are said to have different *absolute configurations*, and we will see later on how these can be assigned.

12.1.5   Flow chart for assigning the type of isomerism

Now we have looked at the different ways in which the structures of molecules can be compared, we can use the flow chart shown in Fig. 12.6 to establish the correct description of the relationship between two molecules.

We shall look into each of these different sorts of isomers in more detail in the remainder of the chapter, but first it is important to be able to draw the molecules in an accurate, unambiguous manner. Box 12.1 on the following page shows the correct way to draw one or more tetrahedral centres, and Box 12.2 on page 525 discusses a number of alternative ways of representing structures.

12.2   The effect of rotations about bonds

The shapes of molecules are not fixed – even a diatomic molecule is constantly vibrating. For larger molecules, many shapes may be obtained simply by rotating around single bonds. The different shapes may also react differently, as we shall see.

Box 12.1 Drawing tetrahedral centres

When trying to understand the three-dimensional chemistry of organic compounds, it is essential to draw the structures in a realistic way. Tetrahedra should be drawn with two bonds in the plane of the paper, one bond coming out of the plane and one going in, as shown by structures **A–C** below. Furthermore, if a line is drawn at a tangent to the two bonds in the plane of the paper (as shown by the red dotted line in the structures below), then the bonds that do not lie in the plane of the paper must be on the opposite side of this line from the bonds that are in the plane. In addition, if a second line is drawn at right angles to the first (as shown by the green dotted line), there must be one bond in each quadrant between the dotted lines.

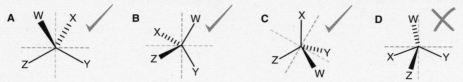

Structures **A–C** are all good representations of tetrahedral structures. In contrast, structure **D** is badly drawn and in fact does not represent a tetrahedral structure at all. Even though it still has two bonds in the plane of the paper, one in and another out, the bond to atom Z has been drawn on the wrong side of the red dotted line, and there is not just one bond in each of the quadrants defined by the dotted lines.

When drawing a number of tetrahedral centres in one molecule, care needs to be taken how the out-of-plane bonds are drawn to ensure that the structures make sense. In structure **E** below, all the bonds coming out of the plane of the paper are to the left of the bonds going into the plane of the paper. This is correct for a molecule viewed slightly from the right. In structure **F**, the bonds coming out of the plane are all to the right of the bonds going into the plane. This is correct for a molecule viewed from the left. Structure **G** is badly drawn and the order of the out-of-plane bonds is not consistent.

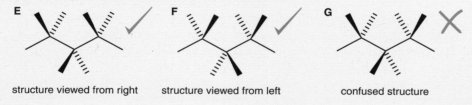

structure viewed from right   structure viewed from left   confused structure

---

12.2.1 Conformation and conformers

Different arrangements arising solely from rotations about single bonds are known as *conformations* e.g. structures **A** and **B** shown in Box 12.2 on the facing page. These two arrangements are particularly important and are known as the staggered and eclipsed conformations.

The conformations of ethane

If we consider a molecule such as ethane, there are essentially an infinite number of different conformations generated by rotating one of the CH_3 groups relative to the other i.e. rotating about the C–C bond. The degree of rotation about this bond is specified by the relevant *dihedral angle*, which in general is the angle between two planes (e.g. Fig. 11.78 on page 502). In the case of ethane, the angle is the one between the plane containing the C–C bond and one of the C–H bonds attached to the first carbon, and the plane containing the C–C bond and one of the C–H bonds attached to the second carbon. The dihedral angle is

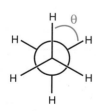

Fig. 12.7 When ethane is viewed along the central C–C bond, the dihedral angle, θ, is the angle between one of the C–H bonds on the front carbon, and one of the C–H bonds on the back carbon.

Box 12.2 Alternative representations of two tetrahedral centres

A number of alternative representations are possible for two tetrahedral centres, each with its own particular merits. Shown in (a) is the usual representation of structure **A** which has two tetrahedral centres. An alternative representation, called a *saw-horse* structure, is shown in (b). This representation shows what **A** would look like when viewed from the left as indicated by the position of the eye. Note that no wedged or hashed bonds are used in this representation, and two of the bonds are aligned vertically.

The staggered conformation

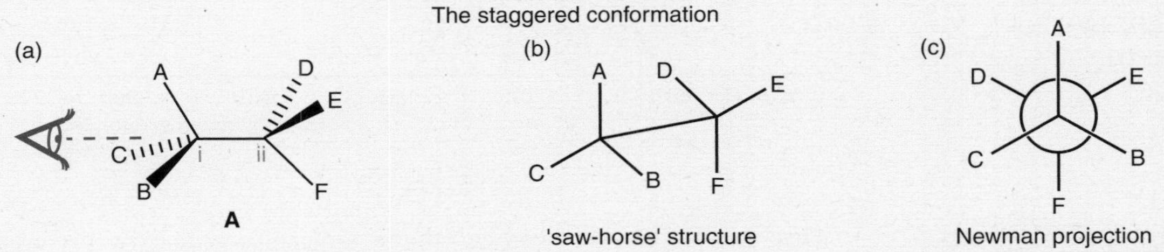

(a)

(b)   A   D   E

'saw-horse' structure

(c)

Newman projection

Another representation of **A**, called a *Newman projection*, is shown in (c). If an observer views **A** along the central C–C bond from the position indicated by the eye, the bond to A appears vertically upwards, the bond to C appears on the left, and the bond to B appears on the right. The carbon atom to which A, B and C are attached (labelled i) is not shown in the Newman projection, but is located where the bonds to A, B and C meet. The carbon atom to which D, E and F are attached (labelled ii) is obscured by carbon i when the structure is viewed in this way. In the Newman projection carbon atom ii is represented by the circular ring. The bonds to D, E, and F all join the ring – they do not meet at the centre. When viewed from the position indicated in (a), the bond to D is on the left, the bond to E is on the right, and the bond to F points vertically down. These are shown in their corresponding positions in the Newman projection.

In **A** none of the bonds to the front carbon overlap with any of the bonds to the rear carbon in this structure. This arrangement is known as the *staggered conformation*.

Rotating the central C–C bond of **A** by 60° gives structure **B**, as shown in (d). This arrangement is called the *eclipsed conformation*. Now when viewed from the position indicated by the eye, the bonds to the substituents on carbon i obscure or eclipse the bonds to the substituents on carbon ii. **B** is shown as a saw-horse structure in (e), and as a Newman projection in (f).

The eclipsed conformation

(d)

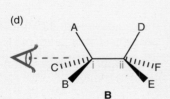

(e)

'saw-horse' structure

(f)

Newman projection

---

most clearly seen in a Newman projection as shown in Fig. 12.7 on the facing page.

A graph showing how the relative energy of ethane varies with the dihedral angle is shown in Fig. 12.8 on the next page. The energy is a minimum in the *staggered* conformation, which is when the dihedral angle is 60°, 180° or 300°. The energy maxima correspond to the *eclipsed* conformation, in which the dihedral angle is 0°, 120° or 240°.

Fig. 12.8 A graph showing how the relative energy of ethane changes as the central C–C bond is rotated. In the staggered conformation the energy is a minimum; in the eclipsed conformation the energy is at a maximum, approximately 12 kJ mol^{-1} higher in energy.

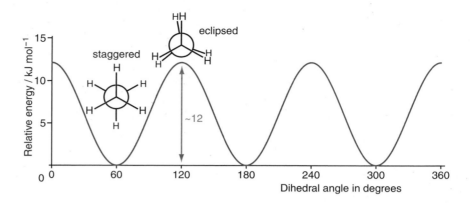

🌐 *Weblink 12.1*

Follow this link to view a real-time version of Fig. 12.8.

This means that on average ethane molecules will exist preferentially in the staggered conformations and only pass through the eclipsed conformations fleetingly. Different conformations which are energy minima are called *conformational isomers* or *conformers* for short. In ethane there are three staggered conformers which are equivalent and have the same energy: they can therefore be described as degenerate.

We might well ask *why* the different conformations have different energy. A simplistic answer would be to say that the eclipsed conformation is higher in energy due to repulsion of the hydrogen atoms as they eclipse one another. Any such repulsion would be due to the electrons in the different C–H bonds repelling each other. In fact, due to the very small size of the hydrogens, there is very little physical interaction between them. The correct explanation lies in the molecular orbitals of the different conformations. Not only is there an unfavourable interaction between the bonds in the eclipsed conformation, but there is actually a favourable interaction in the staggered conformation. This can be thought of as arising from the interactions of the filled C–H σ orbitals on one carbon with the vacant C–H $\sigma^\star$ orbitals on the other.

The conformations of butane

For molecules of the type X–CH$_2$–CH$_2$–Y, the situation is slightly more complicated and the degeneracy of the staggered conformers seen in ethane is reduced. A graph showing how the relative energy of butane varies as the molecule is rotated about the central C–C bond is shown Fig. 12.9 on the facing page.

The highest barrier to rotation is when the two methyl groups are eclipsing each other in conformation **A**. When the methyl groups just eclipse hydrogen atoms, as in the conformations **C** and **E**, the barrier is lower. There are three energy minima or conformers. The lowest, **D**, is where the two methyl groups are arranged opposite each other in the *trans* or *anti* conformation. The other minima, **B** and **D**, are where the two methyl groups are approximately 60° to each other, and are known as the *gauche* conformers.

Note that the two gauche conformers **B** and **F** are in fact enantiomers: they are mirror images of one another and, unless the central C–C bond is rotated, it is not possible to superimpose one on the other. However, since there is such a small energy barrier needed to be overcome in order to convert one conformer to another, it is impossible to isolate different enantiomers of butane. The different conformations interconvert many thousands of times per second.

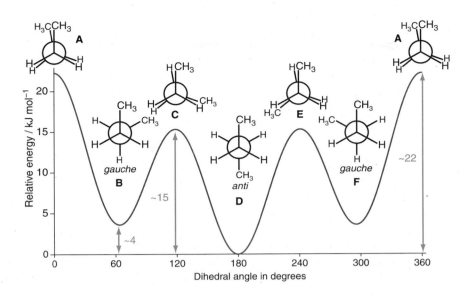

Fig. 12.9 A graph showing how the relative energy of butane changes as the central C–C bond is rotated. The highest energy conformation, **A**, is where the two methyl groups eclipse one another. The other eclipsed conformations, **C** and **E**, in which the methyl groups eclipse smaller hydrogen atoms, are also energy maxima. The lowest energy conformation, **D**, has the two methyl groups opposite one another. The conformations in which the methyl groups are at a dihedral angle of 60° and 300°, **B** and **F**, are also energy minima.

12.2.2 Restricted rotation

Whilst it is relatively easy to rotate about single bonds, as we have already seen in section 4.4.2 on page 153 and Fig. 4.27 on page 154, it is not possible to rotate the two ends of a C=C double bond without breaking the π bond. This means that there is a very large energy barrier to overcome and consequently no such rotation occurs at room temperature.

To break the π bond in ethene in order to allow the two ends to rotate requires approximately 270 kJ mol^{-1}. However, the barrier to rotation can be considerably lower for double bonds which are conjugated and so involved in a delocalized π system. Some examples are shown in Fig. 12.10. In compound **B** there is extensive delocalization between one of the oxygen atoms, the C=C double bond and one of the C=O π bonds. This conjugation weakens the bonding between the two carbons, as can be seen in the resonance structure **B'**.

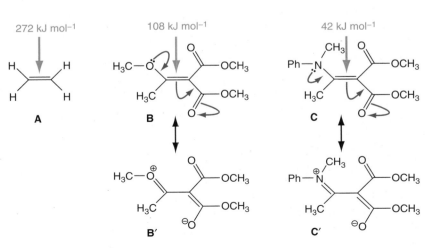

Fig. 12.10 Examples of the energy barrier for rotation about a C=C double bond. In ethene (**A**) the barrier is 272 kJ mol^{-1}, but in **B** and **C** the barriers for rotation about the indicated double bonds are much lower. This is due to the fact that these double bonds are involved in extensive conjugation, as indicated by the blue arrows.

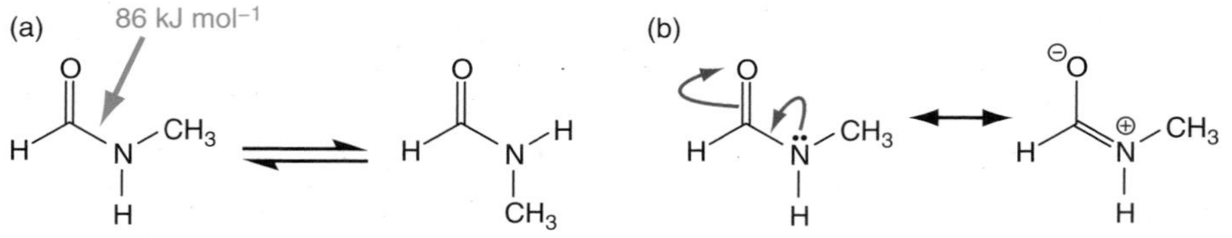

Fig. 12.11 To rotate about the C–N bond in amide shown in (a) requires 86 kJ mol^{-1}. The partial double bond character of the C–N bond is shown by the resonance structures in (b).

A similar situation occurs in **C**, but here it is a nitrogen lone pair that is conjugated with the C=C double bond, rather than an oxygen atom. Since the nitrogen lone pair is higher in energy than the oxygen lone pair, there is even better overlap with this and the π system, and this lowers the barrier needed to rotate about the central C=C bond to just 42 kJ mol^{-1}.

Restricted rotation in amides

We have already seen in section 4.9.3 on page 175 that there is also restricted rotation about the C–N bond in an amide. Whilst we would usually draw this as a single bond, the π MO spread over the oxygen, carbon and nitrogen means there is partial double bond character to the C–N bond. Thus to rotate about the C–N bond indicated in the amide in Fig. 12.11 (a) requires approximately 86 kJ mol^{-1}. The partial double bond character is also revealed in the resonance structures shown in (b).

The important point to understand is that the energy barrier for rotation about a bond varies continuously from around 12 kJ mol^{-1} for the C–C bond in ethane all the way up to almost 300 kJ mol^{-1} for some C=C double bonds. Note that the nominal C=C double bond in compound **C** in Fig. 12.10 on the previous page has a lower barrier to rotation than the nominal C–N single bond in the amide shown in Fig. 12.11.

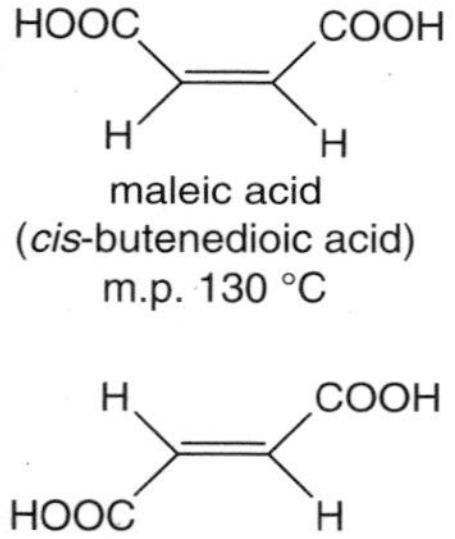

12.3   Isomerism in alkenes

Since there is such a large barrier that needs to be overcome to rotate around a C=C double bond, it is possible to isolate isomers where the two molecules differ due to the arrangement about the double bond. An example is shown in Fig. 12.12. In maleic acid, shown in (a), the two carboxylic acids are on the same side of the double bond, whereas in its isomer, fumaric acid, shown in (b), the carboxylic acid groups are on the opposite sides of the double bond. These two compounds have completely different properties: for example maleic acid melts around 130 °C, whereas fumaric acid sublimes at around 200 °C and melts at 300 °C when heated in a sealed tube.

Fig. 12.12 The isomers maleic acid and fumaric acid differ in the arrangements of the groups about the double bond. Since there is a large barrier to be overcome in order to allow the two to interconvert, the separate compounds can be isolated at room temperature.

12.3.1   *cis-trans* isomerism

The alternative names for maleic acid and fumaric acid are *cis*-butenedioic acid, and *trans*-butenedioic acid. The Latin prefix *cis*- means 'on the same side' and signifies that the two carboxylic acids are on the same side of the double bond. The Latin prefix *trans*- means 'across' and signifies that the two groups are on opposite sides of the double bond, across from one another.

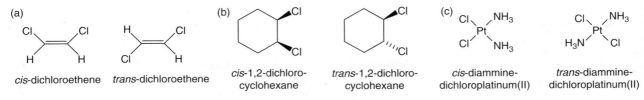

Fig. 12.13 Examples of *cis-trans* isomers. In (a) the two chlorines are either on the same or opposite side of the C=C double bond. In (b) they are either on the same or opposite side of the cyclohexane ring. In the square-planar platinum complex shown in (c) the platinum, two chlorines and two nitrogens are all in the same plane. In the *cis* isomer the two chlorines are on the same side of the square, in the *trans* isomer they are at opposite corners of the square.

This terminology is useful for other sorts of isomers where two groups are on the same or opposite sides in a molecule. Some examples are given in Fig. 12.13. Example (a) shows the *cis* and *trans* isomers of dichloroethene where the two chlorines are either on the same side of the C=C double bond, or on opposite sides. In (b) the two chlorines are either on the same side, above the plane of the six-membered ring (*cis*), or they are on opposite sides (*trans*), with one above the plane of the ring and one below. Finally, example (c) shows two isomers of a square-planar platinum complex. In the *cis* isomer, the two chlorine atoms are on the same side of the square defined by the four groups attached to the central platinum; in the *trans* isomer, they are at opposite corners.

In each pair of isomers in Fig. 12.13 the molecules have the same constitution i.e. it is the same atoms and groups which are connected to each other in the two isomers. For none of the pairs is it the case that one isomer is the non-superimposable mirror image of the other. This means in each case the two isomers are *diastereoisomers*. Sometimes this type of isomerism where one molecule has two groups on the same side and its isomer has the two groups on opposite sides is termed *cis-trans isomerism*.

12.3.2 *E/Z* nomenclature

If there are four different groups attached to one C=C double bond, it it not appropriate to use the prefixes *cis* and *trans*. In such cases a different system of nomenclature is used. As an example, four of the isomers of bromo-chloro-fluoro-iodoethene are shown in Fig. 12.14. Alkenes **W** and **X** have the same

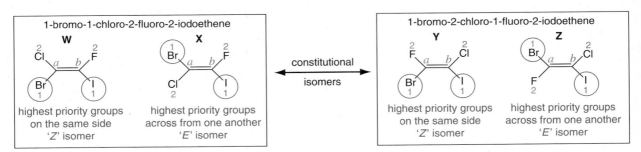

Fig. 12.14 Four of the isomers of $C_2BrClFI$, illustrating how the *E/Z* labels are assigned. The two compounds in the box on the left have the same constitution as each other, but different configurations. Likewise the two compounds in the box on the right have the same constitution, but different configurations.

constitution as each other, as do **Y** and **Z**. However, the two isomers in each box have different configurations. To distinguish between these pairs isomers a priority is assigned to the two groups on each carbon atom in the alkene.

The order of priority for the groups is based on atomic number. For example, in alkene **W**, of the two groups attached to carbon *a*, since bromine has a higher atomic number than chlorine, the bromine is given a higher priority (1) than the chlorine (2). Of the two groups attached to carbon *b*, the iodine is given the higher priority (1) than the fluorine (2). In alkene **W**, the groups on each carbon atom that have the highest priority are on the same side of the double bond. Such an arrangement is termed the (*Z*)-isomer from the German *zusammen* meaning 'together'.

If, as in alkene **X**, the groups with the highest priority from each carbon atom are on opposite sides of the double bond, then the molecule is termed the (*E*)-isomer, from the German *entgegen* meaning 'opposite'.

The Cahn–Ingold–Prelog convention

The assignment of a priority to the different groups according to their atomic numbers illustrates part of the comprehensive system much used in stereochemistry called the Cahn–Ingold–Prelog (CIP) convention. A more elaborate set of rules is needed as the alkenes become more complicated. It is beyond the scope of the book to explain all the intricacies of the system which are only needed for highly specialized examples, but it is important to have at least a basic understanding of the principles which are outlined in Box 12.3 on the next page.

12.4 Enantiomers and chirality

Isomers that are non-superimposable mirror images of one another are known as enantiomers and a molecule that is non-superimposable on its mirror image is said to be *chiral*. The word chiral comes from the Greek word for hand, since left and right hands are also non-superimposable – that is, the right hand will not fit precisely into an exact mould of the left hand. We shall see that different enantiomers have different 'handedness' or *chirality* and this can lead to them having very different properties in some circumstances. First of all though, we shall see how the CIP priority convention provides a useful way of distinguishing between two enantiomers.

12.4.1 Compounds with one stereogenic centre

In Fig. 12.2 on page 521 we saw that a tetrahedral carbon atom with four different groups attached is chiral i.e. there are two isomers which are non-superimposable mirror images. Such a carbon atom with four different groups is sometimes known as a 'chiral centre' or, more generally, a *stereogenic* centre. A simple example of such a molecule is the amino acid alanine, which contains a carbon atom with a hydrogen, a methyl group, a carboxylic acid group and an amine group attached. Figure 12.15 on page 532 shows two different views of the two enantiomers of alanine. All the so-called 'naturally-occurring' amino acids, which make up the proteins in our bodies, have the same general structure as the structure shown in (a) but with different groups replacing the methyl group. Two views of the mirror image structure of naturally occurring alanine are shown in (b).

Box 12.3 The Cahn–Ingold–Prelog priority convention

Consider the alkene shown opposite which has four different groups (a, b, c and d) attached to the two doubly-bonded carbons C(i) and C(ii). We need to assign a priority to each group. Let us first consider the groups a and b which are attached C(i). The two atoms directly attached to C(i) are both carbons, so they have the same priority. We therefore have to move further away and consider the atoms two bonds from C(i) i.e. carbons a_1 and b_1. In order of decreasing atomic number, the atoms directly attached to a_1 are (O, H, H), and those directly attached to b_1 are (C, H, H). This means carbon a_1 has the atom with the highest atomic number directly attached to it (the oxygen) and therefore group a has a higher priority than group b.

Now let us consider groups c and d which are attached to C(ii). The atoms directly attached to C(ii) are both carbon, so as before we have to move out to atoms c_1 and d_1, which are two bonds away. Arranged in decreasing atomic number, the atoms directly attached to c_1 are (C, H, H), and those directly attached to d_1 are also (C, H, H). There is no difference between these, so we have to move out further to atoms c_2 and d_2. The atoms directly attached to c_2 are (C, C, H), whilst those attached d_2 are (C, H, H). In each set the element with the highest atomic number is the same, carbon, but second highest are different: for c_2 the second highest is carbon, whereas for d_2 it is hydrogen. Therefore group c has a higher priority than d.

Now that the priority of the groups has been identified the labels (E) and (Z) can be assigned. When the two highest priority groups are on the same side, we have the (Z)-isomer, but when they are on the opposite side, we have the (E)-isomer; both cases are illustrated opposite.

The CIP system includes many further sub-rules to help distinguish between more complicated examples, and references to these can be found in the further reading section at the end of the chapter.

highest priority substituents a and c on the same side → Z

highest priority substituents a and c on opposite sides → E

Note that it does not matter that later along the carbon chain for substituent d we encounter a chlorine at position d_3. This is because by the time we get to positions c_2 and d_2 there is already a distinction between groups c and d, so the process of assigning priority is complete. What happens beyond this point is not relevant.

The process of determining the priority can also be illustrated using a tree diagram, as shown below.

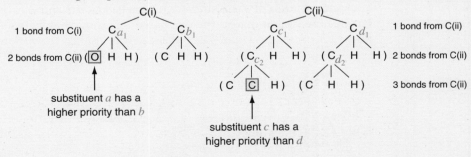

Labelling enantiomers with the *R/S* convention

The two enantiomers are commonly distinguished by the prefixes L- for the natural amino acids, and D- for the 'unnatural' amino acids. This terminology

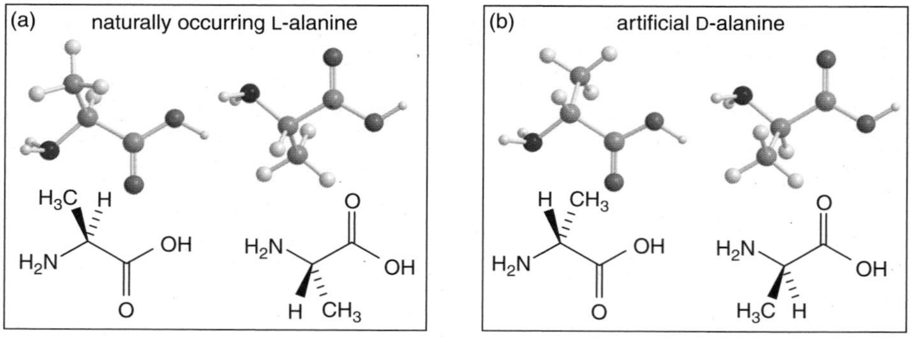

(a) naturally occurring L-alanine

(b) artificial D-alanine

Fig. 12.15 Two views of the naturally-occurring amino acid L-alanine are shown in (a). In both views, the amino group and carboxylic acid group are in the plane of the paper, but whether it is the methyl group or the hydrogen that come out of the plane of the paper depends on which way up the in-plane backbone is drawn. Two views of the artificial isomer D-alanine are shown in (b).

is historical and need not concern us. A systematic nomenclature based on the CIP priority convention is now used to distinguish between the isomers, and this proves to be applicable to many different compounds.

The method is as follows:

(a) Assign a priority following the CIP convention to each of the four groups attached to the stereogenic centre.

(b) Arrange the molecule so it is viewed with the lowest priority group going into the plane of the paper.

(c) If the path from the highest priority group, to the second highest, to the third follows a *clockwise* direction, the prefix (*R*)- is assigned to the centre; if the path follows an *anticlockwise* direction, the prefix (*S*)- is assigned to the centre.

Example one: alanine

Let us first consider how the convention may be applied to the naturally occurring isomer, L-alanine, shown in Fig. 12.16. First of all we need to assign a priority to each of the four groups attached to the chiral carbon in alanine. For this the system used is the one based on atomic numbers and described in Box 12.3 on the previous page where it was applied to working out the *E/Z*

Weblink 12.2

View and rotate a three dimensional structure of L-alanine. It is helpful to do this to be sure that you have the groups arranged correctly when you rotate the hydrogen to the back.

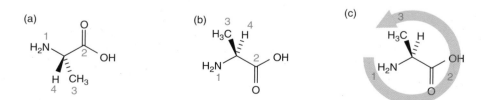

Fig. 12.16 In (a) each group attached to the stereogenic centre in alanine is assigned a priority following the CIP convention. In (b), the structure is rotated so the lowest priority group is going into the plane of the paper. The path from the highest priority group, to the second and third, as shown in (c), moves in an *anticlockwise* direction which means the prefix *S*- is assigned to the centre.

isomers of alkenes. Arranged in order of decreasing atomic number, the atoms directly attached to the stereogenic carbon of alanine are (N, C, C, H).

To distinguish between the two carbon atoms we look at the atoms two bonds from the stereogenic carbon. The atoms attached to the methyl group carbon, are just (H, H, H). The atoms attached to the carboxylic acid carbon are (O, O, O) – note that the double bond to oxygen counts as two oxygens. The carboxylic acid group therefore has a higher priority than the methyl group, so the order of priority is first the –NH$_2$ group, second the –COOH group, third the –CH$_3$ group and finally the H. These are numbered from one to four as shown in Fig. 12.16 (a).

In (b) the molecule is arranged so that when we view it, we look along the C–H bond, i.e. with the lowest priority group going into the background. The path from the highest priority group, to the second, to the third takes us in an anticlockwise direction as shown in (c), and so the prefix (S)- is assigned to the centre.

A mnemonic for remembering the *R/S* convention

A useful aid to help with the *R/S* assignment is to think of a steering wheel. With the column of the steering wheel representing the lowest priority group going into the plane of the paper, if the path from highest priority group to the third highest is in a clockwise direction and the centre is (R)- (from the Latin *rectus* meaning right), this can be remembered by turning the steering wheel in a clockwise direction to turn right, as shown in Fig. 12.17 (a). An anticlockwise path is given the symbol (S)- (from the Latin *sinister* meaning left, or left-hand) and is akin to turning the steering wheel anticlockwise in order to turn left, as shown in (b).

Example two: limonene

As a second example, consider the structure of limonene, shown in Fig. 12.18. First we must identify the centre we wish to assign the configuration of. The carbon marked with the red asterisk in (a) has four different groups attached to it and is a stereogenic centre. The four atoms directly attached to this carbon are (C, C, C, H). These have been arbitrarily labelled *a*, *b*, *c*, and *d* to help us distinguish between the carbons. Clearly, the hydrogen has the lowest priority of these four atoms, but we need to decide between the three carbon atoms.

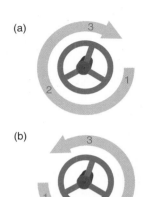

Fig. 12.17 A useful way of remembering how to assign the *R/S* convention to molecules is to think of how you would turn a steering wheel. With the column of the steering wheel representing the lowest priority group going into the background, moving from the highest priority to the lowest in an clockwise path is like turning the wheel to turn right as shown in (a). This configuration is assigned (R)- from the Latin *rectus* meaning 'right'. An anticlockwise path as shown in (b) would be like turning the wheel to turn left. This configuration is assigned (S-) from the Latin *sinister* meaning 'left'.

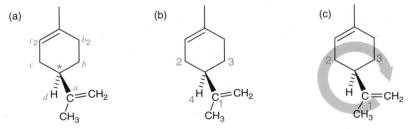

Fig. 12.18 Illustration of assigning the stereogenic centre in limonene. In (a), the four groups attached to the stereogenic centre (marked with an asterisk) are considered to establish the priority order. The priority order is shown in (b). Since the path from the highest priority to the second, to the third takes us in a clockwise direction, as shown in (c), the centre is assigned the prefix (R)-.

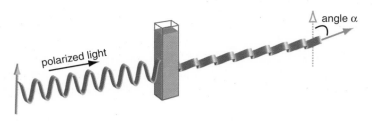

Fig. 12.19 Schematic diagram of a polarimeter which is used to measure the angle α through which the plane of plane-polarized light is rotated by an optically active compound. Plane-polarized light, generating using a polarizing filter, passes through a solution of the test compound and then the angle through which the plane of the polarized light has been rotated is measured using a second polarizing filter.

Since carbon a is part of a C=C double bond which counts as two bonds to carbon, the groups attached to carbon a, at two bonds from the stereogenic centre, are (C, C, C). Attached to b we have (C, H, H) and attached to c we also find (C, H, H). This means carbon a has the highest priority, but we now need to look at the atoms three bonds from the stereogenic centre in order to distinguish between carbons b and c.

Attached to carbon b_2 we have (C, H, H); attached to carbon c_2 we have (C, C, H) which has a higher priority. So now we have a priority order for all four of the groups: $a > c > b > d$. This order is marked on the groups in Fig. 12.18 (b).

We now need to ensure that the lowest priority group (the hydrogen) is pointing into the plane of the paper, which it is. The path from the highest priority group, to the second, to the third takes us in a clockwise direction, as shown in (c), which means the stereogenic centre is assigned the prefix (R)- i.e. the structure shown in Fig. 12.18 is (R)-limonene.

12.4.2 Comparing the properties of enantiomers

Most of the physical properties of two enantiomers are identical – they have the same melting and boiling points, density, refractive index, the same IR spectra and the same NMR spectra. However, if *plane-polarized* light is passed through a solution of a single enantiomer it is found that the plane of polarization of the light is rotated. Furthermore, the two enantiomers rotate the plane of polarization in *opposite* directions. This effect on plane polarized light is the origin of the description of enantiomers as being *optically active*, and of the alternative term *optical isomers* sometimes used for enantiomers.

The rotation of plane-polarized light is measured using an instrument known as a *polarimeter*, which is illustrated schematically in Fig. 12.19. Plane-polarized light, created by passing ordinary light through a polarizing filter, passes through a solution of the compound under test. The angle α through which the plane of the polarized light has been rotated is then measured using a second polarizing filter.

For a standard concentration and a fixed path length through which the light travels, it is found that the *magnitude* of the rotation depends on the particular compound. One enantiomer will rotate the plane of the light in a *clockwise* direction, and the other will rotate in an *anticlockwise* direction. It is not possible to tell from the structure of a compound whether it will cause

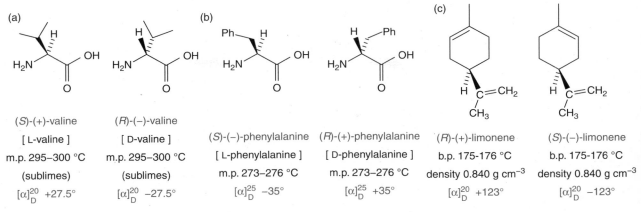

Fig. 12.20 For each pair of enantiomers in examples (a), (b), and (c), one isomer rotates plane-polarized light in a clockwise direction (a positive rotation), the other isomer in an anticlockwise direction (a negative rotation). The other physical properties of each pair of isomers are the same.

a clockwise or an anticlockwise rotation, and furthermore the direction of the rotation has no connection with whether the compound is (R)- or (S)-. All we know is that the two isomers will cause rotations in opposite directions.

The symbol used to specify the degree of optical activity is $[\alpha]_{589}^{20}$ where the superscript 20 indicates the temperature in °C at which the measurement was taken, and the subscript 589 is the wavelength in nm of the light used to make the measurement. Sometimes a subscript 'D' is used in place of the wavelength; this letter refers to the 'sodium D-lines' which are a particularly strong emission, also at a wavelength 589 nm, from a sodium discharge lamp. The concentration and solvent used should also be specified.

Some examples of optically active molecules, along with their $[\alpha]_{589}^{20}$ values, are shown in Fig. 12.20. The first two examples, valine and phenylalanine, are both amino acids. The naturally-occurring (S)-enantiomer of valine rotates plane-polarized light by 27.5°*clockwise*, whereas the (R)-enantiomer rotates the light 27.5°*anticlockwise*. For phenylalanine, the (S)-enantiomer rotates plane-polarized light by 35°*anticlockwise*, whereas the (R)-enantiomer rotates the light by 35°*clockwise*. Note, as mentioned above, that it is not possible to predict from the (R)- or (S)- notation which direction light will be rotated in.

Example (c) shows how plane-polarized light is affected by the two enantiomers of neat liquid limonene. Note for all these pairs of enantiomers, the other physical properties such as the melting point, boiling point and density are the same for each enantiomer.

Terminology

In naming enantiomers the (R)- and (S)- prefixes simply tell us about the absolute configuration of the isomers according to the CIP convention. Sometimes, as shown in the examples in Fig. 12.20, a (+) or (−) is included in the name. This indicates whether the isomer rotates plane-polarized light in a clockwise (+) or anticlockwise (−) direction. In earlier systems of nomenclature, a *d*- was sometimes used if the direction was clockwise and an *l*- if the direction was anticlockwise. In the same way that the (+) and (−) do not correlate directly with (R)- and (S)-, these lowercase *d*- and *l*- do not correlate with the historical uppercase D- and L- used to label the different isomers.

The chemical and biological properties of enantiomers

Whilst, with the exception of the effect on plane-polarized light, the physical properties of enantiomers are identical, this need not be the case with the chemical properties, and is certainly not the case with the biological properties of enantiomers.

In a *chiral environment* the properties of the two isomers can be very different. By 'chiral environment' we mean an environment in which the surrounding molecules are also of one particular chirality. For example, the enzymes in our bodies are all made up of single enantiomers of amino acids and are therefore themselves chiral. As a result, they interact very differently with the two enantiomers of a substance.

As a simple analogy, your feet are chiral objects – your left foot will not fit exactly into a mould of your right foot. Your shoes are an example of a chiral environment; if you put your right foot into your left shoe, this combination has completely different properties from having your left foot in the left shoe. Your socks, however, are not chiral – the left sock is exactly the same as the right sock and it does not matter which foot you put them on.

The naturally occurring amino acids make up all the proteins in our bodies. Their enantiomers have completely different properties in our bodies and cannot be processed by our enzymes in the same way. Some of the 'unnatural' amino acids are even poisonous to us!

Similarly, the two enantiomers of limonene shown in Fig. 12.20 on the preceding page have different properties in our bodies. The (*R*)-isomer occurs in lemons and oranges and is responsible for their characteristic citrus odour. On the other hand, the (*S*)-isomer has an odour reminiscent of pine or turpentine, and is often used to perfume cleaning products.

With regards to the chemical properties of enantiomers, if they are reacting with other chiral reagents, then the two enantiomers may well react in different manners. However, if the reagent they are reacting with is not chiral, there will be no difference in how they react.

12.4.3   Racemic mixtures – racemates

We have seen that if we have two pure enantiomers, then one will rotate plane-polarized light by a certain amount in a clockwise direction, the other by the same amount in an anticlockwise direction. It should not be a surprise to learn that if we have equal amounts of both isomers present, there is no net rotation of light. A mixture of equal amounts of the two enantiomers is known as a *racemic mixture*, or as a *racemate*.

Whenever two achiral compounds react together and form compounds which are chiral, the enantiomers are formed in equal proportions i.e. as a racemic mixture. For example, when sodium borohydride reduces a ketone such as butanone, it is equally likely to approach from either side of the carbonyl, as shown in Fig. 12.21 on the next page. When attacking the ketone as shown in (a), the (*R*)- enantiomer is formed, but when attacking from the opposite face as in (b), the (*S*)- enantiomer is formed. Since both approaches are equally likely, the two isomers are formed in equal amounts and the resulting racemic mixture does not rotate plane-polarized light.

We have seen other examples where a racemic mixture is formed, such as in the S_N1 mechanism where the planar intermediate is attacked from either side as in Fig. 9.88 on page 392.

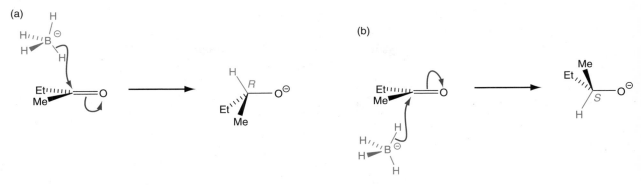

Fig. 12.21 When borohydride attacks the carbonyl of butanone as shown in (a), the (*R*)-enantiomer of the alcohol is formed. When the borohydride attacks from the opposite face, as shown in (b), the (*S*)- enantiomer is formed. Overall, equal amounts of the two isomers are formed, giving a racemic mixture.

12.4.4 Two or more stereogenic centres

When a substance has a number of different stereogenic centres, many isomers are possible; with two different centres, since each centre can be (*R*)- or (*S*)-, four different combinations are possible, as shown in Fig. 12.22 (a). With three different centres, eight combinations are possible, as shown in (b). We say that there are four *stereoisomers* for a compounds with two different stereogenic centres, and eight stereoisomers for a compound with three stereogenic centres. In general, there will be 2^n stereoisomers for a compound with *n* different stereogenic centres.

Each stereoisomer will have a non-superimposable mirror image, or enantiomer. Thus there will be $2^n/2$ or $2^{(n-1)}$ pairs of enantiomers. The pairs of enantiomers are connected by red lines in Fig. 12.22.

Any stereoisomer which is not an enantiomer is called a *diastereoisomer*, or *diastereomer* for short. Any pair within Fig. 12.22 (a) and within (b) that are not connected by a red line are diastereoisomers. Whereas enantiomers have the same properties except when in a chiral environment, diastereoisomers have completely different properties.

An example is given in Fig. 12.23 on the following page which shows the four stereoisomers of 3-amino-2-butanol. There are two stereogenic centres in the molecule, each of which can be (*R*)- or (*S*)-. The top two isomers, labelled (2*S*,3*S*)- and (2*R*,3*R*)-, are enantiomers and hence have the same melting points and rotate plane polarized light by equal amounts but in opposite directions.

(a)
1. A(*R*)–B(*R*) —— 3. A(*S*)–B(*S*)

2. A(*R*)–B(*S*) —— 4. A(*S*)–B(*R*)

(b)
1. A(*R*)–B(*R*)–C(*R*) ——	5. A(*S*)–B(*S*)–C(*S*)
2. A(*R*)–B(*R*)–C(*S*) ——	6. A(*S*)–B(*S*)–C(*R*)
3. A(*R*)–B(*S*)–C(*R*) ——	7. A(*S*)–B(*R*)–C(*S*)
4. A(*S*)–B(*R*)–C(*R*) ——	8. A(*R*)–B(*S*)–C(*S*)

Fig. 12.22 For a compound with two different stereogenic centres A and B, four stereoisomers are possible, as shown in (a). For a compound with three different stereogenic centres, eight stereoisomers are possible. Pairs of enantiomers are shown connected by a red line.

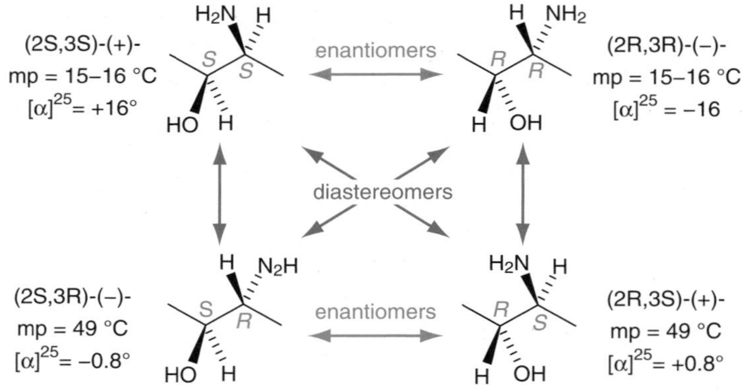

Fig. 12.23 The four stereo-isomers of 3-amino-2-butanol. There are two pairs of enan-tiomers. The isomers from the top row are diastereomers of the isomers on the bottom row, as indicated by the green arrows.

The bottom two isomers, labelled (2S,3R)- and (2R,3S)-, are also enantiomers and so also have the same melting points as each other and rotate plane polarized light by equal amounts but in opposite directions. The isomers from the top row are diastereomers of the isomers from the bottom row and so have different properties from each other.

Meso compounds

It is not necessary for all diastereoisomers to be optically active. Tartaric acid, for example, has just three stereoisomers, two of which are enantiomers, and one an achiral diastereomer. These stereoisomers are shown in Fig. 12.24. The natural form of tartaric acid rotates plane-polarized light in a clockwise direction, while its enantiomer rotates it in an anticlockwise direction. In contrast, the diastereomer, called mesotartaric acid, is superimposable on its mirror image and so is not optically active, and does not rotate plane polarized light at all.

The (2R,3S) isomer of tartaric acid is called a *meso* isomer. The word *meso* comes from the Greek and means 'middle'. This is because the *meso* form of a substance has some symmetry running through the middle of the molecule. As drawn in Fig. 12.24, the structure of mesotartaric acid has a *centre of inversion*. As we saw in section 3.1.3 on page 98, if something has a centre of inversion, it means that if we take any point and go from that point through to the centre of mass and continue in the same direction by the same distance, we come to an equivalent point. This is shown for mesotartaric acid in structure (a) in Fig. 12.25 on the facing page.

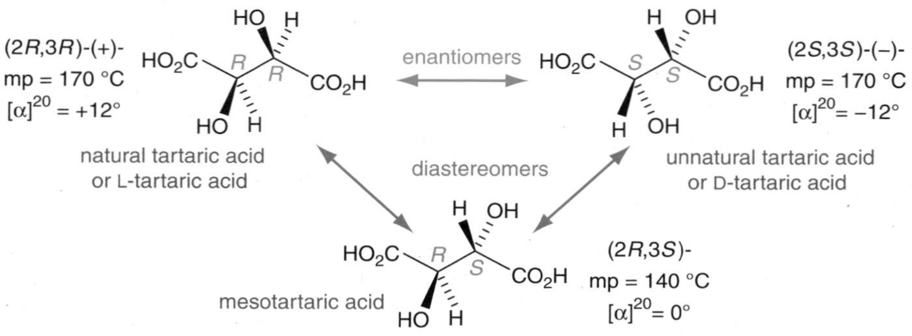

Fig. 12.24 The three stereo-isomers of tartaric acid. Two of the isomers are enantiomers, and the third is a diastereomer with different properties from the other isomers.

It is also possible to arrange the mesotartaric acid so that there is a mirror plane running through the centre of the molecule. By rotating about the central C–C bond, we arrive at the energy-maximum conformation shown in Fig. 12.25 (b). The mirror plane is shown in blue.

Whilst the achiral mesotartaric acid contains either a plane of symmetry, or a centre of inversion, the optically active forms of tartaric acid contain neither. We shall see in the next section that identifying what symmetry is present in a molecule can help us to easily identify whether the molecule is chiral or not.

12.5 Symmetry and chirality

The key point in assessing whether a molecule is chiral or achiral is whether or not it can be superimposed on its mirror image. If it can be superimposed, perhaps after first rotating about some single bonds, then the molecule is achiral. In the previous section, we noted that the achiral mesotartaric acid had either a plane of symmetry or a centre of inversion, depending on the conformation it was in. This is a general result which can be summarized as follows.

A molecule is *achiral* and *can* therefore be superimposed on its mirror image only if its structure has one of the following:

- a plane of symmetry;

- a centre of inversion;

- an improper rotation axis (also known as a rotation–reflection axis).

Molecules that possess only an improper rotation axis but no plane of symmetry or centre of inversion are extremely rare and so we will not consider this type of symmetry any further.

Figure 12.26 on the next page shows some examples of how symmetry can be used to decide whether or not a molecule is chiral. First consider the three isomers **A**, **B** and **C**. **A** has a mirror plane running through the oxygen and the mid point of the C–C bond, as shown by the blue line: the possession of this mirror plane means that the molecule is achiral. Molecules **B** and **C** (which are are diastereomers of **A**) possess neither a mirror plane nor a centre of inversion: they are therefore chiral, and in fact **B** and **C** are non-superimposable mirror images of one another i.e. they are enantiomers.

B and **C** do have rotational symmetry. In each case, rotating the molecule by 180° about the red dashed line brings the molecule into a position which is indistinguishable from the starting position. However, the possession of rotational symmetry does not preclude a molecule from being chiral: the criteria are, as listed above, that a chiral molecule must posses neither a mirror plane nor centre of inversion.

Molecule **D** has two planes of symmetry as indicated by the blue and green lines in the structure. This means it is achiral. There are no planes of symmetry in its diastereoisomer **E**, but there is a centre of inversion, as indicated by the orange dot. Since **E** has a centre of inversion, it too must be achiral.

12.5.1 More exotic chiral compounds

Most of the examples of chiral compounds we have seen so far have contained a chiral centre, or stereogenic centre. Often this might be a carbon atom with

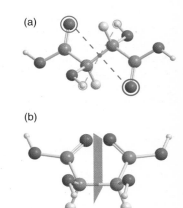

(a)

(b)

Fig. 12.25 The structure of mesotartaric acid in (a) is arranged to show the centre of inversion. Starting at any point, such as the atom circled in green, or the one circled in orange, and moving from that point through the centre of mass and out the other side in the same direction by the same amount, we find an equivalent point. In (b) the structure has been arranged to show the plane of symmetry.

🌐 *Weblink 12.3*

View and rotate about the central C–C bond the three-dimensional structure of mesotartaric acid.

🌐 *Weblink 12.4*

View and rotate in real time the molecules shown in Fig. 12.26.

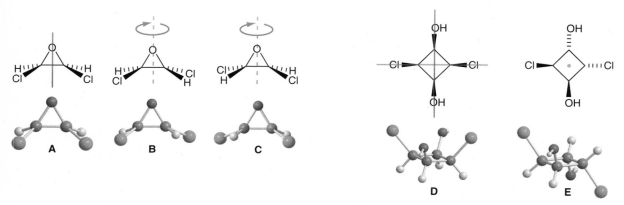

Fig. 12.26 Molecule **A** has a plane of symmetry running through the oxygen and the mid point of the C–C bond as indicated by the blue line: this means that the molecule is achiral. Rotating **B** and **C** by 180° about the red dashed line brings the molecules to an equivalent position i.e. they have rotational symmetry. However, as they have neither a mirror plane nor a centre of inversion, they are chiral. Molecule **D** has two mirror planes, as indicated by the green and blue lines, and so is achiral. Isomer **E** has no mirror planes, but does have a centre of inversion, as indicated by the orange dot. Since **E** has a centre of inversion, it is also achiral.

In allene (CH_2=C=CH_2) the central carbon can be thought of as being *sp* hybridized leaving two *2p* orbitals to form π bonds. These *2p* AOs are at right angles to one another, so the π systems are also oriented in the same way. As a result the CH_2 groups lie in perpendicular planes.

four different groups attached. However, we should point out that there are many examples of chiral compounds with no chiral centre. The key point in trying to decide whether a molecule is chiral or not is to see if it is possible to superimpose the molecule on its mirror image.

Some examples are given in Fig. 12.27 on the facing page. The structure shown in (a) is 1,3-dichloroallene in which the chlorine and hydrogen substituents at one end of the molecule are at right angles to those at the other end. As a result of this geometry, the molecule possesses neither a mirror plane nor a centre of inversion and is therefore chiral; the two isomers which are non-superimposable mirror images of one another are shown in (a)(ii). The molecule does have a rotational axis (by 180°), as shown in (a)(iii), but this has no consequences as to whether the molecule is be chiral or not.

The structure shown in (b) is a substituted biphenyl compound. The framework representation shown in (b)(i) is misleading in that it implies that the molecule is planar, whereas in fact as a result of the large bromine and chlorine groups the two rings are at right angles to one another. Rotation about the central C–C bond is not possible because the substituents are too large and cannot pass one another, as is clear from the space-filling structure shown in (b)(ii). Like allene, the molecule has neither a mirror plane nor a centre of inversion, making it chiral: the two enantiomers are shown in (b)(iii). However, the biphenyl does have a rotational axis, shown in (b)(iv), but this is not relevant in deciding whether or not the molecule is chiral.

Chiral centres with elements other than carbon

It is not necessary for a chiral centre to be a *carbon* with four groups attached. Optically active compounds have been synthesized where the centre is another atom, such as sulfur or phosphorus. Two examples are shown in Fig. 12.28 on the next page. In (a), three different groups are attached to a phosphorus atom. With the lone pair on phosphorus, we would predict the structure to be pyramidal, as shown. This structure has no symmetry and is chiral. A similar

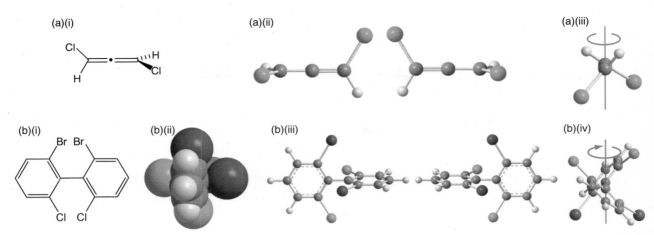

Fig. 12.27 The structure of 1,3-dichloroallene is shown in (a)(i). The molecule has no plane of symmetry and no centre of inversion so it is therefore chiral; the two isomers which are non-superimposable mirror images (enantiomers) are shown in (a)(ii). However, the molecule does possess a rotational axis, as shown in (a)(iii). Due to the effects of the bulky halogen substituents, the substituted biphenyl shown in (b)(i) is not planar but has the two rings at right angles to one another, as illustrated in the space filling representation shown in (b)(ii). Like allene, this molecule is chiral and the two enantiomers are shown in (b)(iii). The biphenyl possesses a rotational axis, as shown in (b)(iv).

case is found with sulfoxides, as shown in (b). With three different groups attached to the sulfur, and a lone pair, these pyramidal molecules also have no symmetry and so are chiral.

Whilst phosphines with three different groups attached can be isolated as enantiomers, this is not possible for amines with three different groups as shown in (c). This is because the amine undergoes a process called *inversion* in which the molecule essentially turns inside-out. During the process, the molecule passes through a planar structure as shown in (c). The nitrogen inversion is usually extremely rapid at room temperature and so even though the two pyramidal structures are chiral, they cannot be isolated in a pure form.

It is thought that the inversion of phosphorus and sulfur does not occur as easily as that for nitrogen since the 3*d* orbitals present in sulfur and phosphorus stabilizes the pyramidal forms more than the planar structures.

> ⊌ *Weblink 12.5*
> View and rotate in real time the molecules shown in Fig. 12.27.

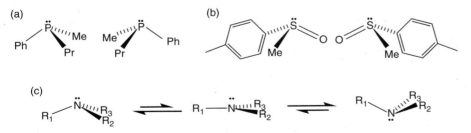

Fig. 12.28 The substituted phosphine in (a), and the sulfoxide in (b) are both chiral, i.e. have non-superimposable mirror images as shown. The amine in (c) rapidly equilibrates between the two pyramidal forms via the planar structure. This means it is not usually possible to isolate the separate enantiomers of an amine.

12.6 The conformation of cyclic molecules

We saw in section 12.2 on page 523 that open-chain molecules such as ethane and butane can adopt different conformations which are related by rotation about single bonds. The conformations which correspond to energy minima, called conformational isomers or conformers, are those which will be populated and are hence are the ones relevant for discussing the spectroscopic and chemical properties of such molecules. Cyclic molecules behave in a similar way and are expected to show conformational flexibility just like their open-chain counterparts.

Rings, especially six-membered rings, are a common feature of many naturally occurring organic compounds such as steroids and sugars. The importance of such compounds has led chemists to make a detailed study of the conformations adopted by six-membered rings, and it is these which we will consider in most detail.

If a ring is formed from three atoms, then there is no choice but for these three to lie in a plane, as is the case for the three-membered *epoxide* ring of structures **A**, **B**, and **C** from Fig. 12.26 on page 540. In this figure, structures **D** and **E** include four-membered rings which we assumed were planar, but in fact it turns out that this is not quite so. Larger rings, with five or more members, are most definitely not planar. We start our discussion by explaining why this is so.

12.6.1 Ring strain

Let us imagine forming various sized rings by joining together CH_2 groups, assuming to start with that the carbon atoms all lie in a plane. It is immediately obvious that there is a problem here since the normal bond angle between two groups attached to a CH_2 is 109° which, in all but one case, does not match the angles within a ring. To form the rings it is therefore going to be necessary to alter the bond angles at the carbons away from their ideal values or make some other accommodation.

Figure 12.29 on the facing page illustrates the way in which nominally tetrahedral carbons can be used construct rings of various sizes. Shown in pink are a series of regular polygons with between three and eight vertices at each of which is placed a tetrahedrally coordinated carbon.

To form a three-membered ring, shown in (a), the bonds of the tetrahedral units must be bent inwards by a considerable degree, as shown by the green arrows, in order to be at the required 60°. This creates considerable strain in the molecule. The bonds in the four-membered ring, shown in (b), do not need to be bent in by so much which means there is less strain in the molecule.

Since the internal angles of a pentagon are 108°, assembling a five-membered ring from tetrahedral units creates a planar ring, essentially free of ring strain, as shown in (c). If we attempted to make larger rings from tetrahedral units, we would find that they would not be planar. Forcing the units into a plane would increase their ring strain as the bond angles would need to increase away from 109° as shown in (d), (e), and (f) for six-, seven, and eight-membered rings. This strain can be avoided by allowing the atoms to move out of the plane.

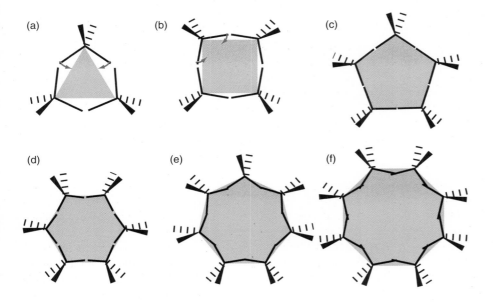

Fig. 12.29 The pink shapes show a series of regular polygons in which a carbon with the usual tetrahedral geometry has been placed at each vertex. For the three-membered ring shown in (a) the C–C–C bond angle at carbon has to be reduced from its optimum value by a considerable amount in order to form the ring. The angles do not have to be reduced by so much for the four-membered ring shown in (b), and do not have to change at all for the five-membered ring in (c). For the larger rings shown in (d), (e), and (f), the bond angle would have to increase to form a planar structure.

Using enthalpies of combustion to measure strain

It is possible to gain an experimental measure of the amount of strain in the different cycloalkanes by looking at the enthalpy of combustion per methylene unit, i.e. per –CH_2– group. The standard enthalpies of combustion of the straight-chain alkanes from hexane, C_6H_{14}, to decane, $C_{10}H_{22}$, are shown in Table 12.1. Across this series, the increase from one alkane to the next homologue is pretty constant at around 659 kJ mol^{-1}. Since there is no strain associated with the straight-chain alkanes, we can conclude that each methylene unit contributes about −659 kJ mol^{-1} to the standard enthalpy of combustion.

Figure 12.30 on the following page shows the enthalpy of combustion per methylene unit for the cycloalkanes from $(CH_2)_3$ to $(CH_2)_{11}$. The red dashed line shows the value per –CH_2– unit in a straight-chain alkane. Where the value per –CH_2– unit in the cycloalkane exceeds this value, it indicates there must be strain in the ring because more energy than expected is released when the ring is broken up.

As expected, the three and four-membered rings cyclopropane and cyclobutane have large amounts of strain. Large rings with ten or more atoms in the ring have no more strain than the straight-chain alkanes. There is some strain in the medium-sized rings with seven, eight, and nine atoms in the ring. This is due to interactions between the atoms from opposite sides of the ring and is known as 'transannular ring strain'.

Table 12.1 Enthalpies of combustion of some straight-chain alkanes

alkane:	hexane		heptane		octane		nonane		decane
$-\Delta_r H°$/ kJ mol^{-1}	4195		4854		5512		6172		6830
difference		659		658		660		658	

Fig. 12.30 The black crosses indicate minus the standard enthalpy of combustion per –CH$_2$– unit in the cycloalkanes from (CH$_2$)$_3$ to (CH$_2$)$_{11}$. The red dashed line indicates the value per –CH$_2$– unit in the straight-chain alkanes. If the value in the cycloalkane exceeds this value, it indicates that there is some strain in the ring. As expected, there is considerable strain in the three and four-membered rings. What is surprising is that there is some strain in the five-membered ring cyclopentane, but none in the six-membered ring cyclohexane.

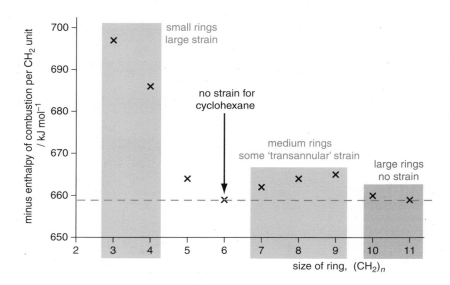

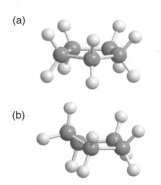

Fig. 12.31 In a planar conformation of cyclopentane, shown in (a), the C–H bonds eclipse the C–H bonds on the neighbouring carbon atoms. These eclipsing interactions can be relieved if the ring puckers slightly, as shown in (b).

Perhaps the real surprise from the graph is that it is the six-membered ring cyclohexane that is the smallest strain-free cyclic hydrocarbon, not the five-membered ring cyclopentane. Whilst *planar* cyclopentane would have very little *angle* strain, all the C–H bonds eclipse the C–H bonds from their neighbouring carbons, as shown in Fig. 12.31 (a). As we have seen for ethane (section 12.2.1 on page 524) these eclipsing interactions are unfavourable. In cyclopentane the eclipsing interactions can be relieved if the ring puckers out-of-plane slightly as shown in (b). This puckering may well increase the angle strain slightly, but the reduction in the eclipsing interactions more than compensates for this. It has been calculated that the *total* strain (angular plus torsional) in the non-planar form is only about 60% of that of the planar version.

Cyclopropane and cyclobutane

In cyclopropane, shown in Fig. 12.32 (a), all of the C–H bonds must eclipse the C–H bonds on the neighbouring carbon atoms and this contributes to the total strain in the molecule. In planar cyclobutane all of the C–H bonds would also be eclipsed. However, the molecule can reduce these eclipsing interactions at the expense of introducing more angle strain by bending slightly, as shown in (b) and (c). This conformation of cyclobutane is called a 'wing' or 'butterfly'

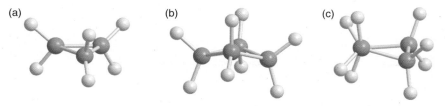

Fig. 12.32 In cyclopropane, shown in (a), the C–H bonds from one carbon eclipse those from the neighbouring carbons. Cyclobutane adopts a 'wing' shape, as shown in (b), in order to minimize eclipsing interactions. Looking down a C–C bond, as in (c), reveals that the C–H bonds are partially staggered.

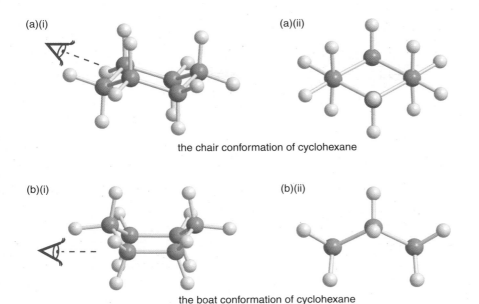

(a)(i)

(a)(ii)

the chair conformation of cyclohexane

(b)(i)

(b)(ii)

the boat conformation of cyclohexane

Fig. 12.33 Two views of the chair conformation of cyclohexane are shown in (a), and two views of the boat conformation are shown in (b). Views (a)(ii) and (b)(ii) are what is seen looking at the chair and boat forms from the position shown by the eye in (a)(i) and (b)(i). In view (a)(ii) of the chair form we see that all the bonds are staggered relative to one another, whereas in view (b)(ii) of the boat form, we see that many bonds eclipse one another.

conformation. The cyclobutane ring can readily flip up and down with a barrier of just 6 kJ mol^{-1} between these two conformations.

12.6.2 The conformations of cyclohexane

The chair and boat forms

Figure 12.30 on the facing page shows that cyclohexane has no ring strain. This is accomplished by having the ring buckle out of the plane so that the tetrahedral angles of 109° are preserved. Such ideal angles are achieved by the two conformations of cyclohexane, the *chair* and *boat* forms.. These structures are shown in Fig. 12.33; two views of the chair conformation are shown in (a) and two views of the boat conformation are shown in (b). Only the chair form is a stable conformer, i.e. a structure corresponding to an energy minimum. It is, in fact, the *lowest* energy conformer. In contrast, the boat form corresponds to an energy *maximum*, and so only has a fleeting existence as cyclohexane interconverts from one form to another.

Why the chair form is an energy minimum but the boat form is an energy maximum may be understood by looking along the line joining two carbons on opposite sides of the ring i.e. from the position of the blue eye in Fig. 12.33 (a)(i) and (b)(i). The structures in (a)(ii) and (b)(ii) show what is seen from this view point. In the chair conformer we see that all the C–H bonds are staggered, whereas in the boat conformation, many are eclipsed and so only the front half of the structure may be seen.

The twist-boat conformation

The eclipsing interactions in the boat conformation may be reduced by twisting the sides of the boat slightly by pushing in the direction of the red arrows shown in (a)(i) from Fig. 12.34 on the following page. The new conformation that results from this movement, known as the *twist-boat*, is an energy-minimum

Weblink 12.6

View and rotate in real time the three-dimensional structures of three-, four- and five-membered rings, and the various conformations of a six-membered ring.

Fig. 12.34 Three views of the boat conformation of cyclo-hexane are shown in (a), and the corresponding views of the twist-boat conformer are shown in (b). The twist-boat conformer may be obtained from the boat form by pushing the carbon atoms in the direction of the red arrows. View (iii) shows that the eclipsing interactions present in the boat form have become staggered in the twist-boat.

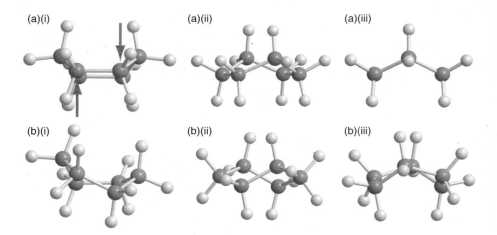

structure but it is not as low in energy as the chair conformer. Three views of the boat conformer are shown in Fig. 12.34 (a), and the corresponding views of the twist-boat are shown in (b). In particular, the end-on view in (a)(iii) shows many eclipsing interactions in the boat conformer which have become staggered in the twist-boat, (b)(iii).

An energy profile for the chair–boat interconversions

Cyclohexane can readily interconvert between the chair and boat conforma-tions. An energy level diagram for this process is shown in Fig. 12.35. The energy barriers indicated are approximate and are measured relative to the energy of the chair conformation. The structures in black only show the carbon atoms – the hydrogens have omitted for clarity.

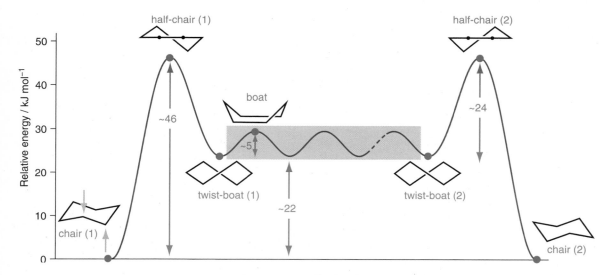

Fig. 12.35 An energy profile for the interconversion between the different conformations of cyclohexane. The energy is plotted relative to the lowest-energy conformation, the chair conformer.

Fig. 12.36 Each carbon atom in the ring has a substituent in an 'upper position', above the approximate plane of the C–C bonds, and a substituent in a 'lower position', below the plane. These are marked with a U and an L. In (a), the blue atoms are in axial positions and alternate in upper and lower positions as we move around the carbon ring. During a ring inversion, if the end carbon atoms are moved in the directions shown by the red arrows, the new chair conformer shown in (c) is formed. In this all the yellow atoms have now become axial, whilst all the blue atoms have become equatorial.

Starting from the chair structure on the left, pushing the two carbon atoms in the directions indicated by the orange arrows initially gives an energy-maximum conformation in which four carbon atoms are more-or-less in the same plane. This structure is called a *half-chair* conformation. Continuing to push the same carbon atoms in the same direction gives rise to a twist-boat structure. The energy barrier that must be overcome in order to convert from the chair conformer to the twist-boat is approximately 46 kJ mol^{-1}.

The region shaded purple in Fig. 12.35 shows the conversion of the cyclohexane ring from one twist-boat structure to another via the energy-maximum boat conformation. The barrier here is only around 5 kJ mol^{-1}, which means this is a fast process. The two twist-boat conformers labelled twist-boat (1) and (2) are actually non-superimposable mirror images of one another. However, it is not possible to isolate the two forms since they interconvert between one another so easily.

The twist-boat can return to the minimum-energy chair conformation via another half-chair, as shown in the right of the scheme. The whole process in which cyclohexane changes from one chair conformation to another via a boat or twist-boat, is known as a *ring-inversion*, or *ring-flip*.

Axial and equatorial substituents in cyclohexane

The chair conformation of cyclohexane has two different environments of hydrogen atoms called *axial* and *equatorial* hydrogens. In Fig. 12.36 (a), the axial hydrogens are shown in blue, and the equatorial hydrogens in yellow. Each carbon atom has one axial hydrogen and one equatorial hydrogen. If we treat the carbon ring as being planar, as suggested by the simple hexagon structure shown in Fig. 12.37, then each carbon has one hydrogen above the carbon plane and one below.

When drawing the three-dimensional conformation of a cyclohexane ring, it is often useful to distinguish those substituents in the 'upper' positions of each carbon i.e. above the carbon plane, and those substituents in the 'lower' positions i.e. below the carbon plane. Moving from one carbon to the next around the ring, the substituents in the 'upper' positions alternate between being axial and equatorial, as do the substituents in the lower positions. This can be seen in Fig. 12.36 (a). The substituents in the upper positions are labelled with a 'U' and those in the lower positions with an 'L'. The substituents in the

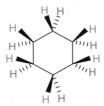

Fig. 12.37 If we treat all the C–C bonds as being in the plane of the paper, then all the hydrogens coloured red lie above this plane, and all the hydrogens coloured green lie below the plane.

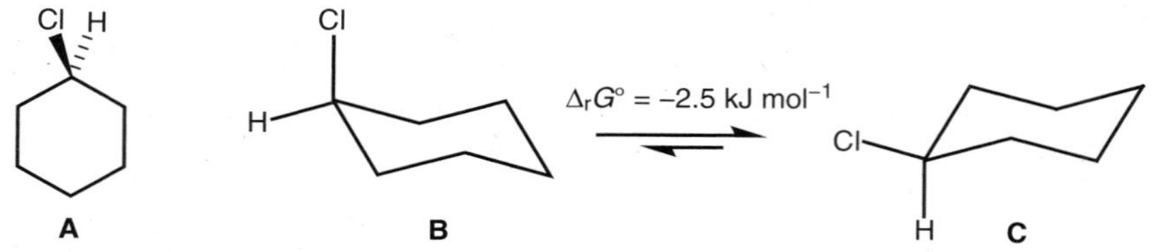

Fig. 12.38 Structure **A** shows the configuration of chlorocyclohexane but it tells us nothing about its conformation. Chlorocyclohexane exists as an equilibrium between the conformer with the chlorine axial, **B**, and the conformation with it equatorial, **C**. The equatorial conformer is 2.5 kJ mol^{-1} lower in Gibbs energy than the axial conformer.

upper positions alternate in being axial (shown in blue) and equatorial (shown in yellow).

During a ring inversion, the environments are interchanged. The substituents in the upper positions on each carbon remain in the upper positions, and those in the lower positions remain in the lower positions but the substituents that were axial become equatorial and those that were equatorial become axial. This is shown in Fig. 12.36 where the chair conformer in (a) flips via a boat conformation (b) to the chair conformation shown in (c). In (a), the atoms shown in blue are all axial, but they all end up equatorial in (c) after the ring-flip. Conversely, the atoms coloured yellow in (a) are all equatorial, but end up axial in (c) after the ring-flip.

For cyclohexane, the ring-inversion occurs so rapidly at room temperature that we only see *one* signal in the ^{1}H NMR spectrum. However, at low temperatures where the ring-inversion is much slower, it might be possible to detect separate signals from the hydrogens in the axial and equatorial environments.

Before we look at what happens when one or more substituents other than hydrogen are included on the ring, it would be worthwhile studying Box 12.4 on the facing page which shows some guidelines to drawing the chair conformation of cyclohexane in a realistic manner.

12.6.3 Substituted cyclohexanes

> The relationship between the standard Gibbs energy change ($\Delta_r G^\circ$) and the equilibrium constant was explored in section 6.9 on page 227.

When one of the hydrogen atoms in cyclohexane is replaced by a different substituent, such as a chlorine atom, two chair forms are possible – one with the substituent equatorial, the other with it axial. These are shown in Fig. 12.38. The framework structure **A** shows the constitution of chlorocyclohexane (i.e. which atoms are joined to which), but tells us nothing about the conformation of the molecule. At room temperature, chlorocyclohexane is in a dynamic equilibrium between the axial conformer, **B**, and the equatorial conformer **C**. For chlorocyclohexane, the standard Gibbs energy change on going from **B** to **C** is about -2.5 kJ mol^{-1}, implying that the equilibrium mixture favours the latter i.e. it is more favourable for the substituent to be equatorial rather than axial.

There are two reasons for the difference in energy between the two conformers. The first is because when the group is axial there are two *gauche* interactions with the C–C bonds in the ring; when it is equatorial, the group is *anti* to the C–C bonds. This is shown in Fig. 12.39 on page 550 where the conformer with the chlorine equatorial is shown in (a), and the conformer with the chlorine axial is shown in (b); Newman projections, seen from the position

---

Box 12.4 Drawing the chair form of cyclohexane

The following are some guidelines for drawing the chair conformation of cyclohexane:

(a) Draw the two C–C bonds connected to the left-most carbon atom on the ring. This should look like a 'V' rotated through 45°.

(b) Draw the middle two C–C bonds, parallel to each other and sloping downwards from left to right.

(c) Draw the two C–C bonds connected to the right-most carbon. These bonds should be parallel to the first two C–C bonds drawn. It should be possible to draw a horizontal line through two pairs of carbons as shown in 3b.

(d) Add the hydrogens on the end carbons first, making sure the angles look tetrahedral. The two axial bonds should be vertical and the two equatorial bonds should be parallel to the C–C bonds indicated.

(e) Add the rest of the axial bonds making sure they alternate up and down.

(f) Add the bonds to the remaining hydrogens. These should form two 'W' shapes for the part of the C–C ring structure highlighted in orange below.

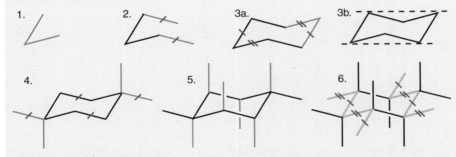

Pairs of parallel bonds are indicated with the blue dashed or double-dashed lines. Note that no wedged or hashed bonds are used to signify bonds coming out of or going into the plane of the paper.

---

of the eye in (a)(i) and (b)(i), are shown in (a)(ii) and (b)(ii). When the chlorine is equatorial, the C–Cl bond is *anti*, i.e. at 180° to two of the C–C bonds in the ring. These bonds are shown highlighted in red.

In contrast, when the chlorine is axial, the C–Cl bond is *gauche*, i.e. at 60°, to two of the C–C bonds. As we have seen in the conformation of butane in section 12.2.1 on page 524, having two groups *anti* to one another is lower in energy than having them *gauche*.

The second reason why the equatorial position is favoured is due to the steric interactions between the axial group and the other axial hydrogens (or other substituents) on the same face of the ring. This is illustrated in Fig. 12.40 on the next page. The unfavourable interactions are between the axial groups shown in (a). In (b) the approximate sizes of the axial chlorine and hydrogens are shown by the dotted surfaces. Some overlap can be seen between the atoms.

Fig. 12.39 The equatorial conformer of chlorocyclohexane is shown in (a) and the axial conformer shown in (b). Views (a)(ii) and (b)(ii) show Newman projections of the structures when viewed from the position shown by the eye. When the chlorine is equatorial, the C–Cl bond is *anti* to two of the C–C bonds from the ring (highlighted in red). When the chlorine is axial, the C–Cl bond is *gauche* to two of the C–C bonds.

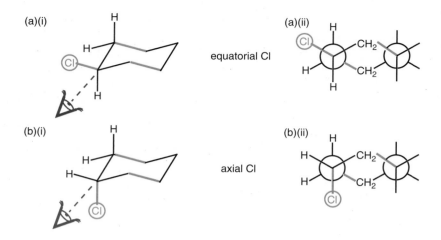

equatorial Cl

axial Cl

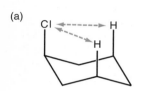

Fig. 12.40 The red arrows in (a) show the interactions between the chlorine and the axial hydrogens. In (b) the sizes of the chlorine and the two axial hydrogen atoms are shown by the dotted surfaces. These surfaces just overlap which indicates that there is repulsion between the atoms.

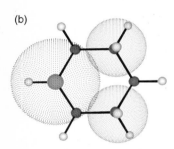

The equilibrium between axial and equatorial substituents

Given that the conversion of the axial to the equatorial conformer of chlorocyclohexane has a $\Delta_r G^\circ$ of -2.5 kJ mol^{-1}, we can use $\Delta_r G^\circ = -RT \ln K$ to compute that the equilibrium constant is 2.74 i.e. [equatorial]/[axial] = 2.74. This corresponds to an equilibrium mix of 73% equatorial chlorocyclohexane and 27% axial.

Since the NMR spectra for the two conformers are different, we might expect to see a total of eight lines in the ^{13}C NMR spectrum: four from the axial conformer and four from the equatorial conformer. However, only *four* lines are seen in the room temperature spectrum, shown in (a) in Fig. 12.41 on the facing page.

The reason that only four lines are seen is that the two conformers are interconverting so rapidly that we see a weighted average of the spectrum from the axial and equatorial isomers. Lowering the temperature slows down the ring flip, and if the temperature is low enough separate signals can then be seen from the two conformers. Under these conditions, we see eight lines in the spectrum, as shown in Fig. 12.41 (b).

The lines from the axial isomer are labelled 1a–4a according to the corresponding carbon atoms, and those from the equatorial isomer are labelled 1b–4b. Since at equilibrium there is less of the axial isomer present than of the equatorial isomer, the peaks due the former are weaker e.g. the peak due to carbon 1a (at 57.1 ppm) is weaker than that due to carbon 1b (at 53.9 ppm).

At room temperature we see a single peak for carbons 1a and 1b, but the shift of this peak is not mid-way between that of 1a and 1b. Rather it is weighted in favour of 1b on account of the fact that there is more of the equatorial isomer present. In fact, in this case the shift of the averaged peak turns out to be little different to that of the dominant peak from the equatorial isomer.

The ^{1}H spectrum of bromocyclohexane is very complex due to the large number of couplings present. However, it is nevertheless useful to think about the number of different environments there are for hydrogen. As is shown in (a) in Fig. 12.42 on the next page, the axial isomer has seven hydrogen environments (labelled 1a–7a), as does the the equatorial isomer (labelled 1b–7b). The rapid interconversion of these two isomers means that at room temperature we only expect there to be seven environments, illustrated on the framework diagram shown in (b).

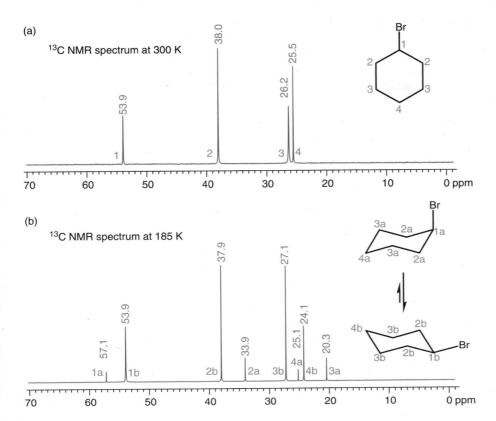

Fig. 12.41 In the room-temperature ^{13}C NMR spectrum of bromocyclohexane, shown in (a), only four peaks are seen, corresponding to the four different environments of carbon. The red numbers show the assignment of each peak to the structure of bromocyclohexane. Only four peaks are seen because the ring is undergoing ring-inversion so rapidly. In the spectrum recorded at 186 K, eight peaks are seen. These correspond to the four different environments of carbon in the conformer with the bromine equatorial, and the four environments in the conformer with the bromine axial. The signals due to the conformer with the bromine axial are smaller since there is less of this conformer at equilibrium.

It is important to realize that the chemical shift of H_2 is the weighted average of the shifts of H_{2a} and H_{2b}, since H_2 can be in either of these positions. It is a common misconception to think that the shift of H_2 is the average of the shifts of H_{2a} and H_{3a}.

The preference of different groups for the equatorial position

The preference of a group for an equatorial position in cyclohexane rather than an axial position is known as the *A-value* for the group. It is defined as minus the value of $\Delta_r G^\circ$ for the equilibrium where an axial substituent becomes equatorial during the ring inversion of cyclohexane. Since this process is invariably favourable, $\Delta_r G^\circ$ is negative and hence the *A*-values are positive.

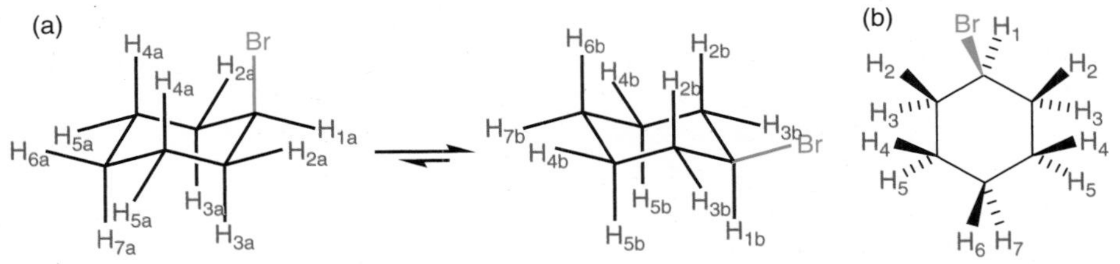

Fig. 12.42 In the axial conformer of bromocyclohexane there are seven distinct environments for hydrogen, labelled 1a–7a, and similarly there are seven (1b–7b) in the equatorial conformer, as shown in (a). At room temperature these conformers interconvert rapidly so only seven averaged environments are seen. These correspond to the seven distinct protons indicated on the framework structure shown in (b).

Table 12.2 The *A*-values for different substituents in cyclohexane

group	*A*-value / kJ mol^{-1}	K_{eq} ([eq]/[ax])	% equatorial
F	1.4	1.8	64
Cl	2.4	2.6	72
OCOCH$_3$	3.2	3.6	78
CO$_2$CH$_3$	5.2	8.2	89
CH$_3$	7.3	19.0	95
Ph	12	127	99.2
t-Bu	20	3205	99.97

Table 12.2 lists the *A*-values for some common groups. Note the particularly large *A*-value of the *t*-butyl group means that *t*-butylcyclohexane essentially exists only with the *t*-butyl group equatorial. We say that this group *locks* the ring and prevents ring flipping.

Substituted cyclohexane rings with more than one substituent

When there is more than one substituent on the ring, it is best to consider *both* chair conformations and see which groups are axial and which are equatorial in each structure. Generally, the conformer with the greater number of groups in the equatorial position will be favoured. However, if a *t*-butyl group is present, this will usually lock the conformation of the cyclohexane ring with the *t*-butyl group equatorial and, if necessary, force other groups to be axial. Two examples are shown in Fig. 12.43.

Two different molecules are shown in (a) and (b); their framework structures are shown in (a)(i) and (b)(i), and the two chair conformations of each are

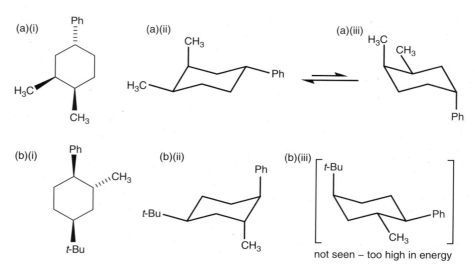

Fig. 12.43 Examples of the conformational preferences of cyclohexanes with more than one substituent. For compound (a) the conformer shown in (a)(ii) is preferred over (a)(iii) as the former has the least number of substituents in axial positions. In (b) one of the substituents is the large *t*-butyl group. This has such a energetic preference for an equatorial position that it essentially locks the ring into conformer (b)(ii), despite the fact that this involves both the phenyl and methyl substituents being axial.

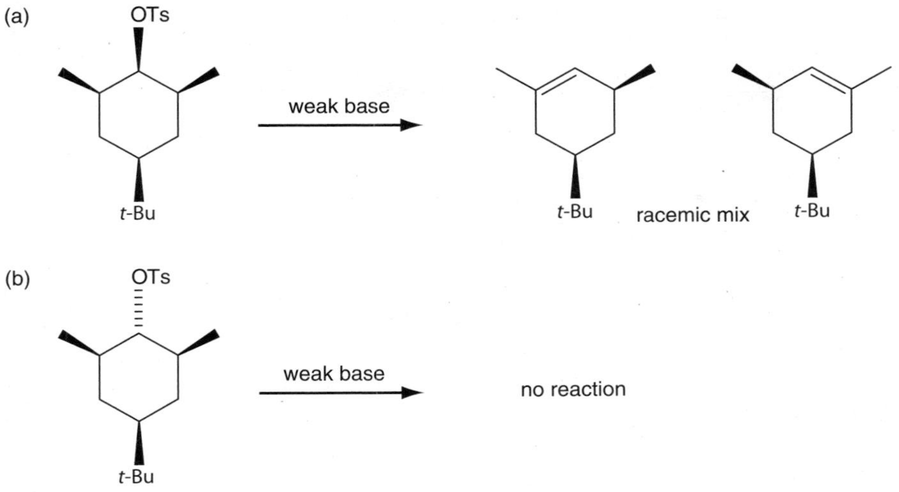

(a)

OTs

t-Bu   t-Bu   racemic mix   t-Bu

weak base →

(b)

OTs

t-Bu

weak base →   no reaction

Fig. 12.44 Illustration of the dramatically different reactivity of two diastereomers (a) and (b). Molecule (a) reacts readily with weak base to give a racemic mixture of the alkenes shown. Molecule (b) has no reaction with weak base.

shown on the right. Conformer (a)(ii) has one axial group and two equatorial groups, and on inversion this gives conformer (a)(iii) which has two axial groups and one equatorial group. Having two axial groups is considerably higher in energy than having just one, so the favourable conformation is that shown in (a)(ii).

In example (b), the framework structure shows a *t*-butyl group in an upper position. Since this group is so large, there is a strong preference for it to be in an equatorial position. Conformer (b)(ii) has the *t*-butyl group equatorial but this means the two other groups – the phenyl and the methyl – must both be in axial positions. The alternative conformer with the *t*-butyl group axial and the phenyl and methyl groups equatorial is higher in energy and is not present to an appreciable extent.

12.7   Moving on

If you look back over this chapter you will see that it is entirely about how you describe the three-dimensional shape of a molecule. The reason why we have devoted so much time to this is that the shape of a molecule has a very strong influence on its reactivity. The example shown in Fig. 12.44 makes the point. Shown in (a) are (b) are two diastereomers of a tosylate which only differ in the orientation of the OTs substituent. Whereas (a) reacts quickly with a weak base to give a racemic mixture of the two alkenes, (b) does not react at all. Why there is such a dramatic difference in the reactivity, and why a racemic mixture is produced in (a) is one of the matters that will be discussed in the following chapter.

FURTHER READING

Basic Organic Stereochemistry E. L. Eliel, S. H. Wilen, and M. P. Doyle, Wiley (2001)
Stereochemistry of Organic Compounds E. L. Eliel and S. H. Wilen, Wiley (1994)

QUESTIONS

12.1 Describe the relationship between the following pairs of structures (e.g. whether they are enantiomers, diastereomers etc.)

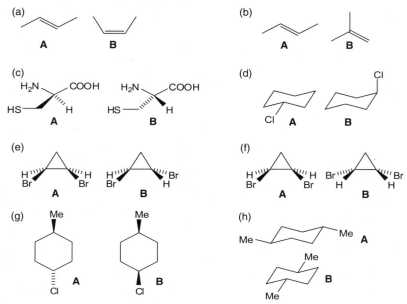

12.2 Redraw each of the following structures in the ways described in Box 12.2 on page 525, indicating the direction in which your views are taken. Describe the relationship between the substituents (other than hydrogen) on the two carbons.

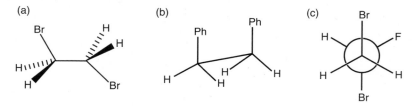

12.3 (a) Sketch (i) a graph of the energy of $BrCH_2CH_2Br$ as a function of the dihedral angle between the two C–Br bonds, and (ii) a similar graph for $BrCH_2CH_3$ as a function of the dihedral angle between the C–Br bond and one of the C–H bonds in the methyl group. Comment on the form of the two graphs.

(b) According to the Boltzmann distribution, if conformation A is higher in energy than conformation B by an amount ΔE the ratio r of the populations of the conformations is given by

$$r = \frac{\text{population of A}}{\text{population of B}}$$
$$= \exp\left(\frac{-\Delta E}{RT}\right),$$

where R is the gas constant (8.3145 J K^{-1} mol^{-1}) and T is the absolute temperature. Use this relationship to calculate the ratio of the populations of the *gauche* and *anti* conformers of butane, the energy profile for which is shown in Fig. 12.9 on page 527, at 298 K and at 398 K.

(c)   If the number of molecules in conformer A is N_A and the number in conformer B is N_B, then it follows that $r = N_A/N_B$. The *fraction* of molecules which are in conformer A is given by $N_A/(N_A + N_B)$. Show that

$$\frac{N_A}{N_A + N_B} = \frac{r}{1 + r}.$$

Hence work out the percentage of butane in the *gauche* and *anti* conformers at 298 K and at 398 K.

2.4   Using the CIP convention (Box 12.3 on page 531) assign *E/Z* labels to the double bonds in the following molecules.

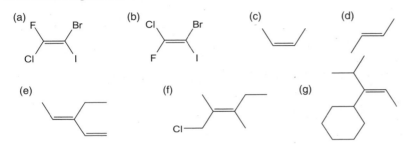

2.5   Using the CIP convention, assign *R/S* labels to the chiral centre or centres in the following molecules.

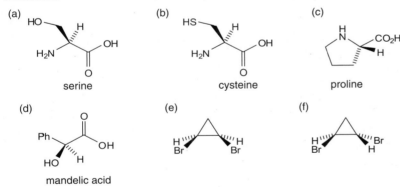

2.6   For each of the following compounds draw all possible stereoisomers, indicating which are enantiomers and which are diastereomers (use wedged and dashed bonds to differentiate the individual isomers). For (a)–(d) assign *R/S* labels to the stereogenic centres. For which of these compounds does a *meso* isomer exist?

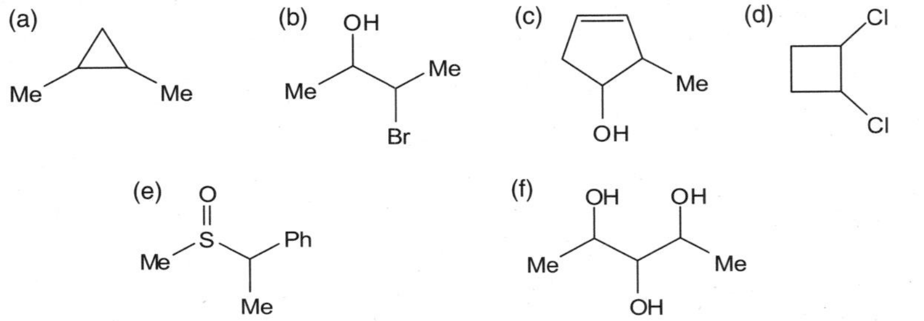

(a)   (b)   OH   Me   Br   (c)   Me   OH   (d)   Cl   Cl

Me   Me   Me

(e)   O   Me   S   Ph   Me   (f)   OH   OH   Me   Me   OH

12.7 Inositol, (a), exists as nine stereoisomers, some of which are shown below.

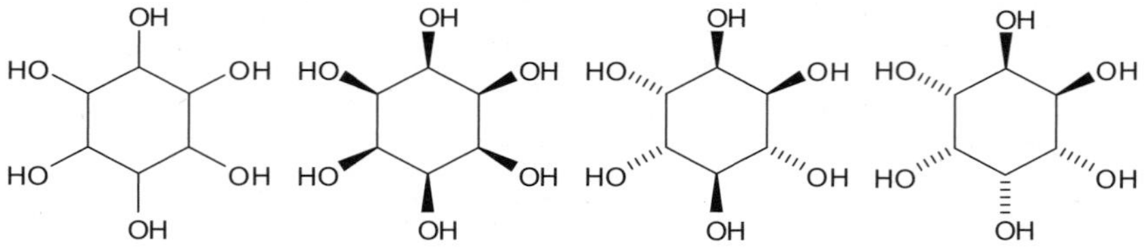

(a)   (b)   (c)   (d)

(a) Using symmetry arguments, explain why isomer (b) is achiral, and why isomer (c) is chiral.

(b) Draw out the structures of all nine stereoisomers (using wedged and dashed bonds), identifying which are chiral.

(c) Draw isomer (d) with the ring in the chair conformation, being careful to place the OH groups in the correct 'up' and 'down' positions. Explain why the structure you have drawn is chiral, and consider what effect a ring flip would have on this structure.

(d) Using your answer to (d), explain why it has not been possible to isolate the enantiomers of isomer (d).

12.8 Explain why the presence of a bulky substituent, such as *t*-Bu, results in a cyclohexane ring being 'locked' in a particular conformation. Draw the lowest energy conformations of (a) and (b), making it clear whether the Cl is axial or equatorial.

12.9 For each of the molecules shown below draw out the two possible chair conformations of the cyclohexane ring, taking care to place the substituents in the correct 'up' or 'down' positions. In each case explain which of the two conformations you would expect to be lower in energy.

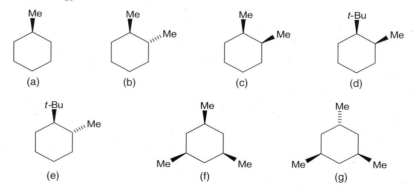

12.10 In the low-temperature spectrum of bromocyclohexane shown in Fig. 12.41 on page 551 the integral of the peak at 20.3 ppm was found to be 0.215 (arbitrary units) and that of the peak at 27.1 ppm was found to be 0.998 (same arbitrary units). Assuming that the integral is proportional to concentration, find the equilibrium constant for the axial ⇌ equatorial equilibrium. Hence find $\Delta_r G°$ for the process and the *A*-value. Compare your value with those in Table 12.2 on page 552. [Be sure to use the correct temperature.]

Organic chemistry 3: reactions of π systems

Key points

- C=C double bonds can be formed in elimination reactions for which there are two principal mechanisms, E1 and E2.

- The E1 mechanism involves loss of an acidic proton from a carbenium ion. The σ conjugation between C–H σ bonding MOs and a vacant orbital in the ion is responsible for increasing the acidity of the protons.

- The E2 mechanism is stereospecific on account of the requirement that the proton and the leaving group be arranged trans co-planar.

- Increasing the number of alkyl substituents on a double bond stabilizes the molecule by σ conjugation.

- Alkenes react by electrophilic addition across the double bond.

- Alkyl substituents increase the reactivity of an alkene.

- For an unsymmetrically substituted C=C bond, the outcome of an addition depends on the electrophile and the details of the mechanism.

- Aldehydes and ketones form enols under acidic conditions and enolate anions under basic conditions. The amount of the enol form increases if the carbonyl is conjugated.

- The conjugation of the oxygen lone pair makes enols and enolates particularly reactive towards electrophiles.

- Aromatic molecules have a continuous, conjugated π system in a ring and contain $4n+2$ electrons π electrons.

- Aromatic molecules usually undergo reactions in which the aromatic system is preserved, typically electrophilic substitution, or addition–elimination reactions.

- Substituents on an aromatic ring have a directing effect towards electrophiles, and can increase or decrease the overall reactivity of the system.

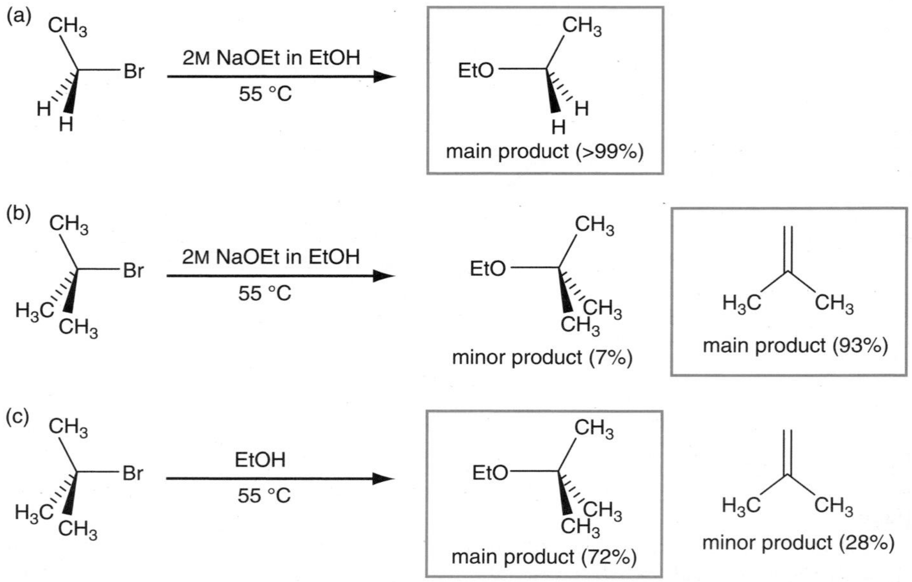

Fig. 13.1 Ethyl bromide (bromoethane) reacts with ethoxide to give the substitution product as shown in (a). Under the same conditions, *t*-butyl bromide gives mainly the alkene isobutene, as shown in (b). In neat ethanol, *t*-butyl bromide gives mainly the substitution product, ethyl *t*-butyl ether, but a significant amount of isobutene is still formed, as shown in (c).

In this, the third and final chapter devoted especially to organic chemistry, we are going to be concerned mainly with reactions in which C=C double bonds are formed and the reactions which such molecules undergo. This topic leads on nicely from where we left off in Chapter 9, since the main way in which C=C double bonds are formed is by elimination reactions which, perhaps surprisingly at first, have quite a lot in common with nucleophilic substitution reactions. We will see that carbenium ions, which were introduced in the discussion of the S_N1 mechanism, play an important role in both the formation of C=C double bonds and their reactions.

Being electron rich, the most characteristic reactions of C=C double bonds are with electrophiles. We will look at typical examples of these reactions and the factors that control which carbon is attacked in an unsymmetrical double bond. The material which was introduced in Chapter 12 will help us to understand the stereochemical requirements and consequences of these and other reactions.

Our attention then turns to the enols and enolates which are formed from aldehydes and ketones under acidic and basic conditions. These species all have an oxygen directly attached to a C=C double bond, and the conjugation of the oxygen lone pair with the double bond makes the latter particularly reactive to electrophiles.

Finally, we will look at aromatic systems, exemplified by benzene. The special stability of the π system in a benzene ring means that the molecule tends to react in such a way that the aromatic ring is preserved. As a result, the reactions of benzene are quite different to those of simple π systems, being mainly electrophilic substitution rather than addition.

Fig. 13.2 The S_N1 mechanism for the reaction between ethanol and *t*-butyl bromide.

13.1 Elimination reactions – the formation of alkenes

In Chapter 9 we looked at nucleophilic substitution and saw how a leaving group, such as bromide ion, can be replaced by a nucleophile. For example, the nucleophile ethoxide reacts with ethyl bromide (bromoethane) by substituting for the bromide to give diethyl ether as shown in Fig. 13.1 (a) on the facing page.

However, under the same conditions, *t*-butyl bromide forms very little of the substitution product ethyl *t*-butyl ether, but instead forms mainly the alkene isobutene, as shown in (b). It *is* possible to form the substitution product as the major product; warming *t*-butyl bromide in ethanol gives mainly the ether, as shown in (c), but a significant proportion of the alkene is still formed.

The formation of isobutene from *t*-butyl bromide is called an *elimination* reaction since the net result is the elimination of HBr from the alkyl halide.

The substitution product is formed from *t*-butyl bromide by the S_N1 mechanism we saw in section 9.7.1 on page 390. First of all the bromide ion leaves to give the carbenium ion, and this then reacts with the alcohol to give the ether as shown in Fig. 13.2. The carbenium ion can also form the alkene and this is the reaction we are now going to look at.

The effects of σ conjugation

It is possible to form the trimethyl carbenium ion from *t*-butyl bromide because of the interaction between the C–H σ bonding MOs with the vacant $2p$ orbital on the carbon. This interaction, known as σ conjugation, lowers the energy of the electrons in the C–H bonding orbitals. MO1 in Fig. 13.3 shows one of the low energy MOs which results from this interaction in which the extensive delocalization across the whole ion can be seen. MO2 in Fig. 13.3 shows the LUMO of the carbenium ion. The main contribution to this orbital is the carbon $2p$ AO, but there is also some out-of-phase contribution from the hydrogen atoms.

The effects of the interaction between the C–H σ bonding MOs and the vacant p orbital are as follows:

- The electrons in the C–H σ bonding MOs are lowered in energy; this stabilizes the ion.

- There is a slight increase in the amount of electron density in between the carbon atoms thus strengthening the C–C bonds.

- Electron density is taken out from the C–H bonds, slightly decreasing the strength of the C–H bonds and making the hydrogens more acidic.

MO2

MO1

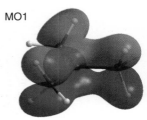

Fig. 13.3 Iso-surface plots of two MOs of the trimethyl carbenium ion. MO1 is a low-energy MO showing how there is delocalization across the whole ion. MO2 is the LUMO, to which the main contributor is the out-of-plane $2p$ AO on the carbon.

Fig. 13.4 In (a) X⁻ acts as a nucleophile and attacks into the vacant p AO of the central carbon atom. In (b) X⁻ acts as a base and removes one of the hydrogens from the carbenium ion to form the alkene. This formation of the C–C π bonding MO from the overlap of the C–H σ bonding MO with the vacant carbon p AO is shown in cartoon form in (c). The same interaction can be seen in the iso-surface plot shown in (d). This is a plot of the HOMO as hydroxide ion approaches the carbenium ion. Some electron density is associated with the oxygen, but the C–C π bonding MO is also clearly beginning to form.

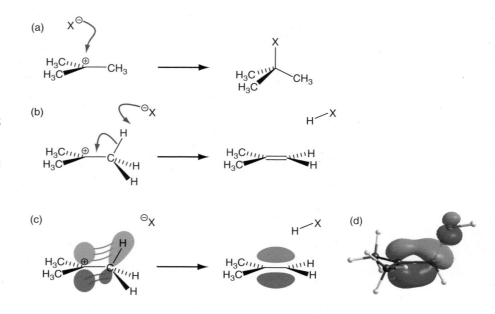

Weblink 13.1

View and rotate the MOs of the trimethyl carbenium ion shown in Fig. 13.3.

Weblink 13.2

View and rotate the HOMO for the OH⁻ / $(CH_3)_3C^+$ interaction shown in Fig. 13.4 (d).

The carbenium ion can react with a nucleophile X⁻ in two ways: either the nucleophile can attack directly into the vacant p AO on the central carbon, or it can attack one of the hydrogens which have been made more acidic because of the σ conjugation. If X⁻ attacks one of the hydrogens, we say it acts as a *base* rather than as a nucleophile. Mechanisms for these two possibilities are shown in (a) and (b) in Fig. 13.4.

Figure 13.4 (c) shows in cartoon form how the interaction between one of the filled C–H σ orbitals and the vacant carbon p AO leads to the formation of the π bonding MO between the two carbons. This interaction can also be seen in the iso-surface plot in (d) which shows the HOMO at an intermediate stage in the reaction as a hydroxide ion approaches the carbenium ion. This single MO shows that some electron density is still associated with the oxygen of the hydroxide ion, but the C–C π bonding MO is clearly beginning to form.

13.1.1 The E1 and E2 elimination mechanisms

We saw that the mechanism for the substitution reaction which proceeds via the carbenium ion is known as an S_N1 mechanism since the rate-limiting step of the reaction is the unimolecular formation of the carbenium ion in the first place. The subsequent reaction of the cation with the nucleophile to form the substitution product is fast.

The alternative reaction of the carbenium ion to form the alkene is also fast compared to the initial formation of the ion. Hence this mechanism for the elimination reaction via the carbenium ion is known as an E1 elimination. Formally, this notation means that it is an elimination reaction in which the rate-limiting step is unimolecular.

We saw in section 9.7.1 on page 390 that there is an alternative substitution mechanism to the S_N1 mechanism, known as the S_N2 mechanism, in which the nucleophile participates directly in the rate-limiting step, forming a new bond

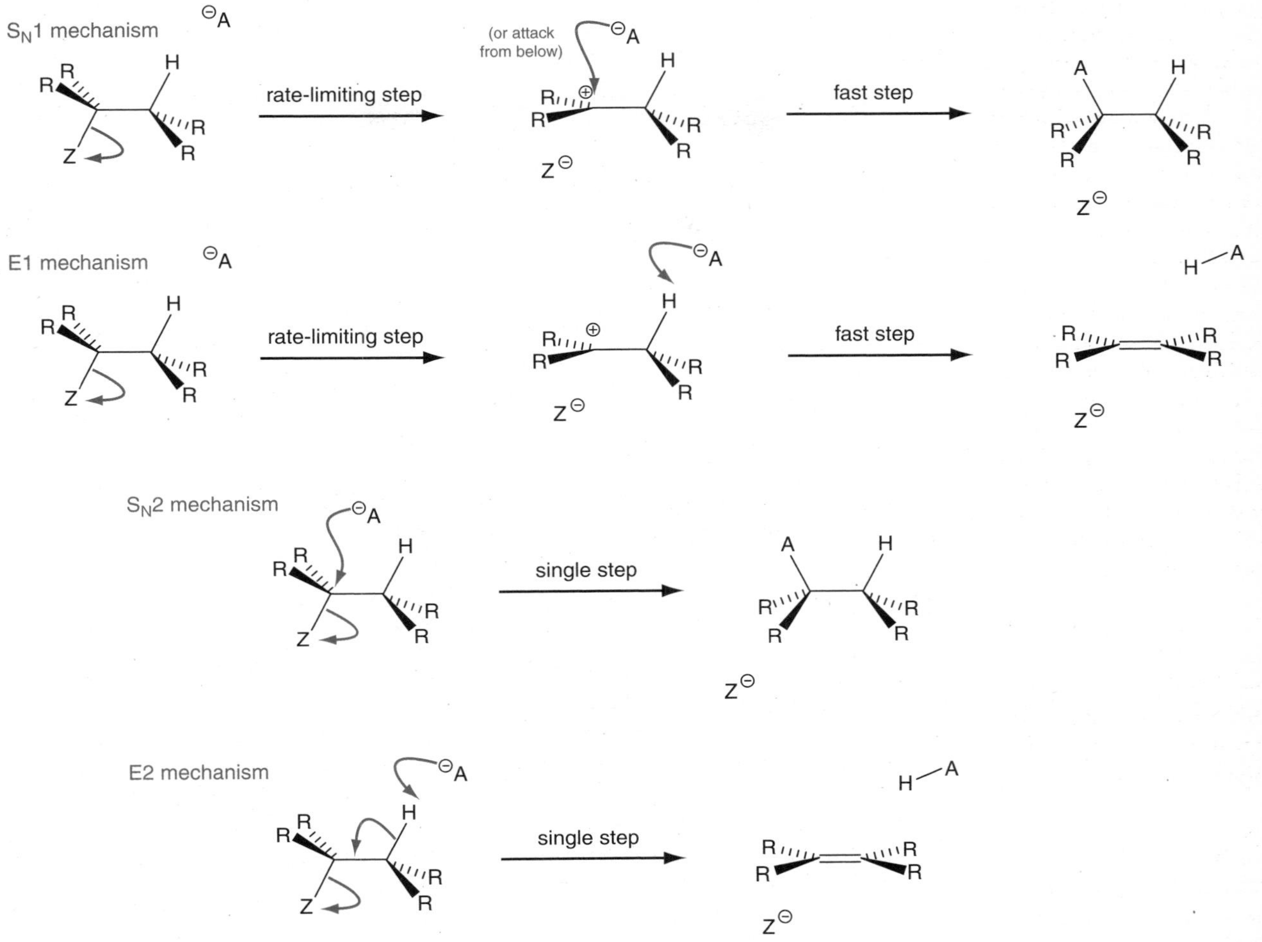

Fig. 13.5 A comparison of the S_N1 and S_N2 mechanisms for nucleophilic substitution with the E1 and E2 elimination mechanisms.

to the carbon as the bond to the leaving group breaks. There is an analogous elimination mechanism, known as the E2 mechanism, where the base removes the proton at the same time as the C=C π MO forms and as the leaving group departs. The S_N1, S_N2, E1, and E2 mechanisms for substitution and elimination are compared in Fig. 13.5.

The trans co-planar or anti-periplanar geometry

In the E1 mechanism we saw how the HOMO of the ion, the filled C–H bonding orbital, interacts with the vacant $2p$ AO of the carbon, the LUMO, as shown in Fig. 13.4 (c) . This interaction increased the acidity of the hydrogens so they could be removed by the base. For an alkyl bromide such as ethyl bromide, C–H σ bonding orbitals can also overlap with the LUMO of the molecule, but this is now the C–Br $\sigma^\star$ MO. This interaction is shown in cartoon form in (a) from Fig. 13.6 on the following page.

In order for one of the C–H σ bonding MOs of ethyl bromide to overlap efficiently with the C–Br $\sigma^\star$ MO, the two orbitals must be in the same plane

Fig. 13.6 One of the filled C–H σ bonding MOs can overlap with the vacant C–Br σ* MO, as shown in cartoon form in (a). The two orbitals must be in the same plane, and the most favourable arrangement is to have them *trans* or *anti* to each other as shown. As a result of the overlap, some electron density is taken from the C–H bond and delocalized between the two carbon atoms. This strengthens the central C–C bond, but weakens the C–H bond and makes the proton slightly more acidic. The C–Br bond is also slightly weakened as a result of the flow of electron density into the antibonding C–Br σ* MO. A base can remove the proton, as shown by the curly arrow mechanism in (b). An iso-surface plot of the HOMO is shown in (c) at the point in the reaction where hydroxide begins to remove the proton.

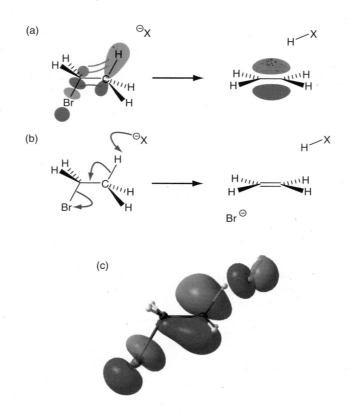

Weblink 13.3

View and rotate the HOMO for the OH⁻/EtBr interaction shown in Fig. 13.6 (c).

i.e. they must be *co-planar*. There are two conformations in which this can be achieved: the *eclipsed* conformation, which is an energy maximum and so not significantly populated (see section 12.2.1 on page 524), and the *staggered* conformation in which the hydrogen and bromine are *trans*, or *anti* to one another with a dihedral angle of 180°. When the hydrogen and the bromine are in the same plane and at 180° to one another, we say they are *trans co-planar* or *anti-periplanar*.

In Fig. 13.6 (a) the ethyl bromide is in a staggered conformation such that the bromine and one of the hydrogens are trans co-planar. There is a favourable interaction between one of the C–H σ bonding MOs with the C–Br σ*. The result of this interaction is that the central C–C bond is slightly strengthened since there is more electron density between the two carbons. However, the C–H bond is slightly weakened, since electron density is withdrawn from it; this also makes the hydrogen slightly more acidic. Furthermore, the flow of electron density into the C–Br σ* MO slightly weakens the C–Br bond.

In the presence of a base, the acidic proton can be removed completely. Figure 13.6 (c) shows an isosurface plot of the HOMO during an intermediate stage as a base (hydroxide) approaches one of the hydrogens of ethyl bromide. Not only can we see electron density associated with both the oxygen of the hydroxide ion and with the bromine atom, but we can also see electron density between the two carbon atoms as the π MO is beginning to form.

We are now in a position to understand the problem posed at the end of the previous chapter where we saw that two different diastereomers behaved

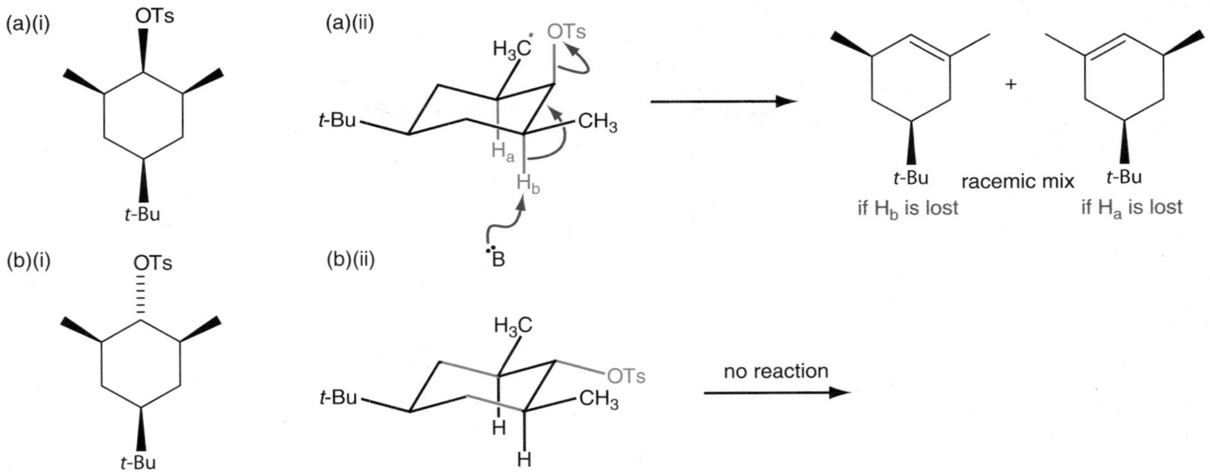

Fig. 13.7 The two structures shown in (a) and (b) are diastereomers in which the bulky *t*-butyl group locks the ring into a chair conformation in which this group is equatorial. For (a) the tosylate leaving group is forced into an axial position, where it is trans co-planar to the two hydrogens coloured red. Either of these can easily be removed by a base to give the two alkenes shown, in equal proportions. In contrast, the tosylate group is fixed in an equatorial position in (b). Only the C–C bonds of the ring shown in red are co-planar to the leaving group. Since there are no hydrogens in the necessary position, no elimination can occur.

very differently in the presence of base: one readily undergoes an elimination reaction, but its isomer does not react. The two isomers are shown again in (a) and (b) in Fig. 13.7, along with their conformational structures. Remember that the large *t*-butyl group locks the ring into a chair conformation with this group in an equatorial position. This means that for the isomer in (a) the tosylate leaving group is forced into an axial position, whereas for the isomer in (b) the tosylate group is forced into an equatorial position.

There are two axial hydrogens which are trans co-planar to the axial tosylate group, as shown by the red bonds in (a)(ii). Either of these two hydrogens can be removed by the base to give the elimination product.

In contrast, when the tosylate is equatorial, there are no hydrogens in the necessary trans co-planar position to allow the elimination to occur. Instead the tosylate is trans co-planar to the two C–C bonds coloured red in the ring shown in (b)(ii).

In these reactions the elimination takes place via the E2 mechanism. No elimination can take place via the E1 mechanism since loss of the tosylate group would leave a secondary carbenium ion which, as we have seen, is not stable enough to be present to any significant extent.

E1 *vs* E2

For *any* elimination reaction to occur, we must have a hydrogen and a good leaving group on adjacent carbon atoms. The E1 mechanism is favoured over the E2 mechanism provided (i) a stable carbenium ion can be formed if the bond to the leaving group breaks and (ii) no strong base is present, since this would make the E2 elimination even more favourable. Since ions are initially formed in the E1 mechanism, a polar solvent, which can help stabilize the ions, may also favour this mechanism.

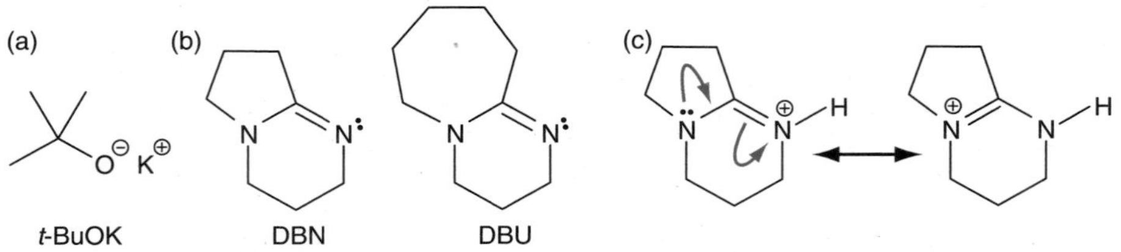

Fig. 13.8 Three bulky bases which do not usually behave as good nucleophiles. Potassium *t*-butoxide, shown in (a), is much larger than its ethoxide equivalent. The two neutral bases DBN and DBU, shown in (b), can both pick up a proton on the doubly-bonded nitrogen to give ions which are stabilized by delocalization as shown in (c).

The E2 mechanism is generally more common and will be favoured when a strong base is present, and for starting materials which cannot form a stable carbenium ion.

Elimination *vs* substitution

We have already seen in Fig. 13.1 on page 558 how substitution and elimination reactions can both occur to different degrees. One way to try to favour the elimination rather than a substitution reaction is to use a stronger base. For example, we saw that *t*-butyl bromide gave mainly the substitution product in ethanol, but mainly the elimination product when the much stronger base ethoxide was present.

The problem is that stronger bases are quite often stronger nucleophiles too, and so if the reactant is not hindered, the base may still act as a nucleophile in a substitution reaction. One way round this is to use a particularly large base such as potassium *t*-butoxide, shown in Fig. 13.8. It is much easier for these large compounds to act as a base by removing one of the peripheral hydrogens than to act as a nucleophile by attacking the less-accessible carbon atom. Two neutral compounds that are particularly good as acting as bases but not as nucleophiles are DBN and DBU, which stand for 1,5-diazabicyclo[4.3.0]non-5-ene and 1,8-diazabicyclo[5.4.0]undec-7-ene. Their structures are shown in (b). Both of these bases pick up a proton on the double-bonded nitrogen, but the resulting formal positive charge is delocalized over both nitrogens as shown by the resonance structures in (c).

13.1.2 Stereospecificity in the E2 mechanism

The requirement that the hydrogen and leaving group be trans co-planar in the E2 mechanism means that different diastereomers give different products. This is because once the hydrogen and leaving group are in the correct orientation for the elimination, there is only one possible product. We therefore say that the reaction is *stereospecific* meaning that the mechanism of the reaction has a particular outcome leading to a specific product. As a result, reactants that differ only in their configuration will be converted to different stereoisomers.

As an example, consider the elimination of HCl from 1,2-diphenyl-1-propyl chloride shown in Fig. 13.9 on the facing page. There are two stereogenic carbon atoms in this molecule, each of which could have an *R*- configuration, or an *S*- configuration. This means there are four stereoisomers possible, which group into two pairs of enantiomers. When treated with ethoxide base at 50 °C, it is found that the (1*R*, 2*R*) isomer gives only the *Z*-alkene, as does

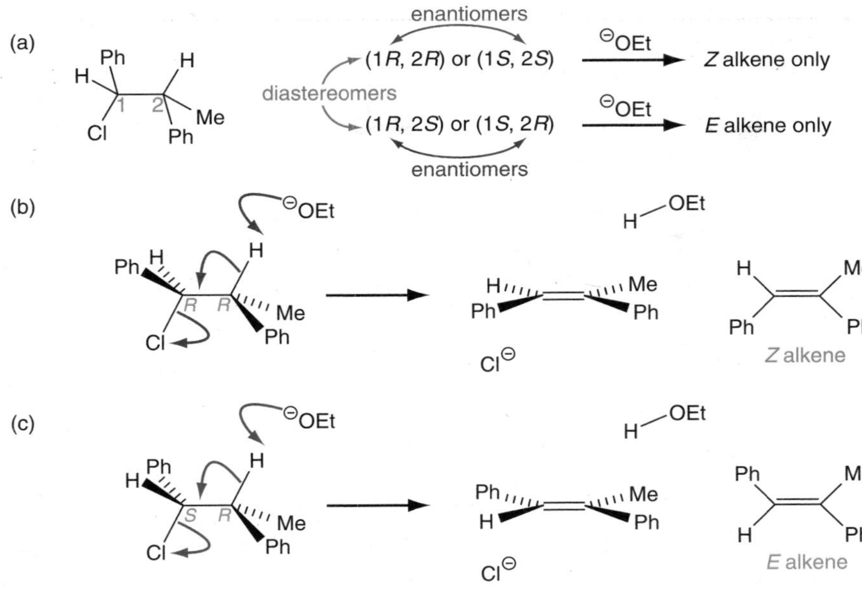

Fig. 13.9 The elimination of HCl from 1,2-diphenyl-1-propyl chloride gives different products depending on which diastereomer reacts, as shown in (a). The results may be understood by inspecting a conformational diagram where the hydrogen and chlorine are arranged trans co-planar for the elimination, as shown in (b) and (c).

its enantiomer the (1S, 2S) form. In contrast, their diastereoisomers, the (1R, 2S) and the (1S, 2R), give exclusively the E-alkene, as shown in (a).

The reason for this stereospecificity is apparent when the structures are drawn out as clear conformational diagrams with the hydrogen and chlorine in the trans co-planar geometry required for the elimination. These are shown in (b) and (c). In the R,R and S,S isomers when the hydrogen and chlorine are trans co-planar the two phenyl groups are on the same side (both coming out of the plane of the paper) and they stay in this arrangement as the alkene is formed. In the S,R and R,S isomers when the hydrogen and chlorine are trans co-planar the phenyl groups are on opposite sides with one coming out of the plane of the paper and the other going in. This leads to the E alkene.

13.1.3 Regioselectivity in elimination reactions

It is sometimes the case that a number of different alkenes can be formed from the same starting material. It might be, for example, that there are a number of alternative hydrogens which could be eliminated together with the same leaving group to form different alkenes. It might even be possible to alter the reaction conditions so as to favour one product rather than another. When a reaction occurs at one particular site in preference to another, we say that the reaction is *regioselective*.

An example is given in Fig. 13.10 on the next page which shows two different alkenes that can be formed from the same alkyl bromide, **A**. The conditions used for the elimination were aqueous *n*-butyl alcohol at 25 °C. Since no strong base was used and since a stable tertiary cation is formed once the bromide ion leaves, this elimination proceeds via the E1 mechanism.

Once the carbenium ion **B** has formed, the solvent can remove a proton to form the alkene. Any one of the hydrogens coloured red, green, or blue could be removed. Removing a red proton gives the alkene **X**, but removing either a blue

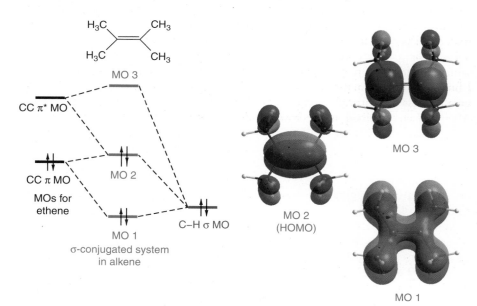

R	% **X**	% **Y**
Me	79	21
Et	71	29
i-Pr	59	41
t-Bu	19	81

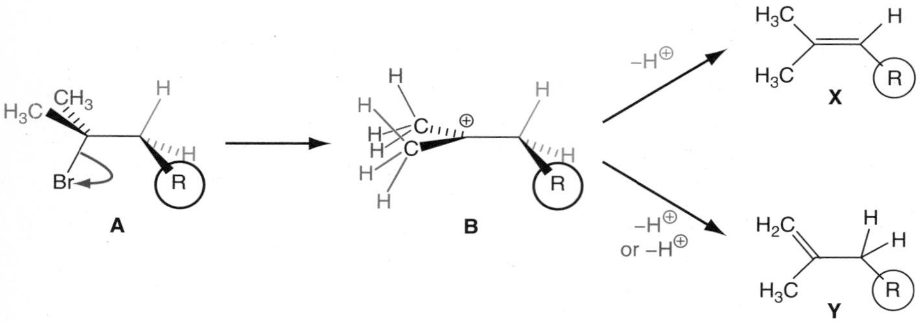

Fig. 13.10 Two different alkenes can be formed from alkyl bromide **A** by the loss of either one of the red hydrogens, or one of the green or blue hydrogens. The size of the R group affects which alkene predominates. It is found that for small R groups the more stable alkene **X** is preferred.

or green proton gives the alkene **Y**. Since there are just two red protons which could be removed to form **X**, but six blue or green protons which could form **Y**, it might be expected that the expected ratio of **X:Y** would be 1:3. However, this is not the observed ratio of the products. In fact when the R group is a methyl group the ratio of **X:Y** is approximately 4:1 i.e. much more of alkene **X** is formed than expected.

Comparing the energies of isomeric alkenes

It turns out that more of alkene **X** is formed than **Y** because **X** is lower in energy than **Y**. As a general rule, a more-substituted alkene is lower in energy than a less-substituted alkene. We have already seen that the more alkyl groups that are attached to the trigonal carbon of a carbenium ion, the lower the energy of the ion due to the more extensive σ-conjugation. The same is true for alkenes. There is no σ-conjugation between any hydrogens directly attached to the C=C

Weblink 13.4

View and rotate the MOs shown in Fig. 13.11 which illustrate the effects of σ conjugation.

Fig. 13.11 The effects of σ conjugation from the methyl groups leads to a net lowering in energy for the alkene and also a slightly raising in energy of the HOMO. The delocalization across the σ and π systems can be seen in the iso-surface plots of the orbitals.

double bond since the hydrogens lie in the nodal plane of the π system. In contrast, the C–H bonds of any alkyl groups attached to the C=C double bond can conjugate with it, thereby lowering the energy of the electrons in the C–H σ bonding MOs.

This interaction between the C–H σ bonding MOs and the π system in 2,3-dimethyl-2-butene is shown in Fig. 13.11 on the facing page and is directly analogous to the interaction between a C–H σ bonding MO and a C=O π system that we saw in Fig. 9.55 on page 375. MO1 is the result of an in-phase interaction between the C–H σ and C=C π orbitals which lowers the energy of the electrons in the σ bonding orbitals. MO2 is essentially just the C=C π bonding MO, but this is raised in energy somewhat due to the out-of-phase interaction with the C–H σ bonding MOs. MO3 is essentially just the C=C $\pi^\star$ MO, but again, this is raised in energy slightly due to an out-of-phase interaction.

In contrast, the isomer of 2,3-dimethyl-2-butene, 1-hexene, which has three hydrogens directly attached to the double bond, only has an interaction with the C–H bonding MOs on one neighbouring carbon. The HOMO for this alkene is shown in Fig. 13.12 where this out-of phase interaction with the C=C π system is clearly visible. Since there is less σ conjugation, 1-hexene is not as low in energy as 2,3-dimethyl-2-butene, and also the HOMO in 1-hexene, the C=C π bonding MO, is not raised as much as it is in its more substituted isomer.

Figure 13.13 gives the standard enthalpies of formation, $\Delta_f H^\circ$, for a series of isomers of hexene. The more negative the value of $\Delta_f H^\circ$, the more stable the molecule. It can be seen that the more substituted the alkene, the lower it is in energy. Notice also that *trans*-2-hexene is slightly lower in energy than *cis*-2-hexene. This is usually explained by saying there are less steric interactions between the two substituents when they are *trans* to one another than when they are *cis*.

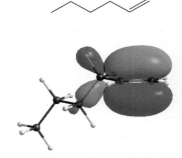

Fig. 13.12 1-Hexene only has σ conjugation from the –CH$_2$– group attached to it. The hydrogens attached to the C=C are in the nodal plane of the π MO and so cannot interact with it.

Thermodynamic product *vs* kinetic product

Returning to the elimination reaction in Fig. 13.10 on the facing page, the more substituted alkene **X** is lower in energy than its isomer **Y** and this is why **X** is formed in preference to **Y**. However, as the size of the substituent R increases, it becomes harder and harder for the base to abstract one of the red protons which are attached to the same carbon as the substituent. When R is a very large *t*-butyl group, it is actually easier to remove one of the blue or green protons to give the less substituted alkene **Y**, even though this is higher in energy than **X**. Alkene **Y** is described as the *kinetic product* of the reaction. It is quicker and

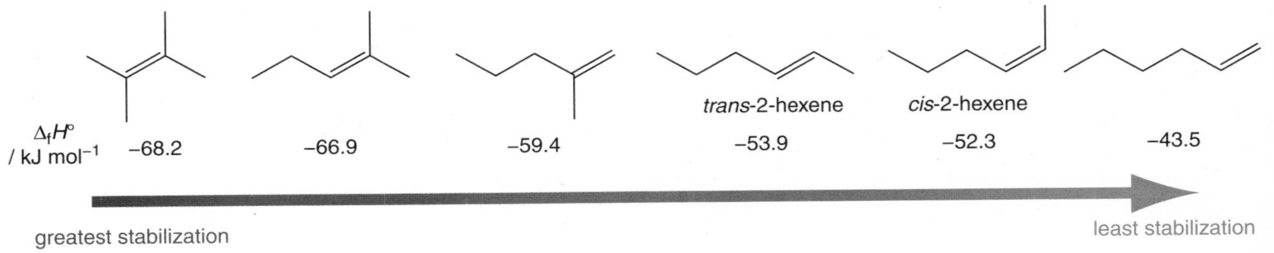

			trans-2-hexene	cis-2-hexene	
$\Delta_f H^\circ$ / kJ mol^{-1} –68.2	–66.9	–59.4	–53.9	–52.3	–43.5

greatest stabilization   least stabilization

Fig. 13.13 The standard enthalpies of formation, $\Delta_f H^\circ$ (in kJ mol^{-1}), for a series of isomeric alkenes. The more substituents there are directly attached to the C=C bond of the alkene, the lower in energy it is, and hence the more negative the value of $\Delta_f H^\circ$.

Fig. 13.14 In this elimination reaction the smaller base ethoxide leads to the formation of alkene **X** which is the most substituted and hence the thermodynamic product. In contrast, the larger base can only remove the more accessible protons giving rise to the less substituted alkene **Y** which is the kinetic product.

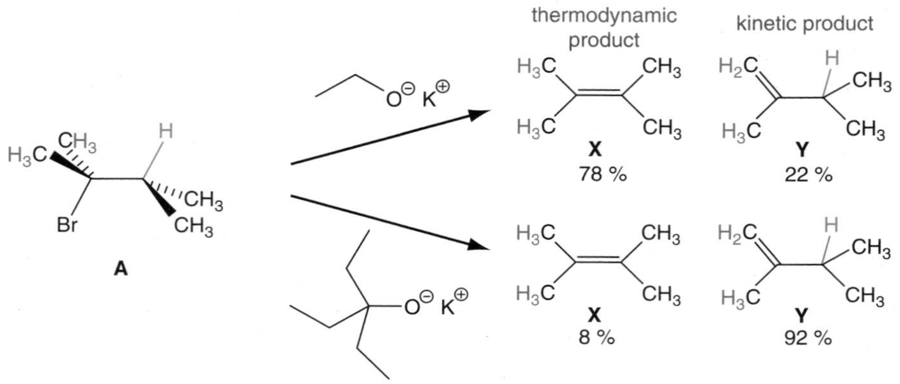

easier to form this alkene than it is to form the more stable alkene **X**, which is described as the *thermodynamic product*.

A similar result is seen for the E2 elimination shown in Fig. 13.14. Since a strong base is present, this reaction proceeds via an E2 mechanism. Two alkene products can be formed by the elimination of HBr from **A**. With the small base potassium ethoxide more of the lower-energy thermodynamic product **X** is formed. When the much larger base Et$_3$COK is used, a proton is removed from the more easily accessible methyl groups to give the kinetic product **Y**.

13.2 Electrophilic addition to alkenes

In the previous section we saw how HBr could be eliminated from an alkyl bromide to form an alkene. In the E1 mechanism a carbenium ion was first formed and then a proton was removed from this ion to give the alkene. The reverse reaction can also occur, where the alkene is first protonated to give the carbenium ion and then a nucleophile attacks the carbenium ion. We have already looked at the orbital interactions for the first part of this reaction in section 8.3.3 on page 321 where we saw how an alkene (ethene) could be protonated to give a carbenium ion. In this reaction, the alkene acts as a nucleophile in attacking the proton. We could also look at it from the other perspective and say that the proton attacks the alkene. In this case the proton is said to act as an *electrophile*, a species which is attracted to areas of high electron density.

This reaction provides a good means of preparing ethyl *t*-butyl ether from isobutene. As shown in Fig. 13.15 on the next page, the alkene is protonated by a strong acid to form the carbenium ion, which then reacts with ethanol to give the desired ether. This reaction is carried out on an industrial scale to prepare ethyl *t*-butyl ether, which is used as an additive in petrol or gasoline.

Protonation of an alkene to give the most stable cation

In this mechanism two different carbenium ions can be formed by protonating the C=C double bond depending on which carbon is protonated, as shown in Fig. 13.16 on the facing page.

However, it is found that only carbenium ion **A** is formed. This is because this ion is considerably lower in energy than ion **B**. In the *t*-butyl carbenium

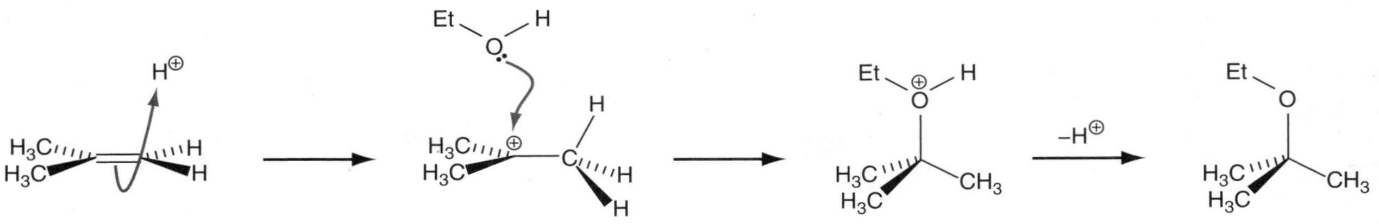

Fig. 13.15 A mechanism for the formation of *t*-butyl ether from isobutene. The alkene is first protonated to give the carbenium ion which is then rapidly attacked by ethanol.

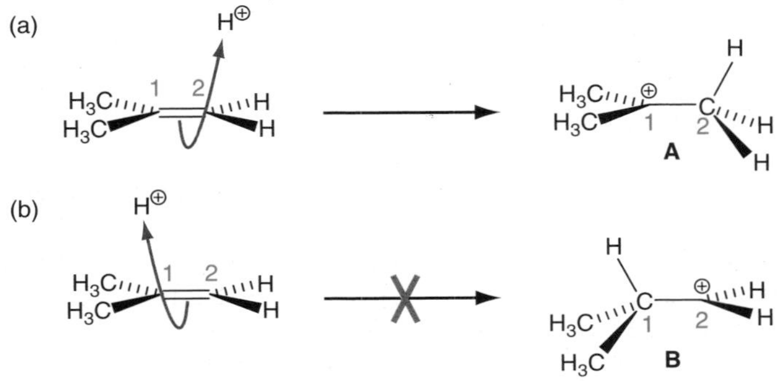

(a)

(b)

Fig. 13.16 Isobutene is preferentially protonated on carbon 2, as shown in (a), since this gives a lower-energy carbenium ion which is stabilized by σ conjugation.

ion **A**, all of the C–H σ bonding MOs can interact with the vacant carbon p orbital, thus lowering the energy of the electrons in these bonds. In contrast, in **B** the two hydrogens directly attached to the carbon with the formal positive charge are in the nodal plane of the p orbital and so have the wrong symmetry to interact with it. In other words, there is far more σ conjugation in **A** than **B**, and it is this that stabilizes the ion.

It also turns out that the effect of the σ conjugation from the methyl groups in isobutene is to put more electron density on carbon 2 of the double bond (the labels are shown in Fig. 13.16). Calculations suggest that carbon 2 has an appreciable negative charge in contrast to carbon 1 which has virtually no charge. Furthermore the HOMO for the alkene, the C=C π bonding MO, has a larger contribution from carbon 2 than from the carbon 1, whereas for the LUMO, the C=C $\pi^\star$ MO, it is the other way round. These two orbitals are shown as iso-surface plots in Fig. 13.17. The greater electron density on the terminal carbon also means that this is the preferred site for protonation.

The general rule is that a proton adds to the carbon of the alkene so as to give the most stable cation. Sometimes this is known as Markovnikov's rule. Once formed, the carbenium ion can react with any suitable nucleophiles that are present. Some examples are given in Fig. 13.18 on the following page. In (a) HCl reacts with the alkene to give a cation stabilized by the phenyl ring. In (b), the cation is a tertiary cation and so is stabilized by σ conjugation. Reactions (c) and (d) are equilibria: the alkene is first protonate by strong acid to give the cation which then reacts with any water present to form the alcohol. The reverse reaction can also take place, the E1 elimination, where the alcohol is first protonated and then leaves to give the carbenium ion, which is finally deprotonated to give the alkene.

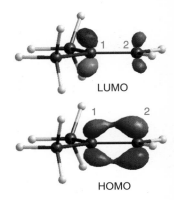

LUMO

HOMO

Fig. 13.17 The HOMO and LUMO of isobutene. The effects of the σ conjugation mean that in the HOMO (the π bonding MO) there is a greater contribution from the right-hand carbon of the double bond (carbon 2 in Fig. 13.16) than from carbon 1. For the LUMO (the $\pi^\star$ MO) it is the other way round.

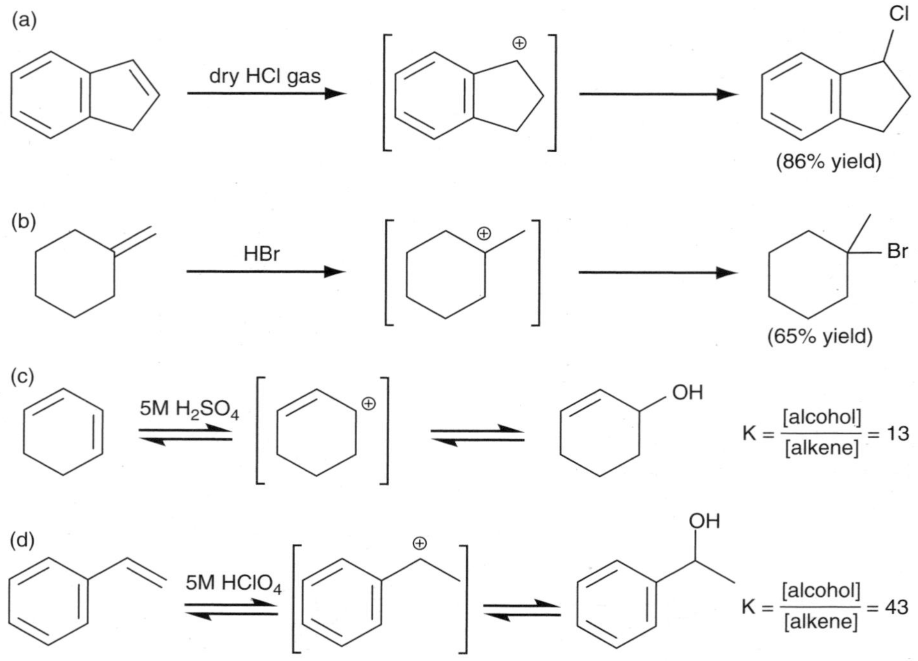

Fig. 13.18 In each reaction, the alkene is first protonated by acid on whichever carbon atom gives the most stable carbenium ion. The carbenium ion then reacts with a nucleophile: halide ion in (a) and (b); water in (c) and (d). In (c) and (d), the alkene is in equilibrium with the alcohol. The reverse reaction is an elimination of H_2O by the E1 mechanism.

(a)

(86% yield)

(b)

(65% yield)

(c)

$K = \dfrac{[\text{alcohol}]}{[\text{alkene}]} = 13$

(d)

$K = \dfrac{[\text{alcohol}]}{[\text{alkene}]} = 43$

13.2.1 Reaction of an alkene with bromine

● **Weblink 13.5**

View and rotate the HOMO and LUMO of isobutene shown in Fig. 13.17.

The high-energy electrons in a π bonding MO can attack other electrophiles. For example, alkenes react with bromine to give a dibromo product in the following way.

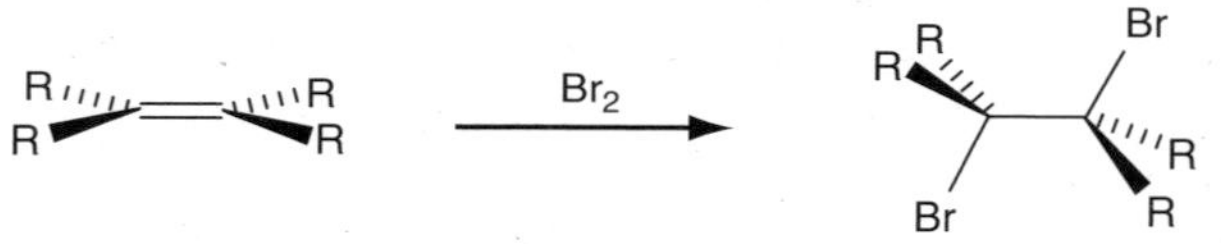

In the first step of the reaction, the high-energy electrons from the C=C π bonding MO attack into the vacant Br–Br $\sigma^\star$ MO. This creates a new C–Br bond and breaks the Br–Br bond. A mechanism for this first step is shown in Fig. 13.19 on the next page. In (a), step (i) shows the electrons from the alkene attacking into the Br–Br $\sigma^\star$ MO and breaking the Br–Br bond. This interaction would leave ion **A**. The LUMO of this ion is the vacant carbon $2p$ AO, and the HOMO is one of the bromine lone pairs. These two orbitals can easily interact to form a three-membered ring as shown in step (ii). This interaction lowers the energy of the bromine non-bonding pair of electrons by making them bonding. The resulting ion **B** is known as a *bromonium ion*, and in it the bromine has two bonds resulting in a formal positive charge.

It may be that this bromonium ion forms in a single step rather than the two suggested in (a). The mechanism in (b) tries to show this, with the electrons from the π system attacking the bromine. This mechanism is slightly unsatisfactory since the convention is that a curly arrow represents the movement of a pair of electrons, whereas two pairs must be used to form the two bonds to bromine in the bromonium ion.

(a)

(i)

A

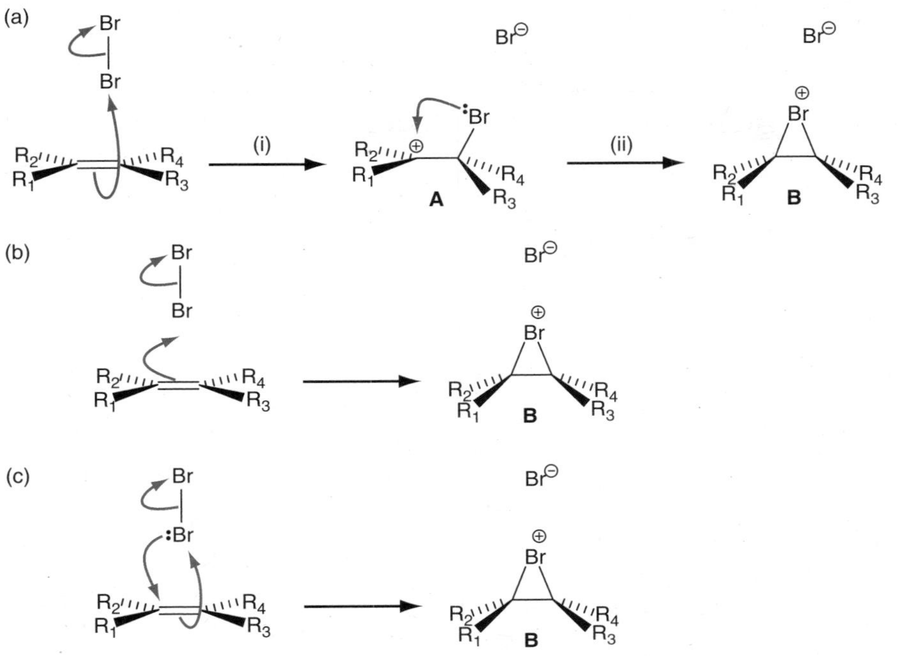

(ii)

B

(b)

B

(c)

B

Fig. 13.19 The first step in the reaction between an alkene and bromine. In step (i) of (a), the electrons from the alkene attack into the Br–Br σ^* to give the cation **A**. This quickly reacts with itself since the bromine lone pair can attack into the vacant $2p$ AO of the carbon atom next to it. This gives the bromonium ion **B**. These two steps probably happen simultaneously and the curly arrow schemes in (b) and (c) are two different ways of trying to show this.

The mechanism shown in (c) shows the correct number of curly arrows to account for all the bonds being formed or broken, but looks a little unusual with electrons both attacking into and leaving from the same bromine atom. It is a matter of personal choice which curly arrow scheme is used.

Since the alkene is flat, the bromine can approach from either face to form the bromonium ion. The approach from different sides gives rise to bromonium ions which are actually enantiomers, as shown in Fig. 13.20. They are formed in equal quantities i.e. as a racemic mixture.

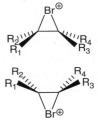

Fig. 13.20 The bromonium ions formed by attacking the alkene on different faces are non-superimposable mirror images, or enantiomers.

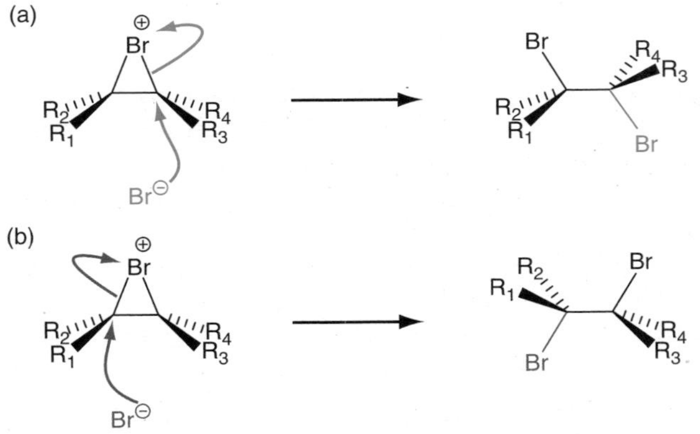

(a)

(b)

Fig. 13.21 In the second step of the reaction, the bromonium ion is attacked by bromide to give the dibromo product. The bromide must attack from directly behind the C–Br bond of the bromonium ion, as shown by the red and blue arrows. It may be easier to attack one carbon than the other depending on the particular substituents. If the bromonium ion is symmetrical, then there is no preference and a racemic mix is formed.

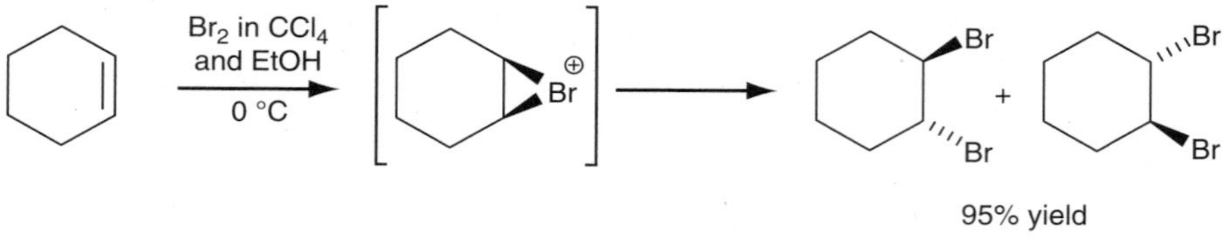

Fig. 13.22 The bromination of cyclohexene gives a racemic mixture of dibromocyclohexane. The bromine substituents end up on opposite sides of the ring.

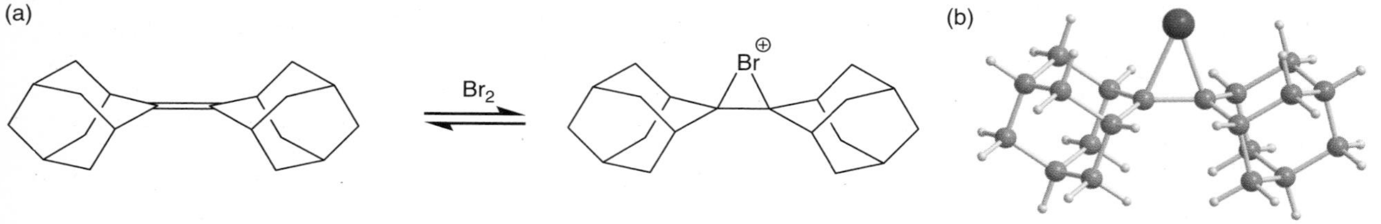

95% yield

(a)

(b)

Fig. 13.23 The alkene shown in (a) is in equilibrium with the bromonium ion, and since the large alkyl groups hinder the approach of the bromide ion the formation of the dibromo compound is inhibited. The crystal structure of the bromonium ion is shown in (b).

Note also that the bromination of an alkene is stereospecific: the groups R_1 and R_3 which start out on the same side of the alkene, remain on the same side of the bromonium ion.

Opening the bromonium ion

Once formed, the bromonium ion is attacked by the bromide ion liberated in the first step. The bromide ion must attack into the LUMO of the bromonium ion which is the C–Br $\sigma^\star$ MO. As we have seen in the S_N2 mechanism, this means the bromide ion must attack from directly behind the C–Br bond.

There are two possible sites of attack, as shown by the red and blue arrows in Fig. 13.21 on the previous page, but in each case in the product the two bromine atoms initially end up *trans* or *anti* to one another with a dihedral angle of 180°. Depending on the nature of the substituents, it may be easier to attack one carbon rather than the other, but if the bromonium ion is symmetrical there will be no preference and so a racemic mixture will be formed.

An example is shown in Fig. 13.22 – the bromination of cyclohexene. The bromonium ion can form on either side of the double bond but the bromide ion must add from the opposite direction. Equal amounts of the two enantiomeric dibromides are formed in which the bromine atoms are on opposite sides of the cyclohexane ring.

It has been possible to isolate a bromonium ion and record its crystal structure, as shown in Fig. 13.23. The extremely large alkyl groups prevent the bromide from attacking into the C–Br $\sigma^\star$ thus inhibiting the formation of the dibromo compound and allowing the bromonium ion to be isolated.

Attacking the bromonium ion with other nucleophiles

Once the bromonium ion has formed other nucleophiles present may attack into the C–Br $\sigma^\star$ orbital before the bromide ion is able to. For example, if the reaction is carried out in water, water will attack the bromonium ion instead of bromide ion and a bromohydrin will be formed which has a bromine atom and hydroxyl group on neighbouring carbon atoms. An example is shown

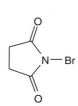

Fig. 13.24
N-bromosuccinimide, NBS, is used as a more convenient source of bromine than toxic liquid bromine. Br$_2$ is formed *in situ* from the reaction between bromide ion and NBS.

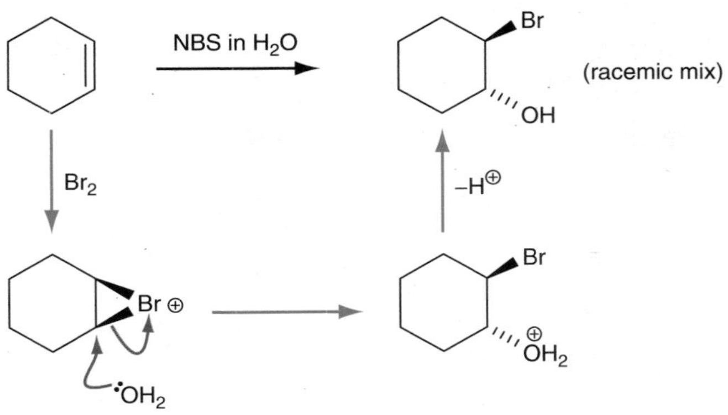

Fig. 13.25 When bromine reacts with an alkene in water, a bromohydrin is formed. Part of the mechanism is shown by the green arrows under the overall scheme. The bromine reacts with the alkene to form a bromonium ion which then reacts with solvent water to give the bromohydrin. In practice N-bromosuccinimide (NBS), whose structure is shown in Fig. 13.24, is used as a source of bromine.

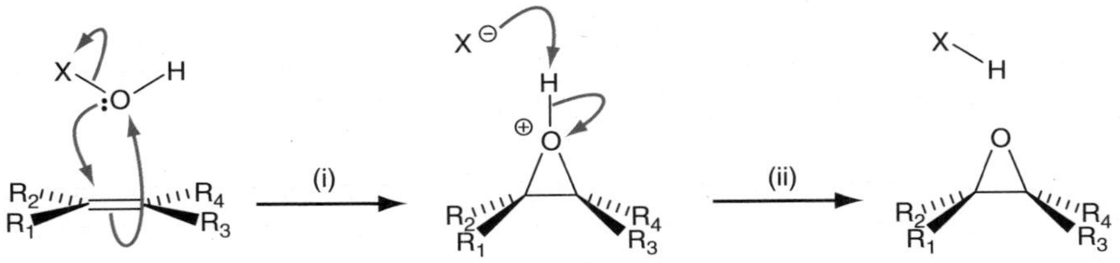

Fig. 13.26 The essence of a mechanism for the formation of an epoxide from an alkene. This mechanism should be compared to the formation of the bromonium ions in Fig. 13.19 on page 571.

in Fig. 13.25 where the overall reaction is shown on the top line and some of the steps from the mechanism are shown below with the green arrows. N-bromosuccinimide (NBS), whose structure is shown in Fig. 13.24 on the preceding page, is used as a source of Br_2 in this reaction. This reagent is much less unpleasant to handle than the extremely toxic liquid bromine.

13.2.2   The formation of epoxides from alkenes

In the previous section the initial reaction between the alkene and bromine gave rise to a three-membered ring which was rapidly opened by a nucelophile. Alkenes can also react to produce a stable three-membered ring containing oxygen in place of bromine; this is called an epoxide.

Figure 13.26 shows the essence of the reaction in such a way that a comparison can be made with the formation of the bromonium ion as shown in Fig. 13.19 on page 571. In step (i) of the reaction the alkene attacks into the O–X $\sigma^\star$ MO, breaking the O–X bond, but forming a bond between carbon and oxygen. At the same time, a lone pair from oxygen forms a C–O bond to the other carbon. These curly arrows are just the same as in Fig. 13.19 (c). In step (ii), the proton is removed from the oxygen to give the neutral epoxide.

The reagent X–OH needs to have a weak O–X bond (i.e. a low-energy $\sigma^\star$ MO) and in addition X^- should be a good leaving group. It is also helpful if X^- is capable of removing a proton in the final step to give the epoxide. The reagent commonly used which fits these criteria is *meta*-chloroperbenzoic acid, *m*-CPBA, whose structure is shown in Fig. 13.27. A mechanism for the formation of an epoxide from an alkene using *m*-CPBA is shown in Fig. 13.28 on the next page. The proton is transferred from the –OH to the carbonyl oxygen in a single step.

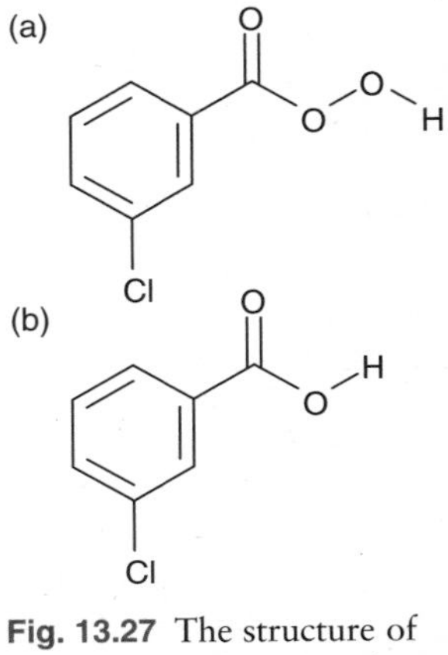

Fig. 13.27 The structure of *m*-chloroperbenzoic acid (*m*-CPBA) is shown in (a) and, for comparison, the structure of *m*-chlorobenzoic acid is shown in (b). The O–O bond in the peracid (a) is particularly weak and has a low energy $\sigma^\star$ MO.

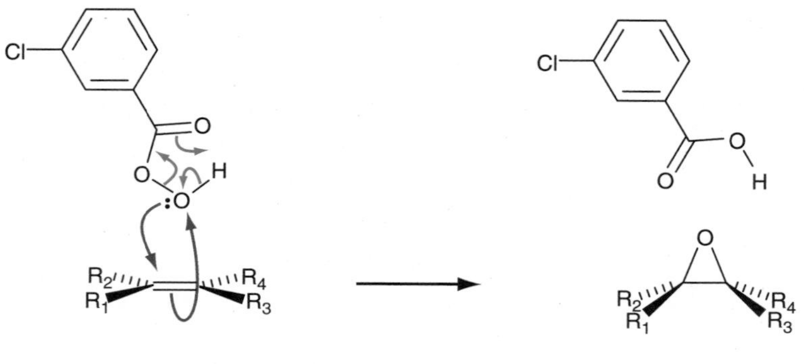

Fig. 13.28 A mechanism for the formation of an epoxide from an alkene using *m*-CPBA.

Notice that the reaction is stereospecific: the groups labelled R_1 and R_3 start out on the same side of the alkene, and remain on the same side in the epoxide. As in the bromination reaction, the *m*-CPBA could attack on either face of the alkene and so a racemic mixture of epoxides is formed.

Formation of epoxides from halohydrins

Epoxides can also be formed from halohydrins, such as the bromohydrin formed in Fig. 13.25 on the preceding page. In the presence of a base such as hydroxide, the alcohol will be deprotonated to give $-O^-$. The lone pair from the oxygen then can attack into the C–Br $\sigma^\star$ MO to give the epoxide as shown below.

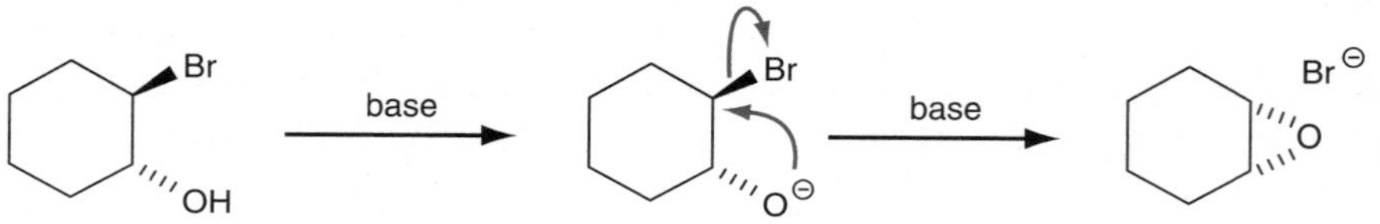

Opening epoxides

Epoxides can be opened up in an analogous way to bromonium ions. Under *basic* conditions, an incoming nucleophile opens up the epoxide in such a way

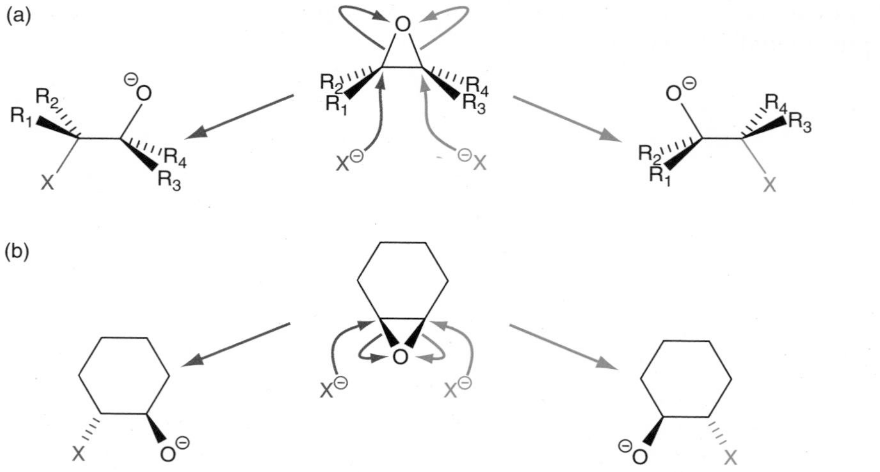

Fig. 13.29 Nucleophiles can open up an epoxide by attacking into the C–O $\sigma^\star$ MO. The mechanism for the general reaction is shown in (a), and for a ring system in (b). In each case, either carbon of the epoxide may be attacked, but if there are large substituents on one carbon and not the other, then the nucleophile may prefer to attack the less-hindered carbon. If the epoxide is symmetrical, as in (b), there will be no preference and a racemic mixture will be formed.

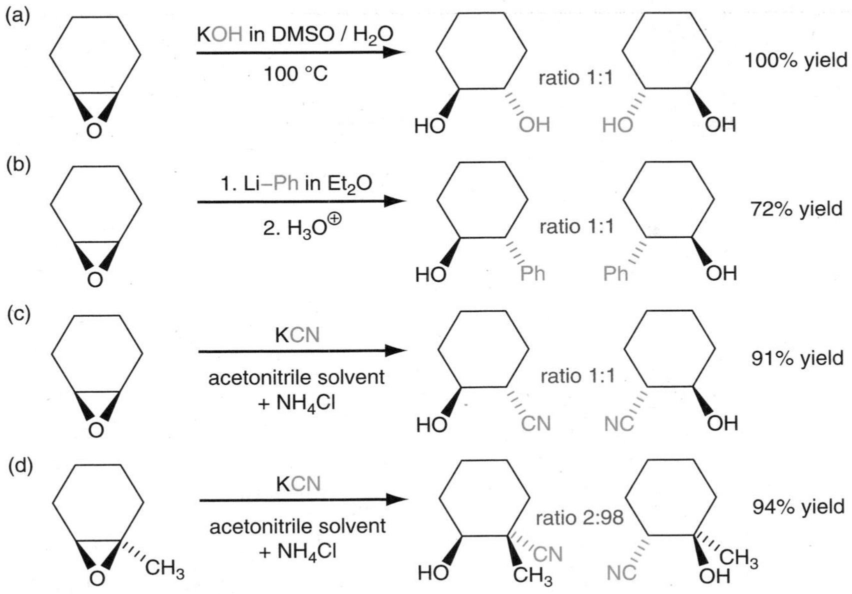

(a) KOH in DMSO / H₂O, 100 °C → ratio 1:1, 100% yield

(b) 1. Li–Ph in Et₂O, 2. H₃O⊕ → ratio 1:1, 72% yield

(c) KCN, acetonitrile solvent + NH₄Cl → ratio 1:1, 91% yield

(d) KCN, acetonitrile solvent + NH₄Cl → ratio 2:98, 94% yield

Fig. 13.30 In each example, the nucleophile attacks the epoxide from the opposite side of the ring to the oxygen, leaving the oxygen and the nucleophile on opposite sides of the cyclohexane ring. In (a), (b), and (c) there is no preference for either site of attack and so a racemic mixture is formed. In (d) the nucleophile attacks the least hindered position in preference to the carbon with the methyl group attached. The ammonium chloride in (c) and (d) is used to protonate the –O⁻ once it has formed.

that the oxygen and the nucleophile end up *trans* or *anti* to one another. This is because the nucleophile must attack into the C–O $\sigma^\star$ MO and therefore approach from the opposite side of the epoxide ring to the C–O bond.

The ring opening of an epoxide with a nucleophile X⁻ under basic conditions is an S_N2 mechanism and is shown in Fig. 13.29 on the preceding page. A general mechanism is shown in (a), and (b) shows how an epoxide in a ring system opens up. Either carbon of the epoxide could be attacked by the nucleophile but large substituents on one carbon may mean attack on the other is favoured. In both cases the nucleophile must approach from behind the C–O bond of the epoxide. If the epoxide is symmetrical, as in (b), then there is no preference between the two alternatives and a racemic mix will be formed. Some examples are given in Fig. 13.30. In (a), (b) and (c), since the epoxide is symmetrical, a racemic mix is formed. In (d), it is easier for the cyanide nucleophile to attack the least hindered carbon and so this product is preferentially formed.

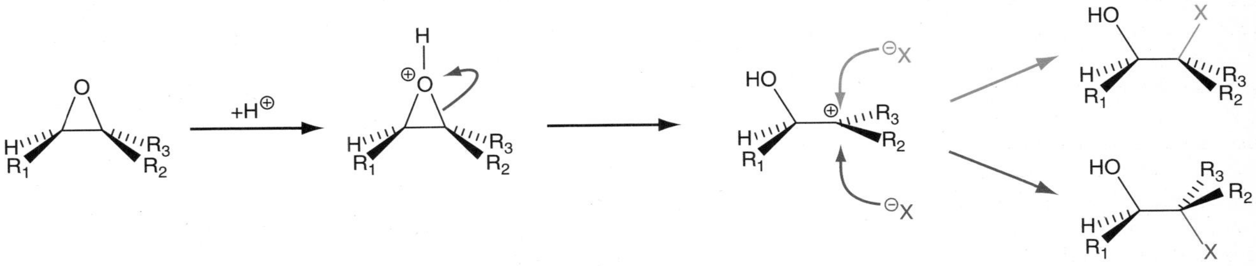

Fig. 13.31 In acid conditions, an epoxide could open via an S_N1 mechanism. The epoxide is first protonated allowing a neutral –OH group to leave and give a carbenium ion. This could be attacked from either direction as shown by the red and blue arrows to give different products.

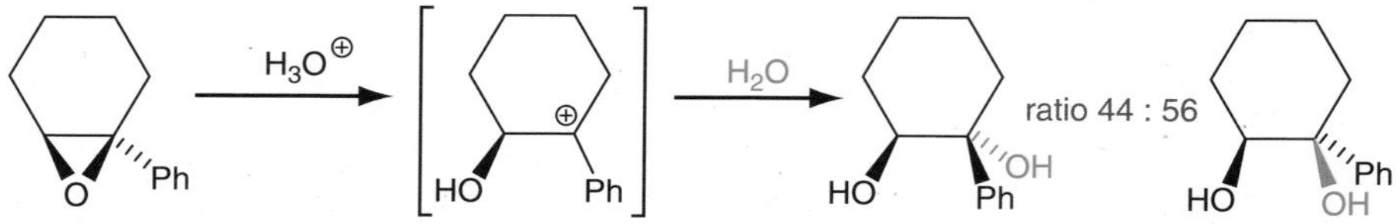

Fig. 13.32 In aqueous acid the epoxide is protonated and then opened up via the S_N1 mechanism. The intermediate cation is stabilized by the phenyl ring and can be attacked on either side by water.

Under *acidic* conditions, the epoxide will first be protonated to give a much better leaving group, neutral –OH as opposed to –O⁻. Under these conditions the ring opening may proceed via an S_N1 mechanism, especially if the resulting cation can be stabilized. The general scheme is shown in Fig. 13.31 on the previous page. If the carbenium ion does form, it could be attacked from either side as shown by the blue and red arrows. However, since the cation has no plane of symmetry, these two routes need not be equally likely and so it is not necessarily the case that equal proportions of the two products will form. Note this mechanism cannot take place unless the epoxide is first protonated. This is because –O⁻ is a poor leaving group which needs a good nucleophile to displace it. An example of an epoxide ring opening in acid conditions is shown in Fig. 13.32.

13.2.3 Hydroboration of alkenes

At the beginning of this section, we saw how it was possible to hydrate an alkene in aqueous acid to give the alcohol. Since the proton adds so as to give the most stable cation, hydration of an unsymmetrical alkene gives one alcohol in preference to the other. For example, in the hydration of styrene, shown in Fig. 13.33, the proton adds to give the most stable cation (Markovnikov's rule) which in turn leads exclusively to alcohol **A**. This alcohol is sometimes known as the *Markovnikov product*. However, what if we wanted to add the water the other way round to give alcohol **B**, as shown in (b)? This alcohol is sometimes known as the *anti-Markovnikov product*. As we will see, it is possible to carry out this reaction, with a very good yield, using a borane as an electrophilic reagent.

Fig. 13.33 In aqueous acid styrene is protonated to give the most stable cation, which can then be attacked by water to give alcohol **A**, the Markovnikov product. The question is how can we add the water to give the isomer **B**, the anti-Markovnikov product?

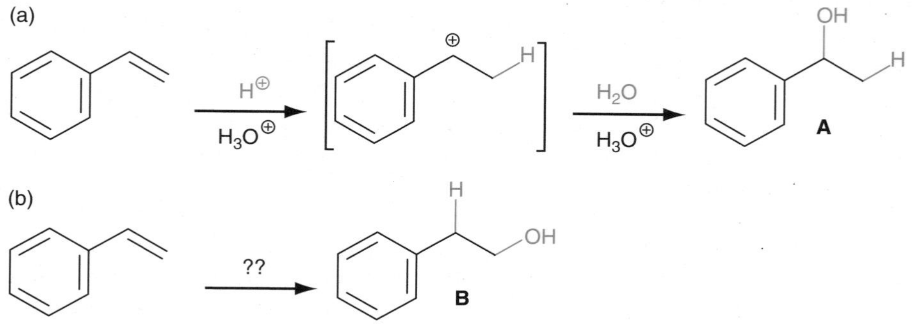

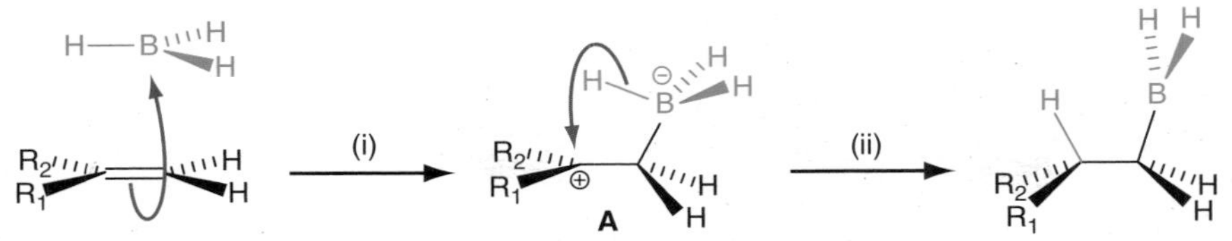

Fig. 13.34 A mechanism for the initial stages of the reaction between BH_3 and an alkene. In step (i) the electrons from the π MO attack into the vacant $2p$ AO on the boron. The resulting adduct is the carbenium ion **A**. This reacts further in step (ii) in which hydride is transferred from the electron-rich boron to the positive carbon.

Step one: the reaction of an alkene with a borane

The simplest borane is BH_3 which actually exists as a dimer (see section 14.2.4 on page 620) but for simplicity here we shall treat it as the monomer. As we saw in Fig. 8.8 on page 318, the LUMO of borane is the vacant $2p$ orbital on the boron. This can readily be attacked by high-energy electrons, such has those in the π bonding MO of an alkene.

To start with, consider the reaction where the electrons from the C=C π bond form a new bond to the boron by attacking into the vacant boron $2p$ orbital, as shown in step (i) of Fig. 13.34. We saw in Fig. 13.17 on page 569 that for isobutene the effect of the σ conjugation means that the HOMO of the alkene has a greater coefficient on the least-substituted carbon of the double bond. It should not be too surprising therefore that it is this least-substituted carbon that preferentially attacks the boron. Furthermore, as the new bond is formed to the boron, one of the carbon atoms of the double bond gains a formal positive charge as electron density is removed from it. The key point is that the initial attack occurs so as to give the most stable carbenium ion, just as we have seen with the protonation reactions at the start of this section.

Once the new carbon–boron bond has formed, the boron in intermediate **A** has four bonds and so has a formal negative charge, just like in the borohydride ion, BH_4^- (see Fig. 8.11 on page 319). The borohydride ion has more electron density associated with the hydrogens than the boron, and it acts like a source of hydride ions, H^-.

Intermediate **A** therefore has a low energy vacant orbital, the empty carbon $2p$ orbital, and high-energy electrons in the B–H σ bonding MOs. The next step is for electrons from one of the B–H bonds to attack into the vacant $2p$ orbital of the carbon, as shown in step (ii).

The overall result is that the C=C bond has broken and a new C–B bond has formed along with a new C–H bond. Both of these two new bonds must form on the same face of the C=C bond: the reaction is therefore stereospecific.

In fact, a full carbenium ion never forms during this reaction and the bond making and breaking reactions occur more-or-less at the same time, as shown by the mechanism shown in (a) from Fig. 13.35 on the next page. There are two HOMO–LUMO interactions taking place at the same time. One interaction is between the HOMO from the alkene (with its greatest coefficient on the more-substituted carbon) and the LUMO of the borane (the vacant $2p$ orbital). This interaction is illustrated in the iso-surface plot, shown in (b), of one of the MOs present at an intermediate stage of the reaction. The second HOMO–LUMO interaction is between the HOMO of the borane (with its greatest coefficient on a hydrogen) and the LUMO of the alkene (with its greatest coefficient on the

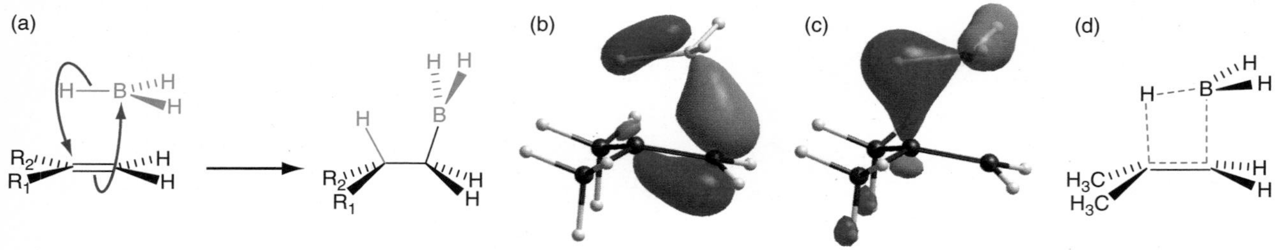

(a)

(b)

(c)

(d)

Fig. 13.35 Shown in (a) is a mechanism for the initial reaction between BH_3 and an alkene. In this mechanism π electrons from the alkene attack into the vacant boron orbital at the same time as the hydrogen from the borane attacks into the $\pi^\star$ MO of the alkene. These two interactions are illustrated by the MOs shown in (b) and (c) for an intermediate stage of the reaction. The bonds being formed and broken are shown by the green dotted lines in (d).

most-substituted carbon). This interaction is illustrated by the MO shown in (c). The bonds being made and broken are shown schematically as dotted green bonds in (d).

The important point here is that the BH_3 adds *as if* the carbenium ion had formed as shown in Fig. 13.34 (a) i.e. the hydrogen adds to whichever carbon atom would best have supported a positive charge. In this reaction, the hydrogen from the borane has added to the alkene in the opposite way that a proton would have added to the alkene.

Further reaction

After this first step has taken place, the boron atom once again has a vacant $2p$ AO. This can be attacked by the π bonding electrons from a second alkene, and then a third alkene, in exactly the same way. The boron atom could therefore ultimately end up with three alkyl groups attached to it rather than the three hydrogen atoms that it started with. The trialkylborane that is formed after three such additions is shown in Fig. 13.36 (a). In order to discuss the next step, in which the boron is replaced by oxygen to give the desired alcohol, we will simplify this structure by writing it as shown in (b).

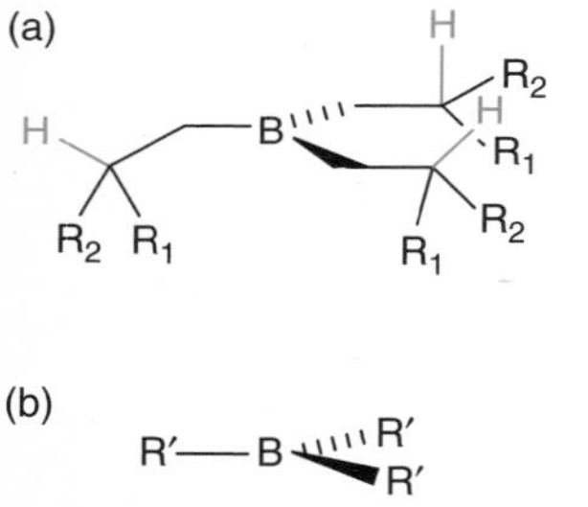

Step two: the removing the borane

To form the alcohol from the trialkylborane a mixture of hydroxide and hydrogen peroxide is used. The pK_a of hydrogen peroxide is 11.65 which means that it is a stronger acid than water which has a pK_a of 15.74. Consequently when sodium hydroxide is dissolved in aqueous hydrogen peroxide, the hydrogen peroxide is deprotonated, giving the HOO^- ion. The HOMO of this ion, a lone pair on the negatively-charged oxygen, attacks into the LUMO of the trialkylborane (the vacant boron $2p$ orbital). A mechanism is given for this in step (i) of Fig. 13.37 on the facing page.

The reason why peroxide is the key nucleophile to use for this conversion of the trialkylborane to the alcohol is apparent by considering the nature of the intermediate **B**. Once again, the boron atom has four bonds and a formal negative charge. This means that there are three electron rich C–B bonds, each of which has more electron density associated with the alkyl groups than with the boron. The LUMO of the ion is the O–O $\sigma^\star$ MO. We have already seen that the O–O bonds are particularly weak, and that one of the oxygens is susceptible

(a)

(b)

Fig. 13.36 If all three of the hydrogens in BH_3 are added to alkenes, the product shown in (a) is formed. For the subsequent stages of the reaction we will simplify this to the structure shown in (b).

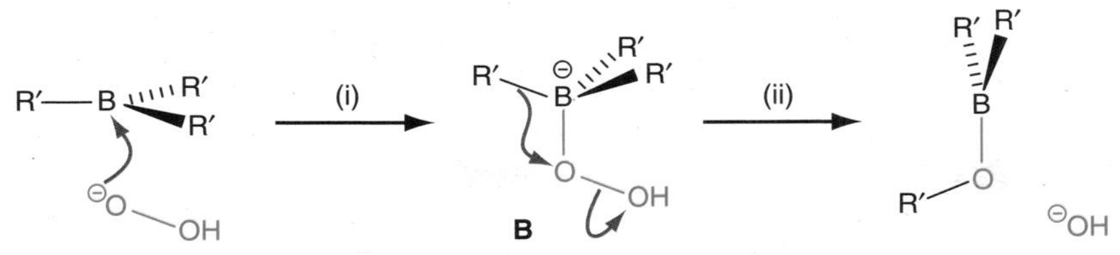

Fig. 13.37 In step (i) the HOO⁻ ion attacks into the vacant *p* orbital on boron. Subsequently the adduct rearranges, as shown in step (ii), by the electron-rich alkyl group attacking into the O–O $\sigma^\star$, thus expelling hydroxide ion.

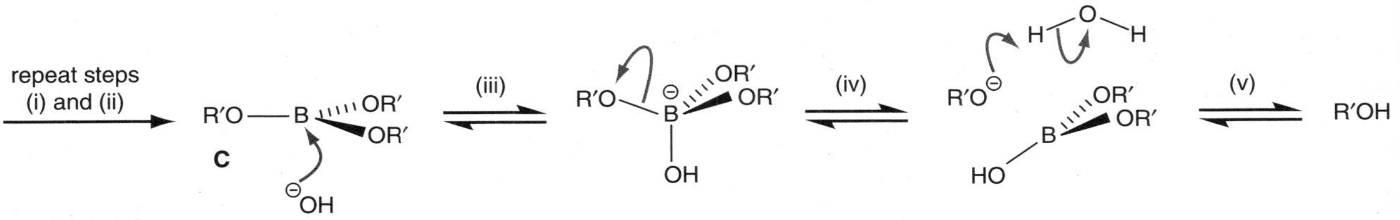

Fig. 13.38 The alkoxide ion, R′O⁻, is released from the boron by being displaced by either an OH⁻ or an HOO⁻ ion. These ions can attack the intermediate **C** to give a tetrahedral anion from which any of the four groups could be lost.

to attack by a nucleophile. This is what happens in step (ii): the high-energy electrons from one of the C–B bonds attack into the O–O $\sigma^\star$ MO and push out hydroxide ion. Hydroxide is not a good leaving group, and will only leave if it displaced by something with higher-energy electrons, as it is here by the high-energy C–B bonding electrons.

The result of this rearrangement is that one of the alkyl groups is now attached to an oxygen rather than to the boron. The boron once again has a vacant 2*p* orbital, and so steps (i) and (ii) can take place again to give **C**, shown in Fig. 13.38.

The vacant *p* orbital in intermediate **C** can then be attacked by another oxygen lone pair either from the HOO⁻ ion, or possibly from some of the hydroxide liberated in step (ii). One of the B–O bonds in the resulting tetrahedral boron ion could then break and it may be the hydroxide ion that falls off again in the reverse of step (iii), or maybe one of the alkoxide ions leaves as R′O⁻, as shown in step (iv). The alkoxide will eventually be protonated by the water to give the desired alcohol, R′OH, and regenerate hydroxide.

Frequently, rather than reacting BH₃ itself with the alkene, a disubstituted borane is used which can deliver just a single hydrogen to the alkene. One such comercially available reagent is called 9-BBN (9-borabicyclo[3.3.1]nonane) whose structure is shown in Fig. 13.39. The great bulk of this reagent helps to deliver the single hydrogen to the least hindered position of the alkene.

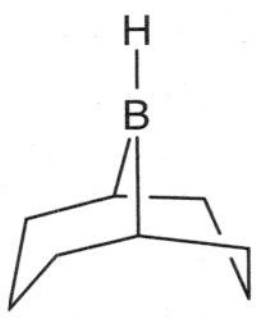

Some examples of the hydroboration of alkenes are shown in Fig. 13.40 on the next page. For each alkene, the anti-Markovnikov product is the favoured one. The selectivity is greater using the bulky 9-BBN rather than BH₃. This is particularly the case in example (c) where there is little selectivity shown with the BH₃, but in contrast the larger 9-BBN shows greater selectivity since it much prefers to bond to the least hindered of the two alkene carbon atoms.

Fig. 13.39 9-BBN is a commonly used reagent which gives excellent yields for the hydroboration of alkenes.

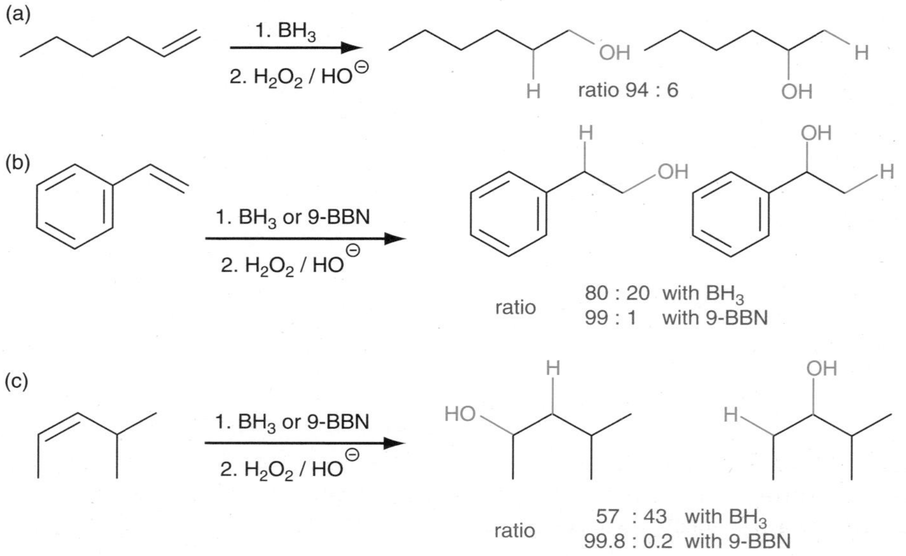

Fig. 13.40 Hydroboration of alkenes gives predominantly the anti-Markovnikov alcohol. Better selectivity is observed when the bulkier 9-BBN is used, especially in example (c) where the large borane only bonds to the least hindered carbon of the alkene.

13.3   Enols and enolates

In the previous sections we have seen how alkyl groups attached to a C=C double bond can raise the energy of the electrons in the π system as illustrated in Fig. 13.11 on page 566. We have also seen in the case of an unsymmetrical alkene, such as isobutene, how there is more electron density on the least-substituted carbon of the double bond, and that this carbon has the greater coefficient in the HOMO.

With a lone pair conjugating into the π system rather than just an alkyl group, the effects are even more pronounced. We have already met such systems briefly in section 4.9.2 on page 174. There we saw how aldehydes and ketones could be deprotonated by treatment with a strong base to form *enolate ions* in which an oxygen lone pair is conjugated with a C=C bond. We are now in a position to understand *why* aldehydes and ketones can be deprotonated in the first place, and to understand some of the reactions of the enolate ions and their neutral counterparts, *enols*.

13.3.1   The formation of enols and enolates

In section 13.1 on page 559 we saw how in a carbenium ion σ conjugation between adjacent C–H bonds and a vacant p orbital led to a net lowering of energy and also increased acidity of the hydrogens as electron density was taken away from the C–H bonds. We have also seen in Chapter 9 that C–H bonds can σ conjugate with neighbouring *carbonyl* groups (see Fig. 9.55 and Fig. 9.56 on page 375). There we were particularly interested in how the CO $\pi^\star$ MO was raised in energy, but now we shall focus on how this interaction also leads to an increased acidity of the hydrogens on the carbon adjacent to the carbonyl bond.

The interaction between a C–H σ bonding MO and the carbonyl π system is shown in cartoon form in (a) from Fig. 13.41 on the facing page. This interaction has the effect of withdrawing some electron density from the C–H bond,

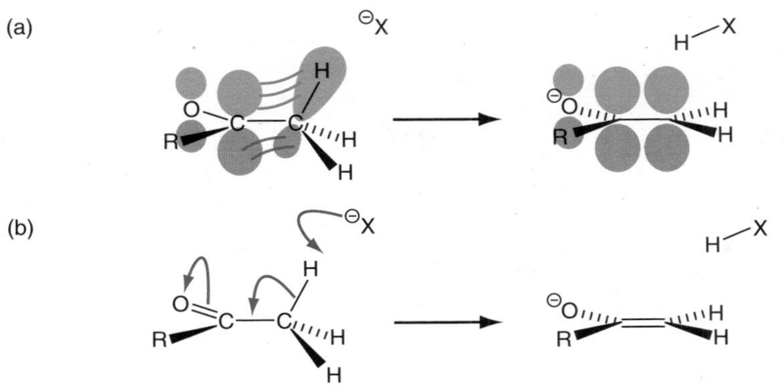

(a)

(b)

Fig. 13.41 The C–H bonding MOs on the α carbon adjacent to a carbonyl group can overlap with the carbonyl π system in the way shown in (a). As a result electron density is withdrawn from the C–H bond, making the hydrogen acidic and liable to be removed by a base. A mechanism for this deprotonation reaction, resulting in the formation of an enolate ion, is shown in (b). The lowest energy π MO of the enolate is shown in cartoon form in (a).

thus making the hydrogen more acidic. A strong base is able to deprotonate one of the hydrogens from the carbon next to the carbonyl carbon (known as the α-carbon) and a mechanism for this is shown in (b).

Deprotonating carbonyl compounds

The approximate pK_a values of a number of carbonyl derivatives are given in Fig. 13.42. We saw in section 6.14.5 on page 249 how we can estimate the equilibrium constant between two acids, or an acid and a base, by looking at the differences in their pK_as. For the equilibrium

$$AH(aq) + X^-(aq) \rightleftharpoons A^-(aq) + XH(aq),$$

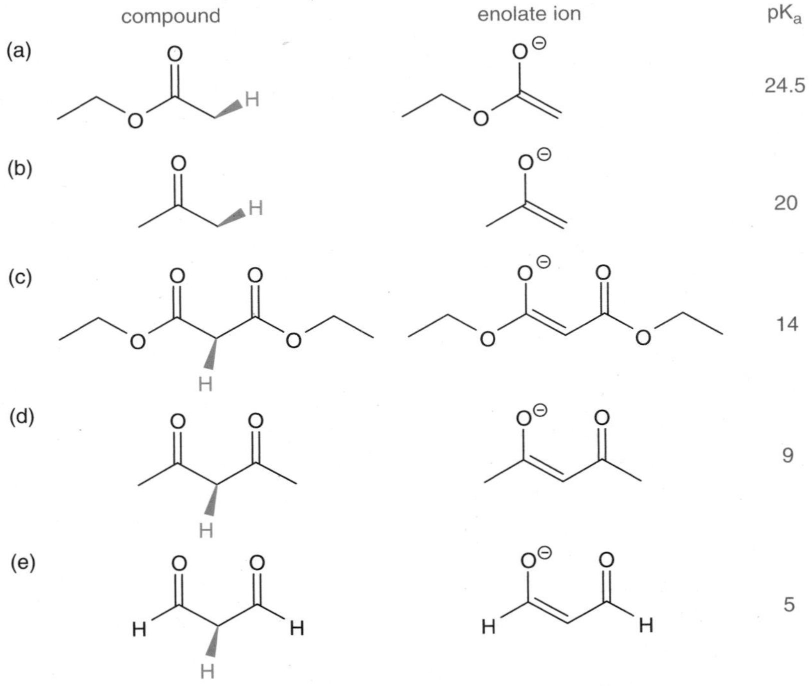

	compound	enolate ion	pK_a
(a)			24.5
(b)			20
(c)			14
(d)			9
(e)			5

Fig. 13.42 The approximate pK_a values of some carbonyl compounds. In each example, the α-hydrogen with is removed is shown in red. For compounds with a pK_a greater than about 16, a strong base is needed to form the enolate ion. For compounds with a pK_a less than this, hydroxide ion can be used. The β-dicarbonyl compounds (c), (d), and (e), which have a CH$_2$ group flanked on either side by a carbonyl group, are much more acidic than the compounds with just the single carbonyl.

the equilibrium constant, K, is given by the equation

$$K = \frac{[A^-][XH]}{[AH][X^-]} = \frac{K_a(AH)}{K_a(XH)} = \frac{10^{-pK_a(AH)}}{10^{-pK_a(XH)}}.$$

This means that the equilibrium constant for the reaction shown below between hydroxide (the conjugate acid of which is water, with a pK_a of 15.74) and propanone (acetone) (pK_a 20) is $10^{-20}/10^{-15.74} = 5 \times 10^{-5}$. This means the equilibrium is well over to the left and the hydroxide deprotonates only 1 part in 20000 of the propanone.

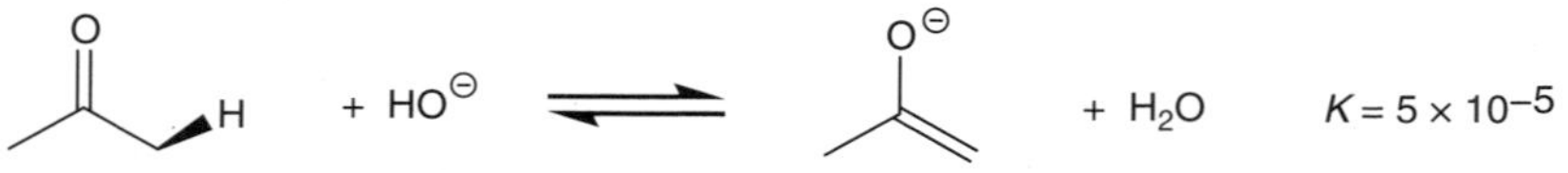

$$K = 5 \times 10^{-5}$$

If the enolate is desired in much greater concentration, a stronger base is needed. A commonly used base is LDA, lithium diisopropylamide, whose structure is given in Fig. 13.43 (a). The pK_a of its conjugate acid diisopropylamine, shown in (b), is around 34 which means that this base can easily deprotonate any α-carbonyl proton. The isopropyl groups increase the size of this base which helps prevent it simply acting as a nucleophile and attacking the carbonyl carbon instead of acting as a base.

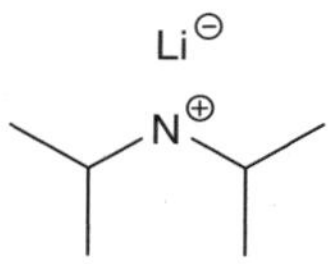

Looking back at Fig. 13.42 on the previous page, we see that when one of the methyl groups of acetone is replaced by an ethoxy group to give an ester, as shown in (a), the α-hydrogen some 10^5 times less acidic. The oxygen is effectively conjugated with the carbonyl system which means less electron density is withdrawn from the α-hydrogens.

If the α-hydrogens can conjugate with *two* carbonyl groups, they become much more acidic and hydroxide can easily remove one to give the enolate ion. Such compounds, which have two carbonyl groups separated by one carbon atom, are known as β-dicarbonyl compounds. In (d), the central hydrogens with a pK_a of 9 are considerably more acidic than the terminal ones on the methyl groups which, as we have seen in propanone, have a pK_a of around 20. In (e) the only hydrogens which are acidic are the central ones. This compound has a pK_a similar to a weak acid such as acetic (ethanoic) acid.

(a)

(b)

The keto–enol equilibrium

We have seen that reasonably strong bases are needed to deprotonate most carbonyl compounds to give enolate ions, but that weaker bases are sufficient to form the enolate ions of β-dicarbonyl compounds. In neutral solutions, a carbonyl compound exists in equilibrium not with the enolate ion, but with the neutral species called an *enol* which has a proton on the oxygen. Its name comes from the fact that it is an alk**ene** with an alc**ohol** attached. The equilibrium is sometimes called the keto–enol equilibrium: that for propanone is shown in Fig. 13.44. As can be seen, the enol is present only in very tiny concentrations

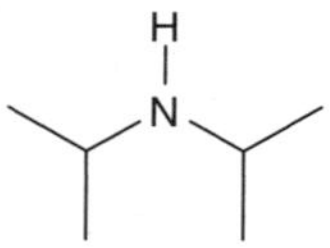

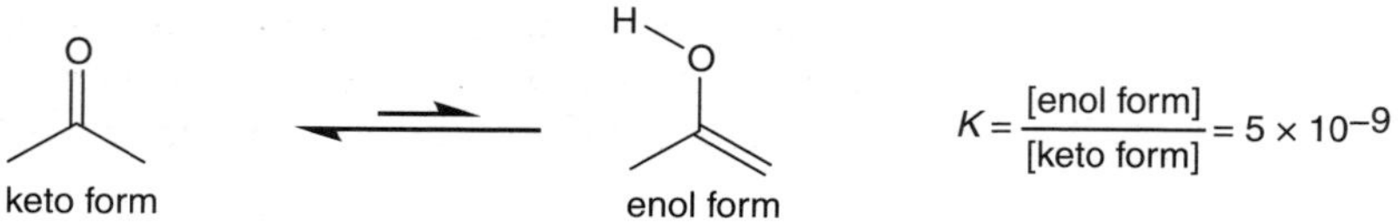

Fig. 13.43 Lithium diisopropylamide, shown in (a), is a commonly used strong base. The pK_a of its conjugate acid, diisopropylamine, shown in (b), is around 33.

Fig. 13.44 The equilibrium between the keto and enol forms of propanone (acetone) lies strongly in favour of the keto form.

keto form

enol form

$$K = \frac{[\text{enol form}]}{[\text{keto form}]} = 5 \times 10^{-9}$$

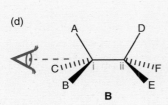

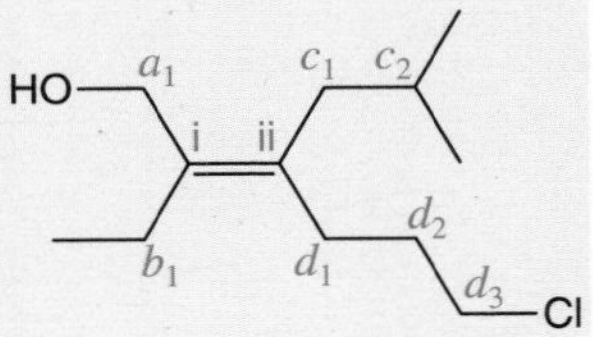

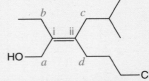

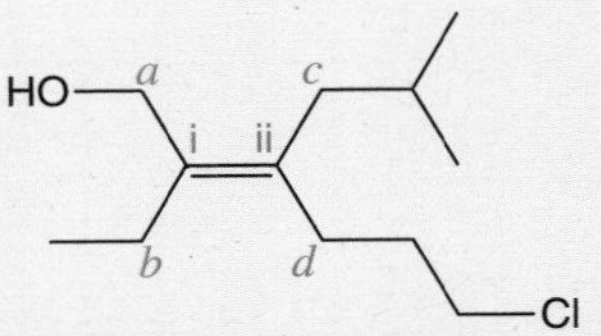

Now that the priority of the groups has been identified the labels (E) and (Z) can be assigned. When the two highest priority groups are on the same side, we have the (Z)-isomer, but when they are on the opposite side, we have the (E)-isomer; both cases are illustrated opposite.

The CIP system includes many further sub-rules to help distinguish between more complicated examples, and references to these can be found in the further reading section at the end of the chapter.

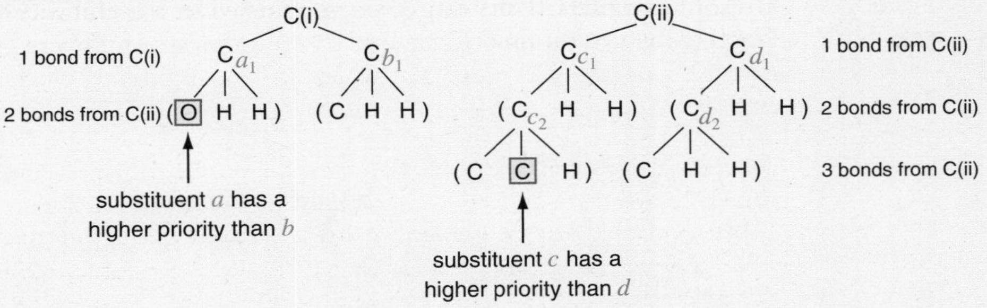

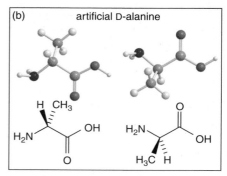

Let us first consider how the convention may be applied to the naturally occurring isomer, L-alanine, shown in Fig. 12.16. First of all we need to assign a priority to each of the four groups attached to the chiral carbon in alanine. For this the system used is the one based on atomic numbers and described in Box 12.3 on the previous page where it was applied to working out the *E/Z*

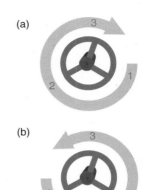

Table 13.1 The percentage keto and enol forms for two β-dicarbonyl compounds

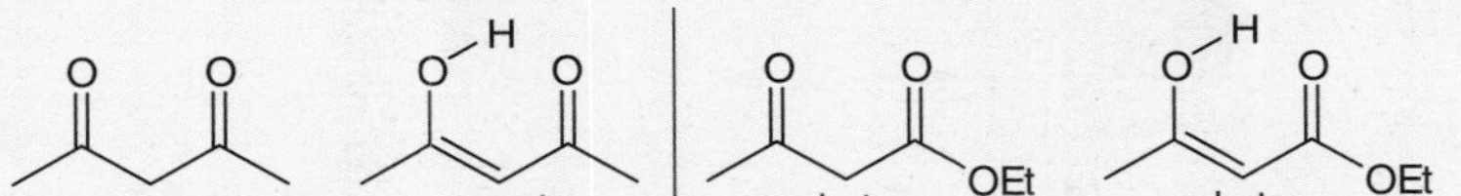

	keto	enol	keto	keto
neat	20	78	94	6
gas phase	2	98	47	53
in CCl₄	5	95	63	37
in H₂O	87	13	99.5	0.5

with the keto form being the more thermodynamically stable form. This is the typical behaviour for simple ketones.

Table 13.1 shows the percentages of keto and enol forms for some β-dicarbonyl compounds. These data show that the proportion of the enol form present at equilibrium is much greater for these β-dicarbonyl compounds that it is for simple ketones. This is because in the enol conjugation is possible over the remaining carbonyl group, the C=C bond, and the –OH group. In contrast, a simple enol, such as that from propanone, has no carbonyl group to extend the conjugation. The position of equilibrium is also strongly dependant on the solvent, and it is notable that solvents which are hydrogen bond donors, such as water, shift the equilibrium in favour of the keto form.

Formation of enols under acidic conditions

With a strong base, a ketone can form the enolate ion as outlined above, but in acidic conditions it is the enol, rather than the enolate, which is formed. In acid, a significant proportion of the carbonyl oxygen will be protonated. Of course, it is possible to remove this proton and re-form the ketone. However, protonating the carbonyl oxygen also means that even more electron density is withdrawn from any C–H bonds on the adjacent carbon atoms that σ conjugate with the π system. As a result, a much weaker base is now able to deprotonate the α-carbon to form the enol. A curly arrow mechanism for this process is shown in Fig. 13.45. After protonation of the carbonyl oxygen in step (i), the weak base water is able to remove one of the protons from the α-carbon to form the enol, as shown in step (ii). Note this reaction is catalyzed by acid: H_3O^+ is used up in step (i), but re-formed in step (ii).

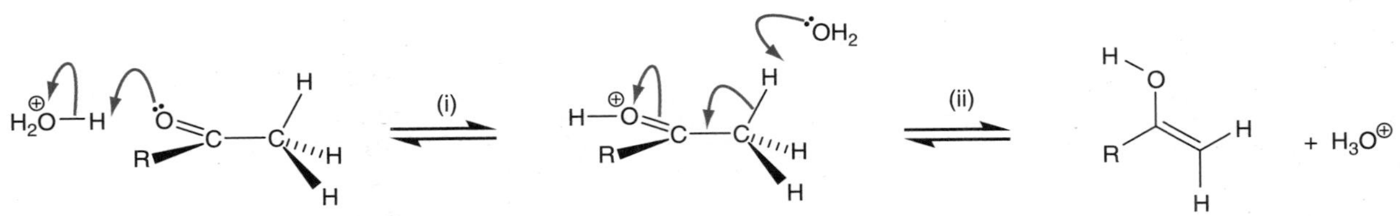

Fig. 13.45 The acid-catalyzed, reversible formation of an enol. Protonation of the carbonyl oxygen makes the protons on the α-carbon more acidic. This means the weak base water is now able to remove one to form the enol.

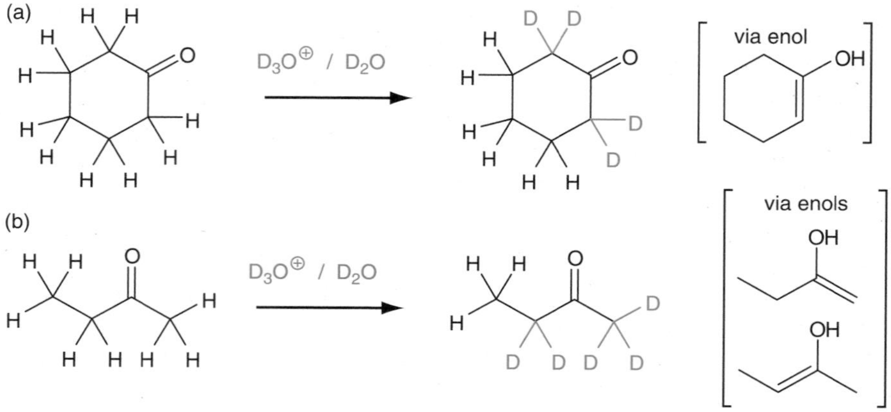

Fig. 13.46 Since the acid-catalyzed formation of an enol is reversible, if a ketone is mixed with D_2O and D_3O^+, any weakly acidic hydrogens will eventually be replaced by deuterium. In (a) four hydrogens can be replaced, going via the enol shown. In (b), five hydrogens can be replaced going via the two enol forms shown.

Proton exchange

Since the acid-catalyzed formation of an enol is completely reversible, if the reaction is carried out in deuterated water (D_2O) and acid, once a proton has been removed, it is far more likely that a deuterium will be added in the reverse step. This means that over time, a number of protons will be replaced by deuterium. Two examples are shown in Fig. 13.46.

13.3.2   Thermodynamic and kinetic control in enolate formation

In Fig. 13.46 (b) since butanone is not symmetrical about the carbonyl bond, two possible enols could be formed in acid, and two enolate ions could be formed in base. These two enols or enolates do not have the same energy since, as we have seen for alkenes, the more substituted double bond has greater σ conjugation and is therefore lower in energy. By careful choice of the reaction conditions it is often possible to favour the production of one of the enolate ions over the other.

Consider the formation of the enolates from 2-methylcyclopentanone using the strong bulky base triphenylmethyllithium, Ph_3CLi, as shown in Fig. 13.47. Removing the proton coloured blue gives the more substituted enolate ion which

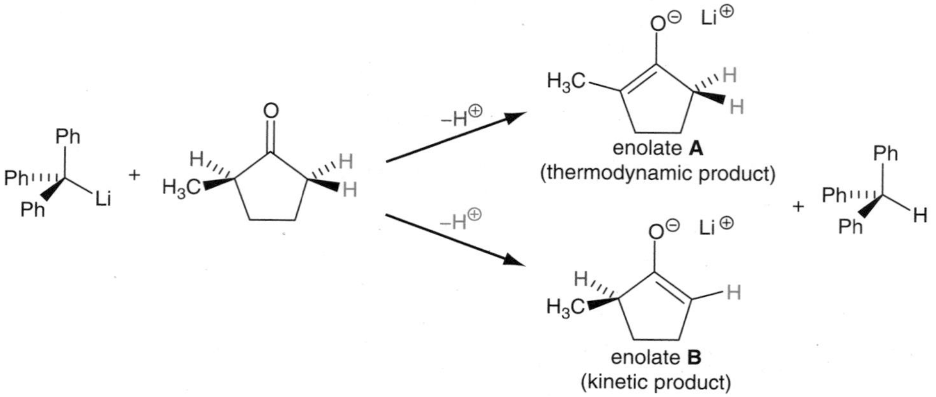

Fig. 13.47 It is easier for a large base to form enolate **B** by removing one of the protons coloured red since these are more accessible. However losing the proton coloured blue gives the more substituted enolate **A**, which is lower in energy.

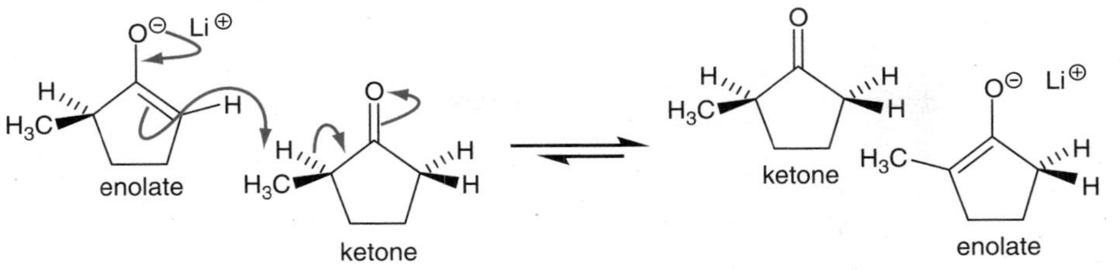

Fig. 13.48 When excess ketone is present, the enolate already formed can deprotonate any excess ketone, but this just gives more ketone and enolate. This proton-swapping eventually allows equilibrium to be reached in which the lowest energy enolate predominates.

is lower in energy: this is the thermodynamic product. However, the protons coloured red are more accessible since the base which is removing them is not obstructed by the methyl group. Since it is easier (quicker) to remove a red proton, the enolate ion that results is the kinetic product.

When an excess of the base is added to the ketone at room temperature, the majority of the enolate formed is the kinetic enolate **B**, with a ratio of **A:B** being 28:72. Once formed, these enolates cannot re-form the ketone since in the presence of excess base there is nothing acidic from which the enolates can pick up a proton. If an electrophile is added for the enolate to react with, the ratio of the two products formed will be the same as the ratio of the two enolates.

If instead of having excess base we have a slight excess of ketone, then under the same conditions (i.e. the same time and temperature), the ratio of the enolates **A:B** becomes 94:6. Now, the lower energy enolate **A** (the thermodynamic product) is the dominant species. This comes about because it is now possible for the enolate to return to the ketone by deprotonating some of the excess ketone present as shown in Fig. 13.48.

This reaction does not increase the concentration of enolate ions present, since all does is convert one enolate to a ketone and vice versa. However, because the reaction is reversible, it allows the enolates and ketone to reach equilibrium. Since enolate **A** is thermodynamically more stable, more of this is present at equilibrium. If an electrophile is added for the enolate to react with once the equilibrium has been established, more product will be formed from enolate **A**.

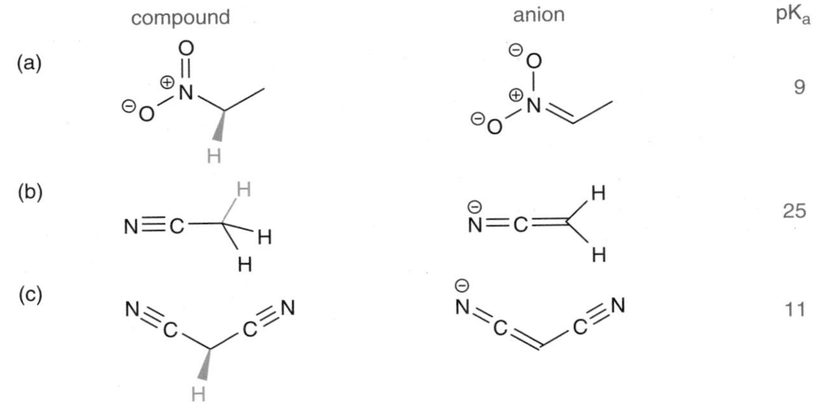

Fig. 13.49 The pK_a of the conjugate acids of some enolate-like compounds. In each case the proton coloured red (or its equivalent) is removed. In each case the negative charge can be stabilized on either an oxygen or a nitrogen.

Fig. 13.50 Iso-surface plots of the HOMOs of enols, enolate ions, and enamines. The HOMOs of the enols and enolate ions of propanone (acetone) are shown in (a), and of 2,4-pentanedione are shown in (b). In (c) the HOMO of an enamine is shown. The HOMO of the enolate ions shows electron density on alternate atoms in the π system, with the greatest electron density on the carbon. The HOMOs for the enol forms and the enamine are similar, but in these there is considerable electron density between the two carbon atoms contributing to the formal C=C double bond.

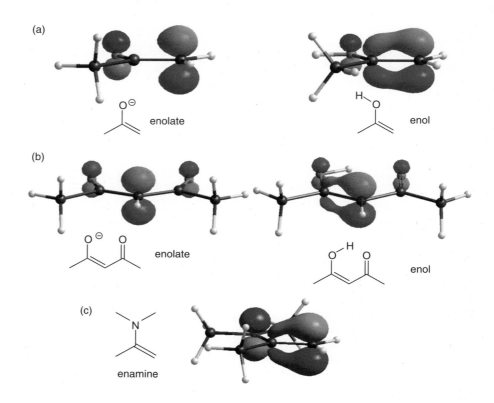

13.3.3 Other enolate-like systems

It is also possible to deprotonate carbons next to π systems other than carbonyls. We have already seen in section 9.4.3 on page 362 how the iminium ions formed from secondary amines and aldehydes or ketones are easily deprotonated to form enamines (see example (b) in Fig. 9.37 on page 363). In these reactions, the $C=NR_2^+$ π system strongly withdraws electron density from any adjacent C–H σ bonds.

Other π systems which also withdraw electron density from any σ-conjugating C–H bonds include the $-NO_2$ group and –CN. As we have seen with carbonyl groups, when a C–H bond σ conjugates to two or more of these groups, the acidity increases still further. Some examples are given in Fig. 13.49 on the preceding page.

13.4 The reactions of enols and enolates

Having seen how enolate ions can be formed in basic conditions, and enols in acidic conditions, we now want to look at how these species react. Like alkenes, these species react with electrophiles, but they are much more reactive than alkenes on account of the conjugation from the oxygen which increases the electron density in the π system.

13.4.1   The form of the HOMO of an enol and enolate ion

We saw how to predict the form of the π MOs for a chain of conjugated atoms in section 4.8 on page 165 and iso-surface plots of the π MOs of the enolate ion of enthanal were shown in Fig. 4.46 on page 175. When considering how enols and enolate ions react towards electrophiles, we are particularly interested in the HOMOs of these species. Iso-surface plots of the HOMOs for the enol and enolate ions of propanone (acetone) and 2,4-pentanedione are shown in Fig. 13.50 on the preceding page.

The HOMOs of the enolate ions show that electron density is concentrated on alternate atoms with a node on the atoms in between, just as predicted by the sine-wave rule we met in section 4.8 on page 165. The form of these non-bonding MOs resulting from the interaction of equivalent p AOs is shown in Fig. 13.51; (a) shows that for three p AOs interacting, (b) for five.

Resonance structures for these enolate ions are shown in Fig. 13.52. These also show that electron density is concentrated on alternate atoms. The structures shown in (a)(ii) and (b)(ii) give a more realistic way of representing the ions, emphasizing the delocalization over the whole π system, and how the charge is concentrated. The numbers in red give the calculated charges. These do not add up to minus one since the charges on the methyl carbons and hydrogens have not been included.

However the MOs shown in Fig. 13.50 on the preceding page are not the result of *equivalent* p AOs interacting since they have both carbon and oxygen AOs participating. The result of this is that the lowest energy π MO, which has all the p AOs in-phase, has a greater contribution from the oxygen atoms, but the HOMOs shown in Fig. 13.50 have a greater contribution on carbon.

There are two ways in which the enolate anions may react to form new bonds – *via* the oxygen or *via* the carbon. For interactions which are mainly electrostatic in nature, the bonds are preferentially formed *via* the oxygen, since this is the atom with the greatest electron density in total. In contrast, for interactions which are dominated by HOMO–LUMO interactions, leading to bonding which is more covalent in nature, bonds are preferentially formed via the carbon, since this atom has the largest coefficient in the HOMO.

The result for enols and enamines is similar but, as can be seen in Fig. 13.50, the HOMOs for these show more delocalization of the electrons, in particular between the two carbons with the formal double bond. Nonetheless, the charge distribution is similar to that in the enolate ion with the greatest electron density on the oxygen atoms, and the carbon still has the greatest coefficient in the HOMO.

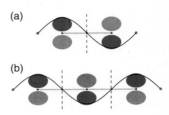

Fig. 13.51 The non-bonding π MOs formed from overlapping (a) three equivalent p AOs, and (b) five equivalent p AOs.

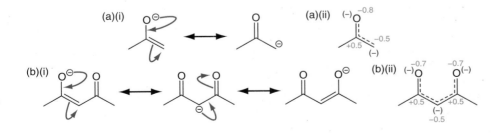

Fig. 13.52 Resonance structures for the enolate from propanone and 2,4-pentanedione are shown in (a)(i) and (b)(i), respectively. The calculated charge distributions on the atoms involved in the π system are shown in (ii). Note that the charges do not sum to −1 as the contributions from the other atoms are not shown.

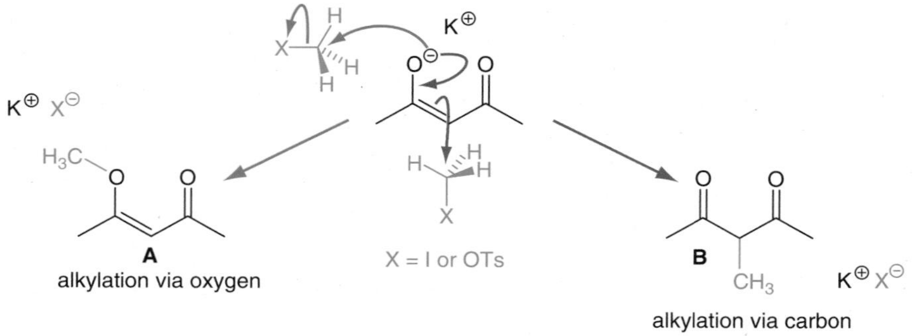

Fig. 13.53 The enolate ion reacts with the methylating agent (shown in red) either via oxygen, to give **A**, or via carbon to give **B**. Reaction conditions which favour ionic interactions favour the reaction via oxygen, whereas conditions which favour HOMO–LUMO interactions favour the reaction via carbon.

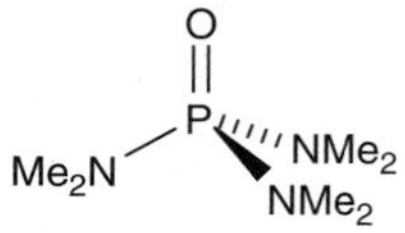

Fig. 13.54 The structure of hexamethylphosphoramide, HMPA.

13.4.2   The alkylation of enolates

As explained above, enolate ions can react via the oxygen or carbon and an example is given in Fig. 13.53 in which the enolate ion reacts to give a product methylated either on oxygen, **A**, or on carbon, **B**. When the reaction is carried out using hexamethylphosphoramide (HMPA) as a solvent and methyl tosylate as the methylating agent, the ratio of **A:B** is 97:3. HMPA, whose structure is shown in Fig. 13.54, is very good at solvating the cation K^+, but not so good at solvating anions. As a result the negative charge of the oxygen is 'exposed'. As it is attached to an electronegative oxygen atom, the methyl carbon in methyl tosylate (MeOTs) has a substantial positive charge. Both these factors help to favour alkylation via oxygen.

However, if the methylating agent is changed to methyl iodide, which has a low-energy C–I $\sigma^\star$ MO to attack into, but much less of a positive charge on the carbon, the ratio of **A:B** swings the other way completely to 3:97. Similarly, if the solvent is changed to an alcohol such as *t*-butyl alcohol which can hydrogen bond to the oxygen and help block this site, then alkylation only occurs on the carbon.

Alkylation via carbon, thereby creating new C–C bonds, is generally of far more use from a synthesis point of view than alkylation on oxygen. Some examples of the alkylation of enolates and enolate-like species are given in Fig. 13.55 on the facing page. In reactions (a), (b), and (c), the base used is sodium amide. This is a strong base which fully deprotonates the α-carbon to give the intermediate anion shown in brackets. In the second step of the reaction, this anion is alkylated on the carbon. Example (d) starts from an enamine, which reacts with the alkylating agent to give the iminium ion as an intermediate. This is hydrolysed in water to form the ketone, as we have seen in section 9.4.3 on page 362.

13.4.3   Bromination of enols and enolate ions

We saw in section 13.2.1 on page 570 how alkenes react with bromine to give 1,2-dibromo products. Enols and enolate ions readily react with bromine but no longer form 1,2-dibromo products since the C=C double bond of the enol or enolate now has far more electron density on one carbon than the other. Furthermore, no bromonium ion is formed in the reaction.

It is possible to prepare the enolate and then add bromine to form the mono-bromide as shown in Fig. 13.56 on the next page. In the first step, the

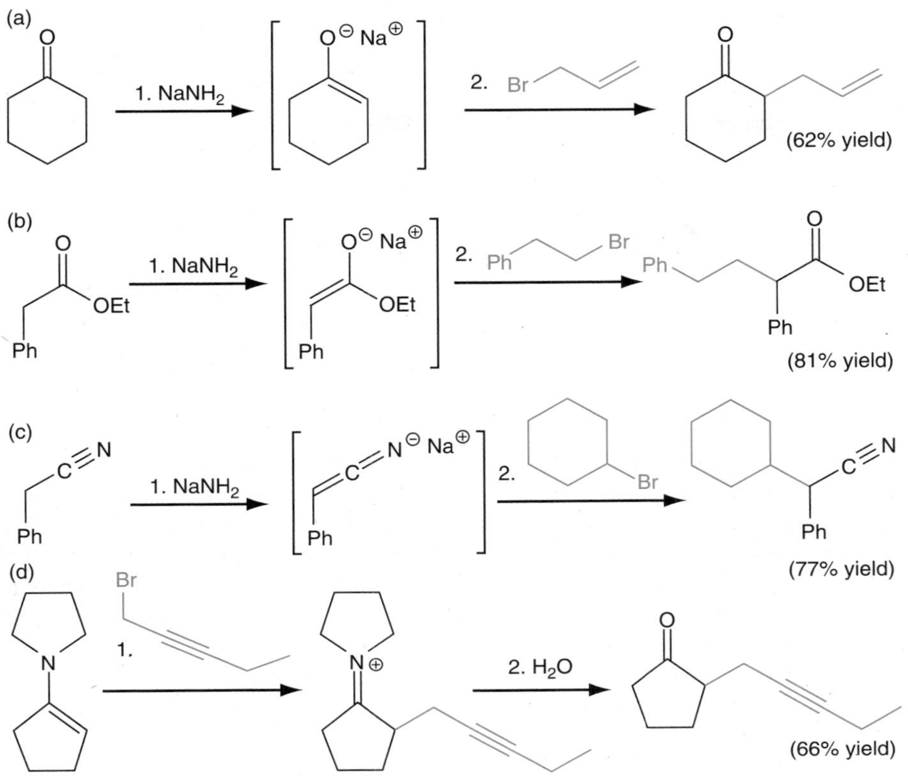

Fig. 13.55 Examples of alkylations of enolates and enolate-like species. In examples (a), (b), and (c) the strong base sodium amide is used to fully deprotonate one of the α carbons to give the anion shown in brackets. This anion then reacts, via the carbon, with the alkylating agent shown in red. In (d), an enamine reacts, again via carbon, with the alkylating agent. This initially gives an iminium ion, which is subsequently hydrolysed in water to give the ketone.

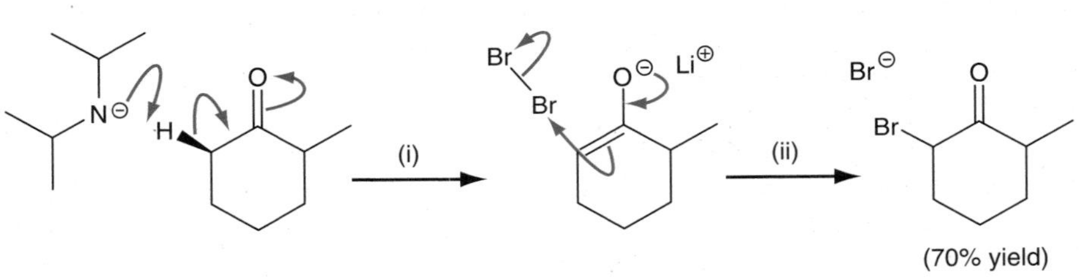

Fig. 13.56 In this reaction, the ketone is first deprotonated to give the kinetically-favoured enolate. In the second step, bromine is added and the enolate ion attacks into the Br–Br $\sigma^\star$ to form the α-bromo ketone.

base LDA is added to the ketone at room temperature. This large base gives mainly the kinetically favoured enolate ion shown. The solution is then cooled to −78 °C before adding bromine in dichloromethane. The enolate ion reacts via the carbon as shown in step (ii) to give the brominated ketone.

Bromination under acidic conditions

In the above reaction, the enolate ion was prepared in the first step before reacting with bromine in the second. However, this method is rather unusual; mono-brominated products are more usually prepared in good yield under mildly acidic conditions by a reaction involving the equilibrium concentration of enol. Two examples are shown in Fig. 13.57 on the following page.

A mechanism for the bromination of propanone (acetone) is shown in Fig. 13.58 on the next page. The key point here is that reaction between the enol and the bromine, step (iii), is extremely rapid but it is the *formation* of the

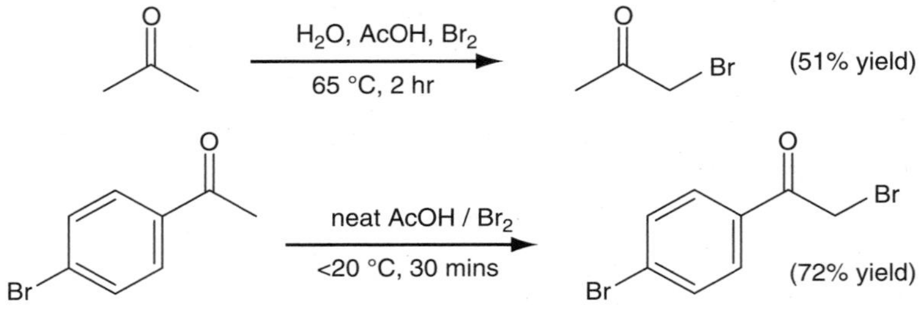

Fig. 13.57 The preparation of α-bromoketones can be carried out in mildly acidic conditions. The reaction proceeds via the enol form.

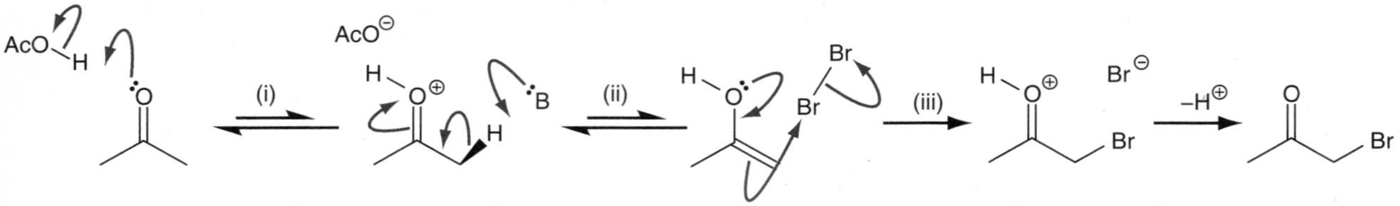

Fig. 13.58 A mechanism for the bromination of acetone (propanone) in dilute acid. Only a small portion of the ketone is ever protonated by the weak acid but this must occur in order to form the enol. In the mechanism above, B could stands for any weak base – either water, or acetate. Once the enol has formed, it reacts very quickly with the bromine, as shown in step (iii).

enol itself that limits the rate of the reaction. The enol is only ever present in very small concentrations and in order for it to form at all, the ketone must first be protonated. The equilibrium for the protonation step with the weak acid is well over to the left favouring the unprotonated propanone. After one bromine has been incorporated into the propanone, the electron-withdrawing nature of the bromine means it is even harder for the weak acid to protonate the ketone and reform more enol. This is why it is possible to obtain a good yield of the mono-brominated product.

Bromination under basic conditions

In contrast to the reaction under acidic conditions, if the bromination is carried out in the presence of a base, such as sodium hydroxide, it is generally not possible to form the mono-brominated product. This is because once one bromine has been incorporated into the structure, the electron-withdrawing nature of the bromine means it is even easier to remove another proton to form an enolate and repeat the process. This is shown in Fig. 13.59 on the facing page. The enolate formed in step (i) is quickly brominated to give the α-bromoketone **C**. This ketone is even easier to deprotonate than propanone itself, forming enolate **D**. This rapidly reacts with bromine to give the dibromoketone **E**. This in turn is even easier to deprotonate than the other ketones and so forms enolate **F**, and hence the tribromoketone, **G**. The reaction does not quite end here, since **G** reacts further with hydroxide to form bromoform.

The formation of bromoform

We saw in section 9.4.1 on page 355 how ketones are in equilibrium with very small proportions of the hydrate. When hydroxide is added to any ketone, it could always attack into the CO $\pi^\star$ MO to give the tetrahedral anion, as shown below.

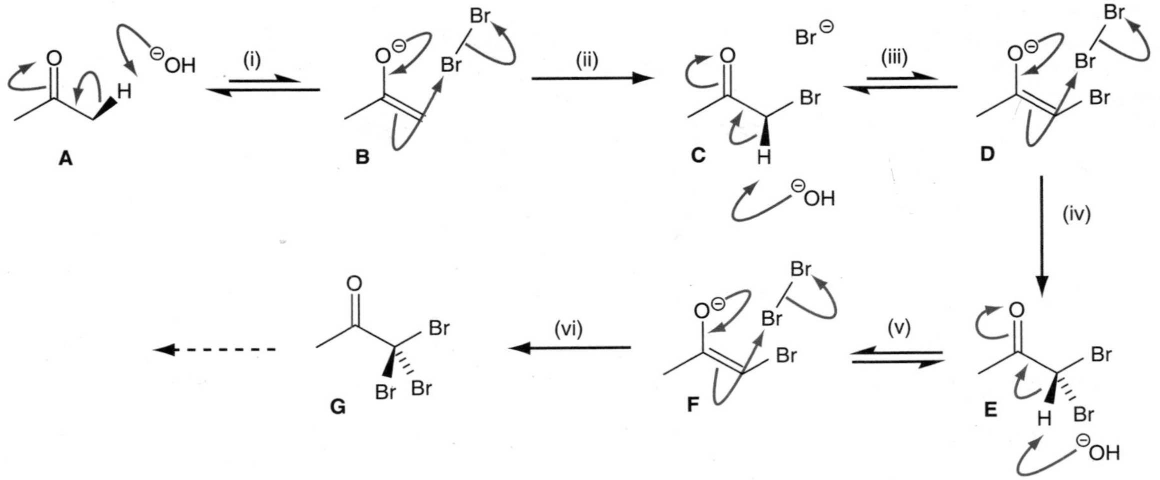

Fig. 13.59 A mechanism for the bromination of acetone in aqueous hydroxide. The incorporation of each bromine atom means the it is even easier to form the next enolate. The tri-bromoacetone **G** undergoes a further reaction with the hydroxide.

Nothing further can happen to this species except for it to re-form the ketone by expelling the only viable leaving group, the hydroxide ion. In alkaline solution this reaction happens all the time but usually does not lead anywhere. However, if hydroxide attacks into the $\pi^\star$ MO of the tribromoketone **G** there is now another viable leaving group – the $[CBr_3]^-$ ion.

The three bromine atoms can stabilize the anion sufficiently for $[CBr_3]^-$ to act as a leaving group, but having done so it quickly picks up a proton to form $CHBr_3$, called bromoform (tribromomethane). The overall reaction is shown in Fig. 13.60. A similar reaction takes place with iodine in place of bromine, and a yellow precipitate of iodoform, CHI_3, forms. This reaction used to be a chemical test for a $-COCH_3$ group before spectroscopic methods became routine. The reaction has also been used for the preparation of carboxylic acids, and an example is given in Fig. 13.61 on the next page.

13.4.4 Larger delocalized systems

As shown in Fig. 13.62 on the following page, when treated with either acid or base it is found that 3-cyclohexen-1-one **A** converts to 2-cyclohexene-1-one

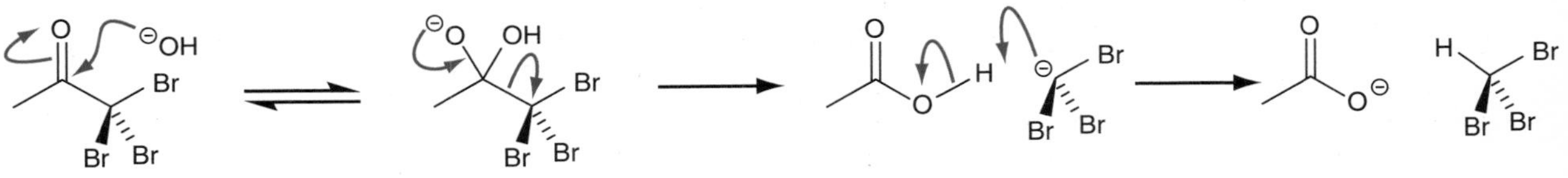

Fig. 13.60 If hydroxide attacks into the $\pi^\star$ MO of the tribromoketone, the $[CBr_3]^-$ ion could leave from the tetrahedral anion instead of hydroxide ion.

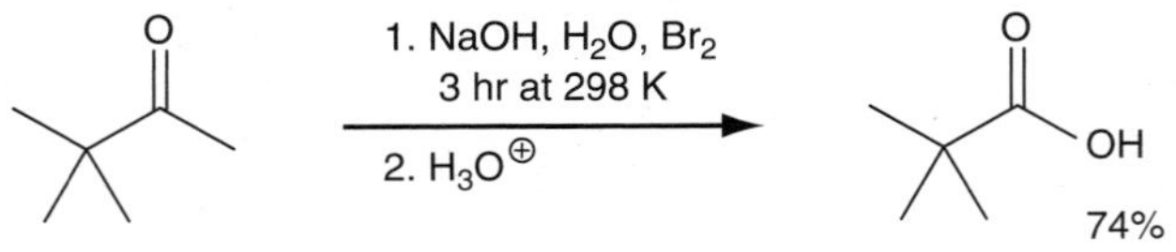

Fig. 13.61 An example of the bromoform reaction used in the synthesis of a carboxylic acid.

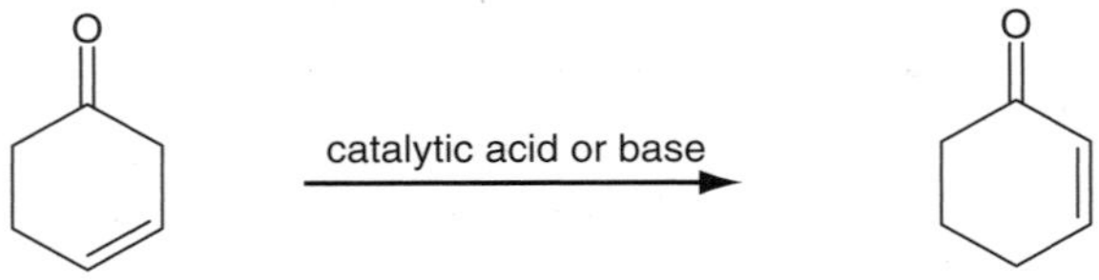

1. NaOH, H₂O, Br₂
   3 hr at 298 K

2. H₃O⊕

74%

Fig. 13.62 On treatment with acid or base, 3-cyclohexen-1-one converts to 2-cyclohexene-1-one.

catalytic acid or base

A   B

B. The conversion takes place via an enol form when in acid, or an enolate ion when in base. Considering the reaction in base, there are two sites which could be deprotonated in **A**, leading to the alternative enolates **C** and **D** shown below.

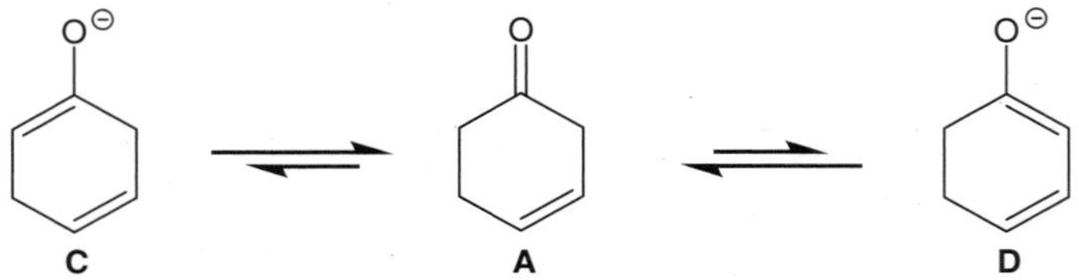

C   A   D

All that can happen with enolate **C** is that it can pick up a proton on the α-carbon and return to **A**. However, there are two possible sites of protonation for enolate **D**. There are five atoms conjugated in the π system of **D**: the oxygen and four carbons. We saw the form of the HOMO for such a system, as predicted using the simple sine wave rule, in (b) of Fig. 13.51 on page 587. An iso-surface plot of the HOMO for enolate **D** is shown in Fig. 13.63 and this broadly agrees with the simplistic picture that the greatest electron density in the HOMO is on the oxygen atom and on alternate carbons. These three sites are where the ion could be protonated and mechanisms for the protonations on carbon are shown below. The overall result is that **D** can reform **A** or **B**, allowing these two ketones to interconvert.

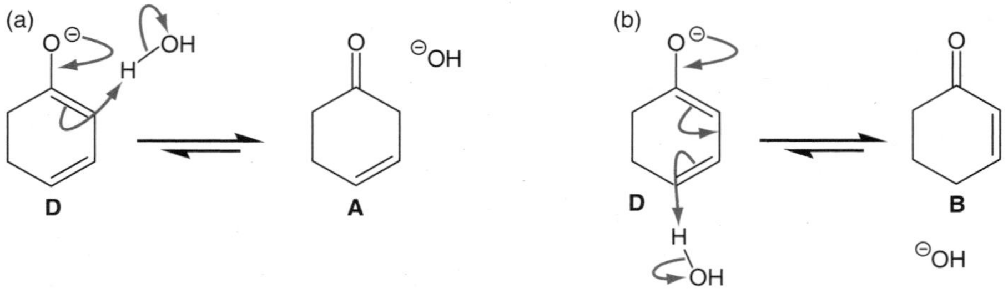

Fig. 13.63 An iso-surface plot of the HOMO of enolate **D**. Electron density is concentrated on the oxygen and on alternate carbons of the π system.

(a)

D   A

(b)

D   B

The analogous mechanism takes place via the enol form if acid is used as a catalyst instead of base. Having seen the mechanism by which ketones **A** and **B** can interconvert via the enolate **D**, we now need to understand *why*

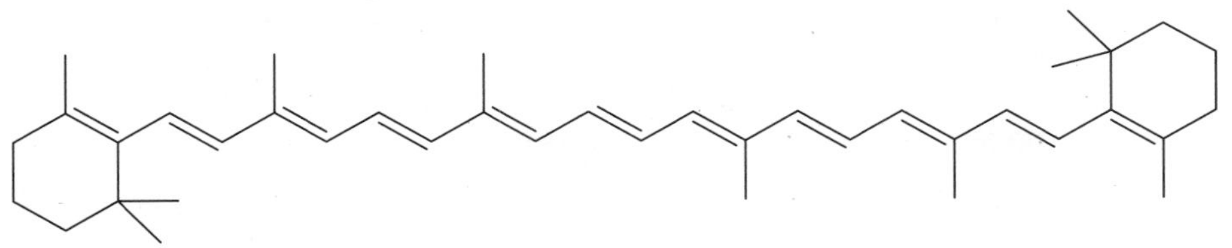

Fig. 13.64 β-carotene is the pigment that gives carrots their colour. All of the C=C bonds are conjugated with each other and this leads to a significant stabilization of the molecule.

ketone **B** is more stable than **A** and hence why 3-cyclohexen-1-one converts to 2-cyclohexene-1-one. The stability is due to the extended π conjugation that is present in **B** where the C=C is conjugated with the C=O. In **A**, there is a $-CH_2-$ group in between these two π bonds which interrupts the conjugation.

The stabilization of having electrons delocalization over a continuous chain of π bonds means that many such systems are found in nature. The pigment in carrots, β-carotene shown in Fig. 13.64, is just one example. When the conjugation extends all round a ring, an even more stable unit is formed, which brings us on to the next section where we look at some reactions of aromatic systems.

13.5   Introduction to aromatic systems

The final section in this chapter is concerned with reactions of the benzene ring, in particular electrophilic substitution reactions in which a hydrogen on the ring is substituted by another group. Despite the fact that benzene can be thought of as having double bonds, its reactivity towards electrophiles is very different to that which we have seen for simple alkenes. For example, whilst cyclohexene reacts rapidly with bromine to form 1,2-dibromocyclohexane as shown in Fig. 13.65 (a), benzene only reacts rather slowly with bromine, and even then only in the presence of a catalyst ($AlCl_3$) as shown in (b). This reaction does not involve the addition of bromine but a *substitution* of one of the hydrogens by a bromine. Note that the conjugated π system of the benzene ring is retained in the product, whereas for the simple double bond the π system is lost.

The special feature of the reactions of benzene is this strong tendency for the delocalized π system to be retained in the products. For this reason, chemists have attributed a special extra 'stability' to the π system of the benzene ring and it is often said that benzene is an especially 'stable' molecule because of this π system. As we will see in the following discussion, MO theory indicates that the kind of conjugated π system possessed by a benzene ring does indeed result in a lowering of the electronic energy.

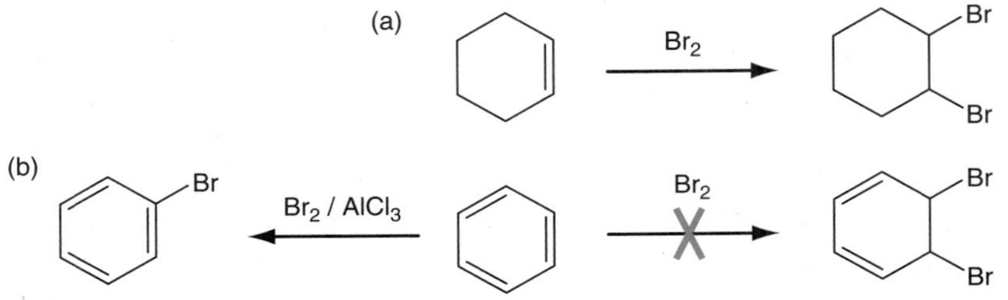

Fig. 13.65 Cyclohexene reacts as in (a) to give the 1,2-dibromo product. Benzene does not undergo an analogous reaction, but instead forms a product where the bromine has substituted for a hydrogen, and then only in the presence of $AlCl_3$ as a catalyst.

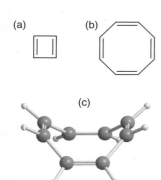

Fig. 13.66 Two examples of ring systems containing alternating double and single bonds. Cyclobutadiene, shown in (a), only has a fleeting existence and has none of the special stability associated with benzene. Cyclooctatetraene, shown in (b), is a well-known molecule, but its reactions and structure are characteristic of alternating single and double bonds. For example, it is not flat but has a 'tub-like' structure as shown in (c).

It is important to realize that not all cyclic compounds with alternating double and single bonds have the special stability which is associated with benzene. Cyclobutadiene, illustrated in Fig. 13.66, only has a fleeting existence, and the structure of cyclooctatetraene is such as to preclude extensive delocalization round the ring. As we will see, the special feature of benzene is that it has *six* electrons in its π system, as opposed to the four electrons in cyclobutadiene and the eight in cyclooctatetraene.

Traditionally, compounds in which the structure is centred on a substituted benzene ring are referred to as *aromatic*. No doubt the name originally arose from the fact that many of these compounds were found to have characteristic odours. The term aromatic is now used to describe any cyclic conjugated π system with a special number of electrons, such as six, which is known to lead to lower electronic energy.

Comparing the π MOs of benzene and cyclobutadiene

Figure 13.67 shows iso-surface plots of the π MOs of benzene and their relative energies. The lowest energy MO is a result of all of the carbon $2p$ AOs combining in-phase; this MO is occupied by two electrons. The two next highest π MOs are degenerate and each has one nodal plane perpendicular to the plane of the ring. These two orbitals are net bonding, and together they can accommodate the remaining four π electrons. This means that all six π electrons in benzene are in bonding MOs.

The next highest energy π MOs are two degenerate antibonding MOs. Each of these has two nodes. The highest energy π MO has three nodes and is a result of each carbon p AO interacting out-of-phase.

Iso-surface plots of the π MOs for cyclobutadiene are shown in Fig. 13.68 on the next page. Two sets of MOs are shown: on the left are the MOs for a geometry in which the carbon atoms are arranged in a square, with all the C–C bond lengths equal, and on the right are the MOs for a geometry in which the carbon atoms are arranged in a rectangle.

Fig. 13.67 The π MOs of benzene. The lowest MO is an in-phase combination of the p orbitals from each carbon. The highest is where each p orbital is out-of-phase with its neighbours. The remaining four orbitals occur as two degenerate pairs. The three orbitals lowest in energy are all bonding, the three highest are antibonding. Only the bonding orbitals are occupied.

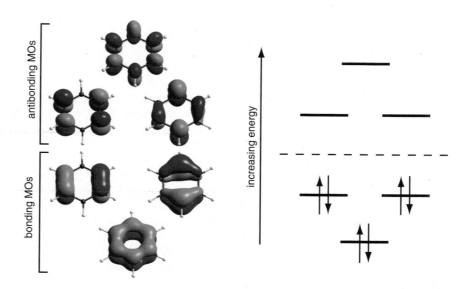

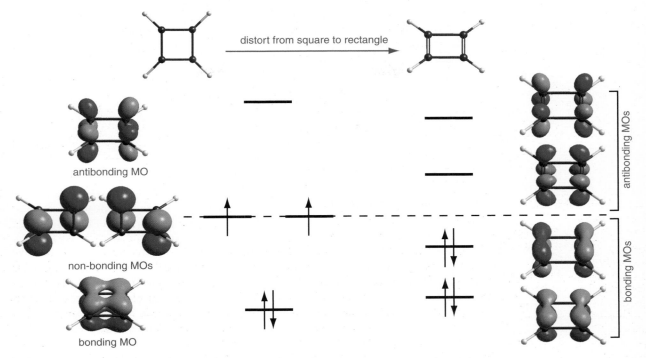

Fig. 13.68 The π MOs for cyclobutadiene with either a square geometry in which each C–C bond length is the same (shown on the left) or with a rectangular geometry in which there are two shorter double bonds and two single bonds (shown on the right). In the square geometry there are two degenerate non-bonding MOs. When the molecule distorts to the rectangular geometry on the right, one of these non-bonding MOs becomes bonding, the other becomes antibonding. The preferred structure is the one on the right with only bonding MOs occupied.

In the case of the square geometry there is one all-in-phase bonding π MO which can hold two electrons, and then two degenerate non-bonding MOs. The key point is, that if the structure distorts from the square to the rectangle by stretching two of the bonds, one of these non-bonding MOs becomes bonding and is lowered in energy, whilst the other becomes antibonding and is raised in energy. Overall since there are only four π electrons to accommodate in the orbitals, the distortion leads to a net lowering in energy.

Each of the two occupied MOs of the distorted cyclobutadiene resembles two C=C π bonding MOs. In the lowest orbital the two π MOs are in-phase with each other, in the highest occupied orbital the two π bonding MOs are out of phase with each other. Similarly, each of the two antibonding orbitals resembles two C=C π* MOs either interacting in-phase, or out-of-phase.

There is a strong contrast between cyclobutadiene on one hand and benzene on the other. In benzene the regular hexagonal structure results in MOs which are delocalized across several atoms. As a result, benzene can be described as having a fully delocalized π system, with identical bonding along each side of the ring.

In contrast, the energy of the π system in cyclobutadiene is lowered by a distortion away from the square geometry, resulting in a structure which is best described as consisting of alternating double and single bonds. Cyclooctate-traene turns out to be similar to cyclobutadiene, having a non-regular structure with alternating single and double bonds.

⚙ *Weblink 13.7*
View and rotate the π MOs of benzene and cyclobutadiene shown in Fig. 13.67 and Fig. 13.68.

All of the π MOs of cyclobutadiene and benzene have a nodal plane in the plane of the ring. This results from the fact these these MOs are constructed from p orbitals which point perpendicular to the plane of the ring.

Hückel's rule

It turns out that if we arrange N $2p$ AOs in a regular cyclic pattern (e.g. equilateral triangle for $N = 3$, square for $N = 4$, pentagon for $N = 5$, hexagon for $N = 6$ etc.) the energies of the resulting MOs fall into a simple pattern. The lowest energy MO is non-degenerate, and then the higher energy MOs all come in degenerate pairs with the exception that if the number of orbitals in the ring is even (as in the case of benzene) the highest energy MO is not degenerate.

The lowest energy MO can accommodate two electrons, and each subsequent pair of degenerate MOs can accommodate four electrons. Therefore, if the total number of electrons is 2, 6, 10 ..., all of the MOs are either fully occupied or empty. On the other hand, if the number of electrons is 4, 8, 12 ..., the highest energy pair of electrons will be assigned to a degenerate pair of orbitals, and so will occupy them singly. This is exactly the situation we had above for cyclobutadiene and so the result will be that, as in cyclobutadiene, a distortion away from the regular geometry will lower the energy, leading to a structure consisting of alternating double and single bonds.

This discussion leads to *Hückel's Rule* which states that a compound with a regular, planar ring-structure that has a continuous π system formed from overlapping p orbitals on adjacent atoms will be aromatic if the number of electrons in the π system is $4n + 2$, where n is a positive integer or zero. By 'aromatic' we mean that the π system is fully delocalized in the way it is for benzene.

In contrast, the regular, planar, ring-structure will *not* be stable with respect to a distortion if the number of electrons in the π system is $4n$, where n is a positive integer. As we have discussed, such molecules are likely to undergo a distortion so as to lower the energy, and as a result the π system will not be fully delocalized in the way it is in benzene but instead will consist of alternating single and double bonds.

Note that it is the number of electrons in the π system which is the important thing, not the number of atoms forming the ring. For example, the cyclopentadienyl anion $[C_5H_5]^-$ has five atoms in the ring but six electrons in the π system which makes it planar and hence aromatic.

13.5.1   Electrophilic addition to aromatic systems

The special stability associated with the π system of benzene governs the way it reacts. In contrast to the reactions of alkenes which usually react by losing the π structure, in benzene there is usually a strong driving force to preserve the cyclic, delocalized π system. We will start by looking at some simple reactions of phenol.

When phenol is mixed with deuterated acid and water i.e. D_3O^+ and D_2O, or when it is mixed with potassium deuteroxide (KOD) and deuterated water, it is found that some (but not all) of the hydrogens attached to the ring are exchanged for deuterium. We will start out by considering the reaction under basic conditions.

The first thing that happens when phenol is mixed with base is that the –OH group is deprotonated to give the phenolate ion, which is an enolate-like ion in the sense that it consists of –O$^-$ attached to a doubly bonded carbon. In the previous section we saw how the HOMO of an extended enolate has electron density on the oxygen and alternate carbon atoms. The same is true

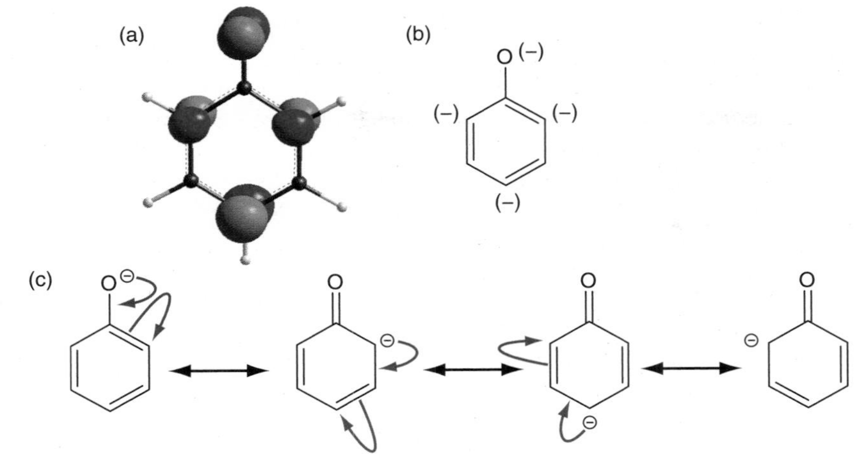

Fig. 13.69 An iso-surface plot of the HOMO of the phenolate ion is shown in (a). This shows that electron density is concentrated on the oxygen and on alternate carbon atoms, as shown in (b). This charge distribution is consistent with the predictions of the resonance structures shown in (c).

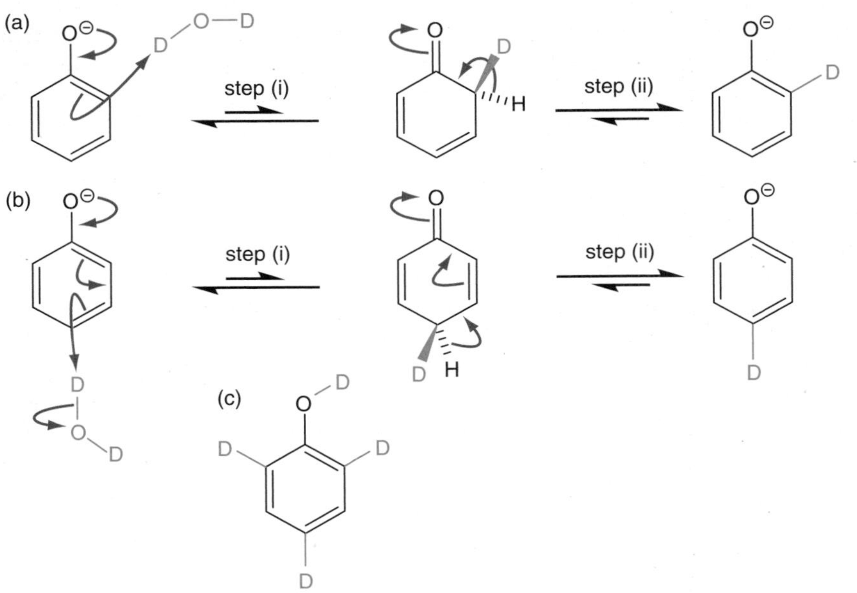

Fig. 13.70 The phenolate ion is preferentially deuterated on the carbon atoms with the greatest electron density in either the ortho positions, as shown in (a), or in the para position, as shown in (b). In both cases, the keto form is an intermediate before either a deuterium is lost in returning to the starting material, or a proton is lost leading to the product. The final product is shown in (c).

for the phenolate ion, as is illustrated by the iso-surface plot of the HOMO shown in Fig. 13.69 (a). Calculations indicate that the positions indicated by the minus signs in (b) have the highest electron density, mirroring the electron distribution from the HOMO. This distribution is also consistent with the resonance structures of the phenolate ion shown in (c). It is these positions of high electron density which become deuterated when the phenolate is mixed with D_2O; a mechanism for the reaction is shown in Fig. 13.70.

In step (i), the phenolate ion is protonated by the water on the positions where there is greatest electron density. This generates the keto form of the original enolate which can then return to the enolate by losing a hydrogen from the carbon to the right of the carbonyl. Either the deuterium which has just been added can be lost (the reverse of step (i)), or the proton attached to the same carbon can be lost, as in step (ii). Notice that both the deuterium and the hydrogen are σ conjugated with the π system, and the oxygen in particular withdraws electron density from the C–H or C–D bonds, increasing their acidity. In addition, loss of H^+ or D^+ from the tetrahedral carbon restores the aromatic π system of the benzene ring, and so we expect this to be a strongly favoured process.

It is usual to refer to the different positions on a benzene ring, relative to a particular substituent, using the terms *ortho*, *meta*, and *para*, as shown in Fig. 13.71. If we number the carbon to which the reference substituent is attached as 1, then the *ortho* positions are at carbons 2 and 6, the *meta* positions are at carbons 3 and 5, and the *para* position is at carbon 4.

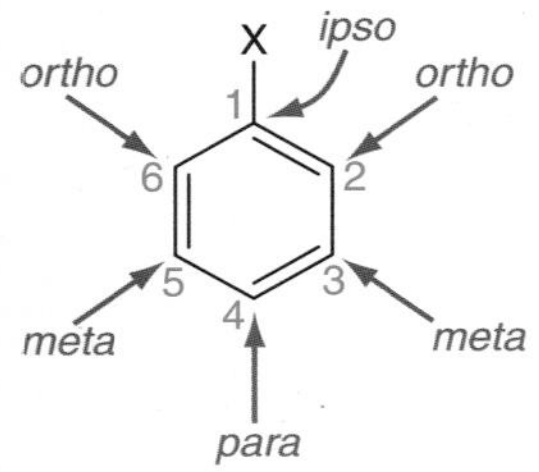

The protonation of a phenolate ion or of phenol takes place preferentially at the *ortho* and *para* positions, that is at carbons 2, 4, and 6. The final product is shown in Fig. 13.70 (c). The reason why these sites are favoured is because these are the sites with the greatest electron density.

Another way of looking at this selectivity is that it is only if the phenol reacts in the *ortho* or *para* positions that the intermediate can be stabilized by the oxygen. We will illustrate this idea by looking at the mechanism of deuterium exchange under acid conditions.

Consider the reaction of phenol with D_3O^+ shown in Fig. 13.72 on the next page. The first step is an electrophilic addition of D^+ to the ring, thereby breaking up the conjugated π system. In (a), the phenol reacts at the *para* position. There are four resonance structures **A1–A4** which can be drawn for the protonated phenol. The first three have a formal positive charge on a carbon, and the last (**A4**) has a formal positive charge on the oxygen. It is this last structure that makes a particularly significant contribution to the overall stabilization of the cation. A similar situation arises if the phenol reacts at the *ortho* positions.

In contrast, when the phenol reacts via a *meta* position, as shown in (b), there is still extensive delocalization over the π system, as shown by the resonance structures **B1–B3**, but it is no longer possible for the oxygen lone pair to participate in the stabilization. As a result, this intermediate is higher in energy and, as we will see, this leads to a slower reaction for substitution at *meta* position.

Fig. 13.71 The different positions in the benzene ring, relative to the substituent X, are referred to by the names indicated. The carbon to which the substituent is attached is sometimes known as the *ipso* position.

Kinetic effects

Figure 13.73 on page 600 shows a schematic energy profile for the reaction of D_3O^+ with phenol. Since the overall reaction is just the substitution of a proton (H) for a deuterium (D) in some position in the benzene ring there is no significant difference in energy between the starting materials and the products. However, as we have just seen, attack at the *ortho* or *para* positions leads to a more stable intermediate than attack at the meta position. Consequently, the activation energy needed to be overcome for the reaction at the *para* position is less than that for reaction at the *meta* position.

This different rate of reaction at different positions is observed for many reactions of substituted benzene rings with electrophiles. Certain substituents speed up the reaction at the *ortho* and *para* positions which leads to more product being formed at these sites. We say that different substituents on the benzene ring have *directing* effects. The oxygen atom in phenol, for example, directs electrophiles to favour attack in the *ortho* and *para* positions.

Table 13.2 Rates of proton exchange relative to benzene for substituted benzene rings

substituent on benzene ring	reaction at ortho position	reaction at para position
– OMe	73000	190 000
–Me	220	450
1,4-dimethyl	1510	
1,3,5-trimethyl	7100 000	
–F	0.14	1.8
–Cl	0.05	0.16

13.5.2 The directing effects of different substituents

We have just seen how the –OH group in phenol directs an electrophile such as a proton to the *ortho* and *para* positions in preference to the *meta* positions. The effects of the conjugation of the oxygen lone pair with the π system of the benzene ring also means that there is more electron density in the π system compared to benzene itself. This means that reactions such as protonation occur more rapidly with phenol than with benzene.

In section 13.2 on page 568 we saw how alkyl substituents on an alkene had a directing effect in the protonation of alkene which resulted in the alkene being protonated so to give the most stable cation. Alkyl groups also have a directing effect when on a benzene ring. Like oxygen, they too lead to electrophilic attack in the *ortho* and *para* positions. However, since they help to stabilize any intermediate only through σ conjugation rather than through the conjugation of a lone pair, their effect is not so strong as that of an oxygen.

Table 13.2 gives the rates of proton exchange under acidic conditions at the *ortho* and *para* postions for various substituted benzene rings. All the rates are relative to the rate of proton exchange for benzene itself. Values are only given for the *ortho* positions for 1,4-dimethylbenzene and 1,3,5-trimethylbenzene since the sites at which these molecules can be protonated are all equivalent and are *ortho* to one of the methyl groups.

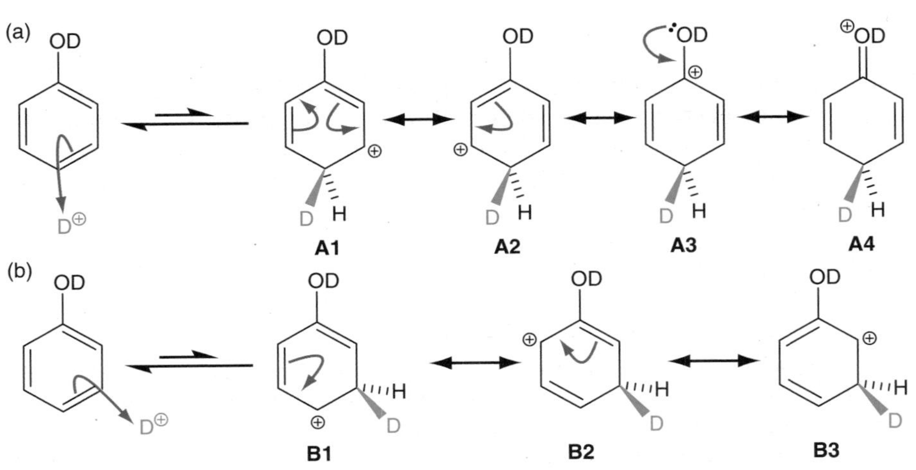

Fig. 13.72 Resonance structures for the intermediate resulting from the deuteration at (a) the *para* position and at (b) the *meta* position. If phenol reacts with an electrophile in the *para* position (or the *ortho* position), then the positive charge can also be stabilized by the oxygen, as shown in the resonance structure **A4**. If the phenol reacts in the *meta* position, whilst there are three resonance structures possible, **B1**, **B2**, and **B3**, none of these have any stabilization from the oxygen.

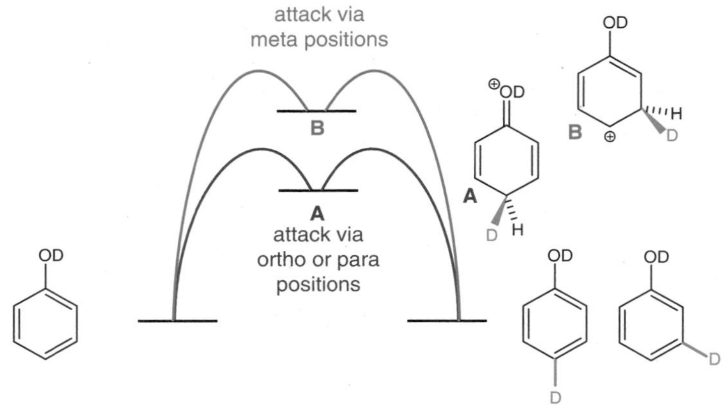

Fig. 13.73 An energy-profile for the deuteration of phenol at the *ortho*, *meta* or *para* positions. The starting materials and products have essentially the same energies, but the intermediates for attack at the *para* or *ortho* positions are lower in energy than that for attack at the *meta* position. Consequently, there is a larger activation energy for reaction at the *meta* position and so this reaction proceeds more slowly than at the other positions.

The first point to note is that in each case, the reaction occurs faster at the *para* position than at the *ortho* position. This is mainly just because the size of the substituent makes it harder for any electrophile to approach the positions on either side of it – the *ortho* positions.

An –OMe group is several hundred times more effective at speeding up the reaction than a single methyl group. However, the effects of the methyl groups are cumulative and with three methyl groups on the ring the reactions is about one hundred times faster than with a single –OMe group.

The halogens tend to slow the reaction down. This is because they are not so good as oxygen at donating electron density via the π system, but they are electronegative and generally reduce the total amount of electron density in the benzene ring. However, the halogens do still generally direct electrophiles to the *ortho* and *para* positions.

Electron-withdrawing groups

Groups which strongly *withdraw* electron density from the π system slow down the rate of reaction with electrophiles, and usually direct electrophilic attack towards the *meta* position. We can consider this in two ways as we did before – either by considering how the substituent affects the electron density in the starting material itself, or by considering the energy of the intermediate once formed by an attack of an electrophile in different positions.

Figure 13.74 shows four resonance structures of acetophenone. The oxygen of the carbonyl group withdraws electron density towards itself and structure **A1** is just the usual resonance structure of the carbonyl that seen before (for

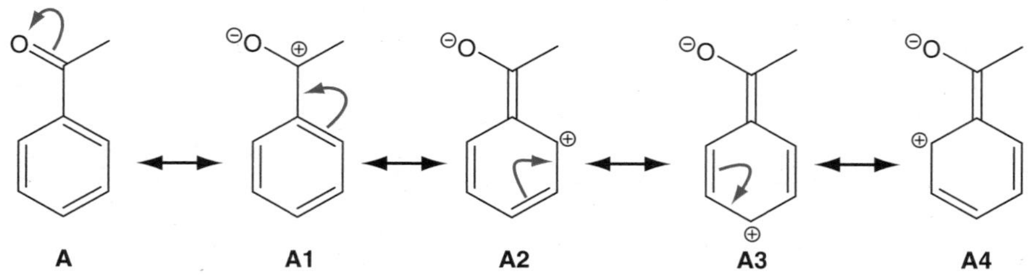

Fig. 13.74 The carbonyl group in acetophenone withdraws electron density from the π system. **A1–A4** are resonance structures showing the specific sites from which most electron density is withdrawn.

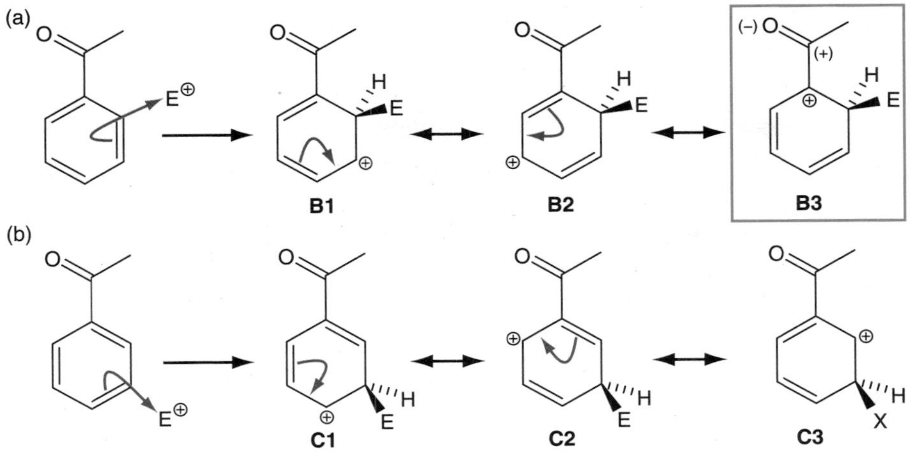

Fig. 13.75 Different resonance structures for the cation formed when an electrophile reacts with acetophenone in (a) the *ortho* position, and (b) the *meta* position. The resonance structure **B3** is particularly high in energy since there are two carbon atoms with significant positive charges on adjacent atoms.

example in Fig. 8.26 on page 331). However, electron density is withdrawn from across the π system in the benzene ring too and resonance structures **A2**, **A3**, and **A4** show this. Together they suggest that more electron density is withdrawn from the *ortho* and *para* positions more than from the *meta*. This conclusion is supported by detailed calculations. If an electrophile were to attack such a system, it would preferentially attack the π system where there is greatest electron density i.e. in the *meta* position.

The reaction between acetophenone and a general electrophile, denoted E⁺, is shown in Fig. 13.75. In (a), the attack is at the *ortho* position. **B1**, **B2**, and **B3** are three resonance structures for the resulting intermediate. One of these structures, **B3**, is particularly unfavourable. This is because in this structure the positive charge is adjacent to the carbonyl carbon, which is itself already significantly positively charged. This destabilizing interaction means that this intermediate is particularly high in energy. A similar result occurs if the electrophile attacks in the *para* position.

In contrast, if the electrophile attacks in the *meta* position as shown in (b), in none of the resonance structures is there a formal positive charge on the carbon adjacent to the carbonyl carbon. Thus the intermediate from attack in the *meta* position is lower in energy than that resulting from attack at the *ortho* or *para* positions.

Since the resulting intermediate during the *meta* attack is lower in energy than that for *ortho* or *para* attack, the former reaction is faster which means that electron withdrawing groups tend to give more *meta* product during electrophilic substitution reactions.

Summary of the directing and activating properties of different substituents

Figure 13.76 on the following page summaries the effects of different substituents towards electrophiles. Since nitrogen has higher energy electrons than oxygen, the nitrogen from an amine is the most effective group for increasing the amount of electron density in the π system of benzene and hence activating it towards electrophilic attack.

The halogens are a complex case. They have lone pairs which could in principle conjugate and increase the electron density in the ring, but these electrons are low in energy, and become increasingly mismatched in size on

Fig. 13.76 A summary of the directing and activating effects of different substituents on a benzene ring.

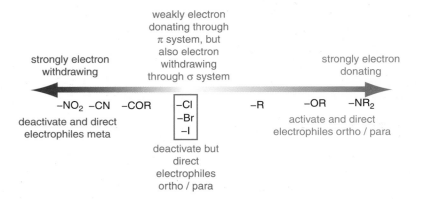

moving from Cl to Br, to I. Being electronegative, the halogens can withdraw electron density from the ring as a whole which deactivates it towards reactions with electrophiles. The experimental observation is that halogens deactivate the ring, but still direct electrophiles towards the *ortho* and *para* positions.

Carbonyl groups and cyanides (nitriles) withdraw electron density and direct electrophiles to the *meta* positions, but the most effective group for doing this is found to be the –NO₂ nitro group. Examples of the directing properties of different groups in specific reactions will be seen in the following sections.

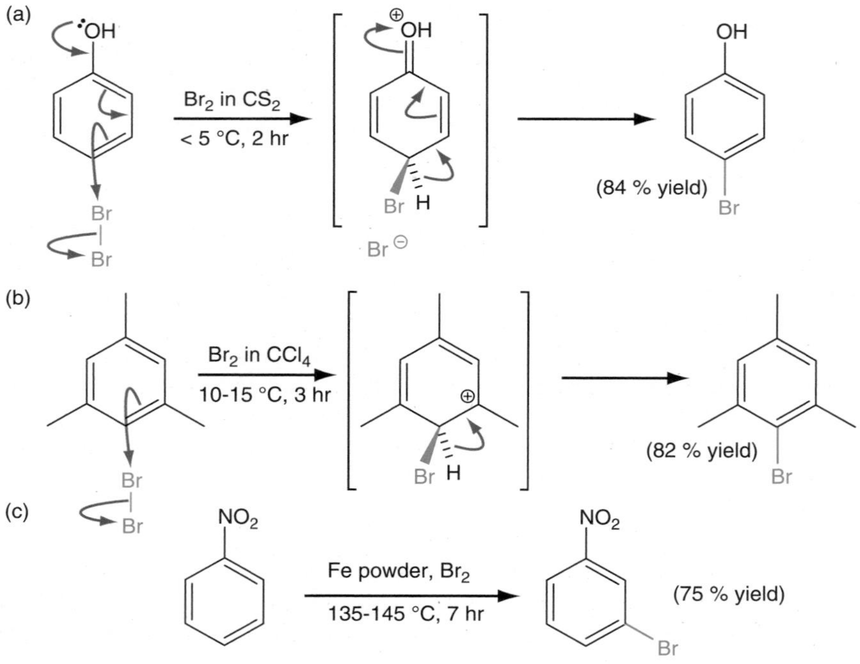

Fig. 13.77 Examples of aromatic bromination. In (a) phenol is brominated preferentially in the *para* position by using a dilute solution of bromine in carbon disulfide. In (b) trimethylbenzene is brominated in good yield. To brominate nitrobenzene, as in (c), an iron catalyst is required, together with a higher temperature and a longer reaction time. The nitro group deactivates the benzene ring by withdrawing electron density and gives predominantly the *meta*-substituted product.

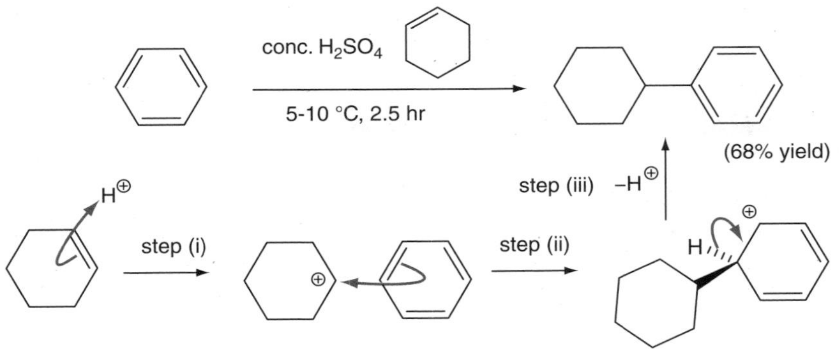

Fig. 13.78 Cyclohexylbenzene may be prepared from cyclohexene and benzene with concentrated sulfuric acid. The cyclohexene is first protonated to give the carbenium ion which is then attacked by the benzene ring. Deprotonation gives the desired product.

13.5.3 Aromatic substitution: bromination

Whilst bromine only reacts with *benzene* in the presence of a catalyst, if there is an activating group on the ring, such as oxygen, bromination is rapid and more than one bromine may be substituted into the ring. For example, phenol reacts to give 2,4-dibromophenol and 2,4,6-tribromophenol. The extent of bromination depends strongly on the solvent and other conditions. For example, in solvent CS_2 and at temperatures around 0 °C *para*-bromophenol can be obtained in good yield, as shown in (a) from Fig. 13.77 on the preceding page.

1,3,5-trimethylbenzene reacts with bromine without any catalyst to give a good yield of the mono-bromo product as shown in (b). In contrast, stronger conditions are needed to brominated nitrobenzene, as shown in (c). The deactivating nature of the $-NO_2$ group means that a catalyst is now needed, together with higher temperatures and longer times. The product is *meta*-bromonitrobenzene.

13.5.4 Aromatic substitution: alkylation

We have seen how enolates are sufficiently reactive that they can be readily alkylated without a catalyst using just an alkyl bromide. In contrast, aromatic systems, even ones with activating groups such as an $-OMe$ group, need a more reactive electrophile or the assistance of a Lewis acid catalyst.

As an example, consider the reaction between benzene and cyclohexene in concentrated sulphuric acid to give cylcohexylbenzene, as shown in Fig. 13.78. In the first step of the reaction, the cyclohexene is protonated to give a reactive carbenium ion. Electrons from the π system of the benzene ring then attack the carbenium ion, and finally a proton is lost to regain the aromaticity of the benzene ring.

In this example, the alkylating agent is the carbenium ion generated by the protonation of the cyclohexene. The reactive carbenium ion necessary for alkylation may also be generated using an alkyl halide and Lewis acid in a reaction known as *Friedel–Crafts alkylation*. A mechanism showing the role of the Lewis acid is shown below.

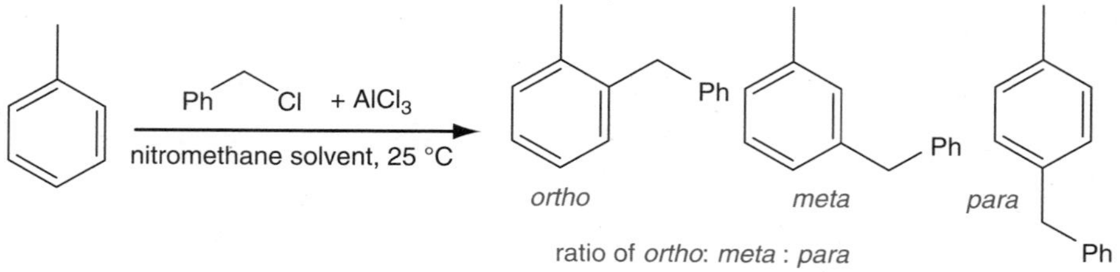

Fig. 13.79 An example of a Friedel–Crafts alkylation. The *ortho/para* directing effect of the methyl group is apparent from the product ratios.

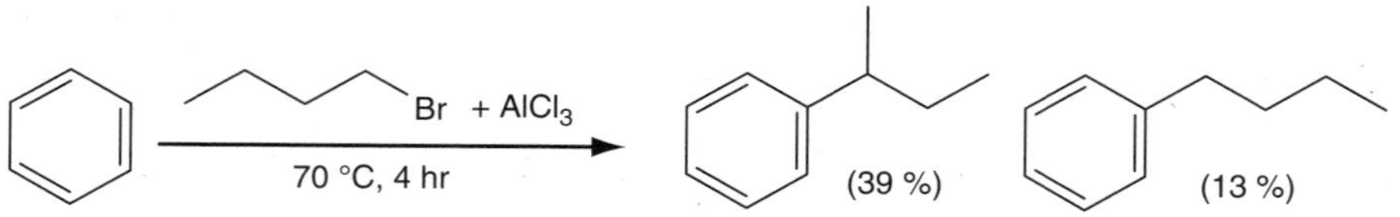

ratio of *ortho*: *meta* : *para*

44 : 4 : 52

Fig. 13.80 In the alkylation of benzene using 1-bromobutane and AlCl₃, more of the branched product **A** is formed than the straight-chain butyl-benzene, **B**.

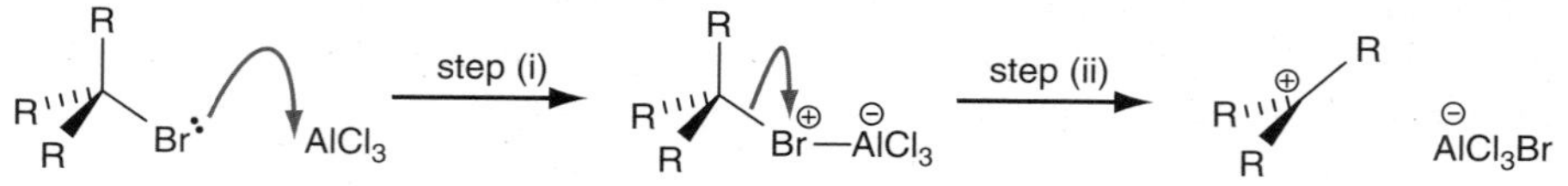

(39 %)   (13 %)

step (i)   step (ii)

In the first step, a lone pair from the alkyl bromide attacks into the vacant *p* orbital of the Lewis acid, the AlCl₃. The bromide gains a formal positive charge and withdraws more electron density from the C–Br bond towards itself, making it easier for it to leave. The C–Br bond may well not break entirely – this depends on the reaction conditions and on the nature of the R groups in the alkyl bromide. An example of a Friedel–Crafts alkylation is given in Fig. 13.79.

Problems with the Friedel–Crafts alkylation

There are two main problems with Friedel–Crafts alkylation: multiple substitution, and rearrangements of the carbenium ion. As we have seen, alkyl groups are activating towards electrophilic substitution reactions. This means that once the alkylation has taken place, the product is actually more reactive than the starting material. Consequently, the product will react in preference to further un-alkylated starting material. For example, any attempt to methylate benzene with methyl chloride and AlCl₃ yields a mixture of mono-, di-, tri-, tetra-, penta-, and hexamethyl benzenes in various proportions depending on the precise conditions used.

The other problem is that the carbenium ion, once formed, can undergo a rearrangement reaction to form a carbenium ion that is lower in energy before it reacts with the aromatic system. An example is given in Fig. 13.80.

If any carbenium ion formed from the bromobutane and AlCl₃, it would be a primary cation, which, as we have seen in section 9.7.1 on page 390, is not as low in energy as a more substituted cation. It is possible for the primary carbenium ion to undergo a rearrangement to give the more stable cation by what amounts to a 'hydride shift' as shown in Fig. 13.81. In this reaction a hydrogen and its bonding electrons appear to move from one carbon to another. It is not the case that a hydride ion H⁻ ever dissociates. In fact it would be more reasonable to think of this reaction as involving the loss of a proton to form the alkene (as we have seen in elimination reactions) and then re-protonation of

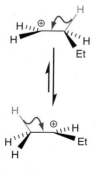

Fig. 13.81 The primary cation above can rearrange to the more stable secondary cation below by a so-called 'hydride-shift'.

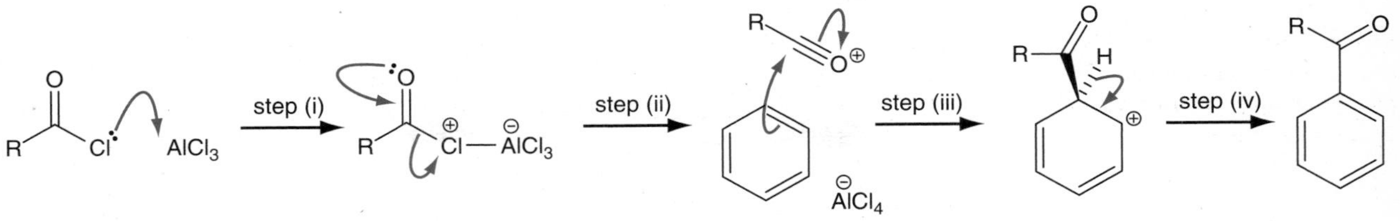

Fig. 13.82 A mechanism for the Friedel–Crafts acylation of benzene from an acyl chloride and Lewis acid, $AlCl_3$.

the alkene to give the more stable carbenium ion. The mechanism is probably somewhere in between these two extremes and is usually represented as shown in Fig. 13.81.

13.5.5 Aromatic substitution: Friedel–Crafts acylation

In contrast to the Friedel–Crafts alkylation of benzene systems where re-arrangements and multiple substitutions are a problem, Friedel–Crafts *acylation* proceed with fewer difficulties. This uses an acyl chloride and Lewis acid rather than an alkyl halide. A general mechanism is given in Fig. 13.82.

In the first step the HOMO of the acyl chloride, the chlorine lone pair, attacks into the vacant p orbital of the $AlCl_3$. This makes the chlorine a much better leaving group, and when it leaves, with assistance from the oxygen lone pair, an acylium ion $R–CO^+$ is generated, as shown in step (ii). In step (iii) the low-energy LUMO of this ion is then attacked by the benzene in the usual manner.

Multiple substitution is not usually a problem with Friedel-Crafts acylation since the product is less reactive than the starting material due to the electron-withdrawing nature of the carbonyl that is conjugated with the π system of the benzene ring. The acylium ion does not undergo any rearrangements since the positive charge is already stabilized by the oxygen.

Two examples of Friedel–Crafts acylation are given in Fig. 13.83. Notice in reaction (a) that only the more reactive acyl chloride reacts – no reaction is

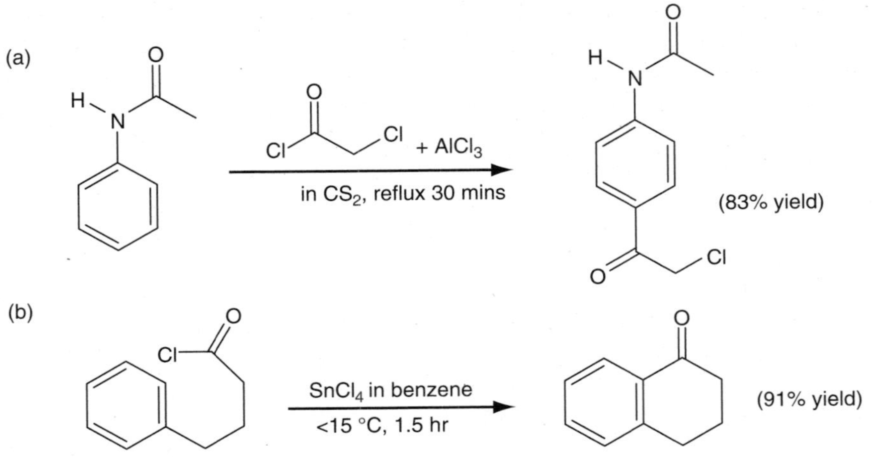

Fig. 13.83 Examples of Friedel–Crafts acylation. In (a) only the acyl chloride reacts with the benzene ring: there is no substitution of the less reactive alkyl chloride. In (b) the acylation leads to a ring closure.

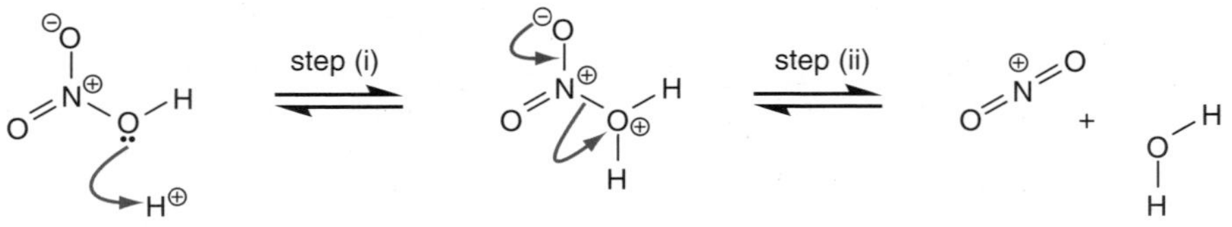

Fig. 13.84 A mechanism for the formation of the nitronium ($[NO_2]^+$) ion from nitric acid. The nitric acid is protonated in step (i) and then water is eliminated in step (ii). Note that the nitronium ion is linear.

seen via the alkyl chloride group in the reagent. The electron-donating nitrogen of the amide directs the acylation to the *para* position. Example (b) is an intramolecular reaction which uses tin(IV) chloride as the Lewis acid catalyst.

13.5.6 Aromatic substitution: nitration

Benzene rings can easily be nitrated in good yields using the classic nitrating mixture of concentrated nitric and sulfuric acids. The actual electrophile that is attacked by the aromatic π system is the nitronium ion $[NO_2]^+$. This is generated in the way shown in Fig. 13.84: first the nitric acid is protonated and then water is lost in a dehydration step. The sulfuric acid acts as both an acid in the first step and a dehydrating agent in the second.

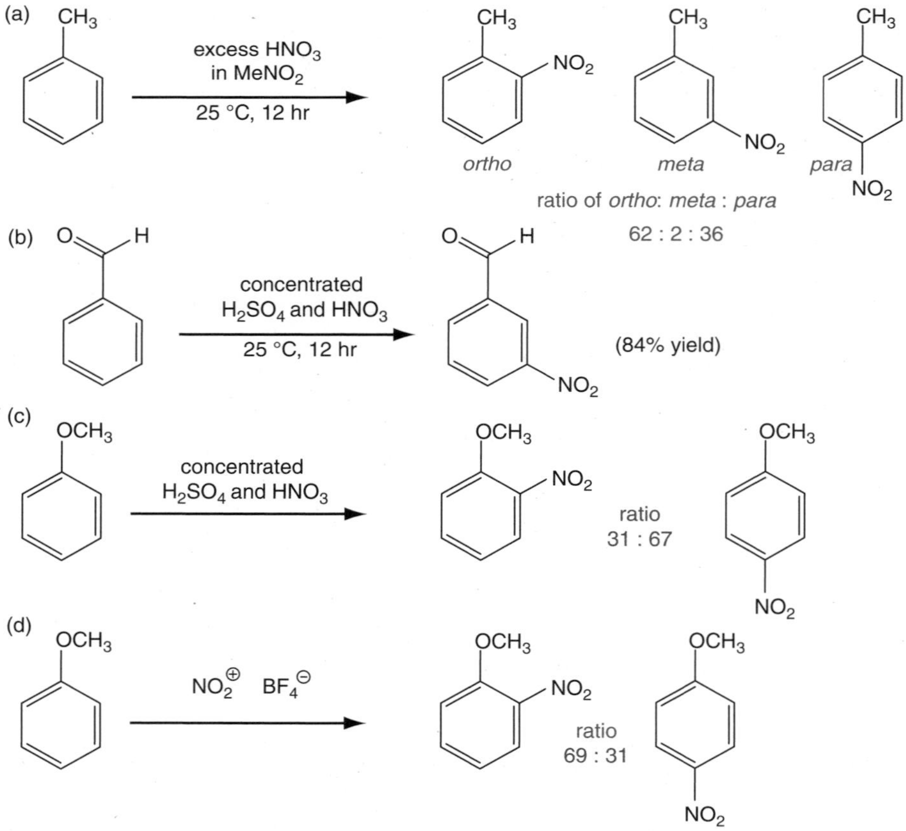

Fig. 13.85 Some examples of aromatic nitration. In (a) nitric acid alone is the nitrating agent, whereas in (b) and (c) the more usual mixture of concentrated sulfuric and nitric acids is used. In (d) the nitronium ion $[NO_2]^+$ is added directly as the salt nitronium tetrafluoroborate.

The nitronium ion is attacked by the aromatic system in the usual manner, with addition being followed by elimination of H^+ to restore the aromatic ring. Some examples are given in Fig. 13.85 on the facing page. In (a), nitric acid alone is used as the nitrating agent, whereas in (b) and (c) the more usual mixture of concentrated sulfuric and nitric acids is used. Note the directing effect of the methyl group in (a) resulting in more of the *ortho* and *para* products being formed. In (b) the carbonyl of the aldehyde withdraws electron density from the ring and directs the nitration primarily to the *meta* position.

In examples (c) and (d) methoxybenzene is nitrated using two different nitrating agents. When concentrated nitric and sulfuric acids are used, more of the *para* product than the *ortho* is formed. This may well be due to the acids hydrogen bonding to the oxygen of the methoxy group, thereby slightly hindering the approach of the electrophile to the *ortho* positions. In (d) nitronium tetrafluoroborate is used as the nitrating agent. This salt is commercially available and has the non-nucleophilic tetrafluoroborate ion as the counterion to the reactive nitronium ion $[NO_2]^+$. The products are still mainly *ortho* and *para* substituted, but now much more of the *ortho* product is formed.

FURTHER READING

Organic Chemistry, Jonathan Clayden, Nick Greeves, Stuart Warren and Peter Wothers, Oxford University Press (2001).
March's Advanced Organic Chemistry, sixth edition, Michael B. Smith and Jerry March, Wiley (2007).

QUESTIONS

13.1   Compounds **A** and **B** both react in aqueous ethanol to give isobutene and *t*-butyl alcohol. **A** and **B** react at different rates, but give exactly the same proportions of the two products. What does this suggest about the mechanism for the reactions?

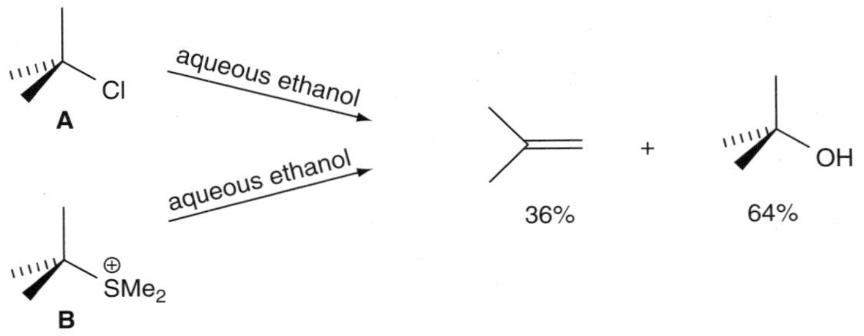

13.2   Both the *cis* and *trans* bromoalkenes shown below react with NaOH to form the alkyne, but the *cis* isomer reacts 2.1×10^5 times faster than the *trans* isomer. Explain why this is so.

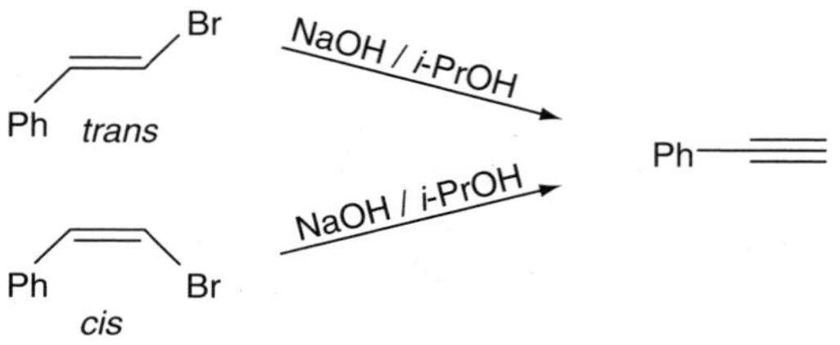

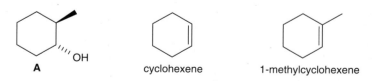

13.3 When 1,2-dimethylcyclohexene, **A**, is treated with dilute acid, diastereoisomers **B** and **C** are formed in approximately equal proportions (together with their enantiomers). No matter how long the mixture is left standing, only around 10% of **A** is found to have reacted. With the aid of a suitable mechanism, account for these observations.

A + enantiomer **B** + enantiomer **C**

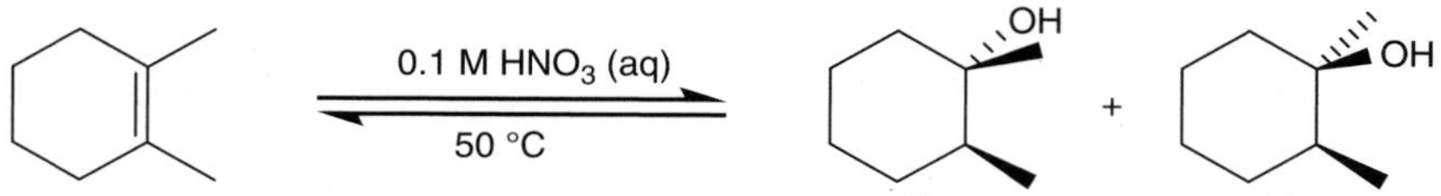

13.4 2-Bromoethyl nitrate, shown below, may be obtained in good yield by bubbling ethene gas through an aqueous solution of bromine and sodium nitrate. Give a mechanism for this reaction.

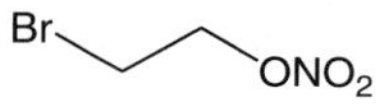

13.5 The diastereoisomers *cis*- and *trans*-but-2-ene both give 2,3-dibromobutane on reaction with bromine. However, one alkene gives only the achiral *meso* form of 2,3-dibromobutane, whilst the other gives only a racemic mixture of the optically active diastereomer. With the aid of clear mechanisms, explain these observations.

13.6 Compound **A**, shown below, may be prepared either from cylohexene via an epoxide, or by the hydroboration of 1-methylcyclohexene. Give mechanisms for each of these reactions and account for the observed stereochemistry.

A

cyclohexene

1-methylcyclohexene

13.7 In the presence of a trace of acid or base, the optically active ketone shown below readily forms a racemic mixture. With the aid of appropriate mechanisms, explain how this occurs.

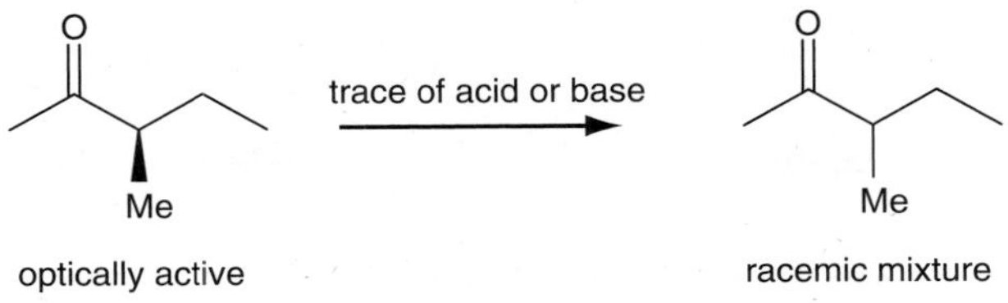

optically active → racemic mixture

13.8 A synthesis of the drug *Amobarbital* (also called amylobarbitone) is shown below.

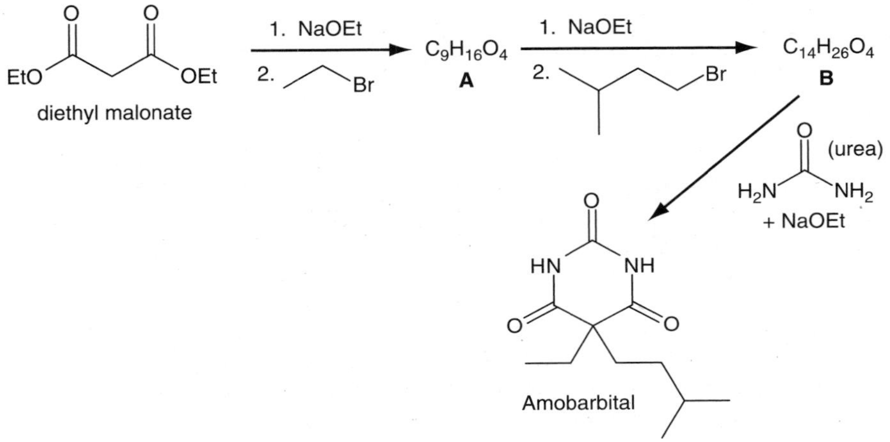

(a) The sodium ethoxide can react with the diethyl malonate either as a nucleophile, or as a base. Give the mechanisms for each of these reactions.

(b) Identify the intermediates **A** and **B**, and give mechanisms for their formation.

(c) Intermediate **B** reacts with urea and NaOEt to form amobarbital. What is the role of the NaOEt? Give a mechanism for the formation of amobarbital.

13.9 Both 5-androstene-3,17-dione, **A**, and 4-androstene-3,17-dione, **B**, may be deprotonated in hydroxide to give a common anion, **X⁻**. The pK_a of **A** to form the anion is 12.7, whereas the pK_a of **B** is 16.1.

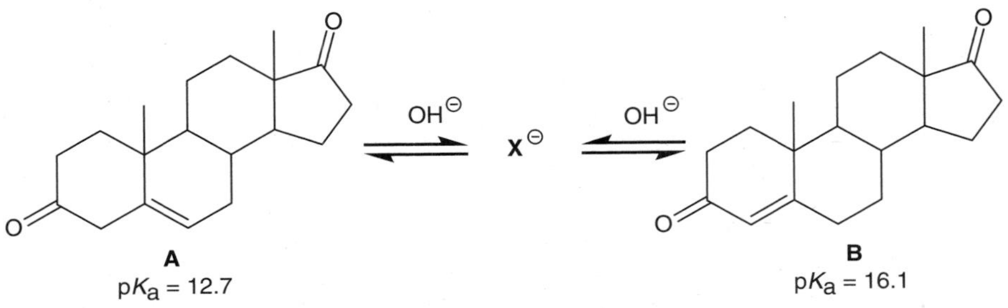

(a) Identify the acidic proton in **A** and in **B** which must be lost to form the common anion **X⁻**. Give a mechanism in each case.

(b) Suggest why the pK_a of **A** is lower than that of **B**, and why both are lower than the pK_a of acetone (around 20).

(c)   Give expressions for $K_a(A)$ and $K_a(B)$, the acidity constants for **A** and **B**, and calculate the values of these equilibrium constants from the given pK_a values.

(d)   Write an expression for the equilibrium constant for $A \rightleftharpoons B$ in terms of $K_a(A)$ and $K_a(B)$, and calculate its value. Comment on your answer.

13.10   In the presence of a catalytic amount of acid or base, 4-methylcyclopent-2-enone, **A**, isomerizes to 3-methylcyclopent-2-enone, **B**. Give a mechanism for this reaction. Why is **B** favoured at equilibrium rather than **A**?

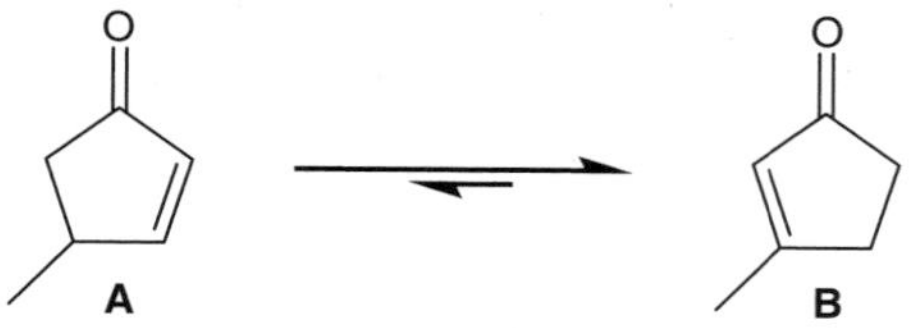

13.11   A synthesis of the drug *Normethadone* is shown below. Give the mechanisms for all the reactions involved in the synthesis.

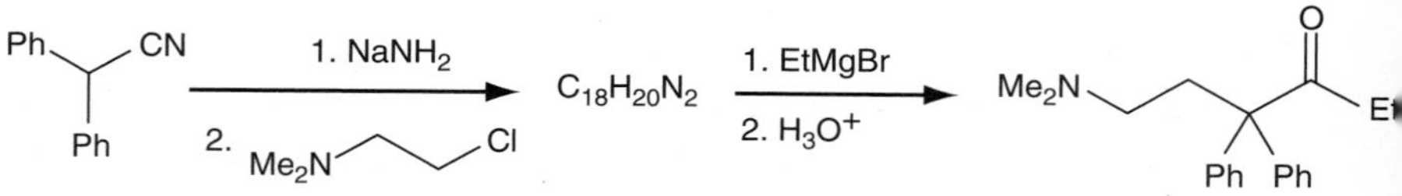

Normethadone

13.12   A synthesis of the bronchodilator *Clenbuterol* is shown below.

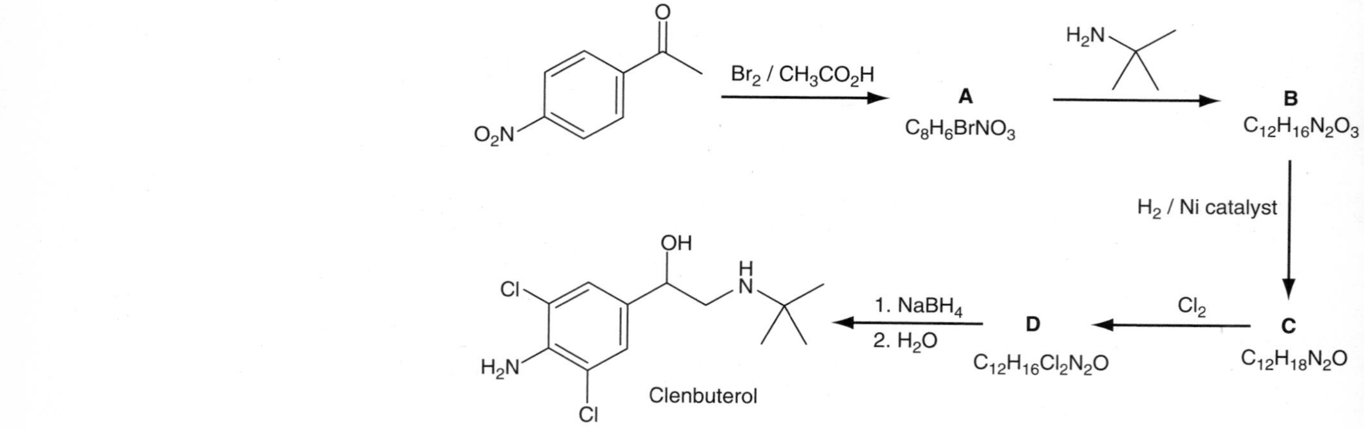

Clenbuterol

(a)   Identify the intermediates **A–D**.

(b)   Give a mechanism for each step of the reaction scheme, except for the catalytic reduction of **B** to **C**.

(c)   How do the two substituents on the benzene ring affect the reaction with chlorine in the formation of **D** from **C**?

Main-group chemistry

Key points

- The main-group elements are those in Groups 1, 2, and 12–18.

- Going *across* the periods we see a trend from metals to non-metals, but for Groups 13–16 we see a trend from non-metals to metals as we go *down* the groups.

- The chemistry of Groups 1 and 2 are dominated by ionic species with oxidation states +1 and +2, respectively.

- The elements from Groups 12–18 show variable oxidation states; within each group the commonly observed oxidation states are often separated by steps of two.

- For a given element the highest oxidation state of (group number minus 10) is often only shown by the fluoride or oxide of that element.

- Towards the bottom of Groups 13–16, an oxidation state of (group number minus 12) becomes increasingly common (the 'inert pair' effect).

- Metal halides may be hydrolysed, depending on the acidity of the aquo ions.

- Hydrolysis of non-metal halides gives acidic solutions.

- Metal oxides tend to be basic or amphoteric, while non-metal oxides tend to be acidic.

This chapter is concerned with the chemistry of what are usually called the 'main-group' elements, which are those in the *s*- and *p*-blocks, i.e. Groups 1, 2 and 13–18, along with Group 12 (see Fig. 14.1 on the next page). Although the main-group elements are very diverse, ranging from the most typical metals, through semi-metals to the most typical non-metals, the reason that they are often considered together is that these elements show a gradual variation in their chemical properties which can largely be understood, or at least rationalized, using a small number of key concepts.

The classification of Group 12 as belonging to the main group is at first sight a little surprising, since this group is part of the *d*-block. The Group 12 elements all have the electronic configuration $nd^{10}(n+1)s^2$, and by the time we reach these elements the *d* electrons have been dropping steadily in energy all the way across the *d*-block, so these electrons are too low in energy and the orbitals too contracted to be involved in bonding. The valence electrons are thus just

Fig. 14.1 The main-group elements are the *s*-block elements in Groups 1 and 2, along with Group 12, and all of the groups in the *p*-block.

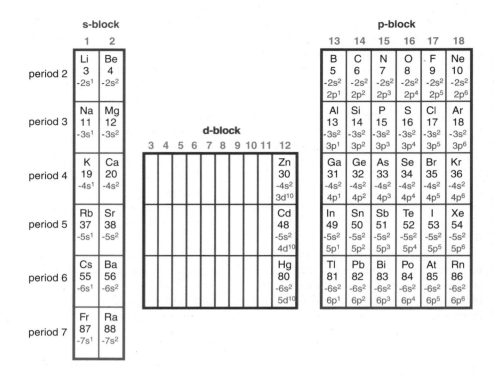

the s^2, which makes the Group 12 elements more analogous to Group 2 then to the transition metals.

This chapter makes extensive use of the ideas introduced in Chapter 7, especially the influence that the energy and size of orbitals has on the type and strength of bonding between the elements. Indeed, we will see that the chemistry of the main-group elements can be understood in terms of how these factors play out as we move across the periods and down the groups. If it has been a while since you studied Chapter 7 you might like to make a quick review of sections 7.1 to 7.5 so as to refresh your understanding of the key concepts.

There are forty-five main-group elements, so one of our problems is avoiding being overwhelmed by the sheer number of different compounds that they form – Greenwood and Earnshaw's *Chemistry of the Elements* devotes over 800 pages to the chemistry of these elements. In this chapter we are only going to focus on the most typical compounds, and especially on the halides and oxides of the elements. In this way, we will keep the topic manageable, but without loosing too much of the essential chemistry of these elements. The structures of the elements themselves have already been discussed in Chapter 7, so this chapter will focus on the compounds formed between these elements.

14.1 Overview

Before we start on the details, it is helpful to have a quick over-view of the chemistry of the main-group elements. The dominant feature is the trend from metals to non-metals as we go across the periods, and the trend from non-metals to metals as we go down the groups which form the *p*-block. These trends are illustrated graphically in Fig. 7.41 on page 295.

Groups 1 and 2 are typical metals in that the valence electrons are in large high-energy orbitals. In the elements, these orbitals overlap to give the delocalized bonding which is so characteristic of metals. The outer electrons are quite readily lost to give cations, so much of the chemistry is dominated by the formation of ionic compounds, or at least compounds with significant ionic character. Group 12 elements, although part of the *d*-block, are usually considered as belonging to the main group since the chemistry of these three elements is more reminiscent of Groups 1 and 2 than of the transition metals.

As we go across the periods the orbitals contract and their energies decrease; this results in a transition from metals to non-metals – a trend most clearly seen in Periods 2, 3 and 4. The non-metals have a tendency to form covalent bonds, especially with one another. Some of these elements, particularly those in Group 17 and at the top of Group 16, also show a tendency to form anions and hence ionic compounds. This reflects the favourable electron affinities of these elements.

As we go down Groups 13 to 16 there is a gradual transition from non-metal to metal, with a number of elements showing intermediate behaviour which is characteristic of semimetals or semiconductors. This transition reflects the increase in both the size and energy of the outer occupied orbitals. We will have a lot more to say about this transition and the consequence it has for the compounds formed by the elements which lie between typical non-metals and metals.

As has already been described in Chapter 7 (page 297), the formation of multiple bonds (via π bonding) is much more common for the non-metals in Period 2 than in the later periods. This reflects the relatively small size of the AOs. The elements in later periods are much more likely to form chains and rings than multiple bonds.

The concept of oxidation state was introduced in section 7.10 on page 303, and the occurrence of different oxidation states is illustrated in Fig. 7.47 on page 305. Group 1 and 2 elements show the oxidation states +1 and +2, respectively, reflecting their tendency to form cations.

In Groups 13 and 14 an oxidation state of the 'group number minus ten' is common, and the same is true for Groups 15 and 17, provided we exclude nitrogen and oxygen. Lower oxidation states are also common. It is interesting to note that the observed oxidation states are often separated by steps of two, e.g. the common oxidation states of sulfur are +6, +4 and +2. We will discuss this point in section 14.2.5 on page 620.

As we descend these groups, there is an increasing tendency for the oxidation state 'group number minus *twelve*' to be favoured e.g. for bismuth in Group 15 the +3 state is far more common than this +5 state. This tendency is often described as being due to the 'inert pair effect', and we will return to this idea later on in this chapter in section 9.4.3 on page 362.

Finally, the non-metals also show negative oxidations states, reflecting their tendency to form anions in ionic compounds with metals. This tendency is most marked for the earlier elements in Groups 15 and 16, together with all of the halogens.

14.2 Key concepts in main-group chemistry

There are a number of key ideas which we will come across again and again in discussing main-group chemistry, so it is convenient to introduce these right away.

14.2.1 Stable compounds

Open any chemistry book, or attend any chemistry lecture, and it will not be long before you find that such-and-such a compound is described as being 'stable' whereas another compound is described as being 'unstable'. As we shall see, whether or not a compound is stable depends very much on the conditions which are prevailing.

Compounds can be unstable in various different ways. The simplest kind of instability is dissociation, in which the molecule falls apart. For example PBr_5 is classed as unstable because at temperatures above 35 °C it readily dissociates into $PBr_3 + Br_2$, whereas PCl_5 is stable as it does not decompose in this way but simply sublimes at 159 °C. Since this kind of dissociation reaction inevitably involves breaking bonds we expect the reaction to be endothermic and thus the extent of dissociation will increase as the temperature rises. For this reason it is sometimes possible to suppress dissociation by keeping the temperature low, and indeed this is the case for PBr_5.

Other compounds are unstable not by dissociation but because they react with other species, notably oxygen and water from the air. Really we should class such compounds as being 'reactive' rather than unstable. For example, the moment that nitric oxide (NO) is exposed to air it reacts with the oxygen to give NO_2, and similarly although sulfur trioxide is indefinitely stable on its own, it reacts instantly with water to give sulfuric acid. Chemists have become adept at doing their reactions under conditions in which oxygen and water are excluded, and this leads to a large number of compounds which are stable under these conditions, but which would not be if exposed to the air.

There are some molecules which are unstable with respect to dimerization or polymerization. Since such reactions are exothermic (bonds are formed), they can be prevented by raising the temperature. This is in contrast to compounds which are unstable with respect to dissociation where low temperatures are required. For example, at temperatures above 720 °C the diatomic species S_2 is the dominant form of sulfur in the gas phase, but as the temperature is lowered the proportion of S_2 is reduced as various polymerization reactions become favoured; by 600 °C, the predominant species is S_8. S_2 is therefore only stable at high temperatures.

Thermodynamic *vs* kinetic stability

When we are thinking about the stability of a compound we need to consider whether or not the reaction which will lead to the decomposition of the compound is feasible in the thermodynamic sense i.e. whether or not the reaction has a significant negative standard Gibbs energy change ($\Delta_r G°$).

For example, the decomposition of SO_3 to SO_2 according to

$$SO_3(g) \longrightarrow SO_2(g) + \tfrac{1}{2}O_2(g)$$

has $\Delta_r G° = +71$ kJ mol^{-1} at 298 K. This large positive value of $\Delta_r G°$ tells us that

⇨ The temperature dependence of equilibrium is discussed in section 17.9 on page 788.

the equilibrium lies very much over to the reactants, so we can conclude that SO_3 is stable with respect to dissociation to SO_2 and O_2.

On the other hand the reaction of SO_3 with water

$$SO_3(g) + H_2O(l) \longrightarrow H_2SO_4(l)$$

has $\Delta_r G^\circ = -82$ kJ mol^{-1} at 298 K, implying that the equilibrium is well over to the products. This provides a ready explanation as to why SO_3 is so reactive to water.

However, these thermodynamic arguments are not always a reliable guide to stability. For example the following two reactions both have significant negative values for $\Delta_r G^\circ$ at 298 K

$$NO(g) \longrightarrow \tfrac{1}{2}N_2(g) + \tfrac{1}{2}O_2(g) \qquad \Delta_r G^\circ = -87 \text{ kJ mol}^{-1}$$
$$N_2H_4(g) \longrightarrow N_2(g) + 2H_2(g) \qquad \Delta_r G^\circ = -51 \text{ kJ mol}^{-1}$$

Despite these favourable $\Delta_r G^\circ$ values, it is found that NO is stable with respect to dissociation into N_2 and O_2, and likewise hydrazine is found to be stable with respect to dissociation into N_2 and H_2.

What is going on here is that although these reactions are favoured thermodynamically, they turn out to be rather slow i.e. they have large activation energies. Hydrazine is said to be *thermodynamically unstable* with respect to dissociation into nitrogen and hydrogen, but *kinetically stable* with respect to this dissociation. It is quite difficult to predict whether or not a reaction will have a high activation energy, but we will see that in some cases it is possible to at least rationalize the occurrence of kinetic stability.

Disproportionation

Some compounds are unstable with respect to *disproportionation*. This is a reaction in which one of the products is a compound in a *higher* oxidation state than the original, and one is in a *lower* oxidation state. For example, AlF_2 decomposes in this way

$$3\,AlF_2(s) \longrightarrow Al(s) + 2\,AlF_3(s)\,.$$
$$\underset{\text{Al(II)}}{} \qquad \underset{\text{Al(0)}}{} \quad \underset{\text{Al(III)}}{}$$

Al in the oxidation state +2 has decomposed into Al metal in oxidation state 0 and Al in oxidation state +3. Similarly, germanium(II) bromide also decomposes by disproportionation to Ge(0) and Ge(IV)

$$2\,GeBr_2(s) \longrightarrow Ge(s) + GeBr_4(s)\,.$$
$$\underset{\text{Ge(II)}}{} \qquad \underset{\text{Ge(0)}}{} \quad \underset{\text{Ge(IV)}}{}$$

This mode of instability is common for intermediate oxidation states.

14.2.2   High oxidation states

In the main group it is observed that the highest oxidation states of the elements are most often exhibited by the fluorides or oxides. For example, SF_6 is a particularly stable compound, but there is no evidence for the existence of SCl_6; similarly only the hexafluorides are known for selenium and tellurium. The oxides of nitrogen are know for oxidation states up up to +5, but no analogous sulfur–nitrogen compounds are known in which the nitrogen is in such a high oxidation state.

In the case of the fluorides we can attribute this propensity to form the highest oxidation states as being a result of a number of factors: (1) the relative weakness of the F–F bond in F_2, (2) the high strength of M–F bonds, (3) the small size of the F atom. As discussed in section 7.6.6 on page 286, the relative weakness of the F–F bond can be attributed to the occupation of the strongly antibonding π MOs. The strong bonds which fluorine forms to other elements can be attributed to the effective overlap with its small orbitals. Finally, the small size of the F atom means that many of them can pack around a central atom at a reasonably short bond length.

As an example, consider the likely decomposition of sulfur(VI) halides according to the following reaction

$$SX_6 \longrightarrow SX_4 + X_2.$$

In this reaction, we lose two S–X bonds and gain one X–X bond. Such a process is clearly more favoured for the chloride than for the fluoride, since we expect the S–Cl bond to be weaker than the S–F bond, and the F–F bond is weaker than the Cl–Cl bond. In addition, the crowding in SCl_6 is greater than in SF_6, again favouring the decomposition of the chloride since this will benefit most from the reduction in crowding.

For oxides perhaps the key point is that oxygen is able to form multiple bonds with the central atom, so that a high formal oxidation state can be achieved with fewer ligands and hence less crowding e.g. SO_3 as compared to SF_6. As we have noted many times before, this ability to form strong multiple bonds is a feature of second-period non-metals such as oxygen.

14.2.3   Hypervalent compounds

Many simple compounds obey the *octet rule*, meaning that there are eight electrons in the valence shell of each atom. In making this count, we imagine that each bond contributes two electrons, even though these two electrons are of course shared between two atoms. In addition, we have to add in any lone pairs. Satisfying the octet rule results in compounds with the 'normal' valency e.g. four for carbon, three for nitrogen and so on.

For hydrogen, the rule is that there should be *two* valence electrons.

For example, in H_2O there are two bonds to the oxygen, each contributing two electrons, and two lone pairs: the total number of electrons in the valence shell of the oxygen is therefore eight. Similarly, in NH_3 there are three bonds to the nitrogen, contributing a total of six electrons, plus one lone pair, bringing the total to eight valence electrons for the nitrogen. A somewhat more complex example is H_3O^+ in which there are three bonds, contributing six electrons and one lone pair, again making a total of eight (recall that this molecule can be thought of as arising when a lone pair from H_2O is donated to H^+).

However, there are many perfectly stable compounds which violate the octet rule. For example, SF_6 has twelve electrons in its valence shell (six bonds and no lone pairs), as does the isoelectronic $[SiF_6]^{2-}$ ion. PF_5 has ten electrons in its valence shell, and IF_7 has fourteen – both compounds are stable and well known. Clearly, it is possible to violate the octet rule and still obtain stable compounds.

Molecules in which the octet rule is violated are termed *hypervalent*. Such compounds are common for elements in Period 3 onwards, but hardly known at all for elements in Period 2. It is also worth noting that hypervalent molecules most often involve bonds to small electronegative atoms, especially fluorine and oxygen.

The octet rule can be rationalized if we recall that a second-period element has available to it one $2s$ and three $2p$ AOs. If these four orbitals overlap with four suitable orbitals on adjacent atoms (ligands), then four bonding MOs and four antibonding MOs will result: this is the maximum number of bonding MOs that can be formed. If the number of ligands, and hence the number of orbitals increases, then more MOs will be formed, but these additional MOs tend to be nonbonding, so occupying them does nothing to increase the strength of bonding.

In CH_4 there are eight valence electrons which nicely occupy all of the bonding MOs. Moving along the period to nitrogen, we can speculate that if a nitrogen atom shared each of its valence electrons with a hydrogen the species NH_5 might be possible. However, this molecule has ten valence electrons, eight of which go into the bonding MOs leaving the remainder to be accommodated in nonbonding MOs. Therefore, in going from CH_4 to NH_5 adding the extra hydrogen does not result in increased bonding since there are no bonding MOs for the extra electrons to occupy In fact, we would expect each bond in NH_5 to be weaker than those in CH_4 since in the former eight electrons are responsible for the bonding to five hydrogens, whereas in the latter the same number of electrons are responsible for bonding to only four hydrogens.

Despite the fact that it has electrons in nonbonding MOs, this description of the bonding in in NH_5 tells us that it would be a bound molecule, in the sense that it would not dissociate into atoms. However it is unlikely to be stable with respect to the dissociation reaction

$$NH_5 \longrightarrow NH_3 + H_2.$$

This is for two reasons, Firstly, the entropy change of the dissociation reaction is favourable, since one molecule go goes to two. Secondly, having five hydrogens around a small atom like nitrogen will be a rather crowded situation, making it impossible for the N–H distance to be small enough to give the best overlap and hence the strongest bonding. So, although the number of bonding electrons in the products is the same as in NH_5, we expect there to be a reduction in energy on dissociation since the bonds in the products are likely to be stronger.

A similar argument applies to the hypothetical species OH_6. This has twelve electrons, only eight of which can be accommodated into bonding MOs, leaving four in nonbonding MOs. Although there is therefore net bonding in this molecule, the extreme crowding caused by having six hydrogens around the small oxygen would undoubtedly result in the O–H bonding being weak, and thus dissociation would be favourable, for example according to the reaction

We are not considering the species NH_4 and OH_5 as these would be very reactive free radicals.

$$OH_6 \longrightarrow OH_2 + 2\,H_2.$$

Similarly, we would expect OH_4 to be unstable with respect to OH_2 and H_2.

If we move to the Third Period we can expect this crowding around the central atom to decrease, simply because the atoms become larger. Despite this, species such as PH_5, SH_4 and SH_6 are not known, presumably once again as a result of their readiness to dissociate. It is only when we move from the hydrides to the *fluorides* that these hypervalent compounds are actually found e.g. the well-known species PF_5, SF_4 and SF_6. It is also interesting to note that

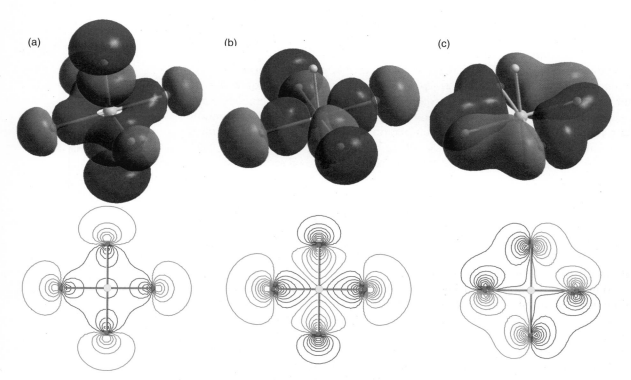

Fig. 14.2 Iso-surface representations and contour plots (take in the xy-plane) of three occupied MOs of SF_6 which, in principle, can contain a contribution from $3d$ AOs. The MO in (a) can have a contribution from the $3d_{z^2}$, whereas (b) and (c) can have contributions from $3d_{x^2-y^2}$ and $3d_{xy}$, respectively. It is clear from the plots that the fluorine $2p$ AOs are the major contributors to all of these MOs.

whereas PH_5 is unknown, PF_2H_3 and PF_3H_2 have been synthesized. One factor which favours the fluorides over the hydrides is that dissociation of the former results in formation of the relatively weak F–F bond, as opposed to the very much stronger H–H bond.

Bonding to oxygen also favours the formation of these hypervalent compounds e.g. SO_3 and $Cl_3P=O$. Typically the oxygen has a multiple bond to the central atom, which means that fewer ligands are needed to give the same oxidation state, thus relieving the crowding. As has been noted many times before, the propensity to form multiple bonds is a feature of Second Period elements such as oxygen.

It has often been proposed that the occurrence of hypervalent molecules in the Third Period is a result of the fact that these elements have more orbitals available to them, and so the number of bonding MOs can be increased above four. Thus whereas nitrogen only has the $2s$ and $2p$, phosphorus has the $3s$, the $3p$ and the five $3d$ AOs available. The $3d$ AOs are considerably higher in energy than the $3s$ and the $3p$, so they are not well matched to the energy of typical ligand orbitals, and as a result we cannot expect them to make a large contribution. Nevertheless, it is the case that the involvement of the $3d$ orbitals would result in some of the nonbonding MOs becoming weakly bonding.

Beguiling though this argument is, the evidence is unfortunately against it. Calculations show that the involvement of the $3d$ AOs in the occupied MOs of molecules such as SF_6 is rather small. This point is illustrated in Fig. 14.2 on the preceding page which shows views of three occupied MOs of SF_6 which can in principle contain contributions from $3d$ AOs on the sulfur. The contour plots show that there is some shift of the electron density towards the sulfur in a way which is suggestive of there being a contribution from a $3d$ AO, but it is clear that the contribution of these AOs is small.

Unfortunately this discussion of hypervalent molecules has left us with no clear-cut answers as to why they are formed in some cases and not in others.

Using hybrid atomic orbitals to describe the bonding in hypervalent molecules

In section 4.4 on page 150 we described how the $2s$ and $2p$ AOs can be combined to give hybrid atomic orbitals (HAOs) which point in various directions. For example, combining the $2s$ and all three $2p$ AOs gives four sp^3 HAOs which point towards the corner of a tetrahedron. These HAOs can be used to describe the bonding on CH_4 by overlapping each HAO with a $1s$ AO on the hydrogen to which the HAO points. Two electrons are placed in the resulting bonding MO to give a two-centre two-electron bond.

To use a similar approach to describing the bonding in PF_5 and SF_6 we need to increase the number of hybrids and also make sure that these are pointing in the right directions. Clearly the s and the three p AOs are insufficient, so it proves necessary to include the d orbitals.

It turns out that to create six hybrids pointing to the corners of an octahedron we need to hybridize the s, the three p and two of the d orbitals together to give six equivalent sp^3d^2 hybrids. For a trigonal bipyramidal arrangement we need to create either sp^3d or spd^3 hybrids.

The bonding in SF_6 can therefore be described as six S–F σ bonds, where each bond consists of two electrons located in the bonding MO formed from the overlap of one of the sp^3d^2 hybrids with an suitable AO on the fluorine to which the HAO points. This is a localized description in terms of two-centre two-electron bonds, and in it all twelve electrons are in bonding MOs. Similarly, the bonding in PF_5 can be described in terms of five two-centre two-electron bonds involving the five hybrids on the phosphorus.

For elements from the second period the $3d$ AOs are far too far away in energy (and far too large) to be able to form hybrids with the $2s$ and $2p$ AOs. We are thus restricted to sp^3 hybrids, which can only account for up to four bonds plus lone pairs. In the third period, the $3d$ AOs are much closer in energy and size to the $3s$ and $3p$ AOs for hybridization to be more reasonable, and so we can use these HAOs to account for the bonding in hypervalent compounds.

Caution over using hybrids to describe the bonding in hypervalent molecules

We have to be very careful with this HAO description of hypervalent molecules. It is emphatically *not* correct to say that molecules such as SF_6 exist *because* it is possible to use $3d$ AOs to form the required HAOs on the sulfur. Rather, all we can say is that in order to describe the bonding in such molecules *in terms of two-centre two-electron bonds* we have to assume that $3d$ AOs are involved in the hybridization.

As we have seen, there is in fact little evidence that $3d$ AOs are much involved in the bonding of molecules such as SF_6, so although the HAO description of the bonding is convenient to use, there is considerable doubt as to whether or not it is realistic.

Fig. 14.3 Iso-surface representations of two of the occupied
MOs from diborane which are
responsible for bonding to the
bridging hydrogen atoms.

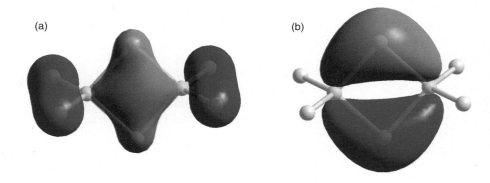

(a)   (b)

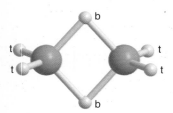

Fig. 14.4 The structure of
diborane, B_2H_6; boron atoms
are shown in yellow and
hydrogen in white. The
molecule has twelve valence
electrons, which is insufficient
to account for the bonds shown
if each is a two-centre
two-electron bond.

14.2.4   Electron deficient compounds

Electron deficient compounds those in which there appear to be insufficient electrons available to account for all of the supposed bonds as two-centre two-electron bonds. The classic example of such a compound is diborane, B_2H_6, whose structure is shown in Fig. 14.4. In this representation atoms which are close enough to be considered to be bonded by the usual critera have been joined up. If each of these bonds is a simple two-centre two-electron bond, then we need a total of sixteen electrons. However, B_2H_6 only has twelve valence electrons, and so can be described as being 'deficient' in electrons.

An MO picture of B_2H_6 reveals, as we must by now expect, that the occupied MOs are spread over several atoms, and so the electrons in them are responsible for bonding more than two atoms together. In fact, it turns out that all twelve electrons are in bonding MOs, and all of the bonding MOs are occupied. The two MOs shown in Fig. 14.3 are particularly interesting as these are the ones responsible for bonding the bridging hydrogen atoms (labelled 'b' in Fig. 14.4).

The MO picture gives a perfectly satisfactory account of the bonding in diborane, but not in terms of two-centre two-electron bonds.

14.2.5   Why common oxidation states are separated by two

Figure 7.47 on page 305 illustrates the fact that the commonly occurring oxidation states for a particular main-group element are often separated by two. For example, the common oxidation states of sulfur are +2, +4 and +6, those of phosphorus are +3 and +5, and those of lead are +2 and +4. It is not that compounds with the oxidation states between these numbers are totally unknown, but just that if they are found they tend to be much less stable.

These common oxidation states all have the property that the electrons in their valence shells are all paired up. For example, SF_4 has four bonded pairs and one lone pair, SF_6 has six bonded pairs, PF_3 has three bonded pairs and one lone pair, and Pb^{2+} ions have the configuration $\ldots 6s^2$. The intervening oxidation states therefore all have at least one unpaired electron e.g. SF_5 has five bonded pairs and one unpaired electron, and PF_4 has four bonded pairs and one unpaired electron. Molecules in these oxidation states are therefore *free radicals*.

Free radicals generally turn out to be rather reactive species, which is probably the reason why such molecules are not regarded as 'stable', but tend only to have a fleeting existence i.e. they are reactive rather than being inherently unstable. It is not that molecules such as SF_5 are fundamentally unstable, in the sense that they will decay by dissociation. Rather, their instability arises from their ready reaction with other species, which often leads to a change in oxidation state.

SF_5 is a good example of this behaviour. This species is thought to exist in the gas phase, but it readily dimerizes to give S_2F_{10} in which there is an S–S bond i.e. F_5S–SF_5. This dimerization can be understood in terms of the orbitals involved. The unpaired electron will be in the HOMO, which is usually a nonbonding MO. If two free radicals come together the HOMOs overlap to form a bonding MO which is then occupied by the two electrons (one from each radical), resulting in a lowering of the energy. Dimerization is thus energetically favoured.

Another common reaction of free radicals is for them to *abstract* an atom from another molecule e.g. H atoms will abstract an H from CH_4 to give H_2 and the radical CH_3. In the case of SF_5, the radical could abstract an F atom from F_2 to give SF_6; such a process is energetically favoured as a weak F–F bond is replaced by a stronger S–F bond. This reaction results in the oxidation state of the sulfur going from 5 to 6, so it is classed as oxidation. It appears that when attempting to prepare sulfur fluorides by fluorinating elemental sulfur this oxidation process accounts for the very low yield of SF_5 (or S_2F_{10}).

If an SF_5 radical abstracts a fluorine atom from another SF_5 we have the reaction

$$SF_5 + SF_5 \longrightarrow SF_4 + SF_6.$$

This is a disproportionation reaction (see page 615) in which two sulfur species in oxidation state 5 go to one is oxidation state 4 and one in oxidation state 6. As the resulting species are *not* free radicals they are much less reactive than SF_5 and so this reaction tends to lead to the destruction of SF_5. As before, such disproportionation reactions are often the mode of instability of these intermediate oxidation states.

Another example of an unstable intermediate oxidation state is the +3 state of sulfur. SF_3 is not known, but sulfur(III) does occur in the dithionite anion, $S_2O_4{}^{2-}$. It is interesting to note that this anion is a dimer, in which the S–S bond can be thought of as arising from the dimerization of the two SOO^- species, which are free radicals. Not surprisingly, this species is unstable with respect to disproportionation, just as for SF_5.

It is important to realize that not all free radicals are unstable. For example NO is a radical (the oxidation state of the nitrogen is 2), and although this species is quite reactive it can nevertheless be prepared in a pure form and stored for long periods. Likewise there are many transition metal compounds which are perfectly stable, despite the metal having unpaired electrons. We will turn to the special properties of these compounds in Chapter 15.

14.2.6 Caution over using the concept of oxidation states

In section 7.10 on page 303 the oxidation state of a particular element in a compound was defined as the charge on that element assuming that the compound consists entirely of ions with closed-shell electronic configurations. For a compound such as NaF the oxidation state of Na is +1, which for this ionic compound matches the charge on the sodium.

S_2F_{10}

dithionite

However, in PF_5 the oxidation state of phosphorus is +5, but we do not of course imagine that phosphorus pentafluoride actually contains P^{5+} ions. The oxidation state is just a formal way of accounting for how electrons are distributed amongst the atoms in a compound. It is not meant to imply that a particular atom necessarily has a charge equal to its oxidation state.

14.3 Hydrolysis of chlorides

Before looking at each group individually, we are going to discuss one particular kind of reaction – the hydrolysis of chlorides – across all of the main-group elements. The reason for doing this is that it will reveal a number of important trends in bonding and reactivity which are very useful for developing a unified view of the chemistry of the main-group elements. If we just look at the elements group by group it is easy to lose sight of the general trends.

The term 'hydrolysis' literally means a splitting apart of water, and this is exactly when happens when BCl_3 comes into contact with water

$$BCl_3(g) + 3\,H_2O(l) \longrightarrow B(OH)_3(aq) + 3\,H^+(aq) + 3\,Cl^-(aq). \tag{14.1}$$

In this reaction the water has indeed been split apart as one if its O–H bonds has been broken, leading to the generation of an acidic proton. The –OH group ends up attached to the boron.

In this reaction the replacement of B–Cl bonds by stronger B–O bands leads to a favourable contribution to the enthalpy change. The reason for this difference in bond strengths is that oxygen is in the second period and so has smaller orbitals than chlorine, and generally smaller orbitals lead to stronger covalent bonds. The hydrolysis of BCl_3 is an example of the general feature that halides of elements which form strong covalent bonds to oxygen will be hydrolysed to give acidic solutions.

We will look at the hydrolysis reaction first for typical ionic chlorides, then for typical covalent chlorides, and then finally for intermediate chlorides in which the bonding is mixed.

14.3.1 Ionic chlorides

When a typical ionic chloride such as NaCl is dissolved in water, the Na^+ and Cl^- ions become separated and each is surrounded by sheath of solvating water molecules. As was described in section 6.12.1 on page 238, between four and eight quite tightly held water molecules form the first layer, as shown rather schematically in Fig. 14.5.

This dissolution of NaCl is not really a hydrolysis reaction as the water molecules are not broken apart in the way they were when BCl_3 reacts with water. However, the water molecules, especially those in the first layer, are affected by their proximity to the sodium ion. There is a strong electrostatic interaction between the dipole on the water and the positive charge on the ion, and this interaction has the effect of weakening the O–H bonds. We can imagine that the Na^+ draws electron density away from the oxygen, and hence away from the O–H bond. As a result this bond is both weakened and further polarized, in the sense that the partial positive charge on the hydrogen is increased.

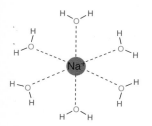

Fig. 14.5 An aqueous Na^+ ion is surrounded by solvating water molecules whose dipoles have a favourable interaction with the ion.

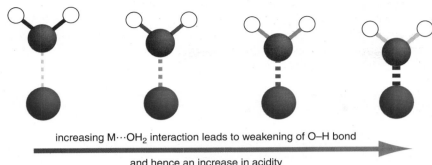

increasing M···OH$_2$ interaction leads to weakening of O–H bond

and hence an increase in acidity

Fig. 14.6 As the interaction between a metal cation (shown in green) and a water molecule increases, the O–H bonds in the water are progressively weakened. They are therefore more likely to dissociate to give H$^+$ and hence an acidic solution. The thickening and darkening dashed bond symbolizes the increasing M–O interaction.

Sodium ions do not appear to cause a significant weakening of the O–H bonds in the water. However, if the sodium is replaced by a smaller ion, or an ion of greater charge, the interaction between the cation and the coordinated water molecules becomes greater. Eventually, the O–H bonds can be sufficiently weakened for some of them to dissociate, thus giving up an acidic proton. Under these circumstances, dissolving an ionic halide in water results in hydrolysis and hence an acidic solution. This trend is indicated schematically in Fig. 14.6.

This drawing of electron density away from the oxygen towards the cation can also be described as the formation of the covalent bond between the metal and the oxygen. Therefore the trend illustrated in Fig. 14.6 can be described as involving an increase in covalent character as we go to the right. The hydrolysis of BCl$_3$ can be seen as this trend taken to an extreme in which there is a such a strong covalent interaction between the boron and the oxygen that the O–H bonds are weakened to the point where they dissociate readily.

Acidity of aquated ions

The tendency of the water molecules coordinated around a cation M^{n+} to become acidic can be described in terms of the following equilibrium

$$[M(H_2O)_6]^{n+} + H_2O \rightleftharpoons [M(H_2O)_5(OH)]^{(n-1)+} + H_3O^+,$$

where we have assumed for simplicity that there are six waters associated with the aquated ion. Just as we did for more conventional acids in section 6.14 on page 245, we can quantify the acidity of $[M(H_2O)_6]^{n+}$ by specifying the K_a, or more usually the pK_a, of the above equilibrium.

$[Na(H_2O)_6]^+$ has a pK_a of 14.5, which is such a high value that the aquated ion is simply not acidic i.e. the above equilibrium is totally to the left. In fact, the pK_a of this aquated ion is not that much different from the pK_a of water itself (15.7), indicating that the Na$^+$ ion is not having much of an effect of the coordinated water molecules.

If we go to a doubly charged cation such as $[Mg(H_2O)_6]^{2+}$ the pK_a falls to 11.4, which means that the aquated ion is a very weak acid. Moving to a triply charged cation such as $[Al(H_2O)_6]^{3+}$ sees to pK_a drop dramatically to 5.0, which is about the same value as ethanoic acid. The aquated aluminium cation is therefore significantly acidic.

This trend in pK_a values as we go from Na$^+$ to Mg^{2+} to Al^{3+} matches exactly what we said above. The higher the charge on the ion, the more strongly the O–H bonds in the water are weakened and polarized, resulting in higher acidity.

It is certainly the case that by the time we reach $[Al(H_2O)_6]^{3+}$ there is significant covalent character to the M–O bond, which is another way of explaining the increased acidity.

This acidity of the aquated ions has some important consequences for the way such solutions behave when the pH of the solution is altered deliberately (for example, by adding a strong acid or alkali). Recall that for the acid dissociation equilibrium

$$AH(aq) + H_2O(l) \rightleftharpoons A^-(aq) + H_3O^+(aq),$$

K_a is defined as

$$K_a = \frac{[H_3O^+]\,[A^-]}{[AH]}.$$

From this we can see that if the concentration of AH and the conjugate base A^- are to be equal, it must follow that $K_a = [H_3O^+]$. Using the usual log scale, this condition is $pK_a = pH$. In other words, if the pH is made equal to the pK_a the concentrations of AH and A^- will be equal.

What this means for the aquated sodium ions, whose pK_a is 14.5, is that we would have to make the solution very basic in order for the amount of $[Na(H_2O)_5(OH)]$ to be significant. On the other hand, for the magnesium ions, a fairly basic solution with pH around 11 will result in equal amounts of $[Mg(H_2O)_6]^{2+}$ and $[Mg(H_2O)_5(OH)]^+$. Raising the pH further will make $[Mg(H_2O)_5(OH)]^+$ the dominant species. In fact this species is itself acidic according to

$$[Mg(H_2O)_5(OH)]^+ + H_2O \rightleftharpoons Mg(H_2O)_4(OH)_2 + H_3O^+.$$

The species on the right, the hydroxide $Mg(OH)_2$ (along with some coordinated water molecules) is uncharged and so precipitates out of solution.

In summary, aquated sodium ions are not significantly acidic and therefore remain in solution even at high pH. In contrast, aquated magnesium ions do have significant acidity, and successive ionization of H^+ from these aquated ions eventually results in the formation of the insoluble hydroxide in moderately high pH.

14.3.2 Covalent chlorides

As we have already seen, BCl_3 hydrolyses to give an acidic solution (Eq. 14.1 on page 622), which is the typical behaviour for covalent halides. In this section we will explore in more detail how such hydrolysis reactions occur, and the reactions which may subsequently follow.

Boron halides

The hydrolysis reactions of BF_3 and BCl_3 are subtly different in way which is revealing about the nature of the reaction. BF_3 has a low-lying LUMO (the out-of-plane $2p$ AO) which makes it ready to act as a Lewis acid with molecules having a suitable HOMO. This is analogous to the behaviour of BH_3 when it acts as a Lewis acid towards H^- to give $[BH_4]^-$, as discussed in section 8.3.1 on page 318.

The initial interaction between BF_3 and H_2O gives a donor–acceptor complex (an adduct) in which the electrons are donated from one of the oxygen lone pairs (the HOMO) into the LUMO of the BF_3.

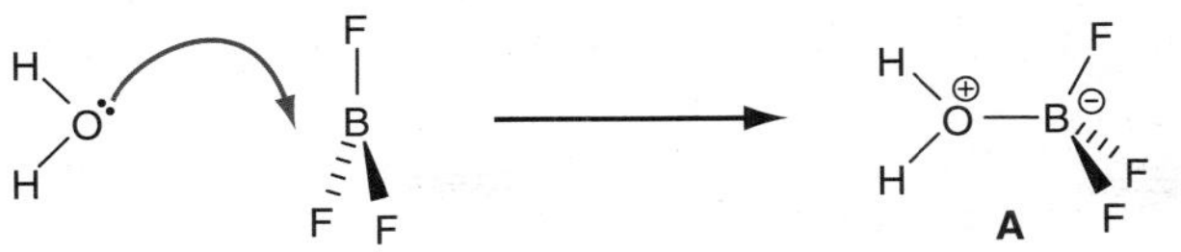

If the amount of water is limited the reaction stops here and it is possible to isolate the adduct **A**.

The adduct is drawn with a formal positive charge on the oxygen since the pair of electrons in the B–O bond comes entirely from the oxygen. This shift of the electrons towards the boron results in a weakening and polarization of the O–H bonds, and it is therefore not surprising to find that the adduct is a strong acid, readily giving up a proton to solvent water

$$H_2O\text{–}BF_3 + H_2O \rightleftharpoons H_3O^+ + [HO\text{–}BF_3]^-.$$

If an excess of water is added, the three fluorides are substituted by OH. We can imagine that the mechanism involves the following steps.

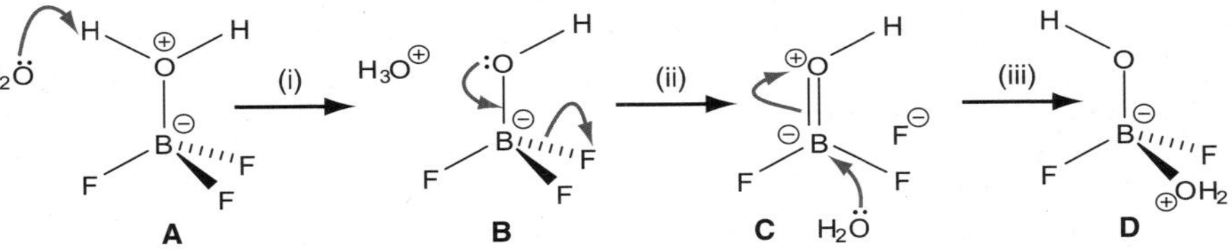

In step (i) the initial adduct **A** donates a proton to solvent water to give H_3O^+. Species **B** has several electronegative atoms attached to a single carbon, and as we have seen in Fig. 8.8 on page 331, this means that a fluoride can easily be expelled by the lone pair on oxygen attacking into the B–F $\sigma^\star$ MO. The resulting species **C** is analogous to a protonated carbonyl group with the carbonyl carbon being replaced by B$^-$. Not surprisingly this species is susceptible to attack by water (see section 8.7 on page 329), as shown in step (iii) to give species **D**.

D can go on to lose a proton just as in step (i), expel a F$^-$ as in step (ii), and then be attacked by water as in step (iii). A further cycle around these three steps gives the tetrahedral species $[B(OH)_4]^-$ according to the overall reaction

$$BF_3(g) + 8\,H_2O(l) \rightleftharpoons 3\,F^-(aq) + [B(OH)_4]^-(aq) + 4\,H_3O^+(aq). \qquad (14.2)$$

The F$^-$ ions also react with further BF$_3$ to give $[BF_4]^-$. Overall, the hydrolysis produces an acidic solution, although the equilibrium is not entirely to the right on account of the fact the the B–F and B–O bonds are comparable in strength.

The hydrolysis of BCl$_3$ proceeds in an analogous was, although the initial adduct $[H_2O\text{–}BCl_3]^-$ has not been isolated, presumably as it reacts too quickly with water. We can attribute this to the fact that BCl$_3$ is a poorer Lewis acid than BF$_3$ on account of its LUMO being higher in energy. In contrast to the fluoride, for the chloride the equilibrium corresponding to Eq. 14.2 is very much to the right, reflecting the greater strength of the B–O bond in comparison to the B–Cl bond.

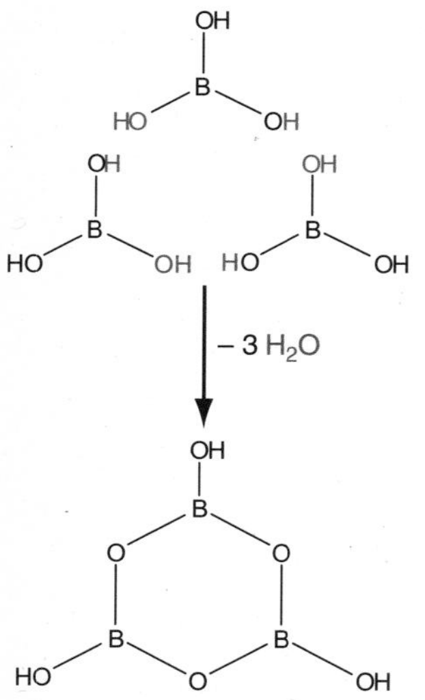

Fig. 14.7 Three $B(OH)_3$ molecules can condense together to give metaboric acid, HBO_2, with the elimination of three water molecules.

Further reactions of the hydrolysis products

The species $[B(OH)_4]^-$ can be thought of as the result of boric acid $B(OH)_3$ reacting with water in the following equilibrium

$$B(OH)_3 + 2\,H_2O \rightleftharpoons [B(OH)_4]^- + H_3O^+.$$

In this reaction we can imagine that water forms a donor–acceptor adduct with $B(OH)_3$. As before, the strong B–O bond makes the hydrogens in the $B-OH_2^+$ group acidic, so one of them can be transferred to water to give H_3O^+. The pK_a value of around 9 indicates that $B(OH)_3$ is a weak acid.

On heating, $B(OH)_3$ eliminates water to form metaboric acid, HBO_2, which (in the α modification) has a cyclic structure derived from the condensation of three $B(OH)_3$ molecules, as illustrated in Fig. 14.7. Further condensation reactions can occur between these six-membered rings, linking them via B–O–B bridges.

An aqueous solution $[B(OH)_4]^-$ is the predominant species at high pH, but as the pH is lowered many other species are formed which can be considered as arising from the condensation of $[B(OH)_4]^-$ with $B(OH)_3$. These species all contain the B_3O_3 six-membered ring illustrated above, with additional OH groups attached to the boron atoms.

Group 14 halides

Chloromethane reacts with water in a nucleophilic displacement reaction following the S_N2 mechanism discussed in section 9.7.1 on page 390. This is the first step shown below

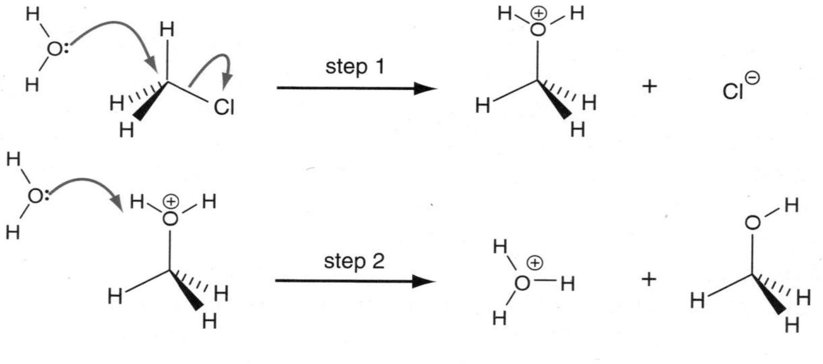

The product of step 1 is protonated methanol, $[CH_3OH_2]^+$, which is very acidic on account of the covalent interaction between the oxygen and the carbon. This species readily gives up a proton to water, as shown in step 2.

Dichloromethane can react in a similar way to give a hydrate or *gem*-diol. As we saw in section 9.4.1 on page 355, hydrates are in equilibrium with the corresponding aldehyde or ketone, the formation of which involves the elimination of water. The hydrate shown here is the favoured form in aqueous solution.

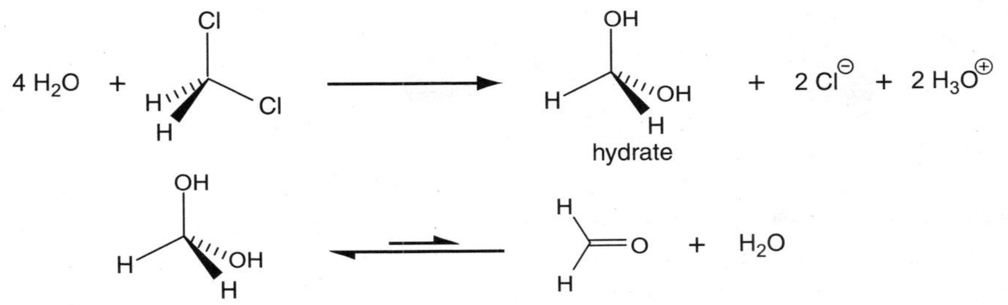

In principle we can imagine trichloromethane (chloroform) being hydrolysed in a similar way, followed by elimination of water to give methanoic acid, and tetrachloromethane resulting in CO_2 after the elimination of two water molecules (see Fig. 9.5 on page 345).

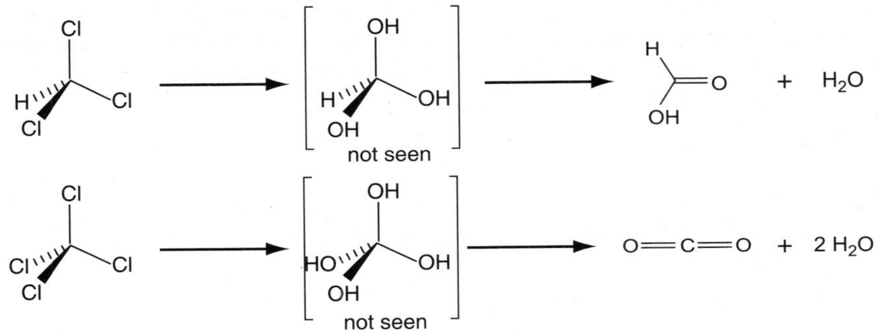

However, although these reactions are all thermodynamically feasible they turn out to be kinetically rather slow and require rather strong conditions to make the reaction go at a reasonable rate. These alkyl chlorides are also immiscible with water, which does not help the situation. We can identify the thermodynamic driving force of the dehydration reactions as the replacement of two C–O single bonds by one C=O double bond. As we have seen, multiple bonding tends to be favoured between second-period elements such as carbon and oxygen.

In contrast to CCl_4, $SiCl_4$ is rapidly hydrolysed by water to give $Si(OH)_4$. As before, replacing Cl by H_2O results in a species containing the acidic $[Si–OH_2]^+$ group which transfers a proton to solvent water. The overall reaction is

$$SiCl_4 + 8\,H_2O \longrightarrow Si(OH)_4 + 4\,H_3O^+ + 4\,Cl^-.$$

In contrast to carbon, multiple bonds to silicon are *not* favoured when compared to single bonds, so rather than eliminating water to give an Si=O group, $Si(OH)_4$ tends to eliminate water *between* two molecules to generate an Si–O–Si linkage. This process can then carry on to give various polymeric silicates based on SiO_4 tetrahedra.

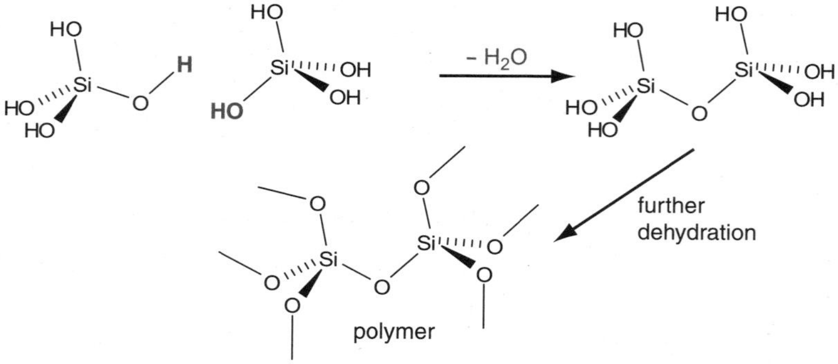

polymer

Group 15

PCl$_5$ reacts readily with water, with all five chlorides being displaced. As with SiCl$_4$, once the water molecules are attached the resulting $-OH_2^+$ groups are very acidic and readily donate a proton to water. The ultimate product of the hydrolysis is phosphoric(V) acid, which can be thought of as arising from the dehydration of the hypothetical species P(OH)$_5$.

Recall that the double headed arrow connects different resonance structures, which are alternative representations of the bonding in a given molecule.

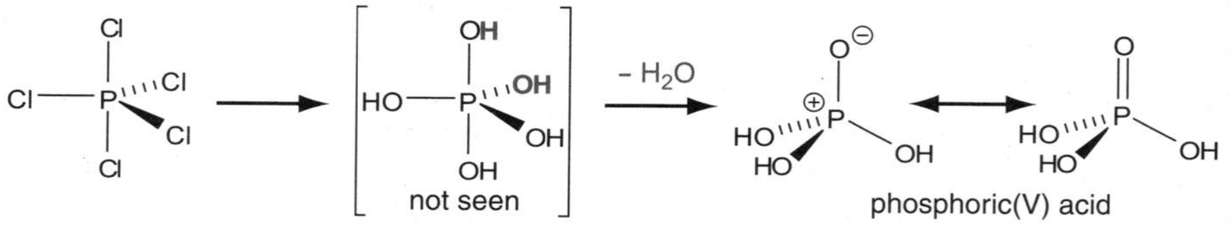

PCl$_3$ hydrolyses easily to give phosphonic acid H$_3$PO$_3$. This acid does not appear to exist as P(OH)$_3$, but in a tautomeric form where one of the H atoms is directly attached to the phosphorus.

Tautomers differ only by the position at which a particular hydrogen is attached. Here one of the H atoms is bonded to O in one tautomer, but to P in the other.

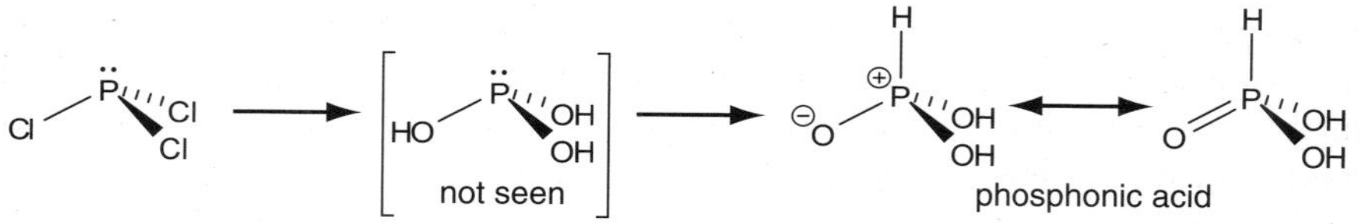

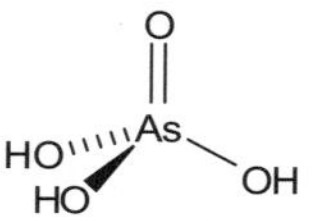

AsCl$_5$ hydrolyses in an analogous fashion to PCl$_5$ to give arsenic(V) acid. Hydrolysis of AsCl$_3$ gives arsenic(III) acid which, in contrast to phosphorus, exists as As(OH)$_3$. This acid does not eliminate water, nor does it exist as a tautomer in which there is an As=O double bond, presumably reflecting the increasing reluctance for As to form multiple bonds.

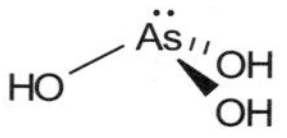

arsenic(V) acid

arsenic(III) acid

Group 16

SCl$_4$ is not known, but SF$_4$ is, and this species is readily hydrolysed in an analogous way to the behaviour of the Group 15 halides. The resulting hypothetical species S(OH)$_4$ can eliminate one water molecule to give sulfuric(IV) acid, and then under suitable conditions another water molecule can be lost to give SO$_2$; this is in direct analogy to the behaviour of CCl$_4$.

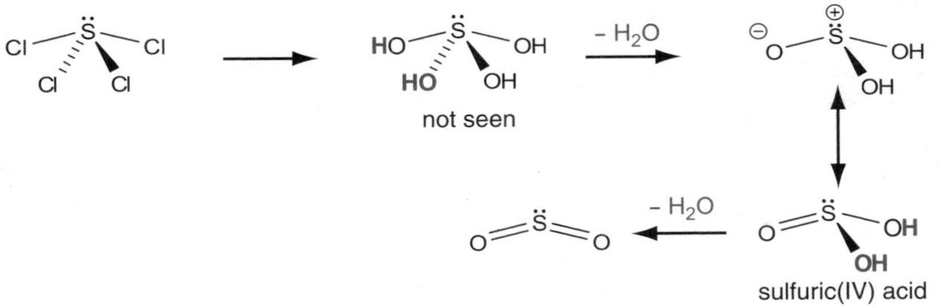

not seen

sulfuric(IV) acid

sulfuric(VI) acid

orthotelluric acid

Like CCl_4, SF_6 is kinetically stable with respect to hydrolysis, but under extreme conditions it will give sulfuric(VI) acid, H_2SO_4, which we recognize as being the result of the elimination of two water molecules from the hypothetical species $S(OH)_6$.

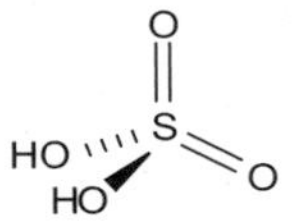

If we move further down the group to tellurium, hydrolysis of TeF_6 gives the octahedral species $Te(OH)_6$. In contrast to sulfur, $Te(OH)_6$ shows no propensity to eliminate water, indicating the reluctance of tellurium to form multiple bonds. Interestingly, $Te(OH)_6$ is hexabasic – meaning that all the hydrogens are acidic.

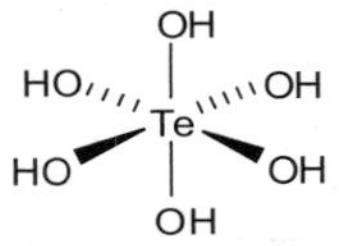

Summary

Hydrolysis of these covalent halides results in species with significant acidity, which is in contrast to the behaviour of typical ionic halides in which hydrolysis results in at best weakly acidic species. For covalent halides, hydrolysis may be followed by elimination of water to give multiply bonded species. Such behaviour is common in Period 2 and is sometimes seen in Period 3, but the propensity to form multiple bonds falls off quite quickly as we go down a group.

14.3.3 Intermediate chlorides

There are many chlorides which fall between the two extremes outlined in the previous sections. These are chlorides in which the bonding has both ionic and covalent characteristics, and which on hydrolysis only give moderately acidic solutions. As these solutions are made deliberately more basic we usually find that an insoluble hydroxide is precipitated, however as the basicity is increased further the precipitate redissolves. This behaviour results in the corresponding oxides (or hydroxides) being classed as *amphoteric*.

Let us take $ZnCl_2$ as an example. At room temperature this is a white solid, but it melts at a modest 275 °C. This, along with some aspects of the crystal structure, indicate that there is a significant covalent contribution to the bonding in the solid.

$ZnCl_2$ dissolves readily in water, and the pK_a of the aquo ion is 9.0, indicating that it is a weak acid. If the pH of the solution is held somewhat below 9, the ion remains in solution as $[Zn(H_2O)_6]^{2+}$. However, as the pH is raised above 9 first one, and then another proton is removed from the aquated ion

$$[Zn(H_2O)_6]^{2+} \xrightarrow{-H^+} [Zn(H_2O)_5(OH)]^+ \xrightarrow{-H^+} Zn(H_2O)_4(OH)_2.$$

The zinc hydroxide, $Zn(OH)_2$ (along with its water of hydration), is insoluble and forms a gelatinous precipitate.

However, the story does not end here. The bonding between the central zinc and the surrounding oxygen atoms has significant covalent character and, as we have seen in the previous section, this increases the acidity of the hydrogens attached to these oxygen atoms. Increasing the pH further can therefore result one or more of these hydrogens being removed

$$Zn(H_2O)_4(OH)_2 \xrightarrow{-H^+} [Zn(H_2O)_3(OH)_3]^- \xrightarrow{-H^+} [Zn(H_2O)_2(OH)_4]^{2-}.$$

In the case of zinc, the species $[Zn(H_2O)_2(OH)_4]^{2-}$ predominates under strongly alkaline conditions (pH 14). These negatively charged *zincate* ions go into solution, so the precipitate of $Zn(OH)_2$ which formed under modestly basic conditions redissolves under very basic conditions.

Looking at it another way, solid $Zn(OH)_2$, or the corresponding oxide ZnO which is easily formed by heating the hydroxide to drive off water, will dissolve in strong acids to give the aquated ion $[Zn(H_2O)_6]^{2+}$, or in strong alkalis to give the zincate ion $[Zn(OH)_4]^{2-}$. An oxide (or hydroxide) which dissolves in both acid and alkali is said to be *amphoteric*.

We can see from our discussion that amphoteric oxides will result from metals whose aquated cations have significant acidity, and in turn this will arise when there is significant covalent character in the interactions between the metal and the surrounding water molecules. Such behaviour is likely to occur for those metals which are near to the border between metals and non-metals.

For example in Group 13 aluminium oxide (Al_2O_3) and gallium oxide (Ga_2O_3) and both amphoteric, as are tin(IV) and lead(IV) oxides from Group 14. In Group 15 arsenic(III) and antimony(III) oxides are also amphoteric.

14.3.4 Overview of the hydrolysis of chlorides

We can now see that there is a steady gradation in the outcome of these hydrolysis reactions.

Ionic chlorides (typically of Group 1 and 2 metals) give simple aquated cations which do not have significant acidity. As we move to chlorides with greater covalent character (often associated with a larger formal charge on the cations), the acidity of the aquated ions increases such that under alkaline conditions a hydroxide may be formed as a precipitate.

If the covalent character is sufficient, further hydrogens may be ionized at even higher pH, resulting in soluble anions. This is the behaviour associated with amphoteric oxides.

Covalent chlorides are hydrolysed to give acidic solutions.

14.4 Oxides

Like the halides, the oxides of main-group elements show a range of behaviour which can be rationalized in much the same way as we have already done for the halides. A useful classification of the oxides is in terms of the acidity or basicity of the solutions they form in water.

Basic oxides

Oxides in which the bonding is predominately ionic can generally be classified as basic. The most typical examples are the Group 1 oxides which dissolve in water to give very basic solutions on account of the reaction of the notional

O^{2-} ion with water to give OH^-. The hydroxides are similarly basic, dissolving in water to give OH^- ions directly.

The oxides (and hydroxides) of the Group 2 metals (with the exception of beryllium) are generally not as soluble as those of the Group 1 metals, but those that do dissolve give strongly basic solutions. The insoluble oxides dissolve in acid solutions, reflecting their basic properties.

Acidic oxides

At the other extreme we have those oxides in which the bonding is primarily covalent; when these dissolve in water they give acidic solutions. A good example is sulfur trioxide SO_3 which is hydrolysed by water to give sulfuric(VI) acid, which is a very strong acid. The oxide is attacked by water and, as before, the hydrogen in the resulting $S-OH_2^+$ is very acidic:

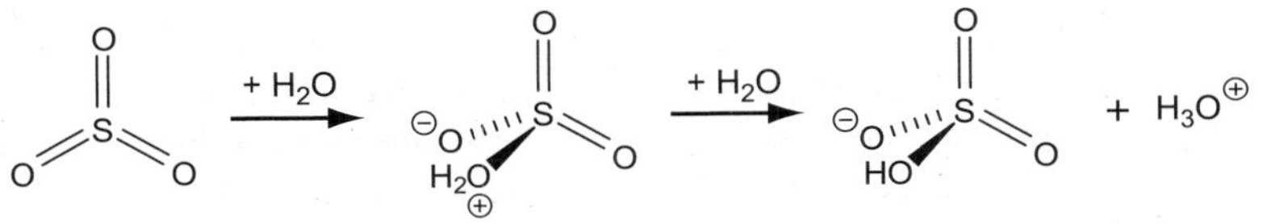

In fact, the remaining hydrogen in HSO_4^- is also acidic, and its dissociation gives the sulfate ion, SO_4^{2-}. CO_2, NO_2 and P_4O_{10} are other examples of acidic oxides.

Even if an acidic oxide does not dissolve readily in water, we would expect such an oxide to dissolve in a basic solution to give a corresponding anion.

Amphoteric oxides

We have already come across these in the previous section – these are oxides (or hydroxides) which dissolve in both acids and in bases. Typically, an amphoteric metal oxide is one in which the cation has high charge or lies near to the border between metals and non metals.

A typical example is aluminium oxide Al_2O_3. This dissolves in strongly acidic solutions to give the aquated ion $[Al(H_2O)_6]^{3+}$ but, as we have seen, on account the high charge of the aluminium and some degree of covalency in the interaction with the oxygen, this species is quite acidic with a pK_a of around 5. Making the solution more basic eventually removes three protons from the aquated ion to give insoluble $Al(OH)_3$. However, this species still has significant acidity so that at high pH a soluble aluminate ion, $[Al(OH)_4]^-$ is formed. Thus, aluminium oxide is soluble both in strong acid and strong base.

14.5   Brief survey of the chemistry of each group

As we explained in the introduction, there it is not possible in just one chapter to describe in detail the chemistry of all of the main-group elements. Rather, we will pick out some key trends in each group and trends between groups, concentrating particularly on halides and oxides, referring back to the key ideas which have already been introduced in this chapter and in Chapter 7.

14.5.1 Group 1

Overview

The elements in Group 1 are archetypical metals whose chemistry is dominated by compounds which can reasonably be described as ionic and containing M^+ ions. This can be attributed to the fact that in these elements the outer electron is in quite a high-energy orbital, and is therefore relatively easily ionized. In addition, the diffuseness of this orbital does not favour covalent bonding.

As expected for such ionic compounds, the oxides and hydroxides are strongly basic.

Halides

All of the Group 1 metal halides with the formula MX are known, and these are invariably high-melting point solids which nevertheless generally dissolve in water i.e. the typical properties of ionic compounds. These halides typically adopt either the NaCl or the CsCl crystal structure, as illustrated in Fig. 5.12 on page 193. Such regular structures, with high coordination numbers are typical of ionic compounds.

In section 6.13.1 on page 241 we discussed the somewhat unusual case of LiF which, unlike most of the other Group 1 halides, is not very soluble. We saw in that section that this insolubility was due to a subtle balance of factors affecting both the enthalpy and entropy change. It is generally observed that ionic compounds in which the ions are of similar size have reduced solubility e.g. LiF is rather insoluble (two small ions), whereas LiI is very soluble (one large and one small ion).

The aquated M^+ ions of Group 1 elements do not exhibit significant acidity.

Oxides

The simple oxides M_2O are all known and are very basic. In addition peroxides such as Na_2O_2 and superoxides such as KO_2 are known. The peroxides can be thought of as ionic compounds containing the O_2^{2-} anion, and the superoxides contain the anion O_2^-.

It is interesting to note that the superoxide of lithium is not known, whereas it is increasingly easy to form the superoxides as we go down Group 1. These compounds tend to be unstable with respect to dissociation to give the peroxide plus oxygen

$$2\,MO_2(s) \longrightarrow M_2O_2(s) + O_2(g).$$

It is possible to argue using lattice enthalpies (see Question 15.4) that the enthalpy change for this process becomes more favourable as the radius of the cation decreases, which offers an explanation for the fact that LiO_2 is not known.

Other features

As we might expect, of all the Group 1 elements lithium is the most likely to participate in covalent bonding since it has the lowest energy and smallest orbitals. There is an extensive and well-studied chemistry of compounds in which lithium is bonded to carbon (termed *organolithium* compounds), in compounds such as methyl lithium $LiCH_3$. We have already seen in section 9.3.2 on page 351 that these compounds are very useful carbon nucleophiles which can be used to create new C–C bonds in organic molecules.

In the solid state organolithium compounds are frequently found as aggregates (e.g ethyl lithium, Fig. 1.14 on page 15), and indeed such structures are often preserved when the compounds are dissolved in organic solvents such as hexane.

14.5.2   Group 2

Overview

In many ways the behaviour of Group 2 elements is very similar to that of Group 1, with the chemistry being dominated by compounds containing the M^{2+} ion. However, beryllium is something of an exception to this general picture. It has the lowest energy orbitals of all of the members of the group and as such has the highest ionization energy and the greatest tendency to form covalent bonds. We will therefore consider it separately to the other elements in the group.

Beryllium

For beryllium, even the simple halides BeX_2 cannot be considered as ionic. For example, solid $BeCl_2$ consists of a chain in which bridging chlorine atoms connecting the metal atoms, as shown in Fig. 14.8. The bonding is covalent in character, and to a reasonable approximation we can imagine that the beryllium atoms are sp^3 hybridized. Part of the interaction holding the structure together is a donation from a lone pair from the chlorine into a vacant orbital on the beryllium.

Beryllium also forms organometallic compounds, such as $Be(CH_3)_2$, which in the solid has a structure analogous to $BeCl_2$ with the Cl atoms being replaced by methyl groups. These compounds are reminiscent of the so-called electron deficient compounds formed by boron.

The oxide BeO is exceptionally stable, the strong bonding being reflected by a melting point of just over 2500 °C. Both the oxide and the hydroxide are amphoteric, which we can interpret either in terms of the strong polarizing effect which the small Be^{2+} ion has on solvating water molecules, or on the significant degree of covalent character in the $Be\cdots O$ interaction.

At high pH, the oxide dissolves to give $[Be(OH)_4]^{2-}$, which is analogous to the species $[Al(OH)_4]^{2-}$ found for aluminium. At low pH the aquated ion $[Be(H_2O)_2]^{2+}$ is formed and, as expected, this species has significant acidity on account of the strong Be–O interaction. In between these two extremes various hydroxo species are formed, such as the cyclic trimer illustrated in Fig. 14.9. This structure further illustrates the propensity for beryllium to form polymeric bridged structures, reflecting the significant degree of covalency in the bonding.

Consistent with the amphoteric nature of BeO, it is found that $BeCl_2$ is hydrolysed in water, the products depending on the pH, just as they do for the dissolution of BeO.

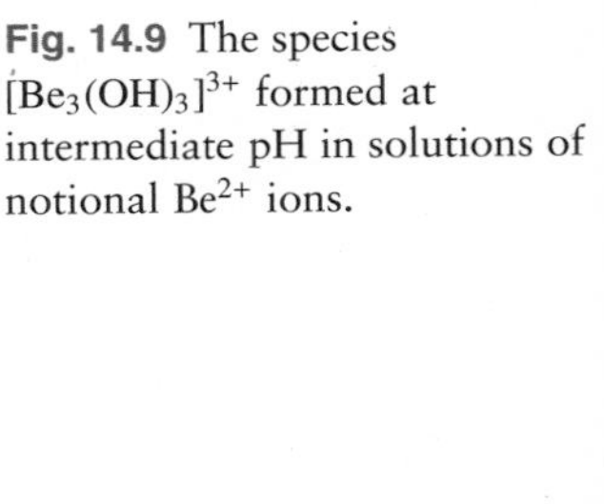

Halides of Mg–Ba

The halides are the remaining elements of Group 2 are all typically ionic in their bonding. In the solid state, the coordination number of the metal ions is six for Mg and eight for the remaining metals, reflecting the increase in the size of the ions. Calcium (Ca), strontium (Sr) and barium (Ba) fluorides all adopt the fluorite structure, illustrated in Fig. 5.12 on page 193. In contrast, the iodides adopt layered structures, as typified by the cadmium iodide structure shown in

Fig. 14.8 Part of the chain structure found in solid $BeCl_2$; the structure continues to the left and right. Beryllium atoms are shown in grey and chlorine atoms in green

Weblink 14.1

View and rotate the structure of solid $BeCl_2$

Fig. 14.9 The species $[Be_3(OH)_3]^{3+}$ formed at intermediate pH in solutions of notional Be^{2+} ions.

Fig. 14.10 on the facing page. The adoption of such layered structures is often taken as evidence of increase covalency in the lattice, as would be expected for the iodides as compared to the fluorides.

One interesting question is why there are no compounds known containing the M^+ ion. In principle, there is no reason why such compounds should not form quite readily, say from the combination of M and X_2 (where X is a halogen) since the first ionization energy of M is much lower than the sum of the first and second. Presumably the problem is that such compounds are unstable with respect to a disproportionation reaction such as

$$2\,MX(s) \longrightarrow M(s) + MX_2(s).$$

The much larger lattice enthalpy of MX_2 than MX appears to be the driving force for this reaction, a point which is explored in *Question 15.6*.

The halides tend to be quite soluble in water, with the exception of the fluorides where the large lattice enthalpies tend to result in these salts being somewhat less soluble. The aquated ions do not show significant acidity.

Oxides of Mg–Ba

The oxides (and hydroxides) are all basic, which fits in with a view of these compounds being primarily ionic, and contrasts with the amphoteric nature of BeO. As we go down the group the hydroxides become more and more basic, which can be rationalized in terms of the increase in ionic character down the group.

Peroxides MO_2 are known for the heavier elements (Ca onwards), and as we go down the group they are more and more readily formed. This parallels the behaviour of the Group 1 metals. The superoxides may exist for Ca, Sr and Ba.

Other features

Like beryllium, there are particularly well characterized organometallic species formed by magnesium. The most important of these are the *Grignard reagents*, with typical composition RMgX (where X is halogen), which, as we have seen in section 9.3.2 on page 351, are widely used reagents in organic chemistry. The structures of these compounds, both in solution and the solid state, is much more complex than the simple formula implies.

14.5.3   Group 12

Overview

The three Group 12 elements, zinc (Zn), cadmium (Cd) and mercury (Hg), all have the electronic configuration $nd^{10}\,(n+1)s^2$. By the time the d sub-shell is full the d electrons have dropped in energy to the point where they cannot be involved in bonding. The valence electrons are therefore just the s^2, and not surprisingly this results in a strong analogy between Groups 12 and 2.

The orbital energies of $4s$ in Zn, $5s$ in Cd and $6s$ in Hg are -8.1 eV -7.7 eV and -8.9 eV, respectively. The increase in energy of going from $4s$ to $5s$ follows the usual trend in which orbital energies increase down a group. The fall in energy on going to Hg is the result of filling the $4f$ shell and relativistic effects which have already been discussed in section 7.2.4 on page 269.

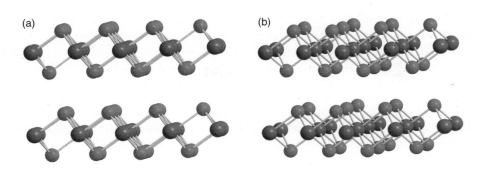

Fig. 14.10 The crystal structure of CdI$_2$; Cd atoms shown in grey, I atoms in purple. View (a) illustrates how the iodine atoms are formed into layers with the cadmium atoms located between the layers. View (b) shows the octahedral coordination of the Cd atoms within the layers of iodine atoms.

The fact that the Hg 6s orbitals are so low in energy means that they match quite well with the energies of the valence orbitals in typical non-metals, e.g. oxygen, fluorine and chlorine. This accounts for the unusually strong covalent bonds between mercury and these elements.

The chemistry of all three elements is dominated by the oxidation state +2. However, mercury also has many compounds containing the $[Hg_2]^{2+}$ species, in which the oxidation state is +1. Again, we can attribute the occurrence of this oxidation state to the unusually low energies of the 6s electrons, making them somewhat harder to ionize. The chemistry of this oxidation state is considered separately.

Halides

All of the MX$_2$ halides are known and occur as solids at room temperature. The fluorides have high melting and boiling points, adopt crystal structures which are characteristic of ionic solids, and are sparingly soluble in water. In these respects the fluorides are very similar to those of Group 2, and can be regarded as ionic compounds.

The other halides are lower melting and have layered crystal structures which are associated with increased covalency. The CdI$_2$ structure, shown in Fig. 14.10, is regarded as the archetype of such structures; the iodine atoms form layers, and the cadmium atoms are located in octahedral sites between these layers. The mercury halides show the greatest degree of covalency: they are soluble in organic solvents, and discrete HgX$_2$ molecules exist in solution.

Oxides

As has already been mentioned, ZnO is amphoteric, dissolving in strong acids to give the aquated Zn^{2+} ion, and in strong bases to give the zincates such as $[Zn(OH)_3]^-$ and $[Zn(OH)_4]^{2-}$. CdO is much more basic than ZnO, only dissolving in very strong bases; HgO is not amphoteric.

Mercury(I)

There are many compounds such as Hg$_2$Cl$_2$ which contain mercury in the +1 oxidation state, and for which this is much evidence that the mercury is present not as Hg^+ but as the dimer $[Hg - Hg]^{2+}$. Corresponding species are known for Zn and Cd, but only under rather exotic circumstances. Mercury(I) compounds tend to be unstable with respect to disproportionation

$$[Hg_2]^{2+}(aq) \longrightarrow Hg^{2+}(aq) + Hg(l).$$

Weblink 14.2

View and rotate the structure of solid CdI$_2$

Box 14.1 Hard and soft Lewis acids and bases

When a Lewis acid and a Lewis base combine to form an adduct, it has been observed that certain combinations are more favoured than others. In an attempt to rationalize this behaviour acids and bases have been categorized as either 'hard' or 'soft'. It is found that the combination of a hard acid with a hard base, or a soft acid with a soft base, is generally more favoured than a hard–soft combination. The following table classifies some commonly encountered species in this way.

hard acids	hard bases	soft acids	soft bases
H^+ Li^+ Mg^{2+}	OH^- F^- Cl^-	Ag^+ Tl^+ Hg^{2+}	I^- S^{2-} R^-

Note that hard acids tend to be small positive ions, possibly with multiple charges, whereas soft acids are larger, more polarizable, positive ions. Hard bases tend to be small electronegative species, whereas soft bases are larger more polarizable anions.

The explanation for why this classification works, and how hardness and softness can be quantified, is the subject of much debate. The argument usually put forward is that in a hard–hard interaction the HOMO of the base and the LUMO of the acid are quite far apart in energy and so do not perturb one another very much. As a result, the interaction between the two species is primarily electrostatic. In contrast, in a soft–soft interaction the HOMO and LUMO are closer in energy, and so they interact more strongly. The result is a significant redistribution of electron density as the bond is formed.

All of the mercury(I) halides are known. Hg_2F_2 is soluble in water, but all of the others are virtually insoluble.

Other features

All of the Group 12 metals form organometallic species of the type RMX (where X is a halogen) and MR_2; the former are analogous to the Grignard reagents. As has been discussed in section 9.6.1 on page 382, the cadmium reagents find use in synthetic organic chemistry where they provide carbon nucleophiles. The MR_2 species are non-polar, soluble in organic solvents, and are generally found as liquids or low-melting solids. These properties point to covalent bonding.

For Zn and Cd the MR_2 species tend to be rather unstable, in the sense that they react quickly with both oxygen and water. For example, the zinc alkyls are spontaneously flammable on contact with air, burning to give ZnO. In contrast, the mercury alkyls HgR_2 are remarkably stable to both air and water – indeed they exist freely in solution in water.

A final feature which is worth noting is the exceptional affinity which mercury has for sulfur, a point well-illustrated by the compound HgS itself which is an insoluble solid, highly resistant to attack by even the strongest acids. It appears that the high toxicity of mercury in part results from its formation of strong bonds to the –SH groups in proteins. This affinity of Hg^{2+} for sulfur can be interpreted using the hard–soft acid base concept, outlined in Box 14.1.

14.5.4 Group 13

Overview

In Group 13 we see for the first time a group containing both metals and a non-metal (boron); as expected, the non-metal occurs at the top of the group, reflecting its lower energy orbitals. The oxidation state +3 is common for all the elements, but as the group is descended the +1 oxidation state starts to become more common. The occurrence of an oxidation state which corresponds to an ion with the configuration $\ldots ns^2$ is referred to as the 'inert pair effect'. Such oxidation states become increasingly common as we descend the groups in the p-block.

In the case of Group 13, the +1 oxidation state is only really common for thallium (Tl), and indeed for this element +1 is preferred over +3. An explanation of this phenomenon is given by Fig. 14.11 which plots the sum of the first three ionization energies for the elements Al–Tl.

The increase from Al to Ga is due to the fact that between these two elements the $3d$ shell has been filled, resulting in a significant increase in the effective nuclear charge. The sum of the ionization energies then drops when we move to indium (In) since the electrons are now being ionized from a shell with a higher principal quantum number. On moving to thallium the ionization energies rises again, partly due to the the filling of the $4f$ shell, and partly due to relativistic effects which lower the energy of the $6s$ electrons. Of course, more energy is still required to form Ga^{3+} than Tl^{3+}, but the gallium ion will be much smaller than that of thallium, so the reduction in energy when a lattice forms is much greater for Ga^{3+} than Tl^{3+}. The difference provides an explanation as to why Tl(III) compounds are relatively more difficult to form, and hence why Tl(I) might be preferred.

Boron stands apart from the rest of the elements in the group, since its bonding is essentially covalent. Boron compounds often have low-lying vacant orbitals making them Lewis acids i.e. electron acceptors.

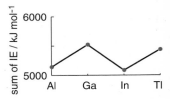

Fig. 14.11 Graph showing the sum of the first three ionization energies of the elements Al–Tl.

Halides

All of the boron halides MX_3 are known, but in contrast to $BeCl_2$ they show no tendency to dimerize. The B–F bond is particularly strong and you will often see this attributed to the presence of some π bonding between the fluorine and the empty p orbital on the boron. However, as was discussed in section 7.9.2 on page 298, the form of the calculated MOs shows that there is in fact rather little π bonding present. These halides are all Lewis acids and readily form adducts with electron donors e.g. with F^- to form $[BF_4]^-$.

The aluminium halides show an interesting gradation of properties. AlF_3 is a high-melting insoluble solid. The structure of the solid is shown in Fig. 14.12 (a): each Al is surrounded by six F atoms in a distorted octahedron, with each F atom being shared by an adjacent octahedron. The $[AlF_6]^{3-}$ unit appears in other solids, such as cryolite Na_3AlF_6. AlF_3 is not a simple ionic compound, but it nevertheless has significant ionic character.

$AlCl_3$ is rather different. As shown in Fig. 14.12 (b), the solid has a layered structure, typically of compounds showing greater covalency; as in AlF_3, the aluminium is six-coordinate. When solid $AlCl_3$ melts at around 190 °C discrete Al_2Cl_6 units, illustrated in Fig. 14.13 on the following page, are formed, and these persist into the gas phase. The bonding can be considered to involve

⊕ *Weblink 14.3*

View and rotate the structures of solid AlF_3 and $AlCl_3$.

Fig. 14.12 The crystal structures of (a) AlF_3 and (b) $AlCl_3$; Al atoms shown in grey, F and Cl atoms shown in lime and green, respectively. In AlF_3 each Al is surrounded by six fluorines in a distorted octahedron; note that adjacent octahedra share F atoms. In contrast, $AlCl_3$ is a layered structure, indicative of greater covalency.

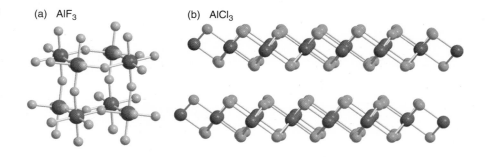

(a) AlF_3 (b) $AlCl_3$

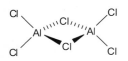

Fig. 14.13 The dimeric unit Al_2Cl_6 found in molten and gas phase aluminium trichloride.

donation of a lone pair from chlorine into an unoccupied MO on the aluminium. It is clear that $AlCl_3$ is not a simple ionic compound. In $AlBr_3$ and AlI_3 such dimers exist in the solid phase, as well as in the liquid and gas. Like AlF_3, the other halides can all act as acceptors.

Thallium(III) fluoride, TlF_3, contrasts with the earlier members of the group. It is a high melting solid and although it cannot be considered to be entirely ionic, there is less evidence of covalency than the earlier halides. In contrast to the other trihalides, it is not an strong acceptor and does not form $[TlF_4]^-$.

Halides with the oxidation state +1 are also known for aluminium and gallium, but these compounds are rather unstable with respect to disproportionation. For indium these halides are easier to prepare, and for thallium the monohalides are formed in preference to the trihalides. The Tl(I) halides can be described as being ionic, as evidenced by the fact that in the solid TlF has a structure which is quite similar to that of NaCl.

Oxides

The oxides show a nice gradation of properties: B_2O_3 is weakly acidic, Al_2O_3 and Ga_2O_3 are amphoteric, In_2O_3 is weakly basic and Tl_2O_3 is basic. This trend follows the increase in metallic character down the group.

In the solid B_2O_3 is amorphous and consists of BO_3 groups joined through the oxygen atoms. It is hydrolysed to $B(OH)_3$, which is a weak acid, which can ionize and polymerize to give a wide range of borate anions, as has already discussed on page 626. The amphoteric nature of Al_2O_3 has already been described in section 14.4 on page 630; gallium oxide behaves similarly.

Tl_2O_3 is, in contrast, weakly basic, but Tl_2O is strongly basic and is hydrolysed by water to give the soluble hydroxide TlOH. In this respect, Tl_2O is reminiscent of the Group 1 oxides.

The amphoteric nature of the Al and Ga oxides implies that the corresponding aquo ions have significant acidity, and indeed it is found that they form a wide range of hydoxy ions, depending on the pH.

Other features

No discussion of Group 13 can pass over the extraordinary range of boron hydrides which are known. The simplest of these, B_2H_6, has already been described in section 14.2.4 on page 620, and it was noted that a key feature of this molecule is that its bonding cannot be described in terms of two-centre two-electron bonds, but rather a more delocalized view has to be taken. There are many more such hydrides with a bewildering range of formulae such as B_4H_{10}

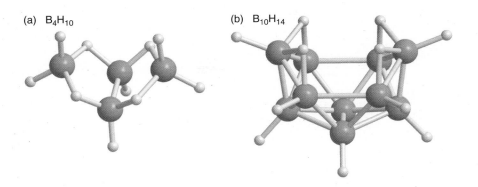

(a) B_4H_{10} (b) $B_{10}H_{14}$

Fig. 14.14 The structures of the boron hydrides (a) B_4H_{10} and (b) $B_{10}H_{14}$; boron atoms are shown in yellow and hydrogen in white. The cylinders joining the atoms indicate close contacts, rather than simple bonds.

and $B_{10}H_{14}$ (illustrated in Fig. 14.14), as well as anions such as $[B_6H_6]^{2-}$ and $[B_{12}H_{12}]^{2-}$. Many of these molecules have hydrogen atoms bridging between boron atoms, as was described for B_2H_6.

The three-dimensional structures adopted by these hydrides can be rationalized by *Wade's Rules*. There are also many related compounds called carboranes in which BH units are replaced by C. Other elements form electron deficient compounds analogous to the boron hydrides, but none have been so well studied or characterised.

Weblink 14.4

View and rotate the structures shown in Fig. 14.14.

14.5.5 Group 14

Overview

This group shows a complete spectrum of behaviour ranging from a typical non-metal (carbon) through semiconductors, silicon (Si) and germanium (Ge), to typical metals, tin (Sn) and lead (Pb). Carbon is typical of the second period in that it forms strong multiple bonds, whereas silicon greatly prefers to form single bonds, particular single bonds to oxygen. In this group we also encounter hypervalent molecules for the first time in the form of $[SiF_6]^{2-}$.

All of the elements show an oxidation state of +4, but as we move towards the bottom of the group the +2 state becomes increasingly common and is indeed the favoured oxidation state for lead. The structure and reactions of carbon-containing compounds is the domain of organic chemistry, so apart from noting that such compounds are typically covalent, we will not discuss them further here.

Halides

All of the carbon and silicon tetrahalides are known; they are discrete molecular species, whose hydrolysis reactions have already been discussed. Silicon can also form the hypervalent anions $[SiF_5]^-$ and $[SiF_6]^{2-}$ (the latter is isoelectronic with the very stable species SF_6).

For germanium, tin and lead both the MX_4 and MX_2 halides are known. GeX_2 tends to be unstable with respect to disproportionation to GeX_4 and Ge, but as we move further down the group the MX_2 halide becomes more stable, being the preferred oxidation state for Pb.

With the exception of SnF_4, the tetrahalides GeX_4 and SnX_4 are either gases or low melting solids, reflecting the fact that they exist as discrete molecules, even in the solid. SnF_4 is something of an exception in that the solid is polymeric, consisting of edge-sharing octahedral SnF_6 units. In the gas phase the dihalides of Ge and Sn are discrete molecules, but in their solid forms they show more complex structures typically consisting of chains created by sharing halogen atoms between two metal atoms.

For lead, PbX_2 is generally more stable than than PbX_4. Only the tetrafluoride is heat stable, the other halides decomposing to $PbX_2 + X_2$. The dihalides are high melting point solids which are sparingly soluble in water – properties rather reminiscent of the Group 2 halides, a comparison further reinforced by the fact that several of the PbX_2 adopt the fluorite crystal structure. These compounds can be regarded are largely ionic.

Oxides

The common oxides of carbon, CO_2 and CO, are molecular and show multiple bonding. In contrast, the common oxide of silicon, with the formula SiO_2, is a giant covalent structure (Fig. 1.2 on page 5), held together by very strong Si–O–Si bonds. As has been discussed before, this behaviour is very typical of the comparison between Periods 2 and 3.

Both CO_2 and SO_2 are acidic. Hydration of 'SiO_2' with various amounts of water results in a bewildering array of polymeric silicates, which have been studied both in the solid state and in solution. The structure of GeO_2 is similar to SiO_2.

Moving further down the group we see the appearance of amphoteric oxides and, as with the halides, the increasing dominance of the +2 oxidation state. SnO_2 is insoluble in water, but dissolves in strong alkalis to give stannates such as $[Sn(OH)_6]^{2-}$ i.e. the oxide is amphoteric. GeO is unstable with respect to disproportionation, but SnO is stable in this regard. This oxide is amphoteric and dissolves in strong alkali to give the species $[Sn(OH)_3]^-$. In the solid state SnO consists of layers of square based pyramidal SnO_4 groups, as illustrated in Fig. 14.15.

The oxides of lead are more complex and include PbO, PbO_2 as well as the mixed valency species Pb_3O_4 and Pb_2O_3. The solid state structures of all these compounds involve sharing of oxygen atoms between two or more metal atoms, continuing the theme of chain and ring formation which runs throughout this group. These oxides are amphoteric.

14.5.6 Group 15

Overview

As with the previous group, in Group 15 we see the transition from non-metal to metal as we descend the group. The +3 and +5 oxidation states are common for the early members of the group, but further down the group the +3 state is preferred and +5 becomes rather uncommon. Hypervalent compounds are well represented in the group, and multiple bonding is seen for Period 2 (nitrogen) and to a lesser extent for Period 3 (phosphorus).

Halides

The trihalides show an interesting gradation in properties. For nitrogen and arsenic (As) they are either gases or low-melting solids, for antimony (Sb) they are solids with modest melting points, and for bismuth they are solids with quite high melting points. All of the nitrogen halides are molecular species, as

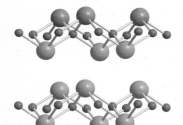

Fig. 14.15 The structure of SnO, which consists of layers of square-based pyramidal SnO_4 units, pointing in opposite directions, and sharing along edges. Tin atoms shown in green, oxygen in red.

Weblink 14.5

View and rotate the structure of solid SnO.

(a)

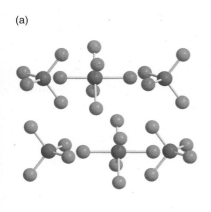

(b)

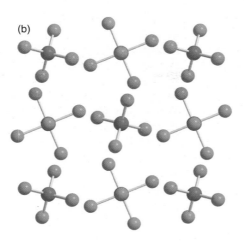

Fig. 14.16 Two views of the solid-state structure of PCl$_5$; phosphorus atoms are in pink, chlorine atoms in green. In (a) the alternating tetrahedral [PCl$_4$]$^+$ and octahedral [PCl$_6$]$^-$ species can be seen. View (b) shows how these two species line up in chains through the solid.

are most of the arsenic and antimony halides. However AsI$_3$, SbI$_3$ and BiI$_3$ have more extended bonding in the solid state. The fluoride of the most metallic element bismuth can be classified as ionic, but the remaining bismuth halides cannot. The trihalides are all hydrolysed by water, and can act as Lewis acids, for example when accepting another halide ion to give [MX$_4$]$^-$.

There are no pentahalides known for nitrogen, but all of the phosphorus pentahalides are known. As is often the case, such hypervalent compounds are not possible for Period 2 elements. These halides are all molecular species, occurring as gases, liquids or low melting solids. Like the trihalides, they are all hydrolysed by water.

Further down the group the pentahalides are confined to the fluoride and chloride for arsenic and antimony, and just the fluoride for bismuth. This is typical behaviour where the highest oxidation state is increasingly confined to the most oxidizing halogens as the group is descended.

As is shown in Fig. 14.16, in the solid state PCl$_5$ exists as equal proportions of the tetrahedral species [PCl$_4$]$^+$ and the octahedral species [PCl$_6$]$^-$, whereas PBr$_5$ exists as PBr$_4^+$ and Br$^-$. AsF$_5$ exists as discrete molecules in the solid state, but solid BiF$_5$ consists of BiF$_6$ octahedra linked through fluorine. This range of structural types reflects subtle changes in the bonding within and between these molecules.

Oxides

Nitrogen forms eight molecular oxides with oxidation states between +1 and +5; of these N$_2$O, NO and NO$_2$ are the most familiar. They are all acidic, for example NO$_2$ is hydrolysed by water to give nitric(V) and nitric (III) acids. Both nitric oxide, NO, and nitrogen dioxide, NO$_2$, are free radicals and as such have a tendency to dimerize to give (NO)$_2$ and N$_2$O$_4$, respectively. However, at room temperature the degree of dimerization is quite low, especially for NO, indicating that these species are unusually stable for free radicals.

Phosphorus also forms a wide range of oxides, but the bonding in them involves P–O–P linkages, rather than P=O double bonds. P$_4$O$_6$ is a low melting solid consisting of discrete molecules whose structure is illustrated in Fig. 14.17 (a); this arrangement of atoms is based on the adamantane structure, shown in Fig. 14.18 on the next page. The oxide is acidic and is hydrolysed by water

Weblink 14.6

View and rotate the structure of solid PCl$_5$.

Fig. 14.17 The molecular structures of (a) P_4O_6 and (b) P_4O_{10}; phosphorus atoms are shown in purple, oxygen atoms in red. Both structures are based on the adamantane fragment.

(a) P_4O_6

(b) P_4O_{10}

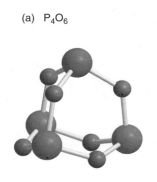

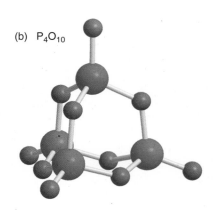

Fig. 14.18 The carbon atoms in the hydrocarbon adamantane ($C_{10}H_6$) form a fragment of the diamond structure.

⚫ *Weblink 14.7*

View and rotate the structures of solid P_4O_6 and P_4O_{10}.

to give phosphoric(III) acid, whose structure has already been discussed on page 628.

The oxide P_4O_{10} also has a structure based on the adamantane framework, and as can be seen from Fig. 14.17 (b); the structure is derived from that of P_4O_6 by adding a terminal oxygen atom to each phosphorus. At high temperatures this oxide forms polymeric structures consisting of rings and sheets, all constructed using P–O–P linkages.

Arsenic(III) oxide and antimony(III) oxide are both amphoteric, but Bi_2O_3 is basic, only dissolving in aqueous acids and not in bases. With bismuth the transition from to a typical metal is therefore complete.

For the higher oxidation state of +5, As_2O_5 is known, but the corresponding oxides for Sb and Bi are not well characterized. As_2O_5 reacts readily with water to give the tri-basic acid H_3AsO_4, which is analogous to phosphoric acid.

Other features

Nitrogen and phosphorus both form a wide range of oxoacids in various oxidation states. For nitrogen nitric(V) acid, HNO_3, and nitric(III) acid (nitrous acid), HNO_2 and the most familiar. For phosphorus the tri-basic phosphoric(V) acid, H_3PO_4, is the most well known.

There is an extensive organic chemistry of phosphorus, for example in the *Wittig reaction*.

14.5.7   Group 16

Overview

The trends in Group 16 are very similar to those seen in Group 15. Oxygen and sulfur are typical non metals, selenium (Se) and tellurium (Te) are semiconductors and the final element polonium (Po) is a metal. Like other Period 2 non-metals, oxygen has a tendency to form multiple bonds, but in the next period the preference for sulfur is for the formation of rings and chains.

Oxygen has a somewhat limited range of oxidation states, but sulfur, and to some extent selenium and tellurium, have a wider range up to a maximum of +6. For polonium, only the +2 and +4 oxidation states are common.

Oxygen has a sufficiently favourable electron affinity, and the resulting anion is sufficiently small, for it to be reasonable propose that the O^{2-} anion is present in oxides such as Na_2O. The same is true for the corresponding sulfides.

However, the presence of such ions becomes increasingly questionable as we move away from the most electropositive metals.

Halides

The only stable halide known for oxygen in OF_2, but sulfur shows a wide range of fluorides from SF_2 to SF_6. These fluorides are molecular and, as has already been discussed on page 628, are hydrolysed by water to give acidic solutions. SF_4 has the somewhat unusual property that it can act as an electron donor to form adducts with species like BF_3, and also as an acceptor to form species such as $[SF_5]^-$. It is thought that the donation is not of the sulfur lone pair but on a lone pair of one of the fluorine ligands. Interestingly, apart from SCl_2, there is little evidence for any halides of sulfur other than these fluorides.

For Se, Te and Po quite a wide range of halides are known. In the highest oxidation state, +6, the only known compounds are SeF_6 and TeF_6, but in the +4 oxidation state all of the MX_4 halides are known, apart from SeI_4. Structurally these halides are quite diverse: some are simple molecules, others show a number of different extended covalent structures in the solid state. There is some evidence for $SeCl_2$ and $TeCl_2$, but $PoBr_2$ and $PoCl_2$ are well characterized and moderately high-melting solids.

Like SF_6, SeF_6 does not react at an appreciable rate with water, but TeF_6 is hydrolysed quickly, to give $Te(OH)_6$, as has already been described on page 629.

Oxides

The two common oxides of sulfur are SO_2 and SO_3, with oxidation states +4 and +6, respectively. These are gaseous species which are hydrolysed readily to give acidic solutions.

For the remaining elements of the group, the dioxides are all well known. SeO_2 is a high melting point solid consisting of corner-linked SeO_3 pyramids. TeO_2 is similarly high melting and also has a polymeric structure in the solid state. However, in contrast to the earlier oxides, TeO_2 is amphoteric. Finally PoO_2 has the fluorite structure which is indicative of significant ionic character. This oxide is also amphoteric, but is more basic that TeO_2.

The monoxides SeO and TeO are only transient, but PoO can be prepared as an easily-oxidized solid. For the oxidation state +6, SeO_3 and TeO_3 are known as solids, the latter being soluble in concentrated aqueous alkali.

Other features

Sulfur has a wide range of oxoacids in which the oxidation state of the sulfur varies from +3 to +6. The most familiar are sulfuric(VI) acid, H_2SO_4, and sulfuric(IV) acid (sulfurous acid), H_2SO_3. Some of the oxoacids have S–S or S–O–S bonds, and the two sulfur atoms can be in different oxidation states, as in disulfurous acid in which the two oxidation states are +5 and +3.

Selenium oxoacids are quite similar to those of sulfur e.g. H_2SeO_4 and H_2SeO_3. However, for oxidation state +6 telluric acid is the tetrabasic $Te(OH)_6$, which is quite different.

disulfurous acid

14.5.8   Group 17

Halides have already been mentioned in the context of the other elements in the main group, so little more needs to be said here. The halogens have sufficiently favourable electron affinities that they readily form anions which appear both in ionic solids and as aquated ions in solution. In addition, there are a large number of molecules in which there is covalent bonding to a halogen, and indeed covalent bonds to fluorine are exceptionally strong.

The halogens also form a series of *interhalogen* compounds of the form XY_n where X and Y are both halogens, with X being the heavier. Compounds with the formulae XY, XY_3, XY_5 and XY_7 are all known, and apart from XY all of these are hypervalent. The high oxidation state compounds tend to become more stable as X becomes heavier and, as usual, are favoured by Y being fluorine. The only example of XY_7 is IF_7.

The halogens form a number of oxoacids such as HOX, HXO_2, HXO_3 and HXO_4. Of these, the most stable are the halate(VII) acids, HXO_4, which are all known apart for X = F. A solution of HIO_4 contains the species $[IO_4]^-$, but in strong acids this species can be protonated to give $[I(OH)_6]^+$.

14.5.9 Group 18

The elements in this group are often called the 'noble gases' precisely because they are so unreactive – indeed for many years it was thought that these elements simply did not form *any* compounds. In terms of bonding theory, the fact that these atoms have filled shells means that the outer orbitals are rather low in energy and also rather contracted. This favours neither the formation of ions nor covalent bonding.

However, as we go down the group these features become less unfavourable, and by the time we reach xenon (Xe) the first ionization energy is not that much different from O_2. This led researchers to believe that, since they had been successful in preparing the ionic compound $[O_2]^+[PtF_6]^-$, ionic compounds of Xe might exist. Their search was fruitful and with some ingenuity they were able to synthesise 'XePtF_6' (which is in fact contains $[XeF^+]$, $[PtF_6]^-$ and $[Pt_2F_{11}]^-$). Soon after a series of fluorides XeF_2, XeF_4 and XeF_6 were synthesised; is is not surprising that the strongly oxidizing fluorine is the ligand. The oxides XeO_3 and XeO_4 are also known.

14.6 Moving on

In just one chapter we cannot possible do justice to the vast range of structures and properties shown by compounds of the main-group elements. However, what we have attempted to do in this chapter is to draw out some themes and ideas which will help you to rationalize and unify the chemistry of these elements so that you can begin to tackle a more detailed study of particular elements of groups.

FURTHER READING

Main Group Chemistry, W. Henderson, Royal Society of Chemistry (2000).
Inorganic Chemistry, fourth edition, Peter Atkins, Tina Overton, Jonathan Rourke, Mark Weller and Fraser Armstrong, Oxford University Press (2006).
Chemistry of the Elements, N. N. Greenwood and A. Earnshaw, second edition, Butterworth–Heinemann (1997).

QUESTIONS

14.1 (a) Discuss why it is that in the gas phase BCl_3 exists as a discrete molecule, whereas $AlCl_3$ forms dimers (illustrated in Fig. 14.13 on page 638).

(b) Explain why B_2H_6 is sometimes described as an 'electron deficient' molecule.

(c) Assuming that the Al is sp^3 hybridized, draw up a description of the bonding in the Al_2Cl_6 dimer. Is this molecule electron deficient?

14.2 Comment on and rationalize the following observations

(a) BF_3 is a gas, whereas the other Group 13 trifluorides are all high-melting solids.

(b) BF_3 and AlF_3 both readily act as Lewis acids toward F^- ions to give $[BF_4]^-$ and $[AlF_4]^-$; however, TlF_3 does not form an analogous adduct with F^-.

(c) GaF and InF are known as unstable gaseous species, but GaI and InI are known as stable solids; all of the thallium(I) halides are known, including TlF.

14.3 (a) Rationalize the trend in the pK_a values of the following three aquo ions

$$[K(H_2O)_6]^+ \; pK_a = 14.5 \quad [Ca(H_2O)_6]^{2+} \; pK_a = 12.8 \quad [Ga(H_2O)_6]^{3+} \; pK_a = 2.6$$

(b) Use these pK_a values to discuss: (i) the nature of the metal-containing species which would be present in aqueous solutions of these ions, (ii) what would happen if such solutions were made progressively more basic.

14.4 One of the pieces of evidence that mercury(I) salts contain the species $[Hg_2]^{2+}$, rather than a simple Hg^+ ion, is that these salts are not paramagnetic (i.e. there are no unpaired electrons). Explain why it is that Hg^+ is paramagnetic whereas $[Hg_2]^{2+}$ is not. [Hint: draw up a simple MO diagram for $[Hg_2]^{2+}$, considering only the $6s$ electrons].

14.5 Comment on the following

(a) AlF_3 is a high melting point solid, whereas SiF_4 is a gas at room temperature.

(b) Silicon has fluorides with four, five and six-fold coordination: SiF_4, $[SiF_5]^-$ and $[SiF_6]^{2-}$, but for carbon only the four-coordinate fluoride CF_4 is known. However, the gas phase species $[CH_5]^+$ has been detected.

(c) The Si–F bond lengths in SiF_4, $[SiF_5]^-$ and $[SiF_6]^{2-}$ are 154 pm, 159 pm, and 169 pm, respectively.

14.6 The strength if the $N{\equiv}N$ triple bond is 946 kJ mol^{-1}, whereas that of $P{\equiv}P$ is 490 kJ mol^{-1}; N–N and P–P single bonds have bond strengths in the range 160 kJ mol^{-1} to 200 kJ mol^{-1}, depending on the compound. Discuss these data and the consequences they have for the kinds of compounds formed by nitrogen and phosphorus.

14.7 If PCl_5 reacts with an excess of water the ultimate product is phosphoric(V) acid, H_3PO_4. However, if equimolar amounts of PCl_5 and water react the compound $POCl_3$ is formed. Describe the likely steps by which these two products might be formed, and explain why limiting the amount of water gives a different product.

14.8 Determine the oxidation state of the sulfur in the following compounds or ions

(a) Na_2S (b) SF_2 (c) S_2F_2 (d) $[SO_3]^{2-}$ (sulfite)   (d) $[SSO_3]^{2-}$ (thiosulfate).

14.9 Discuss the following

(a) The reaction of sulfur with F_2 gives SF_4 and SF_6, but its reaction with Cl_2 gives SCl_2 and S_2Cl_2; there is no evidence for SCl_4 and SCl_6.

(b) The ^{19}F spectrum of SF_6 consists of a single line, whereas that of SF_4 consists of two 1:2:1 triplets (^{32}S has spin zero). [Hint: use VSEPR to predict the structures.]

(c) SO_2 exists as a discrete molecule in which the sulfur is two-fold coordinate; solid $SeO_2(s)$ contains chains of the form $-O-SeO-O-SeO-$ in which the Se is three-fold coordinate; $TeO_2(s)$ has a layered structure in which the Te is four-fold coordinate; $PbO_2(s)$ has a three-dimensional structure similar to the fluorite lattice in which the Pb is eight-fold coordinate.

14.10 Use VSEPR to predict the shapes of the following molecules or ions, and predict the form of the ^{19}F NMR spectrum in each case (ignore any coupling to iodine)

(a) $[IF_2]^+$ (b) $[IF_2]^-$ (c) IF_3 (d) IF_5 (e) IF_7.

Discuss the reasons why it is that IF_4 has not been prepared but $[IF_4]^-$ is well known.

14.11 Explain the following observations

(a) Liquid HF and liquid BF_3 are both very poor conductors of electricity, but a 1:1 mixture of the two liquids is a good conductor.

(b) BF_3 is more resistant to hydrolysis than is BCl_3.

(c) PF_5 is molecular in the solid state, whereas PBr_5 forms an ionic lattice containing $[PBr_4]^+$ and Br^- ions.

(d) The equilibrium constants for the formation of the adducts **A** increase as X is changed from F to Cl and then to Br.

14.12 On careful hydrolysis of PF_3 an intermediate compound **X** is obtained. Accurate mass spectrometry of **X** gives a parent ion peak at 83.9976. The ^{31}P NMR spectrum shows a doublet of doublets with coupling constants 1079 Hz and 756 Hz. The ^1H NMR spectrum shows a very broad peak, and a doublet of doublets with coupling constants 756 Hz and 60 Hz.

Suggest a structure for compound **X** that is consistent with these data, and predict the form of its ^{19}F NMR spectrum.

[Relative atomic masses: ^{31}P 30.9938; ^{19}F 18.9984; ^1H 1.0078; ^{16}O 15.9949. ^{31}P, ^{19}F and ^1H all have spin $I = \frac{1}{2}$]

14.13 The superoxides of Group 1 metals tend to decompose to the peroxide according to the following reaction

$$2MO_2(s) \longrightarrow M_2O_2(s) + O_2(g).$$

The energetics of this reaction can be analysed using the following Hess's Law cycle

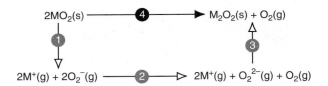

$\Delta_r H°$ for step 1 is *minus twice* the lattice enthalpy of $MO_2(s)$, $\Delta_r H°$ for step 2 is twice the enthalpy of dissociation of the superoxide anion to the peroxide anion, and $\Delta_r H°$ for step 3 is the lattice enthalpy of $M_2O_2(s)$. Our aim is to use this cycle to work out the value of $\Delta_r H°$ for step 4, the decomposition of the superoxide.

(a) Use the Kapustinskii equation, Eq. 5.5 on page 196, to write down expressions the lattice enthalpies needed for steps 1 and 3. Write the radius of the cation as r_+, and assume that of both of the anions O_2^- and O_2^{2-} have the same radius r_- (this is a fair assumption for this rather crude calculation).

(b) $\Delta_r H°$ for step 2 does not change with the metal, so we can simply assume a value, which we will call C. Use this value and your answer to (a) to obtain an expression for $\Delta_r H°$ of step 4.

(c) Carefully explain why your expression predicts that as r_+ increases, the value of $\Delta_r H°$ for step 4 increases. Use this result to rationalize why LiO_2 is not known, but RbO_2 is easily formed.

(d) (Requires calculus) Differentiate your expression for $\Delta_r H°$ of step 4 with respect to r_+, assuming that r_- is constant. Argue that the derivative is positive, and hence leads to the same prediction as in (c) as to the way $\Delta_r H°$ changes with r_+.

(e) Rather than considering $\Delta_r H°$ for step 4, we ought really to consider $\Delta_r G°$ i.e. an entropy term should be included. Discuss whether or not the conclusions of this discussion are likely to be affected by the inclusion of such an entropy term.

14.14 The polyanion I_3^- forms ionic compounds with Group 1 metals, but these compounds tend to decompose to the iodide and iodine according to the following reaction

$$MI_3(s) \longrightarrow MI(s) + I_2(s).$$

Analyse this decomposition using a Hess's Law cycle similar to that in the previous question. Use estimates of the lattice energy to show that this reaction becomes less favoured as the radius of the cation increases. (You should assume that the radius of the I_3^- anion is significantly greater than that of the I^- anion).

14.15 No compounds in which a Group 2 metal is in the oxidation state +1 are known, and it is speculated that this is because such compounds would disproportionate according to

$$2MgCl(s) \longrightarrow Mg(s) + MgCl_2(s),$$

where we have taken MgCl as an example. It is possible to estimate a value for $\Delta_r H°$ for this reaction using the following Hess's Law cycle

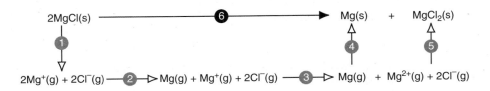

$\Delta_r H°$ for step 1 is *twice minus* the lattice enthalpy of MgCl, $\Delta_r H°$ for step 2 is *minus* the enthalpy of ionization for $Mg(g) \rightarrow Mg^+(g)$, $\Delta_r H°$ for step 3 is the enthalpy of ionization for $Mg^+(g) \rightarrow Mg^{2+}(g)$, $\Delta_r H°$ for step 4 is *minus* the enthalpy of atomization of Mg(s), and $\Delta_r H°$ for step 5 is the lattice enthalpy of $MgCl_2$.

(a) Use the Kapustinskii equation, Eq. 5.5 on page 196, to estimate the lattice enthalpies of MgCl and $MgCl_2$, taking the radius of Mg^+ as 100 pm (a guess based on the radius of Na^+), that of Mg^{2+} as 68 pm, and that of Cl^- as 182 pm.

(b) Given that $\Delta_r H°$ for atomization of Mg(s) is 148 kJ mol^{-1}, $\Delta_r H°$ for $Mg(g) \rightarrow Mg^+(g)$ is 737 kJ mol^{-1}, and $\Delta_r H°$ for $Mg^+(g) \rightarrow Mg^{2+}(g)$ is 1447 kJ mol^{-1}, estimate $\Delta_r H°$ for step 6, the disproportionation of MgCl.

(c) Explain why $\Delta_r S°$ for step 6 is expected to be small.

(d) Do your calculations support the contention that MgCl is unstable with respect to disproportionation? Explain *in words* the origin of this instability.

14.16 High oxidation state metal halides are often unstable with respect to dissociation into a lower oxidation state halide plus the elemental halogen. For example, MX_4 may decompose to MX_2

$$MX_4(s) \longrightarrow MX_2(s) + X_2(g),$$

where M is a metal and X is one of the halogens. This reaction can be analysed using the following Hess's Law cycle

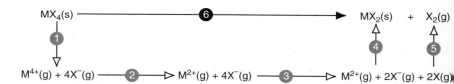

Given the following data, discuss why it is that the higher oxidation state (MX_4) tends to be more stable for the fluoride than the other halides. A quantitative answer is not expected.

The definition of the electron affinity is given in section 7.4.2 on page 278.

	F	Cl	Br	I
electron affinity / kJ mol^{-1}	328	349	325	295
$\Delta_r H°(X_2(g) \rightarrow 2X(g))$ / kJ mol^{-1}	158	243	193	151
$r(X^-)$ / pm	133	182	198	220

14.17 The table below gives the values of $\Delta_r H°$ (in kJ mol^{-1}) for the processes indicated for the cases where M is K or Ca

process	K	Ca
$M(s) \rightarrow M(g)$	90	193
$M(g) \rightarrow M^{2+}(g)$	3470	1735
$M^{2+}(g) + 2Cl^-(g) \rightarrow MCl_2(s)$	-2210	-2226

$\Delta_r H°$ for $Cl_2(g) \rightarrow 2Cl(g)$ is 242 kJ mol^{-1}, and the electron affinity of Cl is 349 kJ mol^{-1}.

Use these data to calculate $\Delta_f H°$ for $KCl_2(s)$ and $CaCl_2(s)$, and hence predict which of these compounds you would expect to form. What is the principle origin of the difference between the values for these two compounds?

Transition metals

Key points

- 4s and 3d electrons are available for bonding in first-row transition metals, although the 3d become increasingly core-like across the row.

- The bonding in transition metal complexes can be understood using a molecular orbital approach.

- In complexes, ligands mainly bond by acting as σ donors. π donation is a weaker effect, and some ligands can act as π acceptors.

- A molecular orbital description of the bonding in a complex helps us to understand its spectroscopic and magnetic properties.

- Quantities such as hydration energies and lattice energies vary across the transition metals in a characteristic way which can be understood using the MO diagram of an octahedral complex.

- A wide range of organometallic complexes exist in which the ligands coordinate through carbon and in which there are often π interactions with conjugated systems on the ligands. In such complexes the metal is usually in a low oxidation state.

- Many organometallic complexes obey the 'eighteen-electron' rule.

- High oxidation states of transition metal ions usually occur as oxoanions in solution.

This chapter is concerned with the *d*-block elements which form Groups 3–11, collectively often referred to as the *transition metals*. Their position within the Periodic Table is illustrated in Fig. 15.1 on the following page. The common feature of these elements is the presence of a partially filled *d* sub-shell, and it is the presence of this partly filled shell which is responsible for most of the special properties which set the transition metals apart from main-group metals. These special properties include:

- The existence of compounds in which a particular element shows a range of oxidation states. In contrast to main-group elements, common oxidation states of transition metals often differ by steps of one.

- The presence of unpaired electrons associated with the metal, giving rise to paramagnetism.

Fig. 15.1 The transition metals are those which form Groups 3 to 11, and which are characterized by the presence of a partly filled *d* sub-shell. In the previous chapter it was explained that Group 12 elements are best considered as being part of the main group, but it is nevertheless common to consider them with the other *d*-block elements. The *d*-block elements in Period 4 are often called the 'first row' transition metals, while those in Periods 5 and 6 are called the second and third rows, respectively.

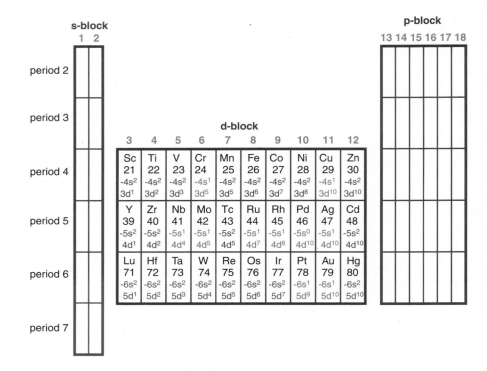

- The formation of coloured compounds and solutions.

- The formation of a large number of *complexes* in which the metal is surrounded by typically between four and six electron-donating ligands.

- The formation of organometallic complexes in which the ligands have π systems e.g. carbon monoxide, small organic molecules with conjugated π systems.

Like the main-group elements, there is a vast amount known about the chemistry of the transition metals, so our task is once again to find the key concepts which will help us to make sense of all this information. Our approach will be based on an understanding of the bonding in the compounds, using the ideas and models developed in Part I.

In order not to be overwhelmed by details, we will focus mainly on transition metal complexes since studying these will give us a good grasp on all of the special properties of the transition metals. We will also confine our attention to the elements from the first row of the transition metals i.e. those in Period 4. Although the second- and third-row transition metals do have quite a lot in common with those of the first row, there are some significant differences which you may well go on to study in a more advanced course.

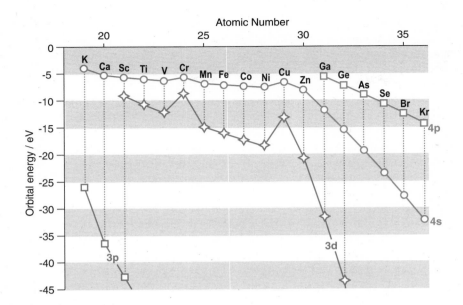

Fig. 15.2 A graph of the atomic orbital energies for the elements of the fourth period. Only the energies of the highest energy occupied orbitals are shown, and the data points are coloured according to the principal quantum number of the orbital. This graph is part of that shown in Fig. 7.2 on page 264.

15.1 Orbital energies and oxidation states

15.1.1 Trends in orbital energies

Figure 15.2 shows the energies of the occupied AOs of the fourth period, including the transition metals Sc–Cu. By the time we reach potassium the $3s$ and $3p$ sub-shells are filled and from this point on these AOs drop rapidly in energy since they are hardly screened by the electrons in the $4s$ and $3d$ AOs. For the transition metals the $3s$ and $3p$ are therefore core electrons and will not concern us further.

The outer electron in potassium occupies the $4s$ AO, rather than the $3d$. This is because the $4s$ orbital penetrates to the nucleus more effectively than does the $3d$, resulting in the energy of the $4s$ being lower than that of the $3d$. The idea that a $2s$ electron is more penetrating than a $2p$ was used in section 2.6.2 on page 69 to explain the order of filling of orbitals in the second period. As is shown in the plots of the RDFs, Fig. 2.45 on page 71, the $2s$ orbital has a subsidiary maximum which lies close in to the nucleus. This results in the electron penetrating the area occupied by the $1s^2$ core electrons, and thus experiencing a larger effective nuclear charge than does the $2p$.

The same argument can be used for the orbitals in the $n = 3$ shell, whose RDFs are shown in Fig. 2.28 on page 56, to explain why the $3s$ is lower in energy than the $3p$, which in turn is lower in energy than the $3d$. In this case both the $3s$ and the $3p$ have subsidiary maxima close in to the nucleus, but the first maximum for the $3s$ is very much closer in. The $3d$ has no such subsidiary maxima, and therefore is much less penetrating than $3s$ or $3p$.

Like the $3s$, the $4s$ is more penetrating than the $4p$ and the $4d$. However, the penetration to the nucleus by the $4s$ is sufficient that, despite being in a higher shell, the $4s$ is lower in energy than the $3d$. Thus, the outer electron in potassium is in the $4s$ rather than the $3d$. At calcium the $4s$ sub-shell is full, so from scandium (Sc) onwards the $3d$ sub-shell is steadily filled.

⇨ The different extent to which electrons screen one another has been discussed section 2.6.4 on page 75.

Looking at Fig. 15.2 we see that for the transition metals the $3d$ AO is *lower* in energy than the $4s$. This can seem rather confusing, as a moment ago we said that the $3d$ was *higher* in energy than the $4s$. The explanation for this apparent paradox is that the electrons in the $4s$ do not screen the $3d$ particularly well since much of the electron density from the $4s$ is further out from the nucleus than that from the $3d$. The $3d$ thus experiences a greater increase in nuclear charge on going from K to Sc than does the $4s$. As a result, by the time we reach Sc the $3d$ has fallen below the $4s$ in energy.

From scandium onwards the energies of both the $4s$ and $3d$ AOs drop steadily: this is simply the result of the increase in nuclear charge not being quite offset by the increase in electron–electron repulsion. Put another way, the electrons in the $3d$ sub-shell do not screen one another particularly well. Although both AOs fall in energy, the $4s$ falls less steeply, which can be explained by noting that this AO is quite well screened by the $3d$ electrons, not least as these are in a lower shell. By the time we reach zinc (Zn) the energy of the $3d$ AOs have dropped to the extent that they are more-or-less core electrons and so are not involved in bonding. This was the justification we made for considering zinc to be a main-group metal. In addition to dropping in energy, the $3d$ AOs also contract, which make them less effective at overlapping with other orbitals in covalent interactions.

The special case of Cr and Cu

These steady trends in orbital energies are interrupted by two 'blips' at chromium (Cr) and copper (Cu). The reason for the blip at Cr is that the ground-state electronic configuration of this atom is $3d^5 4s^1$, rather than $3d^4 4s^2$ as might be expected. The reason for preferring $3d^5 4s^1$ is that this arrangement has six electrons with parallel spins, whereas $3d^4 4s^2$ only has five. As was described in section 2.7.3 on page 82, the larger the number of parallel spins, the greater the favourable exchange interaction. In the case of Cr it turns out that this increase in exchange interaction for $3d^5 4s^1$ is sufficient to favour the configuration over $3d^4 4s^2$.

The energy of a particular orbital depends strongly on which other orbitals are occupied, since the form of the orbitals affects the amount of electron–electron repulsion. Changing the configuration from $3d^3 4s^2$ for vanadium (V) to $3d^5 4s^1$ for Cr therefore changes the electron–electron repulsion in a significant way, and it is to this that we can attribute the sudden change in the energy of the $3d$. A similar argument applies for copper (Cu), for which the lowest energy configuration is $3d^{10} 4s^1$ rather than $3d^9 4s^2$.

Caution in interpreting these orbital energies

⇨ The difficulties of describing the energies of empty orbitals and the way in which orbital energies change with configuration are discussed in section 2.6.5 on page 79.

One other thing can seem rather confusing about Fig. 15.2 is that, since the energy of the $3d$ is *below* that of the $4s$, why then do the electrons not drop down from the $4s$ into the $3d$, thereby lowering the energy? The resolution of this apparent paradox is to understand that what are plotted are the energies of the orbitals for the ground state configurations $3d^n 4s^2$ (with the exception of Cr and Cu). If the configuration was changed to $3d^{n+2}$ the energies of *all* of the orbitals would change since the details of the electron–electron repulsion will change. What we would find is that with the changed orbital energies the configuration $3d^{n+2}$ is higher in energy than $3d^n 4s^2$. It is important to realize that Fig. 15.2 gives the energies of the orbitals when they are occupied in the

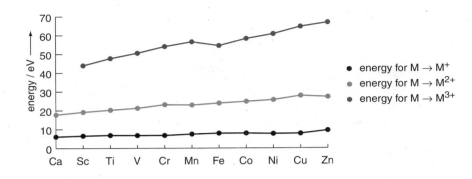

Fig. 15.3 Plot showing the energies needed to ionize one, two and three electrons from a transition metal. The slow rise in the energy needed to form M^+ and M^{2+} reflects the fact that $4s$ electrons are being ionized. To form M^{3+} a $3d$ electron has to be ionized, and this is more sensitive to the increase in nuclear charge, resulting in a steeper rise in the ionization energy.

way that leads to the ground-state configuration; we cannot use this diagram to discuss the energies of orbitals for other configurations.

15.1.2 Ionization

We described in section 2.7.2 on page 81 that, to a good approximation, the ionization energy of an electron is equal to minus its orbital energy. We should therefore expect that the most easily ionized electron from a transition metal will be the $4s$, and this is indeed found to be the case.

Once an electron is removed, the orbital energies of the remaining electrons all decrease as a result of the reduction in electron–electron repulsion i.e. with one less electron, all of the remaining electrons experience a greater net attraction to the nucleus. However, the $3d$ electrons are not well screened by the $4s$, so removing one of the $4s$ electrons does not have a large effect on the energy of the $3d$. In contrast, removing one of the $4s$ electrons has a significant effect on the energy of the other $4s$ electron, so it falls in energy by more than the $3d$. Despite this, it turns out that the remaining $4s$ electron is still higher in energy than the $3d$, so it is this $4s$ electron which is removed to form M^{2+}. The ion will therefore have just d electrons in its outer shell.

Figure 15.3 shows how the energy needed to form M^+, M^{2+} and M^{3+} varies across the transition metals. The first thing we see is that the energy needed to form M^+ and M^{2+} rises rather slowly. This can be explained by noting that the $3d$ electrons form an effective screen for the $4s$. This means that, as the nuclear charge increases by one, the additional electron in the $3d$ largely screens the $4s$ electrons from the increased nuclear charge. As a result, the energy needed to ionize the $4s$ electrons rises rather slowly across the series.

Forming M^{3+} requires significantly more energy, and furthermore this energy increases more sharply along the series when compared to forming M^+ or M^{2+}. The third ionization involves removing a $3d$ electron and since such electrons do not screen one another particularly effectively, much of the increase in the nuclear charge is felt by the $3d$ electrons. As a result, across the series these orbitals drop significantly in energy, causing the ionization energy to rise.

From Fig. 15.3 it can be seen that the ionization energy to form M^{2+} for all of the transition metals is hardly greater than that for Ca, so we can reasonably expect these metals to form ionic compounds containing M^{2+} ions in the same way as Group 2 metals. The formation of M^{3+} requires significantly more energy, but not unfeasibly so. As a result species containing M^{3+} in both the

Fig. 15.4 Three examples of 'classical' transition metal complexes, illustrating the most commonly encountered geometries. These structures have been determined by x-ray diffraction of crystalline materials; the counter ions are not shown.

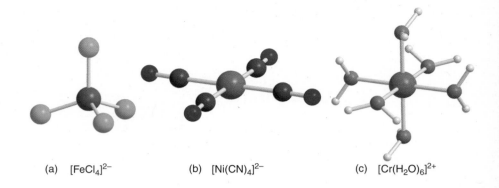

(a) $[FeCl_4]^{2-}$ (b) $[Ni(CN)_4]^{2-}$ (c) $[Cr(H_2O)_6]^{2+}$

solid state and in solution are well-known for the majority of the transition metals. It is only when we come to Cu and Zn that this third ionization energy has become so large as to make the formation of M^{3+} unfeasible.

15.1.3 Oxidation states

Referring to Fig. 7.47 on page 305 we see that for the transition metals up to manganese (Mn) the highest oxidation state is equal to the group number, but that for the later elements the maximum oxidation number falls away, reducing to just +2 by the time we reach zinc. It must be remembered that oxidation state is a formal concept and does not necessarily imply the presence of particular ions e.g. a Mn(VII) compound emphatically does *not* contain Mn^{7+}. Nevertheless, the oxidation state does tell us something about which electrons are available for ionization or to participate in covalent bonding.

The fact that the 3*d* AOs both fall in energy and contract as we go across the transition metals means that they are both harder to ionize and are less likely to be involved in covalent bonding. This trend accounts for why, beyond a certain point (in fact at Mn), the maximum oxidation state starts to fall away.

15.2 Complexes

One of the distinguishing features of transition metals is that they readily form complexes in which the metal is surrounded by a number of ligands. In 'classical' complexes the metal is invariably present as an ion, and the ligands are either negatively charged e.g. halide ions or CN^-, or contain electronegative atoms which make the closest contact with the metal e.g. NH_3 or OH_2. The number of ligands varies with both the metal and type of ligand, but for first-row transition metals the most frequently encountered coordination numbers are four and six. Figure 15.4 gives some examples of such complexes.

Four-coordinate complexes are most often tetrahedral, but square planar complexes are also known, especially for Group 10 (see section 15.7 on page 672). Six-coordinate complexes are usually octahedral, and are also the most commonly encountered: we will have most to say about this kind of complex. It is worth noting that the VSEPR approach, described in section 1.8 on page 18, fails entirely to explain the shapes of these complexes. As we shall see, the shapes are primarily determined by the number of ligands, rather than the number of valence electrons on the metal.

Weblink 15.1

View and rotate the structures shown in Fig. 15.4.

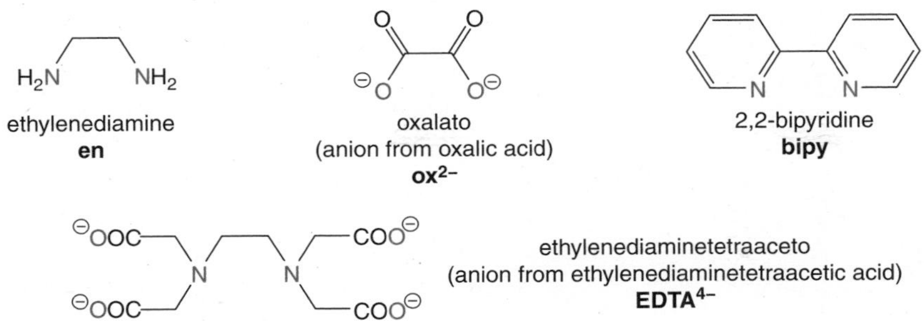

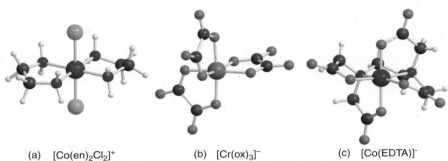

ethylenediamine
en

oxalato
(anion from oxalic acid)
ox^{2-}

2,2-bipyridine
bipy

ethylenediaminetetraaceto
(anion from ethylenediaminetetraacetic acid)
EDTA^{4-}

Fig. 15.5 A selection of polydentate ligands which coordinate to a metal atom through more than one atom (highlighted in blue). The common abbreviation for the ligand is given in bold.

(a) [Co(en)$_2$Cl$_2$]$^+$

(b) [Cr(ox)$_3$]$^-$

(c) [Co(EDTA)]$^-$

Fig. 15.6 Three complexes in which polydentate ligands are involved. Note how the ligand in the EDTA complex has been designed to wrap around the metal facilitating octahedral coordination.

Generally speaking the strength of the interaction between the metal and the ligands is less than that of a full covalent bond between typical non-metals, but not so weak as to be classed as a non-bonded interaction. These complexes retain their structures when dissolved in solution, but in such circumstances one ligand can readily displace another, indicating that there is a low barrier to bond breaking and making.

It is also possible to form complexes in which each ligand makes more than one contact with the metal. Such ligands are described as being *polydentate*, and a selection of them is shown in Fig. 15.5. The three shown at the top are *bidentate* since they have two atoms which can coordinate to the metal. EDTA is *hexadentate*, having a total of six atoms (two nitrogens and four oxygens) which can coordinate to the metal.

Figure 15.6 shows a selection of complexes formed using these polydentate ligands. The ethylenediamine and oxalate ligands have the right shape and size to allow the two coordinating atoms to approach the metal in the appropriate geometry to allow effective interaction with the metal. The ligand EDTA has been designed to be able to wrap around the metal atom and provide all six coordinating atoms.

Forming complexes with polydentate ligands is often advantageous in a thermodynamic sense when compared to forming complexes with monodentate ligands. For example, the reaction

$$[Ni(NH_3)_6]^{2+} + 3\,en \rightleftharpoons [Ni(en)_3]^{2+} + 6\,NH_3$$

in which three bidentate ligands replace six monodentate ligands is found to lie very strongly in favour of the products. At a simple level we can attribute this to the favourable entropy change arising from the fact that the number of

Weblink 15.2

View and rotate the structures shown in Fig. 15.6.

(a)

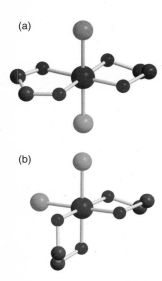

(b)

Fig. 15.7 Two isomers of the complex $[Co(en)_2Cl_2]^+$; for simplicity, the hydrogen atoms are not shown. In (a) the two chlorine atoms are arranged at 180° to one another, whereas in (b) they are at 90°.

🌐 *Weblink 15.3*

View and rotate the structures shown in Fig. 15.7.

---

molecules increases on going from left to right. However, more detailed work shows that there are also contributions arising from the relative strength of the M–L bonds and the relative degree of solvation of the two ligands. The phenomenon in which polydentate ligands show more favourable binding to a metal is called the *chelate effect*.

The ligand EDTA has been designed to provide six coordination sites in just one ligand, and the molecule is sufficiently flexible for these sites to be arranged more or less in an octahedral arrangement. It is not surprising therefore that this hexadentate ligand shows particularly tight binding to transition metal ions.

15.2.1 Isomerism

Complexes can exist in isomers in which the ligands have different spatial arrangements. An example is shown in Fig. 15.7 for the complex $[Co(en)_2Cl_2]^+$. In one isomer the chlorine atoms are at 180° to one another, and in the second isomer the chlorines are at 90° to one another.

Optical isomerism is also possible for complexes, in the same way as for carbon compounds (see section 12.4 on page 530). For example, the two isomers of $[Co(en)_2Cl_2]^+$ shown in Fig. 15.8 are non-superimposable mirror images of one another. This means that we expect such complexes to show optical activity and to be resolvable into enantiomers, which is precisely what is found.

15.3 Bonding in octahedral complexes

To understand why these complexes are formed, and their properties, we need to develop a molecular orbital description of the bonding between the metal and the ligands. The difficulty is that there are rather more atoms and orbitals involved in such complexes than in the relatively simple molecules we have looked at so far. However, the high symmetry of the octahedral environment comes to our aid and makes it possible to construct an MO diagram without too much difficulty.

In Chapter 4 we drew up molecular orbital diagrams for some linear and bent triatomics by using the key idea that only orbitals with the same symmetry can overlap. For these simple molecules it was sufficient to classify the orbitals as being symmetric or antisymmetric according to a mirror plane. As we saw, it was sometimes necessary to combine the atomic orbitals into symmetry orbitals (SOs) which have the desired properties.

Fig. 15.8 The two isomers of the complex $[Co(en)_2Cl_2]^+$ shown here are non-superimposable mirror images of one another i.e. each complex is optically active; hydrogen atoms are not shown.

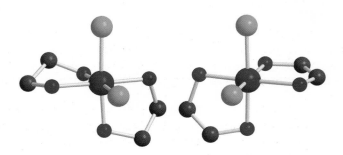

An octahedral complex has much more symmetry than a simple triatomic, and to cope with this we need to use *Group Theory*, which provides a formal method of classifying symmetry. Such a discussion is beyond the level of this text, so the results of such a group theory analysis of the orbitals will simply be presented in a graphical way. From these pictures we will at least be able to appreciate how the symmetry classification works.

The first thing we need to do is to decide which AOs are likely to be involved in the bonding. For the metal, the $3d$ are certainly of appropriate energy, and we should also include the $4s$ and $4p$ AOs. Assuming that the metal is present as an ion, these latter two orbitals will not be occupied and will, as was discussed above, be higher in energy than the $3d$ AOs.

15.3.1 MO diagram for σ-only interactions

To keep things as simple as possible we will start out by assuming that each ligand provides a single σ-type valence orbital directed towards the metal. For a ligand such as a halide ion this orbital might be a p orbital directed towards the metal, for a ligand such as NH_3 or OH_2 the appropriate orbital would be a hybrid containing a lone pair.

Classifying the orbitals according to symmetry

Figure 15.9 on the next page shows how all of these orbitals are classified according to symmetry; for simplicity the ligand σ orbital is simply represented as a sphere, and the metal orbitals are represented in cartoon form. Filled in parts of the orbital indicate that the wavefunction has a positive sign, open parts indicate that the sign is negative. Orbitals with the same symmetry are coloured in the same way.

In orange are shown the metal $4s$ orbital, which is spherical, and the symmetry orbital SO1 in which each ligand orbital appears with the same sign. It is clear from the diagram that SO1 has the correct symmetry to overlap with the $4s$, since all of the overlaps are positive. In the language of group theory these orbitals are classified as having symmetry a_{1g}. Although we cannot explore here the precise meaning of this group theory notation, we will use these designations as convenient labels for the orbitals.

The metal $4p$ AOs are shown in red along with the three symmetry orbitals SO2, SO3 and SO4 which overlap with the three p orbitals. Since the $4p_z$ AO has a positive lobe pointing along $+z$ and a negative lobe pointing along $-z$, it is clear that the correct symmetry orbital will have a ligand orbital with a positive phase located along $+z$ and one with a negative phase located along $-z$. Similar considerations apply to the other p orbitals. Taken together, the group theory designation of these orbitals is t_{1u}.

Recall from section 2.3.4 on page 56 that the $3d$ orbitals come in two groups. The $3d_{z^2}$ and $3d_{x^2-y^2}$ point along the axes, whereas $3d_{xy}$, $3d_{xz}$ and $3d_{yz}$ point between the axes. Given that the ligands are located along the axes, it should not come as a surprise that in an octahedral complex these two groups have different symmetries.

In Fig. 15.9 the $3d_{z^2}$ and $3d_{x^2-y^2}$, along with the matching symmetry orbitals SO5 and SO6, are shown in green. Note how the phases of the ligand orbitals match the phases of the metal orbitals to ensure that there is always positive overlap. These orbitals have the symmetry label e_g.

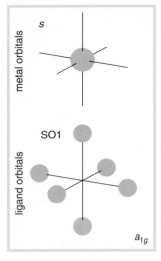

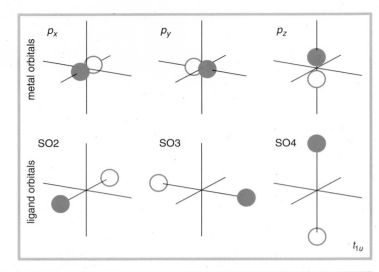

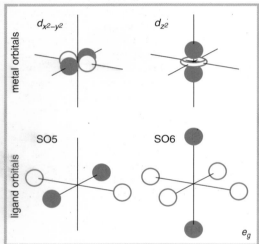

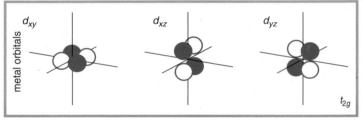

Fig. 15.9 Illustration in cartoon form of the way in which the metal orbitals and the ligand orbitals of an octahedral complex can be classified according to symmetry. Orbitals with the same symmetry have the same colouring. Positive parts are the wavefunction are coloured in, whereas negative parts are left open. The metal orbitals are shown in the upper part of each box, and the symmetry orbitals on the ligands are shown in the lower part. The z-axis is vertical.

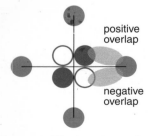

Fig. 15.10 The $3d_{xy}$, $3d_{xz}$ and $3d_{yz}$ orbitals point between the axes and so have no net overlap with a ligand σ orbital, shown in green.

It turns out that there is no combination of the ligand orbitals which has the correct symmetry to overlap with the metal $3d_{xy}$, $3d_{xz}$ and $3d_{yz}$ orbitals (shown in blue). The reason for this is illustrated in Fig. 15.10. For these metal orbitals which point between the axes there is always equal and opposite amounts of positive and negative overlap with a ligand σ orbital, resulting in no net interaction. The group theory designation of these d orbitals is t_{2g}.

Constructing the MO diagram

Now that we have classified the orbitals according to symmetry we can construct the MO diagram using the principle that only orbitals of the same symmetry

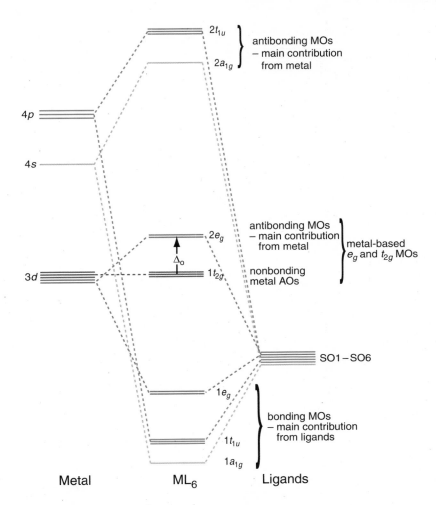

Fig. 15.11 Schematic MO diagram for an octahedral complex ML_6. The orbitals are colour coded according to symmetry in the same way as in Fig. 15.9. The $1a_{1g}$ MO is bonding and results from the interaction between the ligand orbitals and the metal $4s$ AO; similarly the bonding $1t_{1u}$ MOs arise from the interaction between the ligand orbitals and the $4p$ AOs. The $1e_g$ MOs are also bonding, but these result from the interaction between the ligand orbitals and the metal $3d$ AOs. The $1t_{2g}$ are the $3d_{xy}$, $3d_{xz}$ and $3d_{yz}$ nonbonding metal orbitals. The $2e_g$ MOs are the antibonding counterparts of the bonding $1e_g$, and these antibonding MOs are mostly located on the metal. Likewise the $2a_{1g}$ and $2t_{1u}$ are the antibonding counterparts of the bonding $1a_{1g}$ and $1t_{1u}$ MOs. In a typical complex the bonding $1a_{1g}$, $1t_{1u}$ and $1e_g$ MOs are all occupied, whereas the $1t_{2g}$ and $2e_g$ are partially occupied, depending on the number of d electrons present on the metal.

overlap to form MOs. Figure 15.11 shows the resulting schematic diagram; the MOs are coloured according to symmetry in the same way as in Fig. 15.9. Since the ligands are expected to coordinate via electronegative atoms, their orbitals are placed lowest in energy. The $3d$ AOs are placed below the $4s$ and $4p$ since, as was explained above, in a transition metal the ionization of the outer electrons causes the energy of the $3d$ AOs to fall.

The metal $4s$ overlaps with the ligand symmetry orbital SO1 to give a bonding and an antibonding MO (coloured orange). These are labelled $1a_{1g}$ and $2a_{1g}$ respectively, indicating that they are the first and second MOs with this symmetry. Similarly the three metal $4p$ AOs overlap with SO2, SO3 and SO4 (coloured red) to give three bonding MOs, labelled $1t_{1u}$ and three antibonding MOs, labelled $2t_{1u}$. It is important to realize that the label $1t_{1u}$ applies collectively to all three of the bonding MOs.

The metal $3d_{z^2}$ and $3d_{x^2-y^2}$ AOs, and the symmetry orbitals SO5 and SO6 overlap to give the bonding $1e_g$ and antibonding $2e_g$ MOs (coloured green). The label $1e_g$ applies collectively to both the bonding MOs. There are no symmetry orbitals to overlap with the metal $3d_{xy}$, $3d_{xz}$ and $3d_{yz}$ AOs, so these remain nonbonding and have the label $1t_{2g}$ (blue). It should be emphasized that the energy ordering of the MOs shown in the diagram is simply illustrative.

Ligand-based and metal-based MOs

The bonding $1a_{1g}$, $1t_{1u}$ and $1e_g$ MOs are closer in energy to the ligand orbitals than they are to the metal orbitals, and so the major contributor to these MOs will be from the ligand. These MOs are often described as being *ligand based*. On the other hand the major contribution to the antibonding $2e_g$ MOs is from the metal AOs as these are closer in energy to the $2e_g$ than are the ligand orbitals. The nonbonding $1t_{2g}$ are simply metal orbitals. Taken together, the $2e_g$ and the $1t_{2g}$ are described as the *metal-based* orbitals.

In these complexes the ligands are donors and each provides two electrons in the σ-type orbital which is interacting with the metal. These twelve electrons end up in the $1a_{1g}$, $1t_{1u}$ and $1e_g$ bonding MOs of the complex (recall that there are three MOs with the collective label $1t_{1u}$ and two with the label $1e_g$). It is this reduction in energy of the ligand-derived electrons which is responsible for the complex being lower in energy than the separated metal atom and ligands.

Any electrons which come from the metal have to be accommodated in the nonbonding $1t_{2g}$ and antibonding $2e_g$ MOs, and the way that the electrons are arranged in these orbitals turns out, as we shall see, to have a strong influence on the properties of the complex. Much of our attention will therefore be focused on these metal-based orbitals. It is important to realize that the $2e_g$ MOs are antibonding, so electrons in these MOs detract from the overall bonding in the complex.

Since we are invariably going to be discussing complexes formed from transition metal *ions*, the electrons provided by the metal are all from $3d$ orbitals. As a result, it is common to talk about a complex as having such-and-such a number of d electrons, or to refer to it is a d^n complex, meaning that there are n $3d$ electrons. For example, the complex $[CrCl_2(NH_3)_4]^+$ contains Cr^{3+}, which has three d electrons, and so is described as a d^3 complex.

The energy separation between the two metal-based MOs (the $1t_{2g}$ and the $2e_g$) is called Δ_o or Δ_{oct} (the 'o' and 'oct' stand for octahedral). This quantity is often referred to as the *ligand-field splitting*. As we shall see, the size of Δ_o is of considerable significance in determining the properties of complexes.

The stronger the interaction between the metal d orbitals and the ligand orbitals the more antibonding the $2e_g$ becomes, and hence the larger Δ_o becomes. Ligands which can become involved in this kind of bonding are called σ-*donors*, and the strength of the donation is measured by the value of Δ_o: stronger donors give rise to larger values of Δ_o.

15.3.2 Effect of π interactions with the ligand

In constructing the MO diagram shown in Fig. 15.11 on the previous page we have only allowed σ-type interactions between the metal and the ligand. However, in practice there is also likely to be π type interactions between the ligand and the metal, so we now need to consider what effect these will have.

π donor ligands

A ligand such as a halide ion will have p orbitals aligned at right-angles to the M–L direction, and these orbitals can be involved in π interactions with the $3d$ orbitals, as shown in Fig. 15.12. These π interactions are with the $3d_{xy}$, $3d_{xz}$ and $3d_{yz}$ AOs which are not involved in σ interactions. Figure 15.13 on the next page shows the three symmetry orbitals SO7–9 which can overlap with these metal orbitals; they have symmetry t_{2g}.

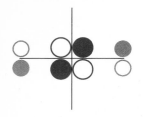

Fig. 15.12 The $3d_{xy}$, $3d_{xz}$ and $3d_{yz}$ orbitals point between the axes and can have a π-type interaction with suitable p orbitals on the ligands.

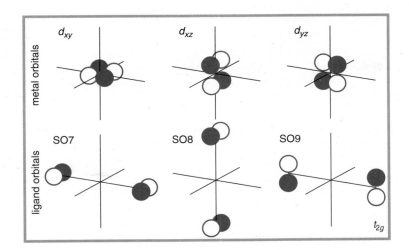

Fig. 15.13 Illustration in cartoon form of the symmetry orbitals, formed from p orbitals on the ligands, which can overlap with the metal $3d_{xy}$, $3d_{xz}$ and $3d_{yz}$ AOs. These SOs have symmetry t_{2g}.

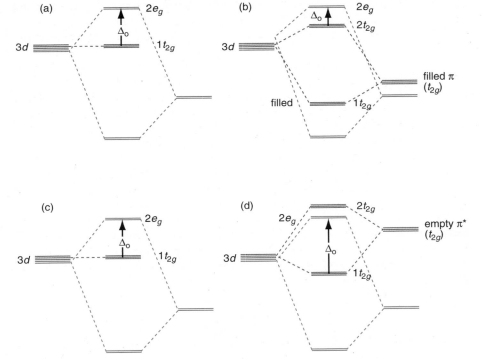

Fig. 15.14 Illustration of the effect on the MO diagram of a complex ML_6 of introducing ligand MOs with t_{2g} symmetry. The diagrams shown in (a) and (c) are for the case where there are only σ interactions between the ligand and metal. Adding a low-lying occupied set of t_{2g} orbitals on the ligand, as in (b), results in the formation of a bonding MO, $1t_{2g}$, and an antibonding MO, $2t_{2g}$. The separation between the two metal-based MOs (the $2t_{2g}$ and the $2e_g$) therefore decreases. Adding a set of high-energy unoccupied t_{2g} ligand orbitals, as shown in (d), also leads to the formation of bonding and antibonding t_{2g} MOs, but this time the separation between the two metal-based MOs (the $1t_{2g}$ and the $2e_g$) is increased.

When these SOs are included the key difference in the MO diagram is that there are now ligand orbitals with the correct symmetry to overlap with the metal t_{2g} orbitals. The result of this interaction is illustrated in Figure 15.14 which shows a series of a partial MO diagrams for the ML_6 complex in which just the t_{2g} and e_g MOs are included.

In (a) we see the situation when there are only σ interactions with the ligand, just as in Fig. 15.11 on page 659; the $1t_{2g}$ MOs are nonbonding. In (b) a set of occupied ligand orbitals with t_{2g} symmetry is included; these AOs are low-lying, just like the σ-type orbitals on the ligand. They interact with the metal t_{2g} AOs to give a bonding $1t_{2g}$ and an antibonding $2t_{2g}$ set of MOs; the $1t_{2g}$ MOs are filled by the electrons from the ligand. The overall result is that the separation Δ_o between the two metal-based sets of orbitals (here that between the $2t_{2g}$ and the $2e_g$) has *decreased*. A ligand which provides a low-lying set of filled π-type orbitals is described as a π *donor*.

π acceptor ligands

If the ligand is a molecule, then as well as having filled π bonding MOs it will also have unoccupied π antibonding MOs lying at higher energies. These MOs can also interact with the t_{2g} metal orbitals in the way shown in Fig. 15.14 (c) and (d). As before (c) is the MO diagram where there are only σ interactions. In (d), a set of $\pi^\star$ MOs with t_{2g} symmetry has been introduced. As these are antibonding they are higher in energy than the π MOs and in this case have been shown as lying above the $3d$ AOs. As before, the metal and ligand t_{2g} orbitals interact to give the bonding $1t_{2g}$ and the antibonding $2t_{2g}$ MOs. This interaction results in an *increase* in the separation between the two metal-based orbitals (the $1t_{2g}$ and the $2e_g$). A ligand which provides such a high energy set of unoccupied π MOs is called a π *acceptor*.

Of course, any ligand which can act as a π acceptor also must have orbitals through which it can act as a π donor. The first effect increases Δ_o, whereas the second decreases Δ_o. Which one of the effects will be dominant is not easy to predict in advance.

Generally speaking the π interaction between ligands and metal is not as strong as the σ interaction. This is thought to be do to the fact that π overlap is particularly sensitive to a mismatch in the size of the orbitals, a point we came across before (on page 297) when considering how π bonding changes on going from the second to the third period.

15.3.3   Factors which affect the value of Δ_o

The value of the ligand-field splitting can be determined from spectroscopic measurements of complexes and, based on a large number of such measurements, it is possible to make some generalizations about how Δ_o varies with the ligand, the metal and the oxidation state of the metal.

The ligand

For a given metal in a particular oxidation state, it is generally found that Δ_o increases along the following series of ligands, called the *spectrochemical series* (where there is any ambiguity, the donor atom is indicated in blue)

increasing Δ_o

I^- Br^- S^{2-} SCN^- Cl^- F^- OH^- H_2O NH_3 PPh_3 CN^- CO

small Δ_o large Δ_o

The stronger the σ interaction, the larger Δ_o will become and the further to the right a ligand will move in this series. The strength of σ donation will be increased by raising the energy of the ligand orbitals so that they provide a better match to the metal d orbitals.

Increasing the strength of the interaction with a π donor *decreases* Δ_o and so moves the ligand to the left in the spectrochemical series. Factors which improve the strength of σ donation will also improve the strength of π donation, but these two effects have the opposite effect on Δ_o. However, as was discussed in the previous section, the σ interaction is usually larger than the π, so the ordering of ligands can usefully be discussed just in terms of their σ donating ability.

In the spectrochemical series ligands which donate through carbon are higher in the series than those which donate through nitrogen, and in turn these are higher than those which donate through oxygen. We can rationalize this order by noting that as we go from carbon to nitrogen to oxygen the ligand orbitals will go *down* in energy and so be a poorer match to the metal $3d$ AOs, resulting in a smaller value of Δ_o.

The metal and its oxidation state

For a given metal and ligand, Δ_o is generally found to increase as the oxidation state on the metal increases, and as we move further along the transition series. We can argue that this increase comes about as a result of the $3d$ AOs falling in energy, thus providing a better match with the ligand orbitals. In addition, the contraction of the AOs may also result in better overlap.

Carbon monoxide as a ligand

Carbon monoxide gives the highest values of Δ_o, and the isoelectronic species CN^- gives similarly high values. As we saw on page 131, the HOMO of CO is a σ type MO which is largely localized on the carbon and points away from the oxygen; the LUMO is the $\pi^\star$ MO to which the carbon is the major contributor. Using its HOMO, carbon monoxide is therefore a σ donor through carbon and, as has been discussed, this makes it a strong σ donor on account of the high energy of the orbital.

The occupied π MOs of carbon monoxide lie lower in energy than the HOMO, resulting in a poorer energy match to the $3d$ AOs. As a result, the π donation is much weaker than the σ donation. In contrast, the empty π^* MO (the LUMO) is higher in energy than the HOMO and so is likely to be reasonably matched to the metal orbitals. As a result, CO is an effective π acceptor.

Taken together, the fact that CO is both a strong σ donor and a π acceptor accounts for the large values of Δ_o which it gives rise to. Similar considerations apply for the isoelectronic CN^- species.

Typical values

Typical values of Δ_o cover the range 7000 cm^{-1} to $30,000$ cm^{-1}. Expressing these values in molar units gives the range as between 80 kJ mol^{-1} and 350 kJ mol^{-1}, which are chemically significant energies.

> For the case where there is only σ donation, the two metal-based orbitals are the $1t_{2g}$ and the $2e_g$. However, as can be seen in Fig. 15.14 on page 661, for π-donors the corresponding MOs have the labels $2t_{2g}$ and the $2e_g$, and for π-acceptors the labels are once again $1t_{2g}$ and the $2e_g$. To avoid confusion, these two sets of MOs which are occupied by the d electrons will simply be referred to as the 'metal-based orbitals', rather than giving their specific labels.

15.4 High-spin and low-spin octahedral complexes

In section 2.7.3 on page 82 it was shown that the energy of the electrons occupying a set of degenerate orbitals is minimized by having the largest number of parallel spins. This is a result of the favourable exchange interaction between

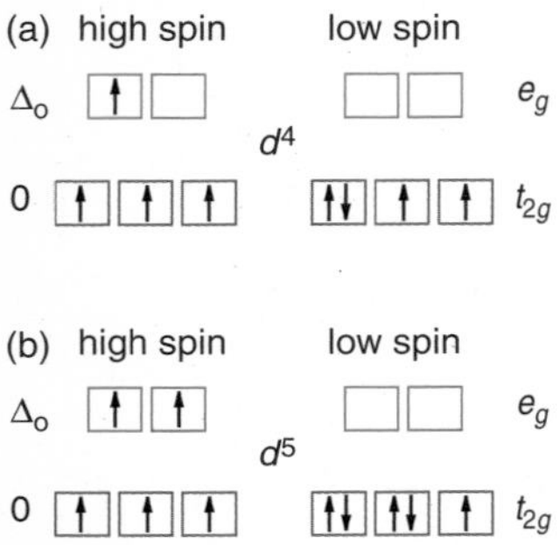

(a) high spin   low spin

Δ_o [↑][ ]   [ ][ ] e_g

d^4

0 [↑][↑][↑]   [↑↓][↑][↑] t_{2g}

(b) high spin   low spin

Δ_o [↑][↑]   [ ][ ] e_g

d^5

0 [↑][↑][↑]   [↑↓][↑↓][↑] t_{2g}

Fig. 15.15 Illustration of the two possible arrangements of (a) four and (b) five electrons in the metal-based t_{2g} and e_g MOs. The high-spin configuration has the greater number of parallel spins, and hence the largest exchange interaction, but in the low-spin configuration all of the electrons are in the lower energy orbital.

It is common to discuss the energy difference between the high- and low-spin configurations by comparing the extra energy needed to place an electron in the higher energy e_g orbitals with the extra electron–electron repulsion which is supposed to arise from pairing up an electron in the lower energy t_{2g} orbitals. We prefer to use an explanation in terms of the exchange energy as such an approach has greater theoretical justification.

such spins. We need to take account of this effect when working out the lowest energy (ground state) arrangement of the electrons in the metal-based t_{2g} and e_g MOs. Recall that it is the d electrons from the metal which are accommodated within these orbitals.

For one, two and three electrons there is no difficulty: each electron can singly occupy one of the three orbitals which make up the t_{2g} set, and the electron spins can all be made parallel in order to maximize the exchange interaction. However, when the fourth electron is added there are two possibilities: either it can pair up with one of the electrons in a t_{2g} MO, or it can go into the higher energy e_g MO, but with its spin parallel to all the others. These two possibilities are illustrated in Fig. 15.15 (a). The arrangement with the greater number of parallel spins is called the *high-spin* configuration, and the one with smaller number of parallel spins is called the *low-spin* configuration.

Which of these arrangements has the lowest energy depends on how the exchange energy compares with the ligand-field splitting Δ_o, and we can work out the energy of the two arrangements in the following way. Suppose that the energy of the t_{2g} MOs is arbitrarily set to zero, making the energy of the e_g MOs Δ_o. The exchange interaction is taken as $-K$ per parallel spin pair. The low-spin configuration $(t_{2g})^4$ has three parallel spins, so there are three possible pairs of parallel spins making the exchange energy $-3K$. All the electrons are in the t_{2g} MOs, which have energy zero, so the total energy of this configuration is $(-3K)$.

The high-spin configuration $(t_{2g})^3 (e_g)^1$ has four parallel spins, so there are six possible pairs of spins making the exchange energy $-6K$. One electron is in the e_g MO, so the total energy is $(\Delta_o - 6K)$. The high-spin configuration will have lower energy than the low spin when

$$\text{energy of high spin} \quad < \quad \text{energy of low spin}$$
$$\Delta_o - 6K \quad < \quad -3K$$
$$\text{hence} \quad \Delta_o \quad < \quad 3K.$$

Therefore, if the ligand-field splitting is less than $3K$ the high-spin configuration is the lower in energy, whereas if Δ_o is greater than $3K$ the low-spin configuration has the lowest overall energy.

For five d electrons there are again two choices illustrated in Fig. 15.15 (b). For the high-spin configuration there are ten possible pairs of parallel spins, giving an exchange energy of $(-10K)$, and two electrons in the e_g giving a contribution of $(2\Delta_o)$; the total energy is therefore $(2\Delta_o - 10K)$. For the low-spin configuration there are three possible pairs of spin-up electrons and one pair of spin-down electrons, giving an exchange contribution of $(-4K)$. The high-spin configuration will have lower energy when

$$2\Delta_o - 10K \quad < \quad -4K$$
$$\text{hence} \quad \Delta_o \quad < \quad 3K.$$

This is the same result as for d^4.

In words, what is going on here is that placing an electron in the e_g is energetically advantageous as it maximizes the favourable exchange interaction, but energetically disadvantageous since the e_g MO is higher in energy. If the ligand-field splitting is large compared to the exchange interaction, then the energy needed to populate the e_g is not compensated for by the exchange interaction, and so the low-spin arrangement is preferred. On the other hand,

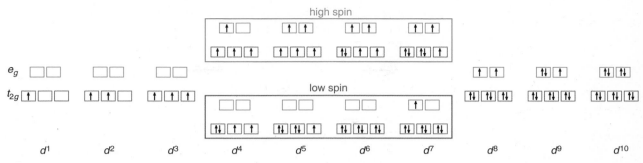

Fig. 15.16 Illustration of the electronic configurations of the d electrons in the metal-based t_{2g} and e_g MOs in an octahedral complex. For between four and seven electrons it is possible that a configuration in which electrons are placed in the e_g MOs before completely filling the t_{2g} MOs will have lower energy on account of the favourable exchange interaction resulting from having more electrons with parallel spins. For d^4 to d^7 we therefore need to consider two possible configurations, termed high spin (shown in the red box) and low spin (shown in the purple box).

if the ligand-field splitting is small compared to the exchange interaction, the extra energy needed to populate the e_g MO is more than compensated for by the favourable exchange interaction, and hence the high-spin arrangement is preferred.

Figure 15.16 illustrates the arrangement of electrons in the metal-derived t_{2g} and e_g MOs for one to ten electrons. For between four and seven electrons a high-spin and a low-spin arrangement is possible, but for other numbers of electrons there is only one plausible configuration for the ground state.

How the energies of the high- and low-spin configurations change depending on the relative size of Δ_o and the exchange term is illustrated in Fig. 15.17 on the following page. In (a) the two contributions to the energy of the electrons are plotted separately. The orbital contribution depends on how many electrons are in the t_{2g} and how many are in the e_g; as before we have assumed that the energy of the t_{2g} MOs are zero. The exchange contribution depends on the number of possible pairs of electrons with parallel spins. Graph (b) shows the total energy, taking into account the orbital and exchange contributions. These graphs have been drawn for the case where Δ_o is small so that the exchange term dominates, and as a result the high-spin configuration has lower energy than the low-spin configuration.

Graphs (c) and (d) are the same as (a) and (b), but plotted for the case where Δ_o is large so that it dominates over the exchange energy. In this case, it is the low-spin configuration which is preferred.

We can expect low-spin complexes to arise when the ligand-field splitting parameter is large which, as we have seen in section 15.3.3 on page 662, will be the case for ligands which are high in the spectrochemical series, higher oxidation state metals, and metals which occur later on in the transition elements.

A good example of this change over between high- and low-spin is provided by Fe^{3+} which has five d electrons. In acidic aqueous solution this ion forms $[Fe(H_2O)_6]^{3+}$, which measurements indicate is a high-spin complex. In contrast the cyano complex $[Fe(CN)_6]^{3-}$ is found to be low spin.

Fig. 15.17 Illustration of how the different contributions to the energy of the high- and low-spin configuration of electrons in the metal-based t_{2g} and e_g MOs changes with the number of electrons. Throughout, blue symbols are for the high-spin configuration and red symbols are for low-spin; black symbols are used for cases where only one configuration is possible. Graph (a) shows separately the two contributions to the energy: that due to the occupation of the orbitals, and that due to the exchange interaction. Graph (b) shows the sum of these two terms. These two graphs have been plotted for the case where the exchange contribution is greater than Δ_o, and so the high-spin configuration has lower energy. Graphs (c) and (d) are the same as (a) and (b) except this time Δ_o is larger than the exchange contribution. Now the low-spin configuration is preferred.

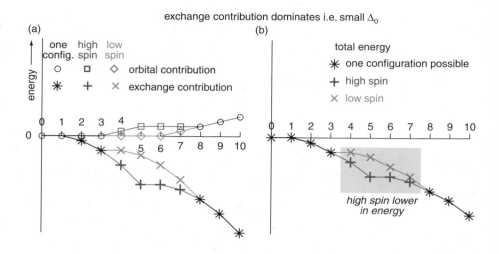

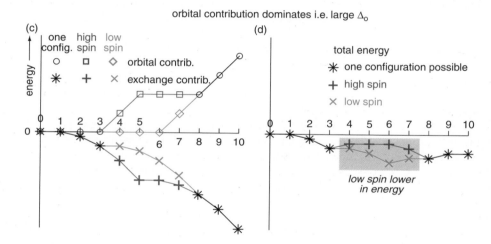

15.5 Magnetic and spectroscopic properties of complexes

The precise arrangement of the electrons in the metal-based orbitals has a strong influence on the magnetic and spectroscopic properties of transition metal complexes. Such measurements are therefore useful ways of probing the details of the electronic structure of such molecules.

15.5.1 Paramagnetism

In discussing the bonding in simple molecules we came across the idea that the presence of electrons with unpaired spins gave rise to the phenomenon of paramagnetism, such as that displayed by molecular oxygen (page 118). A paramagnetic material has the property that it is drawn into a magnetic field, so the presence of paramagnetism is easily detected by comparing the weight of a sample in the presence and absence of a magnetic field: a paramagnetic sample will appear to be heavier when the magnetic field is present. Such measurements are made using the kind of apparatus illustrated in Fig. 15.18 on the facing page.

In fact, there is only a force on a paramagnetic sample when it is placed in a magnetic field *gradient* i.e. a magnetic field which varies across the sample. This is because the sample has lower energy in the region where the magnetic field is stronger and so will have a tendency to move into that region. Therefore, to measure the degree of paramagnetism the sample is placed at the edge of a strong magnet where the gradient will be greatest.

For first-row transition metals, theory indicates that the strength of the paramagnetism (when expressed in appropriate molar units) should be proportional to $\sqrt{n(n+2)}$, where n is the number of unpaired electrons; this is called the *spin-only* approximation. This relationship works quite well in practice and so gives us an experimental way of determining the number of unpaired spins in a compound or complex. Magnetic measurements can easily distinguish between high- and low-spins complexes, since the former have significantly more unpaired spins.

For example, the high-spin d^5 complex $[Fe(H_2O)_6]^{3+}$ has five unpaired spins, whereas the low-spin d^5 complex $[Fe(CN)_6]^{3-}$ has just one (see Fig. 15.16 on page 665 for the configurations). Using the spin-only formula the ratio of the paramagnetism of the two complexes is

$$\frac{\text{paramagnetism of high spin } d^5}{\text{paramagnetism of low spin } d^5} = \frac{\sqrt{5(5+2)}}{\sqrt{1(1+2)}}$$
$$= 3.4;$$

this is a substantial difference which is easily measured.

A substance with no unpaired spins is said to be *diamagnetic*. Such materials are in fact repelled from a magnetic field gradient, thus reducing the apparent weight. However, diamagnetism is usually a much weaker effect than paramagnetism. d^{10} and low-spin d^6 complexes are expected to be diamagnetic.

15.5.2   Spectra

One of the most striking features of many transition metal compounds and complexes is their colour which arises from the fact that these species absorb light in the visible region of the spectrum. Virtually all molecules absorb light at a range of different wavelengths, so the distinctive thing about transition metal compounds is not that they absorb light, but that this absorption happens to take place in the visible region and so is immediately apparent to us.

One way in which light can be absorbed by a molecule is for the energy from a photon to be used to promote an electron from a lower to a higher energy orbital, as illustrated in Fig. 15.19. To be absorbed, the frequency of the photon v must be related to the energy separation ΔE according to $\Delta E = hv$, where h is Planck's constant. The promotion of electrons to different energy orbitals thus causes absorption of light at different frequencies, and if these frequencies correspond to visible light, different colours will be perceived.

It turns out that in transition metal complexes the separation between the metal-based t_{2g} and e_g orbitals often corresponds to the energy of a photon from the visible part of the spectrum. Therefore, the colour of such complexes is associated with the promotion of electrons between these two orbitals. Such processes, and the features the give rise to in the spectrum, and often called *d–d transitions* because the metal d orbitals are the main contributors to the MOs involved.

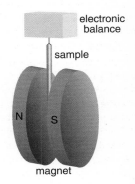

Fig. 15.18 Schematic of the apparatus (a *Gouy balance*) used to measure paramagnetism. A paramagnetic substance is drawn into a magnetic field gradient so by measuring the apparent weight of the sample as a function of the magnetic field strength we can assess the degree of paramagnetism.

⇨ The origin of spectra is discussed in greater detail in section 16.6 on page 719.

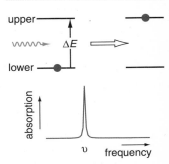

Fig. 15.19 A photon of light, represented here by the red wavy line, can cause an electron to be promoted from a lower energy orbital to a higher one. The energy separation of the two orbitals ΔE must be related to the frequency of the photon v according to $\Delta E = hv$, where h is Planck's constant. In the spectrum we see an absorption at frequency v.

Fig. 15.20 Absorption spectra in the UV/visible region of three hexaquo complexes. The Ti^{3+} species absorbs green light but blue and red pass through, giving a solution of the complex a violet colour. Ni^{2+} absorbs in the blue and red, resulting in a green coloured solution.

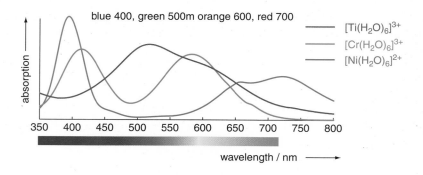

As we have seen in Fig. 15.11 on page 659, the spacing between the metal-based t_{2g} and e_g MOs is defined as Δ_o. This might reasonably lead us to think that measuring the frequency of the light absorbed in such a transition will make it possible to determine the value of Δ_o. If there is just *one* electron in the t_{2g} MOs this is indeed the case. For example, the spectrum of $[Ti(H_2O)_6]^{3+}$, shown in blue in Fig. 15.20, has a maximum in the absorption at a wavelength of 520 nm, which corresponds to $19{,}200$ cm^{-1}. The separation of the t_{2g} and e_g MOs is therefore $19{,}200$ cm^{-1}.

⇨ Wavenumbers (cm^{-1}) are discussed in section 20.2.3 on page 889.

However, if there is more than one electron present in the metal-based orbitals, the relationship between the spectrum and the value of Δ_o is not so straightforward, mainly because the rearrangement of the electrons which accompanies a transition results in a change in the electron–electron repulsion and hence a change in the orbital energies (see section 2.6.5 on page 79). Figure 15.20 also shows spectra of $[Cr(H_2O)_6]^{3+}$ (which has three *d* electrons) and $[Ni(H_2O)_6]^{2+}$ (eight *d* electrons). These spectra are more complicated than that of $[Ti(H_2O)_6]^{3+}$, and it is not possible to equate the photon energy corresponding to one of the bands to the value of Δ_o. Nevertheless, a more sophisticated analysis is capable of yielding a value for this parameter.

Intensities of absorptions

If photons of the correct energy to promote an electron from one orbital to another impinge on a molecule there is a certain rate at which they can be absorbed, and this rate is characteristic of the transition in question. If the rate is high, many photons will be absorbed and the transition is said to be 'strong'; if the rate is low, fewer photons are absorbed and the transition is described as 'weak'. It turns out that these *d–d* transitions are not very strong when compared to other electronic transitions which can occur in molecules. The reason for this is that *d–d* transitions do not entirely obey the spectroscopic *selection rules* which quantum mechanical considerations dictate. Such rules are discussed further in section 16.6.1 on page 720.

Transitions in which electrons are promoted from ligand-based orbitals to metal-based ones (and *vice versa*), called *charge transfer* transitions, often give rise to much stronger absorptions than do *d–d* transitions. Usually these charge-transfer transitions occur in the UV, but in some cases they do cause absorption in the visible part of the spectrum. Such transitions are thought to be the origin of the intense colours in species such as $[MnO_4]^-(aq)$ and $[Cr_2O_7]^{2-}(aq)$.

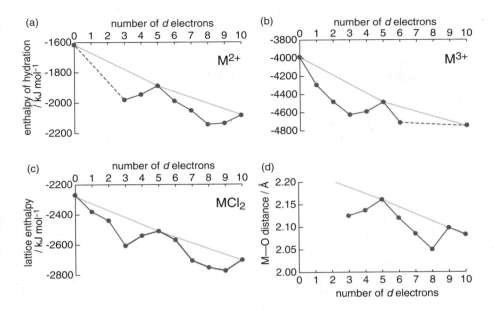

Fig. 15.21 Plots showing how various thermodynamic and structural parameters vary with the number of *d* electrons in a transition metal ion. Graphs (a) and (b) give the standard enthalpies of hydration of M^{2+} and M^{3+}, respectively. Graph (c) gives the lattice enthalpy of the divalent chlorides, and graph (d) gives the average M–O distance in hexaquo complexes. Not all of the data is available to complete these plots; the solid blue lines are of no significance, but are just there as a guide to the eye. The pale green lines connect the data points for d^0, d^5 and d^{10} so as to emphasize the 'double dip' form of the data.

15.5.3 Free radicals

For the majority of oxidation states there are going to be unpaired electrons in the metal-based t_{2g} and e_g MOs, which means that these complexes should be classified as free radicals. In discussing main-group chemistry, we noted that compounds in oxidation states which resulted in free radicals tended to be 'unstable' in the sense that the unpaired electrons made the species very reactive (see section 14.2.5 on page 620).

In contrast, transition metal complexes and compounds do not seem to become very reactive as a result of having unpaired electrons. We can find an explanation for this is the fact that the unpaired electrons are located in orbitals which are either exclusively or primarily metal-based. They are thus buried inside the complex, protected by the surrounding ligands, and in addition the orbitals in which they are located are also quite contracted. Thus, these electrons are simply not as accessible as, for example, the unpaired electrons in an atom or a simple radical such as CH_3.

15.6 Consequences of the splitting of the *d* orbitals

Figure 15.21 shows some plots of experimental data relating to transition metal ions and compounds. Graphs (a) and (b) show the standard enthalpy of hydration of the ions M^{2+} and M^{3+} (i.e. the enthalpy change when the ions are taken from the gas phase into aqueous solution (see section 6.12.1 on page 238), graph (c) shows the lattice enthalpy of the chlorides MCl_2, and graph (d) shows the average M–O distance observed for the complexes $[M(H_2O)_6]^{2+}$.

These data are all plotted against the number of *d* electrons on the metal ion, and they all show a characteristic 'double dip' imposed on a downward trend. If we draw a line (shown in pale green) between the data points for d^0, d^5 and d^{10}, the other data fall beneath this line in two dips: one between d^0 and d^5,

Fig. 15.22 Partial MO diagram for an octahedral complex ML$_6$ with σ donor ligands; only the occupied and partially occupied MOs are shown. The bonding 1e_g MOs are lowered in energy by $-E_\sigma$ relative to the ligand orbitals, and the corresponding antibonding 2e_g MOs are raised in energy by E_σ relative to the metal d orbitals.

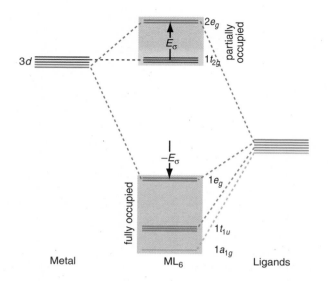

and the second between d^5 and d^{10}. There are many other thermochemical and structural properties of transition metal complexes which behave in this way.

Given that it is the number of d electrons which is varying in these plots, and that in an octahedral complex these electrons are occupying the metal-based t_{2g} and e_g MOs, we expect to be able to explain the form of these graphs in terms of the energetic consequences of the way in which these MOs are occupied.

15.6.1 Binding energy of an octahedral complex

We will start out by considering an octahedral ML$_6$ complex with σ donor ligands, the MO diagram for which is shown in Fig. 15.11 on page 659. Only those orbitals which are going to be occupied need be considered, and these are the ligand-based 1a_{1g}, 1t_{1u} and 1e_g MOs, along with the metal-based 1t_{2g} and 2e_g MOs; Fig. 15.22 is a partial MO diagram showing just these orbitals. Our aim is to work out how the total electronic energy changes when the complex is formed, and how this energy is affected by the number of d electrons and the way in which they occupy the 1t_{2g} and 2e_g MOs.

On forming the complex there is a reduction in energy due to the occupation of the 1a_{1g} and 1t_{1u} MOs. Since these MOs are always fully occupied, regardless of the number of d electrons, we will just write the energy reduction due to all eight electrons in these MOs as $-E_L$.

This leaves the orbitals which have a contribution from the metal d orbitals to consider i.e. the 1e_g, 1t_{2g} and 2e_g. When the complex forms, the bonding 1e_g MOs are lowered in energy relative to the ligand orbitals by an amount which we will call $-E_\sigma$. The corresponding antibonding 2e_g MOs are therefore raised in energy relative to the metal d orbitals by $+E_\sigma$ (in fact they would be raised by somewhat more than this, but this is of no consequence in this simple calculation). The 1t_{2g} MOs are nonbonding, so their energy does not change.

The 1e_g MOs are always fully occupied with four electrons, but the number of electrons in the 2e_g will vary so we will call that n_{e_g}. On forming the complex the total change in energy due to occupation of orbitals which have a contribution from the metal is therefore

$$\begin{aligned}\text{change due to MOs with contributions from M} &= 4 \times (-E_\sigma) + n_{e_g} \times (+E_\sigma) \\ &= E_\sigma(n_{e_g} - 4).\end{aligned}$$

A plot of this energy change against number of *d* electrons is shown in Fig. 15.23, in which it has been assumed that the complex adopts a high-spin configuration. The number of electrons in the $2e_g$ MOs can be read off from Fig. 15.16 on page 665. For example, for d^4 there is one electron in the $2e_g$ so $n_{e_g} = 1$ and hence the stabilization energy is $-3E_\sigma$.

The graph shows that the reduction in energy on forming the complex is greatest when there are no *d* electrons. This makes sense as for this number of electrons the bonding $1e_g$ MOs are full, but there are no electrons in the antibonding $2e_g$. When there are ten *d* electrons the $2e_g$ MOs are full, which cancels out the bonding effect of the $1e_g$; there is then no energy reduction due to the occupation of these MOs.

Recalling that the reduction in energy due to the occupation of the $1a_{1g}$ and $1t_{1u}$ MOs is $-E_L$, the total change in energy of forming the complex is therefore

$$\text{change in energy on forming complex} = -E_L + E_\sigma(n_{e_g} - 4).$$

As we go across the transition metal series the energies of the metal orbitals will decrease (due to the increase in nuclear charge) and hence they will match more closely the energy of the ligand orbitals; as a result there will be a stronger bonding interaction. We therefore expect that both E_L and E_σ will increase in size as we go across the transition metal series. For the moment it will be simplest to keep E_σ fixed and allow E_L to increase, but later on we will lift this restriction.

Figure 15.24 shows how these two contributions to the energy change of forming ML_6 combine as we go across the transition metals i.e. as the number of *d* electrons increase. Graph (a) is exactly the same plot as Fig. 15.23 and shows the reduction in energy due to the occupation of those MOs with a contribution from the metal *d* orbitals. Graph (b) shows the reduction in energy due to the

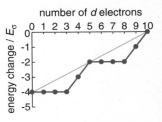

Fig. 15.23 Plot showing the contribution that occupation of those MOs with a contribution from the metal make to the energy change on forming a complex ML_6. A high-spin configuration is assumed, and the energy is plotted in units of E_σ. The pale green line connects the values for d^0, d^5 and d^{10} configurations, just as in Fig. 15.21.

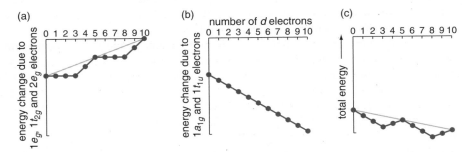

Fig. 15.24 Plots of the different contributions to the change in energy on forming an octahedral complex ML_6. Graph (a) shows the contribution from occupation of the $1e_g$, $1t_{2g}$ and $2e_g$ MOs, all of which involve the metal *d* orbitals (this is the same graph as in Fig. 15.23). Graph (b) shows the contribution from the fully occupied $1a_{1g}$ and $1t_{1u}$ MOs; it has been assumed that there is a steady decrease in the energy of theses orbitals across the transition metals. Graph (c) shows the sum of the energies from (a) and (b); the 'double dip' profile seen in Fig. 15.21 on page 669 is evident. In (a) and (c) the pale green line connects the values for d^0, d^5 and d^{10} configurations, just as in Fig. 15.21.

full ligand-based $1a_{1g}$ and $1t_{1u}$ MOs i.e. $-E_L$. As discussed above, it is assumed that the energies of these MOs falls steadily across the series. Finally (c) shows the sum of the energies in (a) and (b): the 'double dip' seen for the experimental data in Fig. 15.21 on page 669 is immediately apparent.

The exact shape of the plot in (c) depends on how large E_σ is relative to E_L, and how quickly E_L falls as we go across the transition series. The values of these parameters used to plot (c) were chosen so as to reproduce the 'double dip' seen in Fig. 15.21. If we also allow E_σ to increase across the series the line in graph (a) has the same values for d^0 and d^{10}, but now the horizontal portions slope downwards. However, combining this with a suitable steady downward slope as in (b) will still reproduce the 'double dip'.

Historically, this lowering in energy of a complex below the d^0–d^5–d^{10} line is referred to as the *crystal-field stabilization energy* or the *ligand-field stabilization energy*. Our analysis shows that there is nothing mysterious about this drop in energy: it is simply a result of which MOs are occupied, and the energies of these MOs.

15.6.2 Consequences for thermodynamic and structural properties

The data shown in Fig. 15.21 on page 669 can now be interpreted in terms of this discussion of how the binding energy of a complex varies with number of d electrons.

The enthalpies of hydration reflect the energy change when an ion goes from the gas phase into liquid water and, as we have seen, it is typical for the primary hydration sphere of such ions to contain six coordinated water molecules (section 6.12.1 on page 238). Therefore, it is reasonable to expect that the enthalpy of hydration will reflect the reduction in energy on forming a hexaquo complex. This enthalpy change will thus vary with the number of d electrons in a way which mirrors the energy change on forming an octahedral complex i.e. it will have the characteristic double dip shown in Fig. 15.24 (c).

Similar considerations apply to the lattice enthalpies plotted in Fig. 15.21 (c). In the lattice, the coordination around the metal atom is octahedral, so we can expect the energy of forming the lattice in part to reflect the energy of forming an octahedral complex. The variation in the lattice energies as a function of the number of d electrons will therefore follow the plot shown in Fig. 15.24 (c). Of course, the lattice enthalpy also includes contributions from interactions with more distant ions, but nevertheless it is clear from the experimental data that interactions with the closest ions are important.

Finally, the bond-length data shown in Fig. 15.21 (d) can be rationalized by assuming that the stronger the bonding in the complex, the shorter the M–L bonds will be. The strength of bonding is reflected by the energy change shown in Fig. 15.24 (c), and so the bond lengths mirror this variation.

15.7 Tetrahedral and square-planar complexes

The most commonly encountered transition metal complexes are octahedral, but other coordination numbers and geometries are also well known. It is possible to 'force' a complex to adopt a wide range of coordination geometries by using specially designed polydentate ligands. Setting aside these rather special cases, the only other two geometries which are common for simple ligands are tetrahedral and square planar.

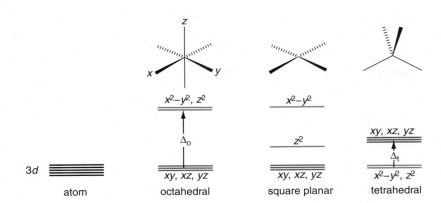

Fig. 15.25 Illustration of the energies of the nonbonding and antibonding MOs, to which the metal $3d$ AOs are the major contributor, for three different geometries. The MOs are labelled according to which $3d$ AO contributes to them, and only σ interactions with the ligands are considered. Simple calculations indicates that $\Delta_t \approx 0.44 \times \Delta_o$, and that in the square-planar geometry the $3d_{z^2}$ is raised by $\frac{1}{3}\Delta_o$.

As we have seen from the octahedral case, it is the populations of the nonbonding and antibonding metal-based orbitals which determine the properties of the complex, so rather than drawing up a full MO diagram we will just concentrate on the energies of these MOs. Figure 15.25 illustrates the relative energies of these MOs for octahedral, square-planar and tetrahedral geometries. Each of the MOs has been labelled according to which particular $3d$ AO contributes to it, and it has been assumed that there are only σ interactions between the ligand and the metal.

For the octahedral case the $3d_{xy}$, $3d_{xz}$ and $3d_{yz}$ MOs are nonbonding (the $1t_{2g}$), whereas the $3d_{x^2-y^2}$ and $3d_{z^2}$ contribute to the antibonding $2e_g$ MOs. Removing the two ligands along the z-axis gives a square planar geometry. Not surprisingly, this does not affect the energy of the $3d_{x^2-y^2}$, as this lies in the xy-plane, but the $3d_{z^2}$ drops in energy as it now has two fewer ligand orbitals to interact with. The other three orbitals remain nonbonding. The pattern of orbitals for the tetrahedral geometry is not obvious from simple considerations, but it turns out that there are two nonbonding and three antibonding MOs – the opposite of the octahedral case.

In the octahedral case the separation between the two sets of orbitals is designated Δ_o, and the equivalent separation in the tetrahedral case is designated Δ_{tet} or Δ_t ('tet' and 't' stand for tetrahedral). Calculations show that, all other things being equal, $\Delta_t = \frac{4}{9}\Delta_o$ i.e. Δ_t is about 44% of the value of Δ_o. This smaller value can be attributed to the fact that in a tetrahedral complex the $3d$ orbitals do not point directly at the ligands, thereby reducing the interaction with the ligand orbitals. Similar calculations show that the $3d_{z^2}$ orbital in the square-planar complex is raised in energy by $\Delta_o/3$.

High- and low-spin complexes

The ligand-field splitting in tetrahedral complexes (Δ_t) is significantly smaller than Δ_o in an octahedral complex, so it is expected to be much more likely that the exchange energy term will dominate, favouring high-spin complexes. Experiment bears this out: no low-spin tetrahedral complexes are known.

For square-planar complexes it is usually considered that low-spin complexes are those in which the lowest *four* MOs are filled completely before filling the most antibonding MO. High-spin complexes are ones in which all five MOs are filled singly before any electrons are paired up.

The pattern of orbitals shown in Fig. 15.25 are for the case where there are only σ interactions with the ligand. If π interactions are present, the $3d_{xy}$, $3d_{xz}$ and $3d_{yz}$ split into a single orbital and a degenerate pair.

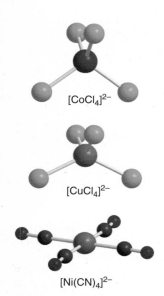

Fig. 15.26 Illustration of the different structures adopted by four-coordinate complexes. $[CoCl_4]^{2-}$ is tetrahedral, $[CuCl_4]^{2-}$ can be thought of as a flattened tetrahedron, and $[Ni(CN)_4]^{2-}$ is square planar.

Weblink 15.4

View and rotate the structures shown in Fig. 15.26.

Structural preferences

The question arises as to which coordination number and geometry is favoured for a particular combination of metal and ligand. Thinking purely in terms of the bonding, ML_6 will always be lower in energy than ML_4 since the interaction between the ligands and the metal is stronger in the former case, as indicated by the fact that $\Delta_o > \Delta_t$. The one exception to this observation is for the d^{10} configuration in which all the metal-based MOs are full: in this case there is no difference between the octahedral and tetrahedral geometry. If the ligand is large, crowding around the metal centre will favour ML_4 over ML_6.

The choice between the square-planar and tetrahedral geometry is more finely balanced. Looking at the arrangement of the MOs depicted in Fig. 15.25 on the previous page, we can see that for d^0, d^1 and d^2 the electrons go into nonbonding MOs in both geometries, and so there is no energy difference between them. However, assuming high-spin configurations, the third and fourth electrons have to go into an antibonding MO in the tetrahedral case, but first into a nonbonding MO and then a less antibonding MO in the square-planar case. For d^3 and d^4 we therefore expect the square-planar geometry to be preferred.

Moving to the configuration d^5 means that in the square-planar case one electron has to go into the most antibonding MO (the one derived from $3d_{x^2-y^2}$). The results in the same total energy as for the tetrahedral case. Carrying on in this way shows that the other configurations for which the square-planar geometry is favoured are d^8 and d^9.

If the square-planar complex is low spin, then this geometry is favoured for all configurations except d^0, d^1, d^2 and d^{10}; the energetic preference is the largest for the d^8 configuration. We must not forget, however, that the simple repulsion between the ligands will favour the tetrahedral geometry, especially if the ligands are large.

Experiment largely confirms these predictions. $[FeCl_4]^-$ (d^5 configuration) and $[CoCl_4]^{2-}$ (d^7 configuration) are both tetrahedral. $[CuCl_4]^{2-}$ (d^9 configuration) adopts a geometry half way between tetrahedral and square planar, whereas $[Ni(CN)_4]^{2-}$ (d^8 configuration) is square planar; the structures are compared in Fig. 15.26. Recall from page 663 that CN^- is a ligand which gives rise to large ligand-field splittings, and so the square-planar complex is likely to be low spin, which favours this geometry strongly over the tetrahedral case.

15.8 Crystal-field theory

From all that we have done so far it is clear that the way in which the electrons occupy the metal-based t_{2g} and e_g orbitals is crucial in determining the thermodynamic, magnetic and spectroscopic properties of octahedral complexes. Historically, the fact that the d orbitals split into these two sets, and the crucial role that the way in which these orbitals are occupied plays, was first recognized using *crystal-field theory*, which is a very simple way of modelling the effect that the ligands have on the metal d orbitals.

In this theory, the ligands are modelled by point negative charges placed around the metal. These charges will repel the electrons on the metal, thereby altering their energies. Electrons in the $3d_{z^2}$ and $3d_{x^2-y^2}$ AOs are affected more than those in the $3d_{xy}$, $3d_{xz}$ and $3d_{yz}$ AOs as the former point towards the negative charges whereas the latter point between them. The d orbitals are thus split into two sets, which can be given the group theory labels t_{2g} (the $3d_{xy}$, $3d_{xz}$ and $3d_{yz}$) and e_g (the $3d_{z^2}$ and $3d_{x^2-y^2}$). For these orbitals, the overall result is effectively the same as we found using the MO treatment, and so our discussion of high- and low-spin complexes, spectra and magnetic properties proceeds as before.

The energy separation between these two sets of d orbitals is Δ_o, called the crystal-field splitting. Although it is clear from the model that the energies of *all* of the d orbitals are raised, it is nevertheless assumed that the t_{2g} set is *lowered* in energy by $\frac{2}{5}\Delta_o$, and the energy of the e_g set is *raised* by $\frac{3}{5}\Delta_o$, as shown in Fig. 15.27. This assumption ensures that the average energy of the $3d$ orbitals is unchanged by the interaction with the ligands.

In this model, occupation of the t_{2g} set of orbitals therefore results in stabilization of the complex, in the sense that in the complex the energy of the $3d$ AOs are lower than they are in the atom. In contrast, occupation of the e_g orbitals destabilizes the complex. By considering the way in which these orbitals are occupied as the number of d electrons increases it is possible to account for the 'double dip' seen in the graphs shown in Fig. 15.21 on page 669.

The crystal field model is appealingly simple, but it is flawed in a number of ways. For example, in this model there is no way of explaining why different ligands give rise to different values of Δ_o, and in particular it is hard to see why a ligand such as carbon monoxide gives rise to the largest Δ_o values. Furthermore, this description of a metal complex has nothing in common with the molecular orbital approach which we have applied systematically to the description of *all* other molecules and reactions. It is for these reasons that, despite its historical importance in the development of an understanding of transition metal complexes, we choose not to discuss crystal-field theory any further.

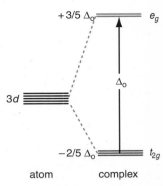

Fig. 15.27 In crystal-field theory the d orbitals are split into a t_{2g} and an e_g set. The energy separation between the two sets is Δ_o, and it is assumed that the t_{2g} set is lowered in energy by $-\frac{2}{5}\Delta_o$ and the e_g set is raised in energy by $+\frac{3}{5}\Delta_o$. As a result of this assumption, the total energy of the d orbitals is unchanged.

15.9 Organometallic complexes

The complexes we have been discussing so far involve a transition metal cation interacting with ligands which are either anions or coordinate to the metal via electronegative atoms; these are called *classical* complexes. There is another extensive and important class of complexes in which the ligands are species such as carbon monoxide, phosphines (PR_3), organic molecules with conjugated π systems (including rings) and alkyl groups. These complexes are usually described as *organometallic*.

Typically, the metal in such complexes has a low formal oxidation state such as 0 or +1, and indeed it is not unusual for the metal to have a negative oxidation state. Such complexes are therefore electron rich on the metal, and broadly speaking this high electron density is stabilised by being partly donated to the ligands. It is for this reason that the ligands seen in such complexes usually have π systems whose empty orbitals can accept electron density i.e π-acceptor ligands.

Figure 15.28 on the next page illustrates some typical organometallic complexes. Complex (a) is Zeise's salt, which is one of the first to be recognised as containing a discrete organic molecule, ethene, as a ligand. Complexes (b) and (c) are just two examples of the very large number of known metal carbonyl

Weblink 15.5

View and rotate the structures shown in Fig. 15.28.

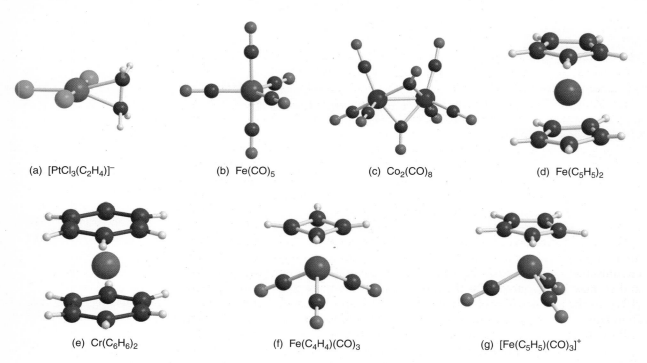

(a) $[PtCl_3(C_2H_4)]^-$ (b) $Fe(CO)_5$ (c) $Co_2(CO)_8$ (d) $Fe(C_5H_5)_2$

(e) $Cr(C_6H_6)_2$ (f) $Fe(C_4H_4)(CO)_3$ (g) $[Fe(C_5H_5)(CO)_3]^+$

Fig. 15.28 Some typical examples of organometallic complexes. In (a) one of the ligands is a discrete ethene molecule. The carbonyl complexes shown in (b) and (c) are representatives of a very large number of such compounds; note that in (c) there are CO molecules bridging between the two metal atoms. In complexes (d)–(e) the ligands included conjugated π systems: the cyclopentadienyl anion ($[C_5H_5]^-$) in (d) and (g), benzene in (e), and cyclobutadiene (C_4H_4) in (f).

complexes. These show a wide range of coordination geometries, and also have a tendency to include several metal atoms, possibly with metal–metal bonds, as in the case of the cobalt species. It is also common for carbonyl groups to form bridges between two or more metal atoms.

In complex (d), known as ferrocene, a conjugated π system acts as the ligand. This complex can be regarded as being formed by two cyclopentadienyl anions, $[C_5H_5]^-$, coordinating to Fe^{2+}. Like benzene, this anion has six electrons in its π system and is therefore regarded as being aromatic. The fact that the ring is 'face on' to the metal clearly indicates that the interaction with the metal involves the π system of the ring. Complexes (e)–(g) are other examples where π systems act as ligands.

Hapticity

The *hapticity* of a ligand is defined as the number of atoms in the ligand which are within bonding distance of the metal. Simple ligands such as CO and NH_3 are therefore monohapto and are designated η^1. The ethene in Zeise's salt, Fig. 15.28 (a), is dihapto (η^2), and the cyclopentadienyl rings in (d) and (g) are pentahapto (η^5).

cyclopentadienyl anion
$[C_5H_5]^-$

15.9.1   The eighteen-electron rule

The structures of many organometallic complexes obey the *eighteen-electron rule*, which turns out to be a very useful guide for understanding which complexes are formed. This rule is probably best regarded as being based on observations rather than theory, but it is possible to construct the following simple argument as to why eighteen is the preferred number of electrons in organometallic complexes.

There are a total of *nine* valence orbitals available on a transition metal: the $4s$, the three $4p$ and the five $3d$ AOs. If suitable orbitals exist of the ligands, these nine metal orbitals can lead to the formation of a maximum of nine bonding MOs and nine antibonding MOs. Filling all of the nine bonding MOs requires *eighteen* electrons, thus this is the optimum number of valence electrons which should be associated with metal–ligand bonding.

If less than eighteen electrons are available, then not all of the bonding MOs are occupied resulting in bonding which is not as strong as it could be. In such circumstances, adding extra ligands increases the number of bonding electrons, and so may be favoured. On the other hand, if there are more than eighteen electrons the excess will probably be located in ligand-based nonbonding MOs. Removing ligands so as to decrease the number of electrons down to eighteen will not reduce the net amount of bonding and so may be a favoured process.

Complexes which can achieve eighteen electrons by losing an electron are likely to be ready to do so i.e. they will be readily oxidized. Likewise, if gaining an extra electron brings the count to eighteen, then the complex may be susceptible to reduction. Finally, we note that complexes obeying the eighteen-electron rule will have no unpaired electrons and so will be diamagnetic.

Electron counting

To apply the eighteen electron rule we need to have a way of counting the number of valence electrons associated with the metal. Rather like assigning an oxidation state or a charge to an atom, counting up the number of valence electrons is a *formal* process based on the application of chemically reasonable rules.

There are actually two commonly used ways of counting the electrons which differ in the way they treat charges; neither approach is entirely satisfactory, but for simple molecules both should yield the same result.

The first method allows the ligands to have their 'usual' charges e.g. a Cl is taken to be Cl^-, CN is taken as CN^-, CO is neutral and cyclopentadienyl is taken as the anion. It is also usual to consider ligands such as H and CH_3 as the anions. The charge on the metal is determined from its oxidation state. Box 15.1 on the following page sets out the rules in full.

The second method treats the ligands and the metal as being neutral for the purposes of electron counting, regardless of the usual charge on the ligands or the oxidation state on the metal. Box 15.2 on the next page sets out the rules.

As an example of using these rules, consider the complex $Fe(CO)_5$. As this is overall neutral, and the ligands are also neutral, both counting schemes proceed in the same way. There are eight electrons from the neutral Fe (configuration $4s^2\,3d^6$), and $5 \times 2 = 10$ from the five CO ligands, giving a total of eighteen electrons. The complex therefore obeys the eighteen-electron rule. $Fe(CO)_4$ would have sixteen electrons, and $Fe(CO)_6$ would have twenty, so the eighteen-electron rule rationalizes why $Fe(CO)_5$ is the observed form of the iron carbonyl. Further examples of electron counting are given in Example 15.1 on page 679.

Box 15.1 Electron counting: scheme 1 (allowing charged ligands)

(a) The ligands are given their usual charges and the charge on the metal is determined by its oxidation state.

(b) If there are any metal–metal bonds, the electrons as divided equally between the two metals e.g. one each for a single bond.

(c) A bridging carbonyl group contributes one electron to each metal.

(d) The number of electrons contributed by particular non-bridging ligand is as below

two electrons	halogen$^-$, H$^-$, alkyl groups (R$^-$), CO, PR$_3$
three electrons	NO (linear coordination)
four electrons	C$_4$H$_4$ 
six electrons	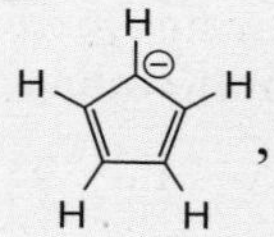 [C$_5$H$_5$]$^-$, C$_6$H$_6$ 

Box 15.2 Electron counting: scheme 2 (metal and all ligands neutral)

(a) For the purposes of electron counting, the metal and the ligands are treated as being neutral.

(b) If there are any metal–metal bonds, the electrons as divided equally between the two metals e.g. one each for a single bond.

(c) A bridging carbonyl group contributes one electron to each metal.

(d) Add one electron if the complex has an overall negative charge; subtract one electron if the complex has an overall positive charge.

(e) The number of electrons contributed by particular non-bridging ligand is as below

one electron	halogen, H, alkyl groups (R)
two electrons	CO, PR$_3$
three electrons	NO (linear coordination)
four electrons	C$_4$H$_4$ 
five electrons	C$_5$H$_5$ 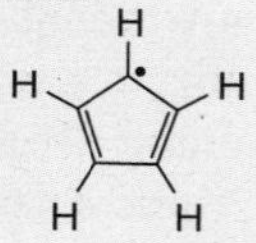
six electrons	C$_6$H$_6$ 

Example 15.1 Electron counting

Let us see if the complexes (c) and (d) in Fig. 15.28 on page 676 obey the eighteen-electron rule.

Complex (c) is neutral and CO is a neutral ligand, so both counting schemes proceed in the same way. Cobalt has the configuration $4s^2 3d^7$ and so contributes nine electrons. Concentrating on one of the cobalt atoms we see that it has three attached CO groups which each contribute two electrons, giving a total of six. The bridging CO groups contribute one electron to each metal atom, so the two bridging COs contribute two electrons to each cobalt. Finally, the Co–Co bond contributes a further electron. So the total is $9 + 6 + 2 + 1 = 18$; the rule is obeyed.

The counting for complex (d) is different in the two schemes. In the charged scheme (Box 15.1) we suppose that the rings are cyclopentadienyl anions $[C_5H_5]^-$ which each contribute six electrons. The complex is neutral, therefore the iron must have a charge of 2+ and so it contributes six electrons. The total is $6 + 6 + 6 = 18$, so the rule is obeyed.

Using the neutral scheme (Box 15.2) the two cyclopentadienyl rings each contribute five electrons and the iron contributes eight (configuration $4s^2 3d^6$). The total is $5 + 5 + 8 = 18$, as before.

15.9.2 Using the eighteen-electron rule

The eighteen-electron rule gives us a way of rationalizing which certain structures occur. For example, while $Fe(CO)_5$ has eighteen electrons, the corresponding manganese pentacarbonyl has seventeen. Having an odd number of electrons makes the species a free radical, and so we might expect it to be rather reactive. If $Mn(CO)_5$ dimerizes by creating a metal–metal bond the electron count rises to eighteen for each Mn (7 from Mn, 10 from the CO and one from the Mn–Mn bond). The reaction of manganese with CO is indeed found to give the dimer $(CO)_5Mn–Mn(CO)_5$, depicted in Fig. 15.29

For nickel the carbonyl $Ni(CO)_4$ is the one formed, which we can rationalize by noting that only four CO ligands are needed to give an eighteen-electron compound. Similarly, for chromium the species formed is $Cr(CO)_6$. For the Group 6 metal, six CO ligands are needed for the compound to achieve eighteen electrons around the metal.

Exceptions to the eighteen-electron rule

Group 9 and 10 metals from the second and third row transition metals are known to form organometallic compounds which adopt a square-planar geometry and have an electron count of sixteen. A good example is Zeise's salt, structure (a) in Fig. 15.28 on page 676, in which the count (using the neutral scheme, Box 15.2) is ten from the Pt, three from the Cl, two from the ethene (the π electrons) and one for the negative charge, giving a total of sixteen. It is interesting to note that metals from Groups 9 and 10 are also the most likely to form square planar classical complexes. Metals from the first row do not seem to form these square planar complexes so readily, but nickel does form the sixteen-electron complex $Ni(PPh)_3$.

$Mn_2(CO)_{10}$

Fig. 15.29 Structure of the manganese carbonyl $(CO)_5Mn–Mn(CO)_5$, which can be regarded as a dimer of $Mn(CO)_5$.

V(CO)$_6$ is a well known species with an electron count of seventeen. In analogy with Mn(CO)$_5$ we might have expected this compound to dimerize, thus increasing the electron count to eighteen for each vanadium atom. However, this is not observed to take place, and it has been proposed that steric crowding around the metal atoms is the reason for this reluctance of V(CO)$_6$ to dimerize.

Why the eighteen-electron rule does not apply to classical complexes

If we look back at the MO diagram for octahedral ML$_6$ with only σ interactions, Fig. 15.11 on page 659, we see that there are six bonding MOs, three nonbonding MOs (the 1t_{2g}) and three moderately antibonding MOs (the 2e_g). Whilst twelve electrons gives the maximum amount of bonding, a further six electrons can be accommodated in the nonbonding MOs without greatly affecting the strength of the bonding, to give a total of eighteen. Further electrons have to go into antibonding MOs, but as these are only weakly antibonding this is not too disadvantageous. Therefore, between twelve and twenty-two electrons can reasonably be accommodated.

The situation alters dramatically if the ligand is changed from those typical of a classical complex, e.g. Cl$^-$ or OH$_2$, to carbon monoxide. As was discussed on page 663, CO is a strong σ donor and as result the 2e_g becomes significantly antibonding. In addition, CO is a π acceptor which will result in the 1t_{2g} orbitals being lowered in energy as they become bonding. Consequently, as shown in Fig. 15.30, there are now *nine* bonding MOs, and above these the next available MOs are significantly antibonding. In such a situation, eighteen electrons is clearly the preferred number. Similar arguments can be made for other geometries.

The sorts of ligands we see in organometallic complexes are strong σ donors through carbon and in addition are often capable of also acting as π acceptors. These are just the sorts of ligands which give rise to an arrangement of MOs in which eighteen bonding electrons is the optimum number.

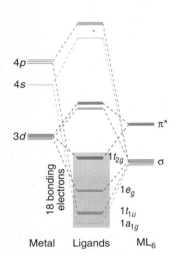

Fig. 15.30 Schematic MO diagram for an octahedral complex in which the ligands are σ donors and π acceptors. The bonding MOs can accommodate eighteen electrons.

15.9.3 IR spectroscopy of carbonyl complexes

As is discussed in detail in section 11.3 on page 461, IR spectroscopy can be used to probe the vibrations of molecules and it is possible to associate the vibrations of specific groups with particular absorptions in the spectrum. For organic molecules, the absorptions due to the carbonyl group are often easily discerned in the spectrum and have been well studied. The absorptions due to C–O vibrations in metal–carbonyl complexes are also rather distinctive and give us a useful guide to both the structure and the bonding of such complexes.

The vibrational frequency of free gaseous carbon monoxide is 2143 cm^{-1}, but it is invariably found that the stretching frequencies of CO ligands in metal complexes occur at *lower* frequencies. For example in Cr(CO)$_6$ the frequency is 2000 cm^{-1}, and in Ni(CO)$_4$ the frequency is 2057 cm^{-1}. The lower frequency for the metal-bound CO means that the C–O bond has been *weakened* by binding to the metal.

These observations are consistent with idea that the bonding between CO and a metal involves overlap between the $\pi^\star$ MOs from the CO with the metal 3d AOs. The resulting bonding MOs are occupied and have a contribution from the CO $\pi^\star$ MOs. Effectively these $\pi^\star$ MOs become partially occupied, and so it is not surprising that the strength of the C–O bonding is weakened. In short, the metal donates electron density into the $\pi^\star$ MOs, thereby weakening the bond.

Further evidence for this interpretation comes from looking at the C–O vibrational frequencies for the following three isoelectronic complexes

$$[V(CO)_6]^- \qquad Cr(CO)_6 \qquad [Mn(CO)_6]^+$$
$$1859 \text{ cm}^{-1} \qquad 1981 \text{ cm}^{-1} \qquad 2101 \text{ cm}^{-1}$$

Going across this series the nuclear charge on the atom increases, so we would expect the energies of the $3d$ AOs to decrease, thereby reducing the interaction with the CO $\pi^\star$ MOs. As a result there is effectively less back-donation into these MOs from the metal, and so the C–O bond is not weakened by as much. This is consistent with the increase in the vibrational frequency.

Carbonyl groups which bridge between two metal atoms are found to have even lower vibrational frequencies than those just attached to one metal. For example, in the cobalt complex shown in Fig. 15.28 on page 676, the vibrational frequencies of the bridging COs are at 1886 cm^{-1} and 1857 cm^{-1}, whereas the other CO groups (described as terminal) have vibrational frequencies in the range 2000–2100 cm^{-1}. Carbonyl groups which are coordinated to three metal atoms have even lower vibrational frequencies in the range 1620–1730 cm^{-1}.

15.10 Aqueous chemistry and oxoanions

Transition metals show a wide range of oxidation states in their compounds and in species which are found in aqueous solution. Low oxidation state metal ions exist in solution as simple aquated species, much as is the case for main-group metals. However, for high oxidation states the species present tend to be *oxoanions*, such as the familiar manganate(VII) $[MnO_4]^-$ and dichromate(VI) $[Cr_2O_7]^{2-}$ ions. For some metals these ions also have a tendency to polymerize e.g. $[V_3O_9]^{3-}$.

The formation of these oxoanions and their polymerization can be understood in terms of the hydrolysis reactions where were first introduced in the discussion of main-group elements, section 14.3 on page 622. Indeed we will discover that there are surprising similarities between the behaviour of certain transition metals and main-group non-metals.

> ⇨ The stabilities of different oxidation states can conveniently be described using oxidation state diagrams, which are discussed in section 19.10 on page 861.

15.10.1 Low oxidation state ions

Transition metal ions in low oxidation states can exist as simple aquated ions in solution e.g. $[Mn(H_2O)_6]^{2+}$ and $[Cr(H_2O)_6]^{3+}$. However, such species show significant acidity and, as has already been discussed, this acidity increases as the charge on the metal increases and as we move along the period. For example, the pK_a of $[Mn(H_2O)_6]^{2+}$ is around 10.5, making it rather feebly acidic, but $[Cr(H_2O)_6]^{3+}$ is much more acidic with a pK_a of around 4.

Increasing the pH of the solution of one of these ions will lead to successive deprotonation of the coordinated water molecules, and eventually to the formation of an insoluble precipitate. In the case of Mn(II) the first species which is formed is $[Mn(H_2O)_5(OH)]^+$, and the further deprotonation gives the insoluble hydroxide, $Mn(OH)_2(H_2O)_4$ (it is likely that some waters of hydration remain associated with the precipitate). If the solution is made strongly alkaline, the precipitate redissolves to give the anion $[Mn(OH)_4(H_2O)_2]^{2-}$ i.e. the hydroxide in amphoteric (see section 14.3.3 on page 629). The formation of this anion can be viewed as resulting from the further deprotonation of the waters of hydration associated with $Mn(OH)_2$.

The hydrolysis of Cr(III) leads to the formation of dimers and higher polymers. Raising to pH first gives the species $[Cr(H_2O)_5(OH)]^{2+}$ which can then eliminate water between itself and the aquo ion $[Cr(H_2O)_6]^{3+}$ to give a dimer connected by a Cr–(OH)–Cr bridge

Fig. 15.31 The structure of the dimer $[Cr_2(H_2O)_8(OH)_2]^{4+}$ which has two Cr–(OH)–Cr bridges between the metal atoms.

This reaction could also be viewed as the OH from $[Cr(H_2O)_5(OH)]^{2+}$ displacing a water from the hexaquo ion. In a subsequent reaction a further water can be eliminated to give a second Cr–(OH)–Cr bridge between the two metal atoms; the structure of the resulting ion (determined by x-ray diffraction) is shown in Fig. 15.31. The Cr–Cr distance in this dimer is just over 300 pm, so there is probably no direct bond between the atoms. Similar condensation reactions involving more chromium centres can take place to give trimers and tetramers, and there is some evidence that related species are present in the insoluble hydroxide which is formed at higher pH.

The formation of dimers (and in some cases higher polymers) in this way is quite common for M^{3+} ions of the transition metals, but is generally not seen for M^{2+}. This presumably reflects the greater acidity of the M^{3+} aquated ions and the stronger interaction between the ligands and the metal.

15.10.2 High oxidation state species

As we have already noted, transition metals tend to occur as oxoanions when in high oxidation states. We will first discuss how this can be rationalized by looking at the case of manganate(VII), and then go on to consider some typical oxoanions of other metals.

The formation of oxoanions

It is of course not reasonable to imagine that a manganese(VII) species contains anything like an 'Mn^{7+}' ion, but it is useful to speculate about what *would* happen if this ion were to be placed in water. Such a highly charged species would have a strongly polarizing effect on the water attracted to it, making the protons very acidic and so easily lost. The result would be manganese with attached OH groups and, by analogy with the behaviour of non-metals (see section 14.3.2 on page 624), we might then expect pairs of these OH groups to eliminate H_2O between them, thus forming an Mn=O bond. How far this process would carry on will depend the relative strength of the Mn–OH and Mn=O bonds, but it is not unreasonable to imagine the tetrahedral $[MnO_4]^-$ ion as being a possible outcome.

Another way of looking at the formation of $[MnO_4]^-$ is to realize that the corresponding Mn(VII) oxide is highly acidic, readily hydrolysing in water according to

$$Mn_2O_7 + 3\,H_2O \longrightarrow 2\,[MnO_4]^- + 2\,H_3O^+.$$

The high-oxidation state oxide thus behaves like a non-metal oxide in that hydrolysis gives acidic solutions. In fact we can draw a direct analogy with Cl_2O_7, in which the oxidation state of chlorine in +7, which hydrolyses in water to give the chlorate(VII) anion $[ClO_4]^-$ (perchlorate).

Manganese is in Group 7 and has the configuration $[Ar]4s^2 3d^5$, and chlorine is in Group 17 with configuration $[Ne]3s^2 3p^5$. Both have seven electrons in their outer shells, so some analogies between their behaviour in oxidation state +7 might be expected. The analogy with the behaviour of the corresponding p-block element will be a constant theme in this section.

The manganate(VII) anion has a regular tetrahedral structure, with an Mn–O distance of 160 pm. This short distance is indicative of multiple bonding, which can arise due to the overlap of p orbitals on the oxygen and d orbitals on the metal.

Chromium oxoanions

The best known chromium oxoanion is dichromate(VI) $[Cr_2O_7]^{2-}$. The simpler monomeric species $[CrO_4]^{2-}$ exists in alkaline solutions, but as the pH is lowered this anion is protonated to give $[HCrO_4]^-$. Elimination of water between two of these anions gives $[Cr_2O_7]^{2-}$

$$2\,[HCrO_4]^- \rightarrow Cr_2O_7^{2-} + H_2O.$$

As is shown in Fig. 15.32, the anion has tetrahedral geometry about the chromium atoms, with a Cr–O–Cr bridge joining the two metal atoms. The Cr–O distance to the non-bridging oxygens is 160 pm, whereas the Cr–O distance to the bridging oxygen is 180 pm. The longer bond in the latter case is presumably indicative of less π bonding.

There is a direct analogy between $[CrO_4]^{2-}$ from Group 6 and $[SO_4]^{2-}$ from Group 16: both Cr and S have a total of six valence electrons. The analogy extends as far as the formation of the disulfate(VI) anion, $[S_2O_7]^{2-}$, whose structure is similar to the dichromate(VI) species. In addition both CrO_3 and SO_3 react with water to give the analogous species H_2CrO_4 and H_2SO_4. Both oxides are highly acidic.

There is some evidence that $[Cr_2O_7]^{2-}$ can go on to form higher polymers such as $[Cr_3O_{10}]^{2-}$ and $[Cr_4O_{13}]^{2-}$, but these species are not favoured in solution. This contrasts with what happens in the case of vanadium.

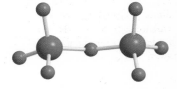

Fig. 15.32 The structure of the dichromate(VI) anion $[Cr_2O_7]^{2-}$, determined by x-ray diffraction.

Vanadium oxoanions

Under strongly alkaline conditions the vanadate(V) oxoanion $[VO_4]^{3-}$ is the favoured species. Not surprisingly, this anion is also formed when vanadium(V) oxide V_2O_5 dissolves in alkai, in accord with the expected acidic nature of the oxide. Like the manganese and chromium oxoanions, $[VO_4]^{3-}$ is tetrahedral.

As the pH is lowered the anion becomes protonated on the oxygen to give $[VO_3(OH)]^{2-}$, and then further protonated to give $[VO_2(OH)_2]^-$. By eliminating water, these species can go on to form a wide range of polyvanadates, for example

$$2\,[VO_3(OH)]^{2-} \longrightarrow [V_2O_7]^{4-} + H_2O$$
$$3\,[VO_2(OH)_2]^- \longrightarrow [V_3O_9]^{3-} + 3\,H_2O$$
$$4\,[VO_2(OH)_2]^- \longrightarrow [V_4O_{12}]^{4-} + 4\,H_2O.$$

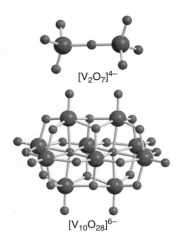

$[V_2O_7]^{4-}$

$[V_{10}O_{28}]^{6-}$

Fig. 15.33 The structures of the polyvanadates $[V_2O_7]^{4-}$ and $[V_{10}O_{28}]^{6-}$, determined by x-ray diffraction (counter ions are not shown).

Weblink 15.6

View and rotate the structures shown in Fig. 15.33.

Some of these polyvanadates can be crystallized and their structures determined by diffraction. The structure of $[V_2O_7]^{4-}$, shown in Fig. 15.33, is analogous to dichromate(VI) except that the V–O–V linkage in linear; the structure can be described as two tetrahedra, joined at an apex.

As the pH is lowered further, yet more of these polyvanadate anions are produced, including species such as $[V_{10}O_{26}(OH)_2]^{4-}$. The deprotonated version of this anion will crystallize with the appropriate cation, allowing the structure shown in Fig. 15.33 to be determined. In this ion the coordination around the vanadium is approximately octahedral, and the structure can be described as ten VO_6 octahedra, sharing along edges.

There are strong analogies between the behaviour of these vanadium(V) species from Group 5 and phosphorus(V) oxoanions in Group 15. The simple phosphate(V) anion $[PO_4]^{3-}$ and the diphosphate(V) species $[P_2O_7]^{4-}$ have direct analogues in vanadium. Under the appropriate conditions phosphate(V) anions also undergo condensation reactions to form many different polyphosphates. In contrast to vanadium, the polyphosphates have a tendency to form chains consisting of PO_4 tetrahedra, connected through their apexes via P–O–P linkages.

15.11 Moving on

One chapter cannot begin to do justice to the diversity and range of transition metal chemistry. However, in this chapter you have been introduced to the key concepts and ideas which are needed for a deeper and more systematic study of the chemistry of these elements. If you go on to take more advanced courses you will find that much is known about the reactivity of transition metal complexes (here we have just concentrated on structures), and that this leads to many practical applications in which particularly organometallic complexes are used to facilitate reactions.

The second- and third-row transition metals have quite a lot in common with those of the first row, but there are significant differences which can be related to the different orbitals available to these later elements. The lanthanide and actinide elements also have some features in common with the transition metals in that they have a partially filled sub-shell (the *f* rather than the *d*). However, the *f* orbitals behave rather differently from the *d*, leading to some distinctive chemistry and physical properties for compounds of the *f*-block elements.

FURTHER READING

Chemistry of the First-Row Transition Metals, Jon McCleverty, Oxford University Press (1999).

Inorganic Chemistry, fourth edition, Peter Atkins, Tina Overton, Jonathan Rourke, Mark Weller and Fraser Armstrong, Oxford University Press (2006).

Chemistry of the Elements, second edition, N. N. Greenwood and A. Earnshaw, Butterworth–Heinemann (1997).

Molecular Symmetry and Group Theory, second edition, Alan Vincent, Wiley (2001).

QUESTIONS

15.1 (a) Using Slater's rules (page 77), determine the effective nuclear charge experienced by the $4s$ and $3d$ electrons in the metals of the first transition series (use the electronic configurations given in Fig. 15.1 on page 650). Compute the energies of the $4s$ and $3d$ orbitals using

$$E_n = -\frac{Z_{\text{eff}}^2 R_H}{n^2},$$

where $R_H = 13.6$ eV and n is the principal quantum number. On the same graph, plot these energies against atomic number.

(b) Comment on the trends that you see in your graph, and compare it with the data given in Fig. 15.2 on page 651.

(c) Repeat the calculations for the ion M^+, assuming that a $4s$ electron has been ionized. Comment on how the orbital energies for M^+ compare with those for M. According to your calculations, which electron will be the easiest to ionize to give M^{2+}?

15.2 Explain or rationalize the following.

(a) The ground-state electronic configuration of potassium is $[\text{Ar}]\,4s^1$ and not $[\text{Ar}]\,3d^1$.

(b) In titanium (Ti) the $3d$ AO is lower in energy than the $4s$, yet the ground-state electronic configuration of Ti is $4s^2\,3d^2$ and not $3d^4$.

(c) The ground-state configuration of copper (Cu) is $4s^1\,3d^{10}$ rather than $4s^2\,3d^9$.

(d) Across the first transition series, the energy of the $4s$ orbital falls more slowly than does the energy of the $3d$.

15.3 Throughout this question assume that the exchange interaction contributes an energy of $-K$ per pair of parallel spins (see section 2.7.3 on page 82).

(a) Show that the exchange contribution to the energy of the configuration $4s^2\,3d^3$ is $-6K$, and that the contribution for the configuration $4s^1\,3d^4$ is $-10K$.

(b) Work out the exchange contribution to the configurations $4s^2\,3d^4$, $4s^1\,3d^5$, $4s^2\,3d^5$ and $4s^1\,3d^6$

(c) By considering the *change* in the exchange contribution on going from $4s^2\,3d^n$ to $4s^1\,3d^{n+1}$, rationalize the ground-state configurations adopted by V, Cr and Mn.

15.4 (a) What do you understand by the *chelate effect*?

(b) For the following reaction

$$[\text{Ni}(\text{NH}_3)_6]^{2+} + 3\,\text{en} \longrightarrow [\text{Ni}(\text{en})_3]^{2+} + 6\,\text{NH}_3$$

it is found that $\Delta_r H^\circ = -12.1$ kJ mol^{-1} and $\Delta_r S^\circ = 185$ J K^{-1} mol^{-1} in aqueous solution at 298 K. Compute the value of $\Delta_r G^\circ$ and hence the value of the equilibrium constant. Does the entropy term make a significant contribution to the value of $\Delta_r G^\circ$?

(c) Discuss the trend in the values of the equilibrium constant for the following reactions (all in aqueous solution)

$$[Co(H_2O)_6]^{2+} + 6\,NH_3 \longrightarrow [Co(NH_3)_6]^{2+} + 6\,H_2O \qquad \lg K = 5.2$$
$$[Co(H_2O)_6]^{2+} + 3\,en \longrightarrow [Co(en)_3]^{2+} + 6\,H_2O \qquad \lg K = 13.9$$
$$[Co(H_2O)_6]^{2+} + (6\text{-en}) \longrightarrow [Co(6\text{-en})]^{2+} + 6\,H_2O \qquad \lg K = 15.8$$

(6-en) is a hexadentate ligand which coordinates through nitrogen.

15.5 (a) Draw two isomers of the octahedral complex $[Co(NH_3)_4Cl_2]^+$. Are either of these isomers optically active?

(b) Draw all the possible isomers of the octahedral complex $[Co(en)_2(NH_3)Cl]^{2+}$. Which of these are optically active?

(c) Draw all the possible isomers of the hypothetical octahedral complex $MA_2B_2C_2$, where A, B and C are three different monodentate ligands. Which, if any, of these isomers are optically active?

15.6 In the spectrochemical series, the halide ions appear in the order

$$\xrightarrow{\text{increasing } \Delta_o}$$

$$I^- \quad Br^- \quad Cl^- \quad F^-.$$

By considering the way in which the *energy* and *size* of the orbitals of the halide ions change along the series, rationalize the order in which these anions appear.

15.7 When carbon monoxide acts as a ligand to a transition metal it invariably bonds through the carbon and not through the oxygen. Discuss why this is so.

15.8 Using the same approach as in section 15.4 on page 663, show that for the d^6 configuration the high-spin arrangement is preferred when

$$2\Delta_o - 10K < -6K.$$

Similarly, show that for the d^7 configuration the high-spin arrangement is preferred when $\Delta_o < 2K$.

15.9 For paramagnetic first-row transition metal complexes the *magnetic moment μ*, in units of the *Bohr magneton* (B.M.), is reasonably well approximated by the spin-only formula

$$\text{magnetic moment in B.M.} = \sqrt{n(n+2)}$$

where n is the number of unpaired electrons. Use this expression to draw up a table of the expected magnetic moments for between one and five unpaired electrons. Then draw up a second table showing the expected magnetic moments for high-spin and for low-spin octahedral complexes with between one and nine d electrons.

Use the following experimentally measured magnetic moments to determine whether the given octahedral complex is high or low spin. [Be aware that the spin-only formula is not expected to give precise agreement with experimental data.]

In the complex $[Co(NO_2)_6]^{4-}$ the ligands are $[NO_2]^-$.

complex	μ / B.M.	complex	μ / B.M.
$[Co(NO_2)_6]^{4-}$	1.9	$[Cr(H_2O)_6]^{3+}$	3.8
$[Cr(H_2O)_6]^{2+}$	4.8	$[Mn(H_2O)_6]^{2+}$	5.9
$[Fe(CN)_6]^{3-}$	2.3	$[Fe(H_2O)_6]^{2+}$	5.3

15.10 Using the approach described in section 15.6.1 on page 670, compute the change in energy due to the occupation of the $1e_g$, $1t_{2g}$ and $2e_g$ MOs as a function of the number of d electrons assuming that a *low-spin* configuration is adopted.

Plot a graph of this energy in the same form as Fig. 15.23 on page 671, and use it to predict how you would expect the energy of a low-spin complex to vary with the number of d electrons.

15.11 This question is about how the energy difference between a square planar and tetrahedral ML_4 complex varies with the number of d electrons. Shown below are partial MO diagrams for the two geometries: *only* those MOs which have a contribution from the metal orbitals are shown. Note that MO3, MO6 and MO8 are triply degenerate; MO7 is doubly degenerate.

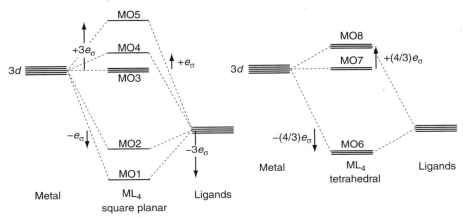

A simplified MO treatment shows that the amount by which each MO is shifted in energy on forming the complex can be expressed in terms of a single energy parameter e_σ. For example, MO1 is shifted down in energy by $3e_\sigma$, whereas MO8 is shifted up in energy by $\frac{4}{3}e_\sigma$. MO3 and MO7 are nonbonding and so do not change energy.

In the square-planar geometry MO1 and MO2 are occupied by two electrons each, contributing an total energy change of $2 \times (-3e_\sigma) + 2 \times (-e_\sigma) = -8e_\sigma$. Similarly in the tetrahedral geometry the triply degenerate MO6 is occupied by six electrons, contributing an total energy change of $6 \times (-\frac{4}{3}e_\sigma) = -8e_\sigma$ i.e. exactly the same amount. In comparing the two geometries, we can therefore ignore these filled orbitals.

The electrons originating from the metal $3d$ AOs can fill MO3, MO4 and MO5 in the square-planar case, and MO7 and MO8 in the tetrahedral case. The precise details of which orbitals are occupied will determine the relative energy of the two possible geometries for ML_4.

Consider the case of four d electrons. Assuming a high-spin configuration, in the square-planar complex three electrons will occupy the three degenerate orbitals MO3, which contribute nothing to the energy change, and one will occupy MO4, which contributes $+e_\sigma$. The energy change is thus $+e_\sigma$. In the tetrahedral case, two electrons occupy MO7 and two occupy MO8, so the energy change is $2 \times 0 + 2 \times (+\frac{4}{3}e_\sigma) = 2.67\,e_\sigma$. It is clear that the square-planar geometry is preferred since the energy increase due to the d electrons is smaller than that for the tetrahedral case.

(a) Repeat this calculation for between one and ten d electrons (assuming a high-spin configuration), and hence identify those configurations for which the square-planar geometry is preferred.

(b) Now consider the case where the square-planar complex is low spin (meaning that MO5 is not occupied until *both* MO3 *and* MO4 are full) but the tetrahedral complex is still high spin. Again identify those configurations for which the square-planar geometry is preferred.

(c) In the light of your calculations, comment on the fact that square-planar complexes are most commonly found for metals with d^8 configurations in conjunction with ligands which give rise to large splitting of the d orbitals.

15.12 Count the number of valence electrons associated with each metal atom in the following complexes (perform the count using the scheme in Box 15.1 on page 678 and then using the scheme in Box 15.2 on page 678).

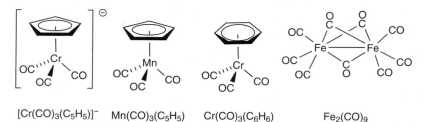

$[Cr(CO)_3(C_5H_5)]^-$ $Mn(CO)_3(C_5H_5)$ $Cr(CO)_3(C_6H_6)$ $Fe_2(CO)_9$

15.13 Determine the number of carbon monoxide ligands which need to be attached to each transition metal atom from the first row in order that the eighteen-electron rule is obeyed by a complex $M(CO)_n$ or $[M(CO)_n]^+$.

15.14 (a) Comment on the fact that no mono-nuclear neutral cobalt carbonyl $Co(CO)_n$ has been detected, but the anion $[Co(CO)_4]^-$ is known.

(b) Reaction of cobalt with carbon monoxide gives the species $Co_2(CO)_8$ which is thought to exist in two structural isomers. The IR spectrum of isomer A shows a number of bands between 2000 cm^{-1} and 2100 cm^{-1}; the spectrum of isomer B shows bands in this same region as well as bands at around 1850 cm^{-1}.

Explain what you can deduce from the IR spectrum about the bonding of the CO groups in these two carbonyl complexes, and hence suggest possible structures for A and B. You should ensure that your structures are consistent with the eighteen-electron rule.

(c) The carbonyl complex $Co_4(CO)_{12}$ is also formed quite readily. Suggest a structure for this in which the eighteen-electron rule is obeyed [hint: start with a tetrahedron of cobalt atoms].

15.15 Discuss the following observations. MnO is a basic oxide, dissolving in aqueous acid to give the species $[Mn(H_2O)_6]^{2+}$. Similarly, Mn_2O_3 is also basic, dissolving in acid to give the aquated ion $[Mn(H_2O)_6]^{3+}$ which tends to be unstable with respect to disproportionation. MnO_2 is an amphoteric oxide; it dissolves in alkali to give the anion $[MnO_3]^{2-}$, but even in strongly acidic solutions there is no evidence for a simple aquo ion '$[Mn(H_2O)_6]^{4+}$'. Mn_2O_7 readily dissolves in and reacts with water to give strongly acidic solutions containing $[MnO_4]^-$.

15.16 Discuss the following observations. Even in the most acidic solutions there is no evidence for the Ti(IV) species '$[Ti(H_2O)_6]^{4+}$'. However, the species $[Ti(H_2O)_5(OH)]^{3+}$, $[Ti(H_2O)_4(OH)_2]^{2+}$ and $[Ti(H_2O)_3(OH)_3]^+$ have all been characterized. Some workers have also suggested the existence of the species $[TiO]^{2+}$ (with associated waters of solvation), but the evidence is not compelling (discuss how this species might be formed). In contrast, there is ample evidence for the existence of the vanadium(IV) species $[VO]^{2+}$. In this ion, the V–O bond length is around 160 pm, indicative of multiple bonding between these atoms.

Quantum mechanics and spectroscopy

Key points

- In quantum mechanics, observables are represented by operators.
- The hamiltonian operator represents the total energy.
- The square of the wavefunction gives the probability density.
- The eigenvalues of the hamiltonian are the energy levels of the system.
- The appearance of quantized energy levels is associated with the constraints placed on the wavefunction by the form of the potential.
- The energy levels and associated wavefunctions are characterized by one or more quantum numbers.
- Wavefunctions are generally oscillatory in character, and show increasing number of nodes as the energy increases.
- The Schrödinger equation for the harmonic oscillator can be solved and the resulting energy levels can be used to model the vibration of a bond.
- The occurrence of spectroscopic transitions are governed by selection rules; the intensities of the resulting spectral lines depend on the populations of the energy levels.
- The Morse potential is a better model for the vibration of a bond than is the harmonic potential.
- The Schrödinger equation for the rigid rotor can be solved and the resulting energy levels can be used to model the rotation of a diatomic.
- From the rotational (microwave) spectrum of a diatomic it is possible to determine the bond length to high accuracy.
- A vibrational transition is accompanied by simultaneous changes in the rotational energy, leading to fine structure in the form of P and R branches in IR spectra.
- Transitions between electronic energy levels give rise to spectra in the UV and visible regions. Such spectra can be interpreted qualitatively in terms of electrons moving to higher energy orbitals.

In Chapter 2 we introduced the idea that the behaviour of electrons in atoms and molecules can be described using the theory known as *quantum mechanics*. Two key ideas emerge from this theory: the first is that the electron is described by a *wavefunction*, and the second is that the energy of the electron is *quantized*.

The wavefunction is important since its square gives the probability density of the electron i.e. it enables us to work out the probability of finding the electron in a particular region of space. This property has been used extensively throughout this book, where the shape and size of atomic and molecular orbitals, which are the wavefunctions for the electrons in atoms and molecules, have been shown to be crucial in determining both shape and reactivity.

Quantization means that the energy can only have certain discrete values, rather than being allowed to vary continuously. This leads to the idea that there are a set of energy levels, or orbitals, available to the electrons. As we have seen, which orbitals are occupied and the energies of these orbitals is of great significance in chemistry.

In this chapter we are going to look at how these wavefunctions and energy levels arise in quantum mechanics, and how they can actually be computed in some simple cases. This will involve setting up and then solving the famous *Schrödinger equation*. We will also see that quantum mechanics is not restricted to the description of electrons in atoms and molecules, but can also be used to describe the motion of molecules in space (translation), as well as the rotation and vibration of molecules.

Each of these kinds of motion has associated with it a set of wavefunctions and energy levels, and it is the transitions between these energy levels which give rise to spectra, a topic which has already been introduced in Chapter 11. In this chapter we will look at the quantum mechanical interpretation of such spectra, and see how they can be used to give detailed information about small molecules.

There are two problems which can arise when we first encounter quantum mechanics. The first is that its predictions are not in accord with our experience of the everyday world, and so can seem counter-intuitive. Our day-to-day experience is of macroscopic objects, which obey the laws of classical physics. We expect energy to vary continuously and we expect objects to have defined positions: in contrast quantum mechanics tells us that energy is quantized, and that we can only talk about the probability of a particle being at a particular position. Strange though the predictions of quantum mechanics seem, they have been tested exhaustively and found to be in excellent agreement with experimental observations. We can therefore be confident that the theory is applicable.

The second problem with quantum mechanics is that once we move away from very simple systems, the mathematics rapidly becomes rather complicated. For example, it is possible to find, with pencil and paper, the wavefunctions and energy levels of a hydrogen atom without too much difficulty. However, for any atom with more than one electron we have to resort to a numerical (computer-based) solution of the problem, and for large molecules we will need to employ the fastest and most powerful computers available to obtain a result in a reasonable time. In this chapter we will only concern ourselves with the very simplest systems, which can be solved with pencil and paper. The solutions we will obtain are nevertheless very useful as they can be used to model, to some degree of approximation, the behaviour of real molecules.

If you have read about quantum mechanics in the popular scientific press you may have obtained the impression that there are problems with this theory, and that it is beset with all sorts of 'philosophical' difficulties. Despite this, quantum mechanics has been shown to work extraordinarily well when it comes to predicting and rationalizing the properties of atoms and molecules. We can therefore be entirely confident in applying quantum mechanics to chemical problems.

16.1 The postulates of quantum mechanics

Like any theory, quantum mechanics starts out by making a number of *postulates* on which the theory is based. A postulate is just a statement that such-and-such is the case. These postulates cannot be proved, but what we can do is compare the predictions which arise from quantum mechanics with experiment. If there is good agreement, which indeed there is, we can say that the theory is consistent with experiment. This does not mean that the theory, or the postulates that it is based on, is correct, but it does give us confidence that the theory is applicable.

Rather than stating all the postulates in one go, we will introduce them throughout this section, discussing and illustrating the meaning of each. There is no escaping the fact that these postulates do seem rather abstract and distant from the practical application of quantum mechanics to molecules, which is what we are really interested in. It is, however, worthwhile discussing these postulates so that we can see what quantum mechanics is based on, rather than simply using it as a recipe.

So far, all the applications of quantum mechanics you have seen have concerned the behaviour of electrons in atoms and molecules. However, quantum mechanics is not restricted to electrons, but can also be used to describe the translational, rotational and vibrational motions of molecules. In the discussion that follows we will simply refer to the object being analysed as the 'system', so as to be as general as possible. The system might be the electrons in an atom, a vibrating bond or a rotating molecule.

The first two postulates we are going to discuss are

- Any property (such as position or energy) of the system can be computed from the wavefunction.

- These properties are represented by mathematical *operators*.

Let us start by discussing operators.

16.1.1 Operators

An operator is a mathematical object which acts on a function to produce a new function:

operator   *acts on*   function   *to give*   new function.

Operators can take a wide variety of forms. For example '$x \times$' is an operator which, in words, means 'multiply by x'. If we apply this to $\sin(Ax)$ we obtain a new function $x \sin (Ax)$

$$\underbrace{x \times}_{\text{operator}} \underbrace{\sin (Ax)}_{\text{function}} = \underbrace{x \sin (Ax)}_{\text{new function}}.$$

A more interesting operator is d/dx, which in words means 'differentiate with respect to x'. For example

$$\underbrace{\frac{d}{dx}}_{\text{operator}} \underbrace{\sin (Ax)}_{\text{function}} = \underbrace{A \cos (Ax)}_{\text{new function}}.$$

It is a postulate of quantum mechanics that each observable quantity is represented by an operator. By observable quantity we mean something which can be measured, and for our purposes there are really only two quantities we are interested in: these are the position (i.e. where the particle is) and the energy. The precise form of the operators which represent these quantities arise from a further postulate of quantum mechanics, a discussion of which is well beyond the level of this text. We will simply have to content ourselves with stating what these operators are.

To keep things as simple as possible we will illustrate the operators by considering a very simple system in which a particle of mass m is constrained to move along the x-axis i.e. a one-dimensional system.

Operator for position

The operator which represents position along the x-axis is simply $x \times$; as we discussed above, this operator means 'multiply by x'. Written more formally we have

$$\hat{x} = x,$$

where the multiplication symbol is taken as being implied. The hat indicates that the object is an operator, so $\hat{x}$ means 'the operator which represents position along the x-axis'.

Operator for energy: the hamiltonian

In classical mechanics the total energy which an object possesses can be divided into two parts: the *kinetic* energy T, and the *potential* energy V

$$E_{\text{tot}} = T + V.$$

Kinetic energy arises from the *motion* of an object, while the potential energy is a contribution which depends on the *position* of the object. For example, if we throw a ball up in the air the kinetic energy depends on its speed, whilst the potential energy depends on the height of the ball above the earth's surface (i.e. the gravitational potential energy).

The operator for kinetic energy along the x-axis, $\hat{T}$, turns out to be

$$\hat{T} = -\frac{\hbar^2}{2m} \frac{d^2}{dx^2},$$

where $\hbar$ is Planck's constant divided by 2π.

The potential energy is a function of the position, so in the one-dimensional case it is simply a function of x and can be written $V(x)$. It turns out that the corresponding operator is just this function

$$\hat{V} = V(x).$$

The reason for this is that the operator for position ($\hat{x}$) is just x, so nothing changes in the expression for $V(x)$ when x is replaced by its operator.

The operator for the *total* energy turns out to be very important in quantum mechanics, so it has its own name and symbol: it is called the *hamiltonian* operator, $\hat{H}$. Just as $E_{tot} = T + V$, $\hat{H}$ is also the sum of the kinetic and potential energy operators:

$$\hat{H} = \hat{T} + \hat{V}.$$

We will frequently come across the hamiltonian in what follows.

16.1.2   Eigenfunctions and eigenvalues

In general, when an operator acts on a function, a new function is generated. However, it is sometimes possible to find a special set of functions which have the property that, when they are acted on by a particular operator, they are *unchanged*, apart from being multiplied by a constant. This property can be expressed as

$$\hat{A}\phi = \lambda \times \phi, \qquad (16.1)$$

where $\hat{A}$ is the operator in question, ϕ is the function it acts on, and λ is the constant. A function ϕ which obeys Eq. 16.1 is said to be an *eigenfunction of the operator $\hat{A}$*, and λ is the associated *eigenvalue*. Equation 16.1 is called the *eigenvalue equation*.

It is important to realize that it is only the eigenfunctions of the operator $\hat{A}$ which obey Eq. 16.1. Some function that is *not* an eigenfunction of $\hat{A}$ will simply be altered by the action of the operator.

As an example, consider the case where the operator is d/dx, so that the eigenvalue equation becomes

$$\frac{d}{dx}\phi(x) = \lambda\,\phi(x).$$

To find the eigenfunctions $\phi(x)$ we have to think of a function whose differential is proportional to itself: the obvious candidate is an exponential, since $d/dx \exp(x) = \exp(x)$.

Let us be a bit more general and try $\phi(x) = B\exp(Cx)$, where B and C are constants, as a possible eigenfunction. To confirm that it is an eigenfunction all we have to do is apply the operator to $\phi(x)$ and see if it is regenerated

$$\begin{aligned}
\frac{d}{dx}\phi(x) &= \frac{d}{dx}[B\exp(Cx)] \\
&= CB\exp(Cx) \\
&= C\phi(x).
\end{aligned}$$

Thus, acting on $\phi(x) = B\exp(Cx)$ with the operator does regenerate the function, multiplied by the constant C. It therefore follows that $\phi(x) = B\exp(Cx)$ is an eigenfunction of the operator d/dx, with eigenvalue C; this is true for all values of B and C.

On the other hand $\phi(x) = B \sin(Cx)$ is *not* an eigenfunction of $\mathrm{d}/\mathrm{d}x$ as the function is not regenerated when the operator acts

$$
\begin{aligned}
\frac{\mathrm{d}}{\mathrm{d}x}\phi(x) &= \frac{\mathrm{d}}{\mathrm{d}x}[B\sin(Cx)] \\
&= CB\cos(Cx) \\
&\neq \mathrm{const.} \times \phi(x).
\end{aligned}
$$

An important point to take away from this discussion is that an operator can have more than one eigenfunction. In the example discussed above we found that $\phi(x) = B\exp(Cx)$ is an eigenfunction for *all* values of B and C.

It is usual only to count an eigenfunction as being distinct if it has a *different* eigenvalue from all the other eigenfunctions. In our example, the eigenvalue depends only on C, and not on the multiplying constant B. Therefore functions with different values of C are counted as distinct eigenfunctions, but those with different values of B are not.

The significance of eigenfunctions and eigenvalues will become apparent as the discussion progresses.

An exception to this are *degenerate* eigenfunctions which have the same eigenvalue but which differ in a more significant way than just the value of a multiplying constant. We return to this point in section 16.4 on page 710.

16.1.3 Measurement in quantum mechanics

We now come to a very strange postulate of quantum mechanics which is concerned with what happens when we make a measurement. We will first state the postulate, and then try to unpick what it means in practice.

- When the value of a particular observable is measured, the result is *always* one of the eigenvalues of the operator which represents that observable.

To give an example, if we take a molecule and measure its energy, then the result we will obtain will *always* be one of the eigenvalues of the operator which represents energy (the hamiltonian operator). Likewise, if we measure the position, the result will be one of the eigenvalues of the operator which represents position.

Note that the postulate does *not* say which eigenvalue will be found: all it says is that we will find *one* of the eigenvalues. For the moment we will leave this idea 'hanging' and come back to it when we look at the significance of the energy eigenvalues.

16.1.4 Wavefunctions

We have already been using the idea that for an electron the square of the wavefunction gives the probability density of the electron: this is an example of the *Born interpretation* of the wavefunction.

Born interpretation

The Born interpretation says that the probability of finding a particle in a small volume $\mathrm{d}V$ (we are using the language of calculus) at position (x, y, z) is given by

probability of being in volume $\mathrm{d}V$ at position $(x, y, z) = [\psi(x, y, z)]^2\,\mathrm{d}V$,

where $\psi(x, y, z)$ is the wavefunction of the particle, and we are assuming that this wavefunction is real. This interpretation means that $[\psi(x, y, z)]^2$ is the *probability density* of the particle, since multiplying this by a volume (dV) gives the probability.

For a one-dimensional wavefunction $\psi(x)$, the Born interpretation is that the probability of finding the particle in the small interval dx at position x is given by

$$\text{probability of being in interval d}x\text{ at position }x = [\psi(x)]^2 \, \mathrm{d}x.$$

Normalization

The *total probability* of finding the particle over all space is found by summing $[\psi(x, y, z)]^2$ dV over all space. In the one-dimensional case this means summing $[\psi(x)]^2$ dx between $x = -\infty$ and $x = +\infty$; this total probability is the integral

$$\int_{-\infty}^{+\infty} [\psi(x)]^2 \, \mathrm{d}x.$$

However, since the particle has to be somewhere, we know that this total probability must be one.

If the total probability predicted by a particular wavefunction is one, the wavefunction is said to be *normalized*

$$\text{normalized} \qquad \int_{-\infty}^{+\infty} [\psi(x)]^2 \, \mathrm{d}x = 1.$$

Some wavefunctions do not conform to this property, and so are not normalized. However we can ensure that such a wavefunction is normalized simply by multiplying it by a constant N, to give $N\psi(x)$, and then adjusting the value of N to ensure that

$$\text{normalization condition} \qquad \int_{-\infty}^{+\infty} N^2 \, [\psi(x)]^2 \, \mathrm{d}x = 1.$$

N is called the *normalization constant*.

Constraints on the wavefunction

If we interpret the square of the wavefunction as being the probability density, then this imposes some constraints on the behaviour of the wavefunction:

(a) The wavefunction can only have a *single value* at any one position. This is because the probability can only have a single value at any one point: you cannot have a probability of 0.2 *and* 0.3 at the same point.

(b) The wavefunction must vary smoothly and cannot suddenly jump from one value to another. If the wavefunction were to jump between values at some point, it would have an infinite number of values at that point, in violation of (1).

(c) The integral of the square of the wavefunction over all space must be finite so that the wavefunction can be normalized. If this integral is infinite, or undefined, the wavefunction cannot be normalized.

Figure 16.1 on the next page shows examples of each of these kinds of unacceptable behaviour in a wavefunction.

The wavefunction can be *complex*, meaning that it has a real and an imaginary part. If this is the case, then the probability density is $\psi^\star(x, y, z)\,\psi(x, y, z)$, where $\psi^\star(x, y, z)$ is the complex conjugate of the wavefunction. All the wavefunctions we will come across are real, in which case $\psi^\star(x, y, z)$ is the same as $\psi(x, y, z)$, so the probability density is simply $[\psi(x, y, z)]^2$.

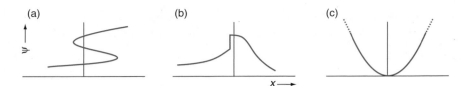

Fig. 16.1 Examples of unacceptable behaviour in the wavefunction. In (a) the wavefunction is multi-valued at some values of x, in (b) there is a sudden jump in the wavefunction, and in (c) the area under the square of the wavefunction (over all space) is infinite. If the square of the wavefunction is interpreted as a probability density, none of these behaviours are acceptable.

16.1.5 Why the eigenvalues and eigenfunctions of the hamiltonian are so important

It is usually the case that the property of the system we are most interested in is its energy. As has been described in section 16.1.3, it is a postulate of quantum mechanics that no matter what the wavefunction of the system is, a measurement of the energy will *always* yield one of the eigenvalues of the operator for energy i.e. one of the eigenvalues of the hamiltonian operator. Therefore, these eigenvalues represent all of the possible results of measuring the energy.

This leads to the idea that the system has available to it a set of *energy levels* which are the eigenvalues of the hamiltonian. Any measurement of the energy will result in one of these eigenvalues, so we say that the system 'occupies one of the energy levels'.

Furthermore, each energy level (eigenvalue) corresponds to a particular eigenfunction of the hamiltonian. By imagining that the system occupies one of these energy levels, we can also imagine that its wavefunction is the corresponding eigenfunction. So, the overall picture is of the system occupying a certain energy level and being described by the corresponding eigenfunction.

We have to be very careful here: we are *not* saying that the wavefunction of the system *is* one of the eigenfunctions of the hamiltonian. All we are saying is that when we measure the energy of the system, the value we obtain is one of the eigenvalues of the hamiltonian, which is associated with a particular eigenfunction of the hamiltonian.

From now on we are going to be concerned exclusively with finding the eigenfunctions and eigenvalues of the hamiltonian operator, since these will tell us about the energy levels which are available to the system. The eigenvalue equation for the hamiltonian is

$$\hat{H}\psi = E\psi, \tag{16.2}$$

where ψ is the eigenfunction and E is the eigenvalue, which is the energy associated with the eigenfunction.

Equation 16.2 is often called the (time-independent) *Schrödinger equation* (SE), so finding the eigenfunctions and eigenvalues of the hamiltonian is also described as 'solving the Schrödinger equation'.

16.1.6 Summary

Some quite subtle ideas have been described in this section, but in the end they boil down to a few key points which we will carry forward to the more practical discussion which follows.

(a) The operator for the total energy is called the hamiltonian $\hat{H}$. It consists of the kinetic energy part $\hat{T}$ and a potential energy part $\hat{V}$: $\hat{H} = \hat{T} + \hat{V}$.

(b) In one dimension

$$\hat{H} = \underbrace{-\frac{\hbar^2}{2m}\frac{d^2}{dx^2}}_{\hat{T}} + \underbrace{V(x)}_{\hat{V}}.$$

(c) The eigenfunctions ϕ and eigenvalues λ of an operator $\hat{A}$ are found by solving the eigenvalue equation

$$\hat{A}\phi = \lambda \times \phi.$$

(d) A measurement of the energy always yields one of the eigenvalues of the hamiltonian operator.

(e) The system can be thought of as having available to it a set of energy levels which are the eigenvalues of the hamiltonian; each level has a corresponding eigenfunction.

(f) The square of the wavefunction gives the probability density of the particle.

16.2 A free particle moving in one dimension

In this section and the next we are going to find the eigenvalues (energy levels) and eigenfunctions of the hamiltonian for a particle moving in one dimension (i.e. along x). This is the very simplest quantum mechanical problem, so it is a good way of starting to get to grips with the theory. The energy levels we will come up with also turn out to be surprisingly useful for modelling various physical systems, despite the idealised nature of the problem.

Some of the mathematics in this section is a little involved, although not fundamentally that complicated. It is not necessary, in the first instance, to understand all of the details, but it is important to appreciate the consequences which are described in words and in Fig. 16.2.

We are first going to consider a 'free' particle, and then in the next section move on to a particle which is constrained to move between certain limits along the axis; this latter arrangement is often called the 'particle in a box'. These two cases make for an interesting comparison, and point the way forward to dealing with more complicated arrangements.

Let us imagine that we have a particle which experiences a potential which is constant, and has the value V_0 i.e. $\hat{V} = V_0$. The hamiltonian operator is therefore

$$\hat{H} = -\frac{\hbar^2}{2m}\frac{d^2}{dx^2} + V_0.$$

We seek the eigenfunctions and eigenvalues of this operator, which means solving the corresponding eigenvalue equation, Eq. 16.2 on the preceding page, i.e.

$$\hat{H}\psi(x) = E\psi(x)$$
$$-\frac{\hbar^2}{2m}\frac{d^2}{dx^2}\psi(x) + V_0\psi(x) = E\psi(x).$$

Rearranging the last line by taking $V_0\psi(x)$ to the right gives

$$-\frac{\hbar^2}{2m}\frac{d^2}{dx^2}\psi(x) = (E - V_0)\psi(x). \tag{16.3}$$

We are going to look at two different cases. The first is when the total energy E exceeds the potential energy V_0, so that $(E - V_0)$ is *positive*. The total energy is equal to the sum of the kinetic and potential energies, and since in classical mechanics the kinetic energy cannot be negative, it follows that the potential energy cannot exceed the total energy. So this first case where $E > V_0$ corresponds to classical mechanics.

The second case we want to look at is when V_0 exceeds E, so that $(E - V_0)$ is *negative*. This is simply not possible in classical mechanics, but it is allowed in quantum mechanics.

16.2.1 The classical case: $E > V_0$

We are simply going to guess a solution to Eq. 16.3 on the preceding page, substitute it into the left- and right-hand sides of this equation, and see if it is indeed a solution. Mathematicians have shown that, for these kinds of equations, a good guess for $\psi(x)$ is

$$\psi(x) = A \cos(kx) + B \sin(kx), \tag{16.4}$$

where A, B and k are constants.

Starting with the left-hand side of Eq. 16.3, we are going to need the second derivative of $\psi(x)$, so let us calculate this first:

$$\frac{d}{dx}[A \cos(kx) + B \sin(kx)] = -kA \sin(kx) + kB \cos(kx)$$

$$\frac{d}{dx}[-kA \sin(kx) + kB \cos(kx)] = -k^2 A \cos(kx) - k^2 B \sin(kx)$$

$$\text{hence} \quad \frac{d^2}{dx^2}[A \cos(kx) + B \sin(kx)] = -k^2 [A \cos(kx) + B \sin(kx)]$$

$$\text{hence} \quad \frac{d^2}{dx^2} \psi(x) = -k^2 \psi(x). \tag{16.5}$$

You can now see why $\psi(x) = A \cos(kx) + B \sin(kx)$ is a good choice, since its second derivative is equal to itself times a constant, which is precisely what Eq. 16.3 requires.

Now that we have an expression for the second derivative, we can evaluate the left-hand side of Eq. 16.3 as

$$\text{LHS} \quad -\frac{\hbar^2}{2m}\frac{d^2}{dx^2} \psi(x) = -\frac{\hbar^2}{2m}(-k^2)\psi(x).$$

The right-hand side is

$$\text{RHS} \quad (E - V_0)\psi(x).$$

$\psi(x)$ cancels between the two sides, and so the LHS and RHS will be equal provided

$$-\frac{\hbar^2}{2m}(-k^2) = (E - V_0).$$

The value of k needed for this to be so is found by rearranging this equation to give

$$k = \sqrt{\frac{2m(E - V_0)}{\hbar^2}}. \tag{16.6}$$

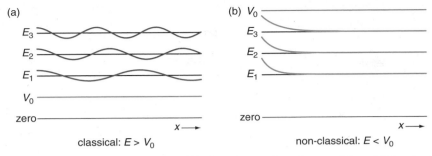

Fig. 16.2 Plots of the eigenfunctions of the hamiltonian for a free particle in (a) the classical, and (b) the non-classical case. In each, three wavefunctions with different energies are shown, where the energies (E_1, E_2 and E_3) are indicated by the horizontal black lines. The potential energy V_0 is shown by the horizontal green line. In the classical case, where the total energy exceeds the potential energy, the wavefunctions (shown in blue) oscillate, with the rate of the oscillation increasing with the energy. In the non-classical case, where the potential energy exceeds the total energy, the wavefunctions (shown in red) decay away. The rate of decay is faster the larger the amount by which V_0 exceeds the total energy. Arbitrarily, it has been assumed that all the wavefunctions start at a positive value on the left-hand side just so their behaviour can be compared. Such a situation is not particularly likely to occur in a physical system.

What we have shown is that $\psi(x) = A\cos(kx) + B\sin(kx)$ is an eigenfunction of the hamiltonian *provided* the constant k is given by Eq. 16.6 on the preceding page. Since $(E - V_0)$ is positive, the calculation of k involves taking the square root of a positive number, so all is well.

The eigenfunction is the sum of a cosine and a sine wave, which oscillate along the x-axis at a rate determined by the constant k. According to Eq. 16.6, the greater the total energy E, the faster the oscillation, as is illustrated in Fig. 16.2 (a). This is a feature we will see over and over again in our solutions to the Schrödinger equation.

Another important feature of the wavefunctions can be seen from Eq. 16.5 on the facing page, which says that as k increases (i.e. as the energy increases) the second derivative of the wavefunction increases. This second derivative is sometimes called the *curvature* of the function as it gives the rate of change of the slope; a function which has a large curvature is therefore one in which the slope changes rapidly. Increasing k results in the cosine and sine functions oscillating more rapidly, which means that their curvature is greater. In general, the more rapidly curving the wavefunction, the higher the energy.

The final point to note here is that the energy is *not* quantized: any value of E is allowed (provided that it exceeds V_0). All that the value of E affects is the size of the constant k.

16.2.2   The non-classical case: $E < V_0$

If the total energy is less than the potential energy, then the problem we are going to have with the trial solution $\psi(x) = A\cos(kx) + B\sin(kx)$ is that since $(E - V_0)$ is negative, the value of k given in Eq. 16.6 on the preceding page will be the square root of a *negative* number. Although we can deal with this by introducing the concept of a complex number, this is not a particularly straightforward path for us to follow: an easier route is to look for a different solution.

For this non-classical case, experience suggest that we should try the solution

$$\psi(x) = C \exp(-k'x). \tag{16.7}$$

The reason for choosing this is that, like sine and cosine, the second derivative of an exponential is itself times a constant. Once we have explored the behaviour of this wavefunction you will see why it is a good choice for the non-classical region.

As before we compute the second derivative of $\psi(x)$ so that we can substitute this into the left-hand side of Eq. 16.3 on page 697.

$$\frac{d}{dx}[C\exp(-k'x)] = -k'C\exp(-k'x)$$

$$\frac{d}{dx}[-k'C\exp(-k'x)] = k'^2C\exp(-k'x)$$

$$\text{hence} \quad \frac{d^2}{dx^2}[C\exp(-k'x)] = k'^2[C\exp(-k'x)]$$

$$\text{hence} \quad \frac{d^2}{dx^2}\psi(x) = k'^2\psi(x).$$

The left-hand side of Eq. 16.3 therefore evaluates to

$$\text{LHS} \quad -\frac{\hbar^2}{2m}\frac{d^2}{dx^2}\psi(x) = -\frac{\hbar^2}{2m}k'^2\,\psi(x).$$

The right-hand side is

$$\text{RHS} \quad (E - V_0)\psi(x).$$

The left- and right-hand sides will be equal provided

$$-\frac{\hbar^2}{2m}k'^2 = (E - V_0).$$

Rearranging this gives k' in terms of the other quantities as

$$k' = \sqrt{\frac{2m(V_0 - E)}{\hbar^2}}. \tag{16.8}$$

With this solution, k' is the square root of a positive number since $(V_0 - E)$ is positive, so all is well. Note that the larger V_0 becomes, the larger k' becomes and hence the more rapidly the eigenfunction $\psi(x) = C\exp(-k'x)$ decays.

The form of the eigenfunctions is very different in the classical and non-classical cases. In the classical case, where the total energy exceeds the potential energy, the eigenfunctions oscillate. The greater the energy, the faster the oscillation. In the non-classical case, where the potential energy exceeds the total energy, the eigenfunctions do not oscillate, but simply decay away. The greater the potential energy, the faster the decay. The contrasting behaviour of the wavefunctions in the two cases is illustrated in Fig. 16.2 on the previous page.

Given that the square of the wavefunction gives the probability density, it follows that the particle has a finite probability of being in a region where the potential energy exceeds the total energy. The particle is said to *penetrate* or *tunnel* into this region. This behaviour is in complete contrast to classical mechanics, where an object simply cannot go into such a region.

This solution is acceptable for positive x as it is a decaying function; for negative x the function increases indefinitely, which is not acceptable behaviour for a wavefunction. For negative x the trial solution ought properly to be written $\psi(x) = C\exp(-k'|x|)$, where $|x|$ is the modulus of x, which is the value of x disregarding the sign. For example, $|-1| = 1$.

It is interesting to consider what happens as the potential approaches infinity. Equation 16.8 on the facing page says that the larger V_0 becomes, the larger k' becomes and hence the more rapidly the wavefunction decays. In the limit that the potential becomes infinite, the wavefunction decays away to zero immediately i.e. the wavefunction is *zero* in a region where the potential is infinite. It follows that there is zero probability of finding the particle in a region with infinite potential energy.

16.3 Particle in a box

The next situation we are going to look at is where the particle, rather than being free to move anywhere along the x-axis, is constrained to a particular region. Such a constraint is created by making the potential energy infinite everywhere outside this region. As has been explained, even a quantum particle cannot go into a region of infinite potential, so the wavefunction is zero in such regions.

Figure 16.3 shows the form of the potential. Everywhere to the left of $x = 0$, and to the right of $x = L$, the potential is infinite; we can thus be sure that the particle cannot go into these regions and so there its wavefunction will be zero. Between $x = 0$ and $x = L$ the potential is finite, and for simplicity we will assume that it is zero. This arrangement is usually described as a *one-dimensional square well* or a *particle in a box*. The latter name is a bit unfortunate as it conjures up a picture of a particle rattling around in box, whereas in fact the particle is only moving along a line.

We already know that the wavefunction is zero outside the region between $x = 0$ and $x = L$, so all we need to concern ourselves with is finding the wavefunction 'inside the box'. Here, since we have decided that $V_0 = 0$, the hamiltonian operator is

$$\hat{H} = -\frac{\hbar^2}{2m}\frac{d^2}{dx^2}.$$

To find the associated eigenfunctions we need to solve the eigenvalue equation for $\hat{H}$ (i.e. the Schrödinger equation)

$$-\frac{\hbar^2}{2m}\frac{d^2}{dx^2}\psi(x) = E\,\psi(x). \tag{16.9}$$

This is very similar that for the free particle, so we can be pretty confident that $\psi(x) = A\cos(kx) + B\sin(kx)$ will be a good guess. We have already computed the second derivative of this function

$$\frac{d^2}{dx^2}[A\cos(kx) + B\sin(kx)] = -k^2[A\cos(kx) + B\sin(kx)]$$

$$\text{hence} \quad \frac{d^2}{dx^2}\psi(x) = -k^2\psi(x),$$

so the left-hand side of Eq. 16.9 is

$$\text{LHS} \quad -\frac{\hbar^2}{2m}\frac{d^2}{dx^2}\psi(x) = -\frac{\hbar^2}{2m}(-k^2)\,\psi(x).$$

The right-hand side is

$$\text{RHS} \quad E\,\psi(x).$$

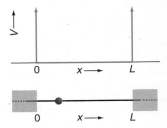

Fig. 16.3 Illustration of the arrangement for the 'particle in a box' problem, in which the particle is constrained to move between $x = 0$ and $x = L$ by making the potential infinite outside this region. The green line on the graph above shows the potential, with the arrows representing the jump to infinity. Beneath is shown another representation in which the regions of infinite potential are shown by the green shading. For simplicity we assume that the potential is zero in the region between $x = 0$ and $x = L$.

The left- and right-hand sides will therefore be equal when

$$-\frac{\hbar^2}{2m}(-k^2) = E.$$

Tidying this up we can express E in terms of k, or *vice versa*

$$E = \frac{\hbar^2 k^2}{2m} \qquad \text{or} \qquad k = \sqrt{\frac{2mE}{\hbar^2}}. \qquad (16.10)$$

In summary, what we have shown is that $\psi(x) = A\cos(kx) + B\sin(kx)$ is an eigenfunction of the hamiltonian (a solution to the Schrödinger equation) provided that the energy E and the constant k are related according to Eq. 16.10. At this stage there is no restriction on the value of the energy E: the system does not show quantization of energy.

16.3.1 Boundary conditions

Recall that we have already decided that outside the region between $x = 0$ and $x = L$ the wavefunction must be zero. As a result of the probabilistic interpretation of the wavefunction (page 695), it is not permitted for the wavefunction to jump from one value to another. Therefore, as $\psi(x) = 0$ to the right of $x = L$, it must be the case that the wavefunction is zero *at* $x = L$. Likewise, the fact that $\psi(x) = 0$ to the left of $x = 0$, means that to avoid a jump in the wavefunction it must be zero *at* $x = 0$. The probabilistic interpretation thus imposes two *boundary conditions*

$$\text{boundary conditions} \qquad \psi(0) = 0 \qquad \psi(L) = 0.$$

We will now show that applying these boundary conditions results in significant restrictions on the wavefunctions, and the appearance of the quantization of energy.

We showed that $\psi(x) = A\cos(kx) + B\sin(kx)$ is an eigenfunction of the hamiltonian, but we now need to make this function compatible with the boundary conditions. First, consider the wavefunction at $x = 0$

$$\begin{aligned} \psi(0) &= A\cos(k \times 0) + B\sin(k \times 0) \\ &= A \times 1 + B \times 0 \\ &= A. \end{aligned}$$

To go to the second line we have used $\cos(0) = 1$ and $\sin(0) = 0$. In order to satisfy the boundary condition $\psi(0) = 0$ the constant A *must be zero*.

Now consider the other boundary condition which relates to the wavefunction at $x = L$; recalling that $A = 0$ we have

$$\psi(L) = B\sin(kL).$$

▷ The key properties of sine and cosine are reviewed in section 20.3 on page 890.

In order to satisfy the boundary condition $\psi(L) = 0$, either B must be zero *or* $\sin(kL)$ must be zero. We reject $B = 0$ as this would make the wavefunction zero everywhere, which is not acceptable since this would give zero probability of finding the particle both inside and outside the box. It must therefore be that $\sin(kL) = 0$, which is the case when (kL) is an integer multiple of π

$$kL = n\pi \qquad n = 0, \pm 1, \pm 2, \ldots$$

Rearranging this gives $k = n\pi/L$, so the wavefunctions $\psi(x) = B\sin(kx)$ become

$$\psi(x) = B\sin\left(\frac{n\pi x}{L}\right).$$

Substituting $k = n\pi/L$ into Eq. 16.10 on the preceding page we find that the energy is

$$E = \frac{\hbar^2 n^2 \pi^2}{2mL^2} \qquad n = 0, \pm 1, \pm 2, \ldots$$

Now we have quantization of energy, since E can only take the discrete set of values which correspond to integer n.

Before carrying on, we need to be a little careful about the values of n we allow. Firstly, if $n = 0$ the wavefunction becomes $\sin(0 \times \pi x/L) = 0$; this is not acceptable since this means that the wavefunction is zero everywhere.

For $n = +1$ and $n = -1$ the wavefunctions are

$$B\sin\left(\frac{\pi x}{L}\right) \qquad \text{and} \qquad B\sin\left(\frac{-\pi x}{L}\right).$$

Given that $\sin(-\theta) \equiv -\sin(\theta)$, the second eigenfunction is just -1 times the first. As we have already noted, simply multiplying an eigenfunction by a constant does not give a distinct eigenfunction. We therefore reject all of the negative integer values of n as not giving distinct eigenfunctions from the corresponding positive values.

In summary, the eigenfunctions and associated energies (eigenvalues) are

$$\psi_n(x) = B\sin\left(\frac{n\pi x}{L}\right) \qquad E_n = \frac{\hbar^2 n^2 \pi^2}{2mL^2} \qquad n = 1, 2, \ldots \qquad (16.11)$$

The subscript n has been added to the eigenfunction and the energy to indicate that they depend on its value. The integer n is called a *quantum number*. The final point to note is that as a result of imposing the boundary conditions we have generated quantized energy levels.

16.3.2 Properties of the wavefunctions

Figure 16.4 on the following page shows two different representations of the four lowest energy eigenfunctions for this particle in a box problem. From looking at these plots it is immediately clear that each wavefunction consists of a whole number of *half* sine waves: $\psi_1(x)$ is one half sine wave, $\psi_2(x)$ is two, $\psi_3(x)$ is three, and so on. This property is simply a consequence of the boundary conditions, which require the wavefunction to go to zero at $x = 0$ and $x = L$. This can only be achieved by fitting whole numbers of half sine waves into the range $x = 0$ to L.

As a consequence of the fact that the number of half sine waves increases as n increases, the number of *nodes*, that is points at which the wavefunction goes to zero, also increases with n. It is not usual to count the boundary conditions at $x = 0$ and $x = L$ as nodes, even though the wavefunction goes to zero at these points. So, the ground state wavefunction $\psi_1(x)$ has no nodes, $\psi_2(x)$ has one node at $x = L/2$, and $\psi_3(x)$ has two nodes at $L/3$ and $2L/3$. It is easy to see that in general the number of nodes is $(n-1)$.

🌐 *Weblink 16.1*

This link takes you to a real-time version of Fig. 16.4 in which you can explore how the energy levels and wavefunctions are affected by changing the width of the box and the mass of the particle.

Fig. 16.4 Plots of the wavefunctions which are solutions to the Schrödinger equation for a particle in a one-dimensional square well (the 'particle in a box'). The four lowest energy wavefunctions with $n = 1$ to $n = 4$ are shown. On the left, the wavefunction is plotted against the variable x in the usual way. On the right, a different representation is used in which the intensity of the shading is proportional to the value of the wavefunction, with red being used for positive values, and blue for negative values. Note that outside the range $x = 0$ to $x = L$ the wavefunction is zero.

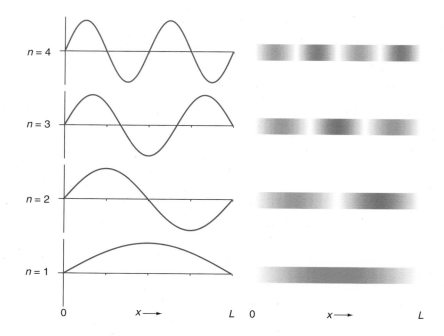

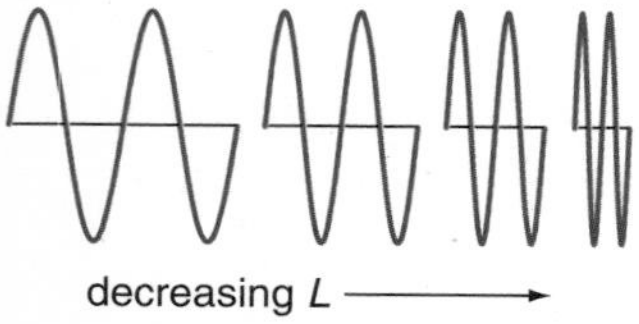

Fig. 16.5 Decreasing the value of L, that is the region over which the particle can move, results in the wavefunction oscillating back and forth more rapidly, as illustrated here for ψ_4. As a result the curvature increases, and so does the energy.

A consequence of the increase in the number of nodes is that the wavefunction has to oscillate back and forth more times, which we can describe as the wavefunction being 'more curved'. Therefore, as n increases the energy increases, the wavefunction oscillates more rapidly and is therefore more curved. This connection between increased curvature and higher energy was noted before on page 699.

From the expression for the energy given in Eq. 16.11 on the previous page we can see that the energy increases as the length L, to which the particle is confined, decreases. To fit the same number of sine waves into a smaller length will require them to be compressed, and so they will oscillate more rapidly, as illustrated in Fig. 16.5. As we have already noted, the resulting increase in curvature is associated with the increase in energy.

In summary, the higher energy the wavefunction, the greater the number of nodes it has and the more rapidly it oscillates. It turns out that all the wavefunctions we will encounter have these same properties, and in what follows we will often make use of them.

16.3.3 The probability density

As we have already described, for this one-dimensional wavefunction, $[\psi(x)]^2 \, dx$ gives the probability of the particle being in a small length dx at position x. Figure 16.6 on the next page shows plots of the squares of the wavefunctions shown in Fig. 16.4. The probability density is also shown in the form of a shaded strip, with darker shading representing higher values. In both representations we can see how the probability density has maxima and minima, with the latter corresponding to the nodes in the wavefunctions.

What behaviour would we expect for a 'classical' particle, that is one obeying Newton's Laws, confined to move along the x-axis between 0 and L?

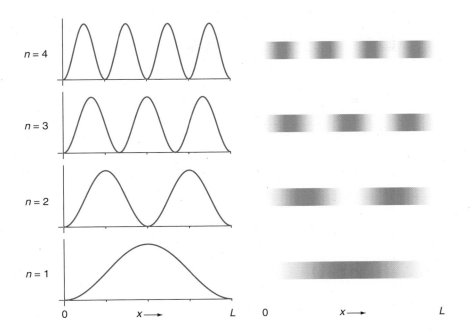

Fig. 16.6 Plots of the squares of the wavefunctions, that is the probability densities, which were shown in Fig. 16.4 on the facing page. On the left, $[\psi(x)]^2$ is plotted against the variable x in the usual way. On the right, a different representation is used in which the intensity of the shading is proportional to the value of $[\psi(x)]^2$.

The first thing to realize is that, as the potential energy is zero, all of the energy which the particle has will be kinetic, so the particle will be moving. When the particle reaches the area of infinite potential energy on the left or right it will 'bounce off the wall' and reverse direction, without loss of energy (an elastic collision). As a result, the particle will be moving back and forth at a constant speed between $x = 0$ and $x = L$. The probability of finding the particle at any point is thus the same i.e. the probability density is constant.

The quantum case is quite different. For the ground state, the probability density is at a maximum at $x = L/2$, and goes to zero at $x = 0$ and $x = L$. For $\psi_2(x)$ there are two maxima at $x = L/4$ and $x = 3L/4$, with the probability density going to zero at $x = 0$, $L/2$ and L. This kind of behaviour of the probability is a feature of the quantum mechanical solutions, and is quite distinct from the expectations of classical mechanics.

When the energy becomes very high, that is when n becomes very large, the probability density will have very many maxima and minima between $x = 0$ and L. Such behaviour becomes more and more like the flat probability density which we expect for a classical particle. This is an example of the *correspondence principle*, which says that at high energies the quantum result must resemble the classical result.

16.3.4 Normalization

As was described in section 16.1.4 on page 694, the sum the square of the wavefunction over all space should come to one, since this is the total probability of finding the particle. In the case of the particle in a box, the wavefunction is zero outside the box, so the total probability is found by integrating $[\psi_n(x)]^2$ between $x = 0$ and $x = L$

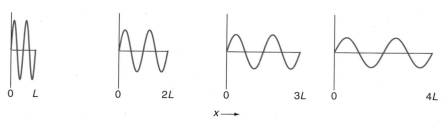

Fig. 16.7 Illustration of how a *normalized* wavefunction (here $\psi_4(x)$) is affected by increasing the width of the region occupied by the particle. The total probability, which is the area under the square of the wavefunction, is one so that as the width of the region occupied increases, the peak amplitude of the wavefunction has to decrease.

$$\int_0^L [\psi_n(x)]^2 \, dx = \int_0^L B^2 \sin^2\left(\frac{n\pi x}{L}\right) dx.$$

The integral is solved by using the identity $\sin^2\theta \equiv \frac{1}{2}[1 - \cos(2\theta)]$ (see section 20.3.4 on page 892)

$$
\begin{aligned}
\int_0^L B^2 \sin^2\left(\frac{n\pi x}{L}\right) dx &= \tfrac{1}{2}B^2 \int_0^L \left[1 - \cos\left(\frac{2n\pi x}{L}\right)\right] dx \\
&= \tfrac{1}{2}B^2 \int_0^L 1 \, dx - \tfrac{1}{2}B^2 \int_0^L \cos\left(\frac{2n\pi x}{L}\right) dx \\
&= \tfrac{1}{2}B^2 \, [x]_0^L - \tfrac{1}{2}B^2 \frac{L}{2n\pi}\left[\sin\left(\frac{2n\pi x}{L}\right)\right]_0^L \\
&= \tfrac{1}{2}B^2 L + 0.
\end{aligned}
$$

To go to the last line we have used the fact that $\sin(2n\pi)$ is zero for any integer value of n.

For the wavefunction to be normalized, we want this integral to be one, which means that

$$\tfrac{1}{2}B^2 L = 1 \qquad \text{or} \qquad B = \sqrt{\frac{2}{L}}.$$

The normalized wavefunctions are therefore

$$\psi_n(x) = \sqrt{\frac{2}{L}} \sin\left(\frac{n\pi x}{L}\right) \qquad E_n = \frac{\hbar^2 n^2 \pi^2}{2mL^2} \qquad n = 1, 2, \ldots \tag{16.12}$$

It is interesting to consider what happens to the wavefunction when the region over which we allow the particle to move is increased i.e. as L is increased. From Eq. 16.12 it is clear that the normalization constant, $\sqrt{2/L}$, decreases as L increases. This is a consequence of the fact that the total area under the square of wavefunction must be one, so as the wavefunction spreads out over a larger range, its peak value must decrease. Figure 16.7 illustrates this idea.

16.3.5 Orthogonality

In quantum mechanics, two wavefunctions are said to be *orthogonal* if the integral of their product over all space is zero. In the case of the particle in a box, the relevant integral is

$$\int_0^L \psi_n(x)\,\psi_m(x)\,\mathrm{d}x.$$

Note that we only need to integrate between $x = 0$ and $x = L$ since we know that the wavefunctions are zero outside this range.

The integral can be worked out by applying the identity $\sin A \sin B \equiv \frac{1}{2}[\cos(A - B) - \cos(A + B)]$ in the following way

▷ These trigonometric identities are discussed in section 20.3.4 on page 892.

$$
\begin{aligned}
\int_0^L \psi_n(x)\,\psi_m(x)\,\mathrm{d}x
&= \frac{2}{L}\int_0^L \sin\left(\frac{n\pi x}{L}\right)\sin\left(\frac{m\pi x}{L}\right)\mathrm{d}x \\
&= \frac{1}{L}\int_0^L \cos\left(\frac{(n-m)\pi x}{L}\right)\mathrm{d}x - \frac{1}{L}\int_0^L \cos\left(\frac{(n+m)\pi x}{L}\right)\mathrm{d}x \\
&= \frac{1}{L}\frac{L}{(n-m)\pi x}\left[\sin\left(\frac{(n-m)\pi x}{L}\right)\right]_0^L - \frac{1}{L}\frac{L}{(n+m)\pi x}\left[\sin\left(\frac{(n+m)\pi x}{L}\right)\right]_0^L \\
&= 0 + 0.
\end{aligned}
$$

The last line follows because $(n \pm m)$ is an integer, and because $\sin(p\pi x/L)$ is zero for integer p when $x = L$.

We have therefore shown that any two eigenfunctions of the particle in a box system are orthogonal to one another, which turns out to be a general property of all the eigenfunctions of any hamiltonian.

16.3.6 Zero-point energy

One of the very unusual features of our quantum mechanical calculations is that the energy of the lowest level, the one with $n = 1$, is *not* zero. In other words, we predict that the particle can never be at rest but must always be moving. This is at complete odds with our experience of the classical world in which there is no problem whatever in a particle having zero energy and so being at rest. This minimum energy that a particle can have is called the *zero-point energy* (ZPE). The existence of such an energy is a particular feature of quantum mechanics.

For the case of the particle in a box, this zero-point energy is E_1, which is $(\hbar^2\pi^2)/(2mL^2)$. We can see from this expression that as the mass increases, or as the length of the region to which the particle is confined increases, the zero-point energy decreases. By the time the mass and the length reach the values associated with everyday objects, the zero-point energy becomes utterly negligible – in other words, it can essentially be taken as zero. However, for very light objects, such as electrons, moving in tiny regions around an atom, the zero-point energy is significant when compared to the absolute energy of the particle.

Uncertainty

The zero-point energy has another rather unusual consequence. As the energy is never zero, the particle is never at rest but is always moving back and forth. Therefore, even in its lowest energy state we cannot say exactly where the particle is. In fact, as we have seen, it is $[\psi_1(x)]^2$ which gives the probability density of this particle, and this is clearly not localized to a particular point but is spread between $x = 0$ and L.

In quantum mechanics this situation is described by saying that there is an *uncertainty* in the position of the particle. In classical mechanics, there is no such uncertainty: if the particle has zero energy, it will not be moving and so we can specify its position exactly.

As the region over which the particle is confined becomes larger (i.e. as L increases), the zero-point energy decreases, but the uncertainty in the position will increase since the particle can be anywhere between $x = 0$ and L. These two quantities – the energy and the uncertainty in the position– seem to have an inverse relationship with one another: as one increases the other decreases.

This behaviour is related to the famous *Heisenberg Uncertainty Principle*, which you may have read about in popular accounts of quantum mechanics. This principle is a very fundamental part of quantum mechanics, and it is no coincidence that our solution to the Schrödinger equation has yielded a result which is consistent with this principle.

16.3.7 Applications of the particle in a box energy levels

The main reason for looking at the particle in a box energy levels is that they are the simplest to derive, and also that they make it clear that it is the boundary conditions that give rise to quantization. However, these energy levels can be used to model, at a simple level, the behaviour of two physical systems of interest: the translational energy levels of a gas and the behaviour of electrons in a metal.

Translational energy levels

If we imagine a gas confined to a container, and assume that the molecules do not interact significantly, then the behaviour of each molecule can be modelled by assuming that it is a particle in a box. The walls of the container are essentially impenetrable, and so represent the infinite potential barriers which confine the particle, but between these walls the molecule is free to move without hindrance. This is the same arrangement as our particle in a box, with the exception that the gas can move in three dimensions, rather than one. For the moment we will set this difference aside (it turns out not to be particularly significant) and simply assume that the molecule moves in one dimension.

The energy levels for a particle in the box given in Eq. 16.12 on page 706 are therefore the energy levels available to each molecule in the container. These energy levels are called the *translational energy levels* since they are associated with the motion (translation) of the molecules within the box.

Let us take a specific example of nitrogen gas confined to a container of dimension 10 cm so that we can get a feel for the energies of these levels. We can work out an expression for the energy of the nth level using Eq. 16.12 by substituting in for the length and the mass. We need to work in SI, so the length of 10 cm must be converted to m, $L = 0.1$ m, and the mass must be in kg. Recalling that one atomic mass unit is 1.6605×10^{-27} kg, the mass of N_2 is $m = 28 \times 1.6605 \times 10^{-27} = 4.65 \times 10^{-26}$ kg. Using $\hbar = 1.055 \times 10^{-34}$ J s, the energy levels are thus

$$
\begin{aligned}
E_n &= \frac{\hbar^2 n^2 \pi^2}{2mL^2} \\
&= \frac{(1.055 \times 10^{-34})^2 \times n^2 \pi^2}{2 \times 4.65 \times 10^{-26} \times (0.1)^2} \\
&= 1.18 \times 10^{-40} \times n^2 \text{ J}.
\end{aligned}
$$

Even on a molecular scale, and even if n is very large, this is a tiny amount of energy.

In section 1.10 on page 26 we introduced the Boltzmann distribution which can be used to predict the population of a particular energy level i.e. how many molecules from a bulk sample will be in that energy level. The key idea that arises from the Boltzmann distribution is that the only levels with significant populations are those with energies less than or comparable to $k_B T$, where k_B is Boltzmann's constant and T is the temperature.

Given that we have already found that the translational energy levels for N_2 in a 10 cm box have energies $1.18 \times 10^{-40} \times n^2$, we can compute the quantum number of the level which has energy equal to $k_B T$; this will give us an idea of how many energy levels are occupied. At 298 K $k_B T = 1.38 \times 10^{-23} \times 298 = 4.11 \times 10^{-21}$ J. The quantum number n_{max} of the level with this energy can be found in the following way

$$
\begin{aligned}
E_{n_{max}} &= k_B T \\
1.18 \times 10^{-40} \times n_{max}^2 &= 4.11 \times 10^{-21} \\
\text{hence } \quad n_{max} &= 5.9 \times 10^9.
\end{aligned}
$$

Therefore for this sample of N_2 gas there are around 6×10^9 translational energy levels which are occupied at 298 K – a very large number indeed. In fact, the number of occupied translational energy levels far far outweighs any other kind of molecular energy level.

In section 6.3.1 on page 206 we described how entropy was related to the way in which the molecules occupy the available energy levels. Since there are so many translational energy levels which can be populated, it is these levels which make the major contribution to the entropy of a gas. A solid, in which the molecules do not have translational motion, therefore has many fewer energy levels available, and hence much lower entropy.

The energies of these translational levels are proportional to $1/(mL^2)$, so increasing the mass or increasing the size of the container will decrease the energies of the levels. As a result, more will have energy less than or comparable to $k_B T$, and as a result the entropy will increase. This is why increasing the molecular mass or expanding the gas increases the entropy.

Electrons in a metal

In section 5.1 on page 180 we described how a picture of the bonding in a metal could be developed by constructing crystal orbitals which extend throughout the solid. An alternative approach is to assume that the outer (valence) electrons are free to move throughout the metal, and then to use the particle in a box as a model for the energy levels of these electrons. The details of how this theory is developed and used is beyond the level of this text, but it is something you may well come across in your more advanced studies of solid state chemistry.

Fig. 16.8 Representations of a number of wavefunctions for the case of a particle confined to a plane; the values of the two quantum numbers n_x and n_y are given in each case. As with Fig. 16.4 on page 704, positive parts of the wavefunction are represented in red and negative in blue, with the intensity of the colour being proportional to the size of the wavefunction. The dashed lines show where the wavefunction is zero (other than at the edges of the plots); these are the equivalent in two dimensions of the nodes seen in the one-dimensional wavefunctions of Fig. 16.4.

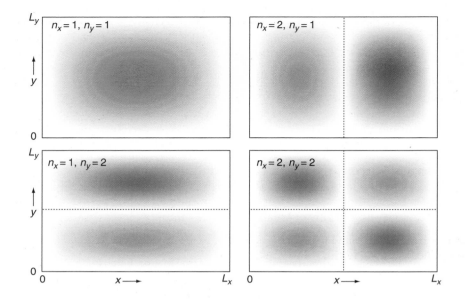

16.4   Particle in a two-dimensional square well

It is relatively easy to extend our treatment of a particle in a box to a two-dimensional case in which we imagine the particle being confined between 0 and L_x along the x-axis, and between 0 and L_y along the y-axis. In other words the particle is confined to a rectangular plane of sides L_x and L_y. The wavefunctions in such a situation are a function of x and y, $\psi(x, y)$. We will not go into the mathematical details of how the Schrödinger equation is solved in such a case, but simply quote the (normalized) wavefunctions and energies

$$\psi_{n_x, n_y}(x, y) = \sqrt{\frac{2}{L_x}} \sin\left(\frac{n_x \pi x}{L_x}\right) \times \sqrt{\frac{2}{L_y}} \sin\left(\frac{n_y \pi y}{L_y}\right)$$

$$E_{n_x, n_y} = \frac{\hbar^2 n_x^2 \pi^2}{2mL_x^2} + \frac{\hbar^2 n_y^2 \pi^2}{2mL_y^2} \qquad n_x = 1, 2, 3, \ldots \qquad n_y = 1, 2, 3, \ldots$$

In this case, there are *two* quantum numbers, n_x and n_y, each of which can be a positive integer. As we had before, both the energy and the form of the wavefunction are determined by the values of these quantum numbers.

It is interesting to compare the two-dimensional wavefunctions and energies with those for the one-dimensional case, Eq. 16.12 on page 706. The two-dimensional wavefunctions are the *product* of a one-dimensional wavefunction which depends on the variable x, the length L_x and the quantum number n_x, and an analogous wavefunction which depends on y, L_y and n_y. In contrast, the energy of the two-dimensional case is the *sum* of two terms which are analogous to those for the one-dimensional case.

Figure 16.8 shows plots of a selection of different two-dimensional wavefunctions. In these plots we are looking down on the xy-plane, and the size of the wavefunction is indicated by shading (red for positive, blue for negative). As a result of the boundary conditions, the wavefunctions go to zero along all four edges of the rectangle.

The wavefunction with $n_x = 1$, $n_y = 1$, $\psi_{1,1}(x, y)$, is positive everywhere, forming a dome. If $n_x = 2$ and $n_y = 1$, $\psi_{2,1}(x, y)$, the wavefunction is zero along the line (shown dashed) which has $x = L/2$ and y any value. This line is the two-dimensional equivalent of the nodes we saw in the one-dimensional wavefunction. Similarly, the wavefunction with $n_x = 1$ and $n_y = 2$, $\psi_{1,2}(x, y)$, has a nodal line along $y = L_y/2$.

The wavefunction with $n_x = 2$, $n_y = 2$, $\psi_{2,2}(x, y)$, has nodal lines along $x = L_x/2$ and along $y = L_y/2$. Just as we saw in the one-dimensional case, the number of nodal lines increases as each quantum number, and hence the energy, increases.

Degeneracy

It is interesting to consider what happens in the two-dimensional case when the dimensions along x and y are equal i.e. $L_x = L_y$. The result is a square, rather than rectangular, plane. In such a case the expression for the energies can be simplified: putting both L_x and L_y equal to L we have

$$
\begin{aligned}
E_{n_x, n_y} &= \frac{n_x^2 \hbar^2 \pi^2}{2mL^2} + \frac{n_y^2 \hbar^2 \pi^2}{2mL^2} \\
&= \frac{\hbar^2 \pi^2}{2mL^2} \left(n_x^2 + n_y^2 \right).
\end{aligned}
$$

Now consider two wavefunctions, one with $n_x = 1$ $n_y = 2$, and one with $n_x = 2$ $n_y = 1$. The energy of these wavefunctions are

$$
E_{1,2} = \frac{\hbar^2 \pi^2}{2mL^2} \left(1^2 + 2^2 \right) = 5 \frac{\hbar^2 \pi^2}{2mL^2}
$$

$$
E_{2,1} = \frac{\hbar^2 \pi^2}{2mL^2} \left(2^2 + 1^2 \right) = 5 \frac{\hbar^2 \pi^2}{2mL^2}.
$$

What we see is that these two *distinct wavefunctions*, with different quantum numbers, have the *same* energy; they are said to be *degenerate*.

This degeneracy is associated with the fact that the two sides of the plane have the same length. In quantum mechanics it turns out that degeneracy is always associated with symmetry: the higher the symmetry, the greater the degeneracy. We see this in its most extreme form in the energy levels of the hydrogen atom. The atom is spherical, giving it high symmetry, which results in the wavefunctions showing considerable degeneracy.

16.5 The harmonic oscillator

The second arrangement for which we will solve the Schrödinger equation is the *harmonic oscillator*. In mechanics, a harmonic oscillator is exemplified by the oscillatory motion that a mass hanging from a spring makes when the mass is displaced and then released, as shown in Fig. 16.9. Like a spring, a chemical bond resists being stretched, and so the vibrations of a bond can be modelled, to a rough approximation, as a harmonic oscillator; this is why we are interested in this system. We will see that we can use the energy levels for the harmonic oscillator to understand the form of the IR spectra of simple molecules.

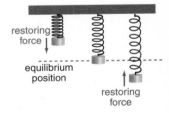

Fig. 16.9 A mass hanging from a spring is an example of a mechanical arrangement in which harmonic oscillations take place. When the mass is displaced there is a restoring force towards the equilibrium position: it is this force which leads to oscillation about the equilibrium position.

The general form of the hamiltonian operator for motion in one dimension is

$$\hat{H} = -\frac{\hbar^2}{2m}\frac{d^2}{dx^2} + V(x),$$

where $V(x)$ is the potential energy, which is a function of the coordinate x. In the case of the particle in a box, $V(x)$ was constant, however this is not going to be the case for the harmonic oscillator. Our first task is to determine the form of $V(x)$, which we do by thinking about a classical harmonic oscillator i.e. one obeying Newton's Laws.

16.5.1   The classical harmonic oscillator

In mechanics harmonic oscillations arise when a mass experiences a force which is: (a) *proportional to the displacement* of the mass from its equilibrium position, and (b) directed towards the equilibrium position. The arrangement is illustrated in Fig. 16.10, where it is assumed that the equilibrium position is at $x = 0$. If the mass is displaced to $x = +a$ and then released, there will be a force which accelerates the mass back *towards* the equilibrium position. The mass will pass through $x = 0$, carry on until it reaches $x = -a$ where it will stop momentarily before accelerating back towards the equilibrium position. In the absence of frictional losses the mass will continue to oscillate back and forth between $x = +a$ and $x = -a$.

Assuming that $x = 0$ is the equilibrium position, the force can be written

$$F(x) = -k_f x,$$

where the minus sign ensures that the force is directed towards the origin, and k_f is the constant of proportion, known as the *force constant*. Since the force constant multiplied by a distance is equal to a force, it follows that the dimensions of k_f must be force per unit length, or N m^{-1} in SI.

In mechanics the potential $V(x)$ and the force $F(x)$ are always related in the following way

$$F(x) = -\frac{dV(x)}{dx}.$$

It therefore follows that $V(x) = \frac{1}{2}k_f x^2$. You can check this by differentiating $V(x)$ with respect to x and then multiplying by -1; this should give $F(x) = -k_f x$, as required. The force and the corresponding potential are shown graphically in Fig. 16.11.

The potential energy function for the harmonic oscillator (HO) is thus

$$V_{HO}(x) = \frac{1}{2}k_f x^2.$$

In fact, an equivalent definition of the harmonic oscillator is a mass experiencing this parabolic potential.

Energy and position during harmonic oscillation

Before we get on with solving the Schrödinger equation for the harmonic oscillator it is useful to think about how a classical harmonic oscillator behaves, so that we can later on compare this to the quantum case.

During the oscillation of the mass about the equilibrium position there is a constant interchange between kinetic and potential energy, as illustrated in

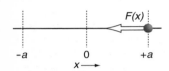

Fig. 16.10 Harmonic oscillation occurs when a mass experiences a force which is proportional to the displacement from the equilibrium position (here $x = 0$) and directed towards that position. If the mass is released at $x = +a$ it will oscillate back and forth between $x = +a$ and $x = -a$.

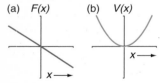

Fig. 16.11 The form of the force, and the corresponding potential, which leads to harmonic oscillations. On the left is shown a graph of the force $F(x) = -k_f x$ which is proportional to the displacement from the origin and is towards that origin. The corresponding potential function $V(x) = \frac{1}{2}k_f x^2$ is shown on the right.

Fig. 16.12. As the particle is released at $x = +a$, the kinetic energy is zero and the potential energy is a maximum. Since $V_{HO}(x) = \frac{1}{2}k_f x^2$, the potential energy at $x = a$, which is the maximum value, is $\frac{1}{2}k_f a^2$. In addition, since the total energy is equal to the sum of the kinetic and potential energies, it follows that the total energy is also $\frac{1}{2}k_f a^2$.

As the particle accelerates towards $x = 0$, the potential energy decreases and the kinetic energy increases, but their sum remains $\frac{1}{2}k_f a^2$. At $x = 0$ the potential energy is zero, so all of the energy is kinetic; this is the maximum in the kinetic energy, so at this point the mass is moving fastest.

As the mass passes beyond $x = 0$ the potential energy increases, and therefore the kinetic energy decreases. At $x = -a$ all of the energy is once again potential, so the mass comes to rest for an instant. After this, the mass accelerates back towards $x = 0$ once more, just as it did when starting from $x = +a$.

Working out the details of classical harmonic motion involves solving a simple second-order differential equation. What we find is that the frequency of the oscillation ω is given by

$$\omega = \sqrt{\frac{k_f}{m}},$$

and that the displacement varies with time according to

$$x(t) = a\cos(\omega t),$$

where a is the displacement at time zero. The frequency ω is in rad s^{-1}, rather than the more familiar units of Hz or s^{-1}; the relation between these units is discussed in section 20.3.6 on page 894.

It follows that the potential and kinetic energies as a function of time, $V(t)$ and $T(t)$, are

$$V(t) = \frac{1}{2}k_f\, a^2 \cos^2(\omega t) \qquad T(t) = \frac{1}{2}k_f\, a^2[1 - \cos^2(\omega t)],$$

where to compute $T(t)$ have have used the fact that the total energy E is $\frac{1}{2}k_f\, a^2$, and that $T(t) = E - V(t)$. The resulting oscillatory interchange of the potential and kinetic energies is illustrated in Fig. 16.13.

16.5.2   Solution of the Schrödinger equation

Now that we know the form of the potential, the hamiltonian operator for the harmonic oscillator can be written explicitly

$$\hat{H}_{HO} = -\frac{\hbar^2}{2m}\frac{d^2}{dx^2} + \frac{1}{2}k_f x^2.$$

To find the associated eigenfunctions and eigenvalues we need to solve $\hat{H}_{HO}\psi(x) = E\psi(x)$, which is

$$-\frac{\hbar^2}{2m}\frac{d^2}{dx^2}\psi(x) + \frac{1}{2}k_f x^2 \times \psi(x) = E\psi(x).$$

You would be right in thinking that this is beginning to look rather complicated, and in fact finding the eigenfunctions $\psi(x)$ which solve this differential equation requires techniques which are well beyond the level of this book. We will therefore simply state the solutions and then go on to see how we can at least make sense of these given what we know about the solutions to the simpler case of the particle in a box.

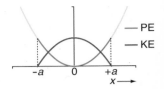

Fig. 16.12 When the mass is released at $x = +a$ its potential energy (PE) is a maximum and its kinetic energy (KE) is zero. As the mass moves towards $x = 0$, the PE decreases and the KE increases, until at $x = 0$ the PE is zero and the KE is a maximum (equal to the original PE). After this, the PE starts to increase and the KE decrease, until at $x = -a$ the PE is once more a maximum and the KE is zero.

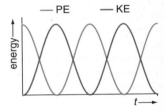

Fig. 16.13 During harmonic motion there is an oscillatory exchange of the potential and kinetic energies, but at all times the sum of these two is constant. Note that when the PE is a maximum, the KE is zero, and *vice versa.*

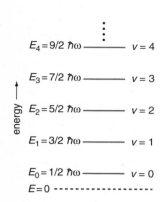

Fig. 16.14 Illustration of the energy levels of the harmonic oscillator. The lowest level has energy $\frac{1}{2}\hbar\omega$, and subsequent levels are spaced by $\hbar\omega$. The (angular) frequency ω is given by $\sqrt{k_f/m}$, as in a classical oscillator.

The energy levels

The eigenvalues, which are the energy levels, turn out to be quantized and take a rather simple form

$$E_v = \left(v + \tfrac{1}{2}\right)\hbar\omega \qquad \omega = \sqrt{\frac{k_f}{m}} \qquad v = 0,\ 1,\ 2,\ \dots. \qquad (16.13)$$

These energy levels are specified by the vibrational quantum number v which takes values 0, 1, 2, …. The energies depend on the (angular) frequency ω, which we recognize as being the same frequency that occurs in the classical harmonic oscillator.

Figure 16.14 shows the energies of the first few levels; note that the spacing between adjacent levels is constant at $\hbar\omega$, and the energy of the lowest level is not zero, but $\frac{1}{2}\hbar\omega$. Like the particle in the box, the harmonic oscillator shows zero-point energy i.e. the energy of the lowest level is not zero. However, in contrast to the particle in a box, for which the energies of the levels vary with the *square* of the quantum number, for the harmonic oscillator the energies are just *linear* in the quantum number. This difference in behaviour arises from the different form of the potential energy in the two cases.

The wavefunctions

For the particle in a box, we were at pains to point out that it was the constraints which the potential places on the wavefunction that give rise to quantized energy levels. In the case of the harmonic oscillator the constraints are not so dramatic, as the potential rises steadily rather than suddenly jumping to infinity. Nevertheless, the harmonic potential does put significant constraints on the form of the wavefunction.

We can understand something about these constraints by thinking about the way a wavefunction must behave. The wavefunction with quantum number v has total energy $(v + \frac{1}{2})\hbar\omega$ which can be separated into a kinetic part $T(x)$ and a potential part $V(x)$. Since $V(x) = \frac{1}{2}k_f x^2$ it follows that

$$\underbrace{(v + \tfrac{1}{2})\hbar\omega}_{\text{total}} = \underbrace{T(x)}_{\text{kinetic}} + \underbrace{\tfrac{1}{2}k_f x^2}_{\text{potential}}. \qquad (16.14)$$

As the displacement x increases from zero, the potential energy increases and so the kinetic energy must decrease, since the sum of these two is equal to the total energy, which is constant. At some point the kinetic energy will be zero, and we can work out where this is by finding the displacement a_v at which $T(x) = 0$ in Eq. 16.14

$$(v + \tfrac{1}{2})\hbar\omega = 0 + \tfrac{1}{2}k_f a_v^2$$

$$\text{hence} \quad a_v = \pm\sqrt{\frac{2(v + \tfrac{1}{2})\hbar\omega}{k_f}}. \qquad (16.15)$$

In a classical harmonic oscillator, the particle cannot go beyond $\pm a_v$, since to do so would result in a negative kinetic energy. However, in the quantum oscillator, no such restriction applies.

There are thus two distinct regions for the quantum oscillator. Between $x = -a_v$ and $x = +a_v$ the potential energy is less than the total energy and so the

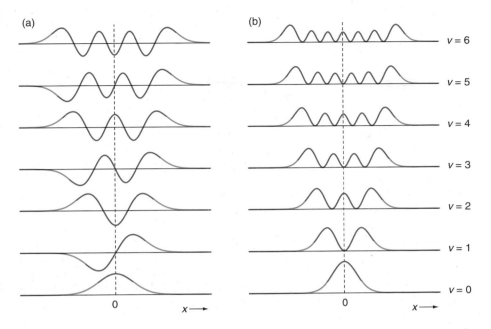

(a)

(b)

$v = 6$

$v = 5$

$v = 4$

$v = 3$

$v = 2$

$v = 1$

$v = 0$

0

$x \longrightarrow$

0

$x \longrightarrow$

Fig. 16.15 Plots of: (a) the seven lowest energy wavefunctions of the harmonic oscillator, and (b) the squares of these wavefunctions; the values of the quantum number v is shown on the right. In the classical region, where the total energy is greater than the potential energy, the wavefunction is plotted in blue; in the non-classical region, where the potential energy exceeds the total energy, the wavefunction is plotted in red. Note that as the energy increases (increasing v) the number of nodes increases, and that these nodes always occur in the classical region. In the non-classical region the wavefunction simply decays away.

kinetic energy is positive: this is called the *classical region*. As we have discussed in section 16.2.1 on page 698, in such a region we expect the wavefunction to oscillate and, as a result, for it to have nodes. The higher the energy, the faster the oscillation, and hence the greater the number of nodes. For the case of a free particle, we found that the wavefunction in such a region was of the form of a sine or cosine. In the case of the harmonic oscillator the wavefunction will be a more complicated function since the potential is not constant, but increases as x increases.

The second region is for x greater than $+a_v$, or less than $-a_v$: here the potential energy exceeds the total energy. This is called the *non-classical region* since a classical particle cannot go into this region. As we have discussed in section 16.2.2 on page 699, in a region where the potential exceeds the total energy we expect that the wavefunction will not oscillate, but will decay away towards zero. The greater the potential becomes, the faster the decay. Since the harmonic potential increases as $|x|$ increases, the wavefunction will decay faster and faster as $|x|$ increases.

In summary, our expectations are that in the classical region the wavefunction will oscillate and have nodes, but in the non-classical region the wavefunction will simply decay to zero. The nature of the oscillations and the rate of the decay depends on the total energy and the exact form of the potential (i.e. the value of k_f). In addition to all this, the wavefunction has to vary smoothly without sudden jumps. Taken together these constraints are sufficient to result in the harmonic oscillator having quantized energy levels.

Figure 16.15 (a) shows plots of the first few wavefunctions for the harmonic oscillator. In the classical region, the wavefunction is shown in blue, and in the non-classical region, it is shown in red. As predicted by Eq. 16.15 on the facing page, the value of a_v, and hence the width of the classical region, increases as the quantum number v increases. The form of these wavefunctions fits with

$|x|$ means the modulus of x, i.e. the value regardless of the sign. For example $|-1| = 1$.

our expectations. In the classical region the wavefunction shows oscillatory behaviour, with the number of nodes increasing as the energy increases. If fact, you can see that the number of nodes is equal to the quantum number v. In the non-classical region (plotted in red), the wavefunction simply decays away to zero: there are no oscillations and no nodes in this region.

The squares of the wavefunctions are plotted in Fig. 16.15 (b). Recall that $\psi^2(x)\,dx$ gives the probability of finding the particle in a small element of length dx at position x. For the lowest energy wavefunction ($v = 0$), the probability is greatest in the centre i.e. at the equilibrium separation. As the energy increases, the probability shows more and more maxima, but the principal maxima are increasingly located at the edges of the classical region.

It is interesting to compare this behaviour with that of a classical harmonic oscillator. As we discussed above, the particle is moving most slowly at the extreme edges of its range, and most quickly as it passes though the equilibrium position. As a result, it is at these edges of the range that we are most likely to find the particle, and we are least likely to find it at the equilibrium position.

This behaviour is in complete contrast to the quantum case for the lowest energy level ($v = 0$), where the greatest probability is at the equilibrium position. However, as the energy increases, the behaviour of the quantum harmonic oscillator begins to resemble the classical oscillator in that the most probable positions move towards the edges. This is another example of the correspondence principle (page 705), which requires that the behaviour of the quantum state becomes classical at high energies.

The final point to note is that the square of the wavefunction is always symmetrical about the equilibrium position $x = 0$ i.e. the probability density at $x = +a$ is exactly the same as at $x = -a$. This is hardly a surprise, since the potential is symmetric about $x = 0$ and so we therefore expect the oscillation to be symmetric about this point. A consequence of this symmetry is that the average position for the oscillating mass is the equilibrium position $x = 0$, since there is equal probability of being at positive and negative values of x.

The wavefunctions themselves behave slightly differently: those with even v (0, 2, 4 ...) are symmetric about $x = 0$, meaning that $\psi_v(x) = \psi_v(-x)$. In contrast, those with those with odd v (1, 3, 5 ...) are antisymmetric about $x = 0$, meaning that $\psi_v(x) = -\psi_v(-x)$.

Representing the energy levels and wavefunctions

Weblink 16.2

This link takes you to a real-time version of Fig. 16.16 in which you can explore how the energy levels and wavefunctions are affected by changing the force constant and the mass.

Figure 16.16 on the next page illustrates a particular way of representing the energy levels, the potential energy function and the wavefunctions which emphasises the relationship between these three things. In (a) the green line is the harmonic potential $V_{HO}(x) = \frac{1}{2}k_f x^2$ plotted as a function of x. On this plot, the energies of the quantum levels are represented by the horizontal black lines, and these are labelled with the value of the quantum number v; the same scale is used for the energy levels and the potential.

The value of x where the line for a given energy level crosses the potential energy curve is the point at which the potential energy is equal to the total energy, resulting in zero kinetic energy; the values a_v at which these crossings occur are given by Eq. 16.15 on page 714. These crossing points separate the classical from the non-classical regions, and the diagram shows clearly how the width of the classical region increases as v, and hence the energy, increases.

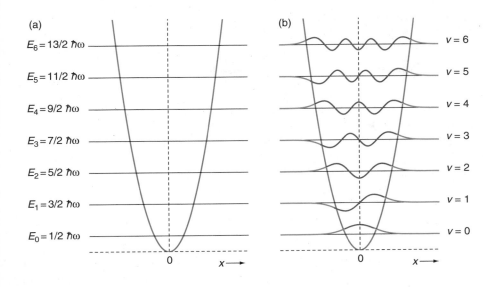

Fig. 16.16 In (a) the potential energy function of the harmonic oscillator $V(x) = \frac{1}{2}k_f x^2$ is plotted in green; imposed on this, and plotted on the *same* vertical scale, are horizontal lines showing the energies of the seven lowest levels. In (b) the wavefunctions corresponding to these energy levels are also plotted, using the line that represents the energy level as the x-axis in each case. The transition from the classical to the non-classical region is where a particular horizontal line crosses the potential energy curve.

A further elaboration in shown in Fig. 16.16 (b). Here the wavefunction corresponding to a particular energy level is plotted using the line which represents the energy level as the x-axis. The nice thing about plotting the wavefunctions in this way is that you can see how the wavefunctions spread out as the width of the potential function increases. In addition, we can see the values of the displacement x at which there is a crossover from the classical to the non-classical regions and how these positions ocur when the potential energy is equal to the total energy.

16.5.3   Mathematical form of the wavefunctions

As has already been mentioned, solving the Schrödinger equation (SE) for the harmonic oscillator involves some mathematics which are well beyond the scope of this text. However, we can get a flavour of what is going on by guessing the form of one of the solutions and then seeing if our guess does indeed solve the SE; this was precisely the approach we used for the particle in a box problem.

The key is to make a good guess in the first place, so it is important to have in mind the attributes we want the wavefunction to have. The ground-state wavefunction must have the following general properties:

(a) It must have no nodes (as in the particle in a box case).

(b) It must decay away smoothly to zero as x becomes large and positive or large and negative.

(c) It must be symmetric (or anti-symmetric) about $x = 0$ on account of the fact that the potential is symmetric about this point.

(d) It must, like all wavefunctions, vary smoothly throughout its range.

Given these constraints, a good choice is the *gaussian* function (see section 20.4.5 on page 900) $\psi(x) = \exp(-Ax^2)$, where A is a constant; this function has all of the properties listed above.

We will not go into the detailed mathematics here, but it can be shown that the particular gaussian function

$$\psi_0(x) = N_0 \exp\left(-\tfrac{1}{2}\beta^2 x^2\right)$$

is a solution to the SE *provided* that $\beta^2 = \sqrt{k_f m / \hbar^2}$; the corresponding energy is $\tfrac{1}{2}\hbar\omega$, so this is the ground state. The reason that this function only solves the SE for a specific value of β^2 is that the value of this constant determines the shape of the wavefunction which has to be just right so that in the non-classical region the rate of decay of the wavefunction matches the increase in the potential energy.

N_0 is a constant, whose value can be found by normalizing the wavefunction, i.e. choosing N_0 such that

$$\int\limits_{-\infty}^{+\infty} N_0^2 \psi_0^2(x)\, \mathrm{d}x = 1.$$

The wavefunction for the next highest energy level presumably has one node, and in addition has to have properties (b)–(d) listed above for the ground state. Our guess for this wavefunction is thus

$$\psi_1(x) = N_1 x \times \exp\left(-\tfrac{1}{2}\beta^2 x^2\right),$$

which has a node at $x = 0$. This does indeed turn out to be the wavefunction with energy $\tfrac{3}{2}\hbar\omega$, provided $\beta^2 = \sqrt{k_f m / \hbar^2}$ as above. We could carry on in this way, but it becomes increasingly difficult to guess at the solutions, and so we should resort to one of the more general and sophisticated ways of finding them.

The general solution

The general form of the wavefunctions for the HO is

$$\psi_v(y) = N_v \times H_v(y) \times \exp\left(-\tfrac{1}{2}y^2\right),$$

where

$$y = \frac{x}{\alpha} \qquad \alpha = \sqrt[4]{\frac{\hbar^2}{mk_f}} \qquad N_v = \sqrt{\frac{1}{2^v v!\,\sqrt{\pi}\,\alpha}}.$$

Table 16.1 Hermite polynomials for $v = 0$ to $v = 4$.

v	$H_v(y)$
0	1
1	$2y$
2	$4y^2 - 2$
3	$8y^3 - 12y$
4	$16y^4 - 48y^2 + 12$

The wavefunctions are written as functions of the variable y, rather than x, but the two variables are related simply via the value of the constant α. N_v is the normalization constant, and $H_v(y)$ is a *Hermite polynomial*. These are polynomials which involve higher and higher powers of the variable y as the quantum number v increases; a few of these polynomials are given in Table 16.1. The third term in $\psi_v(y)$ is the gaussian function $\exp\left(-\tfrac{1}{2}y^2\right)$ we have encountered before.

In general, it is the gaussian term which makes the wavefunctions decay to zero for large positive or negative y. The nodes arise from the Hermite polynomials. For example, $H_1(y)$ is zero when $y = 0$, thus giving one node. Similarly, $H_2(y) = 0$ when $y = \pm\sqrt{1/2}$, giving two symmetrically placed nodes. It is also worth noting that for *even* v the polynomials involve only *even* powers of y, and so are symmetric about $y = 0$. In contrast, for *odd* v the polynomials involve *odd* powers of y, and so are anti-symmetric about $y = 0$.

Let us use this general form of the wavefunction for the case $v = 0$ to find the lowest energy wavefunction. For $v = 0$, $H_0(y) = 1$ so

$$\psi_0(y) = N_0 \exp\left(-\tfrac{1}{2}y^2\right).$$

Given that $y^2 = x^2/\alpha^2$, and $\alpha^2 = \sqrt{(\hbar^2)/(mk_f)}$, it follows that $\psi_0(y)$ can be written in terms of x as follows

$$N_0 \exp\left(-\tfrac{1}{2}y^2\right) = N_0 \exp\left(-\frac{1}{2}\frac{x^2}{\sqrt{(\hbar^2)/(mk_f)}}\right).$$

Tidying this up gives

$$\psi_0(x) = N_0 \exp\left(-\tfrac{1}{2}\sqrt{\frac{mk_f}{\hbar^2}}x^2\right)$$

which is exactly the form of the wavefunction we found above; the constant β^2 we used before is simply the reciprocal of α^2: $\beta^2 = 1/\alpha^2$.

16.6 Spectroscopy and energy levels

In Chapter 11 we introduced the idea that the absorption of electromagnetic radiation (such as light) can cause a molecule to undergo a transition from one energy level to another. The frequency of the radiation which is absorbed depends on the spacing between the energy levels, and since these levels are quantized only photons of very particular energies (frequencies) are absorbed. The result is a *spectrum* in which there are a series of absorption peaks or lines at frequencies which correspond to the spacing of the energy levels. The basic process is illustrated in Fig. 16.17.

In this chapter we will plot spectra as absorption against frequency, so the absorption of photons leads to a peak which points up. In Chapter 11 the IR spectra were plotted as transmission against frequency, which means that an absorption leads to a downward-pointing feature.

Typically a molecule will absorb at a number of different frequencies, thus giving rise to a spectrum in which there are several absorption peaks. The resulting spectra are highly characteristic of the molecule being studied, and it is this property which can be exploited for qualitative structure determination in the way described in Chapter 11.

Now that we are beginning to develop a greater understanding of where the energy levels come from, we are in a position to be more precise about which transitions take place, and how the resulting spectra can be interpreted. The underlying ideas we use to understand spectra are more or less independent of the kind of energy levels involved, so before discussing some particular cases it is helpful to look at the general principles.

Figure 16.18 on the following page shows a typical set of molecular energy levels such as would be obtained by solving the appropriate Schrödinger equation. The levels are labelled with a quantum number p, which takes integer values starting at zero. The lowest energy level, which has $p = 0$, is called the *ground state*; the next highest level ($p = 1$) is sometimes called the *first excited state*, and the next level ($p = 2$) the *second excited state*. The energy of these levels are written E_p.

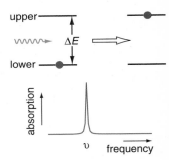

Fig. 16.17 Illustration of the basic process which leads to a spectrum. Absorption of a photon (shown by the red arrow) of the appropriate energy results in a molecule moving from a lower energy level to an upper level. If the photon is to be absorbed its energy must match the separation ΔE of the two levels, which means that the frequency of the radiation v must be such that $\Delta E = hv$. This absorption leads to a peak in the spectrum at frequency v.

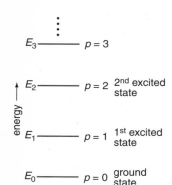

Fig. 16.18 A typical set of molecular energy levels, each of which is labelled with a quantum number p. Transitions between these levels give rise to absorption peaks in the spectrum.

Transitions between these levels can be brought about by the absorption of a photon whose energy matches the energy separation of two of the levels. For example, if the transition is between level p_{lower} and level p_{upper}, the energy separation is

$$E_{p_{upper}} - E_{p_{lower}}.$$

The energy of a photon of frequency υ is $h\upsilon$, where h is Planck's constant, so if the energy of the photon is to match the energy separation of the levels we must have

$$h\upsilon = E_{p_{upper}} - E_{p_{lower}}.$$

The spectrum, which shows the frequencies at which radiation is absorbed, thus reflects the *spacing* between the energy levels. However, it is not necessarily the case that a photon whose energy matches the separation between two levels *will* be absorbed: why this is so, and what determines which photons will be absorbed, are discussed in the next two sections.

16.6.1 Selection rules

Just as quantum mechanics is used to determine the energy levels of a molecule, the same theory can be used to described the interaction of light with a molecule, and hence determine the probability that a photon of the correct energy will be absorbed. This application of quantum mechanics turns out to be rather involved and it is certainly far beyond the level of this text. We will therefore have to content ourselves with describing the results which arise from this theory.

The key idea which arises is that a transition between two particular energy levels will only take place if it is allowed by the appropriate *selection rules*. The application of these selection rules to a given pair of energy levels gives a simple clear-cut result. Either the rules say 'yes', a photon of the correct energy can be absorbed and the transition can take place, or 'no', the transition will not take place. A transition which the selection rules predict can take place is said to be *allowed*; a transition which is predicted not to take place is said to be *forbidden* or *not allowed*.

Selection rules typically come in two types. The first type, sometimes called a *gross selection rule*, refers to some overall property of the molecule. For example, the rule might specify that molecule should possess a permanent dipole moment. The second type of selection rule refers to the quantum numbers of the two energy levels involved. Typically, such a rule would specify that the *difference* in the two quantum numbers should have a particular value. For example, it might be that $p_{upper} - p_{lower}$ has to be 1 (often written $\Delta p = 1$). We will see several examples of both types of selection rules in what follows.

16.6.2 Intensities of spectroscopic transitions

If the transition between levels p_{lower} and p_{upper} is allowed, then a photon of the correct energy can only be absorbed if some of the molecules are actually occupying the lower energy level. If there are no molecules in this level, then there are no molecules available to absorb the photon. The intensity of the transition (i.e. how strongly it appears in the spectrum) thus depends on the *population* of the lower level.

The populations of energy levels are determined by the Boltzmann distribution, discussed in section 1.10.1 on page 27. What we saw there was that it is only levels with energies less than or comparable to $k_B T$, where k_B is Boltzmann's constant, which are populated significantly at temperature T.

Given that the intensity of a transition is proportional to the population of the lower energy level, it therefore follows that in considering whether or not a particular transition is likely to give rise to a measurable absorption of photons, we need to consider the population of the relevant energy levels. This is done by comparing their energies with $k_B T$; we will see several examples of this in due course.

16.6.3   High and low resolution spectroscopy

The spectra of even quite small molecules can be rather complicated, often having many lines closely packed together. Each line has a *width* which is due either to shortcomings in the instrument being used to measure the spectrum, or to some physical process going on in the sample which leads to the energy of a particular level having a spread of values rather than a single, sharply defined value. If we are to disentangle all of the many lines in the spectrum we need to work at the highest possible resolution – meaning that the lines should be as narrow as possible.

Generally speaking such high resolution spectra are only available from gaseous samples in which there are minimal interactions between the molecules. If the sample is a liquid or in solution, the higher density means that there are frequent interactions between the molecules and this leads to broad lines. Under these conditions, all of the closely spaced lines just merge into one another resulting in spectra which consist of broad rather featureless bands. The IR spectra shown in Chapter 11 are typical of such low resolution spectra.

Unless we say otherwise, the spectra referred to in this chapter will be high resolution spectra recorded on gas phase samples.

16.6.4   Other types of transition

So far we have been discussing the process in which the absorption of a photon causes the molecule to move from lower energy level to a higher energy one. This process is known as absorption, or more fully as *stimulated absorption*. It is the most important process when it comes to recording spectra in the microwave and infrared regions, but it is just one of the three processes, illustrated in Fig. 16.19, by which photons interact with molecules.

There are two processes which lead to the emission of photons. The first is called *spontaneous emission*. In this, the molecule drops down from the higher energy level to a lower energy one, and in the process emits a photon whose energy matches the separation between the two levels. Spontaneous emission is only a significant process if it leads to visible or UV photons.

For there to be spontaneous emission the upper energy levels must be populated. In the case of transitions in the visible and UV, these upper levels inevitably have energies very much greater than $k_B T$ above the ground state and so are not populated by thermal motion. However, these levels can be populated by putting in extra energy, for example by passing an electric discharge through a gas. This is precisely how the familiar sodium street lamps work. An electric discharge excites (gaseous) sodium atoms to high energy levels, which then drop down to lower levels, emitting the familiar yellow light.

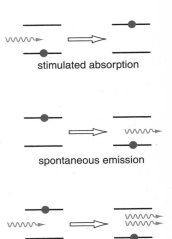

stimulated absorption

spontaneous emission

stimulated emission

Fig. 16.19 In stimulated absorption the absorption of a photon causes the molecule to move to a higher energy level. In spontaneous emission, the molecule drops to a lower energy level and emits a photon whose energy matches the separation of the levels. In stimulated emission, a photon causes the emission of a second photon as the molecule drops down to a lower level.

Stimulated emission is the key process which leads to the generation of *laser light*. In a laser, matters are carefully contrived so that the population of the upper level *exceeds* that of the lower level, making stimulated emission the dominant process. Under the right conditions, this can lead to the intense monochromatic light which is characteristic of a laser.

The second process by which photons can be emitted is called *stimulated emission*. In this process a photon whose energy matches the separation of the two energy levels causes the molecule to drop down from the higher energy level to a lower one, thus emitting a photon. The first photon is not absorbed, so the overall result of the process is an increase in the number of photons.

The rate of stimulated emission depends on the population of the upper energy level. However, due to the Boltzmann distribution this level is less populated than the lower level, so that rate of stimulated absorption is always greater than that of stimulated emission. In effect, this means that for the relatively simple spectra we are interested in we can simply ignore stimulated emission.

16.7 The IR spectrum of a diatomic

The whole point of finding the energy levels of a harmonic oscillator is so that we can use these to help us understand the form of the infrared (IR) spectrum of a diatomic molecule. Typically, it is transitions between *vibrational* energy levels which are responsible for absorptions in the IR part of the spectrum, and it is these energy levels which we can model using a harmonic oscillator.

16.7.1 Vibrational energy levels of a diatomic

In classical mechanics a harmonic oscillator is exemplified by a mass attached to a spring, which is in turn attached to a large immovable object (e.g. the bench). The spring has the property that when it is stretched or compressed there is a force which tries to restore it to its equilibrium length.

The bond between the two atoms in a diatomic behaves, to some degree of approximation, like a spring. The bond has a natural length – the *equilibrium bond length* which you will find quoted in tables of data – which corresponds to the lowest energy arrangement. If the bond is stretched or compressed from this equilibrium position the (potential) energy rises, just as in the harmonic oscillator, and this corresponds to there being a force which attempts to restore the equilibrium separation. Of course, there is no reason to assume that the potential energy function for stretching a bond is the same as that for a harmonic oscillator, but at least for small displacements from equilibrium it is a useful approximation. Later on, we will look at how this approximation fails, and how a better model can be developed.

In the harmonic oscillator we have discussed so far, one end of the spring is attached to an immovable object. This is not quite the right picture for a diatomic in which we have two masses held together by a bond which we are going to treat as a spring, as illustrated in Fig. 16.20. It is however easy to modify the harmonic oscillator energy levels to take account of the fact that there are two masses involved. All we have to do is introduce a quantity known as the *reduced mass* μ defined as

$$\mu = \frac{m_1 m_2}{m_1 + m_2}, \tag{16.16}$$

Fig. 16.20 The vibrational levels of a diatomic can be approximated by the harmonic oscillator energy levels in which the mass is replaced by the reduced mass μ, where $\mu = m_1 m_2/(m_1 + m_2)$. In this model, x is the deviation of the internuclear separation r from the equilibrium bond length r_e: $x = r - r_e$.

where m_1 and m_2 are the masses of the two atoms which form the diatomic.

The energy levels for this modified harmonic oscillator are

$$E_v = \left(v + \tfrac{1}{2}\right)\hbar\omega \qquad \omega = \sqrt{\frac{k_f}{\mu}} \qquad v = 0,\ 1,\ 2,\ \ldots, \qquad (16.17)$$

which are exactly as before (Eq. 16.13 on page 714) but with the mass m replaced by the reduced mass μ. As was discussed on page 713, the harmonic vibrational frequency ω is in units of rad s^{-1}, rather than the more usual Hz or s^{-1}.

16.7.2 Spectrum in the harmonic approximation

There are two selection rules which determine which transitions are allowed between these harmonic oscillator energy levels. The first (the gross selection rule) is that the *dipole moment must change* during the vibration i.e. as the bond length changes. For example, a heteronuclear diatomic such as HCl, ClF or CO obeys this selection rule since it has a permanent dipole moment and we expect that the size of the dipole will change as the separation between the atoms changes, thus altering the details of the electron distribution.

In contrast, a homonuclear diatomic such as H_2 or N_2 has no dipole moment at any separation of the two atoms, and so does not obey the gross selection rule. We therefore expect to see transitions between the vibrational energy levels for heteronuclear diatomics, but not for homonuclear diatomics.

The second selection rule is that the vibrational quantum number v can only change by one. Thus the transition from the ground state with $v = 0$ to the first excited state with $v = 1$ is allowed, but the transition from $v = 0$ to $v = 2$ is not allowed. Similarly, a transition from $v = 1$ to $v = 2$ is allowed, but $v = 1$ to $v = 3$ is not allowed. This selection rule is often written as $\Delta v = 1$, where Δv is the change in the quantum number: $\Delta v = v_{\text{upper}} - v_{\text{lower}}$.

Since it is the lowest level, we can safely assume that the ground state with $v = 0$ will be populated significantly, so for a heteronuclear diatomic we expect to see a transition between $v = 0$ and $v = 1$. The energy change for this transition, and hence the energy of the required photon, can be calculated from Eq. 16.17

$$
\begin{aligned}
E_{0 \to 1} &= E_1 - E_0 \\
&= \tfrac{3}{2}\hbar\omega - \tfrac{1}{2}\hbar\omega \\
&= \hbar\omega.
\end{aligned}
$$

This is the energy of the photon which would need to be absorbed to cause this transition. We can work out the frequency υ of the photon by recalling that $E = h\upsilon$, and also that $\hbar = h/(2\pi)$. Hence the required photon has frequency $\upsilon_{0 \to 1}$ given by

$$
\begin{aligned}
\upsilon_{0 \to 1} &= E_{0 \to 1}/h \\
&= \hbar\omega/h \\
&= \omega/2\pi. \qquad (16.18)
\end{aligned}
$$

Since ω is in rad s^{-1}, $\omega/2\pi$ is a frequency in Hz or s^{-1} (see section 20.3.6 on page 894). The prediction is therefore that there will be an absorption (often simply referred to as a line) in the spectrum at frequency $\omega/(2\pi)$ Hz; in other words the spectrum tells us directly the frequency at which the bond is vibrating, at least to within the harmonic approximation.

Typical values

⇨ More details concerning wavenumbers can be found in section 20.2.3 on page 889.

In the IR spectrum of $^{12}C^{16}O$ there is a strong absorption centred at 2157 cm^{-1}, which we can reasonably assume is due to the $0 \rightarrow 1$ vibrational transition. As was explained in Chapter 11 (Box 11.1 on page 460), it is usual in IR spectroscopy to quote frequencies in wavenumbers (cm^{-1}), a unit which is simply the reciprocal of the wavelength in cm. To convert from wavenumbers to Hz we simply multiply by the speed of light *in cm s^{-1}*

$$\text{frequency in Hz} = \text{wavenumber} \times \text{speed of light in cm s}^{-1}. \tag{16.19}$$

Given that the speed of light is 2.998×10^{10} cm s^{-1}, we can easily work out that 2157 cm^{-1} corresponds to a frequency of 6.467×10^{13} Hz. Using Eq. 16.18 it follows that this frequency is $\omega/(2\pi)$, and hence ω is 4.063×10^{14} rad s^{-1}. From the spectrum we can therefore work out the vibrational frequency of the harmonic oscillator used to model the bond.

From Eq. 16.17 on the preceding page we know that the vibrational frequency is given by $\omega = \sqrt{k_f/\mu}$. It therefore follows that $k_f = \omega^2 \mu$, which we can use to find the value of the force constant. In making the calculation we need to be careful to use SI units for all the quantities, which means that the reduced mass must be in kg.

The reduced mass of $^{12}C^{16}O$ is computed in the following way

$$\begin{aligned}
\mu &= \frac{12.000 \times 15.995}{12.000 + 15.995} \\
&= 6.856 \text{ mass units} \\
&= 6.856 \times 1.6605 \times 10^{-27} \text{ kg} \\
&= 1.138 \times 10^{-26} \text{ kg},
\end{aligned}$$

⇨ More details concerning mass units can be found in section 20.2.3 on page 888.

where to convert from mass units to kg we have used the fact that 1 mass unit is equivalent to 1.6605×10^{-27} kg. Now that we have the reduced mass in kg we can compute the force constant

$$\begin{aligned}
k_f &= \omega^2 \mu \\
&= \left(4.063 \times 10^{14}\right)^2 \times 1.138 \times 10^{-26} \\
&= 1879 \text{ N m}^{-1}.
\end{aligned}$$

We have worked in SI units throughout, so the force constant comes out in N m^{-1}.

It should be emphasized that this value for the force constant has been derived using the assumption that the potential for stretching the bond is harmonic i.e. the potential varies as $\frac{1}{2}k_f x^2$, where x is the extension of the bond from its equilibrium value.

Other transitions

The transition from $v = 1$ to $v = 2$ is also allowed by the selection rule. However, as we shall see, this transition is usually rather weak on account of the very low population of the $v = 1$ level, which is the level from which the transition occurs.

Recalling that the energy of the ground state is $\frac{1}{2}\hbar\omega$, and that of the first excited state is $\frac{3}{2}\hbar\omega$, it follows that the first excited state is $\hbar\omega$ higher in energy than the ground state. In the case of $^{12}C^{16}O$, this energy can be worked out as

$6.626 \times 10^{-34} \times 4.063 \times 10^{14}/2\pi = 4.285 \times 10^{-20}$ J. To determine whether or not this level is occupied we need to know the value of $k_B T$, which at 298 K is $1.381 \times 10^{-23} \times 298 = 4.114 \times 10^{-21}$ J. The first excited state is thus about ten times $k_B T$ above the ground state in energy, in which case we can be confident that the $v = 1$ level will have negligible population, resulting in the $1 \rightarrow 2$ transition being very weak indeed.

If the vibrational frequency of the diatomic is low enough, the $v = 1$ state will have a significant population and so the $1 \rightarrow 2$ transition will be observable. For example $^{127}I^{79}Br$ has a vibrational frequency of 8.06×10^{12} Hz, so the first excited state is 5.34×10^{-21} J above the ground state. This energy is comparable with $k_B T$ which at 298 K, and so the $v = 1$ level is populated significantly. Figure 16.21 shows the comparison between the CO and IBr vibrational energy levels.

However, you can easily work out using the same approach as above that the energy of the $1 \rightarrow 2$ transition is $\hbar\omega$, exactly the same as the $0 \rightarrow 1$ transition. The two transitions thus have exactly the same frequency, and cannot be distinguished. However, we will see shortly in section 16.7.3 that if we use a potential energy function which is more realistic than the harmonic potential, these two transitions are not predicted to have the same frequency.

Working in wavenumbers

Since the frequency scale of an IR spectrum is usually given in wavenumbers (cm^{-1}), it is sometimes convenient to express the energies of the harmonic oscillator levels directly in wavenumbers rather than in joules. Strictly, we should not talk of an 'energy in cm^{-1}', since cm^{-1} is not a unit of energy. Nevertheless, such terminology is used widely in spectroscopy.

To convert an energy in joules to wavenumbers we first divide by Planck's constant to give the frequency in Hz, and then divide by the speed of light in cm s^{-1} to give a quantity in cm^{-1}

$$\text{'energy in cm}^{-1}\text{'} = \frac{\text{energy in joules}}{h \times \text{speed of light in cm s}^{-1}}. \quad (16.20)$$

The harmonic oscillator energy levels are given in Eq. 16.17 on page 723 as $E_v = \left(v + \frac{1}{2}\right)\hbar\omega$. Dividing by $h\tilde{c}$, where $\tilde{c}$ is the speed of light in cm s^{-1}, gives the energies in cm^{-1}, which are denoted $\tilde{E}_v$

$$\begin{aligned}\tilde{E}_v &= \left(v + \tfrac{1}{2}\right)\frac{\hbar\omega}{h\tilde{c}} \\ &= \left(v + \tfrac{1}{2}\right)\frac{\omega}{2\pi\tilde{c}} \\ &= \left(v + \tfrac{1}{2}\right)\tilde{\omega}. \quad (16.21)\end{aligned}$$

On the final line $\tilde{\omega}$ is the vibrational frequency in cm^{-1}, which is related to ω (in rad s^{-1}) by $\tilde{\omega} = \omega/(2\pi\tilde{c})$.

Using Eq. 16.21 the energy of the ground state is $\frac{1}{2}\tilde{\omega}$ and that of the first excited state is $\frac{3}{2}\tilde{\omega}$. The transition between the two therefore has energy $\tilde{\omega}$; this is a nice simple result. Therefore in the case of the spectrum of $^{12}C^{16}O$ the fact that there is a strong absorption at 2157 cm^{-1} tells us immediately that $\tilde{\omega} = 2157$ cm^{-1}.

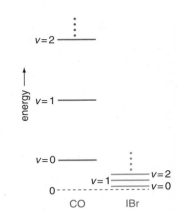

Fig. 16.21 Comparison of the three lowest vibrational energy levels of CO and IBr. The CO levels are much more widely spaced, and as a result the $v = 1$ level has no significant population at 298 K since its energy is about ten times greater than $k_B T$ above the ground state. In contrast, for IBr the $v = 1$ level is populated significantly since this level is within about $k_B T$ of the ground state.

It is usual to add a tilde ˜ to frequencies or 'energies' which are in wavenumbers.

Given that $\tilde{\omega} = \omega/(2\pi\tilde{c})$, we can easily find the value of ω (in rad s^{-1}) in the following way

$$\tilde{\omega} = \frac{\omega}{2\pi\tilde{c}}$$

$$\text{hence } \omega = 2\pi\tilde{c}\tilde{\omega}. \qquad (16.22)$$

In this case we find $\omega = 2\pi \times 2.998 \times 10^{10} \times 2157 = 4.063 \times 10^{14}$ rad s^{-1}, just as before.

16.7.3   Using an anharmonic potential

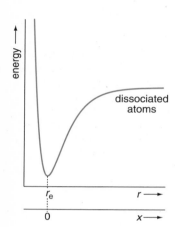

The problem with modelling the vibration of a diatomic using the harmonic oscillator is that it is not very realistic. The harmonic potential is simply $V(x) = \frac{1}{2}k_f x^2$, where x is the deviation of the bond length r from its equilibrium value r_e: $x = r - r_e$. This potential just goes on increasing as the bond is stretched (or compressed) away from the equilibrium position. However, we know from our earlier discussion of the potential energy curve of a diatomic (section 3.1 on page 92) that it is of the form depicted in Fig. 16.22. When the bond is stretched the energy rises at first, but then begins to level off to a value corresponding to the separate non-interacting atoms. What has happened is that at large values of the internuclear separation the atoms have ceased to interact significantly – in other words, the diatomic has *dissociated*. The harmonic potential does *not* predict dissociation, which is the major flaw in using this potential to model the stretching of a bond.

Fig. 16.22 The intermolecular potential for a diatomic has a minimum at the equilibrium position $r = r_e$ (which corresponds to $x = 0$). As r extends beyond this point the potential rises at first but then at large r levels off to a value which corresponds to the energy of the separate atoms i.e. the molecule dissociates.

The exact shape of the potential energy curve can only be found from sophisticated quantum mechanical calculations or from the detailed analysis of spectra. Such curves do not have a simple functional form, and it is not possible to solve the Schrödinger equation 'by hand' for the vibrations in such potential energy curves.

However, we can obtain a feel for the form of the vibrational energy levels associated with such a potential by using another simple model. The most widely used is the *Morse potential*, which is the function

$$V_M(x) = D_e \left[1 - \exp(-ax)\right]^2,$$

where D_e is an adjustable parameter with the dimensions of energy, and a is a second adjustable parameter with dimensions of (length)$^{-1}$. As before, $x = r - r_e$.

When $x = 0$ (i.e. at the equilibrium bond length) the exponential term goes to one, so the square bracket is zero and hence $V_M(0) = 0$. When x becomes large and positive (i.e. separated atoms), the exponential term goes to zero, so the potential tends to D_e. Thus D_e is the energy difference between the diatomic at its equilibrium separation and the separated atoms: this is the *dissociation energy*.

Figure 16.23 on the facing page shows a plot of the Morse potential as a function of x for different values of the parameter a. It can be seen that increasing a results in a 'steeper' potential, but the dissociation energy is unaffected. It is important to realize that the Morse potential is just a *model* for the real potential energy function; there is plenty of evidence that it is not an especially good model, especially at larger bond lengths.

Morse energy levels

The reason that the Morse potential is of particular interest is that it is possible (with rather a lot of effort) to solve the SE for this potential, and hence find the vibrational energy levels. These turn out to be

$$E_v = \left(v + \tfrac{1}{2}\right)\hbar\omega_e - \left(v + \tfrac{1}{2}\right)^2 \hbar\omega_e x_e \qquad v = 0,\ 1,\ 2,\ \dots. \qquad (16.23)$$

As with the harmonic oscillator, the energy levels depend on a quantum number v which takes integer values from zero. The values of the parameters D_e and a in the Morse potential determine the values of the frequency (in rad s^{-1}) ω_e, and the dimensionless quantity x_e, called the *anharmonicity parameter*, according to the following relationships

$$\omega_e x_e = \frac{a^2 \hbar}{2\mu} \qquad \text{and} \qquad \omega_e = a\sqrt{\frac{2D_e}{\mu}}. \qquad (16.24)$$

There are various ways of expressing these relationships, but the ones given above have been chosen because they emphasize that the value of $\omega_e x_e$, which appears in the squared term in the expression for the energy levels, depends only on the parameter a, whereas ω_e depends on both a and D_e. It will be useful later on to rewrite these expressions the other way round i.e. relating a and D_e to ω_e and $\omega_e x_e$

$$a = \sqrt{\frac{2\mu\omega_e x_e}{\hbar}} \qquad \text{and} \qquad D_e = \frac{\omega_e^2 \hbar}{4\omega_e x_e}. \qquad (16.25)$$

The anharmonicity parameter x_e is positive and much less than one, so in Eq. 16.23, and for low values of v, the term in $(v + \tfrac{1}{2})^2$ is much smaller than the term in $(v + \tfrac{1}{2})$. However, as v increases the squared term increases faster than the linear term, and as a result the energy levels get closer and closer together; this is illustrated in Fig. 16.24 where a set of harmonic and Morse energy levels are compared. This uneven spacing of the energy levels has immediate consequences for the IR spectrum.

Using Eq. 16.23 we can compute the energies of the lowest three levels as

$$E_0 = \tfrac{1}{2}\hbar\omega_e - \tfrac{1}{4}\hbar\omega_e x_e \quad E_1 = \tfrac{3}{2}\hbar\omega_e - \tfrac{9}{4}\hbar\omega_e x_e \quad E_2 = \tfrac{5}{2}\hbar\omega_e - \tfrac{25}{4}\hbar\omega_e x_e.$$

The energy of the $0 \to 1$ and $1 \to 2$ transitions are therefore

$$E_{0\to1} = \hbar\omega_e - 2\hbar\omega_e x_e \qquad E_{1\to2} = \hbar\omega_e - 4\hbar\omega_e x_e. \qquad (16.26)$$

In contrast to the case where we used the harmonic oscillator levels, this time we find that the $0 \to 1$ and $1 \to 2$ transitions do not have the same energy (i.e. frequency). Since the Morse energy levels are getting closer together the $1 \to 2$ transition is at a slightly lower energy (frequency), by an amount $2\hbar\omega_e x_e$, than the $0 \to 1$ transition. This means that for molecules where there is some population of the $v = 1$ level, is is possible to see the $1 \to 2$ transition in the spectrum since it does not lie on top of the much stronger $0 \to 1$ transition.

For a harmonic potential the selection rule is that v can only change by one, but this rule does not apply so strictly to an anharmonic potential such as that of the Morse oscillator. As a result, a transition such as $0 \to 2$ will appear in the spectrum, albeit with reduced intensity compared to the 'fully allowed'

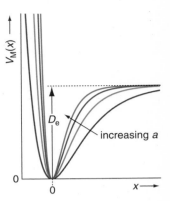

Fig. 16.23 Plots of the Morse potential for a fixed value of the dissociation energy D_e but increasing values of the parameter a. As a increases the potential becomes steeper. x is the displacement from equilibrium: $x = r - r_e$.

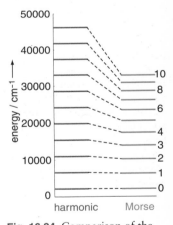

Fig. 16.24 Comparison of the vibrational energy levels for H_2 obtained by modelling the vibration using an harmonic oscillator (blue levels) and a Morse oscillator (red levels); the 'energies' are expressed in wavenumbers, and the vibrational quantum number v is shown on the right. The HO levels are evenly spaced, whereas the Morse levels get progressively closer and closer together.

Fig. 16.25 Shown in (a) are a typical set of wavefunctions for a Morse potential, superimposed on the potential (shown in green) in the same way as in Fig. 16.16 on page 717; the squares of the wavefunctions are shown in (b). In contrast to the harmonic oscillator wavefunctions, those for the Morse oscillator are neither symmetric nor antisymmetric about $x = 0$ (the equilibrium separation), but tend to extend much more to positive values of x then to negative values. As a result the average bond length increases as the vibrational energy increases.

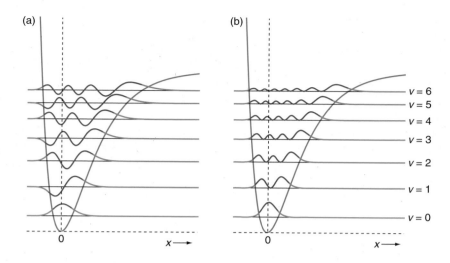

$0 \rightarrow 1$ transition. Using the above energies we can compute the energy of this transition as

$$E_{0\rightarrow2} = 2\hbar\omega_e - 6\hbar\omega_e x_e. \qquad (16.27)$$

This transition is called the *first overtone* as it appears at approximately twice the frequency of the $0 \rightarrow 1$ transition, which is known as the *fundamental*. The transition from $0 \rightarrow 3$ may also be visible in the spectrum: this transition is called the *second overtone*. Absorptions due to the fundamental and these two overtones are clearly visible in the experimental spectrum of CO shown in Fig. 11.17 on page 463.

If we are able to measure the frequency of the $0 \rightarrow 1$ and $0 \rightarrow 2$ transitions, then these data can be manipulated to find the values of ω_e and x_e. In turn, these values can be used to find the Morse parameters D_e and a: an illustration of how this is done is given in Example 16.1 on the next page. It must be remembered that the value of the dissociation energy D_e is only an estimate based on the assumption that the potential is described by a Morse curve: we therefore cannot expect it to be particularly accurate.

Using wavenumbers

As we did for the harmonic oscillator, it is possible to write the 'energies' of the Morse oscillator levels in wavenumbers, giving

$$\tilde{E}_v = \left(v + \tfrac{1}{2}\right)\tilde{\omega}_e - \left(v + \tfrac{1}{2}\right)^2 \tilde{\omega}_e x_e, \qquad (16.29)$$

Weblink 16.3

This link takes you to a real-time version of Fig. 16.25 in which you can explore how the energy levels and wavefunctions are affected by changing the Morse parameters.

where $\tilde{\omega}_e = \omega_e/(2\pi\tilde{c})$ and the value of x_e is unchanged (since it is dimensionless).

Using this expression, the $0 \rightarrow 1$ transition is at $\tilde{\omega}_e - 2\tilde{\omega}_e x_e$ and the $0 \rightarrow 2$ transition is at $2\tilde{\omega}_e - 6\tilde{\omega}_e x_e$.

16.7.4   The Morse oscillator wavefunctions

Figure 16.25 (a) shows a typical set of wavefunctions for the Morse oscillator, imposed on a plot of the potential in the same way as we did for the harmonic

---

Example 16.1 Finding the Morse parameters

In the IR spectrum of CO a strong absorption due to the fundamental was observed at 2143 cm^{-1}, and a weaker absorption was observed at 4260 cm^{-1}. Use these data to find ω_e and $\omega_e x_e$, and hence the Morse parameters a and D_e.

The frequency of the absorption at 4260 cm^{-1} is just a little bit lower than twice that of the fundamental, so it must be the first overtone i.e. $0 \to 2$. The first step is to convert the frequencies in cm^{-1} to energies by multiplying by $h\tilde{c}$

$$
\begin{aligned}
E_{0\to1} &= h \times \tilde{c} \times 2143 \\
&= 6.626 \times 10^{-34} \times 2.998 \times 10^{10} \times 2143 \\
&= 4.257 \times 10^{-20} \text{ J.}
\end{aligned}
$$

Similarly $E_{0\to2} = 8.462 \times 10^{-20}$ J. We now use these known values in the expressions given in Eq. 16.26 and Eq. 16.27 for the energies of these two transitions

$$
\begin{aligned}
E_{0\to1}: \quad 4.257 \times 10^{-20} &= \hbar\omega_e - 2\hbar\omega_e x_e \quad\quad (16.28) \\
E_{0\to2}: \quad 8.462 \times 10^{-20} &= 2\hbar\omega_e - 6\hbar\omega_e x_e.
\end{aligned}
$$

Taking two times $E_{0\to1}$ and subtracting $E_{0\to2}$ eliminates the terms in $\hbar\omega_e$ to give

$$
\begin{aligned}
2 \times E_{0\to1} - E_{0\to2} &= 2\left(\hbar\omega_e - 2\hbar\omega_e x_e\right) - \left(2\hbar\omega_e - 6\hbar\omega_e x_e\right) \\
&= 2\hbar\omega_e x_e.
\end{aligned}
$$

Hence

$$
2\hbar\omega_e x_e = 2 \times \left(4.257 \times 10^{-20}\right) - 8.462 \times 10^{-20},
$$

from which we obtain $\hbar\omega_e x_e = 2.600 \times 10^{-22}$ J. It follows that $\omega_e x_e = 2.465 \times 10^{12}$ rad s^{-1}. Using this value of $\hbar\omega_e x_e$ in the expression for $E_{0\to1}$ in Eq. 16.28 we have

$$
4.257 \times 10^{-20} = \hbar\omega_e - 2 \times 2.600 \times 10^{-22},
$$

hence $\hbar\omega_e = 4.309 \times 10^{-20}$ J, and so $\omega_e = 4.086 \times 10^{14}$ rad s^{-1}.

continued . . .

---

oscillator in Fig. 16.16 on page 717. In their general form the wavefunctions have similarities to the harmonic case: the number of nodes increases as the energy increase, and in the non-classical region the wavefunctions simply decay away. However, in contrast to the harmonic oscillator, the wavefunctions are neither symmetric nor antisymmetric about $x = 0$, the equilibrium separation. This is hardly a surprise, since the Morse potential is not symmetric about this point.

We saw before that for the higher energy harmonic oscillator wavefunctions, the two positions of maximum probability were symmetrically placed at the extreme edges of the classical region. As can be seen from the plots of the squares of the wavefunctions shown in Fig. 16.25 (b), a similar effect occurs for

Example 16.1 Finding the Morse parameters

Now we can use these values of ω_e and $\omega_e x_e$ to find the Morse parameters using Eq. 16.25 on page 727; it will be useful to recall that $\mu = 1.138 \times 10^{-26}$ kg

$$
\begin{aligned}
a &= \sqrt{\frac{2\mu\omega_e x_e}{\hbar}} \\
&= \sqrt{\frac{2 \times 1.138 \times 10^{-26} \times 2.465 \times 10^{12} \times 2\pi}{6.626 \times 10^{-34}}} \\
&= 2.31 \times 10^{10} \text{ m}^{-1},
\end{aligned}
$$

where we have used the fact that $\hbar = h/2\pi$. We expect a to be very large as it is multiplied by x, the deviation of the bond length from its equilibrium value, which is very small.

The dissociation energy is computed using Eq. 16.25

$$
\begin{aligned}
D_e &= \frac{\hbar\omega_e^2}{4\omega_e x_e} \\
&= \frac{6.626 \times 10^{-34} \times \left(4.086 \times 10^{14}\right)^2}{4 \times 2\pi \times 2.465 \times 10^{12}} \\
&= 1.79 \times 10^{-18} \text{ J}.
\end{aligned}
$$

To express this as a molar quantity we multiply by Avogadro's constant to give 1075 kJ mol^{-1}.

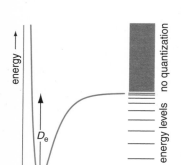

Fig. 16.26 Within the bottom part of the potential energy curve, the potential constrains the atoms on both sides, and as a result there are quantized energy levels. However, once the energy increases beyond the dissociation limit, the atoms are only constrained on one side and as a result there are no quantized energy levels.

the Morse oscillator wavefunctions. However, in this case the probability soon becomes dominated by a *single* maximum at the positive edge of the classical region. As a result, the *average* value of x increases as the energy increases, in contrast to the harmonic oscillator in which the average value of x is always zero. In section 16.13 on page 743 we will see that this increase in the average bond length has a measurable effect on certain kinds of spectra.

16.7.5 Dissociation

As has already been noted, the Morse oscillator energy levels get closer and closer together as the quantum number v increases. There must therefore come a point at which successive energy levels have the same energy (i.e. 'fall on top of one another'), and it is at this point that the molecule has acquired just enough energy to dissociate. The argument is that when the energy is greater than the dissociation energy the atoms are free of the constraining potential and so, as illustrated in Fig. 16.26, we expect there to be no quantized energy levels, just as is the case for a free particle. The point at which two of the Morse energy levels have the same energy thus represents the start of this unquantized region, at which point the energy is equal to the dissociation energy.

A convenient way of finding the highest energy level in the Morse oscillator is to set the the derivative dE_v/dv equal to zero. The derivative can be interpreted as the rate of change of energy with quantum number, and this rate will be zero when two levels have the same energy, which is the point we are trying to find. The derivative is computed as follows

$$\frac{d}{dv}E_v = \frac{d}{dv}\left[\left(v+\tfrac{1}{2}\right)\hbar\omega_e - \left(v+\tfrac{1}{2}\right)^2\hbar\omega_e x_e\right]$$
$$= \hbar\omega_e - 2\left(v+\tfrac{1}{2}\right)\hbar\omega_e x_e.$$

This derivative will be zero when

$$\left(v_{max}+\tfrac{1}{2}\right) = \frac{1}{2x_e},$$

where v_{max} is the quantum number of this highest energy level. Note that its value only depends on the anharmonicity parameter x_e.

We can now substitute this value of $\left(v_{max}+\tfrac{1}{2}\right)$ back into the expression for the energy levels (Eq. 16.23 on page 727) thus obtaining an expression for the energy of this highest level, which will be the dissociation energy

$$E_{v_{max}} = \left(v_{max}+\tfrac{1}{2}\right)\hbar\omega_e - \left(v_{max}+\tfrac{1}{2}\right)^2\hbar\omega_e x_e$$
$$= \frac{1}{2x_e}\hbar\omega_e - \left(\frac{1}{2x_e}\right)^2\hbar\omega_e x_e$$
$$= \frac{\hbar\omega_e}{4x_e} \quad \text{which is sometimes written} \quad \frac{\hbar\omega_e^2}{4\omega_e x_e}.$$

$E_{v_{max}}$ is the dissociation energy D_e. Note that the expression we have just derived for D_e is the same as that given in Eq. 16.25 on page 727.

16.8 Vibrations of larger molecules

The vibrational motion of molecules larger than a diatomic can be analysed using the concept of *normal modes*, which was introduced in section 11.3.4 on page 465. The key idea is that although the motions of the atoms and the stretching of the various bonds in a molecule may look rather complicated, it can always be factored into contributions from a fixed number of normal modes. Each normal mode corresponds to a particular set of motions of the atoms, and these modes can be determined (independent of experiment) for a given molecular geometry. For a molecule containing N atoms there are just $(3N-6)$ normal modes, or $(3N-5)$ if the molecule is linear.

The four normal modes of CO_2 are shown in Fig. 16.27 on the following page. The centre row shows the molecule in its equilibrium geometry, which is linear and symmetric about the central carbon atom, such that the molecule possesses a centre of inversion. For each mode, the blue arrows show the way in which the atoms will move during one half of the vibration; these motions give the distorted structures shown on the lower row. During the other half of the vibration the atoms move in the opposite directions to the blue arrows, giving the distorted structures shown on the upper row. Thus, during the vibration the structure changes continuously between the two extreme positions shown on the top and bottom rows, passing through the equilibrium geometry along the way. Note that the changes in bond lengths or angles shown in this diagram

⊕ *Weblink 16.4*

Follow this link to view animated versions of the normal modes of CO_2.

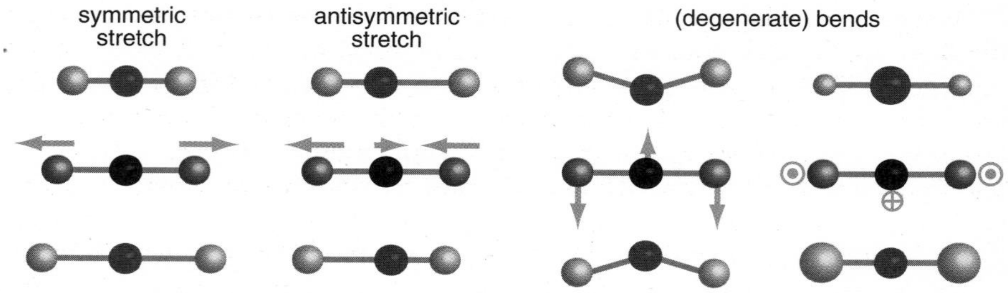

Fig. 16.27 Depiction of the four normal modes of CO_2. The centre row of structures shows the equilibrium geometry, with the upper and lower rows showing the distorted structures which arise during the vibration. The blue arrows show the way in which the atoms move to give the structures shown on the bottom row; the structures shown in the top row result from distortions in the *opposite* directions to those shown by the blue arrows. An arrow coming towards you is indicated by $\odot$, and one going away by $\oplus$. There are two degenerate bends: in one the atoms move in the plane of the paper, whereas in the other they move in a plane coming out of the paper.

have been much exaggerated for clarity: in fact, during a vibration the bond lengths will only change by a few percent of their equilibrium values.

Each normal mode has associated with it a set of energy levels which can be modelled, to the first approximation, using a harmonic oscillator. These energy levels are just the same as we had before

$$E_{v_i} = \left(v_i + \tfrac{1}{2}\right)\hbar\omega_i \qquad v_i = 0,\ 1,\ 2,\ \ldots,$$

where v_i is the quantum number for the ith normal mode, and ω_i is the vibrational frequency of that normal mode. Since several bonds are being stretched in a given normal mode, the vibrational frequency has a complex relationship with the force constants for the individual bonds. In general, each normal mode has a different frequency associated with it, although in some symmetrical molecules it is possible to have physically distinct normal modes which have the same frequency; such modes are said to be *degenerate*. In the case of CO_2 the two bending modes are degenerate, which makes sense since for a linear molecule it clearly makes no difference in which plane the bend takes place.

In the harmonic approximation, the selection rules are the same as for the diatomic: there must be a change in dipole moment when the vibration takes place, and the quantum number can only change by one. For these larger molecules there is an additional rule that a transition is only allowed if the quantum number of a *single normal mode* changes by one.

For example, consider the normal modes of CO_2 depicted in Fig. 16.27. As a result of the particular symmetry of the equilibrium geometry, the molecule has no permanent dipole. However, during the antisymmetric stretch and bending normal modes, this symmetry is broken in such a way that a dipole develops. As a result, transitions between the energy levels associated with these normal modes will be seen in the IR spectrum; the normal modes are said to be *IR active*.

In contrast, for the symmetric stretch normal mode no dipole develops since during the vibration the symmetry of the equilibrium geometry is maintained. Transitions between the energy levels associated with the symmetric stretch normal mode are therefore not allowed: the mode is said to be *IR inactive*.

Just as in the diatomic case, we expect that the strongest transitions will be the $0 \rightarrow 1$. Thus, for each IR active normal mode the simple expectation is that there will be an absorption at frequency $\omega_i/(2\pi)$ Hz. Unfortunately, things are not this simple. As we have seen, the harmonic oscillator approximation is rather poor, so we need to use some kind of anharmonic potential. The strict selection rules break down when we do this, so absorptions due to overtone transitions may be seen.

In addition, the breakdown of the selection rules means that transitions in which the quantum numbers of more that one normal mode change can have significant intensity. For example, in the IR spectrum of CO_2 there is quite a strong absorption at around 3716 cm^{-1}. This is thought to be due to a transition in which the symmetric stretch normal mode goes from the $v = 0$ to $v = 1$ level simultaneous with the antisymmetric stretch normal mode going from $v = 0$ to $v = 1$. There is also an absorption at around 3609 cm^{-1}, in which the antisymmetric stretch goes from $v = 0$ to $v = 1$ simultaneous with the bending mode going from $v = 0$ to $v = 2$.

All in all, the spectrum almost always becomes much more complicated than the 'one absorption per normal mode' which the harmonic approximation predicts.

16.9 Raman spectroscopy

The vibrational energy levels of a molecule can also be probed by a technique known as *Raman spectroscopy*, which is in many ways complementary to IR spectroscopy. Raman spectroscopy uses a completely different kind of interaction between the light and the molecule than is the case for the other kinds of spectroscopy we have discussed so far. Rather than looking at the absorption or emission of photons, in Raman spectroscopy we look at the way in which photons are *scattered*.

Raman spectra are recorded by irradiating the sample with an intense beam of monochromatic light, most conveniently from a laser running in the visible or just beyond the red end of the visible spectrum. The wavelength of the laser light is chosen so that it is *not* absorbed by the molecule, and as a result most of the light passes straight through the sample. However, a small amount of the light is *scattered* by interacting with the molecules in the sample, and it is this scattered light which we detect and analyse.

Most of the scattered photons have exactly the same frequency as the original light from the laser (this is called *Raleigh scattering*). However, a small fraction of the scattered photons are found to be higher or lower in energy than the photons from the laser. What is happening here is that the incident photons are either giving up or gaining some energy from the molecule: this accounts for the shift in the energy of the scattered photons. The basic process is illustrated in Fig. 16.28 on the following page.

Not surprisingly, the amount of energy which can be lost or gained is equal to the difference in energy between two levels of the molecule. If the molecule goes from a level with energy E_{lower} to one with energy E_{upper}, then the energy of the scattered photon is *reduced* by $(E_{upper} - E_{lower})$.

$$h\nu_{Stokes} = h\nu_L - (E_{upper} - E_{lower}).$$

This kind of Raman scattering is called *Stokes scattering*.

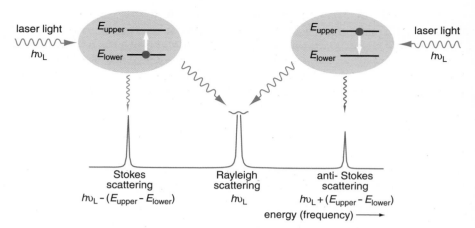

Fig. 16.28 Illustration of how Raman scattering arises. An intense beam of laser light (frequency υ_L) impinges on a molecule, represented by the pale blue oval. Most of the light is scattered without any change in its frequency: this is known as Rayleigh scattering. In the Raman effect the scattering is accompanied by a change in the energy of the molecule. If the molecule moves from a lower level to a higher one, the scattered photon is *lower* in energy by the separation of these two levels: this is Stokes scattering. If the molecule moves from a higher level to a lower one, the photon *gains* an amount of energy equal to this separation: this is anti-Stokes scattering. The *separation* of the Stokes and anti-Stokes lines from the laser line is thus dependent on the spacing of the energy levels.

On the other hand if the molecule drops down from the higher to the lower level, then an amount of energy ($E_{upper} - E_{lower}$) is added to the photon from the laser, and so the energy of the scattered photon is *increased* by this amount

$$h\upsilon_{anti-Stokes} = h\upsilon_L + (E_{upper} - E_{lower}).$$

This kind of Raman scattering is called *anti-Stokes scattering*.

The result of all of this is that the *shift* in energy between the Raman scattered photons and the laser photons is equal to the *difference* in energy between the molecular energy levels. By measuring this shift, we can therefore measure the separation between these levels, just as we do in a conventional spectrum.

In the case of vibrational energy levels, the most intense scattering comes from the process in which the molecule goes from $v = 0$ to $v = 1$, since $v = 0$ is generally the only state which is significantly populated. We therefore expected to see Stokes scattered photons with energy

$$h\upsilon_{Stokes} = h\upsilon_L - \hbar\omega,$$

since $\hbar\omega$ is the energy separation of the $v = 0$ and $v = 1$ states, in the harmonic approximation. These scattered photons will give a line in the Raman spectrum at frequency $\upsilon_L - \omega/(2\pi)$.

Raman scattering is rather weak and the apparatus needed to record the spectra is considerably more complicated than for simple IR spectra. However, the reason that Raman is an interesting technique is that different selection rules apply; in particular, Raman scattering does *not* require there to be a change in the dipole moment during the vibration. The requirement for Raman scattering is that the polarizibility of the molecule must change during the vibration: a discussion of how this can be determined is outside the scope of this text.

Normal modes that are not active in the IR often turn out to be active in Raman spectra. This is why the two techniques are described as being complementary, since together they give us the maximum chance of observing transitions due to all the normal modes. For example, in the case of CO_2 the symmetric stretching mode, which is not IR active, turns out to be Raman active. Its frequency can therefore be measured using Raman spectroscopy.

16.10 Summary of the features of vibrational spectroscopy

Before we move on to look at a different type of energy levels it is useful to summarize the key points about vibrational spectra.

- The vibrational energy levels of a diatomic can be modelled using the harmonic oscillator energy levels.

- The selection rules for transitions between these levels are: (a) there must be a change in the dipole moment during the vibration, (b) the vibrational quantum number can only change by one, $\Delta v = 1$.

- Using the harmonic approximation we predict that we will see a single absorption at frequency $\omega/(2\pi)$, where ω is the vibrational frequency; this transition is between the $v = 0$ and $v = 1$ levels.

- The value of ω can be used to determine the force constant of the bond.

- An anharmonic potential, such as the Morse potential, is a better model for the vibration of a bond.

- The selection rule $\Delta v = 1$ breaks down when the potential is anharmonic, allowing us to see overtones e.g. $v = 0 \rightarrow v = 2$.

- If the frequencies of the fundamental and at least one other transition can be observed, then these data can be used to determine the Morse parameters. This gives an estimate for the dissociation energy.

- The vibrations of larger molecules can be analysed using normal modes.

- Each normal mode can be modelled as a harmonic oscillator and has associated with it a set of energy levels characterized by the vibrational frequency for that mode.

- In the harmonic approximation we expect to see a single absorption at the frequency of each normal mode. Due to anharmonicity, many other transitions, such as overtones and transitions between energy levels from two or more normal modes, are observed.

- Raman spectroscopy is complementary to IR, and is particularly useful as it gives access to data on normal modes which are not IR active.

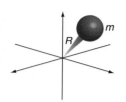

Fig. 16.29 A rigid rotor consists of a mass m free to rotate in any plane but held at a fixed distance R from the origin.

16.11 The rigid rotor

We now come to the third system for which the Schrödinger equation can be solved: the *rigid rotor*. As with the other systems we have discussed, it turns out that the energy levels for this model system can be adapted to describe the motions of molecules – specifically in this case the rotation of a diatomic.

A rigid rotor consists of a mass m which is fixed at a distance R from the origin, but is otherwise free to rotate in any plane and to move between planes; the arrangement is illustrated in Fig. 16.29 In mechanics, such an arrangement is a little difficult to contrive, but a reasonable approximation would be a small sphere moving on the *inside* surface of a larger hollow sphere, or a mass attached by a stiff rod to a central point, about which the rod can pivot freely.

A key quantity needed to describe the rotational motion of an object is its *moment of inertia*, I, which for the case of a rigid rotor is simply

$$I = mR^2.$$

In classical mechanics the rigid rotor has rotational kinetic energy $\frac{1}{2}I\omega^2$, where ω is the angular velocity (in rad s^{-1}) at which the mass is rotating about the origin. This kinetic energy is the rotational analogue of the (linear) kinetic energy $\frac{1}{2}mv^2$ possessed by a mass m moving in a line at velocity v.

16.11.1 Energy levels

As the rotor can move in three dimensions, the corresponding Schrödinger equation is considerably more complicated than the ones we have looked at so far, and its solution involves a level of mathematical complexity which is well beyond the level of this text. We will therefore have to content ourselves with simply stating the solutions to the Schrödinger equation in this case. The energy levels turn out to be

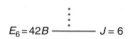

$$E_6 = 42B \underline{\hspace{2cm}} J = 6$$

$$E_5 = 30B \underline{\hspace{2cm}} J = 5$$

$$E_4 = 20B \underline{\hspace{2cm}} J = 4$$

$$E_3 = 12B \underline{\hspace{2cm}} J = 3$$

$$E_2 = 6B \underline{\hspace{2cm}} J = 2$$
$$E_1 = 2B \underline{\hspace{2cm}} J = 1$$
$$E_0 = 0 \underline{\hspace{2cm}} J = 0$$

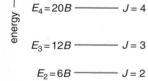

Fig. 16.30 Illustration of the first few energy levels of the rigid rotor. Note that the spacing between the levels increases as J increases.

$$E_J = BJ(J+1) \qquad B = \frac{\hbar^2}{2I} \qquad J = 0, 1, 2, \ldots, \qquad (16.30)$$

where J is the *rotational quantum number* and takes integer values, starting at zero. B is called the *rotational constant*, and its value is related to the moment of inertia in the way stated. Figure 16.30 illustrates the energies of the first few of these levels.

For large J, the energies go as J^2 (as in the particle in a box), and we also note that the separation of the levels increases with J. In contrast to both the particle in a box and the harmonic oscillator, the energy of the lowest level is zero i.e. the rigid rotor does *not* have zero point energy.

A special feature of these energy levels is that they have a *degeneracy* of $(2J + 1)$. Recall from page 711 that degeneracy is where we have distinct wavefunctions which have the same energy. For example, for $J = 1$ there are $(2J + 1) = (2 \times 1 + 1) = 3$ distinct wavefunctions, all with energy $2B$; similarly, for $J = 2$ there are five wavefunctions, all with energy $6B$. This degeneracy has particular consequences when it comes to spectra, which we will discuss in section 16.12.3 on page 740.

16.11.2   Wavefunctions

The wavefunctions which are solutions to the SE for the rigid rotor are most easily expressed in the spherical polar coordinate system which we encountered in section 2.3.3 on page 51 when describing the mathematical form of the hydrogen atomic orbitals. Instead of using x, y and z to describe the position, we use two angles θ and ϕ, which are like a latitude and a longitude, and the distance from the origin r. In the case of the rotor, the distance is fixed, so we only need the two angles.

The wavefunctions for the rigid rotor turn out to be a special set of functions called the *spherical harmonics*, usually given the symbol $Y_{J,M}(\theta, \phi)$. Two quantum numbers are needed to specify the spherical harmonic: the J quantum number is the same as the one which determines the energy in Eq. 16.30 on the preceding page, and the M quantum number takes all possible integer values between $-J$ and $+J$.

This description should start to sound familiar to you, as the same functions, albeit with different labels, form the angular parts of the hydrogen atomic orbitals, introduced in section 2.3.3 on page 51. In that case the labels were l and m_l, with m_l taking all integer values between $-l$ and l. All that we have done is change the labels: the functions are the same.

It is no coincidence that the orbital motion of the electron in hydrogen and the rotational motion of the rigid rotor share the same angular wavefunctions. The only difference between the two situations is that in the hydrogen atom the electron can change its distance from the nucleus, whereas in the rotor the mass is at a fixed distance. This gives the hydrogen atom wavefunctions an additional dependence on the distance r (the radial part of the wavefunction), but the angular part is identical.

The spherical harmonics (which are the wavefunctions) with different values of M but the same value of J all have the same energy since, according to Eq. 16.30, the energy depends only on J. Therefore, these spherical harmonics which only differ in their value of M are degenerate. For a given value of J the quantum number M can take $(2J+1)$ values, and this is the origin of the number of degenerate levels referred to in the previous section.

16.12   The microwave spectrum of a diatomic

In this section we will show how the energy levels of the rigid rotor can be adapted to describe the rotational energy levels of a diatomic and hence the spectrum which arises from transitions between them. Spectra of this type normally occur in the microwave part of the spectrum, and we will see that from such a spectrum it is possible to determine the bond length of a diatomic to high accuracy.

A diatomic is not, at first sight, a rigid rotor since there are two masses involved. However, it can be shown in classical mechanics that the rotation of an object with masses m_1 and m_2 held apart at a distance R is *identical* to a rigid rotor with moment of inertia

$$I = \mu R^2 \qquad \text{where} \quad \mu = \frac{m_1 m_2}{m_1 \times m_2}.$$

You will recognize μ as the reduced mass which we encountered before in section 16.7.1 on page 722.

Fig. 16.31 Idealized form of the rotational spectrum of a diatomic, using the rigid rotor energy levels as a model. The allowed transitions are shown by the arrows connecting the energy levels on the left, and the corresponding peaks (in the same colour) are shown in the spectrum on the right. What we see are a series of lines starting at $2B/h$ for the $J = 0 \rightarrow J = 1$ transition, and then spaced by $2B/h$. The intensities are not shown in a realistic way.

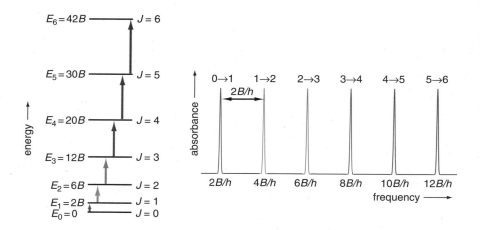

Using this expression for the moment of inertia we can write the energy levels of the diatomic with bond length R as

$$E_J = BJ(J + 1) \qquad B = \frac{\hbar^2}{2I} \qquad I = \mu R^2 \qquad J = 0, 1, 2, \dots \qquad (16.31)$$

In order to predict the spectrum arising from transitions between these energy levels we need to know the selection rules. As in the case of the vibrational spectrum, there are two rules. The first, the gross selection rule, is that the molecule must possess a *permanent* dipole moment. The second is that the quantum number J can only change by one i.e. $\Delta J = 1$. We can therefore expect a spectrum from heteronuclear diatomics, as these have permanent dipole moments, but not from homonuclear diatomics.

16.12.1   Appearance of the spectrum

Since the energy levels get further apart as J increases, the lowest energy (frequency) allowed transition will be that between $J = 0$ and $J = 1$. Using Eq. 16.31 the energies of these two levels are

$$E_0 = 0 \qquad E_1 = 2B,$$

so the energy of the $0 \rightarrow 1$ transition is $2B - 0 = 2B$; the corresponding frequency, $v_{0\rightarrow1}$, is thus $2B/h$.

The next allowed transition is from $J = 1$ to $J = 2$. Given that $E_2 = 6B$, the energy of this transition is $4B$, and the corresponding frequency, $v_{1\rightarrow2}$, is $4B/h$. Carrying on with the $2 \rightarrow 3$ and $3 \rightarrow 4$ transitions we see that a simple pattern develops. The lowest frequency line is at $2B/h$, and subsequent lines are spaced by $2B/h$, as illustrated in Fig. 16.31.

16.12.2   Determination of the bond length

Figure 16.32 on the facing page shows the experimental microwave spectrum of ${}^1H{}^{35}Cl$ gas. What immediately stands out is that there are a series of regularly spaced peaks, precisely as we have just predicted and illustrated in Fig. 16.31.

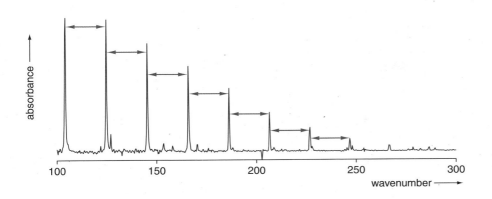

Fig. 16.32 Experimental microwave spectrum of $^1H^{35}Cl$ gas. What is immediately striking is the series of regularly spaced peaks whose separations are indicated by the double-headed blue arrow. Comparison with the idealized spectrum shown in Fig. 16.31 on the facing page allows us to identify this spacing as $2B/h$, and hence determine the value of B.

We can therefore identify this regular spacing as $2B/h$ (in frequency units) and hence determine a value for the rotational constant B. The frequency scale on this spectrum is given in wavenumbers and corresponds to the range 30 GHz to 90 GHz, placing it squarely in the microwave region.

The average spacing between the lines is 20.41 cm^{-1}. This value can be converted to Hz by multiplying by the speed of light in cm s^{-1}, giving the spacing as $20.41 \times 2.998 \times 10^{10} = 6.119 \times 10^{11}$ Hz. As we have shown, the spacing is $2B/h$, so it follows that $B = h \times 6.119 \times 10^{11}/2 = 2.0272 \times 10^{-22}$ J.

Given that $B = \hbar^2/2I$, it follows that $I = \hbar^2/2B$ so we can use the value of B we have just found to determine the moment of inertia I

$$
\begin{aligned}
I &= \frac{\hbar^2}{2B} \\
&= \frac{h^2}{8\pi^2 B} \\
&= \frac{(6.626 \times 10^{-34})^2}{8\pi^2 \times 2.0272 \times 10^{-22}} \\
&= 2.743 \times 10^{-47} \text{ kg m}^2.
\end{aligned}
$$

Recall that $\hbar = h/(2\pi)$.

We have used SI units throughout, and since the moment of inertia is mass × (length)^2 its SI units are kg m^2.

Finally, since $I = \mu R^2$, we can work out the bond length R using $R = \sqrt{I/\mu}$. Given that the mass of ^1H is 1.0078 mass units, and that of ^{35}Cl is 34.9689 mass units, the reduced mass of $^1H^{35}Cl$ is 1.627×10^{-27} kg (recall that 1 mass unit is 1.6605×10^{-27} kg). The calculation of the bond length is therefore

$$
\begin{aligned}
R &= \sqrt{I/\mu} \\
&= \sqrt{2.743 \times 10^{-47} / 1.627 \times 10^{-27}} \\
&= 1.298 \times 10^{-10} \text{ m} \\
&= 129.8 \text{ pm}.
\end{aligned}
$$

We have thus worked out the bond length of $^1H^{35}Cl$ from its microwave spectrum.

The value you will find quoted in data books for the bond length of $^1H^{35}Cl$ is 127.46 pm, which differs from the value we have found by just under 2%. The reason for this discrepancy is that the rigid rotor is only an approximate

Fig. 16.33 Simulated microwave spectrum of $^1H^{35}Cl$ showing the pattern of intensities that are expected. Similar to Fig. 16.31 we see a series of evenly spaces peaks (these have been labelled according to the J-values of the energy levels involved). As J increases the intensities increase at first and then, after reaching a maximum, fall off.

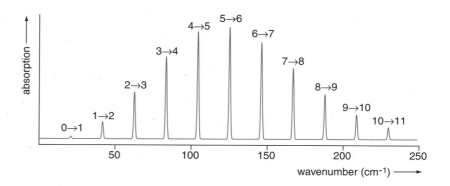

model for the rotational energy levels of a diatomic. For the most accurate work we need to take account of the fact that the bond will stretch as the molecule rotates (*centrifugal distortion*), and that the bond is also vibrating as the molecule rotates. However, it is gratifying that even our very simple analysis gives us a value for the bond length which is within a few percent of the accepted value.

The frequencies of peaks in a microwave spectrum can be measured to very high precision. If these data are analysed with a model which takes account of centrifugal distortion and other effects, the values of the bond lengths which can be obtained are the most precise available from any technique.

By analysing the IR spectrum we were able to determine the vibrational frequency and the force constant of the bond. Now we have seen that by looking at the microwave spectrum we can find the bond length. All this has been done by modelling the energy levels of these molecules using the simple idealized systems of the harmonic oscillator and the rigid rotor. What we see here is that the combination of spectroscopy and quantum mechanics is a powerful tool for determining molecular parameters.

16.12.3 Intensities

In the previous section we found that for $^1H^{35}Cl$ the rotational constant B is 2.0272×10^{-22} J. Recalling that at 298 K k_BT is 4.121×10^{-21} J, we can see that the energies of several of the lower rotational energy levels are going to be less than or comparable to k_BT and so will have significant populations. As a result, we expect to see several transitions in the spectrum, not just the single transition from the ground state as we did in the case of the vibrational spectrum. Indeed, the experimental spectrum of $^1H^{35}Cl$ shown in Fig. 16.32 on the previous page bears this out as there are a number of easily visible transitions.

The experimental spectrum of $^1H^{35}Cl$ shown in Fig. 16.32 does not include the first few lines at low wavenumber, which is why at first glance it looks rather different from the simulated spectrum of Fig. 16.33.

The pattern of intensities of the lines in a microwave spectrum is affected by the fact that the Jth rotational level has a degeneracy of $(2J + 1)$; the simulated spectrum of $^1H^{35}Cl$ shown in Fig. 16.33 is typical of the pattern of intensities which are found. Here we see that as J increases the intensities increase at first before reaching a maximum and then falling off. This is mainly a result of the way in which the populations of the levels vary with J.

Recall from section 1.10 on page 26 that the Boltzmann distribution predicts that the population n_i of level i with energy ε_i is given by

$$n_i = n_0 \exp\left(\frac{-\varepsilon_i}{k_B T}\right).$$

However, this expression is only correct for non-degenerate levels. If the energy levels are degenerate, it needs to be modified to

$$n_i = g_i \times n_0 \exp\left(\frac{-\varepsilon_i}{k_B T}\right),$$

where g_i is the degeneracy of the ith level. Applying this to the case of the rigid rotor energy levels, for which the energies are $BJ(J+1)$ and the degeneracies are $(2J+1)$ we have

$$n_J = \underbrace{(2J+1)}_{\text{degeneracy}} \times n_0 \exp\left(\frac{-BJ(J+1)}{k_B T}\right).$$

The degeneracy term *increases* linearly as J increases, whereas the exponential term *decreases* as J increases. The result is that at first the populations n_J increase on account of the degeneracy term, but as J increases further the decrease of the exponential term outweighs the increase in the degeneracy term. As the result the populations reach a maximum and then decrease. Figure 16.34 shows the typical behaviour of the populations of the rotational levels.

Broadly speaking, the pattern of intensities seen in Fig. 16.33 mirrors this behaviour of the populations of the rotational levels. However, there are some additional more subtle effects involved here which also affect the intensities. As a result, the pattern of the intensities does not follow precisely the variation in the populations of the energy levels. The populations are, however, an important factor.

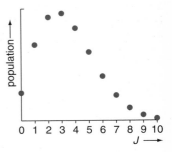

Fig. 16.34 Illustration of how, on account of the degeneracy term, the populations of the rotational levels increase at first, reach a maximum, and then decay. The exact shape of the curve depends on the temperature and the value of the rotational constant.

16.12.4   General form of the spectrum

Working from the energy levels and the selection rule we can determine a general expression for the energies (frequencies) of the transitions which appear in the spectrum. The result is rather a more elegant way of working out the positions of the lines in the spectrum than we have used so far.

Suppose that the lower energy level has quantum number J'' and the upper level has quantum number J'. The energy of the transition between these two is worked out as follows

$$\begin{aligned}
\Delta E_{J'' \to J'} &= E_{J'} - E_{J''} \\
&= BJ'(J'+1) - BJ''(J''+1).
\end{aligned}$$

Since the selection rule is that J can only change by one ($\Delta J = 1$), it follows that J of the upper state must be one greater than J of the lower state i.e. $J' = (J''+1)$. Substituting this in for J' and then tidying up gives

$$\begin{aligned}
\Delta E_{J'' \to (J''+1)} &= B[J''+1]([J''+1]+1) - BJ''(J''+1) \\
&= B\left(J''^2 + 3J'' + 2\right) - B\left(J''^2 + J''\right) \\
&= 2B(J''+1).
\end{aligned}$$

If we put $J'' = 0$ into this expression, the predicted energy of the transition is $2B$, if we put $J'' = 1$ the value is $4B$, and with $J'' = 2$ the value is $6B$. These are exactly the values we found before.

If we divide this energy separation by h to convert it to frequency units we have the following expression for the frequencies of the lines in the spectrum

$$\upsilon(J) = 2(B/h)(J + 1).$$

The notation $\upsilon(J)$ indicates the frequency of the line for which the *lower* level has quantum number J. The double prime is discarded as it is understood (by convention) that in such expressions J is the quantum number of the *lower* level.

16.12.5 Working in wavenumbers

As with the vibrational energy levels, it is sometimes convenient to express the 'energies' of the levels directly in wavenumbers; to do this, we divide the energy in joules by $h\tilde{c}$ to give the following expression for the energies

$$\tilde{E}_J = \tilde{B}J(J + 1) \qquad \tilde{B} = \frac{\hbar^2}{2h\tilde{c}I} \quad \text{or} \quad \tilde{B} = \frac{h}{8\pi^2\tilde{c}I} \qquad J = 0, 1, 2, \ldots \quad (16.32)$$

The rotational constant $\tilde{B}$ is now in cm^{-1}.

Using these energy levels the expression for the frequency, in wavenumbers, of the lines in the spectrum, derived using the approach of the previous section, is rather simple

$$\tilde{\upsilon}(J) = 2\tilde{B}(J + 1).$$

In the case of $^1H^{35}Cl$, the spacing was 20.41 cm^{-1}, so it immediately follows that $\tilde{B} = 20.41/2 = 10.205$ cm^{-1}. From this value we can work out I directly in the following way

$$\tilde{B} = \frac{h}{8\pi^2\tilde{c}I}$$

$$\text{hence} \quad I = \frac{h}{8\pi^2\tilde{c}\tilde{B}}$$

$$= \frac{6.626 \times 10^{-34}}{8\pi^2 \times 2.998 \times 10^{10} \times 10.205}$$

$$= 2.743 \times 10^{-47} \text{ kg m}^2,$$

which is exactly the value we found before.

16.12.6 Larger molecules

In general, the rotational behaviour of any molecule can be characterized by three moments of inertia about three mutually orthogonal axes (i.e. three axes at right angles to one another). These three axes must pass through the centre of mass of the molecule, but can otherwise point in any direction, provided that the axes remain mutually orthogonal.

If the molecule has high symmetry, such as is the case for CH_4 (tetrahedral) or SF_6 (octahedral), the three moments of inertia are all the same and the energy levels are exactly the same as those for a diatomic. If two of the moments of inertia are the same, and one is different, the molecule is classed as a *symmetric top*, and it is possible to solve the SE and find an simple expression for the energy levels. However, if all three moments of inertia are different, there is no simple expression for the energy levels, and so we have to resort to numerical (computer) calculations to find the levels.

In practice, it is usually possible to analyse the microwave spectra of rather small molecules and obtain precise information on their geometry, bond angles and bond lengths. The quality of this information is likely to be more precise that that available from diffraction studies. However, for larger molecules it becomes very difficult to relate the measured spectrum to a detailed molecular structure, and in such cases diffraction is the only viable method for determining precise structures.

▷ Structure determination by X-ray diffraction methods is discussed in section 1.4 on page 10.

16.13   Vibration–rotation spectrum of a diatomic

So far we have analysed the spectrum arising from transitions between the vibrational levels, and between the rotational levels of a diatomic. However, it must surely be the case that the molecule is vibrating and rotating *at the same time*, so how can we talk of these motions are being in some way separate?

What comes to our aid here is that the vibrational and rotational motions have very different energies, and hence take place on very different time scales. A typical vibrational frequency is of the order of 10^{13} Hz, whereas a typical rotational constant is of the order of 10^{10} Hz. So, by the time one rotation is complete, the molecule has vibrated thousands of times.

As a result, when thinking about the rotation of a molecule we can simply assume that the internuclear distance is at its *average* value, rather than worrying about it changing constantly. Similarly, when thinking about the vibrational motion, we need not worry that the molecule will change orientation due to rotation as this is very slow compared to the vibration. This idea that the energy can be separated into contributions from different types of motion which have disparate time scales is called the *Born–Oppenheimer* separation.

Using this separation, we can write the energy of a vibrating–rotating diatomic as

$$\tilde{E}_{v,J} = \tilde{E}_v + \tilde{E}_J$$
$$= \left(v + \tfrac{1}{2}\right)\tilde{\omega} + \tilde{B}J(J+1),$$

where we have chosen to work in wavenumbers. The vibrational part of the energy is $\tilde{E}_v$, which we have approximated using the harmonic oscillator energy levels, specified by the vibrational quantum number v. The rotational energy is $\tilde{E}_J$, which is approximated by the rigid rotor energy levels, specified by the rotational quantum number J.

When the molecule undergoes a vibrational transition, say from $v = 0$ to $v = 1$, it is possible for the rotational energy to change at the same time. Since rotational energies are much smaller than vibrational energies, these changes in rotational energy only alter the energy of the transition by a small amount from that due to the change in vibrational energy. The result is a set of lines which are clustered around the frequency of the vibrational transition and which are known as *rotational fine structure*.

Figure 16.35 on the following page shows the rotational structure associated with the $v = 0 \rightarrow 1$ vibrational transition of $^1\mathrm{H}^{35}\mathrm{Cl}$. The transitions are centred at around 2885 cm^{-1}, but the region plotted is only 500 cm^{-1} wide, hence the

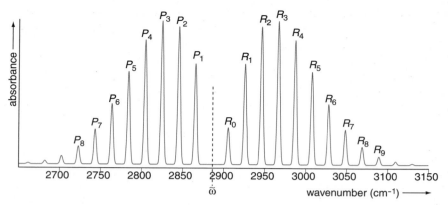

Fig. 16.35 Simulated IR spectrum of $^1H^{35}Cl$ showing the rotational fine structure associated with the vibrational transition from $v = 0$ to $v = 1$. The frequency of the pure vibrational transition (i.e. unaccompanied by any change in rotational energy) is indicated by the dashed line; this frequency is called the *band origin*. To the low frequency side of the band origin there are a series of peaks due to transitions in which the rotational quantum number decreases by one in going from $v = 0$ to $v = 1$: these form the P branch. To the high frequency side are the lines of the R branch in which the rotational quantum number increases by one.

description of these peaks as being 'fine structure'. It is also common to describe the whole set of peaks as a *band*. The many lines which are seen in the band are *all* due to the vibrational transition $v = 0 \rightarrow 1$, but differ by the amount by which the rotational energy changes. We are now going to work out the detailed form of this fine structure.

In respect of the changes in vibrational energy, the selection rules remain the same as we have already described i.e. there must be a change in dipole moment associated with the vibration and, in the harmonic case, the vibrational quantum number can only change by one. For the accompanying change in rotational energy, it turns out that the J quantum number can either *increase* or *decrease* by one i.e. $\Delta J = \pm 1$.

As has already been explained, the most intense vibrational transition is from $v = 0$ to $v = 1$ since it is only the $v = 0$ level which is significantly populated; we will therefore restrict ourselves to considering this transition. To be as general as possible, we will assume that the transition is from an energy level with vibrational quantum number $v = 0$ and rotational quantum number J'', to a level with $v = 1$ and rotational quantum number J'. The 'energy' in wavenumbers of the lower and upper states are therefore

$$\tilde{E}_{0,J''} = \tfrac{1}{2}\tilde{\omega} + \tilde{B}J''(J'' + 1) \quad \text{and} \quad \tilde{E}_{1,J'} = \tfrac{3}{2}\tilde{\omega} + \tilde{B}J'(J' + 1).$$

The frequency of the transition between these two levels is therefore $\tilde{E}_{1,J'} - \tilde{E}_{0,J''}$, which is

$$\tilde{v}(J'' \rightarrow J') = \tilde{\omega} + \tilde{B}J'(J' + 1) - \tilde{B}J''(J'' + 1). \tag{16.33}$$

Recall that when we were just considering the vibrational energy levels we predicted that the transition would occur at frequency $\tilde{\omega}$; this is called the *pure vibrational transition* as it does not involve any contribution from rotational energies. It is also referred to as the *band origin*.

If the J value of the upper level is one greater than that of the lower level i.e. $J' = J'' + 1$, we obtain a set of lines known collectively as the *R branch*. These lines can also be described as having $\Delta J = +1$, where $\Delta J = J' - J''$. If the J value in the upper level is one less than that in the lower level i.e. $J' = J'' - 1$ or $\Delta J = -1$, the set of lines known as the *P branch* is obtained.

We will work out the line positions in the R branch first of all. Putting $J' = J'' + 1$ into Eq. 16.33 on the preceding page gives

$$\begin{aligned}
\tilde{\nu}_{\mathrm{R}}(J'') &= \tilde{\omega} + \tilde{B}\left[J'' + 1\right]\left(\left[J'' + 1\right] + 1\right) - \tilde{B}J''\left(J'' + 1\right) \\
&= \tilde{\omega} + \tilde{B}\left(J''^2 + 3J'' + 2\right) - \tilde{B}\left(J''^2 + J''\right) \\
&= \tilde{\omega} + 2\tilde{B}\left(J'' + 1\right).
\end{aligned}$$

The line has been denoted $\tilde{\nu}_{\mathrm{R}}(J'')$ to indicate that it is the line in the R branch which comes from the rotational level J''; an alternative notation is to label this line $R_{J''}$. As J'' goes through its range of values $0, 1, 2, \ldots$, this expression tells us that there will be a series of lines at

$$\tilde{\nu}_{\mathrm{R}}(0) = \tilde{\omega} + 2\tilde{B} \quad \tilde{\nu}_{\mathrm{R}}(1) = \tilde{\omega} + 4\tilde{B} \quad \tilde{\nu}_{\mathrm{R}}(2) = \tilde{\omega} + 6\tilde{B} \ldots$$

In other words the R branch is a series of lines, spreading to the high-frequency side of the band origin at $\tilde{\omega}$, and separated by $2\tilde{B}$. These lines, labelled R_0, R_1 $\ldots$, can clearly be seen in Fig. 16.35 on the facing page.

For the P branch we have $J' = J'' - 1$ which can be substituted into Eq. 16.33 in the same way. Following through a similar calculation gives as the following general expression for the line in the P branch

$$\tilde{\nu}_{\mathrm{P}}(J'') = \tilde{\omega} - 2\tilde{B}J''.$$

An alternative notation for these lines is $P_{J''}$. Since $J' = J'' - 1$ the lowest value of J'' we can substitute into this expression is $J'' = 1$. We cannot put $J'' = 0$ since this would result in a transition to '$J' = -1$', which does not exist. This expression predicts that in the P branch there will be a series of lines, spreading to the *low* frequency side of the pure vibrational transition, and separated by $2\tilde{B}$; these are clearly visible in Fig. 16.35.

Figure 16.36 on the next page shows the energy levels involved, the allowed transitions between them, and the relationship to the lines in the P and R branches.

From such a band we can determine a value for the rotational constant, since the spacing between successive lines in either the P or R branches, is $2\tilde{B}$. As in the case of the rotational spectrum (the microwave spectrum), this value of B will lead to a value of the bond length. From the position of the band origin we can determine the vibrational frequency.

Intensities

The intensities of the lines in the P and R branches reflect the populations of the rotational energy levels in the $v = 0$ vibrational level. As has already been discussed, these populations at first increase with J, reach a maximum and then fall off. The intensities therefore behave in the same way, as can clearly be seen in Fig. 16.35 on the facing page.

Since as J increases the P and R branch lines spread out from the band origin in opposite directions, the way in which the population of the rotational levels changes with J leads to a 'double hump' which is very characteristic of a band with a P and R branch.

Fig. 16.36 Visualization of how the P and R branches associated with a vibrational transition arise. The lower part of the diagram shows the rotational energy levels, labelled by the quantum number J'', associated with the ground vibrational state which has $v = 0$. In the upper part of the diagram we see the rotational levels associated with the first excited vibrational state $v = 1$; these levels are labelled by the quantum number J'. The separation between the vibrational levels is much greater than that between the rotational levels, as indicated by the dashed lines. Transitions in which the J value increases by one are shown by the red arrows, and lead to successive lines in the R branch; transitions in which J decreases by one are shown by the blue arrows and form the P branch. A schematic spectrum is shown at the foot of the diagram; each peak is shown beneath the arrow from which it arises.

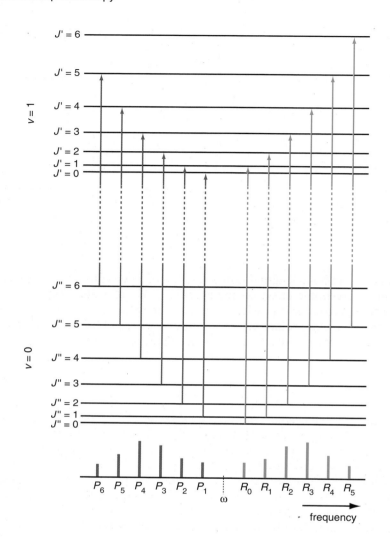

Further complications

Figure 16.37 on the next page shows the experimental IR spectrum of the $v = 0 \rightarrow 1$ transition from $^{12}C^{16}O$; the P and R branch structure can be seen clearly. However, close inspection will show that the spacing of the lines in the branches is not quite constant. In the R branch as J increases the lines are getting closer together, whereas in the P branch they are getting further apart as J increases. This does not agree with the predicted form of the spectrum, for example as shown in Fig. 16.35 on page 744, in which the spacing is constant at $2\tilde{B}$.

In predicting the spectrum we assumed that the rotational constant $\tilde{B}$ is the *same* in the $v = 0$ and $v = 1$ vibrational states. In fact this is not quite true, since, due to the anharmonicity of the potential, the average bond length in the $v = 1$ state is greater than that in the $v = 0$ state (this was described in section 16.7.4 on page 728). As a result the moment of inertia is greater for the $v = 1$ state, and this means that the rotational constant is smaller.

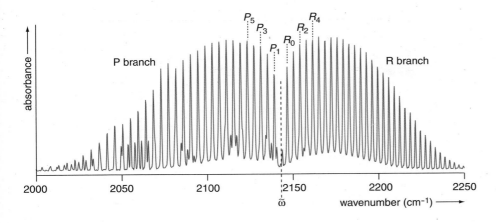

Fig. 16.37 Experimental IR spectrum of $^{12}C^{16}O$ showing the P and R branches associated with the $v = 0 \rightarrow v = 1$ vibrational transition. Broadly speaking the spectrum is of the same form as the simulated spectrum shown in Fig. 16.35 on page 744, although in the experimental spectrum there are some extra peaks due to impurities. Careful inspection will show that the spacing of the lines in the P and R branches is *not* constant.

The variation in bond length is only a fraction of a percent, but it is sufficient to result in a perceptible effect on the spectrum. Using a more sophisticated analysis of the spectrum it is possible to extract the values of the two rotational constants and then extrapolate these back to give the equilibrium bond length, which is the value at the bottom of the potential energy well.

16.14 The hydrogen atom

Of all the cases in which the Schrödinger equation can be solved exactly, the hydrogen atom (or more generally, a one-electron atom) is by far the most important and far-reaching for its consequences in chemistry. We have already devoted the whole of Chapter 2 to the properties of the hydrogen atomic orbitals, which are the solutions to the Schrödinger equation for the hydrogen atom. There is little more to say about these solutions, other than to remind ourselves of their form and the associated energies.

The wavefunctions can be written as product of an angular part $Y_{l,m_l}(\theta, \phi)$ and a radial part $R_{n,l}(r)$

$$\psi_{n,l,m_l}(r, \theta, \phi) = \underbrace{R_{n,l}(r)}_{\text{radial part}} \times \underbrace{Y_{l,m_l}(\theta, \phi)}_{\text{angular part}}.$$

There are three quantum numbers associated with these solutions:

(a) The *principal* quantum number, n, which takes values 1, 2, 3 ...

(b) The *orbital angular momentum* quantum number, l, which takes values from $(n - 1)$ down to 0, in integer steps.

(c) The *magnetic* quantum number, m_l, which takes values from $+l$ to $-l$ in integer steps; there are thus $(2l + 1)$ different values of m_l.

The energy depends *only* on the value of n

$$E_n = -\frac{R_H}{n^2}, \tag{16.34}$$

where R_H is the Rydberg constant (2.180×10^{-18} J). As n increases the number of possible values of l and m_l associated with a given n increases; all of these orbitals which share the same value of n are degenerate, and we can associate this high degeneracy with the high (spherical) symmetry of the problem.

The functions $R_{n,l}(r)$ are called the *associated Laguerre functions* and are responsible for the occurrence of any radial nodes. All of these functions include a term of the form $\exp(-\text{const.} \times r)$ which ensures that the wavefunction goes to zero at large distances: this behaviour is needed to ensure that the integral of the square of the wavefunction is finite.

The angular functions $Y_{l,m_l}(\theta, \phi)$ are the *spherical harmonics* and are responsible for the occurrence of nodal planes or angular nodes. The total number of nodes (radial plus angular) is $(n - 1)$, and so increases as n increase, which in turn means that the number of nodes increases as the energy increases. This is exactly the behaviour we have seen in all of the solutions to the Schrödinger equation.

16.14.1 Spectrum of the hydrogen atom

Since the energy levels of the hydrogen atom depend only on the principal quantum number n, which takes values 1, 2, 3, ..., the resulting spectrum is simple to predict. There is no selection rule for n: any change is allowed.

Hydrogen atoms are usually generated by passing an electric discharge through H_2 gas, resulting in H atoms in a variety of excited states. As these excited states drop down to lower levels, photons are emitted whose energies match the separation of the two levels in the usual way – it is therefore the *emission* spectrum of H atoms which is studied.

If the upper state has principal quantum number n_{upper} and the lower has quantum number n_{lower}, then the energy separation is easily calculated using Eq. 16.34 on the preceding page

$$
\begin{aligned}
E_{n_{upper} \rightarrow n_{lower}} &= E_{n_{upper}} - E_{n_{lower}} \\
&= -\frac{R_H}{n_{upper}^2} - \left(-\frac{R_H}{n_{lower}^2}\right) \\
&= R_H \left(\frac{1}{n_{lower}^2} - \frac{1}{n_{upper}^2}\right).
\end{aligned}
$$

The corresponding frequency is found by dividing by h. It is probably easiest to think about these transitions in terms of the *wavelength* of the photon which is emitted, which can be computed from

$$
\lambda_{n_{upper} \rightarrow n_{lower}} = \frac{hc}{E_{n_{upper} \rightarrow n_{lower}}}.
$$

Table 16.2 lists the wavelengths (in nm) of some emissions expected for the hydrogen atom. The transitions which end up at $n_{lower} = 2$ are particularly important as they occur in the visible part of the spectrum; they are called the *Balmer* series. Those which end up at $n_{lower} = 1$ occur in the UV and are called the *Lyman* series. These two sets of transitions are also illustrated in Fig. 16.38.

Table 16.2 Wavelengths (nm) of emission lines in the spectrum of the H atom

n_{lower}	5	4	3	2	n_{upper}
1	94.9	97.2	103	121	
2	434	486	656		
3	1281	1875			
4	4050				

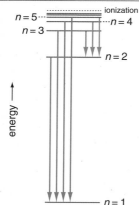

Fig. 16.38 Illustration of the transitions in the hydrogen atom which lead to the Lyman series (green arrows, all ending at $n = 1$) and the Balmer series (red arrows, all ending at $n = 2$).

16.15 Electronic transitions

With the sole exception of one-electron atoms (such as hydrogen), it is not possible to find an exact solution to the Schrödinger equation for the electrons in atoms and molecules. As we noted in Chapter 2, it is the presence of electron–electron repulsion which makes solving the Schrödinger equation so difficult. With modern high-speed computers numerical solutions to the Schrödinger equation can, however, be found.

When thinking about the electronic structure of atoms and molecules we have always used the orbital approximation. As described in section 2.6.1 on page 66, this is an approximation in which we calculate (numerically) the wavefunction for each electron, assuming that it is experiencing some kind of average interaction with all of the other electrons. These one-electron wavefunctions are the atomic or molecular orbitals we have been using so frequently throughout the book.

The orbital approximation leads us to the idea that there are a set of (atomic or molecular) orbitals, which are occupied by the electrons, subject to the usual rule that only two spin-paired electrons can occupy each orbital. We also imagine that there are additional orbitals which are not filled, but to which electrons can be promoted. Just in the same way that the absorption of photons can promote a molecule from one rotational or vibrational level to another, we can also imagine that an appropriately energetic photon could promote an electron from an occupied orbital to an unoccupied one. The spectrum due to such transitions could therefore be interpreted in terms of the energies of the orbitals involved.

Unfortunately, such an interpretation is only very approximate. The difficulty is that when an electron is moved from one orbital to another, the details of the electron-electron repulsion with *all* of the other electrons changes and as a result the orbital energies themselves change. We therefore cannot expect that the energy of the photon will match the separation in energy between two orbitals in the ground state molecule or atom. This problem was encountered before, in section 2.6.5 on page 79, when discussing the relationship between orbital energies and ionization energies.

For small molecules it is possible to compute not just the orbital energies but the overall electronic energy of the whole molecule, both for the ground state and various excited states. Spectra can then be interpreted as arising from transitions between these energy levels of the whole molecule. These transitions are accompanied by simultaneous changes in vibrational and rotational energies, leading to fine structure. The resulting spectra can be very complicated indeed, and their analysis is a very challenging task, but a rewarding one as it reveals many subtle details about the structure and bonding in the molecule.

If we are content with a qualitative description of spectra, then we can return to the idea that absorptions arise due to electrons moving between orbitals, accepting that this is just a rough approximation. Such an approach is frequently used to interpret the spectra of organic molecules, particularly those which have conjugated π bonding; it is to this topic which we now turn.

Typically the spectra of the kind of relatively large organic molecules which absorb in the visible are recorded from solutions, which results in low resolution. All of the vibrational and rotational structure simply merges together to give rather featureless spectra with broad bands.

16.15.1 Spectra of conjugated systems

The energy spacing between molecular orbitals is normally such that transitions between them require photons in the UV or visible region of the spectrum. Promoting an electron from the HOMO to the LUMO, as shown in Fig. 16.39, involves the least amount of energy of any transition, and hence gives rise to an absorption at the longest wavelength. Our discussion will therefore focus on this transition.

Molecules that absorb in the *visible* region appear to our eyes to be coloured, and so are of considerable technological importance as they are used in dyes, inks and paints. The process of photosynthesis and the way in which our eyes are able to detect light both rely on the absorption of visible light by molecules. It is therefore important to understand what properties a molecule must have for it to absorb in the visible region, and how the exact frequency which is absorbed can be altered.

For a simple molecule such as CH_4 the lowest energy transition occurs well into the UV region of the spectrum, at around 130 nm (the absorption is very broad). For ethene, the HOMO→LUMO transition will be between a π and a $\pi^\star$ MO, and we might reasonably expect these MOs to be closer together than the σ and $\sigma^\star$ MOs in CH_4. Even so, the lowest energy absorption in ethene is still well into the UV at around 160 nm. In butadiene, in which there are two conjugated π bonds, the absorption shifts to longer wavelengths, occurring at about 210 nm. This gives us a hint that increasing the amount of conjugation will shift the HOMO→LUMO transition to lower frequencies (and longer wavelengths).

In fact, Fig. 5.1 on page 181 offers a convenient explanation for this effect of increasing the amount of conjugation. This diagram shows the MOs arising from chains of s orbitals of increasing length, but essentially the same picture will arise for a line of out-of-plane $2p$ orbitals forming a conjugated π system. What we see from the picture is that as the number of interacting orbitals increases, the spacing between all of them (and hence the HOMO/LUMO spacing) *decreases*. We therefore expect that the HOMO→LUMO transition will move to lower energies and hence longer wavelengths as the size of the conjugated system increases.

The same conclusion can be reached from an even simpler argument based on the particle-in-a-box energy levels. In a very crude way (and it is indeed crude) we can imagine that the π electrons are confined to a 'box' whose length is from one end of the π system to the other. The electrons can then be thought of as occupying the corresponding particle-in-a-box energy levels, with two spin-paired electrons per level. As we have seen (Eq. 16.11 on page 703), the energies of these levels go as $1/L^2$, where L is the length of the box. Therefore, increasing the length of the box (i.e. increasing the size of the π system) will make the energy levels get closer together, and hence lower the energy of the HOMO→LUMO transition.

A good example of the extent of conjugation needed in order for a molecule to absorb in the visible is β-carotene (whose structure is shown below), one of the molecules responsible for the colour of carrots.

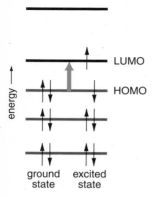

Fig. 16.39 The lowest energy electronic transition usually involves promoting an electron from the HOMO to the LUMO i.e. from the ground electronic state to an excited electronic state.

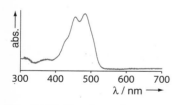

Fig. 16.40 The absorption spectrum of β-carotene in the visible region.

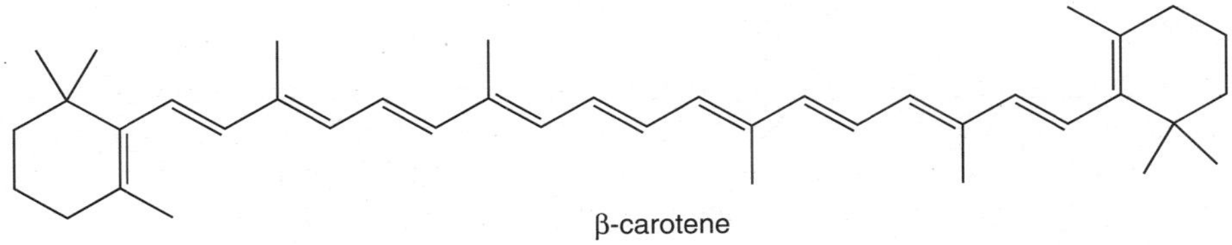

β-carotene

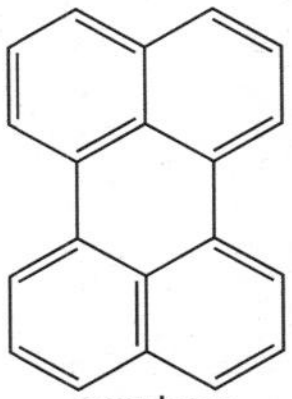

There are twenty-two carbon atoms involved in the conjugated π system. As shown Fig. 16.40 on the facing page, the molecule has a broad absorption centred at around 450 nm, which accounts for its colour.

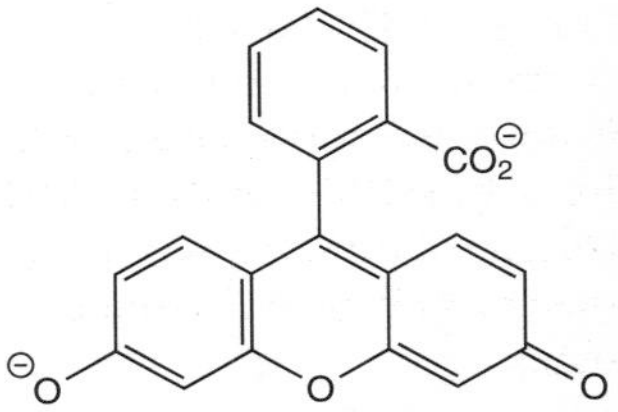

The conjugation does not have to be in the form of a long chain, but can also involve the π bonding systems of benzene rings and other fused ring aromatic molecules. Benzene itself absorbs at around 250 nm, so is colourless, but the more extensive conjugation in a molecule such as perylene (shown opposite) results in a strong absorption at around 425 nm.

Atoms other than carbon can be involved in the conjugated system: for example the dianion of fluorescein has a conjugated system involving both oxygen and carbon. This species has a strong absorption at 500 nm, accounting for its vivid green colour.

16.15.2 Transition metal complexes

One of the particular features of transition metal complexes is their often striking colours. The absorptions which give rise to these colours are usually due to electronic transitions involving the d electrons. Such spectra are discussed in section 15.5.2 on page 667.

perylene

fluorescein dianion

FURTHER READING

Elements of Physical Chemistry, fourth edition, Peter Atkins and Julio de Paula, Oxford University Press (2005).

Physical Chemistry, eighth edition, Peter Atkins and Julio de Paula, Oxford University Press (2006).

Quantum Mechanics 1: Foundations, N. J. B. Green, Oxford University Press (1997).

QUESTIONS

Physical constants: $h = 6.626 \times 10^{-34}$ J s, $k_B = 1.381 \times 10^{-23}$ J K^{-1}, $N_A = 6.022 \times 10^{23}$ mol^{-1}, $c = 2.998 \times 10^{8}$ m s^{-1}, mass unit $(u) = 1.6605 \times 10^{-27}$ kg.

16.1 Show that $\phi_1(x) = A \sin(Bx)$ and $\phi_2(x) = C \cos(Bx)$ are both eigenfunctions of the operator d^2/dx^2, and find the corresponding eigenvalue in each case. Do different values of A and C give different eigenfunctions? Are these two eigenfunctions degenerate?

16.2 State the *Born interpretation* of the wavefunction and use is to explain whether or not the following (one-dimensional) functions are likely to be acceptable wavefunctions. (The dashed lines in (c) imply that the sine function continues indefinitely to positive and negative x.)

(a)   (b)   (c)

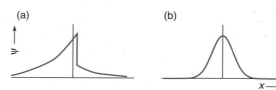 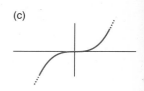

16.3   Imagine a free particle, moving in one dimension, which has total energy E and experiences a fixed potential energy V_0. Explain why it is that in classical mechanics the potential energy cannot exceed the total energy.

Without going into mathematical details, describe the behaviour of a typical eigenfunction of the hamiltonian (the operator for energy) for the classical case where $E > V_0$, and for the non-classical case where $E < V_0$. What happens to the eigenfunction when the potential V_0 becomes very large?

16.4   Consider a particle which experiences the potential shown below in red. The potential is zero in the region (a), equal to V_0 in the region (b), and zero again in the region (c). Also shown in blue is a typical eigenfunction ψ of the hamiltonian; rationalize the behaviour of this eigenfunction. What will happen to ψ as V_0 is increased, and as V_0 is decreased?

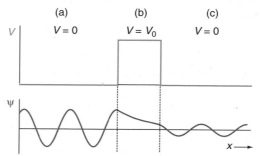

16.5   (a)   Write down the hamiltonian for a particle of mass m experiencing a potential $V_0 = 0$ and moving in one dimension. Show that $\psi(x) = A \cos(kx)$ is an eigenfunction of this hamiltonian, and give an expression for the energy (the eigenvalue) in terms of k.

(b)   Explain the boundary conditions which are imposed on the wavefunction by a potential which is zero between $x = 0$ and $x = L$, and infinite elsewhere.

(c)   Discuss whether or not $\psi(x) = A \cos(kx)$ satisfies these boundary conditions.

16.6   With reference to the wavefunctions and energy levels for the particle in a (one-dimensional) box, explain what you understand by the terms (a) quantum number, (b) normalized, (c) orthogonal, and (d) zero-point energy.

16.7   Why do the number of nodes in a wavefunction increase as its energy increases?

16.8   (a)   With reference to the case of a particle in a two-dimensional square well (described in section 16.4 on page 710), explain what you understand by a *degenerate* energy level.

(b)   Calculate the energies (expressed as multiples of $(\hbar^2 \pi^2)/(2mL^2)$), for all of the levels with n_x in the range 1 to 4, and for n_y in the range 1 to 4. Arrange these in order of increasing total energy. Why are some of the levels degenerate, while others are not?

16.9   In the case of an harmonic oscillator with potential $V(x) = \frac{1}{2} k_f x^2$, what do you understand by the 'classical' and 'non-classical' regions of the oscillation? How does the behaviour of the wavefunction differ between the two regions?

16.10   In the IR spectrum of $^1H^{35}Cl$ there is a strong absorption centred at 2885 cm^{-1}. Assuming that the vibrational energy levels can be approximated by those of the harmonic oscillator, use this observation to (a) determine the frequency of the oscillation (in Hz and in rad s^{-1}), (b) determine the force constant of the bond. (^1H = 1.0078 mass units and ^{35}Cl = 34.9689 mass units).

16.11   The 'energy in wavenumbers' of the energy levels of a Morse oscillator are given by Eq. 16.29 on page 728

$$\tilde{E}_v = \left(v + \tfrac{1}{2}\right)\tilde{\omega}_e - \left(v + \tfrac{1}{2}\right)^2 \tilde{\omega}_e x_e,$$

(a)   Using these energies, show that the wavenumber of the $0 \to 1$ and $0 \to 2$ transitions are given by $\tilde{\omega}_e - 2\tilde{\omega}_e x_e$ and $2\tilde{\omega}_e - 6\tilde{\omega}_e x_e$, respectively.

(b)   In the spectrum of $^1H^{35}Cl$ there is a strong absorption centred at 2885 cm^{-1}, and a weaker absorption centred at 5665 cm^{-1}. Use these data to find values for $\tilde{\omega}_e$ and $\tilde{\omega}_e x_e$ (in cm^{-1}), explaining your method.

(c)   Convert the values of $\tilde{\omega}_e$ and $\tilde{\omega}_e x_e$ from cm^{-1} to rad s^{-1}. Hence determine the values of the Morse parameters a and D_e.

16.12   Using the expression for the energy levels of the Morse oscillator (in cm^{-1}) given in the previous question, show that the value of the vibrational quantum number at which dissociation occurs, v_{max}, is given by

$$v_{max} = \frac{1}{2x_e} - \frac{1}{2}.$$

Hence show that the dissociation energy, expressed in wavenumbers, is

$$\tilde{D}_e = \frac{(\tilde{\omega}_e)^2}{4\tilde{\omega}_e x_e}.$$

Using the values of $\tilde{\omega}_e$ and $\tilde{\omega}_e x_e$ from the previous question, determine $\tilde{D}_e$ (in wavenumbers). Convert your answer to joules and confirm that it is the same as you obtained in the previous question.

16.13   Illustrated below are five of the normal modes of ethyne (acetylene, HCCH). At its equilibrium geometry this molecule has no dipole moment, but for some of the normal modes the distortion during the vibration induces a dipole. Determine which of the normal modes are IR active and explain why modes 4 and 5 are each doubly degenerate.

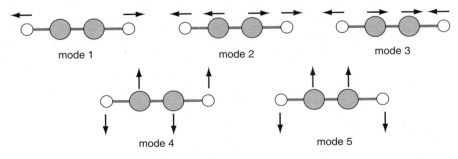

16.14   In a Raman scattering experiment on $^1H_2$ using a green argon ion laser (which emits at a frequency of 20492 cm^{-1}) very strong scattering of light at 20492 cm^{-1} was seen, along with much weaker scattering at 16333 cm^{-1}. Explain the origin of the scattered light at these two different frequencies. Hence determine the vibrational frequency of $^1H_2$ and the corresponding force constant (use the harmonic oscillator energy levels; ^1H = 1.0078 mass units).

16.15 The microwave (rotational) spectrum of $^{12}C^{16}O$ consists of a series of lines spaced by 115.3 GHz (1 GHz $\equiv 10^9$ Hz). Assuming that the rotational energy levels of this molecule can be approximated by those of the rigid rotor, use these data to determine a value for the rotational constant B, the moment of inertia I, and hence the bond length R. Be careful to work in SI units throughout. (^{12}C = 12.000 mass units and ^{16}O = 15.995 mass units.)

16.16 Using the same approach as is described in section 16.13 on page 743, show that the frequencies of the lines in the P branch of the vibration–rotation spectrum of a diatomic are given by

$$\tilde{\nu}_P(J'') = \tilde{\omega} - 2\tilde{B}J''.$$

What range of values can the quantum number J'' take in this expression?

16.17 In the IR spectrum of $^{12}C^{16}O$ there is a strong band at around 2157 cm^{-1} which, on closer inspection, shows P/R branch structure. The spacing between successive lines in the two branches is roughly constant at 3.85 cm^{-1}. Use these data to estimate the bond length in $^{12}C^{16}O$, explaining any approximations you find it necessary to make. (^{12}C = 12.000 mass units and ^{16}O = 15.995 mass units.)

16.18 Why is it that although the rotational level with $J = 0$ has the lowest energy, the first line in the R branch (R_0) is not the most intense?

16.19 Verify that the wavelengths quotes in Table 16.2 on page 748 are correct.

16.20 A very simple model for the electronic structure of a conjugated polyene

is to assume that the π electrons move in a one-dimensional box of length Na, where N is the number of conjugated double bonds and a is the typical length of such a bond. Given that the energy levels of a particle in a box are

$$E_n = \frac{n^2 h^2}{8mL^2} \qquad n = 1, 2, \ldots,$$

the energy levels available to the π electrons are therefore

$$E_n = \frac{n^2 h^2}{8m_e N^2 a^2} \qquad n = 1, 2, \ldots,$$

where m_e is the mass of the electron (9.11×10^{-31} kg).

(a) Given that N conjugated π bonds contribute $2N$ π electrons, and that two electrons occupy each energy level, explain why the highest occupied level has $n = N$, and the lowest unoccupied level has $n = N + 1$.

(b) Show that the energy of the photon needed to cause a transition between the highest occupied and lowest unoccupied levels is given by

$$E_{photon} = \frac{(2N + 1)h^2}{8m_e N^2 a^2},$$

and hence that the corresponding wavelength of the light that will cause this transition is

$$\lambda = \frac{8m_e c N^2 a^2}{(2N + 1)h}.$$

(c) Explain why this expression for λ predicts that as the number of conjugated double bonds increases the wavelength of the absorption also increases. Taking the value of the length a to be 220 pm, find the value of λ for $N = 6$. What region of the spectrum does this correspond to?

Chemical thermodynamics

Key points

- The First Law of Thermodynamics is concerned with the conservation of energy, and can be written $\Delta U = q + w$.
- The internal energy U is a state function, whereas the heat q and the work w are path functions.
- The work done expanding a gas through volume $\mathrm{d}V$ against a pressure p is $-p\,\mathrm{d}V$.
- The work and heat in a reversible process are a maximum.
- ΔU is equal to the heat at constant volume; ΔH is equal to the heat at constant pressure.
- The molar Gibbs energy of a gas in a mixture depends on its partial pressure.
- Heat capacity data can be used to convert tabulated $\Delta_r H^\circ$ and $\Delta_r S^\circ$ values from one temperature to another.
- The temperature dependence of the equilibrium constant is given by the van't Hoff isochore.
- Values of $\Delta_r H^\circ$ and $\Delta_r S^\circ$ can be found from measurements of the temperature dependence of the equilibrium constant.

In this chapter we return to the topic of thermodynamics, first introduced in Chapter 6. In that chapter the aim was to get to the relationships

$$\Delta_r G^\circ = -RT \ln K \quad \text{and} \quad \Delta_r G^\circ = \Delta_r H^\circ - T\Delta_r S^\circ$$

as quickly as possible, since from the chemist's point of view these are the really practical and useful part of thermodynamics. In this chapter, we are going to fill in some of the details which were passed over in our earlier discussion. For example, we will look more carefully at what the internal energy and enthalpy are, and why they are useful to us. We will also prove the relationship $\Delta_r G^\circ = -RT \ln K$, and then go on to show how this can be developed further, for example to discuss the temperature dependence of the equilibrium constant. We will also look at how the tabulated values of standard enthalpy changes and standard entropies at 298 K can be corrected to other temperatures.

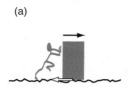 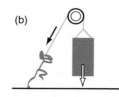 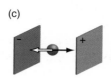

Fig. 17.1 Different situations in which work is done: the force against which the work is done is indicated by the open arrow, and the direction of movement by the solid arrow. In (a) an object moves against a frictional force, in (b) a weight is raised against the force due to gravity, and in (c) a positively charged particle is moved towards a positively charged plate.

The approach we will take in this chapter is more mathematical than that in Chapter 6, and in particular we will use the ideas and techniques from elementary calculus. If you need a refresher on these ideas, you can refer to section 20.5 on page 900 and section 20.6 on page 905. It is all too easy in thermodynamics to get lost in the mathematics and for the topic to seem like the endless manipulation of symbols. We will do our level best to avoid this, always making it clear why a particular manipulation is being performed and what the physical interpretation of the result is.

Our discussion starts with the First Law of Thermodynamics which, although perhaps not so far reaching in its consequences as the Second Law, is nevertheless very important in the application of thermodynamics in chemistry.

17.1   The First Law

The First Law of Thermodynamics is about the conservation of energy. There are many ways in which it can be stated, but we will opt for

> **First Law:** Energy can neither be created nor destroyed but is just transformed from one form into another.

Like all physical laws, the First Law cannot be proved but its validity is established by the fact that the predictions it makes are in accord with experiments and observations.

Energy comes in many different forms, but the ones of interest here are *heat*, *work* and *internal energy*. Of these, heat is already a familiar idea: when energy is transferred from a hotter object to a cooler one, we say that the energy has been transferred as heat. We tend to say that heat 'flows' from the hotter object to a cooler one, but this language is rather dangerous as it makes it sound as if heat is a kind of fluid, which it is not. Rather, heat is the name we give to the process of the transfer of energy from one body to another as a result of a temperature difference.

Work is a concept which is usually introduced in the study of mechanical systems, and the commonest definition is that 'work is done when moving against a force'; various scenarios in which work is done are illustrated in Fig. 17.1. For example, in (a) work is done when a box is slid along the floor: the force is due to friction. Similarly, in (b) work is done when a mass is raised up using a rope and pulley: the force is due to the gravitational attraction the mass experiences. We are going to be particularly interested in the work done when a gas expands. In this situation, the force is that due to the surroundings which are resisting the expansion.

If it has been a while since you have studied Chapter 6, it would be as well to review that chapter before embarking on this one. Sections 6.5 to 6.9 are particularly relevant.

Another kind of work which we will come to later is *electrical work*. This is done when a charge moves in an electric potential. For example, if we imagine the electric potential as being created by two plates with opposite charges, then moving a positively charged species (such as an ion) towards the positive plate results in electrical work being done, as shown in (c). This is because there is a force repelling the positive charge away from the plate.

Internal energy is quite different from heat and work. Whereas heat and work are things that 'happen' to an object, in the sense that work is done or heat flows, internal energy is a property which an object possesses. We can think of it as energy which is stored inside the object, ready to be released in some other form.

At a molecular level, the various ways in which internal energy is stored are illustrated in Fig. 17.2. These stores of energy are the kinetic energy of the molecules due to their thermal agitation, the energy of interaction between the molecules (for example, the interaction between dipoles), and the chemical bonds. This latter source of internal energy is particularly important, since in chemical reactions the breaking and making of bonds often results in significant changes in the internal energy.

The First Law says that these different forms of energy can be interconverted, but that the total amount of energy has to remain constant. For example, if the internal energy is reduced, the resulting energy can appear as heat or be used to do work, provided that the sum of the heat and work energy is equal to the reduction in the internal energy. Similarly, heat supplied to a system or work done on the system can increase its internal energy, but the increase in the internal energy has to be equal to sum of the heat and work energy.

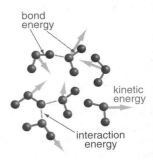

Fig. 17.2 The internal energy is stored as kinetic energy of the molecules, the interaction energy between molecules, and – most importantly – within the chemical bonds.

17.1.1   State functions and path functions

There is an important distinction between internal energy on the one hand, and heat and work on the other: the internal energy is a *state function*, whereas heat and work are *path functions*.

State functions

As the name implies, a state function is one whose value only depends on the *state* of the system. By 'state' we mean those variables which define the physical state and chemical composition of the system e.g. the pressure, the temperature and the amount in moles.

For example, one mole of H_2 gas at 298 K and 1 bar pressure has a particular value of the internal energy which is *always* the same, *regardless* of how the hydrogen was brought to this state. A state function, such as the internal energy, is therefore a property of matter.

Imagine a process in which we go from state A to state B. The internal energy will have a particular value for state A, let us call it U_A, and a particular value U_B for state B. In going from A to B we can therefore talk about the *change* in internal energy, ΔU, given by

$$\Delta U = U_B - U_A.$$

Since internal energy is a state function, the value of ΔU depends only on the initial and final states and *not* on how the transformation between these states is achieved.

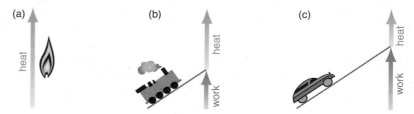

Fig. 17.3 Illustration of how the internal energy change on converting a certain amount of hydrogen and oxygen to water can appear in different combinations of heat and work. In (a) we simply burn the hydrogen, and so all of the change in the internal energy appears as heat. In (b) we burn the hydrogen and use the heat to run a steam engine; as a result some work is done, but as the engine is not 100% efficient, some heat is still produced. If the hydrogen and oxygen are used in a fuel cell to generate electricity and then run an electric motor, we can expect to obtain more work, as shown in (c).

Path functions

In contrast to the internal energy, the amount of heat or work involved in going from A to B does depend on how the transformation is achieved i.e. on the path taken. Hence, heat and work are described as *path functions*.

This is best illustrated with an example. Suppose state A is two moles of H_2 and one mole of O_2, and state B is two moles of H_2O, at some given pressure and temperature. The change in internal energy on going from A to B is $-482\,kJ$ i.e. ΔU for the process

$$\underbrace{2H_2(g) + O_2(g)}_{A} \longrightarrow \underbrace{2H_2O(g)}_{B}.$$

The First Law tells us that this reduction in internal energy can appear as any combination of heat and work, provided the sum of these two is equal to ΔU.

If we simply burnt the hydrogen and oxygen (in a sealed container) then no work will be done, so all of the reduction in internal energy will appear as heat, as shown in Fig. 17.3 (a). Alternatively we could burn the hydrogen but use the heat to create steam to run a steam engine. As a result, we could do some mechanical work. Since steam engines are not 100% efficient, not all of the change in internal energy will be available as work: inevitably, some will appear as heat, as shown in (b).

Another possibility would be to use the hydrogen and oxygen in a *fuel cell* to generate electricity, which in turn could be used to run an electric motor and so do work. Since fuel cells are generally more efficient than steam engines, we can expect that more work will be done in this case, as shown in (c).

This example illustrates that the amount of heat and work involved in going from A to B depends on how the conversion is carried out. There are as many different values of the heat and work as we can devise different ways of going from A to B.

17.1.2   Writing the First Law

For a given process (i.e. going from A to B), the First Law can be written as

$$\text{First Law:} \quad \Delta U = q + w. \tag{17.1}$$

ΔU is the change in the internal energy when we go from A to B, q is the amount of heat involved, and w is the amount of work.

Note that we do not write the heat as 'Δq', as this would imply a 'change in the heat'. Since heat is not a state function, it makes no sense to talk about a *change* in its value. All we can sensibly talk about is an *amount* of heat q associated with some particular process. The same is true of the work.

Often we will want to think about an infinitesimal process in which there is an infinitesimal change dU in the internal energy. We are using the language of calculus to express this infinitesimal change in U as dU in the same way that we write an infinitesimal change in x as dx. For such a change the First Law can be written

$$\text{First Law:} \quad dU = \delta q + \delta w. \qquad (17.2)$$

Here δq is taken to imply an infinitesimal *amount* of heat, and likewise for the work. We do not write 'dq' as this would imply a change in q, which we have already decided is not appropriate.

Sign conventions and the work done *on* or *by* the system

Since energy in the form of heat can flow in to or out of the system, we need to have a convention about which direction corresponds to a positive value of this heat energy. In fact, you are already familiar with this convention, which is that a positive value for the heat means that energy (as heat) is flowing into the system (endothermic), and a negative value means that energy (as heat) is flowing out of the system (exothermic). This means that q is the energy supplied as heat *to the system*.

The convention we will adopt for work mirrors that for heat: w is the work done *on the system*. This means that if a gas is compressed by the action of an external force w is *positive* as work is being done on the gas. On the other hand, if a gas expands by pushing against an external force, w is negative.

Although this choice of the same convention for q and w is consistent, it is not always that convenient to use. We find it more natural to talk about the work done *by* a gas, since we have a tendency to see the system (the gas) as being 'active'. The work done *by* the gas, denoted w', is simply *minus* the work done *on* the the gas

$$w' = -w.$$

We will always be careful to distinguish which kind of work we are talking about, and we will always use w in our calculations involving the First Law, Eq. 17.2.

17.2   Work of gas expansions

The work done when a gas expands against an external force is particularly important to us for two reasons: firstly, it helps us to understand the important concept of a reversible process; secondly, it is crucial in understanding why enthalpy is a useful state function.

The simplest way to think about the work associated with a gas expansion is to imagine that the gas is confined in a cylinder by a piston, as shown in Fig. 17.4; the pressure of the gas inside the cylinder is p_{int}. So we can focus on just the work done by the gas expanding against the external pressure, we imagine that the piston has no mass and moves without friction. The gas inside the cylinder, the piston and the cylinder comprise the system. This system is

Fig. 17.4 Illustration of the arrangement used for thinking about the work done when a gas expands. A gas, shown in green, is confined in a cylinder by a piston, and is at pressure p_{int}. The whole is surrounded by more gas at pressure p_{ext} which exerts a pressure on the open end of the piston.

surrounded by more gas at pressure p_{ext}, which exerts a force on the open end of the piston.

The gas inside the cylinder and the gas surrounding the cylinder both exert a force on the piston. If the internal pressure is greater than the external pressure, the force on the piston due to the gas in the cylinder is greater than the force due to the gas outside. As a result, the piston moves out, and in doing so it pushes against the force due to the external pressure. It is important to realize that although it is the gas in the cylinder which is 'doing the pushing', it is the force due to the gas surrounding the piston which is being pushed against. The work done by the gas in the cylinder therefore depends on this external force, and hence on the *external pressure*.

If the cross-sectional area of the piston is A, then the force on the piston due to the external pressure is $p_{ext} \times A$ (recall that pressure is simply force per unit area). If the piston then moves out a small distance dx, as shown in Fig. 17.5, the work is the force times this distance, which is

$$\text{work done } \textit{by} \text{ the gas} = (p_{ext} \times A)\, dx.$$

We have used the language of calculus to write the small distance as dx i.e. an infinitesimal distance.

As the piston moves out, the volume of the gas inside the piston increases. This volume increase is the cross-sectional area of the piston times the distance the piston has travelled, so in this case $A\, dx$ is the volume increase. Writing this infinitesimal volume increase as dV means that we can write the work done by the gas as

$$\text{work done } \textit{by} \text{ the gas}, \delta w' = p_{ext}\, dV.$$

Note that as this is the work done *by* the gas we have used the symbol w'. The work done *on* the gas is simply the opposite of this

$$\text{work done } \textit{on} \text{ the gas}, \delta w = -p_{ext}\, dV. \tag{17.3}$$

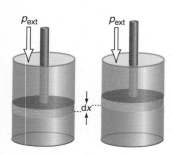

Fig. 17.5 Illustration of how to think about the work done in a gas expansion. The piston moves back by a small distance dx, therefore pushing against the force due to the external pressure. If the cross-sectional area of the piston is A, the force on the piston is $p_{ext}A$, and so the work done when the piston moves is $-p_{ext}A\, dx$.

17.2.1 Expansion against a constant external pressure

Equation 17.3 is the work done (on the gas) when the gas expands by an infinitesimal volume dV. If we want to calculate the work when the gas expands by a finite amount, say from volume V_i to volume V_f, then we will need to integrate the right-hand side of Eq. 17.3 between these two volumes

$$\text{work done } \textit{on} \text{ the gas}, w = \int_{V_i}^{V_f} -p_{ext}\, dV.$$

If we write the increase in volume as ΔV, then the work is $-p_{ext}\Delta V$. An example of using the expression is given in Example 17.1 on the facing page.

If we keep the external pressure constant, then p_{ext} can be taken outside the integral. The steps in the calculation are

Integration is reviewed in section 20.6 on page 905.

$$\text{work done } \textit{on} \text{ the gas } (w) \text{ at const. ext. pressure} = -p_{ext}\int_{V_i}^{V_f} dV$$
$$= -p_{ext}\left[V\right]_{V_i}^{V_f}$$
$$= -p_{ext}\left[V_f - V_i\right]. \tag{17.4}$$

Example 17.1 Work done in a constant pressure expansion

Suppose that 0.1 moles of an ideal gas are confined by a piston into a volume of 1 dm^3, and that the gas is allowed to expand to a volume of 2 dm^3 against a constant external pressure of 1 bar. Calculate the work done on the gas, assuming that the temperature is held constant at 298 K.

All we need is to substitute these values into Eq. 17.4 on the preceding page, being sure to use SI units throughout (SI units are discussed in section 20.2 on page 882). This means that we must put the volume in m^3 (1 dm^3 = 10^{-3} m^3), and the pressure in N m^{-2} (1 bar = 10^5 N m^{-2})

$$
\begin{aligned}
\text{work done on gas, } w \; &= \; -p_{\text{ext}}\,[V_{\text{f}} - V_{\text{i}}] \\
&= \; -10^5 \times \left[2 \times 10^{-3} - 1 \times 10^{-3}\right] \\
&= \; -100 \text{ J}.
\end{aligned}
$$

However, for this expansion to occur it is necessary that the internal pressure is always greater than the external pressure; we need to check that this is the case. We can calculate the initial internal pressure using the ideal gas equation, $pV = nRT$

$$
\begin{aligned}
\text{initial internal pressure} \; &= \; \frac{nRT}{V_{\text{i}}} \\
&= \; \frac{0.1 \times 8.3145 \times 298}{1 \times 10^{-3}} \\
&= \; 2.48 \times 10^5 \text{ N m}^{-2}.
\end{aligned}
$$

When the volume doubles from 1 dm^3 to 2 dm^3, the pressure will halve (since the temperature is constant). So, the final pressure is 1.24×10^5 N m^{-2}. The expansion will take place since at all times the internal pressure is greater than the external pressure.

17.2.2 Expansion doing maximum work

An interesting question to ponder is what is the *maximum* amount of work that can be done in a gas expansion. This discussion is easier to follow if we phrase it in terms of the work done *by* the gas, w'. Since the work done by the gas is $p_{\text{ext}}\,dV$, the maximum amount of work will be done by making the external pressure as large as possible. However, for the expansion to actually take place, the external pressure has to be less than the internal pressure (otherwise the gas will be compressed, not expanded).

Therefore, to obtain the maximum amount of work, the external pressure must be infinitesimally lower then the internal pressure. If this is the case, the gas will expand by a very small amount, in the process lowering the internal pressure infinitesimally. Now the two pressures are equal, and the expansion stops.

To carry on with the expansion we need to lower the external pressure, once more by an infinitesimal amount. As before, the gas will expand until the internal pressure drops to the value of the external pressure. To carry on with the expansion, we need to lower the external pressure again.

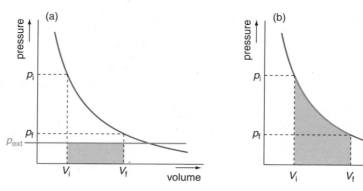

Fig. 17.6 Illustration of the difference between an expansion against a constant external pressure, and a reversible expansion. The blue line gives the relationship between pressure and volume for the gas inside the cylinder, assuming that the temperature is constant; this curve gives the *internal* pressure. The red line in (a), and the red curve in (b), give the *external* pressure in each case. In (a) the expansion from V_i to V_f is against a constant external pressure which must be less than the final pressure, p_f. The work done by the gas in the cylinder is the shaded area. In (b) the external pressure is adjusted to be always just less than the internal pressure. Again, the shaded area is the work done by the gas. Clearly the work done in the reversible case is much greater than that in case (a).

Therefore, to expand the gas doing the maximum amount of work we need to arrange for the external pressure to be infinitesimally less than the internal pressure at all times. Such a process is said to be *reversible* as its direction can be changed by an infinitesimal change in a variable. In this case, increasing the external pressure so that it is infinitesimally greater than the internal pressure will turn the expansion into a compression i.e. reverse its direction. The process is therefore reversible.

This point about the maximum work can also be illustrated graphically, as shown in Fig. 17.6. The blue line shows the relationship between the pressure and volume of the gas inside the cylinder, assuming that the temperature is held constant. Graph (a) depicts an expansion from volume V_i to volume V_f against a constant external pressure, indicated by the red line. The pressure of the gas inside the cylinder corresponding to volumes V_i and V_f can be read off from the blue curve as p_i and p_f. Note that the external pressure needs to be less than the final pressure for the expansion to take place. The work done by the gas in the cylinder is the integral of $p_{ext}\,dV$, which is the area under a graph of p_{ext} against V i.e. under the red line.

Graph (b) shows the same expansion, but this time the external pressure (the red line) is made equal to the internal pressure of the gas (the blue line). The work done by the gas is the area under the red line between the initial and final volumes. It is clear that the work done by the gas in this case is much greater than that in case (a). It is also clear that what is shown in (b) is the maximum work since, throughout the expansion, the external pressure is as large as it can be while maintaining an expansion (i.e. just less that the internal pressure).

We have come to a very important point: when a gas expands it does the *maximum* amount of work if the process is carried out *reversibly*. Although we have shown this for gas expansions, it turns out to be generally true that reversible processes do the maximum amount of work. This important result will be picked up on later in this section.

Expansion against a fixed external pressure is not reversible since the direction of the change cannot be altered by an infinitesimal change in the pressure. Such a process is said to be *irreversible*. Any real process which takes place at a finite speed is inherently irreversible. A reversible process represents an unachievable idealization of a real process.

This discussion has been framed in terms of the work done *by* the gas, w', which we have shown is a maximum in a reversible process. Recall that the work done *on* the gas, w, is just the opposite of the work done by the gas: $w = -w'$. It therefore follows that in a reversible expansion the work done *on* the gas has the largest *negative* value possible.

17.2.3 Reversible isothermal expansion of an ideal gas

If we have n moles of an ideal gas at temperature T and confined to a volume V by our piston, then we can work out the (internal) pressure using the ideal gas equation

$$p_{int} = \frac{nRT}{V}.$$

If this amount of gas is to expand reversibly then the external pressure must be maintained at a value which is essentially equal to the internal pressure at all times i.e. $p_{ext} = p_{int}$. As has been explained, this means that the external pressure must constantly be lowered as the gas expands. If the expansion is from V_i to V_f then the work done on the gas is found from the integral

$$\text{work done on gas, reversible} \quad = \quad \int_{V_i}^{V_f} -p_{ext} \, dV$$

$$= \quad \int_{V_i}^{V_f} -\frac{nRT}{V} \, dV.$$

If we assume that the temperature is held constant (an *isothermal* process, for example by keeping the system in a thermostat), then the factor of T can be taken outside the integral. It is then easy to solve, as follows

$$\text{work done on gas, reversible, isothermal} \quad = \quad -nRT \int_{V_i}^{V_f} \frac{1}{V} \, dV$$

$$= \quad -nRT \left[\ln V \right]_{V_i}^{V_f}$$

$$= \quad -nRT \ln\left(\frac{V_f}{V_i}\right), \qquad (17.5)$$

where we have recalled that the integral of $1/V$ with respect to V is $\ln V$, just as the integral of $1/x$ with respect to x is $\ln x$ (see Table 20.9 on page 906). Example 17.2 on the next page illustrates the use of this final expression.

17.2.4 The heat involved in a gas expansion

Using the First Law

$$\Delta U = q + w,$$

Recall that w is the work done *on* the system.

Example 17.2 Work done in a reversible isothermal expansion of an ideal gas

Suppose that 0.1 moles of an ideal gas are confined by a piston into a volume of 1 dm³, and that the gas is allowed to expand *reversibly* and *isothermally* to a volume of 2 dm³, at 298 K. Calculate the work done on the gas.

All we need is to substitute these values into Eq. 17.5 on the preceding page

$$\text{work done on gas, reversible, isothermal} = -nRT \ln\left(\frac{V_f}{V_i}\right)$$

$$= -0.1 \times 8.3145 \times 298 \times \ln\left(\frac{2}{1}\right)$$

$$= -171 \text{ J}.$$

We do not need to convert the volumes to SI units since we only need their ratio.

Note that the work done on the gas in this reversible expansion is *more negative* than for the same expansion discussed in Example 17.1, which was irreversible. This is in line with our expectations.

---

we can see that if we know w and ΔU we can compute the heat, q. We can use this connection to show that since the work w has its most negative value in a reversible process, the heat will therefore have its most positive value i.e. the heat is a maximum in a reversible process.

The reversible heat

Consider some process which takes us from state A to state B, either by a reversible path or by an irreversible path. Since the internal energy is a state function, ΔU is the same in both cases. However, w and q are not, since these are path functions. We can therefore write for a reversible and an irreversible path

$$\Delta U = q_{rev} + w_{rev} \qquad \Delta U = q_{irrev} + w_{irrev}.$$

Since ΔU is the same in each case the right-hand sides of these two equations can be set equal, and a little rearrangement gives

$$q_{rev} - q_{irrev} = -(w_{rev} - w_{irrev}).$$

For a gas expansion, w_{rev} is more negative than w_{irrev}, so the right-hand side of this equation is positive. It therefore follows that q_{rev} is greater than q_{irrev}. In other words, the *heat* involved in a *reversible* expansion is a *maximum*.

The signs here can get a bit confusing, and it may help with this to think about what is going on in a diagrammatic way, as shown in Fig. 17.7. Here we see the change from A to B, with the work indicated by the green arrows and the heat by the red arrows; ΔU is positive.

In (a) the dashed arrows show w and q for an irreversible process. Since w is negative, q has to be positive in order for $q + w$ to be positive. For the reversible case, shown by the solid arrows, w is more negative so q has to be more positive such that their sum is still equal to ΔU.

In (b), w is positive so now q has to be negative. In the reversible case, w is more positive, so q has to be more negative. The overall conclusion is that in a reversible process the *magnitudes* of the heat and work are a maximum.

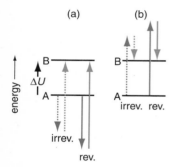

Fig. 17.7 Diagrammatic representation of the relationship between q, w and ΔU for reversible and irreversible processes taking us from state A to state B. The heat and work are represented by the red and green arrows, respectively; dashed arrows are for irreversible processes, and solid arrows are for reversible processes. In (a) w is negative, and since ΔU is positive, q must be positive: for the reversible process, w is more negative, so q must be more positive. In (b) w is positive and so q is negative; for the reversible process q is more negative.

Example 17.3 Heat involved in an isothermal expansion of an ideal gas

Find the heat involved in the expansions described in Example 17.1 and Example 17.2.

Since both of these are isothermal expansions of an ideal gas, it follows that $\Delta U = 0$ and so $q = -w$ (Eq. 17.6).

For the irreversible expansion, which has $w = -100$ J, $q = -w = 100$ J. For the reversible expansion, which has $w = -171$ J, $q = -w = 171$ J. As we expect, the heat for the reversible process is greater in magnitude than that for the irreversible process.

In both cases q is positive, meaning that the process is endothermic. What is happening here is that as the gas expands it is doing work pushing against the external pressure. This energy has to come from somewhere, but since we specified that the temperature must remain constant, the internal energy is fixed and so the energy for the expansion cannot come from this source. Thus, the energy has to be supplied by a flow of heat into the system, which exactly compensates the work done.

Heat for isothermal expansions of an ideal gas

For the special case of an ideal gas, it turns out that the internal energy depends *only* on the temperature and not on the pressure or volume. So, at a particular temperature one mole of an ideal gas has the same internal energy *regardless* of the volume to which it is confined or the pressure it is at.

The proof that U only depends on T for an ideal gas is rather involved, and well beyond the scope of this text. However, at a molecular level we can see that this property is not unexpected since in an ideal gas there are no interactions between the molecules. Therefore, how far apart they are makes no difference to their energy, thus expanding the gas has no effect on the energy.

For an *isothermal* process of an ideal gas we therefore have $\Delta U = 0$, and so from the First Law, $\Delta U = q + w$, we have

$$\text{isothermal, ideal gas} \quad q = -w. \tag{17.6}$$

If we know the work done, we can therefore work out the heat, as illustrated in Example 17.3.

17.2.5   Entropy change for a gas expansion

In Chapter 6 we defined the entropy change in terms of the reversible heat

$$\Delta S = \frac{q_{\text{rev}}}{T}.$$

For the isothermal expansion of an ideal gas we have calculated w_{rev} as

$$w_{\text{rev}} = -nRT \ln\left(\frac{V_{\text{f}}}{V_{\text{i}}}\right)$$

and we have also shown that since $\Delta U = 0$ for such a change, it follows that $q_{\text{rev}} = -w_{\text{rev}}$ i.e.

$$q_{\text{rev}} = nRT \ln\left(\frac{V_{\text{f}}}{V_{\text{i}}}\right).$$

Example 17.4 Entropy change for an isothermal expansion of an ideal gas

Find the entropy change for the (isothermal) expansion described in Example 17.1 on page 761.

All we need to do is to use Eq. 17.7

$$\Delta S = nR \ln\left(\frac{V_f}{V_i}\right)$$

$$= 0.1 \times 8.3145 \times \ln\left(\frac{2}{1}\right)$$

$$= 0.576 \, \text{J K}^{-1}.$$

The same result is also obtained by computing q_{rev}/T, with $q_{rev} = 171$ J, as found in Example 17.3.

In the irreversible expansion described in Example 17.1, the heat is 100 J, so the entropy change of the surroundings is $-q_{sys}/T$ which is -0.336 J K^{-1}. The entropy change of the Universe is therefore $0.576 - 0.336 = 0.24$ J K^{-1}; as expected for this spontaneous expansion against a lower external pressure, the entropy of the Universe increases.

We can immediately use this to work out the entropy change for the isothermal expansion of an ideal gas as

$$\text{isothermal} \qquad \Delta S = nR \ln\left(\frac{V_f}{V_i}\right). \tag{17.7}$$

The important point here is that since entropy is a state function, the value of ΔS is the same for any isothermal expansion between V_i and V_f regardless of how that expansion takes place. Of course, we had to work out q_{rev} in order to find ΔS, since entropy is defined in terms of the reversible heat. However, the resulting expression for ΔS is applicable to both reversible and irreversible processes.

Example 17.4 illustrates the use of this expression to calculate ΔS.

17.3 Internal energy, enthalpy and heat capacity

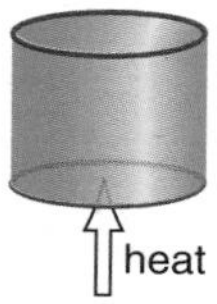

In this section we are going to explore the heat and work involved in two special circumstances: changes taking place under conditions of *constant volume* and under conditions of *constant pressure*. This will enable us to identify the change in internal energy as being the same as the heat under constant volume conditions, and will also lead us to introduce the state function *enthalpy*, which is equal to the heat under constant pressure conditions.

17.3.1 Constant volume conditions

A system that is held under constant volume conditions (e.g. in a sealed container, as illustrated in Fig. 17.8) cannot expand or contract, and so no work can be done against the external pressure i.e. $w = 0$. From the First Law,

Fig. 17.8 When heat is supplied to a system held at constant volume (e.g. a gas in a sealed container), there is no expansion and so no work is done. As a result, all of the heat is converted to internal energy.

$\Delta U = q + w$, it therefore follows that

$$\text{constant volume} \quad \Delta U = q.$$

In words, the heat involved in a constant volume process is equal to the change in internal energy. What is happening here is that when heat is supplied at constant volume it all ends up as internal energy: none is expended in doing work of expansion.

This relationship between q and ΔU gives us a way of measuring ΔU, since all we need to do is to measure the heat given out or absorbed when a process takes place at constant volume. Typically this is done using a *bomb calorimeter*, which is a thick-walled metal container with a tightly fitting lid.

Most often such a calorimeter is used for measuring the heat associated with combustion. The material of interest is placed in the bomb, and then it is pressurized with oxygen. The combustion is initiated using an electrical circuit, and the temperature rise of the bomb is then measured. From a knowledge of the heat capacity of the apparatus we can work back to the heat evolved, and since the volume is constant this heat is equal to ΔU.

Constant volume heat capacity

When heat is supplied to a substance we expect that its temperature will rise in proportion to the amount of heat supplied. The temperature rise ΔT and the amount of heat supplied q are related via the molar heat capacity, C_m

$$q = nC_m\Delta T;$$

we have to multiply by the amount in moles n since C_m is the heat capacity per mole. Under constant volume conditions the heat is equal to the change in internal energy ΔU, and under these conditions it is usual to denote the molar heat capacity as $C_{V,m}$.

$$\text{const. volume} \quad \Delta U = nC_{V,m}\Delta T.$$

The subscript 'V' is there to remind us that it is the heat capacity at constant volume. It will be more convenient to express the internal energy as a molar quantity, U_m in which case

$$\text{const. volume} \quad \Delta U_m = C_{V,m}\Delta T.$$

If the changes in U_m and T are infinitesimal, then this last expression can be written using the language of calculus as

$$\text{const. volume} \quad dU_m = C_{V,m}\, dT,$$

which can be rearranged to

$$\text{const. volume} \quad C_{V,m} = \frac{dU_m}{dT} \quad\quad (17.8)$$

Equation 17.8 is usually taken as the definition of the constant volume heat capacity. It says that $C_{V,m}$ is the rate of change of internal energy with temperature, under conditions of constant volume.

17.3.2 Constant pressure conditions

If we think about a process in which the (external) pressure is constant, then the work term will not be zero as the possibility exists for the system to expand or contract. We showed in section 17.2.1 on page 760 that if the gas expands against a constant external pressure by ΔV, this work term is $-p_{ext}\Delta V$ (Eq. 17.4 on page 760). The First Law then implies

$$\text{const. pressure} \qquad \Delta U = q - p_{ext}\Delta V,$$

which can be rearranged to give the heat as

$$\text{const. pressure} \qquad q = \Delta U + p_{ext}\Delta V. \tag{17.9}$$

The last line shows that the heat under constant pressure conditions is equal to the internal energy change plus an extra term for the work. This can be interpreted in the following way. If heat is supplied to a system at constant pressure, some of the energy goes into the work of expansion, so less is available for increasing the internal energy of the system. This is in contrast to the case of constant volume, where *all* of the heat ends up increasing the internal energy.

We will now go on to show that $(\Delta U + p_{ext}\Delta V)$ is in fact the familiar enthalpy change, ΔH, of the system i.e. the enthalpy change is equal to the heat at constant pressure.

Enthalpy

The enthalpy H is defined in the following way

$$\text{definition} \qquad H = U + pV.$$

Since enthalpy is defined in terms of a state function U, and the variables pressure and volume which describe the state of the system, it follows that H is also a state function.

In order to show that ΔH is equal to the heat at constant pressure we first need to form what is called the *complete differential* of H (if you are not familiar with how this is done, refer to Box 17.1 on the facing page), which is

$$dH = dU + p\,dV + V\,dp. \tag{17.10}$$

This tells us the change in H, dH, when U, p and V all change by the infinitesimal amounts dU, dp and dV, respectively.

If we imagine a process under constant pressure conditions, dp is zero as there is no change in pressure. Thus Eq. 17.10 becomes

$$\text{const. pressure} \qquad dH = dU + p\,dV. \tag{17.11}$$

The First Law, for an infinitesimal change, is $dU = \delta q + \delta w$. For a gas expansion, the work is $-p\,dV$, so the First Law can be written

$$dU = \delta q - p\,dV. \tag{17.12}$$

If we substitute this value of dU into Eq. 17.11, the $p\,dV$ terms cancel to give

$$\text{const. pressure} \qquad dH = \delta q.$$

In other words, the enthalpy change is equal to the heat under *constant pressure* conditions, which is what we set out to prove. For a finite change, this can be written $\Delta H = q_{\text{const. press.}}$.

Box 17.1 The complete differential

We start with the definition of H, $H = U + pV$, and then imagine that the internal energy changes from U to $(U + dU)$, the pressure changes from p to $(p + dp)$, and the volume changes from V to $(V + dV)$. As a result, H must also change to $(H + dH)$, so we have

$$(H + dH) = (U + dU) + (p + dp)(V + dV).$$

Multiplying this out gives

$$H + dH = U + pV + dU + p\,dV + V\,dp + dp\,dV.$$

The last term, $dp\,dV$, is the product of two infinitesimal quantities and so can safely be ignored as it will be negligibly small; we also note that, as $H = U + pV$, the term H on the left and $(U + pV)$ on the right will cancel to give

$$dH = dU + p\,dV + V\,dp. \qquad (17.13)$$

Equation 17.13 is called the *complete differential* of H; it gives the change in H resulting from general changes in U, p and V.

Another way of thinking about how to form the complete differential is just to differentiate both sides of $H = U + pV$ to give

$$dH = dU + d(pV).$$

We recognize that pV is a product of two functions and so needs to be differentiated using the 'product rule', to give

$$dH = dU + p\,dV + V\,dp.$$

Once you get used to it, this is probably the easiest way to form the complete differential.

As is illustrated in Fig. 17.9, when heat is supplied to a constant pressure system, some of the heat is converted to work as the gas expands. As a result, the increase in the internal energy of the system is less than in the constant volume case. This is what Eq. 17.12 says: dU is reduced by an amount $-p\,dV$, which is the work done.

If we rewrite Eq. 17.12 for a finite change we obtain

$$\Delta U = q - p\,\Delta V,$$

which rearranges to $q = \Delta U + p\,\Delta V$. This is the same as Eq. 17.9 on the facing page, assuming that p is the constant external pressure.

Values of the enthalpy change can be determined simply by measuring the heat in a constant pressure process. If we want to measure the enthalpy change associated with combustion, this can be done by using a gaseous sample, pre-mixing it with oxygen, and then letting it burn in an open flame. The heat given off is measured, for example by using it to heat a known volume of water. Such a device is called a *flame calorimeter*.

Fig. 17.9 When heat is supplied to a system held at constant pressure, work is done as the gas expands, so less of the heat is converted to internal energy than in the case of a constant volume process.

Constant pressure heat capacity

When we were thinking about constant volume processes, we saw that the fact that ΔU is the heat at constant volume meant that ΔU and the temperature rise ΔT were related by

$$\text{const. volume} \qquad \Delta U_m = C_{V,m}\,\Delta T.$$

For a constant pressure process, the heat is ΔH, so the analogous expression is

$$\text{const. press.} \qquad \Delta H_m = C_{p,m}\,\Delta T,$$

where $C_{p,m}$ is the *constant pressure* molar heat capacity.

The heat capacity is usually defined as the rate of change of enthalpy with temperature, at constant pressure

$$\text{const. press.} \qquad C_{p,m} = \frac{dH_m}{dT}. \qquad (17.14)$$

The constant pressure heat capacity is greater than that for constant volume. This is because it takes more heat to increase the temperature of a gas at constant pressure than at constant volume since in the former case some of the heat is used to expand the gas rather than it all going into raising the internal energy.

For a monatomic gas, such as helium, $C_{V,m}$ turns out to be $\frac{3}{2}R$ or 12.5 J K^{-1} mol^{-1} (R is the gas constant). More complex molecules have higher values of the heat capacity, so for $H_2(g)$ the value is 20.4 J K^{-1} mol^{-1}, and for $CO_2(g)$ the value is 28.5 J K^{-1} mol^{-1}, both at 298 K.

For an ideal gas it can be shown that $C_{p,m} = C_{V,m} + R$, so for helium gas $C_{p,m} = \frac{5}{2}R$ or 20.8 J K^{-1} mol^{-1}. As we have already mentioned, extensive tabulations of the values of these heat capacities are available. The values are somewhat temperature dependent, as will be discussed in more detail in section 17.8.3 on page 787.

17.3.3 Summary of key points related to the First Law

We have covered quite a lot of ground related to the First Law, so a summary is in order here.

- The First Law is about the conservation of energy, and can be written $\Delta U = q + w$.

- The internal energy U is a state function, but the heat q and work w are path functions.

- The work when a gas expands through a volume dV against a pressure p is $-p\,dV$.

- For a reversible process both the work and heat are a maximum.

- Under constant volume conditions the heat is equal to the internal energy change.

- The enthalpy H is defined as $U + pV$; under constant pressure conditions the heat is equal to the enthalpy change.

17.4 The Gibbs energy

In section 6.7 on page 220 the Gibbs energy was defined as

$$G = H - TS.$$

The useful thing about this state function is that, at constant pressure and temperature, it decreases in a spontaneous process, going to a minimum at equilibrium. We will briefly revisit these properties using the slightly more formal approach we have just been developing.

17.4.1 Gibbs energy and the Universal entropy

From the definition of G the complete differential can be computed as

$$dG = dH - T\, dS - S\, dT.$$

If we then impose the condition of constant temperature, the term $S\, dT$ goes to zero to give

const. temp. $dG = dH - T\, dS.$

Finally, dividing both sides by $-T$ gives

const. temp. $\underbrace{-\dfrac{dG}{T}}_{dS_{\text{univ}}} = \underbrace{-\dfrac{dH}{T}}_{dS_{\text{surr}}} + \underbrace{dS.}_{dS_{\text{sys}}}$

As is indicated by the underbrace, the first term on the right is the entropy change of the surroundings. The argument is that *under constant pressure conditions*, dH is equal to the heat absorbed by the system. Thus, $-dH$ is the heat absorbed by the surroundings, and so $-dH/T$ is the entropy change of the surroundings.

The sum of the two terms on the right is thus the entropy change of the Universe. Since this has to *increase* in a spontaneous process, dS_{univ} must be *positive* and so dG must be *negative*. In other words, the Gibbs energy must decrease in a spontaneous process, which is the conclusion we reached before.

17.4.2 The Master Equations

In order to discuss chemical equilibrium in more detail, we will need to know how the Gibbs energy varies with pressure and temperature. To do this, we will need to develop and use what are often called the thermodynamic *Master Equations*. At first, they seem rather esoteric, but rest assured they are exactly what we need to find how G varies with p and T.

We start with the First Law, written for infinitesimal changes as

$$dU = \delta q + \delta w. \tag{17.15}$$

If the only kind of work is that due to gas expansions, then as we have seen this work is $-p\, dV$, where p is the external pressure. Therefore δw in Eq. 17.15 can be replaced by $-p\, dV$.

We now turn to the δq term. Recall that the definition of entropy is $dS = \delta q_{\text{rev}}/T$, which rearranges to $\delta q_{\text{rev}} = T\, dS$. For a reversible process, we can therefore replace δq in Eq. 17.15 by $T\, dS$ to give

$$dU = T\, dS - p\, dV. \tag{17.16}$$

Now comes the slightly tricky point. Since U is a state function its value is independent of the path taken. So although we came up with Eq. 17.16 on the previous page by thinking about a reversible process, it is equally valid for an irreversible process since the value of dU is independent of the path.

One way of thinking about this is that for a reversible process $T\,dS$ is the heat, and as the internal and external pressures are balanced, $-p\,dV$ is the reversible work. Their sum is therefore equal to the change in internal energy. For an irreversible process $T\,dS$ is *not* the heat, and $-p\,dV$ is *not* the reversible work. However, the *sum* of these two terms is still equal to the change in internal energy. Equation 17.16 on the preceding page is the first of the Master Equations, and is also sometimes called the First and Second Laws combined.

The second Master Equation is found by starting with the definition of the enthalpy, $H = U + pV$, and then forming the complete differential to give

$$dH = dU + p\,dV + V\,dp. \tag{17.17}$$

We now substitute the expression for dU from Eq. 17.16 on the previous page into Eq. 17.17

$$
\begin{aligned}
dH &= dU + p\,dV + V\,dp \\
&= T\,dS - p\,dV + p\,dV + V\,dp \\
&= T\,dS + V\,dp.
\end{aligned}
\tag{17.18}
$$

The last line is the second of the Master Equations.

The process is repeated once more, this time starting with the definition of the Gibbs energy $G = H - TS$. The complete differential is

$$dG = dH - T\,dS - S\,dT. \tag{17.19}$$

We then substitute for dH using Eq. 17.18

$$
\begin{aligned}
dG &= dH - T\,dS - S\,dT \\
&= T\,dS + V\,dp - T\,dS - S\,dT \\
&= V\,dp - S\,dT.
\end{aligned}
\tag{17.20}
$$

The last line is the third of the Master Equations, and the one which we are going to put to immediate use.

17.4.3 How the Gibbs energy varies with pressure, at constant temperature

If we start with the third Master Equation

$$dG = V\,dp - S\,dT,$$

and impose the condition of constant temperature, so that the $S\,dT$ term is zero, we obtain

$$\text{const. temp.} \qquad dG = V\,dp. \tag{17.21}$$

This tells us how the Gibbs energy varies with pressure, but to turn it into a useful form we will need to integrate both sides. We will calculate the definite integral between the limits pressure p_1 and pressure p_2.

The left-hand side is quite straightforward

$$\int_{p_1}^{p_2} dG = \left[G \right]_{p_1}^{p_2}$$

$$= G(p_2) - G(p_1).$$

All we are doing here is saying that the integral of dG is G, just as the integral of dx is x. The integrated function then needs to be evaluated at pressures p_1 and p_2. Since the Gibbs energy varies with pressure, it will be written $G(p)$ to indicate that G is a function of p. So, evaluating the integrated function at p_1 simply gives $G(p_1)$, meaning the value of the Gibbs energy at pressure p_1. Likewise, the Gibbs energy at p_2 is $G(p_2)$.

Integrating the right-hand side of Eq. 17.21 is slightly more difficult. The problem is that the volume V is a function of the pressure p. However, if we assume that we are dealing with n moles of an ideal gas, then the pressure and volume are related via the ideal gas equation $pV = nRT$. Thus, V can be written nRT/p. In addition, we have already supposed that the temperature is constant, so the factor nRT can be taken outside the integral. The integration is now straightforward.

$$\int_{p_1}^{p_2} V dp = nRT \int_{p_1}^{p_2} \frac{1}{p} dp$$

$$= nRT \left[\ln p \right]_{p_1}^{p_2}$$

$$= nRT \ln \left(\frac{p_2}{p_1} \right).$$

Our final result is therefore

const. temp., ideal gas $\qquad G(p_2) - G(p_1) = nRT \ln \left(\dfrac{p_2}{p_1} \right).$

This is usually expressed in a slightly different way. Firstly, by dividing both sides by the amount in moles n, the Gibbs energies become molar quantities, indicated by a subscript 'm'

$$G_m(p_2) - G_m(p_1) = RT \ln \left(\frac{p_2}{p_1} \right).$$

Secondly, the pressure p_1 is taken as the standard pressure p° which is defined as 1 bar (exactly 10^5 N m^{-2}). The molar Gibbs energy at this standard pressure is written G_m°. Finally, pressure p_2 is replaced by the general pressure p to give

const. temp., ideal gas $\qquad G_m(p) = G_m^\circ + RT \ln \left(\dfrac{p}{p^\circ} \right).$ (17.22)

This is a very important equation which we will make much use of in what follows. It tells us how the molar Gibbs energy of an ideal gas varies with the pressure under constant temperature conditions.

A plot of the Gibbs energy as a function of pressure, as predicted by Eq. 17.22, is shown in Fig. 17.10. The equation predicts that, at constant

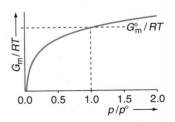

Fig. 17.10 Plot of G_m/RT as a function of p/p° as predicted by Eq. 17.22. We plot G_m/RT rather than G_m itself as the former is dimensionless. When $p = p^\circ$ the molar Gibbs energy is its standard value, G_m°.

temperature, the Gibbs energy increases as the pressure increases. We can understand why this is so if we recall the definition of the Gibbs energy in terms of the enthalpy and entropy: $G = H - TS$. On page 211 it was argued that increasing the volume available to a gas increases its entropy. At constant temperature, the pressure is inversely proportional to the volume, so increasing the pressure results in a decrease in the volume, and hence a reduction in the entropy. Given that $G = H - TS$, a reduction in the entropy results in an increase in the Gibbs energy. The pressure dependence of the Gibbs energy is thus a consequence of the pressure dependence of the entropy.

Mixtures of gases

Equation 17.22 is not quite what we need to discuss chemical equilibrium since it refers to a pure gas, whereas we need to be able to deal with mixtures. The modification needed is very simple since, if the gases are assumed to be ideal, there are no interactions between them. We can therefore assume that the molar Gibbs energy of each gas in the mixture is just the same as it would be if it occupied the whole volume on its own.

Recalling the discussion section 1.9.2 on page 24, the pressure which a gas in a mixture would exert on its own is called the *partial pressure*. We therefore assert that the molar Gibbs energy of gas i in the mixture simply depends on its partial pressure p_i in an analogous way to Eq. 17.22

$$G_{m,i}(p_i) = G^{\circ}_{m,i} + RT \ln\left(\frac{p_i}{p^{\circ}}\right). \tag{17.23}$$

$G_{m,i}(p_i)$ is the molar Gibbs energy of gas i at partial pressure p_i, and $G^{\circ}_{m,i}$ is the molar Gibbs energy of pure gas i at the standard pressure p°. It should be remembered that Eq. 17.23 applies at a given temperature: if the temperature is changed then both $G_{m,i}$ and $G^{\circ}_{m,i}$ will change.

The partial pressure p_i of gas i in a mixture is related to the total pressure p_{tot} and the mole fraction x_i of the gas

$$p_i = x_i p_{tot}.$$

The mole fraction is given by

$$x_i = \frac{n_i}{n_{tot}},$$

where n_i is the amount in moles of i, and n_{tot} is the total amount in moles in the mixture.

17.5 The mixing of ideal gases

As a prelude to discussing chemical equilibrium it is useful to think about mixtures of gases which are not reacting with one another. We know from experience that gases mix unconditionally and completely, so the process of going from separated gases to mixed gases must be accompanied by an increase in the entropy of the Universe or, equivalently, a decrease in the Gibbs energy.

If we restrict our discussion to the mixing of ideal gases, then it is clear that there can be no enthalpy change associated with mixing since there are no interactions between the molecules of such gases. It must therefore be the increase in the entropy of the system (the gases) which is responsible for the mixing.

Let us sharpen up our thinking by analysing the situation depicted in Fig. 17.11. There are two containers separated by a barrier: the first contains n_A moles of an ideal gas A and is at pressure p; the second contains n_B moles of an ideal gas B, also at pressure p. Both gas samples are at the same temperature.

If the barrier between the two containers is removed, then we know that the two gases will mix completely. As the pressure in the two containers was the same to start with, then after mixing the pressure will still be the same. In

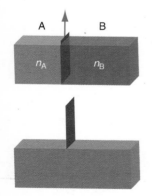

Fig. 17.11 Two (ideal) gases A and B are separated by a partition; there are n_A moles of A and n_B moles of B, and both are at the same pressure p. When the partition is removed, the gases mix spontaneously.

addition, since we have assumed that the gases are ideal, there will be no heat involved in this process.

The fact that the heat is zero tells us that ΔS_{surr} is zero. It therefore follows that the process is spontaneous because ΔS_{sys} is positive. Using our microscopic interpretation of entropy it is easy to see why the entropy increases when the gases mix. On removal of the partition between the two containers, the volume available to each gas increases. We argued in section 6.3.2 on page 210 that increasing the volume decreases the spacing of the energy levels, and so increases W; as a result, the entropy increases.

Equation 17.23 on the preceding page can be used to work out the Gibbs energy change when the two gases mix, and hence to show that there is indeed the expected reduction in its value. Before mixing, the Gibbs energy of the n_A moles of gas A is simply n_A times the molar Gibbs energy of gas A, which is given by Eq. 17.23. Note that, before mixing, the partial pressure is the same as the total pressure p. The Gibbs energy of gas B is computed in the same way, so the total Gibbs energy is given by

$$G_{\text{before mixing}} = \underbrace{n_A\left[G_{m,A}^\circ + RT\,\ln\left(\frac{p}{p^\circ}\right)\right]}_{\text{Gibbs energy of A}} + \underbrace{n_B\left[G_{m,B}^\circ + RT\,\ln\left(\frac{p}{p^\circ}\right)\right]}_{\text{Gibbs energy of B}}.$$

After the gases have mixed the molar Gibbs energy of A depends on its partial pressure, p_A, and likewise that of B depends on its partial pressure. So, the Gibbs energy of the mixture is

$$G_{\text{after mixing}} = \underbrace{n_A\left[G_{m,A}^\circ + RT\,\ln\left(\frac{p_A}{p^\circ}\right)\right]}_{\text{Gibbs energy of A}} + \underbrace{n_B\left[G_{m,B}^\circ + RT\,\ln\left(\frac{p_B}{p^\circ}\right)\right]}_{\text{Gibbs energy of B}}. \qquad (17.24)$$

The change in Gibbs energy on mixing is simply the difference between the Gibbs energy after mixing and the Gibbs energy before mixing

$$\begin{aligned}
\Delta G_{\text{mix}} &= G_{\text{after mixing}} - G_{\text{before mixing}} \\
&= n_A\left[G_{m,A}^\circ + RT\,\ln\left(\frac{p_A}{p^\circ}\right)\right] + n_B\left[G_{m,B}^\circ + RT\,\ln\left(\frac{p_B}{p^\circ}\right)\right] \\
&\quad -n_A\left[G_{m,A}^\circ + RT\,\ln\left(\frac{p}{p^\circ}\right)\right] - n_B\left[G_{m,B}^\circ + RT\,\ln\left(\frac{p}{p^\circ}\right)\right].
\end{aligned}$$

The terms in $G_{m,A}^\circ$ and $G_{m,B}^\circ$ cancel, and then we can tidy up the log terms (using $\ln A - \ln B \equiv \ln(A/B)$) to give

$$\Delta G_{\text{mix}} = n_A RT\,\ln\left(\frac{p_A}{p}\right) + n_B RT\,\ln\left(\frac{p_B}{p}\right). \qquad (17.25)$$

Remember that p was the pressure of the separate gases before mixing, and is also the total pressure after mixing; also the sum of the partial pressures of A and B, p_A and p_B, is equal to the total pressure p. Thus p_A/p is less than one, which means that $\ln(p_A/p)$ must be negative; the same is true for the term $\ln(p_B/p)$. It therefore follows that ΔG_{mix} is *always* negative. We have therefore shown that the mixing of ideal gases is always accompanied by a decrease in the Gibbs energy, and is therefore always spontaneous.

Equation 17.25 can be expressed in a slightly different way if we recognize that the partial pressure of A is determined by its mole fraction (see section 1.9.2 on page 24)

$$p_A = x_A p.$$

It follows that $p_A/p = x_A$, so that Eq. 17.25 on the previous page can be written

$$\Delta G_{mix} = n_A RT \ln(x_A) + n_B RT \ln(x_B).$$

Since, by definition, the mole fraction is less than one, both of the log terms are negative making ΔG_{mix} negative, as we concluded above.

17.6 Chemical equilibrium

In Chapter 6 the very important relationship between the equilibrium constant and the standard Gibbs energy change of the reaction

$$\Delta_r G^\circ = -RT \ln K,$$

was simply stated and then used to discuss chemical equilibrium. We are now in a position to prove this important equation.

The approach we will adopt is to first determine the Gibbs energy of the mixture of reactants and products as a function of the *composition*, that is the ratio of products to reactants. We will show that this Gibbs energy goes to a minimum at a particular composition, which is therefore the equilibrium point. Finally, we will show that the equilibrium composition, and hence the value of the equilibrium constant, is related to $\Delta_r G^\circ$.

To keep things as simple as possible, we will start out by analysing the equilibrium between just two species, such as two isomers. Some examples are shown in Fig. 17.12.

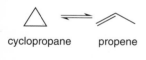

cyclopropane   propene

trans   *cis*

Fig. 17.12 Examples of simple A ⇌ B equilibria in which the two species A and B are isomers of one another.

17.6.1 The equilibrium between two species

In this section we are going to consider the equilibrium between two gaseous species, A and B

$$A(g) \rightleftharpoons B(g)$$

We will call A the reactant and B the product, so the forward reaction is $A(g) \rightarrow B(g)$.

As the stoichiometry is 1:1, the interconversion of A and B does not alter the total amount in moles present, and so the total pressure p remains constant. We will also assume that the reaction is done in a temperature regulated enclosure (a thermostat) so that the temperature is fixed.

If the partial pressures of A and B are p_A and p_B, respectively, and if there are n_A moles of A, and n_B moles of B, then the Gibbs energy of the mixture is given by Eq. 17.24 on the previous page

$$G = n_A \left[G_{m,A}^\circ + RT \ln\left(\frac{p_A}{p^\circ}\right) \right] + n_B \left[G_{m,B}^\circ + RT \ln\left(\frac{p_B}{p^\circ}\right) \right]. \qquad (17.26)$$

By writing the partial pressures in terms of the mole fractions, this expression for G can be rewritten in terms of just the mole fraction of B, x_B, and various

Box 17.2 Gibbs energy of the mixture

We start with Eq. 17.26 on the preceding page, and divide both sides by the total amount in moles, $n_{tot} = n_A + n_B$, which is constant in this equilibrium

$$\frac{G}{n_{tot}} = \frac{n_A}{n_{tot}}\left[G_{m,A}^{\circ} + RT\ \ln\left(\frac{p_A}{p^{\circ}}\right)\right] + \frac{n_B}{n_{tot}}\left[G_{m,B}^{\circ} + RT\ \ln\left(\frac{p_B}{p^{\circ}}\right)\right].$$

We then recognize that n_A/n_{tot} is the mole fraction of A, x_A, and likewise $n_B/n_{tot} = x_B$. Furthermore, p_A can be written as $x_A p$, and p_B as $x_B p$, where p is the total pressure. With these substitutions we have

$$\frac{G}{n_{tot}} = x_A\left[G_{m,A}^{\circ} + RT\ \ln\left(\frac{x_A p}{p^{\circ}}\right)\right] + x_B\left[G_{m,B}^{\circ} + RT\ \ln\left(\frac{x_B p}{p^{\circ}}\right)\right].$$

Finally, as $x_A + x_B = 1$, we can write the equation in terms of x_B by putting $x_A = 1 - x_B$ to give

$$\frac{G}{n_{tot}} = (1 - x_B)\left[G_{m,A}^{\circ} + RT\ \ln\left(\frac{(1 - x_B)p}{p^{\circ}}\right)\right] + x_B\left[G_{m,B}^{\circ} + RT\ \ln\left(\frac{x_B p}{p^{\circ}}\right)\right].$$

Since the pressure p is constant, we can use this expression to compute how the Gibbs energy varies with x_B, for given values of the standard molar Gibbs energies.

constants; the details of how this is done are given in Box 17.2. The resulting expression is

$$\frac{G}{n_{tot}} = (1 - x_B)\left[G_{m,A}^{\circ} + RT\ \ln\left(\frac{(1 - x_B)p}{p^{\circ}}\right)\right] + x_B\left[G_{m,B}^{\circ} + RT\ \ln\left(\frac{x_B p}{p^{\circ}}\right)\right], \quad (17.27)$$

where n_{tot} is the total amount in moles ($= n_A + n_B$), which in this case is constant. G/n_{tot} is the molar Gibbs energy of the reaction mixture.

Figure 17.13 on the following page shows plots of the (molar) Gibbs energy given by Eq. 17.27 as a function of x_B. On the far left of the plot $x_B = 0$, and since $x_A + x_B = 1$ it follows that $x_A = 1$ i.e. only A is present, corresponding to pure reactant. At this point, the value of the Gibbs energy depends only on the pressure, temperature and – most importantly – the standard molar Gibbs energy of A, $G_{m,A}^{\circ}$.

On the far right of the plot we have $x_B = 1$, so only B is present, which corresponds to pure product. At this point, the Gibbs energy depends only on the pressure, temperature and the standard molar Gibbs energy of B, $G_{m,B}^{\circ}$.

In between these two extremes, the Gibbs energy depends on both $G_{m,A}^{\circ}$ and $G_{m,B}^{\circ}$, as well as the mole fraction x_B. The plot shows this variation for three different combinations of the standard molar Gibbs energies. Note that in all the cases the plots form a smooth curve, with a single minimum, indicated by the dashed lines.

Look first at the black curve, which is for the case that $G_{m,A}^{\circ} = G_{m,B}^{\circ}$. If we start with a mixture which has more A than B (i.e. x_B is less than 0.5), as indicated by the green line in Fig. 17.14 on the following page, the graph predicts that increasing x_B will lead to a decrease in the Gibbs energy, and so be

● *Weblink 17.1*

This link takes you to a version of Fig. 17.13 in which you can alter the molar Gibbs energy of the reactant A and the product B and see how this affects the position of equilibrium.

Fig. 17.13 Plots of how the molar Gibbs energy of an interconverting mixture of A and B varies with the mole fraction of B, as predicted by Eq. 17.27. Three curves are given for different combinations of the standard molar Gibbs energies of A and B. In all cases, the curves have a single minimum, whose positions are indicated by the dashed line. The minimum is closest to whichever of A and B has the lowest standard Gibbs energy.

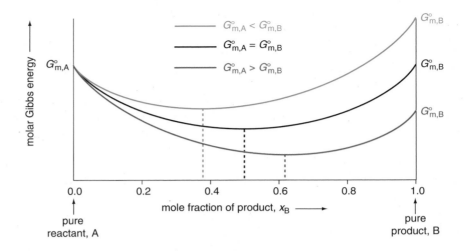

Fig. 17.14 Illustration of how the form of the plot of Gibbs energy against x_B funnels the system towards the equilibrium point. If we start at the mole fraction indicated by the green line, the direction of spontaneous change (i.e. which has a decreasing G) increases x_B taking us towards equilibrium, as shown by the green arrow. Starting from the composition indicated by the purple line, it is a decrease in x_B, indicated by the purple arrow, which takes us to equilibrium.

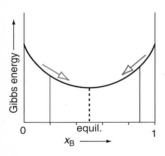

a spontaneous process. The mole fraction of B, x_B, is increased by converting A to B i.e. the reaction proceeding to the right. The proportion of B will go on increasing until we reach the minimum in the plot of G against x_B. At this point no change in the fraction of A or B is allowed as this would result in an increase in the Gibbs energy.

In contrast, if we start with a mixture which has more B than A (i.e. $x_B > 0.5$, the purple line in Fig. 17.14), it is the conversion of B to A, leading to a decrease in x_B, which results in a decrease in the Gibbs energy, and is thus spontaneous. This process will carry on until we reach the minimum in the Gibbs energy, at which point no further change is possible. Therefore, from whatever point we start, the change which is accompanied by a decrease in the Gibbs energy funnels us towards the minimum in G, which corresponds to equilibrium.

The red curve in Fig. 17.13 is for the case where the standard molar Gibbs energy of B is greater than that of A. The general shape of the curve is similar to that when the standard molar Gibbs energies are the same, but the minimum is now at a lower value of x_B (about 0.38). Starting from any value of x_B the shape of the curve funnels the reaction mixture towards the minimum, which is the equilibrium point. In this case, the reactant A is favoured.

The blue curve has $G_{m,A}^\circ > G_{m,B}^\circ$. Again the shape of the curve is much as before, but the minimum is now shifted to a higher value of x_B (about 0.62). At equilibrium, the product B is therefore favoured. The general picture here is that no matter what the values of $G_{m,A}^\circ$ and $G_{m,B}^\circ$ the shape of the curve is such that the composition of the mixture is funnelled towards the minimum.

It turns out that the position of the minimum in the plot of Gibbs energy depends only on the *difference* between $G_{m,A}^\circ$ and $G_{m,B}^\circ$, and not on their absolute values. In the next section, we will prove that this is indeed the case.

17.6.2 Finding the position of equilibrium

We could find the minimum in the curves shown in Fig. 17.13 by using calculus, but there is another way of doing this which is both more convenient and also leads to greater insight into what is going on.

Imagine a mixture containing n_A moles of A and n_B moles of B, with partial pressures p_A and p_B, respectively. The Gibbs energy of this mixture is given by Eq. 17.26 on page 776

$$G = n_A \left[G_{m,A}^{\circ} + RT \, \ln\left(\frac{p_A}{p^{\circ}}\right) \right] + n_B \left[G_{m,B}^{\circ} + RT \, \ln\left(\frac{p_B}{p^{\circ}}\right) \right]. \quad (17.28)$$

Now imagine that there is some conversion of A to B such that the amount of A decreases by dn to $(n_A - dn)$ moles, and as a consequence the amount of B increases by the same infinitesimal amount dn to $(n_B + dn)$ moles. Since the change is infinitesimal, we can assume that the partial pressures of A and B do not change significantly, so after this conversion the Gibbs energy of the mixture is

$$G + dG = (n_A - dn) \left[G_{m,A}^{\circ} + RT \, \ln\left(\frac{p_A}{p^{\circ}}\right) \right] + (n_B + dn) \left[G_{m,B}^{\circ} + RT \, \ln\left(\frac{p_B}{p^{\circ}}\right) \right]. \quad (17.29)$$

We have written this as $G + dG$ to emphasize that the conversion of dn moles of A to B causes an infinitesimal change in the Gibbs energy.

The *change* in Gibbs energy when dn moles of A becomes B is found by subtracting Eq. 17.28 from Eq. 17.29. The terms in n_A and n_B cancel to give

$$dG = dn \left[G_{m,B}^{\circ} + RT \, \ln\left(\frac{p_B}{p^{\circ}}\right) \right] - dn \left[G_{m,A}^{\circ} + RT \, \ln\left(\frac{p_A}{p^{\circ}}\right) \right]. \quad (17.30)$$

It is convenient to define a quantity $\Delta_r G$ as

$$\Delta_r G = \frac{dG}{dn}, \quad (17.31)$$

where the subscript 'r' is to indicate that it is for a reaction. Whereas dG is the change in Gibbs energy, $\Delta_r G$ is the change in Gibbs energy *per mole*.

Taking the dn to the left of Eq. 17.30 and then using Eq. 17.31 we can easily find that

$$\Delta_r G = \left[G_{m,B}^{\circ} + RT \, \ln\left(\frac{p_B}{p^{\circ}}\right) \right] - \left[G_{m,A}^{\circ} + RT \, \ln\left(\frac{p_A}{p^{\circ}}\right) \right]. \quad (17.32)$$

This definition of $\Delta_r G$ as a derivative can be a bit confusing as we are used to thinking about '$\Delta_r G$' as being the *change* in the Gibbs energy, rather than the *derivative* of the Gibbs energy with respect to amount in moles. The following interpretation may be helpful in coming to terms with this.

Imagine that we have a very large amount of the mixture of A and B, combined in the appropriate ratio to make the partial pressures of A and B p_A and p_B, respectively. Now imagine that one mole of A is converted to one mole of B. Since the amounts of A and B are very large, this conversion makes no appreciable difference to the ratio of A to B, so the partial pressures do not change. The Gibbs energy change accompanying this conversion will be the loss of the Gibbs energy of a mole of A, and the gain of the Gibbs energy of a mole of B. Expressed mathematically this is

$$\begin{aligned} \Delta G &= +G_{m,B} - G_{m,A} \\ &= \left[G_{m,B}^{\circ} + RT \, \ln\left(\frac{p_B}{p^{\circ}}\right) \right] - \left[G_{m,A}^{\circ} + RT \, \ln\left(\frac{p_A}{p^{\circ}}\right) \right]. \end{aligned}$$

Comparing this last line with Eq. 17.32 we see that we have precisely the same terms on the right. We can therefore interpret $\Delta_r G$ as the change in Gibbs energy per mole when the reaction proceeds according to the stoichiometric equation under conditions where the ratio of products to reactants do not change.

Returning to Eq. 17.32 on the previous page, this expression can be tidied up by introducing a quantity known as the *standard Gibbs energy change*, $\Delta_r G°$, defined as

$$\Delta_r G° = \underbrace{G°_{m,B}}_{product} - \underbrace{G°_{m,A}}_{reactant}. \tag{17.33}$$

$\Delta_r G°$ is the change in Gibbs energy when one mole of the reactant A, in its standard state, becomes one mole of the product B, also in its standard state. Remember that the standard state means the *pure* substance at a pressure of 1 bar. So, $\Delta_r G°$ is the change in Gibbs energy when one mole of pure A goes to one mole of pure B, at 1 bar.

Inserting Eq. 17.33 into Eq. 17.32 and combining all the log terms gives

$$\Delta_r G = \Delta_r G° + RT \ln \frac{p_B/p°}{p_A/p°}. \tag{17.34}$$

In fact, we could have cancelled the $p°$ terms, but for reasons which will become clearer later we choose not to.

From its definition, Eq. 17.31 on the previous page, we can see that $\Delta_r G$ is the *slope* of a graph of Gibbs energy against amount in moles of A that have become B. For a more general reaction, $\Delta_r G$ can be described as the slope of the graph of Gibbs energy against the *extent* of reaction, where the extent depends on the amount in moles of reactants which have become products. For the simple A $\rightleftharpoons$ B equilibrium the extent is specified by the mole fraction of B (multiplied by the total amount in moles).

It is important to realize that the value of $\Delta_r G$ depends on the relative amount of A and B, since this determines the values of the partial pressures in Eq. 17.34. For a general reaction, we say that $\Delta_r G$ is a function of the *composition* of the reaction mixture, where the composition depends on the ratio of products to reactants.

Figure 17.15 illustrates the interpretation of the value of $\Delta_r G$ as a slope. At the composition indicated by the green line, the slope of the curve, and hence $\Delta_r G$, is negative. This means that converting A to B will result in a decrease in the Gibbs energy, and so be a spontaneous process. On the other hand, at the composition indicated by the purple line, $\Delta_r G$ is positive meaning that converting A to B would result in a increase in the Gibbs energy, which is not allowed. In fact at this composition the reverse process, in which B is converted to A, will lead to a reduction in the Gibbs energy and so be spontaneous. These are the same conclusions we reached before.

At the equilibrium position, the slope, and therefore $\Delta_r G$, is zero. This is the key to the next step in which we take Eq. 17.34 and impose the equilibrium condition $\Delta_r G = 0$ to give

$$0 = \Delta_r G° + RT \ln \frac{p_{B,eq}/p°}{p_{A,eq}/p°}. \tag{17.35}$$

In this equation $p_{A,eq}$ and $p_{B,eq}$ are the *equilibrium* partial pressures of A and B, since the equation is only true at equilibrium.

We now define the *equilibrium constant*, K, for this reaction as

$$K = \frac{p_{B,eq}/p°}{p_{A,eq}/p°}.$$

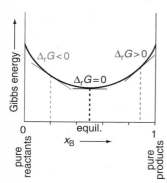

Fig. 17.15 Illustration of how $\Delta_r G$ for the conversion of one mole of A to B is the slope of the graph of Gibbs energy against extent of reaction, here represented by x_B. When $\Delta_r G < 0$ (for example, at the composition shown by the green line), the conversion of A to B is accompanied by a decrease in the Gibbs energy. When $\Delta_r G > 0$ (shown in purple), the conversion of A to B is accompanied by an increase in the Gibbs energy, and so is not allowed; in fact, the reverse process in which B is converted to A is spontaneous. At equilibrium $\Delta_r G = 0$.

As usual, this is of the form 'products over reactants', with the concentrations being expressed as partial pressures. The log term in Eq. 17.35 on the preceding page is therefore $\ln K$, so making this substitution and then rearranging slightly we obtain the by now familiar relationship

$$\Delta_r G^\circ = -RT \ln K. \tag{17.36}$$

This equation says that the value of the equilibrium constant depends on $\Delta_r G^\circ$, which is the change in Gibbs energy when one mole of pure reactant goes to one mole of pure product, both species being in their standard states. The slightly strange thing is that the value of the equilibrium constant, which refers to a mixture of products and reactants, depends on $\Delta_r G^\circ$, which refers to *pure* products and *pure* reactants.

Equation 17.36 explains why it is that the ratio of the concentrations of the products over the reactants is a constant. For a given reaction and at a particular temperature, $\Delta_r G^\circ$ has a particular value, and this determines the value of K via Eq. 17.36. Since K is defined as the ratio of products over reactants, this ratio is fixed for a particular reaction. In a sense, Eq. 17.36 explains why there is such a thing as the equilibrium constant.

17.6.3   The general case

We can now quickly move to the more general reaction than the simple $A(g) \rightleftharpoons B(g)$ equilibrium; we will do this in two stages. Suppose first that the stoichiometry is 1:2 rather than 1:1, so the equilibrium is

$$A(g) \rightleftharpoons 2B(g).$$

This balanced chemical equation, or *stoichiometric* equation, tells us that one mole of A goes to two moles of B. The *stoichiometric coefficient* of A is therefore one, and that of B is two.

Since one mole of A gives *two* moles of B, the expression given in Eq. 17.29 on page 779 is modified to

$$G + dG = \underbrace{(n_A - dn)\left[G^\circ_{m,A} + RT \ln\left(\frac{p_A}{p^\circ}\right)\right]}_{\text{loss of } dn \text{ moles of A}} + \underbrace{(n_B + 2\,dn)\left[G^\circ_{m,B} + RT \ln\left(\frac{p_B}{p^\circ}\right)\right]}_{\text{gain of } 2\,dn \text{ moles of B}}.$$

It follows that $\Delta_r G\ (= dG/dn)$ is given by

$$
\begin{aligned}
\Delta_r G &= 2\left[G^\circ_{m,B} + RT \ln\left(\frac{p_B}{p^\circ}\right)\right] - \left[G^\circ_{m,A} + RT \ln\left(\frac{p_A}{p^\circ}\right)\right] \\
&= \left[2G^\circ_{m,B} + RT \ln\left(\frac{p_B}{p^\circ}\right)^2\right] - \left[G^\circ_{m,A} + RT \ln\left(\frac{p_A}{p^\circ}\right)\right]. \tag{17.37}
\end{aligned}
$$

To go to the second line we have taken the factor of 2 inside the square bracket and used the property that $n \ln A \equiv \ln A^n$.

For this reaction $\Delta_r G^\circ$ is defined as the Gibbs energy change when *one* mole of A becomes *two* moles of B under standard conditions

$$\Delta_r G^\circ = \underbrace{2G^\circ_{m,B}}_{\text{product}} - \underbrace{G^\circ_{m,A}}_{\text{reactant}}.$$

Note that the value of $\Delta_r G°$ takes into account the stoichiometry of the reaction. Using this in Eq. 17.37 on the previous page gives

$$\Delta_r G = \Delta_r G° + RT \ \ln \frac{(p_B/p°)^2}{p_A/p°}.$$

At equilibrium, $\Delta_r G = 0$, and so

$$\Delta_r G° = -RT \ \ln K \quad \text{where} \quad K = \frac{(p_{B,eq}/p°)^2}{p_{A,eq}/p°}.$$

This is similar to the result we obtained before, except that the partial pressure of B is raised to the power of two i.e. to its stoichiometric coefficient. Of course, you already know that in defining an equilibrium constant the concentrations are raised to the appropriate stoichiometric coefficients: now you can see why you need to do this.

If we take a completely general chemical equilibrium

$$a\text{A(g)} + b\text{B(g)} \rightleftharpoons c\text{C(g)} + d\text{D(g)},$$

where $a \ldots d$ are the stoichiometric coefficients in the balanced equation, following through the same line of argument as above gives the same relationship

$$\Delta_r G° = -RT \ \ln K. \tag{17.38}$$

For the general reaction, $\Delta_r G°$ is the change in Gibbs energy when the reaction proceeds from left to right with the stoichiometry indicated by the balanced chemical equation i.e. when a moles of A and b moles of B are converted to c moles of C and d moles of D, with all species under standard conditions (pure, 1 bar pressure)

$$\Delta_r G° = \underbrace{cG°_{m,C} + dG°_{m,D}}_{\text{products}} - \underbrace{aG°_{m,A} - bG°_{m,B}}_{\text{reactants}}.$$

An equilibrium constant for a gas phase reaction in which the concentrations are represented by partial pressures is sometimes denoted K_p.

The equilibrium constant is defined in the familiar way as 'products over reactants', with the partial pressures raised to the appropriate stoichiometric coefficients

$$K = \frac{\left(p_{C,eq}/p°\right)^c \left(p_{D,eq}/p°\right)^d}{\left(p_{A,eq}/p°\right)^a \left(p_{B,eq}/p°\right)^b}. \tag{17.39}$$

17.6.4 Writing the equilibrium constant

In section 6.8.2 on page 224, when we looked at how to write an equilibrium constant, we were most insistent that K had to be written in such a way that it is dimensionless. From the proof we have just completed, you can see that as each partial pressure is divided by the standard pressure, it follows that the equilibrium constant is dimensionless (provided that we use the same units for both pressures).

You may have been told before that equilibrium constants have dimensions, and it is certainly the case that in many books and research papers you will find that they are given dimensions. Strictly speaking, this is not correct.

However, there is a way round this difficulty if you keep in the back of your mind what is going on. If we leave out the standard pressures, then the equilibrium constant for the $A(g) \rightleftharpoons 2B(g)$ equilibrium is

$$K' = \frac{p_{B,eq}^2}{p_{A,eq}}.$$

Writing the equilibrium constant in this way is just about acceptable provided we remember that $p_{A,eq}$ is really a shorthand for $p_{A,eq}/p^\circ$ with $p^\circ = 1$ bar. If we make sure that $p_{A,eq}$ is given in *units of bar*, then the numerical value of $p_{A,eq}$ will be the same as $p_{A,eq}/p^\circ$, and all will be well.

17.7 Equilibria involving other than gases

So far we have only looked at equilibria involving gases, but we will see in this section that the ideas we have developed are quickly and easily adapted to the case where solids and liquids are involved. Reactions involving species in solution need a little more care, but are also readily accommodated.

17.7.1 Solids and liquids

Key in our discussion of equilibria involving gases was the way in which the molar Gibbs energy depends on the partial pressure of a gas, as given by Eq. 17.23 on page 774

$$G_{m,i}(p_i) = G_{m,i}^\circ + RT \ln\left(\frac{p_i}{p^\circ}\right).$$

Recall that $G_{m,i}^\circ$ is the molar Gibbs energy of *pure* substance i under standard conditions.

A solid, being rather incompressible, is hardly affected by external pressure and is *always* present in its pure, unmixed, form. We can therefore assume that its molar Gibbs energy is always equal to its molar standard Gibbs energy

$$\text{solids:} \quad G_{m,i} = G_{m,i}^\circ.$$

The same argument applies to liquids, which are also incompressible and present in their pure form

$$\text{liquids:} \quad G_{m,i} = G_{m,i}^\circ.$$

Of course, under extreme conditions of pressure these approximations will break down, but for the sorts of conditions that the bench chemist is concerned with, they hold well enough.

Let us now see what effect these ideas have on our discussion of equilibrium. We will consider once again the simplest equilibrium, but this time we will make B a gas and A a solid

$$A(s) \rightleftharpoons B(g)$$

Following the same line of argument as before, if we have n_A moles of A and n_B of B, then the Gibbs energy of the mixture is

$$G = n_A G_{m,A}^\circ + n_B\left[G_{m,B}^\circ + RT \ln\left(\frac{p_B}{p^\circ}\right)\right].$$

As has been explained, since A is a *solid* its molar Gibbs energy is simply the standard molar Gibbs energy $G^{\circ}_{m,A}$: there is no pressure term for a solid.

When dn moles of A become B, the Gibbs energy changes to

$$G + dG = (n_A - dn)G^{\circ}_{m,A} + (n_B + dn)\left[G^{\circ}_{m,B} + RT\ \ln\left(\frac{p_B}{p^{\circ}}\right)\right].$$

Therefore $\Delta_r G\ (= dG/dn)$ is

$$\Delta_r G = G^{\circ}_{m,B} + RT\ \ln\left(\frac{p_B}{p^{\circ}}\right) - G^{\circ}_{m,A}.$$

Defining $\Delta_r G^{\circ}$ in the usual way, $\Delta_r G^{\circ} = G^{\circ}_{m,B} - G^{\circ}_{m,A}$, enables us to write

$$\Delta_r G = \Delta_r G^{\circ} + RT\ \ln\left(\frac{p_B}{p^{\circ}}\right).$$

At equilibrium $\Delta_r G = 0$ so it follows that

$$\Delta_r G^{\circ} = -RT\ \ln\left(\frac{p_{B,eq}}{p^{\circ}}\right).$$

The term in the logarithm is the equilibrium constant, $K = p_{B,eq}/p^{\circ}$. No term for A appears in the expression for the equilibrium constant because A is a solid and so is effectively always present in the standard state. However, even though there is no term for A in the expression for K, the properties of A affect the value of $\Delta_r G^{\circ}$. An entirely parallel discussion can be had for the case of equilibria involving liquids.

In section 6.8.2 on page 224 we simply stated that, in writing equilibrium constants, a concentration term is not included for solids or liquids. You can now see why this is the case.

17.7.2   Solution equilibria

Many equilibria we are interested in will involve species which are in solution, for example the dissociation of an acid AH in water

$$AH(aq) + H_2O(l) \rightleftharpoons A^-(aq) + H_3O^+(aq).$$

To use our ideas about Gibbs energy to analyse this equilibrium we need to know how the Gibbs energy of a solution of a species i varies with its concentration. We will denote the concentration of species i using square brackets: $[i]$.

For a gas we had the following relationship

$$G_{m,i}(p_i) = G^{\circ}_{m,i} + RT\ \ln\left(\frac{p_i}{p^{\circ}}\right).$$

More or less proceeding by analogy, we assert that for a solution the relationship is

$$G_{m,i}([i]) = G^{\circ}_{m,i} + RT\ \ln\left(\frac{[i]}{c^{\circ}}\right), \tag{17.40}$$

where c° is the standard concentration, usually taken to be 1 mol dm^{-3}. In this case $G^{\circ}_{m,i}$ is the (molar) Gibbs energy of a solution of i at the standard concentration, c°.

If we accept this assertion, the argument about the approach to equilibrium proceeds exactly as before, with the partial pressures being replaced by concentrations, and the standard pressure by the standard concentration. We will then arrive at $\Delta_r G^\circ = -RT \ln K$, just as before. In the case of the acid dissociation, K is given by

$$K = \frac{([A^-]/c^\circ)\,([H_3O^+]/c^\circ)}{[AH]/c^\circ}. \tag{17.41}$$

An equilibrium constant written in terms of the concentrations of species in solution is sometimes denoted K_c.

Note that, as we have just discussed, there is no term for the liquid water.

The problem with Eq. 17.40 on the preceding page is that it only applies to *ideal* solutions, which are ones in which there are no interactions *between* the solute molecules. In practice such solutions do not exist, but sufficiently dilute solutions may approach this ideal behaviour. For ionic solutes, where the interactions are particularly strong, a solution has to be very dilute before this ideal behaviour is achieved.

There are ways of dealing with the thermodynamic properties of non-ideal (that is, real) solutions, but these are well beyond the scope of this book. We will simply have to assume that any solutions we are dealing with are ideal, so that Eq. 17.40 applies, and hope that this does not cause too many problems.

The units of K

The equilibrium constant defined in Eq. 17.41 is, as we by now must expect, dimensionless. As was explained in section 17.6.4 on page 782, it is nevertheless common for the standard terms (c° or p°) to be missed out, and as a consequence to write equilibrium constants as if they had dimensions. For example, for the acid dissociation we might write

$$K' = \frac{[A^-]\,[H_3O^+]}{[AH]}.$$

K' appears to have units of 'concentration'.

However, if we express the concentrations in mol dm^{-3}, and recall that the standard concentration is 1 mol dm^{-3}, then $[i]$ is really a shorthand for $[i]/c^\circ$ so in fact K' is dimensionless.

17.8   Determination of the standard Gibbs energy change

The way in which $\Delta_r G^\circ$ can be computed from tabulated standard enthalpies of formation and absolute entropies was covered in detail in section 6.9 on page 227. It was mentioned in that section that tabulated data are usually only available at 298 K, and that for the most accurate work it is necessary to correct these data to the temperature in question. In the following two sections we will see how this can be done with the aid of heat capacity data.

17.8.1   Temperature dependence of the enthalpy

The key to working out how the enthalpy varies with temperature is the constant pressure heat capacity $C_{p,\mathrm{m}}$, defined in Eq. 17.14 on page 770

$$C_{p,\mathrm{m}} = \frac{dH_\mathrm{m}}{dT}.$$

By moving the term dT to the left, we are treating the term as if it were a normal variable or function. This is the basis of the method known as the *separation of variables* used to solve differential equations. More details are given in section 20.7.1 on page 910.

In words this says that the rate of change of enthalpy with temperature is given by the heat capacity. To turn this definition into something useful we first need to separate the enthalpy onto one side, and the temperature on the other, to give

$$dH_m = C_{p,m} \, dT. \tag{17.42}$$

Then both sides are integrated between temperatures T_1 and T_2.

The left-hand side is quite straightforward, and we use exactly the same procedure that we employed in section 17.4.3 on page 772 for working out the pressure variation of the Gibbs energy

$$\int_{T_1}^{T_2} dH_m = \left[H_m \right]_{T_1}^{T_2}$$
$$= H_m(T_2) - H_m(T_1).$$

Since the enthalpy varies with temperature, it is written $H_m(T)$; $H_m(T_1)$ is the value of the enthalpy at temperature T_1 and likewise $H_m(T_2)$ is that at T_2.

For the right-hand side we have to make the approximation that the heat capacity is independent of temperature, so that $C_{p,m}$ can be taken outside the integral. Once this is done, the integration is straightforward

$$\int_{T_1}^{T_2} C_{p,m} \, dT = C_{p,m} \int_{T_1}^{T_2} dT$$
$$= C_{p,m} \left[T \right]_{T_1}^{T_2}$$
$$= C_{p,m} \left[T_2 - T_1 \right].$$

The final result is therefore

$$H_m(T_2) - H_m(T_1) = C_{p,m} \left[T_2 - T_1 \right],$$

which is more usefully written

$$H_m(T_2) = H_m(T_1) + C_{p,m} \left[T_2 - T_1 \right].$$

Using this, we can take the value of the enthalpy at temperature T_1 and convert it to the enthalpy at temperature T_2 provided we know the heat capacity (the values of which are tabulated).

In order to find $\Delta_r G^\circ$ we need the value of $\Delta_r H^\circ$, but as we saw in section 6.9 on page 227, $\Delta_r H^\circ$ is just the difference between the molar standard enthalpies of the products and reactants. It therefore follows that the last equation is just as valid for $\Delta_r H^\circ$ values provided that we replace the heat capacity by the *difference* in (standard) molar heat capacities of the products and reactants. The resulting relationship is

$$\Delta_r H^\circ(T_2) = \Delta_r H^\circ(T_1) + \Delta_r C_p^\circ [T_2 - T_1]. \tag{17.43}$$

For the general reaction

$$aA + bB \rightleftharpoons cC + dD,$$

$\Delta_r C_p^\circ$ is given by

$$\Delta_r C_p^\circ = \left[c \times C_{p,m}^\circ(C) + d \times C_{p,m}^\circ(D) \right] - \left[a \times C_{p,m}^\circ(A) + b \times C_{p,m}^\circ(B) \right]. \tag{17.44}$$

$C_{p,m}^\circ(A)$ is the molar heat capacity, measured at constant pressure and under standard conditions, of A, and so on. Equation 17.43 on the preceding page is sometimes known as *Kirchhoff's Law*. Example 17.5 on the following page illustrates the use of this relationship.

17.8.2 Temperature dependence of the entropy

Once again, heat capacities are the key to finding out how the entropy varies with temperature. We start with the definition of entropy, which is $dS = \delta q_{rev}/T$. If we think about a process at constant pressure, then the heat is equal to the enthalpy change, dH. We saw from Eq. 17.42 on the preceding page that dH_m is equal to $C_{p,m}\,dT$, so we use this to write

$$dS_m = \frac{C_{p,m}\,dT}{T}. \tag{17.45}$$

As before, integrating both sides of this equation between temperatures T_1 and T_2 will give us a useful expression for how the entropy varies with the temperature.

The left-hand side is just the same as in the enthalpy case, and evaluates to $S_m(T_2) - S_m(T_1)$. To evaluate the right-hand side we will once again have to assume that the heat capacity does not vary with temperature

$$\int_{T_1}^{T_2} \frac{C_{p,m}}{T}\,dT = C_{p,m} \int_{T_1}^{T_2} \frac{1}{T}\,dT$$

$$= C_{p,m}\Big[\ln T\Big]_{T_1}^{T_2}$$

$$= C_{p,m} \ln\left(\frac{T_2}{T_1}\right).$$

Putting all this together gives

$$S_m(T_2) = S_m(T_1) + C_{p,m} \ln\left(\frac{T_2}{T_1}\right). \tag{17.46}$$

As in the case of the enthalpy, this relationship is easily adapted for $\Delta_r S^\circ$ values to

$$\Delta_r S^\circ(T_2) = \Delta_r S^\circ(T_1) + \Delta_r C_p^\circ \ln\frac{T_2}{T_1}. \tag{17.47}$$

Example 17.5 on the following page illustrates the use of this relationship.

17.8.3 More precise work

For the most precise work it is necessary to take into account the temperature variation of the heat capacity. It is common to tabulate values of the heat capacity by expressing these values in the form of a parametrized equation such as

$$C_{p,m}(T) = A + BT + \frac{C}{T^2}.$$

The parameters (constants) A, B and C are different for different compounds, and it is these values which may be found in data tables. The details of how such parametrized data can be used are explored in the *Questions*.

Example 17.5 Conversion of $\Delta_r H^\circ$ and $\Delta_r S^\circ$ to other temperatures

The following equilibrium was used as an example in Chapter 6

$$CH_4(g) + 2H_2O(g) \longrightarrow CO_2(g) + 4H_2(g).$$

There we found that at 298 K $\Delta_r H^\circ$ was 164.9 kJ mol^{-1}, and that $\Delta_r S^\circ$ was 172.54 J K^{-1} mol^{-1}. The corresponding value of $\Delta_r G^\circ$ is 113.5 kJ mol^{-1}, indicating that to all intents and purposes the reaction does not go at all. In this Example we will compute the value of these parameters at 1000 K.

From data tables the $C_{p,m}^\circ$ values (at 298 K) of $CH_4(g)$, $H_2O(g)$, $CO_2(g)$ and $H_2(g)$ are 35.31, 33.58, 37.11, and 28.82 J K^{-1} mol^{-1}, respectively. Using these, we can compute $\Delta_r C_p^\circ$ as

$$\begin{aligned}
\Delta_r C_p^\circ &= \left[C_{p,m}^\circ(CO_2(g)) + 4 \times C_{p,m}^\circ(H_2(g))\right] - \left[C_{p,m}^\circ(CH_4(g)) + 2 \times C_{p,m}^\circ(H_2O(g))\right] \\
&= [37.11 + 4 \times 28.82] - [35.31 + 2 \times 33.58] \\
&= 49.92 \text{ J K}^{-1} \text{ mol}^{-1}.
\end{aligned}$$

Using this value we can compute $\Delta_r H^\circ$ at 1000 K using Eq. 17.43 on page 786

$$\begin{aligned}
\Delta_r H^\circ(1000) &= \Delta_r H^\circ(298) + \Delta_r C_p^\circ[1000 - 298] \\
&= 164.9 + 49.92 \times 10^{-3} \times [1000 - 298] \\
&= 199.94 \text{ kJ mol}^{-1}.
\end{aligned}$$

Similarly for $\Delta_r S^\circ$ we can use Eq. 17.47 on the preceding page

$$\begin{aligned}
\Delta_r S^\circ(1000) &= \Delta_r S^\circ(298) + \Delta_r C_p^\circ \ln \frac{1000}{298} \\
&= 172.54 + 49.92 \ln \frac{1000}{298} \\
&= 233.0 \text{ J K}^{-1} \text{ mol}^{-1}.
\end{aligned}$$

The values of $\Delta_r H^\circ$ and $\Delta_r S^\circ$ at 1000 K are significantly different from those at 298 K, but of course the temperature difference is substantial. Using the values of $\Delta_r H^\circ$ and $\Delta_r S^\circ$ at 1000 K we find that $\Delta_r G^\circ$ at 1000 K is -33.0 kJ mol^{-1}, indicating that at this high temperature the products are favoured.

If we use the values of $\Delta_r H^\circ$ and $\Delta_r S^\circ$ at 298 K to calculate $\Delta_r G^\circ$ at 1000 K, we find a value of $\Delta_r G^\circ = -7.6$ kJ mol^{-1}, which is significantly different to that found using the correct values of $\Delta_r H^\circ$ and $\Delta_r S^\circ$. Note the standard state *does not* imply a particular temperature, so we can have $\Delta_r G^\circ$ at 298 K, and a different value of $\Delta_r G^\circ$ at 1000 K.

17.9 The temperature dependence of the equilibrium constant

The equilibrium constant is related to the value of $\Delta_r G^\circ$ via

$$\Delta_r G^\circ = -RT \ln K, \tag{17.48}$$

and $\Delta_r G°$ is found using $\Delta_r G° = \Delta_r H° - T\Delta_r S°$. As a result of the factor of T in this expression, $\Delta_r G°$ is generally strongly temperature dependent (except in the special case that $\Delta_r S°$ is small). We saw in the previous section that $\Delta_r H°$ and $\Delta_r S°$ are also temperature dependent, although generally they do not vary as strongly as does $\Delta_r G°$.

In this section we will show that the temperature dependence of the equilibrium constant is determined by the sign and value of $\Delta_r H°$. The proof is quite straightforward: we simply start with Eq. 17.48 on the preceding page and substitute for $\Delta_r G°$ using $\Delta_r G° = \Delta_r H° - T\Delta_r S°$ to give

$$\Delta_r H° - T\Delta_r S° = -RT \ln K.$$

Dividing both sides by $-RT$ and swapping the terms between the left- and right-hand sides gives

$$\ln K = \frac{-\Delta_r H°}{RT} + \frac{\Delta_r S°}{R}. \tag{17.49}$$

If the temperature variation of $\Delta_r H°$ and $\Delta_r S°$ can be ignored, this equation tells us that the way in which K varies with temperature depends on the value of $\Delta_r H°$.

The argument goes as follows. Suppose that the reaction is *endothermic*, meaning that $\Delta_r H°$ is positive so that the term $-\Delta_r H°/(RT)$ in Eq. 17.49 is *negative*. Since the temperature T appears on the bottom of the fraction, as T increases the term $-\Delta_r H°/(RT)$ will become *less negative*. As a result, the value of $\ln K$ and hence the value of K will *increase* as the temperature increases.

For an exothermic reaction $\Delta_r H°$ is negative so $-\Delta_r H°/(RT)$ is positive. Now as T increases $-\Delta_r H°/(RT)$ becomes less positive and so the equilibrium constant decreases. In summary:

- For an endothermic reaction increasing the temperature increases the value of the equilibrium constant, and thus the equilibrium moves towards the products;

- For an exothermic reaction increasing the temperature decreases the value of the equilibrium constant, and thus the equilibrium moves towards the reactants.

These conclusions are in accord with *Le Chatelier's principle* which states that when a system at equilibrium is perturbed, it responds in such a way as to oppose the change. For example, if we have an endothermic reaction, increasing the temperature (by applying heat) is opposed by the reaction taking place so as to absorb the heat. The equilibrium therefore moves to products. In contrast, for an exothermic reaction, the application of heat is opposed by the reverse of the reaction taking place (as this will be endothermic), thus shifting the equilibrium to reactants.

17.9.1 Determination of $\Delta_r H°$ and $\Delta_r S°$ from equilibrium constant data

If we ignore the slight temperature dependence of $\Delta_r H°$ and $\Delta_r S°$, Eq. 17.49 is of the form of a straight line

$$\underbrace{\ln K}_{y} = \underbrace{\frac{-\Delta_r H°}{R}}_{m} \underbrace{\frac{1}{T}}_{x} + \underbrace{\frac{\Delta_r S°}{R}}_{c}.$$

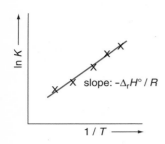

Fig. 17.16 A plot of $\ln K$ against $1/T$ gives a straight line of slope $-\Delta_r H^\circ / R$, assuming that $\Delta_r H^\circ$ does not change significantly over the range of temperatures measured. Shown here is the case for an endothermic reaction with a positive $\Delta_r H^\circ$ value.

Thus, as illustrated in Fig. 17.16, a plot of $\ln K$ against $1/T$ will be a straight line of slope $-\Delta_r H^\circ / R$, provided that $\Delta_r H^\circ$ is temperature independent over the range of temperatures measured.

Why might this be useful? We mentioned in section 17.3 on page 766 that ΔU and ΔH can be measured directly using either a bomb or flame calorimeter. However, for a reaction the measured heat will only be equal to the *standard* enthalpy (or internal energy) change *if* the reaction goes *completely* to products. This is because $\Delta_r H^\circ$ is defined as the enthalpy change when *pure* reactants go to *pure* products.

Typically, such calorimeters are used to measure the heat associated with combustion in pure oxygen, which is likely to be a reaction which goes entirely to products, and so the measured heat can be related to $\Delta_r H^\circ$ or $\Delta_r U^\circ$, as appropriate. However, if the reaction goes to a position of equilibrium in which there are significant amounts of reactants present, then the heat evolved will not be equal to $\Delta_r H^\circ$ for the reaction as there has not been complete conversion to products. Under these circumstances, it is not possible to measure $\Delta_r H^\circ$ by calorimetry.

In contrast, it is often quite straightforward to measure the concentration of reactants and products present at equilibrium (for example, using spectroscopy), and hence determine a value for K. By repeating the measurements over a range of temperatures, we can then determine $\Delta_r H^\circ$ using the graphical method described above.

Determination of $\Delta_r S^\circ$

From Eq. 17.49 on the previous page it follows that in a graph of $\ln K$ against $1/T$ (as in Fig. 17.16) the intercept with the vertical axis at $1/T = 0$ is $\Delta_r S^\circ / R$. The practicality of this method of finding $\Delta_r S^\circ$ is somewhat limited as a long, and potentially inaccurate, extrapolation is often needed to find this intercept.

An alternative approach is to realise that each measurement of K gives a value of $\Delta_r G^\circ$ found using $\Delta_r G^\circ = -RT \ln K$. Then, using the known values of $\Delta_r G^\circ$ and $\Delta_r H^\circ$ (from the slope) in conjunction with $\Delta_r G^\circ = \Delta_r H^\circ - T \Delta_r S^\circ$ we can find $\Delta_r S^\circ$. This can be repeated for each measurement, and so an average value of $\Delta_r S^\circ$ can be determined. An illustration of such an approach is given in Example 17.6 on the next page.

17.9.2   What determines the temperature variation of the equilibrium constant?

It is generally not a good idea to write down things which are just plain wrong, but here we will make an exception. You will, unfortunately, often hear fellow chemists (including many who should know better) proposing the following incorrect piece of logic:

(1) The equilibrium constant is determined by the value of $\Delta_r G^\circ$.

(2) Since $\Delta_r G^\circ = \Delta_r H^\circ - T \Delta_r S^\circ$, the temperature dependence of $\Delta_r G^\circ$ comes from the $T \Delta_r S^\circ$ term.

(3) Therefore the temperature dependence of the equilibrium constant is due to the entropy term. **WRONG.**

The flaw in this argument is that although statement 1 is correct in saying that the value of $\Delta_r G^\circ$ determines the value of K, the statement is *incomplete* as it

---

Example 17.6 Determination of $\Delta_r H°$ and $\Delta_r S°$ from equilibrium constant data

For the reaction

$$N_2O_4(g) \rightleftharpoons 2NO_2(g)$$

The equilibrium constant has been measured as 0.062 at 360 K and 0.17 at 380 K. Estimate values for $\Delta_r H°$ and $\Delta_r S°$ for the reaction.

Equation 17.49 on page 789 can be written at temperatures T_1 and T_2

$$\ln K(T_1) = -\frac{\Delta_r H°}{RT_1} + \frac{\Delta_r S°}{R} \qquad \ln K(T_2) = -\frac{\Delta_r H°}{RT_2} + \frac{\Delta_r S°}{R}$$

Subtracting the first from the second eliminates the term in $\Delta_r S°$ to give

$$\ln K(T_2) - \ln K(T_1) = -\frac{\Delta_r H°}{R}\left[\frac{1}{T_2} - \frac{1}{T_1}\right].$$

This is the same result as would be obtained by integrating the van't Hoff isochore between T_1 and T_2.

All we need to do is to substitute in the values with $T_1 = 360\,\text{K}$ and $T_2 = 380\,\text{K}$

$$\ln 0.17 - \ln 0.062 = -\frac{\Delta_r H°}{8.3145}\left[\frac{1}{380} - \frac{1}{360}\right]$$

$$1.009 = -\frac{\Delta_r H°}{8.3145} \times (-1.462 \times 10^{-4})$$

Hence $\Delta_r H° = 57.4\,\text{kJ mol}^{-1}$.

At 360 K the equilibrium constant is 0.062, so $\Delta_r G°$ is computed as $-RT \ln K = -8.3145 \times 360 \times \ln 0.062$, which is $8.32\,\text{kJ mol}^{-1}$. Hence from $\Delta_r G° = \Delta_r H° - T\Delta_r S°$ we have

$$\Delta_r S° = \frac{\Delta_r H° - \Delta_r G°}{T}$$

$$= \frac{57.4 \times 10^3 - 8.32 \times 10^3}{360}$$

$$= 136\,\text{J K}^{-1}\,\text{mol}^{-1}.$$

A similar calculation at 380 K gives $\Delta_r G° = 5.60\,\text{kJ mol}^{-1}$, and hence $\Delta_r S° = 136\,\text{J K}^{-1}\,\text{mol}^{-1}$. Not surprisingly, with only two data points, both give the same value of $\Delta_r S°$. More precise estimates of both $\Delta_r H°$ and $\Delta_r S°$ could be found if more values of K were available so that a graph could be plotted.

---

fails to recognize that the relationship between $\Delta_r G°$ and K also includes a factor of T: $\Delta_r G° = -RT \ln K$.

If we write $\Delta_r G°$ as $\Delta_r H° - T\Delta_r S°$, $\Delta_r G° = -RT \ln K$ becomes

$$\Delta_r H° - T\Delta_r S° = -RT \ln K.$$

Dividing by $-RT$ and swapping the equation round we have

$$\ln K = \frac{-\Delta_r H°}{RT} + \frac{\Delta_r S°}{R}.$$

It is now clear that, as we have been describing in this section, it is the value of $\Delta_r H^\circ$ which determines the temperature dependence of the equilibrium constant: statement 3 is therefore incorrect. Please do not fall into this trap!

17.10 Determination of absolute entropies

This topic was discussed in section 6.5.1 on page 215, but it is worthwhile revisiting it in order to tidy up a few small points. In section 17.8.2 on page 787, starting from the definition of entropy and assuming constant pressure, it was shown that

$$dS_m(T) = \frac{C_{p,m}(T)\,dT}{T}.$$

The heat capacity has been written as $C_{p,m}(T)$, and the entropy as $S_m(T)$ to remind us that both are functions of temperature.

Integrating both sides between $T = 0$ and $T = T^*$ gives

$$S_m(T^*) - S_m(0) = \int_0^{T^*} \frac{C_{p,m}(T)}{T}\,dT.$$

As has already been discussed on page 216, the entropy at absolute zero is taken to be zero, so the previous equation becomes

$$S_m(T^*) = \int_0^{T^*} \frac{C_{p,m}(T)}{T}\,dT.$$

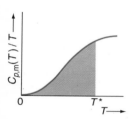

Fig. 17.17 The area under a plot of $C_{p,m}(T)/T$ against T, taken between $T = 0$ and $T = T^*$ (shaded blue), is equal to the absolute molar entropy at T^*. As explained in the text, there are practical problems associated with making measurements very close to absolute zero.

The integral on the right is the *area* under a plot of $C_{p,m}(T)/T$ against T, taken between temperature $T = 0$ and $T = T^*$, as shown in Fig. 17.17. This area has to be found by literally plotting experimental measurements on a graph and measuring the area since the heat capacity will vary significantly over this wide range of temperatures.

It is difficult to make measurements of the heat capacity at very low temperatures. However, at such temperatures it is found that most materials follow the *Debye Law*

$$\text{Debye Law} \qquad C_{p,m}(T) = A\,T^3,$$

where A is a constant characteristic of a particular substance. We can use this law to help us find the area under our $C_{p,m}(T)/T$ *vs* T graph at low temperatures in the following way.

If the lowest accessible temperature is T_0, then a measurement of $C_{p,m}$ is made at this temperature and the values used to determine the constant A in the Debye Law. The integral of $C_{p,m}(T)/T$ between $T = 0$ and $T = T_0$ can then be found using this value

$$\int_0^{T_0} \frac{C_{p,m}(T)}{T}\,dT \;=\; \int_0^{T_0} \frac{A\,T^3}{T}\,dT$$

$$=\; \int_0^{T_0} A\,T^2\,dT$$

$$=\; \tfrac{1}{3}A T_0^3.$$

This quantity, $\frac{1}{3}AT_0^3$, is thus an estimate of the area under the experimentally inaccessible part of the graph between $T = 0$ and $T = T_0$.

If there are any phase changes (e.g. solid → liquid or liquid → gas) in the range 0 to T^* then the entropy of these need to be taken into account. At the temperature of the phase change, T_{pc}, the two phases are in equilibrium, and so the heat associated with this change *is* the reversible heat. This means that the entropy change can be computed directly from its definition as (reversible heat)/(temperature). Assuming that we are working under constant pressure conditions, the heat is equal to the enthalpy change $\Delta_{pc}H$ of the phase transition, so the entropy change is $\Delta_{pc}H/T_{pc}$. Such a term needs to be added in for each phase transition.

FURTHER READING

Thermodynamics of Chemical Processes, Gareth Price, Oxford University Press (1998).
Elements of Physical Chemistry, fourth edition, Peter Atkins and Julio de Paula, Oxford University Press (2005).
Physical Chemistry, eighth edition, Peter Atkins and Julio de Paula, Oxford University Press (2006).

QUESTIONS

The gas constant, R, has the value 8.3145 J K^{-1} mol^{-1}; 1 bar is 10^5 N m^{-2}; 1 atmosphere is 1.01325×10^5 N m^{-2}; 1 Torr is 133.3 N m^{-2}.

17.1 4×10^{-3} moles of an ideal gas are held inside a cylinder by a piston such that the volume of the gas is 10 cm^3; the whole assembly is held in a thermostat at 298 K. Calculate the pressure of the gas in N m^{-2}. [Note that the SI unit of volume is m^3.]

   (a)   Assume that the external pressure is fixed at 1 bar. Explain why the piston moves out when it is released, and why it eventually comes to a stop. What will the pressure of the gas inside the cylinder be when the piston finally stops?

   (b)   Calculate the volume of the gas inside the cylinder when the piston has come to rest, and hence the work for this irreversible expansion.

   (c)   State the change in the internal energy, ΔU, of the gas when it undergoes this isothermal expansion. Hence, using the First Law, calculate the heat associated with the expansion, explaining its sign.

   (d)   Calculate the work associated with reversible isothermal expansion between the same initial and final states as the irreversible expansion described above; hence find the heat. Comment on these values in relation to those for the irreversible expansion.

   (e)   Determine the enthalpy change of the gas in (i) the reversible and (ii) the irreversible expansion.

17.2   A sample of methane gas of mass 4.50 g has volume 12.7 dm^3 at 310 K. It expands isothermally against a constant external pressure of 200 Torr until its volume has increased by 3.3 dm^3. Assuming methane to be a perfect gas, calculate the work and heat associated with the process, along with the change in the internal energy and enthalpy of the gas. [1 Torr = 133.3 N m^{-2}.]

Calculate these same quantities if the expansion is carried out reversibly.

17.3 For the A $\rightleftharpoons$ B equilibrium, sketch a graph showing how the molar Gibbs energy of a mixture of A and B varies as the composition varies from pure reactant A to pure product B for the following cases: (a) $G_{m,A}^\circ = G_{m,B}^\circ$, (b) $G_{m,A}^\circ < G_{m,B}^\circ$. On your graph indicate the equilibrium composition.

Explain why it is that if the proportions of A and B are not at their equilibrium values, there is always a spontaneous process which will result in the composition moving to its equilibrium value, but that once this point is reached, no further change is possible.

17.4 Consider the equilibrium

$$2A(g) \rightleftharpoons B(g).$$

Using the approach illustrated in section 17.6.3 on page 781, show that for this reaction at equilibrium we have

$$\Delta_r G^\circ = -RT \ln K \quad \text{where} \quad K = \frac{p_{B,eq}/p^\circ}{(p_{A,eq}/p^\circ)^2},$$

and

$$\Delta_r G^\circ = G_{m,B}^\circ - 2G_{m,A}^\circ.$$

17.5 This question and the two which follow concern the equilibrium

$$CO(g) + 2 H_2(g) \rightleftharpoons CH_3OH(g),$$

which we are going to investigate as a viable commercial method for the production of methanol. The following data are provided (all at 298 K)

	CO(g)	H$_2$(g)	CH$_3$OH(g)
$\Delta_f H^\circ$ / kJ mol^{-1}	-110.53		-200.66
S_m° / J K^{-1} mol^{-1}	197.67	130.68	239.81
$C_{p,m}^\circ$ / J K^{-1} mol^{-1}	29.14	28.82	43.89

Using these data, determine $\Delta_r H^\circ$, $\Delta_r S^\circ$, $\Delta_r G^\circ$ and hence K, all at 298 K. On the basis of your answer, comment on the viability of the reaction as a method for the production of methanol.

17.6 In practice, it is found that the reaction in the previous question only proceeds at a viable rate at 600 K. Assuming that the values of $\Delta_r H^\circ$ and $\Delta_r S^\circ$ are the same at 600 K as they are at 298 K, find the value of the equilibrium constant at 600 K. Qualitatively, is your answer in accord with Le Chatalier's principle?

17.7 Using the approach described in section 17.8.1 on page 785 and section 17.8.2 on page 787, together with the data given in question 18.7, find the value of $\Delta_r C_p^\circ$ and hence the values of $\Delta_r H^\circ$ and $\Delta_r S^\circ$ at 600 K for the equilibrium

$$CO(g) + 2 H_2(g) \rightleftharpoons CH_3OH(g).$$

Hence compute $\Delta_r G^\circ$ and K at 600 K. How do your values compare with those found in exercise 18.8?

17.8 Consider the equilibrium in which solid calcium carbonate decomposes to the oxide plus carbon dioxide

$$CaCO_3(s) \rightleftharpoons CaO(s) + CO_2(g).$$

Using the approach illustrated in section 17.7.1 on page 783, show that

$$\Delta_r G^\circ = -RT \ln K$$

where p_{CO_2} is the equilibrium pressure of CO_2 and

$$\Delta_r G^\circ = G_m^\circ(CaO) + G_m^\circ(CO_2) - G_m^\circ(CaCO_3) \quad \text{and} \quad K = \frac{p_{CO_2}}{p^\circ}.$$

The standard enthalpies of formation of $CaCO_3(s)$, $CO_2(g)$ and $CaO(s)$ are -1207.6 kJ mol^{-1}, -393.5 kJ mol^{-1} and -634.9 kJ mol^{-1} respectively, and the standard entropies are 91.7 J K^{-1} mol^{-1}, 213.8 J K^{-1} mol^{-1} and 38.1 J K^{-1} mol^{-1} (all at 298 K). Assuming that these values are independent of temperature, compute $\Delta_r H^\circ$, $\Delta_r S^\circ$ and $\Delta_r G^\circ$ at 800 K; hence find the equilibrium pressure of carbon dioxide at this temperature.

17.9 Thermodynamic data, at 298 K, for the reagents and products of the gas phase reaction

$$2HNO_2(g) \rightleftharpoons H_2O(g) + NO(g) + NO_2(g)$$

are given below.

	$\Delta_f H^\circ$ / kJ mol^{-1}	S_m° / J K^{-1} mol^{-1}	$C_{p,m}^\circ$ / J K^{-1} mol^{-1}
$HNO_2(g)$	-79.5	254.0	45.6
$H_2O(g)$	-241.8	188.7	33.6
$NO(g)$	90.2	210.7	29.8
$NO_2(g)$	33.2	240.0	37.2

Calculate $\Delta_r H^\circ$, $\Delta_r S^\circ$ and $\Delta_r G^\circ$ at 298 K. Assuming that the values of $\Delta_r H^\circ$ and $\Delta_r S^\circ$ at 548 K are the same as those at 298 K, calculate $\Delta_r G^\circ$ at 548 K.

Calculate $\Delta_r C_p^\circ$ and, using this value, compute $\Delta_r H^\circ$, $\Delta_r S^\circ$ and $\Delta_r G^\circ$ at 548 K. Compare the two values of $\Delta_r G^\circ$ you have obtained, and comment on what is the major source of the temperature variation of $\Delta_r G^\circ$ for this reaction. Calculate K for the reaction at 298 K and at 548 K.

17.10 The standard molar entropy of N_2 gas at 298 K is 191.6 J K^{-1} mol^{-1}, and its standard molar constant pressure heat capacity, $C_{p,m}^\circ$, at the same temperature is 29.70 J K^{-1} mol^{-1}.

(a) Using Eq. 17.46 on page 787, find the standard molar entropy of N_2 at 398 K.

(b) A better approximation than assuming that $C_{p,m}^\circ$ is constant is to use a parametrized form which includes a temperature dependence. For example

$$C_{p,m}^\circ(T) = A + BT.$$

Using this expression for $C_{p,m}^\circ$, show that integrating Eq. 17.45 on page 787 between T_1 and T_2 gives the following

$$S_m^\circ(T_2) = S_m^\circ(T_1) + A \ln\left(\frac{T_2}{T_1}\right) + B[T_2 - T_1]$$

(c) For N_2 $A = 28.58$ J K^{-1} mol^{-1} and $B = 3.77 \times 10^{-3}$ J K^{-2} mol^{-1}. Using these values in the expression above, calculate the entropy at 398 K. Comment on the difference between your answer and that obtained in (a).

17.11 The equilibrium constant, K, for the reaction

$$COCl_2(g) \rightleftharpoons CO(g) + Cl_2(g)$$

has been measured at a series of temperatures around 700 K as follows

T/K	635.7	670.4	686.0	722.2	760.2
K	0.01950	0.04414	0.07575	0.1971	0.5183

By plotting a graph of $\ln K$ against $1/T$, obtain a value for $\Delta_r H^\circ$ for the above reaction; explain any approximations you have to make. Use your graph to find a value of $\Delta_r G^\circ$ at 700 K, and hence find a value for $\Delta_r S^\circ$ at the same temperature.

17.12 Using the approach illustrated in section 17.9 on page 788 show that the equilibrium constants at T_1 and T_2 are related according to

$$\ln (K(T_2)) - \ln (K(T_1)) = -\frac{\Delta_r H^\circ}{R} \left[\frac{1}{T_2} - \frac{1}{T_1} \right].$$

What assumptions are made in deriving this equation?

For the reaction

$$CO(g) + H_2O(g) \rightleftharpoons H_2(g) + CO_2(g)$$

K is 1.038×10^5 at 298 K and 1.094×10^4 at 350 K. Use the expression above to calculate a value for $\Delta_r H^\circ$. Determine $\Delta_r G^\circ$ at 298 K, and use this value to find $\Delta_r S^\circ$ at this temperature.

17.13 Ketene $O=C=CH_2$ is a reactive gas which can be prepared by the thermal decomposition of propanone (acetone) vapour

$$CH_3COCH_3(g) \rightleftharpoons O=C=CH_2(g) + CH_4(g)$$

Using the approach described in section 6.15.2 on page 255, complete the following table:

line	propanone	$\rightleftharpoons$	ketene	CH₄
1	n_0		0	0
2	$n_0(1 - \alpha)$			
3				

Line 1 gives the initial amount in moles, line 2 gives the amounts in moles at equilibrium, and line 3 gives the mole fractions; α is the fraction of propanone which has decomposed.

Show that the equilibrium constant can be written

$$K = \frac{\alpha^2}{(1 - \alpha)(1 + \alpha)} \frac{p_{eq}}{p^\circ},$$

where p_{eq} is the pressure of the equilibrium mixture. Using this expression, find the value of K which corresponds to 90% decomposition of propanone at a total pressure of 1.2 bar.

At 298 K, $\Delta_r G^\circ$ for this reaction is 42 kJ mol^{-1}, and $\Delta_r H^\circ$ is 81 kJ mol^{-1}. Find the value of the equilibrium constant at 298 K and then, using the relationship below, find the temperature at which there is 90% decomposition of propanone at a total pressure of 1.2 bar.

$$\ln (K(T_2)) - \ln (K(T_1)) = -\frac{\Delta_r H^\circ}{R} \left[\frac{1}{T_2} - \frac{1}{T_1} \right]$$

Chemical kinetics

Key points

- Concentration can be measured as a function of time by using various physical methods such as the measurement of absorbance, conductance or pressure.

- Simple rate laws can be integrated to predict how concentration varies with time.

- Data can be fitted to integrated rate laws in order to determine values of rate constants.

- Rate laws can be simplified by putting one or more reagents in large excess.

- Gas kinetic theory can be used to estimate the collision rate and hence the rate constant; in its simplest form, the theory massively over estimates the rate constant.

- Reactions can be thought of as taking place on a potential energy surface.

- The highest energy point on the pathway between reactants and products is the transition state.

- Using transition state theory, the rate constant is expressed in terms of the enthalpy and entropy of activation; these parameters can be used to probe reaction mechanisms.

In Chapter 10 we looked at how we can describe the rates of reactions using a *rate law*, and how such rate laws can be interpreted in terms of a *mechanism*. In this chapter we are going look at how rate laws are determined. This will involve looking at experimental methods, and also at how the form of the rate law can be determined from data of concentration as a function of time. In addition, we will look at two fundamental theories about the rates of reactions: the *collision theory*, which is based on gas kinetic theory, and the more sophisticated *transition state theory*.

If it has been a while since you studied Chapter 10 it would be as well to review quickly the material in that chapter, especially sections 10.1 to 10.4.

18.1 Measuring concentration

The key part of any experimental study of reaction rates is the measurement of *concentration* as a function of *time*. In this section we will look at a variety of methods which are used to make such measurements. Methods in which the concentration is related to some measurable physical property are generally preferred, as such measurements can usually be made quickly and in an automated way. At the end of this chapter each of these physical techniques is the subject of one or more of the *Questions*.

18.1.1 Traditional chemical analysis

Traditional chemical methods of analysis, such as titration, can be used to measure concentration, but are of restricted applicability in kinetic studies. The main problem is that, since it takes quite a while to complete a titration, we can only study rather slow reactions using this method.

One way in which this limitation can be reduced to some extent is to stop the reaction and then make the chemical analysis. At regular intervals a portion of the reaction mixture is removed (called an *aliquot*), the reaction is stopped by some method, and then each aliquot can be analysed at leisure. How the reaction is stopped, or *quenched*, will depend on the system being studied. One option is to reduce the temperature rapidly, since this will always slow the reaction significantly.

An example of a reaction which has been studied by quenching and then titration is the hydrolysis of ethyl ethanoate by aqueous alkali

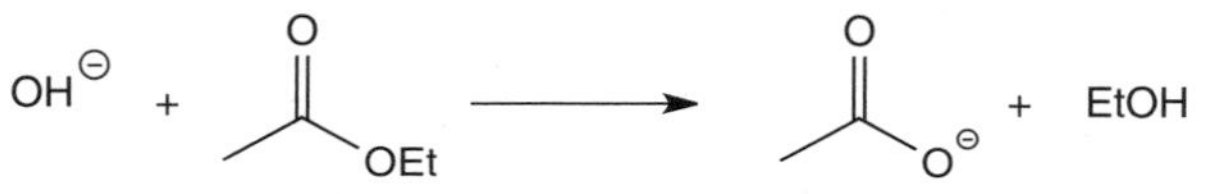

This reaction is rather slow, proceeding over a timescale of tens of minutes. At regular intervals an aliquot of the reaction mixture is withdrawn and mixed with a *known* excess amount of aqueous acid. This neutralizes the alkali, thus effectively stopping the reaction. Then, at leisure, the concentration of the acid is determined by titration. Since the amount of acid added to quench the reaction is known, it is possible to work back to the original concentration of alkali. This method is known as a *back titration*.

18.1.2 Spectroscopic methods

The most widely used method for determining concentration is to measure the attenuation of light (of a suitable wavelength) as it passes through the sample. Using the *Beer–Lambert Law*, the attenuation of the light can be related directly to the concentration of the absorbing species.

You are familiar with the idea that colour is due to the absorption of particular parts of the visible spectrum of light. A solution that appears to be red is so because the molecules are absorbing light from the blue part of the spectrum. Different molecules absorb light at different wavelengths (i.e. of different colours), so even if we have a mixture of species present, as we do in a reaction, it may be possible to find a wavelength at which the species of interest absorbs, and the others do not.

Figure 18.1 shows the basic arrangement for making light absorption measurements using an instrument known as a *spectrophotometer*. The sample (usually a solution) is held in a glass or quartz tube, called a *cuvette*, which has a square cross section. Typically the sides of the cuvette are 1 cm wide, and the total volume is 2 – 3 cm^3. Light of a particular wavelength passes through the solution and then the intensity of the light is measured using a detector, such as a photodiode. The output of the detector is an electrical signal, which can easily be manipulated and logged.

The Beer–Lambert Law gives the relationship between the intensity I_0 of the light entering the cuvette, and the intensity I of the light leaving the cuvette

$$I = I_0 \, 10^{(-\varepsilon c l)}. \tag{18.1}$$

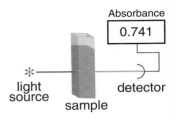

Fig. 18.1 Schematic of a spectrophotometer used to measure light absorption. Light of a given wavelength passes though the sample, which is usually held in a glass or quartz cuvette. Having passed through the sample, and possibly been absorbed, the intensity of the light is measured using a detector. Usually such instruments are arranged so that the size of the detected signal is converted to, and displayed as, an *absorbance*. As explained in the text, this is proportional to the concentration of the absorbing species.

Here c is the concentration and l is the *path length* – the length of sample through which the light has passed. This law says that the intensity decays exponentially with the concentration and the path length, and at a rate determined by the value of the *extinction coefficient*, ε. The extinction coefficient is a function of wavelength, and will only be significant at a wavelength at which the molecule absorbs; different molecules will have different values of ε.

If we take the logarithm to the base 10 of both sides of Eq. 18.1 and do some rearrangement we obtain

$$\lg \frac{I_0}{I} = \varepsilon c l.$$

The quantity $\lg(I_0/I)$ is called the *absorbance*, A, which is therefore directly proportional to the concentration

$$A = \varepsilon c l.$$

The spectrophotometer usually computes and displays the absorbance directly. This is done by first recording the intensity of the light without the sample present (or perhaps just with the solvent in the cuvette): this is I_0. Then the sample is inserted, and the intensity I is measured. Using these two values the instrument then computes, and displays, the absorbance. This is rather convenient, since the absorbance is directly proportional to the concentration.

By making measurements of the absorbance of solutions of known concentration it is possible to find the value of the extinction coefficient ε. Thus, a measurement of absorbance can be directly converted to a concentration, with a minimum of effort. Example 18.1 on the following page illustrates the determination of an extinction coefficient.

Spectrophotometric measurements of this type are very convenient ways of measuring concentration. Not only can they be automated, for example by using a computer to log the values of the absorbance over time, but they can also be made very quickly and without disturbing the sample.

An example of a reaction which has been studied using this approach is the acid catalysed bromination of propanone (acetone) in aqueous solution

$$\text{(reaction scheme)} \quad + \ \mathrm{Br_2} \ \xrightarrow{\mathrm{H_3O^+(aq)}} \ \text{Br} \ + \ \mathrm{Br^{\ominus}}$$

Br$_2$ has a strong absorption at 400 nm (this accounts for its brown colour), a wavelength at which none of the other species absorbs. The consumption

---

Example 18.1 Determination of the extinction coefficient

The absorbance of a solution of Br_2, held in a cuvette with pathlength 1 cm, was measured at a wavelength of 400 nm and as a function of concentration. Use the following data to determine a value of the extinction coefficient.

$[Br_2]$ / mol dm^{-3}	0.002	0.004	0.006	0.008	0.010
absorbance	0.313	0.675	0.990	1.330	1.664

Given that the absorbance A is given by $A = \varepsilon c l$, all we need to do is to plot A as a function of the concentration c.

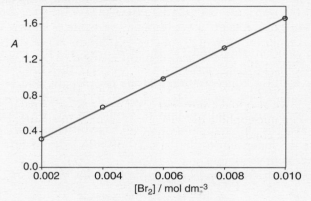

The slope of the best-fit line is 168 mol^{-1} dm^3, which is the value of $\varepsilon \times l$. Given that that pathlength is 1 cm, it follows that $\varepsilon = 168/1 = 168$ mol^{-1} dm^3 cm^{-1}. These rather unusual non-SI units are those traditionally used to quote values of the extinction coefficient.

---

of bromine can therefore be followed by measuring the absorbance at this wavelength.

It is important to make the measurements at a wavelength where only one of the species absorbs, otherwise the data will not reflect accurately the concentration of just a single species. In the visible or UV region the absorption maxima tend to be rather broad and overlapping, so the wavelength needs to be selected with care. The substitution reaction below is a good example of these difficulties

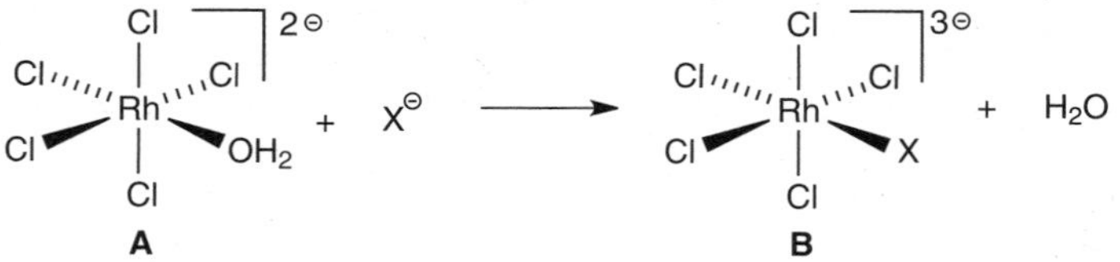

The original rhodium complex **A** has a strong absorbance at 507 nm, and for most of the different incoming ligands X^- which were studied complex **B** does not absorb significantly at this wavelength, so it is suitable for making measurements. However, in the case where the incoming ligand was Br^-, complex **B** also absorbs at 507 nm. In this case, it was necessary to make measurements at 555 nm where the complex **B** absorbs strongly but **A** absorbs hardly at all. This meant following the reaction by monitoring the product, rather than the reactant.

Although it is most common to use absorptions in the UV or visible region to make measurements of the concentration, it is also possible to make measurements in the IR. NMR spectra can also be used, but as this technique is relatively slow and insensitive, it can only be used to monitor rather slow processes.

18.1.3 Conductance measurements

The electrical conductance of a solution depends strongly on the concentration and identity of any ions present, as it is the ions which carry the current. Modern instruments make it straightforward to measure conductance simply by measuring the resistance between two electrodes which are dipped into the solution; a simple example of such a 'dipping electrode' is shown in Fig. 18.2.

Passing a significant current through the solution is not desirable, as this may result in electrolysis and hence chemical changes. Therefore it is usual to measure the conductance using a high-frequency alternating voltage which, as the direction of current flow is constantly changing, results in minimal changes in the solution. We also make sure that the current is very small.

The following nucleophilic displacement reaction, in ethanolic solution, has been studied using conductance measurements

DABCO

The reactants and neutral, but the products are charged, so the conductance increases as the reaction proceeds, and this can be used to measure the rate.

The alkaline hydrolysis of ethyl ethanoate, mentioned at the start of this section, can also be studied using such measurements.

$$CH_3COOEt + OH^- \longrightarrow CH_3COO^- + EtOH.$$

In this case the reaction does not result in a change in the number of ions, but OH^- ions are consumed and ethanoate ions are produced. Different ions contribute differently to the conductance, so for this reaction the conductance changes, even though the number of ions is not changing.

Relating the measured conductance to an absolute concentration is not that straightforward, so such measurements are really only useful as a *relative* measure of concentration. As we will see in the following section, this is not necessarily a problem when it comes to determining certain kinds of rate constant.

18.1.4 Gas pressure

In section 1.9 on page 22 it was shown that, using the ideal gas equation, the pressure can be related directly to the concentration

$$\frac{n}{V} = \frac{p}{RT}.$$

The fraction n/V is the amount in moles per unit volume, which is the concentration. If we have a mixture of gases, the same relationship applies, but it is the *partial pressure* of a particular gas which is related to its concentration.

The *resistance* of a conductor is simply defined as the voltage across the conductor divided by the current through the conductor. The *conductance* is just the reciprocal of the resistance.

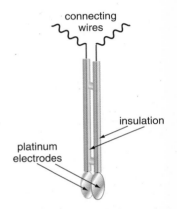

Fig. 18.2 Schematic of a simple electrode assembly which can be dipped into a solution in order to make conductance measurements. Two platinum electrodes, of about 1 cm diameter, are held a few mm apart. The wires which connect the electrodes to the instrument are insulated until they are clear of the solution.

If we have a gas-phase reaction in which there is a change in the *total* amount in moles of gas, and if the reaction takes place in a sealed container, the total pressure will change as the reaction proceeds. For example, in the reaction

$$2NO(g) + O_2(g) \longrightarrow 2NO_2(g)$$

three moles of gas go to two, so as the reaction proceeds in a fixed volume there is a reduction in pressure.

There are many devices which are capable of measuring the pressure rather precisely, varying from simple manometers to electronic pressure gauges. The difficulty it that all we can measure is the *total* pressure, whereas what we want to know is the *partial pressure* (and hence the concentration) of each of the species present. Provided that the overall stoichiometry of the reaction is known it usually possible to relate the change in overall pressure to the required partial pressures, as will be demonstrated for the above reaction.

Suppose that we start out with a 2:1 mixture of NO and O_2, as required by the stoichiometric equation. If the initial partial pressure of O_2 is p_0, then that of NO is $2p_0$, so the initial total pressure is $3p_0$ (recall that the total pressure is the sum of the partial pressures).

Now suppose that the reaction proceeds to some extent, such that the partial pressure of the O_2 falls by Δp; it follows that the partial pressure of the NO will fall by $2\Delta p$, simply because two moles of NO react with one mole of O_2. The partial pressure of the product NO_2 will *rise* from zero to $2\Delta p$, as two moles of NO_2 are produced for each mole of O_2 consumed. The total pressure, p_{tot} is therefore calculated as

$$p_{tot} = \underbrace{2p_0 - 2\Delta p}_{NO} + \underbrace{p_0 - \Delta p}_{O_2} + \underbrace{2\Delta p}_{NO_2}$$
$$= 3p_0 - \Delta p.$$

It therefore follows that $\Delta p = 3p_0 - p_{tot}$. The partial pressure of O_2 is $p_0 - \Delta p$, so using the expression we have just found for Δp we can compute the partial pressure of O_2 in the following way

$$p_{O_2} = p_0 - \Delta p$$
$$= p_0 - (3p_0 - p_{tot})$$
$$= p_{tot} - 2p_0.$$

Finally, recall that the initial total pressure, $p_{tot,init}$ was $3p_0$, so p_0 can be written $p_{tot,init}/3$ in which case

$$p_{O_2} = p_{tot} - \tfrac{2}{3}p_{tot,init}.$$

This expression relates the partial pressure of oxygen, which is what we want to know, to the *measurable* total pressure at the start of the reaction and at some time later.

18.1.5 Electrochemical methods

Electrochemical cells are discussed in detail in Chapter 19.

An electrochemical cell harnesses the energy of redox reactions to produce an electrical current. Such a cell has two electrodes: at one oxidation is taking place and at another reduction is taking place. For example, one electrode might be zinc metal in contact with a solution of Zn^{2+} ions, and the other might be copper metal in contact with a solution of Cu^{2+} ions.

The particularly useful thing about such a cell for our present purposes is that it can be shown that the voltage, or more correctly the potential, produced by the cell depends in a straightforward way on the concentration of the ions in solution. Measuring the potential therefore gives us a way of determining the concentration.

For example, in the oxidation of methanoic acid by bromine

$$HCOOH + Br_2 + 2H_2O \longrightarrow CO_2 + 2Br^- + 2H_3O^+,$$

the concentration of bromine can be followed by setting up a cell of the kind illustrated in Fig. 18.3. One electrode is simply a piece of platinum foil dipping into the solution, and the second electrode is a standard reference electrode (for example a calomel electrode such as that shown in Fig. 19.10 on page 851).

The redox reaction of interest in the solution is

$$Br_2 + 2e^- \longrightarrow 2Br^-.$$

This leads to the development of a potential at the platinum electrode which depends on the concentrations of Br_2 and Br^-. The calomel electrode simply generates a fixed potential, and so the cell potential depends of the concentration of Br_2 and Br^-. The consumption of Br_2 by the reaction can therefore be followed by measuring the potential as a function of time.

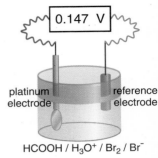

Fig. 18.3 A cell which can be used to measure the kinetics of the oxidation of methanoic acid by bromine. One electrode is simply a piece of platinum foil: at this electrode the redox reaction involves Br_2 and Br^-; the second electrode is simply a reference electrode there to complete the cell. The potential (voltage) produced by the cell depends on the concentration of Br_2 and Br^-, and so changes as the Br_2 is consumed.

18.1.6 Mass spectrometry

We saw in Chapter 11 that mass spectrometry is a very sensitive way of detecting molecules, and is also capable of distinguishing between different molecules on the basis of their masses or fragmentation patterns. In addition, the method can be quantitative, as the rate at which the ions reach the detector is proportional to their concentration in the ionization zone.

Mass spectrometry can therefore be used to follow the progress of a reaction, for example by sampling the reaction mixture at regular intervals. Since the technique is so sensitive and rapid, only a tiny amount of the reaction mixture need be taken, and this can be done quite frequently.

This approach has been used in the study of the gas-phase reaction of ethyl radicals with oxygen. There are two main initial reactions involved here

$$C_2H_5 + O_2 \longrightarrow C_2H_5O_2 \quad \text{and} \quad C_2H_5 + O_2 \longrightarrow C_2H_4 + HO_2.$$

These reactions are thought to be important in the atmospheric oxidation of ethane and also in combustion. By monitoring the reaction mixture with a mass spectrometer it was possible to measure separately the rate of formation of $C_2H_5O_2$ and C_2H_4 as they have different masses.

18.1.7 Fast reactions

If a reaction is very fast, then not only do we need a rapid method of measuring concentration, but we also need a rapid method for mixing the reactants. It is no use if mixing the reactants takes so long that the reaction is all over by the time mixing is complete!

Rather a lot on ingenuity has gone into developing methods for studying fast reactions, but we will look at just two: *continuous flow* and *flash photolysis*.

A free radical is a species with a single unpaired electron. Radicals are often created by breaking a bond e.g. breaking the C–C bond in ethane CH_3–CH_3 creates two methyl radicals, CH_3. Radicals tend to be rather reactive, and hence short lived, as they can readily combine with other radicals, or abstract atoms from stable molecules, creating a different radical e.g.
$$CH_3 + H_2 \rightarrow CH_4 + H$$

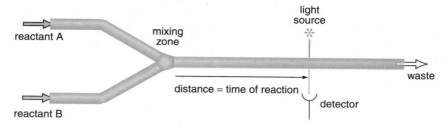

Fig. 18.4 Schematic of a continuous flow apparatus used to study fast reactions. The reactants flow into the mixing zone and then react as they continue to flow down the tube. The time for reaction is therefore proportional to the distance from the mixing zone, but as the reaction mixture is constantly being replenished, measurements can be made at leisure rather than in 'real time'. Usually, light absorption is used to measure concentration.

Continuous flow

Figure 18.4 shows a schematic of a continuous flow apparatus. The key idea is that the reactants (which can be gases or solutions) flow down tubes into a mixing zone and then, having mixed, continue to flow down a tube as they react. If carefully arranged, the mixing can be complete within a few milliseconds, and this sets the upper limit of the rate of reaction which can be studied.

The key thing to understand about the flow method is that different *distances* down the flow tube correspond to different *times* of reaction after mixing. Thus to measure the concentrations after a particular time, all we do is make measurements at a certain distance. As the reactants are constantly being replenished, the measurements do not have to be made in 'real time', which is a great advantage when it comes to studying fast reactions.

A typical flow velocity along the reaction tube is around 10 m s^{-1}, so 1 cm corresponds to 1 ms in time. It is thus possible to study reactions on a ms time scale. As always, we need to have a method for measuring concentration and this is most conveniently done by measuring the absorbance at a suitable wavelength.

A typical reaction in solution which has been studied using a flow method is the complexation reaction between Fe^{2+}(aq) ions and thiocyanate SCN$^-$(aq). The resulting complex absorbs strongly in the visible, whereas the reactants do not, so it is possible to measure the concentration of the products by measuring the absorbance.

A second example of the use of a flow system is in the study in the gas phase of the reactions of oxygen atoms with various small molecules – something of interest in atmospheric chemistry. The oxygen atoms themselves are generated by an electric discharge, and then these flow together with the reactant gases down the tube. Of course, we need to be able to measure the concentration of the oxygen atoms, and this can be done in the following rather ingenious way.

A small amount of nitric oxide, NO, is added to the flow system. An oxygen atom can react with NO to give an electronically excited NO$_2$ molecule

$$O + NO \longrightarrow NO_2^*.$$

This excited molecule drops down to the ground state, emitting a photon in the green part of the spectrum. Finally the NO is regenerated by the following very rapid reaction

$$NO_2 + O \longrightarrow NO + O_2.$$

An electric discharge simply involves passing an electric current through a gas, just as in a neon display sign; usually rather high voltages are needed to set up a discharge. The molecules in the gas through which the discharge passes acquire a great deal of energy, easily enough to dissociate or ionize them. A discharge is therefore used to produce highly energetic species such a free radicals.

As it is regenerated, the amount of NO does not decay down the tube, so the intensity of the green emission is proportional to the concentration of the oxygen atoms. This emission can be measured using a photomultiplier tube, which gives an electrical signal.

Flash photolysis

The idea behind the flash photolysis method is to use an intense flash of light, often from a laser, to create the reactant molecules *in situ*. The reactions of these molecules are then followed by measuring light absorption of the reactants or products. The advantage of this method is that there is no time delay as the reactants are mixed. In addition, since the laser pulse which generates the reactants can be very short (micro- or even nanoseconds), very fast reactions can be studied.

A typical example of using this technique is in studying the recombination reactions of methyl radicals. A pulse of laser light in the UV region was used to generate the CH_3 radicals from propanone (acetone), and then the reactions of these radicals were followed by measuring their absorbance at 216 nm.

Another example, this time in solution, is where a pulse of UV laser light was used to create a carbenium ion

$$Ph_2CHCl \xrightarrow{\text{pulsed UV laser}} Ph_2CH^+.$$

The very fast reactions of the carbenium ion with various nucleophiles were then studied.

18.2 Integrated rate laws

Once we have the basic data of concentration measured as a function of time, the next task is determine the rate law. In general, we expect this to be of the form

$$\text{rate} = k\,[A]^a\,[B]^b \ldots$$

where k is the rate constant, a is the order with respect to species A, and so on.

The immediate difficulty is that our measured data are of concentration as a function of time, whereas the rate law is expressed in terms of the rate i.e. the change in concentration over a time interval. The rate is the *slope* of a graph of concentration against time, and so it can be written as the *derivative* of the concentration with respect to time i.e. $d[A]/dt$. Using this, the rate equation becomes

$$\frac{d[A]}{dt} = -k\,[A]^a\,[B]^b \ldots$$

Since A is a reactant its concentration decreases with time, so the derivative must be negative. This type of equation which includes a derivative is called a *differential equation*. For given values of the orders a, b ... this equation can be solved by integration to give an expression for how the concentration of A is expected to vary with time. The experimentally measured data can then be compared with the predicted variation in the concentration for different values of the rate constant and the orders until the best fit between experiment and theory is obtained.

For a few simple, but nevertheless important, cases it is possible to solve the differential equation by hand. These solutions tell us that a straight line graph will be obtained if the data are plotted in a certain way, and this turns out to be a useful way of assessing whether or not the data are a good fit to the assumed rate law.

For more complex rate laws, a solution by hand is not possible, so it is necessary to resort to a numerical solution by computer. Differential equations of this type come up so frequently in the physical sciences that sophisticated 'off the shelf' software packages are available for solving them.

18.2.1 First-order rate law

The simplest rate law is one which is first order in a single reactant

$$\frac{d[A]}{dt} = -k_{1st}[A]. \tag{18.2}$$

In this expression, the derivative – that is the slope of a plot of [A] against time – is negative since A is a reactant whose concentration *decreases* over time.

This differential equation can be solved using the technique known as the *separation of variables* in which all the terms in one variable (i.e. [A]) are collected on one side, and all those in the other variable (i.e. t) are collected on the other. In making this separation we are allowed to treat 'd[A]' and 'dt' as if they are normal functions. Once the separation has been made, we can integrate each side.

In the case of Eq. 18.2, we take the dt over the the right and bring the term [A] over to the left; k_{1st} is a constant, so it can be left where it is. Making these rearrangements and integrating both sides gives

$$\int \frac{1}{[A]} \, d[A] = \int -k_{1st} \, dt. \tag{18.3}$$

The left-hand side is easy to do since the integral of 1/[A] with respect to [A] is ln[A], just in the same way that the integral of $1/x$ with respect to x is ln x. The right-hand side is also easy since $-k_{1st}$ is a constant and can be taken outside the integral. We are left with the integral of 1 with respect to t, which is simply t. Adding in the constant of integration gives

$$\ln[A](t) = -k_{1st}t + \text{const}. \tag{18.4}$$

where we have written the concentration as [A](t) to remind us that it depends on time.

Graphical analysis

Equation 18.4 is a very nice result, since it says that *if* the reaction is first order, then a plot of ln [A] against time will be a straight line with slope $-k_{1st}$, as is shown in Fig. 18.5 (b). We therefore have a method of checking to see whether or not the data we have conform to a first-order rate law and, if it does, of finding a value for the rate constant.

What we would do would be to plot the graph and then decide whether or not, to within experimental error, it was actually straight. If it is, then we would report that, to within experimental error, the reaction follows a first-order rate law. If the graph is not a straight line, then a different rate law will have to be

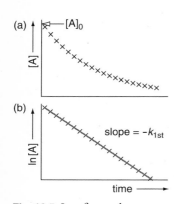

Fig. 18.5 In a first-order reaction the concentration of reactant A drops exponentially with time, as shown in (a). Plotting the log of the concentration against time gives a straight line plot, with slope $-k_{1st}$, as shown in (b).

tried. The nice thing about this approach is that it avoids the need to estimate the rate by drawing tangents to the plot of concentration against time.

This first-order rate law has a very special, and very useful property, which is that it is not necessary to know the absolute concentrations – a relative measure is sufficient. Suppose that we measure some property C_A which is directly proportional to the concentration of A

$$C_A = \alpha[A],$$

it follows that $[A] = C_A/\alpha$. Substituting this into Eq. 18.4 on the preceding page, and doing some rearrangement gives

$$
\begin{aligned}
\ln[A] &= -k_{1st}t + \text{const.} \\
\ln(C_A/\alpha) &= -k_{1st}t + \text{const.} \\
\ln C_A &= -k_{1st}t + \ln\alpha + \text{const.}
\end{aligned}
$$

The last line tells us that a plot of $\ln C_A$ against time will be a straight line with slope $-k_{1st}$, just the same as plotting $\ln[A]$ against time. All that happens is that the intercept changes, but this is not of any interest.

This result is so useful because it is often the case that we can measure something which is proportional to concentration, but determining the constant of proportion is often not so easy. For example, in the oxygen atom flow system we described above, the intensity of the green emission is proportional to [O], but establishing the exact relationship is not at all easy. Another example would be in conductance methods, where the conductance is proportional to the concentration, but again the constant of proportionality might be hard to determine.

It is important to note that this property is unique to a first-order rate law. However, as we shall see, it is often possible to manipulate things in such a way that the rate law is apparently first order.

Explicit form of the rate law

If we return to Eq. 18.3 on the facing page we can remove the constant of integration by computing the definite integral between time $t = 0$ and time $t = t^\star$; the upper limit is taken as $t^\star$ so that we do not get the limit muddled up with the variable t. The concentration at any time t is written $[A](t)$.

$$
\begin{aligned}
\int_{t=0}^{t=t^\star} \frac{1}{[A]}\, d[A] &= \int_{t=0}^{t=t^\star} -k_{1st}\, dt \\
\Big[\ln[A](t)\Big]_{t=0}^{t=t^\star} &= -k_{1st}\Big[t\Big]_{t=0}^{t=t^\star} \\
\ln[A](t^\star) - \ln[A](0) &= -k_{1st}t^\star
\end{aligned}
$$

The concentration at time $t = 0$, $[A](0)$, is written as $[A]_0$, and we can now drop the star from $t^\star$ to write the result in terms of t as

$$\ln\left(\frac{[A](t)}{[A]_0}\right) = -k_{1st}t. \tag{18.5}$$

Taking the exponential of both sides gives

$$\frac{[A](t)}{[A]_0} = \exp(-k_{1st}t) \quad \text{or} \quad [A](t) = [A]_0 \exp(-k_{1st}t). \tag{18.6}$$

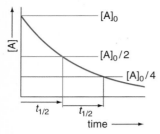

Fig. 18.6 For a first-order process the *half-life*, defined as the time it takes for the concentration to fall to half its initial value, is *independent* of the initial concentration $[A]_0$. Thus it takes the same time to drop from $[A]_0$ to $[A]_0/2$ as it does to drop from $[A]_0/2$ to $[A]_0/4$, and so on. This is a special property of the exponential decay.

This last equation says that the concentration of the reactant A falls exponentially from its initial concentration towards zero i.e. an exponential decay, as is illustrated in Fig. 18.5 (a) on page 806.

The half-life

The *half-life* of a process is the time it takes for the concentration to fall to half its initial value, as illustrated in Fig. 18.6. It is easy to work this out from Eq. 18.5 on the preceding page by setting $[A](t)$ to be $[A]_0/2$

$$-k_{1st}\, t_{\frac{1}{2}} = \ln\left(\frac{[A]_0/2}{[A]_0}\right)$$

$$t_{\frac{1}{2}} = \frac{\ln\frac{1}{2}}{-k_{1st}}$$

$$\text{hence} \quad t_{\frac{1}{2}} = \frac{\ln 2}{k_{1st}},$$

where we have used $\ln\frac{1}{2} = -\ln 2$. The key thing about this result is that the half-life is *independent* of the initial concentration of A, and is therefore the same throughout the reaction. This is a special feature of a first-order reaction.

The concentration of the products

Sometimes it is convenient to think about a reaction in terms of the way in which the concentration of the products increase rather than in terms of the decrease in the concentration of the reactants. For example, if this first-order process produces a product B

$$A \longrightarrow B,$$

then it follows that at any time the amount of B is equal to the amount of A which has reacted. Given that at time zero the concentration of A is $[A]_0$, it follows that at time t the *change* (decrease) in the concentration of A is $[A]_0 - [A](t)$, and so the concentration of B is also $[A]_0 - [A](t)$. Using Eq. 18.6 on the previous page we can therefore write

$$
\begin{aligned}
[B](t) &= [A]_0 - [A](t) \\
&= [A]_0 - [A]_0 \exp(-k_{1st}t) \\
&= [A]_0 \left[1 - \exp(-k_{1st}t)\right].
\end{aligned}
$$

This predicts that the concentration of B rises exponentially from zero toward the initial concentration of A, $[A]_0$.

If we wish to, we can write this final concentration of B as $[B]_\infty$ i.e. the value at very long times, so the last line becomes

$$[B](t) = [B]_\infty \left[1 - \exp(-k_{1st}t)\right].$$

Figure 18.7 (a) shows a plot of [B] as predicted by this equation.

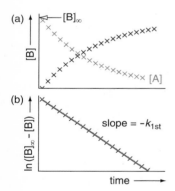

Fig. 18.7 For the first-order reaction A → B the product B increases exponentially to the value $[B]_\infty$ at long times, as shown in (a). For comparison, the behaviour of A is also shown in grey. Graph (b) shows the straight line plot which can be used to determine a value of the rate constant.

This can be rearranged to give a straight line plot in terms of $[B](t)$ as follows

$$[B](t) = [B]_\infty \left[1 - \exp(-k_{1st}t)\right]$$

$$\frac{[B](t)}{[B]_\infty} - 1 = -\exp(-k_{1st}t)$$

$$\frac{[B](t) - [B]_\infty}{[B]_\infty} = -\exp(-k_{1st}t)$$

$$\frac{[B]_\infty - [B](t)}{[B]_\infty} = \exp(-k_{1st}t)$$

$$\ln\left([B]_\infty - [B](t)\right) - \ln\left([B]_\infty\right) = -k_{1st}t,$$

where to go to the last line we have taken logarithms of both sides. It therefore follows that a plot of $\ln\left([B]_\infty - [B](t)\right)$ against t will be a straight line of slope $-k_{1st}$, as shown in Fig. 18.7 (b). As before, the same will be true if we plot not the concentration but any quantity which is proportional to concentration.

18.2.2 Second-order rate law

Another rate law which we can integrate is one which is second order in a single reactant

$$\frac{d[A]}{dt} = -k_{2nd}[A]^2. \tag{18.7}$$

A slightly more complex case is a reaction in which there are two reactants, the order with respect to each is one, making the overall order two i.e. second order

$$\frac{d[A]}{dt} = -k_{2nd}[A][B].$$

If this is for a reaction in which the stoichiometric coefficients of A and B are the same e.g. A + B → products, and if we make the *initial* concentrations of A and B equal, then at all subsequent times their concentrations will also be equal. If this is the case, $[B] = [A]$ so the rate equation reduces to Eq. 18.7.

As before, we can separate the variables ($[A]$ on the left and t on the right), and then integrate to give

$$\int \frac{1}{[A]^2} d[A] = \int -k_{2nd}\, dt$$

$$-\frac{1}{[A]} = -k_{2nd}\, t + \text{const.}$$

$$\frac{1}{[A](t)} = k_{2nd}\, t + \text{const}'.$$

To go to the second line we have used the fact that the integral of $[A]^{-2}$ is $-[A]^{-1}$, and on the last line we have written the concentration as $[A](t)$ to remind ourselves that it is a function of time.

This equation implies that a plot of $1/[A](t)$ against time will be a straight line with slope k_{2nd}, as shown in Fig. 18.8 (b); note that this is a different plot to the one we found for the first-order rate law.

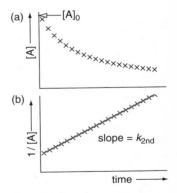

Fig. 18.8 For a second-order process, the concentration of a reactant A drops off as $1/t$ at long times, as shown in (a). If $1/[A]$ is plotted against time, as shown in (b), a straight line with slope k_{2nd} is obtained.

Computing the integral between limits eliminates the constant

$$\int_{[A]_0}^{[A](t)} \frac{1}{[A]^2}\, d[A] \;=\; \int_0^t -k_{2nd}\, dt$$

$$\left[-\frac{1}{[A]}\right]_{[A]_0}^{[A](t)} \;=\; \Big[-k_{2nd}\, t\Big]_0^t$$

$$\frac{1}{[A](t)} - \frac{1}{[A]_0} \;=\; k_{2nd}\, t, \qquad\qquad (18.8)$$

where to go to the last line we have changed the sign of every term. This last equation can be rearranged to give an explicit expression for how the concentration varies with time

$$[A](t) \;=\; \frac{[A]_0}{k_{2nd}\, t\, [A]_0 + 1}.$$

A plot of [A] as a function of time is shown in (a) from Fig. 18.8 on the previous page. At sufficiently long times that the '+1' can be ignored on the bottom of the fraction, the concentration falls off as $1/t$.

The half-life can be found from Eq. 18.8 by setting [A](t) equal to $[A]_0/2$

$$\frac{1}{[A]_0/2} - \frac{1}{[A]_0} \;=\; k_{2nd}\, t_{\frac{1}{2}}$$

$$\text{hence} \quad t_{\frac{1}{2}} \;=\; \frac{1}{k_{2nd}[A]_0}.$$

In contrast to the case of a first-order process, the half-life of a second-order process *does* depend on the concentration, as is illustrated in Fig. 18.9.

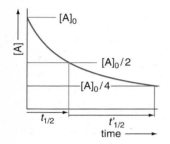

Fig. 18.9 For a second-order process, the time which it takes for the concentration to drop to half its starting value depends on the starting concentration. So, as shown in the plot, it takes much longer for the concentration to drop from $[A]_0/2$ to $[A]_0/4$ than it does for the concentration to fall from $[A]_0$ to $[A]_0/2$.

18.2.3 Zero-order rate law

In Chapter 10 we saw that the bromination of propanone

$$\text{(structure)} + Br_2 \xrightarrow{\;H_3O^+(aq)\;} \text{(structure)} + Br^{\ominus}$$

had the rate law

$$\text{rate of consumption of } Br_2 = k_{obs}\,[\text{propanone}]\,[H_3O^+].$$

Although bromine is consumed in the reaction, its concentration does not appear in the rate law. The rate is said to be *zero order* in bromine, in the sense that, notionally, the bromine concentration is raised to the power of zero.

The simplest possible case of a zero-order rate law for a reactant A is

$$\frac{d[A]}{dt} = -k_0.$$

Recall that $x^0 = 1$, so $[Br]^0 = 1$.

The units of a zero-order rate constant are concentration time^{-1}.

As before, separating and integrating gives

$$\int d[A] \;=\; \int -k_0\, dt$$

$$[A] \;=\; -k_0\, t + \text{const.}$$

This time a simple plot of concentration against time will be a straight line with slope $-k_0$ i.e. the concentration drops linearly with time, as is shown in Fig. 18.10.

Computing the integral between limits eliminates the constant

$$[A](t) = [A]_0 - k_0 \, t,$$

and the half-life is

$$t_{\frac{1}{2}} = \frac{[A]_0}{2k_0}.$$

As with the second-order rate law, the half-life depends on the concentration, a point which is illustrated in Fig. 18.11.

18.2.4   How easy is it to distinguish the order?

In the previous three sections we have shown that for different orders of reaction, a different plot should produce a straight line. In summary these results are

order	straight line plot
0	$[A]$ against t
1	$\ln[A]$ against t
2	$1/[A]$ against t

Figure 18.12 on the following page shows simulated data for zero-, first- and second-order rate processes plotted in these three different ways; for each graph, the blue line is the best-fit straight line. A small amount of scatter has been added to the data in order to make it more like experimental data.

For the zero-order data, shown in the graphs in the first column, the only graph in which the data lie on a straight line is the one where $[A]$ has been plotted against t. The graphs clearly indicate that the order is one.

For the first-order data, shown in the middle column, the log plot produces by far the best fit to a straight line, so on the basis of the graphs we would say that the order is zero. For the second-order data, shown in the right, the reciprocal plot looks to be the best, which is what we expect. However, the plot of $\ln[A]$ against t is certainly not a bad line, and you can imagine that with fewer data points or more scatter, it would be hard to decide whether or not the log plot or the reciprocal plot is the best.

Notice, too, that when the zero- or second-order data is plotted as the log of the concentration against time (the middle row) the data points show a pronounced curvature, visible by the way in which they move from one side of the straight line to the other and then back again. The same behaviour is seen in the other cases where a straight line is not expected.

This simple example shows that, with good quality data, making these straight-line plots is a good way of distinguishing between different orders. However, the more scatter that there is on the data, the more difficult it becomes to decide which is the best. Also, these data are for about three half-lives, so there is a significant change in concentration. If the reaction is followed for a shorter time, the differences between the various plots are much less pronounced.

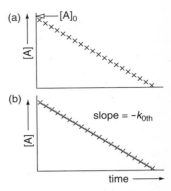

Fig. 18.10 For a zero-order process, the concentration of a reactant A drops off linearly with time, as shown in (a). If a straight line is fitted to such data, as shown in (b), the slope is $-k_0$.

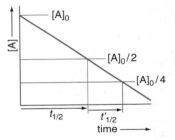

Fig. 18.11 For a zero-order process, the time which it takes for the concentration to drop to half its starting value depends on the starting concentration. So, as shown in the plot, it takes much longer for the concentration to drop from $[A]_0$ to $[A]_0/2$ than it does for the concentration to fall from $[A]_0/2$ to $[A]_0/4$.

Fig. 18.12 Concentration data for a zero-, first- and second-order reaction plotted in different ways; a small amount of scatter has been added to the data. The plot of [A] against t is expected to be a straight line for a zero-order process. For a first-order process the plot of ln [A] against t is expected to be straight, and for a second-order process we need to plot 1/[A] against t. Each set of data is plotted in the three different ways, and the blue line is the best-fit straight line in each case. On the basis of these graphs the orders can be assigned according to which plot gives the best straight line, as indicated by the tick.

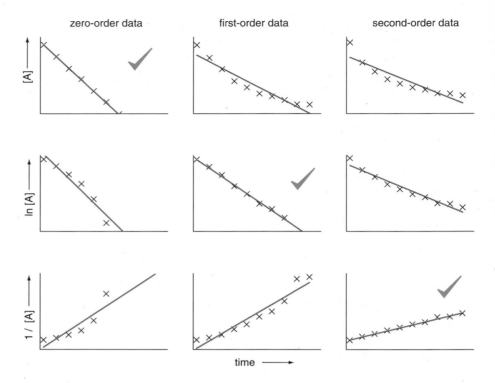

18.2.5 Simplifying rate laws: the isolation method

Suppose that we have a reaction between A and B whose rate law is of the form

$$\frac{d[A]}{dt} = -k\,[A]^a\,[B]^b,$$

where the orders a and b are not known. If we arrange things so that, at the start of the reaction, the amount of B greatly exceeds that of A, then we can assume that the concentration of B will not change significantly from its initial value $[B]_0$ as A is consumed. The rate law can then be written

$$\frac{d[A]}{dt} = -\underbrace{k\,([B]_0)^b}_{k'}\,[A]^a.$$

Under these circumstances, the quantity $k\,([B]_0)^b$ is effectively constant, and so can be written as a new 'rate constant' k'

$$\frac{d[A]}{dt} = -k'\,[A]^a \qquad \text{where} \qquad k' = k\,([B]_0)^b.$$

The rate law is now a great deal simpler, and we can easily try some different (integer) values of a, integrate the rate law, and see if the data fit the predicted concentration dependence. If they do, then we have found the order, and also the value of k'.

The experiment can then be repeated, but this time putting reactant A into large excess, so as to obtain the simplified rate law

$$\frac{d[A]}{dt} = -k''\,[B]^b \qquad \text{where} \qquad k'' = k\,([A]_0)^a.$$

Example 18.2 Using the isolation method

The reaction between bromine and propanone under acidic conditions

$$CH_3COCH_3 + Br_2 \longrightarrow CH_3COCH_2Br + Br^- + H^+$$

is zero order in bromine. Zero-order rate constants for this reaction were measured with the propanone and acid in large *excess* over the bromine, giving the following results

run	1	2	3
$[H^+]$ / mol dm^{-3}	0.15	0.30	0.30
[propanone] / mol dm^{-3}	0.30	0.30	0.60
k_{0th} / mol dm^{-3} s^{-1}	9.90×10^{-7}	1.98×10^{-6}	3.96×10^{-6}

Show that these data are consistent with the rate law

$$\text{rate} = k_{2nd}[\text{propanone}][H^+],$$

and determine a value for the rate constant k_{2nd}.

If propanone and H^+ are in large excess they are effectively constant throughout the reaction. Under these conditions, $k_{2nd}[\text{propanone}][H^+]$ is effectively constant, so the rate law becomes rate $= k_{0th}$, where k_{0th} is a pseudo zero-order rate constant given by $k_{2nd}[\text{propanone}][H^+]$.

Comparing the data for runs 1 and 2 we see that the zero-order rate constant has increased by a factor of $1.98 \times 10^{-6}/9.90 \times 10^{-7} = 2.0$ and the acid concentration has doubled, while the propanone concentration is constant. This implies that $k_{0th} \propto [H^+]^1$.

Similarly from the data for runs 2 and 3 we see that the rate constant increases by a factor of $3.96 \times 10^{-6}/1.98 \times 10^{-6} = 2.0$ when the propanone concentration is doubled, but the acid is held constant. This implies that $k_{0th} \propto [\text{propanone}]^1$. Therefore the data are consistent with the proposed rate law. The second-order rate constant can be found from any one of the runs. Using run 1 we have

$$
\begin{aligned}
k_{2nd} &= k_{0th}/([\text{propanone}][H^+]) \\
&= 9.90 \times 10^{-7}/(0.30 \times 0.15) \\
&= 2.2 \times 10^{-5} \text{ mol}^{-1} \text{ dm}^3 \text{ s}^{-1}.
\end{aligned}
$$

Once again, the data are compared to the predicted time dependence of concentration as a function of time so that the order b can be identified.

Once a and b are known we can go back to the values of k' and k'' and, given that the initial excess concentrations are also known, we can determine the rate constant k. It is important to realize that k' and k'' and not true rate constants, since they depend on concentrations. They are therefore often called *pseudo rate constants*. A typical application of this idea is given in Example 18.2.

This process of simplifying the rate law by putting each reactant, in turn, into excess, is called the *isolation method*. It can, however, lead to some problems. Firstly, putting one reagent in large excess might make the rate inconveniently high. Secondly, there is the possibility that the mechanism of a reaction might be altered by putting a reagent in excess.

18.3   Other methods of analysing kinetic data

As was discussed in Chapter 10 (Fig. 10.2 on page 411), it is possible to convert data of concentration measured as a function of time into *rate* as a function of time, simply by taking the slope of a graph of concentration against time. However, it is not that easy to measure the slope of a curve, especially if there is experimental scatter on the data points.

Once we have obtained this rate data, the order can be determined in a relatively straightforward way. For example, if the rate law is of the form

$$\text{rate} = k[\text{A}]^a,$$

then taking logarithms of both sides gives

$$\ln(\text{rate}) = \ln k + a \ln[\text{A}].$$

Therefore if a plot of $\ln(\text{rate})$ against $\ln[\text{A}]$ is made, the graph will have slope a, the order. This is called the *differential method*.

This is rather nice since, in contrast to the case where we are using integrated rate laws, it was not necessary to test different values of the order a: rather, the value of a is obtained directly. In the case of a more complex rate law, we would simply put all but one reactant into excess, make a series of measurements of rate as a function of the reagent which is not in excess, and hence determine the order with respect to that reagent. The procedure is then repeated, one by one, for the other species.

Half lives can also be used as a way of obtaining an initial estimate of the order. First-order reactions have the unique property that the half-life is independent of the initial concentration, and this can be quite easy to spot just from a table of raw data. For other orders, the behaviour of the half-life is more complex, and it would really be better to try to fit the data to an integrated rate law.

18.4   Collision theory

As was discussed in Chapter 10, the basic picture we have about a bimolecular gas phase reaction is that it takes place when the two molecules collide with sufficient energy to overcome the barrier. What we are going to do in this section is to use gas kinetic theory to estimate the rate of collisions, and then compare this with the experimentally determined rate. What we will find is that, except for reactions between very simple molecules, the collision model dramatically overestimates the rate of reaction. This error is attributed to the fact that the simple theory does not take account of the shape of the molecules and that the chance of reaction depends on the orientation of the colliding molecules.

18.4.1   Estimating the collision rate

If we simply assume that our reacting molecules are hard spheres of a given radius (admittedly rather a dramatic assumption), then it is possible to work out the rate of collisions between two molecules A and B using gas kinetic theory.

Figure 18.13 on the next page is the picture used to work out the collision rate. We concentrate our attention on one B molecule, which is shown in blue,

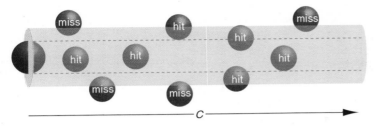

Fig. 18.13 The picture used to compute the number of collisions of one B molecule, shown in blue, with A molecules, shown in red. The A molecules are assumed to be stationary and the B molecule moves forward at velocity c. In one second, the B molecule will hit all of the A molecules any part of which lie within the dashed lines.

and think about its collisions with all the A molecules, a representative selection of which are shown in red. These molecules are all rushing around in different directions, but suppose for the moment that we freeze all of the A molecules. As the B molecule moves through this stationary field of A molecules it is clearly going to hit some of them, and from the figure we can see that the B molecule will hit those A molecules any part of which lies within the dashed lines.

A convenient way of working out which A molecules hit the B molecule is to think about the distance between the centres of the two spheres. As is illustrated in Fig. 18.14, the two molecules collide if the distance between their centres is less than or equal to $(r_A + r_B)$. In Fig. 18.13 the pale green cylinder has radius $(r_A + r_B)$, and so all of the A molecules whose centres are within this cylinder will hit the B molecule.

Now imagine what happens during one second. If the B molecule is moving at speed c, then in one second it will hit all of the A molecules in a cylinder of cross-sectional area $\pi(r_A + r_B)^2$ and length c. The volume of this cylinder is therefore $\pi(r_A + r_B)^2 \times c$, and if the concentration of A molecules is C_A (molecules per unit volume) then the number of A molecules in the cylinder, and hence the number of collisions in one second, is

$$\pi(r_A + r_B)^2 \times c \times C_A.$$

Of course once the B molecule experiences its first collision with an A molecule both will be deflected, and so it is not the case that the B molecule will move steadily forward in the way which is implied by Fig. 18.13. However, when the B molecule starts on its new trajectory it encounters a similar field of stationary A molecules with which it may collide, so the simple calculation gives us essentially the correct result.

In a macroscopic sample different molecules are moving at different speeds, and this can be taken account of by using the *mean* speed to compute the number of collisions. Furthermore, since it is the *relative* motion of A and B which is important in determining the frequency of collisions, we need to use the *mean relative* speed of A and B, $\bar{c}_{rel}$. Finally we need to take into account that there is not just one B molecule, but C_B per unit volume; this means that the total number of collisions is increased by a factor of C_B. The final expression for the total A–B collision rate (per unit volume), Z_{AB}, is therefore

$$Z_{AB} = C_A C_B \pi(r_A + r_B)^2 \bar{c}_{rel}.$$

The area $\pi(r_A + r_B)^2$ is usually called the *collision cross section* and is given the symbol σ. The mean relative speed can be found from gas kinetic theory as

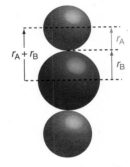

Fig. 18.14 Two spheres of radius r_A and r_B will collide if their centres are separated by a distance of less than or equal to the sum of their radii. The two red spheres are at the maximum distance from the blue sphere which still results in a collision.

The 'bar' in $\bar{c}_{rel}$ is there to indicate that this is the *mean* relative speed. In a sample of gas there is a distribution of speeds, but for our calculation it is sufficiently accurate to use the average or mean speed.

k_B is Boltzmann's constant – not to be confused with various rate constants!

$$\bar{c}_{rel} = \left(\frac{8k_B T}{\pi\mu}\right)^{\frac{1}{2}},$$

where μ is the *reduced mass* of A and B:

$$\mu = \frac{m_A m_{BA}}{m_A + m_B}.$$

Using these, Z_{AB} can be written

$$Z_{AB} = N_A N_B\, \sigma\, \left(\frac{8k_B T}{\pi\mu}\right)^{\frac{1}{2}}.$$

18.4.2   The collision diameter

From Fig. 18.14 on the previous page we saw that the radius of the cylinder containing those A molecules which will collide with the B molecule is $(r_A + r_B)$. If A and B are in fact the same type of molecule, represented by spheres with radius r, then the radius of the cylinder is $2r$, which is the diameter of the sphere. It is therefore common to express the collision cross section σ as πd^2, where d is the *collision diameter*.

For a collision between molecules A and B, with collision diameters d_A and d_B, the collision cross section is

$$\sigma_{AB} = \pi\left(\tfrac{1}{2}d_A + \tfrac{1}{2}d_B\right)^2, \tag{18.9}$$

simply because $r_A = \tfrac{1}{2}d_A$.

18.4.3   Estimating the rate constant

Not all of these collisions will lead to reaction, but only those which are sufficiently energetic to overcome the activation barrier. As was discussed in section 10.4 on page 416, the fraction of collisions with energy greater than E_a is $\exp(-E_a/RT)$, so our estimate of the rate is

$$\text{rate (molecules m}^{-3}\text{ s}^{-1}) = Z_{AB}\exp\left(\frac{-E_a}{RT}\right)$$

$$= C_A C_B\, \sigma\, \left(\frac{8k_B T}{\pi\mu}\right)^{\frac{1}{2}}\exp\left(\frac{-E_a}{RT}\right).$$

Note that since C_A and C_B are in *molecules* per m³, the rate is in molecules m^{-3} s^{-1}. Usually we express the rate in moles dm^{-3} s^{-1}, so to change the rate to these units we need to divide the right-hand side by Avogadro's constant, N_A, and divide again by 10^3 (there are 10^3 dm³ in 1 m³)

$$\text{rate (moles dm}^{-3}\text{ s}^{-1}) = \frac{1}{10^3 N_A}\, C_A C_B\, \sigma\, \left(\frac{8k_B T}{\pi\mu}\right)^{\frac{1}{2}}\exp\left(\frac{-E_a}{RT}\right). \tag{18.10}$$

The reaction we are considering is a simple encounter between A and B, so the rate law is expected to be of the form

$$\text{rate (moles dm}^{-3}\text{s}^{-1}) = k_{2nd}[A][B].$$

To make this comparable with Eq. 18.10 on the preceding page we need to write the concentrations not in moles per dm^3, but in molecules per m^3, which are the units of C_A. The required relationship is $[A] = C_A/(10^3 N_A)$, and similarly for B. Using this the previous expression can be written

$$\text{rate (moles dm}^{-3} \text{ s}^{-1}) = k_{2nd} \frac{C_A}{10^3 \, N_A} \frac{C_B}{10^3 \, N_A}. \tag{18.11}$$

We can now compare the right-hand sides of both Eq. 18.10 and Eq. 18.11 and thus obtain the following expression for k_{2nd}

$$k_{2nd} = 10^3 \, N_A \, \sigma \left(\frac{8k_B T}{\pi\mu}\right)^{\frac{1}{2}} \exp\left(\frac{-E_a}{RT}\right). \tag{18.12}$$

This final equation is our prediction from gas kinetic theory for the second-order rate constant.

18.4.4 Testing the collision theory value for the rate constant

The first thing to notice is that Eq. 18.12 is very similar in form to the Arrhenius expression

$$k_{2nd} = A \, \exp\left(\frac{-E_a}{RT}\right).$$

The terms which multiply the exponential in Eq. 18.12 can be identified as the pre-exponential factor, which we will call A_{coll} to indicate that it is the value predicted by collision theory

$$A_{coll} = 10^3 \, N_A \, \sigma \left(\frac{8k_B T}{\pi\mu}\right)^{\frac{1}{2}}. \tag{18.13}$$

This pre-exponential factor depends on the collision cross section and the mean speed of the molecules, which is turn depends on temperature, varying as $\sqrt{T}$. However, this rather weak dependence would be entirely swamped by the much stronger temperature dependence which comes from the exponential term.

Collision theory does not give any way of estimating the activation energy, so we will simply have to take this value from experiment. Our test of the theory thus reduces to comparing the predictions of Eq. 18.13 with experimental values of the pre-exponential factor. To do this, we need values for the collision cross section, σ. These can be found independently by measurements of bulk properties of a gas, such as the viscosity. A representative selection of values are given in Table 18.1 on the following page.

In Example 18.3 on the next page we see how this expression can be used to compute the pre-exponential factor for the recombination of methyl radicals

$$CH_3 + CH_3 \longrightarrow C_2H_6.$$

The value we find for A_{coll} is 2.5×10^{11} mol^{-1} dm^3 s^{-1}, which can be compared with the experimental value of 2.4×10^{10} mol^{-1} dm^3 s^{-1}. Our predicted value is of the right sort of size, but is around ten times larger than the experimentally determined value.

In a *molecular beam* experiment the reactant molecules are squirted into a vacuum chamber through fine jets, which results in a highly collimated stream of molecules – hence the name molecular beam. The reactant beams are then crossed to allow reaction, and the products are studied using spectroscopic techniques. Although technically difficult, these experiments give unique insight into the details of reactive encounters between molecules.

Example 18.3 Calculation of A using collision theory

We will use Eq. 18.13 to estimate A_{coll} at 298 K for the recombination reaction

$$CH_3 + CH_3 \longrightarrow C_2H_6.$$

As no data is available for the collision cross section of CH_3, we will use the value for CH_4, which is 0.46 nm^2. When we use this value, we have to remember to convert it to m^2; recalling that 1 nm $= 10^{-9}$ m, 0.46 nm^2 is 0.46×10^{-18} m^2.

The mass of CH_3 is 15 amu, and so the reduced mass for the collision of two CH_3 radicals is $15 \times 15/(15 + 15) = 7.5$ amu; we have to remember to convert this to kg by multiplying by the mass of 1 amu, which is 1.6605×10^{-27} kg.

$$
\begin{aligned}
A_{coll} &= 10^3 \, N_A \, \sigma \left(\frac{8 k_B T}{\pi \mu} \right)^{\frac{1}{2}} \\
&= 10^3 \times 6.022 \times 10^{23} \times 0.46 \times 10^{-18} \times \left(\frac{8 \times 1.381 \times 10^{-23} \times 298}{\pi \times 7.5 \times 1.6605 \times 10^{-27}} \right)^{\frac{1}{2}} \\
&= 2.54 \times 10^{11} \; \text{mol}^{-1} \, \text{dm}^3 \, \text{s}^{-1}.
\end{aligned}
$$

Table 18.1 Collision cross sections and diameters

species	σ / nm^2	d / pm
Ar	0.36	340
Xe	0.75	490
H$_2$	0.27	290
Cl$_2$	0.93	540
O$_2$	0.40	360
CO	0.32	320
NO	0.43	370
NH$_3$	0.30	310
CH$_4$	0.46	380
C$_2$H$_4$	0.64	450

Repeating the calculation for

$$NO + Cl_2 \longrightarrow NOCl + Cl,$$

gives A_{coll} as 2.2×10^{11} mol^{-1} dm^3 s^{-1}, which again is significantly larger than the experimental value of 4.0×10^9 mol^{-1} dm^3 s^{-1}. As a final example the reaction

$$H_2 + C_2H_4 \longrightarrow C_2H_6,$$

gives A_{coll} as 4.8×10^{11} mol^{-1} dm^3 s^{-1}, which is some five orders of magnitude larger than the experimental value of 1.2×10^6 mol^{-1} dm^3 s^{-1}.

Calculations similar to these reveal a consistent picture, which is that collision theory overestimates the pre-exponential factor, often by several orders of magnitude. Although there are some uncertainties in the values of the collision cross sections, these are not large enough to account for the very significant discrepancy between the predicted and experimental values.

The underlying problem with simple collision theory is that it treats the molecules as spheres, taking no account of their shape or internal structure. As we have seen in Chapter 8, for a reaction to be successful, the orientation of the two reactants must be appropriate, so that there can be significant overlap of the relevant orbitals. This requirement that the reactants have a suitable orientation in order for there to be a reaction is simply not included at all in the collision model, so it is not surprising that the theory overestimates the rate.

Some very ingenious experiments have been derived to probe this orientation requirement for reaction. For technical reasons the reactions which can be studied at this level of detail using *molecular beams* are a trifle exotic, but nevertheless the results are of interest. The gas phase reaction in which a rubidium atom abstracts an iodine atom from CH_3I

$$Rb + CH_3I \longrightarrow RbI + CH_3,$$

has been studied as a function of the angle between the line of approach of the Rb atom and the C–I bond. Reassuringly for our chemical intuition, collisions in which the Rb approaches the I atom along the line of the C–I bond are successful in leading to reaction, but as the angle of approach increases reaction is less likely. If the Rb atom approaches in the opposite direction (i.e. towards the C), no reaction is seen, and indeed this is also the case for approaches at angles close to this direction.

These detailed studies of the encounters between molecules, and the theory used to describe them, are the field knows as *reaction dynamics*. Such studies, although complex both experimentally and theoretically, give the most detailed picture of what is going on in a chemical reaction at the molecular level.

In the next two sections we will develop a somewhat less ambitious theory about reaction rates which, although it is not as sophisticated as the molecular dynamics approach, nevertheless gives us a useful way of thinking about rate constants.

18.5 Potential energy surfaces

Using modern molecular orbital calculations it is possible to calculate the (total) electronic energy of small molecules in the gas phase with reasonable accuracy. In fact, such an approach can be used to compute the energy of *any* arrangement of atoms, regardless of whether or not the arrangement corresponds to a stable molecule. We could therefore imagine calculating the total energy of the reactants (when they are well separated), and then the energy of all the intermediate arrangements of the atoms as they rearrange themselves into the products. In other words, could determine how the total energy changes as we go from reactants to products.

The arrangement of the atoms in the reactants and products is well defined, but there are essentially an infinite number of ways in which the atoms can be moved in order to go from reactants to products. This is best understood by thinking of a simple example, such as the reaction between an atom A and a diatomic B–C

$$A + B\text{–}C \longrightarrow A\text{–}B + C.$$

As we go from reactants to products the distance r_{AB} has to decrease to the A–B equilibrium bond length, and the distance r_{BC} has to increase as the bond is broken and the atom C leaves. In addition, atom A can approach at any angle θ, as shown in Fig. 18.15. To understand how the energy changes in this reaction we would need to compute the energy for a wide range of different combinations of the two distances and the angle.

The result will be what is called a *potential energy surface* (PES) for the reaction. Such a surface maps out how the energy changes as a function of all of the relevant variables. Of course, it is not a 'surface' in the conventional sense, since that could only represent the energy (the height) as a function of two variables.

Computing the PES for even a simple reaction, such as an atom plus a diatomic, is a formidable task, even for the most powerful computers. This is because there are so many possible arrangements which have to be considered, and for each arrangement a complete calculation of the energy has to be completed. Such surfaces have been computed for $H + H_2$, $F + H_2$ and $H_2 + OH$.

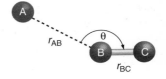

Fig. 18.15 Two distances and an angle are needed to define how A and B–C react to give A–B plus C.

Fig. 18.16 An illustration of the concept of a PES and reaction pathway. A three-dimensional view of the PES is shown, with the height representing the energy. In practice the energy is a function of many more coordinates than the two which we can show here. The products and reactants exist in potential energy minima, and the reaction corresponds to traversing the path between them, illustrated by the white dashed line. The highest point on the pathway, indicated by the orange star, is the *transition state*.

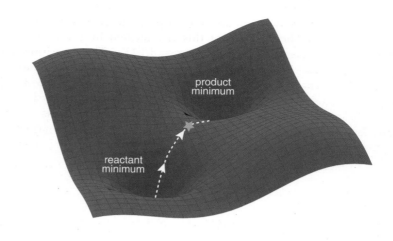

Although we may not have access to complete potential energy surfaces for many reactions, we can discuss the general features of such surfaces and this leads us to some helpful ways of thinking about reactions.

Features of potential energy surfaces

A normal molecule will be at a minimum in the PES. This is because when such a molecule is distorted by a small amount from its equilibrium geometry we expect the energy to rise so that there will be a tendency for the molecule to return to its equilibrium geometry thereby minimizing its energy.

On the surface we expect that there will be a minimum corresponding to the reactants, and another minimum corresponding to the products, as is illustrated in Fig. 18.16. There might be other minima, too, corresponding to alternative products for the reaction. When the reaction proceeds the arrangement of atoms corresponding to the reactants is transformed into an arrangement corresponding to the products. We can envisage this process as following a path on the PES, shown by the dashed white line, from the minimum which corresponds to the reactants to the minimum which corresponds to the products.

As the pathway shown in Fig. 18.16 is traversed, the energy rises at first, reaches a maximum, and then falls as the products are approached. There are of course many pathways which connect the reactants and products, but the one shown here is particularly important as it is the one which involves the least increase in energy, and so is the one by which the greatest number of collisions can result in reactants becoming products.

The dashed line is called the *reaction coordinate*. It is not a single coordinate, such as *x*- or *y*-coordinate of a particular atom, but a complex combination of the many coordinates needed to describe the motion of the atoms as they rearrange themselves into the products. The energy as a function of 'distance' along the reaction coordinate is called the *energy profile*, and it is common to plot this as a simple graph, as shown in Fig. 18.17.

We encountered reaction energy profiles before in section 10.4 on page 416, but using the concept of the PES we can now see more clearly what the reaction coordinate is and why the energy varies in this particular way. From the profile we can also identify the activation energy E_a, which is the energy separation between the reactants and the top of the profile.

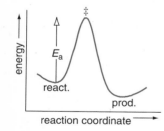

Fig. 18.17 A plot of the energy as a function of the reaction coordinate (i.e. the dashed line in Fig. 18.16) is called the energy profile of the reaction. The activation energy E_a can be identified as the energy separation between the reactants and the top of the profile, which corresponds to the transition state (indicated by ‡).

The arrangement of atoms at the very top of the energy profile is called the *transition state*. The position of this is indicated by the orange star in Fig. 18.16 and the 'double dagger' ‡ in Fig. 18.17. The transition state is not a molecule in the normal sense since, rather than existing at a potential energy minimum, it exists (along the reaction coordinate) at a potential energy maximum. The transition state therefore only has a fleeting existence, but is nevertheless important as, in order to become products, the reactants *must* pass through this arrangement. The transition state plays an important part in the theory of reaction rates, as we shall see in the next section.

18.6 Transition state theory

If the form of the PES for a reaction is known then it is possible to use this to predict the rate constant for the reaction. However, computing the form of the PES is quite a computational challenge, so this approach to predicting the rate constant is not really a practical proposition. We will therefore have to content ourselves with a rather simpler approach, called *transition state theory*, which although it does not predict the actual value of the rate constant, gives us a useful way of thinking about the factors which influence its value. The theory can also be applied to reactions in solution, which makes it especially useful.

The details of this theory are beyond the level of this text, so we will simply quote the final expression, known as the *Eyring equation*, for a second-order rate constant k_{2nd}

$$k_{2nd} = \frac{k_B T}{c^\circ h} \exp\left(\frac{\Delta_r S^{\circ,\ddagger}}{R}\right) \exp\left(\frac{-\Delta_r H^{\circ,\ddagger}}{RT}\right). \qquad (18.14)$$

In this equation, $\Delta_r H^{\circ,\ddagger}$ is the standard enthalpy change in going from the reactants to the transition state, and likewise $\Delta_r S^{\circ,\ddagger}$ is the standard entropy change. These two are related to $\Delta_r G^{\circ,\ddagger}$, the standard Gibbs energy change in going from the reactants to the transition state, in the usual way: $\Delta_r G^{\circ,\ddagger} = \Delta_r H^{\circ,\ddagger} - T\Delta_r S^{\circ,\ddagger}$

If we compare Eq. 18.14 with the Arrhenius expression

$$k_{2nd} = A \exp\left(\frac{-E_a}{RT}\right),$$

we see that the enthalpy of activation can be identified with the activation energy, whereas the value of A depends on the entropy of activation. In fact, it turns out that due to various subtleties of how the activation energy is defined, it is not exactly the same as the enthalpy of activation. The precise relationships turn out to be

bimolecular gas phase:	$E_a = \Delta_r H^{\circ,\ddagger} + 2RT$
unimolecular gas phase:	$E_a = \Delta_r H^{\circ,\ddagger} + RT$
reactions in solution:	$E_a = \Delta_r H^{\circ,\ddagger} + RT.$

Given that a typical value of $\Delta_r H^{\circ,\ddagger}$ is some tens of kJ mol^{-1}, and that RT is 2.5 kJ mol^{-1} at room temperature, these small differences are not very significant, especially as there are often quite large uncertainties associated with the experimental values of these parameters.

In Eq. 18.14, k_B in the Boltzmann constant and c° is the standard concentration.

The expression for a first-order rate constant k_{1st} is identical to Eq. 18.14 except that there is no c° term.

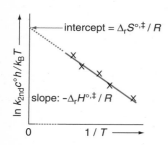

Fig. 18.18 An *Eyring plot* of $\ln([k_{2nd}\,c^{\circ}h]/[k_B T])$ against $1/T$ can be used to determine the values of the activation parameters $\Delta_r H^{\circ,\ddagger}$ and $\Delta_r S^{\circ,\ddagger}$. If the rate constant has only been measured over a small temperature range, a long extrapolation is needed to find the intercept. Any scatter on the data will therefore result in an inaccurate value of $\Delta_r S^{\circ,\ddagger}$.

For a first-order reaction Eq. 18.15 also applies provided we remove the c° term and, of course, replace k_{2nd} with k_{1st}.

18.6.1 Experimental determination of the activation parameters

It is important to realize that transition state theory, as expressed in Eq. 18.14 on the preceding page, does not give us a way of calculating the rate constant since we have no way of calculating the values of $\Delta_r H^{\circ,\ddagger}$ and $\Delta_r S^{\circ,\ddagger}$. Rather the theory is used to *interpret* the values of the rate constant in terms of $\Delta_r H^{\circ,\ddagger}$ and $\Delta_r S^{\circ,\ddagger}$, the values of which must be determined *experimentally*.

We can rearrange Eq. 18.14 into a straight line plot by first taking the fraction $(k_B T)/(c^{\circ}h)$ over to the left-hand side, and then taking logarithms. The steps are

$$k_{2nd} = \frac{k_B T}{c^{\circ}h} \exp\left(\frac{\Delta_r S^{\circ,\ddagger}}{R}\right) \exp\left(\frac{-\Delta_r H^{\circ,\ddagger}}{RT}\right)$$

$$\frac{k_{2nd}\,c^{\circ}h}{k_B T} = \exp\left(\frac{\Delta_r S^{\circ,\ddagger}}{R}\right) \exp\left(\frac{-\Delta_r H^{\circ,\ddagger}}{RT}\right)$$

$$\ln\left(\frac{k_{2nd}\,c^{\circ}h}{k_B T}\right) = \frac{\Delta_r S^{\circ,\ddagger}}{R} - \frac{\Delta_r H^{\circ,\ddagger}}{RT}. \tag{18.15}$$

The last line tells us that a plot of $\ln([k_{2nd}\,c^{\circ}h]/[k_B T])$ against $1/T$ will be a straight line with slope $-\Delta_r H^{\circ,\ddagger}/R$ and intercept (when $1/T = 0$) $\Delta_r S^{\circ,\ddagger}/R$. Such a graph, often called an *Eyring plot*, is illustrated in Fig. 18.18. By measuring k_{2nd} as a function of temperature and plotting the data in this way we have a method of measuring $\Delta_r H^{\circ,\ddagger}$ and $\Delta_r S^{\circ,\ddagger}$, which are collectively called the *activation parameters*.

18.6.2 Interpretation of the activation parameters

Since the enthalpy of activation is a close parallel to the activation energy, its value is interpreted in the same way. For example, a factor which is expected to lower the energy of the transition state will similarly lower the enthalpy of activation. The entropy of activation is a new parameter for us and, as we shall see, its value gives some useful clues as to the form of the transition state.

Gas-phase reactions

For a bimolecular reaction two molecules go to one when the transition state is formed. There is therefore a loss of translational degrees of freedom when the transition state is formed so, by analogy with the discussion in section 6.11.2 on page 237, the process is expected to be accompanied by a significant *decrease* in the entropy. The bimolecular (gas-phase) Diels–Alder reaction between butadiene and ethene

is found to have an entropy of activation of -150 J K^{-1} mol^{-1}, in line with these expectations.

For first-order reactions there is no change in the number of molecules as the transition state is formed, so we do not expect there to be a large change in the entropy.

For example the reverse Diels–Alder reaction

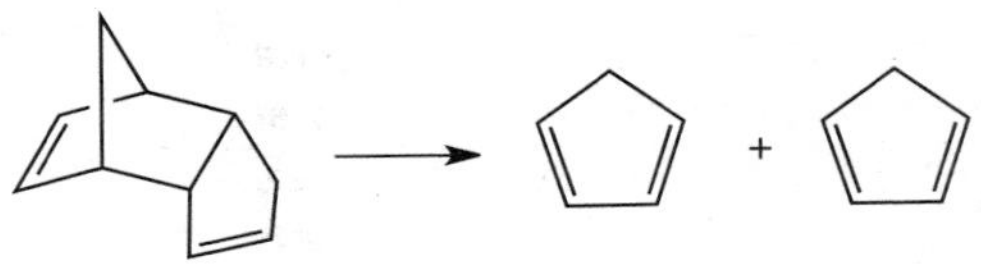

is first order and it found to have $\Delta_r S^{\circ,\ddagger} = -8$ J K^{-1} mol^{-1}. In contrast, the first-order dissociation of ethane to methyl radicals

$$CH_3–CH_3 \longrightarrow CH_3 + CH_3$$

is found to have $\Delta_r S^{\circ,\ddagger} = 58$ J K^{-1} mol^{-1} in the gas phase. This positive value of the entropy change means that the entropy of the transition state is *higher* than that of the reactant, even though both the transition state and the reactant are single molecules. The argument put forward here is that in the transition state the C–C bond is stretched and weakened when compared to the reactant. The transition state therefore has more available energy levels due to its internal motions, and hence has a higher entropy.

In contrast, the first-order isomerization reaction

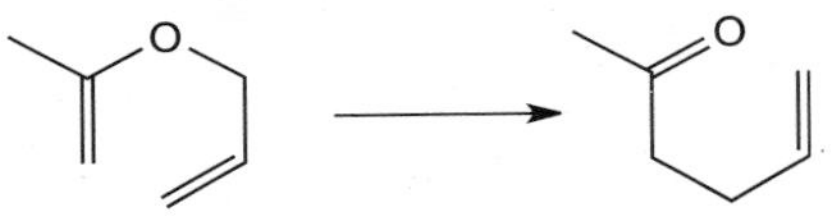

is found to have $\Delta_r S^{\circ,\ddagger} = -32$ J K^{-1} mol^{-1} in the gas phase. This negative value for the activation entropy means that that the transition state has lower entropy than the reactant, and the explanation for this is that this reaction is thought to proceed via a cyclic transition state in which the main atoms form a six-membered ring. The transition state is therefore more 'constrained' than the open-chain reactant, and so has fewer available energy levels and hence lower entropy.

Reactions in solution

The same issues are relevant for reactions in solution as for those in the gas phase. For example, the Diels–Alder reaction

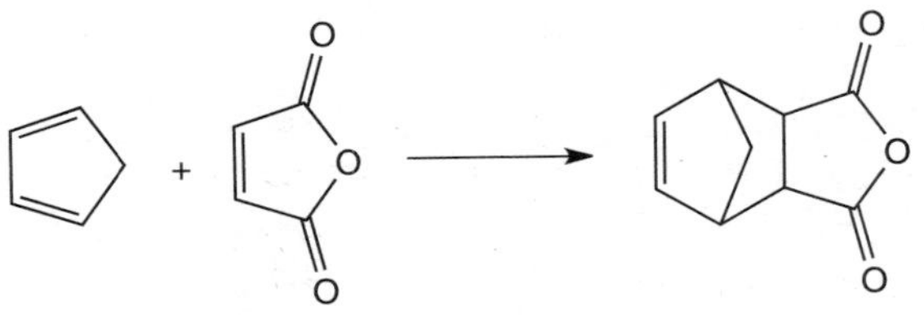

has $\Delta_r S^{\circ,\ddagger} = -150$ J K^{-1} mol^{-1} in dioxan solution – a value very comparable to that for similar reactions in the gas phase.

In polar solvents, reactions involving ionic species are common, and for such reactions some new issues arise. These are well illustrated by the following three reactions substitution reactions

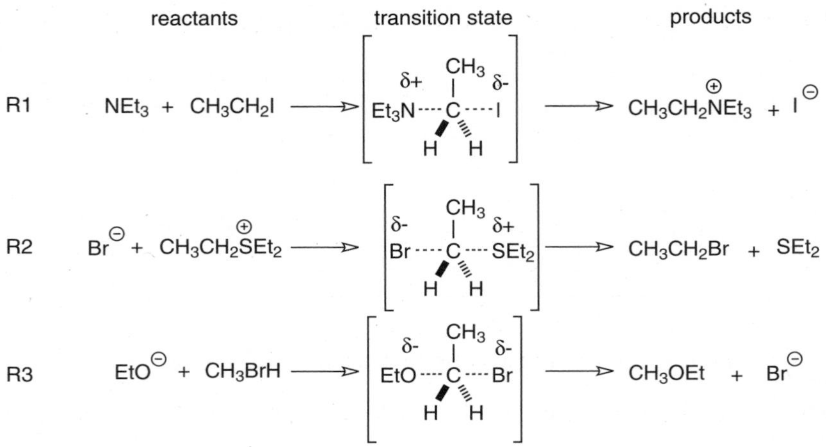

Each of these reactions is thought to take place in a single bimolecular step; the species in the square bracket is an approximate representation of the transition state for the reaction.

For reaction R1 $\Delta_r S^{\circ,\ddagger} = -115$ J K^{-1} mol^{-1}; this negative entropy of activation is just what we expect for a reaction in which two molecules come together to give a transition state. However, for reaction R2 $\Delta_r S^{\circ,\ddagger} = +76$ J K^{-1} mol^{-1}; this is unexpected – how can there be an *increase* in entropy when two molecules go to one?

The explanation for this comes from an understanding of the effect that ion–solvent interactions have on the entropy. As we discussed in section 6.12.3 on page 240, the strong interaction between an ion and a polar solvent is responsible for the overall reduction in entropy when an ion goes from the gas phase into solution. This effect is greater the smaller or more highly charged an ion becomes, as such ions exert a greater restraint on the solvent molecules.

In reaction R2 a *positive* reactant comes together with a *negative* reactant to give a transition state which is overall *neutral* (although it is probably polarized in the way shown). The transition state is therefore much less strongly solvated than the two reactants, and the resulting release of solvent molecules on forming the transition state results in an increase in the entropy. In this case this increase in entropy is sufficient to overcome the decrease in entropy associated with two molecules going to one.

The same idea can be used to understand reaction R3 which has $\Delta_r S^{\circ,\ddagger} = -26$ J K^{-1} mol^{-1}, a value which is much less negative than that for reaction R1. The explanation for this difference is that the negative charge from the reactant EtO$^-$ becomes spread out in the transition state as it moves onto the bromine. The charge is therefore more delocalized in the transition state than it is in the reactant, resulting in a release of the solvent and hence an increase in entropy. The effect is not so strong as in reaction R2 where the transition state is neutral, so $\Delta_r S^{\circ,\ddagger}$ for reaction R3 does not become positive, but it simply less negative than it would have been in the absence of solvation effects.

In reaction R1 the reactants are neutral, but the products are charged, so we therefore expect the transition state to be polarized in the way shown. There is thus an increase in solvation when the transition state is formed, which results in a decrease in entropy. Solvation effects thus contribute to the negative value of $\Delta_r S^{\circ,\ddagger}$ seen for this reaction.

Such reactions can be characterized in the following way:

charge separation: the transition state is formed from neutral reactants and is polarized, as in reaction R1. Such reactions typically have large negative values of $\Delta_r S^{\circ,\ddagger}$.

charge neutralization: the transition state is formed from oppositely charged reactants, leading to an overall reduction in the charge, as in reaction R2. Such reactions typically have positive values of $\Delta_r S^{\circ,\ddagger}$.

charge delocalization: as the transition state is formed the charge which was initially localized on the reactants is spread out, as in reaction R3. Such reactions typically have modest negative values of $\Delta_r S^{\circ,\ddagger}$.

Interpretations such as these are widely used in probing the details of reaction mechanisms.

FURTHER READING

Modern Liquid Phase Kinetics, B. G. Cox, Oxford University Press (1994).
Physical Chemistry, eighth edition, Peter Atkins and Julio de Paula, Oxford University Press (2006).
Reaction Kinetics, second edition, Michael J. Pilling and Paul W. Seakins, Oxford University Press (1995).

QUESTIONS

Physical constants: $R = 8.3145$ J K^{-1} mol^{-1}, $k_B = 1.381 \times 10^{-23}$ J K^{-1}, $h = 6.626 \times 10^{-34}$ J s, $N_A = 6.022 \times 10^{23}$ mol^{-1}, $F = 96{,}485$ C mol^{-1}, mass unit $(u) = 1.6605 \times 10^{-27}$ kg, $1\,\tau = 133.3$ N m^{-2}, 0 °C = 273.15 K.

Many of these questions involve manipulating tables of data and plotting graphs; you will find it convenient to do this using a spreadsheet package, such as EXCEL.

18.1   The alkaline hydrolysis of ethyl ethanoate in aqueous solution is first order in ester and OH$^-$, and so second order overall

$$\frac{d[OH^-]}{dt} = -k_{2nd}[\text{ester}][OH^-].$$

In an experiment to verify this rate law, a reaction mixture was prepared in which the initial concentration of ester and OH$^-$ were both 0.025 mol dm^{-3}. At intervals after the reaction was initiated, a 10 cm^3 aliquot of the reaction mixture was withdrawn and mixed with 10 cm^3 of 0.05 mol dm^{-3} HCl(aq). This amount of acid is sufficient to neutralize any unreacted OH$^-$, thus quenching the reaction. The remaining acid was then titrated with 0.02 mol dm^{-3} NaOH(aq) to the end point in the usual way. This method gave the following data at 0 °C

$t\,/\,s$	vol. of titre $/\,cm^3$	$n_{titre}\,/\,mol$	$n_{aliquot}\,/\,mol$	$[OH^-]\,/\,mol\;dm^{-3}$
120	13.4	2.68×10^{-4}	2.32×10^{-4}	0.0232
300	14.3			
600	15.6			
900	16.4			
1200	17.4			
1500	18.1			
1800	18.5			

Complete the table in the following way (to help you, the first line has been completed).

(a) Work out the amount in moles of OH^- in the volume of NaOH solution added in the titration. Enter this in the column headed n_{titre}.

(b) Work out the amount in moles of H^+ in the acid used to quench the reaction mixture, and hence the amount in moles of *unreacted* OH^- in the aliquot: this is $n_{aliquot}$.

(c) Finally, work out the concentration of OH^- in the aliquot, and enter this into the final column.

Since the initial concentrations of OH^- and ester are equal, the rate law can be simplified to $d[OH^-]/dt = -k_{2nd}[OH^-]^2$. It therefore follows that a plot of $1/[OH^-]$ against t should be a straight line.

Make such a plot and use it to estimate a value for the second-order rate constant, stating the units of your result. (These are real data, and so some scatter is to be expected).

Test the data to see if they fit a first-order rate law (i.e. plot $\ln[OH^-]$ against t).

18.2 The same reaction as in 19.1 can be studied by measuring the conductance as a function of time. Although the number of ions do not change in the course of the reaction, the identity of the ions does, with OH^- ions being gradually replaced by ethanoate ions. The latter are less effective at passing a current, so the conductance *decreases* as the reaction proceeds.

If the conductance at time t is $G(t)$, and at long times (when the reaction has gone to completion) the conductance is G_∞, then it can be shown that the concentration of OH^- is proportional to the difference $(G(t) - G_\infty)$.

A reaction mixture was prepared in which the initial concentration of ethyl ethanoate was 0.25 mol dm^{-3}, and that of NaOH was 0.0025 mol dm^{-3}. The ester is thus in excess so the rate law becomes

$$\frac{d[OH^-]}{dt} = -k_{1st}[OH^-],$$

where the pseudo first-order rate constant k_{1st} is given by $k_{2nd}[ester]$. For a first-order process we expect a plot of $\ln[OH^-]$ against t to be a straight line with slope $-k_{1st}$. As was explained in section 18.2.1 on page 806, for such a first-order process we can also plot the log of any quantity which is *proportional* to concentration. In this case we can therefore plot $\ln(G(t) - G_\infty)$ against time.

The following data were obtained at 0 °C

$t\,/\,s$	60	120	180	240	300	360	420	480
$G\,/\,arb.\;units$	85.5	75.0	65.9	60.3	56.6	52.7	51.0	48.9

At long times the conductance was measured as 45.0. Use these data to plot a suitable graph from which you can determine the pseudo first-order rate constant, and hence determine the second-order rate constant.

18.3 The reaction between propanone and bromine in aqueous acid conditions was studied by measuring the absorbance at 400 nm due to Br_2. The initial concentrations of propanone and acid were both 0.50 mol dm^{-3}, with both reagents being in excess compared to the bromine.

The following data of absorbance at 400 nm as a function of time were obtained at 298 K

t / s	0	60	120	180	240	300	360	420
abs.	0.995	0.964	0.903	0.830	0.772	0.739	0.679	0.605

The path length provided by the cuvette was 1 cm, and the extinction coefficient of Br_2 at 400 nm is 168 mol^{-1} dm^3 cm^{-1}.

(a) Convert the absorbance data to concentrations, and then plot concentration as a function of time. You should find that the graph is a straight line, implying that the reaction is zero-order in bromine.

(b) Determine a value for the (pseudo) zero-order rate constant, stating its units.

(c) Assuming that the reaction is first order in both propanone and acid, determine the second-order rate constant for the overall reaction, stating its units.

18.4 The gas-phase reaction between NO and O_2

$$2NO + O_2 \longrightarrow 2NO_2$$

is thought to have the following rate law, which is overall third order

$$\frac{d[O_2]}{dt} = -k_{3rd}[O_2][NO]^2.$$

If the initial concentration of NO is twice the initial concentration of O_2, then this ratio will persist as the reaction proceeds, in which case the rate law simplifies to

$$\frac{d[O_2]}{dt} = -k'_{3rd}[O_2]^3,$$

where $k'_{3rd} = 4k_{3rd}$.

(a) Integrate this rate law to show that the concentration of O_2 varies with time in the following way

$$\frac{1}{[O_2]^2} = \frac{1}{[O_2]_0^2} + 2k'_{3rd}t,$$

where $[O_2]_0$ is the concentration of O_2 at time zero.

For a gas phase reaction, the partial pressure can be used as a measure of the concentration. As was shown in section 18.1.4 on page 801, for this reaction under the initial conditions described above, the partial pressure of O_2 is given by $p_{tot} - \frac{2}{3}p_{tot,init}$, where p_{tot} is the total pressure, and $p_{tot,init}$ is its initial value.

The following data were obtained at 298 K for a mixture of NO and O_2 in the ratio 2:1

t / s	60	120	180	240	300	360	420	480
p_{tot} / τ	84.0	79.6	77.5	76.2	75.3	74.6	74.0	73.7

The pressure is given in units of Torr (τ), where 1 τ = 133.3 N m^{-2}. The initial pressure was 100.1 τ.

(b) By plotting a suitable graph, show that these data are consistent with the overall third-order rate law given above, and obtain a value of the rate constant, stating its units. [You can either use the pressure directly as a unit of concentration or convert the pressures to concentrations in mol dm^{-3} using the ideal gas law.]

18.5 The oxidation of methanoic acid by bromine

$$HCOOH + Br_2 + 2H_2O \longrightarrow CO_2 + 2Br^- + 2H_3O^+,$$

can be studied by setting up the reaction in an electrochemical cell, as illustrated in Fig. 18.3 on page 803. If it is arranged that in the reaction mixture the bromide ion is in excess, the potential (voltage) E generated by the cell is given by

$$E = C - \frac{RT}{2F} \ln [Br_2],$$

where F is the Faraday constant (96,485 C mol^{-1}) and C is a constant. The consumption of bromine can therefore be followed by measuring the potential as a function of time.

The rate law of this reaction is thought to be first order in bromine. An experiment was designed to investigate the order with respect to methanoic acid (*a*) and acid (*b*)

$$\text{rate} = k[Br_2][HCOOH]^a[H_3O^+]^b.$$

The reaction mixture is set up so that HCOOH and H_3O^+ are in excess, making the rate law pseudo first order

$$\text{rate} = k_{1st}[Br_2] \qquad k_{1st} = k[HCOOH]^a[H_3O^+]^b.$$

The above expression for the cell potential can be rearranged to

$$\ln [Br_2] = \frac{2FC}{RT} - \frac{2FE}{RT}.$$

(a) For a first-order process a plot of $\ln [Br_2]$ against time will have slope $-k_{1st}$. Show that it therefore follows that a plot of E against time will have slope $(k_{1st}RT)/(2F)$.

The following data were obtained at 25 °C for a reaction mixture with the following initial concentrations: $[Br_2] = 3.0 \times 10^{-3}$ mol dm^{-3}, $[HCOOH] = 0.10$ mol dm^{-3}, $[H_3O^+] = 0.05$ mol dm^{-3}

time / s	0	60	120	180	240	300	360
E / V	−0.772	−0.766	−0.761	−0.757	−0.752	−0.745	−0.741

(b) Plot these data as described above, and hence obtain a value of the first-order rate constant.

Similar experiments, with different initial (excess) concentrations of methanoic acid and acid gave the following data (the value indicated by the * is obtained from the graph above)

run	1	2	3	4	5	6
[HCOOH] / mol dm^{-3}	0.10	0.10	0.10	0.12	0.16	0.22
[H$_3$O$^+$] / mol dm^{-3}	0.05	0.12	0.25	0.15	0.15	0.15
$10^3 \times k_{1st}$ / s^{-1}	*	2.77	1.33	2.66	3.55	4.88

Note that $10^3 \times k_{1st} = 2.77$ means $k_{1st} = 2.77 \times 10^{-3}$.

Runs 1, 2 and 3 have the same (excess) concentration of methanoic acid, but different (excess) concentrations of acid. Runs 4, 5 and 6 have the same (excess) concentration of acid, but different (excess) concentrations of methanoic acid.

Taking logarithms of the expression for k_{1st} gives

$$\ln k_{1st} = \ln k + a \ln [HCOOH] + b \ln [H_3O^+].$$

(c)   For runs 1, 2 and 3 plot $\ln k_{1st}$ against $\ln[H_3O^+]$ and hence obtain a value for the order b from the slope of the graph.

(d)   Similarly, use the data from runs 4, 5 and 6 to obtain a value for the order a. Comment on what your values for these orders imply about the mechanism for this reaction.

18.6   Aqueous Fe^{3+} reacts with thiocyanate (SCN^-) to give a complex which absorbs strongly at 460 nm

$$Fe(H_2O)_6^{3+} + SCN^- \longrightarrow Fe(H_2O)_5(SCN)^{2+} + H_2O.$$

The reaction is thought to be first order in each reactant

$$rate = k_{2nd}[Fe(H_2O)_6^{3+}][SCN^-].$$

The reaction was followed using a flow system in which a solution containing Fe^{3+} (concentration 2.0×10^{-4} mol dm^{-3}) was flowed together with a solution of SCN^- (concentration 0.2 or 0.3 mol dm^{-3}, in excess over the Fe^{3+}). The concentration of the complex was monitored at various distances down the flow tube by measuring the absorbance at 460 nm, giving the following data (at 298 K) for two different concentrations of SCN^-. The flow rate was 10 m s^{-1}.

distance / cm	10	20	30	40	50
absorbance ([SCN$^-$] = 0.2 mol dm^{-3})	0.253	0.438	0.576	0.667	0.750
absorbance ([SCN$^-$] = 0.3 mol dm^{-3})	0.344	0.569	0.708	0.802	0.842

Since the thiocyanate is in excess, the reaction is pseudo first order in Fe^{3+} and so, as shown on page 809, the time dependence of the concentration of the *product* (i.e. the complex, denoted B) is given by

$$\ln([B]_\infty - [B](t)) - \ln([B]_\infty) = -k_{1st}t,$$

where $[B]_\infty$ is the concentration of B at long time i.e. when the reaction has gone to completion.

(a)   Assuming that the absorbance at time t, $A(t)$, is proportional to the concentration of B, show that

$$\ln\left(\frac{A_\infty - A(t)}{A_\infty}\right) = -k_{1st}t,$$

where A_∞ is the absorbance when the reaction has gone to completion.

(b)   In this experiment the absorbance at long reaction times was 0.94. Determine the value of the pseudo first-order rate constant (by plotting a suitable graph) for the two different excess concentrations of thiocyanate.

(c)   Hence, determine the value of the second-order rate constant.

[Note that you will have to convert the distances to times, using velocity = distance / time.]

18.7   The rate of the gas phase reaction between the hydroxyl radical and HCl gas

$$OH + HCl \longrightarrow H_2O + Cl$$

has been studied by a combination of flash photolysis and laser induced fluorescence. In this technique, the OH radicals were generated by applying a pulse of UV light from a laser to a small amount of water vapour. Subsequently, the concentration of the OH radicals was monitored by using a laser to promote them to an excited electronic state, and then measuring the intensity of the fluorescence (i.e. the light emitted) from these excited molecules. The intensity of the fluorescence, I_F, is proportional to the concentration of OH radicals. The following data were obtained at 300 K and at 220 K for the indicated concentrations of HCl; the fluorescence intensity I_F is given in arbitrary units.

$t \,/\, \mu s$	0	100	200	300
I_F ([HCl] $= 5 \times 10^{15}$ molec. cm^{-3}, $T = 300$ K)	100	69.5	48.3	33.6
I_F ([HCl] $= 1 \times 10^{16}$ molec. cm^{-3}, $T = 300$ K)	100	48.3	23.3	11.3
I_F ([HCl] $= 1 \times 10^{16}$ molec. cm^{-3}, $T = 220$ K)	100	64.7	41.9	27.1

(a) By plotting a graph, show that in each case the decay of OH follows a first-order process, and hence determine the first-order rate constant.

(b) By comparing the two data sets at 300 K, show that the rate law is first order in HCl, and hence determine the overall second-order rate constant at this temperature. Also determine the second-order rate constant at 220 K. Be sure to state the units of each rate constant.

(c) Using the values of the rate constant at these different temperatures, determine the activation energy and pre-exponential factor for the reaction.

[Data kindly provided by Dr David Husain, Dept. of Chemistry, Cambridge.]

18.8 The gas phase decomposition of ethanal

$$CH_3CHO \longrightarrow CH_4 + CO$$

has the rate law

$$\frac{d[CH_3CHO]}{dt} = -k\,[CH_3CHO]^{\frac{3}{2}}.$$

(a) Integrate this rate law and hence obtain an expression for how the concentration of ethanal varies with time (assume that the concentration of ethanal at time zero is $[CH_3CHO]_0$).

(b) Determine the form a plot which would be expected to give a straight line for data which follows this rate law. How is the rate constant related to the slope of this plot?

(c) Find an expression for the half life of the reaction.

18.9 (a) Using the data in Table 18.1 on page 818, determine the collision cross section for an encounter between NO and Cl_2. Take the masses of N, O and Cl as 14, 16 and 35.5 mass units, respectively.

(b) Following the approach used in Example 18.3 on page 818, use simple collision theory to determine the value of the pre-exponential factor for the reaction between NO and Cl_2 at 298 K.

18.10 Rate constants for the following reaction (in solution)

$$RBr + Cl^- \longrightarrow RCl + Br^- \qquad (R = n\text{Bu})$$

as a function of temperature are given in the following table. Use a graphical method to determine the enthalpy and entropy of activation.

$T \,/\, °C$	25.0	34.6	44.5	55.2	64.8
$10^5 \times k_{2nd} \,/\, \text{mol}^{-1}\,\text{dm}^3\,\text{s}^{-1}$	6.45	16.4	41.0	106	215

Is your value of $\Delta_r S^{\circ,\ddagger}$ consistent with the reaction proceeding via a mechanism in which RBr and Cl^- come together in the rate-limiting step?

18.11 Explain why it is that whereas the bimolecular gas phase Diels–Alder reaction

has $\Delta_r S^{\circ,\ddagger} = -150$ J K^{-1} mol^{-1}, the solution phase bimolecular reaction

$$Cr(H_2O)_6^{3+} + CNS^- \longrightarrow Cr(H_2O)_5(CNS)^{2+} + H_2O$$

has $\Delta_r S^{\circ,\ddagger} = +125$ J K^{-1} mol^{-1}.

Electrochemistry

Key points

- An electrochemical cell is thought of as two half cells in each of which a particular redox reaction is taking place.
- The cell reaction is found by following the conventions set out in section 19.1.1.
- The cell potential is related to the Gibbs energy change of the conventional cell reaction via $\Delta_r G_{cell} = -nFE$.
- The Nernst equation gives the concentration dependence of the cell potential.
- Standard half-cell potentials are measured relative to the standard hydrogen electrode.
- The direction of the spontaneous cell reaction can be found by examining the sign of the cell potential.
- The relative oxidizing or reducing power of different species can be assessed by comparing standard half-cell potentials.
- The thermodynamic properties of ions can be determined with the aid of electrode potentials.
- Electrochemical cells and ion-selective electrodes can be used to measure concentration.

All of our portable electronic gadgets, without which our lives are seemingly impossible, rely on batteries to provide electrical power. What a battery does is to harness the energy from chemical reaction so as to produce an electric current: put another way, it takes the energy output of a chemical reaction and makes it available as electrical work. Some batteries can be recharged, a process in which an electrical current is used to force the chemical reaction in the opposite direction, thus storing up energy. A battery is in fact an example of an *electrochemical cell*.

An enormous amount of technological innovation goes into the design of batteries – for example, to make them as light and small as possible, or to make them suitable for providing continuous current or current in intense bursts. Important though this is, we are going to be interested not in the current which a cell produces, but the voltage (or more precisely the *potential*) which a cell produces when no current is flowing. It will be shown in this chapter that the cell potential is very simply related to the Gibbs energy change of the reaction taking place in the cell. We will also show that the enthalpy and entropy change of the cell reaction can be found from measurements of the cell potential as a function of temperature.

Measuring the cell potential is very simple: attaching a digital voltmeter to the cell enables us to make a precision measurement in seconds. As a result, such measurements give us access to thermodynamic properties in a particularly simple way. Since cell reactions generally involve ions, these measurements of cell potentials are especially useful for finding the standard thermodynamic functions of ions in solution. We will also see that data derived from cells, in the form of standard electrode potentials, are very useful for understanding the relative oxidizing or reducing power of different species.

The cell potential is related in a simple way, given by the *Nernst equation*, to the concentration of the species in the cell. This leads to important practical applications in which cell potentials are used to measure concentrations. Such measurements are convenient and simple to make, and can be adapted for continuous monitoring as well as laboratory based measurements. The chapter closes with a discussion of ion-selective electrodes which are frequently used for making such concentration measurements.

19.1 Electrochemical cells

If you take a strip of metallic zinc, clean the surface with an abrasive, and then dip it into a solution of copper sulfate, you will see that a brown layer of metallic copper is quickly deposited on the surface of the zinc. What has happened in that the zinc metal has *reduced* the copper ions to metallic copper, and in the process the zinc is oxidized to zinc ions

$$Zn(m) + Cu^{2+}(aq) \longrightarrow Zn^{2+}(aq) + Cu(m). \qquad (19.1)$$

We are using 'Zn(m)' to indicate that the physical state of the zinc is the metal. It would be just as acceptable to write Zn(s).

This process takes place by electrons being transferred from the zinc to the copper ions. We can describe what is going on using two chemical (half-) equations in which the electrons appear explicitly

$$\underbrace{Zn(m) \to Zn^{2+}(aq) + 2e^-}_{\text{oxidation}} \qquad \underbrace{Cu^{2+}(aq) + 2e^- \to Cu(m).}_{\text{reduction}}$$

In the first equation the zinc looses electrons, and so is oxidized; in the second, the copper ions gain electrons, and so are reduced. The sum of these two processes is the overall reaction given in Eq. 19.1.

Although electrons are transferred in this reaction, the transfer takes place between nearby atoms and ions. The idea of an electrochemical cell is to arrange things so that although the overall chemical process is the same, these electrons can be 'captured' and made to flow round an external circuit i.e. to generate an electric current.

How this is done is illustrated in Fig. 19.1 on the facing page. The key thing is that the copper and zinc species are kept *completely separate* in what are known as two *half cells*. In the left-hand half cell metallic copper is in contact with a solution of Cu^{2+} ions, and in the right-hand half cell metallic zinc is in contact with a solution of Zn^{2+} ions. For reasons we will describe later, there has to be contact between the two solutions, but we do not want them to mix. Typically this is arranged by having a porous barrier between the solutions.

If a wire is connected between the two metallic electrodes, electrons will flow round this external circuit. At the right-hand electrode the zinc metal dissolves, leaving behind electrons on the electrode i.e. the process $Zn(m) \rightarrow Zn^{2+}(aq) + 2e^-$. These electrons travel round the external circuit and, on arrival at the left-hand electrode, are picked up by copper ions i.e. the process $Cu^{2+}(aq) + 2e^- \rightarrow Cu(m)$. Overall zinc metal dissolves at the right-hand electrode and copper metal is deposited at the left-hand electrode, which is precisely the reaction given in Eq. 19.1 on the preceding page. However, by using the cell the electrons are forced to travel round the external circuit, rather than being transferred directly between the Zn and Cu^{2+}. The cell thus enables us to capture this electron transfer as a current.

As the zinc dissolves to give Zn^{2+} in the right-hand half cell, we will start to build up an excess of positively charged ions. In the left-hand half cell, as the Cu^{2+} ions are taken out of solution, we will be left with an excess of negative ions (remember that there must be negative counter ions, such as $SO_4^{2-}(aq)$, in both solutions). If this charge imbalance were to continue to build up, it would reduce the current to zero, since the electrons would be increasingly reluctant to leave the positively charged right-hand side of the cell.

However, remember that there is a porous barrier between the two solutions. This allows ions to flow from one solution to another, so as to correct any charge imbalance that might build up. For example, if the negative counter ions were to move from the left-hand to the right-hand compartment, then this would correct the imbalance. Such a correction could also be achieved by positive ions moving from the right to the left.

The overall picture of an electrochemical cell is that electrons are generated at one electrode and flow round the external circuit to the other electrode, where they are consumed. At the same time, there is a flow of ions between the two solutions, to as to retain electrical neutrality.

The current which a cell can produce depends on the rate of the electrode reactions and the flow of ions across the porous barrier i.e. it depends on the kinetics of various processes. We will show later on that the *thermodynamic* properties of the constituents of the cell depend on the potential generated by the cell in the *absence of a flow of current*, as shown in Fig. 19.2. It is therefore the cell potential which is of primary interest to us.

Before we look at this relationship in more detail we need to establish how we are going to describe and specify the processes going on in a cell. There are certain conventions about how this is done, and as is always the case, we must obey the accepted conventions otherwise confusion will reign.

19.1.1 Cell conventions

We always think of a cell as consisting of two half cells: one on the left, and one on the right. The reaction which goes on in each half cell will involve electrons in the balanced chemical equation e.g. $Cu^{2+}(aq) + 2e^- \rightarrow Cu(m)$. Regardless of the direction of the actual process, the half cell reactions are *always* specified as *reductions* i.e. with the electrons on the left. For example

$$Cu^{2+}(aq) + 2e^- \rightarrow Cu(m) \qquad Zn^{2+}(aq) + 2e^- \rightarrow Zn(m).$$

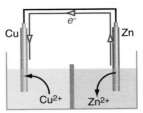

Fig. 19.1 An electrochemical cell which produces an electric current by capturing the electrons in the redox reaction $Zn + Cu^{2+} \rightarrow Zn^{2+} + Cu$. On the right Zn dissolves to give $Zn^{2+}(aq)$ and two electrons. These travel round the external circuit and, at the copper electrode on the left, combine with Cu^{2+} to give Cu. The two solutions are separated by a porous barrier.

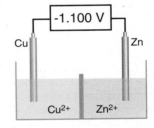

Fig. 19.2 The *thermodynamic* properties of the redox reaction involved in a cell are related to the potential (voltage) produced by the cell under conditions of *zero current flow*. With modern digital voltmeters it is easy to measure this potential without drawing a significant current from the cell. For there to be a measurable potential, there must be contact between the two solutions.

We are not saying that these *are* the reactions which will take place if current is allowed to flow: all we are saying is that the reaction is specified by writing it as a reduction when going from left to right, even though it may be that the actual reaction is the reverse of this. Later on we will see how to work out whether it is oxidation or reduction which is taking place at each electrode.

The four conventions we must adhere to in describing cell reactions and cell potentials. We will first state these conventions and then illustrate the practical implications of each.

(1) The cell is specified, identifying which is the left-, and which is the right-hand half cell.

(2) Each half-cell reaction is written as a reduction.

(3) Having made sure that the half-cell reactions involve the *same* number of electrons, the *conventional cell reaction* is found by taking the *right*-hand half cell reaction and subtracting from it the *left*-hand half-cell reaction i.e. RHS – LHS.

(4) The cell potential is that of the right-hand electrode relative to the left-hand electrode.

These conventions may somewhat obscure at this stage, but as you get used to using them you will see why they are needed. There is an arbitrariness to these conventions which we simply have to live with – after all, that is ultimately what a convention is. For example, you are used to the idea that $\Delta_r H°$ for a reaction refers to 'products – reactants' (rather than the opposite) but this is just a convention, and we could equally well have decided to agree that it is the other way round.

Let us apply these conventions to the cell we have been using as an example, shown in Fig. 19.3; the copper electrode is on the left, and the zinc electrode on the right. The two half-cell reactions, written as reductions are therefore

$$\text{RHS:} \quad Zn^{2+}(aq) + 2e^- \longrightarrow Zn(m) \tag{19.2}$$

$$\text{LHS:} \quad Cu^{2+}(aq) + 2e^- \longrightarrow Cu(m) \tag{19.3}$$

Both half-cell reactions involve two electrons, so we can proceed to work out the conventional cell reaction by subtracting the left-hand side reaction, Eq. 19.3, from the right-hand side reaction, Eq. 19.2. Chemical equations can be added and subtracted in just the same was as algebraic equations, so the subtraction gives

$$Zn^{2+}(aq) + 2e^- - \left[Cu^{2+}(aq) + 2e^- \right] \longrightarrow Zn(m) - Cu(m).$$

The electrons cancel, since we have arranged that they are equal in number on both sides. The Cu^{2+} term can be moved over the the right of the arrow, making it positive, and similarly the Cu term can be moved to the left, also making it positive. The gives the conventional cell reaction as

$$\text{conventional cell reaction:} \quad Zn^{2+}(aq) + Cu(m) \longrightarrow Zn(m) + Cu^{2+}(aq).$$

This reaction is called the *conventional cell reaction* since it is obtained just from the physical arrangement of the electrodes and by the application of the cell conventions. The *actual* reaction which takes place when current is allowed

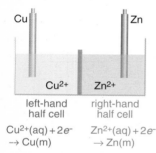

Cu | | Zn

Cu²⁺   Zn²⁺

left-hand half cell   right-hand half cell

$Cu^{2+}(aq) + 2e^-$ $Zn^{2+}(aq) + 2e^-$
$\rightarrow Cu(m)$ $\rightarrow Zn(m)$

conventional cell reaction:
$Zn^{2+}(aq) + Cu(m)$
    $\rightarrow Zn(m) + Cu^{2+}(aq)$

Fig. 19.3 Application of the cell conventions to the cell shown in Fig. 19.2. The cell is thought as as two half cells, one on the right and one on the left. The right- and left-hand half-cell reactions are written as *reductions*, and the conventional cell reaction is found by taking (RHS – LHS).

Example 19.1 Finding the conventional cell reaction

Consider a cell in which the left-hand half cell consists of silver metal in contact with a solution of $Ag^+(aq)$ ions, and the right-hand half cell consists of copper metal in contact with a solution of $Cu^{2+}(aq)$ ions. Write down the two half-cell reactions, as reductions, and hence determine the conventional cell reaction.

The two half-cell reactions, written as reductions are

$$\text{RHS:}\quad Cu^{2+}(aq) + 2e^- \longrightarrow Cu(m)$$
$$\text{LHS:}\quad Ag^+(aq) + e^- \longrightarrow Ag(m).$$

The number of electrons in each reaction is not the same so we must correct this. Either the RHS can be multiplied by $\frac{1}{2}$, or the LHS by 2: both are equally acceptable, but we choose the latter to give

$$\text{RHS:}\quad Cu^{2+}(aq) + 2e^- \longrightarrow Cu(m)$$
$$2 \times \text{LHS:}\quad 2Ag^+(aq) + 2e^- \longrightarrow 2Ag(m).$$

Computing (RHS–LHS) and then tidying up gives the conventional cell reaction as

$$\text{conventional:}\quad Cu^{2+}(aq) + 2Ag(m) \longrightarrow Cu(m) + 2Ag^+(aq).$$

The reaction involves the transfer for *two* electrons. If we had chosen to multiply the RHS by $\frac{1}{2}$, then the conventional reaction would have been

$$\text{conventional:}\quad \tfrac{1}{2}Cu^{2+}(aq) + Ag(m) \longrightarrow \tfrac{1}{2}Cu(m) + Ag^+(aq).$$

This reaction involves the transfer of *one* electron, and is simply half the reaction we found originally. As we shall see, as long as we keep track of the number of electrons involved, it does not matter which of these conventional cell reactions we choose.

to flow may be as the conventional cell reaction is written (left to right), or it may be the opposite (right to left). In order to work out which direction the reaction takes place in we need to inspect the sign of the cell potential, which will be discussed further in section 19.4 on page 846.

Example 19.1 gives a second illustration of finding the conventional cell reaction, but this time for a case where the ions have different charges.

19.1.2 Notation for cells and half-cells

There is a short-hand which is often used for specifying the contents and layout of an electrochemical cell. The electrodes and other species involved are written out on a line, with the left-hand electrode written on the left, and the right-hand electrode on the right. Species which are in *different* phases (i.e. gas, solution or solid) are separated by a vertical line |; species which are in the *same* phase (e.g. both in solution) are separated by a comma. If two solutions are in contact, this is indicated by a double vertical line ||.

The cell illustrated in Fig. 19.2 on page 833 would therefore be written

$$Cu(m) \,|\, Cu^{2+}(aq) \,\|\, Zn^{2+}(aq) \,|\, Zn(m).$$

We will use this notation extensively.

The two species which are involved in a half-cell reaction are said to form a *redox couple*. For example, the half-cell reaction for the silver electrode in contact with $Ag^+(aq)$ ions

$$\underbrace{Ag^+(aq)}_{\text{oxidized form}} + \ e^- \longrightarrow \underbrace{Ag(s)}_{\text{reduced form}},$$

is said to form the Ag^+/Ag redox couple. In specifying the couple, the species are given in the order 'oxidized form'/'reduced form'. In the case of the silver electrode, the oxidized form is the Ag^+ ion, and the reduced form is the Ag metal. Since the conventions require us to write the half-cell reaction as a reduction, it follows that the oxidized form will be on the left, and the reduced form on the right.

19.2 Thermodynamic parameters from cell potentials

In this section we are going to derive the very important relationship between the cell potential E and the Gibbs energy change of the (conventional) cell reaction $\Delta_r G_{\text{cell}}$

$$\Delta_r G_{\text{cell}} = -nFE.$$

In this expression n is the number of electrons involved in the cell reaction, and F is Faraday's constant, which is the charge on a mole of electrons; $F = 96,485 \ C \ mol^{-1}$. Since each electron carries one unit of the fundamental charge, e, the charge on a mole of electrons is found by multiplying this by Avogadro's constant, N_A: $F = N_A \times e$.

19.2.1 Gibbs energy and the reversible work

To prove this relationship between the cell potential and the Gibbs energy change we need to reconsider the First Law (section 17.1 on page 756)

$$dU = \delta q + \delta w,$$

and allow for the fact that other kinds of work than that associated with gas expansions can be done. In the case of an electrochemical cell the other kind of work we are interested in is the *electrical work* done when the potential developed by the cell results in a flow of electrons (a current) around an external circuit. The work done when a charge moves through a potential is given by (charge × potential).

We proceed as we did in section 17.4.2 on page 771 where we derived the Master Equations, but this time an additional work term is included. The derivation starts out by considering a *reversible* process so that δq can be written as $T dS$. The work due to gas expansions is $-p \, dV$, and the reversible additional work will be written $\delta w_{\text{add,rev}}$. The First Law then becomes

$$dU = T dS - p \, dV + \delta w_{\text{add,rev}}.$$

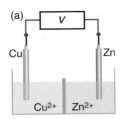
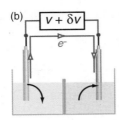
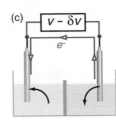

Fig. 19.4 Illustration of the concept of reversibility, as applied to a cell. In (a) a potential V, which is *equal and opposite* to the cell potential, is applied to the cell: as a result, no current flows. If the applied potential is increased by an infinitesimal amount, as shown in (b), an infinitesimal current will flow and redox reactions will occur at the electrodes. On the other hand, if the applied potential is then reduced by an infinitesimal amount, as shown in (c), the current will flow in the opposite direction and the reactions occurring at the electrodes will be in the opposite direction to (b). As the direction of the current flow and the electrode reactions have been changed by an infinitesimal change in the applied potential, the cell is reversible.

Combining this with the complete differential of the enthalpy H (Eq. 17.17 on page 772), $dH = dU + p\,dV + V\,dp$, gives

$$dH = T\,dS + V\,dp + \delta w_{add,rev}.$$

In turn, this is combined with the complete differential of the Gibbs energy G (Eq. 17.19 on page 772), $dG = dH - T\,dS - S\,dT$, to give

$$dG = V\,dp - S\,dT + \delta w_{add,rev}.$$

If we then imagine a process taking place under conditions of constant pressure and temperature, the first two terms in this expression go to zero to give

$$\text{const. temperature and pressure:} \quad dG = \delta w_{add,rev}. \qquad (19.4)$$

In other words, for a *reversible* process taking place under conditions of constant temperature and pressure, the additional work – that is the work other than that due to gas expansions – is equal to the the change in the Gibbs energy, dG.

19.2.2 Reversible electrochemical cells

It is important to realize that it is the *reversible* additional work which is equal to the change in the Gibbs energy. Before we can apply this relationship to an electrochemical cell we need to be clear about what 'reversible' means in such a situation.

If we connect a wire between the two electrodes of a cell, a current will flow, and in the process the concentration of the species in the two half cells will change. This is a spontaneous irreversible process, so Eq. 19.4 does not apply.

We can make the cell reversible by applying across it a potential which is *equal and opposite* to that developed by the cell on its own, as shown in Fig. 19.4 (a). As the two potentials are exactly balanced, no current will flow. If the externally applied potential is then increased infinitesimally (as shown in (b)) the cell potential is now no longer balanced by the applied potential, and so an infinitesimal current will flow in one direction. At the same time the cell reaction will proceed by an infinitesimal amount.

Traditional voltmeters (the ones where a pointer moves over a scale) draw a significant current from the device being measured. In the past, to measure the voltage *without* drawing a current would have required the use of a *potentiometer*, which is a rather elaborate and difficult to use piece of equipment. Now, electronic (digital) voltmeters (often abbreviated as DVM) are readily available: these have high precision and draw very little current from the circuit under test. They are thus ideal for making measurements of the cell potential under reversible conditions.

Now suppose that the applied potential is *decreased* by an infinitesimal amount from its original value, as shown in (c). Again, a current will flow, but this time in the opposite direction, and the reaction will proceed infinitesimally in the opposite direction. What we have here is a system which is reversible in the thermodynamic sense, since a small change in the applied potential causes a reverse in the direction of the cell reaction.

The electrical work done by a cell can therefore only be considered as reversible if the current flow is infinitesimally small. In practice what this means is that the cell potential is measured using a high-impedance voltmeter which draws a negligible current from the cell.

We are now in a position to apply Eq. 19.4 to an electrochemical cell. Suppose that an infinitesimal current is allowed to flow, and that this results in the cell reaction proceeding from left to right in such a way that dz moles of reaction take place. As a result, ndz moles of electrons will have to pass round the external circuit, where n is the number of electrons involved in the cell reaction.

For example, for a cell with the reaction

$$Zn^{2+}(aq) + Cu(m) \longrightarrow Zn(m) + Cu^{2+}(aq),$$

'dz moles of reaction' means that dz moles of Zn^{2+} reacts to give dz moles of Cu^{2+}. For this reaction $n = 2$ since two electrons are needed to convert one Zn^{2+} to one Zn. Thus $2\,dz$ moles of electrons are involved when the reaction proceeds by dz moles.

The electrical charge on this ndz moles of electrons is $-nFdz$. This arises because the charge on a mole of electrons is the Faraday constant, and the electrons are *negatively* charged. If the cell potential is E, then the electrical work done as the electrons pass round the external circuit is given by (charge × potential) which is

$$\delta w_{elec} = -nFE\,dz.$$

We can identify this reversible work as dG, so that $dG = -nFE\,dz$. Taking the dz onto the left gives

$$\frac{dG}{dz} = -nFE.$$

As was shown in section 17.6.2 on page 778, the gradient of G with respect to the amount in moles of reaction is $\Delta_r G$ for the reaction, which in this case is the cell reaction. We therefore have

$$\Delta_r G_{cell} = -nFE. \tag{19.5}$$

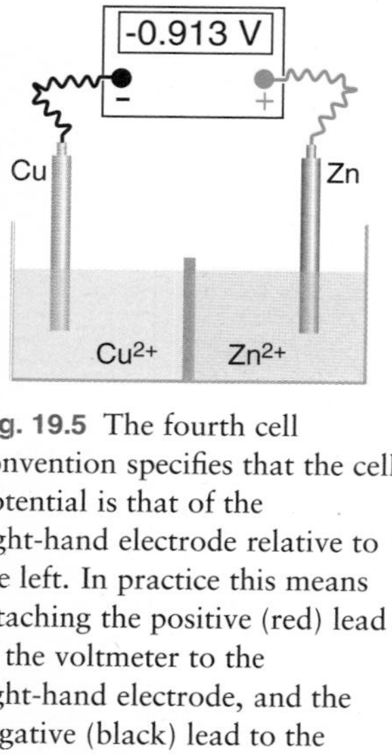

Fig. 19.5 The fourth cell convention specifies that the cell potential is that of the right-hand electrode relative to the left. In practice this means attaching the positive (red) lead of the voltmeter to the right-hand electrode, and the negative (black) lead to the left-hand electrode. The indicated voltage may be positive or negative.

This is an exceptionally important relationship as it enables us to relate the easily measured cell potential to the Gibbs energy change of the cell reaction. There are two important provisos: firstly this relationship applies at constant temperature and pressure; secondly, the cell potential must be measured under reversible conditions. If the temperature is changed then $\Delta_r G_{cell}$ will change, and so therefore will the cell potential i.e. E is a function of the temperature.

The fourth of our cell conventions listed in section 19.1.1 on page 833 is that the cell potential is measured as that of the right-hand electrode relative to the left-hand. What this means is illustrated in Fig. 19.5. The positive lead (usually coloured red) of the voltmeter is attached to the right-hand electrode, and the negative lead (usually black) to the left-hand electrode. The reading, which may be positive or negative, registered by the voltmeter *is* the cell potential. If this potential is then used in Eq. 19.5 on the facing page to determine $\Delta_r G_{cell}$, the reaction to which $\Delta_r G_{cell}$ refers is the *conventional cell reaction*. If we do not follow the cell conventions, we have no way of knowing to which reaction $\Delta_r G_{cell}$ refers.

19.2.3 Enthalpy and entropy changes

Now that we have a relationship between the cell potential and the Gibbs energy change, we can very easily find the associated entropy and enthalpy change. Once again, the Master Equations provide the start we require, specifically in this case we need the third Master Equation, Eq. 17.20 on page 772, $dG = V\,dp - S\,dT$.

If the condition of constant pressure is imposed, then $V\,dp$ goes to zero leaving $dG = -S\,dT$. Taking the dT over to the left gives

$$\text{const. press.}\qquad \frac{dG}{dT} = -S.$$

This relationship applies equally well to changes in these thermodynamic quantities, so for the cell reaction we can write

$$\frac{d\,\Delta_r G_{cell}}{dT} = -\Delta_r S_{cell}.$$

Using Eq. 19.5 on the facing page to write $\Delta_r G_{cell}$ in terms of the cell potential gives

$$\frac{d[-nFE(T)]}{dT} = -\Delta_r S_{cell},$$

where we have written the cell potential as $E(T)$ to remind ourselves that it varies with temperature. This expression which can be rearranged to

$$\frac{dE(T)}{dT} = \frac{\Delta_r S_{cell}}{nF}. \tag{19.6}$$

This relationship tells us that if we measure the cell potential as a function of temperature, and then plot a graph of $E(T)$ against T, the *slope* of the graph will be $\Delta_r S_{cell}/nF$. In this very simple way it is possible to find $\Delta_r S_{cell}$. Having found $\Delta_r G_{cell}$ and $\Delta_r S_{cell}$, it is then possible to find $\Delta_r H_{cell}$ simply by using the definition of the Gibbs energy change: $\Delta_r G_{cell} = \Delta_r H_{cell} - T\Delta_r S_{cell}$.

19.3 The Nernst equation and standard cell potentials

It was mentioned in the introduction that the cell potential depends on the concentration of the species in the cell. Given the relationship $\Delta_r G_{cell} = -nFE$, this concentration dependence should now come as no surprise, since we know that the Gibbs energy of a species depends on its concentration (or partial pressure, if it is a gas), so $\Delta_r G_{cell}$, which is a difference of Gibbs energies, must also depend on the concentrations. The *Nernst equation*, which will be derived in this section, gives the explicit form of this dependence.

Our derivation is very much along the lines of that used in section 17.6.3 on page 781 to find the relationship between the equilibrium constant and the standard Gibbs energy change for a reaction. To make things as general as possible we will assume that the (conventional) cell reaction is

$$aA + bB \longrightarrow cC + dD. \tag{19.7}$$

Here A, B ... are the species involved, and a, b ... are the corresponding stoichiometric coefficients. We will assume that all of these species are in solution, so that their molar Gibbs energies can be written in terms of their concentrations, as was described in section 17.7.2 on page 784

$$G_{m,i} = G_{m,i}^{\circ} + RT \ln([i]/c^{\circ}). \tag{19.8}$$

This expression applies only to *ideal* solutions, which are those in which there are no interactions between the solute molecules. Due to the strong electrostatic interactions between ions, solutions only approach this ideal behaviour at very low concentrations. There are methods of dealing with the thermodynamics of non-ideal solutions, which are briefly discussed in Box 19.1 on the next page, but these are well beyond the level of this book. At this stage we will therefore have to content ourselves with assuming that all solutions are ideal.

Imagine that the cell reaction proceeds by an amount dz according to the conventional cell reaction, Eq. 19.7. This means that the amount in moles of A changes by $-a\,dz$ and that of B changes by $-b\,dz$: these terms are negative as A and B are lost as the reaction proceeds. The amount in moles of C and D change by $+c\,dz$ and $+d\,dz$, respectively: these terms are positive as C and D increase as the reaction proceeds. The overall *change* in the Gibbs energy, dG, is found by multiplying the molar Gibbs energy of each species by the change in its amount in moles

$$dG = c\,dz \times G_{m,C} + d\,dz \times G_{m,D} - a\,dz \times G_{m,A} - b\,dz \times G_{m,B}.$$

Each molar Gibbs energy can then be expressed in terms of the standard molar Gibbs energy and the concentration using Eq. 19.8

$$dG = c\,dz\left[G_{m,C}^{\circ} + RT\ln([C]/c^{\circ})\right] + d\,dz\left[G_{m,D}^{\circ} + RT\ln([D]/c^{\circ})\right]$$
$$-a\,dz\left[G_{m,A}^{\circ} + RT\ln([A]/c^{\circ})\right] - b\,dz\left[G_{m,B}^{\circ} + RT\ln([B]/c^{\circ})\right].$$

This can be tidied up by taking the dz term onto the left-hand side, and multiplying out the brackets to give

$$\frac{dG}{dz} = \underbrace{cG_{m,C}^{\circ} + dG_{m,D}^{\circ} - aG_{m,A}^{\circ} - bG_{m,B}^{\circ}}_{\Delta_r G_{cell}^{\circ}}$$
$$+ cRT\ln([C]/c^{\circ}) + dRT\ln([D]/c^{\circ}) - aRT\ln([A]/c^{\circ}) - bRT\ln([B]/c^{\circ}).$$

The first set of terms on the right we recognise as the standard Gibbs energy change for the cell reaction, $\Delta_r G_{cell}^{\circ}$, and as before the gradient dG/dz is identified as $\Delta_r G_{cell}$. Using these substitutions and bringing all of the log terms together gives

$$\Delta_r G_{cell} = \Delta_r G_{cell}^{\circ} + RT \ln \frac{([C]/c^{\circ})^c\,([D]/c^{\circ})^d}{([A]/c^{\circ})^a\,([B]/c^{\circ})^b}.$$

Now both sides of this equation are divided by $-nF$ to give

$$\frac{-\Delta_r G_{cell}}{nF} = \frac{-\Delta_r G_{cell}^{\circ}}{nF} - \frac{RT}{nF} \ln \frac{([C]/c^{\circ})^c\,([D]/c^{\circ})^d}{([A]/c^{\circ})^a\,([B]/c^{\circ})^b}. \tag{19.9}$$

Box 19.1 Non-ideal solutions and activities

For an *ideal* solution, which is one in which there are no interactions between the solute molecules, the (molar) Gibbs energy depends on concentration according to

$$G_{m,i} = G_{m,i}^{\circ} + RT \ln([i]/c^{\circ}). \tag{19.10}$$

Here $G_{m,i}^{\circ}$ is the molar Gibbs energy of species i when the concentration is equal to the standard concentration c°.

For a non-ideal solution, this concentration dependence is written in terms of a quantity known as the *activity* a_i

$$G_{m,i} = G_{m,i}^{\circ} + RT \ln(a_i). \tag{19.11}$$

In this expression $G_{m,i}^{\circ}$ is the molar Gibbs energy of species i in a solution of concentration c° in which the solute molecules are *not* interacting with one another. This is, of course, an entirely hypothetical construction, but it is used so that $G_{m,i}^{\circ}$ is the *same* for an ideal and a non-ideal solution.

What, then, is the activity? Comparing Eq. 19.10 and Eq. 19.11 we can see that for an ideal solution $a_i = [i]/c^{\circ}$. So, for ideal solutions the activity is simply related to the concentration. This idea is extended to non-ideal solutions by writing the activity as

$$a_i = \gamma_i \times [i]/c^{\circ},$$

where γ_i is the *activity coefficient* of species i. We can immediately see that for an ideal solution the activity coefficient is one. It is also true that the more dilute a solution becomes, the closer it gets to ideality, and hence the closer γ_i gets to one.

The value of the activity, and hence the activity coefficient, is determined by using Eq. 19.11 i.e. we find some property, such as a cell potential, which depends on $G_{m,i}$ and then use the resulting value to find a_i. This may seem like a rather odd approach, but it is taken since in Eq. 19.11 it is $G_{m,i}$ which is the *measurable* quantity.

For dilute solutions of ions there is also a relatively simple theory, the *Debye–Hückel theory*, which relates the activity coefficient to the concentration. Although this theory is not very precise, it does give a useful way of modelling the variation of γ_i with concentration.

When we refer to a solution as having the 'standard concentration', we do not really mean a solution with concentration equal to c° (e.g. 1 mol dm^{-3}), but rather we mean a solution with unit activity i.e. $a_i = 1$. For ionic solutions, unit activity can easily correspond to a concentration very different from c°.

Given that $\Delta_r G_{cell} = -nFE$ we recognise the term on the left as the cell potential, E. By analogy, we define the *standard cell potential* E° in terms of the standard Gibbs energy change for the cell reaction

$$\Delta_r G_{cell}^{\circ} = -nFE^{\circ}. \tag{19.12}$$

This means that the first term on the right of Eq. 19.9 on the facing page can be identified as the standard cell potential, $E°$. This is the potential that the cell will develop when all of the species are present in their standard states i.e. unit concentration for solutes.

It therefore follows that Eq. 19.9 can be written

$$E = E° - \frac{RT}{nF} \ln \frac{([C]/c°)^c ([D]/c°)^d}{([A]/c°)^a ([B]/c°)^b}. \tag{19.13}$$

This is the *Nernst equation* which relates the cell potential to the standard cell potential and the concentration of the species in the cell. The Nernst equation refers to the conventional cell reaction

$$a\text{A} + b\text{B} \longrightarrow c\text{C} + d\text{D}.$$

Note that the fraction inside the log is of the familiar form 'products over reactants', although this is not an equilibrium constant as the contents of the cell are not at equilibrium.

So far we have assumed that all of the species in the cell reaction are present in solution, but this may not always be the case. If a species is present as a gas, the concentration term $[i]/c°$ in the Nernst equation is simply replaced by $p_i/p°$, where p_i is the partial pressure of that gas. If the species is a solid or (pure) liquid, then as we have discussed before, since such species are already in their standard states, they do not contribute a 'concentration' term to the Nernst equation.

For example, for the cell shown in Fig. 19.1 on page 833, the conventional cell reaction is

$$\text{Zn}^{2+}(\text{aq}) + \text{Cu}(\text{m}) \longrightarrow \text{Zn}(\text{m}) + \text{Cu}^{2+}(\text{aq}),$$

so the Nernst equation is

$$E = E° - \frac{RT}{2F} \ln \frac{[\text{Cu}^{2+}]/c°}{[\text{Zn}^{2+}]/c°}. \tag{19.14}$$

No concentration terms are included for the metallic zinc or copper as these are pure solids. We have already established that two electrons are involved in this reaction, so $n = 2$.

19.3.1   The Nernst equation for half cells

Since we think of an electrochemical cell as being composed of two half cells, it is often convenient to express the cell potential as the *difference* of two half-cell potentials

$$E = E_{\text{RHS}} - E_{\text{LHS}},$$

where E_{RHS} is the half-cell potential of the right-hand half cell, likewise E_{LHS} is that of the left-hand half cell. We take 'right minus left' in order to be consistent with the convention that the cell potential is that of the right-hand electrode relative to the left.

Proceeding by analogy with the Nernst equation, if the half-cell reaction (written as a reduction) is

$$p\text{P} + ne^- \longrightarrow q\text{Q},$$

the half-cell potential of the P/Q redox couple is given by

$$E = E°(\text{P/Q}) - \frac{RT}{nF} \ln \frac{([\text{Q}]/c°)^q}{([\text{P}]/c°)^p},$$

where $E^\circ(P/Q)$ is the standard half-cell potential i.e. the potential developed when all species are present in their standard states.

For example, for the cell in Fig. 19.2 on page 833, the half-cell reactions (both written as reductions) are

$$\text{RHS:} \quad Zn^{2+}(aq) + 2e^- \longrightarrow Zn(m)$$
$$\text{LHS:} \quad Cu^{2+}(aq) + 2e^- \longrightarrow Cu(m).$$

The corresponding Nernst equations for the RHS are LHS are

$$\text{RHS:} \quad E_{\text{RHS}} = E^\circ(Zn^{2+}/Zn) - \frac{RT}{2F} \ln \frac{1}{[Zn^{2+}]/c^\circ}$$

$$\text{LHS:} \quad E_{\text{LHS}} = E^\circ(Cu^{2+}/Cu) - \frac{RT}{2F} \ln \frac{1}{[Cu^{2+}]/c^\circ}.$$

Note that there are no concentration terms for the solids as these are present in their standard states. We can find the cell potential by taking RHS − LHS in the usual way to give

$$
\begin{aligned}
E &= E_{\text{RHS}} - E_{\text{LHS}} \\
&= E^\circ(Zn^{2+}/Zn) - \frac{RT}{2F} \ln \frac{1}{[Zn^{2+}]/c^\circ} - \left[E^\circ(Cu^{2+}/Cu) - \frac{RT}{2F} \ln \frac{1}{[Cu^{2+}]/c^\circ} \right] \\
&= E^\circ(Zn^{2+}/Zn) - E^\circ(Cu^{2+}/Cu) - \frac{RT}{2F} \ln \frac{[Cu^{2+}]/c^\circ}{[Zn^{2+}]/c^\circ}.
\end{aligned}
$$

The final expression for the cell potential is the same as that we derived in the previous section (Eq. 19.14 on the preceding page) provided that we identify the standard cell potential as the difference in the standard half-cell potentials

$$E^\circ = E^\circ(Zn^{2+}/Zn) - E^\circ(Cu^{2+}/Cu).$$

In general, the standard cell potential can always be determined from these standard half-cell potentials by taking the usual 'right minus left'

$$E^\circ = E^\circ(\text{RHS}) - E^\circ(\text{LHS}).$$

A somewhat more complex case is illustrated in Example 19.2 on the following page.

19.3.2 Standard half-cell potentials

Just in the same way that it is convenient to tabulate standard enthalpies of formation and standard entropies, in electrochemistry it is useful to tabulate standard half-cell potentials. However, as we can only measure the potential developed by a complete cell i.e. the *difference* between two half-cell potentials, there is no way of measuring the values of individual half-cell potentials.

This difficulty is resolved by arbitrarily setting the potential of one standard half-cell to zero, and then measuring all other half-cell potentials relative to this chosen *reference electrode*. Since any measurable cell potential will always depend on the *difference* of two half-cell potentials, this arbitrarily chosen zero on the scale does not affect the result of any calculation.

Example 19.2 Finding the Nernst equation

Consider the cell we discussed in Example 19.1 on page 835

$$Ag(m)|Ag^+(aq)||Cu^{2+}(aq)|Cu(m)$$

for which the conventional cell reaction, involving two electrons, is

conventional: $Cu^{2+}(aq) + 2Ag(m) \longrightarrow Cu(m) + 2Ag^+(aq).$

From this we can immediately write down the Nernst equation as

$$E = E^\circ - \frac{RT}{2F} \ln \frac{([Ag^+]/c^\circ)^2}{[Cu^{2+}]/c^\circ}$$

Alternatively, we could start with the two half-cell reactions

RHS: $Cu^{2+}(aq) + 2e^- \longrightarrow Cu(m)$
LHS: $Ag^+(aq) + e^- \longrightarrow Ag(m),$

and write down the half-cell Nernst equations for each as

RHS: $E_{RHS} = E^\circ(Cu^{2+}/Cu) - \dfrac{RT}{2F} \ln \dfrac{1}{[Cu^{2+}]/c^\circ}$

LHS: $E_{LHS} = E^\circ(Ag^+/Ag) - \dfrac{RT}{F} \ln \dfrac{1}{[Ag^+]/c^\circ}.$

Note that there are two electrons involved in the RHS half cell, as written, but only one on the LHS. Taking RHS – LHS gives the cell potential as

$$
\begin{aligned}
E &= E^\circ(Cu^{2+}/Cu) - E^\circ(Ag^+/Ag) - \frac{RT}{2F} \ln \frac{1}{[Cu^{2+}]/c^\circ} + \frac{RT}{F} \ln \frac{1}{[Ag^+]/c^\circ} \\
&= E^\circ(Cu^{2+}/Cu) - E^\circ(Ag^+/Ag) - \frac{RT}{2F} \ln \frac{1}{[Cu^{2+}]/c^\circ} + \frac{RT}{2F} \ln \frac{1}{([Ag^+]/c^\circ)^2} \\
&= E^\circ(Cu^{2+}/Cu) - E^\circ(Ag^+/Ag) - \frac{RT}{2F} \ln \frac{([Ag^+]/c^\circ)^2}{[Cu^{2+}]/c^\circ},
\end{aligned}
$$

where to go to the second line we have used $\ln x = \frac{1}{2} \ln x^2$. If we identify $E^\circ(Cu^{2+}/Cu) - E^\circ(Ag^+/Ag)$ as the standard cell potential, then the final line is the same as the Nernst equation we found directly from the cell reaction.

The key point here is that in writing the half-cell Nernst equations, and then subtracting them, we do *not* need to have the same number of electrons in each half-cell reaction.

The chosen reference electrode is the *standard hydrogen electrode* (SHE) for which the half-cell reaction is

$$H^+(aq) + e^- \longrightarrow \tfrac{1}{2}H_2(g).$$

To be entirely precise, the standard state for a solution is unit activity (see Box 19.1 on page 841).

When all of the species are in their standard states (i.e. unit concentration and pressure) this half-cell is taken to produce a potential of zero.

Table 19.1 Standard half-cell potentials in aqueous solutions at 298 K

couple	$E° / V$	couple	$E° / V$
$Li^+ + e^- \rightarrow Li$	−3.040	$Sn^{4+} + 2e^- \rightarrow Sn^{2+}$	+0.151
$K^+ + e^- \rightarrow K$	−2.931	$Cu^{2+} + e^- \rightarrow Cu^+$	+0.161
$Ca^{2+} + 2e^- \rightarrow Ca$	−2.868	$AgCl + e^- \rightarrow Ag + Cl^-$	+0.2223
$Na^+ + e^- \rightarrow Na$	−2.71	$Hg_2Cl_2 + 2e^- \rightarrow 2Hg + 2Cl^-$	+0.2681
$Mg^{2+} + 2e^- \rightarrow Mg$	−2.372	$Cu^{2+} + 2e^- \rightarrow Cu$	+0.3419
$Al^{3+} + 3e^- \rightarrow Al$	−1.662	$O_2 + 2H_2O + 4e^- \rightarrow 4OH^-$	+0.401
$Mn^{2+} + 2e^- \rightarrow Mn$	−1.185	$Cu^+ + e^- \rightarrow Cu$	+0.518
$2H_2O + 2e^- \rightarrow H_2 + 2OH^-$	−0.8277	$Fe^{3+} + e^- \rightarrow Fe^{2+}$	+0.771
$Zn^{2+} + 2e^- \rightarrow Zn$	−0.7618	$Ag^+ + e^- \rightarrow Ag$	+0.7996
$PbO + H_2O + 2e^- \rightarrow Pb + 2OH^-$	−0.580	$Br_2 + 2e^- \rightarrow 2Br^-$	+1.0873
$Fe^{2+} + 2e^- \rightarrow Fe$	−0.447	$Cr_2O_7^{2-} + 14H^+ + 6e^- \rightarrow 2Cr^{3+} + 7H_2O$	+1.232
$Cd^{2+} + 2e^- \rightarrow Cd$	−0.4030	$O_2 + 4H^+ + 4e^- \rightarrow 2H_2O$	+1.229
$Tl^+ + e^- \rightarrow Tl$	−0.336	$Cl_2 + 2e^- \rightarrow 2Cl^-$	+1.358
$Ni^{2+} + 2e^- \rightarrow Ni$	−0.257	$Au^{3+} + 3e^- \rightarrow Au$	+1.498
$AgI + e^- \rightarrow Ag + I^-$	−0.1522	$MnO_4^- + 8H^+ + 5e^- \rightarrow Mn^{2+} + 4H_2O$	+1.507
$Fe^{3+} + 3e^- \rightarrow Fe$	−0.037	$Ce^{4+} + e^- \rightarrow Ce^{3+}$	+1.72
$AgBr + e^- \rightarrow Ag + Br^-$	+0.0713	$F_2 + 2e^- \rightarrow 2F^-$	+2.866

The standard half-cell potential of an electrode is measured by setting up a cell in which the left-hand half cell is the SHE, and the right-hand half cell is the electrode of interest, with all species being in their standard states. The potential of such as cell is

$$E° = E°(RHS) - \underbrace{E°(SHE)}_{= 0.00 \text{ V}}$$

$$= E°(RHS).$$

In this way, tables of standard half-cell electrode potentials can be constructed; a selection of values are given in Table 19.1 (many more can be found in standard reference works – see *Further Reading*). The values in the table are quoted to different numbers of significant figures reflecting the accuracy to which they are known. Note that these standard half-cell potentials cover quite a wide range; how these values can be interpreted and used is discussed later on in this chapter.

Since the standard half-cell potentials are $\Delta_r G°$ values we expect them to be strongly temperature dependent so it is important to use values appropriate to the temperature of interest. Most commonly, standard half-cell potentials are tabulated at 298 K, and for many the value of the temperature coefficient $(dE°/dT)$ is also known.

For the Cu/Zn cell

$$Cu(m) | Cu^{2+}(aq) || Zn^{2+}(aq) | Zn(m)$$

we can easily compute the standard cell potential at 298 K using the data from Table 19.1

$$E° = E°(RHS) - E°(LHS)$$

$$= E°(Zn^{2+}/Zn) - E°(Cu^{2+}/Cu)$$

$$= (-0.7618) - (+0.3419) = -1.1037 \text{ V}.$$

19.4   The spontaneous cell reaction

In section 19.1.1 on page 833 we described how the conventional cell reaction could be determined from the physical arrangement of the cell, but we commented that the *spontaneous cell reaction* – meaning the reaction which takes place when current flows – might not be in the same direction. We are now in a position to determine the direction of the spontaneous cell reaction.

All we have to do is determine the *sign* of the cell potential. If this is *positive*, then it follows from

$$\Delta_r G_{cell} = -nFE$$

that $\Delta_r G_{cell}$ is negative. Thus, the conventional cell reaction is associated with a reduction in the Gibbs energy, and so this reaction will be spontaneous.

On the other hand, if the cell potential is *negative*, $\Delta_r G_{cell}$ is positive. The conventional cell reaction is associated with an increase in Gibbs energy and so is not spontaneous. It follows that the *reverse* of the conventional cell reaction is therefore associated with a reduction in Gibbs energy and so this is the spontaneous reaction.

For the copper–zinc cell we have been using as an example

$$Cu(m) \,|\, Cu^{2+}(aq) \,\|\, Zn^{2+}(aq) \,|\, Zn(m),$$

the conventional reaction is

$$Zn^{2+}(aq) + Cu(m) \longrightarrow Zn(m) + Cu^{2+}(aq).$$

If we set up a cell in which the concentrations of the Zn^{2+} and Cu^{2+} ions are both $1 \ mol \ dm^{-3}$ (i.e. in their standard states), then the cell potential will be the standard cell potential. Using the data from Table 19.1 we have just computed that this standard cell potential is $-1.1037 \ V$ at 298 K.

Since the cell potential is negative, the Gibbs energy change associated with the conventional cell reaction is positive. The spontaneous reaction is therefore the *reverse* of the conventional cell reaction i.e.

conventional, $\Delta_r G$ positive: $Zn^{2+}(aq) + Cu(m) \quad \longrightarrow \quad Zn(m) + Cu^{2+}(aq)$

spontaneous, $\Delta_r G$ negative: $Zn(m) + Cu^{2+}(aq) \quad \longrightarrow \quad Zn^{2+}(aq) + Cu(m).$

This fits in with our expectation that metallic zinc will reduce Cu^{2+} ions to metallic copper. A more complex calculation along the same lines is given in Example 19.3 on the next page.

19.5   Summary

We have now put together the key ideas needed to analyse electrochemical cells, so before we go on to look at applications of these ideas, this is a good point to summarize the key concepts:

- A cell is thought of as being composed of two half cells.

- The conventional cell reaction is deduced by analysing the cell using a set of conventions.

Example 19.3 Finding the spontaneous reaction

For the following cell

$$Cr(m) \,|\, Cr^{3+}(aq) \,\|\, Zn^{2+}(aq) \,|\, Zn(m),$$

find the spontaneous cell reaction when: (a) all of the species are present at the standard concentration; and (b) when the concentration of the Cr^{3+} is 0.01 mol dm^{-3} and that of Zn^{2+} is 1.00 mol dm^{-3} The standard half-cell potentials are $E°(Cr^{3+}/Cr) = -0.74$ V and $E°(Zn^{2+}/Zn) = -0.76$ V at 298 K.

The half-cell reactions are

RHS: $Zn^{2+}(aq) + 2e^- \longrightarrow Zn(m)$ LHS: $Cr^{3+}(aq) + 3e^- \longrightarrow Cr(m)$.

To find the conventional cell reaction we need to adjust the number of electrons to be the same in two half-cell reactions. This can be done by multiplying the RHS by three and the LHS by two to give

RHS: $3Zn^{2+}(aq) + 6e^- \longrightarrow 3Zn(m)$ LHS: $2Cr^{3+}(aq) + 6e^- \longrightarrow 2Cr(m)$.

Now we simply compute RHS – LHS to give the conventional cell reaction as

$$3Zn^{2+}(aq) + 2Cr(m) \longrightarrow 3Zn(m) + 2Cr^{3+}(aq). \qquad (19.15)$$

If the Zn^{2+} and Cr^{3+} are present at the standard concentrations, then the cell potential is the standard cell potential given by

$$E° = E°(Zn^{2+}/Zn) - E°(Cr^{3+}/Cr) = -0.76 - (-0.74) = -0.02 \text{ V}.$$

As the cell potential is negative, the Gibbs energy change is positive and so the spontaneous cell reaction is the *reverse* of the conventional cell reaction i.e. Zn reduces Cr^{3+}.

If the concentrations are not at the standard values, then we need to use the Nernst equation to find the cell potential. From the Eq. 19.15 we can simply deduce the Nernst equation as

$$E = E°(Zn^{2+}/Zn) - E°(Cr^{3+}/Cr) - \frac{RT}{6F} \ln \frac{\left([Cr^{3+}]/c°\right)^2}{\left([Zn^{2+}]/c°\right)^3}.$$

Substituting in $[Cr^{3+}] = 0.01$ mol dm^{-3} and $[Zn^{2+}] = 1.00$ mol dm^{-3}, and evaluating the expression gives $E = +0.02$ V. Now the cell potential is positive, so $\Delta_r G_{cell}$ is *negative*, and the spontaneous reaction is the same as the conventional reaction.

This example illustrates that the direction of the spontaneous cell reaction can be affected by the concentration of the species involved.

- The cell potential is related to the Gibbs energy change of the (conventional) cell reaction via $\Delta_r G_{cell} = -nFE$.

- The Nernst equation gives the concentration dependence of the cell potential.

- The standard half-cell potential is the potential developed by the half cell with all species present in their standard states, measured relative to the standard hydrogen electrode.

- The direction of the spontaneous cell reaction can be found by inspecting the sign of the cell potential.

19.6 Types of half cells

Electrochemical cells give us a very straightforward way of measuring the thermodynamic parameters of the cell reaction. However, in order to take advantage of this, we need to be able to devise a cell in which the reaction of interest will take place. In this section we will look at the variety of different half cells which have been developed in order to maximize the range of reactions which can be studied by electrochemical methods.

19.6.1 Metal/metal ion electrodes

These electrodes consist of a metal in contact with a solution of its ions, such as the copper and zinc electrodes we have already encountered in this chapter. The conventional notation for this half cell (written as the left-hand electrode) and half-cell reaction are

$$M(m) \,|\, M^{n+}(aq) \ldots \qquad M^{n+}(aq) + ne^- \longrightarrow M(m),$$

and the corresponding Nernst equation for the half cell is

$$E = E^\circ(M^{n+}/M) - \frac{RT}{nF} \ln \frac{1}{[M^{n+}]/c^\circ}.$$

Standard half-cell potentials for a great many electrodes of this type are available.

19.6.2 Gas/ion electrodes

These electrodes involve a gaseous species and a solution of a related ion. An inert metal electrode (often made of platinum) is needed in order to create an electrical contact with the species involved. Typically, the electrode is dipped into the solution and the gas is gently bubbled over the surface of the electrode, as is illustrated in Fig. 19.6. Such electrodes are not particularly easy to use experimentally.

For the hydrogen electrode the half-cell reaction involves $H_2(g)$ and a solution of H^+ ions, and in the conventional notation we also include the platinum electrode

$$Pt(m) \,|\, H_2(g) \,|\, H^+(aq) \ldots \qquad 2H^+(aq) + 2e^- \longrightarrow H_2(g).$$

The Nernst equation is

$$E = E^\circ(H^+/H_2) - \frac{RT}{2F} \ln \frac{p_{H_2}/p^\circ}{([H^+]/c^\circ)^2}.$$

Note that the 'concentration' term for hydrogen gas is its partial pressure.

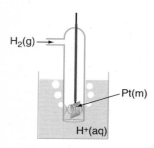

$H_2(g)$

Pt(m)

$H^+(aq)$

Fig. 19.6 A hydrogen electrode is typical of a variety of gas/ion electrodes. Hydrogen gas is bubbled over a platinum electrode which is in contact with a solution of $H^+(aq)$ ions.

Another example of such an electrode is that involving chlorine gas and chloride ions

$$Pt(m)\,|\,Cl_2(g)\,|\,Cl^-(aq)\ldots \qquad Cl_2(g) + 2e^- \longrightarrow 2Cl^-(aq)$$

$$E = E^\circ(Cl_2/Cl^-) - \frac{RT}{2F} \ln \frac{([Cl^-]/c^\circ)^2}{p_{Cl_2}/p^\circ}.$$

Sometimes, the solvent is directly involved in the cell reaction, such as in the half cell formed between hydrogen gas and hydroxide ions, for which the cell reaction is

$$Pt(m)\,|\,H_2(g)\,|\,OH^-(aq)\ldots \qquad 2H_2O(l) + 2e^- \longrightarrow 2OH^-(aq) + H_2(g)$$

$$E = E^\circ(H_2O/H_2) - \frac{RT}{2F} \ln (p_{H_2}/p^\circ)([OH^-]/c^\circ)^2.$$

Note that no term for H_2O appears in the Nernst equation as this is the solvent and so essentially in the pure form. This redox process is best described as the reduction of water to H_2 under alkaline conditions.

19.6.3 Electrodes involving ions in different oxidation states

Transition metals often occur in different oxidation states, and such ions can form a redox couples. As both the oxidized and reduced species are in solution, an inert electrode (typically made of platinum) is needed to make the electrical contact, as shown in Fig. 19.7.

A typical example of such a redox couple is formed from iron in its oxidation states III and II

$$Pt(m)\,|\,Fe^{3+}(aq), Fe^{2+}(aq)\ldots \qquad Fe^{3+}(aq) + e^- \longrightarrow Fe^{2+}(aq)$$

$$E = E^\circ(Fe^{3+}/Fe^{2+}) - \frac{RT}{F} \ln \frac{[Fe^{2+}]/c^\circ}{[Fe^{3+}]/c^\circ}.$$

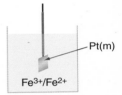

Fe3+/Fe2+   Pt(m)

Fig. 19.7 A half cell in which the oxidized and reduced species (here Fe^{3+} and Fe^{2+}) are both in solution; an inert platinum electrode is needed to make the electrical contact.

A more complex example is the couple formed by the manganese(VII) species, in the form of the manganate(VII) ion MnO_4^-, and manganese(II), as Mn^{2+}. Under acidic conditions the half-cell reaction is

$$MnO_4^-(aq) + 8H^+(aq) + 5e^- \longrightarrow Mn^{2+}(aq) + 4H_2O(l)$$

Note how both solvent water and $H^+(aq)$ ions are involved in the equilibrium. The Nernst equation is

$$E = E^\circ(MnO_4^-/Mn^{2+}) - \frac{RT}{5F} \ln \frac{[Mn^{2+}]/c^\circ}{([H^+]/c^\circ)^8 \left([MnO_4^-]/c^\circ\right)}.$$

19.6.4 Metal/insoluble salt electrodes

These electrodes consist of an layer of an *insoluble* salt coated onto the outside of a metal electrode, as illustrated in Fig. 19.8 on the next page. The commonest example is the silver/silver chloride electrode which has the half-cell reaction

$$Ag(m), AgCl(s)\,|\,Cl^-(aq)\ldots \qquad AgCl(s) + e^- \longrightarrow Ag(m) + Cl^-(aq).$$

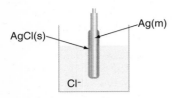

Fig. 19.8 A typical metal insoluble salt electrode. A silver electrode is coated with a *thin* layer of insoluble AgCl; the half cell reaction involves AgCl and the Cl⁻ ions which are in solution. The width of the electrode and the thickness of the layer of AgCl have been exaggerated greatly.

It is very important to realize that it is the anion Cl⁻ which moves between the solid AgCl and the solution: *free Ag⁺ ions are not involved*. The Nernst equation is

$$E = E°(AgCl/Cl⁻) - \frac{RT}{F} \ln([Cl⁻]/c°),$$

there is no concentration term for AgCl as it is present as a solid. The very useful thing about such an electrode is that its potential depends on the concentration of an anion (here Cl⁻). Of course the same is true of the $Cl_2/Cl⁻$ electrode, but the AgCl/Cl⁻ electrode is so much more convenient to use.

There are many of these metal/insoluble salt electrodes. One particularly important one is that involving mercury and (insoluble) mercury(I)chloride (mercurous chloride, traditionally known as *calomel*). The half-cell reaction is

$$Hg(l)\,|\,Hg_2Cl_2(s)\,|\,Cl⁻(aq)\ldots \qquad Hg_2Cl_2(s) + 2e⁻ \longrightarrow 2Hg(l) + 2Cl⁻(aq).$$

$$E = E°(Hg_2Cl_2/Cl⁻) - \frac{RT}{2F} \ln([Cl⁻]/c°)^2.$$

Our final example is a lead/lead(II) oxide electrode i.e. a lead electrode coated in insoluble PbO. Under alkaline conditions the half-cell reaction is

$$Pb(m), PbO(s)\,|\,OH⁻(aq)\ldots \qquad PbO(s) + H_2O(l) + 2e⁻ \longrightarrow Pb(m) + 2OH⁻(aq).$$

The Nernst equation is

$$E = E°(PbO/Pb) - \frac{RT}{2F} \ln([OH⁻]/c°)^2,$$

showing that the electrode is sensitive to the concentration of hydroxide ions.

19.6.5 Liquid junctions

It was noted in section 19.1 on page 832 that for a cell to generate a potential (and a current) there has to be physical contact between the solutions in the two half cells. Unfortunately, there is no way of facilitating this physical contact without at the same time allowing the ions to diffuse between the two solutions. The result of the diffusion is that a potential – called a *liquid junction potential* – builds up across the interface between the two solutions.

This liquid junction potential detracts from the measured cell potential, so that the measured value of E is no longer a reliable measure of $\Delta_r G_{cell}$. The effect is significant, and so it must be eliminated or avoided if our measurements of cell potentials are to lead to reliable and accurate values of $\Delta_r G_{cell}$.

One way in which a junction potential can be avoided is for the two half cells to share the same solution, such as in the cell

$$Pt(m)\,|\,H_2(g)\,|\,HCl(aq)\,|\,AgCl(s), Ag(m).$$

The left-hand half cell is a hydrogen electrode and involves H⁺ from the solution; the right-hand half cell is an Ag/AgCl electrode which involves Cl⁻ from the solution. The cell has no liquid junction.

If the solutions in the two half cells are different, then the liquid junction potential can be minimized by connecting the solutions by a *salt bridge*, as shown in Fig. 19.9 (b). The bridge is a glass tube containing a saturated solution

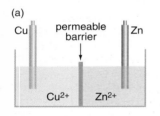

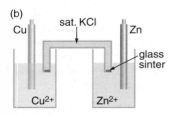

Fig. 19.9 In cell (a) the two solutions are separated by a permeable barrier. Diffusion across this barrier will lead to the formation of a liquid junction potential which will detract from the measured cell potential. The liquid junction potential can be avoided largely by connecting the two solutions by a salt bride, which consists of a tube filled with a saturated solution of KCl, held in place by glass sinters.

of KCl, the ends of which dip into the two solutions. The KCl solution is kept in the tube by closing the ends with a porous barrier, such as a glass sinter. Alternatively, the tube can be filled with a gel containing KCl. For rough work, a filter paper, soaked in saturated KCl, with its ends dipped in the two solutions can also be used as a salt bridge.

There is a particular reason why KCl is used in a salt bridge, which is that the $K^+(aq)$ and $Cl^-(aq)$ ions move at very similar speeds under the influence of an electric potential. For reasons we cannot go into, this means that, provided that the KCl is much more concentrated than the other solutions, the liquid junction potentials developed at each end of the salt bridge are equal and opposite. They therefore cancel one another out.

Sintered glass is formed when roughly powdered glass is partially melted and then cooled before the whole becomes liquid. The result is a material in which the small fragments of glass are joined together, but in which there are quite large pores. These allow liquids to penetrate and, slowly, pass through.

19.6.6 Reference electrodes

As was described in section 19.3.2 on page 843, standard half-cell potentials are measured relative to the standard hydrogen electrode. However, this electrode is very inconvenient to use, and so it is common to use an alternative reference electrode whose potential is reproducible and has been carefully calibrated. By far the commonest such electrode is the *calomel electrode*, whose half cell reaction is

$$Hg_2Cl_2(s) + 2e^- \longrightarrow Hg(l) + 2Cl^-(aq).$$

The physical construction of such an electrode is shown in Fig. 19.10. The (insoluble) Hg_2Cl_2 is held in the bottom of a small tube by a glass sinter, and liquid mercury sits on top of the solid, thereby forming the electrode. The whole assembly is held inside another tube which contains a saturated solution of KCl; saturation is maintained by having some undissolved crystals of KCl in the solution. Finally, this KCl solution is retained in the outer tube by a second glass sinter.

The potential produced by this electrode just depends on the Cl^- concentration, which is constant for a saturated solution (at a given temperature). The KCl solution also forms a salt bridge to the solution into which the calomel electrode is dipped. Overall the electrode is very easy to use: it is simply dipped into the solution which forms the other half cell. If correctly constructed and maintained it produces a well defined potential of +0.2412 V (at 298 K), and so can be used as a reference electrode when making precise measurements of the potentials produced by other half cells.

Another commonly used reference electrode is one based on the $AgCl/Cl^-$ couple (i.e. a silver electrode coated with insoluble AgCl. As with the calomel electrode, the potential of the $AgCl/Cl^-$ electrode depends on the concentration of Cl^-. Therefore if the electrode it to be used as a reference, this concentration must be kept constant, for example in the same way as was described for the calomel electrode. A glass sinter keeps the internal solution in place and provides contact with the solution into which the electrode is dipped.

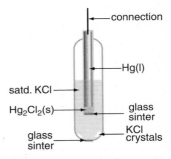

Fig. 19.10 Diagram of a saturated calomel electrode suitable for use as a reference electrode. The Hg_2Cl_2/Hg couple is in contact with a solution of KCl which is kept saturated by the presence of solid KCl; the potential produced by this arrangement is +0.2412 V at 298 K. The KCl solution and the sinter at the bottom of the electrode form a salt bridge to the solution into which this electrode is dipped.

19.7 Assessing redox stability using electrode potentials

One of the very useful things we can do with a table of standard electrode potentials is to use them to assess the relative stability of different species with respect to oxidation and reduction. You may already have come across this idea in the form of the *electrochemical series* in which species are ordered in terms of their oxidizing (or reducing) power. The more positive the standard potential, the more strongly oxidizing a couple is said to be. We are now in a position to explain why this is, and so how this series is constructed.

Let us imagine a cell in which the RHS cell reaction is

$$\text{RHS:} \qquad A_{ox} + n_A e^- \longrightarrow A_{red}.$$

Here A_{ox} is the oxidized species in the couple (e.g. Li^+ or Cl_2), and A_{red} is the corresponding reduced species (e.g. Li or Cl^-). The number of electrons involved is n_A. Similarly, the left-hand side half cell reaction is

$$\text{LHS:} \qquad B_{ox} + n_B e^- \longrightarrow B_{red}.$$

To work out the conventional cell reaction we need to make sure that they number of electrons is the same for each half-cell reaction, which we do by multiplying the RHS reaction by n_B, and the LHS reaction by n_A. Having done this and taken RHS − LHS in the usual way we find the following for the conventional cell reaction

$$\text{conventional:} \qquad n_B A_{ox} + n_A B_{red} \longrightarrow n_B A_{red} + n_A B_{ox}. \qquad (19.16)$$

If we assume that all species are in their standard states, then the cell potential is

$$E = E^\circ(A_{ox}/A_{red}) - E^\circ(B_{ox}/B_{red}),$$

where $E^\circ(A_{ox}/A_{red})$ is the standard half-cell potential of the A_{ox}/A_{red} couple, and likewise $E^\circ(B_{ox}/B_{red})$ is for the B_{ox}/B_{red} couple.

If $E^\circ(A_{ox}/A_{red})$ is *greater* than $E^\circ(B_{ox}/B_{red})$, then the cell potential will be *positive* and so, since $\Delta_r G_{cell} = -nFE$, $\Delta_r G_{cell}$ will be *negative*. This means that the conventional cell reaction will be a spontaneous process i.e. A_{ox} will oxidize B_{red}. In summary

$$\text{If} \quad E^\circ(A_{ox}/A_{red}) > E^\circ(B_{ox}/B_{red}) \quad A_{ox} \text{ oxidizes } B_{red}.$$

For example, consider the following three redox couples, along with their standard half-cell potentials

$$
\begin{array}{ll}
Ce^{4+}(aq) + e^- \longrightarrow Ce^{3+}(aq) & E^\circ = +1.72 \text{ V} \\
Cu^{2+}(aq) + 2e^- \longrightarrow Cu(m) & E^\circ = +0.3419 \text{ V} \\
Al^{3+}(aq) + 3e^- \longrightarrow Al(m) & E^\circ = -1.662 \text{ V}.
\end{array}
$$

Just by glancing at these data we can see that Ce^{4+} is the most strongly oxidizing species as it has the highest potential: it will oxidize Cu to Cu^{2+}, as well as Al to Al^{3+}. Cu^{2+} is sufficiently strongly oxidizing that it will oxidize Al to Al^{3+}, but it will not oxidize Ce^{3+}. Al^{3+} is too weakly oxidizing to oxidize either Cu or Ce^{3+}. Note that it does not matter that the number of electrons involved in each half-cell reaction is different – all we need to consider is the relative size of the half-cell potentials.

If we are interested in reducing power, rather than oxidation, then if A_{red} is to act as a reducing agent, we want the *reverse* of the conventional cell reaction (Eq. 19.16) to be the spontaneous process. For this to be the case, $\Delta_r G_{cell}$ needs to be *positive* and so the cell potential needs to be *negative* i.e. $E^\circ(A_{ox}/A_{red})$ must be *less* than $E^\circ(B_{ox}/B_{red})$. In summary

$$\text{If} \quad E^\circ(A_{ox}/A_{red}) < E^\circ(B_{ox}/B_{red}) \quad A_{red} \text{ reduces } B_{ox}.$$

To see how this works out in practice, consider the following three redox couples

$$Li^+(aq) + e^- \longrightarrow Li(m) \qquad E^\circ = -3.040 \text{ V}$$
$$2H_2O(l) + 2e^- \longrightarrow H_2(g) + 2OH^-(aq) \quad E^\circ = -0.8277 \text{ V}$$
$$Ni^{2+} + 2e^- \longrightarrow Ni(m) \qquad E^\circ = -0.257 \text{ V}.$$

Of these three, the Li^+/Li couple has the smallest (i.e. most negative) potential, so $Li(m)$ will reduce H_2O to H_2, and also Ni^{2+} to Ni. The potential for the $H_2O/H_2,OH^-$ couple is insufficiently negative for H_2 to reduce Li^+, but it can reduce Ni^{2+}. The Ni^{2+}/Ni potential is insufficiently negative for Ni to reduce H_2O (in alkaline conditions) or Li^+.

In summary, for the A_{ox}/A_{red} couple, with the half-cell reaction

$$A_{ox} + n_A e^- \longrightarrow A_{red} :$$

- the *greater* the half-cell potential, the more *strongly oxidizing* A_{ox} becomes;

- the *smaller* (i.e. more negative) the half-cell potential, the more *strongly reducing* A_{red} becomes.

The electrochemical series is simply the redox couples listed in order of standard half-cell potentials, with the most positive at the top and the most negative at the bottom; a selection of couples arranged in this way is shown in Table 19.2. At the top of the table A_{ox} is a strong oxidizing agent, whereas at the bottom of the table A_{red} is a strong reducing agent.

Species in the first column (A_{ox}) will oxidize species in the second column which lie further down the table. Species in the second column (A_{red}) will reduce species in the first column which lie higher up the table.

It is interesting to pause for a moment and consider why all this works, especially why we do not need to worry that different numbers of electrons are involved in the redox couples we are comparing. The reason is due to the relationship between the $\Delta_r G_{cell}$ and the cell potential, $\Delta_r G_{cell} = -nFE$. This can be rewritten as

$$E = \frac{-\Delta_r G_{cell}}{nF},$$

which we can interpret as saying that the cell potential gives us the Gibbs energy change *per electron*. So, by comparing cell potentials, we are really comparing Gibbs energy changes which have been corrected for the number of electrons involved. This is why we do not need to worry about the actual number of electrons involved in the redox couple.

Effect of changing concentrations

In the discussion so far we have assumed that the species involved are all in their standard state i.e. unit concentration and unit pressure. At first sight, this seems like a rather unrealistic assumption. However, we will show here that even if the concentrations deviate significantly from the standard values, our conclusions concerning relative redox stability are barely altered.

Let us take as an example the Zn^{2+}/Zn couple, for which the half-cell reaction is $Zn^{2+}(aq) + 2e^- \longrightarrow Zn(m)$, and the Nernst equation is

$$E = E^\circ(Zn^{2+}/Zn) - \frac{RT}{2F} \ln \frac{1}{[Zn^{2+}]/c^\circ}.$$

Table 19.2 Part of the electrochemical series

A_{ox}	A_{red}	E° / V
A_{ox} is a strong oxidizing agent		
F_2	F^-	+2.866
MnO_4^-	Mn^{2+}	+1.507
$Cr_2O_7^{2-}$	Cr^{3+}	+1.232
O_2	OH^-	+0.401
Sn^{4+}	Sn^{2+}	+0.151
Fe^{3+}	Fe	−0.037
Zn^{2+}	Zn	−0.7618
Mg^{2+}	Mg	−2.372
Li^+	Li	−3.040
A_{red} is a strong reducing agent		

When all of the species are present in their standard states the potential is -0.7618 V. Suppose that the concentration of the Zn^{2+} is reduced to 0.1 mol dm^{-3}, then at 298 K we can compute the term on the right to be -0.03 V. In other words, changing the concentration by a factor of ten only changes cell potential by 30 mV. Looking at the standard half-cell potentials shown in Table 19.1 on page 845, you can see that such a small change in the Zn^{2+}/Zn potential does not shift its position in the table.

Therefore, unless we are comparing redox couples which have very similar standard potentials, the fact that the concentrations we are dealing with are not the standard concentrations will not be of any great significance.

19.8 The limits of stability in aqueous solution

If we attempt to dissolve a powerful oxidizing agent in water then the possibility exists that the solvent itself will be oxidized to give $O_2(g)$. In an acidic solution, the relevant half-cell reaction is

$$O_2(g) + 4H^+(aq) + 4e^- \longrightarrow 2H_2O(l) \qquad E^\circ(O_2, H^+/H_2O) = +1.229 \text{ V}. \quad (19.17)$$

A powerful oxidizing agent will drive this reaction from right to left. In an alkaline solution the half-cell reaction is slightly different in that it involves OH^- rather than H^+

$$O_2(g) + 2H_2O(l) + 4e^- \longrightarrow 4OH^-(aq) \qquad E^\circ(O_2, H_2O/OH^-) = +0.401 \text{ V}. \quad (19.18)$$

Again, a sufficiently powerful oxidizing agent will drive this reaction from right to left.

Under acidic conditions where all species are in their standard states (meaning $[H^+] = 1$ mol dm^{-3}, pH $= 0$) redox couples with E° values *more positive* than $+1.229$ V will oxidize H_2O. For example, we would therefore expect Co^{3+}, for which $E^\circ(Co^{3+}/Co^{2+}) = +1.92$ V, and MnO_4^-, for which $E^\circ(MnO_4^-/Mn^{2+}) = +1.51$ V, to oxidize water.

The other side of the story is that trying to dissolve a powerful reducing agent in water may simply result in the water itself being reduced to H_2. In an acidic solution the relevant half-cell reaction is

$$H^+(aq) + e^- \longrightarrow \tfrac{1}{2}H_2(g) \qquad E^\circ(H^+/H_2) = 0.00 \text{ V}, \quad (19.19)$$

and in an alkaline solution the reaction is

$$2H_2O(l) + 2e^- \longrightarrow H_2(g) + 2OH^- \qquad E^\circ(H_2O/H_2, OH^-) = -0.828 \text{ V}. \quad (19.20)$$

In both cases, sufficiently powerful reducing agents will drive these reactions from left to right.

When all of the species are present in their standard states and at pH 0, redox couples with E° values more negative that 0.00 V will reduce water. For example since $E^\circ(Zn/Zn^{2+}) = -0.76$ V we would expect zinc to reduce water to hydrogen.

The relevant potentials for the oxidation and reduction of water depend on the concentration of H^+, so it is useful to explore this relationship in a quantitative way. For Eq. 19.17 the Nernst equation is

$$E = E^\circ(O_2, H^+/H_2O) - \frac{RT}{4F} \ln\left(\frac{1}{(p_{O_2}/p^\circ)([H^+]/c^\circ)^4}\right).$$

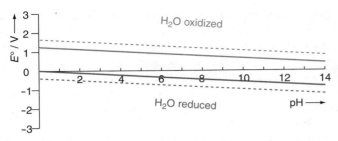

Fig. 19.11 The solid red line gives the half-cell potential for $O_2(g) + 4H^+(aq) + 4e^- \longrightarrow 2H_2O(l)$ as a function of pH, as predicted by Eq. 19.21. In principle, the oxidized species of redox couples whose $E°$ values lie above this line should oxidize water. The solid blue line is the corresponding potential, given by Eq. 19.22, for $H^+(aq) + e^- \longrightarrow \frac{1}{2}H_2(g)$. In principle, the reduced species of redox couples whose $E°$ values lie below this line should reduce water. In practice it seems that, for the redox process to take place at a measurable rate, the potentials need to be about 0.4 V greater in magnitude that these theoretical considerations would indicate; the dashed lines show these extended limits.

If we assume that the oxygen gas is at standard pressure, take the factor of $-\frac{1}{4}$ inside the log, and assume that the concentrations are in mol dm^{-3} so that the $c°$ term can be omitted, we have

$$E = E°(O_2, H^+/H_2O) + \frac{RT}{F} \ln([H^+]).$$

Using the identity $\ln x \equiv \ln 10 \times \log x$, this last expression can be rewritten

$$E = E°(O_2, H^+/H_2O) + \ln 10 \times \frac{RT}{F} \log([H^+]).$$

Finally, recalling that $pH = -\log[H^+]$ we obtain

$$E = E°(O_2, H^+/H_2O) - \ln 10 \times \frac{RT}{F} \times pH. \qquad (19.21)$$

At pH 0 this evaluates to $E°(O_2, H^+/H_2O)$ which is +1.229 V, and at pH 14 it evaluates to +0.401 V. This latter potential is the same as that for Eq. 19.18 on the preceding page, since pH 14 corresponds to $[OH^-] = 1$ mol dm^{-3}. Equation 19.21 thus applies across the full pH range.

For the reduction of water a similar argument starting from the Nernst equation for the half-cell reaction of Eq. 19.19 on the preceding page leads to

$$E = -\ln 10 \times \frac{RT}{F} \times pH. \qquad (19.22)$$

This evaluates to 0.00 V at pH 0, and −0.828 V at pH 14, the latter corresponding to $E°$ for the half-cell of Eq. 19.20 on the facing page.

These half-cell potentials as a function of pH are plotted in Fig. 19.11. The solid red line is the prediction of Eq. 19.21; the oxidized form of couples whose redox potentials lie above this line will oxidize water to O_2. The solid blue line is given by Eq. 19.22; the reduced form of couples whose redox potentials lie below this line will reduce water to H_2. You can see that there is rather a narrow range of redox potentials which are predicted to be stable in water, in the sense that the solvent is neither oxidized nor reduced.

In practice, it is found that for water to be oxidized or reduced by a particular species at a measurable rate the relevant standard half-cell potential

Recall that in aqueous solution the concentrations of OH$^-$ and H$^+$ are related via the ionic product for water, K_w:

$$K_w = [OH^-][H^+].$$

At 298 K, $K_w = 1 \times 10^{-14}$, so taking logarithms to the base 10 gives

$$-14 = \log[OH^-] + \log[H^+].$$

If $[OH^-] = 1$ mol dm^{-3}, $\log[OH^-] = 0$ and so from the above equation it follows that $-\log[H^+] = 14$. This means that the pH of 1 mol dm^{-3} OH$^-$ is 14.

To understand the origin of the over potential we need to look at the rate of the oxidation or reduction processes going on at electrodes or between species in solution. This topic of *dynamic electrochemistry* is outside the scope of this book. Here we are just concerned with *equilibrium* electrochemistry.

needs to be significantly above the red line or below the blue line. A figure of about 0.4 V is often quoted for this *over potential*; the dashed lines plotted in Fig. 19.11 on the preceding page are 0.4 V away from the theoretical lines. The range of potentials which are stable is now considerably wider.

It must be remembered that thermodynamics, which is what all these arguments about cell potentials are based on, only tells us whether or not a reaction is feasible in principle. A feasible reaction may not proceed at a reasonable rate. For example, although for manganate(VII) at pH 0 $E^\circ(MnO_4^-/Mn^{2+}) = +1.51$ V, indicating that manganate(VII) should oxidize water, such solutions are in fact stable for long periods of time.

19.9 Using cell potentials to determine thermodynamic parameters

In this section we will look at two examples where data on cell potentials can be used to find other thermodynamic parameters.

19.9.1 Solubility products

The dissolution of an ionic salt in water can be thought of in terms of an equilibrium between the solid and the aqueous ions. In the case of AgI(s) the equilibrium is

$$AgI(s) \rightleftharpoons Ag^+(aq) + I^-(aq). \tag{19.23}$$

If the salt is sparingly soluble, the equilibrium constant for this dissolution process is known as the *solubility product*, K_{sp}. This equilibrium constant is written

$$K_{sp} = ([Ag^+]/c^\circ)([I^-]/c^\circ);$$

as usual, there is no concentration term for the solid, which is present in its standard state. One way of finding the value of K_{sp} is to determine $\Delta_r G^\circ$ for the reaction given in Eq. 19.23, and then use $\Delta_r G^\circ = -RT \ln K$. We will show that the required $\Delta_r G^\circ$ value can be found using standard cell potentials.

All we need to do is imagine a cell whose conventional cell reaction is Eq. 19.23. The two half cells required are (the E° values at 298 K are also given)

RHS: $AgI(s) + e^- \longrightarrow Ag(m) + I^-(aq)$ $E^\circ(AgI/I^-) = -0.15224$ V
LHS: $Ag^+(aq) + e^- \longrightarrow Ag(m)$ $E^\circ(Ag^+/Ag) = +0.7996$ V.

The right-hand half cell is a metal/insoluble salt electrode (described in section 19.6.4 on page 849) in which a layer of insoluble AgI is coated on to a Ag electrode, and the whole is in contact with a solution of I^-. The left-hand half cell is simply a Ag electrode in contact with a solution of Ag^+.

The two half-cell reactions have the same number of electrons, so we can find the conventional cell reaction by taking RHS – LHS to give

conventional: $AgI(s) \longrightarrow Ag^+(aq) + I^-(aq)$;

this is exactly the same as Eq. 19.23.

The standard cell potential is found by taking $E^\circ_{RHS} - E^\circ_{LHS}$ to give $E^\circ = -0.15224 - (+0.7996) = -0.95184$ V. From this we can easily determine $\Delta_r G^\circ_{cell}$

$$\Delta_r G^\circ_{cell} = -nFE^\circ$$
$$= -1 \times 96,485 \times (-0.95184)$$
$$= 91.8 \text{ kJ mol}^{-1}.$$

This value of $\Delta_r G^\circ_{cell}$ is the same as $\Delta_r G^\circ$ for Eq. 19.23 on the preceding page, so we can compute K_{sp} as $\exp(-\Delta_r G^\circ/RT)$ which gives a value of 8.12×10^{-17}. This is a very small value, which indicates that AgI(s) is indeed sparingly soluble.

Since each AgI which dissolves gives one Ag^+ and one I^-, it follows that $[Ag^+] = [I^-]$. The solubility product can therefore be written as

$$K_{sp} = ([Ag^+]/c^\circ)^2.$$

Using the value of K_{sp} we have just determined, and recalling that $c^\circ = 1$ mol dm^{-3}, we find that

$$[Ag^+] = \sqrt{K_{sp}(c^\circ)^2} = 9.01 \times 10^{-9} \text{ mol dm}^{-3}.$$

This corresponds to just 2.11 μg of solid AgI dissolving in 1 dm^3 of water!

The nice thing about this calculation is that we did not actually have to set up a cell and make any measurements. All we had to do was *imagine* an appropriate cell, and then determine its standard potential from tabulated data. What could be easier?

19.9.2 Thermodynamic parameters for ions

In section 6.12 on page 238 we presented tables of data on the enthalpies and entropies of hydration of ions, and went on to use these in subsequent discussions concerning solubility and acidity. These data are partly derived from electrochemical measurements in a way which we will illustrate in this section.

Imagine a cell constructed using the following half-cells (E° values at 298 K are given)

RHS: $H^+(aq) + e^- \longrightarrow \frac{1}{2}H_2(g)$ $E^\circ(H^+/H_2) = 0.00$ V
LHS: $Ag^+(aq) + e^- \longrightarrow Ag(m)$ $E^\circ(Ag^+/Ag) = +0.7996$ V.

The conventional cell reaction is

$$H^+(aq) + Ag(m) \longrightarrow \frac{1}{2}H_2(g) + Ag^+(aq), \qquad (19.24)$$

and the standard potential of this cell is $0.00 - (+0.7996) = -0.7996$ V. From this we can compute $\Delta_r G^\circ_{cell}$ at 298 K as $-nFE^\circ = -1 \times 96,485 \times (-0.7996) = 77.1$ kJ mol^{-1}.

The standard Gibbs energy change of the cell reaction Eq. 19.24 can be written in terms of the standard Gibbs energies of formation of the reactants and products

$$\Delta_r G^\circ_{cell} = \left[\Delta_f G^\circ(Ag^+(aq)) + \tfrac{1}{2}\Delta_f G^\circ(H_2(g))\right] - \left[\Delta_f G^\circ(H^+(aq)) + \Delta_f G^\circ(Ag(m))\right].$$

Since by definition the standard Gibbs energy of formation of an element in its reference phase is zero, the second and fourth terms are zero, leaving

$$\Delta_r G^\circ_{cell} = \Delta_f G^\circ(Ag^+(aq)) - \Delta_f G^\circ(H^+(aq)).$$

Any measurement on a cell is always going to result in a *difference* between the Gibbs energies of formation of ions. To get round this, we introduce a further convention which is that

convention: $\Delta_f G°(H^+(aq)) = 0$.

This convention just amounts to defining a zero point on our scale, rather in the same way that we define a zero point for cell potentials using the standard hydrogen electrode. Since any measurable quantity will always involve the *difference* of Gibbs energies of formation of ions, this arbitrary choice of zero will have no effect on the outcome.

It therefore follows for our cell that $\Delta_r G°_{cell} = \Delta_f G°(Ag^+(aq))$, and so $\Delta_f G°(Ag^+(aq)) = 77.1 \text{ kJ mol}^{-1}$. The same procedure can be applied to other half-cell potentials, making it possible to determine values for the Gibbs energy of formation of many different ions in a very straightforward way.

To determine the enthalpy and entropy of formation we need to use the approach outlined in section 19.2.3 on page 839 where is was shown that the $\Delta_r S_{cell}$ could be determined from the temperature dependence of the cell potential. For this cell, the reported variation of the standard potential with temperature is

$$\frac{dE°}{dT} = -315 \ \mu\text{V K}^{-1}.$$

Using this value we can find $\Delta_r S°_{cell}$ using Eq. 19.6 on page 839

$$\begin{aligned} \Delta_r S°_{cell} &= nF\frac{dE°}{dT} \\ &= 1 \times 96,458 \times (-315 \times 10^{-6}) \\ &= -30.4 \text{ J K}^{-1} \text{ mol}^{-1}. \end{aligned}$$

Now that we know $\Delta_r G°_{cell}$ and $\Delta_r S°_{cell}$ we can easily compute $\Delta_r H°_{cell}$ using $\Delta_r G°_{cell} = \Delta_r H°_{cell} - T\Delta_r S°_{cell}$; this gives a value of $\Delta_r H°_{cell} = 68.0 \text{ kJ mol}^{-1}$.

This enthalpy change for the cell reaction Eq. 19.24 on the preceding page can be written in terms of standard enthalpies of formation

$$\Delta_r H°_{cell} = \left[\Delta_f H°(Ag^+(aq)) + \tfrac{1}{2}\Delta_f H°(H_2(g))\right] - \left[\Delta_f H°(H^+(aq)) + \Delta_f H°(Ag(m))\right].$$

The second and fourth terms on the right are zero as these are the enthalpies of formation of elements in their reference phases. As with $\Delta_f G°$ values we introduce the convention

convention: $\Delta_f H°(H^+(aq)) = 0$,

so it follows that $\Delta_r H°_{cell} = \Delta_f H°(Ag^+(aq))$. The standard enthalpy of formation of Ag^+ is therefore 68.0 kJ mol^{-1}. Since we have shown that the change in the standard Gibbs energy and enthalpy for the cell reaction are equal to the standard Gibbs energy and enthalpy of formation of $Ag^+(aq)$, it follows that $\Delta_r S°_{cell}$ must be the standard entropy of formation, $\Delta_f S°$, of this species. Just like $\Delta_f H°$, $\Delta_f S°$ refers to the formation of one mole of the species from its elements in their reference phases.

We have therefore determined that for the species $Ag^+(aq)$ at 298 K

$\Delta_f G° = 77.1 \text{ kJ mol}^{-1}$ $\qquad \Delta_f H° = 68.0 \text{ kJ mol}^{-1}$ $\qquad \Delta_f S° = -30.4 \text{ J K}^{-1} \text{ mol}^{-1}$.

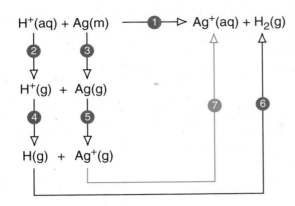

Fig. 19.12 A Hess's Law cycle for determining the standard enthalpy of hydration of Ag^+, step 7 (shown in red). $\Delta_r H°$ for step 1 can be found from an electrochemical cell, as described in the text. The $\Delta_r H°$ values for all the other steps can be found from tabulated data.

These values are determined using the conventions (note that the last of these follows from the first two)

conventions: $\quad \Delta_f G°(H^+(aq)) = 0 \quad \Delta_f H°(H^+(aq)) = 0 \quad \Delta_f S°(H^+(aq)) = 0$

This example shows how simple measurements of cell potentials and their temperature variation gives access to standard Gibbs energies, enthalpies and entropies of formation of aqueous ionic species.

Enthalpies of hydration

Now that we have these values for the enthalpy and entropy of formation, we can go on to work out the corresponding values for hydration, which are the data we presented in section 6.12 on page 238. To do this we return to the conventional cell reaction (Eq. 19.24 on page 857)

$$H^+(aq) + Ag(m) \longrightarrow \tfrac{1}{2}H_2(g) + Ag^+(aq),$$

for which we know $\Delta_r H°_{cell}$, and then dissect it into the Hess' Law cycle shown in Fig. 19.12. This looks rather complicated, but the point of breaking it down in this way is that $\Delta_r H°$ values for steps 1–6 are known, so we can determine $\Delta_r H°$ for step 7, which is the standard enthalpy of hydration of Ag^+. Let us go through the other steps in turn:

(1) This is the cell reaction, for which we know $\Delta_r H°_1 = 77.1 \text{ kJ mol}^{-1}$.

(2) This is the reverse of the hydration of $H^+(g)$: by convention, $\Delta H°_{hyd}(H^+)$ is set to zero (we have more to say about this below).

(3) This is the enthalpy needed to atomize silver metal to gas; the tabulated value is $\Delta_r H°_3 = 284.6 \text{ kJ mol}^{-1}$.

(4) This is the reverse of the ionization of $H(g)$ to $H^+(g)$. The values of these ionization energies are well known, and from tables we can find $\Delta_r H°_4 = -1312 \text{ kJ mol}^{-1}$.

(5) This is the ionization of $Ag(g)$ to $Ag^+(g)$; the tabulated value gives $\Delta_r H°_5 = 731 \text{ kJ mol}^{-1}$.

(6) This is the reverse of the dissociation of $\tfrac{1}{2}H_2(g)$ to $H(g)$; the tabulated value gives $\Delta_r H°_6 = -218 \text{ kJ mol}^{-1}$.

Putting all of this together we have

$$\Delta_r H_1^\circ = \Delta_r H_2^\circ + \Delta_r H_3^\circ + \Delta_r H_4^\circ + \Delta_r H_5^\circ + \Delta_r H_6^\circ + \Delta_r H_7^\circ.$$

Recalling that it is the value of $\Delta_r H_7^\circ$ that we seek, the above equation can be rearranged to

$$
\begin{aligned}
\Delta_r H_7^\circ &= \Delta_r H_1^\circ - \Delta_r H_2^\circ - \Delta_r H_3^\circ - \Delta_r H_4^\circ - \Delta_r H_5^\circ - \Delta_r H_6^\circ \\
&= 77.1 - (0) - (284.6) - (-1312) - (731) - (-218) \\
&= 591.5 \text{ kJ mol}^{-1}.
\end{aligned}
$$

The value of $\Delta H_{hyd}^\circ(Ag^+)$ is therefore 591.5 kJ mol^{-1}.

How can this result be correct? Surely we would expect the process of taking an ion from the gas phase into a polar solvent to be highly *exothermic* on account of the strong ion–solvent interactions. However, the value we have determined indicates that the process is strongly endothermic.

The reason why this value looks so strange is that in step 2 we took the standard enthalpy of hydration of $H^+(g)$ to be zero. Like all of the other conventions we have adopted, this simply sets an arbitrary zero point on our scale, from which we measure all of the other values. So what the value of $\Delta_r H_7^\circ = 591.5$ kJ mol^{-1} really tells us is that the hydration of Ag^+ is 591.5 kJ mol^{-1} *more* endothermic that the hydration of H^+.

Standard hydration enthalpies determined using the convention that $\Delta H_{hyd}^\circ(H^+) = 0$ are called *conventional* hydration enthalpies. So our value should be quoted as

$$\Delta H_{hyd}^\circ(Ag^+)^{conv} = 591.5 \text{ kJ mol}^{-1}.$$

Absolute enthalpies of hydration

If we *do not* adopt the convention $\Delta H_{hyd}^\circ(H^+) = 0$, then in our cycle $\Delta_r H_2^\circ$ is not known so all we can determine is the *sum* of $\Delta_r H_2^\circ$ and $\Delta_r H_7^\circ$

$$\Delta_r H_7^\circ + \Delta_r H_2^\circ = 591.5 \text{ kJ mol}^{-1}.$$

$\Delta_r H_7^\circ$ is the standard enthalpy of hydration of Ag^+, and $\Delta_r H_2^\circ$ is *minus* the standard enthalpy of hydration of H^+

$$\Delta H_{hyd}^\circ(Ag^+) - \Delta H_{hyd}^\circ(H^+) = 591.5 \text{ kJ mol}^{-1}. \tag{19.25}$$

This is the nub of the problem: any experiment will always result in a *difference* of hydration enthalpies.

One way out of this difficulty is to find a value for $\Delta H_{hyd}^\circ(H^+)$ by other means. The details of how this is done are beyond the scope of this text, so we shall have to simply use the accepted value of $\Delta H_{hyd}^\circ(H^+) = -1110$ kJ mol^{-1}. Using this in Eq. 19.25 we find

$$\Delta H_{hyd}^\circ(Ag^+) = -518.5 \text{ kJ mol}^{-1}.$$

This value makes much more sense since, as expected, it is large and negative. A value which has been determined in this way is called an *absolute* enthalpy of hydration.

The values quoted in Table 6.2 on page 239 are absolute enthalpies of hydration, determined precisely in the way demonstrated here for Ag^+. The entropies of hydration are also found using the same cycle, and once again we can either determine conventional values by assuming that $S_m^\circ(H^+) = 0$, or absolute values using the accepted value $S_m^\circ(H^+) = -20.9$ J K^{-1} mol^{-1}.

19.10 Oxidation state diagrams

If we are studying the chemistry of species which show multiple oxidation states, such as transition metals, an *oxidation state diagram* (also known as a *Frost diagram*) can be a very useful aid to our understanding. The general form of such a diagram is illustrated Fig. 19.13. Each point gives the standard Gibbs energy of formation of the ion with the oxidation state indicated by the horizontal scale. The points are connected by a line (which is of no significance). The diagram gives an immediate impression of the relative thermodynamic stabilities of the various oxidation states. For example, in Fig. 19.13 we can immediately see that the +2 oxidation state is the most stable as it has the lowest Gibbs energy of formation. There is much more that we can deduce from these diagrams, as will be explained in this section.

First, we will look at how the diagram is constructed using data from electrochemical potentials. Having done this, we will look at how such diagrams can be interpreted, taking some specific examples from transition metal chemistry.

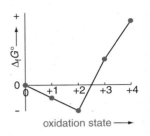

Fig. 19.13 An oxidation state diagram (or Frost diagram) is a plot of the standard Gibbs energy of formation of an ion as a function of its oxidation state. The line joining the points serves as a guide to the eye, but is otherwise of no significance. In some texts, the horizontal scale is plotted with zero on the right, and increasing positive values to the left.

19.10.1 Constructing an oxidation state diagram

We saw in section 19.9.2 on page 857 how it was possible to determine the standard Gibbs energy of formation of an ion from an electrode potential. This approach needs to be generalized somewhat in order to give us the greatest flexibility in using the available standard cell potentials. To do this, consider a cell in which the right-hand side is the couple in which the oxidized form is the ion M^{m+} (i.e. oxidation state $+m$), and the reduced form is the ion $M^{(m-n)+}$ (i.e. oxidation state $+(m-n)$). The left-hand side is the couple H^+/H_2. The half cell reactions are

$$\text{RHS:} \quad M^{m+}(aq) + ne^- \longrightarrow M^{(m-n)+}(aq) \quad E^\circ(M^{m+}/M^{(m-n)+})$$
$$\text{LHS:} \quad nH^+(aq) + ne^- \longrightarrow \tfrac{1}{2}nH_2(g) \quad\;\; E^\circ(H^+/H_2).$$

Note that we have already made sure that both half cells have the same number of electrons. Taking RHS – LHS gives us the conventional cell reaction as

$$M^{m+}(aq) + \tfrac{1}{2}nH_2(g) \longrightarrow M^{(m-n)+}(aq) + nH^+(aq).$$

Since the LHS is the standard hydrogen electrode, for which $E^\circ(H^+/H_2) = 0$, the standard cell potential is simple $E^\circ(M^{m+}/M^{(m-n)+})$. It follows that $\Delta_r G^\circ_{cell} = -nFE^\circ(M^{m+}/M^{(m-n)+})$.

Writing $\Delta_r G^\circ_{cell}$ for the cell reaction in terms of standard Gibbs energies of formation, and recalling that $\Delta_f G^\circ$ for elements in their reference phases and for $H^+(aq)$ are zero by convention, we have

$$\Delta_r G^\circ_{cell} = \Delta_f G^\circ(M^{(m-n)+}) - \Delta_f G^\circ(M^{m+}).$$

Writing $\Delta_r G^\circ_{cell}$ in terms of the cell potential gives

$$-nFE^\circ(M^{m+}/M^{(m-n)+}) = \Delta_f G^\circ(M^{(m-n)+}) - \Delta_f G^\circ(M^{m+}),$$

which can be rearranged to

$$\Delta_f G^\circ(M^{m+}) = \Delta_f G^\circ(M^{(m-n)+}) + nFE^\circ(M^{m+}/M^{(m-n)+}). \qquad (19.26)$$

Example 19.4 Determining the oxidation state diagram for copper

Given the following data (at 298 K), draw up the oxidation diagram for copper.

$$Cu^+(aq) + e^- \longrightarrow Cu(m) \qquad E°(Cu^+/Cu) = +0.518 \text{ V}$$
$$Cu^{2+}(aq) + e^- \longrightarrow Cu^+(aq) \qquad E°(Cu^{2+}/Cu^+) = +0.161 \text{ V}$$
$$Cu^{3+}(aq) + e^- \longrightarrow Cu^{2+}(aq) \qquad E°(Cu^{3+}/Cu^{2+}) = +2.4 \text{ V}.$$

Using Eq. 19.29 with the value for $E°(Cu^+/Cu)$ we find

$$\Delta_f G°(Cu^+)/F = 1 \times E°(Cu^+/Cu) = +0.518 \text{ V}.$$

Using Eq. 19.28 for the Cu^{2+}/Cu^+ couple we have

$$\Delta_f G°(Cu^{2+})/F = \Delta_f G°(Cu^+)/F + 1 \times E°(Cu^{2+}/Cu^+),$$

but we have already determined that $\Delta_f G°(Cu^+)/F = +0.518 \text{ V}$ so

$$\Delta_f G°(Cu^{2+})/F = 0.518 + 0.161 = +0.679 \text{ V}.$$

To find the value for Cu^{3+} we once again use Eq. 19.28, but this time for the Cu^{3+}/Cu^{2+} couple and in conjunction with the known value for Cu^{2+} of $\Delta_f G°(Cu^{2+})/F = +0.679 \text{ V}$

$$\Delta_f G°(Cu^{3+})/F = \Delta_f G°(Cu^{2+})/F + 1 \times E°(Cu^{3+}/Cu^{2+})$$
$$= +0.679 + 2.4 = +3.079 \text{ V}.$$

We have one more point which is $\Delta_f G°$ for Cu(m) which is by definition zero. Using these data, the oxidation state diagram can be plotted; it is shown in Fig. 19.14.

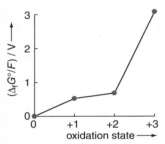

Fig. 19.14 Oxidation state diagram for copper.

For the special case where the reduced form is simply the metal M (which will be when $n = m$), Eq. 19.26 on the preceding page becomes

$$\Delta_f G°(M^{m+}) = mFE°(M^{m+}/M), \qquad (19.27)$$

since $\Delta_f G°$ of M(m) is zero. Using Eqs 19.26 and 19.27 in conjunction with values of standard electrode potentials it is possible to determine the standard Gibbs energies of formation of the various oxidation states of M.

In constructing these oxidation state diagrams it is common to quote the value of $\Delta_f G°(M^{m+})/F$ (which has units of volts) rather than $\Delta_f G°(M^{m+})$ (in kJ mol^{-1}). Equations 19.26 and 19.27 then become

$$\Delta_f G°(M^{m+})/F = \Delta_f G°(M^{(m-n)+})/F + nE°(M^{m+}/M^{(m-n)+}) \qquad (19.28)$$
$$\Delta_f G°(M^{m+})/F = mE°(M^{m+}/M). \qquad (19.29)$$

Example 19.4 shows how these expressions can be used to find values of $\Delta_f G°(M^{m+})/F$ for different oxidation states of copper.

It is common for the higher oxidation states of metals to exist as oxoanions, for example Cr(VI) exists as the dichromate anion $Cr_2O_7^{2-}$ (in an acidic solution). Following the same line of argument as above we imagine the following cell

RHS: $Cr_2O_7^{2-}(aq) + 14H^+(aq) + 6e^- \longrightarrow 2Cr^{3+}(aq) + 7H_2O(l)$ $E^\circ(Cr_2O_7^{2-}/Cr^{3+})$
LHS: $6H^+(aq) + 6e^- \longrightarrow 3H_2(g)$ $E^\circ(H^+/H_2)$.

The conventional cell reaction is therefore

$$Cr_2O_7^{2-}(aq) + 8H^+(aq) + 3H_2(g) \longrightarrow 2Cr^{3+}(aq) + 7H_2O(l),$$

and $\Delta_rG^\circ_{cell}$ for this reaction is

$$\Delta_rG^\circ_{cell} = 2\Delta_fG^\circ(Cr^{3+}) + 7\Delta_fG^\circ(H_2O(l)) - \Delta_fG^\circ(Cr_2O_7^{2-}).$$

As before the H^+ ions do not contribute as their standard Gibbs energy of formation is zero, as is that for $H_2(g)$. Given that $\Delta_rG^\circ_{cell} = -6FE^\circ(Cr_2O_7^{2-}/Cr^{3+})$ the last line can be rearranged to give

$$\begin{aligned} \Delta_fG^\circ(Cr_2O_7^{2-})/F & \\ = 2\Delta_fG^\circ(Cr^{3+})/F &+ 7\Delta_fG^\circ(H_2O(l))/F + 6E^\circ(Cr_2O_7^{2-}/Cr^{3+}). \end{aligned} \tag{19.30}$$

For these oxoanions it is usual to plot the Gibbs energy of formation of the metal-containing species *minus* the Gibbs energy of formation of any H_2O produced in the reaction. The reason for doing this is to take away the (large) influence that the formation of the water molecules has on Δ_fG° for the oxoanion so as to focus on just the contribution from the metal atom. Taking the term for the water to the left of Eq. 19.30 gives

$$\left[\Delta_fG^\circ(Cr_2O_7^{2-}) - 7\Delta_fG^\circ(H_2O(l))\right]/F = 2\Delta_fG^\circ(Cr^{3+})/F + 6E^\circ(Cr_2O_7^{2-}/Cr^{3+}).$$

It is the term on the left which will be plotted on the oxidation state diagram.

Given that $E^\circ(Cr^{3+}/Cr) = -0.744$ V, it follows from Eq. 19.29 on the facing page that $\Delta_fG^\circ(Cr^{3+})/F = 3 \times (-0.744) = -2.232$ V. Putting all of this together with the value of the standard potential $E^\circ(Cr_2O_7^{2-}/Cr^{3+}) = +1.232$ V, we have

$$\left[\Delta_fG^\circ(Cr_2O_7^{2-}) - 7\Delta_fG^\circ(H_2O(l))\right]/F = 2 \times (-2.232) + 6 \times (+1.232) = +2.928V.$$

Since *two* chromium atoms are involved in the Cr(VI) species $Cr_2O_7^{2-}$, the value plotted on the graph is *halved* (i.e. +1.464 V) so that it is the value *per metal atom*.

19.10.2   Interpreting an oxidation state diagram

As a result of the way in which an oxidation state diagram is constructed, the *slope* of a line connecting two points is the redox potential for that couple. For example, in the case of the oxidation state diagram for copper shown in Fig. 19.14 on the preceding page, the slope of the line joining oxidation states 0 and 2 is

$$\text{slope} = \frac{\text{change in } y \text{ coordinate}}{\text{change in } x \text{ coordinate}} = \frac{0.679 - 0.00}{2 - 0} = +0.34 \text{ V}.$$

Thus $E^\circ(Cu^{2+}/Cu) = +0.34$ V, which is indeed the correct value.

If we add some more data, the diagram can be used to compare the redox stability of different species. For example, given that $E^\circ(Zn^{2+}/Zn) = -0.7618$ V, $\Delta_fG^\circ(Zn^{2+})/F = 2 \times (-0.7618) = -1.524$ V. This point can be added to our diagram, along with the obvious one for H^+ which for which the Gibbs energy of formation is zero; the resulting diagram is shown in Fig. 19.15 on the following page.

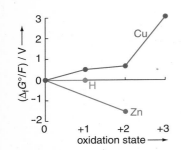

Fig. 19.15 Oxidation state diagram for copper, zinc and hydrogen.

The fact that the $\Delta_f G°$ values for all three oxidation states of copper are positive implies that these copper ions will, in principle, all oxidize H_2 to H^+. This is because, relative to H^+/H_2, there is a reduction in Gibbs energy of the copper species as the oxidation state is reduced.

On the other hand, the negative $\Delta_f G°$ value for Zn^{2+} tells us that in going from Zn^{2+} to Zn there is an increase in Gibbs energy relative to H^+/H_2. So Zn^{2+} will not oxidize H_2. However, the opposite process, in which we go from Zn to Zn^{2+}, results in a reduction in the Gibbs energy relative to hydrogen. As a result, Zn will reduce H^+ to H_2.

Going from Cu^{2+} to Cu^{3+} results in an increase in the Gibbs energy, so to cause this oxidation we need to find an oxidizing agent whose Gibbs energy decreases at a faster rate. It is clear that Zn^{2+} is not the powerful oxidizing agent we are seeking, as its Gibbs energy actually increases as it goes to Zn.

This discussion of redox stability can just as well be formulated using the standard half-cell potentials, as we did in section 19.7 on page 851, so oxidation state diagrams are not particularly useful in this regard.

Disproportionation

Some metal ions are unstable with respect to *disproportionation*, which is a reaction in which an ion oxidizes another ion of *precisely the same type*. The result is that the oxidation state of one ion is increased while that of the other is decreased. For example, Cu(I) disproportionates to Cu(0) (the metal) and Cu(II)

$$2Cu^{2+}(aq) \longrightarrow Cu^{2+}(aq) + Cu(m).$$

$\Delta_r G°$ for this reaction can be expressed in terms of the standard Gibbs energies of formation

$$\Delta_r G° = \Delta_f G°(Cu^{2+}(aq)) + \Delta_f G°(Cu(m)) - 2\Delta_f G°(Cu^+(aq)).$$

If the equilibrium for this reaction is to favour products significantly then $\Delta_r G°$ must be negative, which implies that

$$2\Delta_f G°(Cu^+(aq)) > \Delta_f G°(Cu^{2+}(aq)) + \Delta_f G°(Cu(m)).$$

Dividing each side by $2F$ gives an inequality which we interpret in terms of our oxidation diagram

$$\Delta_f G°(Cu^+(aq))/F > \tfrac{1}{2}\left[\Delta_f G°(Cu^{2+}(aq))/F + \Delta_f G°(Cu(m))/F\right]. \qquad (19.31)$$

This says that for the disproportionation reaction to be feasible, the value of $\Delta_f G°/F$ (the quantity plotted on the vertical axis of an oxidation state diagram) for Cu^+ must be greater than the *average* value for Cu and Cu^{2+}.

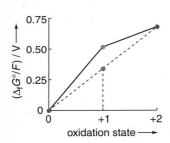

Fig. 19.16 Oxidation state diagram for copper showing why it is that Cu(I) disproportionates to Cu(0) and Cu(II). The dashed line is drawn between the two oxidation states into which it is proposed that the Cu(I) will disproportionate. Where this line crosses the oxidation level of Cu(I) gives the *average* value of $\Delta_f G°/F$ for Cu(II) and Cu(0), indicated by the green dot. As the green dot lies *below* the value of $\Delta_f G°/F$ for Cu(I) (the red dot) it follows that the disproportionation reaction is thermodynamically favoured.

This result can be interpreted graphically on the oxidation state diagram, as shown in Fig. 19.16. The average quantity on the right of Eq. 19.31 can be found by drawing a line (shown dashed) between the $\Delta_f G°/F$ values for Cu(II) and Cu(0). The required average is given by the point where this line crosses the x-coordinate corresponding to Cu(I); the value is shown by the green dot. Since the green dot is *below* the value for $\Delta_f G°(Cu^+(aq))/F$ (the red dot) the inequality of Eq. 19.31 is satisfied. It therefore follows that Cu(I) will disproportionate to Cu(0) and Cu(II). If the green dot had been above the blue dot, then the inequality would not have been satisfied, and Cu(I) would not disproportionate.

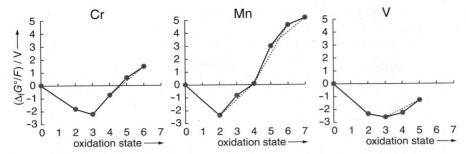

Fig. 19.17 Oxidation state diagrams for Cr, Mn and V under acidic conditions; all of the graphs have been plotted to the same scale. The species are all simple aquo ions unless specified here. For chromium Cr(IV) = CrO_2, Cr(V) is not well characterized, Cr(VI) = $Cr_2O_7^{2-}$; for manganese Mn(IV) = MnO_2, Mn(V) is possibly H_3MnO_4, Mn(VI) = MnO_4^{2-}, Mn(VII) = MnO_4^-; for vanadium V(IV) = VO^{2+}, V(V) = VO_2^{2+}. The dashed lines indicate that Cr(V), Mn(III), Mn(V) and Mn(VI) are all unstable with respect to disproportionation, whereas (for example) V(IV) is not.

Although we have illustrated this construction for copper, it applies generally for any oxidation state diagram. To see if oxidation state (*M*) will disproportionate into (*M* − 1) and (*M* + 1) all we need to do is to draw a line between the $\Delta_f G°/F$ values for (*M* − 1) and (*M* + 1). If the point at which this line crosses the *x*-coordinate for oxidation state (*M*) lies below the value of $\Delta_f G°/F$ for (*M*), then the disproportionation reaction will be favoured.

Some typical oxidation state diagrams

Figure 19.17 shows oxidation state diagrams for three transition metals. For chromium the Cr(III) species has the lowest $\Delta_f G°/F$ value, indicating that redox processes in which this species is formed, either by oxidation or reduction, are likely to be favoured. We can therefore say, with some justification, that 'Cr(III) is the most stable oxidation state'. With its significant positive $\Delta_f G°/F$ value, Cr(VI) can be classed as quite a strong oxidizing agent. We can also see from the dashed line connecting Cr(IV) and Cr(VI) that we would expect Cr(V) to be unstable with respect to disproportionation. The fact that the $\Delta_f G°/F$ value for Cr(III) is significantly negative implies that Cr metal is a reasonable reducing agent in the process in which it forms Cr(III).

For manganese, the species Mn(II) is clearly the most favoured. The high oxidation states are powerful oxidants, as evidenced by the large $\Delta_f G°/F$ values for Mn(V), Mn(VI) and Mn(VII). The dashed line shows that Mn(III) is unstable with respect to disproportionation, which is indeed found to be the case. The diagram also indicates that Mn(VI) and Mn(V) are unstable with respect to disproportionation. However, Mn(IV) is not.

Vanadium behaves rather differently. V(III) is the most stable state, but V(II) and V(IV) are of comparable stability. The dashed line shows that V(IV) is stable with respect to disproportionation. Even the highest oxidation states of vanadium are not particularly strong oxidizing agents.

19.11 Measurement of concentration

As we have seen, cell potentials have a predictable dependence on concentration, so such measurements are often a convenient way measuring concentration, not least since the cell gives an electrical signal which can be displayed or logged by a computer. All we have to do is to devise a cell whose potential responds to the species of interest in a suitable way.

One issue we need to be concerned about is to ensure that the measured cell potential responds only to the species of interest i.e. the electrode needs to be *selective*. A great deal of work has gone into developing such selective electrodes, and we will discuss a few of them here. These electrodes actually work on a somewhat different principle to those we have been discussing so far in that it is the potential developed *across* an interface or membrane which depends on concentration.

19.11.1 Measuring concentration using a cell

In section 18.1.5 on page 802 we described how, in principle, the rate of the oxidation of methanoic acid by bromine

$$HCOOH + Br_2 + 2H_2O \longrightarrow CO_2 + 2Br^- + 2H_3O^+,$$

could be monitored using a cell, which is shown in Fig. 18.3 on page 803. We are now in a position to understand the details of how this works. The left-hand electrode of the cell is formed by the Br_2/Br^- couple (a platinum electrode is needed to make the electrical contact)

$$Br_2(aq) + 2e^- \longrightarrow 2Br^-(aq).$$

The half-cell Nernst equation is

$$E = E^\circ(Br_2/Br^-) - \frac{RT}{2F} \ln \frac{([Br^-]/c^\circ)^2}{[Br_2]/c^\circ}.$$

The right-hand electrode is simply a calomel reference electrode, of the kind depicted in Fig. 19.10 on page 851, which produces a fixed half-cell potential E_{cal}. The cell potential is therefore

$$E = E_{cal} - E^\circ(Br_2/Br^-) + \frac{RT}{2F} \ln \frac{([Br^-]/c^\circ)^2}{[Br_2]/c^\circ}. \tag{19.32}$$

The reaction mixture is made up with an excess concentration of Br^- so that this does not change during the reaction. It is therefore only changes in the concentration of Br_2 which affect the cell potential.

Equation 19.32 can be rearranged to give the following expression for $\ln [Br_2]$

$$\ln ([Br_2]/c^\circ) = \underbrace{\frac{2F}{RT} [E_{cal} - E^\circ(Br_2/Br^-)] + 2 \ln ([Br^-]/c^\circ)}_{\text{constant, } =A} - \frac{2FE}{RT}.$$

For a given temperature and excess concentration of Br^- the term indicated by the underbrace is constant, so we can write

$$\ln [Br_2] = A - \frac{2FE}{RT}. \tag{19.33}$$

(a)

permeable
only to M+

low [M+]   high [M+]

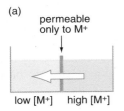

(b)

net +ve   net −ve

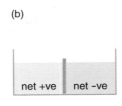

(c)

−0.123 V

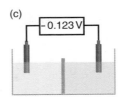

Fig. 19.18 Illustration of how an ion-selective barrier or membrane can give rise to a concentration-dependent junction potential. In (a) two solutions containing the ion M^+ are separated by a barrier only permeable to that ion. Diffusion favours a flow of M^+ from the more concentrated to the less concentrated solution, as indicated by the arrow. However, such a flow leaves an excess of negative counter-ions on the right and an excess of positive charge on the left, as shown in (b). This charge separation opposes the flow and so eventually a situation is reached in which no further flow occurs and the potential across the barrier reaches a steady value. The potential can be measured by creating a cell in which a reference electrode is dipped into each solution, as shown in (c).

In this expression it has been assumed that the concentration is in mol dm^{-3}, so that the $c°$ term can be omitted.

If the reaction is first order in Br_2 then, as explained in section 18.2.1 on page 806, we would expect the concentration as a function of time to obey

$$\ln[Br_2] = -k_{1st}t + \ln[Br_2]_0.$$

Substituting in the expression for $\ln[Br_2]$ from Eq. 19.33 on the preceding page gives

$$A - \frac{2FE}{RT} = -k_{1st}t + \ln[Br_2]_0,$$

which can be rearranged to

$$E = \frac{RTk_{1st}}{2F}t + \text{constant}.$$

If the process is first order, a plot of E against t will be a straight line of slope $(RTk_{1st})/(2F)$; it is not necessary to know the value of the constant. Thus, measurements of the cell potential as a function of time make it possible to determine the rate constant for this reaction.

19.11.2   Ion selective electrodes

In section 19.6.5 on page 850 it was described how a potential, called the liquid junction potential, can arise when two different solutions are in contact. This potential is a nuisance when we are trying to measure cell potentials and relate them to the concentrations using the Nernst equation. However, under some circumstances this potential between two solutions can be used to measure the relative concentrations of the two solutions.

The basic idea is illustrated in Fig. 19.18. Here we have two solutions, with different concentrations, which are separated by a barrier or membrane which has been carefully designed so that it is only permeable to *one* kind of ion. To make the discussion easier, let us assume that this ion is M^+ (although it could just as well be an anion).

If the concentration of M^+ is higher in the right-hand compartment, then there will be a tendency for the M^+ ions to flow into the left-hand compartment in order to equalize the concentrations, as shown in (a). However, as only M^+ ions can pass the barrier, the result of this flow will be that the right-hand compartment will acquire a net negative charge due to the negative counter ions which are 'left behind' when the M^+ ions crossed the barrier. Similarly, the left-hand compartment will acquire a net positive charge, due to the M^+ ions which are arriving; the resulting situation is shown in (b).

This separation of charge which makes the right-hand compartment negative relative to the left *opposes* the flow of the M^+ ions since it is energetically favourable for the *positive* ions to be in the *negatively*-charged right-hand compartment. Eventually the difference in charge between the two compartments will be large enough to stop the flow of M^+ ions. At this point the potential difference between the two compartments reaches a steady values. It can be shown that this potential difference across the barrier $E_{barrier}$ is given by

$$E_{barrier} = \frac{RT}{F} \ln\left(\frac{[M^+]_L/c^\circ}{[M^+]_R/c^\circ}\right)$$

where $[M^+]_R$ is the concentration of M^+ in the right-hand compartment and likewise for the left. This expression tells us that the potential difference across the barrier depends logarithmically on the ratio of the concentrations on either side.

We cannot measure the potential between two solutions as there is nothing to connect to. What we have to do is to create a cell with two electrodes so that there is a measurable potential. Typically, a reference electrode (such as a calomel electrode, with its integral liquid junction) is dipped into each of the compartments, as shown in Fig. 19.18 (c). It is important to realize that these electrodes are just there to create a cell whose potential can be measured. The species M^+ is *not* involved with these two electrodes.

Figure 19.19 shows the practical construction of an ion-selective electrode. Inside the body of the electrode there is a solution containing the ion of interest, and dipping into this is a reference electrode. The bottom of the electrode enclosure is sealed off using the special barrier or membrane which is permeable only to the ion of interest. To make a measurement, this electrode is simply dipped into the solution of interest, along with a second reference electrode (such as a calomel electrode). The potential between these two electrodes is related to the ratio of the concentration of the ion inside the electrode to that in the test solution.

In practice, the potential developed by such an arrangement is measured for a series of solutions of known concentration in order to develop a calibration curve. The concentration of an unknown solution can then be found by measuring the potential and converting it to concentration using this calibration curve.

The absolutely key part in constructing an ion selective electrode is finding the right materials for the barrier or membrane which will allow just one type of ion to pass. Considerable ingenuity has been put into this work, resulting in the availability of electrodes capable of sensing a wide range of different ions. We will look at just two examples of such electrodes.

The fluoride electrode

In this electrode the fluoride-permeable barrier is made from a single crystal of lanthanum fluoride (LaF_3) which has been 'doped' with europium fluoride (EuF_2). Doping means adding a small carefully controlled amount of what is

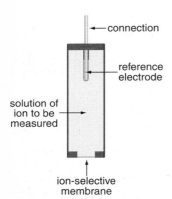

Fig. 19.19 Practical construction of an ion selective electrode. The membrane or barrier at the bottom of the electrode is only permeable to a particular ion. The body contains a solution of this ion, and therefore when the electrode is dipped into a solution containing the same ion a potential is developed across the membrane. This potential is measured between the reference electrode built into the assembly and a second reference electrode (not shown) which is dipped into the solution under test.

essentially an impurity. The result of the doping is that the lanthanum fluoride crystal contains vacancies or holes which are just the right size for F^- ions, thus making the crystal permeable to this ion. Typically Ag/AgCl electrodes are used as the two reference electrodes; the one dipped into the solution under test must have a built-in salt bridge.

Such electrodes are capable of measuring concentrations down to 0.1 ppm, which corresponds to a fluoride ion concentration of 5×10^{-6} mol dm^{-3}. The upper limit is 1900 ppm, which is a concentration of around 0.1 mol dm^{-3}. This is a remarkably wide range.

By using a different crystal for the barrier it is possible to design electrodes of this type which can detect selectively a range of different ions including Br^-, Cl^-, I^-, Cd^{2+}, Cu^{2+}, CN^-, Pb^{2+}, Hg^{2+}, Ag^+, S^{2-} and SCN^-.

The glass electrode

This electrode is selective for H_3O^+ ions and so by measuring their concentration gives a measure of the pH. A knowledge of the pH is vital in large areas of chemistry and biochemistry, so such pH electrodes (as they are often known) are the most widely used of all the selective electrodes.

In this electrode the barrier across which the potential forms is a thin glass membrane. The glass has the special property that the Na^+ ions in the glass can exchange with the H_3O^+ ions in solution. This process, known as *ion exchange*, leads to the development of a potential between the solution and the surface of the glass. The extent of the exchange, and hence the potential developed, depends on the concentration of the H_3O^+ ions,

Ion exchange takes place at the interface between the glass membrane and the outer solution (the solution under test), and also between the membrane and the inner solution (held inside the electrode). The membrane potential thus reflects the difference in the concentration of H_3O^+ ions in the solutions on either side of the membrane.

Figure 19.20 shows the typical construction of a glass electrode. As usual, two reference electrodes are needed to create a cell which will allow the junction potential to be measured. One of these, usually an AgCl/Ag electrode, is built into the glass electrode, and the second reference electrode, which is typically a calomel electrode, is dipped into the solution under test. It is common for this second electrode to be built onto the side of the glass electrode. The result is a simple 'dip and test' assembly which is often referred to as a combination electrode.

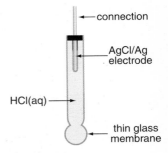

Fig. 19.20 Practical construction of an glass electrode for measuring the concentration of H^+ ions. A potential is built up across the thin glass membrane due to ion exchange between the Na^+ ions in the glass and the $H^+(aq)$ ions is the internal and external solutions. This potential is measured between the AgCl/Ag electrode built into the assembly and a second reference electrode (not shown) which is dipped into the solution under test.

FURTHER READING

Electrode Potentials, Richard G. Compton and Giles H. W. Sanders, Oxford University Press (1996).

QUESTIONS

Physical constants: $R = 8.3145$ J K^{-1} mol^{-1}, $F = 96,485$ C mol^{-1}.

19.1 For each of the following cells: (a) write the cell using the notation described in section 19.1.2 on page 835, (b) determine the conventional cell reaction, stating the number of electrons involved, (c) write down the Nernst equation for the cell, (d) using the data in Table 19.1 on page 845 determine the standard cell potential, (e) hence determine the spontaneous cell reaction when all of the species are present in their standard states.

In the following table it can be assumed that: (i) all ions are in aqueous solution; (ii) H_2O is in its liquid form; (iii) O_2, H_2, and Cl_2 are gases; (iv) the solutions in the right- and left-hand sides are separated by a salt bridge; (v) an inert platinum electrode can be used as required.

	left-hand half-cell reaction	right-hand half-cell reaction
1	$H^+ + e^- \rightarrow \frac{1}{2}H_2$	$Cu^+ + e^- \rightarrow Cu(m)$
2	$H^+ + e^- \rightarrow \frac{1}{2}H_2$	$Fe^{2+} + 2e^- \rightarrow Fe(m)$
3	$Fe^{2+} + 2e^- \rightarrow Fe(m)$	$H^+ + e^- \rightarrow \frac{1}{2}H_2$
4	$Mn^{2+} + 2e^- \rightarrow Mn(m)$	$Au^{3+} + 3e^- \rightarrow Au(m)$
5	$AgCl(s) + e^- \rightarrow Ag(m) + Cl^-$	$H^+ + e^- \rightarrow \frac{1}{2}H_2$
6	$AgI(s) + e^- \rightarrow Ag(m) + I^-$	$Zn^{2+} + 2e^- \rightarrow Zn(m)$
7	$AgCl(s) + e^- \rightarrow Ag(m) + Cl^-$	$Hg_2Cl_2(s) + 2e^- \rightarrow 2Hg(l) + 2Cl^-$
8	$2H_2O + 2e^- \rightarrow H_2 + 2OH^-$	$O_2 + 2H_2O + 4e^- \rightarrow 4OH^-$
9	$O_2 + 4H^+ + 4e^- \rightarrow 2H_2O$	$MnO_4^- + 8H^+ + 5e^- \rightarrow Mn^{2+} + 4H_2O$
10	$2H_2O + 2e^- \rightarrow H_2 + 2OH^-$	$PbO(s) + H_2O + 2e^- \rightarrow Pb(m) + 2OH^-$

19.2 For each of the following cells: (a) write down the right- and left-hand half-cell reactions (as reductions); (b) determine the conventional cell reaction, stating the number of electrons involved, (c) write down the Nernst equation for the cell, (d) using the data in Table 19.1 on page 845 determine the standard cell potential, (e) hence for cells 1–9 hence determine the spontaneous cell reaction when all of the species are present in their standard states.

1 $Zn(m) | Zn^{2+}(aq) \| Mg^{2+}(aq) | Mg(m)$

2 $Pt(m) | H_2(g) | H^+(aq) \| Fe^{3+}(aq) | Fe(m)$

3 $Pt(m) | Sn^{4+}(aq), Sn^{2+}(aq) \| Cu^{2+}(aq) | Cu(m)$

4 $Ag(m), AgI(s) | HI(aq) | H_2(g) | Pt(m)$

5 $Pt(m) | H_2(g) | H^+(aq) | O_2(g) | Pt(m)$

6 $Pt(m) | H_2(g) | OH^-(aq) | O_2(g) | Pt(m)$

7 $Ag(m), AgBr(s) | Br^-(aq), Br_2(aq) | Pt(m)$

8 $Pb(m), PbO(s) | OH^-(aq) | H_2(g) | Pt(m)$

9 $Pt(m) | Cr_2O_7^{2-}(aq), Cr^{3+}(aq), H^+(aq) \| Cl^-(aq) | AgCl(s), Ag(m)$

10 $Ag(m), AgCl(s) | Cl^-(aq, [Cl^-] = c_L) \| Cl^-(aq, [Cl^-] = c_R) | AgCl(s), Ag(m)$

19.3 Devise cells which have the following conventional cell reactions. In each case: (a) write the cell in the way described in section 19.1.2 on page 835; (b) write the right- and left-hand side half-cell reactions (as reductions); (c) determine the cell potential when all of the species are present in their standard states; (d) hence determine the direction of the spontaneous cell reaction when all of the species are present in their standard states.

1    $Zn(m) + Cu^{2+}(aq) \longrightarrow Zn^{2+}(aq) + Cu(m)$

2    $2Ag(m) + Sn^{4+}(aq) \longrightarrow Sn^{2+}(aq) + 2Ag^+(aq)$

3    $AgCl(m) + \frac{1}{2}H_2(g) \longrightarrow H^+(aq) + Cl^-(aq) + Ag(m)$

4    $H^+(aq) + Cl^-(aq) + Ag(m) \longrightarrow AgCl(m) + \frac{1}{2}H_2(g)$

5    $2AgCl(m) + H_2(g) \longrightarrow 2H^+(aq) + 2Cl^-(aq) + 2Ag(m)$

6    $2AgCl(m) + 2Hg(l) \longrightarrow Hg_2Cl_2(m) + 2Ag(m)$

7    $O_2(g) + 4H^+(aq) + 4Cl^-(aq) + 4Ag(m) \longrightarrow 2H_2O(l) + 4AgCl(m)$

8    $H_2O(l) + Pb(m) \longrightarrow H_2(g) + PbO(s)$

9    $AgBr(m) \longrightarrow Ag^+(aq) + Br^-(aq)$

10   $Al(m) + 3Ce^{4+}(aq) \longrightarrow Al^{3+}(aq) + 3Ce^{3+}(aq)$

19.4   Consider the following cell (at 298 K)

$$Fe(m) \,|\, Fe^{2+}(aq, [Fe^{2+}] = c_1) \,\|\, Cd^{2+}(aq, [Cd^{2+}] = c_2) \,|\, Cd(m).$$

  (a)   Determine the conventional cell reaction and hence write down the Nernst equation for the cell in terms of the concentrations c_1 and c_2.

  (b)   Determine the cell potential when all of the species are present in their standard states (i.e. $c_1 = c_2 = 1$ mol dm^{-3}), and hence state the spontaneous cell reaction.

  (c)   By considering the Nernst equation, determine the value of the ratio c_1/c_2 which will reduce the cell potential to zero.

  (d)   Hence state what range of values of c_1/c_2 will lead to the spontaneous cell reaction being the *opposite* of that determined in (b).

19.5   Use the data in Table 19.1 on page 845 to determine whether or not the following processes are thermodynamically feasible when all of the species are present in their standard states and at 298 K (all ions are in aqueous solution).

  (a)   The oxidation of $Cu(m)$ to Cu^{2+} by Fe^{3+} (assume that the Fe^{3+} is reduced to Fe^{2+})

  (b)   The oxidation of Fe^{2+} to Fe^{3+} by $Cl_2(g)$

  (c)   The oxidation of $H_2O(l)$ to $O_2(g)$ by Fe^{3+} (under acidic conditions)

  (d)   The reduction of Cu^{2+} to Cu^+ by $Ag(m)$

  (e)   The reduction of Fe^{3+} to $Fe(m)$ by $Zn(m)$

  (f)   The reduction of Mn^{2+} to $Mn(m)$ by Tl.

19.6   The potential produced by the hydrogen electrode is dependent on concentration of $H^+(aq)$ and the partial pressure of H_2.

  (a)   Given that the half-cell reaction is

$$H^+(aq) + e^- \longrightarrow \tfrac{1}{2}H_2(g) \qquad E^{\circ}(H^+/H_2) = 0.00 \text{ V},$$

write down the Nernst equation for the half-cell potential.

  (b)   Assuming that the partial pressure of H_2 is 1 bar (i.e. the standard pressure) show that the half-cell potential is given by

$$E = \frac{RT}{F} \ln [H^+].$$

  (c)   Hence show that the half-cell potential is related to the pH in the following way [hint $\ln x \equiv \ln 10 \times \log x$]

$$E = -\ln 10 \times \frac{RT}{F} \times \text{pH}.$$

(d) Does it become easier or more difficult for a reducing agent to reduce H^+ to H_2 at the pH increases?

19.7 (a) Devise a cell whose conventional cell reaction is

$$AgCl(s) \longrightarrow Ag^+(aq) + Cl^-(aq).$$

(b) Using tabulated data, determine the standard potential of this cell at 298 K and hence find $\Delta_r G^\circ_{cell}$.

(c) Hence determine the solubility product of AgCl, and the concentration of dissolved Ag^+ ions in a saturated solution of AgCl.

19.8 In this question we will use data on cell potentials to compute some thermodynamic parameters for Mg^{2+} in a way very similar to that illustrated in section 19.9.2 on page 857 for Ag^+. The cell we need to consider in this case is the following

$$\text{RHS:} \quad H^+(aq) + e^- \longrightarrow \tfrac{1}{2}H_2(g)$$
$$\text{LHS:} \quad Mg^{2+}(aq) + 2e^- \longrightarrow Mg(m).$$

At 298 K the standard potential for this cell is +2.360 V, and the temperature dependence of the standard cell potential is $dE^\circ/dT = -1.99 \times 10^{-4}$ V K^{-1}.

(a) Determine the conventional cell reaction.

(b) Determine $\Delta_r G^\circ_{cell}$, $\Delta_r S^\circ_{cell}$ and $\Delta_r H^\circ_{cell}$ (all at 298 K) for this reaction.

(c) Hence, assuming the usual conventions that $\Delta_f G^\circ(H^+(aq)) = 0$, $\Delta_f H^\circ(H^+(aq)) = 0$ and $\Delta_f S^\circ(H^+(aq)) = 0$, find the standard Gibbs energy, enthalpy and entropy of formation of $Mg^{2+}(aq)$.

(d) Dissect the conventional cell reaction found in (a) into a Hess's Law cycle of the same form as that in Fig. 19.12 on page 859.

(e) Use the following data, along with your value of $\Delta_f H^\circ(Mg^{2+}(aq))$, to determine the *absolute* enthalpy of hydration of Mg^{2+}, $\Delta H^\circ_{hyd}(Mg^{2+})$:

$$H^+(g) \rightarrow H^+(aq) \qquad \Delta_r H^\circ = -1110 \text{ kJ mol}^{-1}$$
$$Mg(m) \rightarrow Mg(g) \qquad \Delta_r H^\circ = 147.1 \text{ kJ mol}^{-1}$$
$$H(g) \rightarrow H^+(g) \qquad \Delta_r H^\circ = 1312 \text{ kJ mol}^{-1}$$
$$Mg(g) \rightarrow Mg^{2+}(g) \qquad \Delta_r H^\circ = 1450.7 \text{ kJ mol}^{-1}$$
$$H_2(g) \rightarrow 2H(g) \qquad \Delta_r H^\circ = 456 \text{ kJ mol}^{-1}.$$

[Be very careful to make sure that you use the $\Delta_r H^\circ$ value for the *direction of reaction* specified in the Hess's Law cycle.]

19.9 Use the following data to construct an oxidation state diagram for cobalt. Comment on any features of interest which are revealed by your diagram.

half-cell reaction	E°/V (298K)
$Co^{2+}(aq) + 2e^- \rightarrow Co(m)$	-0.282
$Co^{3+}(aq) + e^- \rightarrow Co^{2+}(aq)$	$+1.92$
$CoO_2(s) + 4H^+ + e^- \rightarrow Co^{3+}(aq) + 2H_2O(l)$	$+1.4$

19.10 Before an ion-selective electrode can be used to make measurements, the potential it generates (when forming a cell with a suitable external reference electrode) needs to be calibrated as a function of the concentration of the ions. This is normally done by measuring the potential for a series of solutions of known concentration and then constructing a calibration curve. For the greatest accuracy, it is best to calibrate the electrode over the range of concentrations likely to be encountered in the measurements.

Low concentrations of ions are usually specified in parts per million (ppm) by weight. For example, if a solution contains 5 ppm of fluoride it means that

$$\frac{\text{mass of } F^-}{\text{mass of solvent}} = 5 \times 10^{-6}.$$

It follows that 1 dm^3 of water, which would weigh 1000 g, will contain 5.0×10^{-3} g of F^-. Given that the RMM of F is 18.998, this corresponds to a concentration of 2.63×10^{-4} mol dm^{-3}. For these very dilute solutions, the concentration expressed in ppm is directly proportional to the concentration in mol dm^{-3}.

Measurements on a series of solutions of known concentration gave the following data

ppm F^-	5	4	3	2	1.5	1	0.5
E / V	0.288	0.282	0.276	0.264	0.258	0.246	0.229

The expected variation of the cell potential with concentration is of the form

$$E = A + B \ln [F^-],$$

where A and B are constants; in this expression, $[F^-]$ can be expressed in ppm. By plotting a suitable straight-line graph, show that these data fit this equation; determine the values of A and B.

The electrode was then used to measure the fluoride concentration in drinking water. Over a series of measurements the average potential was found to be 0.255 V, with a spread of measurements between 0.254 V and 0.256 V. Determine the concentration (in ppm) of F^- in the water, and the likely error on the measurement.

19.11 Manufacturers of glass electrodes, used to measure pH, warn customers that the accuracy of the results may be reduced if the Na^+ concentration in the solution under test is high. Why is this?

19.12 A *concentration cell* is one in which the RHS and LHS electrodes are identical with the only difference between them being the concentration (or pressure) of the species involved. An example of such a cell is once constructed from two $AgCl/Cl^-$ electrodes

$$Ag(m), AgCl(s) \,|\, Cl^-(aq, [Cl^-]_L) \,\|\, Cl^-(aq, [Cl^-]_R) \,|\, AgCl(s), Ag(m),$$

in which the concentration of Cl^- in the left-hand half-cell is $[Cl^-]_L$, and that on the right is $[Cl^-]_R$. In writing the half-cell reactions for such a cell, we need to be careful to keep track of which side of the cell the Cl^- is in. The half cell reactions are therefore written

$$\text{RHS:} \quad AgCl(s) + e^- \longrightarrow Ag(m) + Cl^-(aq)_R$$
$$\text{LHS:} \quad AgCl(s) + e^- \longrightarrow Ag(m) + Cl^-(aq)_L.$$

Note the subscripts 'R' and 'L' indicating which half-cell the Cl^- belongs to.

(a) Determine the conventional cell reaction, and hence show that the cell potential is given by
$$E = \frac{RT}{F} \ln \frac{[Cl^-]_L}{[Cl^-]_R}.$$

(b) Why does the standard half cell potential of the $AgCl/Cl^-$ couple *not* appear in the expression for the cell potential?

(c) Determine the direction of the spontaneous cell reaction when $[Cl^-]_R > [Cl^-]_L$ and when $[Cl^-]_L > [Cl^-]_R$. Rationalize your conclusions.

(d) From the above expression for the cell potential, compute dE/dT [Hint: all but one of the terms on the right do not depend on T]. Hence find $\Delta_r S_{cell}$ and $\Delta_r H_{cell}$. Comment on the results you obtain.

19.13 The potential of the following cell was measured at temperatures of 278, 298 and 318 K as +0.0389, +0.0458 and +0.0527 V respectively

$$Ag(m), AgCl(s) \,|\, KCl(aq, 0.1 \text{ mol dm}^{-3}) \,|\, Hg_2Cl_2(s) \,|\, Hg(l)$$

(a) Write down the half cell reactions of the two electrodes and hence determine the conventional cell reaction. Also, write down the Nernst equation for the cell, and comment on why the cell potential does not depend on the concentration of Cl^-.

(b) Determine $\Delta_r G^\circ_{cell}$ at 298 K.

(c) Using $\Delta_r S^\circ_{cell} = nF(dE^\circ/dT)$, find $\Delta_r S^\circ_{cell}$ and hence $\Delta_r H^\circ_{cell}$ (you will need to plot a graph).

(d) Hence determine the standard enthalpy of formation of $Hg_2Cl_2(s)$ at 298 K, given that at this temperature the standard enthalpy of formation of $AgCl(s)$ is $-126.8 \text{ kJ mol}^{-1}$.

19.14 Consider the cell

$$Pt(m) \,|\, H_2(g) \,|\, H_2SO_4(aq) \,|\, PbSO_4(s) | Pb(m),$$

whose standard potential is found to be -0.356 V at 298 K.

(a) Write down the conventional cell reaction.

(b) Given that the standard half-cell potential of the $Pb^{2+}(aq)/Pb(m)$ electrode is -0.126 V at 298 K, determine the solubility product of $PbSO_4(s)$ at this temperature. What will the concentration of Pb^{2+} ions be in a saturated solution of $PbSO_4$ at 298 K?

(c) Given that the temperature variation of the standard potential of the above cell is given by $dE^\circ/dT = 1.1 \times 10^{-4} \text{ V K}^{-1}$, determine $\Delta_r S^\circ_{cell}$ and hence $\Delta_r H^\circ_{cell}$.

(d) Given that at 298 K the standard enthalpy of formation of $PbSO_4(s)$ is $-918.4 \text{ kJ mol}^{-1}$, determine the standard enthalpy of formation of $SO_4^{2-}(aq)$.

19.15 (a) Describe carefully what is meant by the term *standard electrode potential*. Explain how tables of standard *half-cell* potentials can be drawn up.

(b) An electrochemical cell is constructed from two half cells: on the right a silver wire dips into a 0.001 mol dm^{-3} AgNO$_3$ solution, and the half cell on left is a hydrogen electrode with hydrogen gas at 1 bar pressure and an HCl solution of variable concentration, m. At 298 K the cell potential was found to vary with m in the following way:

m / mol dm^{-3}	0.6	0.8	1.0	1.2	1.4
E / V	0.6375	0.6304	0.6252	0.6214	0.6182

Determine the Nernst equation for the cell and then, *by plotting a suitable straight-line graph*, test how well the above data fit your equation.

(c) From your graph, estimate the standard half-cell potential of the Ag$^+$/Ag couple.

Dimensions, units and some key mathematical ideas

Key points

- Dimensional analysis gives us a way of checking the consistency of an expression.
- The use of SI units provides a consistent and reliable way of calculating quantities.
- A variety of non-SI units remain in common use.
- The exponential function, and the related natural logarithm, appear frequently in the mathematical analysis of chemical processes.
- Differentiation can be used to find the maxima and minima of functions.
- Integration can be thought of in terms of finding the area under a graph.
- The mathematical analysis of some chemical processes leads to differential equations, simple examples of which can be solved straightforwardly.

This chapter is designed to help you with some aspects of the calculations you will encounter in this book, particularly when you attempt the *Questions*. The chances are that you will have encountered quite a lot of the topics presented in this chapter before, either before starting your university course or as part of your continuing study of mathematics during your current course. The first topic – that of *dimensional analysis* – is an exception, since it is less likely that you will have encountered this very useful technique before. This chapter is mainly intended to reinforce things you already know by placing them in a chemical context. It is not a substitute for a proper study of mathematics, which is essential for any scientist.

Mathematics is a subject where the old adage that 'practice makes perfect' definitely applies. You will not gain familiarity and fluency with mathematical techniques just by reading about them – you must practise using them. To this end, throughout the chapter you will find short self-test exercises which we recommend very strongly that you work through. The answers to these exercises will be found at the end of the chapter (starting on page 912), along with further exercises for you to try.

20.1 Dimensional analysis

The following four equations are all used to calculate the energy in a particular circumstance

$$E = \tfrac{1}{2}mv^2 \tag{20.1}$$

$$E = mc^2 \tag{20.2}$$

$$E = mgh \tag{20.3}$$

$$E = -\frac{m_e e^4}{32\pi^2 \varepsilon_0^2 \hbar^2}. \tag{20.4}$$

Let us go through these one by one. Equation 20.1 is the kinetic energy of a mass m moving at a velocity v. Equation 20.2 is Einstein's celebrated equation giving the energy equivalent of a mass m; c is the speed of light. Equation 20.3 is the gravitational potential energy of a mass m at height h above the surface of the earth; g is the acceleration due to gravity. Finally, Eq. 20.4 is the energy of the $1s$ orbital in a hydrogen atom; we will not go into what all the symbols stand for right now, but will return to this later on.

These expressions are all ways of calculating an energy, but they do all look rather different, especially the last two. How can we be sure that they are all correct? Of course, you can look all of these expressions up in a text book as they are standard results, but suppose that you have just derived an expression for an energy – for example in the course of solving a problem – how can you check that it is correct?

This is where *dimensional analysis* comes in. This technique is a way of checking any expression to see if it has the 'correct dimensions'. Using this approach we will be able to show that Equations 20.1–20.4 not only all have the same dimensions but also have the the correct dimensions to be expressions for energy. This approach is extremely useful for checking a calculation, since if the dimensions of an expression are not correct, the expression simply cannot be correct.

Unfortunately, the converse is not true. An expression which has the right dimensions is not necessarily correct, but we can be sure that it might be correct!

20.1.1 The basic dimensions

Any physical quantity (e.g. energy, length, velocity) always has some units associated with it. For example, energy can be given in joules (J), length in metres (m), and velocity in metres per second (m s^{-1}). In dimensional analysis, rather than using specific units such as metres or kilograms, we use a basic set of units or *dimensions*, which are listed in Table 20.1. It turns out that *any* unit can be expressed in terms of these dimensions.

To work out the dimensions of some quantity, all we need to do is look at its definition and then substitute in the basic dimensions from the table. For example, the area of a square is found by multiplying together the length of each side. A length has dimension L, so the dimensions of area are $L \times L = L^2$. Similarly, the volume of a cube has dimensions $L \times L \times L = L^3$. A slightly more complex example is velocity, which is defined as (distance travelled)/(time); the dimensions of distance are L and of time are T, so velocity has the dimensions

Table 20.1 Dimensions

quantity	symbol
length	L
mass	M
time	T
temperature	K
amount of substance	N
electric current	J
luminous intensity	I

L/T. This is usually written

$$[\text{velocity}] = L\,T^{-1},$$

where the square brackets indicate the quantity whose dimensions are given on the right.

Another example is acceleration, which is the rate of change of velocity i.e. (change in velocity)/(time). The dimensions of velocity are $L\,T^{-1}$ and of time are T, so those of acceleration are $L\,T^{-1}\,/\,T$

$$[\text{acceleration}] = L\,T^{-2}.$$

Once we have worked out the dimensions of these basic quantities we can move on to some more complex cases. Equation 20.1 on the facing page is an expression for the kinetic energy: the mass m has dimensions M, and we have already worked out that the velocity v has dimensions $L\,T^{-1}$. So the dimensions of $\frac{1}{2}mv^2$ can be worked out as

$$\begin{aligned}[\tfrac{1}{2}mv^2] &= M \times \left(L\,T^{-1}\right)^2 \\ &= M\,L^2\,T^{-2}.\end{aligned}$$

Note that the factor of a half is just a number and so has no dimensions. The dimensions of energy are thus $M\,L^2\,T^{-2}$. In Eq. 20.2 c is also a velocity, so the dimensional analysis is the same as for Eq. 20.1.

Finally, in Eq. 20.3, g is an acceleration, so has dimensions $L\,T^{-2}$, and h is a height with dimensions L. The dimensions of mgh are therefore

$$\begin{aligned}[mgh] &= M \times L\,T^{-2} \times L \\ &= M\,L^2\,T^{-2}.\end{aligned}$$

As expected, the dimensions are those for energy. Dimensional analysis confirms that Eqs. 20.1–20.3 must all be expressions for energy, since they have the dimensions of energy.

20.1.2 Working from the units

A slightly different way of finding the dimensions of a quantity is to work from its known units; sometimes, this can be easier than working from the formal definition. For example, velocity can be given in m s^{-1}, which is (length) $\times$ (time)$^{-1}$. Substituting in the symbols for these quantities gives the dimensions as $L\,T^{-1}$. Similarly, acceleration can be given in m s^{-2}, which is (length) $\times$ (time)$^{-2}$: this translates to dimensions of $L\,T^{-2}$.

As a final example, we will consider the dimensions of the molar heat capacity, which typically has units of J K^{-1} mol^{-1}. Joule is a unit of energy, with dimensions $M\,L^2\,T^{-2}$, kelvin is a unit of temperature which we see from Table 20.1 has dimensions K, and mole is a measure of the amount of substance, which from the table has dimensions N

$$[\text{molar heat capacity}] = M\,L^2\,T^{-2} \times K^{-1} \times N^{-1} = M\,L^2\,T^{-2}\,K^{-1}\,N^{-1}.$$

Self Test 20.1 will give you some practice at finding dimensions using the ideas presented so far.

Self Test 20.1 Finding dimensions: 1

Determine the dimensions of the following quantities or expressions:

(a) Momentum, define as mass × velocity.

(b) The quantity $p^2/(2m)$, where p is the momentum and m is the mass.

(c) Density, defined as mass per unit volume.

(d) Force, defined as mass × acceleration.

(e) Pressure, defined as force per unit area.

(f) Molar entropy, whose units are $J \, K^{-1} \, mol^{-1}$.

(g) The molar enthalpy change, whose units are $kJ \, mol^{-1}$.

20.1.3 Dimensions in equations

If we have an equation or expression in which several terms are added together, such as in

$$A = B + C,$$

then the dimensions of all the terms *must be the same*. It makes no sense to add an energy to a velocity and expect to obtain a distance!

For example, consider the well known expression in dynamics

$$v = u + at,$$

where v is the velocity at time t, u is the initial velocity and a is the acceleration. Clearly u and v must have the dimensions of velocity, which we have found to be $L T^{-1}$. The dimensions of at are $(L T^{-2}) \times T$, which is $L T^{-1}$ i.e. velocity. Thus everything is consistent since all three terms have the same dimensions.

20.1.4 Electrical and electrostatic quantities

Table 20.2 Dimensions of electrical quantities

quantity	dimensions
current	J
voltage	$M L^2 T^{-3} J^{-1}$
charge	$T J$
capacitance	$T^4 J^2 L^{-2} M^{-1}$

Looking back at Table 20.1 on page 876 you will see that one of the basic dimensions is electric current, given the symbol J. All of the other electrical quantities, such as voltage (potential), charge and capacitance, involve J along with the other dimensions. We will not derive all of these here, not least as most of them are not relevant to chemistry, but simply summarize them in Table 20.2.

20.1.5 The dimensions of fundamental constants

More complex expressions, such as Eq. 20.4 on page 876, often include various fundamental constants, so to complete a dimensional analysis of such an expression we need to know the dimensions of these constants. Sometimes, these are quite easy to work out. For example, the Boltzmann constant k_B has units $J \, K^{-1}$. Given that we already know that the dimensions of energy (joules) are $M L^2 T^{-2}$, it follows that the dimensions of k_B are $M L^2 T^{-2} K^{-1}$.

We can put this to immediate use by checking the dimensions of the expression given by kinetic theory for the mean speed of the molecules in a gas $\bar{c}$

$$\bar{c} = \left(\frac{8 k_B T}{\pi m} \right)^{\frac{1}{2}},$$

where m is the mass and T the temperature. Noting that the number '8' and the constant π are dimensionless (i.e. have no dimensions), we now know the dimensions of all the quantities so we can compute the overall dimensions of the expression on the right.

$$\left[\left(\frac{8k_\text{B}T}{\pi m}\right)^{\frac{1}{2}}\right] = \left(\frac{(M\,L^2\,T^{-2}\,K^{-1})(K)}{M}\right)^{\frac{1}{2}}$$

$$= \left(L^2\,T^{-2}\right)^{\frac{1}{2}}$$

$$= L\,T^{-1}.$$

The square braces [...] around an expression mean 'take the dimensions of the expression'.

To go to the second line we have cancelled the M between the top and bottom of the fraction, and have also used $K^{-1}K = 1$. The dimensions of this expression are therefore $L\,T^{-1}$ which are the dimensions of velocity, as required.

The product $k_\text{B}T$, where k_B is the Boltzmann constant and T the temperature, appears very often when dealing with molecular energy levels. Its dimensions are

$$[k_\text{B}T] = (M\,L^2\,T^{-2}\,K^{-1})(K)$$

$$= M\,L^2\,T^{-2},$$

which are those of energy i.e. $k_\text{B}T$ has the dimensions of energy.

This product $k_\text{B}T$ appears in the Boltzmann distribution, introduced in section 1.10.1 on page 27, which gives the population of energy level i, n_i, in terms of its energy ε_i and the population n_0 of the lowest level (Eq. 1.4 on page 27)

$$n_i = n_0 \exp\left(\frac{-\varepsilon_i}{k_\text{B}T}\right). \tag{20.5}$$

Since $k_\text{B}T$ has the dimensions of energy, the dimensions of the ratio $\varepsilon_i/(k_\text{B}T)$ are

$$\left[\frac{\varepsilon_i}{k_\text{B}T}\right] = \frac{M\,L^2\,T^{-2}}{M\,L^2\,T^{-2}} = 1.$$

The ratio therefore has no dimensions: it is said to be *dimensionless*. In retrospect this is obvious, since $k_\text{B}T$ and ε_i both have the same dimensions of energy.

Taking the exponential of a dimensionless quantity results in quantity which is also dimensionless, so the exponential term on the right-hand side of Eq. 20.5 is dimensionless. The populations n_i are just numbers, and so are also dimensionless, so the dimensions of both sides of the equation are the same.

The gas constant R is simply $N_\text{A}k_\text{B}$, where N_A is Avogadro's constant, which has dimensions N^{-1}. It therefore follows that the dimensions of R are $M\,L^2\,T^{-2}\,K^{-1}\,N^{-1}$. The product RT therefore has dimensions $M\,L^2\,T^{-2}\,N^{-1}$, which is energy per amount of substance i.e. energy per mole. RT is thus the molar analogue of $k_\text{B}T$.

The product RT appears very often. For example, in the Arrhenius law (section 10.3 on page 415, Eq. 10.2), which gives the temperature dependence of the rate constant k in terms of a pre-exponential factor A and an activation energy E_a

$$k = A \exp\left(\frac{-E_\text{a}}{RT}\right).$$

As discussed in section 20.4.4 on page 899, strictly speaking we can only take the exponential of a dimensionless quantity, giving a dimensionless result.

Table 20.3 Dimensions of selected physical constants

name	symbol	SI units	dimensions
speed of light	c	m s^{-1}	$L\,T^{-1}$
Planck's constant	h	J s	$M\,L^2\,T^{-1}$
Avogadro's constant	N_A	mol^{-1}	N^{-1}
Elementary charge	e	C	$T\,J$
Boltzmann constant	k_B	J K^{-1}	$M\,L^2\,T^{-2}\,K^{-1}$
Faraday constant	F	C mol^{-1}	$T\,J\,N^{-1}$
Gas constant	R	J K^{-1} mol^{-1}	$M\,L^2\,T^{-2}\,K^{-1}\,N^{-1}$
Permittivity of vacuum	ε_0	F m^{-1}	$T^4\,J^2\,L^{-3}\,M^{-1}$

C stands for coulomb and F for farad.

E_a is given in energy per mole, and so has the same dimensions as RT, making the ratio E_a/RT dimensionless. It therefore follows that k_r and A must have the same dimensions.

You can practice some more calculations along these lines in Self Test 20.2.

Sometimes it is relatively easy to find the dimensions of a physical constant, as we did for Boltzmann's constant, but for others it is not quite so straightforward. Table 20.3 gives the dimensions of a selection of physical constants which you may come across in your calculations.

Two units from this table may be unfamiliar to you: C stands for coulomb, the SI unit of electrical change; F stands for farad, the SI unit of capacitance. The dimensions of these quantities are given in Table 20.2 on page 878.

20.1.6 Further examples

Using the information given in Table 20.3 we can finally check the dimensions of Eq. 20.4 on page 876, which is the expression for the energy of the $1s$ AO in hydrogen

$$E = -\frac{m_e e^4}{32\pi^2 \varepsilon_0^2\, \hbar^2}.$$

In this equation m_e is the mass of the electron, e is the charge on the electron (the elementary charge), $\hbar$ is Planck's constant divided by 2π, and ε_0 is the permittivity of a vacuum. Using the information from the table, and recalling that numbers (including π) do not have dimensions, the calculation is

$$\left[-\frac{m_e e^4}{32\pi^2 \varepsilon_0^2\, \hbar^2}\right] = \frac{M \times (T\,J)^4}{(T^4\,J^2\,L^{-3}\,M^{-1})^2 \times (M\,L^2\,T^{-1})^2}$$

$$= \frac{M \times \left(T^4\,J^4\right)}{(T^8\,J^4\,L^{-6}\,M^{-2}) \times (M^2\,L^4\,T^{-2})}$$

$$= \frac{M\,T^4\,J^4}{T^6\,J^4\,L^{-2}}$$

$$= M\,L^2\,T^{-2}.$$

After quite a lot of computation, we see that the expression does indeed have the dimensions of energy.

Self Test 20.2 Finding dimensions: 2

Using the results you have already found as necessary, determine the dimensions of the following quantities or expressions:

(a) The gas constant, R, whose units are J K^{-1} mol^{-1}.

(b) An energy E expressed in J mol^{-1}.

(c) The ratio $E/(RT)$.

(d) Planck's constant h, whose units are J s.

(e) Planck's constant divided by 2π, usually called 'h-cross' and written $\hbar$.

(f) The product $h\upsilon$, where υ is a frequency with dimensions T^{-1}.

(g) The expression

$$\frac{h^2}{8ma^2},$$

where a is a length.

Which of these expressions are energies or energies per amount of substance?

When we are looking at electrochemical cells, the Nernst equation for the cell potential is of great importance (Eq. 19.13 on page 842)

$$E = E^\circ - \frac{RT}{nF} \ln \frac{([C]/c^\circ)^c \, ([D]/c^\circ)^d}{([A]/c^\circ)^a \, ([B]/c^\circ)^b}.$$

The term inside the logarithm is dimensionless since each concentration $[i]$ is divided by the standard concentration c°. E and E° are cell potentials (i.e. voltages), and n is the number of electrons involved in the stoichiometric equation.

Let us check the dimensions of RT/nF, recalling that R is the gas constant and F is the Faraday constant. Referring to Table 20.3 on the facing page the calculation is

$$
\begin{aligned}
\left[\frac{RT}{nF}\right] &= \frac{\left(M L^2 T^{-2} K^{-1} N^{-1}\right) \times K}{T J N^{-1}} \\
&= \frac{M L^2 T^{-2} N^{-1}}{T J N^{-1}} \\
&= M L^2 T^{-3} J^{-1}.
\end{aligned}
$$

Comparison with Table 20.2 on page 878 reveals that these are the dimensions of voltage. So in the Nernst equation the term on the left and both terms on the right have the dimensions of voltage, which is as expected.

Some further examples for you to try are given in Self Test 20.3.

20.1.7 Atomic units

Quantum mechanical calculations are considerably simplified by adopting what are called *atomic units*, which is a system of units in which the length is expressed as a multiple of the Bohr radius, charge is expressed as a multiple of the fundamental change, and mass is expressed as a multiple of the mass

---

Self Test 20.3 Finding dimensions: 3

You should use any results you have already found, along with information from the tables, in answering the following:

(a) The energy of interaction between charges q_1 and q_2 separated by a distance r is given by the expression below. Confirm that it has the dimensions of energy.

$$E = \frac{q_1 \, q_2}{4\pi\varepsilon_0 r}.$$

(b) The energy stored by a capacitor of capacitance C charged to a voltage V is given by $\frac{1}{2}CV^2$. Confirm that this expression has the dimensions of energy.

(c) Find the dimensions of the molar entropy S_m°, given in J K^{-1} mol^{-1}.

(d) The change in the molar entropy when a gas expands from volume V_1 to V_2 is given by the expression below. Confirm that the terms on the left and right have the same dimensions.

$$\Delta S_m = R \ln \frac{V_2}{V_1}.$$

(e) The ideal (perfect) gas equation relates the pressure p, the volume V, the amount in moles n and the temperature T

$$pV = nRT.$$

Find the dimensions of the left- and right-hand sides.

---

of the electron. In this system of units, energy turns out to be a multiple of a quantity known as the *hartree*. Although this all sounds rather complicated, it really does simplify the calculations and manipulations.

One point to watch out for is that an expression written in these atomic units will *not* have the correct dimensions since a consequence of using these units is that many of the physical constants 'disappear' from the equations.

20.2 Units

Suppose that we want to compute the kinetic energy using

$$E = \tfrac{1}{2}mv^2.$$

The question arises as to which *units* we should use for the velocity and the mass. If we put the mass in kg we will obtain one value for the energy, whereas is we put the mass in grams we will obtain another. How do we decide which units to use?

The solution to this difficulty is to use SI units for *all the quantities* in your calculation. If you do this, an energy will *always* come out in joules, a force in newtons and a velocity in m s^{-1} regardless of how the quantities are computed.

Table 20.4 Basic SI units

quantity	unit (symbol)
length	metre (m)
mass	kilogram (kg)
time	second (s)
temperature	kelvin (K)
amount	mole (mol)
electric current	ampere (A)
luminous intensity	candela (cd)

Table 20.5 Derived SI units

quantity	symbol	derived SI unit (symbol)	unit in terms of basic SI units
force	F	newton (N)	$kg\ m\ s^{-2}$
pressure	p	pascal (Pa) or N m^{-2}	$kg\ m^{-1}\ s^{-2}$
frequency	f or v	hertz (Hz) or s^{-1}	s^{-1}
charge	q	coulomb (C)	$A\ s$
energy	various	joule (J)	$kg\ m^2\ s^{-2}$
voltage or potential	V or E	volt (V)	$kg\ m^2\ s^{-3}\ A^{-1}$

The abbreviation SI comes from the full title (in French) *Système International d'Unités*.

The basic SI units are given in Table 20.4 on the facing page: these map one-for-one onto the basic dimensions which were given in Table 20.1 on page 876. All other SI units can be derived from these basic units. So, for example, velocity, which is (distance travelled)/(time) is given in m s^{-1}, and acceleration, which is (change in velocity)/(time) is given in m s^{-2}.

We can work out the units of a quantity from its dimensions, since there is a one-to-one correspondence between each dimension and a particular SI unit. For example, velocity has dimensions $L\,T^{-1}$, which translates to an SI unit of m s^{-1}. Similarly, the fact that energy has the dimensions $M\,L^2\,T^{-2}$ means that its SI unit is kg m^2 s^{-2}.

However, we rarely quote energies in units of 'kg m^2 s^{-2}' but instead use joules. In fact, 1 J is *exactly* the same as 1 kg m^2 s^{-2}. For commonly measured quantities, such as energy, force and pressure, there are a set of units which are derived from the basic SI units but which are more practical to use, not least as they are shorter. A selection of these derived units are listed in Table 20.5.

The four expressions given in Eqs. 20.1–20.4 on page 876 are all for energy. In each case if we use the correct SI unit for all the quantities on the right we can be confident that the computed energy will come out in joules. This simplicity saves us from a great deal of worry and trouble. You can work through some examples of assigning SI units in Self Test 20.4 .

20.2.1 Decimal prefixes

In SI units the bond length in N_2 is 1.10×10^{-10} m. The problem with this is that it is both rather a mouthful to say, and also rather long to write out. What is more, since we have to read both the mantissa (the 1.10) and the power of 10 in order to comprehend the number, it can be quite difficult to make comparisons when data are presented in this format.

For these reasons, it is common to add to the unit a prefix which indicates the power of ten. For example the prefix 'p' (pronounced pico) implies an exponent of 10^{-12}; using this the above bond length can be expressed as 110 pm (the unit is spoken as 'pico metres'). Written in this way, the number is easier to comprehend, and shorter to read and pronounce.

These prefixes are available for each power of 10, but with the exception of 10^{-2} and 10^{-1} it is common only to use those that are multiples of 10^3; those prefixes frequently encountered in chemistry are given in Table 20.6.

Table 20.6 Decimal prefixes

prefix	symbol	multiple
pico	p	10^{-12}
nano	n	10^{-9}
micro	μ	10^{-6}
milli	m	10^{-3}
centi	c	10^{-2}
deci	d	10^{-1}
kilo	k	10^{3}
mega	M	10^{6}
giga	G	10^{9}

Self Test 20.4 Finding SI units

Determine the correct SI units of the the following quantities or expressions (use derived units where it leads to simplification):

(a) Momentum, define as mass × velocity.

(b) The quantity $p^2/(2m)$, where p is the momentum and m is the mass.

(c) Density, defined as mass per unit volume.

(d) Electric field strength, defined as voltage per unit length.

(e) The tensile stress in a rod, defined as the force per unit cross-sectional area.

To convert 1.10×10^{-10} m to pm we *divide* by the multiple, which is 10^{-12}

$$1.10 \times 10^{-10} \text{ m is equivalent to } \frac{1.10 \times 10^{-10}}{10^{-12}} = 110 \text{ pm.}$$

Alternatively we could express the length in nm, for which the multiple is 10^{-9}

$$1.10 \times 10^{-10} \text{ m is equivalent to } \frac{1.10 \times 10^{-10}}{10^{-9}} = 0.110 \text{ nm.}$$

Similarly, inconveniently large numbers can be scaled down by using those prefixes which correspond to positive powers of 10. For example, the energy needed to convert NaCl(s) to gaseous Na^+ and Cl^- ions is 7.88×10^5 J mol^{-1} which is conveniently expressed in kJ mol^{-1} as follows

$$7.88 \times 10^5 \text{ J mol}^{-1} \text{ is equivalent to } \frac{7.88 \times 10^5}{10^3} = 788 \text{ kJ mol}^{-1}.$$

Once you get the hang of this you can see that converting from m to pm just means adding 12 to the power of 10, and converting from J to kJ just means subtracting 3 from the power of 10.

While it is permissible to prefix SI units in this way, you have to be careful to take account of the prefix when you use the quantity to make a calculation. For example, suppose that we want to determine the value of

$$\frac{\Delta_r H^\circ}{R},$$

where $\Delta_r H^\circ$ is the molar enthalpy change of the reaction and R is the gas constant. We almost always quote $\Delta_r H^\circ$ values in kJ mol^{-1}, but this must be converted to J mol^{-1} *before* the value is used in this expression. To take a specific example, if $\Delta_r H^\circ$ is 100 kJ mol^{-1} then the calculation is

$$\frac{\Delta_r H^\circ}{R} = \frac{100 \times 10^3}{8.3145} = 12027 \text{ K.}$$

If we had simply computed 100/8.3145 we would not have obtained the correct answer.

A second example concerns computing the standard Gibbs energy change $\Delta_r G^\circ$ from the standard enthalpy and entropy changes, $\Delta_r H^\circ$ and $\Delta_r S^\circ$, using

Self Test 20.5 Manipulating prefixes

Express the following quantities using a convenient prefix:

(a) The ionic radius of Na^+, 1.00×10^{-10} m.

(b) The average C–H bond length, 1.08×10^{-10} m.

(c) A wavelength corresponding to red light, 7.80×10^{-7} m.

(d) The frequency of a typical FM radio transmitter 9.4×10^7 Hz.

Express the following quantities in the form $n.nn \times 10^m$:

(e) The enthalpy of combustion of $CH_4(g)$: -891 kJ mol^{-1}.

(f) The spacing between adjacent ions in $NaCl(s)$ of 0.282 nm.

(g) The wavelength corresponding to UV light of 350 nm.

(h) The frequency of 2 GHz used by some radar.

$\Delta_r G^\circ = \Delta_r H^\circ - T\Delta_r S^\circ$. We have to take account of the fact that whereas $\Delta_r H^\circ$ is usually quoted in kJ mol^{-1}, $\Delta_r S^\circ$ is quoted in J K^{-1} mol^{-1}. Suppose that $\Delta_r H^\circ = 100$ kJ mol^{-1} and $\Delta_r S^\circ = 50$ J K^{-1} mol^{-1}, then at 298 K the calculation of $\Delta_r G^\circ$ is

$$\begin{aligned} \Delta_r G^\circ &= \Delta_r H^\circ - T\Delta_r S^\circ \\ &= 100 \times 10^3 - 298 \times 50 \\ &= 8.51 \times 10^4 \text{ J mol}^{-1}. \end{aligned}$$

Note that on the second line the value of $\Delta_r H^\circ$ was converted from kJ mol^{-1} to J mol^{-1} by multiplying by 10^3. Since everything is in SI, we can be sure that the answer will be in J mol^{-1}; this can be expressed as 85.1 kJ mol^{-1} simply by dividing by 10^3.

Alternatively, as $\Delta_r G^\circ$ is usually quoted in kJ mol^{-1}, we can make the whole calculation in kJ by converting $\Delta_r S^\circ$ from J K^{-1} mol^{-1} to kJ K^{-1} mol^{-1}; this is achieved by multiplying the given value of $\Delta_r S^\circ$ by 10^{-3}.

$$\begin{aligned} \Delta_r G^\circ &= \Delta_r H^\circ - T\Delta_r S^\circ \\ &= 100 - 298 \times 50 \times 10^{-3} \\ &= 85.1 \text{ kJ mol}^{-1}. \end{aligned}$$

You can work through some examples of manipulating prefixes in Self Test 20.5 .

20.2.2 Concentrations

In SI, the 'proper' unit for concentration would presumably be mol m^{-3}, but as you know concentrations are invariably expressed in mol dm^{-3}. A *length* of 1 dm is equivalent to 0.1 m, so a *volume* of 1 dm^3 is equivalent to $0.1 \times 0.1 \times 0.1 = (0.1)^3$ m^3 which is 10^{-3} m^3. Expressed the other way round, 1 m^3 corresponds to 10^3 dm^3. The 'deci' prefix in dm^3 means 0.1 metres, not 0.1 cubic metres: this is an important point to appreciate if errors are to be avoided.

A volume of 1 dm^3 is also called a litre, and is the same as 1000 cm^3. Since laboratory glassware is commonly marked in cm^3, we need to be adept at converting back and forth between dm^3, cm^3 and sometimes m^3. We give here a few relevant examples of such calculations.

(a) What is the concentration in $mol\ m^{-3}$ of a solution with concentration $0.1\ mol\ dm^{-3}$?

Since there are $10^3\ dm^3$ in $1\ m^3$, the concentration in $mol\ m^{-3}$ must be 10^3 times *greater* than that in $mol\ dm^{-3}$ i.e. the concentration is $10^3 \times 0.1 = 100\ mol\ m^{-3}$.

(b) How many moles are there in $15\ cm^3$ of a solution with concentration $0.1\ mol\ dm^{-3}$?

This concentration means that there are 0.1 moles in $1\ dm^3$ or, equivalently, in $1000\ cm^3$. Therefore, in $15\ cm^3$ the number of moles will be

$$\underbrace{\frac{15}{1000}}_{\text{fraction of 1 dm}^3} \times \underbrace{0.1}_{\text{concentration in mol dm}^{-3}} = 1.5 \times 10^{-3}\ mol.$$

(c) What volume (in cm^3) of a solution with concentration $0.15\ mol\ dm^{-3}$ will contain $2.5 \times 10^{-3}\ mol$?

The easiest way to approach this is first to recall that concentration is defined as (amount in moles)/volume

$$\text{concentration in mol dm}^{-3} = \frac{\text{amount in moles}}{\text{volume in dm}^3}.$$

Rearranging this gives

$$\text{volume in dm}^3 = \frac{\text{amount in moles}}{\text{concentration in mol dm}^{-3}}.$$

Finally, multiplying by 1000 converts the volume from dm^3 to cm^3

$$\text{volume in cm}^3 = 1000 \times \frac{\text{amount in moles}}{\text{concentration in mol dm}^{-3}}.$$

Therefore the required volume is

$$\text{volume in cm}^3 = 1000 \times \frac{2.5 \times 10^{-3}}{0.15} = 16.67\ cm^3.$$

It is advisable to cross-check the answer by working out the number of moles in $16.67\ cm^3$ of a $0.15\ mol\ dm^{-3}$ solution. This is $(16.67/1000) \times 0.15 = 2.5 \times 10^{-3}\ mol$, confirming that our answer is correct.

In Self Test 20.6 you can try out some more calculations along these lines.

20.2.3   Non-SI units

Always using SI units gives us a consistent approach to calculating and reporting the values of physical quantities. However, in practice there are quite a lot of non-SI units which remain in widespread use. Sometimes these units have survived from earlier attempts to harmonize the system of units, but more often than not they are units which people have felt were convenient and practical, despite not fitting into the SI scheme. It is essential that you recognize what these

---

Self Test 20.6 Concentrations

(a) What is the concentration, in mol m^{-3}, of a solution with concentration 1.5 mol dm^{-3}?

(b) What is the concentration, in mol dm^{-3}, of a solution with concentration 0.1 mol m^{-3}?

(c) How many moles are there in 25 cm^3 of a solution with concentration 0.2 mol dm^{-3}?

(d) What volume of a solution with concentration 0.2 mol dm^{-3} contains 0.01 mol?

---

various non-SI units are, and know how to convert them to SI before making your calculations.

In this section we will consider a selection of these non-SI units that you are likely to come across, and it will also be explained how quantities in these units can be converted to SI.

Length: the angstrom (Å)

One angstrom (symbol Å) is equivalent to 10^{-10} m. The conversion to metres is therefore

$$\text{length in m} = \text{length in Å} \times 10^{-10}.$$

For example, the average C–H bond length of 1.08 Å is equivalent to 1.08×10^{-10} m or 108 pm.

Angstroms are widely used when quoting values of quantities such as bond lengths, lattice spacings, ionic radii and atomic radii. No doubt the continued popularity for using this unit for such quantities is due to that fact that the values turn out to be between 0.5 and 5 – numbers which can easily be grasped.

Energy: the calorie (cal)

Originally one calorie was defined as the heat needed to raise 1 g of water through 1 °C. It is now defined in terms of the SI unit of energy, the joule: 1 cal is equivalent to exactly 4.1840 J. A kcal is 1000 calories. The conversion is therefore

$$\text{energy in J} = \text{energy in cal} \times 4.1840.$$

For example 50 kcal mol^{-1} is $50 \times 4.1840 = 209$ kJ mol^{-1}.

For some reason, kcal mol^{-1} are still commonly used by North American scientists to quote the values of quantities such as $\Delta_r H°$ and activation energies. Entropies are sometimes also quoted in cal K^{-1} mol^{-1} (sometimes called an entropy unit, e.u.). Conversion to J K^{-1} mol^{-1} is achieved simply by multiplying by 4.1840.

Energy: the electronvolt (eV)

The electronvolt (eV) is commonly used to quote the energies of electrons in atoms and molecules. One electronvolt is the energy acquired when an electron is accelerated through a potential of one volt. Since energy is charge × potential, the conversion from eV to joules simply involves the charge on the electron energy in J = energy in eV × charge on the electron in C,

where the charge on the electron has the value 1.602×10^{-19} C.

For example, the energy of the $1s$ electron in a hydrogen atom is -13.61 eV. The energy in joules is therefore $-13.61 \times 1.602 \times 10^{-19} = 2.181 \times 10^{-18}$ J. If we want the energy in J mol^{-1}, then we multiply by the Avogadro constant to give $-2.181 \times 10^{-18} \times 6.022 \times 10^{23} = -1.313 \times 10^{6}$ J mol^{-1} or -1313 kJ mol^{-1}.

Pressure

The SI unit for pressure is N m^{-2} or Pa, but it is rare to find a pressure gauge actually calibrated in Pa. Various other units are in common use.

Atmospheric pressure varies constantly, but the standard atmospheric pressure has been defined as 101 325 Pa. It is common to quote pressures in terms of this standard atmospheric pressure, given the symbol atm. The conversion is

$$\text{pressure in Pa} = \text{pressure in atmospheres} \times 101\,325\,.$$

For example, the pressure of the air inside a full tank which a scuba diver might use is around 170 atm, which is 1.72×10^{7} Pa or 17.2 MPa.

One bar pressure is defined as *exactly* 10^{5} Pa. This unit plays an important role in thermodynamics since the standard state is defined as being at 1 bar pressure. The conversion is

$$\text{pressure in Pa} = \text{pressure in bar} \times 10^{5}\,.$$

The other place where you will commonly encounter bar is on weather maps, where the 'isobars' – contours are equal atmospheric pressure – are quoted in millibar (mbar). A typical value of 1026 mbar corresponds to $1026 \times 10^{-3} \times 10^{5} = 1.026 \times 10^{5}$ Pa.

Pressure can be measured in terms of the height of liquid it can support: this is the principle on which early barometers worked. It was common to use mercury as the liquid, on account of its high density, and as a result pressure was quoted in terms of the height of a mercury column. Standard atmospheric pressure can support a mercury column of height 760 mm. The pressure is therefore quoted as '760 mm Hg'. It follows that 1 mm Hg corresponds to a pressure of $101325/760 = 133.3$ Pa.

An alternative name for mm Hg is 'Torr', the symbol for which is τ. The conversion is

$$\text{pressure in Pa} = (\text{pressure in mm Hg } or \ \tau) \times 133.3$$

For example a standard laboratory rotary pump can reduce the pressure to around $10^{-3}\ \tau$, which corresponds to 0.13 Pa or 1.3×10^{-6} atm.

Mass unit

The relative formula mass (relative molecular mass) is determined using a scale based on the mass of one mole of ^{12}C being exactly 12 g. For example, the RMM of ^{12}C^{16}O is $12.00 + 15.99 = 27.99$ g mol^{-1}.

Sometimes we want to know the *actual* mass (in kg) of a single molecule or atom. We can find this by dividing the RMM by Avogadro's constant N_A, and then multiplying by 0.001 to convert the result from g to kg

$$\text{mass of one molecule in kg} = \frac{\text{RMM}}{N_A} \times 0.001\,.$$

For example, using this expression we can compute the mass of one molecule of $^{12}C^{16}O$ as 4.648×10^{-26} kg.

An alternative, and entirely equivalent, way of looking at this is to define the mass of one atom of ^{12}C as exactly 12 *atomic mass units*, usually given the symbol u. With this definition, the tables of relative atomic masses can be interpreted as the masses in atomic mass units. For example the mass of $^{12}C^{16}O$ is $12.00 + 15.99 = 27.99$ u.

One atomic mass unit is equivalent to 1.6606×10^{-27} kg, so to convert from mass units to kg we simply multiply by this number

$$\text{mass of one molecule in kg} = (\text{mass in atomic mass units}) \times (1.6606 \times 10^{-27}).$$

Using this, the mass of $^{12}C^{16}O$ is $27.99 \times 1.6606 \times 10^{-27} = 4.648 \times 10^{-26}$ kg.

1.6606×10^{-27} is $0.001 / N_A$.

Wavenumber (cm^{-1})

Wavenumber is a unit used to specify the frequency of an electromagnetic wave e.g. of light. At first sight it is rather a curious unit: the wavenumber is the *reciprocal* of the *wavelength in cm*

$$\text{wavenumber} = \frac{1}{\text{wavelength in cm}};$$

the units of wavenumber are thus cm^{-1} (usually spoken as 'wavenumbers' or 'centimetres to the minus one').

Recall that the wavelength λ and frequency v of an electromagnetic wave are related by $c = \lambda v$, where c is the speed of light. It therefore follows that

$$\frac{1}{\lambda} = \frac{v}{c},$$

which tells us that $1/\lambda$ is proportional to the frequency v, with the constant of proportionality being $1/c$. Since the wavenumber is the reciprocal of wavelength, it too is proportional to the frequency.

To convert the wavenumber to frequency in Hz (or s^{-1}), $c = \lambda v$ is rearranged to $v = c/\lambda$. Recalling that wavenumber in $1/\lambda$, it follows that

$$\text{frequency in Hz or } s^{-1} = \text{wavenumber} \times \text{speed on light in cm } s^{-1}.$$

It is very important to note that in this conversion the speed of light must be in cm s^{-1} and *not* in m s^{-1}; $\tilde{c} = 2.998 \times 10^{10}$ cm s^{-1}.

It is usual to denote a quantity in wavenumbers by adding a tilde ˜ over the symbol e.g. $\tilde{\omega}$. To avoid confusion, it is also helpful to distinguish the speed of light in cm s^{-1} by also adding a tilde: $\tilde{c}$. Using this nomenclature, the conversion to Hz can be written

$$v = \tilde{\omega} \times \tilde{c},$$

where v is in Hz, $\tilde{\omega}$ in cm^{-1}, and $\tilde{c}$ in cm s^{-1}.

Wavenumbers are often used to specify the frequency of absorptions in IR spectra, which correspond to molecular vibrations. For example, a typical absorption due to the vibration of the C–O carbonyl bond in a ketone appears at 1715 cm^{-1}. This corresponds to a frequency of $1715 \times 2.998 \times 10^{10} = 5.14 \times 10^{13}$ Hz; note the use of the speed of light in cm s^{-1}.

Sometimes we will want to convert the frequency in wavenumbers to the corresponding energy. Recall that a photon of frequency v has energy $E = hv$,

---

Self Test 20.7 Conversion to SI

Convert the following to SI and, if appropriate, also give your answer using a suitable prefix:

(a) The ionic radius of Cs^+, 1.68 Å.

(b) The activation energy for $EtONa + MeI \rightarrow EtOMe + NaI$, 19.5 kcal mol^{-1}.

(c) A pressure of 150 τ.

(d) A pressure of 0.15 atm.

(e) A pressure of 0.15 bar.

(f) The IR absorption due to the CO, 2100 cm^{-1} (convert to Hz and then to J).

(g) The ionization energy of He^+, 54.44 eV (convert to J and kJ mol^{-1}).

---

where h is Planck's constant; $h = 6.626 \times 10^{-34}$ J s^{-1}. Given that $\upsilon = \tilde{\omega}\tilde{c}$, it follows that

$$E = h \times \tilde{\omega} \times \tilde{c}.$$

For example a photon with wavenumber 1715 has energy

$$E = 6.626 \times 10^{-34} \times 1715 \times 2.998 \times 10^{10} = 3.41 \times 10^{-20} \text{ J}.$$

Since the energy is directly proportional to the wavenumber, it is not uncommon to refer to the 'energy of wavenumbers'. Strictly speaking this is incorrect as the dimensions of wavenumber are L^{-1}, whereas those of energy are $M L^2 T^{-2}$. Nevertheless, since the wavenumber is proportional to the energy, this loose usage is acceptable.

You can practice converting to SI units using the problems in Self Test 20.7.

20.3 Trigonometric functions

The trigonometric functions (principally sine and cosine) appear frequently in the wavefunctions which are solutions to the Schrödinger equation. A particularly important feature of these solutions is their periodic behaviour which derives from the trigonometric functions.

20.3.1 Measuring angles

An angle can be measured in degrees, where a complete rotation (i.e. going all the way round a circle) corresponds to 360°. Although this is the measure of angle which we usually encounter first, for more advanced applications it is neither the most convenient nor the most natural measure to adopt. Rather, angles are best specified in *radians* (symbol rad).

If we imagine a sector cut from a circle, as shown in Fig. 20.1, then the angle in radians is the ratio of the length of the arc to the radius. Since the circumference of a circle is $2\pi r$, it follows that 360° (a full rotation) is 2π radians.

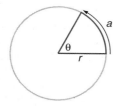

Fig. 20.1 The size of an angle in radians is given by the ratio of the length of the arc a to the radius r.

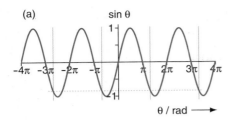

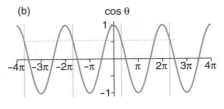

θ / rad ⟶

Fig. 20.2 Plots of (a) sin θ and (b) cos θ. The pale green lines are separated by 2π and demonstrate that the functions repeat (i.e. have the same value) each 2π rad.

Thus a quarter rotation, 90°, is π/2 rad, and a half rotation, 180°, is π rad. The conversion between radians and degrees is simply a scaling

$$\text{angle in radians} = \frac{2\pi}{360} \times \text{angle in degrees}$$

$$\text{angle in degrees} = \frac{360}{2\pi} \times \text{angle in radians}.$$

Strictly speaking, an angle in radians is a dimensionless quantity, since it is the ratio of two lengths. However, in order to keep track of what is going on it is common to give the unit of an angle as rad.

20.3.2 Sine and cosine

The sine and cosine of an angle θ can be defined in terms of the sides of a right-angle triangle, as shown in Fig. 20.3. Figure 20.2 shows a plot of sin θ and cos θ for θ in the range −4π to +4π rad. Both functions vary between −1 and +1 and are said to be *periodic* with period 2π. This means that if you select any angle θ and then add 2π, to give an angle (θ + 2π), the functions have exactly the same value at these two angles. Expressed mathematically this periodic behaviour is

$$\sin(\theta + 2\pi) = \sin(\theta) \qquad \cos(\theta + 2\pi) = \cos(\theta).$$

In fact these functions repeat when any *whole* multiple of 2π is either added to or subtracted from the angle:

$$\sin(\theta + 2n\pi) = \sin(\theta) \qquad \cos(\theta + 2n\pi) = \cos(\theta) \qquad n = 0, \pm1, \pm2, \dots.$$

The pale green lines in Fig. 20.2 are separated by 2π radians; note how at each point where these lines cross the curve, the function has the *same* value, indicated by the dashed line. This periodic behaviour is the most important feature of these trigonometric functions.

The functions have some other important properties which we will describe for angles in the range 0 to 2π. Any feature we mention will of course repeat indefinitely every 2π on account of the periodic nature of the function.

Three features are of particular interest: these are the angles at which the function is zero, the location of any maxima and the location of any minima; these are illustrated in Fig. 20.4. In the case of sin θ, there are zero crossings at angles of 0 and π (180°); the zero crossing at 2π (360°) is just the periodic repeat of the one at 0. In contrast, for cos θ there are zero crossings at π/2 (90°) and 3π/2 (270°).

When it comes to minima and maxima, sin θ has a maximum of +1 at θ = π/2, and a minimum of −1 at θ = 3π/2. For cos θ, the maximum of +1 is at θ = 0 and the minimum of −1 is at θ = π.

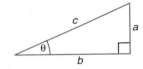

Fig. 20.3 In a right-angle triangle, sin θ is defined as a/c and cos θ as b/c.

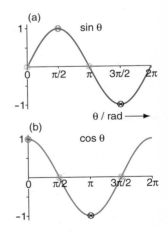

Fig. 20.4 Plots of (a) sin θ and (b) cos θ showing the location of the maxima (green ⊗), minima (black ⊗) and zero-crossings (orange ⊗) in the range 0 to 2π. The value of the function at θ = 2π is the same as that at θ = 0, therefore any extrema or zero crossing at this point is simply a repeat of that at θ = 0 and so is not marked.

It is evident from these descriptions, and from the graphs shown in Fig. 20.2 on the previous page, that the sine function is just the cosine function shifted by $\pi/2$ to the right i.e. $\sin(\theta) = \cos(\theta - \pi/2)$. Alternatively, the cosine function can be thought of as the sine function shifted by $\pi/2$ to the left i.e. $\cos(\theta) = \sin(\theta + \pi/2)$. Shifting a function in this way is called a *phase shift*. So, $\sin\theta$ can be described as $\cos\theta$ shifted in phase by $-\pi/2$.

The final property of these two functions we should note is their symmetry about $\theta = 0$. From Fig. 20.2 we can see that the value of $\cos\theta$ is the same at $+\theta$ and $-\theta$; in other words the function is *symmetric* about $\theta = 0$. Such a function that has the property $f(-\theta) \equiv f(\theta)$ is said to be *even*.

In contrast, the sine function is *antisymmetric* about $\theta = 0$, meaning that its value at angle $-\theta$ is *minus* its value at $+\theta$. A function that has the property $f(-\theta) \equiv -f(\theta)$ is said to be *odd*. On account of the periodicity of these functions, these symmetry properties we have discussed in reference to $\theta = 0$ reoccur at $\theta = 2n\pi$ ($n = 0, \pm1, \pm2, \ldots$). Furthermore, because of the $\pi/2$ phase shift which relates sine and cosine, $\sin\theta$ is symmetric (even) about the angle $\pi/2$, and $\cos\theta$ is antisymmetric (odd) about the same angle.

20.3.3 Multiple angles

Figure 20.5 shows plots of $\sin(\theta)$, $\sin(2\theta)$ and $\sin(3\theta)$, plotted in the range $\theta = 0$ to 2π. We can describe the differences between these plots by noting that, in the range $\theta = 0$ to 2π, $\sin(\theta)$ goes through *one* complete sine wave, $\sin(2\theta)$ goes through *two* complete sine waves, and $\sin(3\theta)$ goes through *three*. By 'complete sine wave' we mean $\sin(\theta)$ in the range $\theta = 0$ to 2π. In a similar way $\cos(\theta)$, $\cos(2\theta)$ and $\cos(3\theta)$, when plotted in the range $\theta = 0$ to 2π, can be described as one, two and three complete cosine waves.

We described $\sin(\theta)$ as having a period of 2π in the sense that the function repeats every whole multiple of 2π. The function $\sin(2\theta)$ has a period of π, since each time θ increases (or decreases) by a multiple of π, 2θ changes by 2π and so the function repeats. Similarly, $\sin(3\theta)$ has a period of $2\pi/3$. The general result is that $\sin(M\theta)$ has a period of $2\pi/M$; the same is true for $\cos(M\theta)$.

The periodicity of these multiple angle functions can be thought of in terms of the occurrence of the zero crossings. As we described above, $\sin(\theta)$ has zero crossings at $\theta = 0$, $\theta = \pi$, $\theta = 2\pi$ and $\theta = 3\pi$. It therefore follows that $\sin(2\theta)$ has zero crossings at $2\theta = 0$, $2\theta = \pi$, $2\theta = 2\pi$ and $2\theta = 3\pi$ i.e. at $\theta = 0$, $\theta = \pi/2$, $\theta = \pi$ and $\theta = 3\pi/2$.

In other words, whereas the zero crossings are spaced by π in $\sin(\theta)$, they are spaced by $\pi/2$ in $\sin(2\theta)$. Similarly the zero crossings in $\sin(3\theta)$ are spaced by $\pi/3$.

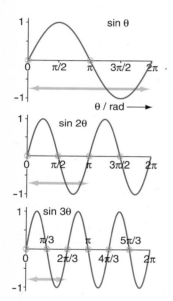

Fig. 20.5 Plots of $\sin(\theta)$, $\sin(2\theta)$ and $\sin(3\theta)$ in the range $\theta = 0 \to 2\pi$. For $\sin(\theta)$ the period, indicated by the green arrow, is 2π, for $\sin(2\theta)$ the period is π, and for $\sin(3\theta)$ the period is $2\pi/3$. The zero crossings are indicated by $\otimes$; these are spaced by π, $\pi/2$ and $\pi/3$ for $\sin(\theta)$, $\sin(2\theta)$ and $\sin(3\theta)$, respectively.

20.3.4 Trigonometric identities

Products of sine and cosine functions, and powers of these functions, quite often arise in quantum mechanical calculations. There are a set of identities which can be useful for manipulating such products and powers, and we simply summarize these here, without proof. The first set relate to products of sines and cosines:

$$\cos A \cos B \equiv \tfrac{1}{2}[\cos(A+B) + \cos(A-B)]$$
$$\sin A \sin B \equiv \tfrac{1}{2}[\cos(A-B) - \cos(A+B)]$$
$$\sin A \cos B \equiv \tfrac{1}{2}[\sin(A+B) + \sin(A-B)]$$
$$\cos A \sin B \equiv \tfrac{1}{2}[\sin(A+B) - \sin(A-B)].$$

The fourth of these is the same as the third with A and B swapped, but it is included for convenience. These identities can be manipulated to bring $\sin(A \pm B)$ or $\cos(A \pm B)$ to the left-hand side, giving the identities

$$\sin(A \pm B) \equiv \sin A \cos B \pm \cos A \sin B$$
$$\cos(A \pm B) \equiv \cos A \cos B \mp \sin A \sin B.$$

Of special interest is the case when $A = B$ for which the above identities become (for the positive sign)

$$\sin(2A) \equiv 2\sin A \cos A$$
$$\cos(2A) \equiv \cos^2 A - \sin^2 A.$$

From the definition of sine and cosine it follows that (see Fig. 20.6)

$$\cos^2 A + \sin^2 A \equiv 1,$$

which can be rearranged to

$$\cos^2 A \equiv 1 - \sin^2 A \quad \text{and} \quad \sin^2 A \equiv 1 - \cos^2 A.$$

These can be used to rewrite $\cos(2A) \equiv \cos^2 A - \sin^2 A$ in two different ways

$$\cos(2A) \equiv 2\cos^2 A - 1 \quad \text{or} \quad \cos(2A) \equiv 1 - 2\sin^2 A.$$

These identities can be used to express $\sin^2 A$ and $\cos^2 A$ in terms of $\cos(2A)$:

$$\cos^2 A \equiv \tfrac{1}{2}[1 + \cos(2A)] \quad \text{and} \quad \sin^2 A \equiv \tfrac{1}{2}[1 - \cos(2A)].$$

These identities will be particularly useful to us in our quantum mechanical calculations.

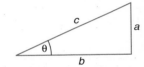

Fig. 20.6 In a right-angle triangle, $a^2 + b^2 = c^2$. From the definitions of $\sin\theta$ and $\cos\theta$ it follows that $\sin\theta = a/c$ and $\cos\theta = b/c$. Thus $\cos^2\theta + \sin^2\theta = b^2/c^2 + a^2/c^2 = (b^2 + a^2)/c^2 = 1$.

20.3.5 Definitions and units

The function $\sin\theta$ can be defined in terms of the (infinite) power series

$$\sin\theta = \theta - \frac{\theta^3}{3!} + \frac{\theta^5}{5!} - \frac{\theta^7}{7!} + \cdots,$$

where $N!$ (N factorial) is defined as $N \times (N-1) \times (N-2)\ldots \times 1$. Dimensional analysis tells us that each term on the right must have the same dimensions, and as each involves ever higher powers of θ this can only be achieved if θ is dimensionless. As a result, $\sin\theta$ is *dimensionless*. Recall from section 20.3.1 on page 890 that an angle expressed in radians is dimensionless, so taking the sine of an angle in radians is dimensionally consistent.

Self Test 20.8 Trigonometric functions

(a) Convert angles of 45° and 60° to radians.

(b) Convert angles of $2\pi/3$ rad and $5\pi/4$ rad to degrees.

(c) Sketch $\sin(4\theta)$ in the range $\theta = 0$ to 2π; what is the period (in radians, expressed as a fraction of π)?

(d) Identify the angles (in radians, expressed in terms of π) at which zero crossings occur in $\sin(4\theta)$ in the range $0 \to 2\pi$.

(e) Which cosine function will have a period of $2\pi/3$ radians?

(f) Which cosine function will have zero crossings at $\pi/4$, $3\pi/4$, $5\pi/4$ and $7\pi/4$ radians?

Similarly, $\cos\theta$ can be defined in terms of the power series

$$\cos\theta = 1 - \frac{\theta^2}{2!} + \frac{\theta^4}{4!} - \frac{\theta^6}{6!} - \cdots.$$

As before, each term must be dimensionless, making $\cos\theta$ dimensionless.

Self Test 20.8 will give you some practice with these various aspects of trigonometric functions.

20.3.6 Angular frequency

The speed at which an object rotates can be described by specifying the *period* or *frequency* of the rotation. The period T_r is the time that it takes for a complete rotation (360°): this has dimensions T, and its SI unit will be seconds. The frequency v is the reciprocal of the period

$$\text{frequency} = \frac{1}{\text{period}} \qquad v = \frac{1}{T_r}.$$

Frequency has dimensions T^{-1}, and SI unit of s^{-1} or Hz.

The angle through which an object has rotated is most naturally measured in radians. This leads to the idea of specifying the rate of rotation using the *angular frequency* ω, which is the rate of change of angle (in radians) with time i.e. (change in angle)/(interval of time). Although an angle measured in radians is dimensionless, it is common to use units of rad s^{-1} for angular frequency so as to make it clear that it is different from the frequency v, which is in s^{-1}. Since a complete revolution is 2π radians, the relationship between angular frequency and frequency is

$$\omega = 2\pi \times v.$$

For example, a frequency of 100 Hz corresponds to an angular frequency of 628 rad s^{-1}.

Consider a point moving round a circular path of radius r and at (constant) angular frequency ω, as shown in Fig. 20.7. As a result of the way angular frequency is defined, after a time t the point rotates through an angle θ given by

$$\theta = \omega \times t.$$

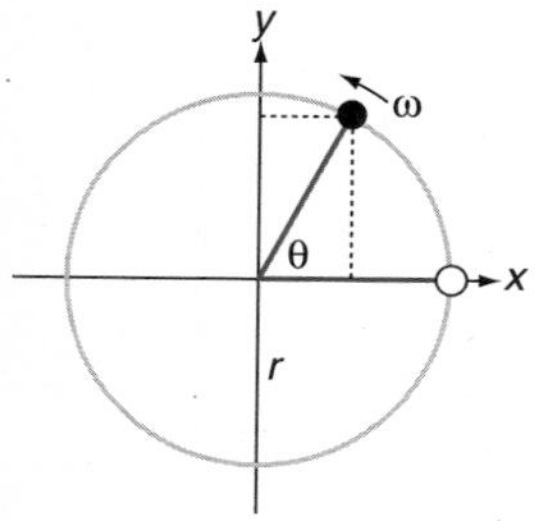

Fig. 20.7 A point moving round a circular path of radius r with angular velocity ω rotates through an angle $\theta = \omega t$ after time t. If the point starts on the x-axis at time $t = 0$, then after time t the x- and y-components (indicated by the dashed lines) are $x(t) = r\cos(\omega t)$ and $y(t) = r\sin(\omega t)$.

If the point starts on the x-axis, then simple geometry tells us that the x coordinate of the point is $r\cos\theta$, and the y component is $r\sin\theta$. The time dependence of these two components can therefore be written

$$x(t) = r\cos(\omega t) \qquad y(t) = r\sin(\omega t).$$

We therefore see that a rotation is conveniently described using trigonometric functions and an angular frequency.

In these expressions for $x(t)$ and $y(t)$, ω is in rad s^{-1} and t is in s. The product ωt is therefore in radians, so the sine or cosine can be computed without further ado. If we put the frequency in Hz (s^{-1}), υt will *not* be in radians, and it will therefore not be permissible to take its sine or cosine. Rather, we would first have to convert υ to an angular frequency by multiplying by 2π; then the sine or cosine can be computed e.g. $x(t) = r\cos(2\pi\upsilon t)$.

20.4 The exponential function

The exponential function, written e^{Ax} or $\exp(Ax)$ (A is a constant, x is the variable), appears very frequently in the analysis of all sorts of physical and chemical problems. In chemistry it occurs in the wavefunctions for the hydrogen atom (the atomic orbitals), in the Boltzmann distribution, in the relationship between the equilibrium constant and the standard Gibbs energy change, in the Arrhenius expression for the temperature dependence of the rate constant, and in the time dependence of the concentration in a first-order reaction – the exponential function is truly ubiquitous.

In this section we will look at the important properties of the exponential function, and the related natural logarithm ($\ln(x)$). Box 20.1 on the following page revises the properties of powers, which are relevant to this discussion of the exponential function.

20.4.1 Introducing the exponential function

The exponential of x is the number e raised to the power of x: e^x. The number e is the limit of the (infinite) power series

$$e = 1 + \frac{1}{2!} + \frac{1}{3!} + \frac{1}{4!} + \frac{1}{5!} + \dots .$$

The decimal value of e is $2.71828\dots$; e is sometimes referred to as the base of natural logarithms.

The e^x is defined in terms of the power series

$$e^x = 1 + \frac{x}{1!} + \frac{x^2}{2!} + \frac{x^3}{3!} + \frac{x^4}{4!} + \dots . \tag{20.6}$$

The most important property of e^x is that its derivative is itself $d(e^x)/dx = e^x$, and it is for this reason that this function appears in the solution to so many mathematical problems.

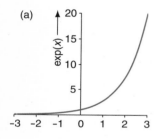

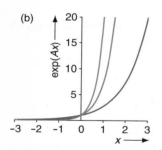

Fig. 20.8 Graph (a) is a plot of exp(x): for positive x the function increases indefinitely, whereas for negative x the function decreases towards zero, only reaching zero when $x = -\infty$. Graph (b) is a plot of exp(Ax) for A = 1 (blue), A = 2 (green) and A = 3 (red). As A increases the function climbs more steeply for positive x, and decays more rapidly for negative x. Regardless of the value of A, exp(Ax) = 1 when x = 0.

Box 20.1 Powers

Raising a number A to an integer power simply implies repeated multiplication

$$A^n = \underbrace{A \times A \times A \dots}_{n \text{ times}} \qquad \text{for integer } n.$$

However, the properties of A^n we are about to discuss apply for *any* value of n i.e. they are not restricted to integers. These properties are

$$A^{-n} \equiv \frac{1}{A^n} \qquad \frac{1}{A^{-n}} \equiv A^n$$

$$A^n \times A^m \equiv A^{n+m} \qquad \frac{A^n}{A^m} \equiv A^{n-m}$$

$$\sqrt{A^n} \equiv A^{n/2} \qquad \sqrt[p]{A^n} \equiv A^{n/p}$$

$$(A^n)^p \equiv A^{pn} \qquad A^0 \equiv 1$$

Some of these are redundant – for example the first property and the one to its right are the same – but they are included for convenience. Self Test 20.9 will give you some practice in using these properties.

Self Test 20.9 Manipulating powers

Simplify the following:

(a) $A^2 \times A^3 \times A^{-1}$ and $\dfrac{A^2 \times A^3}{A^2}$

(b) $\dfrac{A^2 \times A^{-3}}{A^{-2}}$ and $A^n \times A^{-n}$

(c) $\sqrt{A^5}$ and $\left(A^6\right)^{\frac{1}{3}}$

Graph of exp(x) and exp(Ax)

Figure 20.8 (a) shows a graph of e^x for both positive and negative values of x. For positive x the function is always positive and climbs steeply without limit. The slope is always positive, and increases as x increases. When $x = 0$, e^x takes the value 1. For negative values of x the function decays towards zero as x becomes more negative, only reaching 0 in the limit that $x = -\infty$. The slope is still positive, and decreases as x becomes more negative.

As is shown in Fig. 20.8 (b), exp(Ax) (A is a positive constant) behaves similarly except that the larger A becomes, the more rapidly the function rises for positive x, and the more rapidly it decays for negative x. Regardless of the value of A, exp(A × 0) = 1.

Although the rising part of the exponential function (the part for positive x) does occur as the solution to some physical problems, for chemistry it is the decaying part of the exponential (i.e. for negative x) which is of the greatest interest. Therefore from now on we will consider the function exp(−Ax) for *positive* values of x and for positive A i.e. only the decaying part, as illustrated in Fig. 20.9. To make things a little more general we will include a further constant B which scales the function: $B \exp(-Ax)$.

The form of the exponential decay

One of the very special features of this decaying exponential is that for a given increase in x, the function decreases by a particular fraction *regardless of the initial value of x*. This property is most commonly encountered in the form of the half life, but we will demonstrate it here for a more general case.

At some arbitrary value of x, the function has the value $B \exp(-Ax)$. We seek the value of x, which we will denote x', at which the function has fallen to a fraction α of its original value i.e.

$$B \exp(-Ax') = \alpha \times B \exp(-Ax).$$

The constant B cancels, and taking $\exp(-Ax)$ over to the left gives

$$\frac{\exp(-Ax')}{\exp(-Ax)} = \alpha.$$

Using the usual rules for how powers behave on division (Box 20.1 on the preceding page) we have

$$\exp(-A[x' - x]) = \alpha.$$

This can be simplified by noting that $[x' - x]$ is the increase in the value of x when the function falls to a fraction α of its original value; if we let this increase in x be δ, then

$$\exp(-A\delta) = \alpha.$$

What this tells us is that the increase in x required for the function to fall to a fraction α of its initial value depends *only* on the constant A, and is *independent of the initial value of x* we choose; this property is illustrated in Fig. 20.10. If we take the natural logarithm of both sides we have

$$-A\delta = \ln\alpha,$$

so it follows that $\delta = -(\ln\alpha)/A$. Using this we can compute the increase in x needed for a given fractional fall in the value of the function.

20.4.2 Natural logarithms

The natural logarithm of x, written $\ln(x)$, is defined in the following way

$$\text{if } x = e^a \quad \text{then} \quad \ln(x) = a. \tag{20.7}$$

In words $\ln(x)$ is the number to which e must be raised in order to give the value x. The natural logarithm is sometimes called the logarithm to the base e since it involves raising e to a power.

Since $e^0 = 1$, it follows that $\ln(1) = 0$. Also, since e^a is *always* positive, $\ln(x)$ is only defined for *positive* values of x.

Sometimes $\ln(x)$ is written without the bracket i.e. $\ln x$. Either way of writing the function are acceptable, although sometimes it is useful to include the bracket to avoid ambiguity.

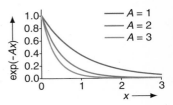

Fig. 20.9 Plot of $\exp(-Ax)$ for positive x i.e. the decaying part of the exponential function. As A increases, the decay becomes more rapid. Regardless of the value of A, $\exp(-Ax) = 1$ when $x = 0$.

> Logarithms are considered in the next section.

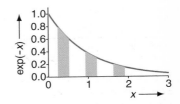

Fig. 20.10 A special property of $\exp(-Ax)$ is that the increase in x needed for the function to decay to a given fraction of its initial value is *independent* of the initial value chosen. In the plot, each pale green block shows the increase in x needed for the function to decay to $\frac{3}{4}$ of its initial value; these blocks are all of the same width.

Manipulations with logarithms

Suppose that

$$y = e^a \quad \text{and} \quad z = e^b,$$

which means that $\ln(y) = a$ and $\ln(z) = b$. The product $y \times z$ can be computed as

$$\begin{aligned} y \times z &= e^a \times e^b \\ &= e^{a+b} \end{aligned}$$

From the last line, and the definition of the logarithm (Eq. 20.7 on the previous page), it follows that

$$\ln(y \times z) = a + b.$$

Recalling that $\ln(y) = a$ and $\ln(z) = b$, this equation can be written

$$\ln(y \times z) = \ln(y) + \ln(z).$$

This is an important practical property of logarithms: the *logarithm of a product* of two numbers is equal to the *sum of the logarithms* of the two numbers. This property is independent of the base used. It is easy to show using a similar line of argument that

$$\ln\left(\frac{y}{z}\right) = \ln(y) - \ln(z).$$

If we set $y = 1$ and recall that $\ln(1) = 0$ it follows that

$$\ln\left(\frac{1}{z}\right) = -\ln(z).$$

Logarithms are also useful for dealing with powers. The argument proceeds along the following lines

$$\text{if} \quad x = e^a \quad \text{then it follows that} \quad x^n = e^{n \times a}.$$

From the definition of logarithms, Eq. 20.7, the previous line implies

$$\ln(x) = a \quad \text{and} \quad \ln(x^n) = (n \times a).$$

It therefore follows that

$$\ln(x^n) = n \times \ln(x);$$

this applies to all values of n i.e. positive, negative and non-integer.

If we want to go from the logarithm back to the original number, then all we do is to raise e to the logarithm

$$e^{\ln(x)} = x.$$

This follows simply from the definition of a logarithm, Eq. 20.7. If $x = \exp(a)$, then by definition $\ln(x) = a$; substituting this value for a into $x = \exp(a)$ gives $x = \exp(\ln(x))$.

Table 20.7 summarizes the useful properties of logarithms.

Table 20.7 Properties of logarithms

$$\ln(x \times y) = \ln(x) + \ln(y)$$

$$\ln\left(\frac{x}{y}\right) = \ln(x) - \ln(y)$$

$$\ln\left(\frac{1}{x}\right) = -\ln(x)$$

$$\ln(x^n) = n\ln(x)$$

$$\exp(\ln(x)) = x$$

20.4.3 Logarithms to the base 10

The other commonly encountered type of logarithm are those to the base 10, denoted $\lg(x)$. These are defined in an analogous way to natural logarithms (Eq. 20.7 on page 897) but the logarithm to the base 10 of a number is the power to which 10 has to be raised to equal that number

$$\text{if } x = 10^a \quad \text{then} \quad \lg(x) = a.$$

Sometimes the base is indicated explicitly, so natural logarithms can be denoted $\log_e(x)$ and logarithms to the base 10 can be denoted $\log_{10}(x)$. All of the properties of natural logarithms listed in Table 20.7 on the facing page apply to logarithms to the base 10, except for the last line which becomes $10^{\lg(x)} = x$.

The two types of logarithms are related as follows

$$
\begin{aligned}
\ln(x) &= \ln(10) \times \lg(x) \\
&= 2.303 \times \lg(x).
\end{aligned}
\tag{20.8}
$$

20.4.4 Dimensions

If you look back at the definition of e^x, Eq. 20.6 on page 895, and recall that in such an equation all of the terms which are added together must have the same dimensions, it is clear that x *must be dimensionless*. Only by satisfying this condition can x^n have the same dimensions for all values of n. It therefore follows that e^x is *dimensionless*.

Furthermore, the definition of a logarithm (Eq. 20.7)

$$\text{if } x = e^a \quad \text{then} \quad \ln(x) = a,$$

along with the fact that a and x are dimensionless, implies that $\ln(x)$ is also dimensionless. In other words, if we are to be dimensionally consistent we can only take the logarithm or exponential of a dimensionless quantity.

Let us look at two examples to see how this works out. On page 879 we argued that in the Arrhenius equation

$$k = A \exp\left(\frac{-E_a}{RT}\right),$$

the fraction $E_a/(RT)$ is dimensionless. It it thus dimensionally acceptable to take the exponential of this quantity.

As a second example consider a first-order process for which we showed (section 18.2.1 on page 806) that the concentration of reactant A, $[A](t)$, varies with time according to

$$\ln\left(\frac{[A](t)}{[A]_0}\right) = -k_{1st}t,\tag{20.9}$$

where t is time, k_{1st} is the rate constant, and $[A]_0$ is the concentration at time t. The ratio $[A](t)/[A]_0$ is dimensionless, so we are taking the logarithm of a dimensionless quantity, as required.

Using the properties of logarithms we can rewrite Eq. 20.9 as

$$\ln([A](t)) - \ln([A]_0) = -k_{1st}t.\tag{20.10}$$

It now appears that we are taking the logarithm of a quantity which has dimensions, specifically in this case the dimensions of concentration. However,

---

Self Test 20.10 Exponentials and logarithms

Simplify the following:

(a) Sketch $2e^{-x}$ in the range $x = 0$ to $x = +5$; on the same graph, also sketch the function $2e^{-2x}$.

(b) Consider the exponential decay $A_0 e^{-kt}$, where t is the time. In what interval of time will the function drop to one half of its original value?

(c) Express the following as a sum of logarithms: (i) $\ln(a \times x)$, (ii) $\ln(a \times x^2)$, (iii) $\ln(b \times \sqrt{y})$, (iv) $\ln(b/z^3)$.

(d) Is the expression $\exp(-k_{1st}t)$, where k_{1st} has units s^{-1} and t has units s, dimensionally consistent?

---

because Eq. 20.10 on the preceding page is in fact an identity with Eq. 20.9, we can be sure that there is nothing dimensionally inconsistent about Eq. 20.10 on the previous page. You will often find that although it appears that you are about to take the logarithm or exponential of a dimensioned quantity, some rearrangement will ensure that the exponential or logarithm is of a dimensionless quantity.

Self Test 20.10 gives you some practice in handling exponential and logarithms.

20.4.5 Gaussian function

The gaussian function,

$$g(x) = \exp(-Ax^2),$$

where A is a *positive* constant, occurs in solutions to the Schrödinger equation and in the mathematical analysis of statistical problems. Since x appears as the *square*, the function is symmetric about $x = 0$ i.e. $g(-x) = g(x)$, which is in contrast to the function $\exp(Ax)$. When $x = 0$ the function takes the value one, and it tails away to zero as x becomes large and positive or large and negative. The function is plotted in Fig. 20.11 for a series of values of A; as A increases, the decay becomes more rapid.

A slightly more general version of the gaussian is

$$g(x) = \exp(-A[x - x_0]^2),$$

the centre of which is located at $x = x_0$ rather than $x = 0$.

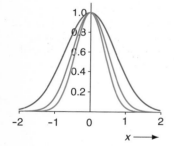

Fig. 20.11 Plots of the gaussian function $\exp(-Ax^2)$ for $A = 1$ (blue), $A = 2$ (green), and $A = 3$ (red). In contrast to the exponential, the gaussian is symmetric about $x = 0$.

20.5 Calculus: differentiation

Calculus is an exceptionally important tool for the mathematical analysis of phenomena in all parts of science, and in developing mathematical models of such phenomena. The chapters concerned with thermodynamics, chemical kinetics and quantum mechanics make extensive use of calculus. In this section we will simply state the key results and methods which will be used in this book; no proofs will be given.

Table 20.8 Derivatives of some commonly encountered functions

$f(x)$	$\mathrm{d}f(x)/\mathrm{d}x$	comments
A	0	A is a constant throughout this table
x^n	nx^{n-1}	n can be positive or negative
$\sin(Ax)$	$A\cos(Ax)$	
$\cos(Ax)$	$-A\sin(Ax)$	
$\exp(Ax)$	$A\exp(Ax)$	
$\ln(x)$	$1/x$	
$\ln(Ax)$	$1/x$	recall that $\ln(Ax) = \ln A + \ln(x)$

Generally speaking throughout this section we will use x as the variable and it can also be assumed that all *upper case* symbols represent constants i.e. values which are not dependent on the variables. It is important to realize that changing the variable from x to y or θ or p does not affect the result: x is just an *example* of a variable.

20.5.1 Introducing differentiation

Differentiating a function $f(x)$ with respect to x gives a function $g(x)$; the process is written

$$g(x) = \frac{\mathrm{d}f(x)}{\mathrm{d}x}.$$

The function $g(x)$ is called the *derivative of $f(x)$ with respect to x*. As shown in Fig. 20.12, the derivative is interpreted as the *slope* of a graph of $f(x)$ against x. Therefore, $g(x_1)$ is the slope at $x = x_1$, likewise $g(x_2)$ is the slope at $x = x_2$.

Figure 20.13 illustrates that at a maximum or a minimum in the function $f(x)$, the slope, and hence the derivative, is zero

$$\text{at maximum or minimum} \qquad \frac{\mathrm{d}f(x)}{\mathrm{d}x} = 0.$$

Although there are systematic methods for determining whether a particular value of x at which the derivative is zero is a maximum or a minimum, in physical problems it is usually not necessary to use these since it will be obvious from the context whether we have found a maximum or a minimum.

Table 20.8 lists the derivatives of some commonly encountered functions. In using this table it is important to recall that

$$\frac{\mathrm{d}(B \times f(x))}{\mathrm{d}x} = B \times \frac{\mathrm{d}f(x)}{\mathrm{d}x} \qquad \text{where } B \text{ is a constant.}$$

In words, multiplying a function by a constant simply results in the derivative being multiplied by the same constant.

An alternative way of writing the derivative is

$$\frac{\mathrm{d}}{\mathrm{d}x} f(x);$$

this is often used if $f(x)$ is inconveniently large to write alongside the 'd' on the top of the fraction.

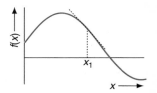

Fig. 20.12 The derivative of a function $f(x)$ with respect to x gives the slope of a graph of $f(x)$ against x. For example, at x_1, the slope is $\mathrm{d}f(x)/\mathrm{d}x$, evaluated at $x = x_1$.

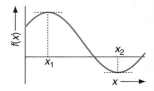

Fig. 20.13 At a maximum or minimum in a function the slope, and hence the derivative, is zero e.g. in this case $\mathrm{d}f(x)/\mathrm{d}x$, evaluated at $x = x_1$ and at $x = x_2$, is zero.

Self Test 20.11 Derivatives: 1

In the following problems all *upper case* symbols are constants.

(a) Find the derivative with respect to x of $3x^2 + Ax^{-1} + \dfrac{2}{x^2}$.

(b) Find the derivative with respect to θ of $A\cos(2\theta)$.

(c) Compute

$$\frac{\mathrm{d}}{\mathrm{d}p}\left(\frac{pV}{RT}\right).$$

(d) Find the derivative with respect to t of (i) $B\ln(At)$ and (ii) $B\ln(t)$.

(e) Find the derivative with respect to r of (i) $B\exp(-Ar)$ and (ii) $B\exp(Ar)$.

(f) Find the *second* derivative with respect to θ of $A\cos(2\theta)$.

20.5.2 Second derivative

The second derivative with respect to x of a function is denoted as

$$\frac{\mathrm{d}^2 f(x)}{\mathrm{d}x^2} \quad \text{or} \quad \frac{\mathrm{d}^2}{\mathrm{d}x^2} f(x),$$

and is found by taking the derivative twice in succession

$$\frac{\mathrm{d}^2 f(x)}{\mathrm{d}x^2} = \frac{\mathrm{d}}{\mathrm{d}x}\frac{\mathrm{d}f(x)}{\mathrm{d}x}.$$

For example, the second derivative of $A\sin(Bx)$ is

$$
\begin{aligned}
\frac{\mathrm{d}^2}{\mathrm{d}x^2} A\sin(Bx) &= \frac{\mathrm{d}}{\mathrm{d}x}\frac{\mathrm{d}}{\mathrm{d}x} A\sin(Bx) \\
&= \frac{\mathrm{d}}{\mathrm{d}x} AB\cos(Bx) \\
&= -AB^2 \sin(Bx).
\end{aligned}
$$

Self Test 20.11 will give you some practice at computing some derivatives.

20.5.3 Chain rule

The differentiation of more complex functions, such as

$$f(x) = \exp(Ax^2)$$

can be achieved by making use of the *chain rule*. There are quite a lot of different ways of describing how this is used, so if you do not find the explanation we give here familiar, do not worry.

The key idea is to find a substitution which transforms the function we want to differentiate into one we can differentiate straightforwardly using the standard results, such as those given in Table 20.8 on the preceding page. In the case of $f(x) = \exp(Ax^2)$ the substitution $y = x^2$ achieves what we want, since this gives $f(y) = \exp(Ay)$ which is easy to differentiate

$$\frac{\mathrm{d}}{\mathrm{d}y} \exp(Ay) = A\exp(Ay).$$

Self Test 20.12 Derivatives: 2

In the following problems all *upper case* symbols are constants.

(a) Differentiate $\exp(By^2)$ with respect to y, using the substitution $x = y^2$.

(b) Differentiate $\cos(At^2)$ with respect to t, using the substitution $y = t^2$.

(c) Differentiate the following function with respect to θ, using the substitution $y = \sin(B\theta)$

$$\frac{A}{\sin(B\theta)}.$$

However, we want $\mathrm{d}f(x)/\mathrm{d}x$, not $\mathrm{d}f(y)/\mathrm{d}y$. Fortunately, there is a general relationship between these, which is

$$\frac{\mathrm{d}f(x)}{\mathrm{d}x} = \frac{\mathrm{d}f(y)}{\mathrm{d}y}\frac{\mathrm{d}y}{\mathrm{d}x}. \tag{20.11}$$

In a crude (and certainly non-mathematical) way you can imagine that on the right the 'dy' terms cancel to give $\mathrm{d}f(x)/\mathrm{d}x$.

Using Eq. 20.11 to differentiate $f(x) = \exp(Ax^2)$ with the substitution $y = x^2$ proceeds as follows

$$\frac{\mathrm{d}f(x)}{\mathrm{d}x} = \frac{\mathrm{d}f(y)}{\mathrm{d}y}\frac{\mathrm{d}y}{\mathrm{d}x}$$

$$\frac{\mathrm{d}}{\mathrm{d}x}\exp(Ax^2) = \frac{\mathrm{d}}{\mathrm{d}y}\exp(Ay) \times \frac{\mathrm{d}}{\mathrm{d}x}x^2$$

$$= A\exp(Ay) \times 2x$$

$$= A\exp(Ax^2) \times 2x.$$

To go to the last line, we have substituted back $y = x^2$.

Once you get used to it you can calculate the derivatives of these kinds of functions 'in your head' without going through all these formalities. Self Test 20.12 will give you some practice at using the chain rule.

20.5.4 Differentiation of a product or quotient of two functions

If the function we wish to differentiate is a *product* of two functions, then the *product rule* must be used to compute the derivative. Suppose that $f(x) = u(x) \times v(x)$, where $u(x)$ and $v(x)$ are both functions of x. The product rule gives the derivative of $f(x)$ with respect to x as

$$\frac{\mathrm{d}[u(x) \times v(x)]}{\mathrm{d}x} = u(x)\frac{\mathrm{d}v(x)}{\mathrm{d}x} + v(x)\frac{\mathrm{d}u(x)}{\mathrm{d}x}. \tag{20.12}$$

The *quotient rule* is used for the case $f(x) = u(x)/v(x)$

$$\frac{\mathrm{d}}{\mathrm{d}x}\frac{u(x)}{v(x)} = \frac{u(x)\frac{\mathrm{d}v(x)}{\mathrm{d}x} - v(x)\frac{\mathrm{d}u(x)}{\mathrm{d}x}}{[v(x)]^2}. \tag{20.13}$$

It is a common error to assume that $\mathrm{d}/\mathrm{d}x\,[u(x) \times v(x)] = \mathrm{d}u(x)/\mathrm{d}x \times \mathrm{d}v(x)/\mathrm{d}x$.

Self Test 20.13 Derivatives: 3

In the following problems all *upper case* symbols are constants.

(a) Differentiate $Ar \exp(-Br)$ with respect to r.

(b) Differentiate $\cos(At) \exp(-Bt)$ with respect to t.

(c) Differentiate $x \exp(-Ax^2)$ with respect to x; you should recall that $d/dx \exp(-Ax^2) = -2Ax \exp(-Ax^2)$.

As an example, let us consider the function

$$f(x) = \underbrace{Ax^2}_{u(x)} \times \underbrace{\exp(-Bx)}_{v(x)},$$

where $u(x)$ and $v(x)$ have been identified by the braces. Application of Eq. 20.12 on the preceding page gives the following

$$
\begin{aligned}
\frac{d[u(x) \times v(x)]}{dx} &= u(x)\frac{dv(x)}{dx} + v(x)\frac{du(x)}{dx} \\
&= Ax^2 \frac{d}{dx}\exp(-Bx) + \exp(-Bx)\frac{d}{dx}Ax^2 \\
&= Ax^2 \underbrace{[-B\exp(-Bx)]}_{dv(x)/dx} + \exp(-Bx)\underbrace{[2Ax]}_{du(x)/dx} \\
&= -ABx^2 \exp(-Bx) + 2Ax \exp(-Bx).
\end{aligned}
$$

It is worth noting that a quotient can always be expressed as a product using a negative power, so you can use either the quotient or the product rule, as you wish

$$\frac{d}{dx}\frac{u(x)}{v(x)} \equiv \frac{d}{dx} u(x) \times [v(x)]^{-1}.$$

Self Test 20.13 will give you some practice at using the product rule.

20.5.5 Finding minima and maxima

It was mentioned in section 20.5.1 on page 901 that the positions of minima and maxima (collectively called extrema) can be found by setting the derivative equal to zero. We will look at an example of this which will turn out to be useful in the quantum mechanical treatment of the hydrogen atom. The function whose extrema we seek is

$$f(x) = x^2 \exp(-2x). \tag{20.14}$$

Applying the rule for differentiating a product, section 20.5.4, gives

$$\frac{df(x)}{dx} = 2x \exp(-2x) - 2x^2 \exp(-2x).$$

To find the extrema we set this equal to zero

$$
\begin{aligned}
2x \exp(-2x) - 2x^2 \exp(-2x) &= 0 \\
2x(1 - x)\exp(-2x) &= 0,
\end{aligned}
$$

where to go to the second line we have realized $2x$ and $\exp(-2x)$ are factors.

This last line is zero if *any* of the three terms are zero i.e.

$$2x = 0 \qquad (1-x) = 0 \qquad \exp(-2x) = 0.$$

The first simply says that there is a maximum or minimum at $x = 0$, and the second says that there is another at $x = 1$. Recalling that $\exp(-Ax)$ goes to zero when $x = +\infty$, the third tells us that there is another at $x = +\infty$.

Our task is now to identify the type of extrema that these three values indicate. First we note that the original function, Eq. 20.14 on the facing page, is positive for all values of x since both the exponential and x^2 are always positive. We also not that at $x = 0$ the function is zero. Therefore, $x = 0$ must correspond to a minimum.

As x increases, x^2 increases, but $\exp(-2x)$ decreases. One of the properties of an exponential is that, for large enough x, it always wins out over any simple power of x. It therefore follows that as x goes to infinity, the exponential term will drive $f(x)$ down to zero. Given that: (a) there is a minimum at $x = 0$, (b) the function is always positive, and (c) the function goes to zero at $x = \infty$, then there is no choice other than for $x = 1$ to correspond to a maximum. A plot of the function is shown in Fig. 20.14.

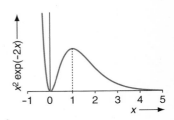

Fig. 20.14 Plot of $f(x) = x^2 \exp(-2x)$ showing the minima at $x = 0$, the maxima at $x = 1$, and the way in which the derivative tends to zero as $x \to +\infty$.

20.5.6 How dimensions are affected by differentiation

The derivative of a function $f(x)$ with respect to x can be interpreted as the slope of a graph of $f(x)$ against x. Since the slope is (change in $f(x)$)/(change in x), the dimensions of the slope must be

$$\frac{\text{dimensions of } f(x)}{\text{dimensions of } x}.$$

For example, the rate of change of concentration of a species $[i]$ is given by the derivative of $[i]$ with respect to time, $d[i]/dt$. The dimensions of concentration are NL^{-3} (i.e. amount per unit volume), and those of time are T, so the dimensions of the derivative are

$$\left[\frac{d[i]}{dt}\right] = \frac{NL^{-3}}{T} = NL^{-3}T^{-1}.$$

20.6 Calculus: integration

20.6.1 Indefinite integral

Integration is best viewed as the opposite of differentiation. So if

$$\frac{df(x)}{dx} = g(x)$$

then

$$\int g(x)\,dx = f(x) + \text{const.}$$

Recall that the evaluation of such an *indefinite* integral always results in the addition of a constant term. We always have to integrate with respect to a

Self Test 20.14 Integrals: 1

In the following problems all *upper case* symbols are constants. Compute the following indefinite integrals, checking your answers by differentiating them to confirm that the original function is regenerated:

(a) $\int Ay^3\,dy$ (b) $\int B\exp(-Cx)\,dx$

(c) $\int \sin(2\theta)\,d\theta$ (d) $\int \dfrac{1}{Br}\,dr$

(e) $\int \dfrac{C}{y^3}\,dy$ (f) $\int dz$

particular variable: here the variable of integration is x, as indicated by the dx to the right of the integral sign. Table 20.9 lists the integrals of some commonly encountered functions; to save space, the constant is omitted. You can always check an integral by differentiating it and seeing if you get back to the original function.

Any multiplying constants or terms which do not depend on the variable with respect to which the integral is being calculated can be taken outside the integral. For example, the integral of $A\cos(B\theta)$ with respect to θ (A and B are constants) is

$$\int A\cos(B\theta)\,d\theta \;=\; A\int \cos(B\theta)\,d\theta$$

$$=\; A\times\frac{1}{B}\sin(B\theta) + \text{const.}$$

One type of integral which comes up quite often in physical problems is

$$\int A\,dx,$$

where A is a constant. This is very easy to do, but sometimes its simplicity makes it hard. Referring to the first line of Table 20.9 we see that the integral evaluates to $Ax + \text{const.}$ If $A = 1$ the integral looks rather peculiar

$$\int 1\,dx,$$

but the first line of the table tells us that with $A = 1$ the integral evaluates to $x + \text{const.}$ If we omit the '1', which we can do as multiplying by 1 has no effect, the integral looks even more peculiar

$$\int dx,$$

but it is still the same integral, evaluating to $x + \text{const.}$

Self Test 20.14 will give you some practice at computing integrals.

Table 20.9 Indefinite integrals of some commonly encountered functions

$f(x)$	$\int f(x)\,dx$
A	Ax
$x^n\ (n \neq -1)$	$\dfrac{1}{n+1}x^{n+1}$
$\dfrac{1}{x}$	$\ln(x)$
$\sin(Ax)$	$-\dfrac{1}{A}\cos(Ax)$
$\cos(Ax)$	$\dfrac{1}{A}\sin(Ax)$
$\exp(Ax)$	$\dfrac{1}{A}\exp(Ax)$

20.6.2 Integral between limits (definite integral)

A *definite* integral is taken between *limits* i.e. between two values of the variable of integration: for example

$$\int_a^b g(x)\,dx = [f(x)]_a^b$$
$$= f(b) - f(a).$$

The notation $[f(x)]_a^b$ means evaluate $f(x)$ at $x = b$ and subtract from it $f(x)$ at $x = a$. In contrast to the indefinite integral, there is no constant of integration as this cancels when you compute the difference $f(b) - f(a)$.

The definite integral

$$\int_a^b g(x)\,dx$$

can be interpreted as the *area* between $x = a$ and $x = b$ under a graph of $g(x)$ against x, as illustrated in Fig. 20.15. The area is interpreted as a signed quantity: if $g(x)$ is negative between the limits, then the area (the integral) is also negative. This idea is also illustrated in Fig. 20.15.

As an example let us compute the definite integral of $\sin\theta$ with respect to θ and between limits $\theta = \theta_1$ and $\theta = \theta_2$. The steps are

$$\int_{\theta_1}^{\theta_2} \sin\theta\,d\theta = [-\cos\theta]_{\theta_1}^{\theta_2}$$
$$= -\cos\theta_2 + \cos\theta_1.$$

If $\theta_1 = 0$ and $\theta_2 = \pi$, the integral evaluates to $(-\cos\pi + 1) = 2$; if $\theta_1 = 0$ and $\theta_2 = 2\pi$, the integral evaluates to $(-\cos 2\pi + 1) = 0$. Finally if $\theta_1 = \pi$ and $\theta_2 = 2\pi$, the integral evaluates to $(-\cos 2\pi + \cos\pi) = -2$.

When the limits are $0 \rightarrow \pi$ the integral is positive, since it corresponds to the area under the first half of a sine wave, where the function is positive. In contrast, when the limits are $\pi \rightarrow 2\pi$ the integral is negative, since the corresponding area under the second half of a sine wave is negative; note that these integrals are equal and opposite. It is not surprising therefore that when the limits are $0 \rightarrow 2\pi$ the integral is zero, since these limits cover a complete sine wave, which has a positive area between 0 and π, and an equal negative area between π and 2π.

Self Test 20.15 will give you some practice at computing definite integrals.

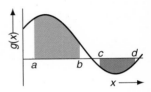

Fig. 20.15 The definite integral of $g(x)$ between limits x_1 and x_2 can be interpreted as the *area* under a graph of $g(x)$ against x between x_1 and x_2. The area is a *signed* quantity: e.g. the definite integral between a and b is positive (shaded red), whereas that between c and d is negative (shaded blue), since the function is negative in this interval.

20.6.3 Integration by parts

Unlike differentiation, there is no simple 'formula' for the integral of a *product* of two functions. However, if the function to be integrated can be written as

$$\underbrace{u(x)}_{\text{function 1}} \times \underbrace{\frac{dv(x)}{dx}}_{\text{function 2}},$$

i.e. one of the functions can be thought of as a derivative of another function, the following rule applies

$$\int_A^B u(x)\frac{dv(x)}{dx}\,dx = \left[u(x) \times v(x)\right]_A^B - \int_A^B \frac{du(x)}{dx} \times v(x)\,dx. \tag{20.15}$$

This is called *integration by parts*.

Self Test 20.15 Integrals: 2

In the following problems all *upper case* symbols are constants. Compute the following definite integrals, and illustrate them on suitable graphs.

(a) $\displaystyle\int_0^A x\,dx$

(b) $\displaystyle\int_{-A}^A x\,dx$

(c) $\displaystyle\int_0^\pi \cos(\theta)\,d\theta$

(d) $\displaystyle\int_{-\pi/2}^{\pi/2} \cos(\phi)\,d\phi$

(e) $\displaystyle\int_0^\infty \exp(-x)\,dx$ (note $e^{-\infty} = 0$)

(f) $\displaystyle\int_{z_1}^{z_2} dz$

At first sight, this does not look very helpful as one integral is expressed in terms of another. However, by a careful choice of $u(x)$ and $v(x)$ it is sometimes possible to use this rule to express the integral we want to evaluate in terms of a simpler integral which we can evaluate more easily.

It is easiest to see how this works by looking at an example

$$\int_0^{2\pi} \underbrace{x}_{u(x)}\ \underbrace{\cos x}_{dv(x)/dx}\ dx$$

If we break up the integrand as shown, then since $dv(x)/dx = \cos x$ it follows that $v(x) = \sin x$. Furthermore, since $u(x) = x$, $du(x)/dx = 1$; using these results in Eq. 20.15 on the preceding page we can complete the integration as follows

$$\int_0^{2\pi} \underbrace{x}_{u(x)}\ \underbrace{\cos x}_{dv(x)/dx}\ dx \;=\; \Big[x\sin x\Big]_0^{2\pi} - \int_0^{2\pi} 1 \times (\sin x)\,dx$$

$$= \; 0 - \Big[-\cos x\Big]_0^{2\pi}$$

$$= \; -[-1 - (-1)]$$

$$= \; 0.$$

The 'trick' with integration by parts is to choose $u(x)$ and $dv(x)/dx$ correctly, so that the integral on the right of Eq. 20.15 is simpler than the one on the left. If you make a 'bad' choice, then you will find that the problem is getting more difficult, and if this happens it is a sure sign that you should go back to the start and try an alternative tack.

Self Test 20.16 will give you some practice with integration by parts.

20.6.4 Standard integrals

There are some integrals which require special techniques to evaluate, or which are very laborious to work out (e.g. requiring repeated integration by parts). To save you effort, the results of a number of these integrals, which occur commonly in quantum mechanics, are listed below.

Self Test 20.16 Integration by parts

Compute the given integrals using integration by parts using the indicated choices for $u(x)$ and $dv(x)/dx$. For (c) you will find that the result from part (a) will be useful.

(a) $\displaystyle\int_0^{2\pi} x \sin x \, dx$ $\qquad u(x) = x$ $\qquad \dfrac{dv(x)}{dx} = \sin x$

(b) $\displaystyle\int_0^{\infty} x \exp(-x) \, dx$ $\qquad u(x) = x$ $\qquad \dfrac{dv(x)}{dx} = \exp(-x)$ $\qquad$ (note $e^{-\infty} = 0$)

(c) $\displaystyle\int_0^{2\pi} x^2 \cos x \, dx$ $\qquad u(x) = x^2$ $\qquad \dfrac{dv(x)}{dx} = \cos x$

$$\int_{-\infty}^{\infty} \exp(-Ax^2) \, dx \;=\; \sqrt{\frac{\pi}{A}} \quad (A > 0) \tag{20.16}$$

$$\int_{-\infty}^{\infty} x^2 \exp(-Ax^2) \, dx \;=\; \sqrt{\frac{\pi}{A}} \, \frac{1}{2A} \quad (A > 0) \tag{20.17}$$

$$\int_{-\infty}^{\infty} x^4 \exp(-Ax^2) \, dx \;=\; 3 \times \sqrt{\frac{\pi}{A}} \, \frac{1}{4A^2} \quad (A > 0) \tag{20.18}$$

$$\int_{0}^{\infty} r^n \exp(-Ar) \, dr \;=\; \frac{n!}{A^{n+1}} \quad (n \geq 0; A > 0) \tag{20.19}$$

Spotting that the integral you want to compute is one of these standard forms will save a great deal of time!

20.6.5 How dimensions are affected by integration

The integral of a function $f(x)$ with respect to x can be interpreted as the area under a graph of $f(x)$ plotted against x. The dimensions of the area, and hence of the integral, are

$$\text{dimensions of } f(x) \times \text{dimensions of } x.$$

For example consider the integral

$$\int p \, dV,$$

where p is the pressure of a gas, and V is the volume. The dimensions of pressure are $M L^{-1} T^{-2}$ and those of volume are L^3, so the dimensions of the integral are

$$\left[\int p \, dV \right] = M L^{-1} T^{-2} \times L^3 = M L^2 T^{-2},$$

which we recognize as the dimensions of energy.

20.7   Differential equations

A differential equation is one in which both a function and its derivative appear e.g.

$$\frac{df(x)}{dx} = -Af(x), \tag{20.20}$$

where $f(x)$ is some (as yet unspecified) function of x and A is a constant. Differential equations are ubiquitous in the mathematical description of physical phenomena.

It is usually the case that we want to know the form of the function (e.g. $f(x)$ in the above equation). In some relatively simple cases this is not too difficult to do by hand, and we will look at one example here which is particularly relevant to chemistry.

20.7.1   Solution by separation of variables

Equation 20.20 can be solved by using the method known as the *separation of variables* in which all of the terms involving $f(x)$ are take onto one side, and all those involving x are taken onto the other. Having done this, both sides are integrated. In making this separation we treat 'd(f)x' and 'dx' as if there were ordinary variables or functions; the justification for this is beyond the scope of this discussion.

Taking $f(x)$ from Eq. 20.20 over to the left and dx over to the right, and then integrating both sides gives

$$\int \frac{1}{f(x)} \, df(x) = \int -A \, dx.$$

Given that the integral of $1/x$ with respect to x is $\ln(x)$, it follows that the integral of $1/f(x)$ with respect to $f(x)$ is $\ln(f(x))$. The integral on the right evaluates to $-Ax$, so the final result of integrating both sides is

$$\ln(f(x)) = -Ax + \text{const.} \tag{20.21}$$

Note that since the integrals are indefinite a constant of integration is needed.

If we want to be cautious we can differentiate both sides of Eq. 20.21 with respect to x to check that we get back to the original equation. To differentiate the left-hand side we use the substitution $y = f(x)$ and the chain rule (section 20.5.3 on page 902)

$$\frac{d\ln(f(x))}{dx} = \frac{d\ln(y)}{dx}$$
$$= \frac{d\ln(y)}{dy} \frac{dy}{dx}$$
$$= \frac{1}{y} \frac{dy}{dx}$$
$$= \frac{1}{f(x)} \frac{df(x)}{dx}.$$

To go to the last line we have used $y = f(x)$. The right-hand side of Eq. 20.21 differentiates of $-A$. The overall result of differentiating Eq. 20.21 is therefore

$$\frac{1}{f(x)}\frac{\mathrm{d}f(x)}{\mathrm{d}x} = -A \qquad \text{which rearranges to} \qquad \frac{\mathrm{d}f(x)}{\mathrm{d}x} = -Af(x).$$

This is indeed where we started from, Eq. 20.20.

Returning to our solution, Eq. 20.21 on the preceding page, we can determine the value of the constant if we know the value of $f(x)$ for some particular value of x. These known values form what is called a *boundary condition*; such conditions often arise from a physical constraint which applies to the system being modelled.

To be entirely general, we will assume that the boundary condition is that, at $x = x_0$, $f(x)$ has the value f_0. Putting these values into Eq. 20.21 we have

$$\ln(f_0) = -Ax_0 + \text{const.},$$

hence is follows that const. $= \ln(f_0) + Ax_0$. Putting this back into Eq. 20.21 and doing some rearrangement gives

$$
\begin{aligned}
\ln(f(x)) &= -Ax + \ln(f_0) + Ax_0 \\
\ln(f(x)) - \ln(f_0) &= A[x_0 - x] \\
\ln(f(x)/f_0) &= A[x_0 - x] \\
f(x)/f_0 &= \exp(A[x_0 - x]) \\
f(x) &= f_0 \exp(A[x_0 - x]). \qquad (20.22)
\end{aligned}
$$

Equation 20.22 is a general solution to the differential equation given in Eq. 20.20.

As an example of such a differential equation, consider a mass m, moving with velocity v, which experiences a retarding force which is proportional to the velocity (Fig. 20.16). The force F is given by $F = -kv$, where k is a constant and the minus sign is there to ensure that the force retards the mass. Recall that by definition force is mass × acceleration, and that acceleration is $\mathrm{d}v/\mathrm{d}t$, where t is the time. Using these substitutions $F = -kv$ can be rearranged to

$$m\frac{\mathrm{d}v(t)}{\mathrm{d}t} = -kv(t),$$

where the velocity has been written as $v(t)$ to remind us that it depends on time. Taking the m over to the right gives

$$\frac{\mathrm{d}v(t)}{\mathrm{d}t} = -\frac{k}{m}v(t),$$

which is of the form of Eq. 20.20 on the facing page with the substitutions

$$x \to t \quad f(x) \to v(t) \quad A \to \frac{k}{m}.$$

We can therefore immediately apply the general solution Eq. 20.21 on the preceding page

$$\ln(v(t)) = -\frac{k}{m}t + \text{const.}$$

If at $t = 0$ the velocity is v_0, then we can eliminate the constant using Eq. 20.22 with $f_0 \to v_0$ and $x_0 \to 0$ to give

$$v(t) = v_0 \exp\left(-\frac{k}{m}t\right).$$

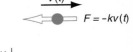

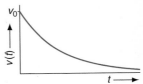

Fig. 20.16 A particle moving at velocity $v(t)$ experiences a retarding force (i.e. a force in the opposite direction to that in which the particle is moving) which is proportional to the velocity; this force can be written $F = -kv(t)$, where k is a constant. By solving the corresponding differential equation we can show that the velocity decreases exponentially from its initial value at a rate determined by k.

Self Test 20.17 Differential equations

A first-order reaction has the following rate equation

$$\frac{d[A](t)}{dt} = -k_{1st}[A](t), \tag{20.23}$$

where $[A](t)$ is the concentration of reactant A at time t, and k_{1st} is the first-order rate constant. At time $t = 0$, the concentration of A is $[A]_0$.

(a) By comparison with the differential equation given in Eq. 20.20 on page 910, and using the solution given in Eq. 20.22 on the preceding page, deduce an expression for $[A](t)$.

(b) Solve Eq. 20.23 by separating the variables ($[A](t)$ to the left and t to the right), integrating and imposing the boundary condition.

This says that the velocity decreases exponentially over time at a rate given by (k/m), as illustrated in Fig. 20.16.

Self Test 20.17 will give you some practice with this sort of approach.

FURTHER READING

Foundations of Science Mathematics, Deviderjit Singh Sivia and S. G. Rawlings, Oxford University Press (1999).
Maths for Chemistry, Paul Monk, Oxford University Press (2006)
Foundations of Physics for Chemists, Grant Ritchie and Devinder Sivia, Oxford University Press (2000).

ANSWERS TO SELF TEST EXERCISES

20.1 (a) MLT^{-1}, (b) ML^2T^{-2}, (c) ML^{-3}, (d) MLT^{-2}, (e) $ML^{-1}T^{-2}$, (f) $ML^2T^{-2}K^{-1}N^{-1}$, (g) $ML^2T^{-2}N^{-1}$.

20.2 (a) $ML^2T^{-2}K^{-1}N^{-1}$, (b) $ML^2T^{-2}N^{-1}$, (c) dimensionless, (d) ML^2T^{-1}, (e) ML^2T^{-1}, (f) ML^2T^{-2}, (g) ML^2T^{-2}.
(b), (f) and (g) are energies.

20.3 (c) $ML^2T^{-2}K^{-1}N^{-1}$, (d) both have the same dimensions as in (c), (e) both have dimensions ML^2T^{-2} i.e. energy.

20.4 (a) kg m s^{-1}, (b) J, (c) kg m^{-3}, (d) V m^{-1}, (e) N m^{-2}.

20.5 (a) 100 pm or 0.100 nm, (b) 108 pm or 0.108 nm, (c) 780 nm or 0.780 μm, (d) 94 MHz, (e) -8.91×10^5 J mol^{-1}, (f) 2.82×10^{-10} m (g) 3.50×10^{-7} m (h) 2.00×10^9 Hz.

20.6 (a) 1.5×10^3 mol m^{-3}, (b) 1.0×10^{-4} mol dm^{-3}, (c) 5×10^{-3} mol, (d) 50 cm^3.

20.7 (a) 1.68×10^{-10} m³ or 168 pm, (b) 8.16×10^4 J mol⁻¹ or 81.6 kJ mol⁻¹, (c) 2.0×10^4 Pa or 20 kPa, (d) 1.52×10^4 Pa or 15.2 kPa, (e) 1.50×10^4 Pa or 15.0 kPa, (f) 6.30×10^{13} Hz and 4.17×10^{-20} J, (g) 8.72×10^{-18} J or 5252 kJ mol⁻¹.

20.8 (a) $\pi/4$ and $\pi/3$, (b) $120°$ and $225°$, (c) $\pi/2$, (d) $0, \pi/4, \pi/2, 3\pi/4, \pi, 5\pi/4, 3\pi/2, 7\pi/4$, (e) $\cos(3\theta)$, (f) $\cos(2\theta)$.

20.9 (a) A^4 and A^3, (b) A and 1, (c) $A^{5/2}$ (or $A^{2.5}$) and A^2.

20.10 (b) The fraction α is $\frac{1}{2}$, so the time is $-\ln(\frac{1}{2})/k$ which is $\ln(2)/k$, (c) (i) $\ln a + \ln x$ (ii) $\ln a + 2\ln x$ (iii) $\ln b + \frac{1}{2}\ln y$ (iv) $\ln b - 3\ln z$, (d) Yes, since $k_{1st}t$ is dimensionless.

20.11 (a) $6x - Ax^{-2} - 4x^{-3}$, (b) $-2A\sin(2\theta)$, (c) $V/(RT)$, (d) (i) B/t (ii) B/t, (e) (i) $-AB\exp(-Ar)$ (ii) $AB\exp(Ar)$, (f) $-4A\cos 2\theta$.

20.12 (a) $2By\exp(By^2)$, (b) $-2At\sin(At^2)$, (c) $-AB[\sin(B\theta)]^{-2}\cos(B\theta)$.

20.13 (a) $A\exp(-Br) - ABr\exp(-Br)$, (b) $-A\sin(At)\exp(-Bt) - B\cos(At)\exp(-Bt)$, (c) $\exp(-Ax^2) + x[-2Ax\exp(-Ax^2)] \equiv \exp(-Ax^2) - 2Ax^2\exp(-Ax^2)$.

20.14 (all + const.) (a) $\frac{1}{4}Ay^4$, (b) $-(B/C)\exp(-Cx)$, (c) $-\frac{1}{2}\cos(2\theta)$, (d) $(1/B)\ln r$, (e) $-\frac{1}{2}Cy^{-2}$, (f) z.

20.15 (a) $\frac{1}{2}A^2$, (b) 0, (c) 0, (d) 2, (e) 1, (f) $(z_2 - z_1)$.

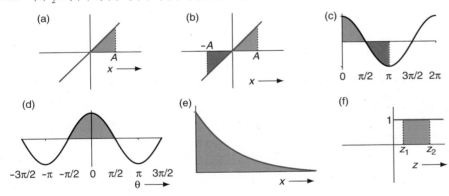

20.16 (a) -2π, (b) 1, (c) applying integration by parts gives another integral which is of the form of that in (a), so we can just reuse the result from (a): the final answer of 4π.

20.17 (a) $[A](t) = [A]_0\exp(-k_{1st}t)$, (b) Integral $\int([A](t))^{-1}\,d[A](t) = \int -k_{1st}\,dt$; result of integral $\ln([A](t)) = -k_{1st}t + \text{const}$; $\text{const} = \ln([A]_0)$.

QUESTIONS

Physical constants: $R = 8.3145$ J K⁻¹ mol⁻¹, $h = 6.626 \times 10^{-34}$ J s, $k_B = 1.381 \times 10^{-23}$ J K⁻¹, $N_A = 6.022 \times 10^{23}$ mol⁻¹, $F = 96,485$ C mol⁻¹, $c = 2.998 \times 10^8$ m s⁻¹.

20.1 The so-called 'SUVAT' equations apply to an object experiencing a constant acceleration a and give various relationships between the distance travelled s, the initial velocity u, the final velocity v, and the time t. Determine the dimensions of each term in these equations, and hence show that they are dimensionally consistent.

$$v = u + at \quad s = ut + \tfrac{1}{2}at^2 \quad s = vt - \tfrac{1}{2}at^2 \quad v^2 = u^2 + 2as \quad s = \tfrac{1}{2}(u + v)t$$

20.2 A spherical particle of radius r moving with velocity v through a viscous medium (e.g. a liquid) experiences a retarding force F which can, under some circumstances, be approximated by the Stokes' Law

$$F = 6\pi\eta rv,$$

where η is the viscosity of the medium. Determine the dimensions of η, and hence its SI unit. [Hint: rearrange the expression to give $\eta = \ldots$, and then use the known dimensions of all the quantities then on the right.]

20.3 A first-order reaction has the rate law

$$\frac{d[A]}{dt} = -k_{1st}[A],$$

where $[A]$ is the concentration of a reactant, t is time and k_{1st} is the rate constant. Assuming that $[A]$ is given in amount of substance per unit volume, determine the dimensions of k_{1st}. Hence, state the SI unit of k_{1st}.

20.4 A second-order reaction has the rate law

$$\frac{d[A]}{dt} = -k_{2nd}[A]^2,$$

where $[A]$ is the concentration of a reactant, t is time and k_{2nd} is the rate constant. Assuming that $[A]$ is given in amount of substance per unit volume, determine the dimensions of k_{2nd}. If the concentration is expressed in mol dm^{-3}, what will be the units of k_{2nd}?

20.5 The work done when a force F moves a distance x is $F \times x$. The work done when a charge q moves through a potential (voltage) V is $q \times V$. The work done when a surface is increased in area by A is given by $A \times \gamma$, where γ is the surface tension (SI unit N m^{-1}).

Show that each of these work terms has the dimensions of energy.

20.6 (a) The rotational kinetic energy of an object with *moment of inertia* I rotating at angular frequency ω is given by $\tfrac{1}{2}I\omega^2$. Given that I has dimensions ML^2, confirm that $\tfrac{1}{2}I\omega^2$ has the dimensions of energy. [Hint: ω could be given in rad s^{-1}; recall that radians are dimensionless.]

(b) For a diatomic molecule, I is given by

$$I = \frac{m_1 m_2}{m_1 + m_2}R^2,$$

where m_1 and m_2 are the masses of the two atoms and R is the bond length. In quantum mechanics the energy E_J of a rotating diatomic is given by

$$E_J = BJ(J + 1) \qquad \text{where } B = \frac{\hbar^2}{2I};$$

in this expression J is a dimensionless quantum number which takes integer values, and $\hbar$ is Planck's constant divided by 2π: $\hbar = h/2\pi$.

Determine the dimensions of I and B; hence show that E_J has the dimensions of energy.

20.7 In quantum mechanics, a simple model for the energy of a vibrating bond is

$$E_v = (v + \tfrac{1}{2})\hbar\omega \qquad \text{where } \omega = \sqrt{\frac{k_f}{m}};$$

in this expression v is a dimensionless quantum number which takes integer values, k_f is the force constant (SI unit N m^{-1}), m is the mass, and $\hbar$ is Planck's constant divided by 2π: $\hbar = h/2\pi$.

Determine the dimensions of k_f and ω; hence show that E_v has the dimensions of energy.

20.8 Convert the following to SI, using an prefix in your answer, where appropriate.

(a) The collision cross-section of O_2, 40 Å^2.

(b) The entropy change when one mole of liquid water is vaporized at 373 K, 26.05 e.u.

(c) The vapour pressure of water at its triple point, 4.58 τ.

(d) The vibrational frequency of the bond in H_2, 4401 cm^{-1}.

(e) The orbital energy of a $1s$ electron in He, -25 eV.

(f) The mass of one molecule of $^{19}F_2$, given that the mass of ^{19}F is 18.998 u.

20.9 (a) What volume, in cm^3, of a solution of concentration 0.15 mol dm^{-3} contains the same amount in moles as 13.5 cm^3 of a solution of concentration 1.2 mol dm^{-3}?

(b) 1.50 g of NaCl is dissolved in water and the solution made up to a total volume of 50 cm^3. What is the concentration, in mol dm^{-3} and in mol m^{-3}, of the resulting solution. [RFM for NaCl is 58.35 g mol^{-1}.]

(c) How much water needs to be added to 10 cm^3 of a solution of concentration 1.00 mol dm^{-3} to give a solution of concentration 0.15 mol dm^{-3}?

20.10 In quantum mechanics, the energy of a rotating diatomic is given by

$$E_J = BJ(J + 1) \qquad \text{where } B = \frac{\hbar^2}{2I},$$

where all of the symbols are defined in question 22.7. If SI units are used for $\hbar$ and I, then the energy E_J will of course be in joules.

(a) Explain why the energy expressed in cm^{-1}, $\tilde{E}$ and the energy expressed in joules, E, are related by
$$E = h \times \tilde{c} \times \tilde{E},$$
where $\tilde{c}$ is the speed on light in cm s^{-1}.

(b) Hence show that the rotational energy, expressed in cm^{-1}, is given by

$$\tilde{E}_J = \frac{B}{h\tilde{c}} J(J + 1).$$

(c) By substituting in $\hbar^2/2I$ for B, go on to show that the rotation energy in cm^{-1} can be written

$$\tilde{E}_J = \frac{h}{8\pi^2 \tilde{c} I} J(J + 1).$$

20.11 Sketch $\cos(3\alpha)$ in the range $\alpha = 0$ to 2π. What is the period of this function? At what values of α (expressed in terms of π) does the function go to zero? Classify the function as even or odd about the value $\alpha = \pi$.

20.12 (a) Given that

$$\sin(A + B) \equiv \sin A \cos B + \cos A \sin B \qquad \cos(A + B) \equiv \cos A \cos B - \sin A \sin B.$$

show that

$$\sin(2A) \equiv 2 \sin A \cos B \qquad \cos(2A) \equiv \cos^2 A - \sin^2 A.$$

(b) Using $\cos^2 A + \sin^2 A = 1$, show that $\cos(2A) \equiv \cos^2 A - \sin^2 A$ can be rewritten in two different ways

$$\cos(2A) \equiv 2\cos^2 A - 1 \quad \text{or} \quad \cos(2A) \equiv 1 - 2\sin^2 A.$$

(c) Show that these last two identities can be rearranged to give

$$\cos^2 A \equiv \tfrac{1}{2}[1 + \cos(2A)] \quad \text{and} \quad \sin^2 A \equiv \tfrac{1}{2}[1 - \cos(2A)].$$

20.13 The concentration of a reactant A in a first-order reaction varies with times as follows

$$[A](t) = [A]_0 \exp(-k_{1st}t),$$

where $[A](t)$ is the concentration at time t, $[A]_0$ is the concentration at time zero, and k_{1st} is the rate constant. Show that the half life t_{half}, which is the time taken for the concentration to fall to half its initial value, is given by $t_{half} = \ln(2)/k_{1st}$.

20.14 (a) The Boltzmann distribution gives the population of energy level i, n_i, in terms of its energy ε_i and the population n_0 of the lowest level

$$n_i = n_0 \exp\left(\frac{-\varepsilon_i}{k_B T}\right).$$

Rearrange this to find expressions for $\ln(n_i)$ and $\ln(n_i/n_0)$.

(b) The standard Gibbs energy change $\Delta_r G°$ and the equilibrium constant K are related by $\Delta_r G° = -RT \ln K$. Rearrange this to find expressions for $\ln K$ and for K.

20.15 Compute the first and second derivatives of $f(t) = A \exp(-Bt^2)$ with respect to t (A and B are constants). Sketch a graph of $f(t)$ against t and, by interpreting the first derivative as the slope, explain how your expression for $df(t)/dt$ is consistent with the plot.

20.16 Differentiate the function $f(r) = r^4 \exp(-r)$ with respect to r and hence show that there are extrema at $r = 0$, $r = 4$ and $r = +\infty$. Explain why $r = 4$ must correspond to a maximum.

20.17 (a) Using the identity $\sin^2 \theta \equiv \tfrac{1}{2}[1 - \cos(2\theta)]$, show that

$$\int_0^A \sin^2\left(\frac{\pi x}{A}\right) dx$$

can be written

$$\tfrac{1}{2}\int_0^A dx - \tfrac{1}{2}\int_0^A \cos\left(\frac{2\pi x}{A}\right) dx.$$

(b) Evaluate both integrals and show that together they come to $A/2$.

20.18 A second-order reaction has the following rate equation

$$\frac{d[A](t)}{dt} = -k_{2nd}[A]^2(t),$$

where $[A](t)$ is the concentration of reactant A at time t, and k_{2nd} is the second-order rate constant.

(a) Separate the variables, taking the terms in $[A](t)$ to the left, and those in t to the right.

(b) Integrate both sides of the resulting equation.

(c) Use the fact that at time $t = 0$ the concentration of A is $[A]_0$ to determine the constant of integration.

Index

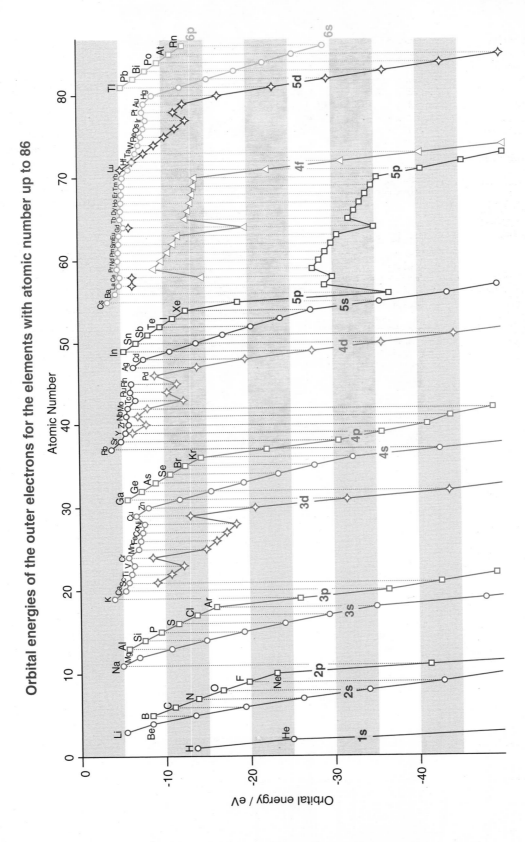

Orbital energies of the outer electrons for the elements with atomic number up to 86